Contents

In March 2020 the national lockdown, caused by the Coronavirus pandemic, forced the National Garden Scheme to close all 3,700 gardens for the first time in its 93-year history. The impact on the charity, which exists to raise money for nursing and health charities was, as for so many, a major crisis.

However, National Garden Scheme garden owners are as resilient as they are generous. Many decided that if they could not physically welcome visitors they would do so virtually, by making short videos of their gardens. The videos were part of a campaign the charity launched called Help Support our Nurses, with a request for donations to make up for the loss of garden income. I was delighted to contribute a video to the campaign, which raised some £250,000. I was also thrilled to hear that gardens started re-opening in June, and by the end of September, ticket sales had raised £450,000. Not only that, but I was greatly impressed to learn that the first Great British Garden Party, despite all the restrictions, produced very nearly £50,000. This is a wonderful tribute to the determination and kindness of so many garden owners, to whom I can only offer my heartfelt thanks.

Now, as we embark on a new season, just looking through this 2021 Garden Visitor's Handbook at the array of wonderful gardens gives one a great sense of optimism. As always, the National Garden Scheme is offering a uniquely affordable way for people to enjoy gardens of all shapes and sizes, as well as the intrinsic benefits to our health and wellbeing that come from visiting a garden.

2020 confirmed the positivity gardens bring to people's lives. Once again, the generous people who open for the National Garden Scheme are inviting everyone to share their gardens and gain that reassuring feeling of relaxation, and connectivity with Nature that is so good for us all, while continuing to provide gardening inspiration for people to take home. Above all, they will be raising funds for the charities that the National Garden Scheme supports, whose time of need has never been greater.

Chairman's message

I am very proud to have been appointed Chairman of the National Garden Scheme and I would like to first acknowledge the superb contribution of my predecessor, Martin McMillan OBE, during his six years in the post – and indeed during nine years before that as a Trustee.

I have been a Trustee since 2014 and Charles and I have opened our garden in Clapham for many years. Both levels of involvement have given me plenty of opportunity to witness what a remarkable charity I am now fortunate enough to be leading.

I took office at the end of undoubtedly the most challenging year in the charity's long history and yet, as we show on page 5, thanks to the wonderful support and resourcefulness of our garden owners we were still able to donate nearly £2.9 million to our beneficiaries. And the year was one of innovation for the National Garden Scheme in many different ways, innovation that will strengthen our armoury this year and beyond.

I am thrilled that our garden owners registered with enthusiasm to open their gardens in 2021. As a result, with as many gardens as in previous years and with the prospect of calmer times I am confident we can look forward to a positive and rewarding year.

On the occasions that we opened our garden in 2020 we were particularly struck by the visible happiness and relaxation that being able to spend time in a garden clearly brought to our visitors. I am sure those feelings will continue to be prevalent through 2021. The personal nature of a visit and the variety of gardens that open, together make the National Garden Scheme offering completely unique – especially if you add in the charitable use of the funds raised and the wonderful organisations that we support. If you are an established visitor we look forward to welcoming you back, if you are new to the National Garden Scheme we are certain you will find visiting gardens to be an enjoyable and memorable experience (not just for the tea).

Rupert Tyler

Discover the nation's best gardens

The National Garden Scheme gives visitors unique access to exceptional private gardens across England and Wales and raises impressive amounts of money for nursing and health charities through admissions, teas and cake.

Thanks to the generosity of garden owners, volunteers and visitors we have donated over £60 million to nursing and health charities since we were founded in 1927. Last year, despite the challenges we all faced, we made total annual donations of £2.88 million from our 2019 season.

2020 confirmed the National Garden Scheme's extraordinary dual impact, both within the gardens that open and through the donations made with the funds raised at their open days. As well as their unrivalled horticultural delights and quality, the gardens offer visitors enjoyment, relaxation and confirmation that having access to a garden is good for everybody's health and wellbeing, especially those without their own garden or who have particular health needs.

Originally established to raise funds for district nurses, we are now the most significant charitable funder of nursing in the UK and our beneficiaries include Macmillan Cancer Support, Marie Curie, Hospice UK and The Queen's Nursing Institute. Year by year, the cumulative impact of the grants we are able to give grows incrementally, helping our beneficiaries make their own important contribution to the nation's health and care, and strengthening the partnerships that we have with them all.

With over 3,600 gardens opening across England and Wales in 2021, we hope you enjoy exploring our beautiful gardens this year. Each inspiring space not only provides a unique glimpse into hidden horticultural delights but also the chance to support the health and wellbeing of thousands.

© Judi Lion

YOUR GARDEN VISITS HELP CHANGE LIVES

In 2020 the National Garden Scheme donated £2.88 million from funds raised in our gardens. These donations provided critical support to nursing and health charities in a year of crisis, and supported charities doing amazing work in gardens and health.

Nursing and health

Macmillan Cancer Support £425,000	**Marie Curie** £425,000
Hospice UK £425,000	**The Queen's Nursing Institute** £370,000

Carers Trust £340,000	**Parkinson's UK** £157,500	**Mind** £80,000

Gardens and health

Maggie's £100,000	**ABF The Soldier's Charity** £80,000	**Horatio's Garden** £75,000
Well Halton £75,000	**Greenfingers Charity** £15,000	**Patchworking Garden Project** £20,000
Community garden projects £97,210	**Support for gardeners** £200,000	

Thank you

To find out more, visit **ngs.org.uk/beneficiaries**

⊕ Investec

A partner for all seasons

After every long, cold winter, gardens return to bloom, reminding us that all things come and go.

We are collectively emerging from a difficult past year, in which the enjoyment of outdoor spaces acted as a much-needed tonic to our physical and mental health. And as we enter a hopefully more positive period, these spaces will blossom in celebration with us.

As wealth managers, we are more accustomed to volatility than most. We pursue growth for our clients through rises and falls, and know that difficult times are, more often than not, followed by good.

Through both, we're delighted to sponsor the National Garden Scheme, facilitating access to exquisite gardens, and supporting their incredible fundraising for nursing and health charities.

With Investment Your Capital is at Risk.

Know where life can take you.

investecwin.co.uk

Some tips on using your Handbook

This book lists all the gardens opening for the National Garden Scheme between January 2021 and early 2022. It is divided up into county sections, each including a map, calendar of opening dates and details of each garden, listed alphabetically.

Symbols explained

NEW Gardens opening for the first time this year or re-opening after a long break.

◆ Garden also opens on non-National Garden Scheme days. (Gardens which carry this symbol contribute to the National Garden Scheme either by opening on a specific day(s) and/or by giving a guaranteed contribution.)

♿ Wheelchair access to at least the main features of the garden.

🐕 Dogs on short leads welcome.

❋ Plants usually for sale.

NPC Plant Heritage National Plant Collection.

🛏 Gardens that offer accommodation.

☕ Refreshments are available, normally at a charge, subject to Covid-19 guidelines.

🪑 Picnics welcome.

D Garden designed by a Fellow, Member, Pre-registered Member, or Student of The Society of Garden Designers.

🚌 Garden accessible to coaches. Coach sizes vary so please contact garden owner or County Organiser in advance to check details.

Group Visits Group Organisers may contact the County Organiser or a garden owner direct to organise a group visit to a particular county or garden. Otherwise contact the National Garden Scheme office on 01483 211535.

Children must be accompanied by an adult.

Photography is at the discretion of the garden owner; please check first. Photographs must not be used for sale or reproduction without prior permission of the owner.

Funds raised In most cases all funds raised at our open gardens comes to the National Garden Scheme. However, there are some instances where income from teas or a percentage of admissions is given to another charity.

Toilet facilities are not always available at gardens.

If you cannot find the information you require from a garden or County Organiser, call the National Garden Scheme office on 01483 211535.

There are many ways to book and enjoy a visit to our gardens...

Cashless payments and online booking

Pre-booking available

Many gardens plan to accept card payments in 2021. Look for the cashless symbol in the garden listing on our website.

For over 3,000 gardens you can purchase tickets online in advance, as well as purchasing them on the gate on the day. Look for the Pre-booking available symbol in the garden listing.

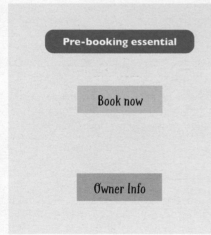

The **Pre-booking essential** symbol in the garden listing means you need to pre-book your visit. You will also see either:

A book now button which allows you to book your tickets through the National Garden Scheme

OR

If there is no book now button, you will need to book your tickets direct with the Garden Owner. Click on the Owner Info tab to find their contact details

Special Events

Some gardens offer Special Events which may include a tour, an exclusive visit or a 3 course meal. These must be pre-booked, and sell out fast. Visit **ngs.org.uk/special-events/** to book your ticket and sign up to receive the latest updates.

Open by arrangement

Over 1,100 gardens invite you to visit By Arrangement – so that you can visit on a date to suit you. Contact the Garden Owner to discuss availability and book your visit.

COBRA

One of the UK's largest range of lawnmowers

Create a lawn that is the envy of your neighbours with a new lawnmower from Cobra. At the heart of these powerful, stylish mowers is a choice of either electric, cordless or petrol engines powered by Briggs & Stratton, Honda Kohler and more.

Cobra have over 65 lawnmowers in their extensive range including the 'T3' award winning Cobra MX3440V 40V cordless mower and a new range of powerful twin 40V cordless models for 2021. Whatever your gardening needs, Cobra has a lawnmower for you.

Promo prices start from just £89.99 inc VAT

Model Shown: Cobra MX51S80V RRP: £559.99 inc vat

POWERED BY BRIGGS & STRATTON

Expertly Powered By

COBRA KOHLER

For your nearest dealer visit: **www.cobragarden.co.uk** or call: **0115 986 6646**   *Promotional prices only at participating dealers*

The impact of our donations to nursing and health

Since our inception the National Garden Scheme has developed an efficient fundraising formula which allows us to donate 80% of money raised at our open gardens to nursing and health beneficiaries. Even with the closure of many gardens during the 2020 lockdowns we were still able to donate £2.58 million to key nursing and health charities without which the NHS would have struggled during the coronavirus crisis.

Our beneficiaries are major contributors to the key areas of community health and care including:

- Nursing people at home so they do not need to be in hospital

- End of life and palliative care for patients and support for their families

- Specialist nursing and care for health conditions such as cancer and Parkinson's

Here are some of the incredible things our donations have supported in 2020:

127 New Queen's Nurses received the prestigious title in 2020 bringing the total number to 1,420

158 Hospices across the UK supported with the supply of PPE to protect frontline staff during the pandemic

300 Additional people living with Parkinson's able to access the support they need to manage their condition

124 Inpatients supported by the Y Bwthyn NGS Macmillan Specialist Palliative Care Unit in Wales

18,130 Unpaid carers supported through funding for Carers Trust

Traditionally...
The Gardener's Glasshouse

Griffin has been designing bespoke glasshouses since the early 1960s. With a history of innovating glasshouse design for the commercial sector, Griffin continues to be the choice of many estate managers and discerning professional gardeners today, with glasshouses installed in the UK and worldwide.

Each Griffin glasshouse is individually created for you, tailored to your individual requirements and unique surroundings whilst being both a pleasure for you to use and an enhancing feature of your property. Manufactured from powder coated aluminium in any colour of your choice, your glasshouse will have all the appeal of a traditional wooden structure but without the maintenance issues.

Unreservedly distinct *Glasshouses, Greenhouses* and *Orangeries*

For more information please call or visit our website.

GRIFFIN GLASSHOUSES
GREENHOUSES OF DISTINCTION
www.griffinglasshouses.com
01962 772512

Partnering National Garden Scheme with our select range of free standing glasshouses

Gardens and health: people and communities

We have championed and supported the concept that access to gardens and practical gardening are good for people's health and wellbeing since 2016. Last year, we published a report *Gardens and Coronavirus 2020: the importance of gardens and outdoor space during lockdown*, which illustrated just how beneficial our gardens and green spaces were during that time. Most crucially it shone a spotlight on how even the tiniest glimpse of nature provided hope to those shielding, isolated or alone.

We continue to make substantial donations to specific charities for health-related garden projects, supporting Horatio's Garden to build eleven gardens for NHS spinal injury units and Maggie's to add to their portfolio of gardens at their cancer support centres.

We also support a broad spectrum of people and communities. From those looking for a change of career and getting into horticulture for the first time to supporting garden projects that help rehabilitate ex-servicemen or reduce social isolation, our funding provides a positive introduction to gardening. Our support for Perennial, also meant continued support for gardeners, horticulturalists and their families who have fallen on tough times.

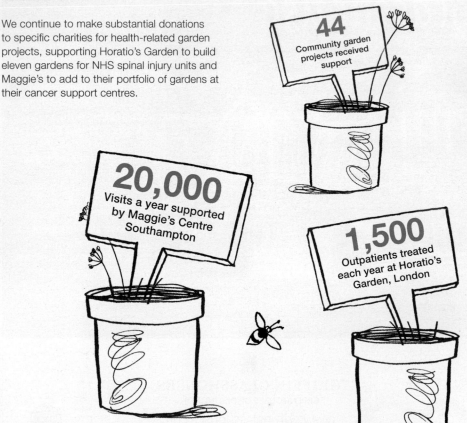

44
Community garden projects received support

20,000
Visits a year supported by Maggie's Centre Southampton

1,500
Outpatients treated each year at Horatio's Garden, London

Opposite page: Patient and nurse in Horatio's Garden, Salisbury, © Russell Sach

Our rich diversity of gardens

National Garden Scheme Chief Executive, George Plumptre looks as how the variety of gardens that open is increasing steadily

The National Garden Scheme has long been renowned for the quality of its gardens. People know that when they follow the yellow arrows their destination will be rewarding and often a wonderful surprise. But in recent years our portfolio of gardens has also assumed greater diversity, balancing lawns and borders at rural old rectories with urban tropical extravaganzas, or wildlife-focused sanctuaries where the planting is for pollinators.

Certain types of garden encourage this diversity, in particular gardens that open as a group, community gardens and allotments. Groups have been a major feature of the National Garden Scheme for many years and in 2021 there will be 250 opening. Involving a number of gardens in the same village or in close proximity in a town or city, they encourage different members of a local community to take part. They also have the scope to include gardens that might not be able to open on their own and this in itself adds a rich vein of diversity to the portfolio.

More than 20 community gardens will open in 2021; many of them have a strong emphasis on individual and community health and all of them introduce new people to the joys of gardening in an often life-changing way.

In Birmingham's historic Bournville Village, one of the group of gardens that opens is Masefield Community Garden whose origins typify the determination to improve a local environment that so many of these gardens have: 'Until 2012, the overgrown site was home to vandalised lock-up garages. Volunteers have developed an eco-friendly garden comprising raised beds, mixed borders, a forest garden (permaculture), an orchard, a wildlife pond, two greenhouses (one made from plastic bottles) and a polytunnel.'

In 2021 35 groups of allotments will open for the National Garden Scheme. They have become increasingly popular with visitors who love meeting all the different people involved and the mixture of ornamental flower and practical fruit and veg gardening – often heading home with a bag of freshly grown produce or a jar of home-made jam. Joining the group for the first time in 2021 will be Oswestry Gateacre Allotments and Gardens Association. Filling two sites in the heart of the Shropshire town, they have become a vital and much-loved part of the urban landscape as allotments so often do.

Left: The Therapy Garden
Opposite page: Ordnance House & Maggie's Manchester

Oswestry is also home to one of a special and growing group of gardens that open, all created by or part of one of the National Garden Scheme's beneficiary charities. In 2019 the charity Horatio's Garden, which creates gardens for spinal injury centres, opened its garden at Oswestry and it will be one of four Horatio's Gardens opening in 2021. Similar solace and respite that these gardens offer to patients can be found in the seven gardens of Maggie's Centres that look after people living with cancer, and in the 21 hospice gardens that open in recognition of the National Garden Scheme's support of Hospice UK.

We will perpetuate the maxim that has always applied to our gardens, that they should have 'quality, interest and character'. But we also want to represent our country's uniquely rich diversity of gardens and the people that create and look after them. The restrictions and isolation that the 2020 pandemic inflicted confirmed the vital importance of access to a garden or outdoor green space. By opening a wide range of gardens in a variety of settings we will champion this access and, we hope, introduce more and more people to it.

Below: Oswestry Allotments & Wicor Primary Community School

The National Garden Scheme are looking for new gardens

© Matthew Bruce

Join a community of like-minded individuals passionate about gardens and open your garden to raise money for vital nursing and health charities.

Our garden portfolio is diverse: from gorgeous country estates to tropical urban extravaganzas, gardens planted for wildlife and allotments. So, whatever its size or style, if your garden has quality, character and interest we'd love to hear from you.

Call us on 01483 211 535 or email hello@ngs.org.uk

NATIONAL open GARDEN SCHEME

Join the Great British Garden Party 2021

Join our Great British Garden Party by hosting your own event. A garden party, cake sale or a community plant and produce sale – whatever you choose to host you can help us to raise funds for vital nursing and health charities.

The lockdown in 2020 highlighted just how important gardens and public spaces are to our health and wellbeing and a source of joy. So whatever kind of garden or outdoor space you have, however big or small, tidy or untidy we want you to celebrate it with friends and family by having fun and raising funds. Our main event will run from Saturday September 4th – Sunday September 12th 2021, but you can host a Great British Garden Party whenever it's best for you.

Joining the Great British Garden Party this year could not be simpler. Sign up on our website to register your interest (ngs.org.uk/gardenparty) and we will share additional information and materials including invitations, posters, menu and party inspiration to help you on your way.

Whether you choose to host an al fresco lunch, an afternoon tea party, prosecco by candlelight or a plant sale in your front garden, everyone can join in and help raise funds for the frontline nursing and health charities that are at the heart of the National Garden Scheme

Despite the challenges that last year brought in terms of restrictions to socialising and the rule of six, our first ever Great British Garden Party fundraiser illustrated just how resourceful and innovative people can be. From National Garden Scheme President, Mary Berry to garden owners, visitors, friends and family, small events took place up and down the country, from intimate tea parties to cake sales, and plant and produce stalls in the front garden. Together you raised an amazing £40,000 for the Help Support Our Nurses appeal. With your help we know we can raise even more in 2021.

The Most Beautiful Pots
& Garden Furniture in The World

Works of *Art* to enhance your *Garden*

Great DISCOUNTs for National Garden Scheme supporters

Help open gardens for the future
with a gift in your will

A love of gardens is often inherited from our parents or grandparents. If you love gardens and garden visiting, you can help inspire that passion in future generations by leaving a gift in your will to the National Garden Scheme.

For more information or to speak to our friendly team call **01483 211535**, email **giftinwill@ngs.org.uk** or visit **ngs.org.uk/giftinwill**

Royal Botanic Gardens

Kew

Discover beauty
Visit the world's greatest
botanic garden

Open daily*

⊖ Kew Gardens
⇌ Kew Bridge

*except 24 and 25 December

Image © RBG Kew / Jeff Eden

ENJOY EXCLUSIVE DISCOUNTS VIA THE NATIONAL GARDEN SCHEME

on select Honda Lawn and Garden products at store.honda.co.uk

Treat your garden to Honda's durable, quality and reliable power tools for the best cut and finish, season after season, year after year. **To claim your discount, please contact: lee.stern@honda-eu.com**

HONDA | ENGINEERING FOR *Life*

store.honda.co.uk

BEDFORDSHIRE

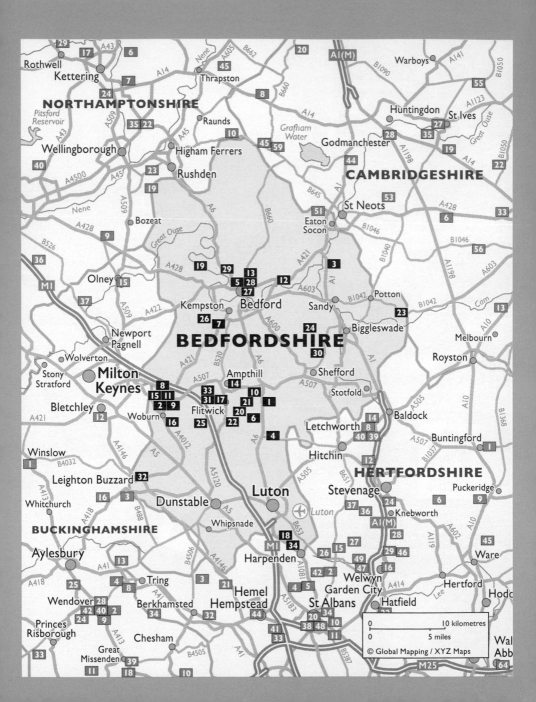

The Birthplace of John Bunyan, it is little wonder the county of Bedfordshire inspired the author of Pilgrim's Progress.

Dear Visitor, we are delighted to welcome you again in January to the National Garden Scheme in Bedfordshire. More than ever our garden owners are eager to share their passion and enthusiasm for plants and design.

We begin the season in January with the National Collection of Snowdrops followed by spring bulbs in March and April. You can enjoy beautiful plantings and inspirational designs at numerous gardens throughout the summer, some of which have been featured in national magazines and on television. There are themed gardens such as the Alpine garden, the Japanese garden and the Texan garden as well as several walled gardens. A trio of 18th century gardens designed by Capability Brown, Southill Park, Luton Hoo and The Walled Garden at Luton Hoo are also open for the National Garden Scheme in Bedfordshire. We will conclude the year at the end of October with an enchanting, Diwali themed evening of lanterns and diya lamps.

Please note that most of our garden openings require tickets to be purchased through the on-line booking system. We look forward to welcoming you

Volunteers

County Organiser
Indi Jackson
01525 713798
indi.jackson@ngs.org.uk

County Treasurer
Colin Davies
01525 712721
colin.davies@ngs.org.uk

Press Officer
Hannah Sardar
07590 385911
hannah.sardar@ngs.org.uk

Facebook & Twitter
Hannah Sardar
(as above)

Booklet Co-ordinator
Indi Jackson
(as above)

Photography
Venetia Barrington
07787 668027
venetiajanesgarden@gmail.com

Talks
Kate Gardner
07725 307803
kate.gardener@ngs.org.uk

Assistant County Organisers
Geoff & Davina Barrett
geoffanddean@gmail.com

Ann Davies
annie1davies1@gmail.com

Julie Neilson
julieneilson@outlook.com

f @bedfordshire.ngs

Left: 40 Leighton Street

OPENING DATES

All entries subject to change. For latest information check www.ngs.org.uk

Extended openings are shown at the beginning of the month.

Map locator numbers are shown to the right of each garden name.

January

Saturday 9th
127 Stoke Road 32

February

Snowdrop Festival

Sunday 28th
◆ King's Arms Garden 14
127 Stoke Road 32

March

Sunday 21st
The Old Rectory,
Westoning 22

April

Sunday 25th
Steppingley Village
Gardens 31

May

Sunday 9th
The Old Rectory,
Wrestlingworth 23

Saturday 15th
Secret Garden 29

Sunday 16th
NEW Church Farm 6

Sunday 23rd
The Old Rectory,
Wrestlingworth 23

Saturday 29th
Orchard Grange 24

Sunday 30th
Steppingley Village
Gardens 31

Monday 31st
◆ The Manor House,
Stevington 19
Steppingley Village
Gardens 31

June

**Daily from
Wednesday 9th
to Sunday 13th**
Ash Trees 2

Saturday 5th
NEW Bedford Heights 5
22 Elmsdale Road 7

Sunday 6th
Ash Trees 2
22 Elmsdale Road 7
Southill Park 30

Saturday 12th
192 Kimbolton Road 13
NEW 69 Mill Lane 20
NEW 70 Mill Lane 21

Sunday 13th
NEW 69 Mill Lane 20
NEW 70 Mill Lane 21
Royal Oak Cottage 26

Saturday 19th
NEW Hollington Farm 10

Sunday 20th
NEW Barton le Clay
Gardens 4
NEW Hollington Farm 10

Sunday 27th
The Old Rectory,
Westoning 22

July

Saturday 3rd
22 Elmsdale Road 7
Glade House 9
Hollydale, Woburn
Lane 11

Sunday 4th
22 Elmsdale Road 7
Glade House 9
Hollydale, Woburn
Lane 11

Saturday 10th
4 St Andrews Road 27

Sunday 11th
Lindy Lea 17

Saturday 17th
NEW Rads End Farm 25

Sunday 18th
NEW Rads End Farm 25
Royal Oak Cottage 26

Sunday 25th
Luton Hoo Hotel
Golf & Spa 18

August

Friday 6th
◆ The Walled Garden 34

Saturday 7th
1a St Augustine's
Road 28

Wednesday 18th
Flaxbourne Farm 8

Sunday 29th
NEW Church Farm 6

September

Sunday 5th
Flaxbourne Farm 8

Saturday 11th
192 Kimbolton Road 13

Sunday 12th
Howbury Hall Garden 12

Saturday 18th
40 Leighton Street 16

October

Saturday 30th
Townsend Farmhouse 33

By Arrangement

Arrange a personalised garden visit on a date to suit you. See individual garden entries for full details.

10 Alder Wynd 1
1c Bakers Lane 3
22 Elmsdale Road 7
NEW Lake End House 15
40 Leighton Street 16
Royal Oak Cottage 26
4 St Andrews Road 27
1a St Augustine's
Road 28
Secret Garden 29
127 Stoke Road 32

The Old Rectory, Westoning

THE GARDENS

❶ 10 ALDER WYND

Silsoe, Bedford, MK45 4GQ.
David & Frances
Hampson, 01525 861356,
mail@davidhampson.com. *From
Barton Road, turn into Obelisk Way,
R at the school & first L into Alder
Wynd.* Visits by arrangement June
to Sept for groups of up to 10.
Admission includes refreshments.
Adm £8, chd free.
Finalist in the Gardeners World 2017
'Small Garden' category. 10m x 11m
of exotic lush plantings of bananas,
gingers, tree ferns, cannas, hostas,
bamboos, herbaceous perennials and
tetrapanax. Created from scratch in
autumn 2014, it is courtyard in style
and formed around a series of raised
beds and oak structures. There is a
decked area adjacent to the house
and a pond with water feature.

❷ ASH TREES

Green Lane, Aspley Guise, Milton
Keynes, MK17 8EN. John &
Teresa. *Yellow arrow from the village
square. Green Lane off Wood Lane.
Some on-road parking in Wood
Lane.* Disabled or mobility issue
parking only at house. Evening
opening Sun 6 June (4.30-7).
Adm £5, chd free. Live music,
supper additional charge. Daily
Wed 9 June to Sun 13 June
(10.00-4.00). Adm £7.50, chd free
incl refreshments. Pre-booking
essential, please visit www.ngs.
org.uk for information & booking.
Secluded garden, walled in with
borrowed tree-scape and hidden in
a private lane. Herbaceous borders,
shrubs, trees, bulbs, and fruit
including apricot and peach. There
is a unique garden arch sculpture
created for a Hampton Court show
garden. Seats in quiet spots amongst
the plants. Children and wheelchairs
welcome as are dogs on leads.
Garden is flat with reasonable access
for wheels and walkers.

❸ 1C BAKERS LANE

Tempsford, Sandy,
SG19 2BJ. Juliet & David
Pennington, 01767 640482,
juliet.pennington01@gmail.com. *E
side of the A1, Bakers Lane approx
300 yards down Station Road on L.*

Visits by arrangement Feb to Oct
for groups of up to 20. Adm £5,
chd free.
This is a garden for all seasons with
interesting and unusual plants. It
comes to life with a winter border
planted with colourful Cornus and
evergreen shrubs underplanted
with snowdrops and hellebores with
gravel areas carpeted with miniature
cyclamen. It develops through the
seasons culminating in early autumn
with a spectacular show of sun-loving
herbaceous plants.

GROUP OPENING

❹ NEW BARTON LE CLAY GARDENS

Manor Road, Barton-Le-Clay,
Bedford, MK45 4NR. David Pilcher.
*Old A6 through Barton-le-Clay
Village, Manor Rd is off Bedford Rd.
Parking in paddock.* Sun 20 June
(2-5). Combined adm £6, chd free.
Home-made teas. Pre-booking
essential, please visit www.ngs.
org.uk for information & booking.

GIFLA HOUSE
Jim & Rosemary Bottoms.

THE MANOR HOUSE
David Pilcher.

WAYSIDE COTTAGE
Nigel Barrett.

Three beautifully landscaped gardens
with picturesque streams, bridges
and waterfalls over a natural river. At
the Manor House, colourful stream
side planting includes an abundance
of arum lilies and a sunken garden
with lily pond. The garden at Gifla
House has decorative bridges, mixed
borders and trees. A well stocked
pond with a fountain and waterfalls
and a variety of outbuildings nestled
within the old walled garden create a
tranquil scene at Wayside Cottage.

❺ NEW BEDFORD HEIGHTS

Brickhill Drive, Bedford,
MK41 7PH. Graham A. Pavey,
bedfordheights.co.uk/. *Park in
main carpark and not at Travelodge.*
Sat 5 June (10-2). Adm £5. Light
refreshments in The Graze
Bistro. Pre-booking essential,
please visit www.ngs.org.uk for
information & booking.
Originally built by Texas Instruments,

a Texan theme runs throughout
the building and the gardens. The
entrance simulates an arroyo or dry
river bed, and is filled with succulents
and cacti, plus a mixture of grasses
and herbaceous plants, many of
which are Texan natives. There are
also 3 courtyard gardens, where
less hardy plants thrive in the frost
free environment. Gardens designed
by Graham A Pavey. No wheelchair
access to two of the courtyards
but can be viewed from a viewing
platform.

❻ NEW CHURCH FARM

Church Road, Pulloxhill, Bedford,
MK45 5HD. Keith & Sue Miles.
*Yellow signs from village to Church
Road, towards parish church.* Sun
16 May, Sun 29 Aug (2-5). Adm £5,
chd free. Home-made teas. Pre-
booking essential, please visit
www.ngs.org.uk for information
& booking.
Mixed colourful planting to the rear
of the house including a long sunny
border of drought tolerant plants and
large topiary subjects. Beyond the
farmyard a small kitchen garden and
wild flower orchard incorporating a
formal rose walk and views to the
wider countryside. Access to the
wild flower orchard includes sloping
ground. Livestock may be present in
adjoining fields so sorry, no dogs.

❼ 22 ELMSDALE ROAD

Wootton, Bedford, MK40 0JN. Roy
& Dianne Richards, 07733 222495,
roy.richards60@ntlworld.com.
*Follow signs to Wootton, turn R at
The Cock PH, follow to Elmsdale
Rd.* Sat 5, Sun 6 June, Sat 3, Sun
4 July (12-5). Adm £5, chd free.
Home-made teas. Pre-booking
essential, please visit www.ngs.
org.uk for information & booking.
Visits also by arrangement Mar to
Oct for groups of 10 to 20.
Topiary garden greets visitors before
they enter a genuine Japanese Feng
Shui garden including bonsai. Large
collection of Japanese plants, Koi
pond, lily pond and a Japanese Tea
House. The garden was created from
scratch by the owners about 20 years
ago and has many interesting features
including Japanese lanterns and the
Kneeling Archer terracotta soldier
from China.

Orchard Grange

8 FLAXBOURNE FARM

Salford Road, Aspley Guise, MK17 8HZ. Paul Linden, www.flaxbournegardens.com. *From village centre take Salford Road, go over the railway line and 250 yards on L.* **Wed 18 Aug, Sun 5 Sept (12-5). Adm £5, chd free. Light refreshments. Pre-booking essential, please visit www.ngs.org.uk for information & booking.**

An entertaining and fun garden, lovingly developed with numerous water features, a windmill, romantic bridges, a small moated castle, lily pond, herbaceous borders and a Greek temple ruin. Children will enjoy exploring the intriguing nooks and crannies, discovering a grotto, crocodiles and a crow's nest. There is also a large Roman arched stone gateway.

9 GLADE HOUSE

Spinney Lane, Aspley Guise, Milton Keynes, MK17 8JT. Alex Ballance & Lindsay Walker. *From village square, turn S up Woburn Lane, L into Spinney Lane, R into Village Hall car park. Blue Badge parking ONLY outside individual gardens.* **Sat 3, Sun 4 July (1-5). Combined adm with Hollydale, Woburn Lane £5, chd free. Home-made teas in Village Hall. Pre-booking essential, please visit www.ngs.org.uk for information & booking.**

A colourful, multi-level garden of about half an acre, around a pretty thatched house. Divided into different areas by perennial and shrub borders and banks, it also has a sunny terrace, fish pond and woodland areas with shade loving plants. Over 400 varieties of perennials, shrubs and herbs, many of them unusual, plus several seating areas from where you can enjoy varied vistas within the garden.

GREYWALLS

See Northamptonshire

10 NEW HOLLINGTON FARM

Flitton Hill, Bedford, MK45 2BE. Mr John & Mrs Susan Rickatson. *Off A507 between Clophill & Ampthill- take Silsoe/ Flitton then bear R & follow yellow signs.* **Sat 19, Sun 20 June (12-5.30). Adm £6, chd free. Home-made teas. Pre-booking essential, please visit www.ngs.org.uk for information & booking.**

Two acre country garden. Semi formal near house with small parterre, pergola, pond and borders. Planting is massed perennials, shrubs and roses. Outer areas are wilder with mature trees. An option to walk through a wild flower farm meadow with views over Mid Beds. A good variety of plants for sale. Some steps and slopes.

11 HOLLYDALE, WOBURN LANE

Aspley Guise, MK17 8JH. Mrs Gill Cockle. *From village square, turn L to Woburn Lane.* **Sat 3, Sun 4 July (1-5). Combined adm with Glade House £5, chd free. Home-made teas in Village Hall. Pre-booking essential, please visit www.ngs.org.uk for information & booking.**

Set in ⅓ acre on three levels around Grade II Listed Georgian house, this classically styled garden is a series of rooms - flower gardens, leisure area for entertaining, vegetable garden, herb garden, fruit tree orchard and chicken pen. There are some steep steps but lots of seating around the garden. The large front garden with specimen shrubs and flower borders is wheelchair accessible. During your visit to our garden, you can try tasting some of our prize winning local honey. If you enjoy it, our honey is available to buy and for every jar sold, we donate £1 to the NGS charities.

12 HOWBURY HALL GARDEN

Howbury Hall Estate, Renhold, Bedford, MK41 0JB. Julian Polhill & Lucy Copeman, www.howburyfarmflowers.co.uk. *Leave A421 at A428/Gt Barford exit towards Bedford. Entrance to house & gardens ½m on R. Parking in field.* **Sun 12 Sept (2-5). Adm £5, chd free. Home-made teas. Pre-booking essential, please visit www.ngs.org.uk for information & booking.**

A late Victorian garden designed with mature trees, sweeping lawns and herbaceous borders. The large walled garden is a working garden, where one half is dedicated to growing a large variety of vegetables whilst the other is run as a cut flower business.

In the woodland area, walking towards the large pond, the outside of a disused ice house can be seen. Gravel paths and lawns may be difficult for smaller wheels.

♿ ☕

13 192 KIMBOLTON ROAD
Bedford, MK41 8DP. Tricia Atkinson. *On B660 between Brickhill Drive & Avon Drive nr pedestrian crossing.* Sat 12 June, Sat 11 Sept (2-5). Adm £5, chd free. Pre-booking essential, please visit www.ngs.org.uk for information & booking.
A third of an acre cottage garden specialising in many varieties of roses. There are also vegetable and soft fruit patches and an orchard. The garden includes over 90 roses. Wheelchair access with care. Some gravel.

♿ ❉ ☕

14 ◆ KING'S ARMS GARDEN
Brinsmade Road, Ampthill, Bedford, MK45 2PP. Ampthill Town Council. *Free parking in town centre. Entrance opp Old Market Place, down King's Arms Yard.* For NGS: Sun 28 Feb (2-4). Adm £4, chd free. Pre-booking essential, please visit www.ngs.org.uk for information & booking.
Small woodland garden of about 1½ acres created by plantsman, the late William Nourish. Trees, shrubs, bulbs and many interesting collections throughout the year. Since 1987, the garden has been maintained by 'The Friends of the Garden' on behalf of Ampthill Town Council. Mass plantings of snowdrops and early spring plants in February and beautiful autumn colours in October.

♿ 🐖 ❉ ☕

15 NEW LAKE END HOUSE
Mill Lane, Woburn Sands, Milton Keynes, MK17 8SP. Mr & Mrs G Barrett, 07831 110959, Geoffanddean@gmail.com. *Directions advised upon arrangement.* Visits by arrangement for groups of 20+. Groups are invited to bring their own picnics. Adm £5, chd free.
A large 3 acre lake created on a brown field site, provides sanctuary to a pair of black swans and large number of wild waterfowl. Beds adjacent to lake are planted with colourful perennials, shrubs and trees for year round interest. Larger garden sited above the house, with Japanese style ponds, bridges, tea house and

rock formations, is further enhanced by several cloud pruned trees and shrubs. Owners, Davina and Geoff are also the inspirational couple who created the Flaxbourne fun garden which opens for the NGS separately.

🐖 🚐

16 40 LEIGHTON STREET
Woburn, MK17 9PH. Ron & Rita Chidley, 01525 290802, ritachidley@gmail.com. *500yds from centre of village L side of road. Very limited on road parking outside house.* Sat 18 Sept (12-5). Adm £5, chd free. Home-made teas. Pre-booking essential, please visit www.ngs.org.uk for information & booking. Visits also by arrangement May to Aug for groups of 5 to 10.
Large cottage style garden in three parts. There are several 'rooms' with interesting features to explore and many quiet seating areas. There are perennials, climbers, vegetables, shrubs, trees and two ponds with fish. In late summer and autumn there is an abundance of colour from dahlias, fuchsias and grasses and late flowering roses. Many pots of unusual and tender plants.

❉ ☕

17 LINDY LEA
Ampthill Road, Steppingley, Bedford, MK45 1AB. Roy & Linda Collins. *Next to Steppingley Hospital. Park in hospital car park.* Sun 11 July (2-5). Adm £4, chd free. Light refreshments.
Set within an acre, this garden is a haven for wildlife with two water features, cottage garden style perennial plantings, variety of shrubs and mature trees. There is also a vegetable garden and a sunny terrace furnished with pots and climbers. Some paths may be difficult to negotiate.

♿ ❉ ☕

18 LUTON HOO HOTEL GOLF & SPA
The Mansion House, Luton Hoo, Luton, LU1 3TQ. Luton Hoo Hotel Golf & Spa, www.lutonhoo.co.uk. *Approx 1m from J10 M1, take London Rd A1081 signed Harpenden for approx ½m - entrance on L for Luton Hoo Hotel Golf & Spa.* Sun 25 July (12-5). Adm £5, chd free. Pre-booking essential, please visit www.ngs.org.uk for information & booking. Visitors wishing to

have lunch/formal afternoon tea at the hotel must book in advance directly with the hotel.
The gardens and parkland designed by Capability Brown are of national historic significance and lie in a conservation area. Main features - lakes, woodland and pleasure grounds, Victorian grass tennis court and late C19 sunken rockery. Italianate garden with herbaceous borders and topiary garden. Gravel and grass paths.

♿ 🐖 🚐 🚉 ☕

19 ◆ THE MANOR HOUSE, STEVINGTON
Church Road, Stevington, Bedford, MK43 7QB. Kathy Brown, www.kathybrownsgarden.com. *Off A428 through Bromham. If using Sat Nav please enter entire address, not just postcode.* For NGS: Mon 31 May (12-5). Adm £7, chd £3. Home-made teas. Pre-booking essential, please visit www.ngs. org.uk for information & booking. Refreshments in aid of Sue Ryder. For other opening times and information, please visit garden website.
The Manor House Garden has many different rooms, including six art inspired gardens. Early roses and wisteria along with several different Clematis montana will festoon the walls and pergolas; foxgloves, poppies and peonies providing colour down below. Elsewhere avenues of white stemmed birches, gingko, eucalypts, and metasequoia will lead the eye to further parts of the garden. 85% wheelchair access. Disabled WC.

♿ ☕

20 NEW 69 MILL LANE
Greenfield, Bedford, MK45 5DG. Pat Rishton. *On right hand side of Mill lane near the mill.* Sat 12, Sun 13 June (1-5). Combined adm with 70 Mill Lane £5, chd free. Pre-booking essential, please visit www.ngs.org.uk for information & booking.
Deceptively long back garden, recently re-loved from neglect with most of the planting less than three years old. Includes an old English Rose border with perennials, a white border in shade and a hot border in full sun. The waterlily pond leads to the fruit area then onto an arch with Rambling Rector screening a woodland area. All planting has to survive deer, rabbits and a granddaughter.

♿ ☕

21 NEW **70 MILL LANE**
Greenfield, Bedford, MK45 5DF.
Lesley Arthur. *At the end of Mill
Lane, on the left side.* Sat 12,
Sun 13 June (1-5). Combined
adm with 69 Mill Lane £5, chd
free. Pre-booking essential,
please visit www.ngs.org.uk for
information & booking.
This is a split level garden, near to
where the old Greenfield mill was
situated. The planting is cottage
style, with a small pond. The banks
of the River Flit, which flows through
the garden, are left to allow nettles
and comfrey to grow for the benefit
of wildlife. There are two fruit trees,
a small vegetable plot and a second
pond on the other side of the river.
✿ ☕

22 **THE OLD RECTORY,
WESTONING**
Church Road, Westoning,
MK45 5JW. Ann & Colin Davies.
*Off A5120, ¼ m up Church Rd,
next to church.* Sun 21 Mar, Sun
27 June (2-5.30). Adm £5, chd
free. Cream teas at C14 church
next door. Pre-booking essential,
please visit www.ngs.org.uk for
information & booking.
Ancient box and yew hedges
surround the colour co-ordinated
beds of this 2 acre garden. Spring is
greeted by hellebores and daffodils
with magnolias blooming in profusion.
A rose garden and herbaceous
borders complement the flowering
trees and shrubs of summer with
a stunning display of poppies and
cornflowers in the meadow. Come

and enjoy the sights and smells of a
traditional English garden. Wheelchair
access generally good.
& ✿ ☕

23 **THE OLD RECTORY,
WRESTLINGWORTH**
Church Lane, Wrestlingworth,
Sandy, SG19 2EU. Josephine Hoy.
*The Old Rectory is at the top of
Church Lane, which is well signed,
behind the church.* Sun 9, Sun
23 May (2-6). Adm £6, chd free.
Home-made teas.
4 acre garden full of colour and
interest. The owner has a free style
of gardening sensitive to wildlife.
Beds overflowing with tulips, alliums,
bearded iris, peonies, poppies,
geraniums and much more. Beautiful
mature trees and many more planted
in the last 30 years. Includes a large
selection of betulas. Gravel gardens,
box hedging and clipped balls,
woodland garden and wild flower
meadows. Wheelchair access may be
restricted to grass paths.
& 🐕 ✿ ☕

24 **ORCHARD GRANGE**
Old Warden, Biggleswade,
SG18 9HB. Robert & Victoria
Diggle. *In Old Warden village, NE of
entrance to Shuttleworth College.
Parking in village hall car park.* Sat
29 May (2.30-5.30). Adm £5, chd
free. Home-made teas. Pre-
booking essential, please visit
www.ngs.org.uk for information
& booking.
Orchard Grange is a late Georgian
rectory in the pretty estate village of
Old Warden. A yew-edged formal
garden with alliums and roses sits
next to a gravel garden with drought-
resistant planting designed to make
the most of local conditions. A small
conservatory houses a collection of
pelargoniums and succulents. There
are extensive lawns, a mulberry
avenue, croquet lawn and a huge
beech tree. Some parts of the garden
may not be wheelchair accessible.
& 🐕 ✿ ☕

PREACHERS PASSING
See Cambridgeshire

25 NEW **RADS END FARM**
Higher Rads End, Eversholt,
Milton Keynes, MK17 9ED. Carolyn
Howell. *If using satnav you may
be told you have 'reached your
destination' at a T junction, but you
haven't. Turn L there. L round S*

Hollydale

© Venetia Jane

Lake End House

bend around a brick wall is our gate. Park in field opposite gates. Sat 17, Sun 18 July (2-5). Adm £5, chd free. Home-made teas. Pre-booking essential, please visit www.ngs.org.uk for information & booking.
A country garden with lawns and herbaceous borders surrounding the house. There is also a veg and fruit garden, water garden, bog garden and rough areas. Pergolas and sitting areas with quiet spots. Specimen trees of mulberry and medlar. Some gravel paths.

26 ROYAL OAK COTTAGE
46 Wood End Road, Kempston Rural, Bedford, MK43 9BB. Teresa & Mike Clarke, teresaclarke@idnet.com. From Wood End Lane turn L at the Cross Keys, no 46 is on the L after 400yds. Sun 13 June (12-5); Sun 18 July (12-5.30). Adm £5, chd free. Home-made teas. Pre-booking essential, please visit www.ngs.org.uk for information & booking.
Visits also by arrangement May to Aug for groups of 10 to 30. Groups may request a gardening talk.
¾ acre garden created on the land of an old beer house, with wildlife in mind. There are old orchard trees, herbaceous beds, a small oriental garden and a vegetable plot. A blue and white border surrounds an old renovated piggery building. Agapanthus and blue hydrangeas are a patio speciality. Shallow ponds

support newts and other amphibians. The wild flower meadow is Austrian scythed in August.

27 4 ST ANDREWS ROAD
Bedford, MK40 2LJ. David Meghen, 07710 404323, davidmeghen@aol.com. Off Park Avenue. Sat 10 July (10-4). Adm £3.50, chd free. Pre-booking essential, please visit www.ngs.org.uk for information & booking.
Visits also by arrangement for groups of up to 10.
An elegant, multi-functional urban garden set within a semi-formal layout featuring clipped beech hedges softened by colourful shrubs and perennials and a rose pergola. Focal points have been created to be enjoyed from both within the garden as well from the house.

28 1A ST AUGUSTINE'S ROAD
Bedford, MK40 2NB. Chris Bamforth Damp, 01234 353730/01234 353465, chrisdamp@mac.com. St Augustine's Rd is on L off Kimbolton Rd as you leave the centre of Bedford. Sat 7 Aug (12-4.30). Adm £5, chd free. Home-made teas.
Visits also by arrangement July to Sept for groups of 10 to 30.
A colourful town garden with herbaceous borders, climbers, a greenhouse and pond. Planted in cottage garden style with traditional

flowers, the borders overflow with late summer annuals and perennials including salvias and rudbeckia. The pretty terrace next to the house is lined with ferns and hostas. The owners also make home-made chutneys and preserves which can be purchased on the day. The garden is wheelchair accessible.

29 SECRET GARDEN
4 George Street, Clapham, Bedford, MK41 6AZ. Graham Bolton, 07746 864247, bolton_graham@hotmail.com. Clapham Village High St. R into Mount Pleasant Rd then L into George St. 1st white Bungalow on R. Sat 15 May (12.30-5.30). Adm £3.50, chd free. Tea. Pre-booking essential, please visit www.ngs.org.uk for information & booking.
Visits also by arrangement Apr & May for groups of 10 to 20.
Profiled in the RHS The Garden, alpine lovers can see a wide collection of alpines in two small scree gardens, front and back of bungalow plus pans. Planting also incl dwarf salix, rhododendrons, daphnes, acers, conifers, pines, hellebores and epimediums. Two small borders of herbaceous salvias, lavenders and potentillas. Alpine greenhouse with rare varieties and cold frames with plants for sale.

30 SOUTHILL PARK

Southill, nr Biggleswade, SG18 9LL. Mr & Mrs Charles Whitbread. *In the village of Southill.* Sun 6 June (2-5). Adm £5, chd free. Cream teas. Refreshments in aid of another charity.

Southill Park first opened its gates to NGS visitors in 1927. A large garden with mature trees and flowering shrubs, herbaceous borders, a formal rose garden, sunken garden, ponds and kitchen garden. It is on the south side of the 1795 Palladian house. The parkland was designed by Lancelot 'Capability' Brown. A large conservatory houses the tropical collection and a new fernery was added in 2019.

STAPLOE GARDENS

See Cambridgeshire

GROUP OPENING

31 STEPPINGLEY VILLAGE GARDENS

Steppingley, Bedford, MK45 5AT. *Follow signs to Steppingley, pick up yellow signs from village centre.* Sun 25 Apr, Sun 30, Mon 31 May (2-5). Combined adm £6, chd free. Home-made teas at Townsend Farmhouse. Pre-booking essential, please visit www.ngs.org.uk for information & booking. Classic car show on Monday 31st May.

MIDDLE BARN

Bruce & Pauline Henninger. Open on Sun 30, Mon 31 May

NEW 37 RECTORY ROAD

Bill & Julie Neilson. Open on all dates

TOP BARN

Tim & Nicky Kemp. Open on Sun 30, Mon 31 May

TOWNSEND FARMHOUSE

Hugh & Indi Jackson. Open on all dates (See separate entry)

WEST OAK

John & Sally Eilbeck. Open on all dates

Steppingley is a picturesque Bedfordshire village on the Greensand ridge, close to Ampthill, Flitwick and Woburn. Although a few older buildings survive, most of Steppingley was built by the 7th Duke of Bedford between 1840 and 1872. Five gardens in the village offer an interesting mix of planting styles and design to include pretty courtyards, cottage garden style perennial

The Manor House, Barton Le Clay Gardens

© Venetia Jane

Townsend Farmhouse

borders, ponds, a Victorian well, glass houses, an orchard, vegetable gardens, a herb garden, wild life havens and country views. Livestock include chickens, ducks and fish.

32 127 STOKE ROAD

Linslade, Leighton Buzzard, LU7 2SR. Steve Owen, 07973 318383. *Opposite the turn for Rothschild Road.* Sat 9 Jan, Sun 28 Feb (10-4). Adm £5, chd free. Pre-booking essential, please visit www.ngs.org.uk for information & booking. **Visits also by arrangement in Jan for groups of up to 20. Donation to Plant Heritage.**

The garden houses the National Collection of Galanthus (snowdrop), with nearly 2,000 different varieties. Most are grown in the garden setting but some specialised ones can be seen on display racks. There is a Japanese themed garden and a raised alpine bed. Other plants grown include daphnes and a collection of over 100 maple trees. The alpine house contains a collection of Primula allionii.

NPC

33 TOWNSEND FARMHOUSE

Rectory Road, Steppingley, Bedford, MK45 5AT. Hugh & Indi Jackson. *Follow directions to Steppingley village and pick up yellow signs from village centre.* Evening opening Sat 30 Oct (5.30-8.30). Adm £10, chd £2. Entry includes light refreshment. Pre-booking essential, please visit www.ngs.org.uk for information & booking. Opening with Steppingley Village Gardens on Sun 25 Apr, Sun 30, Mon 31 May. Country garden with tree lined driveway and perennial borders. Pretty cobbled courtyard with a glass house and a Victorian well 30 metres deep, viewed through a glass top.'A Spectacular evening of lanterns, diya lamps and flower rangoli. Admission includes a light snack. Bring a torch to evening opening. Date may change if poor weather.

34 ◆ THE WALLED GARDEN

Luton, LU1 4LF. Luton Hoo Estate, www.lutonhooestate.co.uk. *From A1081 turn at West Hyde Road After 100 metres turn L through black gates, follow signs to Walled Garden.* For NGS: Fri 6 Aug (11-4). Adm £5, chd free. Light refreshments at Woodyard Coffee Shop. Pre-booking essential, please visit www.ngs.org.uk for information & booking. For other opening times and information, please visit garden website.

The 5 acre Luton Hoo Estate Walled Garden was designed by Capability Brown and established by the notorious Lord Bute in the late 1760s. The Walled Garden now offers a unique opportunity to see conservation and restoration combined with outstanding volunteer involvement in action. The garden continues to be restored, repaired and reimagined for the enjoyment of all. Volunteer garden and local history experts are on hand to explain and expand on what you see. An amazing cactus collection and original restored prop houses. Exhibition of Victorian tools. Walled Garden Shop with ever changing seasonal produce, including our outstanding Estate honey.

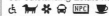

The National Garden Scheme searches the length and breadth of England and Wales for the very best private gardens

BERKSHIRE

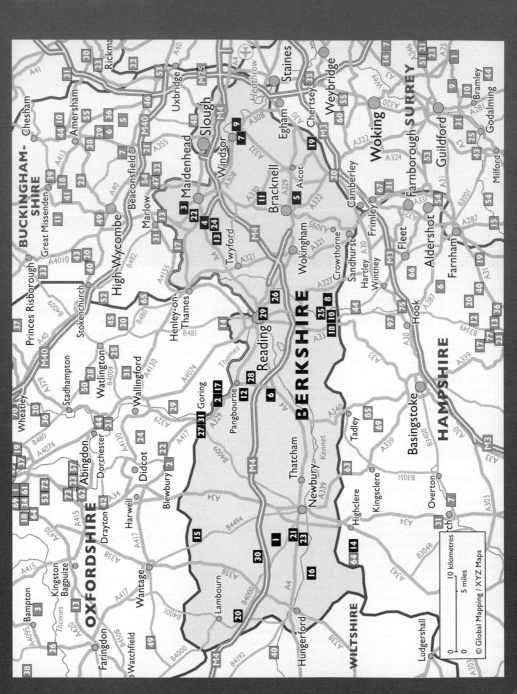

The Royal County of Berkshire offers a wonderful mix of natural beauty and historic landmarks that are reflected in the portfolio of gardens opening for the National Garden Scheme.

Our gardens come in all shapes, sizes and styles, from urban oases and community plots, to village gems and country house landscapes. Last year, many of our garden owners were very disappointed at having to close their gates due to Covid-19. They are all looking forward to welcoming you back in 2021 to share uplifting moments, conversations and cake in their beautiful gardens. Amongst them are the staff and volunteers at Eton College, and we hope you will come to help us celebrate their 60th anniversary opening on 12th June.

We are delighted to welcome 6 gardens opening for the first time this year. They range from community gardens at Jeallotts Hill near Maidenhead and Watlington House in Reading, to private gardens such as Lower Bowden Manor, 7 The Knapp, Thurle Grange and Wynders. All are generously opening to raise funds for the nursing and health charities we support.

In addition to our open days, 14 gardens offer opportunities for private visits 'By Arrangement' and you can contact the garden openers directly to organise them. Also, if you are involved in a club or society and would like to know more about the talks we offer (in person or on-line), please contact Angela at angela. oconnell@icloud.com.

So while some gardens may capture your interest due to their designers or their historic setting, most have evolved thanks to the efforts of their enthusiastic owners. We think they all offer moments of inspiration and look forward to welcoming you at a garden soon.

Volunteers

County Organiser
Heather Skinner
01189 737197
heather.skinner@ngs.org.uk

County Treasurer
Hugh Priestley
01189 744349 Fri – Mon
hughpriestley@aol.com

Booklet Co-ordinator
Heather Skinner
(as above)

Talks & Group Visits
Angela O'Connell
01252 668645
angela.oconnell@icloud.com

Assistant County Organisers
Claire Fletcher
claire.fletcher@ngs.org.uk

Carolyn Foster
01628 624635
candrfoster@btinternet.com

Angela O'Connell
(as above)

Graham O'Connell
01252 668645
graham.oconnell22@gmail.com

Rebecca Thomas
01491 628302
rebecca.thomas@ngs.org.uk

 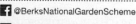 @BerksNationalGardenScheme
@BerksNGS

© Adam Jepson

Left: The Tithe Barn

OPENING DATES

All entries subject to change. For latest information check **www.ngs.org.uk**

Map locator numbers are shown to the right of each garden name.

February

Snowdrop Festival

Wednesday 3rd
◆ Welford Park   30

March

Saturday 6th
Stubbings House   24

Sunday 7th
Stubbings House   24

April

Sunday 18th
The Old Rectory,
    Farnborough   15

Wednesday 21st
Rooksnest   20

Sunday 25th
Rookwood Farm
    House   21

Wednesday 28th
Malverleys   14

May

Saturday 15th
Stubbings House   24
NEW Thurle Grange   27
NEW Wynders   31

Sunday 16th
The Old Rectory,
    Farnborough   15
Stubbings House   24
NEW Thurle Grange   27
NEW Wynders   31

Sunday 23rd
Church Farm House   2
The Old Rectory,
    Lower Basildon   17

Monday 31st
Rookwood Farm
    House   21

June

Sunday 6th
Stockcross House   23

Saturday 12th
Eton College Gardens   7

Sunday 13th
Stockcross House   23
The Tithe Barn   28

Sunday 20th
Deepwood Stud Farm   4

Wednesday 23rd
The Old Rectory,
    Farnborough   15
Rooksnest   20

Sunday 27th
Swallowfield Village
    Gardens   25

July

Sunday 4th
Farley Hill Place
    Gardens   8
St Timothee   22
NEW Watlington House,
    Geoff Hill Memorial
    Garden   29

Wednesday 7th
Lower Lovetts Farm   13

Thursday 15th
Malverleys   14

Saturday 17th
NEW Jealott's Hill
    Community
    Landshare   11

Sunday 18th
Deepwood Stud Farm   4

August

Sunday 29th
Stockcross House   23

Monday 30th
Rookwood Farm
    House   21

September

Sunday 5th
Stockcross House   23

By Arrangement

Arrange a personalised garden visit on a date to suit you. See individual garden entries for full details.

Boxford House   1
Compton Elms   3
Deepwood Stud Farm   4
Devonia   5
Farley Hill Place Gardens   8
Handpost   10
NEW Lower Bowden
    Manor   12
The Old Rectory Inkpen   16
The Priory   18
Priory House   19
Rooksnest   20
Rookwood Farm House   21
St Timothee   22
NEW 7 The Knapp   26

Rooksnest

THE GARDENS

1 BOXFORD HOUSE

Boxford, Newbury, RG20 8DP.
Tammy Darvell, Head
Gardener, 07802 883084,
tammydarvell@hotmail.com. *4m NW
of Newbury. Directions will be provided
on booking.* **Visits by arrangement
May to Sept for groups of 10 to 15.
Adm £10, chd free. Guided tour &
refreshments included.**
Beautiful large family garden
extensively developed over the past
7 yrs. Emphasis on roses and scent
throughout the 5 acre main garden. Old
and new orchards, laburnum tunnel,
formal and colourful herbaceous
borders. Handsome formal terraces,
pond, water features and garden
woodland areas. Inviting cottage
garden and productive vegetable
gardens. Partial wheelchair access,
gravel paths and sloping lawns.

2 CHURCH FARM HOUSE

Church Lane, Lower Basildon,
Reading, RG8 9NH. Meredith &
Max Green. *2m NW of Pangbourne
on A329. Into Lower Basildon,
200yds past petrol station, turn R
down lane to large barn & follow
signs to field parking.* **Sun 23 May
(2-5). Combined adm with The Old
Rectory, Lower Basildon £6, chd
free. Home-made teas at adjacent
St Bartholomew's Church.**
10 acres of organic garden with
formal and natural planting set in
Thames Valley, with lovely views
across Chiltern Hills. Formal lawn and
pond, herbaceous borders, vegetable
plots, fruit cage, large greenhouse,
wildflower meadow, young woodland,
paddock with small coppices and
layered hedges. Partial wheelchair
access over hard rolled chippings and
flat grassland, but some gravel and
uneven paths.

3 COMPTON ELMS

Marlow Road, Pinkneys Green,
Maidenhead, SL6 6NR. Alison
Kellett, kellettaj@gmail.com.
*Situated at the end of a gravel road
located opp & in between the Arbour
& Golden Ball pubs on the A308.*
**Visits by arrangement Mar & Apr
for groups of 10 to 30. Adm £7.50,
chd free. Light refreshments by
prior request.**
A delightful spring garden set
in a sunken woodland, lovingly
recovered from clay pit workings.
The atmospheric garden is filled with
snowdrops, primroses, hellebores and
fritillaria, interspersed with anemone
and narcissi under a canopy of ash
and beech.

4 DEEPWOOD STUD FARM

Henley Road, Stubbings, nr
Maidenhead, SL6 6QW. Mr &
Mrs E Goodwin, 01628 822684,
ed.goodwin@deepwood.co.
*2m W of Maidenhead. M4 J8/9
take A404M N. 2nd exit for A4 to
Maidenhead. L at 1st r'about on
A4130 Henley, approx 1m on R.* **Sun
20 June, Sun 18 July (2-5). Adm
£5, chd free. Home-made teas.
Visits also by arrangement Mar to
Oct for groups of 10+.**
4 acres of formal and informal
gardens within a stud farm, so great
roses! Small lake with Monet style
bridge and 3 further water features.
Several neo-classical follies and
statues. Walled garden with windows
cut in to admire the views and horses.
Woodland walk and enough hanging
baskets to decorate a pub! Partial
wheelchair access.

5 DEVONIA

Broad Lane, Bracknell,
RG12 9BH. Andrew Radgick,
aradgick@btinternet.com. *1m S of
Bracknell. From A322 Horse & Groom
r'about take exit into Broad Lane, over
2 r'abouts. 3rd house on L after railway
bridge.* **Visits by arrangement Aug &
Sept for groups of 5 to 20. Adm £5,
chd free.**
Inspirational urban garden, planted for
all season interest with many plants
that survive regardless of the weather,
both current and with climate change
in mind. Contains traditional borders,
prairie planting and areas with foliage
effects. With three different soil types
and many aspects, this ⅓ acre
garden offers lots of ideas. Particularly
interesting for gardening groups, with
optional guided tour.

6 ◆ ENGLEFIELD HOUSE GARDEN

Englefield, Theale, Reading,
RG7 5EN. Mr & Mrs Richard
Benyon, 01189 302221,
peter.carson@englefield.co.uk,
www.englefieldestate.co.uk. *6m W
of Reading. M4 J12. Take A4 towards
Theale. 2nd r'about take A340 to
Pangbourne. After ⅙m entrance
on the L.* **For opening times and
information, please phone, email or
visit garden website.**
The 12 acre garden descends
dramatically from the hill above the
historic house through woodland
where mature native trees mix with
rhododendrons and camellias. Drifts
of spring and summer planting are
followed by striking autumn colour.
Stone balustrades enclose the lower
terrace, with wide lawns, roses and
mixed borders. A stream linked by
a series of shallow pools meanders
through the woodland. Open every
Monday from Apr-Sept (10am-6pm)
and Oct-Mar (10am-4pm). Please
check the Englefield Estate website
for any changes before travelling.
Wheelchair access to some parts of
the garden.

7 ETON COLLEGE GARDENS

Eton, nr Windsor, SL4 6DB. Eton
College. *½m N of Windsor. Parking
signed off B3022, Slough Rd. Walk
from car park across playing fields
to entry. Follow signs for tickets &
maps which are sold at gazebo near
entrance.* **Sat 12 June (2-5). Adm
£6, chd free. Home-made teas in
the Fellows Garden.**
A rare chance to visit a group of
central College gardens surrounded
by historic school buildings, including
Luxmoore's garden on a small island
in the Thames reached across two
attractive bridges. Also an opportunity
to explore the fascinating Eton
College Natural History Museum,
the Museum of Eton Life and a small
group of other private gardens.
Plant Sale at Warre House. Sorry,
the gardens are not suitable for
wheelchairs due to gravel, steps and
uneven ground.

National Garden Scheme
gardens are identified by their
yellow road signs and posters.
You can expect a garden
of quality, character and
interest, a warm welcome and
plenty of home-made cakes!

8 FARLEY HILL PLACE GARDENS

Church Road, Farley Hill, Reading, RG7 1TZ. Tony & Margaret Finch, 01189 762544, tony.finch7@btinternet.com. *From M4 J11, take A33 S to Basingstoke. At T-lights turn L for Spencers Wood, B3349. Continue 2m, turn L through Swallowfield towards Farley Hill. Garden ½ m on R.* **Sun 4 July (2-5). Adm £5, chd free. Home-made teas. Visits also by arrangement May to July for groups of 10+.**
A 4 acre, C18 cottage garden. 1½ acre walled garden with yr-round interest and colour. Well stocked herbaceous borders, large productive vegetable areas with herb garden, dahlia and cutting flower beds. Enjoy wandering around the garden, with spontaneous singing from a Barber's Shop Quartet. Victorian glasshouse recently renovated and small nursery. Plants, lovely cut flowers and produce for sale. Partial wheelchair access.

SPECIAL EVENT

9 ◆ FROGMORE HOUSE & GARDEN

Windsor, SL4 1LU. Her Majesty The Queen. *1m SE of Windsor. Enter via Park Street gate into Long Walk to car parking. Visiting the garden involves 10-15 min walk from Long Walk entrance.* **Opening date and times to be confirmed, please visit www.ngs.org.uk for more information.**
The private royal garden at Frogmore House on the Crown Estate at Windsor. This beautiful landscaped garden, set in 35 acres with notable trees, lawns, flowering shrubs and C18 lake, is rich in history. It is largely the creation of Queen Charlotte, who in the 1790s introduced over 4,000 trees and shrubs to create a model picturesque landscape. The historic plantings, including tulip trees and redwoods, along with Queen Victoria's Tea House, remain key features of the garden today. The Royal Mausoleum is closed due to long term restoration. Tickets for the garden and also to visit the house may be available on the day, please check NGS website for updates. Wheelchair access is difficult due to gravel paths. Please note Windsor traffic may be halted around 11am for Guard change.

10 HANDPOST

Basingstoke Road, Swallowfield, Reading, RG7 1PU. Faith Ramsay, 07801 239937, faith@mycountrygarden.co.uk, www.mycountrygarden.co.uk. *From M4 J11, take A33 S. At 1st T-lights turn L on B3349 Basingstoke Rd. Follow road for 2¾ m, garden on L, opp Barge Lane.* **Visits by arrangement May to Sept for groups of 10 to 30. Adm £6, chd free. Home-made teas.**
4 acre designer's garden with many areas of interest. Features incl two lovely long herbaceous borders attractively and densely planted in six colour sections, a formal rose garden, an old orchard with a grass meadow, pretty pond and peaceful wooded area. Large variety of plants, trees and a productive fruit and vegetable patch. Some gravel areas, but largely accessible.

11 NEW JEALOTT'S HILL COMMUNITY LANDSHARE

Wellers Lane, at junction of Penfurzen Lane, Warfield, Bracknell, RG42 6BQ. Jealott's Hill Community Landshare, www.jealottshilllandshare.org.uk. *Enter Wellers Lane off the A330, 1st gateway on R at junction with Penfurzen Lane.* **Sat 17 July (1-4). Adm £4.50, chd free.**
Jealott's Hill Community Landshare is an inspirational 6 acre multi-purpose community garden, tended by the local community. The site offers various horticultural activities as well as creating a haven for wildlife, amongst a developed sensory garden, vineyard, a 450 tree orchard, wildlife pond and crop areas. Wheelchair access to most of the site from level car park, please phone 07867 695931 with any queries. Disabled WC.

12 NEW LOWER BOWDEN MANOR

Bowden Green, Pangbourne, RG8 8JL. Juliette & Robert Cox-Nicol, 07552 217872, robert.cox-nicol@orange.fr. *1½ m W of Pangbourne. Directions provided on booking.* **Visits by arrangement Apr to Sept for groups of 5 to 30. Adm £8, chd free. Home-made teas by prior request.**
A 3 acre designer's garden where structure predominates. Stunning views. Specimen trees show

contrasting bark and foliage. A marble 'Pan' plays to a pond with boulders and boulder-shaped evergreens. Versailles planters with standard topiaries line a rill. A stumpery leads to the orchard's carpet of daffodils and later a wave of white hydrangeas. Some gravel, but most areas accessible. Dogs welcome on leads.

SPECIAL EVENT

13 LOWER LOVETTS FARM

Knowl Hill Common, Knowl Hill, RG10 9YE. Richard Sandford. *5m W of Maidenhead. Turn S off A4 at Knowl Hill Church into Knowl Hill Common. Past pub & across common to T-junction. Turn L down dead end lane.* **Wed 7 July (10.30-4). Adm £12. Pre-booking essential, please visit www.ngs.org.uk/special-events for information & booking. There will be two time slots 10.30am-12.30pm & 2pm-4pm.**
Enjoy a 'Talk & Walk' in this fascinating large modern organic kitchen garden (60 x 30 metres) with a wildflower meadow and herbaceous border. Wide variety of vegetables and fruit grown for yr-round home consumption and nutritional value. Flowers grown for eating or herbal teas. Lots of interesting growing techniques and tips. Before you walk around the garden, join Richard Sandford for a talk to learn about this amazing garden and how his plant-based diet influences his growing methods. Sorry, no children, teas or WC. Visitors welcome to bring refreshments.

SPECIAL EVENT

14 MALVERLEYS

Fullers Lane, East End, Newbury, RG20 0AA. *A34 S of Newbury, exit signed for Highclere. Follow A343 for ½ m, turn R to Woolton Hill. Pass school & turn L to East End. After 1m R at village green, then after 100 metres, R onto Fullers Lane.* **Wed 28 Apr, Thur 15 July (10.30-3). Adm £15, chd free. Pre-booking essential, please visit www.ngs.org.uk/special-events for information & booking. Visits are by guided tours at 10.30am,1pm & 3pm with tea, coffee & cake included.**
10 acres of dynamic gardens which have been developed over the last 9 yrs to incl magnificent mixed borders and a series of contrasting yew

Lower Bowden Manor

hedged rooms, hosting flame borders, a cool garden, a pond garden and new stumpery. A vegetable garden with striking fruit cages sit within a walled garden, also encompassing a white garden. Meadows open out to views over the parkland. Due to steps and uneven paths, the garden is not suitable for wheelchairs.

---

15 THE OLD RECTORY, FARNBOROUGH

nr Wantage, OX12 8NX. Mr & Mrs Michael Todhunter, 01488 638298. *4m SE of Wantage. Take B4494 Wantage-Newbury road, after 4m turn E at sign for Farnborough. Approx 1m to village, Old Rectory on L.* **Sun 18 Apr, Sun 16 May (2-5.30); Wed 23 June (11-5). Adm £5, chd free. Home-made teas. Donation to Farnborough PCC.**
In a series of immaculately tended garden rooms, incl herbaceous borders, arboretum, secret garden, roses, vegetables and bog garden, there is an explosion of rare and interesting plants, beautifully combined for colour and texture.

Regional Finalist, The English Garden's The Nation's Favourite Gardens 2019. With stunning views across the countryside, it is the perfect setting for the 1749 rectory (not open), once home of John Betjeman, in memory of whom John Piper created a window in the local church. Access over some steep slopes and gravel paths.

---

16 THE OLD RECTORY INKPEN

Lower Green, Inkpen, RG17 9DS. Mrs C McKeon, 01488 668793, claremckeon@gmail.com. *4m SE of Hungerford. From centre of Kintbury at the Xrds, take Inkpen Rd. After ½m turn R, then go approx 3m (passing Crown & Garter Pub, then Inkpen Village Hall on L). Nr St Michaels Church, follow car park signs.* **Visits by arrangement May to Sept for groups of 10 to 20. Adm £10, chd free. Light refreshments included.**
On a gentle hillside with lovely countryside views, the Old Rectory offers a peaceful setting for this pretty 2 acre garden. Enjoy strolling through the formal and walled gardens, herbaceous

borders, pleached lime walk and wildflower meadow (some slopes).

---

17 THE OLD RECTORY, LOWER BASILDON

Church Lane, Lower Basildon, Reading, RG8 9NH. Charlie & Alison Laing. *2m NW of Pangbourne on A329. Into Lower Basildon, 200yds past petrol station, turn R down lane to church & follow signs to field parking.* **Sun 23 May (2-5). Combined adm with Church Farm House £6, chd free. Home-made teas at adjacent St Bartholomew's Church.**
Recently remodelled 2 acre garden based around mature trees (notably cedar, magnolia and mulberry) with a pond and crinkle crankle wall. New plantings incl fruit trees, white border, rose and peony beds, herbaceous long border and substantial kitchen garden. In AONB the garden also benefits from proximity to listed church (site of memorial to Jethro Tull) and Thames river walks. Wheelchair access over level site, but grass is uneven and some gravel paths.

18 THE PRIORY

Beech Hill, RG7 2BJ. Mr & Mrs C Carter, 07957 151534, ljcbryan@gmail.com. *5m S of Reading. M4 J11, A33 S to Basingstoke. At T-lights, L to Spencers Wood. After 1½m turn R for Beech Hill. After approx 1½m, L into Wood Lane, R down Priory Drive.* **Visits by arrangement May & June for groups of 10 to 30. Adm £10, chd free. Tea, coffee & cake included.**

Extensive gardens in grounds of former C12 French Priory (not open), rebuilt 1648. The mature gardens are in a very attractive setting beside the River Loddon. Large formal walled garden with espalier fruit trees, lawns, mixed and replanted herbaceous borders, vegetables and roses. Woodland, fine trees, lake and Italian style water garden. A lovely garden for group visits.

19 PRIORY HOUSE

Priory Road, Sunningdale, Ascot, SL5 9RQ. Mrs J Leigh, 07973 746979, rio4jen@gmail.com. *Approx 4½m SE of Ascot. Take turning opp Waitrose, Ridgemount Rd. Turn 1st L into Priory Rd. Priory House located at end of road. Parking for 15 cars.* **Visits by arrangement June to Aug for groups of 10 to 30. Adm £10. Tea & home-made cake included.**

3 acre garden designed in 1930s by Percy Cane with yr-round interest, colour and fragrance. Ornamental pond, large lawn with shrub borders, yew hedges and rare trees. Lavish borders, vegetable garden, rose garden, rhododendrons, azaleas, daphne, gunnera and Lysichiton americanus by stream. Mature camellias, magnolias, fine shrubs and conifers.

20 ROOKSNEST

Ermin Street, Lambourn Woodlands, RG17 7SB. 07787 085565, gardens@rooksnest.net. *2m S of Lambourn on B4000. From M4 J14, take A338 Wantage Rd, turn 1st L onto B4000 (Ermin St). Rooksnest signed after 3m.* **Wed 21 Apr, Wed 23 June (11-4). Adm £5, chd free. Light refreshments. Last entry 3.30pm. Visits also by arrangement Apr to June for groups of 20 to 30.**

Approx 10 acre, exceptionally fine traditional English garden. Rose garden (redesigned 2017), herbaceous garden, pond garden, herb garden, fruit, vegetable and cutting garden, and glasshouses. Many specimen trees and fine shrubs, orchard and terraces. Garden mostly designed by Arabella Lennox-Boyd. Most areas have step-free wheelchair access, although surface consists of gravel and mowed grass.

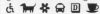

Wynders

21 ROOKWOOD FARM HOUSE

Stockcross, Newbury, RG20 8JX. The Hon Rupert & Charlotte Digby, 01488 608676, charlotte@rookwoodhouse.co.uk, www.rookwoodhouse.co.uk. *3m W of Newbury. M4 J13, A34(S). After 3m exit for A4(W) to Hungerford. At 2nd r'about take B4000 towards Stockcross, after approx ¾ m turn R into Rookwood.* **Sun 25 Apr, Mon 31 May, Mon 30 Aug (11-5). Adm £5, chd free. Home-made teas & light refreshments. Visits also by arrangement Apr to Sept for groups of 10+.**
This exciting valley garden, a work in progress, has elements all visitors can enjoy. A rose covered pergola, fabulous tulips, giant alliums, and a recently developed jungle garden with cannas, bananas and echiums. A kitchen garden features a parterre of raised beds, which along with a bog garden and colour themed herbaceous planting, all make Rookwood well worth a visit. WC available.

22 ST TIMOTHEE

Darlings Lane, Pinkneys Green, Maidenhead, SL6 6PA. Sarah & Sal Pajwani, 07976 892667, pajwanisarah@gmail.com. *1m N of Maidenhead. M4 J8/9 to A404M. 3rd exit onto A4 to Maidenhead. L at 1st r'about to A4130 Henley Rd. After ½ m turn R onto Pinkneys Drive. At Pinkneys Arms Pub, turn L into Lee Lane, follow NGS signs.* **Sun 4 July (11-4). Adm £5, chd free. Home-made teas. Visits also by arrangement May to Aug for groups of 10+.**
Colour themed borders planted for yr-round interest with a wide range of attractive grasses and perennials, all set within the 2 acre plot of a 1930s house.

23 STOCKCROSS HOUSE

Church Road, Stockcross, Newbury, RG20 8LP. Susan & Edward Vandyk. *3m W of Newbury. M4 J13, A34(S). After 3m exit A4(W) to Hungerford. At 2nd r'about take B4000, 1m to Stockcross, 2nd L into Church Rd.* **Sun 6, Sun 13 June, Sun 29 Aug, Sun 5 Sept (12-5). Adm £5, chd free. Home-made teas.**
Romantic 2 acre garden set around a former rectory (not open) designed to create vistas of the impressive church tower of St. John's. Deep herbaceous borders with emphasis on naturalistic planting and colour combinations. Large orangery, wisteria and clematis clad pergola. Naturalistic pond on lower level and small stumpery. Folly and reflecting pond, croquet lawn and long rose clad wall. Plants from garden for sale. Partial wheelchair access with some gravelled areas.

24 STUBBINGS HOUSE

Stubbings Lane, Henley Road, Maidenhead, SL6 6QL. Mr & Mrs D Good, www.stubbingsnursery.co.uk. *From A404(M) W of Maidenhead, exit at A4 r'about & follow signs to Maidenhead. At the small r'about turn L towards Stubbings. Take the next L onto Stubbings Lane.* **Sat 6, Sun 7 Mar, Sat 15, Sun 16 May (10-4). Adm £3.50, chd free. Light refreshments in onsite Café.**
Parkland garden accessed via adjacent retail nursery. Set around C18 Grade II listed house (not open), home to Queen Wilhelmina of Netherlands in WW2. Large lawn with ha-ha and woodland walks. Notable trees incl historic cedars and araucaria. March brings an abundance of daffodils and in May a 60 metre wall of wisteria. Attractions incl a C18 icehouse and access to adjacent NT woodland. Wheelchair access to a level site with firm gravel paths.

GROUP OPENING

25 SWALLOWFIELD VILLAGE GARDENS

The Street, Swallowfield, RG7 1QY. *5m S of Reading. From M4 J11 take A33 S. At 1st T-lights turn L on B3349 signed Swallowfield. In the village follow signs for parking. Purchase tickets & map in Swallowfield Medical Practice car park, opp Crown Pub.* **Sun 27 June (2-5.30). Combined adm £7, chd free. Home-made teas at Brambles.**

5 BEEHIVE COTTAGE
Ray Tormey.

BRAMBLES
Sarah & Martyn Dadds.

5 CURLYS WAY
Carolyn & Gary Clark.

THE FIRS
Harmi Kandohla & Mark Binns.

LAMBS FARMHOUSE
Eva Koskuba.

LODDON LOWER FARM
Talal & Nour Chamsi-Pasha.

NORKETT COTTAGE
Jenny Spencer.

PRIMROSE COTTAGE
Hilda Phillips.

WESSEX HOUSE
Val Payne.

This year Swallowfield is offering at least 9 gardens to visit. A number are in the village itself or just outside, others are nearby, so there is a mix of walking to some, with those in different directions needing a car or bicycle to reach them comfortably. Whilst each provides its own character and interest, they all nestle in countryside by the Whitewater, Blackwater and Loddon rivers with an abundance of wildlife and lovely views. The garden owners, many of whom are members of the local Horticultural Society, are always happy to chat and share their enthusiasm and experience. Plants for sale. Wheelchair access to many gardens, some have slopes and uneven ground.

26 NEW 7 THE KNAPP

Earley, Reading, RG6 7DD. Mrs Ann McKie, 07881 451708, annmckie@hotmail.com. *Directions & parking details will be provided on booking.* **Visits by arrangement May & June for groups of 5 to 20. Adm £7, chd free. Home-made teas included.**
A mature garden, surrounded by trees and divided into rooms with lawns, mixed borders, gravel area with box hedging, pond and an area with natural planting where beehives are situated. The garden provides interest most of the year, but the highlights are the roses, particularly rambling roses covering sheds, arches and climbing into the trees. Wheelchair access across an area of gravel.

Online booking is available for many of our gardens, at ngs.org.uk

27 NEW **THURLE GRANGE**
Rectory Road, Streatley, Reading,
RG8 9QH. David Juster. *From
Streatley, N on A329 Wallingford Rd.
Fork L onto A417 Wantage Rd, then
fork L again along Rectory Rd. Past
golf course, downhill & after stables
Thurle Grange is on R. Parking
beyond on the R.* **Sat 15, Sun 16
May (10-4). Combined adm with
Wynders £6, chd free. Home-
made teas at Wynders.**
Recently developed 1 acre garden
set around attractive country house
(c1900, not open) with lovely views
across peaceful valley. Centred upon
a splendid Catalpa tree, with a mix
of formal and informal successional
planting for yr-long interest. Features
incl rose filled parterre, wildflower
area, yew avenue and richly planted
herbaceous borders. Wheelchair
access over gravel.
♿ ☕

28 **THE TITHE BARN**
Tidmarsh, RG8 8ER. Frances
Wakefield. *1m S of Pangbourne, off
A340. In Tidmarsh, turn by side of
Greyhound Pub, over bridge, R into
Mill Corner field for car park. Short
walk over field to garden.*

Sun 13 June (2-5). Adm £4, chd
free. Home-made teas.
This is a delightful ¼ acre village
garden within high brick walls around
The Tithe Barn (not open) dating
from 1760. Formally laid out with
parterres of box and yew. There
are roses, hostas and lavender as
well as interesting vintage pots and
containers.
❄ ☕

29 NEW **WATLINGTON HOUSE,
GEOFF HILL MEMORIAL
GARDEN**
44 Watlington Street, Reading,
RG1 4RJ. Watlington House Trust,
www.watlingtonhouse.org.uk.
*Limited car parking on site via South
St. Public car parks at Oracle &
Queen's Rd, approx 10 mins walk.*
**Sun 4 July (10.30-5). Adm £3.50,
chd free. Light refreshments.**
An attractive walled garden at
Watlington House, a Grade II *
property, recreated since 2012 using
archival research on site of previous
car park. Clipped panel-pleached
hornbeam, box hedging and pruned
fruit trees give formal structure within
a quadrangle design. Featuring
a shade walk, knot garden, and

colourful herbaceous planting, it
is a calm oasis in a bustling town.
Beautiful flint and brick walled
surround. House dating back to
Medieval times with Jacobean and
Georgian facades. Wheelchair access
via gravel topped paths, but no
access to buildings and WC.
❄ ☕

30 ♦ **WELFORD PARK**
Welford, Newbury, RG20 8HU.
Mrs J H Puxley, 01488 608691,
snowdrops@welfordpark.co.uk,
www.welfordpark.co.uk. *6m NW
of Newbury. M4 J13, A34(S). After
3m exit for A4(W) to Hungerford. At
2nd r'about take B4000, after 4m
turn R signed Welford. Entrance on
Newbury-Lambourn road.* **For NGS:
Wed 3 Feb (11-4). Adm £8, chd
£4. Light refreshments. For other
opening times and information,
please phone, email or visit garden
website.**
One of the finest natural snowdrop
woodlands in the country, approx 4
acres, along with a wonderful display
of hellebores throughout the garden
and winter flowering shrubs. This is
an NGS 1927 pioneer garden on the
River Lambourn set around Queen
Anne House (not open). Also the
stunning setting for Great British Bake
Off 2014 - 2019. Dogs welcome on
leads. Coach parties please book in
advance.
♿ 🐕 🚗 ☕

31 NEW **WYNDERS**
Rectory Road, Streatley, Reading,
RG8 9QA. Marcus & Emma
Francis. *From Streatley, N on A329
Wallingford Rd. Fork L onto A417
Wantage Rd, then fork L again along
Rectory Rd. After the golf course,
down hill & after stables, Wynders is
on L & parking to the R.* **Sat 15, Sun
16 May (10-4). Combined adm
with Thurle Grange £6, chd free.
Home-made teas.**
Magnificent views combined with
generous and varied planting
across ¾ acre surrounded by open
countryside. Grass borders, formal
garden, orchard, wildflower meadow,
rose gardens, shrub borders, ferns
and vegetable plots. Classic cars
on display for non-gardeners and a
playground for children. Something
for everyone. Partial wheelchair
access over gravel drive, please call
07920 712571 in advance for special
parking.
♿ ❄ ☕

Thurle Grange

The Old Rectory, Lower Basildon

BUCKINGHAMSHIRE

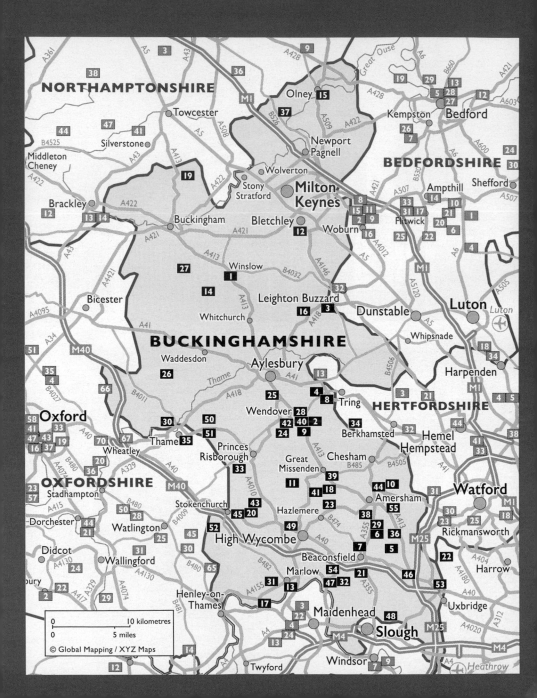

NORTHAMPTONSHIRE

BEDFORDSHIRE

BUCKINGHAMSHIRE

OXFORDSHIRE

HERTFORDSHIRE

Oxford

Bicester

Brackley

Buckingham

Bletchley

Milton Keynes

Stony Stratford

Wolverton

Newport Pagnell

Olney

Towcester

Silverstone

Middleton Cheney

Winslow

Whitchurch

Leighton Buzzard

Dunstable

Luton

Whipsnade

Harpenden

Hemel Hempstead

Watford

Rickmansworth

Harrow

Uxbridge

Slough

Windsor

Maidenhead

Marlow

High Wycombe

Beaconsfield

Hazlemere

Amersham

Chesham

Great Missenden

Berkhamsted

Tring

Aylesbury

Wendover

Princes Risborough

Thame

Wheatley

Stokenchurch

Watlington

Dorchester

Stadhampton

Didcot

Wallingford

Henley-on-Thames

Twyford

Waddesdon

Kempston

Bedford

Ampthill

Shefford

Flitwick

Woburn

Brackley

0   10 kilometres
0   5 miles

© Global Mapping / XYZ Maps

Buckinghamshire has a beautiful and varied landscape; edged by the River Thames to the south, crossed by the Chiltern Hills, and with the Vale of Aylesbury stretching to the north.

This year Buckinghamshire will hold six group openings, many of which can be found in villages of thatched or brick and flint cottages.

Many Buckinghamshire gardens have been used as locations for films and television, with the Pinewood Studios nearby and excellent proximity to London.

We also boast historical gardens including Ascott, Cowper and Newton Museum Gardens, Hall Barn, and Stoke Poges Memorial Gardens (Grade I listed).

Most of our gardens offer homemade tea and cakes to round off a lovely afternoon, visitors can leave knowing they have enjoyed a wonderful visit and helped raise money for nursing and health charities at the same time.

Volunteers

County Organiser
Maggie Bateson
01494 866265
maggiebateson@gmail.com

County Treasurer
Tim Hart
01494 837328
timgc.hart@btinternet.com

Publicity
Sandra Wetherall
01494 862264
sandracwetherall@gmail.com

Talks
Janice Cross
01494 728291
janice.cross@ngs.org.uk

Booklet Co-ordinator
Maggie Bateson
(as above)

Assistant County Organisers
Janice Cross
(as above)

Judy Hart
01494 837328
judy.elgood@gmail.com

Mhairi Sharpley
01494 782870
mhairisharpley@btinternet.com

Stella Vaines
07711 420621
stella@bakersclose.com

📘 @BucksNGS
🐦 @BucksNGS
📷 @national_garden_scheme_bucks

© Ellen Rooney

Left: Little Missenden Gardens

OPENING DATES

All entries subject to change. For latest information check www.ngs.org.uk

Extended openings are shown at the beginning of the month.

Map locator numbers are shown to the right of each garden name.

March

Wednesday 24th
Montana 34

April

Sunday 4th
Overstroud Cottage 39

Sunday 11th
Long Crendon
　Gardens 30

Wednesday 21st
Montana 34

Sunday 25th
Aston Clinton Gardens 4
Orchard House 38

May

Sunday 2nd
Lindengate 28
Overstroud Cottage 39

Monday 3rd
Turn End 50

Thursday 6th
Bowers Farm 6

Tuesday 11th
Red Kites 43

Wednesday 19th
Montana 34

Sunday 23rd
Higher Denham
　Gardens 22

Sunday 30th
The White House 53

Monday 31st
◆ Ascott 3
The Claydons 14
Glebe Farm 19
The Plough 42

June

**Every Sunday from
Sunday 13th**
11 The Paddocks 40

Friday 4th
The Walled Garden 52

Sunday 6th
Cublington Gardens 16
Overstroud Cottage 39

Saturday 12th
◆ Cowper & Newton
　Museum Gardens 15
Old Park Barn 37
Woodside 55

Sunday 13th
126 Church Green
　Road 12
◆ Cowper & Newton
　Museum Gardens 15
Long Crendon
　Gardens 30
The Manor House 33
Old Park Barn 37
◆ Stoke Poges
　Memorial Gardens 48
Woodside 55

Tuesday 15th
126 Church Green
　Road 12

Saturday 19th
Acer Corner 2

Sunday 20th
Acer Corner 2
Aston Clinton Gardens 4
18 Brownswood Road 7
Canal Cottage 8

Wednesday 23rd
Montana 34

Thursday 24th
Lords Wood 31

Friday 25th
11 The Paddocks 40

Saturday 26th
St Michaels Convent 46

Sunday 27th
Chiltern Forage Farm 11
Overstroud Cottage 39
Tythrop Park 51

July

Saturday 3rd
Little Missenden
　Gardens 29

Sunday 4th
Little Missenden
　Gardens 29
11 The Paddocks 40
Robin Hill 45

Saturday 10th
Old Park Barn 37

Sunday 11th
Fressingwood 18
Old Park Barn 37

Tuesday 13th
Red Kites 43

Saturday 17th
8 Claremont Road 13

Sunday 18th
8 Claremont Road 13
Higher Denham
　Gardens 22

Wednesday 21st
Montana 34

Saturday 24th
8 Claremont Road 13

Sunday 25th
8 Claremont Road 13

August

Wednesday 4th
Danesfield House 17

Sunday 8th
NEW Howe Farm
　Flowers 26

Sunday 29th
Horatio's Garden 25
◆ Nether Winchendon
　House 35

Monday 30th
◆ Ascott 3

September

Sunday 5th
Overstroud Cottage 39

Sunday 12th
Lindengate 28

Saturday 25th
Acer Corner 2

Sunday 26th
Acer Corner 2

By Arrangement

Arrange a personalised garden visit on a date to suit you. See individual garden entries for full details.

Abbots House 1
Acer Corner 2
Beech House 5
Cedar House 9
Chesham Bois House 10
Glebe Farm 19
NEW 11a Green Lane 20
Hall Barn 21
Hollytrees 23
Homelands 24
Kingsbridge Farm 27
Lindengate 28
Magnolia House 32
Montana 34
North Down 36
Old Park Barn 37
Overstroud Cottage 39
11 The Paddocks 40
Peterley Corner
　Cottage 41
Red Kites 43
Rivendell 44
Robin Hill 45
The Shades 47
NEW 1 Talbot Avenue 49
20 Whitepit Lane 54
Wind in the Willows,
　Higher Denham
　Gardens 22

Online booking is available for many of our gardens, at ngs.org.uk

THE GARDENS

1 ABBOTS HOUSE

10 Church Street, Winslow, MK18 3AN. Mrs Jane Rennie, 01296 712326, jane@renniemail.com. *9m N of Aylesbury. A413 into Winslow. From town centre take Horn St & R into Church St, L fork at top. Entrance 20 metres on L, in Church St, not Church Walk. Parking in town centre & adjacent streets.* **Visits by arrangement Apr to Aug for groups of 5 to 20. Adm £4, chd free. Light refreshments.**
Behind red brick walls a ³/₄ acre garden on four different levels, each with unique planting and atmosphere. Lower lawn with white wisteria arbour and pond, upper lawn with rose pergola and woodland, pool area with grasses, Victorian kitchen garden and wild meadow. Spring bulbs in wild areas and woodland is a major feature in April, remaining areas peak in June/July. Late spring bulbs, water feature and many pots. Experimental wild areas. Some sculptures. Please visit www.ngs.org.uk for pop up openings. Partial wheelchair access, garden levels accessed by steps. Guide dogs and medical-aid dogs only.

2 ACER CORNER

10 Manor Road, Wendover, HP22 6HQ. Jo Naiman, 07958 319234, jo@acercorner.com, www.acercorner.com. *3m S of Aylesbury. Follow A413 into Wendover. L at clock tower r'about into Aylesbury Rd. R at next r'about into Wharf Rd, continue past schools on L, garden on R.* **Sat 19, Sun 20 June, Sat 25, Sun 26 Sept (2-5). Adm £3.50, chd free. Home-made teas. Visits also by arrangement May to Oct for groups of up to 20.**
Garden designer's garden with Japanese influence and large collection of Japanese maples. The enclosed front garden is Japanese in style. Back garden is divided into three areas; patio area recently redesigned in the Japanese style; densely planted area with many acers and roses, also a corner which incl a productive greenhouse and interesting planting.

3 ◆ ASCOTT

Ascott, Wing, Leighton Buzzard, LU7 0PP. The National Trust, 01296 688242, info@ascottestate.co.uk, www.ascottestate.co.uk. *2m SW of Leighton Buzzard, 8m NE of Aylesbury. Via A418. Buses: 150 Aylesbury - Milton Keynes, 100 Aylesbury & Milton Keynes.* **For NGS: Mon 31 May, Mon 30 Aug (1-5.30). Adm £6, chd £3. Light refreshments. (NT members are required to pay to enter the gardens on NGS days.)** For other opening times and information, please phone, email or visit garden website.
Combining Victorian formality with early C20 natural style and recent plantings to lead it into the C21, with a completed garden designed by Jacques and Peter Wirtz, and also a Richard Long sculpture. Terraced lawns with specimen and ornamental trees, panoramic views to the Chilterns. Naturalised bulbs, mirror image herbaceous borders, and impressive topiary incl box and yew sundial. Ascott House is closed on NGS Days. Please contact the Estate Office to reserve a wheelchair.

GROUP OPENING

4 ASTON CLINTON GARDENS

Green End Street, Aston Clinton, Aylesbury, HP22 5JE. *3m E of Aylesbury. From Aylesbury take A41 E. At large r'about, continue straight (signed Aston Clinton). Continue onto London Rd. L at The Bell Pub, parking on Green End St & side roads.* **Sun 25 Apr, Sun 20 June (2-5). Combined adm £4, chd free. Home-made teas at The Lantern Cottage. Also open Canal Cottage (20 June only).**

101 GREEN END STREET
Sue Lipscomb.

THE LANTERN COTTAGE
Jacki Connell.

These two lovely cottage gardens, one well established and the other having recently undergone a radical redesign by its new owner, share a basis of seasonal interest underpinned by evergreens and perennial planting. At The Lantern Cottage, spring hellebores and an abundance of tulips give way to an early summer display of roses, peonies, bearded iris,

alliums and climbers including various clematis, wisteria and akebia. A wide selection of salvias and herbaceous perennials, mostly raised from seed and cuttings. Pelargoniums provide yr-round colour in the conservatory and the greenhouse is always full! At 101, the new owner took up residence in autumn 2015, quickly establishing raised beds for vegetable production, a number of fruit trees and a variety of soft fruits. There is also a wildlife pond, an arbour overlooking the Victorian greenhouse, plus herbaceous borders, ornamental grasses surrounding a red kite sculpture and varied container planting. Wildlife is encouraged to visit.

5 BEECH HOUSE

Long Wood Drive, Jordans, Beaconsfield, HP9 2SS. Sue & Ray Edwards, 01494 875580, raychessmad@hotmail.com. *From A40, L to Seer Green & Jordans for approx 1m, turn into Jordans Way on R, Long Wood Drive 1st L. From A413, turn into Chalfont St Giles, straight ahead until L signed Jordans, 1st L Jordans Way.* **Visits by arrangement Mar to Sept. Adm £4, chd free.**
2 acre plantsman's garden built up over the last 33 yrs with a wide range of plants in a variety of habitats to provide enjoyment all yr. Many bulbs, perennials, shrubs, roses and grasses provide continuous show. Trees planted for their foliage, ornamental bark and autumn display. Two flowering meadows are always a popular feature. Wheelchair access dependent upon weather conditions.

6 BOWERS FARM

Magpie Lane, Coleshill, HP7 0LU. John & Linda Daly. *2m S of Amersham & 3m N of Beaconsfield. Magpie Lane is off A355, by Harte & Magpies Pub.* **Thur 6 May (12.30-5). Adm £5, chd free. Light refreshments.**
An established 5 acre garden that has undergone significant renovation over the past 5-6 yrs with trees, perennials, annuals and bulbs; creating a variety of planting spaces including herbaceous beds, ponds, kitchen and cutting garden, new bulb meadow, small woodland and open parkland areas.

7 18 BROWNSWOOD ROAD

Beaconsfield, HP9 2NU. John & Bernadette Thompson. *From New Town turn R into Ledborough Lane, L into Sandleswood Rd, 2nd R into Brownswood Rd.* **Sun 20 June (2-5). Adm £4, chd free.**
A plant filled garden designed by Barbara Hunt. A harmonious arrangement of arcs and circles introduces a rhythm that leads through the garden. Sweeping box curves, gravel beds, brick edging and lush planting. A restrained use of purples and reds dazzle against a grey and green background. There has been considerable replanning and replanting during the winter.

D

8 CANAL COTTAGE

11 Wharf Row, Buckland Road, Buckland, Aylesbury, HP22 5LJ. Angela Hale. *4m E of Aylesbury. On the A41 exit at Aston Clinton/ Wendover. Down Lower Ikneild Way onto Buckland Rd. Park along road & walk over white bridge. Garden is signed.* **Sun 20 June (12-5). Adm £3, chd free. Home-made teas at The Lantern Cottage. Also open Aston Clinton Gardens.**
A peaceful garden celebrates what can be achieved in a long narrow strip behind a cottage (not open). Recently landscaped to create rooms and feeding places for wildlife. Open deep borders of perennials welcome the visitor. Shaped lawns lead you through shady rose arbour undergrown with a variety of ferns. Final room is a pond, seating area with Mediterranean terracing and many pots.

9 CEDAR HOUSE

Bacombe Lane, Wendover, HP22 6EQ. Sarah Nicholson, 01296 622131, sarahhnicholson@btinternet.com. *5m SE Aylesbury. From Gt Missenden take A413 into Wendover. Take 1st L before row of cottages, house at top of lane. Parking for no more than 10 cars.* **Visits by arrangement Feb to Sept for groups of 10 to 20. Adm £5, chd free. Home-made teas.**
A plantsman's garden in the Chiltern Hills with a great variety of trees, shrubs and plants. A sloping lawn leads to a natural swimming pond with wildflowers including native orchids. A lodge greenhouse and a good collection of half-hardy plants

in pots. Local artist's sculptures can be viewed. Picnics welcome on prior request. Wheelchair access over gentle sloping lawn.

& ✿ ☕ 🏕

10 CHESHAM BOIS HOUSE

85 Bois Lane, Chesham Bois, Amersham, HP6 6DF. Julia Plaistowe, 01494 726476, plaistowejulia@gmail.com, cheshamboishouse.co.uk. *1m N of Amersham-on-the-Hill. Follow Sycamore Rd (main shopping centre road of Amersham), which becomes Bois Lane. Do not use SatNav once in lane as you will be led astray.* **Visits by arrangement Mar to Aug. Adm £4.50, chd free.**
3 acre beautiful garden with primroses, daffodils and hellebores in early spring. Interesting for most of the year with lovely herbaceous borders, rill with small ornamental canal, walled garden, old orchard with wildlife pond, and handsome trees of which some are topiaried. It is a peaceful oasis. Wheelchair access with gravel in front of the house.

& ✿ 🚐 ☕

11 CHILTERN FORAGE FARM

Speen, Princes Risborough, HP27 0SU. Emma Plunket, www.plunketgardens.com. *Directions will be provided in email confirmation when ticket has been purchased. Parking in field, suitable footwear is recommended for uneven ground.* **Sun 27 June. Adm £5, chd free. Pre-booking essential, please email emma@ plunketgardens.com for information & booking. Light refreshments.**
Pre-booked tour at 2.30pm and 4.00pm for limited numbers only, by owners of this new project under development in stunning AONB setting. 8 acres of pasture being restored to native hay meadows and planted with fruit trees, soft fruit and perennial vegetables. Creation of wildlife habitats with native planting, dead hedges and green manure.

D ☕

12 126 CHURCH GREEN ROAD

Bletchley, Milton Keynes, MK3 6DD. David & Janice Hale. *13m E of Buckingham, 11m N of Leighton Buzzard. Turn R off B4034 into Church Green Rd, take L turn at mini-r'about.* **Sun 13 June (2-6); Tue 15 June (2-5.30). Adm £4, chd**

free. Home-made teas.
A gentle sloping mature garden of ½ acre is a plant lover's delight, which incl a small formal garden, shady areas and mixed borders of shrubs, perennials and roses. Features incl a thatched wendy house, pergola, formal pond, wildlife pond, productive fruit and vegetable garden, two greenhouses and patio.

🐕 ✿ ☕ 🏕

13 8 CLAREMONT ROAD

Marlow, SL7 1BW. Andi Gallagher. *No parking at garden, but short walk from all town car parks. From town centre walk S on High St, L on Institute Rd, L on Beaufort Gardens & R on Claremont Rd.* **Sat 17, Sun 18, Sat 24, Sun 25 July (2-5). Adm £4, chd free.**
Paintings, prints and pots to see and buy in this small town garden owned by an artist gardener. The unusual house was built in 2015. Gravel paths divide the rectangular beds filled with herbaceous perennials, grasses and ferns. A cow trough water feature and owner's ceramics add surprise. A gate leads to a deliberate wild area with fruit trees and art studio with garden related paintings.

GROUP OPENING

14 THE CLAYDONS

East Botolph and Middle Claydon, MK18 2ND. *1½ m SW Winslow. In Winslow turn R off High St, by the Bell Pub & follow NT signs towards Claydon House & The Claydons.* **Mon 31 May (2-6). Combined adm £5, chd free. Home-made teas in village hall.**

CLAYDON COTTAGE
Mr & Mrs Tony Evans.

THE OLD RECTORY
Mrs Jane Meisl.

THE OLD VICARAGE
Nigel & Esther Turnbull.

Three small villages, originally part of the Claydon Estate with typical north Buckinghamshire cottages and two C13 churches. Claydon Cottage, a pretty thatched cottage (not open) with an unconventional garden. The Old Vicarage, a large garden on clay with mixed borders, scented garden, dell, shrub roses, vegetables and a natural clay pond. Small meadow area and planting to encourage wildlife, beehives. A free children's quiz. The

Old Rectory is a large garden with a wildflower meadow, herbaceous borders, a woodland walk and cloud hedging. Wheelchair access via gravel drive at The Old Vicarage.

15 ✦ COWPER & NEWTON MUSEUM GARDENS

Orchard Side, Market Place, Olney, MK46 4AJ. Anne Kempson, 01234 711833, house-manager@cowperandnewtonmuseum.org.uk, www.cowperandnewtonmuseum.org.uk. *5m N of Newport Pagnell. 12m S of Wellingborough. On A509. Please park in public car park in East St.* **For NGS: Sat 12, Sun 13 June (10.30-4.30). Adm £3.50, chd free. Home-made teas. For other opening times and information, please phone, email or visit garden website.**

The tranquil Flower Garden of C18 poet William Cowper, who said 'Gardening was of all employments, that in which I succeeded best', has plants introduced prior to his death in 1800, many mentioned in his writings. The Summer House Garden with Cowper's 'verse manufactory', now a Victorian Kitchen Garden, has new and heritage vegetables organically grown, also a herb border and medicinal plant bed. Features incl lacemaking demonstrations and local artists painting live art on both days. Georgian dancers on Sun. Wheelchair access on mostly hard paths.

GROUP OPENING

16 CUBLINGTON GARDENS

Cublington, Leighton Buzzard, LU7 0LF. *5m SE Winslow, 5m NE Aylesbury. From Aylesbury take A413 Buckingham Rd. After 4m, at Whitchurch, turn R to Cublington.* **Sun 6 June (2-6). Combined adm £5, chd free. Home-made teas in Biggs Pavilion, Orchard Ground.**

CHERRY COTTAGE, 3 THE WALLED GARDEN
Gwyneira Waters.

LARKSPUR HOUSE
Mr & Mrs S Jenkins.

OLD MANOR COTTAGE
Mr & Mrs J Packer.

1 STEWKLEY ROAD
Tom & Helen Gadsby, www.structuredgrowth.co.uk.

A group of diverse gardens in this attractive Buckinghamshire village listed as a conservation area. Cherry Cottage is adapted for wheelchair gardening with raised beds and artificial grass. Larkspur House garden uses a variety of plants and hard landscaping to create distinct areas. Through an Art Nouveau inspired gate there is a large orchard and wildflower meadow. 1 Stewkley Road has a strong focus on home-grown food with an idyllic organic kitchen garden, small orchard and courtyard garden. Old Manor Cottage, a cottage garden with low hedges enclosing secluded seating areas. Partial wheelchair access to some gardens.

17 DANESFIELD HOUSE

Henley Road, Marlow, SL7 2EY. Danesfield House Hotel, 01628 891010, amoorin@danesfieldhouse.co.uk, www.danesfieldhouse.co.uk. *3m from Marlow. On the A4155 between Marlow & Henley-on-Thames. Signed on the LH side Danesfield House Hotel & Spa.* **Wed 4 Aug (10-4). Adm £4.50, chd free. Pre-booking essential for lunch & afternoon tea.**

The gardens at Danesfield were completed in 1901 by Robert Hudson, the Sunlight Soap magnate who built the house. Since the house opened as a hotel in 1991, the gardens have been admired by several thousand guests each yr. However, in 2009 it was discovered that the gardens contained outstanding examples of pulhamite in both the formal gardens and the waterfall areas. The 100 yr old topiary is also outstanding. Part of the grounds incl an Iron Age fort. Wheelchair access on gravel paths throughout garden.

98 High Street, Long Crendon Gardens

18 FRESSINGWOOD

Hare Lane, Little Kingshill, Great Missenden, HP16 0EF. John & Maggie Bateson. *1m S of Gt Missenden, 4m W of Amersham. From the A413 at Chiltern Hospital, turn L signed Gt & Lt Kingshill. Take 1st L into Nags Head Lane. Turn R under railway bridge, then L into New Rd & continue to Hare Lane.* **Sun 11 July (2-5.30). Adm £4.50, chd free. Home-made teas.**
Thoughtfully designed and structured garden with yr-round colour and many interesting features. Including herbaceous borders, a shrubbery with ferns, hostas, grasses and hellebores. Small formal garden, pergolas with roses and clematis. A variety of topiary and a landscaped terrace. A central feature area incorporating water with grasses. Large bonsai collection.

19 GLEBE FARM

Lillingstone Lovell, Buckingham, MK18 5BB. Mr David Hilliard, 01280 860384, thehilliards@talk21.com, www.glebefarmbarn.co.uk. *Off A413, 5m N of Buckingham & 2m S of Whittlebury. From A5 at Potterspury, turn off A5 & follow signs to Lillingstone Lovell.* **Mon 31 May (1.30-5). Adm £4, chd free. Home-made teas. Visits also by arrangement in June for groups of 5 to 10.**
A large cottage garden with an exuberance of colourful planting and winding gravel paths, amongst lawns and herbaceous borders on two levels. Ponds, a wishing well, vegetable beds, a knot garden, a small walled garden and an old tractor feature. Everything combines to make a beautiful garden full of surprises.

20 NEW 11A GREEN LANE

Radnage, High Wycombe, HP14 4DJ. Ms Jo Dudley, 07710 484434, jo5456@hotmail.com. *Turn off A40 Stokenchurch to Radnage & follow signs to village hall. Park at village hall (use postcode HP14 4DF). Walk through the gap in the hedge to 11a Green Lane opp.* **Visits by arrangement Mar to Sept for groups of 5 to 30. Adm £4, chd free. Light refreshments.**
Yr-round interest in this ⅔ acre of relaxed cottage garden planting with a nod to prairie style, augmented by local artist sculptures. White border, hot border, wildflower meadow and woodland walk. Productive no-dig vegetable garden with greenhouses and fruit cage. A double pond water feature with bog garden. Lots of places to sit and enjoy. Some steps and slopes.

21 HALL BARN

Windsor End, Beaconsfield, HP9 2SG. Mrs Farncombe, garden@thefarncombes.com. *½m S of Beaconsfield. Lodge gate 300yds S of St Mary & All Saints' Church in Old Town centre. Please do not use SatNav.* **Visits by arrangement Feb to Oct. Adm £5, chd free. Home-made teas provided for groups of 10+.**
Historical landscaped garden laid out between 1680-1730 for the poet Edmund Waller and his descendants. Features 300 year old cloud formation yew hedges, formal lake and vistas ending with classical buildings and statues. Wooded walks around the grove offer respite from the heat on sunny days. One of the original NGS garden openings of 1927. Gravel paths, but certain areas can be accessed by car for those with limited mobility.

GROUP OPENING

22 HIGHER DENHAM GARDENS

Higher Denham, UB9 5EA. *6m E of Beaconsfield. Turn off A412, approx ½m N of junction with A40 into Old Rectory Lane. After 1m enter Higher Denham straight ahead. Tickets available for all gardens at community hall, 70yds into the village.* **Sun 23 May, Sun 18 July (2-5). Combined adm £6, chd free. Home-made teas in village hall. Donation to Higher Denham Community CIO (Garden Upkeep Fund).**

9 LOWER ROAD
Mrs Patricia Davidson.
Open on Sun 18 July

NEW 33B LOWER ROAD
Mr & Mrs J Hughes.
Open on Sun 18 July

5 SIDE ROAD
Jane Blyth.
Open on Sun 23 May

Canal Cottage

WIND IN THE WILLOWS
Ron James, 07740 177038,
r.james@company-doc.co.uk.
Open on all dates
Visits also by arrangement Mar to Sept for groups of 10+.

At least 4 gardens will open in 2021 in the delightful Misbourne chalk stream valley, 2 will open in May and 3 in July. Wind in the Willows has over 350 shrubs and trees, informal woodland and wild gardens incl riverside and bog plantings and a collection of 80 hostas and 12 striped roses in 3 acres. A new water lily pond has been created within the river. 'Really different' was a typical visitor comment. The garden at 5 Side Road is medium sized with lawns, borders and shrubs, and many features which children will love. This year a big conifer has been removed and the area replanted for greater interest. Opening in July, 9 Lower Road is a small garden backing onto the river. Recently professionally designed on concentric circles, it is now maturing and has new shrubs. 33b Lower Road is a mid-sized garden crammed with flowering plants, many of which are uncommon and incorporating a few veg. In May the owner of Wind in the Willows will lead optional guided tours of the garden starting at 2.30pm and 4pm. Tours generally last approx 1 hour. Partial wheelchair access to each garden.

23 HOLLYTREES
Parish Piece, Holmer Green, High Wycombe, HP15 6SP.
Brian Fisher, 07751 720060, brian@the-fishers.org.uk. *From Amersham (3m) or High Wycombe (4m). Follow the A404, signed exit to Holmer Green (Earl Howe Rd). After approx ½ m continue over Xrd by pond & Parish Piece is 2nd road on L.* **Visits by arrangement Aug & Sept for groups of 10 to 20. Adm £3.50, chd free. Light refreshments.**
A garden that reflects the enthusiasm of a highly qualified, respected and widely travelled plantsman who loves passing on his knowledge. Many unusual hardy plants combine in a unique micro-climate with exotic, tender species such as oleander, to reach a colourful peak in late summer. Also features three beautifully displayed collections; rocks from around the world, old gardening tools, wall nails and curiosities, and West Indian treasures which completes a fascinating experience. Wheelchair access with paved areas all around the garden.

24 HOMELANDS
Springs Lane, Ellesborough, Aylesbury, HP17 0XD. Jean & Tony Young, 01296 622306, young.ellesborough@gmail.com. *6m SE of Aylesbury. On the B4010 between Wendover & Princes Risborough. Springs Lane is between village hall at Butlers Cross & the church. Narrow lane with an uneven surface.* **Visits by arrangement Apr to Aug. Adm £4, chd free. Light refreshments.**
Secluded ¾ acre garden on difficult chalk, adjoining open countryside. Designed to be enjoyed from many seating positions. Progress from semi-formal to wildflower meadow and wildlife pond. Deep borders with all season interest, and gravel beds with exotic late summer and autumn planting.

25 HORATIO'S GARDEN
National Spinal Injuries Centre (NSIC), Stoke Mandeville Hospital, Mandeville Road, Stoke Mandeville, Aylesbury, HP21 8AL. Jacqui Martin-Lof, www.horatiosgarden.org.uk. *The closest car park to Horatio's Garden at Stoke Mandeville Hospital is Car Park B, opp Asda. Free parking on open day.* **Sun 29 Aug (10.30-4). Adm £5, chd free. Home-made teas in the Garden Room.**
Opened in Sept 2018, Horatio's Garden at the National Spinal Injuries Centre, Stoke Mandeville Hospital is designed by Joe Swift. The fully accessible garden for patients with spinal injuries has been part funded by the NGS. The beautiful space is cleverly designed to bring the sights, sounds and scents of nature into the heart of the NHS. Everything is high quality and carefully designed to bring benefit to patients who often have lengthy stays in hospital. The garden features a contemporary garden room, designed by Andrew Wells as well as a stunning Griffin Glasshouse. We also have a wonderful wildflower meadow. Please come along and meet the Head Gardener and volunteer team and taste our delicious tea and home-made cake! The garden is fully accessible, having been designed specifically for patients in wheelchairs or hospital beds.

26 NEW HOWE FARM FLOWERS
Ashendon Road, Dorton, Aylesbury, HP18 0NY.
Mrs Amber Partner, www.howefarmflowers.com. *If using SatNav, please make sure it takes you to Dorton & not the nearby village of Westcott.* **Sun 8 Aug (10-2). Adm £4, chd free. Home-made teas.**
Howe Farm Flowers is a flower farm located in the beautiful Buckinghamshire countryside. We currently grow on a 1 acre plot and in 2 large polytunnels with plans for further expansion. We use the 'no-dig' method and avoid using chemicals and pesticides. Our growing season is from Mar-Nov. We are committed to providing beautiful British blooms for everyone to enjoy. Gravel path easily accessible by wheelchairs leading to the flower field.

27 KINGSBRIDGE FARM
Steeple Claydon, MK18 2EJ. Mr & Mrs T Aldous, 01296 730224. *3m S of Buckingham. Halfway between Padbury & Steeple Claydon. Xrds signed to Kingsbridge Only.* **Visits by arrangement Mar to July. Adm £6, chd free.**
Stunning and exceptional 6 acre garden imaginatively created over last 30 yrs. Main lawn is enclosed by softly, curving, colour themed herbaceous borders, and many roses and shrubs interestingly planted with cleverly created landscaping features. Clipped topiary yews, pleached hornbeams lead out to the ha-ha and countryside beyond. A natural stream with bog plants and nesting kingfishers, meanders serenely through woodland gardens with many walks. A garden always evolving, to visit again and again.

The National Garden Scheme was set up in 1927 to raise funds for district nurses. We remain strongly committed to nursing and today our beneficiaries include Macmillan Cancer Support, Marie Curie, Hospice UK and The Queen's Nursing Institute

Nether Winchenden House

28 LINDENGATE

The Old Allotment Site, Dobbies Garden Centre, Aylesbury Road, Wendover, HP22 6BD. Lindengate Charity, 01296 622443, info@lindengate.org.uk, www.lindengate.org.uk. *4m SE of Aylesbury on A413. Turn into Dobbies Garden Centre, Lindengate on LH-side.* **Sun 2 May, Sun 12 Sept (10-12.30). Adm £4, chd free. Visits also by arrangement May to Sept.**

Lindengate's mission statement is to 'Foster an improved state of mental health and wellbeing through the healing power of nature and horticulture'. The charity's 5 acre site, has been developed into a series of wild spaces in synergy with more formalised gardens and is described as an oasis in a busy world. It successfully supports many people on their road to recovery. The garden has a sensory garden which incl a log wall, stumpery and various sensory experiences including water ball and rill. We will be selling products and produce from the gardens incl cut flowers, fruit and veg and crafted items. Accessible with welfare facilities and a fully accessible path throughout.

  占 ☕ ⛟

GROUP OPENING

29 LITTLE MISSENDEN GARDENS

Amersham, HP7 0RD. *On A413 between Great Missenden & Old Amersham. Please park in signed car parks.* **Sat 3, Sun 4 July (2-6). Combined adm £7, chd free. Home-made teas.**

A variety of gardens set in this attractive Chiltern village in an area of outstanding natural beauty. You can start off at one end of the village and wander through stopping off halfway for tea at the beautiful Anglo-Saxon church built in 975. The church has recently received a lottery grant for restoration and many medieval wall paintings have been found. Tours will be given of these discoveries. The gardens reflect different style houses including several old cottages, a Mill House and a more modern house (houses not open). There are herbaceous borders, shrubs, trees, old fashioned roses, hostas, topiary, koi and lily ponds, kitchen gardens, play areas for children and the River Misbourne runs through a few. Some gardens are highly colourful

and others just green and peaceful. Beekeeper at Hollydyke House. Partial wheelchair access to some gardens due to gravel paths and steps.

  占 ☕ ⛟

GROUP OPENING

30 LONG CRENDON GARDENS

Long Crendon, HP18 9AN. *2m N of Thame. On the B4011 Thame-Bicester road, a NGS yellow arrow sign will indicate R turn down High St, then continue down to St. Mary's Church Hall. Maps will be provided.* **Sun 11 Apr, Sun 13 June (2-6). Combined adm £6, chd free. Home-made teas.**

BAKER'S CLOSE
Mr & Mrs Peter Vaines.
Open on Sun 11 Apr

BARRY'S CLOSE
Mr & Mrs Richard Salmon.
Open on Sun 11 Apr

NEW **BRINDLES**
Sarah Chapman.
Open on Sun 11 Apr

COP CLOSE
Sandra & Tony Phipkin.
Open on Sun 13 June

25 ELM TREES
Carol & Mike Price.
Open on Sun 13 June

NEW **98 HIGH STREET**
Richard & Alex Rogers,
www.richardrogersdesigns.com.
Open on Sun 13 June
D

LOPEMEAD FARM
Wendy Thompson & Bryony Rixon.
Open on Sun 13 June

NEW **MANOR FARM BUNGALOW**
Ms Tracey Russell.
Open on Sun 13 June

MANOR HOUSE
Mr & Mrs West.
Open on Sun 11 Apr

TOMPSONS FARM
Mr & Mrs T Moynihan.
Open on Sun 11 Apr

Five gardens will open on Sun 11 Apr. Barry's Close, spring flowering trees, borders and water garden. Baker's Close, 1000s of daffodils, tulips and narcissi, shrubs and wild area. Manor House, a large garden with views to the Chilterns, two ornamental lakes,

a variety of spring bulbs and shrubs. Brindles, an organic garden with natural swimming pool, beehives and a developing wildflower area. Tompsons Farm, a large woodland garden with lake and newly planted borders. Five gardens will also open on Sun 13 June. A formal courtyard garden at Lopemead Farm with raised beds, vegetables and flower borders. Cop Close with mixed borders, vegetable and cutting garden. 25 Elm Trees, featuring herbaceous borders, roses and clematis. Manor Farm Bungalow, a biodiverse mixed garden using natural materials, wildflower area, exotic hot house and many occupied bird boxes. 98 High Street, a dramatic, contemporary cottage garden created by garden designer Richard Rogers. Partial wheelchair access to some gardens.

  占 ☕ ⛟

31 LORDS WOOD

Frieth Road, Marlow Common, SL7 2QS. Mr & Mrs Messum, www.messums.com. *1½m NW Marlow. From Marlow turn off the A4155 at Platts Garage into Oxford Rd & Chalkpit Lane towards Frieth for 1½m, 100yds past the Marlow Common road turn L down a made up bridlepath & follow parking signs.* **Thur 24 June (11-4). Adm £6, chd free. Home-made teas.**

'An outpost of Old Bloomsbury in Marlow Woods' was how diarist Frances Partridge described Lords Wood. James and Alex Strachey entertained many of the Bloomsbury Group including Lytton Strachey and Dora Carrington. The 5 acres surrounding the house (not open) showcase sculpture, water features, extensive mature borders, flower and herb gardens, orchard, and woodland walks with spectacular views over the Chilterns. Partial wheelchair access; gravel paths, steep slopes and open water.

  占 �car ☕

In our first year 609 private gardens opened their gates to all, for the modest sum of one shilling. Today the National Garden Scheme retains that combination of inclusivity and affordability

32 MAGNOLIA HOUSE

Grange Drive, Wooburn Green, Wooburn, HP10 0QD. Elaine & Alan Ford, 01628 525818, lanforddesigns@gmail.com. *On A4094 2m SW of A40 between Bourne End & Wooburn. From Wooburn Church, direction Maidenhead, Grange Drive is on L before r'about. From Bourne End, L at 2 mini-r'abouts, then 1st R.* **Visits by arrangement Feb to Aug. Light refreshments. Combined visit with The Shades may be possible.**
½ acre garden with mature trees incl large magnolia. Cacti, fernery, stream, ponds, greenhouses, aviaries, 10,000 snowdrops, hellebores, bluebells and over 60 varieties of hosta. Child friendly. Constantly being changed and updated. Partial wheelchair access.

33 THE MANOR HOUSE

Off Perry Lane, Bledlow, nr Princes Risborough, HP27 9PB. Lord & Lady Carrington, www.carington.co.uk/gardens/. *9m NW of High Wycombe, 3m SW of Princes Risborough. ½m off B4009 in middle of Bledlow village. For SatNav use postcode HP27 9PA.* **Sun 13 June (2-4.30). Adm £6, chd free. Tea.**
8 acres of gardens revitalised during 2019/20. Highlights incl the walled kitchen garden criss-crossed by paths with vegetables, fruit, herbs and flowers. The sculpture garden, the replanted Granary garden with fountain and borders, as well as the individual paved gardens and parterres divided by yew hedges and more. The Lyde water garden formed out of old cress beds fed by numerous springs. Partial wheelchair access via steps or sloped grass to enter gardens.

34 MONTANA

Shire Lane, Cholesbury, HP23 6NA. Diana Garner, 01494 758347, montana@cholesbury.net. *3m NW of Chesham. From Wigginton turn R after Champneys, 2nd R onto Shire Lane. From Cholesbury common, turn on Cholesbury Rd by cricket club, take 1st L onto Shire Lane. Montana is ½m down Shire Lane on LH-side.* **Weds 24 Mar; 21 Apr; 19 May; 23 June; 21 July (11-3). Adm £4, chd free. Pre-booking essential, please**

phone 01494 758347 or email montana@cholesbury.net for information & booking. **Visits also by arrangement Mar to July for groups of up to 30.**
A peaceful large country garden planted with rare trees, unusual flowering shrubs and perennials under planted with thousands of bulbs; kitchen garden edged by sweet peas; shade loving plants and small meadow. A gate leads to a mixed deciduous wood with level paths, a fernery planted in an old clay pit and an avenue of daffodils and acers. Lots of seats to enjoy the atmosphere. Small apiary. An un-manicured garden high in the Chiltern Hills. Surrounding fields have been permanent pasture for more than 100 yrs. Picnics welcome. The majority of paths in the garden and wood are wheelchair friendly.

35 ◆ NETHER WINCHENDON HOUSE

Nether Winchendon, Thame, Aylesbury, HP18 0DY. Mr Robert Spencer Bernard, 01844 290101, Contactus@ netherwinchendonhouse.com, www.nwhouse.co.uk. *6m SW of Aylesbury, 6m from Thame. Approx 4m from Thame on A418, turn 1st L to Cuddington, turn L at Xrds, downhill turn R & R again to parking by house.* **For NGS: Sun 29 Aug (2-5.30). Adm £4, chd free. Home-made teas at church. For other opening times and information, please phone, email or visit garden website.**
Nether Winchendon House has fine and rare trees, set in a stunning landscape surrounded by parkland with 7 acres of lawned grounds running down to the River Thame. A Founder NGS Member (1927). Enchanting and romantic Mediaeval and Tudor House, one of the most romantic of the historic houses of England and Grade I listed. Picturesque small village with an interesting church. Unfenced riverbank.

36 NORTH DOWN

Dodds Lane, Chalfont St Giles, HP8 4EL. Merida Saunders, 01494 872928. *4m SE of Amersham, 4m NE of Beaconsfield. Opp the green in centre of village, at Costa turn into UpCorner onto Silver Hill. At top of hill fork R into Dodds Lane.*

North Down is 7th on L. **Visits by arrangement May to Sept for groups of up to 30. Adm £4.50, chd free. Light refreshments.**
A passion for gardening is evident in this plantswomans lovely ¾ acre garden, which has evolved over the yrs with scenic effect in mind. Colourful and interesting throughout the yr. Large grassed areas with island beds of mixed perennials, shrubs and some unusual plants. Variety of rhododendrons, azaleas, acers and clematis. Displays of sempervivum, alpines, grasses and ferns. Small patio and water feature, greenhouse and an Italianate front patio to owner's design. You can see in the village Milton's Cottage and garden where John Milton wrote Paradise Lost.

37 OLD PARK BARN

Dag Lane, Stoke Goldington, MK16 8NY. Emily & James Chua, 01908 551092, emilychua51@yahoo.com. *4m N of Newport Pagnell on B526. Park on High St. A short walk up Dag Lane. Disabled parking for two cars near garden via Orchard Way.* **Sat 12, Sun 13 June, Sat 10, Sun 11 July (2-5). Adm £5, chd free. Home-made teas. Visits also by arrangement June & July for groups of 10+.**
A garden of almost 3 acres made from a rough field over 20 yrs ago. Near the house (not open) a series of terraces cut into the sloping site create the formal garden with long and cross vistas, lawns and deep borders. The aim is to provide interest throughout the yr with naturalistic planting and views borrowed from the surrounding countryside. Beyond is a wildlife pond, meadow and woodland garden. Partial wheelchair access.

38 ORCHARD HOUSE

Tower Road, Coleshill, Amersham, HP7 0LB. Mr & Mrs Douglas Livesey. *From Amersham Old Town take the A355 to Beaconsfield. Appox ¾m along this road at top of hill, take 1st R into Tower Rd. Parking in cricket club grounds.* **Sun 25 Apr (1-4.30). Adm £5, chd free. Home-made teas in the barn.**
The 5 acre garden incl several wooded areas with eco bug hotels for wildlife. Two ponds with wildflower planting, large avenues of silver birches, a bog garden with board

walk and a wildflower meadow. There is a cut flower garden and a dramatic collection of spring bulbs set amongst an acer glade. Wheelchair access with sloping lawn in rear garden.

    🔥 🐛 🍵

39 OVERSTROUD COTTAGE

The Dell, Frith Hill, Gt Missenden, HP16 9QE. Mr & Mrs Jonathan Brooke, 01494 862701, susanmbrooke@outlook.com. ½ m E Gt Missenden. Turn E off A413 at Gt Missenden onto B485 Frith Hill to Chesham Rd. White Gothic cottage set back in lay-by 100yds uphill on L. Parking on R at church. **Suns 4 Apr; 2 May; 6, 27 June; 5 Sept (2-5). Adm £4, chd free. Cream teas at parish church. Visits also by arrangement Apr to Sept for groups of 20 to 30.**

Artistic chalk garden on two levels. Collection of C17/C18 plants including auriculas, hellebores, bulbs, pulmonarias, peonies, geraniums, dahlias, herbs and succulents. Many antique species and rambling roses. Potager and lily pond. Blue and white ribbon border. Cottage was once C17 favour house for Missenden Abbey. Features incl a garden studio with painting exhibition (share of flower painting proceeds to NGS).

    🌸 🍵

40 11 THE PADDOCKS

Wendover, HP22 6HE. Mr & Mrs E Rye, 01296 623870, pam.rye@talktalk.net. 5m from Aylesbury on A413. From Aylesbury turn L at mini-r'about onto Wharf Rd. From Gt Missenden turn L at the Clock Tower, then R at mini-r'about onto Wharf Rd. **Every Sun 13 June to 4 July (11-5). Adm £4, chd free. Evening opening Fri 25 June (6-8). Adm £5, chd free. Wine. Visits also by arrangement June & July for groups of 10 to 30. Donation to Bonnie People in South Africa.**

Small peaceful garden with mixed borders of colourful herbaceous perennials, a special show of David Austin roses and a large variety of spectacular named Blackmore and Langdon delphiniums. A tremendous variety of colour in a small area. The White Garden with a peaceful and shady arbour, and The Magic of Moonlight created for the BBC.

    🔥 🌸 🍵

41 PETERLEY CORNER COTTAGE

Perks Lane, Prestwood, Great Missenden, HP16 0JH. Dawn Philipps, 01494 862198, dawn.philipps@googlemail.com. Turn into Perks Lane from Wycombe Rd (A4128), Peterley Corner Cottage is the 3rd house on the L. **Visits by arrangement May to Aug for groups of 10 to 30. Adm £5, chd free. Light refreshments.**

A 3 acre mature garden, incl an acre of wildflowers and indigenous trees. Surrounded by tall hedges and a wood, the garden has evolved over the last 30 yrs. There are many specimen trees and mature roses incl a Paul's Himalaya Musk and a Kiftsgate. A large herbaceous border runs alongside the formal lawns with other borders like heathers and shrubs. The most recent addition is a potager.

    🔥 🍵

42 THE PLOUGH

Chalkshire Road, Terrick, Aylesbury, HP17 0TJ. John & Sue Stewart. 2m W of Wendover. Entrance to garden & car park signed off B4009 Nash Lee Rd. 200yds E of Terrick r'about. Access to garden from field car park. **Mon 31 May (1-5). Adm £4, chd free. Home-made teas.**

Formal organic garden with open views to the Chiltern countryside. Designed as a series of outdoor rooms around a listed former C18 inn (not open), incl border, parterre, vegetable and fruit gardens, and a newly planted orchard. Delicious home-made teas in our barn and adjacent entrance courtyard. Jams and apple juice for sale, made with fruits from the garden.

    🐛 🌸 🍵

43 RED KITES

46 Haw Lane, Bledlow Ridge, HP14 4JJ. Mag & Les Terry, 01494 481474, lesterry747@gmail.com. 4m S of Princes Risborough. Off A4010 halfway between Princes Risborough & West Wycombe. At Hearing Dogs sign in Saunderton turn into Haw Lane, then ¾ m on L up the hill. **Tue 11 May, Tue 13 July (2-5). Adm £5, chd free. Home-made teas. Visits also by arrangement May to Sept for groups of 20+.**

This much admired 1½ acre Chiltern hillside garden is planted for yr-round interest and is lovingly maintained with mixed and herbaceous borders, wildflower orchard, established pond,

vegetable garden, managed woodland area and a lovely hidden garden. Many climbers used throughout the garden which changes significantly through the seasons. Sit and enjoy the superb views from the top terrace.

44 RIVENDELL

13 The Leys, Amersham, HP6 5NP. Janice & Mike Cross, 01494 728291, janice.cross@ngs.org.uk. Off A416. From Amersham take A416 N towards Chesham. The Leys is on L ½ m after Boot & Slipper Pub. **Visits by arrangement Apr to Aug for groups of 20 to 30. Adm £10, chd free. Home-made teas included. Pre-payment is required.**

An established, but always evolving, south-facing garden with different linked areas. A woodland area, gravel area with grasses and pond, bug hotels, fruit and vegetables, herbaceous beds containing a wide variety of shrubs, bulbs and perennials designed to encourage pollinators, surround a circular lawn with a rose and clematis arbour. Many of the plants are propagated for sale.

    🌸 🍵

45 ROBIN HILL

Water End, Stokenchurch, High Wycombe, HP14 3XQ. Caroline Renshaw & Stuart Yates, 07957 394134, Info@cazrenshawdesigns.co.uk. 2m from M40 J5 Stokenchurch. Turn off A40 just S of Stokenchurch towards Radnage, then 1st R to Waterend & then follow signs. Limited parking. **Sun 4 July (10-4.30). Adm £4.50, chd free. Home-made teas. Visits also by arrangement Apr to Oct for groups of up to 10.**

1½ acre informal country garden at the start of The Chiltern Hills, open to the views, over its own wildflower meadow. The garden is full of planting with shrubs, perennials and grass borders, new and established trees, and lots of places to sit and enjoy the views. Wind your way through the paths in the meadow and you can also visit the chickens in the large cherry orchard. Wheelchair access over mainly flat and lawned garden, but no paths.

    🔥 🌸 🅳 🍵

The Walled Garden

46 ST MICHAELS CONVENT
Vicarage Way, Gerrards Cross,
SL9 8AT. Sisters of the Church.
*15mins walk from Gerrards Cross
station. 10mins from East Common
buses. Limited parking at convent.*
**Sat 26 June (2-4.30). Adm by
donation. Tea.**
Recently acquired garden, having
been neglected for many years,
is now being developed by the
Community as a place for quiet,
reflection and to gaze upon beauty.
Includes a walled garden with
vegetables, labyrinth and beehives, a
shady woodland dell and a recently
built chapel. Colourful borders, beds
and mature majestic trees. Come and
see how the garden is developing and
growing!

47 THE SHADES
High Wycombe, HP10 0QD.
Pauline & Maurice Kirkpatrick,
01628 522540. *On A4094 2m SW
of A40 between Bourne End &
Wooburn. From Wooburn Church,
direction Maidenhead, Grange Drive
is on L before r'about. From Bourne*

End, L at 2 mini-r'abouts, then 1st R.
**Visits by arrangement Feb to Aug.
Combined visit with Magnolia
House. Light refreshments at
Magnolia House.**
The Shades drive is approached
through mature trees, areas of
shade loving plants, beds of shrubs,
60 various roses and herbaceous
plants. The rear garden with natural
well, surrounded by plants, shrubs
and acers. A green slate water
feature and scree garden with alpine
plants completes the garden. Partial
wheelchair access.

48 ◆ STOKE POGES
MEMORIAL GARDENS
Church Lane, Stoke
Poges, Slough, SL2 4NZ.
Buckinghamshire Council,
01753 523744, memorial.
gardens@buckinghamshire.gov.
uk, www.southbucks.gov.uk/
stokepogesmemorialgardens.
*1m N of Slough, 4m S of Gerrards
Cross. Follow signs to Stoke Poges
& from there to the Memorial
Gardens. Car park opp main*

entrance, disabled visitor parking
*in the gardens. Weekend disabled
access through churchyard.* **For
NGS: Sun 13 June (1-4.30). Adm
£4.50, chd free. Home-made
teas. For other opening times and
information, please phone, email or
visit garden website.**
Unique 22 acre Grade I registered
garden constructed 1934-9.
Rock and water gardens, sunken
colonnade, rose garden, 500
individual gated gardens, beautiful
mature trees and newly landscaped
areas. Guided tours every hour. Guide
dogs only.

49 NEW 1 TALBOT AVENUE
Downley, High Wycombe,
HP13 5HZ. Mr Alan Mayes,
01494 451044,
alan.mayes2@btopenworld.com.
*From Downley T-lights off West
Wycombe Rd, take Plomer Hill turn
off, then 2nd L into Westover Rd,
then 2nd L into Talbot Ave.* **Visits
by arrangement Mar to Sept for
groups of up to 10. Adm £4, chd
free. Tea.**

A Japanese garden, shielded from the top garden level by Shoji screen. A winding path leads you over a traditional Japanese bridge by a pond and waterfall, and invites you through a moongate to reveal a purpose-built tea house, all surrounded by traditional Japanese planting including maples, cherry blossom trees, azaleas and rhododendrons. Ornamental grasses and bamboo complement the hard landscaping with feature cloud tree and checkerboard garden path.

50 TURN END
Townside, Haddenham, Aylesbury, HP17 8BG. Margaret & Peter Aldington, turnendgarden@gmail.com, www.turnend.org.uk. *3m NE of Thame, 5m SW of Aylesbury. Exit A418 to Haddenham. Turn at Rising Sun Pub into Townside. Street parking, very limited. Please park with consideration for residents. See Turn End website for parking info for this event.* **Mon 3 May (2-5). Adm £4.50, chd free. Home-made teas.** Grade II registered series of garden rooms, each with a different planting style enveloping architect's own Grade II* listed house (not open). Dry garden, formal box garden, sunken gardens, mixed borders around curving lawn, all framed by ancient walls and mature trees. Bulbs, irises, wisteria, roses, ferns and climbers. Courtyards with pools, pergolas, secluded seating and Victorian Coach House. Open artist's studio with displays and demonstrations.

51 TYTHROP PARK
Kingsey, HP17 8LT. Nick & Chrissie Wheeler. *2m E of Thame, 4m NW of Princes Risborough. Via A4129, at T-junction in Kingsey turn towards Haddenham, take L turn on bend. Parking in field on L.* **Sun 27 June (2-5.30). Adm £7, chd free. Home-made teas.** 10 acres of garden surrounds a C17 Grade I listed manor house (not open). This large and varied garden blends traditional and contemporary styles, featuring pool borders rich in grasses with a green and white theme, walled kitchen/cutting garden with large greenhouse at its heart, box parterre, deep mixed borders, water feature, rose garden, wildflower meadow, and many old trees and shrubs.

52 THE WALLED GARDEN
Wormsley, Stokenchurch, High Wycombe, HP14 3YE. Wormsley Estate. *Leave M40 at J5. Turn towards Ibstone. Entrance to estate is ¼m on R. NB: 20mph speed limit on estate. Please do not drive on grass verges.* **Fri 4 June (10-3). Adm £7, chd free. Pre-booking essential, please visit www.ngs.org.uk/ special-events for information & booking. Home-made teas.** The Walled Garden at Wormsley Estate is a 2 acre garden providing flowers, vegetables and tranquil contemplative space for the family. For many yrs the garden was neglected until Sir Paul Getty purchased the estate in the mid-1980s. In 1991 the garden was redesigned by the renowned garden designer Penelope Hobhouse. The garden has changed over the yrs, but remains true to the original brief. Wheelchair access to grounds, please ensure you confirm upon booking.

53 THE WHITE HOUSE
Village Road, Denham Village, UB9 5BE. Mr & Mrs P G Courtenay-Luck. *3m NW of Uxbridge, 7m E of Beaconsfield. Signed from A40 or A412. Parking in village road. The White House is in centre of village, opp St Mary's Church.* **Sun 30 May (2-5). Adm £6, chd free. Cream teas.** Well established 6 acre formal garden in picturesque setting. Mature trees and hedges with River Misbourne meandering through lawns. Shrubberies, flower beds, rockery, rose garden and orchard. Large walled garden. Herb garden, vegetable plot and Victorian greenhouses. Wheelchair access with gravel entrance and path to gardens.

54 20 WHITEPIT LANE
Flackwell Heath, High Wycombe, HP10 9HS. Trevor Jones, 01628 524876, trevorol4969@gmail.com. *¾m E Hackwell Heath, 4m High Wycombe. From M40 J3 (west bound exit only) take 1st L, 300yds T-junction turn R, 300yds turn L uphill to centre of Flackwell Heath, turn L ¾m, garden 75yds on R past mini-r'about.* **Visits by arrangement May to Aug for groups up to 6. Adm £7, chd free.** **Cream teas included.** The front garden has a seaside type landscape with timber groynes and a rock pool. The rear garden is long and thin with colour from Apr to Oct, containing many unusual plants, with steps and bridges over two ponds. Amongst the plants are a large Banana, an Albizia, Grevillea's, Callistemon's Abutilon, Salvia's and several Alstroemeria's. The beds are filled with mixed shrubs and herbaceous plants. Many rare and unusual plants and a garden railway.

55 WOODSIDE
23 Willow Lane, Amersham, HP7 9DW. Elin & Graham Stone. *On A413, 1m SE from Old Amersham, between Barley Lane & Finch Lane. The garden is the last but one, on the RH side. Limited parking, please park with consideration to neighbours.* **Sat 12, Sun 13 June (1-5.30). Adm £3.50, chd free. Pre-booking essential, please phone 07592 367434 or email glsimagesuk@gmail.com for information & booking. Light refreshments & delicious home-made cakes.** A small cottage garden, created on s-facing slope, integrating a circular lawn, lily pond, gravel paths and steps to a curved clematis and rose pergola. Rose beds and abundant humped borders. In contrast, a secret naturally arching woodland path winds past a stumpery, a shade area and wildlife hedging leading to a kitchen garden and bee hotel. Hidden seating areas abound. Artist's Studio. Features incl dragon flies, wildlife habitat, fernery, seasonal potted decorative plants, and a very effective 8 monthly composting process.

OPENING DATES

All entries subject to change. For latest information check www.ngs.org.uk

Extended openings are shown at the beginning of the month.

Map locator numbers are shown to the right of each garden name.

February

Snowdrop Festival

Saturday 13th
Caldrees Manor 5
NEW Landwade Hall 31

Sunday 14th
NEW Landwade Hall 31

Sunday 21st
Clover Cottage 10

Sunday 28th
Clover Cottage 10

March

Sunday 7th
Clover Cottage 10

Sunday 28th
Kirtling Tower 30

April

Sunday 4th
Netherhall Manor 40

Sunday 18th
◆ Docwra's Manor 13

Monday 26th
Staploe Gardens 51

Thursday 29th
28 Houghton Road 27

Friday 30th
28 Houghton Road 27

Online booking is available for many of our gardens, at ngs.org.uk

May

Every Friday and Saturday from Friday 14th
23A Perry Road 44

Every Sunday from Sunday 16th
23A Perry Road 44

Sunday 2nd
Barton Gardens 2
Chaucer Road Gardens 9
Netherhall Manor 40

Monday 3rd
Chaucer Road Gardens 9

Friday 7th
28 Houghton Road 27

Saturday 8th
28 Houghton Road 27
NEW Milton Hall 38

Sunday 9th
◆ Ferrar House 20

Saturday 15th
28 Houghton Road 27

Sunday 16th
28 Houghton Road 27

Sunday 23rd
Catworth, Molesworth & Brington Gardens 8
College Farm 11
28 Houghton Road 27

Monday 24th
28 Houghton Road 27

Sunday 30th
Cambourne Gardens 6
Cottage Garden 12
Elm House 16
NEW Gifu 21
NEW Hinxton Gardens 25
Island Hall 28
NEW 38 Kingston Street 29
Sutton Gardens 52

June

Every Friday and Saturday
23A Perry Road 44

Every Sunday
23A Perry Road 44

Wednesday 2nd
28 Houghton Road 27

Thursday 3rd
28 Houghton Road 27

Saturday 5th
NEW Fenstanton 19
NEW 34 Millington Road 37
Staploe Gardens 51
Twin Tarns 55

Sunday 6th
Duxford Gardens 14
Ely Open Gardens 17
NEW Fenstanton 19
Highsett Cambridge 24
The Old Rectory 43
Staploe Gardens 51
Twin Tarns 55

Wednesday 9th
Wild Rose Cottage 57

Thursday 10th
28 Houghton Road 27

Friday 11th
28 Houghton Road 27

Saturday 12th
NEW Abbots Barn 1
Beaver Lodge 3
Little Oak 32

Sunday 13th
NEW Abbots Barn 1
Beaver Lodge 3
Clover Cottage 10
Little Oak 32
◆ Mary Challis Garden 36

Sunday 20th
Green End Farm 22

Sunday 27th
Castor House 7
Kirtling Tower 30

July

Every Friday and Saturday
23A Perry Road 44

Every Sunday
23A Perry Road 44

Saturday 3rd
NEW Bramley Cottage 4
38 Norfolk Terrace Garden 42
Trinity Hall - Wychfield Site 54
Wrights Farm 59

Sunday 4th
NEW Bramley Cottage 4

Green End Farm 22
38 Norfolk Terrace Garden 42
Sawston Gardens 48
The Six Houses 50
NEW Two Toft Gardens 56
Wrights Farm 59

Thursday 15th
NEW Preachers Passing 45

Sunday 18th
Madingley Hall 33
NEW Silver Birches 49

Saturday 24th
38 Norfolk Terrace Garden 42

Sunday 25th
38 Norfolk Terrace Garden 42

August

Every Friday and Saturday
23A Perry Road 44

Every Sunday
23A Perry Road 44

Sunday 1st
◆ Elgood's Brewery Gardens 15
Netherhall Manor 40

Saturday 7th
NEW Top Farm 53

Sunday 8th
Netherhall Manor 40
NEW Top Farm 53

Saturday 14th
Beaver Lodge 3
NEW Preachers Passing 45

Sunday 15th
Beaver Lodge 3
Robynet House 47

Sunday 22nd
Castor House 7

Saturday 28th
The Night Garden 41

Sunday 29th
The Night Garden 41

Monday 30th
The Night Garden 41

All entries are
subject to change.
For the latest
information check
ngs.org.uk

The Manor House, Fenstanton

THE GARDENS

1 NEW ABBOTS BARN

Southorpe, Stamford, PE9 3BX.
Carl & Vanessa Brown. *5m SE of Stamford, 8m W of Peterborough. On entering Southorpe Village from A47, 1st house on left.* **Sat 12, Sun 13 June (1-6). Adm £5, chd free. Home-made teas. Vegans catered for.**
2-acre garden surrounding converted Georgian barn. Large family garden with herbaceous borders and wildflower meadow, allotment-sized vegetable plot, orchard, mediterranean gravel garden with mulberry tree, large glasshouse, 20 year-old woodland with mown grass paths, beach garden, wildlife pond, large courtyard garden with garden-sized croquet lawn, cutting flower garden, gravel/pergola walkways.

We maintain the garden using organic methods, providing a rich wildlife habitat. Children of all ages welcome.

GROUP OPENING

2 BARTON GARDENS

High Street, Barton, Cambridge, CB23 7BG. *3½m SW of Cambridge. Barton is on A603 Cambridge to Sandy Rd, ½m from J12 M11.* **Sun 2 May (1-5). Combined adm £5, chd free.**

FARM COTTAGE
Dr R M Belbin.

114 HIGH STREET
Meta & Hugh Greenfield.

11 KINGS GROVE
Mrs Judith Bowen.

THE SIX HOUSES
Perennial.
(See separate entry)

Varied group of large and small gardens reflecting different approaches to gardening. Farm Cottage: large landscaped cottage garden with herbaceous beds and themed woodland walk. 114 High Street: small cottage garden with an unusual layout comprising several areas incl vegetables, fruit and a secret garden. The Six Houses: Six unique gardens filled with charm, each beautifully tended by resident retired professional gardeners. 11 Kings Grove: a garden developed from a wilderness since 1992 with a lawn, flowers and shrub area and a fruit area. Some gardens have gravel paths.

Twin Tarns

❸ BEAVER LODGE

45 Henson Road, March, PE15 8BA. Mr & Mrs Nielson-Bom, 07455 495592, beaverbom@gmail.com. *A141 to Wisbech rd into March, L into Westwood Ave, at ond, R into Henson Rd, property on R opp school playground.* **Sat 12, Sun 13 June, Sat 14, Sun 15 Aug (10-4). Adm £3, chd free. Home-made teas. Visits also by arrangement June to Sept for groups of up to 30.**

An oriental garden with more than 120 large and small bonsai trees, different acers, pagodas, oriental statues, water features and pond with koi carp, creating a peaceful and relaxing atmosphere. The garden is divided into different rooms, one with a Mediterranean feel including tree ferns, lemon trees, bougainvilleas and a great variety of plants and water fountain. Plenty of sitting areas.

❹ NEW BRAMLEY COTTAGE

Barton Road, Wisbech St Mary, Wisbech, PE13 4RP. Jim & Mel Wakefield, 01945 410554, mclaniewright001@btinternet.com. *Coming towards Wisbech St Mary from Wisbech, 1st house on R following a sweeping RH bend.* **Sat 3, Sun 4 July (10-5). Adm £5, chd free. Light refreshments. Visits also by arrangement May to Sept for groups of 10+. Arranged visits evenings and weekends only with no machinery on display.**

This garden has been built over the last thirteen years from a blank canvas. It is an adult garden with structures of a rose arch and a wisteria and laburnum arch. It has various quirky pieces that we like. The garden can also be viewed from a platform giving an aerial view of the garden and surrounding fields. On display also will be a selection of vintage horticultural machinery and hand tools.

❺ CALDREES MANOR

2 Abbey Street, Ickleton, Saffron Walden, CB10 1SS. *In the centre of Ickleton on Abbey Street. Parking for 30 cars at The Village Hall. From M11 J10, A505 East, then through Duxford village, signed Ickleton. From Saffron Walden, via Gt Chesterford.* **Sat 13 Feb, Sat 16 Oct (10.30-4). Adm £6, chd free. Home-made teas.**

C19 Manor House with extensive formal gardens, lakes and streams, over 150 varieties of Acer palmatum,

many specimen trees, a Japanese garden, orchard, woodland walks, wildlife garden and wild flower meadows as well as snowdrops in late winter. The garden has been designed to peak in Spring and Autumn. Some gravel, and a few steps.

GROUP OPENING

❻ CAMBOURNE GARDENS

Great Cambourne, CB23 6AH. *8m W of Cambridge on A428. From A428: take Cambourne junction into Great Cambourne. From B1198, enter village at Lower Cambourne & drive through to Great Cambourne. Follow NGS signs via either route to start at any garden.* **Sun 30 May (11-5). Combined adm £6, chd free. Home-made teas. Coffee and biscotti at 43 Monkfield Lane.**

13 FENBRIDGE
Lucinda & Tony Williams.

14 GRANARY WAY
Jackie Hutchinson.

88 GREENHAZE LANE
Darren & Irette Murray.

8 LANGATE GREEN
Steve & Julie Friend.

5 MAYFIELD WAY
Debbie & Mike Perry.

14 MILLER WAY
Geoff Warmington.

43 MONKFIELD LANE
Tony & Penny Miles.

A unique and inspiring modern group, all created from new build in just a few years. This selection of seven demonstrates how imagination and gardening skill can be combined in a short time to create great effects from unpromising and awkward beginnings. The grouping includes foliage gardens with collections of carnivorous pitcher plants and hosta; a suntrap garden for play, socialising and colour; gardens with ponds, a vegetable plot and many other beautiful borders showing their owners' creativity and love of growing fine plants well. Cambourne is one of Cambridgeshire's newest communities, and this grouping showcases the happy, vibrant place our village has become.

❼ CASTOR HOUSE

Castor, Peterborough, PE5 7AX. Ian & Claire Winfrey, ian@winfrey.co.uk, www.castorhousegardens.co.uk. *4m W of Peterborough. House on main Peterborough Rd in Castor. Parking in paddock off Water Lane.* **Sun 27 June, Sun 22 Aug (2-5). Adm £7.50, chd free. Home-made teas. Visits also by arrangement June to Sept for groups of 20+.**

12 acres of gardens and woodland on a slope, terraced and redesigned 2010. Italianate spring fed ponds and stream gardens. Potager with greenhouse and exotic borders. Willow arbour and woodland garden. Peony and prunus walk. Rose and cottage gardens, 'Hot' double border, stumpery. Orchard in walled gardens, autumn cyclamen. Only partial access for wheelchairs due to sloping nature of the garden.

GROUP OPENING

❽ CATWORTH, MOLESWORTH & BRINGTON GARDENS

Molesworth, Huntingdon, PE28 0QD. *10m W of Huntingdon. A14 W for Molesworth & Brington exit at J16 onto B660.* **Sun 23 May (2-6). Combined adm £5, chd free. Home-made teas at Molesworth House and Yew Tree Cottage.**

32 HIGH STREET
Colin Small.

MOLESWORTH HOUSE
John Prentis.

YEW TREE COTTAGE
Christine & Don Eggleston.

Molesworth House is an old rectory garden with everything that you'd both expect and hope for, given its Victorian past. There are surprising corners to this traditional take on a happy and relaxed, yet also formal garden. Yew Tree Cottage, informal garden approx 1 acre, complements the C16 building (not open) and comprises flower beds, lawns, vegetable patch, boggy garden, copses and orchard. Plants in pots and hanging baskets. High Street, Catworth is a long narrow garden with many rare plants including ferns, herbaceous borders, woodland area and wildlife pond. Partial wheelchair access.

GROUP OPENING

9 CHAUCER ROAD GARDENS
Cambridge, CB2 7EB. *1m S of Cambridge. Off Trumpington Rd (A1309), nr Brooklands Ave junction. Parking available at MRC Psychology Dept on Chaucer Rd.* **Sun 2, Mon 3 May (2-5). Combined adm £7, chd free. Home-made teas at Upwater Lodge.**

11 CHAUCER ROAD
Mark & Jigs Hill.

12 CHAUCER ROAD
Mr & Mrs Bradley.

16 CHAUCER ROAD
Mrs V Albutt.

UPWATER LODGE
Mr & Mrs George Pearson, 07890 080303, jmp@pearson.co.uk. **Visits also by arrangement Apr to Sept.**

11 Chaucer Road is a ¾ acre Edwardian garden that has changed rapidly over the ensuing 110 yrs. A rock garden with pond and large weeping Japanese maple dates from about 1930. 12 Chaucer Road is a half acre town garden with mature trees, Japanese themed garden and unusually for this part of the world, some impressive ericaceous planting. 16 Chaucer Road is an artist's garden, full of surprises, and attractions for children. Upwater Lodge is an Edwardian academic's house with 7 acres of grounds. It has mature trees, fine lawns, old wisterias, and colourful borders. There is a small, pretty potager with a selection of fruits, and a well maintained grass tennis court. A network of paths through a wooded area lead down to a dyke, water meadows and a small flock of rare breed sheep. Enjoy a walk by the river and watch the punts go by. Buy home-made teas and sit in the garden or take them down to enjoy a lazy afternoon with ducks, geese, swans and heron on the riverbank. Cakes made with garden fruit where possible. Swings and climbing ropes. Plant stall possible but please email to check. Some gravel areas and grassy paths with fairly gentle slopes.

&

10 CLOVER COTTAGE
50 Streetly End, West Wickham, CB21 4RP. Mr Paul & Mrs Shirley Shadford, 01223 893122, shirleyshadford@live.co.uk. *3m from Linton, 3m from Haverhill & 2m from Balsham. From Horseheath turn L, from Balsham turn R, thatched cottage opposite triangle of grass next to old windmill.* **Sun 21 Feb, Sun 28 Feb, Sun 7 Mar (2-4). Sun 13 June (12-5). Light refreshments. In June also open Mary Challis Garden. Adm £3.50, chd free. Visits also by arrangement Feb to June.**
In winter find a flowering cherry tree, borders of snowdrops, aconites, iris reticulata, hellebores and miniature narcissus throughout the packed small garden which has inspiring ideas on use of space. Pond and arbour, raised beds of fruit and vegetables. In summer arches of roses and clematis, hardy geraniums, delightful borders of English roses and herbaceous plants. Snowdrops, hellebores and spring flowering bulbs for sale for the snowdrop festival, plants also for sale in June. Access not suitable for wheelchair users, prams, pushchair and wheeled walkers. No Dogs

&

11 COLLEGE FARM
Station Road, Haddenham, Ely, CB6 3XD. Sheila & Jeremy Waller, www.primaveragallery.co.uk. *From Stretham & Wilburton, at Xrds in Haddenham, turn R. Pass the church & exactly at the bottom of the hill, turn L down narrow drive, with a mill wheel on R of the drive.* **Sun 23 May (2-5). Adm £7.50, chd £5.**
40 acres around an intact Victorian farm. Walks, galleries, flower and sculpture cattle yard. Roses, wildflowers, water plants, foxgloves and plantings of trees, hedges and fruit trees, amongst ponds and through meadows, add colour and structure. Splendid fen views, lovely water features and ancient ridge and furrow pasture land. Original farm buildings, outside galleries and inside galleries, Interesting (not seeded) wildflowers like Jack go to bed at noon and Salsify. Wheelchair access is possible around the garden near the house, but not through the gallery, farm buildings, milking parlour and many of the walks.

&

12 COTTAGE GARDEN
79 Sedgwick Street, Cambridge, CB1 3AL. Rosie Wilson, 07805 443818, liccycat@icloud.com. *Cambridge City. Off Mill Rd, south of the railway bridge, now only open for bus and bicycle traffic.* **Sun 30 May (2-6). Adm £3, chd free. Home-made teas. Also open 38 Kingston Street. Visits also by arrangement May to Aug for groups of up to 10.**
Small, long, narrow, and planted in the cottage garden style with over 40 roses some on arches and growing through trees. Particularly planned to encourage wildlife with small pond, mature trees and shrubs. Perennials and some unusual plants interspersed with sculptures. Some narrow paths.

&

13 ♦ DOCWRA'S MANOR
2 Meldreth Road, Shepreth, Royston, SG8 6PS. Mrs Faith Raven, 01763 260677, faithraven@btinternet.com, www.docwrasmanorgarden.co.uk. *8m S of Cambridge. ½m W of A10. Garden is opp the War Memorial in Shepreth. King's Cross-Cambridge train stop 5 min walk.* **For NGS: Sun 18 Apr (2-5). Adm £5, chd free. Home-made teas. For other opening times and information, please phone, email or visit garden website.**
2½ acres of choice plants in a series of enclosed gardens. Tulips and Judas trees. Opened for the NGS for more than 50 yrs. The garden is featured in great detail in a book published 2013 'The Gardens of England' edited by George Plumptre. Unusual plants. Wheelchair access to most parts of the garden, gravel paths. Guide dogs only.

&

GROUP OPENING

14 DUXFORD GARDENS
Duxford, CB22 4RP. *Most gardens are close to the centre of the village of Duxford. S of the A505 between M11 J10 & Sawston.* **Sun 6 June (2-6). Combined adm £7, chd free. Home-made teas at one of the churches in the village (tea and cake) and at Bustlers Cottage (for cream teas if fine). Venues will be clearly signed.**

BUSTLERS COTTAGE
John & Jenny Marks.

DUXFORD MILL
Mrs Frankie Bridgwood.

2 GREEN STREET
Mr Bruce Crockford.

5 GREEN STREET
Jenny Shaw.

16 ICKLETON ROAD
Claire James.

NEW KINGS HEAD HOUSE
Russell Waller.

NEW 26 PETERSFIELD ROAD
Robert & Josephine Smit.

31 ST PETER'S STREET
Mr David Baker.

Eight village gardens of different sizes and characters. In size the gardens range from the charming small cottage garden at 2 Green Street to Duxford Mill, 11 acres on the River Cam. The sunny aspect at 16 Ickleton Road has interesting and unusual planting, as does 31 St Peters Street also with its rockery in front of the house. Bustlers Cottage has an acre of cottage garden including vegetable garden and newly planted hedges. 5 Green Street's walled garden contains lawn, pergolas, various beds and vegetables. The innovative green walls in the garden at 26 Petersfield Rd are vibrant with colour, and Kings Head House, new to the NGS, has a long herbaceous border, a hydrangea border and a wide variety of roses. Some gravel paths and a few steps, mostly avoidable.

 ◆ ELGOOD'S BREWERY GARDENS
North Brink, Wisbech, PE13 1LW.
Elgood & Sons Ltd, 01945 583160,
info@elgoods-brewery.co.uk,
www.elgoods-brewery.co.uk.
1m W of town centre. Leave A47 towards Wisbech Centre. Cross river to North Brink. Follow river & brown signs to brewery & car park beyond. **For NGS: Sun 1 Aug (11.30-4.30). Adm £5, chd free. Light refreshments. For other opening times and information, please phone, email or visit garden website.**
Approx 4 acres of peaceful garden featuring 250 yr old specimen trees providing a framework to lawns, lake, rockery, herb garden and maze. Wheelchair access to Visitor Centre and most areas of the garden.

Norfolk Terrace Garden

© Howard Rice

16 ELM HOUSE

Main Road, Elm, Wisbech, PE14 0AB. Mrs Diana Bullard. *2½m SW of Wisbech. From A1101 take B1101, signed Elm, Friday Bridge. Elm House is ⅓m on L, well signed.* **Sun 30 May (1-5). Adm £4, chd free. Home-made teas.** Walled garden with arboretum, many rare trees and shrubs, mixed perennials C17 house (not open). New 3 acre flower meadow. Children welcome. Guide dogs only please. Some uneven ground in flower meadow otherwise most paths suitable.

GROUP OPENING

17 ELY OPEN GARDENS

Ely, CB7 4DL. *14m N of Cambridge. Parking at Barton Rd car park, Tower Rd, (adjacent to 42 Cambridge Rd); the Grange Council Offices; St Mary's St. These within easy reach of most of the gardens. A map given at first garden visited.* **Sun 6 June (12-6). Combined adm £8, chd free.**

BISHOP OF HUNTINGDON'S GARDEN
Dagmar, Bishop of Huntingdon.

THE BISHOP'S HOUSE
The Bishop of Ely.

42 CAMBRIDGE ROAD
Mr & Mrs J & C Switsur.

12 CHAPEL STREET
Ken & Linda Ellis.

38 CHAPEL STREET
Peter & Julia Williams, 01353 659161, peterrcwilliams@btinternet.com.
Visits also by arrangement Mar to Oct for groups of up to 20.

NEW 9 CHIEFS STREET
Amanda Seamark.

5B DOWNHAM ROAD
Mr Christopher Cain.

A delightful and varied group of gardens in an historic Cathedral city: The Bishop's House adjoins Ely Cathedral and has mixed planting, including a formal rose garden, bordered by wisteria. There is a local artist displaying and selling work. The Bishop of Huntingdon's garden is for family and entertaining. Lawns, herbaceous borders, an orchard and more! 9 Chiefs Street is new this year. Divided into areas, it has developing borders, an all weather play area, mini allotment and chickens. 12 Chapel Street, a small town garden, reflecting the owners varied gardening interests, from alpines to herbaceous and vegetables all linked with a railway! 38, Chapel St has year round interest and is bursting with unusual and interesting planting. 42 Cambridge Road, a secluded town garden has colourful herbaceous borders, roses, shrubs and trees. 5b Downham Road shows how to rise to the challenge of a small interestingly shaped area. Careful planting has created a tranquil area. Wheelchair access to areas of most gardens.

18 FENLEIGH

Inkerson Fen, off Common Rd, Throckenholt, PE12 0QY. Jeff & Barbara Stalker, 07800575100, barbara.stalker@ngs.org.uk. *Half way between Gedney Hill & Parson Drove. Postcode for Sat Navs. PE12 0QY, please email for further directions.* **Visits by arrangement May to Sept for groups of up to 30. Adm £3, chd free. Tea.** A fish pond dominates this large garden surrounded with planting. Patio with pots and raised beds, undercover BBQ area. Small wooded area, poly tunnels and corners of the garden for wildlife.

GROUP OPENING

19 NEW FENSTANTON

Huntingdon, PE28 9JW. *Public car park next to the Church, a two minute walk away from both gardens.* **Sat 5, Sun 6 June (10-4). Combined adm £5, chd free. Light refreshments.**

NEW 5 CHURCH LANE
Mrs Jan Stone.

Old Farm Cottage, Staploe Gardens

THE MANOR HOUSE
Lynda Symonds & Nigel Ferrier.

Two delightful gardens. The Manor House was once the home of Capability Brown. Its formal garden is designed in a series of rooms with pleached limes, an avenue of Himalayan Birch and interesting cottage borders set in just over a third of an acre. Featured on C4 TV with Alan Titchmarsh and in the book: 'East Anglia's Secret Gardens'.Nearby in Church Lane, the new opening at number 5 presents a chance to see the small, enclosed, 'secret' garden hidden behind this house, one of the oldest in the village. The garden is packed with a wide variety of plants with a focus on hostas, succulents and ferns. Extra interest is provided by two ponds, one raised and one at ground level for encouraging wildlife.

🐾🐄☕

20 ◆ FERRAR HOUSE
Little Gidding, Huntingdon, PE28 5RJ. Mrs Susan Capp, 01832 293383, info@ferrarhouse.co.uk, www.ferrarhouse.co.uk. *Take Mill Rd from Great Gidding (turn at Fox & Hounds) then after 1m turn R down single track lane. Car Park at Ferrar House.* **For NGS: Sun 9 May (10.30-5). Adm £4, chd free. Home-made teas. For other opening times and Information, please phone, email or visit garden website.**
A peaceful garden of a Retreat House with beautiful uninterrupted views across meadows and farm land. Adjacent to the historic Church of St John's, it was here that a small religious community was formed in the C17. The poet T. S. Eliot visited in 1936 and it inspired the 4th of his Quartets named Little Gidding. Lawn and walled flower beds with a walled vegetable garden. Games on the lawn. Traditional wooden garden games, tea and home-made cakes, plant stall. History talks in Church. WC accessible at Ferrar House.

♿🐄❄🚍⛲☕

21 NEW GIFU
3 Birds Close, Ickleton, Saffron Walden, CB10 1SU. Cali Holberry. *Birds Close is very narrow please try to park elsewhere in the village.* **Sun 30 May (2-5). Adm £4, chd free. Also open Hinxton Gardens.**
Artist and plant lover's ornamental bee and wildlife friendly garden,

including a small woodland with seat. South facing, dry garden with grasses, dry planting and red nut tree. Main garden with mature bamboos and small deep pond overlooked by a kitchen garden with herb and salad beds. Most paths suitable for wheelchairs. Some small ramps.

♿

22 GREEN END FARM
Over Road, Longstanton, Cambridge, CB24 3DW. Sylvia Newman, www.sngardendesign.co.uk. *From A14 take direction of Longstanton At r'about, take 2nd exit At next r'about, turn L (this shows a dead end on the sign) We're 200 metres on L.* **Sun 20 June, Sun 4 July (11-3). Adm £5, chd free. Home-made teas. and refreshments.**
A developing garden that's beginning to blend well with the farm. An interesting combination of new and established spaces executed with a design eye. An established orchard with beehives; two wildlife ponds. An outside kitchen, productive kitchen and cutting garden. Doves, chickens and sheep complete the picture!

♿❄☕

23 NEW 6 HEMINGFORD CRESCENT
Stanground, Peterborough, PE2 8LL. Michael & Nick Mitchell, 07880 871763, michaelandnick64@gmail.com. *S side of city centre. 10 mins drive from A1M. Take Whittlesey Rd B1092. Coming out of town look for Apple Green petrol stn on L. Turn into Coneygree Rd. Follow rd round, Hemingford Crescent is 6th turning on L.* **Visits by arrangement June to Aug for groups of up to 20. Adm £5, chd free. Light refreshments.**
Medium sized city garden. Inspired by our travels to Morocco, Egypt and India. Mixture of exotic planting, shrubs and perennials. Various seating areas including an outside dining area and an enclosed moroccan/egyptian room. Raised beds and pond area. Michael is an artist and his work is available for viewing. Partial wheelchair access. No dogs.

☕

GROUP OPENING

24 HIGHSETT CAMBRIDGE
Cambridge, CB2 1NZ. Alice Fleet. *Centre of Cambridge. Via Station Rd, Tenison Rd, 1st L Tenison*

Ave, entrance ahead. SatNav CB2 1NZ Regret no parking available in Highsett.* **Sun 6 June (2-5). Combined adm £6, chd free. Home-made teas at 82 Highsett.**
9 delightful town gardens will be open to view in this well known housing complex within Central Cambridge. Set in large communal grounds with fine specimen trees, many 70+ years old, and lawns. A haven for children and wildlife. Architect Eric Lyons planned the whole estate in the late 1950's with a mixture of flats, small houses and large town houses, the very ethos of tranquil living space for all generations. Several of the Open Gardens have been skilfully modernised by garden designers. Most paths accessible into the gardens.

♿🐾☕

GROUP OPENING

25 NEW HINXTON GARDENS
High Street, Hinxton, Saffron Walden, CB10 1QY. *Most gardens are on or close to Hinxton High Street. Accessible via A1301.* **Sun 30 May (2-5). Combined adm £5, chd free. Home-made teas in Hinxton Village Hall. Also open Gifu.**

NEW **2 HALL FARM BARNS**
Ivan Yardley.

NEW **25 HIGH STREET**
Steve Trudgill.

NEW **85 HIGH STREET**
Mike Boagey.

NEW **MANOR HOUSE**
Sara Varey.

NEW **MILLER'S COTTAGE**
Sue & Chris Elliott.

NEW **THE OAK HOUSE**
Jane Chater.

A group of six village gardens of varying sizes including cottage gardens, with traditional planting, a garden belonging to a C15 Manor House with planting appropriate to the era. Millers Cottage, next to the old water mill, with narrow riverside garden and planting appropriate to its position, a barn conversion garden with tropical dry planting, and a family garden with lawns, shrubs and perennials. Delicious teas and large selection of plants for sale. Some gravel paths, a few steps, narrow entrances, but mainly navigable.

♿❄🚍☕

26 HORSESHOE FARM

Chatteris Road, Somersham, Huntingdon, PE28 3DR. Neil & Claire Callan, 01354 693546, nccallan@yahoo.co.uk. *9m NE of St Ives, Cambs. Easy access from the A14. Situated on E side of B1050, 4m N of Somersham Village. Parking for 8 cars in the drive.* **Visits by arrangement May to July. Any number of visitors up to 20. Adm £4, chd free. Home-made teas.**
This ¾ acre plant-lovers' garden has a large pond with summer-house and decking, bog garden, alpine troughs, mixed rainbow island beds with over 30 varieties of bearded irises, water features, a small hazel woodland area, wildlife meadow, secret corners and a lookout tower for wide Fenland views and bird watching.

✿ 🚗 ☕

27 28 HOUGHTON ROAD

St Ives, PE27 6RH. Julie Pepper. *On A1123, western edge of St Ives. Garden on the L as you enter St Ives from Huntingdon.* **Thur 29, Fri 30 Apr, Fri 7, Sat 8, Sat 15, Sun 16, Sun 23, Mon 24 May, Wed 2, Thur 3, Thur 10, Fri 11 June (11-5). Adm £3.50, chd free. Pre-booking essential, please visit www.ngs. org.uk for information & booking. Light refreshments. Gluten free options available.**
Step into the peace and tranquility of Acer heaven, created by designer Julie. Among more than 60 Acers, of which there are 45 varieties, you will also find a patio area set out as a room for outside dining and relaxation. There is much use of dramatic accent planting, punctuated with sculpture, artifacts, water feature, seating areas, bonsai, hostas, trees, ferns and many other interesting plants.

☕ 🪑

28 ISLAND HALL

Godmanchester, PE29 2BA. Christopher & Grace Vane Percy, www.islandhall.com. *1m S of Huntingdon (A1). 15m NW of Cambridge (A14). In centre of Godmanchester next to free Mill Yard car park.* **Sun 30 May (10.30-4.30). Adm £5, chd free. Home-made teas in the Fisherman's Lodge on the island set within the grounds.**
3-acre grounds. Tranquil riverside setting with mature trees. Chinese bridge over Saxon mill race to embowered island with wild flowers. Garden restored in 1983 to mid C18

formal design, with box hedging, clipped hornbeams, parterres, topiary, good vistas over borrowed landscape and C18 wrought iron and stone urns. The ornamental island has been replanted with Princeton elms (ulmus americana). Mid C18 mansion (not open).

♿ 🐕 ✿ ☕ ☕

29 NEW 38 KINGSTON STREET

Cambridge, CB1 2NU. Wendy & Clive Chapman. *Central Cambridge. Off Mill Road, north of railway bridge which is now only open for bus & bicycle traffic.* **Sun 30 May (2-6). Adm £3, chd free. Also open Cottage Garden.**
A tiny courtyard garden with contrasts between deep shade and sunlit areas. Raised beds and pots of various sizes with a range of perennial and annual planting. Small step into garden.

♿

30 KIRTLING TOWER

Newmarket Road, Kirtling, Newmarket, CB8 9PA. The Lord & Lady Fairhaven. *6m SE of Newmarket. From Newmarket head towards village of Saxon Street, through village to Kirtling, turn L at war memorial, signed to Upend, entrance is signed on the L.* **Sun 28 Mar, Sun 27 June (11-4). Adm £5, chd free. Light refreshments at Church.**
Surrounded by a moat, formal gardens and parkland. In the spring there are swathes of daffodils, narcissi, crocus, muscari, chionodoxa and tulips. Closer to the house, vast lawn areas Secret and Cutting Gardens. In the summer the Walled Garden has superb herbaceous borders with anthemis, hemerocallis, geraniums and delphiniums. The Victorian Garden is filled with peonies. Views of surrounding countryside. A Classic car display will be in attendance. The Arcadia Recorder Group will be playing in the walled garden. A variety of plant and craft stalls on both dates as well as a display of Stonework from the Fairhaven Stoneyard. Selection of delicious hot and cold food, sandwiches, cakes, tea and coffee. Many of the paths and routes around the garden are grass - they are accessible by wheelchairs, but can be hard work if wet.

♿ 🐕 ✿ 🚗 ☕ ☕

31 NEW LANDWADE HALL

Landwade Road, Exning, Newmarket, CB8 7NH. Mr Simon Gibson. *Landwade Hall is reached by talking the turning off the A142 r'about towards Snailwell. Disabled parking in drive.* **Sat 13, Sun 14 Feb (10-3). Last entry (2.30). Adm £7.50, chd free. Hot soup.**
4 acre garden with small C15 church. Moat with foundations of the original house. 100 years of snowdrops, aconites and daffodils.A delight in early spring. Wheelchair access to grassed areas (approx 50%) may be hard work on damp ground

♿ 🚗 ☕

32 LITTLE OAK

66 Station Road, Willingham, Cambridge, CB24 5HG. Mr & Mrs Eileen Hughes, www.littleoak.org.uk/garden. *4m N of A14 J29 near Cambridge. Easy to find on the main road in the village.* **Sat 12, Sun 13 June (1-5). Adm £4.50, chd free. Home-made teas.**
Michael and Eileen welcome you to our 1 acre garden. Featuring a 50' Laburnum Walk with perennial borders, ponds, Cottage Garden, fruit cage, kitchen garden/greenhouses/growing tunnels, orchard, Mediterranean and rose garden, coppice and chickens. Teas and home-made cakes. Picnics welcome. Weather permitting, Michael will demonstrate woodturning using a pole lathe throughout afternoon openings. Main garden is wheelchair accessible. Driveway parking for disabled use only.

♿ ✿ ☕ 🪑

33 MADINGLEY HALL

Cambridge, CB23 8AQ. University of Cambridge, 01223 746222, reservations@madingleyhall.co.uk, www.madingleyhall.co.uk. *4m W of Cambridge. 1m from M11 J13. Located in the centre of Madingley village. Entrance adjacent to mini roundabout.* **Sun 18 July (2.30-5.30). Adm £5, chd free. Home-made teas.**
C16 Hall (not open) set in 8 acres of attractive grounds landscaped by Capability Brown. Features incl landscaped walled garden with hazel walk, alpine bed, medicinal border and rose pergola. Historic meadow, topiary, mature trees and wide variety of hardy plants.

✿ 🏠 ☕

34 NEW 67 MAIN ST

Yaxley, Peterborough, PE7 3LZ.
Mrs Karen Woods, 07787 864426,
karenandstevewoods@yahoo.co.uk.
S of Peterborough, approx 3m from
J16 on A1. On entering Yaxley, turn r
into Dovecote Lane. At the bottom,
turn L onto Main St. Find us approx
1m on R. **Visits by arrangement
Apr to Dec. Adm £4, chd free.
Tea.**
1 acre working organic cottage
garden providing something for the
kitchen and vase all year round.
Seasonal cut flowers grown in
raised beds and traditional borders.
Allotment style vegetable garden.
Paddock with fruit trees and views
over the fen to the rear. Chickens for
eggs. We enjoy working alongside the
wildlife that we share our garden with.

35 ♦ THE MANOR, HEMINGFORD GREY

Hemingford Grey, PE28 9BN.
Mrs D S Boston, 01480 463134,
diana_boston@hotmail.com,
www.greenknowe.co.uk. 4m E of
Huntingdon. Off A1307 Entrance
to garden by small gate off river
towpath. Limited parking on verge
halfway up drive. Parking near house
also limited. Otherwise please park
in village/. **For opening times and
information, please phone, email or
visit garden website.**
Garden designed by author Lucy
Boston, surrounds 12th Century
manor house on which Green Knowe
books based (house open by appt).
3 acre 'cottage' garden with topiary,
snowdrops, old roses, extensive
collection of irises incl Dykes Medal
winners and Cedric Morris varieties,
herbaceous borders with mainly
scented plants. Meadow with mown
paths. Enclosed by river, moat and
wilderness. Late May splendid show
of irises followed by the old roses.
Care is taken with the planting to
start the year with a large variety
of snowdrops and to extend the
flowering season through to the
first frosts with colour from unusual
annuals. The garden is interesting
even in winter with the topiary.
Gravel paths but wheelchairs are
encouraged to go on the lawns.

36 ♦ MARY CHALLIS GARDEN

68 High Street, Sawston,
Cambridge, CB22 3BG. A M
Challis Trust Ltd, 01223 834511,
chair@challistrust.org.uk,

www.challistrust.org.uk. 5m SE of
Cambridge. Passageway between
60 High St & 66 High St (Billsons
Opticians). **For NGS: Sun 13 June
(1-5). Adm £3, chd free. Home-
made teas. Also open Clover
Cottage. For other opening times
and information, please phone,
email or visit garden website.**
Gifted to Sawston in 2006, this
2 acre garden is maintained by
volunteers to benefit wildlife and for
the local community. Winter/spring
long border, drifts of crocuses,
snowdrops and aconites in spring;
very colourful summer flowerbeds.
Woodland, specimen trees, pond,
raised vegetable beds, orchard,
vine house, wildflower meadow and
beehives. Paths and lawns accessible
by wheelchair from car-park.

37 NEW 34 MILLINGTON ROAD

Cambridge, CB3 9HP. Charlotte
Forbes. Just off the Barton Road in
Newnham. Coming into Cambridge
on Barton Road, turn R onto
Millington Road (just before St Marks
church), proceed to very end, house
is the penultimate one on R. **Sat 5
June (2-5). Adm £3.50, chd free.
Light refreshments.**
A small organic city garden with a
romantic country feel belonging to
garden designer Charlotte Forbes.
The garden includes an informal
back garden with large circular beds
filled with perennials and self-seeded
annuals/biennials, a herb garden,
a shady seating area, a rose arch,
newly planted prairie beds, a small
cutting bed for annual flowers, raised
dahlia beds and a patio with pots.

38 NEW MILTON HALL

Milton Park, Milton, Peterborough,
PE6 7AG. Lady Isabella Naylor-
Leyland. Please use entrance off
A47, shared with Peterborough
(Milton) Golf Club. **Sat 8 May
(10.30-5.30). Adm £7, chd free.
Cream teas.**
20 acres of pleasure grounds laid
out by Humphrey Repton in 1791
including lake, mature trees, extensive
lawns, hard gravel paths, historic
orangery. Enclosed walled Italian
garden and kitchen garden. Teas
served in orangery all day. Garden
level, with hard gravel surfaced paths.

39 MUSIC MAZE AND GARDEN

Troy Green, 2A Nine Chimneys
Lane, Balsham, Cambridge,
CB21 4ES. Mr & Mrs Jim &
Hilary Potter, 01223 891211,
hppotter@btinternet.com,
www.balshammaze.org.uk. In centre
of Balsham just off High St. 3m E of
A11, 12m S of Newmarket & 10m SE
of Cambridge. Car parking in the High
St or the Church car park. 2 disabled
spaces in Nine Chimneys Lane. **Visits
by arrangement Apr to Oct for
groups of 10 to 20. Tour by Garden
Owners, usually 1-2 hours. Tea.**
Two acres of garden with spring
bulbs, mixed borders, raised
vegetable beds, gravel garden, large
duck pond, wildflower meadow, an
orchard, modern sculptures and
human sundial. A yew Music Maze
was planted in 1993 with golden yew
(Taxus elegantissima) in the shape of
a treble clef, overlooked by a viewing
hill. Mature and new trees. In the
maze there are over 1500 trees. Two
paved areas form the shape of French
horns which enclose an alpine garden
and a mobile fountain. Large pond.
Children's swings and sandpit. Hard
paths around formal garden area. Ltd
wheelchair access to grassland areas.
Fine in dry weather with a good driver.

40 NETHERHALL MANOR

Tanners Lane, Soham, CB7 5AB.
Timothy Clark, 01353 720269. 6m
Ely, 6m Newmarket. Enter Soham
from Newmarket, Tanners Lane
2nd R 100yds after cemetery. Enter
Soham from Ely, Tanners Lane 2nd L
after War Memorial. **Sun 4 Apr, Sun
2 May, Sun 1, Sun 8 Aug (2-5).
Adm £3, chd free. Home-made
teas. Visits also by arrangement
Mar to Aug for groups of 10 to 30.**
Elegant garden 'touched with
antiquity'. Good Gardens Guide 2000.
Unusual garden appealing to those
with historical interest in individual
collections of plant groups: March-old
primroses, daffodils, Victorian double
flowered hyacinths & first garden
hellebore hybrids. May-old English
tulips, Crown Imperials. Aug-Victorian
pelargonium, heliotrope, calceolaria,
dahlias. Author-Margery Fish's Country
Gardening & Mary McMurtrie's Country
Garden Flowers Historic Plants 1500-
1900. The only bed of English tulips on
display in the country. Author's books
for sale. Featured on Gardeners' World
three times. Flat garden with two
optional steps. Lawns.

41 THE NIGHT GARDEN

37 Honeyhill, Paston, Peterborough, PE4 7DR. Andrea Connor, 07801 987905, andrea.connors@ntlworld.com, www.gardenofsanctuary.co.uk. *From A47 Soke Parkway at J19. Take exit N on Topmoor Way. R at next r'about (3rd exit) to Paston Ridings, over the speed humps. Take 4th L to Honeyhill. 1st R into car park. No.37 at top L corner.* **Evening opening Sat 28, Sun 29, Mon 30 Aug (7.30-10.30). Adm £5, chd free. Light refreshments in garden, or house if weather inclement. Tea, coffee, wine, cake and biscuits, £3 donation appreciated. Visits also by arrangement July to Sept for groups of 10 to 20.**

Small town garden with shrubs, roses, clematis, bedding plants and Ash tree for shade, seclusion and privacy. Many Salvias for 'wellbeing' and other colourful flowers. Arches and trellis give height, water adds tranquility. At dusk, the essential features of the garden are the solar-powered lighting which gradually transforms the space into an enchanting magical world. Sit and enjoy the gradual transformation from seating areas at the ends of the garden. Night time garden lighting brings an unusual dimension, transforming the experience, hence suggested evening visiting times.

42 38 NORFOLK TERRACE GARDEN

Cambridge, CB1 2NG. John Tordoff & Maurice Reeve. *Central Cambridge. A603 East Rd turn R into St Matthews St to Norfolk St, L into Blossom St & Norfolk Terrace is at the end.* **Sat 3, Sun 4, Sat 24, Sun 25 July (12-6). Adm £3, chd free. Home-made teas.**

A small, paved courtyard garden in Moroccan style. Masses of colour, backed by oriental arches. An ornamental pool decorated in patterned tiles. The garden won third prize in the 2018 Gardener's World national competition. It is also included in the book 'The Secret Gardens of East Anglia'. The owners' previous, London garden, was named by BBC Gardeners' World as 'Best Small Garden in Britain'. There will also be a display of recent paintings by John Tordoff and handmade books by Maurice Reeve.

43 THE OLD RECTORY

312 Main Road, Parson Drove, Wisbech, PE13 4LF. Helen Roberts, 01945 700415, yogahelen@talk21.com. *SW of Wisbech. From Peterborough on A47 follow signs to Parson Drove L after Thorney Toll. From Wisbech follow the B1166 through Leverington Common.* **Sun 6 June (11-4). Adm £4.50, chd free. Home-made teas. Visits also by arrangement May to Aug for groups of 10+.**

After parking in our field walk through the wildflower meadow and paddocks, past the long pond and then enter the main walled cottage style garden under our new moon gate. No hills but lovely open Fen views then maybe on your way out have a putt on our golf hole?

44 23A PERRY ROAD

Buckden, St Neots, PE19 5XG. David & Valerie Bunnage, 01480 810553, d.bunnage@btinternet.com. *5m S of Huntingdon on A1. From A1 Buckden roundabout take B661, Perry Rd approx 300yds on left.* **Every Fri and Sat 14 May to 5**

Abbots Barn

Sept (2-5). Every Sun 16 May to 5 Sept (10-4). Adm £4, chd free. Visits also by arrangement May to Sept.

Approx 1 acre garden consisting of many garden designs including Japanese interlinked by gravel paths. Large selection of acers, pines, rare and unusual shrubs. Also interesting features, a quirky garden. Plantsmans garden for all seasons, small bog garden. WC. Coaches welcome.

45 NEW PREACHERS PASSING

55 Station Road, Tilbrook, Huntingdon, PE28 0JT. Mr & Mrs Keith & Rosamund Nancekievill. *Tilbrook is 4½m S of J16 on A14. Station Road in Tilbrook can be accessed from the B645 or B660. Preachers Passing faces the small bridge over the River Til at a sharp bend in Station Road with All Saints church behind it.* **Thur 15 July, Sat 14 Aug (10.30-5). Adm £5, chd free.**

¾ acre garden fits into its pastoral setting. Near the house, parterre, courtyard and terrace offer formality; but beyond, prairie planting leads to a wild life pond, rock gardens, meadow, stumpery, rose garden and copse. Enjoy different views from arbour, honeysuckle-covered swing or scattered benches. Deciduous trees and perennials give changing colour. Here are open spaces and hidden places.

46 QUERCUS

Bradley Road, Burrough Green, Newmarket, CB8 9NH. Dulce Threlfall, 01638 508470, thedulcethrelfall@gmail.com. *5m S of Newmarket. First house on L entering BG from Newmarket immediately after 30 speed limit.* **Visits by arrangement Apr to Sept for groups of 10 to 30. Adm £10, chd free. Home-made teas.**

There are 7 different areas in this renovated, replanted, redesigned and evolving 1.5 acre garden. Mirrored mixed borders, roses, shade plants, vegetables, orchard, interesting trees, greenhouse. Brand new for 2021 a large natural swimming pool finished in October 2020. Many lovely aquatic plants and amphibians. Wheelchair access to all garden.

47 ROBYNET HOUSE

28 Green Street, Duxford, Cambridge, CB22 4RG. Gordon Lister. *From A505 just E of M11, lm S turn L onto St John's Street & R into Green St by St John's Church.* **Sun 15 Aug (2-6). Adm £5, chd free.**

This garden, partly created from a field, has been gardened by the owner since 1986. Over 30 different varieties of dahlias, 3 main varieties of hibiscus, several acers, and possibly the healthiest horse chestnut in the area. Many 'rooms' of different character and atmosphere, a stream and pond and a formal fountain. The dry garden contains a rock from the family's original plant nursery in Bute. Some gravel paths, and a few steps which can be circumnavigated.

GROUP OPENING

48 SAWSTON GARDENS

Sawston, Cambridge, CB22 3HY. *5m SE of Cambridge. Midway between Saffron Walden & Cambridge on A1301.* **Sun 4 July (1-5). Combined adm £5, chd free. Cream teas at the Sweet Tea cafe in the High St.**

BROOK HOUSE
Mr & Mrs Ian & Mia Devereux.

11 MILL LANE
Tim & Rosie Phillips.

22 ST MARY'S ROAD
Ann & Mike Redshaw.

3 gardens in this large South Cambs. village. Brook House has many delightful features in 1½ acre grounds of a grade II listed house, including a pond with expansive patio and pergolas, and a large walled kitchen garden with greenhouse, raised beds and herbaceous borders. 11 Mill Lane is an attractive C16/C19 house with an impressive semi circular front lawn edged with roses and mixed borders, and a secluded sun-dappled garden at the rear. A large wildlife-friendly family garden at 22 St Mary's Road has colour-themed borders, pond, fruit trees, gravel beds, raised salad beds and beehives, and good views over SSSI meadows. Enjoy a cream tea at Sweet Tea cafe on the High Street. Most parts of the gardens are accessible by wheelchair.

49 NEW SILVER BIRCHES

77 North Street, Burwell, Cambridge, CB25 0BB. Mr & Mrs Jeremy Lander. *In village very close to The Anchor Pub. Coming from Cambridge take The Causeway off B1102 carry on 0.5m N of Co-op, or use Toyse Lane off B1102. Park in Toyse Lane, or on North St N of Toyse Lane junction.* **Sun 18 July (12-5.30). Adm £4, chd free. Home-made teas.**

A 400' long garden on the edge of the Cambridgeshire Fens, divided into separate 'compartments': an informal flower garden with seating, lawns and small pond, a wooded area with silver birches, coppiced hazel, and two hammocks. A gate then leads to an orchard with free range chickens. Beyond is a vegetable garden, and finally a wooden jetty onto the 'Catchwater Drain', a medieval watercourse. Approx half the garden is wheelchair accessible.

50 THE SIX HOUSES

30-45 Comberton Road, Barton, CB23 7BA. Perennial. Sun 4 July (12-5). Combined adm with Two Toft Gardens £5, chd free. Opening with Barton Gardens on Sun 2 May. Six unique gardens filled with charm, each beautifully tended in a variety of cottage styles by resident retired professional gardeners, beneficiaries of the charity Perennial. Further communal garden of almost two acres features a collection of choice trees including maple, ginkgo and cherry, underplanted with naturalised bulbs. Several newly planted herbaceous beds. Many plants labelled.

> "The support of Hospice UK and the National Garden Scheme has been invaluable to hospice nurses across the country whilst we've been battling the coronavirus crisis, helping hospices such as Derian House to continue providing vital end of life and respite care to 400 children and young adults from across the North West. Thank you."
> – Katie Turner, Perinatal Nurse at Derian House Children's Hospice

GROUP OPENING

51 STAPLOE GARDENS

Staploe, St Neots, PE19 5JA. caroline-falling@hotmail.co.uk. *Great North Rd in western part of St Neots. At r'about just N of the Coop store, exit westwards on Duloe Rd. Follow this under the A1, through the village of Duloe & on to Staploe.* **Mon 26 Apr (1-5). Sat 5, Sun 6 June (1-5). Home-made teas in June only. Sun 17 Oct (1-5). Combined adm £5, chd free. Visits also by arrangement Apr to Oct for groups of up to 30.**

FALLING WATER HOUSE
Caroline Kent, 07702 707880

OLD FARM COTTAGE
Sir Graham & Lady Fry

Old Farm Cottage: flower garden surrounding thatched house (not open), with 3 acres of orchard, grassland, woodland and pond maintained for wildlife. Ginkgo, loquat and manuka trees grown from seed, and wildflower meadow. Cherry blossom brightens the spring, and in autumn spindles, ginkgos and maples provide vivid pinks and yellows. Falling Water House: a mature woodland garden, partly reclaimed from farmland 10yrs ago, is constructed around several century old trees incl three Wellingtonia. Kitchen garden potager, courtyard and herbaceous borders, planted to attract bees and wildlife, through which meandering paths have created hidden vistas. Old Farm Cottage has uneven ground and one steep slope.

GROUP OPENING

52 SUTTON GARDENS

Sutton, Ely, CB6 2QQ. *6m W of Ely. From A142 turn L at r'about on to B1381 to Earith. Parking at Brooklands Centre 1 km on R.* **Sun 30 May (1-6). Combined adm £7, chd free. Home-made teas at The Burystead.**

THE BURYSTEAD
Sarah Cleverdon & Stephen Tebboth.

NEW 8 HADDOCK'S RISE
Mr William England.

61 HIGH STREET

Ms Kate Travers & Mr Jon Megginson, 01353 778427, Jonmegginson1@gmail.com. **Visits also by arrangement.**

63 HIGH STREET

Ms Ruth & Mr Arthur Brown.

THE OLD BAPTIST CHAPEL

Janet Porter & Steve Newton.

19 THE ROW

Alistair & Jane Huck.

NEW 89 THE ROW

Andrew Thompson.

The Old Baptist Chapel is a new garden created around an C18 building converted to a family home. 61 The High Street - a tiny shaded garden with a host of features: interesting trees, herb and vegetable plots, two greenhouses and boxes for birds, bats and bugs. 63 The High Street - a sloping sunny garden with spectacular views over the fens. Borders are in the 'cottage garden' style with many traditional favourites. 19 The Row is a long ¾ acre mature garden. Summer house, shed with green roof, pond, meadow, sculpture, unusual trees, shrubs, perennials, fruit and vegetables. 89 The Row has a variety of planting: mature trees, shrubs, fernery, a small vegetable plot and herbaceous plants with an emphasis on scent. 8 Haddocks Rise - a new, small but packed urban garden: still maturing with trees, roses and shrubs giving year round colour. The Burystead - a ½ acre walled courtyard garden of formal design within a large country garden against the backdrop of a restored C16 thatched barn.

53 NEW TOP FARM

High Street, Toseland, St Neots, PE19 6RX. Mandy & Steve Hill. *1½m from A428; 9m from Black Cat r'about on A1; 5½m from St Neots. Leave A1 & drive towards Cambridge along A428. Turn L at yellow sign opposite a car wash. Or, from St Neots, drive towards Great Paxton. Turn R at yellow sign.* **Sat 7, Sun 8 Aug (11-4). Adm £4, chd free. Home-made teas.**
Planted from scratch in 2016, on heavy clay, the garden now comprises a number of different areas. The main garden has long herbaceous borders in different colour schemes either side of a gravel path with rose covered

pergolas. Other areas include a large vegetable garden and cut flower bed, a natural pond with informal planting for shade, and a paddock with fruit trees and wildflowers.

54 TRINITY HALL - WYCHFIELD SITE

Storey's Way, Cambridge, CB3 0DZ. The Master & Fellows, www.trinhall.cam.ac.uk/about/gardens/. *1m NW of city centre. Turn into Storey's Way from Madingley Rd (A1303) & follow the yellow NGS signs. Limited on-road parking is available.* **Sat 3 July (11-3). Adm £4, chd free. Home-made teas. Hot and cold beverages, selection of home-made cakes in the cricket pavilion.**
A beautiful, large garden that complements the interesting and varied architecture. The Edwardian Wychfield House and its associated gardens contrast with the recent, contemporary development located off Storey's Way. Majestic trees, roses and herbaceous perennials, shady under-storey woodland planting and established lawns, work together to provide an inspiring garden. Plant sale. Some gravel paths.

55 TWIN TARNS

6 Pinfold Lane, Somersham, PE28 3EQ. Michael & Frances Robinson, 07938 174536, mikerobinson987@btinternet.com. *Easy access from the A14. 4m NE of St Ives. Turn onto Church St. Pinfold Lane is next to the church. Please park on Church Street as access is narrow & limited.* **Sat 5, Sun 6 June (1-6). Adm £4, chd free. Visits also by arrangement May to Sept for groups of up to 30.**
One-acre wildlife garden with formal borders, kitchen garden and ponds, large rockery, mini woodland, wildflower meadow (June/July). Topiary, rose walk, greenhouses. Character oak bridge, veranda and tree-house. Adjacent to C13 village church.

Online booking is available for many of our gardens, at ngs.org.uk

26 Petersfield Road

GROUP OPENING

56 NEW TWO TOFT GARDENS
Comberton Road, Toft,
Cambridge, CB23 2RY. *8m W of
Cambridge. Opp each other on
main road through Toft (B1046).
What3words app - bags.holidays.
grafted.* **Sun 4 July (12-5).
Combined adm with The Six
Houses £5, chd free.**

NEW 21 COMBERTON ROAD
Mr Sheppard.

NEW OLD FARM
Kay Brown.

Two contrasting gardens. The
sheltered south facing garden at Old
Farm is in a colourful cottage style
brimming with perennials, vegetables
and flower cutting beds. The newly
designed garden opposite at No.
21, owned by a college gardener,
is creatively set out in four zones
with planting to reflect history, food,
sunshine and much more besides.
&

57 WILD ROSE COTTAGE
Church Walk, Lode,
Cambridge, CB25 9EX. Mrs
Joy Martin, 01223 811132,
joymartin123@outlook.com. *From
A14 take the rd towards Burwell turn
L into Lode & park on L after 150
metres. Walk straight on between
thatched cottages to the archway
of Wild Rose Cottage.* **Evening
opening Wed 9 June (6-9). Adm
£10, chd free. Wine or home made
elderflower is incl in the adm.
Visits also by arrangement Apr to
Sept.**
A real cottage garden overflowing
with plants. Gardens within gardens
of abundant vegetation, roses
climbing through trees, laburnum
tunnel, a daisy spiral in the summer,
circular vegetable garden and wildlife
pond. Described by one visitor as
a garden to write poetry in! It is a
truly wild and loved garden where
flowers in the vegetable circle are
not pulled up! Geese, a dog, circular
vegetable garden, wild life pond, and
wild romantic garden. Food and drink
available at 'The Shed'. Lots of little
path ways on grass and gravel so not
very easy for a wheelchair.
🍵

58 THE WINDMILL
10 Cambridge Road, Impington,
CB24 9NU. Pippa & Steve
Temple, 07775 446443,
mill.impington@ntlworld.com,
www.impingtonmill.org. *2½m N of
Cambridge. Off A14 at J32, B1049
to Histon, L into Cambridge Rd at
T-lights, follow Cambridge Rd round
to R, the Windmill is approx 400yds
on L.* **Visits by arrangement Apr
to Oct. Adm £5, chd free. Light
refreshments incl coffee/tea/wine
and nibbles by arrangement.**
A previously romantic wilderness of 1½
acres surrounding windmill, now filled
with bulbs, perennial beds, pergolas,
bog gardens, grass bed and herb
bank. Secret paths and wild areas with
thuggish roses maintain the romance.
Millstone seating area in smouldering
borders contrasts with the pastel
colours of the remainder of the garden.
Also 'Pond Life' seat, 'Tree God' and
amazing compost area! The Windmill -
an C18 smock on C19 tower on C17
base on C16 foundations - is being
restored. Guide dogs only.
& ✿ ⊜ 🍵

59 WRIGHTS FARM
Tilbrook Road, Kimbolton,
Huntingdon, PE28 0JW. Russell &
Hetty Dean. *13m SW of Huntingdon.
¼m W of Kimbolton on the B645.
Do not follow sat nav.* **Sat 3, Sun 4
July (10.30-5). Adm £5, chd free.
Home-made teas.**
A new garden started in 2013, set
within 4 acres. Varied borders with
bee and butterfly friendly planting.
Formal walled vegetable garden
with raised beds, potting shed
and greenhouses. Mediterranean
style courtyard with dry garden.
Paved entrance area with plenty of
pots. Tropical hot border. Riverside
paths and seating areas with open
countryside views.
✿ 🍵

CHESHIRE & WIRRAL

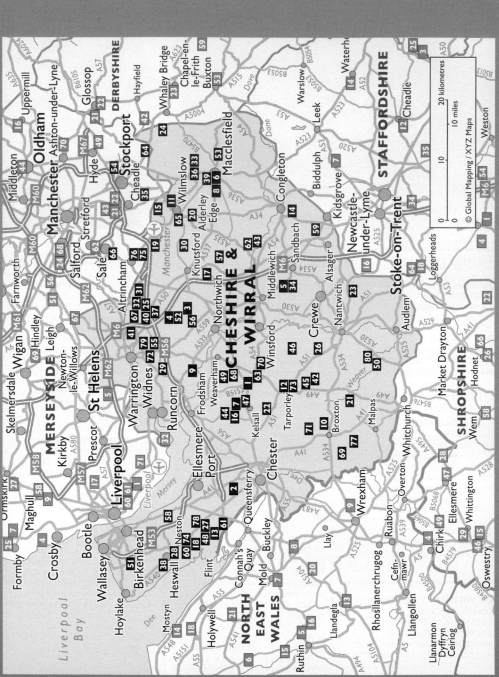

The area of Cheshire and Wirral comprises what are now the four administrative regions of West Cheshire and Chester, East Cheshire, Warrington and Wirral, together with gardens in the south of Greater Manchester, Trafford and Stockport.

The perception of the area is that of a fertile county dominated by the Cheshire Plain, but to the extreme west it enjoys a mild maritime climate, with gardens often sitting on sandstone and sandy soils and enjoying mildly acidic conditions.

A large sandstone ridge also rises out of the landscape, running some 30-odd miles from north to south. Many gardens grow ericaceous-loving plants, although in some areas, the slightly acidic soil is quite clayey. But the soil is rarely too extreme to prevent the growing of a wide range of plants, both woody and herbaceous.

As one travels east and the region rises up the foothills of the Pennine range, the seasons become somewhat harsher, with spring starting a few weeks later than in the coastal region.

As well as being home to one of the RHS's major shows, the region's gardens include two National Garden Scheme 'founder' gardens in Arley Hall and Peover Hall, as well as the University of Liverpool Botanic Garden at Ness.

Below: 181a Higher Lane

f @National Garden Scheme Cheshire & Wirral   🐦 CheshireWirrNGS

Volunteers

County Organiser
Janet Bashforth
01925 349895
jan.bashforth@ngs.org.uk

County Treasurer
Andrew Collin
01513 393614
andrewcollin@btinternet.com

Booklet Co-ordinator
John Hinde
0151 353 0032
johnhinde059@gmail.com

Assistant County Organisers
Graham Beech
01625 402946
graham@dhgb.co.uk

Sue Bryant
0161 928 3819
suewestlakebryant@btinternet.com

Jean Davies
01606 892383
mrsjeandavies@gmail.com

Linda Enderby
07949 496747
linda.enderby@ngs.org.uk

Sandra Fairclough
0151 342 4645
sandrafairclough51@gmail.com

Juliet Hill
01829 732804
t.hill573@btinternet.com

Romy Holmes
01829 732053
romy@bowmerecottage.co.uk

Richard Goodyear
01270 528944
goodyear.pickford@btinternet.com

Mike Porter
01925 753488
porters@mikeandgailporter.co.uk

OPENING DATES

All entries subject to change. For latest information check www.ngs.org.uk

Map locator numbers are shown to the right of each garden name.

February

Snowdrop Festival

Saturday 13th
Briarfield 13

Sunday 14th
Briarfield 13

Sunday 28th
Bucklow Farm 17

March

Sunday 21st
◆ Abbeywood Gardens 1

April

Saturday 3rd
Poulton Hall 58

Sunday 4th
Poulton Hall 58

Sunday 11th
Briarfield 13
Parm Place 56

Sunday 18th
Long Acre 42

Sunday 25th
NEW Ash Bank 5
NEW Hill Farm 34

May

Saturday 1st
64 Carr Wood 19

Sunday 2nd
All Fours Farm 3
Laskey Farm 41
10 Statham Avenue 67

Monday 3rd
All Fours Farm 3

Framley 27
Laskey Farm 41
10 Statham Avenue 67

Thursday 6th
◆ Cholmondeley Castle
Gardens 21

Saturday 8th
Brooke Cottage 15
◆ Lane End Cottage
Gardens 40

Sunday 9th
Brooke Cottage 15
◆ Lane End Cottage
Gardens 40

Saturday 15th
The Old Parsonage 52

Sunday 16th
The Old Parsonage 52
◆ Stonyford Cottage 68

Sunday 23rd
Hall Lane Farm 29
Manley Knoll 44
Sandymere 63

Saturday 29th
NEW Ash Bank 5
Cheriton 20
NEW Hill Farm 34
◆ Mount Pleasant 47

Sunday 30th
NEW Ash Bank 5
Cheriton 20
NEW Hill Farm 34
73 Hill Top Avenue 35
Laskey Farm 41
◆ Mount Pleasant 47
15 Park Crescent 55
Rowley House 62
10 Statham Avenue 67
NEW Tiresford 73

Monday 31st
Laskey Farm 41
10 Statham Avenue 67

June

Saturday 5th
◆ Peover Hall
Gardens 57

Sunday 6th
24 Old Greasby Road 51
◆ Peover Hall
Gardens 57

Wednesday 9th
10 Statham Avenue 67

Saturday 12th
Drake Carr 24
The Homestead 37
Oakfield Villa 50
NEW One House
Walled Garden 53
Thorncar 72
Wren's Nest 80

Sunday 13th
Bucklow Farm 17
Drake Carr 24
60 Kennedy Avenue 39
NEW One House
Walled Garden 53

Wednesday 16th
10 Statham Avenue 67

Friday 18th
Twin Gates 74

Saturday 19th
Ashmead 6
NEW Brook Farm 14
18 Highfield Road 33
Laskey Farm 41
Willaston Grange 78

Sunday 20th
Ashmead 6
18 Highfield Road 33
Laskey Farm 41
Long Acre 42
Tattenhall Hall 71

Saturday 26th
All Fours Farm 3
61 Birtles Road 8

Sunday 27th
All Fours Farm 3
61 Birtles Road 8
Burton Village
Gardens 18

July

Saturday 3rd
◆ Bluebell Cottage
Gardens 9
181a Higher Lane 32
8a Warwick Drive 76

Sunday 4th
Ashton Grange 7
◆ Bluebell Cottage
Gardens 9
181a Higher Lane 32
Parm Place 56
34 Stanley Mount 66
8a Warwick Drive 76

Thursday 8th
5 Cobbs Lane 23

Saturday 10th
5 Cobbs Lane 23
Inglewood 38
10 Statham Avenue 67
Stretton Old Hall 69

Sunday 11th
The Homestead 37
Inglewood 38
Manley Knoll 44
Rowley House 62
10 Statham Avenue 67
Stretton Old Hall 69

Saturday 17th
NEW 5 Gayton Lane 28

Sunday 18th
NEW 5 Gayton Lane 28
Norley Bank Farm 49

Saturday 24th
Field House 25
Milford House Farm 46

Sunday 25th
Field House 25
The Firs 26
NEW Rose Brae 60

Saturday 31st
Laskey Farm 41
21 Scafell Close 64
10 Statham Avenue 67

August

Sunday 1st
◆ Arley Hall & Gardens 4
Laskey Farm 41
Norley Bank Farm 49
21 Scafell Close 64
10 Statham Avenue 67

Saturday 7th
Thorncar 72

Sunday 8th
The Firs 26
15 Park Crescent 55

Sunday 29th
Laskey Farm 41

Monday 30th
Laskey Farm 41

September

Friday 3rd
◆ Ness Botanic
Gardens 48

THE GARDENS

❶ ◆ ABBEYWOOD GARDENS
Chester Road, Delamere, Northwich, CW8 2HS. The Rowlinson Family, 01606 889477, info@abbeywoodestate.co.uk, www.abbeywoodestate.co.uk. *11m E of Chester. On the A556 facing Delamere Church.* For NGS: Sun 21 Mar, Sun 19 Sept (10-4). Adm £6, chd free. Light refreshments. Restaurant in Garden. For other opening times and information, please phone, email or visit garden website.
Superb setting near Delamere Forest. Total area 45 acres incl mature woodland, new woodland and new arboretum all with connecting pathways. Approx 4½ acres of gardens surrounding large Edwardian House. Vegetable garden, exotic garden, chapel garden, pool garden, woodland garden, lawned area with beds.

❷ ADSWOOD
Townfield Lane, Mollington, CH1 6LB. Ken & Helen Black, 01244 851327, keneblack@outlook.com, www.kenblackclematis.com. *3m N of Chester. From Wirral take A540 towards Chester. Cross A55 at r'about, past Wheatsheaf PH, turn L into Overwood Lane. At T junction turn R into Townfield Lane. Parking will be signed.* Visits by arrangement Apr to July for groups of up to 20. Adm £6, chd free. Home-made teas.
A cottage garden with several borders packed with a wide range of bulbs, perennials, climbers and English roses. The garden is planted with over 100 varieties of clematis with many more grown in containers. There is a small woodland garden, wildlife pond and a raised ornamental fish pond. There are several seating areas including a garden pavilion. There are no steps but the front drive and some of the garden paths are gravelled and side access to the garden is quite narrow.

❸ ALL FOURS FARM
Colliers Lane, Aston by Budworth, Northwich, CW9 6NF. Mr & Mrs Evans, 01565 733286. *M6 J19, take A556 towards Northwich. Turn immed R, past The Windmill Pub. Turn R after approx 1m, follow rd, garden on L after approx 2m. We're happy to allow direct access for drop off & collection for those with limited mobility.* Sun 2, Mon 3 May, Sat 26, Sun 27 June (10-4). Adm £5, chd free. Home-made teas. Visits also by arrangement May & June for groups of 20+.
A traditional and well established country garden with a wide range of roses, hardy shrubs, bulbs, perennials and annuals. You will also find a small vegetable garden, pond and greenhouse as well as vintage machinery and original features from its days as a working farm. The garden is adjacent to the family's traditional rose nursery. The majority of the garden is accessible by wheelchair.

> "I love the National Garden Scheme which has been the most brilliant supporter of Queen's Nurses like me. It was founded by the Queen's Nursing Institute which makes me very proud. As we battle Coronavirus on the front line in the community, knowing we have their support is a real comfort." – Liz Alderton, Queen's Nurse

4 ◆ ARLEY HALL & GARDENS

Arley, Northwich, CW9 6NA.
Viscount Ashbrook,
01565 777353, enquiries@
arleyhallandgardens.com,
www.arleyhallandgardens.com.
*10m from Warrington. Signed from
J9 & 10 (M56) & J19 & 20 (M6) (20
min from Tatton Park, 40 min to
Manchester). Please follow the brown
tourist signs.* **For NGS: Sun 1 Aug
(10-4.30). Adm £9, chd £4. All
refreshments available.** **For other
opening times and information,
please phone, email or visit
garden website.**
Within Arley's 8 acres of formal garden,
there are many different areas, each
with its own distinctive character.
Beyond the Chapel is The Grove,
a well established arboretum and a
Woodland Walk of about another 6
or 7 acres. Gardens are wheelchair
accessible, however parts of the estate
have cobbles which can prove a little
difficult for manual wheelchairs.

5 NEW ASH BANK

Mill Lane, Moston, Sandbach,
CW11 3PS. Mr & Mrs Alan Holder,
07736 108770, anne@ashbank.net.
*In a quiet country lane just outside
Sandbach. Take A533 from
Sandbach towards Middlewich.
Past The Fox pub on L & take next L
onto Mill Lane. At T junction over the
canal turn L & Ashbank is on your L.*
**Sun 25 Apr, Sat 29, Sun 30 May
(10.30-6). Adm £4, chd free. Light
refreshments. Also open Hill
Farm. Visits also by arrangement
Apr & May for groups of 5 to 10.**
Ashbank's garden is a large country
garden grown and tended specifically
to attract wildlife and to increase
the biodiversity of the Cheshire
countryside. We have an ornamental
Koi pond, stream as well as a natural
wildlife pond and bog garden with
jetty. Rose garden herbaceous border,
Cottage garden, wildflower meadow,
Mediterranean garden and Veg plot all
provide nectar for the insects.

6 ASHMEAD

2 Bramhall Way, off Gritstone
Drive, Macclesfield, SK10 3SH.
Peter & Penelope McDermott. *1m
W of Macclesfield. Turn onto Pavilion
Way, off Victoria Rd , then immed L
onto Gritstone Drive. Bramhall Way
first on R.* **Sat 19, Sun 20 June
(1-5). Adm £4, chd free. Home-
made teas.**
⅛ acre suburban cottage garden,
featuring plant packed mixed borders,
rock gardens, kitchen garden, island
beds, water feature, pond. The
garden demonstrates how small
spaces can be planted to maximum
effect to create all round interest.
Extensive range of plants favoured
for colours, texture and scent. Pots
used in a creative way to extend and
enhance borders.

7 ASHTON GRANGE

Grange Road, Ashton Hayes,
Chester, CH3 8AE. Martin &
Kate Slack, 01829 759172,
kateslack1@icloud.com. *8m E of
Chester. Grange Rd is a single track
road off B5393, by the village sign at
N end of Ashton Hayes.* **Sun 4 July
(12-4). Adm £6, chd free. Visits
also by arrangement May to Aug
for groups of 10 to 30.**
Ashton Grange has a traditional
country house garden with sweeping
lawns, herbaceous borders, kitchen
garden and unusual trees. You can
wander through mature woodland,
see tree carvings, visit the wildflower
meadow with lovely views and sit
by a large wildlife pond. The garden
and woodland extend to 9 acres and
there will be access to a new three
acre native woodland planted in
2018. Picnics welcome. Wheelchair
access to most parts of the garden
but woodland paths could be difficult.
Limited parking for wheelchair users.

8 61 BIRTLES ROAD

Macclesfield, SK10 3JG. Kate
& Graham Tyson. *Close to
Macclesfield Leisure Centre &
Macclesfield Hospital. Follow NGS
signs - B5087 Prestbury Rd onto
Priory Lane, R Birtles Rd, or from
Fallibroome Rd, L onto Priory Lane,
L Birtles Rd.* **Sat 26, Sun 27 June
(10.30-3.30). Adm £4, chd free.
Home-made teas.**
A South facing suburban garden
created to give year round interest
and colour. Well stocked herbaceous
borders, Rose garden, island beds
and a pond creating a habitat for frogs
and newts all add to an ever changing
colour palette. We look forward to
greeting you! There is plenty of seating
for homemade cake and afternoon
tea! Gazebo provision for inclement
weather! Wheelchairs will be able to
access onto the patio there are steps
down onto the garden.

9 ◆ BLUEBELL COTTAGE GARDENS

Lodge Lane, Dutton, WA4 4HP.
Sue Beesley, 01928 713718,
info@bluebellcottage.co.uk,
www.bluebellcottage.co.uk. *5m
NW of Northwich. From M56 (J10)
take A49 to Whitchurch. After 3m
turn R at T-lights towards Runcorn/
Dutton on A533. Then 1st L. Signed
with brown tourism signs from A533.*
**For NGS: Sat 3, Sun 4 July (10-5).
Adm £5, chd free. Home-made
teas.** **For other opening times and
information, please phone, email or
visit garden website.**
South facing country garden wrapped
around a cottage on a quiet rural lane
in the heart of Cheshire. Packed with
thousands of rare and familiar hardy
herbaceous perennials, shrubs and
trees. Unusual plants available at
adjacent nursery. The opening dates
coincide with the peak of flowering
in the herbaceous borders. Regional
Finalist, The English Garden's The
Nation's Favourite Gardens 2019.
Some gravel paths. Wheelchair
access to 90% of garden. WC is not
fully wheelchair accessible.

10 BOLESWORTH CASTLE

Tattenhall, CH3 9HQ. Mrs
Anthony Barbour, 01829 782210,
dcb@bolesworth.com,
www.bolesworth.com. *8m S of
Chester on A41. Enter through
Broxton Lodge in Bolesworth Hill
Road using this postcode: CH3 9HN.*
**Visits by arrangement Apr to
June for groups of 10+. Adm £5,
chd free.**
The Spring garden on The Rock
Walk above Castle, planted with
superb collection of Rhododendrons,
Azaleas, Camellias and specimen
trees in the 90's, is undergoing
restoration and development with
exciting new planting. Well planted
shrub/herbaceous borders around
Castle. Unusual and rare trees
planted over a period of 25 years by
Anthony Barbour now at their best
(Autumn colour in October). Regret no
wheelchair access.

11 BOLLIN HOUSE

Hollies Lane, Wilmslow,
SK9 2BW. Angela Ferguson &
Gerry Lemon, 07828 207492,
fergusonang@doctors.org.uk.
*From Wilmslow past Station &
proceed to T-junction. Turn L onto
Adlington Rd. Proceed for ½m,*

Bowmere Cottage

© Joa Wainwright

then turn R into Hollies Lane (just after One Oak Lane). Drive to the end of Hollies Lane and follow yellow signage. Park on Hollies L, or Browns L (other side Adlington Rd). Visits by arrangement May to July for groups of 10 to 30. Parking for 5/6 cars at the house. Adm £5, chd free. Home-made teas. Tea and cake or scones, cold drinks also available.

There are 3 parts to the garden, an established perennial formal garden and the wildflower meadow. The garden has deep herbaceous borders and an orchard. A new landscaped area has been added with a central water feature. The meadow with it's annual and perennial wildflower areas (with mown pathways and benches) attracts lots of butterflies and humming insects on a sunny day. Bollin House is in an idyllic location with the garden, orchard and meadow flowing into the Bollin Valley. The meadow is a combination of perennial and annual wildflower areas. Ramps to gravel lined paths to most of the garden. Some narrow paths through borders. Mown pathways in the meadow.

&

12 BOWMERE COTTAGE
5 Bowmere Road, Tarporley, CW6 0BS. Romy & Tom Holmes, 01829 732053, romy@bowmerecottage.co.uk. 10m E of Chester. From Tarporley High St (old A49) take Eaton Rd signed Eaton. After 100 metres take R fork into Bowmere Rd, Garden 100 metres on LH-side. Visits by arrangement June & July. Adm £5, chd free. Home-made teas.

A colourful and relaxing one acre country style garden around a Grade II listed house. Lawns are surrounded by three well stocked herbaceous and shrub borders and rose covered pergolas. There are two plant filled courtyard gardens and a small vegetable garden. Shrubs and rambling roses, clematis, hardy geraniums and a wide range of mostly hardy plants make this a very traditional English garden

&

13 BRIARFIELD
The Rake, Burton, Neston, CH64 5TL. Liz Carter, 07711 813732, carter.burton@btinternet.com. 9m NW of Chester. Turn off A540 at

Willaston-Burton Xrds T-lights & follow rd for 1m to Burton village centre. Sat 13, Sun 14 Feb (11-4). Sun 11 Apr, Sun 12 Sept (11-5). Home-made teas in St Nicholas Church, just along the lane. Adm £5, chd free. Refreshments available after 12 noon in aid of Claire House Children's Hospice. Opening with Burton Village Gardens on Sun 27 June. Visits also by arrangement Feb to Sept.

Tucked under the S-facing side of Burton Wood the garden is home to many specialist and unusual plants, some available in plant sale. This 2 acre garden is on two sites, a couple of minutes along an unmade lane. Shrubs, colourful herbaceous, bulbs, alpines and water features compete for attention as you wander through four distinctly different gardens. Always changing, Liz can't resist a new plant! Rare and unusual plants sold by appointment (70% to ngs) from the drive each Friday from 9am to 5pm.

&

14 NEW BROOK FARM

Newcastle Road, Astbury, Congleton, CW12 4RL. Mrs Emma Ingham. *Follow A34 until you reach Astbury Garden Centre, Watery Lane is opposite & we are opposite the small red post box approx 100 metres down the lane.* Sat 19 June (12-4). Adm £4, chd free. Home-made teas.

Established country gardens set in over an acre of lawns and borders packed with all seasonal plants. With large pond and brook running through the property.

15 BROOKE COTTAGE

Church Road, Handforth, SK9 3LT. Barry & Melanie Davy. *1m N of Wilmslow. Centre of Handforth, behind Health Centre. Turn off Wilmslow Rd at St Chads, follow Church Rd round to R. Garden last on L. Parking in Health Centre car park.* Sat 8, Sun 9 May (12-5). Adm £4.50, chd free. Home-made teas.

A chance to see this plant-filled garden in Spring. Shady woodland area of ferns, azaleas, rhododendrons, camellias, magnolia, erythroniums, trilliums, arisaema and blue poppies. Patio with hostas, daylilies, small pond. Borders with grasses, perennials, euphorbia, alliums and tulips. Anthriscus, aquilegia, persicaria and astrantia create meadow effect popular with insects. Featured in RHS magazine.

16 BROOKLANDS

Smithy Lane, Mouldsworth, CH3 8AR. Barbara & Brian Russell-Moore, 01928 740413, wbrm1@netscape.co.uk. *1½m N of Tarvin. 5½m S of Frodsham. Smithy Lane is off B5393 via A54 Tarvin/ Kelsall or the A56 Frodsham/ Helsby rd.* Visits by arrangement May to Sept for groups of 5 to 30. Cakes using eggs from our own hens. Adm £5, chd free.

A lovely country style, ¾ acre garden with backdrop of mature trees and shrubs. The planting is based around azaleas, rhododendrons, mixed shrub and herbaceous borders. There is a small vegetable garden, supported by a greenhouse and hens providing eggs for all the afternoon tea cakes!! Repeat visitors will notice significant changes following damage and tree loss caused by 'Storm Doris' etc.

17 BUCKLOW FARM

Pinfold Lane, Plumley, Knutsford, WA16 9RP. Dawn & Peter Freeman. *2m S of Knutsford. M6 J19, A556 Chester. L at 2nd set of T-lights. In 1¼m, L at concealed Xrds. 1st R. From Knutsford A5033, L at Sudlow Lane. becomes Pinfold Lane.* Sun 28 Feb (1-4). Adm £3.50, chd free. Light refreshments. Sun 13 June (2-5). Adm £4.50, chd free. Cream teas. Mulled Wine in February. Cream teas in June. 2022: Sun 27 Feb. Donation to Knutsford First Responders.

Country garden with shrubs, perennial borders, rambling roses, herb garden, vegetable patch, meadow, wildlife pond/water feature and alpines. Landscaped and planted over the last 30yrs with recorded changes. Free range hens. Carpet of snowdrops and spring bulbs. Leaf, stem and berries to show colour in autumn and winter. Featured in Cheshire Life. Cobbled yard from car park, but wheelchairs can be dropped off near gate.

GROUP OPENING

18 BURTON VILLAGE GARDENS

Burton, Neston, CH64 5SJ. *9m NW of Chester. Turn off A540 at Willaston-Burton Xrds T-lights & follow rd for 1m to Burton. Maps given to visitors. Buy your ticket at first garden.* Sun 27 June (11-5). Combined adm £5, chd free. Home-made teas in the Sports and Social Club behind the village hall. Drinks and biscuits in Burton Manor Glasshouse.

BRIARFIELD
Liz Carter.
(See separate entry)

♦ BURTON MANOR WALLED GARDEN
Burton Manor Gardens Ltd, 0151 336 6154, www.burtonmanorgardens.org.uk.

TRUSTWOOD
Peter & Lin Friend, lin@trustwoodbnb.uk, , www.trustwoodbnb.uk.
Visits also by arrangement May to July for groups of up to 20. Adm incl refreshments.

Burton is a medieval village built on sandstone overlooking the Dee estuary. Three gardens are open. Trustwood is a country wildlife garden with fruit, flowers and vegetables in raised beds at the front; at the back a more formal garden blends into the wood where the hens live. Briarfield's sheltered site, on the south side of Burton Wood (NT), is home to many specialist and unusual plants, some available in the plant sale at the house. The 1½ acre main garden invites exploration not only for its huge variety of plants but also for the imaginative use of ceramic sculptures. Period planting with a splendid vegetable garden surrounds the restored Edwardian glasshouse in Burton Manor's walled garden. Plants for sale at two gardens. Well signed free car parks. Maps available. All within walking distance. Limited disabled parking at each garden. Briarfield is unsuitable for wheelchairs.

---

19 64 CARR WOOD

Hale Barns, Altrincham, WA15 0EP. Mrs John Booth & Mr David Booth. *10m S of Manchester city centre. 2m from J6 M56: Take A538 to Hale Barns. L at 'triangle' by church into Wicker Lane & L at mini r'about into Chapel Lane & 1st R into Carr Wood.* Sat 1 May (2-6). Adm £5, chd free. Home-made teas.

Two-thirds acre landscaped, S-facing garden overlooking Bollin Valley laid out in 1959 by Clibrans of Altrincham, royal warrant holders. Gently sloping lawn, woodland walk, seating areas and terrace, extensive mixed shrub and plant borders. Partial wheelchair access and ample parking on Carr Wood. Wheelchair access to terrace overlooking main garden.

20 CHERITON

34 Congleton Road, Alderley Edge, SK9 7AB. David & Jo Mottershead. *400yds S of Alderley Edge on R of Congleton Road. Park on road.* Sat 29, Sun 30 May (11-5). Adm £5, chd £2. Light refreshments. Selection of teas, coffee, wine, biscuits and cakes. SW-facing, 1-acre garden with views on a fine day to the Clwydian Range. Garden of mature rhododendron, magnolias and wisteria with early clematis, hellebores, spring bulbs and a range of unusual herbaceous plants and young specimen trees. The third year of the new and previously open garden with significant and exciting changes made. Large free-standing Wisteria. Interesting planted flowing

Tom Makin, 07790 610586, s_makingardens@yahoo.co.uk. *8m SE of Chester, 3m W of Tarporley. A51 from Chester towards Tarporley. 1m after Tarvin turn off, at bus shelter, turn L into Willington Rd. After community centre, 2nd L into Well Lane. Third house.* Visits by arrangement May to Sept for groups of 5+. Coaches drop off then park on A51 lay-bys. Groups max 50. Adm £6, chd free. Home-made teas. Home grown organic fruits used in jams & vegan cakes. Gluten free available by prior arrangement.

2 acre organic, wildlife friendly, gold award winning cottage garden. Orchard, 3 wildlife ponds, perennial wildflower meadow, fruit and vegetable areas, badger sett, rose pergola & verandah, gazebo, shepherd's hut, summer house, barn owl and many other nest and bat boxes. Drought tolerant gravel garden and shade garden. New wild areas. Year round interest. Run on vegan principles. 'Frogwatch' charity volunteers transport migrating amphibians to the safety of these ponds when they are found on the roads in early spring. Butterflies, bees, dragonflies etc are plentiful in their season. The garden is a clear example of how an oasis for wildlife can be beautiful. Gravel paths may be difficult to use but most areas are flat and comprise grass paths or lawn.

Twin Gates

23 5 COBBS LANE

Hough, Crewe, CW2 5JN. David & Linda Race. *4m S of Crewe. M6 J16 A500 Nantwich. At r'about 1st exit. Next r'about 2nd exit Hough. 1m turn L into Cobbs Lane. From W take A51 Nantwich bypass to A500 r'about. 3rd exit Shavington. 3m passing White Hart PH. R.* Thur 8, Sat 10 July (11-5). Adm £5, chd free. Home-made teas at Village Hall 300 metres up Cobbs Lane. A plant person's ⅔ acre garden with island beds, wide cottage style herbaceous borders with bark paths. A large variety of hardy and some unusual perennials. Interesting features, shrubs, grasses and trees, with places to sit and enjoy the surroundings. A water feature runs to a small pond, wildlife friendly garden containing a woodland area. Finalists in Daily Mail Garden Competition, Best Kept Garden winners.

water feature with ponds. Front garden now planted as a stunning winter garden, with pruned white barked birches underplanted with heuchera and hellebore. Wheelchair access to rear garden only via a short flight (4) of low steps. No WC facilities for wheelchair users.

21 ♦ CHOLMONDELEY CASTLE GARDENS

Cholmondeley, Malpas, SY14 8AH. The Cholmondeley Gardens Trust, 01829 720383, office@cholmondeleycastle.co.uk, www.cholmondeleycastle.com. *4m NE of Malpas Sat Nav SY14 8ET. Signed from A41 Chester-Whitchurch rd & A49 Whitchurch-Tarporley rd.* For NGS: Thur 6 May (10-4.30). Adm £8.50, chd £4. For other opening times and information, please phone, email or visit garden website.

70 acres of romantically landscaped gardens with fine views and eye-catching water features, which still manages to retain its intimacy. Beautiful mature trees form a background to millions of spring bulbs and superb plant collections including magnolias, rhododendrons, camellias. Magnificent magnolias. One of the finest features of the gardens are its trees, many of which are rare and unusual, Cholmondeley Gardens is home to over 40 county champion trees. 100m long double mixed herbaceous border and extended Rose Garden with 250 roses. Light lunches and home-made teas available in Tea Room located in the heart of the Gardens. Partial wheelchair access.

22 CLEMLEY HOUSE

Well Lane, Duddon Common, Tarporley, CW6 0HG. Sue &

24 DRAKE CARR

Mudhurst Lane, Higher Disley, SK12 2AN. Alan & Joan Morris. *8m SE of Stockport, 12m NE of Macclesfield. From A6 in Disley centre turn into Buxton Old Rd, go up hill 1m & turn R into Mudhurst Lane. After ⅓m park on lay-by or grass verge. No parking at garden. Approx 150 metre walk.* Sat 12, Sun 13 June (11-5). Adm £4, chd free. Home-made teas. Home-made gluten-free cakes available. ½ acre cottage garden in beautiful rural setting with natural stream running into large wildlife pond containing many native species. Surrounding C17 stone cottage, the garden, containing herbaceous borders, shrubs and veg plot, is on several levels divided by grassed areas, slopes and steps. This blends into boarded walk through bog garden and mature wooded area into a small wildflower meadow. Not suitable for wheelchairs.

25 FIELD HOUSE

Crouchley Lane, Lymm, WA13 0TQ. Emma Aspinall, emmalaspinall@gmail.com. *1m S/ SE of Lymm Village. From Lymm Dam Head E along Church Rd (A56) past The Church Green Pub. Take next R into Crouchley Lane for approx 1m garden on R. From M56 J7 follow A56 Lymm Turn L into Crouchley Lane.* Sat 24, Sun 25 July (12.30-5). Adm £5, chd free.

Visits also by arrangement July & Aug for groups of 10 to 30. An evolving garden designed and tended by local garden designer, Emma Aspinall. and family. Set in peaceful countryside on an exposed plot extending to approx 0.4 acres together with a one acre paddock. Stable block with sedum roof. Well stocked herbaceous borders, vegetable/cut flower garden, fruit trees, orangery, water features and small rose/scented garden, horse and hens. Some gravel paths.

26 THE FIRS

Old Chester Road, Barbridge, Nantwich, CW5 6AY. Richard & Valerie Goodyear, 07775 924929, Goodyear.Pickford@btinternet. com. *3m N of Nantwich on A51. After entering Barbridge turn R at Xrds after 100 metres. The Firs is 2nd house on L.* Sun 25 July, Sun 8 Aug (1.30-5.30). Adm £4.50, chd free. Home-made teas. Visits also by arrangement May to Sept for groups of 10 to 30. Canalside garden set idyllically by a wide section of the Shropshire Union Canal with long frontage. Garden alongside canal with varied trees, shrubs and herbaceous beds, with some wild areas. All leading down to an observatory at the end of the garden. Usually have nesting friendly swans with cygnets May to September that occupy a small area of the garden. Wheelchair access to

some or all of the garden. The extent of access depends on the type of wheelchair.

27 FRAMLEY

Hadlow Road, Willaston, Neston, CH64 2US. Mrs Sally Reader, 07496 015259, sllyreader@yahoo.co.uk. *½m S of Willaston village centre. From Willaston Green, proceed along Hadlow Rd, crossing the Wirral Way. Framley is the next house on R.* Mon 3 May (10-4). Adm £5, chd free. Home-made teas. Visits also by arrangement May & June for groups of 5 to 20. This 5 acre garden holds many hidden gems. Comprising extensive mature wooded areas, underplanted with a variety of interesting and unusual woodland plants - all at their very best in spring. A selection of deep seasonal borders surround a mystical sunken garden, planted to suit its challenging conditions. Wide lawns and sandstone paths invite you to discover what lies around every corner. Please phone ahead for parking instructions for wheelchair users - access around much of the garden although the woodland paths may be challenging.

28 NEW 5 GAYTON LANE

Gayton, Wirral, CH60 3SH. Stuart & Maggie Watson. *From the 'Devon Doorway' r'about in Heswall follow the A540 towards Chester & Gayton Lane is 1st turning on R. No. 5 is 3rd entrance on L.* Sat 17, Sun 18 July (11-5). Adm £4, chd free. Light refreshments. ⅓ acre suburban garden stocked with a variety of shrubs and perennials. Roses and clematis are accompanied by a large selection of colourful pots. You may even find a few hidden fairy houses.

29 HALL LANE FARM

Hall Lane, Daresbury, Warrington, WA4 4AF. Sir Michael & Lady Beverley Bibby. *1m from J11 of M56. Leave M56 at J11, head towards Warrington take 2nd R turn into Daresbury village go around sharp bend then take L into Daresbury Lane. Entrance on L after 100 yds.* Sun 23 May (12-5). Adm £5, chd free. Home-made teas. The 2 acres of private formal garden

Sandymere

originally designed by Arabella Lennox-Boyd are arranged in a 'gardens within gardens' style to create a series of enclosed spaces each with their own character and style. The gardens also include a vegetable garden, orchard, Koi pond, as well as lawns and a tree house.

30 HIGHER DAM HEAD FARM

Damson Lane, Mobberley, WA16 7HY. Richard & Alex Ellison, 01565 873544, alex.ellison@talk21.com. *2m N E of Knutsford. 4m W of Alderley Edge. Turn off B5085 into Mill Lane at sign for Roebuck Inn. Damson Lane runs alongside the Roebuck car park and property lies 150yds further on R.* **Visits by arrangement May to Sept for groups of 10+. Adm £5, chd free. Home-made teas.**
Two acre country garden with box lined courtyards, large terracotta pots, deep herbaceous borders, topiary, pergola, orchard, walled vegetable garden, large glass house with stoned fruit. Natural pond with oak framed summerhouse and waterside deck. Newly planted area of silver birch and grasses in modern style. Plenty of areas to sit.

31 NEW 213 HIGHER LANE

Lymm, WA13 0HN. Mark Stevenson, 07470 715007, mjs.stevenson@btinternet.com. *From M56 use J7 follow the signs for Lymm A56 the garden is approx 300 metres from the Jolly Thresher pub towards, Lymm on R.* **Visits by arrangement May to Oct for groups of up to 20. Adm by donation. Picnics are most welcome.**
Established on an acre of grounds, segregated into various planting schemes. Herbaceous borders, specimen rhododendrons, hydrangeas, ferns and grasses, gingers and subtropical plants. along with so much more. A pond complete with fish and waterfalls surrounded by Japanese acers with far reaching views across fields and woodland. Many rare and unusual plants and shrubs with wooded pathways. 70% of the garden is accessible to people in wheelchairs.

32 181A HIGHER LANE

Higher Lane, Lymm, WA13 0RF. Melanie Farrow. *M6 signed Lymm approx 2m T-junction turn R. Follow rd approx 2m house on R. From M56 one straight rd. Follow Lymm approx. ¾ m house on R ½ m after T-lights.* **Sat 3 July (12-5); Sun 4 July (12-4). Adm £4, chd free. Pre-booking essential, please phone 07803 079495 or email mguest@tektura. com for information & booking. Cream teas. Home-made cakes, teas, coffees and juices.**
A structured garden, backing on to open fields with garden room. Clipped box and yew hedges and Pleached trees. Herbaceous borders filled with poppies, foxgloves, ammi, iris, grasses and lots more, four separate dahlia beds mid to late summer. Vegetable garden and greenhouse. Various seating areas.

33 18 HIGHFIELD ROAD

Bollington, Macclesfield, SK10 5LR. Mrs Melita Turner. *3m N of Macclesfield. A523 to Stockport. Turn R at B5090 r'about signed Bollington. Pass under viaduct. Take next R (by Library) up Hurst Lane. Turn R into Highfield Rd. Property on L. Park on wider road just past it.* **Sat 19, Sun 20 June (10-4.30). Adm £4.50, chd free. Home-made teas.**
This small terraced garden packed with plants was designed by Melita and has evolved over the past 13yrs. This plantswoman is a plantaholic and RHS Certificate holder. An attempt has been made to combine formality through structural planting with a more casual look influenced by the style of Christopher LLoyd. Refreshments available. Steep step on to top tier at front. Steps up to higher tiers at rear.

34 NEW HILL FARM

Mill Lane, Moston, Sandbach, CW11 3PS. Mr & Mrs Chris & Richard House, 01270 526264, housecr2002@yahoo.co.uk. *2m NW of Sandbach. From Sandbach Town Centre take A533 towards Middlewich. After Fox Pub take next L (Mill Lane) to a canal bridge & turn L. Hill Farm 400m on L.* **Sun 25 Apr, Sat 29, Sun 30 May (11-5). Adm £5, chd free. Light refreshments. Also open Ash Bank. Visits also by arrangement Apr to Sept for groups of up to 10.**
The garden extends to approximately ½ acre and is made up of a series of

gardens including a formal courtyard with a pond, a vegetable garden with. south facing wall. An orchard and meadow were established about 3 years ago. A principal feature is a woodland garden which supports a rich variety of woodland plants. This was extended in April 2020 to include a pond and grass/ herbaceous borders. A level garden, the majority of which can be accessed by wheelchair.

35 73 HILL TOP AVENUE

Cheadle Hulme, Stockport, SK8 7HZ. Mrs Elaine Land, 0161 486 0055. *4m S of Stockport. Leave A34 (new bypass) at r'about signed Cheadle Hulme (B5094). 2nd turn L into Gillbent Rd signed Cheadle Hulme Sports Centre. At end, small r'about, R into Church Rd. Garden 2nd rd on L. From Stockport or Bramhall turn R/L into Church Rd by Church Inn. Garden 1st rd on R.* **Sun 30 May (2-6). Adm £3.50, chd free. Tea. Visits also by arrangement May to Aug for groups of 5+.**
⅙ acre plantswoman's garden. Well stocked with a wide range of sun-loving herbaceous plants, shrub and climbing roses, many clematis varieties, pond and damp area, shade-loving woodland plants and some unusual trees and shrubs, in an originally designed, long narrow garden.

36 HILLTOP

Flash Lane, Prestbury, SK10 4ED. Martin Gardner, 07768 337525, hughmartingardner@gmail.com, www.yourhilltopwedding.com. *2m N of Macclesfield. A523 to Stockport. Turn R at B5090 r'about signed Bollington, after ½ m turn L at Cock & Pheasant Pub into Flash Lane. At bottom of lane turn R into cul de sac. Hilltop Country House signed on R.* **Visits by arrangement May to Aug for groups of up to 20. Monday, Tuesday or Wednesday only. Adm £6, chd free. Tea / Coffee and a cake £2 each.**
Interesting country garden of approx 4 acres. Woodland walk, parterre, herb garden, herbaceous borders, dry stone walled terracing, lily ponds with waterfall. Wisteria clad 1693 house (not open). ancient trees, orchard, magnificent views to Pennines and to West. Easy parking. Partial wheelchair access, disabled WC.

37 THE HOMESTEAD

2 Fanners Lane, High Legh, Knutsford, WA16 0RZ. Janet Bashforth, 01925 349895, janbash43@sky.com. *J20 M6/J9 M56 at Lymm interchange take A50 for Knutsford, after 1m turn R into Heath Lane then 1st R into Fanners Lane. Follow parking signs.* **Sat 12 June, Sun 11 July (11-4.30). Adm £4, chd free. Visits also by arrangement May to Aug for groups of 10 to 20.**

This compact gem of a garden has been created over the last 5 in years by a keen gardener and plants woman. Enter past groups of Liquidambar and White Stemmed Birch, visit shaded nooks with their own distinctive planting. Past topiary and obelisks covered with many varieties of clematis and roses, enjoy the colours of the hot border, a decorative greenhouse and pond complete the picture. Further on there is a small pond with water lilies and Iris. Many types of roses and clematis adorn the fencing along the paths and into the trees. Winner Period Living magazine's Best Garden 2020. Featured in April 2020 edition of Gardens Answers.

38 INGLEWOOD

4 Birchmere, Heswall, CH60 6TN. Colin & Sandra Fairclough. *6m S of Birkenhead. From A540 Devon Doorway/Clegg Arms r'about go through Heswall. ¼m after Tesco, R into Quarry Rd East, 2nd L into Tower Rd North & L into Birchmere.* **Sat 10, Sun 11 July (1-4). Adm £5, chd free. Home-made teas.**

Beautiful ½ acre garden with stream, large koi pond, 'beach' with grasses, wildlife pond and bog area. Brimming with shrubs, bulbs, acers, conifers, rhododendrons, herbaceous plants and hosta border. Interesting features including hand cart, antique mangle, wood carvings, bug hotel and Indian dog gates leading to a secret garden. Lots of seating to enjoy refreshments. Live music may be available.

39 60 KENNEDY AVENUE

Macclesfield, SK10 3DE. Bill North. *5min NW of Macclesfield Town Centre. Take A537 Cumberland St, 3rd r'bout (West Park) take B5087 Prestbury Rd. Take 5th turn on L into Kennedy Ave, last house on L before Brampton Ave opp Belong Care Home.* **Sun 13**

June (12.30-5). Adm £4.50, chd free. Home-made teas. Provided by East Cheshire Hospice.

Small suburban garden which featured in 2017 August edition of Amateur Gardening magazine, designed to provide Al Fresco dining and relaxed entertaining, also providing relaxing Cottage Garden tranquillity which for 2021 features new plantings as well as 65 hanging baskets. Wheelchair access will require a carer to guide up the drive and will only provide access to part of the garden.

40 ◆ LANE END COTTAGE GARDENS

Old Cherry Lane, Lymm, WA13 0TA. Imogen & Richard Sawyer, 01925 752618, imogen@laneendcottagegardens.co.uk, www.laneendcottagegardens.co.uk. *1m SW of Lymm. J20 M6/J9 M56/A50 Lymm interchange. Take B5158 signed Lymm. Turn R 100 metres into Cherry Corner, turn immed R into Old Cherry Lane.* **For NGS: Sat 8, Sun 9 May, Sat 11, Sun 12 Sept (10-5). Adm £4, chd free. Light refreshments. For other opening times and information, please phone, email or visit garden website.**

Formerly a nursery, this 1 acre cottage garden is densely planted for all year round colour with many unusual plants. Features include deep mixed borders, scented shrub roses, ponds, herb garden, walled orchard with trained fruit, shady woodland walk, sunny formal courtyard, vegetable garden and chickens. Teas may be available. Plants for sale. For other opening times see garden website. Wheelchair accessible WC. Park by garden entrance. Flat garden. Some bark paths may be difficult if wet.

41 LASKEY FARM

Laskey Lane, Thelwall, Warrington, WA4 2TF. Howard & Wendy Platt, 07740 804825, howardplatt@lockergroup.com, www.laskeyfarm.com. *2m From M6/M56. From M56/M6 follow directions to Lymm. At T-junction turn L onto the A56 in Warrington direction. Turn R onto Lymm Rd. Turn R onto Laskey Lane.* **Sun 2, Mon 3, Sun 30, Mon 31 May (11-4), also open 10 Statham Avenue. Sat 19, Sun 20 June (11-4). Sat 31 July, Sun 1 Aug (11-4), also**

open 10 Statham Avenue. Sun 29, Mon 30 Aug (11-4). Adm £5, chd £1. Home-made teas. **Visits also by arrangement May to Aug for groups of 10+. Coaches and parties of up to 50. Can also open along with 10 Statham Ave (a very different garden) 2 mins away**

1½ acre garden including herbaceous and rose borders, vegetable area, a greenhouse, parterre and a maze showcasing grasses and prairie style planting Interconnected pools for wildlife, specimen koi and terrapins form an unusual water garden which features a swimming pond There is a tree house plus a number of birds and animals. Most areas of the garden may be accessed by wheelchair.

42 LONG ACRE

Wyche Lane, Bunbury, CW6 9PS. Margaret & Michael Bourne, 01829 260944, mjbourne249@tiscali.co.uk. *3½m SE of Tarporley. In Bunbury village, turn into Wyche Lane by Nags Head Pub car park, garden 400yds on L.* **Sun 18 Apr, Sun 20 June (2-5). Adm £5, chd free. Home-made teas. Visits also by arrangement Apr to June for groups of 10+. Donation to St Boniface Church Flower Fund and Bunbury Village Hall.**

Plantswoman's garden of approx 1 acre with unusual and rare plants and trees including Kentucky Coffee Tree, Scadiopitys, Kalapanax Picta and others, pool garden, exotic conservatory with bananas, anthuriums and medinilla, herbaceous, greenhouses with Clivia in Spring and Disa Orchids in Summer. Spring garden with camellias, magnolias, bulbs; roses and lilies in summer.

43 ◆ THE LOVELL QUINTA ARBORETUM

Swettenham, CW12 2LD. Tatton Garden Society, 01565 831981, admin@tattongardensociety.org.uk, www.lovellquintaarboretum.co.uk. *4m NW of Congleton. Turn off A54 N 2m W of Congleton or turn E off A535 at Twemlow Green, NE of Holmes Chapel. Follow signs to Swettenham. Park at Swettenham Arms. Do not follow Sat Nav.* **For NGS: Sun 10 Oct (1-4). Adm £5, chd free. See above. For other**

Abbeywood Gardens

© Joe Wainwright

opening times and information, please phone, email or visit garden website.
This 28-acre arboretum has been established since 1960s and contains around 2,500 trees and shrubs, some very rare. Incl National Collections of Pinus and Fraxinus. A large selection of oak, a private collection of hebes plus autumn flowering, fruiting and colourful trees and shrubs. A lake and way-marked walks. Autumn colour, winter walk and spring bulbs. Refreshments at the Swettenham Arms during licenced hours or by arrangement. Care required but wheelchairs can access much of the arboretum on the mown paths.

44 MANLEY KNOLL
Manley Road, Manley, WA6 9DX. Mr & Mrs James Timpson, www.manleyknoll.com. *3m N of Tarvin. On B5393, via Ashton & Mouldsworth. 3m S of Frodsham, via Alvanley.* **Sun 23 May, Sun 11 July (12-5). Adm £5, chd free. Light refreshments.**
Arts and Crafts garden created early 1900s. Covering 6 acres, divided into different rooms encompassing parterres, clipped yew hedging, ornamental ponds and herbaceous borders. Banks of rhododendron and azaleas frame a far-reaching view of the Cheshire Plain. Also a magical quarry/folly garden with waterfall and woodland walks.

45 MAYFIELD HOUSE
Moss Lane, Bunbury Heath, Tarporley, CW6 9SY. Mr & Mrs J France Hayhurst, jeanniefh@me.com. *Mayfield House is off A49 on Tarporley/ Whitchurch Rd. Moss Lane is opposite School Lane which leads to Bunbury village. There's a yellow speed camera (30 mph) across the road from the house.* **Visits by arrangement Apr to Sept for groups of 10 to 30. Light refreshments.**
A thoroughly English mature garden with a wealth of colour and variety throughout the year. A background of fine trees, defined areas bordered by mixed hedging, masses of rhododendrons, azaleas, camellias, hydrangeas and colourful shrubs. Clematis and wisteria festoon the walls in early Summer. Easy access and random seating areas. Small lake with an island and broad lawns with glades of foxgloves. Garden statuary, large pond with fish, swimming pool, close to pretty villages and very good gastro-pubs and some beautiful gardens nearby. Gravel drive. Very wide wrought iron gates lead to the garden and it is entirely navigable.

"Unfortunately, cancer was not in lockdown in 2020. The continued support of our long-standing and valued partner, the National Garden Scheme is more important than ever."
Macmillan

46 MILFORD HOUSE FARM

Long Lane, Wettenhall, Winsford, CW7 4DN. Chris & Heather Pope, 07887 760930, hclp@btinternet.com. *3m Et of Tarporley. From A51 at Alpraham turn into Long Lane. Proceed for 2½m to St David's Church on L. Milford House Farm is 150 metres further on L Park at Church. Limited mobility park at house.* **Sat 24 July (12-5). Adm £4, chd free. Home-made teas at St. David's Church 150 metres from garden on flat lane. Visits also by arrangement July & Aug for groups of 5 to 20. Adm incl refreshments. Accessible WC available.**

A large country garden created over 19 years with lawn, herbaceous borders, trees, perennials and annuals. Modern walled garden for vegetables, fruit, greenhouse, flowers, shrubs and various pots. Orchard with fruit, native and ornamental trees, large wildlife pond with toads and newts. Gardened on organic lines to increase biodiversity and attract wildlife. All areas are flat. Parking. Good views over borders, lawn, orchard. Teas in aid of Motor Neurone Disease. Walled garden has gravel and flags. Church fully accessible for refreshments/ WC.

&. 🐄 ✳ ☕

47 ◆ MOUNT PLEASANT

Yeld Lane, Kelsall, CW6 0TB. Dave Darlington & Louise Worthington, 01829 751592, louisedarlington@btinternet.com, www.mountpleasantgardens.co.uk. *8m E of Chester. Off A54 at T-lights into Kelsall. Turn into Yeld Lane opp Farmers Arms Pub, 200yds on L. Do not follow SatNav directions.* **For NGS: Sat 29, Sun 30 May, Sat 4, Sun 5 Sept (11-4). Adm £7, chd £5. Light refreshments. For other opening times and information, please phone, email or visit garden website.**

10 acres of landscaped garden and woodland started in 1994 with impressive views over the Cheshire countryside. Steeply terraced in places. Specimen trees, rhododendrons, azaleas, conifers, mixed and herbaceous borders; 4 ponds, formal and wildlife. Vegetable garden, stumpery with tree ferns, sculptures, wildflower meadow and Japanese garden. Bog garden, tropical garden. Sculpture trail. Sculpture Exhibition. Please ring prior to visit for wheelchair access.

✳ 🚗 ☕

48 ◆ NESS BOTANIC GARDENS

Neston Road, Ness, Neston, CH64 4AY. The University of Liverpool, 0151 795 6300, nessgdns@liverpool.ac.uk, www.liverpool.ac.uk/ness-gardens. *10m NW of Chester. Off A540. M53 J4, follow signs M56 & A5117 (signed N Wales). Turn onto A540 follow signs for Hoylake. Ness Gardens is signed locally.* **For NGS: Fri 3 Sept (10-4.30). Adm £7.50, chd £3.50. Light refreshments in our Café, based in our Visitor Centre. For other opening times and information, please phone, email or visit garden website.**

Looking out over the dramatic Dee Estuary from a lofty perch of the Wirral peninsula, Ness Botanic Gardens boasts 64 spectacular acres of landscaped and natural gardens overflowing with horticultural treasures. With a delightfully peaceful atmosphere, a wide array of events taking place, plus a cafe and gorgeous open spaces it's a great fun-filled day out for all. National Collections of Sorbus and Betula. Herbaceous borders, Rock Garden, Mediterranean Bank, Potager and conservation area. Wheelchairs are available for hire [donations gratefully accepted] but advance booking is highly recommended.

&. ✳ 🚗 NPC ☕

49 NORLEY BANK FARM

Cow Lane, Norley, Frodsham, WA6 8PJ. Margaret & Neil Holding, 07828913961, neil.holding@hotmail.com. *From the Tigers Head pub in the centre of Norley village keep the pub on L, carry straight on through the village for approx 300 metres. Cow Lane is on the R.* **Sun 18 July, Sun 1 Aug (11-5). Adm £4.50, chd free. Home-made teas. Visits also by arrangement July & Aug for groups of 10 to 30. Homemade teas available.**

A garden well stocked with perennials, annuals and shrubs that wraps around this traditional Cheshire farmhouse. There is an enclosed cut flower and nursery garden surrounding a greenhouse and just beyond the house a vegetable garden. Wander past this to a large wildlife pond with a backdrop of further planting. All set in about 2 acres. Free range hens, donkeys and Coloured Ryeland sheep. Access in the main is available for wheelchairs although not the toilet. There are

some stone steps and a small number of narrow paths.

🐄 ✳ ☕ 🍴

50 OAKFIELD VILLA

Nantwich Road, Wrenbury, Nantwich, CW5 8EL. Carolyn & Jack Kennedy. *6m S of Nantwich & 6m N of Whitchurch. Garden on main rd through village next to Dairy Farm. Limited parking outside house. Parking available in Community Centre 2 minutes walk away.* **Sat 12 June (10-4.30). Combined adm with Wren's Nest £7, chd free.**

Romantic S-facing garden of densely planted borders and creative planting in containers, incl climbing roses, clematis and hydrangeas. Divided by screens into 'rooms'. Pergola clothed in beautiful climbers provides relaxed sheltered seating area and there is a small water feature. Small front garden, mainly hydrangeas and clematis. Refreshments at Café in the village, booking preferred. Some gravelled areas and small lawn area.

&. ☕

51 24 OLD GREASBY ROAD

Upton, Wirral, CH49 6LT. Lesley Whorton & Jon Price, 07905 775750, wlesley@hotmail.co.uk. *Approx 1m from J2A M53 (Upton Bypass). M53 J2; follow Upton sign. At r'about (J2A) straight on to Upton Bypass. At 2nd r'about, turn L by Upton Cricket Club. 24 Old Greasby Rd on L.* **Sun 6 June (11-4). Adm £4, chd free. Home-made teas. Visits also by arrangement June to Aug for groups of 10 to 30.**

A multi-interest and surprising suburban garden. Both front and rear gardens incorporate innovative features designed for climbing and rambling roses, clematis, under-planted with cottage garden plants with a very productive kitchen garden. Unfortunately, due to narrow access and gravel paths, there is no wheelchair access.

✳ ☕

52 THE OLD PARSONAGE

Arley Green, Northwich, CW9 6LZ. The Hon Rowland & Mrs Flower, www.arleyhallandgardens.com. *5m NNE of Northwich. 3m NNE of Great Budworth. M6 J19 & 20 & M56 J10. Follow signs to Arley Hall & Gardens. From Arley Hall notices to Old Parsonage which lies across park at Arley Green (approx 1m).* **Sat**

15, Sun 16 May (2-5). Adm £5, chd free. Home-made teas.
2-acre garden in attractive and secretive rural setting in secluded part of Arley Estate, with ancient yew hedges, herbaceous and mixed borders, shrub roses, climbers, leading to woodland garden and unfenced pond with gunnera and water plants. Rhododendrons, azaleas, meconopsis, cardiocrinum, some interesting and unusual trees. Wheelchair access over mown grass, some slopes and bumps and rougher grass further away from the house.

53 NEW ONE HOUSE WALLED GARDEN

Rainow, SK11 0AD. Louise Baylis.
2½m NE of Macclesfield. Just off A537 Macclesfield to Buxton rd. 2½m from Macclesfield stn. **Sat 12, Sun 13 June (10-5). Adm £4, chd free. Home-made teas at One House Lodge Buxton New Road, Rainow.**
An historic early C18 walled kitchen garden, hidden for 60yrs and restored by volunteers. This romantic and atmospheric garden has a wide range of vegetables, flowers and old tools. There is an orchard with friendly pigs, a wildlife area and pond, and a traditional greenhouse with ornamental and edible crops. There is a free rural life exhibition next to the car park.

54 39 OSBORNE STREET

Bredbury, Stockport, SK6 2DA. Geoff & Heather Hoyle, www.youtube.com/user/Dahliaholic.
1½m E of Stockport, just off B6104. Follow signs for Lower Bredbury/ Bredbury Hall. Leave M60 J27 (from S & W) or J25 (from N & E). Osborne St is adjacent to pelican crossing on B6104. **Sat 4, Sun 5 Sept (11-4). Adm £4, chd free. Home-made teas. Teas, coffees, and cakes.**
This dahliaholic's garden contains over 400 dahlias in 150+ varieties, many of exhibition standard. Shapely lawns are surrounded by deep flower beds that are crammed with dahlias of all shapes, sizes and colours, and complemented by climbers, soft perennials and bedding plants. An absolute riot of early autumn colour. The garden comprises two separate areas, both crammed with very colourful flowers. The dahlias range in height from 18 inches to 8 feet tall, and are in a wide variety of shapes and colours. They are interspersed with salvias, fuchsias, argyranthemums, and bedding plants. The garden is on YouTube: search for Dahliaholic.

Rode Hall

55 15 PARK CRESCENT

Appleton, Warrington, WA4 5JJ. Linda & Mark Enderby, 07949 496 747, lmaenderby@outlook.com.
2½m S of Warrington. From M56 J10 take A49 towards Warrington for 1½m. At 2nd set of lights turn R into Lyons Lane , then 1st R into Park Crescent. No.15 is last house on R. **Sun 30 May, Sun 8 Aug (11.30-5). Adm £4, chd free. Light refreshments. Prosecco will be available by donation. Visits also by arrangement May to Aug for groups of 10 to 30.**
An abundant garden containing many unusual plants, trees and a mini orchard. A cascade, ponds and planting encourage wildlife. There are raised beds, many roses in various forms and herbaceous borders. The garden has been split into distinct areas on different levels each with their own vista drawing one through the garden. A Chinese model railway will also be on show. A cascade, 2 ponds and large rambling roses, plus other roses and clematis. Some unusual plants and a good example of planting under a lime tree in clay soil and shade.

56 PARM PLACE
High Street, Great Budworth, CW9 6HF. Jane Fairclough, 01606 891131, janefair@btinternet.com. *3m N of Northwich. Great Budworth on E side of A559 between Northwich & Warrington, 4m from J10 M56, also 4m from J19 M6. Parm Place is W of village on S side of High St.* **Sun 11 Apr, Sun 4 July (1-5). Adm £4, chd free. Visits also by arrangement Apr to Aug for groups of 10 to 30. Donation to Great Ormond Street Hospital.**
Well-stocked ½ acre plantswoman's garden with stunning views towards S Cheshire. Curving lawns, parterre, shrubs, colour co-ordinated herbaceous borders, roses, water features, rockery, gravel bed with some grasses. Fruit and vegetable plots. In spring large collection of bulbs and flowers, camellias, hellebores and blossom.

57 ◆ PEOVER HALL GARDENS
Over Peover, Knutsford, WA16 9HW. Brooks Family, 07836 219128, bookings@peoverhall.com, www.peoverhall.com. *4m S of Knutsford. Do not rely on SATNAV. From A50/Holmes Chapel Rd at Whipping Stocks Pub turn onto Stocks Lane. Follow - R onto Grotto Ln ¼m turn R onto Goostrey Ln. Main entrance on R on bend - through white gates.* **For NGS: Sat 5, Sun 6 June (2-5). Adm £5, chd free. Home-made teas in the Park House Tea Room. For other opening times and information, please phone, email or visit garden website.**
The extensive formal gardens to Peover Hall feature a series of 'garden rooms' filled with clipped box, water garden, Romanesque loggia, warm brick walls, unusual doors, secret passageways, beautiful topiary work and walled gardens, C19 dell and rockery, rhododendrons and pleached limes. With Peover Hall a Grade 2* listed Elizabethan family house dating from 1585 providing a fine backdrop. The Grade I listed Carolean Stables which are of significant architectural importance will be open to view. Partial wheelchair access to garden - wheelchair users please ask the car-park attendant for parking on hard standing rather than on the grass.

58 POULTON HALL
Poulton Lancelyn, Bebington, Wirral, CH63 9LN. The Poulton Hall Estate Trust & Poulton Hall Walled Garden Charitable Trust, www.poultonhall.co.uk. *2m S of Bebington. From M53, J4 towards Bebington; at T-lights R along Poulton Rd; house 1m on R.* **Sat 3, Sun 4 Apr (2-5). Adm £6, chd free. Home-made teas.**
3 acres; lawns fronting house, wildflower meadow. Surprise approach to walled garden, with reminders of Roger Lancelyn Green's

Higher Dam Head Farm

retellings, Excalibur, Robin Hood and Jabberwocky. Scented sundial garden for the visually impaired. Memorial sculpture for Richard Lancelyn Green by Sue Sharples. Rose, nursery rhyme, witch, herb and oriental gardens and now Memories Reading room. There are often choirs or orchestral music in the garden. Childrens toys and play house. Afternoon teas are served by a separate charity, usually St Johns Hospice, with any surplus going to their funds. Level gravel paths. Separate wheelchair access (not across parking field). Disabled WC.

& ⛤ ✿ ☕

59 ◆ RODE HALL
Church Lane, Scholar Green, ST7 3QP. Randle & Amanda Baker Wilbraham, 01270 873237, enquiries@rodehall.co.uk, www.rodehall.co.uk. *5m SW of Congleton. Between Scholar Green (A34) & Rode Heath (A50).* **For opening times and information, please phone, email or visit garden website.**

Nesfield's terrace and rose garden with stunning view over Humphry Repton's landscape is a feature of Rode, as is the woodland garden with terraced rock garden and grotto. Other attractions incl the walk to the lake with a view of Birthday Island complete with heronry, restored ice house, working two-acre walled kitchen garden and Italian garden. Snowdrop Walks in Feb, Tue-Sat (11-4). Bluebell Walks in May. Summer: Weds and Bank Holiday Mons until end of Sep (11-5). Courtyard Kitchen offering wide variety of refreshments. Partial wheelchair access, some steep areas with gravel and woodchip paths, access to WC, kitchen garden and tearooms.

⛤ ✿ 🚌 ☕

60 NEW ROSE BRAE
Earle Drive, Parkgate, Neston, CH64 6RY. Mr Joe & Mrs Carole Rae. *11m NW of Chester. From A540 take B5134 at the Hinderton Arms towards Neston. Turn R at the T-lights in Neston & L at the Cross onto Parkgate Road, B5135. Earle Drive is ½m on R.* **Sun 25 July (1-4.30). Adm £5, chd free. Light refreshments.**

An all-season, half acre garden comprising bulbs, trees, shrubs, perenials and climbers. In July roses, clematis, hydrangeas, phlox and agapanthus predominate together with ferns, grasses, flowering trees,

topiary, an unmown bulb lawn and a formal pool with water lilies.

☕

61 ROSEWOOD
Old Hall Lane, Puddington, Neston, CH64 5SP. Mr & Mrs C E J Brabin, 0151 353 1193, angela.brabin@btinternet.com. *8m N of Chester. From A540 turn down Puddington Lane, 1½m. Park by village green. Walk 30yds to Old Hall Lane, turn L through archway into garden.* **Visits by arrangement for groups of up to 30. Adm £3, chd free. Tea.**

All yr garden; thousands of snowdrops in Feb, Camellias in autumn, winter and spring. Rhododendrons in April/May and unusual flowering trees from March to June. Autumn Cyclamen in quantity from Aug to Nov. Perhaps the greatest delight to owners are two large Cornus capitata, flowering in June. Bees kept in the garden. Honey sometimes available.

& ✿ 🚌 ☕

62 ROWLEY HOUSE
Forty Acre Lane, Kermincham, Holmes Chapel, CW4 8DX. Tim & Juliet Foden. *3m ENE from Holmes Chapel. J18 M6 to Holmes Chapel, from Holmes Chapel take A535 (Macclesfield) Take R turn in Twemlow (Swettenham) at Yellow Broom restaurant. Rowley House ½m on L.* **Sun 30 May, Sun 11 July (11-4.30). Adm £5, chd free. Home-made teas.**

Our aim is to give nature a home and create a place of beauty. There is a formal courtyard garden and informal gardens featuring rare trees, and herbaceous borders, a pond with swamp cypress and woodland walk with maples, rhododendrons, ferns and shade-loving plants. Beyond the garden there are wildflower meadows, natural ponds and a wood with ancient oaks. Also wood sculptures by Andy Burgess. Many unusual plants and trees, also many areas dedicated to wildlife.

⛤ ☕

63 SANDYMERE
Middlewich Road, Cotebrook, CW6 9EH. Sir John Timpson. *5m N of Tarporley. On A54 approx 300yds W of T-lights at Xrds of A49/A54.* **Sun 23 May (12-5). Adm £6, chd free. Home-made teas.**

16 landscaped acres of beautiful

Cheshire countryside with terraces, walled garden, extensive woodland walks and an amazing hosta garden. Turn each corner and you find another gem with lots of different water features including a new rill built in 2014, which links the main lawn to the hostas. Partial wheelchair access.

& ✿ 🚌 ☕

64 21 SCAFELL CLOSE
High Lane, Stockport, SK6 8JA. Lesley & Dean Stafford, 01663 763015, lesley.stafford@live.co.uk. *High Lane is on A6 SE of Stockport towards Buxton. From A6 take Russell Ave then Kirkfell Drive. Scafell Close on R.* **Sat 31 July, Sun 1 Aug (1-4.30). Adm £4, chd free. Light refreshments. tea/coffee and cakes. Visits also by arrangement July & Aug for groups of 10+. £7.50 Adm incl Cream Teas.**

⅓ acre landscaped suburban garden. Colour themed annuals border the lawn featuring the Kinder Ram statue in a heather garden, passing into vegtables, soft fruits and fruit trees. Returning perennial pathway leads to the fishpond and secret terraced garden with modern water feature and patio planting. Finally visit the blue front garden. Refreshments in aid of Cancer Research UK. Partial wheelchair access.

& ⛤ ☕

65 68 SOUTH OAK LANE
Wilmslow, SK9 6AT. Caroline Melliar-Smith, 01625 528147, caroline.ms@btinternet.com. *¾m SW of Wilmslow. From M56 (J6) take A538 (Wilmslow) R into Buckingham Rd. From centre of Wilmslow turn R onto B5086, 1st R into Gravel Lane, 4th R into South Oak Lane.* **Visits by arrangement May to July for groups of up to 20. Adm £5, chd £2.**

With year-round colour, scent and interest, this attractive, narrow, hedged cottage garden has evolved over the years into 5 natural 'rooms'. The owners passion for unusual plants, is reflected in the variety of shrubs, trees, flower borders and pond, creating havens for wildlife. Share this garden with its' varied history from the 1890's. Some rare and unusual hardy, herbaceous and shade loving plants and shrubs.

✿

66 34 STANLEY MOUNT

Sale, M33 4AE. Debbie & Steve Bedford. *1m from J7 M60 heading S on A56 (Washway Rd.) towards Altrincham. From A56 Altrincham N towards Sale (Washway Rd) Turning for Stanley Mount off Washway Rd, no.34 on R.* **Sun 4 July (1-5). Adm £4, chd free. Home-made teas.** Informal small suburban garden designed and created from scratch. Mature trees, hedges, shrubs and climbers provide the framework. Planted areas feature herbaceous perennials, grasses, bulbs, clematis, hostas, ferns and groundcover. Seated areas, creative pots, productive greenhouse, cacti, alpines and a shed hotel.

❀ ☕

67 10 STATHAM AVENUE

Lymm, WA13 9NH. Mike & Gail Porter, 01925 753488, porters@mikeandgailporter.co.uk. *Approx 1m from J20 M6 /M56 interchange. From M/way follow B5158 to Lymm. Take A56 Booth's Hill Rd, L towards Warrington, turn R on to Barsbank Lane, pass under Bridgewater canal, after 50 metres turn R on to Statham Ave. No 10 is 100 metres on R.* **Sun 2, Mon 3, Sun 30, Mon 31 May (12-5), also open Laskey Farm. Wed 9, Wed 16 June, Sat 10, Sun 11 July (12-5). Sat 31 July, Sun 1 Aug (12-5), also open Laskey Farm. Adm £5, chd free. Home-made teas. Enjoy home-made cakes or try Gail's famous meringues with fresh fruit and cream. Visits also by arrangement May to Aug for groups of 10 to 30. Can also open along with the beautiful garden at Laskey Farm (2 mins away).** Peaceful, pastel shades in early summer. Beautifully structured ¼ acre south facing garden carefully terraced and planted rising to the Bridgewater towpath. Hazel arch opens to clay paved courtyard with peach trees. Rose pillars lead to varied herbaceous beds and quiet shaded areas bordered by fuchsias, azaleas and rhododendrons. Wide variety of plants and shrubs. Interesting garden buildings. A treasure hunt/quiz to keep the children occupied. Delicious refreshments to satisfy the grown ups. Short gravel driveway and some steps to access the rear garden. The rear garden is sloping.

🐃 ❀ 🚗 ☕

68 ◆ STONYFORD COTTAGE

Stonyford Lane, Oakmere, CW8 2TF. Janet & Tony Overland, 01606 888970, info@stonyfordcottagegardens.co.uk, www.stonyfordcottagegardens.co.uk. *5m SW of Northwich. From Northwich take A556 towards Chester. ¾m past A49 junction turn R into Stonyford Lane. Entrance ½m on L.* **For NGS: Sun 16 May (12-4.30). Adm £4.50, chd free. Light refreshments. For other opening times and information, please phone, email or visit garden website.** Set around a tranquil pool this Monet style landscape has a wealth of moisture loving plants, incl iris and candelabra primulas. Drier areas feature unusual perennials and rarer trees and shrubs. Woodland paths meander through shade and bog plantings, along boarded walks, across wild natural areas with views over the pool to the cottage gardens. Unusual plants available at the adjacent nursery. Open Tues - Fri, Apr - Oct 10-5pm. Plant Nursery. Some gravel paths.

♿ 🐃 ❀ 🚗 ☕

69 STRETTON OLD HALL

Stretton, Tilston, Malpas, SY14 7JA. Stephen Gore Head Gardener, 07800 602225, Strettonoldhallgardens@hotmail.com. *5m N of Malpas. From Chester follow A41, Broxton r'about follow signs for Stretton Water Mill, turn L at Cock o Barton. 2m on L.* **Sat 10, Sun 11 July (10-4). Adm £7, chd free. Home-made teas. Visits also by arrangement Mar to Oct for groups of 10+.** 5 acre Cheshire countryside garden with a planting style best described as controlled exuberance with a definite emphasis upon perennials, colour, form and scale. Divided into several discrete and individual gardens incl stunning herbaceous borders, scree garden, walled kitchen garden and glass house. wildflower meadows, wildlife walk around the lake with breathtaking vistas in every direction. Gravel paths.

♿ 🚗 ☕ 📷

70 SUNNYSIDE FARM

Shop Lane, Little Budworth, CW6 9HA. Mike & Joan Smeethe, 01829 760618, joan.ann.smeethe@btinternet.com. *4m NE of Tarporley. On A54 1m E of T-lights at A54/A49 junction. Turn off A54 into Shop Lane opp Shrewsbury Arms Pub. After 30yds take farm*

track on R & continue to end. Parking at house. **Visits by arrangement May to Aug. Adm £5, chd free. Home-made teas.** A plant lover's and beekeeper's atmospheric country garden. Wildlife and child friendly the five acres encompass oak woodland, wildflower meadow, pool garden, potager and fruit garden, orchard with beehives, long border with naturalistic planting and cottage garden. With a wealth of unusual plants and interesting colour schemes this is a constantly evolving garden. Always something new. Wheelchair access to most parts of the garden.

♿ 🐃 ❀ ☕ 🚗

71 TATTENHALL HALL

High Street, Tattenhall, Chester, CH3 9PX. Jen & Nick Benefield, Chris Evered & Jannie Hollins, 01829 770654, janniehollins@gmail.com. *8m S of Chester on A41. Turn L to Tattenhall, through village, turn R at Letters pub, past war memorial on L through Sandstone pillared gates. Park on rd or in village car park.* **Sun 20 June (2-5). Adm £5, chd free. Home-made teas. Visits also by arrangement Feb to Oct.** Plant enthusiasts' garden around Jacobean house (not open). 4½ acres, wildflower meadows, interesting trees,large pond, stream, walled garden, colour themed borders, succession planting, spinney walk with shade plants, yew terrace overlooking meadow, views to hills. Glasshouse and vegetable garden. Wildlife friendly sometimes untidy garden, interest throughout the year, continuing to develop. Partial wheelchair access due to gravel paths, cobbles and some steps.

♿ 🐃 ❀ 🚗 ☕

72 THORNCAR

Windmill Lane, Appleton, Warrington, WA4 5JN. Mrs Kath Carey, 01925 267633, john.carey516@btinternet.com. *South Warrington. From M56 J10 take A49 towards Warrington for 1½m. At 2nd set of T-lights turn L into Quarry Lane, as the rd swings R it becomes Windmill Lane. Thorncar is 4th house on R.* **Sat 12 June, Sat 7 Aug (12.30-4). Adm £2, chd free. Light refreshments. Visits also by arrangement Mar to Oct for groups of 5 to 20.** One third acre plantwoman's suburban garden planted since 2011 within the framework of part of an older garden to provide year round interest. In August the hydrangeas, fuchsias, Eucryphia 'Nymansay' and in the 'in the face' yellow of a hardy Calceolaria bring colour following on from daffodils in April, self seeded perfoliate alexandras in early May, peonies in June and day lilies in July. Refreshments provided by WI. Plants for St Josephs Family Centre, Warrington.

&. ✿ ☕

73 NEW **TIRESFORD**
Tarporley, CW6 9LY. Susanna Posnett. *Tiresford is on A49 Tarporley Bypass. It is adjacent to the farm on R leaving Tarporley in the direction of Four Lane Ends T-lights.* Sun 30 May (2-5.30). Adm £5, chd free. Home-made teas. Ice-Cream.
Established garden from the 1930's undergoing a major restoration project to reinstate it to its former glory. Overlooking both Beeston and Peckforton Castles, it was last opened for the National Garden Scheme 25 years ago. Work has focused on hard landscaping but planting has begun on the herbaceous borders, the rhododendron woodland walk and the sunken garden. This is a work in progress. The House has been transformed into a stylish 6 Bedroom B&B. There is parking for wheelchair users next to the house. There is a disabled loo in the house which wheelchair users will have access to.

&. ✿ ☕

74 **TWIN GATES**
3 Earle Drive, Parkgate, Neston, CH64 6RY. John Hinde & Lilian Baker. *Approx ½ m N of Neston town centre. From Neston (Tesco/ Brown Horse PH), N towards Methodist Church. At church (Brewers Arms on L), take L fork into Park St, becomes Leighton Rd. Continue for approx ½ m, L into Earle Drive.* Evening opening Fri 18 June (6-9). Adm £6, chd free. Adm incl one glass of wine or soft drink. Additional glasses by donation.
Garden developed over last decade and now looking mature. Mixed herbaceous and shrub borders, with some choice small trees all set in fully stocked sinuous borders. In June, peonies and roses come to the fore, but there are plenty of other genus, too.

&. ✿ ☕ ⌆

76 **8A WARWICK DRIVE**
Hale, Altrincham, WA15 9EA. Gill & Chris Turner, 01619 803258, gillandchris_turner@hotmail.com. *M56 J6, take A538 to Altrincham. Turn L at 2nd T-lights into Park Rd. Take 2nd R into Bower Rd & turn L immed into Warwick Drive. Garden is 200yds on L, on corner of Lindop Rd.* Sat 3, Sun 4 July (1-5). Adm £4, chd free. Home-made teas at 21 Warwick Drive. Visits also by arrangement May to Aug for groups of 20 to 30. Conducted tour.
This is a small suburban garden which is constantly evolving. Recent changes include replacing an old mixed hedge with yew, making room for more planting areas, more sunlight and, of course, more plants! We design and maintain the garden ourselves for all year round interest but the front herbaceous border provides an explosion of colour in July.

&. ✿ ☕ ☕

77 **THE WELL HOUSE**
Wet Lane, Tilston, Malpas, SY14 7DP. Mrs S H French-Greenslade, 01829 250332. *3m NW of Malpas. On A41, 1st R after Broxton r'about, L on Malpas Rd through Tilston. House on L.* Visits by arrangement Feb to Sept for groups of 5+. Adm by donation. Pre-booked refreshments for small groups only.
1-acre cottage garden, bridge over natural stream, spring bulbs, perennials, herbs and shrubs. Triple ponds. Adjoining ¼ -acre field made into wildflower meadow; first seeding late 2003. Large bog area of kingcups and ragged robin. February for snowdrop walk. Victorian parlour and collector's items on show only. Not suitable for wheelchairs. Dogs on leads only.

✿ 🚐 ☕

78 **WILLASTON GRANGE**
Hadlow Road, Willaston, Neston, CH64 2UN. Mr & Mrs M Mitchell. *On A540 (Chester High Rd) take the B5151 into Willaston. From the Village Green in the centre of Willaston turn onto Hadlow Rd, Willaston Grange is ½ m on your L.* Sat 19 June (12-4.30). Adm £5, chd free.
Willaston Grange is an Arts and Craft property. Twelve years ago, the house and gardens had been derelict for three years. After months of detailed planning, the significant restoration work began. The Gardens extend to

6 acres with a small lake, wide range of mature and rare tree species, herbaceous border, woodland and vegetable gardens, orchard and magical treehouse. The fully restored Arts and Crafts house provides the perfect backdrop for a summer visit, along with afternoon tea and live music. Most areas accessible by wheelchair.

✿ ✿ ☕ ⌆

79 NEW **WOODSIDE**
6 Sunbury Gardens, Appleton, Warrington, WA4 5QE. Mr Peter & Mrs Margaret Wilkinson, 01925 269919, marg.wilkinson43@gmail.com. *M56, J10 r'about take A49 signed Warrington to T- lights just past Golf club. Turn R into Lyons Lane. R at next r'about into Longwood Rd. Sunbury Gardens is second rd on L.* Visits by arrangement June to Aug for groups of 5 to 10. Adm £5, chd free. Light refreshments.
Woodside garden is positioned adjacent to a deciduous woodland area. This acts as a backdrop to a small suburban garden on 2 levels of beds packed with shrubs interspersed with perennials to create an artistic colour blend. Many clematis on trellis and obelisks give height to this flagged garden. More design is achieved with large pots of Bananas, Palms and Tree ferns and Hostas and flowers.

✿ ☕

80 **WREN'S NEST**
Wrenbury Heath Road, Wrenbury, Nantwich, CW5 8EQ. Sue & Dave Clarke, 07855398803, wrenburysue@gmail.com. *12m from M6 J16. From Nantwich signs for A530 to Whitchurch, reaching Sound school turn 1st R Wrenbury Heath Rd, across the Xrds & bungalow is on L, telegraph pole outside.* Sat 12 June (10.30-4). Combined adm with Oakfield Villa £7, chd free. Visits also by arrangement in June for groups of 10 to 30.
Set in a semi-rural area, this bungalow has a Cottage Garden Style of lush planting and is 80ft x 45ft. The garden is packed with unusual and traditional perennials and shrubs incl over 100 hardy geraniums, campanulas, crocosmias, iris and alpine troughs. Plants for sale. Plant Heritage National Collection of Hardy Geraniums sylvaticum and renardii.

✿ 🚐 NPC

CORNWALL

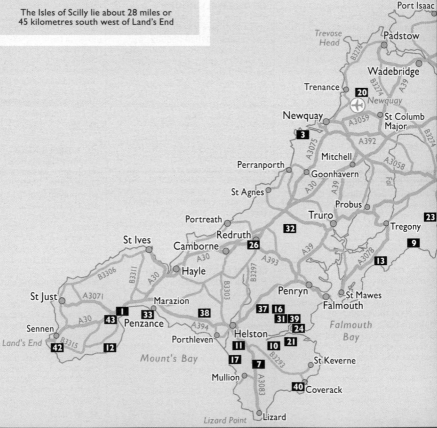

ISLES OF SCILLY

Tresco
St Martin's
Bryher
Hugh Town
St Mary's
St Agnes

The Isles of Scilly lie about 28 miles or
45 kilometres south west of Land's End

Port Isaac
Padstow
Trevose
Head
Wadebridge
Trenance
20 Newquay
Newquay
St Columb
Major
3
Perranporth
Mitchell
Goonhavern
St Agnes
Probus
Portreath
Truro
Tregony **23**
Redruth
32
Camborne
26
9
St Ives
13
Hayle
Penryn
37 **16**
St Mawes
St Just
Marazion
31 **39**
Falmouth
1
38
24
Falmouth
43
33 Penzance
Helston
21
Bay
Sennen
10
Land's End
12
Porthleven
11
42
Mount's Bay
17
7
St Keverne
Mullion
40 Coverack
Lizard Point
Lizard

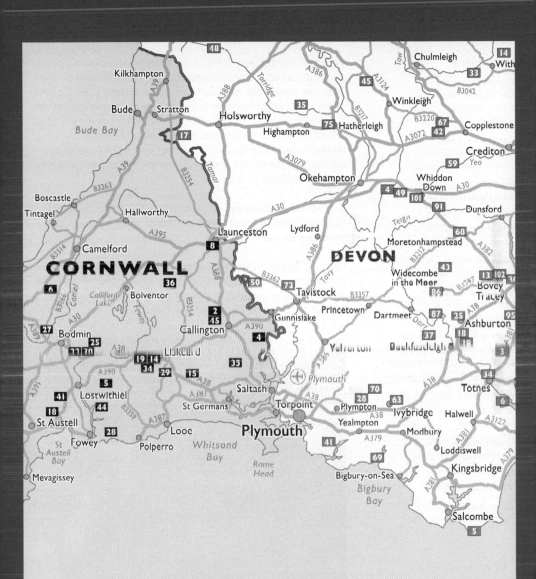

© Global Mapping / XYZ Maps

Volunteers

County Organiser
Christopher Harvey Clark
01872 530165
suffree2012@gmail.com

County Treasurer
Theresa Elgar 07830 946065
theresa.elgar@ngs.org.uk

Publicity
Sara Gadd 07814 885141
sara@gartendesign.co.uk

Emma Skilton 07772 542143
emma24nyrorganic@gmail.com

Claire Woodbine 01208 821339
cwoodbine@btinternet.com

Booklet Co-ordinator
Peter Stanley 01326 565868
stanley.m2@sky.com

Assistant County Organisers
Ginnie Clotworthy 01208 872612
ginnieclotworthy@hotmail.co.uk

Theresa Elgar 07830 946065
theresa.elgar@ngs.org.uk

Alison O'Connor 01726 882460
tregoose@tregoose.co.uk

Libby Pidcock 01208 821303
libby@cardinhambb.co.uk

Marion Stanley 01326 565868
stanley.m2@sky.com

Claire Woodbine 01208 821339
cwoodbine@btinternet.com

 @CornwallNGS
 @ngs_cornwall

Cornwall has some of the most beautiful natural landscapes to be found anywhere in the world.

Here, you will discover some of the country's most extraordinary gardens, a spectacular coastline, internationally famous surfing beaches, windswept moors and countless historic sites. Cornish gardens reflect this huge variety of environments particularly well.

A host of National Collections of magnolias, camellias, rhododendrons and azaleas, as well as exotic Mediterranean semitropical plants and an abundance of other plants flourish in our acid soils and mild climate.

Surrounded by the warm currents of the Gulf Stream, with our warm damp air in summer and mild moist winters, germination continues all year.

Cornwall boasts an impressive variety of beautiful gardens. These range from coastal-protected positions to exposed cliff-top sites, moorland water gardens, Japanese gardens and the world famous tropical biomes of the Eden Project.

Right: Trenarth

OPENING DATES

All entries subject to change. For latest information check **www.ngs.org.uk**

Extended openings are shown at the beginning of the month.

Map locator numbers are shown to the right of each garden name.

April

Monday 12th
◆ Pencarrow   27

Thursday 22nd
◆ Trewidden Garden   43

Saturday 24th
Ash Barn   4
◆ Chygurno   12

Sunday 25th
Ash Barn   4
◆ Chygurno   12
Trebartha Estate Garden and Country Garden at Lemarne   36

May

Sunday 2nd
East Down Barn   15
Navas Hill House   24
NEW Trebarvah Woon   37

Tuesday 4th
Pinsla Garden   30

Wednesday 5th
Pinsla Garden   30

Sunday 9th
◆ Boconnoc   5
Penmilder   29
South Lea   35

Wednesday 12th
South Bosent   34

Thursday 13th
South Bosent   34

Sunday 16th
Alverton Cottage   1
◆ The Japanese Garden   20

Monday 17th
◆ The Japanese Garden   20

Tuesday 18th
Pinsla Garden   30

Wednesday 19th
Pinsla Garden   30

Sunday 23rd
Anvil Cottage   2
Windmills   45

Sunday 30th
Caervallack   10
The Lodge   22
NEW The Old School House   25

Monday 31st
Bokelly   6

June

Tuesday 1st
Pinsla Garden   30

Wednesday 2nd
Kestle Barton   21
Pinsla Garden   30

Tuesday 8th
NEW ◆ Caerhays Castle   9

Wednesday 9th
Gardens Cottage   18
Kestle Barton   21

Thursday 10th
Gardens Cottage   18

Saturday 12th
◆ St Michael's Mount   33

Tuesday 15th
Pinsla Garden   30

Wednesday 16th
Kestle Barton   21
Pinsla Garden   30
South Bosent   34

Thursday 17th
South Bosent   34

Sunday 20th
Crugsillick Manor   13
Trenarth   39
NEW Trethew   41
Trevilley   42

Wednesday 23rd
NEW Dove Cottage   14
NEW 9 Higman Close   19
Kestle Barton   21

Sunday 27th
NEW Dove Cottage   14
NEW 9 Higman Close   19

Tuesday 29th
Pinsla Garden   30

Wednesday 30th
Kestle Barton   21
Pinsla Garden   30

July

Saturday 3rd
◆ Roseland House   32

Sunday 4th
◆ Roseland House   32
South Lea   35

Tuesday 6th
Pinsla Garden   30

Wednesday 7th
NEW Dove Cottage   14
Gardens Cottage   18
NEW 9 Higman Close   19
Kestle Barton   21
Pinsla Garden   30

Thursday 8th
Gardens Cottage   18

Sunday 11th
NEW Dove Cottage   14
NEW 9 Higman Close   19
Parkhenver   26
NEW Treprenn Vean   40

Wednesday 14th
Kestle Barton   21
South Bosent   34

Thursday 15th
South Bosent   34

Saturday 17th
◆ Chygurno   12

Sunday 18th
◆ Chygurno   12
Tregonning   38

Tuesday 20th
Pinsla Garden   30

Wednesday 21st
NEW Dove Cottage   14
NEW 9 Higman Close   19
Kestle Barton   21
Pinsla Garden   30

Sunday 25th
Byeways   8
NEW Dove Cottage   14
NEW 9 Higman Close   19

Wednesday 28th
Kestle Barton   21

August

Tuesday 3rd
Pinsla Garden   30

Wednesday 4th
NEW Dove Cottage   14
Gardens Cottage   18
NEW 9 Higman Close   19
Kestle Barton   21
Pinsla Garden   30

Thursday 5th
Gardens Cottage   18

Sunday 8th
Anvil Cottage   2
NEW Dove Cottage   14
NEW 9 Higman Close   19
Windmills   45

Wednesday 11th
Kestle Barton   21
South Bosent   34

Thursday 12th
◆ Bonython Manor   7
South Bosent   34

Tuesday 17th
Pinsla Garden   30

Wednesday 18th
NEW Dove Cottage   14
NEW 9 Higman Close   19
Kestle Barton   21
Pinsla Garden   30

Sunday 22nd
NEW Dove Cottage   14
NEW 9 Higman Close   19

Wednesday 25th
Kestle Barton   21

Tuesday 31st
Pinsla Garden   30

September

Wednesday 1st
Gardens Cottage   18
Kestle Barton   21
Pinsla Garden   30

Thursday 2nd
Gardens Cottage   18

Wednesday 8th
Kestle Barton   21

Wednesday 15th
Kestle Barton   21
South Bosent   34

By Arrangement

Arrange a personalised garden visit on a date to suit you. See individual garden entries for full details.

There are brilliant plant sales at many gardens. Look out for the symbol in the garden description – and don't forget to bring a bag to carry your plants home in

---

THE GARDENS

1 ALVERTON COTTAGE

Alverton Road, Penzance, TR18 4TG. David & Lizzie Puddifoot. *Next door to YMCA. About 600 metres from Penlee car park travelling towards A30. Morrab Gardens are also close to car park.* **Sun 16 May (2-5). Adm £4, chd free. Tea and home-made cakes.**
Alverton Cottage is a grade II listed Regency house. The garden is modest in size though large for a Penzance garden, S-facing and sheltered by mature trees incl elms. Very large monkey-puzzle tree and holm oak. Laid out in 1860s, we have added a succulent area, fernery and have cleared and replanted since 2019. In nearby Morrab Sub-tropical Gardens the 'Friends' will also welcome visitors. Off road disabled parking but wheelchair to garden terrace only.

2 ANVIL COTTAGE

South Hill, PL17 7LP. Geoff & Barbara Clemerson, 01579 362623, gcclemerson@gmail.com. *3m NW of Callington. Head N on A388 from Callington centre. After ½m L onto South Hill Rd (signed South Hill), straight on for 3m. Gardens on R just before St Sampson's Church.* **Sun 23 May, Sun 8 Aug (1-5). Combined adm with Windmills £5, chd free. Home-made teas. Gluten free refreshments available. Visits also by arrangement May to Aug for groups of 5 to 30 except during June. Donation to Cornwall Air Ambulance.**
During lockdown, we took the

opportunity to revamp some of the garden. Regular visitors will spot that some overgrown areas have been thinned, but it remains a plantsman's garden. Winding paths lead through a series of themed rooms with familiar, rare and unusual plants. Steps lead up to a raised viewpoint looking towards Caradon Hill and Bodmin Moor, and then into a formal rose garden. Only partial wheelchair access due to steps.

3 ARUNDELL

West Pentire, Crantock, TR8 5SE. Brenda & David Eyles, 01637 831916, david@davideyles.com. *1m W of Crantock. From A3075 take signs to Crantock. At junction in village keep straight on to West Pentire (1m). Park in field (signed) or public car parks at W Pentire.* **Visits by arrangement 10 May to 28 Aug (daily excl Friday (2-5)) for groups of up to 20. Adm £5, chd free. Cream teas & home-made biscuits.**
A garden where no garden should be! - on windswept NT headland between 2 fantastic beaches. 1 acre packed with design and plant interest round old farm cottage. Front: cottage garden. Side: Mediterranean courtyard. Rear: rockery and shrubbery leading to stumpery and fernery and on to stream and pond, Cornish Corner, herbaceous borders, Beth Chatto dry garden, and exotic garden. Wheelchair access from public car park with entrance via rear gate. 14 shallow steps in centre of garden useable with care.

4 ASH BARN

Callington, PL17 8BP. Mary Martin. *3m E of Callington. Leaving Callington on A 390 (Tavistock direction) take 1st R signed Harrowbarrow. Turn R at T junction in village then 1m down hill to Glamorgan Mill parking on L.* **Sat 24, Sun 25 Apr (10-5). Adm £5, chd free. Home-made teas.**
Terraced 3 acre woodland garden, full of magnolias, camellias and fruit trees, under-planted with old roses, shrubs and Tamar Valley narcissi. Intensive herbaceous planting provides interest for most of the year, and contrasts with box and holly topiary. Narrow, uneven paths and steps reflect the wild wood atmosphere. There is also a quarry edge, so no dogs or unsupervised children. Regret no wheelchair access.

5 ◆ BOCONNOC

Lostwithiel, PL22 0RG. Elizabeth Fortescue, 01208 872507, office@boconnoc.com, www.boconnoc.com. *Off A390 between Liskeard & Lostwithiel. From East Taphouse follow signs to Boconnoc. (SatNav does not work well in this area).* **For NGS: Sun 9 May (2-5). Adm £5, chd free. For other opening times and information, please phone, email or visit garden website.**
20 acres surrounded by parkland and woods with magnificent trees, flowering shrubs and stunning views. The gardens are set amongst mature trees which provide the backcloth for exotic spring flowering shrubs, woodland plants, with newly-planted magnolias and a fine collection of hydrangeas. Bathhouse built in 1804, woodland gardens, obelisk built in

1771, house dating from Domesday, deer park, C15 church.

 🐕 🐷 🍴 🛏 ☕

6 BOKELLY
St Kew, Bodmin, PL30 3DY. Toby & Henrietta Courtauld. *Follow road to Trelill from St Kew Highway, leaving Red Lion pub on L. Garden on L after approx 1m.* **Mon 31 May (1.30-5.30). Adm £5, chd free. Cream teas.**
The 7-acre garden surrounds a beautiful C15 barn and lichened stone outbuildings. It has been expanded and rejuvenated with new plantings to suit the varied levels and soil types, incl woodland planting, herbaceous border, little orchard and vegetable plot, and cut flower beds (a particular passion of garden designer Henrietta).
☕

7 ◆ BONYTHON MANOR
Cury Cross Lanes, Helston, TR12 7BA. Mr & Mrs Richard Nathan, 01326 240550, sbonython@gmail.com, www.bonythonmanor.co.uk. *5m S of Helston. On main A3083 Helston to Lizard Rd. Turn L at Cury Cross Lanes (Wheel Inn). Entrance 300yds on R.* **For NGS: Thur 12 Aug (2-4.30). Adm £9, chd £2. Tea, coffee, fruit juices and home-made cakes. For other opening times and information, please phone, email or visit garden website.**
Magnificent 20 acre colour garden incl sweeping hydrangea drive to Georgian manor (not open). Herbaceous walled garden, potager with vegetables and picking flowers; 3 lakes in valley planted with ornamental grasses, perennials and South African flowers. A 'must see' for all seasons colour.
♿ 🐕 ✳ 🍴 🛏 ☕

8 BYEWAYS
Dunheved Road, Launceston, PL15 9JE. Tony Reddicliffe. *Launceston town centre. 100yds from multi-storey car park past offices of Cornish & Devon Post into Dunheved Rd, 3rd bungalow on R.* **Sun 25 July (1-5). Adm £5, chd free. Home-made teas.**
Small town garden developed over 11yrs by enthusiastic amateur gardeners. Herbaceous borders, rockery. Tropicals incl bananas, gingers and senecio. Stream and water features. Roof garden. Japanese inspired tea house and courtyard with bridge. Fig tree and Pawlonia flank

Caerhays Castle

area giving secluded seating. Living pergola. Wild flower planting.
🐕 ✳ ☕

SPECIAL EVENT

9 NEW ◆ CAERHAYS CASTLE
Caerhays, St Austell, PL26 6LY. Charles Williams, www.caerhays.co.uk. *Please follow Brown Signs to Caerhays Estate.* **For NGS: Tue 8 June (9.30-2.30). Adm £120. Pre-booking essential, please visit www.ngs.org.uk/special-events for information & booking. £120.00 per person Includes: teas & coffees, buffet lunch with wine, private lecture on Styrax/Stewartia, private garden tour with Charles Williams. For other opening times and information, please visit garden website.**
Caerhays is a woodland garden covering well over 140 acres, full of magnolias, camellias, and rhododendrons. It has been designated as being of outstanding importance by Kew Gardens and are holders of one of the UK's National Magnolia Collections. The owner Charles Williams will host a private day, including a lecture on Styrax/Stewartia and a private garden tour followed by a buffet lunch. Caerhays gardens are situated on a tall hillside and, consequently, the routes to the top of the gardens and back down towards the castle are steep.
🐕 ✳ 🚗 🍴 🛏 ☕

10 CAERVALLACK
St Martin, Helston, TR12 6DF. Matt Robinson & Louise McClary, 01326 221130, mat@build-art.co.uk. *5m SE of Helston. Go through Mawgan village, over 2 bridges, past Gear Farm shop; go past turning on L, garden next farmhouse on L.* **Sun 30 May (1.30-4.30). Adm £5, chd free. Tea & home-made cakes; scones, clotted cream & jam. Visits also by arrangement for groups of 5 to 30 any time of the year. Refreshments possible for large groups.**
Romantic garden arranged into rooms, the collaboration between an artist and an architect. Colour and form of plants against architectural experiments in cob, concrete and shaped hedges and topiary. Grade II listed farmhouse and orchard. Roses and wisteria a speciality. 25 years in the making. Arts & Crafts cob walls and paving; Heroic 54ft pedestrian footbridge, 5 sided meditation studio, cast concrete ponds and amphitheatre; coppice and wild flower meadow; mature orchard and veg plot. Partial wheelchair access but if you can negotiate the opening 2 steps, some of front garden will be accessible. Brick, grass and gravel paths.
✳ 🍴 🛏 ☕

11 CARMINOWE VALLEY GARDEN

Tangies, Gunwalloe, TR12 7PU. Mr & Mrs Peter Stanley, 01326 565868, stanley.m2@sky.com. *3m SW of Helston. A3083 Helston-Lizard rd. R opp main gate to Culdrose. 1m downhill, garden on R.* **Visits by arrangement Apr & May. Adm £5, chd free.**
Overlooking the beautiful Carminowe Valley towards Loe Pool this abundant garden combines native oak woodland, babbling brook and large natural pond with more formal areas. Wild flowers, mown pathways, shrubberies, orchard. Enclosed cottage garden, spring colours and roses early summer provide huge contrast. Gravel paths, slopes.

12 ◆ CHYGURNO

Lamorna, TR19 6XH. Dr & Mrs Robert Moule, 01736 732153, rmoule010@btinternet.com. *4m S of Penzance. Off B3315. Follow signs for The Lamorna Cove Hotel. Garden is at top of hill, past Hotel on L.* **For NGS: Sat 24, Sun 25 Apr, Sat 17, Sun 18 July (2-5). Adm £5, chd free. For other opening times and information, please phone or email.**
Beautiful, unique, 3 acre cliffside garden overlooking Lamorna Cove. Planting started in 1998, mainly S-hemisphere shrubs and exotics with hydrangeas, camellias and rhododendrons. Woodland area with tree ferns set against large granite outcrops. Garden terraced with steep steps and paths. Plenty of benches so you can take a rest and enjoy the wonderful views.

13 CRUGSILLICK MANOR

Ruan High Lanes, Truro, TR2 5LJ. Dr Alison Agnew & Mr Brian Yule, 07538 218201, alisonagnew@icloud.com. *On Roseland Peninsula. Turn off A390 Truro-St Austell rd onto A3078 towards St Mawes. Approx 5m after Tregony turn 1st L after Ruan High Lanes towards Veryan, garden is 200yds on R. Limited parking.* **Sun 20 June (11-6). Adm £5, chd free. Tea, coffee, soft drinks, cakes and light lunches. Visits also by arrangement May to Sept for groups of 10+. Refreshments by prior arrangement.**
2 acre garden, substantially re-

landscaped and planted, mostly over last 8 yrs. To the side of the C17/C18 house, a wooded bank drops down to walled kitchen garden and hot garden. In front, sweeping yew hedges and paths define oval lawns and broad mixed borders. On a lower terrace, the focus is a large pond and the planting is predominantly exotic flowering trees and shrubs. Partial wheelchair access. Garden is on several levels connected by fairly steep sloping gravel paths.

14 NEW DOVE COTTAGE

Lantoom, Dobwalls, Liskeard, PL14 4LR. Becky Martin, 07871 368227, beckymartin1@hotmail.com. *Do not use postcode for SatNav. In Dobwalls at double mini r'about take road to Duloe, About 20m after end-of-speed-limit sign turn L at football club. Follow concrete track.* **Wed 23, Sun 27 June, Wed 7, Sun 11, Wed 21, Sun 25 July, Wed 4, Sun 8, Wed 18, Sun 22 Aug (2-6). Combined adm with 9 Higman Close £6, chd free. Pre-booking essential, please visit www.ngs. org.uk for information & booking. Visits also by arrangement June to Aug for groups of up to 10.**
Two very different, modestly-sized enclosed gardens. The three owners have between them over 100 years of professional horticultural experience. Replanted from scratch in 2017 showing what can be achieved in 4 years. Separate areas with different types of planting. Emphasis on colour. Tropical deck, sunroom, lush foliage area. Steps, gravel and narrow paths. Unsuitable for wheeelchairs or buggies.

15 EAST DOWN BARN

Menheniot, Liskeard, PL14 3QU. David & Shelley Lockett, 07803 159662. *S side of village near cricket ground. Turn off A38 at Hayloft restaurant/railway station junction and head towards Menheniot village. Follow NGS signs from sharp L hand bend as you enter village.* **Sun 2 May (1-5). Adm £3.50, chd free. Home-made teas. Visits also by arrangement Apr & May for groups of 20+.**
Garden laid down between 1986-1991 with the conversion of the barn into a home and covers almost ½ acre of East sloping land with stream running North - South acting as the

Easterly boundary. 3 terraces before garden starts to level out at the stream. Garden won awards in the early years under the stewardship of the original owners. Steep slopes.

16 ETHNEVAS COTTAGE

Constantine, Falmouth, TR11 5PY. Lyn Watson & Ray Chun, 01326 340076. *6m SW of Falmouth. Nearest main rds A39, A394. Follow signs for Constantine. At lower village sign, at bottom of winding hill, turn off on private lane. Garden ¾m up hill.* **Visits by arrangement Apr to June for groups of 10 to 20. Adm £5, chd free.**
Isolated granite cottage in 2 acres. Intimate flower and vegetable garden. Bridge over stream to large pond and primrose path through semi-wild bog area. Hillside with grass paths among native and exotic trees. Many camellias and rhododendrons. Mixed shrubs and herbaceous beds, wild flower glade, spring bulbs. A garden of discovery of hidden delights.

17 GARDEN COTTAGE

Gunwalloe, Helston, TR12 7QB. Dan & Beth Tarling, 01326 241906, beth@gunwalloe.com, www.gunwalloecottages.co.uk. *Just beyond Halzephron Inn at Gunwalloe. Cream cottage with green windows.* **Visits by arrangement Apr to Sept. Adm £4, chd free.**
Coastal cottage garden. Small garden with traditional cottage flowers, vegetable garden, greenhouse and meadow with far reaching views. Instagram: seaview_gunwalloe. Featured in Country Living magazine. Gravel paths and a few steps.

18 GARDENS COTTAGE

Prideaux, St Blazey, PL24 2SS. Sue & Roger Paine, 07786 367610, sue.newton@btinternet.com. *1m from railway Xing on A390 in St Blazey. Turn into Prideaux Rd opp Gulf petrol station on A390 in St Blazey (signed Luxulyan). Proceed ½m. Turn R (signed Luxulyan Valley and Prideaux) and follow signs.* **Wed 9, Thur 10 June, Wed 7, Thur 8 July, Wed 4, Thur 5 Aug, Wed 1, Thur 2 Sept (2-5). Adm £5, chd free. Home-made teas. Visits also by arrangement May to Aug for groups of up to 30.**

Set in tranquil location on edge of Luxulyan Valley, work commenced on creating a 1½ acre garden from scratch in winter 2015. The aim has been to create a garden that's sympathetic to its surrounding landscape, has yr-round interest with lots of colour, is productive and simply feels good to be in. Check our Facebook page for special events - @ gardenscottageprideaux.

🐕 ✤ ☕

19 **NEW** **9 HIGMAN CLOSE**
Dobwalls, Liskeard,
PL14 4LW. Jim Stephens &
Sue Martin, 01579 321074,
J.l.stephens@btinternet.com. *3m
W of Liskeard. From double mini-
r'about in Dobwalls village take road
to Duloe, take 2nd R onto Treheath
Road, then 1st R into Higman Close.*
**Wed 23, Sun 27 June, Wed 7, Sun
11, Wed 21, Sun 25 July, Wed
4, Sun 8, Wed 18, Sun 22 Aug
(2-6). Combined adm with Dove
Cottage £6, chd free. Pre-booking
essential, please visit www.ngs.
org.uk for information & booking.
Visits also by arrangement June to
Aug for groups of up to 10.**
Two very different, modestly sized enclosed gardens. The three owners have between them over a hundred years of professional horticultural experience. A restless, constantly changing, unapologetically busy garden, full of good plants but also colour, scent and surprises. Sue's glasshouse is overflowing with cacti and succulents, some over 35 years old.

✤

20 ✦ **THE JAPANESE GARDEN**
St Mawgan, TR8 4ET. Natalie Hore & Stuart Ellison, 01637 860116, info@japanesegarden.co.uk, www.japanesegarden.co.uk. *6m East of Newquay. St Mawgan village is directly below Newquay Airport. Follow brown and white road signs on A3059 and B3276.* **For NGS: Sun 16, Mon 17 May (10-6). Adm £5, chd £2.50. Refreshments available in village 2-3 min walk from garden. For other opening times and information, please phone, email or visit garden website.**
Discover an oasis of tranquillity in a Japanese-style Cornish garden, set in approx 1 acre. Spectacular Japanese maples and azaleas, symbolic teahouse, koi pond, bamboo grove, stroll woodland, zen and

moss gardens. A place created for contemplation and meditation. Adm free to gift shop, bonsai and plant areas. Featured in BBC2 Big Dreams Small Spaces. 90% wheelchair accessible, with gravel paths.

♿ ✤

21 **KESTLE BARTON**
Manaccan, Helston, TR12 6HU.
Karen Townsend, 01326 231811,
info@kestlebarton.co.uk,
www.kestlebarton.co.uk. *A3083
Helston - Lizard, 2m L B3293 for St
Keverne. 2m L Helford, Newtown St
Martin, R then L for Helford. 1m L at
Xrds to Kestle Barton, R after several
hundred yds, follow signs. Large
coaches cannot reach Kestle Barton.*
**Every Wed 2 June to 29 Sept
(10.30-5). Adm by donation. Tea.**
A delightful garden near Frenchmans Creek, on the Lizard, which is the setting for Kestle Barton Gallery; wild flower meadow, Cornish orchard with named varieties and a formal garden with prairie planting in blocks by James Alexander Sinclair. It is a riot of colour in summer and continues to delight well into late summer. Parking, honesty box tea hut. Art Gallery. Good wheelchair access and

reasonably accessible loo. Dogs on leads welcome.

♿ 🐕 🚗 ☕ ⛱

22 **THE LODGE**
Fletchersbridge, Bodmin,
PL30 4AN. Mr Tony Ryde. *2m
E of Bodmin. From A38 at Glynn
Crematorium r'about take rd towards
Cardinham and continue through
hamlet of Fletchersbridge. Pass
Stable Art on R, garden 1st R over
river bridge.* **Sun 30 May (11-5.30).
Combined adm with The Old
School House, Cardinham £8,
chd free.**
3-acre riverside garden some 20 yrs old specialising in trees and shrubs chosen for their flowers, foliage and form, embracing a Gothic lodge remodelled in 2016, once part of the Glynn estate. Late May sees the cornus varieties at their best, with viburnums, rhododendrons, wisteria, clematis, roses and early perennials, and featuring a water garden with ponds, waterfalls and sculptures. Wheelchair access to gravelled areas around house and along 2 sides of garden.

♿ 🐕 ⛱

Alverton Cottage

23 ◆ THE LOST GARDENS OF HELIGAN
Pentewan, St Austell, PL26 6EN. Heligan Gardens Ltd, 01726 845100, info@heligan.com, www.heligan.com. *5m S of St Austell. From St Austell take B3273 signed Mevagissey, follow signs.* **For opening times and information, please phone, email or visit garden website.**
Lose yourself in the mysterious world of The Lost Gardens where an exotic sub-tropical jungle, atmospheric Victorian pleasure grounds, an interactive wildlife project and the finest productive gardens in Britain all await your discovery. Wheelchair access to Northern gardens. Armchair tour shows video of unreachable areas. Wheelchairs available at reception free of charge.

24 NAVAS HILL HOUSE
Bosanath Valley, Mawnan Smith, Falmouth, TR11 5LL. Aline & Richard Turner. *1½m from Trebah & Glendurgan Gdns. Head for Mawnan Smith, pass Trebah and Glendurgan Gdns then follow yellow signs. Don't follow SatNav which suggests*

you turn R before Mawnan Smith - congestion alert! **Sun 2 May (2-5). Adm £5, chd free. Home-made teas. 'All you can eat' £5.00.**
8½-acre elevated valley garden with paddocks, woodland, kitchen garden and ornamental areas. The ornamental garden consists of 2 plantsman areas with specialist trees and shrubs, walled rose garden, water features and rockery. Young and established wooded areas with bluebells, camellia walks and young large leafed rhododendrons. Seating areas with views across wooded valley. Partial wheelchair access, some gravel and grass paths.

25 NEW THE OLD SCHOOL HOUSE
Averys Green, Cardinham, Bodmin, PL30 4EA. Mrs Libby Pidcock. *Edge of Cardinham village. A30 to Cardinham. Xrds in village towards church on R. Pass cemetery, and tennis court on L, 3rd house on R after tennis court. B & B sign outside. Parking Parish Hall opp church.* **Sun 30 May (11-5). Combined adm with The Lodge £8, chd free.**

Treprenn Vean

Cottage garden created in the grounds of a Victorian school by a passionate and enthusiastic gardener over 35yrs. Not so much by design, but led by a love of plants and trying anything: herbaceous, annuals, bulbs, shrubs, trees, fruit cage, espalier apples, cut flower patch and some veg. Wild flower area, wildlife ponds, bird feeding stations, jungle corner, quirky containers. Plenty of seating. Sadly gravel paths not wheelchair friendly.

26 PARKHENVER
Penventon, Redruth, TR15 3AA. Dr & Mrs David Quill Smart. *Important: To enter gardens, please use main entrance of Roman Catholic Church - about 100m to town end of our main gate (white balustrading). Leave by Exit lane.* **Sun 11 July (1-5.30). Adm £5, chd free. Light refreshments on lawn and in marquee. Wine available.**
Gardens created in about 1850. Most features of house and gardens now complete. This year, we have cleared the 'slips', the area outside the walled garden. Good use of bark chippings have supressed many weeds. Woodland walks and a walk through the 'slips' now available. Other areas around the estate are now clear. Restored greenhouses. Some features are, we think, unique, such as the style of the fountain which faces the main house. It is interesting that this small estate is within the Redruth Town boundary. Wheelchair access to some gravelled areas may be difficult.

27 ◆ PENCARROW
Washaway, Bodmin, PL30 3AG. Molesworth-St Aubyn family, 01208 841369, info@pencarrow.co.uk, www.pencarrow.co.uk. *4m NW of Bodmin. Signed off A389 & B3266. Free parking.* **For NGS: Mon 12 Apr (10-5.30). Adm £6.95, chd free. Cream teas. For other opening times and information, please phone, email or visit garden website.**
50 acres of tranquil, family-owned Grade II* listed gardens. Superb specimen conifers, azaleas, magnolias and camellias galore. 700 varieties of rhododendron give a blaze of spring colour; blue hydrangeas line the mile-long carriage drive throughout the summer. Discover the

Iron Age hill fort, lake, Italian gardens and granite rockery. Dogs welcome, café and children's play area. Gravel paths, some steep slopes.

28 PENDOWER HOUSE

Lanteglos-by-Fowey, PL23 1NJ. Mr Roger Lamb, 01726 870884, rl@rogerlamb.com. *Near Polruan off B3359 towards Bodinnick. 2m from Fowey using the Bodinnick ferry. Please ask for directions when arranging to visit the garden. Do not use SatNav.* **Visits by arrangement May & June for groups of up to 20. Adm £5, chd free. Home-made teas. Other refreshments available by prior arrangement.**
Set in the heart of Daphne du Maurier country in its own valley this established garden surrounding a Georgian rectory is now undergoing a revival having been wild and neglected for some years. It has formal herbaceous terraces, a cottage garden, orchard, ponds, streams and a C19 shrub garden with a fine collection of azaleas, camellias and rhododendrons plus rare mature specimen trees. House and garden available for filming and photography. Sadly difficult wheelchair access to much of this garden.

29 PENMILDER

Lodge Hill, Liskeard, PL14 4EL. Chris & Amanda Deegan. *1m S of Liskeard town centre. On B254 (Duloe road). Entrance on L directly opp 'Santa Trees'.* **Sun 9 May (10.30-4.30). Adm £5, chd free. Light refreshments.**
This is a gently sloping, S-facing garden of approx 2½ acres. Lawns with mature borders and a lily pond with plenty of wildlife. There are also natural wooded areas and an apple orchard which is particularly pretty in spring with daffodils, bluebells and primroses. Sadly it is not suitable for wheelchair users due to gravel paths.

30 PINSLA GARDEN

Glynn, Nr. Cardinham, Bodmin, PL30 4AY. Mark & Claire Woodbine, www.pinslagarden.wordpress.com. *3½m E of Bodmin. From A30 or Bodmin take A38 towards Plymouth, 1st L to Cardinham & Fletchers Bridge, 2m on R.* **Tue 4, Wed 5, Tue 18, Wed 19 May, Tue 1, Wed 2,**

Tue 15, Wed 16, Tue 29, Wed 30 June, Tue 6, Wed 7, Tue 20, Wed 21 July, Tue 3, Wed 4, Tue 17, Wed 18, Tue 31 Aug, Wed 1 Sept (9-5). Adm £4, chd free.
Surround yourself with deep nature. Pinsla is a tranquil cottage garden buzzing with insects enjoying the sheltered sunny edge of a wild wood. Lose yourself in a colourful tapestry of naturalistic planting. There are lots of unusual planting combinations, cloud pruning, intricate paths, garden art and a stone circle. Sorry, no teas but you are welcome to bring your own thermos. Partial wheelchair access as some paths are narrow and bumpy.

31 ◆ POTAGER GARDEN

High Cross, Constantine, Falmouth, TR11 5RF. Mr Mark Harris, 01326 341258, enquiries@potagergarden.org, www.potagergarden.org. *5m SW of Falmouth. From Falmouth, follow signs to Constantine. From Helston, drive through Constantine and continue towards Falmouth.* **For opening times and information, please phone, email or visit garden website.**
Potager has emerged from the bramble choked wilderness of an abandoned plant nursery. With mature trees which were once nursery stock and lush herbaceous planting interspersed with fruit and vegetables Potager Garden aims to demonstrate the beauty of productive organic gardening. There are games to play, hammocks to laze in and boule and badminton to enjoy.

32 ◆ ROSELAND HOUSE

Chacewater, TR4 8QB. Mr & Mrs Pridham, 01872 560451, charlie@roselandhouse.co.uk, www.roselandhouse.co.uk. *4m W of Truro. At Truro end of main st. Park in village car park (100yds) or on surrounding rds.* **For NGS: Sat 3, Sun 4 July (1-5). Adm £5, chd free. Home-made teas. For other opening times and information, please phone, email or visit garden website.**
The 1-acre garden is a mass of summer colour when the National Collection of Clematis viticella is at its peak in July, other climbing plants abound lending both foliage and scent, the conservatory and greenhouses are also full of unusual and interesting plants. Two ponds

and a Victorian conservatory. Some slopes.

33 ◆ ST MICHAEL'S MOUNT

Marazion, TR17 0HS. James & Mary St Levan, 01736 710507, mail@stmichaelsmount.co.uk, www.stmichaelsmount.co.uk. *2½m E of Penzance. ½m from shore at Marazion by Causeway; otherwise by motor boat.* **For NGS: Sat 12 June (10.30-5). Adm £8.50, chd £4. Pre-booking essential, please phone 01736 710507 or email mail@stmichaelsmount. co.uk for information & booking. For other opening times and information, please phone, email or visit garden website.**
Infuse your senses with colour and scent in the unique sub-tropical gardens basking in the mild climate and salty breeze. Clinging to granite slopes the terraced beds tier steeply to the ocean's edge, boasting tender exotics from places such as Mexico, the Canary Islands and South Africa. Laundry lawn, mackerel bank, pill box, gun emplacement, tiered terraces, well, tortoise lawn. Walled gardens, seagull seat. Garden lawn can be accessed with wheelchairs although further exploration is limited due to steps and steepness.

34 SOUTH BOSENT

Liskeard, PL14 4LX. Adrienne Lloyd & Trish Wilson. *2½m W of Liskeard. From r'about at junction of A390 and A38 take turning to Dobwalls. At mini-r'about R to Duloe, after 1¼m at X-rds turn R. Garden on L after ¼m.* **Wed 12, Thur 13 May, Wed 16, Thur 17 June, Wed 14, Thur 15 July, Wed 11, Thur 12 Aug, Wed 15, Thur 16 Sept (2-5.30). Adm £5, chd free. Home-made teas.**
This garden is an example of work in progress currently being developed from farmland. The aim is to create a combination of interesting plants coupled with habitat for wildlife over a total of 9.5 acres. There are several garden areas, woodland gardens, a meadow, ponds of varying sizes, incl new rill and waterfall. In spring, the bluebell wood trail runs alongside the stream. Regret no wheelchair access to bluebell wood due to steps.

National Garden Scheme gardens are identified by their yellow road signs and posters. You can expect a garden of quality, character and interest, a warm welcome and plenty of home-made cakes!

35 SOUTH LEA

Pillaton, Saltash, PL12 6QS. Viv & Tony Laurillard, 01579 350629, tonyandviv.laurillard@gmail.com. *Pillaton, opp Weary Friar PH. 4m S of Callington. Signed from r'abouts on A388 at St Mellion and Hatt, and on A38 at Landrake. Roadside parking. Please do not park in PH car park.* **Sun 9 May, Sun 4 July (1-5). Adm £5, chd free. Home-made teas. Visits also by arrangement May to July for groups of 10+.**

In the front a path winds through interesting landscaping with a small pond. Tropical beds by front door with palms, cannas, etc. The back garden, with views over the valley, is a pretty picture in May with spring bulbs and clematis, whilst in July the herbaceous borders are a riot of colour. Lawns are separated by a fair sized fish pond and the small woodland area is enchanting in spring. Plenty of seating. Due to steps, wheelchair access to front dry garden and rear terrace, from which much of garden can be viewed.

 ♿ 🐕 🚗 ☕

36 TREBARTHA ESTATE GARDEN AND COUNTRY GARDEN AT LEMARNE

Trebartha, nr Launceston, PL15 7PD. The Latham Family. *6m SW of Launceston. North Hill, SW of Launceston nr junction of B3254 & B3257. No coaches.* **Sun 25 Apr, Sun 10 Oct (2-5). Adm £8, chd free. Home-made teas.**

Historic landscape gardens featuring streams, cascades, rocks and woodlands, incl fine trees, bluebells, ornamental walled garden and private modern country garden at Lemarne. Ongoing development of C19 American garden and C18 fish ponds. Allow at least 1 hour for a circular walk. Some steep and rough paths, which can be slippery when wet. Stout footwear advised. October opening for autumn colour.

🐕 ☕

37 NEW TREBARVAH WOON

Seworgan, Constantine, Falmouth, TR11 5QN. Paul & Vicky Bryant. *Access from A394. From Falmouth take L signed Constantine and Gweek after Edgecumbe, turning R to Seworgan after 1½m. From Helston take R marked Seworgan and Laity after Manhay.* **Sun 2 May (11-5). Adm £5, chd free. Light refreshments. Tea/coffee, cake/scones.**

A steeply sloping upper valley garden of just over 1 acre. The Helford stream forms a long lower boundary some of which is managed. The site is terraced and planted at several levels incl small cottage garden by conservatory. Azaleas, camellias, rhododendrons and magnolias enjoy the poor acid soil. The garden also contains vegetable beds, a wild area, and a small wood. Not suitable for wheelchairs.

🐕 ❀ ☕

38 TREGONNING

Carleen, Breage, Helston, TR13 9QU. Andrew & Kathryn Eaton, 01736 761840, alfeaton@aol.com, Tregonninggarden.co.uk. *1m S of Godolphin Cross. From Xrds in centre of Godolphin Cross head S towards Carleen. In ½m at fork signed Breage 1¼ turn R up narrow lane marked no through road. After ½m parking on L opp Tregonning Farm.* **Sun 18 July (1-5). Adm £5, chd free. Home-made teas. Visits also by arrangement Mar to Oct for groups of up to 20. By arrangement adm price £6 incl tea or coffee and biscuits.**

Located 300ft up NE side of Tregonning Hill this small (1¼ acre) maturing garden will hopefully inspire those thinking of making a garden from nothing more than a pond and copse of trees (in 2009). With the ever present challenge of storm force winds, garden offers yr-round interest and a self-sufficient vegetable and soft fruit paddock. Sculpted grass meadow, with panoramic views from Carn Brea to Helston. A section of the garden is designed in the form of a plant (incorporating a deck, leaf shaped beds, stream and large pond). Front cottage garden, spring garden, Mediterranean patio. Carp pond/fernery. Packed vegetable garden. Spring garden not accessible to wheelchairs. See us on Facebook - Tregonninggarden.

♿ 🐕 ❀ 🛏 ☕

39 TRENARTH

High Cross, Constantine, Falmouth, TR11 5JN. Lucie Nottingham, 01326 340444, lmnottingham@btinternet.com, www.trenarthgardens.com. *6m SW of Falmouth. Main rd A39/A394 Truro to Helston, follow Constantine signs. High X garage turn L for Mawnan, 30yds on R down dead end lane, Trenarth is ½m at end of lane.* **Sun 20 June (2-5). Adm £5, chd free. Great teas provided by local groups. Visits also by arrangement Feb to Nov for groups of up to 30.**

4 acres round C17 farmhouse in peaceful pastoral setting. Yr-round interest. Emphasis on tender, unusual plants, structure and form. C16 courtyard, listed garden walls, yew rooms, vegetable garden, traditional potting shed, orchard, woodland area with children's interest, palm garden, with dierama show. Circular walk down ancient green lane via animal pond to Trenarth Bridge, returning through woods. Abundant wildlife. Bees in tree bole, lesser horseshoe bat colony, swallows, wild flowers and butterflies. Family friendly, children's play area, the Wolery, and plenty of room to run, jump and climb.

🐕 ❀ 🚗 ☕ 🍽

40 NEW TREPRENN VEAN

Pednavounder, Coverack, Helston, TR12 6SE. Mike & Jill Newell. *Pednavounder near Coverack. Head out of Helston on A3083 towards Lizard and take B3293 to Coverack and St Keverne. After about 7m turn R at Zoar Garage and follow the road for about 2m.* **Sun 11 July (1-5). Adm £4, chd free. Home-made teas in neighbour's garden through a linking gate.**

The ⅓ of an acre garden has been created by the owners since late 2014. Rural setting about ½m from the sea. Large pond with bridge over it which forms the major axis of the garden. To the right the planting is more tropical, to the left more cottage garden. Productive veg growing area and fruit cage with other fruit trees within the garden. Garden is flat, only 1 step up onto bridge. Alternative routes to see all of the garden.

♿ 🐕 ❀ ☕

41 NEW TRETHEW

Lanlivery, nr Bodmin, PL30 5BZ. Ginnie & Giles Clotworthy. *3 m W of Lostwithiel. On rd between Lanlivery and Luxulyan. Signed from both villages and A390. Do not use*

Satnav. **Sun 20 June (12-6). Adm £5, chd free. Home-made teas.** Series of profusely planted and colourful areas surrounding an ancient Cornish farmhouse. Features incl terracing with pergola, gazebo and herbaceous borders within yew hedges, all overlooking orchard with roses and pond beyond. Magnificent views.

42 TREVILLEY
Sennen, Penzance, TR19 7AH. Patrick Gale & Aidan Hicks, 01736 871808. *For walkers, Trevilley lies on footpath from Trevescan to Polgigga and Nanjizal. Trevilley is up a signposted unmade track just outside Trevescan. If you get to the white house, confusingly called Trevilley Farmhouse, you need to reverse!* **Sun 20 June (1-5). Adm £6, chd free. Light refreshments. Pimms as well as tea and cake. Visits also by arrangement June to Aug for groups of 5 to 30.** Eccentric, romantic and constantly evolving garden, as befits the intense creativity of its owners, carved out of an expanse of concrete farmyard over 20 yrs. Incl elaborate network of decorative cobbling, pools, container garden, veg garden, shade garden, the largely subtropical mowhay garden and both owner's studios but arguably its glory is the westernmost walled rose garden in England. Plant stall. Dogs are welcome on leads and there's direct access to fields where they can let off steam. The garden visit can form the climax of enjoying the circular coast path walk from Land's End to Nanjizal and back across the fields.

43 ◆ TREWIDDEN GARDEN
Buryas Bridge, Penzance, TR20 8TT. Mr Alverne Bolitho - Richard Morton, Head Gardener, 01736 364275/363021, contact@trewiddengarden.co.uk, www.trewiddengarden.co.uk. *2m W of Penzance. Entry on A30 just before Buryas Bridge. Postcode for SatNav is TR19 6AU.* **For NGS: Thur 22 Apr (10.30-5.30). Adm £8, chd free. For other opening times and information, please phone, email or visit garden website.** Historic Victorian garden with magnolias, camellias and magnificent tree ferns planted within ancient tin workings. Tender, rare and unusual exotic plantings create a riot of colour throughout the season.

Water features, specimen trees and artefacts from Cornwall's tin industry provide a wide range of interest for all.

44 WAYE COTTAGE
Lerryn, nr Lostwithiel, PL22 0QQ. Malcolm & Jennifer Bell, 01208 872119, lerrynbells@gmail.com. *4m S of Lostwithiel along north river bank. Parking usually available at property or in village car park. Garden 10 minute level stroll along riverbank/stepping stones.* **Visits by arrangement Apr to Sept for groups of up to 20. The garden is open most days. Please ring first - best after 6pm. Adm £5, chd free. Refreshments/light lunches available in shop and pub on village green.** An enchanting cottage garden on the footprint of an old market garden - many interesting plants, enticing paths, secluded seats and stunning river views. New grass garden. 'Magical! The perfect place for a botanical recharge and horticultural inspiration.' Reproduced courtesy of Cornwall Life magazine. Garden groups very welcome. Pretty, tidal village. Lovely walks. Good pub and

cafe. Sadly the garden is steep with too many steps for disabled access.

45 WINDMILLS
South Hill, Callington, PL17 7LP. Mr & Mrs Peter Tunnicliffe, tunnicliffesue@gmail.com. *3m NW of Callington. Head N from Callington A388, after about ½m turn L onto South Hill Rd (signed South Hill). Straight on for 3m, gardens on R just before church.* **Sun 23 May, Sun 8 Aug (1-5). Combined adm with Anvil Cottage £5, chd free. Home-made teas. Gluten free cakes available. Visits also by arrangement May to Aug (excl June) for groups of 5 to 30. Donation to Cornwall Air Ambulance.** Next to medieval church and on the site of an old rectory and there are still signs in places of that long gone building. A garden full of surprises, formal paths and steps lead up from the flower beds to extensive vegetable and soft fruit area. More paths lead to a pond, past a pergola, and down into large lawns with trees and shrubs and chickens. Partial wheelchair access.

Trebartha Estate Garden and Country Garden at Lemarne

OPENING DATES

All entries subject to change. For latest information check www.ngs.org.uk

Extended openings are shown at the beginning of the month.

Map locator numbers are shown to the right of each garden name.

February

Snowdrop Festival

Every Saturday from Saturday 20th
Summerdale House   42

March

Every Saturday
Summerdale House   42

Sunday 21st
◆ Dora's Field   8
◆ High Close Estate & Arboretum   17
◆ Holehird Gardens   18
◆ Rydal Hall   40

April

Every Saturday
Summerdale House   42

May

Every Saturday
Summerdale House   42

Sunday 2nd
Low Fell West   30

Monday 3rd
Low Fell West   30

Sunday 9th
Chapelside   3
NEW Lea Cottage   27
◆ Rydal Hall   40

Sunday 16th
Matson Ground   33

Friday 21st
◆ Holker Hall Gardens   19

Saturday 29th
Galesyke   12
Langholme Mill   25

Sunday 30th
Chapelside   3
Galesyke   12
Grange over Sands Hidden Gardens   14
Langholme Mill   25

Monday 31st
Langholme Mill   25

June

Every Saturday to Saturday 19th
Summerdale House   42

Saturday 5th
Coombe Eden   6
Hazel Cottage   16

Sunday 6th
Coombe Eden   6
Hazel Cottage   16
Park House   36
Quarry Hill House   38
◆ Rydal Hall   40
Tithe Barn   43
Yewbarrow House   48

Sunday 13th
Hayton Village Gardens   15
NEW Little Crag   28
NEW Low Crag   29
8 Oxenholme Road   35

Saturday 19th
NEW Whetstone Croft & Cottage   44

Sunday 20th
Chapelside   3
NEW Eden Mount Gardens   9
NEW Whetstone Croft & Cottage   44

Wednesday 23rd
Church View   5

Sunday 27th
Askham Hall   1
Fernhill Coach House   11
Grange over Sands Hidden Gardens   14
Ivy House   22
Park House   36

July

Saturday 3rd
NEW Keerside   23

Sunday 4th
NEW Keerside   23
Yewbarrow House   48

Thursday 8th
◆ Holehird Gardens   18
Larch Cottage Nurseries   26

Sunday 11th
Chapelside   3
Winton Park   46

Saturday 17th
NEW Keerside   23

Sunday 18th
Holme Meadow   20
NEW Keerside   23
◆ Rydal Hall   40

Sunday 25th
Grange over Sands Hidden Gardens   14

Saturday 31st
Coombe Eden   6
Hazel Cottage   16

August

Sunday 1st
Coombe Eden   6
Hazel Cottage   16
NEW Newlands   34
Yewbarrow House   48

Sunday 8th
NEW Beulah   2
HPB Merlewood   21
NEW Mains House   32

Thursday 12th
Larch Cottage Nurseries   26

Saturday 14th
Woodend House   47

Sunday 15th
Grange Fell Allotments   13
Woodend House   47

Sunday 22nd
Fell Yeat   10

Sunday 29th
Grange over Sands Hidden Gardens   14
Kinrara   24

September

Sunday 5th
Yewbarrow House   48

Wednesday 15th
Church View   5

Thursday 16th
Larch Cottage Nurseries   26

Sunday 26th
Grange over Sands Hidden Gardens   14

October

Sunday 17th
Low Fell West   30

By Arrangement

Arrange a personalised garden visit on a date to suit you. See individual garden entries for full details.

Chapelside   3
Cherry Cottage   4
Church View   5
Crumble Cottages   7
Elder Cottage, Grange over Sands Hidden Gardens   14
Fernhill Coach House   11
Galesyke   12
Grange Fell Allotments   13
Holme Meadow   20
Ivy House   22
Langholme Mill   25
Low Fell West   30
Lower Rowell Farm & Cottage   31
Matson Ground   33
Pear Tree Cottage   37
Quarry Hill House   38
Rose Croft   39
Sprint Mill   41
Windy Hall   45
Woodend House   47
Yewbarrow House   48

THE GARDENS

1 ASKHAM HALL

Askham, Penrith, CA10 2PF.
Charles Lowther, 01931 712350,
enquiries@askhamhall.co.uk,
www.askhamhall.co.uk. *5m S of
Penrith. Turn off A6 for Lowther &
Askham.* **Sun 27 June (11-6). Adm
£3, chd free. Light refreshments.
Donation to Askham and Lowther
Churches.**
Askham Hall is a Pele Tower
incorporating C14, C16 and early C18
elements in a courtyard plan. Opened
in 2013 with luxury accommodation,
a restaurant, cafe and wedding barn.
Splendid formal garden with terraces
of herbaceous borders and topiary,
dating back to C17. Meadow area
with trees and pond, kitchen gardens
and animal trails. Combined with
Summer Fair. Cafe serving tea, coffee,
light lunches, cake. Wood-fired pizza
oven, BBQ and bar in the courtyard.
Partial wheelchair access but not all
areas due to steps.

2 NEW BEULAH

Pooley Bridge, Penrith, CA10 2NG.
Mr & Mrs M Macinnes. *½m outside
the village of Pooley Bridge on
Ullswater. From M6 (J40), A66 W
onto A592 to Ullswater. At the lake,
turn L into Pooley Bridge; through
village towards Tirril. Parking signed
at 1st L turn.* **Sun 8 Aug (1-5).
Combined adm with Mains House
£5, chd free. Home-made teas at
Mains House CA10 2NG.**
Beech hedged country garden
with ha-ha giving stunning views
towards Ullswater and the fells. Mixed
perennial borders; formal box edged
parterre with flowers, herbs, fruits
and vegetables. Small rose garden.
Orchard and shrubbery. Interesting
shrubs and trees. Varied garden
sculptures.

♿ ☕

3 CHAPELSIDE

Mungrisdale, Penrith,
CA11 0XR. Tricia & Robin
Acland, 017687 79672,
rtacland@gmail.com. *12m W of
Penrith. On A66 take minor rd N
signed Mungrisdale. After 2m, sharp
bends, garden on L immed after tiny
church on R. Use church car park
at foot of our short drive. On C2C
Reivers 71, 10 cycle routes.* **Sun 9,
Sun 30 May, Sun 20 June, Sun 11
July (1-5). Adm £4, chd free. Visits
also by arrangement Apr to Sept.**
1 acre windy garden below fell round
C18 farmhouse and outbuildings.
Fine views. Tiny stream, large pond.
Herbaceous, gravel, alpine, damp
and shade areas, bulbs in grass.
Wide range of plants, many unusual.
Relaxed planting regime. Run on
organic lines. Art constructions in
and out, local stone used creatively.
Featured in leading magazines and
several books.

🐄 ✿ ☕

4 CHERRY COTTAGE

Crosby Moor, Crosby-On-Eden,
Carlisle, CA6 4QX. Mr & Mrs John
& Lesley Connolly, 01228 573614,
jtlaconnolly@aol.com. *Off
A689 midway between Carlisle
& Brampton. Take turn signed
'Wallhead'. Cherry Cottage is on the
corner of junction on R. Park in the
lane.* **Visits by arrangement Apr to
Sept for groups of 5 to 20. Adm
£4, chd free. Light refreshments.**
Relaxed country garden surrounding
an C18 cottage on an approx.
⅓ acre site. It has a wide range
of habitats including herbaceous
borders, wildlife pond, bog garden
and shady woodland. The vegetable
area is gradually changing over to cut
flower beds. It has 2 greenhouses, a
potting shed and a summer house.
Various seating areas connected by
grass and gravel paths. Gravel paths
not suitable for wheelchairs.

5 CHURCH VIEW

Bongate, Appleby-in-
Westmorland, CA16 6UN.
Mrs H Holmes, 01768 351397,
engcougars@btinternet.com, www.
sites.google.com/site/engcougars/
church-view. *0.4m SE of Appleby
town centre. A66 W take B6542 for
2m St Michael's Church on L garden
opp. A66 E take B6542 & continue
to Royal Oak Inn, garden next door,
opp church.* **Wed 23 June, Wed 15
Sept (1-4.30). Adm £4, chd free.
Visits also by arrangement May to
Sept for groups of up to 30.**
It's all about the plants! Less than
½ acre of garden but with layers
of texture, colour and interest in
abundance, this is a garden for
plantaholics. Plant combinations
are at the heart of the design. With
self-contained vistas and maximum
use of planting space, the garden
photographs very well and has been
a subject for many local and national
publications and photographers over
the last decade. Partial wheelchair
though main garden is on a sloping
site with gravel paths.

6 COOMBE EDEN

Armathwaite, Carlisle, CA4 9PQ.
Belinda & Mike. *8m SE of Carlisle.
Turn off A6 just South of High Hesket
signed Armathwaite. Continue to
bottom of hill where garden can
be found on R turn for Lazonby.*
**Sat 5, Sun 6 June, Sat 31 July,
Sun 1 Aug (12-5). Combined
adm with Hazel Cottage £5, chd
free. Home-made teas at Hazel
Cottage garden.**
A garden of traditional and
contemporary beds. Steep banks
down to stream with pretty Japanese
style bridge. Large Rhododendrons
give a breathtaking show - end
of May. New formal garden and
vegetable areas added 2020. ½
garden on flat with some path, mainly
grass, rest of garden steep banks
and steps.

🐄 ☕

7 CRUMBLE COTTAGES

Beckside, Cartmel, LA11 7SP.
Sarah Byrne & Stewart
Cowe, 015395 34405,
sarah@crumblecottages.co.uk,
www.crumblecottages.co.uk.
*http://crumblecottages.co.uk/
contact-us/. Car sat nav brings you
directly here however mobile phone
sat navs take you up the wrong lane.*
**Visits by arrangement Apr to Oct.
Adm £5, chd £3.**
1½ acre water garden built just
over 5 years ago to improve the
biodiversity of the area. Grade 2 listed
wall with 7 beeboles. Walled kitchen
garden. Ornamental cottage style
planting around the house with yew
hedges, topiary, and box hedging.
wildflower meadow. Cut flower
and butterfly borders. Wheelchair
access ornamental areas by house
and outlying areas of garden are
accessible but only by a wheelchair
that can cope with uneven ground.

♿ 🏠

*All entries are subject
to change. For the latest
information check
ngs.org.uk*

8 ◆ DORA'S FIELD
Rydal, Ambleside,
LA22 9LX. National Trust,
www.nationaltrust.org.uk. 1½ m
N of Ambleside. Follow A591 from
Ambleside to Rydal. Dora's Field is
next to St Mary's Church. **For NGS:
Sun 21 Mar (11-4). Combined
adm with High Close Estate &
Arboretum by donation. For other
opening times and information,
please visit garden website.**
Named for Dora, the daughter
of the poet William Wordsworth.
Wordsworth planned to build a
house on the land but, after her
early death, he planted the area with
daffodils in her memory. Now known
as Dora's field the area is renowned
for its spring display of daffodils and
bluebells. 21 March; Wordsworth's
Daffodil Legacy. Old School Tea
Room Cafe located at neighbouring
Rydal Hall.

GROUP OPENING

9 NEW EDEN MOUNT GARDENS
Eden Mount, Grange-Over-Sands,
LA11 6BZ. Dr Ian & Pauline Wilson.
Please park on Eden Mount Road as
there is no parking near the gardens.
**Sun 20 June (11-4). Combined
adm £5, chd free.**

NEW **THE COTTAGE**
Dr Ian and Pauline Wilson.

NEW **EDEN MOUNT**
Mr & Mrs John Barker.

NEW **EDENHURST**
Mr & Mrs P Bowe.

NEW **GROVE HOUSE**
Mr & Mrs Melissa Harrington.

NEW **2 MOUNT EDEN**
Mr & Mrs P Dutta.

The Cottage - Hidden from view with
wide gravel paths and a number
of steps. Lovely changing views
over Morecambe Bay. Eden Mount
- Compact garden. Mostly hand
landscaped with planted borders,
a range of container plants and
pebble art feature. Edenhurst -
Sloping front garden with large patio.
Evergreen trees and two topiary trees.
Limestone rockeries and walkways,
3 pools, a waterfall, stream, fountain
and fish. Small rear garden, large
variety of plants, flowers and roses.
Grove House - Victorian layout of
front terrace with limestone above
a sloping lawn and hedged kitchen

side lawn. Collection of David Austen
roses, cottage garden borders,
orchard border, wild area and stone
circle garden feature. Planted by Tom
and Sarah Westley and the current
owners to re-establish traditional
English garden planting, incl 3000
daffodils, it is a work in progress. 2
Mount Eden – L shaped garden with
mixed planting, wildlife ponds and
lawns. Interesting features provide
year interest.

10 FELL YEAT
Casterton, Kirkby Lonsdale,
LA6 2JW. Mrs A E Benson. 1m E
of Casterton Village. On the rd to
Bull Pot. Leave A65 at Devils Bridge,
follow A683 for 1m, take the R fork
to High Casterton at golf course,
straight on L, ¼ m from no-through-rd
sign. **Sun 22 Aug (1-5). Adm £4,
chd free.**
1 acre country garden with mixed
planting, incl unusual trees, shrubs
and some topiary. Small woodland
garden and wooded glades. 2 ponds
which encourage wildlife, including
newts and dragonflies. Explore the
fernery and maturing stumpery and
new grotto house with rocks and
ferns. Large collection of hydrangeas.
A garden to explore, with several
arbours where you can sit and relax.
Adjoining nursery specialising in
ferns, hostas, hydrangeas and many
unusual plants. Partial wheelchair
access.

11 FERNHILL COACH HOUSE
Bleacragg Road, Witherslack,
Grange-Over-Sands,
LA11 6RX. Adele & Mike
Walford, 015395 52102,
mwandaj@btinternet.com. Country
road, ½ m beyond Halecat & Abi and
Tom's Garden Plants. From A590
turn N to Witherslack. Follow brown
signs to Halecat, continue on lane for
½ m. Fernhill on L. Rail; Grange-over-
sands; 5m, bus X6. 2m walk, NCR
70. Roads & parking are not suitable
for coaches. **Sun 27 June (10-5).
Adm £4, chd free. Home-made
teas. Visits also by arrangement
June to Sept fr groups of up to 20.
Not suitable for coach parties.**
Exuberant cottage garden; mixed
borders, lots of vegetables,
greenhouse, polytunnel. Orchard,
hens, bees. Small ponds and lots
of roses. We propagate cottage

garden plants and grow heritage
and northern fruit trees. Witherslack
Orchard Group apple juice, damson
and apple juice. Numerous apple
trees for sale. Sale of refreshments
supports our community shop which
is managed largely by volunteers and
is a registered charity. Wheelchair
access to flower garden only. Paths
are uneven and on sloping ground.

12 GALESYKE
Wasdale, CA20 1ET. Christine &
Mike McKinley, 01946 726267,
mckinley2112@sky.com. From
Gosforth, follow signs to Nether
Wasdale & then to Lake, approx 5m.
From Santon Bridge follow signs to
Wasdale then to Lake, approx 2¼ m.
**Sat 29, Sun 30 May (10-5). Adm
£4, chd free. Cream teas. Visits
also by arrangement May to Sept.
Cream Teas by arrangement only.**
4 acre woodland garden with
spectacular views of the Wasdale
fells. The R lrt runs through the garden
and both banks are landscaped,
you can cross over the river via
a picturesque, mini, suspension
bridge. The garden has an impressive
collection of rhododendrons and
azaleas that light up the woodlands in
springtime.

**13 GRANGE FELL
ALLOTMENTS**
Fell Road, Grange-Over-
Sands, LA11 6HB. Mr Bruno
Gouillon, 01539 532317,
brunog45@hotmail.com. Opposite
Grange Fell Golf Club. Rail 1.3
m, Bus 1CR X6, NCR 70. **Sun 15
Aug (11-4). Adm £5, chd free.
Light refreshments. Visits also
by arrangement Apr to Aug for
groups of 10 to 30.**
The allotments are managed by
Grange Town Council. Opened
in 2010, 30 plots are now rented
out and offer a wide selection of
gardening styles and techniques. The
majority of plots grow a mixture of
vegetables, fruit trees and flowers.
There are a few communal areas
where local fruit tree varieties have
been donated by plot holders with
herbaceous borders and annuals.

*Online booking is available
for many of our gardens,
at ngs.org.uk*

GROUP OPENING

14 GRANGE OVER SANDS HIDDEN GARDENS

Grange-Over-Sands, LA11 7AF.
Bruno Gouillon. *Off Kents Bank Road, 3 gardens on Cart Lane then 1 garden up Carter Road for Shrublands Rail 1.4m; Bus X6; NCR 70.* Sun 30 May (11-3); Sun 27 June, Sun 25 July, Sun 29 Aug (11-4); Sun 26 Sept (11-3). Combined adm £5, chd free. Tea.

21 CART LANE
Veronica Cameron.

ELDER COTTAGE
Bruno Gouillon & Andrew Fairey, 01539 532317, brunog45@hotmail.com.
Visits also by arrangement Mar to Oct.

NEW HAWTHORNE COTTAGE
Mr & Mrs Carroll & John Ashton.

SHRUBLANDS
Jon & Avril Trevorrow.

4 very different gardens hidden down narrow lanes off the road south out of Grange. Off Kents Bank Road, 3 gardens on Cart Lane all back onto the railway embankment, providing shelter from the wind but also creating a frost pocket. 21 is a series of rooms designed to create an element of surplus with fruit and vegetables in raised beds. Elder cottage is an organised riot of fruit trees, vegetables, shrubby perennials and herbaceous plants. Productive and peaceful. Hawthorne Cottage has been redesigned and replanted over the last 2 years to create a garden with colour and interest. Up the hill on Carter Road for Shrublands, a ³⁄₄ acre garden situated on a hillside overlooking Morecambe Bay. Access to Shrublands and 21 Cart Lane only. Elder Cottage only be seen from the road side. Hawthorne Cottage only seen from side gate.

♿ 🐴 ✿ 🚌 ☕ 🍷

GROUP OPENING

15 HAYTON VILLAGE GARDENS

Hayton, Brampton, CA0 9HR.
5m E of M6 J43 at Carlisle, ¹⁄₂ m S of A69. 3m W of Brampton. signed to Hayton off A69. Narrow roads: Please park courteously L side only. If able park less centrally leaving space for less able nr pub.

Sun 13 June (12-6). Combined adm £5, chd free. Home-made teas at Hayton Village Primary School with live music, and also cold refreshments in one of the gardens - outside if suitable. **Donation to Hayton Village Primary School.**

ARNWOOD
Joanne Reeves-Brown.

ASH TREE FARM
Mr & Mrs J Dowling.

BECK COTTAGE
Fiona Cox.

CHESTNUT COTTAGE
Mr Barry Brian.

CURLEW COTTAGE
Frances & David Scales.

HAYTON C OF E PRIMARY SCHOOL
Hayton C of E Primary School, www.hayton.cumbria.sch.uk.

KINRARA
Tim & Alison Brown.
(See separate entry)

MILLBROOK
M & J Carruthers.

THE PADDOCK
Phil & Louise Jones.

TOWNHEAD COTTAGE
Chris & Pam Haynes.

WEST GARTH COTTAGE
Debbie Jenkins, www.westgarth-cottage-gardens.co.uk.

NEW WHITE HOUSE COTTAGE
Mr Chris Potts.

A valley of gardens of varied size and styles all within ¹⁄₂ m, mostly of old stone cottages. Smaller and larger cottage gardens, courtyards and containers, steep wooded slopes, lawns, exuberant borders, frogs, pools, colour and texture throughout. Home-made teas and live music at the school where the children annually create gardens within the main school garden. Other refreshments usually in a garden and outside if suitable. Gardens additional to those listed also generally open or visible. Popular Scarecrow Trail also open over the same route this year. Facebook: Hayton Open Gardens. Two of the gardens: Westgarth Cottage and Kinrara are designed by artist/s and an architect with multiple garden design experience. Accessibility from full to none. Note that gardens are spread out & some walking needed to see all. Come early or late for easier parking.

♿ ✿ ☕

Beulah

▮6 HAZEL COTTAGE

Armathwaite, Carlisle, CA4 9PG.
Mr D Ryland & Mr J Thexton.
8m SE of Carlisle. Turn off A6 just S of High Hesket signed Armathwaite, after 2m house facing you at T-junction 1¼ m walk from Armathwaite railway station. **Sat 5, Sun 6 June, Sat 31 July, Sun 1 Aug (12-5). Combined adm with Coombe Eden £5, chd free. Home-made teas at Hazel Cottage.**
Flower arrangers and plantsman's garden. Extending to approx 5 acres. Includes mature herbaceous borders, pergola, ponds and planting of disused railway siding providing home to wildlife. Many variegated and unusual plants. Varied areas, planted for all seasons, S facing, some gentle slopes, ever changing. Abandoned railway cutting now a woodland walk. Various animals around garden. Only partial access for wheelchair users, small steps to WC area. Main garden planted on gentle slope.

▮7 ◆ HIGH CLOSE ESTATE & ARBORETUM

Loughrigg, Ambleside, LA22 9HH.
National Trust, 015394 37623,
neil.winder@nationaltrust.org.
uk, www.nationaltrust.org.uk. *10 min NW from Ambleside. Ambleside (A593) to Skelwith Bridge signed for High Close, turn R & head up hill until you see a white painted stone sign to 'Langdale', turn L, High Close on L.* **For NGS: Sun 21 Mar (11-4). Combined adm with Dora's Field by donation. For other opening times and information, please phone, email or visit garden website.**
Originally planted in 1866 by Edward Wheatley-Balme, High Close was designed in the fashion of the day using many of the recently discovered 'exotic' conifers and evergreen shrubs coming into Britain from America. Today the garden works in partnership with the Royal Botanic Gardens Edinburgh and International Conifer Conservation Programme, preserving endangered Conifers species. Guided walk at 11am, tree trail and chat to our volunteers. Small cafe in house part of the YHA (please check with YHA for opening times).

▮8 ◆ HOLEHIRD GARDENS

Patterdale Road, Windermere, LA23 1NP. Lakeland Horticultural Society, 015394 46008, enquiries@ holehirdgardens.org.uk, www.holehirdgardens.org.uk. *1m N of Windermere. On A592, Windermere to Patterdale rd.* **For NGS: Sun 21 Mar, Thur 8 July (10-5). Adm £5, chd free. Self-service hot drinks available. For other opening times and information, please phone, email or visit garden website. Donation to Plant Heritage.**
Run by volunteers of the Lakeland Horticultural Society to promote knowledge of gardening in Lakeland conditions. On fellside overlooking Windermere, the 12 acres provide interest year round. 4 National Collections (*astilbe*, daboecia, *polystichum*, meconopsis). Lakeland

Elder Cottage

© Richard Jakobson

Collection of hydrangeas. Walled garden has colourful mixed borders and island beds. Alpine beds and display houses. Many interesting trees and shrubs. Wheelchair access limited to walled garden and beds accessible from drive.

ᕕ ✿ NPC ☕

19 ◆ HOLKER HALL GARDENS
Cark-in-Cartmel, Grange-over-Sands, LA11 7PL. The Cavendish Family, 015395 58328, info@holker.co.uk, www.holker.co.uk. *4m W of Grange-over-Sands. 12m W of M6 (J36) Follow brown tourist signs. Rail 1m, NCR 700 ½ m.* **For NGS: Fri 21 May (10.30-4). Adm by donation. NGS will receive all donations made from guided tours and the garden entrance income. For other opening times and information, please phone, email or visit garden website.**
25 acres of romantic gardens, with peaceful arboretum, inspirational formal gardens, flowering meadow and Labyrinth. Summer brings voluptuous mixed borders and bedding. Discover unusually large rhododendrons, magnolias and azaleas, and the National Collection of Styracaceae. Discover our The Pagan Grove, designed by Kim Wilkle. We have a range of accessible routes for wheelchairs. There are also wheelchairs available to hire.

ᕕ ⚘ ✿ 🚌 NPC ☕

20 HOLME MEADOW
1 Holme Meadow, Cumwhinton, Carlisle, CA4 8DR. John & Anne Mallinson, 01228 560330, jwai.mallinson@btopenworld.com. *2m S of Carlisle. From M6 J42 take B6263 to Cumwhinton, in village take 1st L then bear R at Lowther Arms, Holme Meadow is immed on R.* **Sun 18 July (11-4). Adm £3.50, chd free. Light refreshments. Visits also by arrangement June to Aug for groups of 10 to 30.**
Village garden developed and landscaped from scratch by owners. Incl shrubbery, perennial beds supplemented by annuals; summerhouse, pergola and trellis with climbers, slate beds and water feature, ornamental copse, pond, wildflower meadow and kitchen garden. Designed, planted and maintained to be wildlife friendly.

✿ ☕

21 HPB MERLEWOOD
Windermere Road, Grange-Over-Sands, LA11 6JT. Marion St Quinton Site Manager. *Merlewood is on B5271 between Grange over Sands & Lindale. Rail 1m; Bus X6, NCR 70 (1m).* **Sun 8 Aug (10.30-3.30). Adm £5, chd free. Home-made teas.**
Extensive and varied gardens with a dramatic view to Morecambe Bay from the terrace. Newly planted formal and terraced gardens surround the house. A woodland including a nature trail for children. The wood on the limestone crags contain a variety of tree species many of which are from the original Victorian planting. The rockery is work in progress! Some areas are not accessible due to steps.

ᕕ ✿ ☕

22 IVY HOUSE
Cumwhitton, CA8 9EX. Martin Johns & Ian Forrest, 01228 561851, martinjohns193@btinternet.com. *6m E of Carlisle. At the bridge at Warwick Bridge on A69 take turning to Great Corby & Cumwhitton. Through Great Corby & woodland until you reach a T junction Turn R.* **Sun 27 June (1-5). Adm £4, chd free. Home-made teas in Cumwhitton village hall. Visits also by arrangement Apr to Sept.**
Approx 2 acres of sloping fell-side garden with meandering paths leading to a series of 'rooms': pond, fern garden, gravel garden with assorted grasses, vegetable and herb garden. Copse with meadow leading down to beck. Trees, shrubs, ferns, bamboos and herbaceous perennials planted with emphasis on variety of texture and colour. WC available. Steep slopes.

⚘ 🚌 ☕

23 NEW KEERSIDE
Arkholme, Carnforth, LA6 1AP. Geoffrey Ford. *7m from J35 or J36 of M6. From Burton in Kendal take Dalton Lane at southern end of village. After approx 3m turn R signed Arkholme & Docker. Continue for approx 1m. Garden is on R.* **Sat 3, Sun 4, Sat 17, Sun 18 July (10-4). Adm £4, chd free. Home-made teas.**
This ½ acre rural garden is constantly evolving, after 8 years' hard work. An abundance of roses ramble over trees and an unusual pergola is clad in more roses and honeysuckle. A new bog garden adds to the recent stumpery and wildlife pond.

Herbaceous borders; well-mown lawns; and a vista which follows the R Keer towards Morecambe Bay complete the picture.

ᕕ ✿ ☕

24 KINRARA
Hayton, Brampton, CA8 9HR. Tim & Alison Brown. *Less than 1 mile south of the A69, 5 miles east of J43 of M6. 3 miles west of Brampton. Centre of village diagonally opposite The Stone Inn (meals, WCs). Immediately west of the 'Walnut Field' village green. Parking on road - please park considerately on one side of the road only.* **Sun 29 Aug (1-6). Adm £5. Home-made teas. Cold drinks too if warm weather. Opening with Hayton Village Gardens on Sun 13 June.**
Densely planted, mainly evergreen gardens around C18 stone house. Only the front garden open in June Group opening. All year interest from rhododendrons & camellias to perennials, roses, clematis & bulbs. Hostas, ferns & grasses, climbers, topiary & a few exotics. Slope/rocks create terraces with pools. Newly opened back garden designed as a 'jungle' with bamboo, bananas, Gunnera & tree ferns. Limited accessibility and not suitable for young children.

✿ ☕

25 LANGHOLME MILL
Woodgate, Lowick Green, Ulverston, LA12 8ES. Judith & Graham Sanderson, 01229 885215, judith@themill.biz. *7m NW of Ulverston. West on A590. At Greenodd, North on A5092 towards Broughton. Langholme Mill is approx 3m along this rd on L as rd divides on the hill.* **Sat 29, Sun 30, Mon 31 May (11-5). Adm £4, chd free. Home-made teas. Visits also by arrangement Apr to Oct for groups of up to 30.**
Approx 1 acre of mature woodland garden with meandering lakeland slate paths surrounding the mill race stream which can be crossed by a variety of bridges. The garden hosts well established bamboo, rhododendrons, hostas, acers and astilbes and a large variety of country flowers. There is a surprise round every corner! Home-made teas £3, large variety of delicious home-made scones, cakes, tea and coffee. Accessible by a side gate entrance from the road side.

ᕕ ⚘ 🚌 🏠 ☕

Mains House

26 LARCH COTTAGE NURSERIES

Melkinthorpe, Penrith, CA10 2DR. Peter Stott, www.larchcottage.co.uk. *From N leave M6 J40 take A6 S. From S leave M6 J39 take A6 N signed off A6.* **Thur 8 July, Thur 12 Aug, Thur 16 Sept (1-4). Adm £4, chd free. Tea in La Casa Verde - Perfect food in fantastic surroundings!.** We are pleased to open our gardens and chapel for 3 days this year in support of the NGS. The gardens include lawns, flowing perennial borders, rare and unusual shrubs, trees, small orchard and a kitchen garden. A natural stream runs into a small lake - a haven for wildlife and birds. At the head of the lake stands a small frescoed chapel designed and built by Peter for family use. Larch Cottage has a Japanese Dry garden, ponds and Italianesque columned garden specifically for shade plants, the Italianesque tumbled down walls are draped in greenery acting as a backdrop for the borders filled with stock plants. Newly designed and constructed lower gardens and chapel. Accessible to wheelchair users although the paths are rocky in places.

  ♿ ✿ ☕ 🖤

27 NEW LEA COTTAGE

Red Bank Road, Grasmere, Ambleside, LA22 9PY. Tony & Marie Reynolds. *1m W of Grasmere village on Red Bank Road. Park in Grasmere village public car parks. There is NO visitor parking at Lea Cottage. Visitors wishing to see the garden must park in Grasmere near the garden centre & walk the 1m to the cottage along Red Bank Road in a West direction.* **Sun 9 May (12.30-4.30). Adm £4, chd free.** Intimate and secluded Arts & Crafts garden, subdivided into 'rooms' by yew hedges, stone walls & gabions using stone from garden. Inspired by its location in Wordsworth's 'loveliest spot that man hath found'. Designed to be both productive & beautiful by its architect owner. Steep steps, hidden behind a high wall and stone, potentially slippery, steps within garden at level changes. Regret no wheelchair access to garden.

🐕

28 NEW LITTLE CRAG

Crook, Kendal, LA8 8LE. Simon & Georgina Higgins. *Turn off the B5284 opposite the Sun Inn. Passing Ellerbeck farm on R continue for ¼m. Take the second drive on L & Little Crag is 1st on R.* Sun 13 June (10-5). Combined adm with Low Crag £6, chd free. Light refreshments at Low Crag next door.

A beautiful well-established cottage garden designed in terraces and slopes over different levels. Riotous with mature shrubs, evergreen topiary, roses, peonies and herbaceous planting, all jostling for attention. Accommodates a yurt and quirky additions of art and statuary, all of which add to its characterful charm. Steep slopes and steps. Not suitable for wheelchair access.

29 NEW LOW CRAG

Crook, Kendal, LA8 8LE. Chris Dodd & Liz Jolley. *Turn off B5284 opp Sun Inn. Pass Ellerbeck farm on R, continue for ¼m. Take the 3rd drive on L. Steep drive to large, flat yard. Note - Satnav takes you to Ellerbeck Farm.* Sun 13 June (10-5). Combined adm with Little Crag £6, chd free. Light refreshments.

A relaxed, wildlife-friendly 2-acre farmhouse garden, designed and maintained on organic principles. Transitions from a formal garden, featuring yew hedges and herbaceous planting, through to a wildlife pond, arboretum, small vegetable garden, and an orchard, and out to meadows and the greater landscape. Long views with seating and viewpoints throughout. Opens with Little Crag next door. Some steep slopes and steps.

30 LOW FELL WEST

Crosthwaite, Kendal, LA8 8JG. Barbie & John Handley, 015395 68297, barbie@handleyfamily.co.uk. *4½m S of Bowness. Off A5074, turn W just S of Damson Dene Hotel. Follow lane for ½m.* Sun 2 May (1-5). Home-made teas. Mon 3 May (10.30-1.30); Sun 17 Oct (12-4). Light refreshments. Adm £5, chd free. Visits also by arrangement for groups of up to 30. Lane unsuitable for coaches. Coach parking in lay-by half a mile away.

This 2 acre woodland garden in the tranquil Winster Valley has extensive views to the Pennines. The four season garden, restored since 2003, incl expanses of rock planted sympathetically with grasses, unusual trees and shrubs, climaxing for autumn colour. There are native hedges and areas of plant rich meadows. A woodland area houses a gypsy caravan and there is direct access to Cumbria Wildlife Trust's Barkbooth Reserve of Oak woodland, bluebells and open fellside. Wheelchair access to much of the garden, but some rough paths, steep slopes.

🦽 🐕 🌸 ☕

31 LOWER ROWELL FARM & COTTAGE

Milnthorpe, LA7 7LU. John & Mavis Robinson & Julie & Andy Welton, 015395 62270. *Approx 2m from Milnthorpe, 2m from Crooklands. Signed to Rowell off B6385, Milnthorpe to Crooklands Rd. Garden ½m up lane on L.* Visits by arrangement Feb to June for groups of 10 to 30. Refreshments by arrangement.

Approx 1¼ acre garden with views to Farleton Knott and Lakeland hills. Unusual trees and shrubs, plus perennial borders; architectural pruning; retro greenhouse; polytunnel with tropical plants; cottage gravel garden and vegetable plot. Fabulous display of snowdrops in spring followed by other spring flowers, with colour most of the year. Wildlife ponds and 2 friendly pet hens.

🦽 🌸 🚗 ☕

32 NEW MAINS HOUSE

Pooley Bridge, CA10 2NG. Bea Ray. *Mains House is ¼m out of Pooley Bridge, at the B5320 junction with the Howtown road. Limited parking available in the yard.* Sun 8 Aug (1-5). Combined adm with Beulah £5, chd free. Home-made teas at house.

New garden built over the last five years by the owners, with herbaceous borders, terraced lawn, new trees and meadow with views to the fells. Cut flower garden, curved beds, sloping lawns and gravel and cut grass paths to explore. Only partial wheelchair access due to steps to the terraced beds. The rest of the garden is accessible over sloping lawns.

🦽 ☕

33 MATSON GROUND

Windermere, LA23 2NH. Matson Ground Estate Co Ltd, 07831 831918, sam@matsonground.co.uk. *⅔m E of Bowness. Turn N off B5284 signed Heathwaite. From E 100yds after Windermere Golf Club, from W 400yds after Windy Hall Rd. Rail 2½m; Bus 1m, 6, 599, 755, 800; NCR 6 (1m).* Sun 16 May (1-5). Adm £6, chd free. Home-made teas. Visits also by arrangement for groups of 5+. 2 hour garden tour with Head Gardener included in all group visits.

2 acres of mature, south facing gardens. A good mix of formal and informal planting including topiary features, herbaceous and shrub borders, wildflower areas, stream leading to a large pond and developing arboretum. Rose garden, rockery, topiary terrace borders, ha-ha. Productive, walled kitchen garden c 1862, a wide assortment of fruit, vegetables, cut flowers, cobnuts and herbs. Greenhouse.

🦽 🐕 🌸 🚗 🚙 ☕

34 NEW NEWLANDS

Cumrew, Brampton, CA8 9DD. Mr & Mrs Colin & Teresa Clark. *On the edge of the Pennines, in village of Cumrew. Take track by telephone box, then 2nd on the L. Approach the village via B6413 or use satnav postcode. On arrival in Cumrew, park tactfully on village street & walk up gravelled lane by red telephone box. House is last (2nd) on L.* Sun 1 Aug (10-4). Adm £4, chd free.

A newish and developing garden with spectacular views, nestled on the edge of the Pennines. Garden areas include a small, young but well stocked orchard with quince, medlar, pears, damson, plum, apples and cherry; long borders filled with colourful perennials, a pond, shrubbery, ferns, alpines, and ericaceous beds, gravel garden and a well stocked vegetable garden with greenhouse. Newly established pond, gravel garden. Mostly level. Access to the garden is via the gravel area. Lawns and solid paths are easily accessible thereafter. Single step can be bypassed.

🦽 🌸

35 8 OXENHOLME ROAD
Kendal, LA9 7NJ. Mr & Mrs John & Frances Davenport. *SE Kendal. From A65 (Burton Rd, Kendal/Kirkby Lonsdale) take B6254 (Oxenholme Rd). No.8 is 1st house on L beyond red post box.* **Sun 13 June (9.30-5.30). Adm £4, chd free. Tea.**
Artist and potters garden of approx ½ acre of mixed planting designed for year-round interest, incl two linked small ponds. Roses, grasses and colour themed borders surround the house, with a gravel garden at the front, as well as a number of woodland plant areas, vegetable and fruit areas and sitting spaces. John is a ceramic artist and Frances is a painter. Paintings and pots are a feature of the garden display. Garden essentially level, but access to WC is up steps.

36 PARK HOUSE
Barbon, Kirkby Lonsdale, LA6 2LG. Mr & Mrs P Pattison. *2½ m N of Kirkby Lonsdale. Off A683 Kirkby Lonsdale to Sedburgh rd. Follow signs into Barbon Village.* **Sun 6, Sun 27 June (10.30-4). Adm £5, chd free. Cream teas.**
Romantic Manor House (not open). Extensive vistas. Formal tranquil pond encased in yew hedging. Meadow with meandering pathways, water garden filled with bulbs and ferns. Formal lawn, gravel pathways, cottage borders with hues of soft pinks and purples, shady border, kitchen garden. An evolving garden to follow.

37 PEAR TREE COTTAGE
Dalton, Burton-in-Kendal, LA6 1NN. Linda & Alec Greening, 01524 781624, greening@ngs.org.uk. *5m from J35 & J36 of M6. From northern end of Burton-in-Kendal (A6070) turn E into Vicarage Lane & continue approx 1m.* **Visits by arrangement June & July for groups of 10+. Adm £4.50, chd free. Refreshments by prior arrangement.**
⅓ acre cottage garden in a delightful rural setting. A peaceful and relaxing garden, harmonising with its environment and incorporating many different planting areas, from packed herbaceous borders and rambling roses, to wildlife pond, bog garden, rock garden and gravel garden. A plantsperson's delight, incl over 200 different ferns, and many other rare and unusual plants.

38 QUARRY HILL HOUSE
Boltongate, Mealsgate, Wigton, CA7 1AE. Mr Charles Woodhouse & Mrs Philippa Irving, 01697 371225/ 07785 934 377, cfwoodhouse@btinternet.com. *1/3m W of Boltongate, 6m SSW of Wigton, 13m NW of Keswick, 10m E Cockermouth. M6 J41, direction Wigton, through Hesket Newmarket, Caldbeck, & Boltongate on B5299, entrance gates Quarry Hill House drive on R. On A595 Carlisle to Cockermouth turn L for Boltongate, Ireby.* **Sun 6 June (1-5.30). Adm £5, chd free. Home-made teas. Visits also by arrangement May to July. Donation to Cumbria Community Foundation--Quarry Hill Grassroots Fund.**
3 acre parkland setting country house (not open) woodland garden with marvellous views of Skiddaw, Binsey and the Northern Fells and also the Solway and Scotland. Trees, some very old and many specimen, shrubs, herbaceous borders, potager vegetable garden.

39 ROSE CROFT
Levens, Kendal, LA8 8PH. Enid Fraser, 07976 977018, enidfraser123@btinternet.com. *Approx 4m from J36. M6 J36 take A590 toward Barrow. R turn Levens, L at pub, follow rd to garden on L. From A6, into Levens, past shop, bear L, over Xrds, downhill. L turn signed 'PV Dobson'. Garden on R after Dobsons.* **Visits by arrangement June to Sept for groups of up to 30. Adm £4, chd free. Light refreshments.**
Gardening on a steep slope with wildlife in mind. Naturalistic plantings, perennials and grasses pouring away from the top terrace, shrubs, rose arches and silver birches for supporting structure. August sees the peak of this colourful drama: set against the views that dominate the westerly scene beyond the summer house, past the sown wildflowers and lawn which fold into the garden-bounding stream. Refreshments at Hare and Hounds pub approx 6 min walk, 1 min drive. Sizergh Castle less than 10 mins drive.

40 ◆ RYDAL HALL
Ambleside, LA22 9LX. Diocese of Carlisle, 01539 432050, gardens@rydalhall.org, www.rydalhall.org. *E from Ambleside. E from A591 at Rydal signed Rydal Hall. Bus 555, 599, X8, X55; NCR 6.* **For NGS: Sun 21 Mar, Sun 9 May, Sun 6 June, Sun 18 July (9-4). Adm by donation. Light refreshments in tea shop on site. For other opening times and information, please phone, email or visit garden website.**
Forty acres of Park, Woodland and Gardens to explore. The Formal Thomas Mawson Garden has fine examples of herbaceous planting, seasonal displays and magnificent views of the Lakeland Fells. Enjoy the peaceful atmosphere created in the Quiet Garden with informal planting around the pond 'and stunning views of the waterfalls from The Grot, the UK's first viewing station.'. Partial Wheelchair access, top terrace only.

41 SPRINT MILL
Burneside, Kendal, LA8 9AQ. Edward & Romola Acland, 01539 725168, mail@sprintmill.uk. *2m N of Kendal. From Burneside follow signs towards Skelsmergh for ½ m, then L into drive of Sprint Mill. Or from A6, about 1m N of Kendal follow signs towards Burneside, then R into drive.* **Visits by arrangement Apr to Oct for groups of up to 30. Adm £4, chd free. Light refreshments available and can be varied according to need.**
Unorthodox organically run garden, the wild and natural alongside provision of owners' fruit, vegetables and firewood. Idyllic riverside setting, 5 acres to explore including wooded riverbank with hand-crafted seats. Large vegetable and soft fruit area, following no-dig and permaculture principles. Hand-tools prevail. Historic water mill with original turbine. The 3-storey building houses owner's art studio and personal museum, incl collection of old hand tools associated with rural crafts. Goats, hens, ducks, rope swing, very family-friendly. Short walk to our flower-rich hay meadows. Access for wheelchairs to some parts of both garden and mill.

42 SUMMERDALE HOUSE

Cow Brow, Nook, Lupton, LA6 1PE. David & Gail Sheals, www.summerdalegardenplants.co.uk. *7m S of Kendal, 5m W of Kirkby Lonsdale. From J36 M6 take A65 towards Kirkby Lonsdale, at Nook take R turn Farleton. Location not signed on highway. Detailed directions available on our website.* **Every Sat 20 Feb to 19 June (11-4.30). Adm £5, chd free.**
1½ acre part-walled country garden set around C18 former vicarage. Several defined areas have been created by hedges, each with its own theme and linked by intricate cobbled pathways. Relaxed natural planting in a formal structure. Rural setting with fine views across to Farleton Fell. Large collections of auricula, primulas and snowdrops. Adjoining RHS Gold Medal winning nursery. During adverse weather, please check garden website for opening information. Not suitable for wheelchairs, many steps.

43 TITHE BARN

Laversdale, Irthington, Carlisle, CA6 4PJ. Mr & Mrs Gordon & Christine Davidson, 01228 573090, christinedavidson7@sky.com. *8m N E of Carlisle. Laversdale, N. Cumbria, ½ m from Carlisle Lake District Airport. From 6071 turn for Laversdale, from M6 J44 follow A689 Hexham/Brampton/Airport. Follow NGS signs.* **Sun 6 June (1-5). Adm £4, chd free. Home-made teas.**
Set on a slight incline, the thatched property has stunning views of the Lake District, Pennines and Scottish Border hills. Planting follows the cottage garden style, the surrounding walls, arches, grottos and quirky features have all been designed and created by the owners. There is a peaceful sitting glade beside a rill and pond. This property also offers self-catering accommodation (sleeps 2) within the grounds of the garden. Wheelchairs unable to view the sloped south boundary area down steep steps.

44 NEW WHETSTONE CROFT & COTTAGE

Woodland, Broughton-In-Furness, LA20 6AE. John & Elaine Hudson, Rob Wilson & Iain Speak. *On the A593, approx 2m N of Broughton-in-Furness. Follow NGS signs through yard to park.* **Sat 19, Sun 20 June (12-5). Adm £5, chd free. Home-made teas.**
Two different, tranquil, south-facing cottage gardens overlooking the Woodland valley. Herbaceous borders bursting with cottage-garden favourites, mature flowering shrubs, azaleas, herb beds, stoop garden, courtyard and shrubbery surrounding C18 farmhouse and barns. Easy access to ten acres of renowned traditional hay meadows including eyebright, orchids, yellow rattle and oxeye-daisies. Stunning views of the woodland fells. Restricted access due to gravel, slate and bark paths and some steep steps.

Low Crag

Whetstone Croft and Cottage

15 Aug (11-4.30). Adm £4, chd free. Home-made teas. **Visits also by arrangement Apr to Sept for groups of up to 20.**
A small interesting garden tucked away in a small hamlet. Meandering gravel paths lead around the garden with imaginative, colourful planting and quirky features. Take a look around a productive, organic potager, wildlife pond, mini spring and summer meadows and sit in the pretty summerhouse. Designed to be beautiful throughout the year and wildlife-friendly. Plant sale, home-made teas, mini-quiz for children. Live Blues and Jazz in the garden. The gravel drive and paths are difficult for wheelchairs but more mobile visitors can access the main seating areas in the rear garden.

48 YEWBARROW HOUSE
Hampsfell Road, Grange-over-Sands, LA11 6BE. Jonathan & Margaret Denby, 015395 32469, jonathan@bestlakesbreaks.co.uk, www.yewbarrowhouse.co.uk. ¼m from town centre. Proceed along Hampsfell Rd passing a house called Yewbarrow to brow of hill then turn L onto a lane signed 'Charney Wood/ Yewbarrow Wood' & sharp L again. Rail 0.7m, Bus X6, NCR 70. **Sun 6 June, Sun 4 July, Sun 1 Aug, Sun 5 Sept (11-4). Adm £5, chd free. Home-made teas. Visits also by arrangement May to Sept for groups of 10+.**
'More Cornwall than Cumbria' according to Country Life, a colourful 4 acre garden filled with exotic and rare plants, with dramatic views over the Morecambe Bay. Outstanding features include the Orangery; the Japanese garden with infinity pool, the Italian terraces and the restored Victorian kitchen garden. Dahlias, cannas and colourful exotica are a speciality. Find us on youTube. Ltd wheelchair access owing to the number of steps.

45 WINDY HALL
Crook Road, Windermere, LA23 3JA. Diane Hewitt & David Kinsman, 015394 46238, dhewitt.kinsman@gmail.com, windy-hall.co.uk. ½m S of Bowness-on-Windermere. On western end of B5284, pink house up Linthwaite House Hotel driveway. Rail 2.6m; Bus 1m, 6, 599, 755, 800; NCR 6 (1m). **Visits by arrangement Apr to July for groups of up to 30. Adm £5, chd free. Light refreshments.**
"Paradise!". "I was bowled over by the ecologically intelligent approach you and Diane take and the exquisitely planted back garden or hill with rare species and subspecies so elegantly placed where they will flourish. It was truly superb." "The garden left a lasting impression on all. It is beautiful, exciting and charming. It nestles so comfortably in its landscape and totally belongs there.". Mosses and Ferns, rare Hebridean sheep, exotic waterfowl. Teas and home-made cake (ad lib) for any size group and Light Lunches for groups of 8+.

46 WINTON PARK
Appleby Road, Kirkby Stephen, CA17 4PG. Mr Anthony Kilvington, www.wintonparkgardens.co.uk. 2m N of Kirkby Stephen. On A685 turn L signed Gt Musgrave/Warcop (B6259). After approx 1m turn L as signed. **Sun 11 July (11-5). Adm £6, chd free. Light refreshments.**
5 acre country garden bordered by the banks of the R Eden with stunning views. Many fine conifers, acers and rhododendrons, herbaceous borders, hostas, ferns, grasses, heathers and several hundred roses. Four formal ponds plus rock pool. Partial wheelchair access.

47 WOODEND HOUSE
Woodend, Egremont, CA22 2TA. Grainne & Richard Jakobson, 019468 13017, gmjakobson22@gmail.com. 2m S of Whitehaven. Take the A595 from Whitehaven towards Egremont. On leaving Bigrigg take 1st turn L. Go down hill, garden at bottom on R opp Woodend Farm. **Sat 14, Sun**

The National Garden Scheme searches the length and breadth of England and Wales for the very best private gardens

Low Fell West

OPENING DATES

All entries subject to change. For latest information check www.ngs.org.uk

Map locator numbers are shown to the right of each garden name.

February

Snowdrop Festival

Saturday 13th
The Dower House   18

Sunday 14th
The Dower House   18

Sunday 21st
10 Chestnut Way   13

March

Sunday 21st
◆ Cascades Gardens   12

Saturday 27th
Chevin Brae   14

April

Monday 5th
◆ The Burrows
  Gardens   8

Sunday 11th
334 Belper Road   6

Sunday 25th
◆ Cascades Gardens   12
The Paddock   46
15 Windmill Lane   66

May

Saturday 1st
NEW Ashbourne Road
  and District
  Allotments Ltd   2

Sunday 2nd
12 Ansell Road   1
Barlborough Gardens   5

Monday 3rd
12 Ansell Road   1

Sunday 9th
Littleover Lane
  Allotments   33

Sunday 16th
Fir Croft   20
27 Wash Green   61

Tuesday 18th
◆ Thornbridge Hall
  Gardens   55

Saturday 22nd
12 Water Lane   62

Sunday 23rd
Fir Croft   20
Highfield House   24
12 Water Lane   62

Sunday 30th
12 Ansell Road   1
Rectory House   49
Tilford House   56

Monday 31st
12 Ansell Road   1
Askew Cottage   3
◆ The Burrows
  Gardens   8
10 Chestnut Way   13
Repton Allotments   51
◆ Tissington Hall   57

June

Saturday 5th
The Holly Tree   30

Sunday 6th
Fir Croft   20
Highfields House   25
The Holly Tree   30
Walton Cottage   60

Saturday 12th
◆ Melbourne Hall
  Gardens   39

Sunday 13th
334 Belper Road   6

Hollies Farm Plant
  Centre   29
◆ Melbourne Hall
  Gardens   39
◆ Meynell Langley
  Trials Garden   40
13 Westfield Road   63

Wednesday 16th
◆ Bluebell Arboretum
  and Nursery   7

Saturday 19th
12 Ansell Road   1
◆ Calke Abbey   10
Holmlea   31

Sunday 20th
12 Ansell Road   1
◆ The Burrows
  Gardens   8
Hill Cottage   26
Holmlea   31

Friday 25th
330 Old Road   44

Saturday 26th
Elmton Gardens   19
NEW Mastin Moor
  Gardens and
  Allotments   37
330 Old Road   44

Sunday 27th
Elmton Gardens   19
High Roost   22
58A Main Street   36
Old Shoulder of Mutton -
  7 Main Street   45

July

Saturday 3rd
NEW Avalon   4
Barlborough Gardens   5

Sunday 4th
Barlborough Gardens   5
8 Curzon Lane   16
The Lilies   32
Littleover Lane
  Allotments   33

Tuesday 6th
◆ Renishaw Hall &
  Gardens   50

Wednesday 7th
Tilford House   56

Saturday 10th
New Mills School   42

Sunday 11th
NEW Candlemas
  Cottage   11
◆ Cascades Gardens   12
10 Chestnut Way   13
◆ Meynell Langley
  Trials Garden   40
New Mills School   42
22 Pinfold Close   48

Wednesday 14th
◆ Bluebell Arboretum
  and Nursery   7
NEW Spingles   53

Thursday 15th
NEW Spingles   53

Saturday 17th
NEW Longford Hall
  Farm   34

Sunday 18th
◆ The Burrows
  Gardens   8
8 Curzon Lane   16
58A Main Street   36
Moorfields   41
15 Windmill Lane   66

Saturday 24th
12 Ansell Road   1
NEW Ashbourne Road
  and District
  Allotments Ltd   2
Stanton in Peak
  Gardens   54

Sunday 25th
12 Ansell Road   1
The Paddock   46
Stanton in Peak
  Gardens   54

Saturday 31st
Byways   9
9 Main Street   35
NEW Valdona   59

August

Sunday 1st
Byways   9
8 Curzon Lane   16
Hollies Farm Plant
  Centre   29
9 Main Street   35

Saturday 7th
NEW Valdona   59
26 Windmill Rise   67

Sunday 8th
Barlborough Gardens   5

Ashbourne Road and District Allotments

THE GARDENS

1 12 ANSELL ROAD

Ecclesall, Sheffield, S11 7PE.
Dave Darwent, 01142 665881,
dave@poptasticdave.co.uk. *Approx
3m SW of City Centre. Travel to
Ringinglow Rd (88 bus), then Edale
Rd (opp Ecclesall C of E Primary
School). 3rd R - Ansell Rd. No 12 on
L ¾ way down, solar panel on roof.*
Sun 2, Mon 3, Sun 30, Mon 31
May, Sat 19, Sun 20 June, Sat 24,
Sun 25 July (11.30-6). Adm £3.50,
chd free. Light refreshments.
Visits also by arrangement May
to Aug for groups of up to 20. By
Arrangement admission fee incl tea
and selection of home-made cakes.
Now in its 93rd year since being
created by my grandparents, this is
a suburban mixed productive and
flower garden retaining many original
plants and features as well as the
original layout. A book documenting
the history of the garden has been
published and is on sale to raise
further funds for charity. Original
rustic pergola with 90+ year old
roses. Then and now pictures of
the garden in 1929 and 1950's vs
present. Map of landmarks up to
55m away which can be seen from
garden. 7 water features. Wide variety
of unique-recipe homemade cakes
with take-away service available. New
winter garden.

2 NEW ASHBOURNE ROAD AND DISTRICT ALLOTMENTS LTD

Mackworth Road, Derby,
DE22 3BL. Elaine Crick,
www.araa.org.uk. *The site sits
between main roads into Derby city.
From Kedleston Rd turn onto Cowley
St which becomes Mackworth
Rd. From Ashbourne Rd turn
onto Merchant St which becomes
Mackworth Rd.* Sat 1 May, Sat 24
July (10.30-3.30). Adm £5, chd
free. Home-made teas.
This beautiful allotment site is nestled
quietly between two main roads
leading into Derby city. It boasts of
being over 100 years old with plots
of all differing shapes and sizes. It
houses a giant cockerel made from
sheet metal and recycled tools, as
well as a Growing Academy, packed
poly tunnels, as well as Starter
Plots, a Centenary Allotment and is
the 'home' of Radio Derby's Potty

Plotters! The site will have plots for
you to venture on to & plot holders
to chat to. There are 12 Starter Plots
to inspect & main plots filled with
flowers, vegetables & fruits in varying
stages of growth & production. A
giant cockerel will be standing guard
as you view the bees & allotmenty
things to buy! The main grass paths
are flat and accessible with care.

3 ASKEW COTTAGE

23 Milton Road, Repton,
Derby, DE65 6FZ. Louise
Hardwick, 07970411748,
louise.hardwick@hotmail.co.uk,
www.hardwickgardendesign.co.uk.
*6m S of Derby. From A38/A50 junction
S of Derby. Follow signs to Willington
then Repton on B5008. In Repton turn
1st L then bear sharp R into Milton Rd.*
Mon 31 May, Thur 9 Sept (1.30-6).
Adm £4, chd free. Also open 10
Chestnut Way. Refreshments
available at 10 Chestnut Way. Visits
also by arrangement Apr to Oct for
groups of 5 to 20.
The owner, a professional garden
designer, has used curved paths
and structural beech, box and yew
hedges to make several different
areas, each with a distinct feel. The
planting within the borders changes
as areas are altered and new
combinations tried. Features include
a circle of meadow grass set within a
cloud box hedge, trained apple trees
and a small wildlife pool.

4 NEW AVALON

Mansfield Road, Heath,
Chesterfield, S44 5SG. Elizabeth &
Mervyn Blackwell. *1m from J29 M1
motorway. 5m from Chesterfield. 1m
from J29 M1. From Chesterfield take
the A617 take the B6039 Temple
Normanton. Follow the signs for
Heath. From the M1 take the A6175
to Clay Cross. Turn R into Heath.*
Sat 3 July (1-6). Adm £4, chd free.
Light refreshments.
Delightful cottage garden overlooking
farmland and paddocks of the
Chatsworth estate. An eclectic mix
of mature trees, intimate seating and
dining areas, form the framework for
prolific, interesting planted borders and
areas of potted Mediterranean and
jungle planting. A small mature raised
fishpond and a quirky water feature
add extra interest. The pottager leads
onto a large swath of grass.

GROUP OPENING

5 BARLBOROUGH GARDENS

Chesterfield Road, Barlborough,
Chesterfield, S43 4TR. Christine
Sanderson, 07956 203184,
christine.r.sanderson@uwclub.
net, www.facebook.com/
barlboroughgardens. *7m NE of
Chesterfield. Off A619 midway
between Chesterfield & Worksop.
½m E M1, J30. Follow signs for
Barlborough then yellow NGS signs.
Parking available in village.* Sun 2
May (1-5). Combined adm £5,
chd free. Sat 3, Sun 4 July (1-6).
Combined adm £6, chd free. Sun
8 Aug (1-5). Combined adm £5,
chd free. Refreshments available
at 'The Hollies' on all dates
listed plus chefs canapes & cool
drinks at 'Raiswells House' at the
August opening.

CLARENDON

Neil & Lorraine Jones.
Open on Sun 2 May, Sat 3, Sun
4 July

GOOSE COTTAGE

Mick & Barbara Housley.
Open on Sat 3, Sun 4 July

THE HOLLIES

Vernon & Christine Sanderson.
Open on all dates
(See separate entry)

LINDWAY

Thomas & Margaret Pettinger.
Open on all dates

RAISWELLS HOUSE

Mr Andrew & Mrs Rosie Dale.
Open on Sun 8 Aug

NEW 19 WEST VIEW

Mrs Pat Cunningham.
Open on Sat 3 July

WOODSIDE HOUSE

Tricia & Adrian Murray-Leslie.
Open on Sat 3, Sun 4 July

Barlborough is an attractive historic
village and a range of interesting
buildings can be seen all around
the village centre. The village is
situated close to Renishaw Hall for
possible combined visit. Map detailing
location of all the gardens is issued
with admission ticket, which can be
purchased at any of the gardens
listed. For more information visit our
Facebook page - see details above.
Barlborough Gardens celebrates 10
years of opening for National Garden
Scheme in 2021. The Hollies incl
an area influenced by the Majorelle

Garden in Marrakech. This makeover was achieved using upcycled items together with appropriate planting. It was featured in an episode of TVs "Love Your Garden". Partial wheelchair access at The Hollies.

6 334 BELPER ROAD

Stanley Common, DE7 6FY. Gill & Colin Hancock, 01159 301061, gillandcolin@tiscali.co.uk. *7m N of Derby. 3m W of Ilkeston. On A609, ¾m from Rose & Crown Xrds (A608). Please park in field up farm drive or Working Men's Club rear car park if wet.* **Sun 11 Apr, Sun 13 June (12-5). Adm £4, chd free. Home-made teas. Home-made soup available in April. Visits also by arrangement Feb to June for groups of 5+.**

Beautiful country garden with many attractive features incl a laburnum tunnel, rose and wisteria domes, old workmen's hut, wild life pond and much more. Take a scenic walk through the ten acres of woodland and glades to a ½ acre lake and see Snowdrops and Hellebores in February, cowslips in April and wild orchids in May. Now for 2021 Rose meadow. Plenty of seating to enjoy home-made cakes. Children welcome with plenty of activities to keep them entertained. Paths round wood and lake not suitable for wheelchairs.

7 ◆ BLUEBELL ARBORETUM AND NURSERY

Annwell Lane, Smisby, Ashby de la Zouch, LE65 2TA. Robert & Suzette Vernon, 01530 413700, sales@bluebellnursery.com, www.bluebellnursery.com. *1m NW of Ashby-de-la-Zouch. Arboretum is clearly signed in Annwell Lane (follow brown signs), ¼m S, through village of Smisby off B5006, between Ticknall & Ashby-de-la-Zouch. Free parking.* **For NGS: Wed 16 June, Wed 14 July, Wed 11 Aug, Wed 15 Sept (9-5). Adm £5, chd free. For other opening times and information, please phone, email or visit garden website.**

Beautiful 9 acre woodland garden with a large collection of rare trees and shrubs. Interest throughout the yr with spring flowers, cool leafy areas in summer and sensational autumn colour. Many information posters describing the more obscure plants. Adjacent specialist tree and shrub nursery. Please be aware this is not a wood full of bluebells, despite the name. The woodland garden is fully labelled and the staff can answer questions or talk at length about any of the trees or shrubs on display. Rare trees and shrubs. Educational signs. Woodland. Arboretum. Please wear sturdy, waterproof footwear during or after wet weather. Full wheelchair access in dry, warm weather however grass paths can become wet and inaccessible in snow or after rain.

8 ◆ THE BURROWS GARDENS

Burrows Lane, Brailsford, Ashbourne, DE6 3BU. Mrs N M Dalton, 01335 360745, enquiries@burrowsgardens.com, www.burrowsgardens.com. *5m SE of Ashbourne; 5m NW of Derby.* *A52 from Derby: turn L opp sign for Wild Park Leisure 1m before village of Brailsford, ¼m & at grass triangle head straight over through wrought iron gates.* **For NGS: Mon 5 Apr, Mon 31 May, Sun 20 June, Sun 18 July, Mon 30 Aug (11-4). Adm £5, chd free. Home-made teas. For other opening times and information, please phone, email or visit garden website.**

Immaculate lawns show off exotic rare plants and trees, mixing with old favourites in this outstanding garden. A huge variety of styles from temple to Cornish, Italian and English, gloriously designed and displayed. This is a must see garden. Visit our website for more information. Most of garden accessible to wheelchairs.

Melbourne Hall Gardens

© Andrea Jones

9 BYWAYS
7A Brookfield Avenue, Brookside, Chesterfield, S40 3NX. Terry & Eileen Kelly, 07414827813, telkel1@aol.com. *1½m W of Chesterfield. Follow A619 from Chesterfield towards Baslow. Brookfield Ave is 2nd R after Brookfield School. Please park on Chatsworth Rd (A619).* Sat 31 July, Sun 1 Aug (11.30-4.30). Adm £3.50, chd free. Home-made teas. incl gluten free options. Tea or coffee refills and squash for children are free. Visits also by arrangement July & Aug for groups of 10 to 30. Donation to Ashgate Hospice.
Previous winners of the Best Back Garden over 80sqm, Best Front Garden, Best Container Garden and Best Hanging Basket in Chesterfield in Bloom. Well established perennial borders incl helenium, monardas, phlox, grasses, acers, giving a very colourful display. Rockery and many planters containing acers, pelargoniums, hostas & fuchsias. Also a greenhouse with a display of regal pelargoniums.

10 ◆ CALKE ABBEY
Ticknall, DE73 7LE. National Trust, 01332 863822, calkeabbey@nationaltrust.org.uk, www.nationaltrust.org.uk/calke. *10m S of Derby. On A514 at Ticknall between Swadlincote & Melbourne. For Sat Nav use DE73 7JF.* For NGS: Sat 19 June, Sat 18 Sept (10-5). Adm £5, chd £2.50. Light refreshments at the main visitor facilities. For other opening times and information, please phone, email or visit garden website.
With peeling paintwork and overgrown courtyards, Calke Abbey tells the story of the dramatic decline of a country-house estate. The large kitchen garden, impressive collection of glasshouses, garden buildings and tunnels hint at the work of past gardeners, while today the flower garden, herbaceous borders and unique auricula theatre providing stunning displays all year. Restaurant at main visitor facilities for light refreshments and locally sourced food. House is also open by timed ticket only. Electric buggy available for those with mobility problems.

11 NEW ◆ CANDLEMAS COTTAGE
Alport Lane, Youlgrave, Bakewell, DE45 1WN. Jane Ide. *3½m S of Bakewell. From A6: pass playing fields on L, then Youlgrave Garage on R. From A515: through village & straight on at church/George pub Xrds. Blue house directly opp Youlgrave Primary School.* Sun 11 July (11.30-4.30). Adm £3.50, chd free. Home-made teas.
A newly established plantswoman's contemporary cottage garden in the heart of the Peak National Park, featuring rose beds and wisteria arch, winter garden, green garden, long mixed border, specimen trees, kitchen garden and pollinator's patch. Gardened to maximise benefits to wildlife and humans, with carefully sited seating areas in shade and sunshine.

12 ◆ CASCADES GARDENS
Clatterway, Bonsall, Matlock, DE4 2AH. Alan & Alesia Clements, 07967337404, alan.clements@ cascadesgardens.com, www.cascadesgardens.com. *5m SW of Matlock. From Cromford A6 T-lights turn towards Wirksworth. Turn R along Via Gellia, signed Buxton & Bonsall. After 1m turn R up hill towards Bonsall. Garden entrance at top of hill. Park in village car park.* For NGS: Sun 21 Mar, Sun 25 Apr, Sun 11 July, Sun 15 Aug (12-4). Adm £7, chd £3. Home-made teas. For other opening times and information, please phone, email or visit garden website.
The Meditation Garden: Fascinating 4 acre peaceful garden in spectacular natural surroundings with woodland, cliffs, stream, pond and ruined corn mill. Inspired by Japanese gardens and Buddhist philosophy, secluded garden rooms for relaxation and reflection. Beautiful landscape with a wide collection of unusual perennials, conifers, shrubs and trees. Nursery. Plants for sale. Hellebore month in March. Mostly wheelchair accessible. Gravel paths, some steep slopes.

13 10 CHESTNUT WAY
Repton, DE65 6FQ. Robert & Pauline Little, 01283 702267, rlittleq@gmail.com, www.littlegarden.org.uk. *6m S of Derby. From A38/A50, S of Derby, follow signs to Willington, then Repton. In Repton turn R at r'about. Chestnut*

Way is ¼m up hill, on L. Sun 21 Feb (11-3); Mon 31 May, Sun 11 July, Thur 9 Sept (12-5). Adm £4, chd free. Home-made teas. Home-made soup available in Feb. Visits also by arrangement Feb to Oct for groups of 5+. Admission price incl refreshments & tour.
A large garden full of interesting and unusual plants designed to have colour and interest throughout the year. Storm damage to some large trees has given yet more opportunities for new planting. Many benches to sit and soak up the atmosphere enjoying our unusual sculptures. Overflowing borders and a surprise round every corner. Excellent plant stall, especially good in Spring. Special interest in viticella clematis, organic vegetables and composting. Level garden, good solid paths to main areas. Some grass/bark paths.

14 CHEVIN BRAE
Milford, Belper, DE56 0QH. Dr David Moreton, 07778004374, davidmoretonchevinbrae@gmail. com. *1½m S of Belper. Coming from S on A6 turn L at Strutt Arms & cont up Chevin Rd. Park on Chevin Rd. After 300 yds follow arrow to L up Morrells Lane. After 300 yds Chevin Brae on L with silver garage.* Sat 27 Mar, Sat 21 Aug (1-5). Adm £3, chd free. Home-made teas. Visits also by arrangement Feb to Oct for groups of up to 30.
A large garden, with swathes of daffodils in the orchard a spring feature. Extensive wild flower planting along edge of wood features aconites, snowdrops, wood anemones, fritillaries and dog tooth violets. Other parts of garden will have hellebores and early camelias. During the summer the extensive flower borders and rose trellises give much colour. Large kitchen garden and fruit cages. Tea and home-made cakes, pastries and biscuits, many of which feature fruit and jam from the garden. Please note that the front drive and steps are steep.

15 COXBENCH HALL
Alfreton Road, Coxbench, Derby, DE21 5BB. Mr Brian Ballin, 01332 880200, office@coxbench-hall.co.uk, www.coxbench-hall.co.uk. *4m N of Derby close to A38. After passing thru Little Eaton, turn L onto Alfreton Rd for 1m, Coxbench Hall is on L next to*

Fox & Hounds PH between Little Eaton & Holbrook. From A38, take Kilburn turn & go towards Little Eaton. **Sun 12 Sept (2.30-4.30). Adm £5, chd free. Light refreshments. incl diabetic and gluten free cakes. Visits also by arrangement for groups of up to 20.** Formerly the ancestral home of the Meynell family, the gardens reflect the Georgian house standing in 4½ acres of grounds most of which is accessible and wheelchair friendly. The garden has 2 fishponds connected by a stream, a sensory garden for the sight impaired, a short woodland walk through shrubbery, rockery, raised vegetable beds, an orchard and seasonal displays in the mainly lawned areas. As a Residential Home for the Elderly, our Gardens are developed to inspire our residents from a number of sensory perspectives - different colours, textures and fragrances of plants, growing vegetables next to the C18 potting shed. There is also a veteran (500 - 800 yr old) Yew tree. Most of garden is lawned or block paved incl a block paved path around the edges of the main lawn. Regret no wheelchair access to woodland area.

🔥 🐕 ☕ ♨

16 8 CURZON LANE
Alvaston, Derby, DE24 8QS. John & Marian Gray, 01332 601596, maz@curzongarden.com, www.curzongarden.com. *2m SE of Derby city centre. From city centre take A6 (London Rd) towards Alvaston. Curzon Lane on L, approx ½m before Alvaston shops.* **Sun 4, Sun 18 July, Sun 1 Aug (12-5). Adm £3, chd free. Light refreshments. Visits also by arrangement July & Aug for groups of 10 to 30.** Mature garden with lawns, borders packed full of perennials, shrubs and small trees, tropical planting and hot border. Ornamental and wildlife ponds, greenhouse with different varieties of tomato, cucumber, peppers and chillies. Well stocked vegetable plot. Gravel area and large patio with container planting.

🐕 �֍ ☕ ♨

17 NEW DOVECOTE COTTAGE
Stony Houghton, Mansfield, NG19 8TR. Rachel Hayes, 01623 811472, rachelhayes_@hotmail.com. *Between Rotherham Rd & Water Lane opp the phone box.* **Visits by arrangement May to Aug for groups of up to 20. Adm £3.50,**

chd free.
Wildlife friendly, mature garden set in a rural location. A variety of evergreens, foliage plants and cottage garden favourites with Spring and Autumn highlights. The garden is on different levels with steps and slopes and is not suitable for wheelchairs. Two local cafes within 5 mins drive. Dogs welcome. Limited parking.

🐕

18 THE DOWER HOUSE
Church Square, Melbourne, DE73 8JH. William & Griselda Kerr, 01332 864756, griseldakerr@btinternet.com. *6m S of Derby. 5m W of exit 23A M1. 4m N of exit 13 M42. When in Church Square, turn R just before the church by a blue sign giving church service times. Gates are then 50 yds ahead.* **Sat 13, Sun 14 Feb (10-4); Sat 14, Sun 15 Aug (10-5). Adm £4, chd free. Light refreshments. Visits also by arrangement Jan to Nov for groups of 10 to 30. Coffee or tea is usually served in the dining room - max seating 26.** Beautiful view of Melbourne Pool from balustraded terrace running length of 1829 house. Garden drops steeply by paths and steps to lawn with herbaceous borders and bank of shrubs. Numerous paths lead to different areas of the garden, providing many different planting opportunities incl a bog garden, glade, shrubbery, grasses, herb and vegetable garden, rose tunnel, orchard and small woodland. Hidden paths and different areas for children to explore and various animals such as a bronze crocodile and stone dragon to find. They should be supervised by an adult at all times due to the proximity of water. If it is very cold in February, there will be a log fire in the cottage. Wheelchair access to top half of the garden only. Shoes with a good grip are highly recommended as slopes are steep. No parking within 50 yards.

�֍ 🐕 ✳ 🚗 ♨

GROUP OPENING

19 ELMTON GARDENS
Elmton, Worksop, S80 4LS. *2m from Creswell, 3m from Clowne, 5m from J30, M1. From M1 J30 take A616 to Newark. Follow approx 4m. Turn R at Elmton signpost. At junction turn R, the village centre is in ½m.* **Sat 26, Sun 27 June (12-5). Combined adm**

£5, chd free. Cream teas at the Old Schoolroom next to the church. Food also available all day at the Elm Tree Inn.

NEW THE BARN
Mrs Anne Merrick.

ELM TREE COTTAGE
Mark and Linda Hopkinson.

ELM TREE FARM
Angie & Tim Caulton.

PEAR TREE COTTAGE
Geoff & Janet Cutts.

PINFOLD
Nikki Kirsop, Barry Davies.

Elmton is a lovely little village situated on a stretch of rare unimproved Magnesian limestone grassland with quaking grass, bee orchids and harebells all set in the middle of attractive, rolling farm land. It has a pub, a church and a village green with award winning wildlife conservation area and newly restored Pinfold. Garden opening coincides with Elmton Festival and Well Dressing celebrations (three well dressings on display). There are two exhibitions to view, both with a local theme. Cream teas are served in the old School Room. The five very colourful but different open gardens have wonderful views. They show a range of gardening styles, themed beds and have a commitment to fruit and vegetable growing. Elmton received a gold award and was voted best small village for the 7th time and best wildlife and conservation area in the East Midlands in Bloom competition in 2019.

✤ 🐕 ✳ 🚗 ♨

20 FIR CROFT
Froggatt Road, Calver,
S32 3ZD. Dr S B Furness,
www.alpineplantcentre.co.uk. *4m
N of Bakewell. At junction of B6001
with A625 (formerly B6054), adjacent
to Froggat Edge Garage.* Sun 16,
Sun 23 May, Sun 6 June (2-5).
Adm by donation.
Massive scree with many varieties.
Plantsman's garden; rockeries;
water garden and nursery; extensive
collection (over 3000 varieties)
of alpines; conifers; over 800
sempervivums, 500 saxifrages and
350 primulas. Many new varieties not
seen anywhere else in the UK. Huge
new tufa wall planted with many rare
Alpines and sempervivums.

21 GAMESLEY FOLD COTTAGE
Gamesley Fold, Glossop, SK13 6JJ.
Mrs G M Carr, 01457 867856,
gcarr@gamesleyfold.co.uk,
www.gamesleyfold.co.uk. *2m W of
Glossop. Off A626 Glossop/Marple
Rd nr Charlesworth. Turn down lane
directly opp St. Margaret's School,
white cottage at the bottom. Car
parking in the adjacent field if weather
is dry.* Visits by arrangement May
to Aug. Adm £4, chd free. Home-
made teas.
Old fashioned cottage garden with
rhododendrons, herbaceous borders

with candelabra primulas, cottage
garden perennial flowers and herbs
also a plant nursery selling a wide
variety of these plus wild flowers.
Small ornamental fish pond and
an orchard. Lovely views of the
surrounding countryside and plenty
of seats available to relax and enjoy
tea and cakes. Garden planted to be
in keeping with the great age of the
house, 1650. Gravel & grass, not very
good wheelchair access.

22 HIGH ROOST
27 Storthmeadow Road,
Simmondley, Glossop,
SK13 6UZ. Peter & Christina
Harris, 01457 863888,
harrispeter448@gmail.com. *³⁄₄m
SW of Glossop. From Glossop
A57 to M/CL at 2nd r'about, up
Simmondley Ln nr top R turn. From
Marple A626 to Glossop, in Chworth
R up Town Ln past Hare & Hound PH
2nd L.* Sun 27 June (12-4). Adm
£3, chd free. Light refreshments.
Visits also by arrangement May
to July for groups of up to 30.
Donation to Donkey Sanctuary.
Garden on terraced slopes, views
over fields and hills. Winding paths,
archways and steps explore different
garden rooms packed with plants,
designed to attract wildlife. Alpine
bed, gravel gardens; vegetable

garden, water features, statuary,
troughs and planters. A garden
which needs exploring to discover
its secrets tucked away in hidden
corners. Craft stall, children's garden
quiz and lucky dip.

23 HIGHER CROSSINGS
Crossings Road, Chapel-
en-le-Frith, High Peak,
SK23 9RX. Malcolm & Christine
Hoskins, 01298 812970,
malcolm275@btinternet.com. *Turn
off B5470 N from Chapel-en-le-Frith
on Crossings Rd signed Whitehough/
Chinley. Higher Crossings is 2nd
house on R beyond 1st Xrds. Park
best before crossroads on Crossings
Rd or L on Eccles Rd.* Visits by
arrangement May to July for
groups of 10 to 20. Adm £4, chd
free. Light refreshments.
Nearly 2 acres of formal terraced
country garden, sweeping lawns
and magnificent Peak District views.
Rhododendrons, acers, azaleas,
hostas, herbaceous borders, Japanese
garden. Mature specimen trees and
shrubs leading through a dell. Beautiful
stone terrace and sitting areas. Garden
gate leading into meadow. Tea and
biscuits or home-made cakes available
(not included in admission price).
Wheelchair access around the house
and upper terrace and gravel garden.

St Oswalds Crescent

24 HIGHFIELD HOUSE
Wingfield Road, Oakerthorpe, Alfreton, DE55 7AP. Paul & Ruth Peat and Janet & Brian Costall, 01773 521342, highfieldhouseopengardens@hotmail.co.uk, www.highfieldhouse.weebly.com. *Rear of Alfreton Golf Club. A615 Alfreton-Matlock Rd.* Sun 23 May (10.30-5). Adm £3, chd free. Home-made teas. Visits also by arrangement Feb to July for groups of 10+. Please note the entry price includes refreshments.
Lovely country garden of approx 1 acre, incorporating a shady garden, woodland, pond, laburnum tunnel, orchard, herbaceous borders and vegetable garden. Fabulous AGA baked cakes and lunches. Groups welcome by appointment - (incl 16th-24th February for Snowdrops with afternoon tea or lunch inside by the fire). Lovely walk to Derbyshire Wildlife Trust nature reserve to see Orchids in June. Some steps, slopes and gravel areas.

25 HIGHFIELDS HOUSE
Shields Lane, Roston, Ashbourne, DE6 2EF. Sarah Pennell. *6m SW of Ashbourne. From Ashbourne: Follow A515 S, after 3m turn R onto B5033. After 2m turn L at grass triangle with blue flower container and follow signs. Turn L at Roston Inn & follow signs.* Sun 6 June, Sun 5 Sept (1-5). Adm £3, chd free. Home-made teas.
A countryside garden with extensive views. Colourful mixed borders, patio, small pond, greenhouse and veg plot with fruit trees. Summerhouse area and replanted wildlife bank. poly-tunnel and a short walk through field to the brook and willow area. Many quiet areas to sit and new shade/woodland border. Partial wheelchair access. Garden crafts. Some gravel paths.

26 HILL COTTAGE
Ashover Road, Littlemoor, Ashover, nr Chesterfield, S45 0BL. Jane Tomlinson and Tim Walls. *Littlemoor. 1.8m from Ashover village, 6.3m from Chesterfield and 6.1m from Matlock. Hill Cottage is on Ashover Rd (also known as Stubben Edge Lane). Opp the end of Eastwood Lane.* Sun 20 June (10.30-4). Adm £3.50, chd free. Home-made teas.

Hill Cottage is a lovely example of an English country cottage garden. Whilst small, the garden has full, colourful and fragrant mixed borders with hostas and roses in pots along with a heart shaped lawn. A greenhouse full of chillies and scented pelargoniums and a small veg patch. Views over a pastoral landscape to Ogston reservoir. A wide variety of perennials and annuals grown in pots and containers. There are several steps.

27 HILLSIDE
286 Handley Road, New Whittington, Chesterfield, S43 2ET. Mr E J Lee, 01246 454960, eric.lee5@btinternet.com. *3m N of Chesterfield. Between B6056 & B6052.* Visits by arrangement Apr to Sept. Adm £4, chd free. Tea.
Part of the ⅓ of an acre site, slopes steeply but there are handrails for all the steps. New items are a Japanese feature showcasing Japanese plants and a tree trail with 60 named trees, both small and large There is a Himalayan bed with 40 species grown from wild collected seeds and an Asian area with bamboos, acers and a Chinese border. Other attractions are pools, streams and bog gardens.

28 THE HOLLIES
87 Clowne Road, Barlborough, Chesterfield, S43 4EH. Vernon & Christine Sanderson, 07956 203184, wwwfacebook.com/barlboroughgardens. *7m NE of Chesterfield. Off A619 midway between Chesterfield & Worksop. ½m F M1, J30. Follow signs for Barlborough then yellow NGS signs. Parking available along Clowne Road.* Sun 12 Sept (11-5). Adm £3, chd free. Home-made teas. The 'Cool Café' cake menu will include a number of Autumnal home-made cakes. Opening with Barlborough Gardens on Sun 2 May, Sat 3, Sun 4 July, Sun 8 Aug.
The Hollies maximises the unusual garden layout and includes a shade area, patio garden with a Moroccan corner, cottage border plus a fruit and vegetable plot. A wide selection of home-made cakes on offer including gluten free and vegan options, plus tea, coffee and cold drinks. Home-made tombola stall. Extensive view across arable farmland. The garden includes an area influenced by the Majorelle Garden in Marrakech.

This recent makeover was achieved using upcycled items together with appropriate planting. This garden project was featured on an episode of "Love Your Garden". Partial wheelchair access.

29 HOLLIES FARM PLANT CENTRE
Uppertown, Bonsall, Matlock, DE4 2AW. Robert & Linda Wells, www.holliesfarmplantcentre.co.uk. *From Cromford turn R off A5012 up The Clatterway. Keep R past Fountain Tearoom to village cross, take L up High St, then 2nd L onto Abel Lane. Garden straight ahead.* Sun 13 June, Sun 1 Aug (11-3). Adm £3.50, chd free. Home-made teas.
The best selection in Derbyshire with advice and personal attention from Robert and Linda Wells at their family run business. Enjoy a visit to remember In our beautiful display garden - set within glorious Peak District countryside. Huge variety of hardy perennials incl the rare and unusual. Vast selection of traditional garden favourites. Award winning hanging baskets. Ponds, herbaceous borders and glorious views. Plenty of parking available.

30 THE HOLLY TREE
21 Hackney Road, Hackney, Matlock, DE4 2PX. Carl Hodgkinson. *½m NW of Matlock, off A6. Take A6 NW past bus stn & 1st R up Dimple Rd. At T-junction, turn R & immed L, for Farley & Hackney. Take 1st L onto Hackney Rd. Continue ¾m.* Sat 5, Sun 6 June (11-4.30). Adm £3.50, chd free. Home-made teas.
In excess of 1½ acres and set on a steeply sloping S-facing site, sheltering behind a high retaining wall and including a small arboretum, bog garden, herbaceous borders, pond, vegetables, fruits, apiary and chickens. Extensively terraced with many paths and steps. Spectacular views across the Derwent valley.

Online booking is available for many of our gardens, at ngs.org.uk

31 HOLMLEA

Derby Road, Ambergate, Belper, DE56 2EJ. Bill & Tracy Reid. *On the A6 in Ambergate - between Belper & Matlock. Bungalow set slightly back from the road between petrol station & St Anne's Church. Additional parking at The Hurt Arms overflow car park.* Sat 19, Sun 20 June, Sat 11, Sun 12 Sept (11-4). Adm £4, chd free. Light refreshments.

A garden with something for everyone to enjoy. Set beside the River Derwent the garden comprises a formal garden, a large vegetable plot, fruit trees, greenhouses, herbaceous borders, and a gravel garden. A path from the vegetable garden leads you to a dramatic wooden arch down steps to the stunning riverside walk and canal lock water feature. Partial wheelchair access. Most of the garden is level and easy to access, except the lower garden and riverside walk. Some grass paths.

32 THE LILIES

Griffe Grange Valley, Grangemill, Matlock, DE4 4BW. Chris & Bridget Sheppard, www.thelilies.com. *On A5012 Via Gellia Rd 4m N Cromford. 1st house on R after junction with B5023 to Middleton. From Grange Mill, 1st house on L after IKO Grangemill (Formerly Prospect Quarry).* Sun 4 July, Sun 15 Aug (10.30-4.30). Adm £4, chd free. Home-made teas. Light lunches served 11.30am to 2pm. Tea, coffee and cake served all day.

One acre garden gradually restored over the past 15yrs situated at the top of a wooded valley, surrounded by wildflower meadow and ash woodland. Area adjacent to house with seasonal planting and containers. Mixed shrubs and perennial borders many raised from seed. 3 ponds, vegetable plot, barn conversion with separate cottage style garden. Natural garden with stream developed from old mill pond. Walks in large wildflower meadow and ash woodland both SSSI's. Handspinning demonstration and natural dyeing display using materials from the garden and wool from sheep in the meadow. Locally made crafts for sale. Partial wheelchair access. Steep slope from car park, limestone chippings at entrance, some boggy areas if wet.

33 LITTLEOVER LANE ALLOTMENTS

19 Littleover Lane, Normanton, Derby, DE23 6JF. Mr David Kenyon, 07745227230, davidkenyon@tinyworld.co.uk, littleoverlaneallotments.org.uk. *On Littleover Lane, opp the junction with Foremark Avenue. Just off the Derby Outer Ring Road (A5111). At the Normanton Park r'about turn into Stenson Rd then R into Littleover Lane. The Main Gates are on the L as you travel down the rd.* Sun 9 May, Sun 4 July, Sun 12 Sept (11-3). Adm £5, chd free. Light refreshments. Visits also by arrangement Apr to Sept for groups of 10 to 30.

A large private allotment site in SW Derby. About 170 plots, many cultivated to a high standard. If you are interested in growing your own you should find help, advice and inspiration. Many plot holders practice organic methods and some cultivate unusual and heritage vegetables. On site disabled parking available. All avenues have been stoned but site is on a slope and extensive (12 acres) so some areas may not be accessible.

34 [NEW] LONGFORD HALL FARM

Longford, Ashbourne, DE6 3DS. Liz Wolfenden, Longfordhallfarmholidaycottages. com. *Use the drive which has Longford Hall Farm sign at the entrance off Long Lane.* Sat 17 July (1.30-5). Adm £5, chd free. Home-made teas.

The front garden is mainly shrubs Hydrangeas Roses Ferns Grasses and Hostas beds. These surround a modern pond and fountain. There are on going projects in this garden. The walled garden has large traditional herbaceous perennial borders, landscaped rill and small orchard. Most of the garden is on the level.

35 9 MAIN STREET

Horsley Woodhouse, Derby, DE7 6AU. Ms Alison Napier, 01332 881629, ibhillib@btinternet.com. *3m SW of Heanor. 6m N of Derby. Turn off A608 Derby to Heanor Rd at Smalley, towards Belper, (A609). Garden on A609, 1m from Smalley turning.* Sat 31 July, Sun 1 Aug (1.30-4.30). Adm £3.50, chd free. Cream teas.

Water provided and soft drinks available to purchase. Visits also by arrangement Apr to Sept for groups of 5+.

⅓ acre hilltop garden overlooking lovely farmland view. Terracing, borders, lawns and pergola create space for an informal layout with planting for colour effect. Features incl large wildlife garden with water lilies, bog garden and small formal pool. Emphasis on carefully selected herbaceous perennials mixed with shrubs and old fashioned roses. Gravel garden for sun loving plants and scree garden, both developed from former drive. Please ensure you bring your own carrier bags for the plant stall. Wide collection of home grown plants for sale. All parts of the garden accessible to wheelchairs. Wheelchair adapted WC.

36 58A MAIN STREET

Rosliston, DE12 8JW. Paul Marbrow, 01283 362780, paulmarbrow@hotmail.co.uk. *Rosliston. if exiting the M42, J11 onto the A444 to Overseal follow signs Linton then Rosliston. From A38, exit to Walton on Trent, then follow Rosliston signs.* Sun 27 June, Sun 18 July (11.30-5.30). Adm £4, chd free. Light refreshments. Visits also by arrangement June to Sept for groups of 5 to 30.

Large ½ acre garden formed from a once open field over the last few years. The garden has developed into themed areas and is ongoing changing and maturing. Japanese, arid beach, bamboo grove with ferns etc. The one hundred square metre indoor garden for cacti, exotic and tender plants, are now becoming established as a mini Eden. There are also vegetable and fruit gardens. Areas are easily accessible, although some are only for the sure of foot, advice signs will be situated on non suitable routes.

37 [NEW] MASTIN MOOR GARDENS AND ALLOTMENTS

Worksop Road, Mastin Moor, Chesterfield, S43 3DN. Mastin Moor Gardens and Allotments CIO, www. mastinmoorgardensandallotments. com. *Immed below the Bolsover Rd Xrds in Mastin Moor. Access is easiest by using the layby on Bolsover Rd approx. 100 metres from the Crossroad junction with*

the A619 Worksop Rd. Sat 26 June (11-5). Adm £4, chd free. Light refreshments.
This public access (open 24/7) three acre site in Mastin Moor Derbyshire has 'open plan' allotments and gardens. Established in the early 21st century the gardens incl an arboretum, pond, bog garden, orchard, picnic area and sensory garden in addition to some 25 allotments and memorials. There are over one mile of paths around the site. A composting toilet is available. A wide variety of trees, shrubs, wildflowers and fruiting trees as well as flower and herb displays in addition to open allotments. The site is fully available for wheelchair and motor scooter access although some paths may be subject to the effects of wet weather at times.

& ♞ ❀ ㋡ ♨ ㅠ

38 NEW MEADOW COTTAGE
1 Russell Square, Hulland Ward, Ashbourne, DE6 3EA. Michael Halls, 01335 372064, mvaehalls@gmail.com. *5m E of Ashbourne. There is a lane between & on the same side as the two garages in Hulland Ward. Meadow Cottage is just down the lane to the R.* Visits by arrangement Mar to Sept for groups of up to 30. Adm £6, chd free. Home-made teas.
This garden was created from a field from 1995. It faces south and is on a sloping site. There is a wooded area, a wild flower orchard, herbaceous and vegetable areas. The soil is heavy clay so roses do well here. At 750' above sea level, growing seasons are later and shorter. The lawns are all original meadow. There are bees and hens too. The access to the garden is over a gravel area and the slope is significant down the garden but can be mastered with care.

& ❀ ㋡

39 ◆ MELBOURNE HALL GARDENS
Church Square, Melbourne, Derby, DE73 8EN. Melbourne Gardens Charity, 01332 862502, info@melbournehall.com, www.melbournehallgardens.com. *6m S of Derby. At Melbourne Market Place turn into Church St, go down to Church Sq. Garden entrance across visitor centre next to Melbourne Hall tea room.* For NGS: Sat 12, Sun 13 June (1.30-5.30). Adm £6, chd £5. Light refreshments in Melbourne Hall

Tearooms and locally. For other opening times and information, please phone, email or visit garden website.
A 17 acre historic garden with an abundance of rare trees and shrubs. Woodland and waterside planting with extensive herbaceous borders. Meconopsis, candelabra primulas, various Styrax and Cornus kousa. Other garden features incl Bakewells wrought iron arbour, a yew tunnel and fine C18 statuary and water features. 300yr old trees, waterside planting, feature hedges and herbaceous borders. Fine statuary and stonework. Don't forget to visit our pigs, alpacas, goats and various other animals in their garden enclosures. Gravel paths, uneven surface in places, some steep slopes.

& ❀ ㋡ ㋡

40 ◆ MEYNELL LANGLEY TRIALS GARDEN
Lodge Lane (off Flagshaw Lane), Kirk Langley, Ashbourne, DE6 4NT. Robert & Karen Walker, 01332 824358, enquiries@meynell-langley-gardens.co.uk, www.meynell-langley-gardens.co.uk. *4m W of Derby, nr Kedleston Hall. Head W out of Derby on A52. At Kirk Langley turn R onto Flagshaw Lane (signed to Kedleston Hall) then R onto Lodge Lane. Follow Meynell Langley Gardens signs.* For NGS: Sun 13 June, Sun 11 July, Sun 15 Aug, Sun 12 Sept (10-4). Adm £5, chd free. For other opening times and information, please phone, email or visit garden website.
Completely re-designed during 2020 with new glasshouse and patio area incorporating water rills and small ponds. New wildlife and fish ponds also added. Displays and trials of new and existing varieties of bedding plants, herbaceous perennials and vegetable plants grown at the adjacent nursery. Over 180 hanging baskets and floral displays. Adjacent tea rooms serving lunches and refreshments daily. Level ground and firm grass and some hard paths.

& ♞ ❀ ㋡ ㋡

41 MOORFIELDS
261 Chesterfield Road, Temple Normanton, Chesterfield, S42 5DE. Peter, Janet & Stephen Wright. *4m SE of Chesterfield. From Chesterfield take A617 for 2m, turn on to B6039 through Temple Normanton, taking R fork signed Tibshelf, B6039. Garden ¼ m on R. Limited parking.* Sun 18

July (1-5). Adm £3.50, chd free. Light refreshments.
Two adjoining gardens each planted for seasonal colour. The larger one has mature, mixed island beds and borders, a gravel garden to the front, a small wild flower area, large wildlife pond, orchard and soft fruit, vegetable garden. The smaller gardens of No. 257 feature herbaceous borders and shrubs. Extensive views across to mid Derbyshire.

❀ ㋡

42 NEW MILLS SCHOOL
Church Lane, New Mills, High Peak, SK22 4NR. Mr Craig Pickering, 07833 373593, cpickering@newmillsschool.co.uk, www.newmillsschool.co.uk. *12m NNW of Buxton. From A6 take A6105 signed New Mills, Hayfield. At C of E Church turn L onto Church Lane. School on L. Parking on site.* Sat 10 July (1-5); Sun 11 July (2-5). Adm £4, chd free. Light refreshments in School Library. Visits also by arrangement July & Aug for groups of 10 to 20.
Mixed herbaceous perennials/shrub borders, with mature trees and lawns and gravel border situated in the semi rural setting of the High Peak incl a Grade II listed building with 4 themed quads. The school was awarded highly commended in the School Garden 2019 RHS Tatton Show and won the Best High School Garden and the People's Choice Award. Come and see exhibits from this. Hot and Cold Beverages and a selection of sandwiches, cream teas and home-made cakes are available. Ramps allow wheelchair access to most of outside, flower beds and into Grade II listed building and library.

& ❀ ㋡ ㋡

"I love the National Garden Scheme which has been the most brilliant supporter of Queen's Nurses like me. It was founded by the Queen's Nursing Institute which makes me very proud. As we battle Coronavirus on the front line in the community, knowing we have their support is a real comfort." – Liz Alderton, Queen's Nurse

43 ◆ OLD ENGLISH WALLED GARDEN, ELVASTON CASTLE COUNTRY PARK

Borrowash Road, Elvaston, Derby, DE72 3EP. Derbyshire County Council, 01629 533870, www.derbyshire.gov.uk/elvaston. *4m E of Derby. Signed from A52 & A50. Car parking charge applies.* For NGS: Sun 8 Aug (12-4). Adm £2.50, chd free. Light refreshments. For other opening times and information, please phone or visit garden website. Come and discover the beauty of the Old English walled garden at Elvaston Castle. Take in the peaceful atmosphere and enjoy the scents and colours of all the varieties of trees, shrubs and plants. Summer bedding and large herbaceous borders. After your visit to the walled garden take time to walk around the wider estate featuring romantic topiary gardens, lake, woodland and nature reserve. Estate gardeners on hand during the day. Delicious home-made cakes available.

&. 🐑 ❀ ☕

44 330 OLD ROAD

Brampton, Chesterfield, S40 3QH. Christine Stubbs & Julia Stubbs. *Approx 1½m from town centre. 50 yds from junc with Storrs Rd. 1st house next to grazing field; on-road parking available adjacent to tree-lined roadside stone wall.* Fri 25, Sat 26 June (10.30-5). Adm £3, chd free. Light refreshments. Deceptive ⅓ acre plot of mature trees, landscaped lawns, orchard and cottage style planting. Unusual perennials, species groups such as astrantia, lychnis, thalictrum, heuchera and 40+ clematis. Acers, actea, hosta, ferns and acanthus lie within this interesting garden. Through a hidden gate, another smaller plot of similar planting, with delphinium, helenium, Echinacea, acers and hosta. Winner Chesterfield in Bloom Best Large Garden 2017. Wheelchair access on terrace area only, for views and refreshments - all welcome. Steps with handrail down to main garden. WC available.

❀ 🚗 ☕

45 OLD SHOULDER OF MUTTON - 7 MAIN STREET

Walton on Trent, DE12 8LY. Sarah & Mark Smith, mark_and_sarah@live.co.uk. *Off the A38 at the Barton/Walton junction. Parking available at* The White Swan pub. Garden *2 min walk from there.* Sun 27 June (11-4). Adm £4, chd free. Light refreshments. Home baked sausage rolls back by popular demand. Visits also by arrangement in July for groups of 5 to 20.
Delightful herbaceous borders fill this formerly neglected pub garden. From the topiary courtyard garden, step up past pleached limes to the cottage garden idyll scented by delicate roses. The abundant herbaceous borders with careful colour arrangements delight the eye. Walk past the lush shade border to explore the pond and beyond. Only the Courtyard is accessible to wheelchairs as access to the main garden is via 7 steps.

❀ ☕

46 THE PADDOCK

12 Manknell Rd, Whittington Moor, Chesterfield, S41 8LZ. Mel & Wendy Taylor, 01246 451001, debijt9276@gmail.com. *2m N of Chesterfield. Whittington Moor just off A61 between Sheffield & Chesterfield. Parking available at Victoria Working Mens Club, garden signed from here.* Sun 25 Apr, Sun 25 July (11-5). Adm £3.50, chd free. Cream teas. Visits also by arrangement Apr to Aug.
½ acre garden incorporating small formal garden, stream and koi filled pond. Stone path over bridge, up some steps, past small copse, across the stream at the top and back down again. Past herbaceous border towards a pergola where cream teas can be enjoyed.

&. 🐑 ❀ ☕

47 PARK HALL

Walton Back Lane, Walton, Chesterfield, S42 7LT. Kim Staniforth, 07785784439, kim.staniforth@btinternet.com. *2m SW of Chesterfield centre. From town on A619 L into Somersall Lane. On A632 R into Acorn Ridge. Park on field side only of Walton Back Lane.* Visits by arrangement Apr to July for groups of 10+. Minimum group charge £50. Adm £6, chd free. Light refreshments. Donation to Bluebell Wood Childrens Hospice.
Romantic 2 acre plantsmans garden, in a stunningly beautiful setting surrounding C17 house (not open) 4 main rooms, terraced garden, parkland area with forest trees, croquet lawn, sunken garden

with arbours, pergolas, pleached hedge, topiary, statuary, roses, rhododendrons, camellias, several water features. Runner-up in Daily Telegraph Great British Gardens Competition 2018. Two steps down to gain access to garden.

&. ❀ 🚗 ☕ 🛖

48 22 PINFOLD CLOSE

Repton, DE65 6FR. Mr & Mrs O Jowett. *6m S of Derby. From A38, A50 J, S of Derby follow signs to Willington then Repton. Off Repton High St TL into Pinfold Lane, Pinfold Close 1st L.* Sun 11 July (11-5). Adm £3, chd free.
Small garden with an interest in tropical plants. palms, gingers, cannas and bananas. Mainly foliage plants. Changes over the last year incl construction of a Safari Breeze Hut and repositioning of pond. Not suitable for wheelchairs.

🐑

49 RECTORY HOUSE

Kedleston, Derby, DE22 5JJ. Helene Viscountess Scarsdale. *5m NW Derby. A52 from Derby turn R Kedleston sign. Drive to village turn R. Brick house standing back from rd on sharp corner.* Sun 30 May (2-5). Adm £5, chd free. Home-made teas.
The garden is next to Kedleston Park and is of C18 origin. Many established rare trees and shrubs also rhododendrons, azaleas and unusual roses. Large natural pond with amusing frog fountain. Primulas, gunneras, darmeras and lots of moisture loving plants. The winding paths go through trees and past wild flowers and grasses. New woodland with rare plants. An atmospheric garden. A sphere of Cumbrian slate built by Jo Smith, Kirkcudbrightshire. Lily fountain in tea courtyard area. Uneven paths.

❀ ☕

50 ◆ RENISHAW HALL & GARDENS

Renishaw, Sheffield, S21 3WB. Alexandra Hayward, 01246 432310, enquiries@renishaw-hall.co.uk, www.renishaw-hall.co.uk. *10m from Sheffield city centre. By car: Renishaw Hall only 3m from J30 on M1, well signed from junction r'about.* For NGS: Tue 6 July (10.30-4). Adm £8, chd £3.50. Light refreshments at The Cafe. For other opening times and

Mastin Moor Gardens and Allotments

information, please phone, email or visit garden website.

Renishaw Hall and Gardens boasts 7 acres of stunning gardens created by Sir George Sitwell in 1885. The Italianate gardens feature various rooms with extravagant herbaceous borders. Rose gardens, rare trees and shrubs, National Collection of Yuccas, sculptures, woodland walks and lakes create a magical and engaging garden experience. Wheelchair route around garden.

51 REPTON ALLOTMENTS
Monsom Lane, Repton, Derby, DE65 6FX. Mr A Topping. *From Willington, L at r'about in Repton then L into Monsom Lane, at bottom of hill.* Mon 31 May (1-5). Adm £2, chd free. Refreshments available at 10 Chestnut Way.

Set on the edge of Repton with lovely views over the Derbyshire countryside. There are about 20 plots growing many types of vegetables and flowers too, a Community Garden is a new addition and now includes a large communal polytunnel. Mainly grass paths.

♿

52 NEW 40 ST OSWALD'S CRESCENT
Ashbourne, DE6 1FS. Anne McSkimming, 01335 342235, anneofashbourne@hotmail.com. *From centre of Ashbourne, Shaw Croft car park, go along Park St. L into Park Ave then 2nd R into St Oswald's Crescent. Follow rd til you see signs. Park on rd.* Visits by arrangement

May & July for groups up to 6. Adm £4, chd free. Light refreshments. £1 for tea/coffee and biscuits. Small enclosed town garden, developed by a plantswoman, from scratch 9 years ago. Trees, lots of shrubs, perennials, climbers. Some unusual specimens. Wide and narrow beds. Plenty of colour and interest throughout the season. A few steps lead up to a level garden. Can be slippery in places.

53 NEW SPINGLES
1 Nunbrook Grove, Buxton, SK17 7AU. Sue & Geoff Ashby. *All main roads bring you into town centre. 5 mins walk from Springs shopping centre c/park. Signed from side of Aldi, under r/way bridge, 2 L turns to Nunbrook Grove. Ltd on st parking near house.* Wed 14, Thur 15 July (11-5). Adm £3, chd free. Light refreshments. Gluten free options available.

We invite you to visit our beautiful gardens created by us from a blank canvas. Front garden a mix of white yellow and purple Back garden a haven of pastel colours with Roses & Clematis and much more.Various seating around the garden so come and relax with some home baking, inc gluten free, tea/coffee. We also sell plants. Our profits from refreshments and plants to Dogs Trust. 2 steps to garden. Our garden is also open for the Buxton Garden Trail held in June. Two shallow steps to access rear garden.

GROUP OPENING

54 STANTON IN PEAK GARDENS
Stanton-In-The-Peak, Matlock, DE4 2LR. *At the top of the hill in Stanton in Peak, on the rd to Birchover. Stanton in Peak is 5m south of Bakewell. Turn off the A6 at Rowsley, or at the B5056 & follow signs up the hill.* Sat 24, Sun 25 July (1-5). Combined adm £4.50, chd free. Home-made teas at 2 Haddon View. At Woodend a pop-up pub serves real ale from the barrel.

2 HADDON VIEW
Steve Tompkins.

HARE HATCH COTTAGE
Bill Chandler.

WOODEND COTTAGE
Will Chandler.

Stanton in Peak is a hillside, stone village with glorious views, and is a Conservation Area in the Peak District National Park. Three gardens are open. Steve's garden at 2 Haddon View is at the top of the village and is 1/10th acre crammed with plants. Follow the winding path up the garden with a few steps. There are semi flowering in the greenhouse, lots of patio pots, herbaceous borders, three wildlife ponds with red, pink and white water lilies, a koi pond, rhododendrons and a summerhouse. Tea and home-made cakes served. Just down the hill is Woodend where Will has constructed charming roadside follies on a strip of raised land along the road. The hidden, rear garden has diverse planting, and an extended vegetable plot. All with breath-taking views. A pop-up 'pub' will have draught beer from a local brewery. Nearby is Hare Hatch where Bill's garden wraps around the cottage. This is carefully designed to get the best of every planting opportunity, with a little bit of everything!

55 ◆ THORNBRIDGE HALL GARDENS
Ashford in the Water, DE45 1NZ. Jim & Emma Harrison, 07500 698795, gardeners@thornbridgehall.co.uk, www.thornbridgehall.co.uk. *2m NW of Bakewell. From Bakewell take A6, signed Buxton. After 2m, R onto A6020. ½m turn L, signed Thornbridge Hall.* For NGS: Tue 18 May (9-5). Adm £7, chd free. Light refreshments. For other opening times and information, please phone, email or visit garden website.
A stunning C19, 14 acre garden, set in the heart of the Peak District overlooking rolling Derbyshire countryside. Designed to create a vision of 1000 shades of green, the garden has many distinct areas. These incl koi lake and water garden, Italian garden with statuary, grottos and temples, 100ft herbaceous border, kitchen garden, scented terrace, hot border and refurbished glasshouses. Contains statuary from Clumber Park, Sydnope Hall and Chatsworth. Tea, coffee, sandwiches and cakes available. Gravel paths, steep slopes, steps.

56 TILFORD HOUSE
Hognaston, Ashbourne, DE6 1PW. Mr & Mrs P R Gardner, 01335 373001, peter.gardner@ngs.org.uk. *5m NE of Ashbourne. A517 Belper to Ashbourne. At Hulland Ward follow signs to Hognaston. Downhill (2m) to bridge. Roadside parking 100 metres.* Sun 30 May, Wed 7 July (1-5). Adm £4, chd free. Home-made teas. Visits also by arrangement May to July for groups of 5 to 30.
A 1½ acre streamside English country garden. Woodland, wildlife areas, alpine beds and ponds lie alongside colourful seasonal planting. Fruit cage and raised beds for vegetables. New beds and borders continually under development in this plantsman's garden. Sit and relax with tea and cake whilst listening to the continuous sounds of the birds.

57 ◆ TISSINGTON HALL
Tissington, Ashbourne, DE6 1RA. Sir Richard & Lady FitzHerbert, 01335 352200, tisshall@dircon.co.uk, www.tissingtonhall.co.uk. *4m N of Ashbourne. E of A515 on Ashbourne to Buxton Rd in centre of the beautiful Estate Village of Tissington.* For NGS: Mon 31 May, Mon 23, Mon 30 Aug (12-3). Adm £6, chd free. Cream teas at Award winning Herbert's Fine English Tearooms. For other opening times and information, please phone, email or visit garden website.
Large garden celebrating over 80yrs in the NGS, with stunning rose garden on west terrace, herbaceous borders and 5 acres of grounds. Refreshments available at the award winning Herberts Fine English Tearooms in village (Tel 01335 350501). Edward & Vintage Sweetshop, Andrew Holmes Butchers and Onawick Candle workshop also open in the village. Wheelchair access advice from ticket seller. Please seek staff and we shall park you nearer the gardens.

58 TREETOPS HOSPICE CARE
Derby Road, Risley, Derby, DE72 3SS. Treetops Hospice Care, www.treetopshospice.org.uk. *On main rd, B5010 in centre of Risley village. From J25 M1 take rd signed to Risley. Turn L at T-lights, Treetops Hospice Care on L approx ½m through village. From Borrowash direction Treetops is on R just after church.* Sat 4 Sept (11-3). Adm £3, chd free. Light refreshments.
The 12 acre site began as a Spring garden over 12 years ago following an appeal for daffodil bulbs. A team of 10 volunteers help to create all-year round colour with woodland planting, herbaceous/perennial borders, rose garden, wildlife meadow and pond and small fruit orchard. Raised wheelchair walkways allow access through some woodland areas with a circular walk along bark chipped paths. There are plants and homemade preserves stalls. Only partial wheelchair access through some of the woodland areas.

59 NEW VALDONA
Wheston, Tideswell, Buxton, SK17 8JA. Mrs Janet Walker. *Wheston, near Tideswell, Derbyshire (Peak District). From Tideswell village (Market Square), take the road pointing to Wheston up the hill. Valdona is a bungalow on the RHS, approx 1m from Tideswell.* Sat 31 July (2-5); Sat 7 Aug (10.30-1.30). Adm £3, chd free. Light refreshments.
A newly developed garden. Previously rough grass and trees. Now a landscaped garden, 1200 feet above sea level in a very exposed location. The garden is cottage style at the back and mainly lawn at the front, with views over the Peak District. There are three raised beds down one side for vegetables. There are also a number of fruit bushes. Great view over the Peak District.

60 WALTON COTTAGE
Matlock Road, Walton, Chesterfield, S42 7LG. Neil & Julie Brown, 07968 128313, jmbrown.waltoncottage@gmail.com. *3m from Chesterfield. 6½m from Matlock. On A632 Matlock to Chesterfield main rd. From Chesterfield 1m after junction nr garage with T-lights. From Matlock 1m after B5057 junction.* Sun 6 June (11-4). Adm £4, chd free. Home-made teas. Visits also by arrangement in June for groups of up to 30.
A large garden that has both formal and informal areas, incl woodland, orchard, kitchen garden and sweeping lawns with views over Chesterfield.

61 27 WASH GREEN
Wirksworth, Matlock, DE4 4FD. Mr Paul & Mrs Kathy Harvey, 01629 822218, pandkharvey@btinternet.com. *⅓m E of Wirksworth centre. From Wirksworth centre, follow B5035 towards Whatstandwell. Cauldwell St leads over railway bridge to Wash Green, 200 metres up steep hill on L. Park in town or uphill from garden entry.* Sun 16 May, Sun 29 Aug (11-4). Adm £4, chd free. Light refreshments. Visits also by arrangement Apr to Oct for groups of up to 30.

Secluded 1 acre garden, with outstanding views over Wirksworth and the Ecclesbourne valley. Inner enclosed area has a topiary garden, pergola and lawn surrounded by mixed borders, with paved seating area. The larger part of the garden has an open sweep of grass, with large borders, beds, areas of woodland and orchard. A quarter of the plot is a productive fruit and vegetable garden with polytunnel. Drop off at property entry for wheelchair access. The inner garden has flat paths with good views of whole garden.

62 12 WATER LANE

Middleton, Matlock, DE4 4LY. Hildegard Wiesehofer, 07809 883393, wiesehofer@btinternet.com. *Approx 2½ m SW of Matlock. 1½ m NW of Wirksworth. From Derby: at A6 & B5023 intersection take rd to Wirksworth. R to Middleton. Follow NGS signs. From Ashbourne take Matlock rd & follow signs. Park on main rd. Limited parking in Water Ln.* Sat 22 May (11-5); Sun 23 May (10-5). Adm £3.50, chd free Home-made teas. Gluten free options available. Visits also by arrangement May to Aug for groups of 10 to 20.
A small, eclectic hillside garden on different levels, created as a series of rooms. Each room has been designed to capture the stunning views over Derbyshire and Nottinghamshire and incl a woodland walk, ponds, eastern and infinity garden. Emphasis is on holistic and organic principles. Featured in Derbyshire Life, Practical Gardener. Glorious views and short distance from High Peak Trail, Middleton Top and Engine House. Very rare specimen of a limestone vaulted ceiling, so we are told.

63 13 WESTFIELD ROAD

Swadlincote, DE11 0BG. Val & Dave Booth, 01283 221167 or 07891 436632, valerie.booth1955@gmail.com. *5m E of Burton-on-Trent, off A511. Take A511 from Burton-on-Trent. Follow signs for Swadlincote. Turn R into Springfield Rd, take 3rd R into Westfield Rd.* Sun 13 June, Sun 8 Aug (1-5). Adm £3.50, chd free. Home-made teas. Visits also by arrangement May to Aug for

groups of 10 to 30.
A deceptive country-style garden in Swadlincote, a real gem. The garden is on 2 levels of approx ½ acre. Packed herbaceous borders designed for colour. Shrubs, baskets and tubs. Lots of roses. Greenhouses, raised-bed vegetable area, fruit trees and 2 ponds. Free range chicken area. Plenty of seating to relax and take in the wonderful planting from 2 passionate gardeners. The small top garden is accessible for wheelchairs, the lower garden can be accessed via a slope instead of the 7 steps.

64 WHARFEDALE

34 Broadway, Duffield, Belper, DE56 4BU. Roger & Sue Roberts, 01332 841905, rogerroberts34@outlook.com, www.garden34.co.uk. *4m N of Derby. Turn onto B5023 to Wirksworth (Broadway) off A6 at T-lights midway between Belper & Derby.* Visits by arrangement May to Sept for groups of up to 30. Adm £4.50, chd free. Home-made teas.
Garden design and plant enthusiast with over 600 varieties and rare specimens. Eclectic yet replaceable. 12 distinct areas incl naturalistic, choice shrub and single colour borders. Italian walled garden. Woodland with pond and walkway. Japanese landscape with stream, moon gate and pavilion. Front cottage garden with winter shrubs. Stone, wire and wood sculptures. Fully labelled. Comfortable seating around the garden. Close to Kedleston Hall and Derwent Valley World Heritage Site.

65 WILD IN THE COUNTRY

Hawkhill Road, Eyam, Hope Valley, S32 5QQ. Mrs Gill Bagshawe, www.wildinthecountryflowers. co.uk. *In Eyam, follow signs to public car park. Located next to Eyam Museum & opp public car park on Hawkhill Rd.* Sat 14 Aug (11-4). Adm £2.50, chd free.
A rectangular plot devoted totally to growing flowers and foliage for cutting. Sweet pea, rose, larkspur, cornflower, nigella, ammi. All the florist's favourites can be found here. There is a tea room, a village pub and several cafes in the village to enjoy refreshments.

66 15 WINDMILL LANE

Ashbourne, DE6 1EY. Jean Ross & Chris Duncan. *Take A515 (Buxton Rd) from the market place and at the top of the hill turn R into Windmill Lane; house by 4th tree on L.* Sun 25 Apr, Sun 18 July (1-5). Adm £3, chd free.
Landscaped and densely-planted town garden providing ideas others may wish to develop. Using a limited palette (mainly white, pink, mauve & burgundy) and emphasising leaf shape and colour, the key aims in establishing this garden were all-year interest; low maintenance; no grass; attraction of pollinators; and growing some soft fruit, peas & beans. Extensive views. A few steps.

67 26 WINDMILL RISE

Belper, DE56 1GQ. Kathy Fairweather. *From Belper Market Place take Chesterfield Rd towards Heage. Top of hill, 1st R Marsh Lane, 1st R Windmill Lane, 1st R Windmill Rise - limited parking on Windmill Rise - disabled mainly.* Sat 7, Sun 8 Aug (11-4.30). Adm £3.50, chd free. Light refreshments.
Behind a deceptively ordinary façade, lies a real surprise. A lush oasis, much larger than expected, with an amazing collection of rare and unusual plants. A truly plant lovers' organic garden divided into sections: woodland, Japanese, secret garden, cottage, edible, ponds and small stream. Many seating areas, incl a new summer house in which to enjoy a variety of refreshments. Home made delicious cakes and light lunches available.

Volunteers

**County Organisers
& Central Devon**
Edward & Miranda Allhusen
01647 440296
Miranda@allhusen.co.uk

County Treasurer
Nigel Hall 01884 38812
nigel.hall@ngs.org.uk

Publicity
Brian Mackness 01626 356004
brianmackness@clara.co.uk

Cath Pettyfer 01837 89024
cathpettyfer@gmail.com

Paul Vincent 01803 722227
paul.vincent@ngs.org.uk

Booklet Co-ordinator
Edward Allhusen 01647 440296
edward@allhusen.co.uk

Talks
Julia Tremlett 01392 832671
jandjtremlett@hotmail.co.uk

Assistant County Organisers

East Devon
Penny Walmsley 01404 831375
walyp_uk@yahoo.co.uk

Exeter
Jenny Phillips 01392 254076
jennypips25@hotmail.co.uk

Exmoor
Anna Whinney 01598 760217
annawhinney@yahoo.co.uk

North Devon
Jo Hynes 01805 804265
hynesjo@gmail.com

North East Devon
Jill Hall 01884 38812
jill22hall@gmail.com

Plymouth
Maria Ashurst 01752 351396
maria.ashurst@ngs.org.uk

South Devon
Sally Vincent 01803 722227
sallyvincent14@gmail.com

Torbay
Michelle Fairley 01626 879576
grosvenorgreengardens@gmail.com

West Devon
Alex Meads 01822 615558
ameads2015@outlook.com

Devon is a county of great contrasts in geography and climate, and therefore also in gardening.

The rugged north coast has terraces clinging precariously to hillsides so steep that the faint-hearted would never contemplate making a garden there. But here, and on the rolling hills and deep valleys of Exmoor, despite a constant battle with the elements, National Garden Scheme gardeners create remarkable results by choosing hardy plants that withstand the high winds and salty air.

In the south, in peaceful wooded estuaries and tucked into warm valleys, gardens grow bananas, palms and fruit usually associated with the Mediterranean.

Between these two terrains is a third: Dartmoor, 365 square miles of rugged moorland rising to 2000 feet, presents its own horticultural demands. Typically, here too are many National Garden Scheme gardens.

In idyllic villages scattered throughout this very large county, in gardens large and small, in single manors and in village groups within thriving communities – gardeners pursue their passion.

Below: Hole's Meadow

 @Devon NGS   @DevonNGS   @ngsdevon

OPENING DATES

All entries subject to change. For latest information check www.ngs.org.uk

Extended openings are shown at the beginning of the month.

Map locator numbers are shown to the right of each garden name.

February

Snowdrop Festival

Friday 5th
Higher Cherubeer · 45

Friday 12th
Higher Cherubeer · 45

Saturday 13th
The Mount, Delamore · 70

Sunday 14th
The Mount, Delamore · 70

Saturday 20th
Higher Cherubeer · 45

Sunday 28th
East Worlington House · 33

March

Sunday 7th
East Worlington House · 33

Saturday 13th
Haldon Grange · 38

Sunday 14th
Bickham House · 9
Haldon Grange · 38

Saturday 20th
Haldon Grange · 38

Sunday 21st
Haldon Grange · 38

Monday 22nd
NEW Houndspool · 51

Tuesday 23rd
NEW Houndspool · 51

Saturday 27th
Haldon Grange · 38
Monkscroft · 67

Sunday 28th
Haldon Grange · 38
Heathercombe · 43
Monkscroft · 67

April

Thursday 1st
Holbrook Garden · 47

Friday 2nd
Holbrook Garden · 47

Saturday 3rd
Haldon Grange · 38
Holbrook Garden · 47

Sunday 4th
Andrew's Corner · 4
Haldon Grange · 38
High Garden · 44
Higher Cherubeer · 45
Holbrook Garden · 47
Kia-Ora Farm & Gardens · 54

Monday 5th
Andrew's Corner · 4
Bickham House · 9
Haldon Grange · 38
Higher Cherubeer · 45
Kia-Ora Farm & Gardens · 54

Saturday 10th
Byes Reach · 21
Haldon Grange · 38

Sunday 11th
Byes Reach · 21
Haldon Grange · 38

Monday 12th
NEW Houndspool · 51

Tuesday 13th
NEW Houndspool · 51

Wednesday 14th
Haldon Grange · 38

Friday 16th
Sidbury Manor · 82

Saturday 17th
Haldon Grange · 38

Sunday 18th
Haldon Grange · 38
◆ Hotel Endsleigh · 50
Kia-Ora Farm & Gardens · 54
Sidbury Manor · 82
Upper Gorwell House · 97

Saturday 24th
Haldon Grange · 38
South Wood Farm · 85

Sunday 25th
Haldon Grange · 38
Shapcott Barton Knowstone Estate · 80
South Wood Farm · 85

May

Saturday 1st
Byes Reach · 21
Greatcombe · 37
Haldon Grange · 38
Holbrook Garden · 47
Musbury Barton · 71
Torview · 95

Sunday 2nd
Andrew's Corner · 4
Byes Reach · 21
Chevithorne Barton · 23
Greatcombe · 37
Hayne · 42
Haldon Grange · 38
High Garden · 44
Holbrook Garden · 47
Kia-Ora Farm & Gardens · 54
Mothecombe House · 69
Musbury Barton · 71
Torview · 95

Monday 3rd
Andrew's Corner · 4
Byes Reach · 21
Greatcombe · 37
Haldon Grange · 38
Holbrook Garden · 47
Kia-Ora Farm & Gardens · 54
Torview · 95

Wednesday 5th
Haldon Grange · 38

Saturday 8th
NEW Bradford Tracey House · 14
Gardens at Lake Farm · 35
Haldon Grange · 38
Spitchwick Manor · 87

Sunday 9th
Bickham House · 9
NEW Bradford Tracey House · 14
Gardens at Lake Farm · 35
Haldon Grange · 38
Heathercombe · 43
Spitchwick Manor · 87

Saturday 15th
Haldon Grange · 38
Kentlands · 53
The Old Vicarage · 74
The Walled Garden, Lindridge · 99

Sunday 16th
Haldon Grange · 38
Heathercombe · 43
Kentlands · 53
Kia-Ora Farm & Gardens · 54
Middle Well · 66
The Old Vicarage · 74
Upper Gorwell House · 97

Tuesday 18th
Avenue Cottage · 6

Wednesday 19th
Avenue Cottage · 6
Haldon Grange · 38

Friday 21st
Moretonhampstead Gardens · 68

Saturday 22nd
NEW Blundell's School Gardening Project · 11
Haldon Grange · 38
Heathercombe · 43
Kilmington (Shute Road) Gardens · 55
Moretonhampstead Gardens · 68

Sunday 23rd
Andrew's Corner · 4
Bickham House · 9
NEW Blundell's School Gardening Project · 11
Haldon Grange · 38
Heathercombe · 43
Kilmington (Shute Road) Gardens · 55
Moretonhampstead Gardens · 68

Tuesday 25th
Heathercombe · 43

Wednesday 26th
Heathercombe · 43

Thursday 27th
Heathercombe · 43

Friday 28th
Heathercombe · 43
Little Ash Bungalow · 60

Saturday 29th
Brocton Cottage · 18
Goren Farm · 36

Cleave Hill

Bradford Tracey House

THE GARDENS

GROUP OPENING

❶ ABBOTSKERSWELL GARDENS

Abbotskerswell, TQ12 5PN. *2m SW of Newton Abbot town centre. A381 Newton Abbot/Totnes rd. Sharp L turn from NA, R from Totnes. Field parking at Fairfield. Maps available at all gardens and at Church House.* **Sat 5, Sun 6 June (1-5). Combined adm £6, chd free. Home-made teas at Church House. Teas available from 2pm. Maps and tickets from 1pm.**

ABBOTSFORD
Wendy & Phil Grierson.

ABBOTSKERSWELL ALLOTMENTS
Tasha Mundy.

1 ABBOTSWELL COTTAGES
Jane Taylor.

BRIAR COTTAGE
Peggy & David Munden.

FAIRFIELD
Brian Mackness.

 THE POTTERY
Mr D & Mrs B Dubash.

7 WILTON WAY
Mr & Mrs Cindy & Vernon Stunt.

10 WILTON WAY
Mrs Margaret Crompton.

16 WILTON WAY
Katy & Chris Yates.

For 2021, Abbotskerswell offers 8 gardens plus the village allotments, ranging from very small to large they offer a wide range of planting styles and innovative landscaping. Cottage gardens, terracing, wild flower areas, a wild garden and specialist plants. Changes to some gardens from 2019. Ideas for every type and size of garden. Visitors are welcome to picnic in the field or arboretum at Fairfield. Sales of plants, garden produce, jams and chutneys and other creative crafts. Disabled access to 3 gardens. Partial access to most others.

❷ 32 ALLENSTYLE DRIVE

Yelland, Barnstaple, EX31 3DZ. Steve & Dawn Morgan, 01271 861433, fourhungrycats@aol.com, www.devonsubtropicalgarden.rocks. *5m W of Barnstaple. From Barnstaple take B3233 towards Instow. Through Bickington & Fremington. L at Yelland sign into Allenstyle Rd. 1st R into Allenstyle Dr. Light blue bungalow. From Bideford go past Instow on B3233.* **Sun 29 Aug, Sun 5, Sun 12 Sept (12-5). Adm £4, chd free. Light refreshments. Visits also by arrangement Aug & Sept for groups of up to 30.**
Our garden is 30m x 15m and is packed full of all our favourite plants. See our huge bananas, cannas, colocasias, delicate and scented tropical passion flowers, prairie planting, a wildlife pond and 2 large greenhouses. Relax and inhale the heady scents of the ginger lilies and rest awhile in the many seating areas. Partial wheelchair access due to narrow gravel paths.

❄ 🍵

❸ AM BROOK MEADOW

Torbryan, Ipplepen, Newton Abbot, TQ12 5UP. Jennie & Jethro Marles. *5m from Newton Abbot on A381. Leaving A381 at Causeway Cross go through Ipplepen village, heading towards Broadhempston. Stay on Orley Rd for ¾m. At Poole Cross Turn L signed Totnes, then 1st L into Am Brook Meadow.* **Sat 24, Sun 25 July (2-6.30). Adm £5, chd free.**
Country garden developed over past 14 years to encourage wildlife. Perennial native wildflower meadows, large ponds with ducks and swans, streams and wild areas covering 10 acres are accessible by gravel and grass pathways. Formal courtyard garden with water features and herbaceous borders and prairie-style planting together with poultry and bees close by. Wheelchair access to most gravel path areas is good, but grass pathways in larger wildflower meadow are weather dependent.

🚲 ❄ 🪑

❹ ANDREW'S CORNER

Skaigh Lane, Belstone, EX20 1RD. Robin & Edwina Hill, 01837 840332, edwinarobinhill@outlook.com, www.andrewscorner.garden. *3m E of Okehampton. Signed to Belstone from A30. In village turn L, signed Skaigh. Follow NGS signs. Garden approx ½m on R. Visitors may be dropped off at house, parking in nearby field.* **Sun 4, Mon 5 Apr, Sun 2, Mon 3, Sun 23, Sun 30, Mon 31 May (2-5). Adm £5, chd free. Home-made teas. Visits also by arrangement Feb to Oct.**
Join us as we enter our 50th year of opening and take a walk on the wild side in this tranquil moorland garden. April openings highlight magnolias, trillium and the lovely erythroniums. Early May the maples, rhododendrons and unusual shrubs provide interest, late May brings the flowering davidia, cornus, embothrium and the spectacular blue poppies. Live music 23rd May celebrates 1st opening in 1972. Wheelchair access difficult when wet.

♿ 🐕 ❄ 🍵

❺ ASH PARK

East Prawle, Kingsbridge, TQ7 2BX. Chris & Cathryn Vanderspar. *Ash Park, East Prawle South Devon. Take A379 Kingsbridge to Dartmouth, at Frogmore after PH R to East Prawle, after 1.1m l, in 1.4m at Cousins Cross bear R (middle of 3 rds). In village head to Prawle Point.* **Sat 18, Sun 19 Sept (11-5). Adm £5, chd free. Home-made teas.**
In a stunning location, with 180° view of the sea, Ash Park nestles at the foot of the escarpment, with 3½ acres of sub-tropical gardens, paths to explore, woodland glades, ponds and hidden seating areas. In Sept, cannas, hydrangeas, ginger lilies, salvias and dahlias should be at their best. Partial access for wheelchairs. Access is possible up a drive to the teas, but the garden itself has no hard paths and the lawns are sloping.

♿ 🐕 ❄ 🍵

❻ AVENUE COTTAGE

Ashprington, Totnes, TQ9 7UT. Mr Richard Pitts & Mr David Sykes, 01803 732769, richard.pitts@btinternet.com, www.avenuecottage.com. *3m SW of Totnes. A381 Totnes to Kingsbridge for 1m; L for Ashprington, into village then L by PH. Garden ¼m on R after Sharpham Estate sign.* **Tue 18, Wed 19 May, Tue 20, Wed 21 July (10-6). Adm £4, chd free. Pre-booking essential, please visit www.ngs.org.uk for information & booking. Home-made teas. Visits also by arrangement Apr to Oct for groups of up to 20.**
11 acres of mature and young trees and shrubs. Once part of an C18 landscape, the neglected garden has been cleared and replanted over the last 30 yrs. Good views of Sharpham House and R Dart. Azaleas and hydrangeas are a feature.

🐕 🛏 🍵

25 Elmfield Road, Barnstaple World Gardens

7 NEW **BACKSWOOD FARM**
Bickleigh, Tiverton, EX16 8RA.
Andrew Hughes, 01884 855005,
andrew@backswood.co.uk. *2m
SE of Tiverton. Take A396 off A361.
Turn L to Butterleigh opp Tesco's, L
at mini r'about, after 150 yds turn R
up Exeter hill for 2m then 1st R. 500
yds turn 1st R into Farm entrance.*
**Sat 17, Sun 18 July (11.30-5). Adm
£5, chd free. Home-made teas
in the Sheep Barn. Also open
Shutelake.**
This newly created 2 acre nature
garden provides many uniquely
designed homes for wildlife. Wander
through the flower meadow visiting
individually designed gardens,
structures, water features and ponds.
Seating areas afford stunning views
towards Exmoor & Dartmoor. Both
native and herbaceous plants have
been chosen to benefit insect and
bird life. A Photographic exhibition will
be open each day. Home-made teas
undercover.

GROUP OPENING

8 NEW **BARNSTAPLE WORLD
GARDENS**
Anne Crescent, Little Elche,
Barnstaple, EX31 3AF. *31 Anne
Cres L off Old Torrington Rd into
Phillips Ave then follow signs: 21
Becklake Close from A3125 turn
into Roundswell on Westermoor
Way then signs: 25 Elmfield Rd from
B3233 L at Bickington PO.* **Sun 20
June (11-4). Combined adm £5,
chd free. Light refreshments.
Japanese snacks. teas and
cakes.**

> **31 ANNE CRESCENT**
> Mr Gavin Hendry.
>
> **21 BECKLAKE CLOSE**
> Karen & Steve Moss.
>
> **25 ELMFIELD ROAD**
> Nigel & Carol Oates.

Explore three very different spaces,
one focusing on everything Japanese,
one on tropical plants, and one to
interest the plantsman. 31 Anne
Crescent: North Devon's Little Elche.
Small urban L-shaped tropical garden
complete with wide variety of palms,
agaves, bananas, cacti, tree ferns
and delightful pond loaded with
a wide variety of colourful koi. 21
Becklake Close: All things Japanese.
25 Elmfield Road: Front: mixed

borders and gravel area containing hardy palms and Mount Etna Broom. Rear: L shaped patio leading to lawn edged by mixed borders of unusual plants. The overall impression is of a colourful but calm place, enhanced by the sound of the stream trickling past the end of the garden. A garden to attract those interested in plants with a difference.

⓿ BICKHAM HOUSE
Kenn, Exeter, EX6 7XL. Julia Tremlett, 01392 832671, jandjtremlett@hotmail.com. *6m S of Exeter, 1m off A38. Leave A38 at Kennford Services, follow signs to Kenn, 1st R in village, follow lane for ³⁄₄m to end of no through rd. Only use SatNav once you are in the village of Kenn.* **Sun 14 Mar, Mon 5 Apr, Sun 9, Sun 23 May, Sun 6, Sun 20 June, Sun 11, Sun 25 July, Sun 15, Sun 22 Aug (1-5). Adm £6, chd free. Home-made teas. Visits also by arrangement Mar to Aug.** 6 acres with lawns, borders, mature trees. Formal parterre with lily pond. Walled garden with colourful profusion of vegetables and flowers. Palm tree avenue leading to millennium summerhouse. Late summer colour with dahlias, crocosmia, agapanthus etc. Cactus and succulent greenhouse. Pelargonium collection. Lakeside walk. WC, disabled access.

⓾ ◆ BLACKPOOL GARDENS
Dartmouth, TQ6 0RG. Sir Geoffrey Newman, 01803 771801, beach@blackpoolsands.co.uk, www.blackpoolsands.co.uk. *3m SW of Dartmouth. From Dartmouth follow brown signs to Blackpool Sands on A379. Entry tickets, parking, toilets and refreshments available at Blackpool Sands. Sorry, no dogs permitted.* **For information, please phone, email or visit garden website.**
Carefully restored C19 subtropical plantsman's garden with collection of mature and newly planted tender and unusual trees, shrubs and carpet of spring flowers. Paths and steps lead gradually uphill and above the Captain's seat offering fine coastal views. Recent plantings follow the S hemisphere theme with callistemons, pittosporums, acacias and buddlejas. Gardens open 1st Apr - 30th Sept

(10-4pm) weather permitting. Admission: adult £5, children free. Group visits by arrangement.

⓫ NEW BLUNDELL'S SCHOOL GARDENING PROJECT
Blundells Road, Tiverton, EX16 4DN. Harry Flower. *From Tiverton: Continue up Blundells Rd, take 1st R into School. Follow parking signs over mini r'about and park on signed field.* **Sat 22, Sun 23 May (1-4.30). Adm £3.50, chd free. Cream teas.**
Pupils of Blundell's School have created an area of vegetable production to supply the school kitchens. They have created a wildlife pond and an area of wild flowers with an outdoor seating area. Also a chance to see inside the Headmaster's Garden and other areas of the grounds. Pupils will be on hand to talk about the different projects they are undertaking in the garden. Fairly steep gravel and grass paths.

⓬ BOCOMBE MILL COTTAGE
Bocombe, Parkham, Bideford, EX39 5PH. Mr Chris Butler & Mr David Burrows, 01237 451293, www.bocombe.co.uk. *6m E of Clovelly, 9m SW of Bideford. From A39 just outside Horns Cross village, turn to Foxdown. At Xrds follow signs for parking.* **Visits by arrangement Apr to July for groups of 10+. Guided Tour for groups. Adm £5, chd £1. Ploughmans lunches & traditional home-made cakes & cream teas.**
12 flower gardens and many unique features that punctuate an undulating landscape of 5 acres in a wooded valley. Streams, 3 bog gardens - raised walkway, 12 water features, 3 large pools. White pergola. Hillside orchard. Soft fruit and kitchen gardens. Wild meadow, a wildlife haven. Short flower garden walk or longer circular walk, boots suggested. Garden plan incl 80+ specimen trees. Real hermit in the hermitage with adjoining shell grotto. Goats on hillside. Japanese pavilion. Garden kaleidoscope. New gated stone archway to hermitage walk. All organic.

GROUP OPENING

⓭ BOVEY TRACEY GARDENS
Bovey Tracey, TQ13 9NA. *6m N of Newton Abbot. Take A382 to Bovey Tracey. Car parking at town car parks or on some roads for 4 near-central gardens, on other roads for 3 gardens towards Brimley and for St Peter's Close.* **Sat 19, Sun 20 June (1.30-5.30). Combined adm £6, chd free. Tea at Gleam Tor, Wine at Ashwell.**

ASHWELL
TQ13 9EJ. Jeanette Pearce.

NEW FOOTLANDS
TQ13 9JX. Jon & Helen Elliott.

GLEAM TOR
TQ13 9DH. Gillian & Colin Liddy.

GREEN HEDGES
TQ13 9LZ. Alan & Linda Jackson.

PARKE VIEW
TQ13 9AD. Peter & Judy Hall.

2 REDWOODS
TQ13 9YG. Mrs Julia Mooney.

11 ST PETER'S CLOSE
TQ13 9ES. Pauline & Keith Gregory.

23 STORRS CLOSE
TQ13 9HR. Roger Clark & Chie Nakatani.

Bovey Tracey nestles in the Dartmoor foothills. Its gardens range from small - 23 Storrs Close, a plantsman's garden full of trees, shrubs, herbaceous and bulbous plants, with many rare examples from China and Japan, and colourful 11 St Peter's Close with its mini railway - to large, the romantic Parke View with meandering stone walls and colour themed borders, and Ashwell on a steep stone walled slope, with a vineyard, orchard, vegetables, mixed borders, wild flower areas and views. Green Hedges is packed full of interest – colourful borders, shade plants, organic vegetables, greenhouses and a stream. There's lots to enjoy at Gleam Tor - long herbaceous border, white garden, wild flower meadow, prairie planting, interesting 'memory patio' (and legendary cakes for tea), at Footlands, newly redesigned, unusual and colourful shrubs and conifers, with perennials, grasses and young trees and 2 Redwoods, mature trees, Dartmoor leat, fernery, sunny gravel garden and acid loving shrubs. Partial wheelchair access, none at Storrs Close, St Peter's Close, Ashwell or Green Hedges.

"The support of Hospice UK and the National Garden Scheme has been invaluable to hospice nurses across the country whilst we've been battling the coronavirus crisis, helping hospices such as Derian House to continue providing vital end of life and respite care to 400 children and young adults from across the North West. Thank you." – Katie Turner, Perinatal Nurse at Derian House Children's Hospice

14 NEW ▶ BRADFORD TRACEY HOUSE
Witheridge, Tiverton, EX16 8QG. Elizabeth Wilkinson. *20 minutes from Tiverton. Postcode will get you to thatched lodge at bottom of drive which has 2 bouncing hares on the top.* **Sat 8, Sun 9 May (1.30-5.30). Adm £5, chd free. Home-made cakes including gluten free and low sugar served in lovely coach house with lashings of top notch tea! Cakes and cookie boxes to take away.**
A pleasure garden set around a Regency hunting lodge combining flowers with grasses, shrubs and huge trees in a natural and joyful space. It is planted for productivity and sustainability giving harvests of wonderful flowers, fruits, herbs and vegetables (and weeds!). There are beautiful views over the lake, forest walks, deep blowsy borders, an ancient wisteria and an oriental treehouse garden. Mostly grass, gravel and stepping stones so not suitable for wheelchairs but disabled parking up at the house.
✿ ☕

15 BRAMBLE TORRE
Dittisham, nr Dartmouth, TQ6 0HZ. Paul & Sally Vincent, www.rainingsideways.com. *¾m from Dittisham. Leave A3122 at Sportsman's Arms. Drop down into village, at Red Lion turn L to Cornworthy. Continue ¾m, Bramble Torre straight ahead. Follow signs to Car Park.* **Fri 11, Sat 12, Sun 13 June (2-5). Adm £5, chd free. Cream teas.**
Set in 30 acres of farmland in the beautiful South Hams, the 3 acre garden follows a rambling stream through a steep valley: lily pond, herbaceous borders, roses, camellias, lawns and shrubs, a formal herb and vegetable garden. All are dominated by a huge embothrium glowing scarlet in late spring against a sometimes blue sky! Well behaved dogs on leads welcome. Partial wheelchair access, parts of garden very steep and uneven. Tea area with wheelchair access and excellent garden view.
♿ 🐾 ✿ ☕

GROUP OPENING

16 BRENDON GARDENS
Brendon, Lynton, EX35 6PU. 01598 741343, lalindevon@yahoo.co.uk. *1m S of A39 North Devon coast rd between Porlock and Lynton.* **Sat 12, Sun 13 June, Sat 31 July, Sun 1 Aug (12-5). Combined adm £5, chd free. Light refreshments. Higher Tippacott Farm serves light lunches, home-made cakes & cream teas. WC. Visits also by arrangement May to Sept.**

BRENDON HOUSE
Pat Young & Martin Longhurst.

HALL FARM
Karen Wall.

HIGHER TIPPACOTT FARM
Angela & Malcolm Percival.

NEW **LOWER WILSHAM COTTAGE**
Jane Glover.

Stunning part of Exmoor National Park. All gardens have lovely views. Excellent walking along river and between gardens; map online or pick up from gardens. Brendon Hse: C18 in idyllic village location. Established front garden, kitchen garden and greenhouse. Emphasis on recycling and gardening in harmony with wildlife. Hall Farm: C16 longhouse set in 2 acres of tranquil mature gardens, with lake and wild area beyond. Rheas, chickens, rare-breed cattle and black bees. H. Tippacott Farm: 950ft alt. on moor, overlooking own idyllic valley with stream and pond. Sunny levels of planting and lawns. Organic. L.Wilsham Cottage: Medium-sized cottage garden, small fernery, ponds and raised flower beds with wide variety of interesting and unusual plants. Plants, produce, books and bric-a-brac for sale.
🐄 ✿ ☕

17 THE BRIDGE MILL
Mill Rd, Bridgerule, Holsworthy, EX22 7EL. Rosie & Alan Beat, 01288 381341, rosie@thebridgemill.org.uk, www.thebridgemill.org.uk. *In Bridgerule village on R Tamar between Bude and Holsworthy. Between the chapel by river bridge and church at top of hill. Garden is at bottom of hill opp Short and Abbott agricultural engineers. See website for detailed directions.* **Visits by arrangement May & June for groups of 20+. Adm £4, chd free. Home-made teas. Refreshments in garden with friendly ducks if fine or in stable if wet! Plenty of dry seating.**
1-acre organic gardens around mill house and restored working water mill. Small cottage garden; herb garden with medicinal and dye plants; productive fruit and vegetable garden, and wild woodland and water garden by mill. 16 acre smallholding: lake and riverside walks, wildflower meadows, friendly livestock, sheep, ducks and hens. The historic water mill was restored to working order in April 2012 and in 2017 was awarded a plaque by the Society for the Protection of Ancient Buildings. For other opening times and information, please see website. Wheelchair access to some of gardens. WC with access for wheelchairs.
♿ ✿ 🚗 ☕ ⛱

18 BROCTON COTTAGE
Pear Tree, Ashburton, Newton Abbot, TQ13 7QZ. Mrs Naomi Hindley, 01364 654902, hindley1@clara.co.uk. *¼m from A38. Leave A38 at Princetown turning and turn R at end of slip road. Parking is restricted so if possible please park by Shell garage and walk up road to thatched house.* **Sat 29, Sun 30, Mon 31 May (12-4). Adm £4, chd free. Home-made teas. Lunches available at Dartmoor Lodge Hotel and Furzleigh Hotel nearby, also in Ashburton town centre. Visits also by arrangement May to Oct for groups of 5 to 20.**

The Old School House

1.3 acres recovered from neglect, still being developed after 10 years. Orchard, woodland, ponds and productive areas linked to established herbaceous borders and shrubberies. Views over Devon countryside. Wheelchair friendly. Dogs on leads only please. There is wheelchair access to the main terrace, one of the greenhouses and to the wildlife pond at the bottom of the garden.

19 BULLAND FARM

Ashburton, Newton Abbot, TQ13 7NG. S & L Middleton. *1m from A38. Exit A38 at Peartree Cross near Ashburton and follow signs directing initially towards Landscove. Turning R halfway up hill, follow NGS signs. From Totnes, NGS signs will direct you from A384.* **Sat 3, Sun 4 July (10.30-3.30). Adm £5, chd free. Light refreshments.**

Once an old cider orchard that had grown wild, over the last 10 yrs it has been transformed into a beautiful and productive garden. Set over 8 acres, it makes the most of wonderful views over rolling rural countryside. Designed with wildlife in mind it encompasses formal areas, prairie planting, boardwalk water garden, woodland trail, wildflower meadows and terraced vegetable garden. Steps and steep slopes although the central path of the garden is suitable for wheelchairs which allows great views over the countryside and garden.

20 ◆ BURROW FARM GARDENS

Dalwood, Axminster, EX13 7ET. Mary & John Benger, 01404 831285, enquiries@ burrowfarmgardens.co.uk, www.burrowfarmgardens.co.uk. *3½ m W of Axminster. From A35 turn N at Taunton Xrds then follow brown signs.*
Beautiful 13 acre garden with unusual trees, shrubs and herbaceous plants. Traditional summerhouse looks towards lake and ancient oak woodland with rhododendrons and azaleas. Early spring interest and superb autumn colour. The more formal Millennium garden features a rill Anniversary garden featuring late summer perennials and grasses. A photographer's dream. Open daily 1st Apr – 31 Oct (10am – 6pm). Adult £9 Child £2 Season ticket £25. Café, nursery and gift shop. Various events incl spring and summer plant fair and open air theatre held at garden each yr. Visit events page on Burrow Farm Gardens website for more details.

21 BYES REACH

26 Coulsdon Road, Sidmouth, EX10 9JP. Lynette Talbot & Peter Endersby, 01395 578081, latalbot01@gmail.com. *From Exeter on A3052 turn R at lights onto Sidford Rd A375. Turn ½m L into Coulsdon Rd.* **Sat 10, Sun 11 Apr, Sat 1, Sun 2, Mon 3 May (1-5.30). Adm £3, chd free. Home-made teas in parkland through back gate or within garden. Gluten free options for cakes. Opening with Sidmouth Gardens on Sat 29, Sun 30, Mon 31 May, Sat 28, Sun 29, Mon 30 Aug. Visits also by arrangement May to Aug for groups of 5 to 20.**
¼ acre garden with colour themed hot border at front. At rear of house: potager style vegetable garden, seasonal colour-themed beds, studio and greenhouse, 20m fruit arched walkway, rill, ferns, rockery, collection of hostas, pond. Seating in secluded niches with views to garden. Easy access.

22 ◆ CADHAY

Ottery St Mary, EX11 1QT. Rupert Thistlethwayte, 01404 813511, jayne@cadhay.org.uk, www.cadhay.org.uk. *1m NW of Ottery St Mary. On B3176 between Ottery St Mary and Fairmile. From E exit A30 at Iron Bridge. From W exit A30 at Patteson's Cross, follow brown signs for Cadhay.* **For NGS: Sun 30, Mon 31 May, Sun 29, Mon 30 Aug (2-5). Adm £5, chd £1. Cream teas. For other opening times and information, please phone, email or visit garden website.**
Tranquil 2 acre setting for Tudor manor house. 2 medieval fish ponds surrounded by rhododendrons, gunnera, hostas and flag iris. Roses, clematis, lilies and hellebores surround walled water garden. 120ft herbaceous border walk informally planted. Magnificent display of dahlias throughout. Walled kitchen gardens have been turned into allotments and old apple store is now tea room. Gravel paths.

23 CHEVITHORNE BARTON

Chevithorne, Tiverton, EX16 7QB. Chris McDonald (Head Gardener), chevithornebarton.co.uk. *3m NE of Tiverton. Follow yellow signs from A361, A396 or Sampford Peverell.* **Sun 2, Sun 30 May, Sun 4 July, Sun 29 Aug (2.30-5.30). Adm £5,**
chd free. Home-made teas.
Newly planted areas complement walled garden, summer borders and Robinsonian inspired woodland of rare trees and shrubs. In spring, garden features a large collection of magnolias, camellias, and rhododendrons with grass paths meandering through a sea of bluebells. Home to National Collection of Quercus (Oaks) comprising over 440 different taxa. From time to time within the gardens are a flock of Jacob sheep and rare breed woodland pigs. Lanning Roper's favourite garden (Country Life 1969). Recently planted collection of birches and witch hazels. Orchard of West country apples and recently added mazzards.

24 CLEAVE HILL

Membury, Axminster, EX13 7AJ. Andy & Penny Pritchard, 01404 881437, penny@tonybenger.com. *4m NW of Axminster. From Membury Village, follow rd down valley. 1st R after Lea Hill B&B, last house on drive, approx 1m.* **Sat 26, Sun 27 June (11-5). Adm £4.50, chd free. Light refreshments. Visits also by arrangement for groups of 5+.**
Artistic garden in pretty village situated on edge of Blackdown Hills. Cottage style garden, planted to provide all season structure, texture and colour. Designed around pretty thatched house and old stone barns. Wonderful views, attractive vegetable garden and orchard, wild flower meadow.

25 COOMBE MEADOW

Ashburton, Newton Abbot, TQ13 7HU. Angela Patterson & Mike Walker, 07775 627237, coombemeadow2016@gmail.com. *Opp Waterleat. Enter Ashburton town centre, turn onto North St. After Victoria Inn, bear R at junction, do not cross bridge to Buckland in the Moor. Follow signs.* **Visits by arrangement Mar to Sept for groups of 5 to 20. Parking for up to 8 cars. Adm £4, chd free. Light refreshments. Tea, coffee, sandwiches and cake.**
This lost garden, by a beautiful Dartmoor stream was overgrown and neglected. Since 2016 its mature magnolias, camellias, azaleas and other trees have been rescued. Paths and ponds cleared, archaeology
preserved, borders created, bridges and terraces repaired and over 5000 bulbs planted. The meadows contain wild daffodils and bluebells, so in spring the garden is a kaleidoscope of colour. Wheelchair access to front and rear patios only.

26 THE CROFT

Yarnscombe, Barnstaple, EX31 3LW. Sam & Margaret Jewell, 01769 560535. *8m S of Barnstaple, 10m SE of Bideford, 12m W of South Molton, 4m NE of Torrington. From A377, turn W opp Chapelton railway stn. Follow Yarnscombe signs for 3m. From B3232, ¼m N of Huntshaw Cross TV mast, turn E and follow Yarnscombe signs for 2m. Parking in village hall car park.* **Sun 20 June, Sun 18 July, Sun 15 Aug (2-6). Adm £4, chd free. Home-made teas. Visits also by arrangement June to Sept. Refreshments by arrangement, min 3 people. Donation to North Devon Animal Ambulance.**
1 acre plantswoman's garden featuring exotic Japanese garden with tea house, koi carp pond and cascading stream, tropical garden with exotic shrubs and perennials, herbaceous borders and shrubs, bog garden with collection of irises, astilbes and moisture-loving plants. New beds around duck pond and bog area, large collection of rare and unusual plants. See our garden on Facebook under 'The Croft Gardens Yarnscombe'. Not all areas wheelchair accessible.

27 NEW DECOY ALLOTMENT FIELD

Bladon Close, Newton Abbot, TQ12 1WA. Newton Abbot Town Council, nadcaa.org.uk/. *From Sainsburys at Penn Inn go past Keyberry Hotel, next L into Deer Park Rd, then R into Bladon Close, entrance to allotment field at top of the cul-de-sac, which is next to Decoy Primary School.* **Sat 12, Sun 13 June (1-4). Combined adm with 37 Kingskerswell Rd £5, chd free. Home-made teas. Cream teas will be available at 'The Hub'.**
Within 2 acres are hundreds of allotments, each of them individual to their owners. Beautiful wildlife areas, unusual plants, diverse habitats for flora and fauna and a shop with garden plants and honey from local bees. Many different nationalities

and people with different abilities are involved in working on the allotments. Also open is the nearby garden of one of our allotment holders. Some of the paths are uneven on the field with steep inclines in places. The other garden nearby does not have wheelchair access.

28 NEW DERRYDOWN
Sparkwell, Plymouth, PL7 5DF. Peter & Ann Tremain, 07940 543707, anntremain1942@gmail.com. *From Treby Arms Sparkwell 100yds on take RH fork. 2nd property on R, a bungalow on the corner.* **Visits by arrangement Apr to Sept for groups of 4 to 25. Adm £5, chd free. Cream teas.** 2017 new garden, ½ acre with raised beds, quirky fish pond. Pagoda with kiwi and grape vines. Water harvesting and rill. Lawn surrounded by hydrangeas and herbaceous. 2020 new plot with a 50ft circular lawn, frog pond, berry hedge, wild meadows, flowers, moon gates, double glazed greenhouses. Rain water harvesting and upcycling. Panoramic views. Children's quiz. Dogs on leads.

29 ◆ DOCTON MILL
Lymebridge, Hartland, EX39 6EA. Lana & John Borrett, 01237 441369, docton.mill@btconnect.com, www.doctonmill.co.uk. *8m W of Clovelly. Follow brown tourist signs on A39 nr Clovelly.* **For NGS: Sun 13 June (10-5). Adm £4.50, chd free. Cream teas and light lunches available all day. For other opening times and information, please phone, email or visit garden website.** Situated in stunning valley location. Garden surrounds original mill pond and the microclimate created within the wooded valley enables tender species to flourish. Recent planting of herbaceous, stream and summer garden give variety through the season. Regret not suitable for wheelchairs.

30 DUNLEY HOUSE
Bovey Tracey, TQ13 9PW. Mr & Mrs F Gilbert. *2m E of Bovey Tracey on rd to Hennock. From A38 going W turn off slip rd R towards Chudleigh Knighton on B3344, in village follow yellow signs to Dunley House. From A38 eastwards turn off on Chudleigh K slip rd L and follow signs.* **Sat 5, Sun 6 June, Sat 16, Sun 17 Oct (2-5). Adm £5, chd free. Home-made teas.** 9 acre garden set among mature oaks, sequoiadendrons and a huge liquidambar started from a wilderness in mid eighties. Rhododendrons, camellias and over 40 different magnolias. Arboretum, walled garden with borders and fruit and vegetables, rose garden and new enclosed garden with lily pond. Large pond renovated 2016 with new plantings. Woodland walk around perimeter of property.

"Unfortunately, cancer was not in lockdown in 2020. The continued support of our long-standing and valued partner, the National Garden Scheme is more important than ever." Macmillan

Harbour Lights

GROUP OPENING

EAST CORNWORTHY GARDENS

East Cornworthy, Totnes, TQ9 7HG. *Leave A3122 at Sportsman's Arms to Dittisham. At Red Lion L to Totnes. Continue 1m following signs to East Cornworthy.* **Sat 17, Sun 18 July (2-5). Combined adm £5, chd free. Cream teas.**

BLACKNESS BARN
Andrew & Karen Davis.

BROOK
John & Michelle Pain.

RIVENDALE FARM
Marina Pusey.

SANFORD HOUSE
Anne Mitchell.

TOAD HALL
Denis & Jacky Kerslake.

East Cornworthy is a small hamlet nestling in a valley just outside the village of Dittisham near the River Dart where 5 beautiful gardens will be opening their gates in July. Surrounded by rolling hills, with Dartmoor in the distance, you can enjoy woodland walks, streams, ponds as well as interesting planting and magnificent mixed borders in sun and shade in these very different gardens. A Devonshire cream tea and a plant stall will complete the afternoon. Partial wheelchair access. Parts of East Cornworthy are steep.
✿ ☕

32 EAST WOODLANDS FARMHOUSE

Alverdiscott, Newton Tracey, Barnstaple, EX31 3PP. Ed & Heather Holt. *5m NE of Great Torrington, 5m S of Barnstaple, off B3232. From Great Torrington turn R into single track rd before Alverdiscott; and from Barnstaple turn L after Alverdiscott. 1m down rd R fork at Y-junction.* **Sat 26, Sun 27 June (2-5). Adm £5, chd free. Home-made teas. Gluten free cakes available.**
East Woodlands is a beautiful RHS inspired and designed garden full of rooms packed with plants, shrubs and trees. Enjoy the spectacular bamboos, flowing grasses, colourful roses, Mediterranean, Japanese, cottage and bog gardens (unfenced pond), all set in an acre looking out over N Devon countryside. Occasional live music. Lots of seating areas and vintage crockery teas served. Plants for sale. Partial wheelchair access.
♿ ✿ ☕

33 EAST WORLINGTON HOUSE

East Worlington, Witheridge, Crediton, EX17 4TS. Barnabas & Campie Hurst-Bannister. *In centre of East Worlington, 2m W of Witheridge. From Witheridge Square R to East Worlington. After 1½m R at T-junction in Drayford, then L to Worlington. After ½m L at T-junction. 200 yds on L. Parking nearby, disabled parking at house.* **Sun 28 Feb, Sun 7 Mar (1.30-5). Adm £4, chd free. Cream teas in thatched parish hall next to house.**
Thousands of crocuses. In 2 acre garden, set in lovely position with views down valley to Little Dart river, these spectacular crocuses have spread over many years through the garden and into the neighbouring churchyard. Dogs on leads please.
♿ 🐕 ✿ ☕

Oakwood

34 FOXHOLE COMMUNITY GARDEN

Dartington, Totnes, TQ9 6EB. Zoe Jong, www. foxholecommunitygarden.org.uk. *On the Dartington Estate near the Foxhole Centre at Old School Farm.* **Thur 24, Sun 27 June (10-3). Adm £4, chd free. Teas, coffee, juice and home-made cakes.** Beautiful community garden and orchard on the Dartington Estate. Since 2016 it has been developed to provide a garden space for all abilities. Nature trail and garden crafts for children, talks and walks run on organic, no-dig low maintenance principles. Raised veg beds, orchard, herb, wildlife, wildflower, cutting flower, pond and potager planting areas. Full of colour, produce and wildlife. Parking directly outside the garden, main area of the garden accessible by wheelchair as is the toilet.

GROUP OPENING

35 GARDENS AT LAKE FARM

Sheepwash, Beaworthy, EX21 5PF. Richard Coward, 01409 231677, coward.richard@sky.com. *1.3m N of Sheepwash. A3072 to Highampton. Through Sheepwash. L on track signed Lake Farm. /A386- S of Merton take road to Petrockstow. Up hill opp, eventually L. After 350yds turn R down track 'Lake Farm'. SatNav error!* **Sat 8, Sun 9 May, Sat 17, Sun 18 July (11-5). Combined adm £5, chd free. Tea. Visits also by arrangement Apr to Sept for groups of up to 20.**

LAKE FARMHOUSE
Erica Fisher, www.mudandblossom.com.

MUSSELBROOK COTTAGE GARDEN
Richard Coward, 01409 231677, coward.richard@sky.com. **Visits also by arrangement Apr to Sept for groups of up to 20. Parking for 8 cars. Guided tours £25.**

2 gardens on opp sides of a track. Lake Farmhouse: A plantaholics garden with themed areas: rose garden, hosta, peony and hydrangea borders, large cut flower garden with dahlias, large productive kitchen garden, raised beds and containers. Many plants grown from seeds,

cuttings and divisions. Informal, cottage style planting, Musselbrook Cottage: 1 acre naturalistic/wildlife/ plantsman's garden of all season interest. Many rare/unusual plants on sloping site. 11 ponds (koi, orfe, rudd, dragonflies, lilies, aquatics). Stream, Japanese/Mediterranean gardens, oriental features. Wildflower meadow, clock golf. 1000s of bulbs. Hundreds of ericaceous plants incl acers, rhododendrons, camellias, hydrangeas and magnolias. Grasses, dierama, crocosmia. Wildlife haven. Aquatic nursery incl waterlilies.

36 GOREN FARM

Broadhayes, Stockland, Honiton, EX14 9EN. Julian Pady, www.goren.co.uk/pages/open-days. *6m E of Honiton, 6m W of Axminster. Go to Stockland television mast. Head 100 metres N signed from Ridge Cross, head E towards ridge.* **Every Sat and Sun 29 May to 13 June (10-5). Home-made teas. Evening openings Mon 31 May to Sat 31 July (5-10). Adm £5, chd free. The farm has a licenced premises to serve cider and refreshments.** Wander through 50 acres of natural species rich wild flower meadows. Easy access footpaths. Dozens of varieties of wild flowers and grasses. Thousands of orchids from early June and butterflies July. Stunning views of Blackdown Hills. Georgian house and walled gardens. Species information signs and picnic tables around the fields. Farm café and shop selling seeds and home grown produce. Partial wheelchair access to meadows. Dogs welcome on a lead only, please clean up after your pet.

37 GREATCOMBE

Holne, Newton Abbot, TQ13 7SP. Robbie & Sarah Richardson. *Michelcombe, Holne, TQ13 7SP. 4m NW Ashburton via Holne Bridge and Holne Village. 4m NE Buckfastleigh via Scorriton. Narrow lanes. Large car park adjacent to garden.* **Sat 1, Sun 2, Mon 3, Sat 29, Sun 30, Mon 31 May, Fri 25, Sat 26, Sun 27 June, Fri 23, Sat 24, Sun 25 July, Fri 6, Sat 7, Sun 8, Fri 20, Sat 21, Sun 22 Aug (1-5). Adm £5, chd free. Home-made teas.** Let's try again!! We thought we'd let our visitors describe the garden. "A truly beautiful, tranquil garden with

unusual plants and exciting use of colour. A garden to visit again and again! Fantastic Cream Tea by the stream!", "Enchanting, magical place. One of our favourites which has given us much inspiration. A real asset to the NGS. Fabulous home-made cake and scones too!". Artist's Studio featuring brightly coloured acrylic paintings, prints and cards all available to purchase along with ornamental metal plant supports in all sizes and shapes and 'Made by Robbie' metal artefacts. Regret only partial wheelchair access.

38 HALDON GRANGE

Dunchideock, Exeter, EX6 7YE. Ted Phythian, 01392 832349. *5m SW of Exeter. From A30 through Ide Village to Dunchideock 5m. L to Lord Haldon, Haldon Grange is next L. From A38 (S) turn L on top of Haldon Hill follow Dunchideock signs, R at village centre to Lord Haldon.* **Every Sat and Sun 13 Mar to 13 June (1-5). Mon 5, Wed 14 Apr, Mon 3, Wed 5, Wed 19, Mon 31 May (1-5). Adm £5, chd free. Home-made teas. Visits also by arrangement Mar to July for groups of 10+.** Peaceful, well established 12 acre garden some dating back to 1770's. This hidden gem boasts one of the largest collections of rhododendrons, azaleas, magnolias and camellias. Interspersed with mature and rare trees and complimented by a lake and cascading ponds. 5 acre arboretum, large lilac circle, wisteria pergola with views over Exeter and Woodbury complete this family run treasure. Wheelchair access to main parts of garden.

39 HALSCOMBE FARM

Halscombe Lane, Ide, Exeter, EX2 9TQ. Prof J Rawlings. *From Exeter go through Ide to mini r'about take 2nd exit and continue to L turn into Halscombe Lane.* **Sat 26, Sun 27 June (2-5). Adm £4, chd free. Home-made teas. Donation to The Friends of Exeter Cathedral.** Farmhouse garden created over last 8 yrs. Large collection of old roses and peonies, long and colourful mixed borders, productive fruit cage and vegetable garden all set within a wonderful borrowed landscape.

40 HARBOUR LIGHTS

Horns Cross, Bideford, EX39 5DW. Brian & Faith Butler, 01237 451627, brian.nfu@gmail.com, harbourlightsgarden.org. *8m W of Bideford, 3m E of Clovelly. On main A39, so easy to find and access, between Bideford and Clovelly, halfway between Hoops Inn and Bucks Cross. There will be union jack flag, and yellow arrow signs at the entrance.* **Sat 19, Sun 20 June (11-5.30). Adm £4, chd free. Cream teas, cakes, light lunches, wine all available. Visits also by arrangement June to Aug for groups of 10+.**
½ acre colourful garden with Lundy views. A garden of wit, humour, unusual ideas, installation art, puzzles, volcano and many surprises. Water features, shrubs, foliage area, grasses in an unusual setting, fernery, bonsai and polytunnel, plus masses of colourful plants. You will never have seen a garden like this! Free leaflet. We like our visitors to leave with a smile! Child friendly. A 'must visit' interactive garden. Intriguing artwork of various kinds, original plantings and ideas.

41 THE HAVEN

Wembury Road, Hollacombe, Wembury, South Hams, PL9 0DQ. Mrs S Norton & Mr J Norton, 01752 862149, suenorton1@hotmail.co.uk. *20mins from Plymouth city centre. Use A379 Plymouth to Kingsbridge Rd. At Elburton r'about follow signs to Wembury. Parking on roadside. Bus stop nearby on Wembury Rd. Route 48 from Plymouth.* **Visits by arrangement Mar to May for groups of 5 to 20. Adm £4, chd free. Cream teas.**
½ acre sloping plantsman's garden in South Hams AONB. Tearoom and seating areas. 2 ponds. Substantial collection of large flowering Asiatic and hybrid tree magnolias. Large collection of camellias including camellia reticulata. Rare dwarf, weeping and slow growing conifers. Daphnes, early azaleas and rhododendrons, spring bulbs and hellebores. Wheelchair access to top part of garden.

42 HAYNE

Zeal Monachorum, Crediton, EX17 6DE. Tim & Milla Herniman, www.haynedevon.co.uk. *Located ½ m S of Zeal Monachorum. From Zeal Monachorum, keeping church on L, drive through village. Continue on this road for ⅓ m, garden drive is 1st entrance on R.* **Sun 2 May (2-6). Adm £5, chd free. Home-made teas.**
In the magical walled garden exciting new planting blends with the beautiful tree peonies, mature wisteria purple and white and rambling wild roses in combination with a more modern Piet Oudolf style perennial planting surrounding the recently renovated grade II* farm buildings. Extensive veg and cut flower planting adds to the scene ... magic, mystery and soul by the spadeful! Live jazz band. Disabled WC. Wheelchair access to walled garden through orchard.

43 HEATHERCOMBE

Manaton, nr Bovey Tracey, TQ13 9XE. Claude & Margaret Pike Woodlands Trust, 01626 354404, gardens@pike.me.uk, www.heathercombe.com. *7m NW of Bovey Tracey. From Bovey Tracey take scenic B3387 to Haytor/ Widecombe. 1.7m past Haytor Rocks (before Widecombe hill) turn R to Hound Tor and Manaton. 1.4m past Hound Tor turn L at Heatree Cross to Heathercombe.* **Sun 28 Mar, Sun 9, Sun 16, Sat 22, Sun 23 May (1.30-5.30). Daily Tue 25 May to Sun 30 May (1.30-5.30). Daily Tue 1 June to Sun 6 June (1.30-5.30). Sat 12, Sun 13 June (11-5.30). Daily Tue 15 June to Sun 20 June (11-5.30). Daily Tue 22 June to Sun 27 June (11-5.30). Daily Tue 29 June to Sun 4 July (11-5.30). Adm £5, chd free. Home-made teas. Some days may be self service. Please purchase tickets on NGS website if possible. Visits also by arrangement Apr to Oct. Donation to Rowcroft Hospice.**
Tranquil secluded valley with streams running through woods, ponds and lake - 30 acres of spring/summer interest with many sculptures and new developments - daffodils, extensive bluebells, large displays of rhododendrons, many unusual specimen trees, cottage gardens, orchard, wild flower meadow, bog/ fern/woodland gardens and woodland walks. 2 miles of mainly level sandy paths with many benches. Disabled reserved parking close to tea room & toilet.

44 HIGH GARDEN

Chiverstone Lane, Kenton, EX6 8NJ. Chris & Sharon Britton, www.highgardennurserykenton. wordpress.com. *5m S of Exeter on A379 Dawlish Rd. Leaving Kenton towards Exeter, L into Chiverstone Lane, 50yds along lane. Entrance clearly marked at High Garden. Phone for directions 01626 899106.* **Sun 4 Apr, Sun 2 May, Sun 4 July, Sun 1 Aug, Sun 5 Sept (12-5). Adm £4, chd free. Delicious home-made cakes and usually cream teas are available in tea room.**
Stunning garden of over 4 acres. Huge range of trees, shrubs, perennials, grasses, climbers and exotics planted over past 13 yrs. Great use of foliage to give texture and substance as well as offset the floral display. 70 metre summer herbaceous border. Over 40 individual mixed beds surrounded by meandering grass walkways. Exciting new formal plantings. Teas not for NGS charities. Selection of interesting plants available at attached nursery. 10% of plant sales to NGS. For other opening times and information please phone, email or visit the garden website. Slightly sloping site but the few steps can be avoided.

45 HIGHER CHERUBEER

Dolton, Winkleigh, EX19 8PP. Jo & Tom Hynes, 01805 804265, hynesjo@gmail.com. *2m E of Dolton. From A3124 turn S towards Stafford Moor Fisheries, take 1st R, garden 500m on L.* **Fri 5, Fri 12, Sat 20 Feb, Sun 4, Mon 5 Apr (2-5). Adm £5, chd free. Home-made teas. 2022: Fri 4, Fri 11, Sat 19 Feb. Visits also by arrangement Feb to Oct (excl Aug) for groups of 10+.**
1¾ acre country garden with gravelled courtyard and paths, raised beds, alpine house, lawns, herbaceous borders, woodland beds with naturalised cyclamen and snowdrops, kitchen garden with large greenhouse and orchard. Winter openings for National Collection of cyclamen species, hellebores and over 400 snowdrop varieties.

46 HIGHER ORCHARD COTTAGE
Aptor, Marldon, Paignton,
TQ3 1SQ. Mrs Jenny Saunders,
01803 551221. *1m SW of Marldon.
A380 Torquay to Paignton. At
Churscombe Cross r'about R for
Marldon, L towards Berry Pomeroy,
take 2nd R into Farthing Lane. Follow
for exactly 1m. Turn R at NGS sign
Follow signs for parking.* **Sat 5, Sun
6 June (11-5). Adm £4, chd free.
Home-made teas. Refreshments
from 2pm provided by Marldon
Garden Club if weather permits.
Outside seating only. Visits also
by arrangement Apr to Sept for
groups of up to 10. The number of
car parking spaces is limited to 4
at any one time.**
2 acre garden with generous colourful
herbaceous borders, wildlife pond,
productive vegetable beds and
grass path walks through wild flower
meadows in lovely countryside.
Sculpture and art installations by
local artists add excitement at every
turn. Often the featured artists
are in residence ready to chat to
visitors about their work. Only area
immediately at house is accessible for
wheelchairs. All paths through garden
are on gently sloping grass.

47 HOLBROOK GARDEN
Sampford Shrubs, Sampford
Peverell, EX16 7EN. Martin
Hughes-Jones & Susan Proud,
holbrookgarden.com. *1m NW from
M5 J27. From M5 J27 follow signs to
Tiverton Parkway. At top of slip rd off
A361 follow brown sign to Holbrook
Garden 07741 192915.* **Thur 1, Fri
2, Sat 3, Sun 4 Apr, Sat 1, Sun 2,
Mon 3 May, Tue 1, Wed 2, Thur
3 June, Thur 1, Fri 2, Sat 3 July,
Sun 1, Mon 2, Tue 3 Aug, Wed 1,
Thur 2, Fri 3 Sept (11-5). Adm £5,
chd free. Tea, coffee, home-made
biscuits and cakes.**
Ask our visitors - "Heaven for bumble
bees", "sun shining through layers of
plants", "an immersive experience",
"inspiring", "exciting plants", "magical,
artistic, free flowing paradise",
"cleverly planted but looks so natural".
Productive vegetable garden and
polytunnel. 2 acres of respite for
people and wildlife alike but if you
like 'neat and tidy' then this garden
may not be for you. Coaches by
arrangement only, see holbrookgarden.
com. Donation to MSF UK (Medecin
sans Frontieres). Partial access for
wheelchairs and buggies.

48 HOLE FARM
Woolsery, Bideford, EX39 5RF.
Heather Alford. *11m SW of
Bideford. Follow directions for
Woolfardisworthy, signed from A39
at Bucks Cross. From village follow
NGS signs from school for approx
2m.* **Sun 18 July, Sun 12 Sept (2-
6). Adm £5, chd free. Home-made
teas in converted barn. Room to
sit and have a cup of tea even if
its raining!**
3 acres of exciting gardens with
established waterfall, ponds,
vegetable and bog garden. Terraces
and features incl round house have all
been created using natural stone from
original farm quarry. Peaceful walks
through Culm grassland and water
meadows border R Torridge and host
a range of wildlife. Home to a herd of
pedigree native Devon cattle.

49 HOLE'S MEADOW
South Zeal, Okehampton,
EX20 2JS. Fi & Paul
Reddaway, 07850 305040,
holesmeadow@gmail.com,
holesmeadow.com. *4½ m from
Okehampton on B3260, 4m from
Whiddon Down. Signed from main
street. Half way between the King's
Arms and Oxenham Arms and opp
village hall. A minute's fairly level walk
along private path.* **Sun 27 June,
Sun 25 July (11-5). Adm £4, chd
free. Light refreshments. Visits
also by arrangement June & July.
Donation to Plant Heritage.**
Set below Dartmoor's Cawsand
(Cosdon) Beacon, in 2 acre burgage
plot. Planting focuses on gardening
for pollinators, incl Plant Heritage
National Plant Collections of Monarda
and Nepeta. Bottom half of garden
incl an orchard, ornamental trees
and a maturing native woodland area
interspersed with pathways.

50 ◆ HOTEL ENDSLEIGH
Milton Abbot, Tavistock,
PL19 0PQ. Olga Polizzi,
01822 870000,
info@hotelendsleigh.com,
www.hotelendsleigh.com/garden.
*7m NW of Tavistock, midway
between Tavistock and Launceston.
From Tavistock, take B3362 to
Launceston. 7m to Milton Abbot then
1st L, opp school. From Launceston
& A30, B3362 to Tavistock. At
Milton Abbot turn R opp school.*
**For NGS: Sun 18 Apr, Sun 18 July
(11-4). Adm £8, chd free. For other**
opening times and information,
please phone, email or visit garden
website.
200 year old Repton-designed garden
in 3 parts; formal garden around
house, picturesque dell with pleasure
dairy and rockery and arboretum.
Gardens were laid out in 1814 and
have been carefully renovated over
last 14yrs. Bordering River Tamar, it
is a hidden oasis of plants and views.
Hotel was built in 1810 by Sir Jeffry
Wyattville for the 6th Duchess of
Bedford in the romantic cottage Orné
style. Plant Nursery adjoins hotel's
108 acres. Partial wheelchair access.

51 NEW HOUNDSPOOL
Ashcombe Road, Dawlish,
EX7 0QP. Mr & Mrs Edward
Bourne. *Situated 3m from Dawlish
centre, 3-4m from A380. From
A380 exit at Ashcombe Cross
(2.8m) or Great Haldon Café (3.8m),
follow signs to Ashcombe. Garden
is adjacent to Whetman Plants
International nursery.* **Mon 22, Tue
23 Mar, Mon 12, Tue 13 Apr, Sat
31 July, Sun 1 Aug, Mon 13, Tue
14 Sept (2-5). Adm £4, chd free.
Light refreshments.**
Formerly a market garden, now a
private pleasure garden developed
over past 40 yrs. The garden is
very much a work in progress, the
owners aiming to make it as labour
saving as possible indulging in their
love of trees, shrubs, herbaceous,
water, fruit/vegetables and flowers,
providing interest all yr round. Dogs
on leads. Children welcome. If social
restrictions apply please feel free to
wander around at your leisure. Gravel
paths may cause a little difficulty for
some wheelchairs but motorised
wheelchairs should have no problem
accessing most of the garden.

The National Garden Scheme
was set up in 1927 to raise funds
for district nurses. We remain
strongly committed to nursing
and today our beneficiaries
include Macmillan Cancer
Support, Marie Curie, Hospice
UK and The Queen's Nursing
Institute

52 KENTISBEARE HOUSE

Kentisbeare, Cullompton, EX15 2BR. Nicholas & Sarah Allan. *2m E of M5 J28 (Cullompton). Turn off A373 at Post Cross signed Kentisbeare. After ½ m past cricket field and main drive on R, entrance to carpark is through next gate on R.* **Sat 12 June (2-5); Sun 13 June (11-5). Adm £5, chd free. Home-made teas.**
Surrounding the listed former Kentisbeare rectory, the gardens have been redesigned and planted by the present owners in recent years with various planting themes that complement the surrounding countryside. Formal beds, lake walk, kitchen garden and glasshouse, recently established wildflower meadow, orchard. Diverse and interesting collection of trees, shrubs and woodland plants.

& ✳ D ☕ ♥

53 KENTLANDS

Whitestone, Exeter, EX4 2JR. David & Gill Oakey. *NW of Exeter mid way between Exwick and Tedburn St Mary. Follow NGS signs from centre of Whitestone village.* **Sat 15, Sun 16 May, Sun 25 July (11-5); Sun 12 Sept (11-4). Adm £4, chd free. Home-made teas.**
Our 2 acre tucked away garden is S-facing with distant views across Exeter towards W Devon and Sidmouth. Garden was started in 2010 and is still developing. Planting is mainly perennials with some shrubs, large salvia collection, orchids and alpines, productive vegetable garden with polytunnel, fruit cage and fruit trees. Sloping garden.

& 🐄 ✳ ☕ ♥

54 KIA-ORA FARM & GARDENS

Knowle Lane, Cullompton, EX15 1PZ. Mrs M B Disney, www.kia-orafarm.co.uk. *On W side of Cullompton and 6m SE of Tiverton. M5 J28, through town centre to r'about, 3rd exit R, top of Swallow Way turn L into Knowle Lane, garden beside Cullompton Rugby Club.* **Sun 4, Mon 5, Sun 18 Apr, Sun 2, Mon 3, Sun 16, Sun 30, Mon 31 May, Sun 13, Sun 27 June, Sun 11, Sun 25 July, Sun 8, Sun 22, Sun 29, Mon 30 Aug, Sun 12, Sun 26 Sept (2-5.30). Adm £4, chd free. Home-made teas at Kia-ora, inside or outside depending on personal preference and the weather! Teas and sales not for NGS charities.**
Charming, peaceful 10 acre garden with lawns, lakes and ponds. Water features with swans, ducks and other wildlife. Mature trees, shrubs, rhododendrons, azaleas, heathers, roses, herbaceous borders and rockeries. Nursery avenue, novelty crazy golf. Stroll leisurely around and finish by sitting back, enjoying a traditional home-made Devonshire cream tea or choose from the wide selection of cakes!

& ✳ 🚗 ☕ ♥

GROUP OPENING

55 KILMINGTON (SHUTE ROAD) GARDENS

Kilmington, Axminster, EX13 7ST. www.kilmingtonvillage.com. *1½ m W of Axminster. Signed off A35.* **Sat 22, Sun 23 May (1.30-5). Combined adm £6, chd free. Home-made teas at Breach.**

BETTY'S GROUND
Mary-Anne Driscoll.

BREACH
Judith Chapman & BJ Lewis, 01297 35159, jachapman16@btinternet.com. **Visits also by arrangement Apr to Oct for groups of 10 to 30. Refreshments provided by arrangement.**

SPINNEY TWO
Paul & Celia Dunsford.

Set in rural E Devon in AONB yet easily accessed from A35, 3 gardens just under 1 mile apart. Spinney Two: ½ acre garden planted for yr-round colour, foliage and texture. Mature oaks and beech. Spring bulbs, hellebores, shrubs; azaleas, camellias, cornus, pieris, skimmias, viburnums. Roses, acers, flowering trees, clematis and other climbers. Vegetable patch. Breach: 3+ acres with woodland, partially underplanted with rhododendrons, camellia and hydrangea; rose bed, variety of trees and shrubs, wild flower area, colourful borders, vegetable garden and fruit trees. Two ponds, one in bog garden. Betty's Ground, Haddon Corner: 1½ acres, a third of which was designed and replanted 4 yrs ago. Remaining two thirds has been restored, replanted in places and continues to evolve. Good selection of mature trees incl beautiful Wisteria walk. Beds both formal and relaxed incl woodland area all united by repeated perennial planting.

& 🐄 ✳ ☕

56 NEW 37 KINGSKERSWELL RD

Kingskerswell Road, Newton Abbot, TQ12 1DQ. Ferai Akyol & Heather Mather. *Entrance via rear of 37 Kingskerswell Rd. One minute's walk from Decoy allotments. A small garden, open in conjunction with the allotments.* **Sat 12, Sun 13 June (1-4). Combined adm with Decoy Allotment Field £5.**
Small town garden created for peace and relaxation. It is a garden for entertaining family and friends as well as bringing nature to your home. Not suitable for wheelchair access.

57 LANGUARD PLACE

Middle Warberry Road, Torquay, TQ1 1RS. Alison & Steve Dockray. *From Babbacombe Rd opp St Matthias church, turn into Higher Warberry Rd, then L into Middle Warberry Rd. Approx 800yds up hill to junction on R. Languard Place is on the corner.* **Sat 12, Sun 13 June (1-5). Adm £4, chd free. Teas, coffees and cakes available in the garden.**
This small level S-facing cottage style organic garden is a plantsman's delight. Enjoy the herbaceous borders packed with roses, clematis and many unusual perennials. Archways and paths lead to a separate Japanese style area with natural wildlife pond and gazebo, acers and bamboos. There's a productive greenhouse, vegetable and fruit areas and a wisteria arch. Limited wheelchair access.

✳ ☕

58 LEE FORD

Knowle Village, Budleigh Salterton, EX9 7AJ. Mr & Mrs N Lindsay-Fynn, 01395 445894, crescent@leeford.co.uk, www.leeford.co.uk/. *3½ m E of Exmouth. For SatNav use postcode EX9 6AL.* **Visits by arrangement Apr to Sept for groups of 10 to 20. Adm £7, chd free. Light refreshments. Numbers and special dietary requests must be pre-booked. Donation to Lindsay-Fynn Trust.**
Extensive, formal and woodland garden, largely developed in 1950s, but recently much extended with mass displays of camellias, rhododendrons and azaleas, incl many rare varieties. Traditional walled garden filled with fruit and vegetables, herb garden, bog garden, rose garden, hydrangea collection, greenhouses. Ornamental

conservatory with collection of pot plants. Lee Ford has direct access to the Pedestrian route and National Cycle Network route 2 which follows the old railway line that linked Exmouth to Budleigh Salterton. Garden is ideal destination for cycle clubs or rambling groups. Formal gardens are lawn with gravel paths. Moderately steep slope to woodland garden on tarmac with gravel paths in woodland.

♿ 🚗 ☕

59 LEWIS COTTAGE

Spreyton, nr Crediton, EX17 5AA. Mr & Mrs M Pell & Mr R Orton, 07773 785939, rworton@mac.com, www.lewiscottageplants.co.uk. *5m NE of Spreyton, 8m W of Crediton. A377 to Cred. Follow signs for Coleford & Colebrooke then follow NGS signs after New Inn PH. From A30, slip road signed M/hampstead. Turn R to Spreyton then over A30 flyover. Then follow NGS signs.* **Sat 29, Sun 30, Mon 31 May, Sat 26, Sun 27 June (11-5). Adm £4.50, chd free. Light refreshments. Visits also by arrangement May to Sept for groups of 10 to 30.**

4 acre garden located on SW-facing slope in rural Mid Devon. Evolved primarily over last 27 yrs, harnessing and working with the natural landscape. Using informal planting and natural formal structures to create a garden that reflects the souls of those who garden in it, it is an incredibly personal space that is a joy to share. Spring camassia cricket pitch, rose garden, large natural dew pond, woodland walks, bog garden, hornbeam rondel, winter garden, hot and cool herbaceous borders, fruit and veg garden, picking garden, outdoor poetry reading room and plant nursery selling plants mostly propagated from the garden. Wheelchairs/motorised buggies not advised due to garden being on a slope (though many have successfully tried!).

🐕 ✿ ☕

Online booking is available for many of our gardens, at ngs.org.uk

Breach, Kilmington (Shute Road) Gardens

21 Woodland Avenue, Teignmouth Gardens

60 LITTLE ASH BUNGALOW

Fenny Bridges, Honiton, EX14 3BL. Helen & Brian Brown, 01404 850941, helenlittleash@hotmail.com, www.facebook.com/littleashgarden. *3m W of Honiton. Leave A30 at 1st turn off from Honiton 1m, Patteson's Cross from Exeter ½m and follow NGS signs.* **Fri 28 May, Sun 15 Aug (1-5). Adm £4.50, chd free. Light refreshments. Visits also by arrangement May to Sept for groups of 10+.**
Country garden of 1½ acres, packed with different and unusual herbaceous perennials, trees, shrubs and bamboos. Designed for yr-round interest, wildlife and owners' pleasure. Naturalistic planting in colour coordinated mixed borders, highlighted by metal sculptures, providing foreground to the view. Natural stream, pond and damp woodland area, mini wildlife meadows and raised gravel/alpine garden. Grass paths.

61 LITTLE WEBBERY

Webbery, Bideford, EX39 4PS. Mr & Mrs J A Yewdall, 01271 858206, jyewdall1@gmail.com. *2m E of Bideford. From Bideford (East the Water) along Alverdiscott Rd, or from Barnstaple to Torrington on B3232.*

Take rd to Bideford at Alverdiscott, pass through Stoney Cross. **Visits by arrangement May to July for groups of up to 30. Adm £4.50, chd free.**
Approx 3 acres in valley setting with pond, lake, mature trees, 2 ha-has and large mature raised border. Large walled kitchen garden with yew and box hedging incl rose garden, lawns with shrubs and rose and clematis trellises. Vegetables and greenhouse and adj traditional cottage garden. Partial wheelchair access.

62 NEW LITTLEFIELD

Parsonage Way, Woodbury, Exeter, EX5 1HY. Bruno Dalbiez & Caryn Vanstone. *Please park cars in Woodbury village and follow signs to narrow driveway shared with Summer Lodge. Access to driveway from Parsonage Way, opp the stone cross on junction with Pound Lane.* **Sat 19, Sun 20 June (11-5). Adm £4, chd £1. Light refreshments.**
½ acre eco-garden. Rescued from derelict land in 2009/10, only 13m wide, 150m long, divided into herbaceous, shrub and tree planting, with large veg area, fruit and orchard with chickens and beehive. Plantsperson's garden - stocked with large range of varieties, in colourful combinations. Sculptures, unusual ironwork, wildlife pond all

add charm and interest. Managed using permaculture techniques. The entire garden can be accessed in a wheelchair, but be aware that most paths are gravel and can be soft.

63 ◆ LUKESLAND

Harford, Ivybridge, PL21 0JF. Mrs R Howell & Mr & Mrs J Howell, 01752 691749, lorna.lukesland@gmail.com, www.lukesland.co.uk. *10m E of Plymouth. Turn off A38 at Ivybridge. 1½m N on Harford rd, E side of Erme valley. Beware of using SatNavs as these can be very misleading.* **For opening times and information, please phone, email or visit garden website.**
24 acres of flowering shrubs, wild flowers and rare trees with pinetum in Dartmoor National Park. Beautiful setting of small valley around Addicombe Brook with lakes, numerous waterfalls and pools. Extensive and impressive collections of camellias, rhododendrons, azaleas and acers; also spectacular Magnolia campbellii and huge Davidia involucrata. Superb spring and autumn colour. Children's trail. Open Suns, Weds and BH (11-5) 14 March - 13 June and 3 Oct - 14 November. Adm £6, under 16s free. Group discount for parties of 20+. Group tours available by appointment. Partial wheelchair access in garden. Accessible café and WC.

64 MARSHALL FARM

Ide, nr Exeter, EX2 9TN. Jenny Tuckett. *Between Ide and Dunchideock. Drive through Ide to top of village r'about, straight on for 1½m. Turn R onto concrete drive, parking in farmyard at rear of property.* **Fri 11, Sun 13 June (1-5). Adm £4, chd free. Home-made teas. Donation to Spinal Injuries Association.**
Garden approached along lane lined with home grown lime, oak and chestnut trees. A country garden created approx 1967. One acre featuring wild flower gardens, gravel beds, pond, parterre garden and a vegetable and cutting garden. New wildlife ponds set in old orchard. Stunning views of Woodbury, Sidmouth gap and Haldon. Partial wheelchair access.

65 ◆ MARWOOD HILL GARDEN
Marwood, EX31 4EB. Dr J
A Snowdon, 01271 342528,
info@marwoodhillgarden.co.uk,
www.marwoodhillgarden.co.uk.
*4m N of Barnstaple. Signed from
A361 & B3230. Look out for brown
signs. See website for map and
directions. www.marwoodhillgarden.
co.uk. Coach & Car park.* **For NGS:
Fri 11 June (10-4.30). Adm £7,
chd free. Garden Tea Room offers
selection of light refreshments
throughout the day, all home-
made or locally sourced delicious
food to suit most tastes. For other
opening times and information,
please phone, email or visit garden
website.**
Marwood Hill is a very special private
garden covering an area of 20
acres with lakes and set in a valley
tucked away in N Devon. From early
spring snowdrops through to late
autumn there is always a colourful
surprise around every turn. National
Collections of astilbe, iris ensata
and tulbaghia, large collections
of camellia, rhododendron and
magnolia. Winner of MacLaren Cup
at rhododendron and camellia show
RHS Rosemoor. Partial wheelchair
access.

66 MIDDLE WELL
Waddeton Road, Stoke Gabriel,
Totnes, TQ9 6RL. Neil & Pamela
Millward, 01803 782981,
neilandpamela@talktalk.net. *A385
Totnes towards Paignton. R at
Parkers Arms in Collaton St. Mary. L
at Four Cross.* **Sun 16 May (11-5).
Adm £5, chd free. Home-made
teas. Soup lunch available. Visits
also by arrangement Apr to Oct for
groups of up to 30. Large coaches
must park 300m away.**
Tranquil 2 acre garden plus woodland
and streams contain a wealth of
interesting plants chosen for colour,
form and long season of interest.
Many seating places from which to
enjoy the vistas. Interesting structural
features (rill, summerhouse, pergola,
cobbling, slate bridge). Heady
mix of exciting perennials, shrubs,
bulbs, climbers and specimen trees.
Vegetable garden. Child friendly.
Produce and books. Featured in The
English Garden Nov 2020. Mostly
accessible by wheelchair.

67 MONKSCROFT
Zeal Monachorum, Crediton,
EX17 6DG. Mr & Mrs Ken & Jane
Hogg. *Lane opp Church. Parking in
farmyard.* **Sat 27, Sun 28 Mar (12-
5). Adm £3.50, chd free. Home-
made teas.**
Pretty, medium sized garden of oldest
cottage in village. Packed with spring
colours, primroses, daffodils, tulips,
magnolias and camellias. Views to far
hills. New exotic garden. Also tranquil
fishing lake with daffodils and wild
flowers in beautiful setting, home to
resident kingfisher. Steep walk to lake
approx 20mins, or 5mins by car. Dogs
on leads welcome. WC at lake.

GROUP OPENING

**68 MORETONHAMPSTEAD
GARDENS**
Moretonhampstead, TQ13 8PW.
*12m W of Exeter, 12m N of Newton
Abbot. Signs from the Xrd of
A382 and B3212. On E slopes of
Dartmoor National Park. Parking
at both gardens.* **Fri 21, Sat 22,
Sun 23 May, Sat 4, Sun 5 Sept
(1-5). Combined adm £6, chd
free. Home-made teas at both
gardens.**

MARDON
Graham & Mary Wilson.

SUTTON MEAD
Edward & Miranda
Allhusen, 01647 440296,
miranda@allhusen.co.uk.
Visits also by arrangement Apr
to Oct.

2 large gardens on edge of moorland
town. One in a wooded valley, the
other higher up with magnificent
views of Dartmoor. Both have mature
orchards and yr-round vegetable
gardens. Substantial rhododendron,
azalea and tree planting, croquet
lawns, summer colour and woodland
walks through hydrangeas and
acers. Mardon: 4 acres based on
its original Edwardian design. Long
herbaceous border and formal
granite terraces, stunning grasses.
Fernery and colourful bog garden
beside stream fed pond with its
thatched boathouse. Arboretum.
Sutton Mead: also 4 acres, shrub
lined drive. Lawns surrounding granite
lined pond with seat at water's edge.
Unusual planting, dahlias, grasses,
bog garden, rill fed round pond,
secluded seating and gothic concrete

greenhouse. Sedum roofed summer
house. Enjoy the views as you wander
through the woods. Dogs on leads
welcome, plant sale. Teas are a must.
Partial wheelchair access.

69 MOTHECOMBE HOUSE
Mothecombe, Holbeton,
Plymouth, PL8 1LA. Mr & Mrs J
Mildmay-White, www.flete.co.uk.
*12m E of Plymouth. From A379
between Yealmpton and Modbury
turn S for Holbeton. Continue
2m to Mothecombe.* **Sun 2 May
(11-5). Adm £6, chd free. Home-
made teas. Lunches at The
Schoolhouse, Mothecombe
village.**
Queen Anne house (not open) with
Lutyens additions and terraces set in
private estate hamlet. Walled pleasure
gardens with planting of lavenders
and bee friendly plants, borders and
Lutyens courtyard. Orchard with
spring bulbs, unusual shrubs and
trees, camellia walk. Autumn garden,
streams, bog garden and pond.
Bluebell woods. Yr-round interest.
Sandy beach at bottom of garden,
unusual shaped large liriodendron
tulipifera. Gravel paths, two slopes.

70 THE MOUNT, DELAMORE
Cornwood, Ivybridge, PL21 9QP.
Mr & Mrs Gavin Dollard. *Delamore
Park PL21 9QP. Please park in car
park for Delamore Park Offices not
in village. From Ivybridge turn L at
Xrds in Cornwood village keep PH on
L, follow wall on R to sharp R bend,
turn R.* **Sat 13, Sun 14 Feb (10.30-
3.30). Adm £4.50, chd free. Village
pub now community owned and
open. Excellent food.**
Welcome one of the first signs of
spring by wandering through swathes
of thousands of snowdrops in this
lovely wood. Closer to the village
than to Delamore gardens (open
only in May for the Sculpture and Art
Exhibition), paths meander through
a sea of these lovely plants, some of
which are unique to Delamore and
which were sold as posies to Covent
Garden market as late as 2002. Main
house and garden open for sculpture
exhibition every day in May. Mainly
rough paths/woodland tracks so
difficult wheelchair access.

71 MUSBURY BARTON
Musbury, Axminster, EX13 8BB.
Lt Col Anthony Drake. *3m S of Axminster off A358. Turn E into village, follow yellow arrows. Garden next to church, parking for 12 cars, otherwise park on road in village.* **Sat 1, Sun 2 May, Sat 26, Sun 27 June (1.30-5). Adm £5, chd free. Home-made teas. Tea proceeds to Musbury Church.**
6 acres. Extensive areas of well established trees and shrubs, many rare or unusual. Over 2000 roses spread round the garden. Stream from the top to the bottom. Lots of steps and bridges. Always interesting never perfect.

72 OAKWOOD
Orchard Court, Lamerton, Tavistock, PL19 8SF. Karen & Rod Dreher, 07813 435987, k.dreher@hotmail.co.uk. *At Blacksmith's Arms PH in Lamerton on Tavistock to Launceston Road, turn into village, follow road for ¼ m cross bridge, turn L into lane, follow NGS signs.* **Visits by arrangement Mar to Sept for groups of up to 20. Excluding period 21 July to 31 Aug. Adm £5, chd free. Home-made teas. Please request when booking if required with preferences and any dietary requirements.**
Created in 2008, this SW facing gently sloping garden provides all year round interest with winding paths through well stocked borders and rockery, water features, laburnum and wisteria tunnel. Specimen trees, intimate garden rooms taking advantage of the landscape. Soft fruit and vegetable areas with small potting shed; wilder area of informal lawn and trees with shrubs and spring bulbs. Wheelchair users can access around 60% of flower garden. Some steps. Grassland area sloping.

73 NEW THE OLD SCHOOL HOUSE
Ashcombe, Dawlish, EX7 0QB. Vanessa Hurley. *From A38 at Kennford take A380 towards Torquay. Turn off onto B3192 signed to Teignmouth, take 1st exit off r'about follow windy road for 1½ m 1st R after church and follow signs.* **Sat 31 July (1-5); Sun 1 Aug (1.30-5). Adm £4, chd free. Light refreshments. Tea, coffee, cold drinks, home-made cakes for sale.**
The Old School House is part of Ashcombe Estate. Created by Queen's Nurse Vanessa and husband Chaz. It is an artistic palette of colour with a hint of the theatrical. Set in over ½ acre there is a variety of plants shrubs, vines, bananas and trees attracting insects and wildlife plus an interesting array of quirky recycled materials. Dawlish water stream circles around the pretty garden. There are gravelled areas to entrances and some uneven ground. Some disabled parking outside house.

74 THE OLD VICARAGE
West Anstey, South Molton, EX36 3PE. Tuck & Juliet Moss, 01398 341604, julietm@onetel.com. *9m E of South Molton. From S Molton go E on B3227 to Jubilee Inn. From Tiverton r'about take A396 7m to B3227 then L to Jubilee Inn. Follow NGS signs to garden.* **Sat 15, Sun 16 May, Sat 14, Sun 15 Aug (12-5). Adm £5, chd free. Cream teas. Visits also by arrangement Apr to Aug for groups of up to 30.**
Croquet lawn leads to multi-level garden overlooking 3 large ponds with winding paths, climbing roses and overviews. Brook with waterfall flows through garden past fascinating summerhouse built by owner. Benched deck overhangs first pond. Features rhododendrons, azaleas and primulas in spring and large collection of wonderful hydrangeas in Aug. A wall fountain is mounted on handsome, traditional dry wall above house. Access by path through kitchen garden. A number of smaller standing stones echoing local Devon tradition.

75 PANGKOR HOUSE
Runnon Moor Lane, Hatherleigh, EX20 3PL. Sally & John Ingram. *8m NW of Okehampton. Follow Runnon Moor Lane for the whole length (approx 1 mile). At fork at end turn R onto unmade track (follow Pangkor House sign). Pangkor House is ¼ m uphill on R.* **Sat 24, Sun 25 July (1.30-5). Adm £5, chd free. Home-made teas.**
4 acre naturalistic garden with spectacular views over Dartmoor. Freeform lawn with a series of grass paths with lovely vista views. Quirky and highly personal with sculptures and random artefacts hidden within the borders. Wild meadow, wild

garden, Zen garden, courtyard garden and pond area with waterfall. Access to much of garden is via mown grass paths with a medium slope, i.e. wheelchair accessible if the ground is reasonably firm and dry.

76 ◆ PLANT WORLD
St Marychurch Road, Newton Abbot, TQ12 4SE. Ray Brown, 01803 872939, info@plant-world-seeds.com, www.plant-world-gardens.co.uk. *2m SE of Newton Abbot. 1½ m from Penn Inn turn-off on A380. Follow brown tourist signs at end of A380 dual carriageway from Exeter.*
The 4 acres of landscape gardens with fabulous views have been called Devon's 'Little Outdoor Eden'. Representing each of the five continents, they offer an extensive collection of rare and exotic plants from around the world. Superb mature cottage garden and Mediterranean garden will delight the visitor. Attractive viewpoint café, picnic area and shop. **Open April 1st to end of Sept (9.30-5.00).** Wheelchair access to café and nursery only.

77 POUNDS
Hemyock, Cullompton, EX15 3QS. Diana Elliott, 01823 680802, shillingscottage@yahoo.co.uk, www.poundsfarm.co.uk. *8m N of Honiton. M5 J26. From ornate village pump, near pub and church, turn up rd signed Dunkeswell Abbey. Entrance ½ m on R. Park in field. Short walk up to garden on R.* **Sat 12 June (1-5). Mon 14 June (1-5), also open Regency House. Sat 11, Mon 13 Sept (1-5). Adm £5, chd free.**
Cottage garden of lawns, colourful borders and roses, set within low flint walls with distant views. Slate paths lead through an acer grove to a swimming pool, amid scented borders. Beyond lies a traditional ridge and furrow orchard, with a rose hedge, where apple, pear, plum and cherries grow among ornamental trees. Further on, an area of raised beds combine vegetables with flowers for cutting. Some steps, but most of the garden accessible via sloping grass, concrete, slate or gravel paths.

78 REGENCY HOUSE

Hemyock, EX15 3RQ. Mrs Jenny Parsons, 01823 680238, jenny.parsons@btinternet.com, www.regencyhousehemyock.co.uk. *8m N of Honiton. M5 J26. From Catherine Wheel pub and church in Hemyock take Dunkeswell-Honiton Rd. Entrance ½ m on R. Please do not drive on the long, uncut grass alongside the drive. Disabled parking (only) at house.* **Mon 14 June (2-6). Adm £6, chd free. Pre-booking essential, please visit www.ngs. org.uk for information & booking. Home-made teas on terrace also garden at Pounds open same day. Regency House will open on various other dates but pre-booking will be essential; check ngs.org.uk for details. Visits also by arrangement June to Oct for groups of 10 to 30.**

5 acre plantsman's garden approached across private ford. Many interesting and unusual trees and shrubs. Visitors can try their hand at identifying plants with the plant list or have a game of croquet. Plenty of space to eat your own picnic. Walled vegetable and fruit garden, lake, ponds, bog plantings and sweeping lawns. Horses, Dexter cattle and Jacob sheep. Gently sloping gravel paths give wheelchair access to the walled garden, lawns, borders and terrace, where teas are served.

79 SAMLINGSTEAD

Near Roadway Corner, Woolacombe, EX34 7HL. Roland & Marion Grzybek, 01271 870886, roland135@msn.com. *1m outside Woolacombe. Stay on A361 road all the way to Woolacombe. Passing through town head up Chalacombe Hill, L at T-junction, garden 150metres on L.* **Sun 11 July (10-3.30). Adm £4, chd free. Cream teas in 'The Swallows' a purpose built out-building. Hot sausage rolls and cakes will also be available. Visits also by arrangement Mar to Sept.**

Garden is within 2 mins of N Devon coastline and Woolacombe AONB. 6 distinct areas; cottage garden at front, patio garden to one side, swallows garden at rear, meadow garden, orchard and field (500m walk with newly planted hedgerow). Slightly sloping ground so whilst wheelchair access is available to most parts of garden certain areas may require assistance.

80 SHAPCOTT BARTON KNOWSTONE ESTATE

(East Knowstone Manor), East Knowstone, South Molton, EX36 4EE. Anita Allen, 01398 341664. *13m NW of Tiverton. J27 M5 take Tiverton exit. 6½ m to r'about take exit South Molton 10m on A361. Turn R signed Knowstone. Leave A361 travel ¼ m to Roachhill through hamlet turn L at Wiston Cross, entrance on L ¼ m.* **Sun 25 Apr, Sun 20 June, Sun 18, Sun 25 July, Sun 8 Aug (10.30-4). Adm £5, chd free. Visits also by arrangement Apr to Aug. Donation to Cats Protection.**

Large garden of 200 acre estate around ancient historic manor house. Wildlife garden. Restored old fish ponds, stream and woodland rich in bird life. Unusual fruit orchard. Scented historic narcissi bulbs in Apr, roses in June, astilbes and phlox early July. Flowering burst July/Aug of National Plant Collections Leucanthemum superbum (shasta daisies) and Buddleja davidii. Large collection of hydrangea species. Large kitchen garden. Dowsing lessons and History of House talks. Only partial wheelchair access, steep slopes.

 NPC

81 SHUTELAKE

Butterleigh, Cullompton, EX15 1PG. Jill & Nigel Hall, 01884 38812, jill22hall@gmail.com. *3m W of Cullompton; 3m S of Tiverton. Between Tiverton & Cullompton, Follow signs for Silverton from Butterleigh village. Take L fork 100yds after entrance to Pound Farm. Car park sign on L after 150yds.* **Sat 29, Sun 30 May, Sat 17, Sun 18 July (2-5). Adm £5, chd free. Home-made teas. Also V/VG/GF Cakes. Visits also by arrangement May to Sept for groups of 10 to 20.**

Hidden in the Devon countryside is a terraced garden surrounding an ancient farmhouse. Roses and Wisteria come early followed by rich colours in the border. Plenty of variety with a natural pond teaming with wildlife, a woodland walk beside stream and sculptures. Plenty of spots to relax and enjoy great teas. Alphabet trail to amuse all ages.

82 SIDBURY MANOR

Sidmouth, EX10 0QE. Lady Cave, www.sidburymanor.com. *1m NW of Sidbury. Sidbury village is on A375, S of Honiton, N of Sidmouth.* **Fri 16, Sun 18 Apr (2-5). Adm £5, chd free. Cream teas.**

Built in 1870s this Victorian manor house built by owner's family and set within E Devon AONB comes complete with 20 acres of garden incl substantial walled gardens, extensive arboretum containing many fine trees and shrubs, a number of champion trees, and areas devoted to magnolias, rhododendrons and camellias. Partial wheelchair access.

GROUP OPENING

83 SIDMOUTH GARDENS

Woolbrook Park, Sidmouth, EX10 9DX. *For Rowan Bank. From Exeter on A3052 10m. R at Woolbrook Rd. In ½ m R at St Francis Church. Directions will be available for other gardens. 2 gardens will be open in May and 3 gardens in August.* **Sat 29, Sun 30, Mon 31 May (1-5.30). Combined adm £5, chd free. Sat 28, Sun 29, Mon 30 Aug (1-5.30). Combined adm £6, chd free. Home-made teas. Gluten free, lactose free cakes available**

BYES REACH

Lynette Talbot & Peter Endersby. Open on all dates
(See separate entry)

FAIRPARK

Knowle Drive, Sidmouth, EX10 8HP. Helen & Ian Crackston. Open on Sat 28, Sun 29, Mon 30 Aug

ROWAN BANK

44 Woolbrook Park, Sidmouth, EX10 9DX. Barbara Mence. Open on all dates

Situated on Jurassic Coast World Heritage Site, Sidmouth has fine beaches, beautiful gardens and magnificent coastal views. 3 contrasting ¼ acre gardens 1m apart. Byes Reach: Potager style vegetable garden, colour-themed herbaceous borders, rill, ferns, rockery, hostas. Spring colour of fruit blossom, spring bulbs - tulips, erythroniums and alliums. 20m arched walkway. Seating in secluded niches each with views of garden. Front hot border. Fairpark lies behind a 12ft red brick wall, terraces, rockery, impressionist palette of colour, texture, many acers, small woodland and greenhouse, grasses, raised beds and willow sculptures. Seating areas to enjoy home-made treats. Rowan Bank is NW facing, sloping, generously planted with trees, shrubs, perennials, bulbs. Steps lead to wide zigzag path leading to woodland and Mexican pine, Seats and summerhouse. No wheelchair access at Rowan Bank and Fairpark.

🐐 ✳ ☕

84 SILVER STREET FARM
Prescott, Uffculme, Cullompton, EX15 3BA. Alasdair & Tor Cameron, www.camerongardens.co.uk. Signs will guide you from A38 between M5 and Wellington. **Sat 11 Sept (1-5). Adm £5, chd free. Tea and cakes available.**
A plantsman's garden in rural setting, alive with scent, colour and dynamic planting. Roses, herbs and perennials, enormous herbaceous borders with meandering paths, an eclectic collection of plants and shrubs. Though designed this is a home and haven to people who love to nurture and be playful with their plants. Embracing the agricultural setting it encourages birds and insects and crafts a magical family space. Seating areas.

🚻 🚗 D ☕

85 SOUTH WOOD FARM
Cotleigh, Honiton, EX14 9HU. Professor Clive Potter, Southwoodfarmgarden@gmail.com. 3m NE of Honiton. From Honiton head N on A30, take 1st R past Otter Valley Field Kitchen layby. Follow for 1m. Go straight over Xrds and take first L. Entrance after 1m on R. **Sat 24, Sun 25 Apr, Sat 18, Sun 19 Sept (11-3). Adm £6, chd free. Pre-booking essential, please visit www.ngs. org.uk for information & booking. Light refreshments. Visits also by arrangement Apr to Sept for groups of 10 to 30.**
Designed by renowned Arne Maynard around C17 thatched farmhouse, country garden exemplifying how contemporary design can be integrated into a traditional setting. Herbaceous borders, roses, yew topiary, knot garden, wildflower meadows, orchards, lean-to greenhouses and a mouthwatering kitchen garden create an unforgettable sense of place. Rare opportunity to visit spring garden in all its glory. In Apr: 5000 bulbs incl 2500 tulips and hundreds of camassias

in flower meadow. Regional Finalist, The English Garden's The Nation's Favourite Gardens 2019. Gravel pathways, cobbles and steps.

🚻 🐐 ☕

86 SOUTHCOMBE BARN
Southcombe Barn, Widecombe-in-the-Moor, Newton Abbot, TQ13 7TU. Tom Dixon, www.southcombebarn.com. 6m W of Bovey Tracey. B3387 from Bovey Tracey after village church take rd SW for 400yds then sharp R signed Southcombe, after 200yds pass C17 farmhouse and park on L. **Sat 12, Sun 13 June (10-4). Adm £5, chd free. Cream teas.**
New owners are enjoying the challenge of a Dartmoor garden. 2 tea lawn terraces with a super colourful rockery between 3 acre garden of crazy colourful flower meadow and flowering trees with mown grass paths running through and around and alongside the stream. Art Gallery will be open. Wildlife. If it is a very hardy wheelchair it can bump it's way as far as the tea terraces.

🐐 ☕

87 SPITCHWICK MANOR
Poundsgate, Newton Abbot, TQ13 7PB. Mr & Mrs P Simpson. 4m NW of Ashburton. Princetown rd from Ashburton through Poundsgate, 1st R at Lodge. From Princetown L at Poundsgate sign. Past Lodge. Park after 300yds at Xrds. **Sat 8, Sun 9 May (11-4.30). Adm £5, chd free. Home-made teas.**
6½-acre garden with extensive beautiful views. Mature garden undergoing refreshment. A variety of different areas; lower walled garden with glasshouses, formal rose garden with fountain, camellia walk with small leat and secret garden with Lady Ashburton's plunge pool built 1763. 2.6 acre vegetable garden sheltered by high granite walls housing 9 allotments and lily pond. Mostly wheelchair access.

🚻 🐐 ✳ ☕

88 SPRINGFIELD HOUSE
Seaton Road, Colyford, EX24 6QW. Wendy Pountney, 01297 552481. Starting on A3052 coast rd, at Colyford PO take Seaton Rd. House 500m on L. Ample parking in field. **Sat 29 May, Sat 3 July, Sat 7 Aug (10.30-5). Adm £4, chd free. Home-made teas. Visits also by arrangement May**

to Sept for groups of 5 to 30. Refreshments on request.
1 acre garden of mainly fairly new planting. Numerous beds, majority of plants from cuttings and seed keeping cost to minimum, full of colour spring to autumn. Vegetable garden, fruit cage and orchard with ducks and chickens. Large formal pond. Wonderful views over R Axe and bird sanctuary, which is well worth a visit, path leads from the garden. Featured in Amateur Gardening magazine.

🚻 🐐 ✳ ☕

89 SQUIRRELS
98 Barton Road, Torquay, TQ2 7NS. Graham & Carol Starkie. 5m S of Newton Abbot. From Newton Abbot take A380 to Torquay. After ASDA store on L, turn L at T-lights up Old Woods Hill. 1st L into Barton Rd. Bungalow 200yds on L. Also could turn by B&Q. Parking nearby. **Sat 24, Sun 25 July (2-5). Adm £5, chd free. Light refreshments.**
Plantsman's small town environmental garden, landscaped with small ponds and 7ft waterfall. Interlinked through abutilons to Japanese, Italianate, Spanish, tropical areas. Specialising in fruit incl peaches, figs, kiwi. Tender plants incl bananas, tree fern, brugmansia, lantanas, oleanders. Collection of fuchsia, dahlias, abutilons, bougainvillea. Enviromental and Superclass Winners. 27 cleverly hidden rain water storage containers. Advice on free electric from solar panels and solar hot water heating and fruit pruning. 3 sculptures. Many topiary birds and balls. Huge 20ft Torbay palm. 9ft geranium. 15ft abutilons. New Moroccan and Spanish courtyard with tender succulents etc. Regret no wheelchair access. Conservatory for shelter and seating.

✳ 🚗 ☕

90 STONE FARM
Alverdiscott Rd, Bideford, EX39 4PN. Mr & Mrs Ray Auvray. 1½m from Bideford towards Alverdiscott. From Bideford cross river using Old Bridge and turn L onto Barnstaple Rd. 2nd R onto Manteo Way and 1st L at mini r'about. **Sat 29, Sun 30 May, Sat 26, Sun 27 June, Sat 24, Sun 25 July, Sat 28, Sun 29 Aug (2-5). Adm £4, chd free. Home-made teas.**
1 acre country garden with striking herbaceous borders, dry stone wall

terracing, white garden, hot garden and dahlia beds. Extensive fully organic vegetable gardens with polytunnels and newly created ¼ acre walled garden, together with an orchard with traditional apples and pears. Farm walks to see our herd of pedigree Red Ruby Devon cattle, pigs and sheep flock will be arranged. Some gravel paths but wheelchair access to whole garden with some help.

& ⴲ ✿ ☕

91 ◆ **STONE LANE GARDENS**
Stone Farm, Chagford,
**TQ13 8JU. Stone Lane Gardens
Charitable Trust, 01647 231311,
admin@stonelanegardens.com,
www.stonelanegardens.com.**
*Halfway between Chagford and
Whiddon Down, close to A382. 2.3m
from Chagford, 1.5m from Castle
Drogo, 2.5m from A30 Whiddon
Down via Long Lane.* **For NGS: Sat
9, Sat 16 Oct (10-6). Adm £6, chd
£3. Home-made teas. For other
opening times and information,
please phone, email or visit garden
website.**
Outstanding and unusual 5-acre arboretum and water garden on edge of Dartmoor National Park. Our birch have lovely colourful pooling bark, from dark brown, reds, orange, pink and white. Interesting under-planting. Meadow walks and lovely views. Sculpture exhibition situated in the garden. Near by: NT Castle Drogo and garden. RHS Partner Garden with National Collection of Birch and Alder. Tea Room. Partial wheelchair access to the gardens.

& ⴲ ✿ [NPC] ☕

92 STONELANDS HOUSE
Stonelands Bridge, Dawlish,
**EX7 9BL. Mr Kerim Derhalli
(Owner) Mr Saul Walker (Head
Gardener), 07815 807832,
saulwalkerstonelands@outlook.
com.** *Outskirts of NW Dawlish. From
A380 take junction for B3192 and
follow signs for Teignmouth, after
2m L at Xrds onto Luscombe Hill,
further 2m main gate on L.* **Visits
by arrangement Apr to July for
groups of 10 to 20. Adm £7, chd
free.**
Beautiful 12 acre pleasure garden surrounding late C18 property designed by John Nash. Mature specimen trees, shrubs and rhododendrons, large formal lawn, recently landscaped herbaceous beds, vegetable garden, woodland garden, orchard with wild-

Sidmouth Gardens

flower meadow and river walk. An atmospheric and delightful horticultural secret! Wheelchair access to lower area of gardens, paths through woodland, meadow and riverside walk may be unsuitable.

& ⴲ [D] 🏕

GROUP OPENING

93 TEIGNMOUTH GARDENS
Cliff Road, Teignmouth, TQ14 8TW.
*½m from Teignmouth town centre.
5m E of Newton Abbot. 11m S
of Exeter. Purchase ticket for all
gardens at first garden visited, a map
will be provided showing location of
gardens and parking.* **Sat 19, Sun
20 June (1-5). Combined adm
£6, chd free. Home-made teas at
High Tor, Cliff Road.**

BERRY COTTAGE
Alan & Irene Ward.

**GROSVENOR GREEN
GARDENS**
Michelle & Neal Fairley.

26 HAZELDOWN ROAD
Mrs Ann Sadler.

HIGH TOR
Gill Treweek.

LOWER COOMBE COTTAGE
Tim & Tracy Armstrong.

THE ORANGERY
Teignmouth Town Council,
teignmouthorangery.wordpress.
com/.

NEW **SEA VISTA**
Mrs S A Williams.

65 TEIGNMOUTH ROAD
Mr Terry Rogers.

NEW **21 WOODLAND AVENUE**
Larissa Letwyn.

6 YANNON TERRACE
Stuart Barker & Grahame Flynn,
mail@stuartbarker.info, ,
No6yannon.co.uk.
🛏

Popular coastal town Teignmouth has 3 new gardens joining the group this yr and visitors can view the Orangery, a beautifully restored glasshouse built in 1842. 21 Woodland Ave: large exotic clifftop garden overlooking sea with pond, palm trees and subtropical plants incl echiums, agaves, puyas and ginger lilies. Sea Vista: stunning sea views with Mediterranean court yard and Asian inspired planting. Gros Green Gdns: ⅓ acre with fruit, veg and large greenhouse. 6 Yannon Terrace: shrubs and perennials for yr round interest incl summer hot bed. Lower Coombe Cottage: country garden with Bitton Brook running through. 26 Hazeldown Rd: manicured garden with clipped topiary and large koi carp pond. Berry Cott: planted to provide a haven for wildlife. High Tor: home-made teas served overlooking sea in $\frac{1}{2}$ acre bee and butterfly friendly garden. 65 Teignmouth Rd: Coastal garden enjoying beautiful views over Lyme Bay with colour themed borders. Partial wheelchair access at some gardens.

& ✽ ☕

GROUP OPENING

94 NEW TEIGNMOUTH SEPTEMBER GARDENS
Bitton Park Road, Teignmouth, TQ14 9DF. *A379 to The Orangery Bitton Park Road, follow yellow signs to Coombe Vale Road to Coombe Ave. [(steps - see gate post with 1-8 Coombe Ave sign) or via level lane from Westbrook Ave.)] and Coombe Lane.* **Sat 11, Sun 12 Sept (1-5). Combined adm £5, chd free. Home-made teas at Bitton House (former home of Sir Edward Pellew - Hornblower series), adjacent to The Orangery.**

7 COOMBE AVENUE
Stewart and Pat Henchie

LOWER COOMBE COTTAGE
Tim & Tracy Armstrong.

THE ORANGERY
Teignmouth Town Council,
teignmouthorangery.wordpress.
com/.

7 Coombe Ave: Small sheltered plantmans garden, developed from 1998. South and West facing, on 2 levels surrounding Edwardian semi. Wide range of hardy and exotic plants with small glasshouse. Lower Coombe Cottage: Traditional country garden of just under $\frac{1}{4}$ acre. S-facing front garden incl range of unusual trees, shrubs and newly-planted roses leading to greenhouse and potager garden. Approached through C17 cottage's garden room, rear garden offers a Mediterranean-style courtyard with small vegetable patch and steps down to Bitton Brook. Teignmouth Orangery: Lying in the grounds of Bitton House, is believed to have been built in 1842. It was beautifully restored in 1985 and now houses a wide range of plants from different climate zones including brugmansia,

bougainvillea, gloriosa and strelitzia. Wheelchair access at Bitton House. Wheelchair users will find 7 Coombe Ave and Lower Coombe Cottage difficult because of steps.

🐄 ✽ 🚗 ☕

95 TORVIEW
44 Highweek Village, Newton Abbot, TQ12 1QQ. Ms Penny Hammond, penny.hammond2@btinternet.com. *On N of Newton Abbot accessed via A38. From Plymouth: A38 to Goodstone, A383 past Hele Park, L onto Mile End Rd. From Exeter: A38 to Drumbridges then A382 past Forches X, R signed Highweek. R at top of hill. Locally take Highweek signs.* **Sat 1, Sun 2, Mon 3 May (12-5). Adm £5, chd free. Home-made teas. Visits also by arrangement Apr to Sept for groups of 10 to 20.**
Run by two semi-retired horticulturalists: Mediterranean formal front garden with wisteria-clad Georgian house, small alpine house. Rear courtyard with tree ferns, pots/ troughs, lean-to 7m conservatory with tender plants and climbers. Steps to 30x20m walled garden - flowers, vegetables and trained fruit. Shade tunnel of woodlanders. Many rare/ unusual plants. Rear garden up 7 steps, pebble areas in front garden.

✽ ☕

96 TREETOPS
Broadclyst, Exeter, EX5 3DT. Geoffrey & Margaret Gould. *From Exeter direction on B3181 drive through Broadclyst past school, next turn R. Follow signs.* **Sat 12, Sun 13 June (1-5). Adm £4, chd free. Home-made teas.**
Previously an orchard this 1-acre cottage garden, which is still evolving, is bordered by a forest and set within a beautiful borrowed landscape. Incorporating an avenue of Olivia Austin roses and an old restored brick path surrounded by borders featuring traditional and unusual cottage garden plants and pond.

& 🐄 ☕

97 UPPER GORWELL HOUSE
Goodleigh Rd, Barnstaple, EX32 7JP. Dr J A Marston, www.gorwellhousegarden.co.uk. *$\frac{3}{4}$m E of Barnstaple centre on Bratton Fleming rd. Drive entrance between 2 lodges on L coming*

uphill (Bear Street) approx $\frac{3}{4}$m from Barnstaple centre. Take R fork at end of long drive. New garden entrance to R of house up steep slope. **Sun 18 Apr, Sun 16 May, Sun 20 June, Sun 18 July, Sun 19 Sept (2-6). Adm £5, chd free. Cream teas provided by Goodleigh W.I.**
Created mostly since 1979, this 4 acre garden overlooking the Taw estuary has a benign microclimate which allows many rare and tender plants to grow and thrive, both in the open and in walled garden. Several strategically placed follies complement the enclosures and vistas within the garden. Mostly wheelchair access but some very steep slopes at first to get into garden.

& 🐄 ✽ ☕ 🪑

98 VENN CROSS ENGINE HOUSE
Venn Cross, Waterrow, Taunton, TA4 2BE. Kevin & Samantha Anning, 01398 361392, venncross@btinternet.com. *Devon/Somerset border. Located on the B3227 between Bampton and Wiveliscombe.* **Sat 24, Sun 25 July, Sat 28, Sun 29 Aug (2-5.30). Adm £5, chd free. Visits also by arrangement May to Sept.**
Former GWR goods yard. 4 acres of formal and less formal gardens of interest to gardeners and railway enthusiasts alike. An acre of orchid rich wild flower meadow. Areas of mass-planted candelabra primulas start the summer with many large sweeping herbaceous borders bursting into colour as summer progresses. Ponds, streams, sculptural features, some railwayana and woodland walk. Vegetable beds. Wheelchair access to main areas (difficult if wet).

& 🐄 ✽ 🚗

99 THE WALLED GARDEN, LINDRIDGE
Humber, Teignmouth, TQ14 9TE. William & Surya Patterson. *12m S of Exeter. From B3192 pass Teignmouth golf course, follow signs for Bishopsteignton. After 1m at Rowden Cross, R to Lindridge. Pass main entrance to Lindridge Park on L, ahead 400m, on L through white gates.* **Sat 15 May, Sat 12 June, Sat 10 July, Sat 14 Aug (10-4). Adm £5, chd free. Pre-booking essential, please visit www.ngs. org.uk for information & booking.**
1 acre historical walled former kitchen garden of the Lindridge Park Estate,

fallow for over 50 years. Now half way through a 10-yr renovation plan, the garden has a geometric layout, lawns, mixed borders, juvenile hedges, trained fruit trees, wild flower meadow, ponds, woodland area, set in attractive countryside with far reaching views. Managed on an organic basis to provide a haven for wildlife with appropriate planting and ponds. Sloping site with gravel paths and some steps. Limited accessible parking.

Decoy Allotment Field

GROUP OPENING

100 NEW **WEST CLYST BARNYARD GARDENS**
West Clyst, Exeter, EX1 3TR.
From Pinhoe take B3181 towards Broadclyst. At Westclyst T-lights continue straight past speed camera 1st R onto Private Road over M5 bridge, R into West Clyst Barnyard. **Sat 5, Sun 6 June (1-5). Combined adm £5, chd free. Home-made teas.**

NEW **2 WEST CLYST BARNYARD**
Mark & Gill McIlroy

NEW **3 WEST CLYST BARNYARD**
Adam & Sarah Hemmings.

NEW **6 WEST CLYST BARNYARD**
Alan & Toni Coulson.

7 WEST CLYST BARNYARD
Malcolm & Ethel Hillier.

These gardens have been planted in farmland around a converted barnyard of a medieval farm. There are 4 gardens with a wild flower meadow, wildlife ponds, bog garden, David Austin roses, magnolias and many trees and shrubs. Cars may be driven to the gate of No 7 for disabled access to the gardens but then please park in the car park.

101 NEW **WHIDDON GOYLE**
Whiddon Down, Okehampton, EX20 2QJ. Mr &Mrs Lethbridge.
From Whiddon Down, follow signs for Okehampton, take 2nd exit at r'about. Whiddon Goyle is on L, signed. **Sat 5, Sun 6 June (11-4). Adm £4, chd free. Home-made teas.**
Whiddon Goyle enjoys stunning views over Dartmoor and sits 1000 ft

above sea level. Built in 1930's and cleverly designed to protect its 2 acre garden against the Dartmoor weather. It enjoys many features including a rockery, croquet lawn, rose garden herbaceous borders, ponds, small vegetable and flower plot along with a pair of majestic monkey puzzle trees. Access is via a gravelled driveway on a slope.

102 **WHITSTONE FARM**
Whitstone Lane, Bovey Tracey, TQ13 9NA. Katie & Alan Bunn, 01626 832258, klbbovey@gmail.com. *½m N of Bovey Tracey. From A382 turn towards hospital (sign opp golf range), after ⅓m L at swinging sign 'Private road leading to Whitstone'. Follow NGS signs.* **Visits by arrangement Mar to Sept for groups of up to 30. Small coaches/mini buses can access. Adm £5, chd free. Tea and home-made cakes, gluten free option. Donation to Plant Heritage.**
Nearly 4 acres of steep hillside garden with stunning views of Haytor and Dartmoor. Bluebells throughout the garden in spring. Arboretum planted 40 yrs ago, over 200 trees from all over the world incl magnolias,

camellias, acers, alders, betula, davidias and sorbus. Always colour in the garden and wonderful tree bark. Major plantings of rhododendrons and cornus. Late flowering eucryphia (National Collection) and hydrangeas. Display of architectural and metal sculptures and ornaments. Partial access to lower terraces for wheelchair users.

Open your garden with the National Garden Scheme and join a community of like-minded individuals, all passionate about gardens, and raise money for nursing and health charities. Big or small, if your garden has quality, character and interest we'd love to hear from you. Call us on 01483 211535 or email hello@ngs.org.uk

Volunteers

County Organiser
Alison Wright 01935 83652
alison.wright@ngs.org.uk

County Treasurer
Richard Smedley 01202 528286
richard@carter-coley.co.uk

Publicity
Clare Arber 07939 071806
clare.arber@ngs.org.uk

Social Media
Alison Wright (as above)

Booklet Editor
Judith Hussey 01258 474673
judithhussey@hotmail.com

Booklet Distributor
Alison Wright (as above)

Photographer
Jane Terry 07968 800075
darwineurope@aol.com

Assistant County Organisers

Central East
Trish Neale 01425 403565
trish.neale@ngs.org.uk

North East
Alexandra Davies 01747 860351
alex@theparishhouse.co.uk

North East Central
Caroline Renner 01747 811140
croftfarm12@gmail.com

**North East/Ferndown/
Christchurch**
Mary Angus 01202 872789
mary@gladestock.co.uk

North West & Central
Annie Dove 01300 345450
anniedove1@btinternet.com

South East/Poole/Bournemouth
Position vacant. For details please
contact Alison Wright (as above)

South Central/East
Helen Hardy 01929 471379
helen.hardy@ngs.org.uk

South Central/West
Fiona Johnston 07702 077200
fiona.johnston@ngs.org.uk

South West/Lyme Regis
Debbie Bell 01297 444833
debbie@debbiebell.co.uk

South West/Beaminster
Christine Corson 01308 863923
christine.corson@ngs.org.uk

West Central
Alison Wright (as above)

Dorset is not on the way to anywhere. We have no cathedral and no motorways. The county has been inhabited forever and the constantly varying landscape is dotted with prehistoric earthworks and ancient monuments, bordered to the south by the magnificent Jurassic Coast.

Discover our cosy villages with their thatched cottages, churches and pubs. Small historic towns including Dorchester, Blandford, Sherborne, Shaftesbury and Weymouth are scattered throughout, with Bournemouth and Poole to the east being the main centres of population.

Amongst all this, we offer the visitor a wonderfully diverse collection of gardens, found in both towns and deep countryside. They are well planted and vary in size, topography and content. In between the larger ones are the tiniest, all beautifully presented by the generous garden owners who open for the National Garden Scheme. Most of the county's loveliest gardens in their romantic settings also support us.

Each garden rewards the visitor with originality and brings joy, even on the rainiest day! They are never very far away from an excellent meal and comfortable bed.

So do come, discover and explore what the gardens of Dorset have to offer with the added bonus of that welcome cup of tea and that irresistible slice of cake, or a scone laden with clotted cream and strawberry jam!

f @NGS.Dorset
y @DorsetNGS
o @dorset_national_garden_scheme

Above: Farrs

OPENING DATES

All entries subject to change. For latest information check **www.ngs.org.uk**

Extended openings are shown at the beginning of the month.

Map locator numbers are shown to the right of each garden name.

February

Snowdrop Festival

Every day from Sunday 14th to Sunday 21st
Lawsbrook	52

Friday 12th
The Old Vicarage	72

Sunday 14th
The Old Vicarage	72

Sunday 21st
Herons Mead	34

Saturday 27th
Manor Farm, Hampreston	59

Sunday 28th
Manor Farm, Hampreston	59

March

Wednesday 3rd
The Mill House	65

Thursday 4th
The Mill House	65

Sunday 14th
Frankham Farm	28
Ivy House Garden	44

Sunday 21st
Herons Mead	34
The Old Vicarage	72

Saturday 27th
Knowle Cottage	51

Sunday 28th
Ivy House Garden	44

April

Sunday 4th
Herons Mead	34
Ivy House Garden	44

Monday 5th
◆ Edmondsham House	20
Ivy House Garden	44

Wednesday 7th
◆ Edmondsham House	20
Knitson Old Farmhouse	49

Saturday 10th
Chideock Manor	11
◆ Cranborne Manor Garden	15

Sunday 11th
Chideock Manor	11

Wednesday 14th
◆ Edmondsham House	20
Knitson Old Farmhouse	49

Sunday 18th
Broomhill	7
Frankham Farm	28
Ivy House Garden	44
The Old Vicarage	72

Wednesday 21st
◆ Edmondsham House	20
Horn Park	43

Saturday 24th
17 Lower Golf Links Road	58
NEW The Pines	76

Sunday 25th
17 Lower Golf Links Road	58

Wednesday 28th
◆ Edmondsham House	20

May

Sunday 2nd
NEW Dorchester Gardens	18
Gillans	30
Herons Mead	34
Holworth Farmhouse	42
Ivy House Garden	44

The Manor House, Beaminster	62
The Old Rectory, Litton Cheney	68
NEW The Pines	76
Wolverhollow	99

Monday 3rd
Holworth Farmhouse	42
Ivy House Garden	44
The Manor House, Beaminster	62
Wolverhollow	99

Wednesday 5th
Knitson Old Farmhouse	49
The Old Rectory, Litton Cheney	68

Sunday 9th
22 Avon Avenue	3
NEW Falconers	23
Mayfield	64
The Old Rectory, Pulham	70

Wednesday 12th
Deans Court	17
Knitson Old Farmhouse	49
Mayfield	64

Thursday 13th
The Old Rectory, Pulham	70

Saturday 15th
NEW Little Benville House	54

Sunday 16th
NEW Falconers	23
NEW Little Benville House	54
Wincombe Park	98

Wednesday 19th
Wincombe Park	98

Friday 21st
NEW Little Benville House	54

Saturday 22nd
Edwardstowe	21
17 Lower Golf Links Road	58
Pilsdon View	74
NEW Rosebank	83
Well Cottage	94

Sunday 23rd
Edwardstowe	21
NEW Higher Brimley Coombe Farm	35

17 Lower Golf Links Road	58
Manor Farm, Hampreston	59
Manor House Farm	61
Mayfield	64
The Old Vicarage	72
Pilsdon View	74
NEW Rosebank	83
Well Cottage	94

Monday 24th
Pilsdon View	74
Well Cottage	94

Tuesday 25th
NEW Higher Brimley Coombe Farm	35
◆ Keyneston Mill	46

Saturday 29th
NEW Manor Farm, Stourton Caundle	60
NEW Wagtails	93

Sunday 30th
Annalal's Gallery	2
Holworth Farmhouse	42
NEW Manor Farm, Stourton Caundle	60
The Manor House, Beaminster	62
2 Pyes Plot	79
Staddlestones	89
NEW Utopia	92
NEW Wagtails	93

Monday 31st
Holworth Farmhouse	42
The Manor House, Beaminster	62
Mayfield	64
Staddlestones	89

June

Wednesday 2nd
Knitson Old Farmhouse	49
Old Down House	67

Friday 4th
24 Carlton Road North	8

Saturday 5th
24 Carlton Road North	8
The Grange	31
Knowle Cottage	51
Old Down House	67

Sunday 6th
24 Carlton Road North	8
Frankham Farm	28
The Grange	31

Mayfield 64
Old Down House 67

Monday 7th
24 Carlton Road North 8

Wednesday 9th
Knitson Old
Farmhouse 49

Friday 11th
NEW Lewell Lodge 53
17 Lower Golf Links
Road 58

Saturday 12th
NEW Cattistock
Gardens 9
Chideock Manor 11
17 Lower Golf Links
Road 58
Philipston House 73

Sunday 13th
22 Avon Avenue 3
NEW Cattistock
Gardens 9
Chideock Manor 11
NEW Dorchester
Gardens 18
The Old Rectory, Litton
Cheney 68

Tuesday 15th
NEW Encombe House 22

Wednesday 16th
The Old School House 71
Stable Court 88

Saturday 19th
The Hollow, Blandford
Forum 38
Stable Court 88

Sunday 20th
Cliff Cottage 12
Cliff Lodge 13
East End Farm 19
NEW 5 Fosters
Meadows 27
The Hollow, Blandford
Forum 38

Online booking is available for many of our gardens, at ngs.org.uk

Little Cliff 55
Manor House Farm 61
The Manor House,
Beaminster 62
The Old School House 71
25 Richmond Park
Avenue 82
Western Gardens 95

Tuesday 22nd
NEW The Dairy House 16
◆ Holme for Gardens 40
◆ Littlebredy Walled
Gardens 56
NEW Yardes Cottage 100

Wednesday 23rd
NEW The Dairy House 16
Deans Court 17
Farrs 24
The Hollow, Blandford
Forum 38
Horn Park 43

Thursday 24th
Rampisham Gardens 81
NEW Yardes Cottage 100

Friday 25th
◆ Knoll Gardens 50
NEW Lewell Lodge 53

Saturday 26th
The Manor House,
Beaminster 62
Philipston House 73
NEW White House 96
NEW Yardes Cottage 100

Sunday 27th
Annalal's Gallery 2
Grove House 32
Hanford School 33
6 Hillcrest Road 36
Semley Grange 84
NEW White House 96
NEW Yardes Cottage 100

Tuesday 29th
◆ Littlebredy Walled
Gardens 56

July

Every Wednesday
The Hollow, Swanage 39

Friday 2nd
NEW Stafford House 90

Saturday 3rd
16 Chapel Rise 10
NEW Julia's House
Children's Hospice 45

Lower Abbotts Wootton
Farm 57

Sunday 4th
16 Chapel Rise 10
Holworth Farmhouse 42
NEW Julia's House
Children's Hospice 45
Lower Abbotts Wootton
Farm 57
25 Richmond Park
Avenue 82

Wednesday 7th
Knitson Old
Farmhouse 49
Lower Abbotts
Wootton Farm 57

Saturday 10th
◆ Cranborne Manor
Garden 15
NEW White House 96

Sunday 11th
22 Avon Avenue 3
Broomhill 7
NEW Dorchester
Gardens 18
NEW White House 96

Wednesday 14th
Knitson Old
Farmhouse 49

Sunday 18th
Hilltop 37
Manor Farm,
Hampreston 59

Saturday 24th
Edwardstowe 21

Sunday 25th
Annalal's Gallery 2
Black Shed 4
Edwardstowe 21
Hilltop 37

August

Every Wednesday
The Hollow, Swanage 39

Sunday 1st
Hilltop 37
The Old Rectory,
Pulham 70
25 Richmond Park
Avenue 82
Wolverhollow 99

Monday 2nd
Wolverhollow 99

Wednesday 4th
Knitson Old
Farmhouse 49

Thursday 5th
The Old Rectory,
Pulham 70

Saturday 7th
10 Brookdale Close 6

Sunday 8th
Annalal's Gallery 2
10 Brookdale Close 6
6 Hillcrest Road 36
Western Gardens 95

Wednesday 11th
Knitson Old
Farmhouse 49

Thursday 12th
Broomhill 7

Sunday 15th
22 Avon Avenue 3
Hilltop 37
Manor Farm,
Hampreston 59
The Old Vicarage 72

Saturday 21st
10 Brookdale Close 6

Sunday 22nd
Brook View Care Home 5
10 Brookdale Close 6
Hilltop 37

Wednesday 25th
Brook View Care Home 5

Sunday 29th
Black Shed 4

September

Wednesday 1st
Knitson Old
Farmhouse 49

Friday 3rd
NEW 1 Pine Walk 75

Saturday 4th
NEW 1 Pine Walk 75

Sunday 5th
NEW 1 Pine Walk 75

Tuesday 7th
◆ Holme for Gardens 40

Wednesday 8th
Knitson Old
Farmhouse 49

THE GARDENS

1 ◆ ABBOTSBURY GARDENS
Abbotsbury, Weymouth, DT3 4LA.
Ilchester Estates, 01305 871387,
info@abbotsbury-tourism.co.uk,
www.abbotsburygardens.co.uk.
*8m W of Weymouth. From B3157
Weymouth-Bridport, 200yds W of
Abbotsbury village.* For opening
times and information, please
phone, email or visit garden
website.
30 acres, started in 1700 and
considerably extended in C19. Much
recent replanting. The maritime
micro-climate enables Mediterranean
and southern hemisphere garden to
grow rare and tender plants. National
Collection of Hoherias (flowering
Aug in NZ garden). Woodland valley
with ponds, stream and hillside walk
to view the Jurassic Coast. Open
all year except for Christmas week.
Featured in Country Life and on
Countrywise and Gardeners' World.
Partial wheelchair access, some very
steep paths and rolled gravel but we
have a selected wheelchair route with
sections of tarmac hard surface.
& 🐕 ✿ ♿ NPC 🍵

2 ANNALAL'S GALLERY
25 Millhams Street,
Christchurch, BH23 1DN.
Anna & Lal Sims, 01202 567585,
anna.sims@ntlworld.com,
www.annasims.co.uk. *Town centre.
Park in Saxon Square PCP - exit
to Millhams St via alley at side of
church.* Sun 30 May, Sun 27 June,
Sun 25 July, Sun 8 Aug, Sun 12
Sept, Sun 5, Sun 12 Dec (2-4).
Adm £3, chd free. Visits also by
arrangement May to Dec.
Enchanting 150 yr-old cottage, home
of two Royal Academy artists. 32ft
x 12½ ft garden on 3 patio levels.
Pencil gate leads to colourful scented
Victorian walled garden. Sculptures
and paintings hide among the
flowers and shrubs. Unusual studio
and garden room. Not suitable for
wheelchairs; not suitable for dogs.

3 22 AVON AVENUE
Ringwood, BH24 2BH. Terry &
Dawn Heaver. *Past Ringwood from
E A31 turn L after garage, L again
into Matchams Ln, Avon Castle 1m
on L. A31 from west turn R into
Boundary Ln, then L into Matchams
Ln, Avon Ave ½m on R.* Sun 9 May,
Sun 13 June, Sun 11 July, Sun 15
Aug, Sun 12 Sept (12-5). Adm £5,
chd free. Home-made teas.

Japanese themed water garden
featuring granite sculptures, ponds,
waterfalls, azaleas, rhododendrons
cloud topiary and a collection of
goldfish and water lilies. Children only
under parental supervision, due to
large, deep water pond.
✿ 🍵

4 BLACK SHED

Blackmarsh Farm, Dodds Cross, Sherborne, DT9 4JX. Paul & Helen Stickland, blackshedflowers.blogspot.com/. *On A30 just E of Sherborne. From Sherborne, follow A30 towards Shaftesbury. Black Shed approx 1m E at Blackmarsh Farm, on L, next to The Toy Barn. Large car park shared with The Toy Barn.* **Sun 25 July, Sun 29 Aug (1-5). Adm £5, chd free. Home-made teas.**
Over 200 colourful and productive flower beds growing a sophisticated selection of cut flowers and foliage to supply florists and the public, for weddings, events and occasions throughout the seasons. Traditional garden favourites, delphiniums, larkspur, foxgloves, scabious and dahlias alongside more unusual perennials, foliage plants and grasses, creating a stunning and unique display. A warm welcome and generous advice on creating your own cut flower garden is offered. Easy access from gravel car park. Wide grass pathways enabling access for wheelchairs. Gently sloping site.

5 BROOK VIEW CARE HOME

Riverside Road, West Moors, Ferndown, BH22 0LQ. Charles Hubberstey, www.brookviewcare.co.uk. *Past village shops, L into Riverside Road, Brook View Care Home is on R after 100 metres. Parking onsite or nearby roads.* **Sun 22, Wed 25 Aug (11-5). Adm £3.50, chd free. Teas in morning. Cream teas in afternoon. Payment by donation.**
Our colourful and vibrant garden is spread over two main areas, one warm and sunny, the other cooler and shadier. A cool fountain area, games lawn and mixed borders, then walking past our greenhouse leads to the fruit and vegetable gardens. Produce is eagerly used by the kitchen, and residents will help out with the production of the bedding plants, all expertly managed by our gardener.

6 10 BROOKDALE CLOSE

Broadstone, BH18 9AA. Michael & Sylvia Cooper, 01202 693280, Michaelcooper6744@hotmail.co.uk. *Located just 100yds from centre of Broadstone, Brookdale Close is on Higher Blandford Rd, with additional parking in next road, Fairview Cres.* **Sat 7, Sun 8, Sat 21, Sun**

22 Aug (2-5). Adm £4, chd free. Home-made teas. Visits also by arrangement Aug to Dec.
A little piece of paradise with the 'WOW' factor, our 70ft x 50ft garden is centred around a wildlife pond and tumbling waterfall. A rich kaleidoscope of colour combining both tropical and cottage garden, with tree ferns, bananas, grasses, beautiful perennials and stunning dahlia display. Featured on Radio Solent, in Amateur Gardening magazine, Lewis Manning Cancer Trust Best Dorset Garden 2020. Only partial wheelchair access.

7 BROOMHILL

Rampisham, Dorchester, DT2 0PU. Mr & Mrs D Parry, 01935 83266, carol.parry2@btopenworld.com. *11m NW of Dorchester. From Dorchester A37 Yeovil, 9m L Evershot. From Yeovil A37 Dorchester, 7m R Evershot. Follow signs. From Crewkerne A356, 1½ m after Rampisham Garage L Rampisham. Follow signs.* **Sun 18 Apr, Sun 11 July, Thur 12 Aug (2-5). Adm £5, chd free. Home-made teas. Opening with Rampisham Gardens on Thur 24 June. Visits also by arrangement May to Aug for groups of 10+. Morning coffee, light lunches and teas.**
Once a farmyard now a delightful, tranquil garden set in 2 acres. Island beds and borders planted with shrubs, roses, masses of unusual perennials and choice annuals to give vibrancy and colour into the autumn. Orchard, veg garden. Lawns and paths lead to less formal area with large wildlife pond, meadow, shaded areas, bog garden and late summer border. Gravel entrance, the rest is grass, some gentle slopes.

8 24 CARLTON ROAD NORTH

Weymouth, DT4 7PY. Anne & Rob Tracey, 01305 786121, mellie_52@hotmail.com. *8m S of Dorchester. A354 from Dorchester, almost opp Rembrandt Hotel R into Carlton Road North. From Town Centre follow esplanade towards A354 Dorchester, L into Carlton Road North.* **Fri 4, Sat 5, Sun 6, Mon 7 June (2-5). Adm £3, chd free. Home-made teas. Visits also by arrangement Mar to Sept for groups of up to 10.**
Town garden near the sea. Long garden on several levels. Steps and

narrow sloping paths lead to beds and borders filled with trees, shrubs and herbaceous plants incl many unusual varieties. A garden which continues to evolve and reflect an interest in texture, shape and colour. Wildlife is encouraged. Raised beds in front garden create a space for vegetable growing.

GROUP OPENING

9 NEW CATTISTOCK GARDENS

Cattistock, Dorchester, DT2 0JJ. www.cattistockvillage.co.uk/the-great-outdoors/open-gardens/. *8m NW of Dorchester. From Dorchester: N on Maumbury Rd/B3147 towards Great Western Rd/B3144. At r'about 2nd exit onto A37. L then L onto W End, destination on L. Parking in field below the kennels in village.* **Sat 12, Sun 13 June (1-5). Combined adm £8, chd free. Admission covers both days. Sandwich lunches, tea/coffee and cake refreshments available throughout afternoon in village hall.**
From elegant walled gardens to perfectly-formed courtyard spaces this pretty West Dorset village has something for everyone, with plenty of high quality, well stocked gardens, showcasing different planting styles from classic cottage summer perennials to the pub garden community vegetable patch. Many of the gardens have views out to the beautiful Dorset countryside. Flower festival in church, open to visitors. The village community adventure playground will keep the children amused and the village pub will be open. Popular plant stall and raffle. Ample parking is available. Dogs on leads welcome. Toilets in village hall. Limited wheelchair access to some gardens.

10 16 CHAPEL RISE

Ringwood, BH24 2BL. Angela Ward. *From E A31, past Ringwood L after garage, immediate L Hurn Ln, 1½ m L Avon Ave, R to Egmont Drv, R Chapel Rise. A31 from W turn R into Boundary Ln then L Matchams Ln Avon Ave ½ m R.* **Sat 3, Sun 4 July (12-5). Adm £4, chd free. Home-made teas.**
Garden of ¾ acre set to lawn with shrubs and herbaceous borders surrounded by yew hedge. Water feature, bamboos and various

wooden and stone sculptures by owner's late husband. Steep short drive to garden, thereafter only grass paths.

1 CHIDEOCK MANOR
Chideock, Bridport, DT6 6LF. Mr & Mrs Howard Coates. *2m W of Bridport on A35. In centre of village turn N at church. The Manor is ¼ m along this rd on R.* **Sat 10, Sun 11 Apr, Sat 12, Sun 13 June (2-5). Adm £6, chd free. Home-made teas.**
6/7 acres of formal and informal gardens. Bog garden beside stream and series of ponds. Yew hedges and mature trees. Lime and crab apple walks, herbaceous borders, colourful rose and clematis arches, fernery and nuttery. Walled vegetable garden and orchard. Woodland and lakeside walks. Fine views. Partial wheelchair access.

2 CLIFF COTTAGE
West Cliff, West Bay, Bridport, DT6 4HS. Mr Asit Acharya. *Across harbour bridge in West Bay, 2nd exit at r'about, along Forty Foot Way and uphill to very top of West Walks.* **Sun 20 June (1-5). Adm £5, chd free. Also open Cliff Lodge.**
One acre cottage garden, originally 3 small workers' dwellings built around 1750 and set on top of West Cliff. The garden was overgrown when the present owner moved in 6 yrs ago and is still a work in progress. Some mature borders, now areas of planting, small orchard. Relaxed, informal planting. Shepherds hut and viewing deck with views towards East Cliff and Portland.

3 CLIFF LODGE
West Cliff, West Bay, Bridport, DT6 4HS. Mrs Dawn Gibson. *At top of West Cliff private estate in West Bay. See directions on Cliff Cottage.* **Sun 20 June (1-5). Adm £5, chd free. Home-made teas. Also open Cliff Cottage.**
⅓ acre mature garden with good rural views. Mature camellias, hydrangeas, silver birch, old bramley apple tree with rambling fragrant roses planted through, fruit cage, geraniums, roses, lavenders, grasses, thatched summerhouse.

4 COTTESMORE FARM
Newmans Lane, West Moors, Ferndown, BH22 0LW. Paul & Valerie Guppy, 07413 925372, paulguppy@googlemail.com. *1m N of West Moors. Off B3072 Bournemouth to Verwood rd. Car parking in owner's field.* **Visits by arrangement July & Aug for groups of up to 30. Adm £5, chd free. Home-made teas.**
Gardens of over an acre, created from scratch over 20 yrs. Wander through a plantsman's tropical paradise of giant gunneras, bananas, towering bamboos and over 100 palm trees, into a floral extravaganza. Large borders and sweeping island beds overflowing with phlox, heliopsis, helenium and much more combine to drown you in scent and colour. Wheelchair access to garden to avoid 2 lots of steps, please ask on arrival to make use of level route through main drive gate.

5 ◆ CRANBORNE MANOR GARDEN
Cranborne, BH21 5PP. Viscount Cranborne, 01725 517289, info@cranborne.co.uk, www.cranborne.co.uk. *10m N of Wimborne on B3078. Enter garden via Cranborne Garden Centre, on L as you enter top of village of Cranborne. Contactless payment at gate.* **For NGS: Sat 10 Apr, Sat 10 July (10-4). Adm £6.50, chd £1. Light refreshments. Please see our website for further details www.cranborne.co.uk. For other opening times and information, please phone, email or visit garden website.**
Beautiful and historic garden laid out in C17 by John Tradescant and enlarged in C20, featuring several gardens surrounded by walls and yew hedges: blue and white garden, cottage style and mount gardens, water and wild garden. Many interesting plants, with fine trees and avenues. Mostly wheelchair access.

29 Fleet Street

16 NEW **THE DAIRY HOUSE**
Melbury Osmond, Dorchester, DT2 0LS. Angela & Graham Chidgey, 01935 83619, angelachidgey@gmail.com. *6m S of Yeovil. A37, turn W signed Melbury Osmond. 1m into village, L at Xrds, park at village hall. Follow signs to garden, 5 min walk.* **Tue 22 June (2-5.30). Tea. Evening opening Wed 23 June (5-7.30). Wine. Adm £4, chd free. Visits also by arrangement for groups of up to 10.**
1 acre garden with 4 separate rooms, with views to fields, once upon a time it was a farmyard. Mature trees, incl English elm planted in 2002. Subdued colours of silver, blue, pale pink and white. A painter's work in progress. An orchard with wildflowers, pots (many of which came with the owners when they moved here from Tuscany) full of scented geraniums and an outside library.

🐑 ✿ ☕

17 **DEANS COURT**
Deans Court Lane, Wimborne Minster, BH21 1EE. Sir William Hanham, 01202 849314, info@deanscourt.org, www.deanscourt.org. *Pedestrian Entrance (no parking) - Deans Court Lane, Wimborne (BH21 1EE). Coach parties welcome but no parking for coaches available at Deans Court - please park in the town or contact Garden Owner for more information. Other vehicle entrance (free parking) on A349,Poole Rd, Wimborne (BH21 1QF) Please see extended description with regard to disabled access.* **Wed 12 May, Wed 23 June (11-5). Adm £5, chd free. Light refreshments at our Deans Court Café on Deans Court Lane. Donation to Friends of Victoria Hospital, Wimborne.**
13 acres of peaceful, partly wild gardens in ancient monastic setting with mature specimen trees, Saxon fish pond, herb garden, orchard and apiary beside R Allen close to town centre. First Soil Association accredited garden, within C18 serpentine walls. The Permaculture system has been introduced here with chemical free produce supplying the Deans Court café (open) nearby. With regard to disabled access, please contact us in advance of visiting to help us understand your specific needs regarding accessibility, and work out the best plan. Follow signs within grounds for disabled parking

closer to gardens. Deeper gravel on some paths. See website extended description re access.

♿ 🐎 🚗 🚌 ☕

GROUP OPENING

18 NEW **DORCHESTER GARDENS**
Fordington, Dorchester, DT1 1ED. *In the vicinity of Salisbury Fields, St Georges Church and Fordington Green, accessed by a lane leading to the large park, Salisbury Fields, between 10 &12 South Walks Rd. Follow signs to parking.* **Sun 2 May, Sun 13 June, Sun 11 July (2-5). Combined adm £4, chd free.**

> NEW **6 SOUTH WALKS ROAD**
> Paul Cairnes.

> NEW **8 SOUTH WALKS ROAD**
> Mrs Margaret Somerville.

> NEW **18 SOUTH WALKS ROAD**
> Sandra Manfield.

18 South Walks Road: Tiny Victorian front garden with olive trees and bulbs: back garden, a corridor of stunning Japanese acers and exotic trees leading into park. 6 South Walks Rd: Large mid C20 walled garden, with ongoing restoration. Paeonies, magnolias, climbing and shrub roses, Wisteria sinensis and spring bulbs. 8 South Walks Road: Small sheltered garden. Upper level Heptacodium miconoides; lower courtyard garden with mature climbers, incl Hydrangea seemanii. Contact: Margaret Somerville, msomerville162@gmail.com. Cream teas at the Bean on the Green. Follow signs.

🐑 🚗 ☕

19 **EAST END FARM**
Barkers Hill, Semley, Shaftesbury, SP7 9BJ. Celia & Piers Petrie. *Semley. From A350 take exit to Semley and continue to Church. Take turning next to church towards Tisbury then 1st R to Barkers Hill. Continue along lane (1m) continue up hill East End Farm on R.* **Sun 20 June (2-5). Adm £5, chd free. Home-made teas.**
Exquisite wildflower meadow started 20 yrs ago. 2 acres on spectacular site with views to Pyt House. Orchids, yellow rattle, grass vetchling, ragged robin, agrimony, meadow cranesbill, various vetches, amongst many other species. Very pretty and relaxed

garden with wild areas. Wheelchair access to garden only but views from garden to wildflower meadow.

♿ ☕

20 ◆ **EDMONDSHAM HOUSE**
Edmondsham, Wimborne, BH21 5RE. Mrs Julia Smith, 01725 517207, julia.edmondsham@homeuser.net. *9m NE of Wimborne. 9m W of Ringwood. Between Cranborne & Verwood. Edmondsham off B3081.* Wheelchair access and disabled parking at West front door. **For NGS: Mon 5 Apr (2-5). Every Wed 7 Apr to 28 Apr (2-5). Every Wed 6 Oct to 27 Oct (2-5). Adm £2.50, chd £0.50. Tea, coffee, cake and soft drinks 3.30 to 4.00 p.m. in Edmondsham House, Weds only. For other opening times and information, please phone or email. Donation to Prama Care.**
6 acres of mature gardens, grounds, views, trees, rare shrubs, spring bulbs and shaped hedges surrounding C16/C18 house, giving much to explore incl C12 church adjacent to garden. Large Victorian walled garden is productive and managed organically (since 1984) using 'no dig' vegetable beds. Wide herbaceous borders planted for seasonal colour. Traditional potting shed, cob wall, sunken greenhouse. Coaches by appointment only. Some grass and gravel paths.

♿ ✿ 🚗 ☕

21 **EDWARDSTOWE**
50-52 Bimport, Shaftesbury, SP7 8BA. Mike & Louise Madgwick. *Park in town's main car park. Walk along Bimport (B3091) 500mts, Edwardstowe last house on L.* **Sat 22, Sun 23 May (10.30-5), also open Rosebank. Sat 24, Sun 25 July (10.30-5). Adm £4, chd free.**
Parts of the garden were extensively remodelled during 2018, a new greenhouse and potting shed have been added, along with changing pathways and the vegetable garden layout. Long borders have been replanted in places with trees managed letting in more light.

🐑 ✿

SPECIAL EVENT

22 NEW ENCOMBE HOUSE

Kingston, Corfe Castle, Wareham, BH20 5LW. James & Arabella Gaggero, 01935 83652, alison.wright@ngs.org.uk. *Drive through Corfe turn R to Kingston. Turn R at The Scott Arms. Drive past church, 300yds 2nd turning L. Turn onto tarmac driveway signed Encombe. Follow signs for parking.* **Tue 15 June (11-3.30). Adm £120. Pre-booking essential, please visit www.ngs.org.uk/special-events for information & booking. Light refreshments. Coffee/tea on arrival and light buffet lunch.**
The historic Encombe House and Estate is nestled within a unique stunning valley in the Purbeck hills. The garden has been extensively redeveloped since 2009, with a modern, sympathetic design for the garden by Tom Stuart-Smith. The main garden to the south of the house includes large sweeping borders filled with grasses and perennials, alongside extensive lawns, lake and deep herbaceous beds. Partial wheelchair access only (gravel paths and steps).

23 NEW FALCONERS

89 High Street, Lytchett Matravers, Poole, BH16 6BJ. Hazel & David Dent. *6m W from Poole. Garden is past Village Hall at end of High Street, on L.* **Sun 9, Sun 16 May (11-4). Adm £3.50, chd free. Home-made teas.**
Behind the gate of 150 yr old Falconers Cottage lies a ¼ acre mature garden with some interesting plants. The characterful cottage and garden have been cared for and enhanced by the current custodians. As you wander round the many aspects, incl pond, herbaceous bed, climbers and vegetable plot you will find restful areas to sit and enjoy this garden in spring. Regret unsuitable for wheelchairs.

SPECIAL EVENT

24 FARRS

3, Whitcombe Rd, Beaminster, DT8 3NB. John Makepeace, 01308 862204, info@johnmakepeacefurniture.com, www.johnmakepeacefurniture.com. *Southern edge of Beaminster. On B3163. Car parking on site only for disabled. Enter through garden door in the wall adjacent to Museum. Park in the Square or side streets.* **Wed 23 June, Wed 15 Sept (2.30-4.30). Adm £40, chd free. Pre-booking essential, please visit www.ngs.org.uk/special-events for information & booking. Cream teas in the house. Visits also by arrangement May to Sept for groups of 10 to 30. Group visits by prior arrangement.**
Enjoy several distinctive walled gardens, rolling lawns, sculpture and giant topiary around one of Beaminster's historic town houses. John's inspirational grasses garden, Jennie's riotous potager with a cleft oak fruit cage. Glasshouse, straw bale studio, geese in orchard. Remarkable trees, planked and seasoning in open sided barn for future furniture commissions. House also open on NGS days with furniture by John Makepeace, and paintings, sculpture and applied arts by living artists. Talk on design at 2.30pm each opening. Jennie Makepeace will give plant talks at 3.30pm. Some gravel paths, alternative wheelchair route through orchard.

25 NEW 29 FLEET STREET

Beaminster, DT8 3EF. Mrs Jane Pinkster, 01308 863923, christine.corson@ngs.org.uk. *Short walk from 21A The Square and central car park.* **Visits by arrangement May to Sept for groups of 5 to 10. Combined opening with 21A The Square and Shadrack House. Adm £5, chd free.**
A walled town garden 72 ft sq, totally restored and replanted since 2010 with a mass of roses, clematis, penstemons and unusual perennials and shrubs. Maintained by owner with occasional outside help. 3rd time of opening. Easy wheelchair access but all gravel.

26 ♦ FORDE ABBEY GARDENS

Forde Abbey, Chard, TA20 4LU. Mr & Mrs Julian Kennard, 01460 221290, info@fordeabbey.co.uk, www.fordeabbey.co.uk. *4m SE of Chard. Signed off A30 Chard-Crewkerne and A358 Chard-Axminster. Also from Broadwindsor B3164.* **For opening times and information, please phone, email or visit garden website www.fordeabbey.co.uk.** 30 acres of fine shrubs, magnificent specimen trees, ponds, herbaceous borders, rockery, bog garden containing superb collection of Asiatic primulas, Ionic temple, working walled kitchen garden supplying the tearoom. Centenary fountain, England's highest powered fountain. Gardens open from Feb and house from 1 April Tues to Fri incl Suns & BH Mons but please see our website for updated information. Please ask at reception for best wheelchair route. Wheelchairs available to borrow/hire, advance booking 01460 221699.

27 NEW 5 FOSTERS MEADOWS

Wintorborno Whitoohuroh, Blandford Forum, DT11 0DW. John & Ann Somerville. *6m SW of Blandford Forum. From A354 turn into Whatcombe Lane by tel box, approx 150yds, R over bridge into Fosters Meadows (no parking, please park sensitively in village).* **Sun 20 June (11-4). Adm £4, chd free. Light refreshments.**
5 year old medium-sized level garden with gravel paths linking loose cottage-style planting. Well established plants providing structure and rich floral colour, incl persicaria, Japanese anemones, lilies and crocosmia. Small potager, patio and pergola with trelliswork filled with climbing roses, Clematis montana and Clematis armandii. No steps, gravel paths, easy access. Seating available. Some driveway parking.

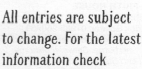

All entries are subject to change. For the latest information check ngs.org.uk

The Manor House, Beaminster

© Val Corbett

28 FRANKHAM FARM

Ryme Intrinseca, Sherborne, DT9 6JT. Susan Ross MBE, 07594 427365, neilandsusanross@gmail.com. *3m S of Yeovil. Just off A37 - turn next to Hamish's signed to Ryme Intrinseca, go over small bridge and up hill, drive is on L.* **Sun 14 Mar, Sun 18 Apr, Sun 6 June, Sun 10 Oct (12.30-5). Adm £6, chd free. Please see website for more information.** 3½ acre garden, created since 1960 by the late Jo Earle for yr-round interest. This large and lovely garden is filled with a wide variety of well grown plants, roses, unusual labelled shrubs and trees. Productive vegetable garden. Clematis and other climbers. Spring bulbs through to autumn colour, particularly oaks. Modern toilets incl disabled. Sorry, no dogs.

29 FRITH HOUSE

Stalbridge, DT10 2SD. Mr & Mrs Patrick Sclater, 01963 250809, rosalynsclater@btinternet.com. *5m E of Sherborne. Between Milborne Port and Stalbridge. From A30 1m, follow sign to Stalbridge. From Stalbridge 2m and turn W by PO.* **Visits by arrangement Apr to July for groups of 10+. Adm £5, chd**

free. **Home-made teas.**
Approached down long drive with fine views. 4 acres of garden around Edwardian house and self-contained hamlet. Range of mature trees, lakes and flower borders. House terrace edged by rose border and featuring Lutyensesque wall fountain and game larder. Well stocked kitchen gardens. Woodland walks with masses of bluebells in spring. Garden with pretty walks set amidst working farm.

30 GILLANS

Minterne Parva, nr Dorchester, DT2 7AP. Robert & Sabina ffrench Blake. *N of Cerne Abbas off A352, plenty of parking. Dogs on leads. Signed off main road - 200 yards to gate on L.* **Sun 2 May (2-6). Adm £5, chd free. Home-made teas.**
Hidden 5 acre garden set in deep valley beside ponds and River Cerne; surrounded by ancient trees with borrowed landscape. Spring bulbs and blossom with species magnolia and cornus. Carpets of primroses around camellias and rhododendrons. Steep paths, stout footwear recommended! Partially replanted over past 6yrs. Formal beds, pots, climbing roses, wisteria and productive veg garden surround

house. Wheelchair access around house.

31 THE GRANGE

Burton Street, Marnhull, Sturminster Newton, DT10 1PS. Francesca Pratt. *Marnhall. 8m SW from Shaftesbury take A30 and B3092, to Marnhull turn R into Sodom lane. From S B3092 to Marnhall turn L into Church St - follow NGS signs to The Grange.* **Sat 5, Sun 6 June (2-5). Adm £5, chd free. Home-made teas.**
Newly laid out garden, having lost a main feature - its 300 year old copper beech tree in 2019. The terrace has been expanded and borders replanted, flower beds surrounding main lawn have been re-configured and a number of areas of the woodland have been newly planted. Herb garden has also been expanded. In 2021, everything will look very different. Wheelchair access to lawn and woodland garden - 6 steps down to terrace.

32 GROVE HOUSE

Semley, Shaftesbury, SP7 9AP. Judy & Peter Williamson. *A350 turn to Semley. Grove House approx ½m on L just before railway bridge.*

From Semley village leave pub on R go under railway bridge (approx ⅓m) Grove House 2nd on R. **Sun 27 June (12-5). Adm £5, chd free. Home-made teas. Also open Semley Grange.**
Classic English garden divided into rooms. Yew hedges and topiary. Rose garden. Large herbaceous border. Late summer hot garden with grasses. Lawns. Mown walks through specimen trees.

Ġ ⚘ ♿

33 HANFORD SCHOOL
Child Okeford, Blandford Forum, DT11 8HN. Rory & Georgina Johnston. From Blandford take A350 to Shaftesbury. Approx 2m after Stourpaine take L turn for Hanford. From Shaftesbury take A350 to Poole. After Iwerne Courtney take next R to Hanford. **Sun 27 June (2-5). Adm £5, chd free.**
Perhaps the only school in England with a working kitchen garden growing quantities of seasonal vegetables, fruit and flowers for the table. The rolling lawns host sports matches, gymnastics, dance and plays while ancient cedars look on. The stable clock chimes the hour and the chapel presides over it all. Teas in Great Hall (think Hogwarts). What a place to go to school or visit. Several steps/ramp to main house. No wheelchair access to WC.

Ġ ♞ ⚘

34 HERONS MEAD
East Burton Road, East Burton, Wool, BH20 6HF. Ron & Angela Millington, 01929 463872, ronamillington@btinternet.com. 6m W of Wareham on A352. Approaching Wool from Wareham, turn R just before level Xing into East Burton Rd. Herons Mead ¾m on L. **Sun 21 Feb, Sun 21 Mar, Sun 4 Apr, Sun 2 May, Sun 19 Sept (2-5). Adm £3.50, chd free. Home-made teas. Visits also by arrangement Feb to Sept for groups of 10 to 30.**
½ acre plantlover's garden full of interest from spring (bulbs, many hellebores, pulmonaria, fritillaries) through abundant summer perennials, old roses scrambling through trees and late seasonal exuberant plants amongst swathes of tall grasses. Wildlife pond and plants to attract bees, butterflies, etc. Tiny woodland. Cacti. Small wheelchairs can gain partial access - as far as the tea house!

Ġ ♞ ✿ ⚘

35 NEW HIGHER BRIMLEY COOMBE FARM
Stoke Abbott, Beaminster, DT8 3JZ. Linda & Will Bowditch. 1m W of Stoke Abbott. On B3162, 2m S of Broadwinsor, 5.7m N of Bridport. **Sun 23, Tue 25 May, Thur 9, Sun 12 Sept (2-5). Adm £4, chd free. Cream teas.**
This is a new garden of just over an acre, planted over the last 5 yrs to give yr round interest. Rose and herbaceous borders surround the house and new prairie styled planting blends into the orchard where mown paths wind through the long grass. It is an open site at 500' on Lewesdon Hill with stunning views across Marshwood Vale to the sea. Garden to front is grass and flat, other areas are sloped so help with wheelchair may be required.

Ġ ⚘

36 6 HILLCREST ROAD
Weymouth, DT4 9JP. Helen & Michael Toft. 6 Hillcrest Road. Via Weymouth take A345 towards Portland, L at Rylands Lane, last L at bottom of hill 3rd Bungalow on R. Via Bridport take B13156 to W'mth follow signs to Portland, L at A345, R at Rylands Lane. **Sun 27 June, Sun 8 Aug (1-6). Adm £4, chd free. Tea and home-made cakes, gluten free catered for.**
A suburban garden with the sea at the end, allows the cultivation of some exotic plants inc Strelizia, banana and Cobaea scandens. Fish Pond, mature fruit trees, veg patch and soft fruit. Hot bed nearest the house is balanced by white bed at bottom of lawn. A mixed border of perennials and annuals runs down one side. Other beds with colour themes.

♞ ✿ ⚘

37 HILLTOP
Woodville, Stour Provost, Gillingham, SP8 5LY. Josse & Brian Emerson, www.hilltopgarden.co.uk. 7m N of Sturminster Newton, 5m W of Shaftesbury. On B3092 turn E at Stour Provost Xrds, signed Woodville. After 1¼m thatched cottage on R. On A30, 4m W of Shaftesbury, turn S opp Kings Arms. 2nd turning on R signed Woodville, 100 yds on L. **Sun 18, Sun 25 July, Sun 1, Sun 15, Sun 22 Aug (2-6). Adm £3.50, chd free. Home-made teas.**
Summer at Hilltop is a gorgeous riot of colour and scent, the old thatched cottage barely visible amongst the flowers. Unusual annuals and perennials grow alongside the traditional and familiar, boldly combining to make a spectacular display, which attracts an abundance of wildlife. Always something new, the unique, gothic garden too a great success.

♞ ✿ 🚗 ⚘

38 THE HOLLOW, BLANDFORD FORUM
Tower Hill, Iwerne Minster, Blandford Forum, DT11 8NJ. Sue Le Prevost. Between Blandford and Shaftesbury. Follow signs on A350 to Iwerne Minster. Turn off at Talbot Inn, continue straight to The Chalk, bear R along Watery Lane for parking in Parish Field on R. 5 min uphill walk to house. **Sat 19, Sun 20, Wed 23 June (2-5). Adm £3.50, chd free. Cream teas. Home-made cakes and gluten-free available.**
Hillside cottage garden built on chalk, about ⅓ acre with an interesting variety of plants in borders that line the numerous sloping pathways. Water features for wildlife and well placed seating areas to sit back and enjoy the views. Productive fruit and vegetable garden in converted paddock with raised beds and greenhouses. A high maintenance garden which is constantly evolving. Use of different methods to plant steep banks. Slopes, narrow gravel paths and steep steps so regret not suitable for wheelchairs or limited mobility.

✿ 🚗 ⚘

39 THE HOLLOW, SWANAGE
25 Newton Road, Swanage, BH19 2EA. Suzanne Nutbeem, 01929 423662, gdnsuzanne@gmail.com. ½m S of Swanage town centre. From town follow signs to Durlston Country Park. At top of hill turn R at red postbox into Bon Accord Rd. 4th turn R into Newton Rd. **Every Wed 7 July to 25 Aug (2-5.30). Adm £4, chd free. Visits also by arrangement July & Aug for groups of up to 30.**
Come and wander round a dramatic sunken former stone quarry, a surprising garden at the top of a hill above the seaside town of Swanage. Stone terraces, with many unusual shrubs and grasses, form a beautiful pattern of colour and foliage attracting happy butterflies and bees. Pieces of mediaeval London Bridge lurk in the walls. Steps have elegant handrails. WC available. Exceptionally wide range of plants including cacti and airplants.

♞

40 ◆ HOLME FOR GARDENS
West Holme Farm,
Wareham, BH20 6AQ. Simon
Goldsack, 01929 554716,
simon@holmefg.co.uk,
www.holmefg.co.uk. *Easy to find
on the B3070 road to Lulworth 2m
out of Wareham.* **For NGS: Tue 22
June, Tue 7 Sept (10-4.30). Adm
£5, chd £1. Light refreshments
in The Orchard Café. For other
opening times and information,
please phone, email or visit garden
website.**
Extensive formal and informal gardens
strongly influenced by Hidcote Manor
and The Laskett. The garden is made
up of distinct rooms separated by
hedges and taller planting. Extensive
collection of trees, shrubs, perennials
and annuals sourced from across the
UK. Spectacular wildflower meadows.
Grass amphitheatre, Holme henge
garden, lavender avenue, cutting
garden, pear tunnel, hot borders, white
borders, ornamental grasses, unusual
trees and shrubs. Grass paths are kept
in good order and soil is well drained
so wheelchair access is reasonable
except immediately after heavy rain.

41 22 HOLT ROAD
Branksome, Poole, BH12 1JQ.
Alan & Sylvia Lloyd, 01202 387509,
alan.lloyd22@ntlworld.com. *2½ m
W of Bournemouth Square, 3m E
of Poole Civic Centre. From Alder
Rd turn into Winston Ave, 3rd R
into Guest Ave, 2nd R into Holt
Rd, at end of cul de sac. Park in
Holt Rd or in Guest Ave.* **Visits by
arrangement Mar to Aug for
groups of 20+. Adm £3.50, chd
free. Home-made teas.**
¾ acre walled garden for all
seasons. Garden seating throughout
the diverse planting areas, incl
Mediterranean courtyard garden
and wisteria pergola. Walk up slope
beside rill and bog garden to raised
bed vegetable garden. Return
through shrubbery and rockery back
to waterfall cascading into a pebble
beach. Partial wheelchair access.

42 HOLWORTH FARMHOUSE
Holworth, Dorchester,
DT2 8NH. Anthony & Philippa
Bush, 01305 852242,
bushinarcadia@yahoo.co.uk. *7m E
of Dorchester. 1m S of A352. Follow
signs to Holworth. Through farmyard
with duckpond on R. 1st L after
200yds of rough track. Ignore No*

Access signs. **Sun 2, Mon 3, Sun
30, Mon 31 May, Sun 4 July (2-5).
Adm £5, chd free. Home-made
teas. Visits also by arrangement
May to Sept.**
Escape briefly from the cares and
pressures of the world. Visit this
unusual garden tucked away in an
area of extraordinary peace and
tranquility chosen by the Monks
of Milton Abbey. The garden
was created 35 years ago and is
constantly evolving. New projects
every year. Many mature and unusual
trees and shrubs, numerous borders
with seats to ponder and reflect.
Vegetables, water and wild spaces.
Beautiful unspoilt views. Many birds
and butterflies. Only partial wheelchair
access.

43 HORN PARK
Tunnel Rd, Beaminster,
DT8 3HB. Mr & Mrs David
Ashcroft, 01308 862212,
angieashcroft@btinternet.com.
*1½ m N of Beaminster. On A3066
from Beaminster, L before tunnel
(see signs).* **Wed 21 Apr, Wed 23
June (2.30-4.30). Adm £4.50, chd
free. Home-made teas. Visits also
by arrangement Apr to Oct for
groups of up to 30.**
Large plantsman's garden with
magnificent views over Dorset
countryside towards the sea. Many
rare and mature plants and shrubs
in terrraced, herbaceous, rock and
water gardens. Woodland garden
and walks in bluebell woods. Good
amount of spring interest with
magnolia, rhododendron and bulbs
which are followed by roses and
herbaceous planting, wildflower
meadow with 164 varieties incl
orchids. Only partial access for
wheelchair users, gravel paths and
some steep slopes.

44 IVY HOUSE GARDEN
Piddletrenthide, DT2 7QF.
Bridget Bowen, 07586 377675,
beepeebee66@icloud.com. *9m N
of Dorchester. On B3143. In middle
of Piddletrenthide village, opp Village
Stores near Piddle Inn.* **Sun 14, Sun
28 Mar, Sun 4, Mon 5, Sun 18 Apr,
Sun 2, Mon 3 May (2-5). Adm £5,
chd free. Home-made teas. Visits
also by arrangement Mar to May
for groups of 10+. Garden not
suitable for wheelchair users.**
A steep and challenging ½ acre
garden with fine views, set on

S-facing site in the beautiful Piddle
Valley. Wildlife-friendly garden with
mixed borders, ponds, propagating
area, large vegetable garden, fruit
cage, greenhouses and polytunnel,
chickens and bees, plus a nearby
allotment. Daffodils, tulips and
hellebores in quantity for spring
openings. Come prepared for steep
terrain and a warm welcome! We
hope to hold a live music event in the
garden sometime during the last w/e.
Run on organic lines with plants to
attract birds, bees and other insects.
Insect-friendly plants usually for sale.
Honey and hive products available
and, weather permitting, observation
hive of honey bees in courtyard.
Beekeeper present to answer queries!

**45 NEW JULIA'S HOUSE
CHILDREN'S HOSPICE**
135 Springdale Road, Broadstone,
BH18 9BP. Liz Thompson. *From
Broadstone turn L at T-lights on
Lower Blandford Rd. ½ m on L.
Parking not available on site.* **Sat
3, Sun 4 July (10-3.30). Adm £4,
chd free. Proceeds from plant
sale and refreshments to Julia's
House.**
Julia's House Children's Hospice
provides respite care for children with
life-threatening or limiting conditions.
Our garden has been designed so
the children can experience and enjoy
different sensory elements and is
cared for by our volunteer gardeners.
Explore how our children use the
garden. Tours of the hospice may be
available.

46 ◆ KEYNESTON MILL
Tarrant Keyneston, Blandford
Forum, DT11 9HZ. Julia &
David Bridger, 01258 786022,
events@keynestonmill.com,
www.keynestonmill.com. *From
Blandford or Wimborne take B3082.
Turn into Tarrant Keynston village and
continue right through to Xrds, we
are straight ahead.* **For NGS: Tue
25 May, Tue 14 Sept (2-5.30). Adm
£5, chd free. Tea and home-made
cake on NGS days. For other
opening times and information,
please phone, email or visit garden
website.**
Keyneston Mill is the creative home
of Parterre Fragrances - a 50 acre
working estate dedicated to fragrant
and aromatic plants. Enjoy the
gardens, each compartment featuring
plants from a different perfume family

eg. floral, fern and citrus. Enjoy a walk around the river meadow and the perfume crop fields where we grow the ingredients for our perfumes, and see the exhibition and distillery. Open all yr, please see above website for details. Compacted gravel paths in floral garden, lawns elsewhere. Wheelchair access to bistro-café and WCs. Dogs on leads welcome in the gardens.

& 🐕 ✿ ☕

47 ♦ KINGSTON LACY

Wimborne Minster, DH21 4EA. National Trust, 01202 883402, kingstonlacy@nationaltrust.org.uk, www.nationaltrust.org.uk/kingston-lacy. *2½ m W of Wimborne Minster. On Wimborne-Blandford rd B3082.* **For opening times and information, please phone, email or visit garden website.**
35 acres of formal garden, incorporating parterre and sunk garden planted with Edwardian schemes during spring and summer. 5 acre kitchen garden and allotments. Victorian fernery containing over 35 varieties. Rose garden, mixed herbaceous borders, vast formal lawns and Japanese garden restored to Henrietta Bankes' creation of 1910. 2 National Collections: Convallaria and Anemone nemorosa. Snowdrops, blossom, bluebells, autumn colour and Christmas light display. Deep gravel on some paths but lawns suitable for wheelchairs. Slope to visitor reception and South lawn. Dogs allowed in some areas of woodland.

& ✿ 🚌 NPC ☕

48 ♦ KINGSTON MAURWARD GARDENS AND ANIMAL PARK

Kingston Maurward, Dorchester, DT2 8PY. Kingston Maurward College, 01305 215003, events@kmc.ac.uk, *www.morekmc.com. 1m E of Dorchester. Off A35. Follow brown Tourist Information signs.* **For opening times and information, please phone, email or visit garden website.**
Stepping into the grounds you will be greeted with 35 impressive acres of formal gardens. During the late spring and summer months, our National Collection of penstemons and salvias display a lustrous rainbow of purples, pinks, blues and whites, leading you on through the ample hedges and stonework balustrades. An added treat is the Elizabethan walled garden, offering a new vision of enchantment. Open early Jan to mid Dec. Hours will vary in winter depending on conditions - check garden website or call before visiting. Partial wheelchair access only, gravel paths, steps and steep slopes. Map provided at entry, highlighting the most suitable routes.

& 🚌 ☕

17 Lower Golf Links Road

Yardes Cottage

49 KNITSON OLD FARMHOUSE
Corfe Castle, Wareham, BH20 5JB.
Rachel Helfer, 01929 421681,
rjehelfer@gmail.com. *Purbeck.
Between Corfe Castle and Swanage.
Follow the A351 3m E from Corfe
Castle. Turn L signed Knitson.
After 1m fork R. We are on L after
¼ m.* **Wed 7, Wed 14 Apr, Wed
5, Wed 12 May, Wed 2, Wed 9
June, Wed 7, Wed 14 July, Wed 4,
Wed 11 Aug, Wed 1, Wed 8 Sept
(12-5). Adm £4, chd free. Light
refreshments. Sweet and savoury
snacks and hot/cold drinks. Visits
also by arrangement Mar to Nov
for groups of up to 30. Various
refreshments and/or talks can be
organised.**
Mature cottage garden nestled at
base of chalk downland. Herbaceous
borders, rockeries, climbers and
shrubs, evolved and designed over
55yrs for yr-round colour and interest.
Large wildlife friendly kitchen garden
for self sufficiency. Historical stone
artefacts are used in the garden
design. We have used a lot of
local stone in the design and have
interesting old stones and stone baths
around the garden. Ancient trees
and shrubs are retained as integral
to garden design. Over 20 varieties
of fruits and many vegetables all yr
round are the basis for a sustainable
lifestyle. Garden is on a slope, main
lawn and tea area are level but there
are uneven, sloping paths.

50 ♦ KNOLL GARDENS
Hampreston, Wimborne,
BH21 7ND. Mr Neil Lucas,
01202 873931,
enquiries@knollgardens.co.uk,
www.knollgardens.co.uk. *2½ m
W of Ferndown. ETB brown signs
from all directions approaching
Wimborne. Large car park.* **For
NGS: Fri 25 June, Fri 24 Sept
(10-5). Adm £6.95, chd £4.95.
Light refreshments. Self catering
drinks and pre-wrapped cakes.
For other opening times and
information, please phone, email or
visit garden website.**
A naturalistic and calming garden,
renowned for its whispering
ornamental grasses, it surprises
and delights with an abundance of
show-stopping flowering perennials.
A stunning backdrop of trees and
shrubs add drama to this wildlife
and environmentally-friendly garden
that works with nature. An onsite
nursery, expert advice on hand,
provides opportunity to replicate
a little bit of Knoll at home. Year
round event programme includes
grass masterclasses, naturalistic
design workshops and creative
courses. Events are bookable
online at knollgardens.co.uk The
garden's charity the Knoll Gardens
Foundation researches and promotes
sustainable, wildlife friendly gardening:
knollgardensfoundation.org. Some
slopes. Various surfaces incl gravel,
paving, grass and bark.

51 KNOWLE COTTAGE
No.1, Shorts Lane, Beaminster,
DT8 3BD. Claire & Guy Fender.
*Near St Mary's church, Beaminster.
Park in main square of Beaminster or
in main town car park and walk down*

Church Lane and R onto single track Shorts Lane. Entry is through brown gates after 1st cottage on L. **Sat 27 Mar, Sat 5 June (11-3.30). Adm £4.50, chd free.**

Large 1½ acre garden with 35m long S-facing herbaceous border with yr round colour. Formal rose garden within a circular floral planting is bound on 3 sides by lavender. Small orchard and vegetables in raised beds in adjacent walled area, with whole garden leading to small stream, and bridge to pasture. Beds accessed from level grass, slope not suitable for wheelchairs. Picnickers welcome. Limited outside seating available. No dogs.

&♿ ☕ 🏕

52 LAWSBROOK
Brodham Way, Shillingstone, DT11 0TE. Clive Nelson, 07771 658846, cne70bl@aol.com, www.facebook.com/Lawsbrook. 5m NW of Blandford. To Shillingstone on A357. Turn up Gunn Lane - up lane (past Wessex Avenue on L & Everetts Lane on R) then turn R at next opportunity as road bends to L. Lawsbrook 250m on R. **Daily Sun 14 Feb to Sun 21 Feb (10-4), Sun 31 Oct, Sun 7 Nov (10-4). Adm £4, chd free. Light refreshments. Visits also by arrangement Feb to Nov.**

Open for over 10 yrs for NGS. Garden incl over 130 different tree species spread over 6 acres. Native species and many unusual specimens incl Dawn redwood, Damyio oak and Wollemi pine. Large garden, raised bed vegetable and flower garden. Wildlife, stream and meadow. You can be assured of a relaxed and friendly day out. Children's activities, all day teas/cakes and dogs are very welcome. Large scale snowdrop days in Feb and intense autumn hues in Nov. Enough space and interest for everyone. Gravel path at entrance, grass paths over whole garden.

&♿ 🐕 🌼 🚗 ☕ 🏕

53 NEW LEWELL LODGE
West Knighton, Dorchester, DT2 8RP. Rose & Charles Joly. 3m SE of Dorchester. Turn off A35 onto A352 towards Wareham, take West Stafford bypass, turn R for West Knighton at T junction. ¾m turn R up drive after village sign. **Fri 11, Fri 25 June (1-5). Adm £6, chd free. Home-made teas.**

Elegant 2 acre classic English garden designed by present owners over

last 25 yrs, surrounding Gothic Revival house. Double herbaceous borders enclosed by yew hedges with old fashioned roses. Shrub beds edged with box and large pyramided hornbeam hedge. Crab apple tunnel, box parterre and pleached hornbeam avenue. Large walled garden, many mature trees and woodland walk. Garden is level but parking is on gravel and there are gravel pathways.

&♿ ☕

54 NEW LITTLE BENVILLE HOUSE
Benville Lane, Corscombe, Dorchester, DT2 0NN. Jo & Gavin Bacon. 2½m (6mins) from Evershot village on Benville Lane. Benville may be approached from A37, via Evershot village. House is on L ½m after Benville Bridge. Alternatively from A356, Dorchester to Crewkerne road, 1m down on R. **Sat 15, Sun 16 May (9.30-5). Adm £8, chd £4. Home-made teas/juice on kitchen terrace or in Barn if wet. Egg sandwiches with local Evershot Bakery bread at lunch time. Evening opening Fri 21 May (5.30-9). Adm £12, chd £0. Light refreshments and wine on kitchen terrace or in Barn if wet.**

Contemporary garden, with landscape interventions by Harris Bugg Studio within a varied ecological AONB and historic landscape off Benville Lane, mentioned in Thomas Hardy's Tess. Within the curtilage there are new herbaceous borders, woodland planting, walled vegetable and cutting garden, cloud pruned topiary, ha-ha, ornamental and productive trees and moat which is a listed Ancient Monument. Bring a tennis racquet and appropriate footwear and try the tennis court and enjoy the view.

🌼 ☕

55 LITTLE CLIFF
Sidmouth Road, Lyme Regis, DT7 3EQ. Mrs Debbie Bell. Edge of Lyme Regis. Turn off A35 onto B3165 to Lyme Regis. Through Uplyme to mini r'about by Travis Perkins. 3rd exit on R, up to fork and L down Sidmouth Rd following NGS arrows from mini r'about. Garden on R. **Sun 20 June (1-5). Adm £4, chd free. Home-made teas.**

S-facing seaward, Little Cliff looks out over spectacular views of Lyme Bay. Spacious garden sloping down hillside through series of garden rooms where visual treats unfold.

Vibrant herbaceous borders incl hot garden and bog garden intermingled with mature specimen trees, shrubs and wall climbers. Jungle walkway at bottom of garden, Indian influenced pavilion leading to white borders and vegetable garden. New hidden garden amongst the palms and rich colours of the hot garden, and new Mediterranean terrace garden at top by house. Steep slopes.

🌼 ☕

56 ◆ LITTLEBREDY WALLED GARDENS
Littlebredy, DT2 9HL. The Walled Garden Workshop, 01305 898055, secretary@wgw.org.uk, www.littlebredy.com. 8m W of Dorchester. 10m E of Bridport. 1½m S of A35. NGS days: park on village green then walk 300yds. For the less mobile (and on normal open days) use gardens car park. **For NGS: Tue 22, Tue 29 June (2-5.30). Adm £5, chd free. Home-made teas. For other opening times and information, please phone, email or visit garden website.**

1 acre walled garden on S-facing slopes of Bride River Valley. Herbaceous borders, riverside rose walk, lavender beds and potager vegetable and cut flower gardens. Original Victorian glasshouses, one under renovation. Partial wheelchair access, some steep grass slopes. For disabled parking please follow signs to main entrance.

&♿ 🐕 🌼 ☕

57 LOWER ABBOTTS WOOTTON FARM
Whitchurch Canonicorum, Bridport, DT6 6NL. Johnny & Clare Trenchard. 6m W of Bridport. Well signed from A35 at Morecombe Lake (2m) and Bottle Inn at Marshwood on B3165 (1.5m). Some disabled off-road parking. **Sat 3, Sun 4, Wed 7 July (2-5). Adm £5, chd free. No teas on Wed 7 July, but picnickers welcome.**

Sculptor owner reflects her creative flair in garden form, shape and colour. New open gravel garden contrasts with main garden consisting of lawns, borders and garden rooms, making a perfect setting for sculptures. The naturally edged pond provides a tranquil moment of calm, but beware of being led down the garden path by the running hares! Partial wheelchair access.

🐕 🌼 ☕ 🏕

58 17 LOWER GOLF LINKS ROAD
Broadstone, Poole, BH18 8BQ.
Dr & Mrs Nicholas Dunn. *½m N of Broadstone centre. Approaching from Gravel Hill, along Dunyeats Rd, Lower Golf Links Road is 2nd turn on R, past the Middle School.* **Sat 24, Sun 25 Apr, Sat 22, Sun 23 May, Fri 11, Sat 12 June (1-5). Adm £4, chd free. Home-made teas.**
Town garden of ²/₃ acre, created in a heathland suburb. Originally mainly acid-loving plants, now, after much clearance and soil enrichment, plants for all seasons. Well organised vegetable garden with brick borders to beds, chickens, fruit trees and pond as well as borders. Something of interest all yr round, but particularly impressive in spring and early summer. Camellias, azaleas and rhododendrons; roses and pergola leading to pond. Gravel drive and stone steps. Garden on a slight slope.
&. ✿ 🚗 🍵 🎋

59 MANOR FARM, HAMPRESTON
Wimborne, BH21 7LX. Guy & Anne Trehane, 01202 574223, anne.trehane@live.co.uk. *2½m E of Wimborne, 2½m W of Ferndown. From Canford Bottom r'about on A31, take exit B3073 Ham Lane. ½m turn R at Hampreston Xrds. House at bottom of village.* **Sat 27 Feb (10-1); Sun 28 Feb (1-4); Sun 23 May, Sun 18 July, Sun 15 Aug (1-5). Adm £5, chd free. Soup also available at Feb openings. Visits also by arrangement Feb to Aug for groups of 10 to 30.**
Traditional farmhouse garden designed and cared for by 3 generations of the Trehane family through over 100yrs of farming and gardening at Hampreston. Garden is noted for its herbaceous borders and rose beds within box and yew hedges. Mature shrubbery, water and bog garden. Open for hellebores and snowdrops in Feb. Dorset Hardy Plant Society sales at openings. Hellebores for sale in Feb.
&. ✿ 🍵 🍵

60 NEW MANOR FARM, STOURTON CAUNDLE
Stourton Caundle, DT10 2JW.
Mr & Mrs O S L Simon. *6m E of Sherborne, 4 m W of Sturminster Newton. From Sherborne take A3030. At Bishops Caundle, L signed Stourton Caundle. After 1½m, L opp Trooper Inn in middle*

of village. **Sat 29, Sun 30 May (2-5.30). Combined adm with Wagtails £10, chd free. Home-made teas.**
C17 farmhouse and barns with walled garden in middle of village. Mature trees, shrubberies, herbaceous borders, lakes and vegetable garden. Lovingly created over last 50 yrs by current owners. Wheelchair access to lower areas of garden, steps to top areas of garden.
&. 🍵

61 MANOR HOUSE FARM
Ibberton, Blandford Forum, DT11 0EN. Fiona Closier, www.instagram.com/thefloristwithin/?hl=en. *Off A357 between Blandford Forum and Sturminster Newton. From A357 at Shillingstone take road to Okeford Fitzpaine. Follow signs for Belchalwell and Ibberton. Continue through Belchalwell to Ibberton.* **Sun 23 May, Sun 20 June (2-5). Adm £5, chd free. Home-made teas.**
Set in the lee of Bulbarrow Hill, this 1½ acre partially terraced garden has formal yew and beech hedging interspersed with topiary, lawn, and many abundantly filled herbaceous borders. Native trees, 2 spring-fed ponds attracting wildlife, and walled kitchen garden also give many areas of interest. Orchard area left to meadow with spring bulbs. Partial wheelchair access in dry weather due to steps and steep grass areas. No ground floor level disabled toilet.
&. ✿ 🚗 🍵

62 THE MANOR HOUSE, BEAMINSTER
North St, Beaminster, DT8 3DZ.
Christine Wood. *200yds N of town square. Park in the square or public car park, 5 mins walk along North St from the square. Limited disabled parking on site.* **Sun 2, Mon 3, Sun 30, Mon 31 May, Sun 20, Sat 26 June (11-5). Adm £6, chd free. Home-made teas in Coach House garden. Sorry, no teas on 2/3 May, do bring a picnic.**
Set in heart of Beaminster, 16 acres of stunning parkland with mature specimen trees, lake and waterfall, The Manor House looks forward to welcoming visitors after a 3 year break from opening it's grounds. This peaceful garden is a haven for wildlife with a woodland walk, wildflower meadow and walled garden 'serendipity'. Ornamental ducks, black swans, pigmy goats, chickens

and guinea pigs. Partial wheelchair access.
&. 🐎 ✿ 🍵 🎋

63 ◆ MAPPERTON GARDENS
Mapperton, Beaminster, DT8 3NR. The Earl & Countess of Sandwich, 01308 862645, office@mapperton.com, www.mapperton.com. *6m N of Bridport. Off A356/A3066. 2m SE of Beaminster off B3163.* **For opening times and information, please phone, email or visit garden website.**
Terraced valley gardens surrounding Tudor/Jacobean manor house. On upper levels, walled croquet lawn, orangery and Italianate formal garden with fountains, topiary and grottoes. Below, C17 summerhouse and fishponds. Lower garden with shrubs and rare trees, leading to woodland and spring gardens. Partial wheelchair access (lawn and upper levels).
&. ✿ 🚗 🍵

64 MAYFIELD
4 Walford Close, Wimborne Minster, BH21 1PH. Mr & Mrs Terry Wheeler, 01202 849838, terpau@talktalk.net. *½m N of Wimborne Town Centre. B3078 out of Wimborne, R into Burts Hill, 1st L into Walford Close.* **Sun 9, Wed 12, Sun 23, Mon 31 May, Sun 6 June (2-5). Adm £3.50, chd free. Home-made teas. Visits also by arrangement May & June for groups of 5 to 30.**
Town garden of approx ¼ acre. Front: formal hard landscaping planted with drought-resistant shrubs and perennials. Back garden contrasts with a seductive series of garden rooms containing herbaceous perennial beds separated by winding grass paths and rustic arches. Pond, vegetable beds and greenhouses containing succulents and vines. Garden access is across a pea-shingle drive. If this is manageable, wheelchairs can access the back garden provided they are no wider than 65cms.
&. 🐎 ✿ 🚗 🍵

65 THE MILL HOUSE
Crook Hill, Netherbury, DT6 5LX.
Mike & Shirley Bennett, 01308 488252, enquiries@ themillhousenetherbury.com, www.themillhousenetherbury.com. *1m S of Beaminster. Turn R off A3066 Beaminster to Bridport rd*

signed Netherbury. Car park in field at Xrds at bottom of hill. **Wed 3, Thur 4 Mar (11-2). Adm £6, chd £3.**

6½ acres of garden next to R Brit, incl mill house, mill stream and pond. Extensive formal garden with beds, lawns, terraced areas, stunning walled garden and vegetable patch. Explore wild areas and orchard. Rare and interesting trees incl magnolias, conifers, eucalyptus and fruit trees. The Mill House garden is a haven for wildlife - you may be lucky to see a kingfisher or two. Children to be supervised at all times due to river and mill pond. Partial wheelchair access due to some steps.

   ♿ 🍴 🏕

66 ◆ MINTERNE GARDEN
Minterne House, Minterne Magna, Dorchester, DT2 7AU. Lord & Lady Digby, 01300 341370, enquiries@minterne.co.uk, www.minterne.co.uk. *2m N of Cerne Abbas. On A352 Dorchester-Sherborne rd.* **For opening times and information, please phone, email or visit garden website.**
As seen on BBC Gardeners' World and voted one of the 10 prettiest gardens in England by The Times. Famed for their display of rhododendrons, azaleas, Japanese cherries and magnolias in April/May. Small lakes, streams and cascades offer new vistas at each turn around the 1m horseshoe shaped gardens covering 23 acres. The season ends with spectacular autumn colour. Snowdrops in Feb. Spring bulbs, blossom and bluebells in April. Garden at its peak in April/May with historic rhododendron collection, magnolias and azaleas. Over 200 acers provide spectacular autumn colour in Sept/Oct. Regret unsuitable for wheelchairs.

   🐕 🚌 🍵

67 OLD DOWN HOUSE
Horton, Wimborne, BH21 7HL. Dr & Mrs Colin Davidson, 07765 404248, pipdavidson59@gmail.com. *7½m N of Wimborne. Horton Inn at junction of B3078 with Horton Rd, pick up yellow signs leading up through North Farm. No garden access from Matterley Drove. 5min walk to garden down farm track.* **Wed 2, Sat 5, Sun 6 June (2-5). Adm £3.50, chd free. Home-made teas in comfortable garden room if weather inclement. Visits also**

by arrangement May to July.
Nestled down a farm track, this ¾ acre garden on chalk, surrounds C18 farmhouse. Stunning views over Horton Tower and farmland. Cottage garden planting with formal elements. Climbing roses clothe pergola and house walls along with stunning wisteria sinensis and banksian rose. Part walled potager, well stocked. Chickens. Hardy Plant Society plant stall. Not suitable for wheelchairs.

   🌸 🍵

68 THE OLD RECTORY, LITTON CHENEY
Litton Cheney, Dorchester, DT2 9AH. Richard & Emily Cave, 01308 482266, emilycave@rosacheney.com. *9m W of Dorchester. 1m S of A35, 6m E of Bridport. Small village in the beautiful Bride Valley. Park in village and follow signs.* **Sun 2, Wed 5 May, Sun 13 June (2-5.30). Adm £6, chd free. Home-made teas. Visits also by arrangement May to Oct.**
Steep paths lead to beguiling 4 acres of natural woodland with many springs, streams, 2 pools one a natural swimming pool planted with native plants. Front garden with pleached crabtree border, topiary and soft planting incl tulips, peonies, roses and verbascums. Walled garden with informal planting, kitchen garden, orchard and 350 rose bushes for a cut flower business. Formal front garden designed by Arne Maynard. Not suitable for wheelchairs.

   🐕 🍵 🏕

69 THE OLD RECTORY, MANSTON
Manston, Sturminster Newton, DT10 1EX. Andrew & Judith Hussey, 01258 474673, judithhussey@hotmail.com. *6m S of Shaftesbury, 2½m N of Sturminster Newton. From Shaftesbury, take B3091. On reaching Manston, past Plough Inn, L for Child Okeford on R-hand bend. Old Rectory last house on L.* **Visits by arrangement May to Sept for groups of 5+. Adm £6, chd free. Home-made teas.**
Beautifully restored 5 acre garden. S-facing wall with 120ft herbaceous border edged by old brick path. Enclosed yew hedge flower garden. Wildflower meadow marked with mown paths and young plantation of mixed hardwoods. Well maintained walled Victorian kitchen garden with new picking flower section. Large

new greenhouse also installed. Knot garden now well established.

   ♿ 🌸 🚌 🍵

70 THE OLD RECTORY, PULHAM
Dorchester, DT2 7EA. Mr & Mrs N Elliott, 01258 817595, gilly.elliott@hotmail.com. *13m N of Dorchester. 8m SE of Sherborne. On B3143 turn E at Xrds in Pulham. Signed Cannings Court.* **Sun 9, Thur 13 May, Sun 1, Thur 5 Aug (2-5). Adm £6, chd free. Home-made teas. Visits also by arrangement May to Sept for groups of 10+.**
4 acres formal and informal gardens surround C18 rectory, splendid views. Yew pyramid allées and hedges, circular herbaceous borders with late summer colour. Exuberantly planted terrace, purple and white beds. Box parterres, mature trees, pond, fernery, ha-ha, pleached hornbeam circle. 10 acres woodland walks. Flourishing extended bog garden with islands; awash with primulas and irises in May. Interesting plants for sale. Mostly wheelchair access.

   ♿ 🐕 🌸 🚌 🍵

71 THE OLD SCHOOL HOUSE
The Street, Sutton Waldron, Blandford Forum, DT11 8NZ. David Milanes. *Turn into Sutton Waldron from A350, continue for 300 yds, 1st house on L in The Street. Entrance past house through gates in wall.* **Wed 16, Sun 20 June (2-5). Adm £3.50, chd free. Home-made teas.**
Small village garden laid out in last 7 yrs with planting of hedges into rooms incl orchard, secret garden and pergola walkway. Recent addition of raised beds for growing vegetables. Strong framework of existing large trees, beds are mostly planted with roses and herbaceous plants. Pleached hornbeam screen. A designer's garden with interesting semi-tender plants close to house. Level lawns.

   ♿ 🐕 🌸 🍵

72 THE OLD VICARAGE
East Orchard, Shaftesbury, SP7 0BA. Miss Tina Wright, 01747 811744, tina_lon@msn.com. *4½m S of Shaftesbury, 3½m N of Sturminster Newton. Between 90 degree bend and lay-by with defibrillator red phone box. Parking is on opp corner towards Hartgrove.* **Fri 12, Sun 14 Feb (1.30-4.30); Sun**

Manor House Farm

© Rachael Brewer

21 Mar, Sun 18 Apr, Sun 23 May, Sun 15 Aug (2-5). Adm £4, chd free. Teas will be inside if raining, with wood stove if very cold. Visits also by arrangement. Can give a talk about bats if wanted. Advice when booking. Teas £4.
1.7 acre wildlife garden with lots of different snowdrops and other winter flowering shrubs. Followed by other bulbs incl hundreds of narcissi and tulips, followed by camassias and alliums. A wonderful stream meanders down to a pond and there are lovely reflections from the swimming pond. Tree viewing platform allows you to look over garden and to the wider area. Grotto, old Victorian man pushing his lawn mower his owner purchased brand new in 1866. Tree viewing platform. First swimming pond built in Dorset. Not suitable for wheelchairs if very wet.

73 PHILIPSTON HOUSE
Winterborne Clenston, Blandford Forum, DT11 0NR. Mark & Ana Hudson. *Off road between Winterborne Whitechurch and Winterborne Stickland. 2 km N of Winterborne Whitechurch and 1 km S of Crown Inn in Winterborne Stickland. Park in signed track/field off road, near Bourne Farm Cottage. Enter garden from field.* **Sat 12, Sat 26 June (2-6). Adm £5, chd free. Cream tea and/or cake will be served for a £5 charge - please bring £5 cash. Social-distanced seating will be provided.**
Charming 2 acre garden with lovely views in the Clenston Valley. Many unusual trees, rambling roses, wisteria, mixed borders and shrubs. Rose parterre, walled garden, swimming pool garden, vegetable garden. Winterbourne stream with bridge over to wooded shady area

with cedar-wood pavilion. Orchard with mown paths planted with spring bulbs. Sorry, no toilet available. No dogs except on leads, and please clear up. Wheelchair access good providing it is dry. Plants for sale on at least one of the open days.

74 PILSDON VIEW
Junction Butts Lane and Pitman's Lane, Ryall, Bridport, DT6 6EH. D Lloyd. *5m W of Bridport, through Chideok. From E through Morecombelake A35. Take the Ryall turning opp Felicity's farm shop on A35. Garden 3/4 m on L at junction Butts Lane/Pitmans Lane. High hedge with PO box in wall.* **Sat 22, Sun 23, Mon 24 May (12-5). Combined adm with Well Cottage £7, chd free. Teas and tickets at Well Cottage.**
Started over 25 yrs ago, the hard

landscaping provides different levels with breathtaking views over the Marshwood Vale towards Pilsdon Pen. Mature copper beech and evolving garden gives all yr round interest. Water features with wildlife add to the essence of the garden. We have now added a rose garden. 2nd in the large garden contest Melplash Show 2019. Partial wheelchair access.

75 NEW 1 PINE WALK
Lyme Regis, DT7 3LA. Mrs Erika Savory. *At end of Pine Walk through Holmbush Car Park at top of Cobb Road. Please park in Holmbush Car Park.* **Fri 3, Sat 4, Sun 5 Sept (11-4). Adm £5, chd £1. Light refreshments. Also several cafés and hotel in Lyme Regis.**
Unconventional ½ acre, multi level garden above Lyme Bay, adjoining NT's Ware Cliffs. Abundantly planted with an exotic range of shrubs, cannas, gingers and magnificent ferns. Apart from a rose and hydrangea collection, planting reflects owner's love of Southern Africa incl staggering succulents and late summer colour explosion featuring drifts of salvias, dahlias, asters, grasses and rudbeckia.

76 NEW THE PINES
15 Longacre Drive, Ferndown, BH22 9EE. Ian Gallimore. *½ m from centre of Forndown. R off A348 towards Poole, Longacre drive is almost opp M&S Foodhall.* **Sat 24 Apr, Sun 2 May (1-5). Adm £4, chd free. Light refreshments in aid of Cats Protection.**
Suburban garden. Heavily planted with mostly perennials, trees and exotic type plants to give a secluded and private feel. 2 ponds, seating areas and outdoor kitchen/BBQ area. Plants for sale.

77 ◆ PRIEST'S HOUSE MUSEUM & GARDEN
23-27 High Street, Wimborne, BH21 1HR. Priest's House Museum Trust, 01202 882533, museum@priest-house.co.uk, www.priest-house.co.uk. *Wimborne town centre. Wimborne is just off A31. From W take B3078, from E take B3073 towards town centre. From Poole and Bournemouth enter town from S*

on A341. **For opening times and information, please phone, email or visit garden website.**
Discover a real gem tucked away in the centre of Wimborne. The walled garden, with its path leading from the back door to the mill stream, is 100 metres long. Colourful herbaceous borders and old varieties of apple and pear trees line the path further down. The garden is sheltered by brick walls, which mark ancient property boundaries and a medieval burgage plot. Groups are asked to pre-book. Wheelchair access throughout the ground floor, garden and tearoom.

78 PUGIN HALL
Rampisham, nr Dorchester, DT2 0PR. Mr & Mrs Tim Wright, 01935 83652, wright.alison68@yahoo.com. *Near centre of village. NW of Dorchester. From Dorchester A37 Yeovil, 9m L Evershot, follow signs. From Crewkerne A356, take 1st L to Rampisham, Pugin Hall is on L after ½ m.* **Visits by arrangement Apr to Sept. Adm £6, chd £3. Home-made teas at Pugin Hall, refreshments such as canapes and early evening wine can be provided if required. A special double garden visit can be organised.**
Pugin Hall was once Rampisham Rectory, designed in 1847 by Augustus Pugin, who also helped to design the interior of the Houses of Parliament. A grade 1 listed building, it is surrounded by 4½ acres of garden, including a large front lawn with rhododendrons and perennial borders, a walled vegetable, fruit and cut flower garden, orchard and beyond the river Frome a woodland walk. Partial wheelchair access, gravel driveway and steps to terraced lawns.

79 2 PYES PLOT
St. James Road, Netherbury, Bridport, DT6 5LP. Sarah Porter Martyn Lock. *2m SW of Beaminster. Turn off A3066. Go over R Brit into centre of village. L into St James Rd, signed to Waytown, R corner Hingsdon Lane.* **Sun 30 May (1-5). Adm £4, chd free. Light refreshments.**
Small but perfectly formed front and back courtyard garden, created from new in 2007. Cream walls and black paintwork make a striking

framework for softer planting. Climbing plants, foliage and running water feature enhance the tranquil feel to this space, which uses every inch creatively.

80 Q
113 Bridport Road, Dorchester, DT1 2NH. Heather & Chris Robinson, 01305 263088, hmrobinson45@gmail.com. *Approx 300m W of Dorset County Hospital. From Top o' Town r'about head W towards Dorset County Hospital, Q 300 metres further on from Hospital on R.* **Visits by arrangement Mar to July for groups of 10 to 30. Adm £3.50, chd free. Can provide light refreshment if requested early.**
Q is essentially all things to all men, a modern cottage town garden, divided into rooms, with many facets, jam packed with shrubs, trees, climbers and herbaceous plants. Gazebo, statutes, water, bonsai and topiary. Planting reflects the owners' many and varied interest incl over 100 clematis plus many varied and unusual spring bulbs from early snowdrops. Herbaceous plants incl hellebores, clematis, daphne. Vegetable garden and fruit trees dotted around the garden. Small number of paths available for wheelchair users.

GROUP OPENING

81 RAMPISHAM GARDENS
Dorchester, DT2 0PU. *11m NW of Dorchester. From Dorchester A37 Yeovil, 9m L to Evershot. From Yeovil A37 Dorchester, 7m R Evershot, follow signs. From Crewkerne A356, 1½ m after Rampisham Garage L to Rampisham.* **Thur 24 June (1.30-5). Combined adm £8, chd free. Home-made teas at Broomhill.**

BROOMHILL
Mr & Mrs D Parry.
(See separate entry)

THE CURATAGE
Mr & Mrs Tim Hill.

ROSEMARY COTTAGE
Lady Ford.

This beautiful historic village hosts a wide variety of gardens with 3 hidden gems. Easy walking distance between gardens. Broomhill: A delightful, tranquil garden set in 2 acres. Island beds and borders are planted with shrubs, roses, masses of unusual perennials and choice annuals to give vibrancy and colour. Less formal area with large wildlife pond, meadow, shaded areas, bog garden and late summer border. The Curatage: A small cottage style garden growing flowers, fruit and vegetables. Eating area, small pond, wild grass circle, and Wendy house! Rosemary Cottage: An informal, 1½ acre, country garden, with stream, pond, and bog garden. Borders of shrub roses and perennials give way to an area of wildflowers and grasses. Since allowing the wild areas to flourish many flowers are returning, incl orchids, and insect life is thriving. C15 church with modifications in the 1840s by Pugin. At Rosemary Cottage grassy slopes and stone steps make wheelchair access to most of the garden difficult.

82 25 RICHMOND PARK AVENUE
Bournemouth, BH8 9DL. **Barbara Hutchinson & Mike Roberts, 01202 531072, barbarahutchinson@tiscali.co.uk.** *2½ m NE Bournemouth Town Centre. From T-lights at junction with Alma Rd and Richmond Park Rd, head N on B3063 Charminster Rd, 2nd turning on R into Richmond Park Ave.* **Sun 20 June, Sun 4 July, Sun 1 Aug (2-5). Adm £4, chd free. Home-made teas. Visits also by arrangement June & July for groups of 10 to 30. Home-made teas can be pre-booked £2.50 each.**
Beautifully designed town garden with pergola leading to ivy canopy over raised decking. Cascading waterfall connects 2 wildlife ponds enhanced with domed acers. Circular lawn with colourful herbaceous border planted to attract bees and butterflies. Fragrant S-facing courtyard garden at front, sparkling with vibrant colour and Mediterranean planting. As featured in Amateur Gardening & Garden News. Partial wheelchair access.

83 NEW ROSEBANK
8 Love Lane, Shaftesbury, SP7 8BG. **Nigel & Shouanna Hawkins.** *Park in town's main car park. Walk along Bimport, take 3rd L opp hospital sign into Magdalene Lane. Continue past hospital into Love Lane. Reserved disabled parking in drive.* **Sat 22, Sun 23 May (10.30-4.30). Adm £3.50, chd free. Also open Edwardstowe.**
Delightful small garden living up to its name, 106 rose plants have been bedded in during the last 4 yrs, transforming the front garden into box-edged flower beds with rose arches and peonies. The back garden is in cottage style, planted with many varieties of herbaceous perennials and T-roses. A pergola shades an alpine rockery and a herbaceous border. Wheelchair access around garden. Slope near entrance.

84 SEMLEY GRANGE
Semley, Shaftesbury, SP7 9AP. **Mr & Mrs Reid Scott.** *From A350 take turning to Semley continue along road and under railway bridge. Take 1st R up Sem Hill. Semley Grange is 1st on L - parking on green.* **Sun 27 June (12-5.30). Adm £5, chd free. Tea. Also open Grove House.**
Large garden recreated in last 10 yrs. Herbaceous border, lawn and wildflower meadow intersected by paths and planted with numerous bulbs. The garden has been greatly expanded by introducing many standard weeping roses, new mixed borders, pergolas and raised beds for dahlias, underplanted with alliums. Numerous fruit and ornamental trees introduced over last 10 yrs.

85 SHADRACK HOUSE
Shadrack Street, Beaminster, DT8 3BE. **Mr & Mrs Hugh Lindsay, 01308 863923, christine.corson@ngs.org.uk.** *Centre of Beaminster. Within 5 mins walk of 21A The Square.* **Visits by arrangement May to Sept for groups of 5 to 10. Opening together with 29 Fleet Street and 21A the Square. Adm £5, chd free.**
Delightful small mature garden, hidden behind high walls, with abundance of roses, clematis, shrubs, and climbers galore. Entrance through garden gates below the house, with views over to the church. Beaminster Festival end of June, beginning of July. Sadly not very wheelchair friendly, steepish steps to enter.

86 ◆ SHERBORNE CASTLE
New Rd, Sherborne, DT9 5NR. **Mr E Wingfield Digby, www.sherbornecastle.com.** *½ m E of Sherborne. On New Rd B3145. Follow brown signs from A30 & A352.* **For opening times and information, please visit garden website.**
40+ acres. Grade I Capability Brown garden with magnificent vistas across surrounding landscape, incl lake and views to ruined castle. Herbaceous planting, notable trees, mixed ornamental planting and managed wilderness are linked together with lawn and pathways. Short and long walks available. Partial wheelchair access, gravel paths, steep slopes, steps.

87 NEW 21A THE SQUARE
Beaminster, DT8 3AU. **Christine Corson, 01308 863923, christine.corson@ngs.org.uk.** *The Square Beaminster. 6m N of Bridport. 6m S of Crewkerne on B3162. 5 mins walk from Public Car Park.* **Visits by arrangement May to Sept for groups of 5 to 10. Combined with 29 Fleet Street and Shadrack House. Adm £5, chd free.**
3 town gardens, within walking distance of each other, one 4 yrs old, started from scratch, the other 2 well established, both with lovely views and both quite large for town gardens. 21A is a plantaholic and flower arranger's garden which says all! There is something to pick all yr round. Wide collection of shrubs, trees, roses and borders. 3rd prize at Melplash Show in 2019 for medium gardens, sadly no show or competitions in 2020! Very wheelchair friendly, house and garden flat, and ramp from house into garden.

"The annual donation from the National Garden Scheme to Perennial is the cornerstone of our fundraising activities and encourages many of our donors". Perennial

Ivy House Garden

88 STABLE COURT

Chalmington, Dorchester, DT2 0HB. Jenny & James Shanahan, 01300 321345, jennyshanahan8@gmail.com. *From A37 travel towards Cattistock & Chalmington for 1½m. R at triangle to Chalmington. ½m house on R, red letter box at gate. From Cattistock take 1st turn R then next L at triangle.* **Wed 16, Sat 19 June (2-5.30). Adm £5.50, chd free. Home-made teas. Visits also by arrangement May to Sept for groups of up to 20.**
This exuberant garden was begun in 2010. Extending to about 1½ acres, it is naturalistic in style with a shrubbery, gravel garden, wild garden and pond where many trees have been planted. Overflowing with roses scrambling up trees, over hedges and walls. More formal garden closer to house with lawns and herbaceous borders. Lovely views over Dorset countryside. Exhibition of paintings in studio. Gravel path partway around garden, otherwise the paths are grass, not suitable for wheelchairs in wet weather.

♿ 🐕 ❀ ☕

89 STADDLESTONES

14 Witchampton Mill, Witchampton, Wimborne, BH21 5DE. Annette & Richard Lockwood, 01258 841405, richardglockwood@yahoo.co.uk. *5m N of Wimborne off B3078. Follow signs through village and park in sports field, 7 min walk to garden, limited disabled parking near garden.* **Sun 30, Mon 31 May (2-5). Adm £4.50, chd free. Cream teas. Visits also by arrangement Apr to Sept for groups of up to 30.**
A beautiful setting for a cottage garden with colour themed borders, pleached limes and hidden gems, leading over chalk stream to shady area which has some unusual plants. Plenty of areas just to sit and enjoy the wildlife. Wire bird sculptures by local artist. Wheelchair access to 1st half of garden.

♿ ❀ ☕

SPECIAL EVENT

90 NEW STAFFORD HOUSE

West Stafford, Dorchester, DT2 8AD. Lord & Lady Fellowes. *2m E of Dorchester in Frome Valley. Follow signs to West Stafford from the West, 1st house on L before you get to the village, green park railings and gate* **Fri 2 July (10-12). Adm £50. Pre-booking essential, please visit www.ngs.org.uk/ special-events for information & booking. Home-made elevenses including cake and biscuits in a gardening theme.**
The gardens at Stafford House include a river walk and tree planting in the style of early-19th century Picturesque. Humphrey Repton prepared landscape proposals for the garden and the designs were later implemented, they were also included in his famous Red Books. Lord & Lady Fellowes will host a private morning in their garden, including special home-made elevenses with Julian and Emma under the turkey oak tree planted in 1633 on the main terrace.

☕

91 ✦ UPTON COUNTRY PARK

Upton, Poole, BH17 7BJ. BCP Council, 01202 262753, uptoncountrypark@ bcpcouncil.gov.uk, www.uptoncountrypark.com. *3m W of Poole town centre. On S side of A35/A3049. Follow brown signs.* **For information, please phone, email or visit garden website.**
Over 65 hectares of award winning parkland incl formal gardens, walled garden, woodland and shoreline. Maritime micro-climate offers a wonderful collection of unusual trees, vintage camellias and stunning roses. Home to Upton House, Grade II* listed Georgian mansion. Regular special events, art gallery and tea rooms. Car parking pay + display (cash or card), free entry to the park. Open 8am - 6pm (winter) and 8am - 9pm (summer). www.facebook.com/ uptoncountrypark. Easy access for wheelchair users.

♿ 🐕 ❀ 🚐 ☕ 🪑

92 NEW UTOPIA

Tincleton, Dorchester, DT2 8QP. Mr Nick & Mrs Sharon Spiller. *Take signs to Tincleton from Dorchester, Puddletown. Pick up garden signs in the village.* **Sun 30 May (2-5). Adm £4.50, chd £3. Light refreshments.**
Approximately ½ acre of secluded, peaceful garden made up of several rooms inspired by different themes. Inspiration is taken from Mediterranean and Italian gardens, Woodland space, water gardens and the traditional country vegetable garden. Seating is scattered throughout to enable you to sit and enjoy the different spaces and take advantage of both sun and shade. The garden, unfortunately, due to the layout and terrain, is not suitable for pushchairs and is not accessible for wheelchairs.

❀ ☕

93 NEW WAGTAILS

Stourton Caundle, Sturminster Newton, DT10 2JW. Sally & Nick Reynolds. *6m E of Sherborne. From Sherborne take A3030. At Bishops Caundle, L signed Stourton Caundle. After 1½m in village on R as descending hill, 5 houses before Trooper Inn.* **Sat 29, Sun 30 May (2-5.30). Combined adm with Manor Farm, Stourton Caundle £10, chd free.**
Contemporary Arne Maynard inspired garden of almost 3 acres, with wildflower meadows, orchards, kitchen garden and lawns, linked by a sweeping mown pathway, studded with topiary, divided by box, yew and beech hedging. Landscaped and planted over last 10 yrs by current owners, and still a work in progress. Wheelchair access over majority of garden and orchards over lawn grade paths. Gentle slopes.

♿ ☕

94 WELL COTTAGE

Ryall, Bridport, DT6 6EJ. John & Heather Coley, 01297 489066, jfrcoley@btinternet.com. *Less than 1m N of A35 from Morcombelake. From E: R by farm shop in Morcombelake. Garden 0.9m on R. From W: L entering Morcombelake, immed R by village hall and L on Pitmans Lane to T junc. Turn R, Well Cott on L. Parking on site and nearby.* **Sat 22, Sun 23, Mon 24 May (12-5). Combined adm with Pilsdon View £7, chd free. Home-made teas at Well Cottage along with the ticket desk. Visits also by arrangement Apr to Sept for groups of up to 30.**
1+acre garden brought back to life since 2012. There is now much more light after some trees were taken down and new areas have been cultivated. A number of distinct areas, some quite surprising but most still enjoy wonderful views over Marshwood Vale. Heather's textile art studio will be open to view. Wheelchair access possible but there are a few hard surface paths, slopes and steps.

 ☕

95 WESTERN GARDENS
24A Western Ave, Branksome
Park, Poole, BH13 7AN. Mr
Peter Jackson, 01202 708388,
pjbranpark@gmail.com. *3m W of
Bournemouth. From S end Wessex
Way (A338) at gyratory take The
Avenue, 2nd exit. At T-lights turn R
into Western Rd then at bottom of
hill L. At church turn R into Western
Ave.* **Sun 20 June, Sun 8 Aug (2-5).
Adm £4.50, chd free. Home-made
teas. Visits also by arrangement
Apr to Sept for groups of 20+.
Please discuss availability of
refreshments when booking.**
'This secluded and magical 1-acre
garden captures the spirit of warmer
climes and begs for repeated visits'
(Gardening Which?). Created over
40 yrs it offers enormous variety
with rose, Mediterranean courtyard
and woodland gardens, herbaceous
borders and cherry tree and camellia
walk. Lush foliage and vibrant flowers
give yr-round colour and interest
enhanced by sculpture and topiary.
Plants and home-made jams and
chutneys for sale. Wheelchair access
to ¾ garden.

96 NEW WHITE HOUSE
Newtown, Witchampton,
Wimborne, BH21 5AU.
Mr Tim Read, 01258 840438,
tim.read@catralex.com. *5m N of
Wimborne off B3078. Travel through
village of Witchampton towards
Newtown for 800m. Pass Crichel
House's castellated gates on L.
White House is a modern house
sitting back from road on L after
further 300m.* **Sat 26, Sun 27 June,
Sat 10, Sun 11 July (12-4.30). Adm
£4, chd free. Light refreshments.
Visits also by arrangement May to
Sept for groups of up to 10.**
1½ acre garden set on different
levels, with a Mediterranean feel,
planted to encourage wildlife and
pollinators. wildflower border,
pond surrounded by moisture
loving plants, prairie planting of
grasses and perennials, orchard.
Chainsaw sculptures of Birds of Prey.
Reasonable wheelchair access to all
but the top level of the garden.

**97 ◆ WIMBORNE MODEL
TOWN & GARDENS**
16 King Street, Wimborne,
BH21 1DY. Wimborne Minster
Model Town Ltd Registered
Charity No 298116, 01202 881924,

info@wimborne-modeltown.com,
www.wimborne-modeltown.com.
*2 mins walk from Wimborne Town
Centre & the Minster Church.
Follow Wimborne signs from A31;
from Poole/Bournemouth follow
Wimborne signs on A341; from
N follow Wimborne signs B3082/
B3078. Public parking opp in King
Street Car Park.* **For opening times
and information, please phone,
email or visit garden website.**
Attractive garden surrounding
intriguing model town buildings, incl
original 1950s miniature buildings
of Wimborne. Herbaceous borders,
rockery, perennials, shrubs and rare
trees. Miniature river system incl bog
garden and other water features.
Sensory area incl vegetable garden,
grasses, a seasonally fragrant and
colourful border, wind and water
features. Plentiful seating. Open 27
March - 31 October (10am - 5pm).
Seniors discount; groups welcome.
Tea room, shop, crazy golf, miniature
dolls' house collection, 00 gauge
model railway, Wendy Street play
area. Wheelchair available.

98 WINCOMBE PARK
Shaftesbury, SP7 9AB. John &
Phoebe Fortescue, 01747 852161,
pacfortescue@gmail.com,
www.wincombepark.com. *2m N
of Shaftesbury. A350 Shaftesbury
to Warminster, past Wincombe
Business Park, 1st R signed
Wincombe & Donhead St Mary.
¾m on R.* **Sun 16, Wed 19 May
(2-5). Adm £6, chd free. Home-
made teas. dairy and gluten free
options available. Visits also
by arrangement Apr to June for
groups of 10+.**
Extensive mature garden with
sweeping panoramic views from lawn
over parkland to lake and enchanting
woods through which you can
wander amongst bluebells. Garden is
a riot of colour in spring with azaleas,
camellias and rhododendrons in
flower amongst shrubs and unusual
trees. Beautiful walled kitchen garden.
Partial wheelchair access only, slopes
and gravel paths.

99 WOLVERHOLLOW
Elsdons Lane, Monkton Wyld,
DT6 6DA. Mr & Mrs D Wiscombe,
01297 560610. *4m N of Lyme Regis.
4m NW of Charmouth. Monkton
Wyld is signed from A35 approx
4m NW of Charmouth off dual

carriageway. Wolverhollow next to
church.* **Sun 2, Mon 3 May, Sun 1,
Mon 2 Aug (11.30-4.30). Adm £4,
chd free. Home-made teas. Visits
also by arrangement Feb to Oct.**
Over 1 acre of informal mature garden
on different levels. Lawns lead past
borders and rockeries down to a
shady lower garden. Numerous
paths take you past a variety of
uncommon shrubs and plants. A
managed meadow has an abundance
of primulas growing close to stream.
A garden not to be missed! Cabin in
meadow area from which vintage,
retro and other lovely things can be
purchased.

100 NEW YARDES COTTAGE
Dewlish, Dorchester,
DT2 7LT. Christine & Ross
Robertson, 07774 855152,
christine.m.robertson@hotmail.
com. *9m NE of Dorchester. From
A35 Puddletown/A354 junction,
take B3142 and immediately turn
R onto Long Lane. Follow the road
around the bend to R, then turn R
to Dewlish. Follow yellow NGS signs
through village to garden.* **Tue 22,
Thur 24, Sat 26, Sun 27 June
(1-5). Adm £5, chd free. Pre-
booking essential, please visit
www.ngs.org.uk for information
& booking. Home-made teas.
Refreshments in aid of All Saints
Parish Church, Dewlish. Visits
also by arrangement June & July
for groups of up to 10. Parking
available for up to 5 cars; also
suitable for a minibus.**
Country cottage garden of 1½ acres
bordering the Devil's Brook, having
a wealth of different planting areas
incl woodland, stream, small lake,
extensive lawns fringed with formal
herbaceous borders, vegetable and
soft fruit areas. Much of the planting
encourages insect life and supports
our bees and chickens. Yardes
Cottage Honey for sale. From lower
parking area there is flat paving to
upper sections of garden around
house only.

---

*The National Garden Scheme
searches the length and breadth
of England and Wales for the
very best private gardens*

ESSEX

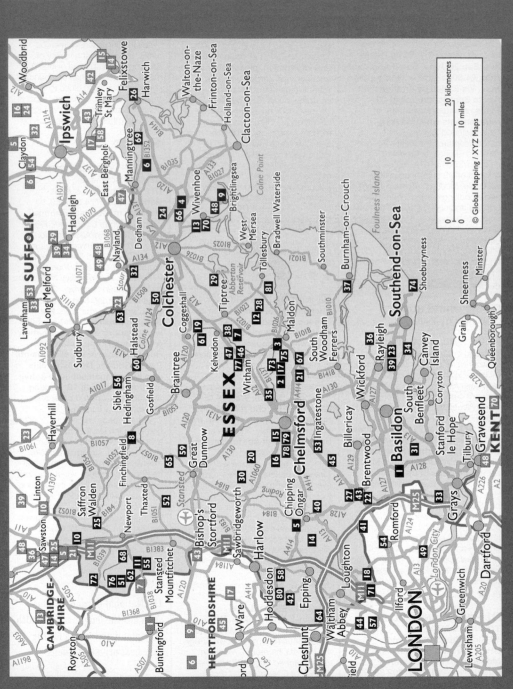

Close to London but with its own unique character, Essex is perhaps England's best kept secret - with a beautiful coastline, rolling countryside and exquisite villages. It is a little known fact that over seventy per cent of Essex is rural. There are wide horizons, ancient woodlands and hamlets pierced by flint church spires.

A perfect mix of villages, small towns, coast line and riverside settings enables Essex to provide some of the most varied garden styles in the country.

Despite its proximity to London, Essex is a largely rural county, its rich, fertile soil and long hours of sunshine make it possible to grow a wide variety of plants and our garden owners make the most of this opportunity. We have some grand country estates, small town gardens and much in between.

Our year starts with snowdrop gardens in February, runs through tulip gardens in the spring and makes the most of the high summer months with gardens full of roses and colourful perennials. Our visiting season extends into late summer when many gardens feature more exotic and tropical planting schemes making September a good time to explore.

Recent months have shown that, more than ever, visiting a garden provides huge benefits to mind and body. And whether you wish to visit as part of a group or club, with family, friends or on your own, you will be sure of a warm welcome. Our garden owners are usually on hand to talk about their garden to ensure you leave with a little more knowledge and lots of inspiration.

We look forward to sharing our gardens with you.

Volunteers

County Organiser
Susan Copeland
01799 550553
susan.copeland@ngs.org.uk

County Treasurer
Richard Steers
07392 426490
steers123@aol.com

Publicity & Social Media Coordinator
Debbie Thomson 07759 226579
debbie.thomson@ngs.org.uk

Booklet Co-ordinator and Publicity Assistant
Doug Copeland
01799 550553
doug.copeland@ngs.org.uk

Assistant County Organisers
Tricia Brett 01255 870415
tricia.brett@ngs.org.uk

Avril & Roger Cole-Jones
01245 225726
randacj@gmail.com

David Cox 01245 222165
david.cox@ngs.org.uk

Lesley Gamblin 07801 445299
lesley.gamblin@ngs.org.uk

Linda & Frank Jewson
01992 714047
linda.jewson@ngs.org.uk

Talks
Ed Falrey 07780 685634
ed@faireyassociates.co.uk

County Photographer
Caroline Cassell 07973 551196
caroline.cassell@ngs.org.uk

Left: Long House Plants

 @EssexNGS @EssexNGS @essexngs

OPENING DATES

All entries subject to change. For latest information check **www.ngs.org.uk**

Extended openings are shown at the beginning of the month.

Map locator numbers are shown to the right of each garden name.

January

Sunday 24th
◆ Green Island 24

February

Snowdrop Festival

Friday 12th
Horkesley Hall 32

Sunday 14th
Longyard Cottage 42

Wednesday 17th
Dragons 16

Sunday 21st
NEW Grove Lodge 25

Wednesday 24th
Ulting Wick 73

March

Saturday 20th
◆ Beth Chatto's Plants & Gardens 4

Sunday 21st
Heath House 29

April

Every Thursday from Thursday 22nd
Barnards Farm 1

Every Thursday and Friday
Feeringbury Manor 19

Saturday 17th
Loxley House 43

Sunday 18th
30 Glenwood Avenue 23

Friday 23rd
◆ Beeleigh Abbey Gardens 3
Wycke Farm 81

Sunday 25th
NEW Heyrons 30
Ulting Wick 73
Wycke Farm 81

Wednesday 28th
Furzelea 21
Writtle University College 79

Friday 30th
Scrips House 61
Ulting Wick 73

May

Every Thursday
Barnards Farm 1

Every Thursday and Friday
Feeringbury Manor 19

Sunday 2nd
Chestnut Cottage 11
Furzelea 21
◆ Green Island 24
Wickets 76

Monday 3rd
NEW Grove Lodge 25

Friday 7th
NEW Bassetts 2

Saturday 8th
NEW 39 Wakering Avenue 74

Sunday 9th
NEW Bassetts 2
The Old Vicarage 52

Saturday 15th
NEW 14 Clifton Terrace 13

Sunday 16th
NEW 14 Clifton Terrace 13
1 Whitehouse Cottages 75

Thursday 20th
Dragons 16

Friday 28th
8 Dene Court 15

Sunday 30th
NEW Bradfield Gardens 6
NEW The Gates 22
NEW Sheepcote Green Gardens 62
Silver Birches 64
Tudor Roost 70

Monday 31st
Fairwinds 18
Rookwoods 56
79 Royston Avenue, E4 57
Tudor Roost 70

June

Every Thursday
Barnards Farm 1

Every day from Monday 7th to Sunday 13th
Elwy Lodge 17

Every Thursday and Friday
Feeringbury Manor 19

Thursday 3rd
8 Dene Court 15

Saturday 5th
Isabella's Garden 34
NEW 38 Juniper Road 35

Sunday 6th
Furzelea 21
Isabella's Garden 34
NEW 38 Juniper Road 35
Miraflores 47
NEW Stocksmead 68
Two Cottages 72

Wednesday 9th
Oak Farm 50

Friday 11th
Heath House 29

Sunday 13th
NEW Chesterford Village Gardens 10
NEW Clunes House 14
Heath House 29
61 Humber Avenue 33
Miraflores 47
Peacocks 53
37 Turpins Lane 71
1 Whitehouse Cottages 75

Wednesday 16th
Heath House 29

Friday 18th
NEW Chilterns 12

Saturday 19th
NEW Chilterns 12

Sunday 20th
Fudlers Hall 20
The Old Vicarage 52
Two Cottages 72

Saturday 26th
Keeway 37
18 Pettits Boulevard, RM1 54
Spring Cottage 66
NEW Witham Town Gardens 77

Sunday 27th
Barnards Farm 1
Havendell 28
9 Malyon Road 46
Miraflores 47
Old Bell Cottage 51
18 Pettits Boulevard, RM1 54
Silver Birches 64
Spring Cottage 66
Wickets 76
Writtle Gardens 78
Wychwood 80

Tuesday 29th
8 Dene Court 15

Wednesday 30th
Keeway 37

July

Every Thursday
Barnards Farm 1

Every Thursday and Friday
Feeringbury Manor 19

Friday 2nd
Scrips House 61

Saturday 3rd
9 Malyon Road 46

Sunday 4th
Fudlers Hall 20
262 Hatch Road 27
Hilldrop 31
9 Malyon Road 46
37 Turpins Lane 71

Tuesday 6th
Braxted Park Estate 7

Wednesday 7th
Long House Plants 41
Oak Farm 50

Isabella's Garden

Online booking is
available for many of our
gardens, at ngs.org.uk

THE GARDENS

◻ BARNARDS FARM

Brentwood Road, West Horndon, CM13 3LX. Bernard & Sylvia Holmes & The Christabella Charitable Trust, 01268 454075, vanessa@barnardsfarm.eu, www.barnardsfarm.eu. *5m S of Brentwood. On A128 1½ m S of A127 Halfway House flyover. From Junction continue south on A128 under the railway bridge. Garden on R just past bridge.* **Every Thur 22 Apr to 26 Aug (11-4.30). Adm £7.50, chd free. Sun 27 June (1-5); Sun 5 Sept (2-5). Adm £10, chd free. Due to Covid-19 restrictions, refreshments may only be available on Sun 5 Sept. Visits also by arrangement Jan to Nov for groups of 30+. Donation to St Francis Church.**
So much to explore! Climb the Belvedere for the wider view and take the train for a woodland adventure. Spring bulbs and blossom, summer beds and borders, ponds, lakes and streams, walled vegetable plot. 'Japanese garden', sculptures grand and quirky enhance and delight. See the new Talia-May Avenue, new for 2021. Barnards Miniature Railway rides (BMR) :Separate charges apply. Sunday extras: Bernard's Sculpture tour 2.30pm Car collection. 1920s Cycle shop. Archery. Model T Ford Rides. Collect loyalty points on Thur visits and earn a free Sun or Thur entry. Season Tickets available Aviators welcome (PPO). Picnics welcome Wheelchair accessible WC Golf Buggy tours available.

◻ NEW BASSETTS

Bassetts Lane, Little Baddow, Chelmsford, CM3 4BZ. Mrs Margaret Chalmers. *1m down Tofts Chase which becomes Bassetts Lane. Big yellow house on L, wooden gates, red brick wall. From Spring Elms Lane go down Bassetts Lane ½ m. House on R.* **Fri 7, Sun 9 May (10.30-5). Adm £5, chd free.**
Two acre garden, tennis court, swimming pool surrounding an early C17 house with plants for all year round interest set on gently sloping ground with lovely distant views of the Essex countryside. Shrub borders and mature ornamental trees, an orchard and two natural ponds. Many places to sit and relax.

No refreshments but bring a picnic and enjoy the views ! Please check wheelchair access with garden owner 01245 226768.

◻ ◆ BEELEIGH ABBEY GARDENS

Abbey Turning, Beeleigh, Maldon, CM9 6LL. Christopher & Catherine Foyle, 07506 867122, www. visitmaldon.co.uk/beeleigh-abbey. *1m NW of Maldon. Leaving Maldon via London Road take 1st R after Cemetery into Abbey Turning.* **For NGS: Fri 23 Apr (10.30-4.30). Adm £6.50, chd £2.50. Light refreshments. For other opening times and information, please phone or visit garden website.**
3 acres of secluded gardens in rural historic setting. Mature trees surround variety of planting and water features, woodland walks under planted with bulbs leading to tidal river, cottage garden, kitchen garden, orchard, wildflower meadow, rose garden, wisteria walk, magnolia trees, lawn with 85yd long herbaceous border. Scenic backdrop of remains of C12 abbey incorporated into private house (not open). Refreshments incl Hot and Cold Drinks, Cakes and rolls. Gravel paths, some gentle slopes and some steps. Large WC with ramp and handlebars.

◻ ◆ BETH CHATTO'S PLANTS & GARDENS

Elmstead Market, Colchester, CO7 7DB. Beth Chatto's Plants & Gardens, 01206822007, office@bethchatto.co.uk, www.bethchatto.co.uk. *¼ m E of Elmstead Market. On A133 Colchester to Clacton Rd in village of Elmstead Market.* **For NGS: Sat 20 Mar (10-5). Adm £6.95, chd £1.50. Wed 29 Sept (10-5). Adm £8.95, chd £1.50. Light refreshments. For other opening times and information, please phone, email or visit garden website.**
Internationally famous gardens, including dry, damp, shade, reservoir and woodland areas. The result of over 60 years of hard work and application of the huge body of plant knowledge possessed by Beth Chatto and her husband Andrew. Visitors cannot fail to be affected by the peace and beauty of the garden. Beth is renowned internationally for her books, her gardens and her influence on

the world of gardening and plants. At present, prebooking your arrival time is required. Please visit www. bethchatto.co.uk/visit for up to date visiting details. Picnic area available in the adjacent field. Disabled WC & parking. Wheelchair access around all of the gardens - on gravel or grass (concrete in Nursery, Welcome Centre & Gardener's shop areas).

◻ BLAKE HALL

Bobbingworth, CM5 0DG. Mr & Mrs H Capel Cure, www.blakehall.co.uk. *10m W of Chelmsford. Just off A414 between Four Wantz r'about in Ongar & Talbot r'about in North Weald. Signed on A414.* **For opening times and information, please visit www. ngs.org.uk**
25 acres of mature gardens within the historic setting of Blake Hall (not open). Arboretum with broad variety of specimen trees. Spectacular rambling roses clamber up ancient trees. Traditional formal rose garden and herbaceous border. Sweeping lawns. Teas served from Essex Barn. Some gravel paths.

GROUP OPENING

◻ NEW BRADFIELD GARDENS

Heath Road, Bradfield, Manningtree, CO11 2UZ. *3m E of Manningtree. 9m W of Harwich. Take A137 from Manningtree Station, turn L opp garage. Take 1st R towards Clacton. At Radio Mast turn L into Bradfield. Continue through village. Gardens opposite Primary School.* **Sun 30 May, Sun 11 July (11-4.30). Combined adm £6, chd free. Home-made teas at Chippins. Delicious home-made cakes!.**

CHIPPINS
Kit & Ceri Leese, 01255 870730, ceriandkit@gmail.com.
Visits also by arrangement May to July for groups of 5 to 30.
NEW 2 HOPE COTTAGES
Mr Martin Ford.

Two beautiful, but very contrasting gardens in the village of Bradfield, bordering on the Stour Estuary in north east Essex. Chippins an artist's and plantaholics' paradise, packed with interest. Spring heralds hostas, aquilegia and irises. Includes meandering stream, wildlife pond and Horace the Huge! Summer hosts a colour explosion with daylilies and rambling roses. The front garden has an abundance of tubs and hanging baskets. Also includes an exotic border with cannas, dahlias and cacti. Studio in conservatory with paintings and etching press. 2 Hope Cottages, a new garden opening for the first time for the NGS is a tranquil, country style garden overlooking the serene, open countryside. The beautifully planted garden reflects the owner's love of trees. In early summer the rich variety of roses and peonies give the garden a soft, romantic feel. Followed later with splashes of vibrant colour from dahlias, salvias, agapanthus and sweet peas. Chippins - Kit is a landscape artist and printmaker, pictures always on display. Afternoon tea with delicious home made cakes is also available for small parties (minimum of 5) on specific days if booked in advance, May to July.

  ♻ ☕

---

7 BRAXTED PARK ESTATE
Braxted Park Road, Great Braxted, Witham, CM8 3EN. Mr Duncan & Mrs Nicky Clark, 01621 892305, office@braxtedpark.com, www.braxtedpark.com. *A12 north,*

by-pass Witham,. Turn L to Rivenhall & Silver End by pub called the Fox (now closed) At T junction turn R to Gt Braxted & Witham. Follow brown sign to Braxted Pk (NOT Braxted Golf Course). Down drive to Car Park. **Tue 6 July (10-4). Adm £6, chd free. Light refreshments in the Walled Garden Pavilion. Light lunches and afternoon tea available. Visits also by arrangement Mar to Sept for groups of 10+. Refreshments available by arrangement when booking.**

Idyllic Braxted Park, a prestigious luxury events venue, welcomes gardeners to enjoy the tranquil surroundings. Extensive plantings of perennials, mixed borders of unusual shrubs and roses abound. Historic Walled Garden featuring individually themed gardens radiating from the central fountain and parasol mulberry trees, rarely seen in the UK. Themes incl a Black and White Garden, Italian Garden and English Garden, each containing a wealth of design planting inspiration. Head Gardener, Andrea Cooper, will be on hand to answer questions and guided tours for a further donation of £5pp payable upon the day. An extensive new native meadow has been created between house and lakes. A walk around the lake is not to be missed.

Guide dogs only.
  ♻ ❄ 🚗 ☕

---

8 NEW BRICK HOUSE
The Green, Finchingfield, CM7 4JS. Mr & Mrs Graham & Susan Tobbell. *9m NW of Braintree. Brick House in the centre of the picturesque village of Finchingfield, at the Xrds of the B1053 & B1057 in rural north-west Essex.* **Fri 23 July (1-5). Adm £6, chd free. Home-made teas.**

This recently renovated period property and garden covers 1½ acres, with a brook running through the middle. The garden features contemporary sculptures, and several distinct planting styles; most notably a hidden scented cottage-style garden, a crinkle-crankle walled area with an oriental feel and some rather glorious late-summer herbaceous borders inspired by the New Perennial Movement. Modern marble sculptures by Paul Vanstone add drama and contrast to the soft landscaping. Garden is professionally designed to be a calming space. Wheelchair access to where the garden can be viewed but only partial wheelchair access to the garden itself.

🐕 ☕

Scrips House

GROUP OPENING

9 BRIGHTLINGSEA GARDENS
Brightlingsea, CO7 0JF. *Approx 10m SE of Colchester. Go to Thorrington on B1027 & head S for approx 3m towards Brightlingsea on B1029. Group gardens are either side of Church Rd. just beyond Autosmith Garage.* **Sun 18 July (10-4.30). Combined adm £8, chd free. Light refreshments at 44 Church Road.**

44 CHURCH ROAD
Mandy Livingstone & Steven Nicholson.
77 CHURCH ROAD
Mr & Mrs Mick & Gill Tokley.
SANDY HOOK
Mr & Mrs Peter & Elaine Sedwell.

Three very interesting and contrasting gardens to enjoy in the unique and ancient maritime town of Brightlingsea. Sandy Hook has traditional borders, a raised rose bed, dahlias, penstemons and salvias, with a small woodland stream and a sheltered "White Garden". 44 Church Road has a traditional country garden, with gentle colours, mature trees, evergreen shrubs, a water feature and a wisteria covered pergola. The garden offers various places to relax and contemplate the different features of this quintessentially English garden. At 77 Church Road we have a colourful garden with an eclectic mix of trees, shrubs and perennials in borders alongside a range of hard landscaping. The garden is filled with plants and features and you can wander through pathways to hidden corners offering peaceful and tranquil seating. Come to Brightlingsea and make a day of it. At Brightlingsea the spectacular award winning floral displays adorning the centre are not to be missed. There is much to see along the harbour, the marina with bracing walks along the promenade. Brightlingsea is blessed with plenty of watering holes and places to eat. Access to main border garden only at Sandy Hook. At No 77 the side passage is partially obstructed by a gas meter.
♿ ❄ ⛾

GROUP OPENING

10 NEW CHESTERFORD VILLAGE GARDENS
High Street, Great Chesterford, Saffron Walden, CB10 1TZ. *4m NW Saffron Walden. Parking & ticket sales at either Gt Chesterford Community Centre CB10 1NS or Manor Cottage, Lt Chesterford CB10 1TZ.* **Sun 13 June (1-5). Combined adm £8, chd free. Home-made teas at Great Chesterford Community Centre from 2-4pm.**

NEW BANK COTTAGE
Nicola Kearton.
NEW BISHOP'S HOUSE
Amanda & Matt Bonass.
NEW GELDARDS
Liz & Mark Gamble.
NEW MANOR COTTAGE
Krista & David Bagley.
NEW WEARNS FOLLY
Lorna & Howard Rolfe.

Bank Cottage has a pretty cottage garden with mixed borders, clipped hedges, many roses and trim lawns. Also featuring a gravel garden and raised vegetable beds framed by espaliered fruit trees. Bishops House has a beautiful 9 acre garden dating back to 19C. Abundant perennial borders, velvet lawns, mature trees and walled kitchen garden. A bridge across the river leads to a natural meadow. Geldards is a varied garden: a formal parterre with ballerina roses, mixed borders with cottage style planting and raised veg beds. A small orchard, wildlife pond and woodland area complete the garden. Manor Cottage has a pretty cottage garden with views across the river. It boasts mature trees, colourful borders and summerhouse. Raised veg beds, fruit trees and Mediterranean inspired patio also feature. Wearns Folly showcases several styles: hot and cool borders, a rose/jasmine covered pergola leading to a small pond, lawns and old rose terrace. A gate leads to a well stocked kitchen garden. Village orchard comprising well over 100 different fruit trees (labelled) and village allotments available for viewing at Gt Chesterford Community Centre. Partial access at Bank Cottage.
♿ ⛾

11 CHESTNUT COTTAGE
Middle Street, Clavering, CB11 4QL. Carol & Mike Wilkinson. *At the bottom of Middle Street, Chestnut Cottage faces you across the ford (River Stort) - walk across the footbridge. Parking in Middle Street.* **Sun 2 May (2-5). Adm £4, chd free. Also open Wickets.**
Cottage garden with sweeping lawns and wide flower borders, opposite Clavering Ford. Lots of pathways and seats to sit and contemplate. Stunning views over the medieval heart of the village. Come and see 'The Little House', built in 1760 and reputedly the smallest thatched cottage in England.
❄ 🚌 ⛾

12 NEW CHILTERNS
Chelmer Close, Little Totham, CM9 8JN. Mrs Jane Kynaston. *3m S of Tiptree. From A12 take Rivenall/ Silverend junction towards Braxted/ Tiptree. At T junction turn R then immediately L signed Little Totham. At The Swan pub turn R onto Post Office Lane 100yds Chelmer Close on L.* **Fri 18, Sat 19 June (11-4). Adm £3.50, chd free. Light refreshments and cream teas.**
Chilterns has always been a family garden, now with a 'grown up' feel that has a friendly welcoming atmosphere, full of plants that give something special to the visitor. Plants are here for their beauty, their perfume, their pollen and nectar for our beehive. Learning which plants work where seems to have now created a serendipitous, restful and reflective space. Informal cottage style garden where ex battery hens roam free. Small exhibition of arts, crafts, cards and paintings for sale. Mediterranean gravel area, please check with owner re wheelchair access 07743 906788.
❄ ⛾

13 NEW 14 CLIFTON TERRACE
Wivenhoe, Colchester, CO7 9DY. Mr & Mrs Michael & Liz Taylor Jones. *Clifton Terrace turning on R off Wivenhoe High St opp Greyhound Pub, beside Concord estate agents. On R of Clifton Terrace is the public car park and WC. Clifton Terrace ends in our front garden situated just above the station platform.* **Sat 15, Sun 16 May (12-5). Adm £4, chd free. Home-made teas.**
A very long, narrow garden alongside the railway station. Near house:

herbaceous border, shrubs, lawn and roses. Slope down to station footpath: ground cover shrubs. Beyond the lawn: vegetable garden and small orchard, tree house in walnut tree and swing. Continue to part of old coppiced Wivenhoe Woods. Beyond a wildlife pond, natural woodland continues to rhododendrons, bamboos and a second pond.

14 NEW CLUNES HOUSE
Mill Lane, Toot Hill, Ongar, CM5 9SF. Dr & Mrs Hugh & Elaine Taylor, 07704230155, elaine_taylor_clunes@outlook.com. *4m from J7 M11 & 4m from the Towns of Ongar & Epping. The garden is at the end of Mill Lane in Toot Hill.* Sun 13 June, Sun 11 July (10.30-4.30). Adm £6, chd free. Home-made teas. Visits also by arrangement May to Sept.
This is a beautiful, and interesting traditional country garden with wonderful views over the Essex countryside. A pergola with roses, clematis and wisteria leads into a woodland walk and pond surrounded by herbaceous borders and sculpture. A path through the wildlife meadow

takes you into the orchard and a small holding with pigs and sheep, ending at the kitchen garden and cut flower beds. Original World War II air raid shelter. The garden is adjacent to the 'Essex Way' long distance walking trail. Beautiful views over the countryside and in the distance London rooftops.

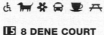

15 8 DENE COURT
Chignall Road, Chelmsford, CM1 2JQ. Mrs Sheila Chapman, 01245 266156. *W of Chelmsford (Parkway). Take A1060 Roxwell Rd for 1m. Turn R at T-lights into Chignall Rd. Dene Court 3rd exit on R. Parking in Chignall Rd.* Fri 28 May, Thur 3, Tue 29 June, Fri 9 July (2-5). Tue 20 July (2-5), also open Dragons. Wed 28 July, Tue 10 Aug (2-5). Wed 18 Aug (2-5), also open Dragons. Tue 31 Aug (2-5). Sun 5 Sept (2-5), also open Dragons. Adm £3.50, chd free. Visits also by arrangement May to Sept for groups of up to 30.
Beautifully maintained and designed compact garden (250sq yds). Owner is well-known RHS gold medal-winning exhibitor (now retired). Circular lawn, long pergola and walls

festooned with roses and climbers. Large selection of unusual clematis. Densely-planted colour coordinated perennials add interest from May to Sept in this immaculate garden.

16 DRAGONS
Boyton Cross, Chelmsford, CM1 4LS. Mrs Margot Grice, 01245 248651, margot@snowdragons.co.uk. *3m W of Chelmsford. On A1060. ½m W of The Hare Pub.* Wed 17 Feb (11-3). Light refreshments. Thur 20 May (2-5). Tea. Tue 20 July, Wed 18 Aug, Sun 5 Sept (2-5). Tea. Also open 8 Dene Court. Adm £4, chd free. Visits also by arrangement Feb to Oct for groups of 10+.
A plantswoman's ¾ acre garden, planted to encourage wildlife. Sumptuous colour-themed borders with striking plant combinations, featuring specimen plants, fernery, clematis and grasses. Meandering paths lead to ponds, patio, scree garden and small vegetable garden. Two summerhouses, one overlooking stream and farmland.

Geldards, Chesterford Village Gardens

17 ELWY LODGE

West Bowers Rd, Woodham Walter, CM9 6RZ. David & Laura Cox. *Just outside Woodham Walter village. From Chelmsford, A414 to Danbury. L at 2nd mini r'about into Little Baddow Rd. From Colchester, A12 to Hatfield Peverel, L onto B1019. Follow NGS signs.* **Daily Mon 7 June to Sun 13 June (12-5). Adm £6, chd free. Light refreshments.**

Tucked away in the gentle Essex countryside, the glorious garden at Elwy Lodge offers a welcome that is hard to beat. Join the growing numbers of visitors who regularly return each year to sit and savour the different sights this exceptional garden has to offer. Plenty of seating throughout the garden. A secluded chamomile-scented lower garden with raised veg. beds and fruit trees leads to a delightful summer house, where visitors relax and enjoy the peaceful surroundings and amazing views. Gravel drive, sloping uneven lawn in parts.

18 FAIRWINDS

Chapel Lane, Chigwell Row, IG7 6JJ. Sue & David Coates, 07731 796467, scoates@forest.org.uk. *2m SE of Chigwell. Grange Hill Tube, turn R at exit, 10 mins walk uphill. Car: Nr M25 J26 & N Circular Waterworks r'about. Follow signs for Chigwell. Fork R for Manor/Lambourne Rd. Park in Lodge Close Car Park.* **Mon 31 May (2-5). Adm £4.50, chd free. Home-made teas. Free refills of tea/coffee Soya milk available. Some green/fruit teas and decaf on request. Vegan cake (limited amount available). Visits also by arrangement May to Sept for groups of 5 to 20.**

Gravelled front garden and three differently styled back garden spaces with planting changes every year. Meander, sit, relax and enjoy. NEW this year: Wildflower meadow seeded autumn 2019, extension to Long Border and fresh planting in woodland. Beyond the rustic fence, lies the wildlife pond and vegetable plot. Planting influenced by Beth Chatto, Penelope Hobhouse and Christopher Lloyd. Happy hens. Happy insects in bee house, bug houses, log piles, meadow and sampling the spring pollen. Newts in pond. There be dragons a plenty!! Space for 2 disabled cars to park by the house. Wood chip paths in woodland area may require assistance.

19 FEERINGBURY MANOR

Coggeshall Road, Feering, CO5 9RB. Mr & Mrs Giles Coode-Adams, 01376 561946, seca@btinternet.com, *Between Feering & Coggeshall on Coggeshall Rd, 1m from Feering village.* **Every Thur and Fri 1 Apr to 30 July (10-4). Every Thur and Fri 9 Sept to 22 Oct (10-4). Adm £5, chd free. Visits also by arrangement. Donation to Feering Church.**

There is always plenty to see in this peaceful 10 acre garden with two ponds and river Blackwater. Jewelled lawn in early April then spectacular tulips and blossom lead on to a huge number of different and colourful plants, many unusual, culminating in a purple explosion of michaelmas daisies in late Sept. Wonderful sculpture by Ben Coode-Adams. No wheelchair access to arboretum, steep slope.

20 FUDLERS HALL

Fox Road, Mashbury, CM1 4TJ. Mr & Mrs A J Meacock, 01245 231335. *7m NW of Chelmsford. Chelmsford take A1060, R into Chignal Rd. ½ m L to Chignal St James approx 5m, 2nd R into Fox Rd signed Gt Waltham. Fudlers from Gt Waltham. Take Barrack Lane for 3m.* **Sun 20 June, Sun 4 July (2-5). Adm £6, chd free. Cream teas. Visits also by arrangement June & July.**

An award winning, romantic 2 acre garden surrounding C17 farmhouse with lovely pastoral views, across the Chelmer Valley. Old walls divide garden into many rooms, each having a different character, featuring long herbaceous borders, ropes and pergolas festooned with rambling old fashioned roses. Enjoy the vibrant hot border in late summer. Yew hedged kitchen garden. Ample seating. Wonderful views across Chelmer Valley. 2 new flower beds, many roses. 500year old yew tree. gravel farmyard and 30foot path to gardens, all of which is level lawn.

21 FURZELEA

Bicknacre Road, Danbury, CM3 4JR. Avril & Roger Cole-Jones, 01245 225726, randacj@gmail.com. *4m E of Chelmsford, 4m W of Maldon A414 to Danbury. At village centre S into Mayes Lane. Then 1st R. Go past Cricketers Pub, L on to Bicknacre Rd Use NT carpark immed on L Garden 50m on R. Extra parking 200 m past on L. Use NT Common paths from carparks.* **Wed 28 Apr (2-5); Sun 2 May, Sun 6 June, Sun 18 July, Sun 12 Sept (11-5); Wed 15 Sept (2-5). Adm £5, chd free. Home-made teas. Visits also by arrangement Apr to Oct. Min 15.**

Country garden of nearly one acre has evolved overtime to showcase the different seasons by titillating the senses with colour,scent,texture and form. Paths ,steps and archways lead you onwards through intimate spaces and flowing lawns. Clipped box hedges edge flower beds of tulips in Spring,roses,hemerocallis and other herbaceous plants in Summer,dahlias,grasses,salvias, and exotics in Autumn. Black and White garden plus exotics and many unusual plants add to the Visitors interest. International garden tour visitors voted us the best garden in their recent tours of East Anglia. Opp Danbury Common (NT), short walk to Danbury Country Park and Lakes and short drive to RHS Hyde Hall. Only partial wheelchair access, steps and gravel paths and drive.

22 NEW THE GATES

London Road Cemetery, London Road, Brentwood, CM14 4QW. Ms Mary Yiannoullou, www.frontlinepartnership.org. *At the rear of the cemetery. Enter the Cemetery (opp Tesco Express). Drive to the rear of the cemetery where you will see our site on L. Limited parking within but plenty in the surrounding rds.* **Sun 30 May, Sun 22 Aug (11-3.30). Adm £3.50, chd free. Tea, coffee and cold drinks, with a variety of home-made cakes.**

'The Gates' horticultural project offers local citizens, including vulnerable adults, an opportunity to develop new skills within a horticultural setting. Greenhouses and allotments for raising bedding plants, educational workshops and growing fruit and vegetables. Walks through the Woodland Dell and Sensory area.

Refreshments in the Tea House. Large variety of plants and produce for sale. Enjoy the mosaic garden, visit the apiary and relax in the various seating areas around the site. All areas are accessed via slopes and ramps.

23 30 GLENWOOD AVENUE
Leigh-On-Sea, SS9 5EB.
Joan Squibb, 07543 031772,
squibb44@gmail.com. *Follow A127 towards Southend. Past Rayleigh Weir. At Progress Rd T-lights turn L. At next T-lights turn L down Rayleigh Rd A1015. Past shops turn 2nd L into Glenwood Ave. Garden halfway down on R.* **Sun 18 Apr (11-4). Adm £6, chd free. Entrance fee includes refreshments. Visits also by arrangement Apr to Sept for groups of 5 to 20.**
Nestled next to the busy A127 lies a beautifully transformed town garden. Home-made raised beds with tulips in spring, dahlias and roses in the summer. From a corridor, you emerge into an open garden giving a vista of colour, inspiration and peaceful harmony, with many different scents to savour. Hanging baskets bloom in the fruit trees and along the fences. A paved area allows one to enjoy the view of the garden as does a deck at the back where herbs and some vegetables live alongside the flowers. A peaceful vista, to sit in and restore the batteries. Vegetables are also grown in raised beds.

24 ♦ GREEN ISLAND
Park Road, Ardleigh, CO7 7SP.
Fiona Edmond, 01206 230455,
greenislandgardens@gmail.com,
www.greenislandgardens.co.uk.
3m NE of Colchester. From Ardleigh village centre, take B1029 towards Great Bromley. Park Rd is 2nd on R after level Xing. Garden is last on L. **For NGS: Sun 24 Jan, Sun 2 May, Sat 9 Oct (10-5). Adm £8, chd £2.50. Home-made cakes, scones, sandwiches and baguettes, Cream teas or full afternoon teas to order Picnics only in car park area. 2022: Sun 23 Jan. For other opening times and information, please phone, email or visit garden website. Donation to Plant Heritage.**
A garden for all seasons, highlights incl bluebells, azaleas, autumn colour, winter hamamelis and snowdrops. A plantsman's paradise. Carved within 20 acre mature woodland are

huge island beds, Japanese garden, terrace, gravel, seaside and water gardens, all packed with rare and unusual plants. Bluebells Bazaar weekend 1 & 2 May, Autumn colour weekend 9 & 10 Oct, Snowdrops 23 Jan 2022. National collections of Hamamelis cvs and Camellias (autumn and winter flowering). January-Hamamelis, February-snowdrops, May-bluebells, azaleas, acers and rhododendrons. Summer-water gardens, island beds and tree lilies. Oct/Nov-Stunning autumn colour. Light lunches, home-made cakes and cream teas. Flat and easy walking /pushing wheelchairs. Ramps at entrance and tearoom. Disabled parking and WC.

25 NEW GROVE LODGE
3, Chater's Hill, Saffron Walden, CB10 2AB. Chris Shennan, 01799522271. *Approx 10 mins walk from town centre. Facing The Common on E side, about 100 yds from the turf maze. Note: Chater's Hill is one way.* **Sun 21 Feb, Mon 3 May (2-5). Adm £5, chd free. Home-made teas. Tea and cake will be provided, subject to demand. Visits also by arrangement Jan to Oct for groups of up to 10.**
A large walled garden close to the town centre with unusually high biodiversity (e.g. 16 species of butterfly recorded), close to the turf maze and Norman castle. Semi-woodland on light free-draining chalk soil allows bulbs, hellebores and winter-flowering shrubs to thrive. Two ponds, topiary, orchard and small vegetable garden blend some formality with informal areas where wildlife thrives. A profusion of snowdrops, other bulbs and spring blossom. Wheelchair access via fairly steep drive leading to terrace from which the garden may be viewed.

GROUP OPENING

26 HARWICH GARDENS
Harwich, CO12 3NH. *Centre of Old Harwich. Car park on Wellington Rd, CO12 3DT within 50 metres of St Helens Green. Also street parking available.* **Sun 25 July (11-4). Combined adm £5, chd free. Home-made teas at 8 St Helens Green. Soup available between 11 and 1. Teas, coffees and cakes**

available all day.

63 CHURCH STREET
Sue & Richard Watts.

QUAYSIDE COURT
Dave Burton.

8 ST HELENS GREEN
Frances Vincent.

NEW 63 WEST STREET
Mr & Mrs Turnbull.

Four different gardens all within walking distance in the historical town of Harwich. 8 St Helens Green, just 100m from the sea. A small town garden, with an abundance of dahlias, and hydrangeas mixed with perennials. 63 Church Street (access via Cow Lane/Kings Head Street) is a long, narrow walled courtyard with shrubs, climbers, perennials and veg packed into the garden and architectural features including an Elizabethan window with original glass. Brought back to life the garden at 63 West Street was started in 2016 from a mess of rubble to a beautiful garden. Containing shrubs, perennials and annuals, the owner has brought plants from previous gardens to help stock this garden. Many scented plants. Pots also feature highly. Lovely at this moment in time, it is ever evolving. Views of historic buildings Quayside Court unusually boasts a sunken garden hidden from view at the end of the car park which features a pond, veg patch, roses and climbers.

27 262 HATCH ROAD
Pilgrims Hatch, Brentwood, CM15 9QR. Mike & Liz Thomas. *2m N of Brentwood town centre. On A128 N toward Ongar turn R onto Doddinghurst Rd at mini-r'about (to Brentwood Centre) After the Centre turn next L into Hatch Rd. Garden 4th on R.* **Sun 4, Sun 18 July (11.30-4.30). Adm £5, chd free. Home-made teas. and cakes.**
A formal frontage with lavender. An eclectic rear garden of around an acre divided into 'rooms' with themed borders, several ponds, three green houses, fruit and vegetable plots and oriental garden. There is also a secret white garden, spring and summer wildflower meadows, Yin and Yang borders, a folly and an exotic area. There is plenty of seating to enjoy the views and a cup of tea and cake.

28 HAVENDELL

Beckingham Street, Tolleshunt Major, CM9 8LJ. Malcolm & Val. *5m E of Maldon 3m W of Tiptree. From B1022 take Loamy Hill rd. At the Xrds L into Witham Rd. Follow NGS signs.* **Sun 27 June (11-4.30). Adm £4, chd free. Home-made teas.**
This beautiful and tranquil garden will amaze and inspire you as you walk among the herbaceous and shrub borders. Meandering paths and seating areas entice you to take time out in a garden that surrounds you with nature in all its splendour. The many scented roses allowing all of your senses to be aroused. Exotic sub- tropical plants and over 50 hostas. WC available.

29 HEATH HOUSE

Crayes Green, Layer Breton, CO2 0PN. Geoff & Julia Russell Grant. *7m SW of Colchester off B1022. Going S ½m after Heckfordbridge go L to Birch, past school & church spire, after 0.7m go L. Garden 50 metres, next but one to Hare & Hounds. Parking on street.* **Sun 21 Mar (2-4). Adm £5, chd free. Evening opening Fri 11 June (6-8). Adm £10. Wine. Sun 13,**

Wed 16 June (12.30-5.30). Adm £5, chd free. Home-made teas.
2 acre spring and summer garden of surprises. Some areas ablaze, others calming. Formal ponds with box, yew, lavender, nepeta, and many different hostas, leading gently into lyrical colour-co-ordinated beds, rambling roses and fragrant Philadelphus. On to lawns bordered by stunning imaginative planting, expanding to an oak-bordered perennial wildflower meadow with a large pond; a wildlife haven. Starting with the sparkle of a formal fountain, drifting through lyrical island plantings to an oak-lined wildflower meadow and pond. Wander through many 'garden rooms' and relaxing spaces. Some onsite parking. Most areas accessible on grass paths.

 🚲 ✿ ☕

30 NEW HEYRONS

High Easter, Chelmsford, CM1 4QN. Mr Richard Wollaston. *½m outside High Easter towards Good Easter. From Chelmsford - A1060 to Good Easter. Follow road to High Easter. ½m before High Easter on L. From Dunmow R at Barnston to High Easter - ½m on R. Black barn & red brick buttresses.* **Sun 25 Apr (11-5). Adm £5, chd**

free. Light refreshments.
An informal garden of four parts created for family enjoyment within and around an ancient restored farmyard. An Essex barn and red brick farm buildings surround an intensely planted walled garden. A terraced area with mature trees, herbaceous and shady beds leads up to a rose garden. Beyond is an open area with grass tennis court, big skies, a small meadow and fine views over Essex countryside. The driveway is gravelled and there is a small curb and some shallow steps down into the farmyard, but a ramp will be in place on the steps.

 🚲 ✿ ☕

31 HILLDROP

Laindon Road (B1007), Horndon-On-The-Hill, Stanford-le-Hope, SS17 8QB. John Little & Fiona Crummay, www.grassroofcompany.co.uk. *Just N of Horndon on the Hill. 80m E of junction B1007 & Lower Dunton Rd.* **Sun 4 July (11-7). Adm £5, chd free. Light refreshments. veggie/ vegan snacks.**
Since building our turf roof house in 1995 we have trialled waste materials in our 4 acre garden to mimic brownfield habitat, one of the most undervalued

Clunes House

places for wildlife. These are now our most beautiful and diverse habitats for plants and insects. Several green roofs and lots of ideas to create biodiverse habitats, especially for solitary bees. Views across the Thames Estuary to Kent. The 4 acre garden features a self build timber house with green roof, 6 other green roofs, 3 ponds, brownfield landscapes to create plant diversity and solitary bee habitats. New area created this year, using local sand and gravel from A13 widening work to trial climate change plant species. Some wheelchair friendly paths and grass paths that may be accessible depending on rabbit damage. Garden is on a slight slope.

32 HORKESLEY HALL

Little Horkesley, Colchester, CO6 4DB. Mr & Mrs Johnny Eddis, 07808 599290, pollyeddis@hotmail.com, www.airbnb.co.uk/rooms/10354093. *6m N of Colchester City Centre. 2m W of A134, 10 mins from A12. At the grass triangle with tree in middle, turn into Little Horkesley Church car park to the very far end - access is via low double black gates at the far end.* Fri 12 Feb (12-4). Adm £6, chd free. Light refreshments in Little Horkesley Church (right next to house) if weather is poor. Visits also by arrangement Feb to Oct for groups of 10+. Excellent parking for coaches and cars. Teas held in church if poor weather.

Magical setting with 8 acres of romantic garden surrounding classical house. Parkland setting with major 20' high sculpture of balancing stones. Exceptional trees inc largest Gingko outside Kew. New planting of over 50 varieties of Iris from disbanded National Collection. Walled garden, snowdrops, hellebores, wildflower garden, superb hydrangeas, jungle walk, roses, charming enclosed swimming pool garden to relax with teas. Excellent Plant Stall. Soups, teas, cream teas, cakes, meringues, ice creams available depending on the season. Partial wheelchair access to some areas, gravel paths and slopes but easy access to tea area with lovely views over lake and garden.

33 61 HUMBER AVENUE

South Ockendon, RM15 5JW. Mr & Mrs Kasia & Greg Purton-Dmowski, 07711 721629, kasia.purtondmowski@gmail.com.

M25 - J30/31 or A13 - exit to Lakeside. J31/M25 to Thurrock Services, or Grays from A13, at r'about take exit onto Ship Ln to Aveley, turn R then 2nd exit at r'about onto Stifford Rd/B1335, take Foyle Dr to Humber Ave. Sun 13 June, Sun 25 July (12-5). Adm £4, chd free. Home-made teas. Gluten Free refreshments available. Visits also by arrangement Apr to Oct for groups of 10+.

A suburban garden close to Nature Rocorvo, 30m x 14m space featuring an old cherry tree from the orchard of the famous Belhus Mansion. The garden features abundant large borders planted heavily with herbaceous plants, small trees, roses, lilies, and ornamental grasses with oriental senses mixed with a traditional English feel with shady areas with hostas, tree ferns and Japanese style plants. The owners are interested in design and are members of Sogetsu International School, original arrangements will be on display during open day.

34 ISABELLA'S GARDEN

42 Theobalds Road, Leigh-On-Sea, SS9 2NE. Mrs Elizabeth Isabella Ling-Locke, 01702714424, ling_locke@yahoo.co.uk. *Take A13 towards Southend on Sea As you pass welcome to Leigh-on-Sea sign, turn R at T-lights onto Thames Drive then L onto Western Road, carry on 0.6m then R onto Theobalds Rd.* Sat 5, Sun 6 June, Sat 17, Sun 18 July (11.30-5). Adm £4, chd free. Home-made teas. Visits also by arrangement June to Sept for groups of 5 to 20.

This enchanting town garden is bursting with a profusion of colour from early Spring through to the Autumn months. Roses, clematis, agapanthus, herbaceous plants, alpines and pots with unusual succulents fill every corner of this garden. There is a wildlife pond. Lilies and water features as well as other garden ornaments which are to be found hiding within the shrubbery and throughout the garden. This garden is situated in the town of Leigh on Sea, and just a 5 min walk from the cockle sheds of old Leigh and Leigh railway station. Cakes made by Broadway Belles, the local WI. Most of the garden is accessible, there are steps near to the house.

35 NEW 38 JUNIPER ROAD

Boreham, Chelmsford, CM3 3DX. Sharon Rose. *Opposite Boreham Primary School.* Sat 5, Sun 6 June (9-5). Adm £3.50, chd free. Cream teas.

End of terrace house situated within the heart of Boreham Village. Three distinctly different gardens in one, with a colour palette ranging from hot reds/oranges to the softer pinks and lilacs with a little bit of the jungle thrown in to spice up the mix. Built on a budget with lots of recycling ideas from waterfalls to sundials.

36 KAMALA

262 Main Road, Hawkwell, Hockley, SS5 4NW. Karen Mann, 07976 272999, karenmann10@hotmail.com. *3m NE of Rayleigh. From A127 at Rayleigh Weir take B1013 towards Hockley. Garden on L after White Hart Pub & village green.* Sat 31 July, Sun 22 Aug (12-5). Adm £4, chd free. Home-made teas. Visits also by arrangement in Aug for groups of 10+.

Come and enjoy spectacular herbaceous borders which sing with colour as displays of salvia are surpassed by Dahlia drifts. Gingers, brugmansia, various bananas, bamboos and canna add an exotic note. Grasses sway above the blooms, giving movement. Rest awhile in the rose clad pergola while listening to the two Amazon Parrots in the aviary. The garden also features a completely new, RHS accredited Dahlia named "Jake Mann". Trees and shrubs include Acers, Catalpa aurea, Cercis 'Forest Pansy' and a large unusual Sinocalycanthus (Chinese Allspice) a stunning, rare plant with fantastic flowers.

"Unfortunately, cancer was not in lockdown in 2020. The continued support of our long-standing and valued partner, the National Garden Scheme is more important than ever."
Macmillan

37 KEEWAY
Ferry Road, Creeksea, nr Burnham-on-Crouch, CM0 8PL. John & Sue Ketteley, 01621 782083, sueketteley@hotmail.com. *2m W of Burnham-on-Crouch. B1010 to Burnham on Crouch. At town sign take 1st R into Ferry Rd signed Creeksea & Burnham Golf Club & follow NGS signs.* **Sat 26, Wed 30 June (2-5). Adm £5, chd free. Home-made teas. Visits also by arrangement June & July for groups of 10 to 30. Adm for groups includes home-made tea and cake.**
Large, mature country garden with stunning views over the River Crouch. Formal terraces surround the house with steps leading to sweeping lawns, mixed borders packed full of bulbs and perennials, formal rose and herb garden with interesting water feature. Further afield there are wilder areas, paddocks and lake. A productive greenhouse, vegetable and cutting gardens complete the picture.
&. ✿ ☕

38 KELVEDON HALL
Kelvedon, Colchester, CO5 9BN. Mr & Mrs Jack Inglis, v_inglis@btinternet.com. *Take Maldon Rd direction Great Braxted from Kelvedon High St. Go over R Blackwater bridge & bridge over A12 At T-junction turn R onto Kelvedon Rd. Take 1st L, single gravel road, oak tree on corner.* **Visits by arrangement Apr to June for groups of 20+. Adm £7.50, chd free. Home-made teas in the Courtyard Garden by the house or for larger numbers in the Pool Walled Garden - weather permitting. Teas, coffees and cakes.**
Varied 6 acre garden surrounding a gorgeous C18 house. A blend of formal and informal spaces interspersed with modern sculpture. Pleached hornbeam and yew and box topiary provide structure. A courtyard walled garden juxtaposes a modern walled pool garden, both providing season long displays. Herbaceous borders offset an abundance of roses around the house. Lily covered ponds with a wet garden. Topiary, sculpture, tulips and roses. Wheelchair access not ideal as there is a lot of gravel.
✿ Ⓓ ☕

39 NEW LITTLE HAVENS HOSPICE
Daws Heath Road, Hadleigh, Benfleet, SS7 2LH. Jane Hopkins. *Off A127 Rayleigh Weir. From A127 take A129 Rayleigh/Hadleigh. At the r'about take exit onto A129 towards Hadleigh. At next r'about take 1st exit onto Daws Heath Rd, & Little Havens is on L.* **Sun 11 July (10-4). Adm £4, chd free. Pre-booking essential, please visit www.ngs. org.uk for information & booking. Home-made teas.**
The gardens and surrounding areas are an important part of the care Little Havens provides for the children and their families who use the hospice. The gardens are a beautiful and peaceful retreat and offers plenty of play opportunities. There are gently sloping paths which allow children to access all parts of the garden including wheelchair users. Potting area and vegetable garden. Most of our garden is wheelchair accessible.
&. ✿ ☕

40 LITTLE MYLES
Ongar Road, Stondon Massey, CM15 0LD. Judy & Adrian Cowan. *1½ m SE of Chipping Ongar. Off A128 at Stag Pub, Marden Ash, towards Stondon Massey. Over bridge, 1st house on R after 'S' bend. 400yds Ongar side of Stondon Church.* **Fri 9, Wed 14 July (2-5). Adm £5, chd free. Home-made teas.**
A 3 acre romantic garden that has shifted its emphasis to providing an abundance of nectar-rich flowers for struggling bees and butterflies. The Herb garden full of vipers bugloss and roses. Meandering paths past full borders, themed hidden gardens, hornbeam pergola, sculptures and tranquil benches. Hand painted beach & jungle mural. Crafts and handmade herbal cosmetics for sale. Explorers sheet and map for children. New expanded flower meadow full of colourful nectar-rich flowers for bees and butterflies. Gravel paths. No disabled WC available.
&. ✿ ☕

41 LONG HOUSE PLANTS
Church Road, Noak Hill, Romford, RM4 1LD. Tim Carter, 01708 371719, tim@thelonghouse.net, www.longhouse-plants.co.uk. *3½ m NW of J28 M25. J28 M25 take A1023 Brentwood. At 1st T-lights, turn L to South Weald after 0.8m turn L at T junction. After 1.6m turn L, over M25 after ½ m turn R into Church Rd, nursery opp church.* **Wed 7 July, Wed 4 Aug, Wed 8 Sept (11-4). Adm £6, chd £3. Home-made teas. Visits also by arrangement June to Oct for groups of 30+.**
A beautiful garden - yes, but one with a purpose. Long House Plants has been producing home grown plants for more than 10 years - here is a chance to see where it all begins! With wide paths and plenty of seats carefully placed to enjoy the plants and views. It has been thoughtfully designed so that the collections of plants look great together through all seasons. Longhouse Plants garden was featured on BBC Gardeners World in October 2020 presented with Joe Swift. Disabled Car Parking. Paths are suitable for wheelchairs and mobility scooters. Disabled toilet in nursery. Two small cobbled areas not suitable.
&. ✿ 🚗 ☕

42 LONGYARD COTTAGE
Betts Lane, Nazeing, Waltham Abbey, EN9 2DA. Jackie & John Copping, 07780 802863, Nigella11@btopenworld.com. *Opposite red telephone box.* **Sun 14 Feb (11-4). Adm £4, chd free. Tea. Visits also by arrangement Apr to Sept for groups of 5 to 30.**
Longyard Cottage is an interesting ¾ of an acre garden situated within yards of an SSSI site, a C11 Church, a myriad of footpaths and a fine display of snowdrops and spring bulbs. This conceptual garden is based on a 'journey' and depicted through the use of paths which take you through 3 distinct areas, with its own characteristics. It's a tactile garden with which you can engage or simply sit and relax.
✿ ☕

43 LOXLEY HOUSE
49 Robin Hood Road, Brentwood, CM15 9EL. Robert & Helen Smith. *1m N of Brentwood town centre. On A128 N towards Ongar turn R onto Doddinghurst Rd at mini r'about. Take the 1st rd on L into Robin Hood Rd. 2 houses before the bend on L.* **Sat 17 Apr, Sat 17 July (11-3). Adm £4, chd free. Home-made teas.**
On entering the rear garden you will be surprised and delighted by this town garden. A colourful patio with pots and containers. Steps up onto a lawn with new circular beds surrounded by hedges, herbaceous

borders, trees and climbers. 2 water features, one a Japanese theme and another with ferns in a quiet seating area. The garden is planted to offer colour throughout the seasons.

44 16 MAIDA WAY, E4
Chingford, E4 7JL. Clare & Steve Francis. *1m from Chingford town centre off Kings Head Hill. Maida Way is a cul-de-sac off Maida Avenue that can be accessed via Kings Head Hill or Sewardstone Rd.* **Sat 24 July (2-7). Adm £7.50, chd free. Wine. Adm includes a glass of wine and selection of canapes.** There are three distinct areas to the garden. A walled patio with raised beds of ferns, shrubs,climbers, hostas and pots. Steps up to middle garden with a large koi pond, seating area, with acers, shrubs, herbaceous plants, annuals, and grasses. Numerous retro artefacts creatively upcycled. The third area is a kitchen garden with raised beds of fruit trees, bushes and canes; vegetables and herbs.

45 NEW MALTINGS COTTAGE
High Street, Ingatestone, CM4 9EZ. Richard & Susan Martin, 07747 765512. *Leave A12 on B1002 exit to Ingatestone. Maltings Cottage is on High Street B1002. Parking at Station Car Park CM4 0BW 8 mins walk. Disabled parking please contact owner to reserve space on driveway.* **Sat 18 Sept (12.30-4.30). Adm £4, chd free. Home-made teas.** The garden echoes the look and feel of a traditional cottage garden reflecting the character of the property. Rear courtyard surrounded by large structural plants including banana plants, dahlias, cannas and Japanese anemones. Dry and Shade area beds. Front garden has a modern slant with pleached hornbeams and wildflower meadow planting using cosmos, zinnia, echinacea and lupin.

46 9 MALYON ROAD
Witham, CM8 1DF. Maureen & Stephen Hicks. *Car park at bottom of High St opp Swan Pub, 5 min walk from car park cross High St by pedestrian crossing onto River Walk, follow path, take 1st R turn into Luard Way, Malyon Rd straight ahead.* **Sun 27 June, Sat 3, Sun 4 July (10.30-5). Adm £4, chd free. Home-made teas. Opening with**

Witham Town Gardens on Sat 26 June. Large town garden with mature trees and shrubs made up of a series of garden rooms. Flower beds, pond, summerhouse and greenhouse giving all year interest. Plenty of places to sit and relax with paths that take you on a tour of the garden. A quiet hidden place not expected in a busy town. Our garden is a hidden oasis, where you can escape from the hustle and bustle of busy life. On the edge of town, close to the Witham's rambling Town River Walk. Wheelchair access, but there is a step down into part of the garden.

47 MIRAFLORES
Witham, CM8 2LJ. Yvonne & Danny Owen, 07976 603863, danny@dannyowen.co.uk. *Please DO NOT go via Rowan Way, go to Forest Rd as there is rear access only to garden. Access is ONLY by rear gate at the top The Spinney, off of FOREST RD, CM8 2TP. Please follow yellow signs. PLEASE do not park in The Spinney as this is a private parking area.* **Sun 6, Sun 13, Sun 27 June (2-5). Adm £4, chd free. Home-made teas. Visits also by arrangement in June for groups of 10 to 30.** An award-winning, medium-sized garden described as a "Little Bit of Heaven'. A blaze of colour with roses, clematis, pergola, rose arch, triple fountain with box hedging and deep herbaceous borders. See our 'Folly', exuberant and cascading hanging baskets... find our Secret Door. Featured in Garden Answers, Amateur Gardening and Essex Life. Tranquil seating areas incl new summerhouse for 2020. Cakes (incl gluten free), savouries, teas, coffee, soft drinks, herbal teas are available. No access for wheelchair users as step up to garden.

48 MOVERONS
Brightlingsea, CO7 0SB. Lesley & Payne Gunfield, lesleyorrock@me.com, www.moverons.co.uk. *7m SE of Colchester. At old church turn R signed Moverons Farm. Follow lane & garden signs for approx 1m. Beware some SatNavs take you the wrong side of the river.* **Open garden with Sculpture Exhibition. Visits by arrangement for groups of 10+. Light refreshments.** Tranquil 4 acre garden in touch with its surroundings and enjoying

stunning estuary views. A wide variety of planting in mixed borders to suit different growing conditions and provide all year colour. Courtyard, large natural ponds, sculptures and barn for rainy day teas! We're busy redeveloping the reflection pool area. Magnificent trees some over 300yrs old give this garden real presence. Most of the garden is accessible by wheelchair via grass and gravel paths, There are some steps and bark paths.

49 23 NEW ROAD
Dagenham, RM10 9NH. John Seaman, 07504 712818, echinopsis100@gmail.com. *Leave A13 at the junction signed for A1306 - Dagenham & Hornchurch. Go around the r'about following it off at the 4th exit. Then follow the 2nd r'about straight onwards.* **Sat 14, Sun 15 Aug (12-5). Adm £4, chd free. Visits also by arrangement for groups of 5 to 20.** This is an exotic garden, themed on the foothills of the Himalayas in India. This garden will be like no other you may have seen before despite its modest size. It is filled with exquisite tropical plants including Oanna, Gingers and Bananas. A feast for the eyes filled with good ideas for vertical space. An extravaganza of plants in all shapes, colour and form.

50 OAK FARM
Vernons Road, Wakes Colne, CO6 2AH. Ann & Peter Chillingworth. *Vernons Road off A1124 between Ford Street & Chappel. From Colchester, 3rd R after Ford St; Oak Farm is 200 metres up lane on R. From Halstead, 2nd L after viaduct in Chappel.* **Wed 9 June, Wed 7 July, Wed 8 Sept (2-5). Adm £4, chd free. Home-made teas.** Farmhouse garden of about 1 acre on an exposed site with extensive views south and west across the Colne Valley. Shrubberies, lawned borders and shady spots where people can enjoy refreshments. Look out for the secret garden. On-going projects include the rose avenue, Mediterranean bed and prairie garden. Although there are some steps in places, access can be gained to most of the garden.

Blake Hall

51 OLD BELL COTTAGE

Old Bell Cottage, Langley Upper Green, Saffron Walden, CB11 4RU. Richard Vallance, 01799 550474, r.vallance1234@gmail.com. *On rd to & from Clavering, opp the children's playground, house is painted red.* **Sun 27 June (2-5). Combined adm with Wickets £7.50, chd free. Home-made teas at Wickets. Individual garden adm £5. Visits also by arrangement May to July.**

Cottage garden, successfully incorporating part of a former field, with various herbaceous beds. Well stocked with spring bulbs, Alliums, Lupins and Delphiniums. Tree planting so as not to obscure views across valley. Greenhouse, patio area and cabinets for hardening off. Raised beds for vegetables and in fruit cage, to overcome lack of drainage in lower garden. Sunken BBQ area. Natural pond. The only limitation for wheelchair access is the need to travel over mown lawns.

&. 🐾 ☕

52 THE OLD VICARAGE

Church End, Broxted, CM6 2BU. Ruth & Adam Tidball. *3m S of Thaxted. From Thaxted on the B1051, take Broxted turning. Just before next junction, turn R into field for parking.* **Sun 9 May, Sun 20 June (1-5). Adm £5, chd free. Light refreshments.**

Victorian former vicarage surrounded by three acres of formal gardens, one-acre woodland garden and seven-acre hay meadow. Deep mixed borders. Rose garden with recently added perennial underplanting. Many rare and special plants and trees. Alpine and succulent rockeries, series of ponds providing habitat for ornamental goldfish and native wildlife. Views across the meadow to church and beyond.

&. 🐾 ❄ ☕

53 PEACOCKS

Main Road, Margaretting, CM4 9HY. Phil Torr, 07802 472382, phil.torr@btinternet.com. *Margaretting Village Centre. From village xrds go 75yds in the direction of Ingatestone, entrance gates will be found on L set back 50 feet from road frontage.* **Sun 13 June (11-4). Adm £5, chd free. Pre-booking essential, please visit www.ngs. org.uk for information & booking. Visits also by arrangement Mar to July for groups of 20+. Donation to**

St Francis Hospice.

5 acre garden, mature native and specimen trees. Restored horticultural buildings. Series of garden rooms including walled Paradise Gardens, Garden of Reconciliation, Alhambra fusion. Long herbaceous/mixed border. Temple of Antheia on the banks of a lily lake. Large areas for wildlife incl woodland walk, nuttery and orchard. Traditionally managed wildflower meadow. Sunken dell with waterfall. Narrow gauge railway trips on some days. Display of old Margaretting postcards. Garden sculpture. Most of garden wheelchair accessible.

&. 🚗 ☕ 🧺 🍽

54 18 PETTITS BOULEVARD, RM1

Rise Park, Romford, RM1 4PL. Peter & Lynn Nutley. *From M25 take A12 towards London, turn R at Pettits Lane junction then R again into Pettits Blvd or Romford Stn then 103 or 499 bus to Romford Fire Stn and follow NGS signs.* **Sat 26, Sun 27 June, Sat 4, Sun 5 Sept (1-5). Adm £3.50, chd free. Home-made teas.**

The garden is 80ft x 23ft on three levels with an ornamental pond, patio area with shrubs and perennials, many in pots. Large eucalyptus tree leads to a woodland themed area with many ferns and hostas. There are agricultural implements and garden ornaments giving a unique and quirky feel to the garden. There are also tranquil seating areas situated throughout.

&. 🐾 ☕

55 PIERCEWEBBS

40 Pelham Road, Clavering, Saffron Walden, CB11 4PQ. Mrs J William-Powlett, 01799 550809, jwp@william-powlett.net. *7m N of Bishops Stortford on B1038. Turn W onto B1383 at Newport.* **Visits by arrangement May to Sept for groups of 10 to 20. Home-made teas.**

Formal old walled garden, shrubs, lawns, ha-ha, yew with topiary, stilt hedge, pond and trellised rose garden. Landscaped field walk. Extensive views over countryside. Wendy House; field walk with swing for adults and children. Wheelchair access over gravel drive or via a shallow step. Ramp provided with advance warning.

&. 🐾 🚗 ☕ 🍽

56 ROOKWOODS

Yeldham Road, Sible Hedingham, CO9 3QG. Peter & Sandra Robinson, 07770 957111, sandy1989@btinternet.com, www.rookwoodsgarden.com. *8m NW of Halstead. Entering Sible Hedingham from direction of Haverhill on A1017 take 1st R just after 30mph sign. Coming through SH from the Braintree direction turn L just before the 40mph leaving the village.* **Mon 31 May (10.30-4.30). Adm £7, chd free. Home-made teas. Visits also by arrangement Apr to Sept.**

Rookwoods is a tranquil garden where you can enjoy a variety of mature trees; herbaceous borders; a shrubbery; pleached hornbeam rooms; a work-in-progress wildflower bed; a buttercup meadow, and an ancient oak wood. There is no need to walk far, you can come and linger over tea, under a dreamy wisteria canopy while enjoying views across the garden. Beehive. Wildflower garden. Meadow to walk through leading to Ancient Oak Wood. Mature Foxglove tree. Terrace with views across garden. Gravel drive.

&. ❄ 🚗 ☕ 🍽

57 79 ROYSTON AVENUE, E4

Chingford, E4 9DE. Paul & Christine Lidbury. *From A406 Crooked Billet r'about take A112 towards Chingford. Continue ½m, across T-lights (Morrison's), Royston Ave is 3rd turn on R. Bus 97, 158, 215, 357 (Leonard Rd or Ainslie Wood Rd stops).* **Mon 31 May (12-5). Adm £3.50, chd free. Home-made teas.**

A 55 x 19ft urban garden with over 600 different varieties of plants - a great many in containers, including collections of Hostas, Acers, Heucheras and Sempervivums. An oasis for local birds and wildlife with two small ponds and insect habitats with further interest provided by ornaments, sculptures and artwork, where words of gardening wisdom abound. Not suitable for young children. Tea/coffee and home-made cakes. Quality home-brewed beer tastings available.

❄ ☕

All entries are subject to change. For the latest information check ngs.org.uk

Heyrons

58 69 RUNDELLS - THE SECRET GARDEN

Harlow, CM18 7HD. Mr & Mrs K Naunton, 01279 303471, k_naunton@hotmail.com. *3m from J7 M11. A414 exit T-lights take L exit Southern Way, mini r'about 1st exit Trotters Rd leading into Commonside Rd, after shops on left 3rd L into Rundells.* **Sat 17 July (2-5). Adm £3, chd free. Home-made teas. Visits also by arrangement Apr to Sept for groups of 5 to 30.**
Featured on Alan Titchmarsh's first 'Love Your Garden' series ('The Secret Garden') 69, Rundells is a very colourful, small town garden packed with a wide variety of shrubs, perennials, herbaceous and bedding plants in over 200 assorted containers. Hard landscaping on different levels has a summer house, various seating areas and water features. Steep steps. Access to adjacent allotment open to view. Various small secluded seating areas. A small fairy garden has been added to give interest for younger visitors. The garden is next to a large allotment and this is open to view with lots of interesting features including a bee apiary. Honey and other produce for sale (conditions permitting). Cakes, tea, coffee and soft drinks available.

❄ ☕ ▭

59 ST HELENS

High Street, Stebbing, CM6 3SE. Stephen & Joan Bazlinton, 01371 856495, jbazlinton@gmail.com. *3m E of Great Dunmow. Leave Gt Dunmow on B1256. Take 1st L to Stebbing, at T-junction turn L into High St, garden 2nd on R.* **Visits by arrangement Apr to July. Adm £8, chd free. Home-made teas. Donation to Dentaid.**
A garden established over 40 years from a bog-ridden, cricket-bat plantation into a gently sloping woodland garden. Springs and ponds divide the garden into different areas with various shrubs and perennial planting. Partial wheelchair access.

♿ ☕ ▭

60 SANDY LODGE

Howe Drive, Hedingham Road, Halstead, CO9 2QL. Emma & Rick Rengasamy. *8m NE of Braintree. Turn off Hedingham Rd into Ashlong Grove. Howe Drive is on L. Please park in Ashlong Grove & walk up Howe Drive.* **Sun 19 Sept (11-5). Adm £4.50, chd free. Home-made teas.**
A stunning garden, with amazing views over Halstead, there is something to see every day. $3/4$ acre created over the last 5 years, enter into our gravel Bee Border, wander in our Winter Wedding border and the Woodlands Walk. Across the lawn you find double borders with flowing prairie planting. Lots of seating and viewing spots. Featured in the Garden Gate is Open https://thegardengateisopen.blog/. Wheelchair access restricted due to large amount of gravel.

🐕 ☕

61 SCRIPS HOUSE

Cut Hedge Lane, Coggeshall, CO6 1RL. Mr & Mrs James & Sophie Bardrick. *From A12 N/S take Kelvedon/Feering exit. Turn L/R passed Kelvedon stn. Continue for 2m turn L into Scrips Rd. From A120 take Coggeshall exit. Drive through village towards Kelvedon. Turn R up Scrips Rd.* **Fri 30 Apr, Fri 2 July (10-3). Adm £6, chd free. Home-made teas.**
A large garden with further paddocks. Includes a mature woodland with a memorial avenue of fastigiate oaks. The garden is divided into smaller areas including a walled pool garden and a Spring garden. A pleached lime walk leads to a white garden flanked by two 30m herbaceous borders. There is an ornamental pond with ducks, an ample vegetable garden and fruit cage, an orchard and a chicken run. Most areas accessible with a fairly strong wheelchair pusher.

♿ ❄ ☕ ▭

GROUP OPENING

62 NEW SHEEPCOTE GREEN GARDENS

Sheepcote Green, Clavering, Saffron Walden, CB11 4SJ. *Midway between Clavering & Langley. Take lane signed Sheepcote Green on Clavering to Langley Lower Green road. Park on the green at the end of the lane.* **Sun 30 May (1-5).**

Combined adm £8, chd free. Home-made teas.

NEW APPLETREE COTTAGE
Judy & Adrian Wilson-Smith.

APRIL COTTAGE
Anne & Neil Harris.

NEW SHEEPCOTE GREEN HOUSE
Jilly & Ross McNaughton, www.thehabitatgarden.co.uk.

NEW WAGGON AND HORSES
Jenny & Peter Milledge.

Four quite different - yet equally charming - gardens of varying size, offering inspiration and delight! The first three gardens neighbour one another, while the Waggon and Horses is just a short stroll up the leafy footpath. April Cottage boasts a well-established, colourful cottage garden, filmed in 2020 with Alan Titchmarsh for the Love Your Garden Series. Appletree Cottage is a compact cottage garden of fragrant roses, clematis, honeysuckle with annuals and perennial beds. Sheepcote Green House is an atmospheric, old country garden, set in the three acre grounds of a former farmhouse, currently undergoing a gentle reawakening with wildlife in mind. The Waggon and Horses is a ¼ acre garden offering rural views and mixed planting incorporating rhubarb and soft fruit amongst flowers grown for perfume and pollinators. All enjoy an idyllic setting in a hamlet in the ancient countryside of North West Essex. Teas served on the lawn at Sheepcote Green House. Plant sale on the green. Partial wheelchair access due to gravel, narrow paths and uneven surfaces.

& ❀ ☕ ☕

63 SHRUBS FARM
Lamarsh, Bures, CO8 5EA. Mr & Mrs Robert Erith, 01787 227520, bob@shrubsfarm.co.uk, www.shrubsfarm.co.uk. 1¼ m from Bures. On rd to Lamarsh, the drive is signed to Shrubs Farm. Visits by arrangement Apr to Oct for groups of 5+. Home-made teas £6 per head. Adm £7, chd free. Wine & canapes by arrangement.
2 acres with shrub borders, lawns, roses and trees. 50 acres parkland with wildflower paths and woodland trails. Over 60 species of oak. Superb 10m views over Stour valley. Ancient coppice and pollards incl largest goat

(pussy) willow (*Salix caprea*) in England. Wollemi and Norfolk pines, and banana trees. Full size black rhinoceros. Display of Bronze Age burial urns. Large grass maze. Guided Tour to incl garden, park and ancient woodland, historic items including Bronze Age Burial Urns and painting of the Stour valley 200 years ago. Restored C18 Essex barn is available for refreshment by prior arrangement. Tea proceeds to Lamarsh Church. Some ground may be boggy in wet weather.

& ☕ ☕ ☕

64 SILVER BIRCHES
Quendon Drive, Waltham Abbey, EN9 1LG. Frank & Linda Jewson, 01992 714047, linda.jewson@ngs.org. M25, J26 to Waltham Abbey. At T-lights by McD turn R to r'about. Take 2nd exit to next r'about. Take 3rd exit (A112) to T-lights. L to Monkswood Av follow ngs signage. Sun 30 May, Sun 27 June, Sun 12 Sept (11.30-5). Adm £5, chd free. Home-made teas. Visits also by arrangement May to Sept for groups of 10 to 20.
The garden boasts three lawns on the two levels. This surprisingly secluded garden has many mixed borders packed with colour. Mature shrubs and trees create a short woodland walk. Crystal clear water flows through a shady area of the garden. Wheelchairs are able to visit the garden, but there are areas which would not be suitable.

& ❀ ☕

65 SNARES HILL COTTAGE
Duck End, Stebbing, CM6 3RY. Pete & Liz Stabler, 01371 856565, petestabler@gmail.com. Between Dunmow & Bardfield. On B1057 from Great Dunmow to Great Bardfield, ½ m after Bran End on L. Visits by arrangement Apr to Sept for groups of 5+. Adm £5, chd free. Home-made teas.
A 'quintessential English Garden' - Gardeners World. Our quirky 1½ acre garden has surprises round every corner and many interesting sculptures. A natural swimming pool is bordered by romantic flower beds, herb garden and Victorian folly. A bog garden borders woods and leads to silver birch copse, beach garden and 'Roman' temple. Classic cars. Shepherds Hut. Not wheelchair friendly as it is a hilly garden with some steep slopes.

☕ ☕ ☕

66 SPRING COTTAGE
Chapel Lane, Elmstead Market, CO7 7AG. Mr & Mrs Roger & Sharon Sciachettano. 3m from Colchester. Overlooking village green North of A133 through Elmstead Market. Parking limited adjacent to cottage, village car park nearby on South side of A133. Sat 26, Sun 27 June (1.30-4.30). Adm £4, chd free. Home-made teas on Elmstead Market Green, in front of the cottage.
From Acteas to Zauschenerias and Aressima to Zobra grass we hope our large variety of plants will please. Our award winning garden features a range of styles and habitats e.g. woodland dell, stumpery, Mediterranean area, perennial borders and pond. Our C17 thatched cottage and garden show case a number of plants found at the world famous Beth Chatto gardens ½ m down the road. In early June our 17 rose varieties take centre stage and several clematis will be at their best. We can promise colour and scent to delight the senses and a range of unusual species to please the plant enthusiast. Refreshments provided by the WI served on the village green in front of Spring Cottage.

❀ ☕

67 NEW 2 SPRING COTTAGES
Conduit Lane, Woodham Mortimer, CM9 6SZ. Sharon & Michael Cox, 07841867908, coxmichael1958@gmail.com. From Danbury continue on A414 towards Maldon for 1½ m until you come to Oak Corner r'about, 1st L continuing on A414. 200 yards take 1st R into Conduit Lane. Fri 16, Sun 18 July (10-5). Adm £4, chd free. Cream teas and home-made cakes available. Visits also by arrangement in July for groups of up to 20.
A small pretty terraced cottage garden, featuring many different levels, each area exhibiting good use of space and varied planting. Plenty of seating in a relaxed quiet setting. The garden has been developed to encourage wildlife, with an abundance of bee, butterfly, bird houses water feature and lilly pond.

Online booking is available for many of our gardens, at ngs.org.uk

68 NEW STOCKSMEAD

Wicken Road, Arkesden, CB11 4EY. Mr & Mrs Paul & Louise Kimberley. *5m W of Saffron Walden. 3m W from Newport via B1038 through Wicken Bonhunt. Enter Wicken Road from the village we are the last house on R. Conversely if you enter Wicken Road from Wicken Bonhunt we are the 1st house on L as you enter the village.* Sun 6 June (11-4). Adm £5, chd free. Home-made teas.

Set in nearly one acre, Stocksmead is formed of four distinct garden areas, a Mediterranean inspired gravel garden, a colour gravel garden, a traditional perennial garden and a structural evergreen garden. With far reaching views over the Essex countryside, there is something of interest for everyone.

69 STRANDLANDS

off Rectory Road, Wrabness, Manningtree, CO11 2TX. Jenny & David Edmunds, 01255 886260, strandlands@outlook.com. *1km along farm track from the corner of Rectory Rd. If using a SatNav, the post code will leave you at the corner of Rectory Rd. Turn onto a farm track, signed to Woodcutters Cottage & Strandlands, & continue for 1km.* Visits by arrangement May & June for groups of 10 to 20. Adm £5, chd free. Tea, coffee & a slice of home-made cake.

Cottage surrounded by 4 acres of land bordering beautiful and unspoilt Stour Estuary. One acre of decorative garden: formal courtyard with yew, box and perovskia hedges, lily pond, summerhouse and greenhouse; 2 large island beds, secret 'moon garden', madly and vividly planted 'Madison' garden, 3 acres of wildlife meadows with groups of native trees, large wildlife pond, also riverside bird hide. View the Stour Estuary from our own bird hide. Grayson Perry's 'A House for Essex' can be seen just one field away from Strandlands. Mostly accessible and flat although parking area is gravelled.

70 TUDOR ROOST

18 Frere Way, Fingringhoe, Colchester, CO5 7BP. Chris & Linda Pegden, 01206 729831. *5m S of Colchester. In Fingringhoe by Whalebone PH follow sign to Ballast Quay, after ½ m turn R into Brook Hall Rd, then 1st L into Frere Way.* Sun 30, Mon 31 May, Sat 17, Sun 18 July (2-5). Adm £4, chd free. Home-made teas. Large conservatory to sit in if inclement weather.

An unexpected hidden colourful ¼-acre garden. Well manicured grassy paths wind round island beds and ponds. Densely planted subtropical area with architectural and exotic plants - cannas, bananas, palms, agapanthus, agaves and tree ferns surround a colourful gazebo. Garden planted to provide yr-round colour and encourage wildlife. Many peaceful seating areas. Within 1m of Fingringhoe Wick Nature Reserve.

71 37 TURPINS LANE

Chigwell, IG8 8AZ. Fabrice Aru & Martin Thurston, 0208 5050 739, martin.thurston@talktalk.net. *Between Woodford & Epping. Tube: Chigwell, 2m from North Circular Rd at Woodford, follow the signs for Chigwell (A113) through Woodford Bridge into Manor Rd & turn L, Bus 275 & W14.* Sun 13 June, Sun 4 July (11-6.30). Adm £4, chd free. Visits also by arrangement May to Oct for groups of up to 10.

An unexpected hidden, magical, part-walled garden showing how much can be achieved in a small space. An oasis of calm with densely planted rich, lush foliage, tree ferns, hostas, topiary and an abundance of well maintained shrubs complemented by a small pond and 3 water features designed for year round interest. Featured on BBC Gardeners' World and ITV Good Morning Britain.

72 TWO COTTAGES

Church Road, Chrishall, SG8 8QT. Michelle Thomas, 07581 745130, mrsdthomas@btinternet.com. *7m W of Saffron Walden. Continue on B1039, take R turn to Chrishall. Bury Lane leading into Church Rd. Two Cottages is 1st L with an old planted boat on the bank.* Sun 6, Sun 20 June (12-4). Adm £5, chd free. Home-made teas. Visits also by arrangement May to July for groups of up to 30.

Magic lurks within this charming 1½ acre garden, evolved over 30 yrs. Many treasures hidden amongst a variety of planting. Over 215 roses showcased in island borders. Meandering lawn paths lead through wisteria walkway to find three miniature Shetland ponies keen to show off to guests. Tranquil seating areas, teas, cakes and gifts for sale in 'The Shed' shop. The perfect place to buy unusual gifts. Large smoking dragon, once on display at Hampton Court garden show. Superb views over countryside to village church.

73 ULTING WICK

Crouchmans Farm Road, Maldon, CM9 6QX. Mr & Mrs B Burrough, 01245 380216, philippa. burrough@btinternet.com, www.ultingwickgarden.co.uk. *3m NW of Maldon. Take R turning to Ulting off B1019 as you exit Hatfield Peverel by a green. Garden on R after 2 M.* Wed 24 Feb (11.30-2). Adm £20. Light refreshments. Sun 25 Apr (11-5). Adm £6, chd free. Light refreshments. Fri 30 Apr, Fri 9 July, Mon 30 Aug, Fri 3 Sept (2-5). Adm £6, chd free. Home-made teas. Pre-booking essential, please visit www.ngs.org.uk for information & booking. Visits also by arrangement Jan to Nov. Winter Walks Talks priced at £20 pp incl homemade soup & rolls cake & tea. Donation to All Saints Ulting Church.

Listed black barns provide backdrop for vibrant and exuberant planting in 8 acres. Thousands of colourful tulips, flowing innovative spring planting, herbaceous borders, pond, mature weeping willows, kitchen garden, dramatic late summer beds with zingy, tender, exotic plant combinations. Drought tolerant perennial and mini annual wildflower meadows. Woodland. Many plants propagated in-house. Lots of unusual plants for sale. Beautiful dog walks along the R Chelmer from the garden. Some gravel around the house but main areas of interest are accessible for wheelchairs.

74 NEW 39 WAKERING AVENUE

Shoeburyness, Southend-On-Sea, SS3 9BE. Wendy Adlington. *Shoeburyness 3m E of Southend Town Centre on A13. Roadside parking available on Wakering Avenue and surrounding roads.* Sat 8 May (11-4). Adm £4, chd free. Home-made teas. Vegan and Gluten Free Cakes available.

A small well-established town garden with an interesting use of space creating several garden rooms. Seating areas capture the sun as it moves across the garden. Lush

2 Hope Cottages, Bradfield Gardens

green foliago of palm and cordyline contrast with established ceanothus and silver birch. Wisteria clambers over the pergola while arum and calla lilies, Californian poppies, nigella and bluebells fill the pots and beds.

75 1 WHITEHOUSE COTTAGES
Blue Mill Lane, Woodham Walter, CM9 6LR. Mrs Shelley Rand, 07799 848772, shelley@special-p co uk. *In between Maldon & Danbury, short drive from A12. From A414 Danbury, turn L at The Anchor & continue into the village. Directly after the white village gates at the far end of the village, turn R into Blue Mill Lane.* **Sun 16 May, Sun 13 June, Sun 25 July (10-3.30). Adm £4, chd free. Home-made teas. Visits also by arrangement May to Aug for groups of up to 30.**
Nestled betwixt farmland in rural Essex, is our small secret garden,

that has a wonderful charm and serenity to it. Set in 3½ acres, mostly paddocks, a little plot of loveliness wraps around our Victorian cottage, and roses smother the porch in June. A meandering lawn takes you through beds and borders softly planted with a cottage feel, a haven for wildlife and people alike. Dean Harris a local blacksmith will have a pop-up forge on-site making and selling metal plant accessories on the day of your visit. Parking available a short walk up the lane near I he Cats pub. Unsuitable for wheelchairs unless you're intrepid.

76 WICKETS
Langley Upper Green, CB11 4RY. Susan & Doug Copeland. *7m W of Saffron Walden, 10m N of Bishops Stortford. At Newport take B1038 After 3m turn R at Clavering, signed Langley. Upper Green is 3m further on. At cricket green turn R. House 200m on R.* **Sun 2 May (2-5). Adm**

£5, chd free. Sun 27 June (2-5). Combined adm with Old Bell Cottage £7.50, chd free. Home-made teas. Individual garden adm £5 on 27 June.
Rejuvenate your spirits in this wonderfully floral garden with fine rural views. Billowing Borders inc narcissus and tulips, roses and summer colour. wildflower meadow with shepherd's hut, orchard and small prairie/pond area. Lily pond with 'Monet' bridge sheltered by silver birch. New Mediterranean courtyard. Parterre enclosed by espalier apples. Lots of places to sit and enjoy the rural views. Perhaps the ultimate romantic garden? Gravel drive.

The National Garden Scheme searches the length and breadth of England and Wales for the very best private gardens

Bassetts

GROUP OPENING

77 NEW WITHAM TOWN GARDENS

Witham, CM8 1NB. 01376 514931, whitechat@sky.com. *Witham is on B1038 approx 8m from Chelmsford & 10m from Colchester. A map will be available at both gardens with locations and parking details.* Sat 26 June (11-4.30). Combined adm £6, chd free. Home-made teas.

9 MALYON ROAD
Maureen & Stephen Hicks. (See separate entry)

NEW 13 STEVENS ROAD
CM8 1NB. Robin & Isobel Norton.

A warm welcome awaits you at our two contrasting town gardens in Witham. 13 Stevens Road is an established garden full of plants including many Salvia, Sanguisorba and Thalictrum. From the lawn the garden slopes down to a secluded shaded area and stream. The paved patio is a welcoming area to sit and enjoy home-made cakes, tea and soft drinks as well as an opportunity to purchase home made jam. Malyon Road is a large town garden with mature trees and shrubs laid out in a series of garden rooms. Flower beds, pond and summerhouse give year round interest. Plenty of places to sit and relax with paths taking you on a tour of the garden. Partial wheelchair access at Stevens Road and Malyon Road. Plant Sales at Malyon Road.

&. ✿ ☕ ▆

"I love the National Garden Scheme which has been the most brilliant supporter of Queen's Nurses like me. It was founded by the Queen's Nursing Institute which makes me very proud. As we battle Coronavirus on the front line in the community, knowing we have their support is a real comfort." - Liz Alderton, Queen's Nurse

---

GROUP OPENING

78 WRITTLE GARDENS

Chelmsford, CM1 3NA. *Writtle can be approached from 3 directions. From the A1060, A1016 & A414 follow the yellow signs to the Village. A local map showing the gardens open in the village will be available at each house.* Sun 27 June (1-6). Combined adm £6, chd free.

8 THE GREEN
CM1 3DU. Andrea Johnson.

53 LONG BRANDOCKS
CM1 3JL. Roger & Margaret Barker.

65 ONGAR ROAD
CM1 3NA. Doug & Jean Pinkney.

40 ST JOHNS ROAD
CM1 3EB. Catherine Eubanks.

Four contrasting, colourful and interesting gardens to enjoy in the delightful village of Writtle. 8 The Green offers creatively planted borders and a tapestry of colour, texture and form, with perennials, shrubs, ornamental trees, annuals and alpines, and a south-facing summerhouse. The garden at 65 Ongar Road will transport visitors to tropical destinations, with its colour and "summer living" features. 40 St Johns Road is a relatively new garden, now in its 8th year and beginning to mature. The front is designed to echo the country hedgerows of the local area and the back is a tranquil haven in a modern Italian style. 53 Long Brandocks is a plantsman's garden with a wealth of unusual shrubs, clematis, daylilies, and some exotic herbaceous plants. There is also a selection of trees including catalpas and acers. No refreshments are offered at these gardens, but Writtle offers a number of pubs and cafes, including the renowned Tiptree Tea Room in Lordship Rd. The ancient and traditional village of Writtle, with its delightful Norman church, village green and pond, dates back to pre-Roman times, and was featured in the Doomsday Book.

79 WRITTLE UNIVERSITY COLLEGE

Writtle, CM1 3RR. Writtle University College, www.writtle.ac.uk. *4m W of Chelmsford. On A414, nr Writtle village.* Wed 28 Apr (10.30-3.30). Adm £5, chd free. Light refreshments in The Garden Room (main campus) & The Lordship tea room (Lordship campus).

Writtle University college has 15 acres of informal lawns with naturalized bulbs and wildflowers. Large tree collection, mixed shrubs, herbaceous border, dry/ Mediterranean borders, seasonal bedding and landscaped glasshouses . All gardens have been designed and built by our students studying a wide range of horticultural courses. Some gravel, however majority of areas accessible to all.

&. ✿ ☕ ▆

80 WYCHWOOD

Epping Road, Roydon, CM19 5DW. Mrs Madeleine Paine. *At Tylers Cross r'about head in the direction of Roydon. Garden on R approx 400 metres from the r'about. Parking is available in Redrick's nursery next to garden.* Sun 27 June (12-5.30). Adm £4, chd free. Home-made teas.

A garden approx. $\frac{3}{4}$ acre with a large pond, attracting much wildlife, as well as the owners resident ducks. Free ranging chickens roam in the shrubbery and budgerigar aviary. There are numerous features incl. vegetable and fruit plot, mixed shrub and herbaceous borders, 1920's summer house and Scandinavian cabin. Old fashioned style roses are a particular feature of the garden. Lake fully stocked with fish and inhabited by resident and wild ducks. Sit in the Norwegian hut and enjoy your refreshments.

✿ ☕ ▆

81 WYCKE FARM

Pages Lane, Tolleshunt D'Arcy, CM9 8AB. Nancy & Anthony Seabrook. *5m E of Maldon, 10m SW of Colchester. B1023 from Tolleshunt D'Arcy 1m towards Tollesbury. Turn R into Pages Lane. Follow for 1m to Wycke Farm.* Fri 23, Sun 25 Apr (11-5). Adm £5, chd free. Home-made teas.

Large cottage style farmhouse garden situated in the peaceful Essex countryside with mature trees, mixed borders, vegetables, greenhouses and a small flock of sheep. Developed from a neglected state over 15 years ago with a fine view of the Blackwater Estuary. Walk to estuary, 1200 metres. Some gravel and grass paths.

&. ✿ ☕ ▆

GLOUCESTERSHIRE

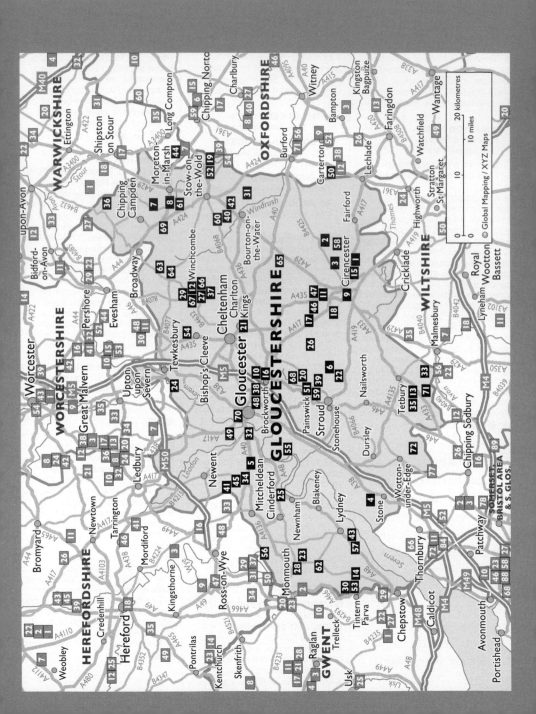

Gloucestershire is one of the most beautiful counties in England, spanning as it does a large part of the area known as the Cotswolds as well as the Forest of Dean and Wye and Severn Valleys.

The Cotswolds is an expanse of gently sloping green hills, wooded valleys and ancient, picturesque towns and villages; it is designated as an area of Outstanding Natural Beauty, and its quintessentially English charm attracts many visitors.

Like the county itself many of the gardens that open for the National Garden Scheme are simply quite outstanding. There are significant gardens which open for the public as well, such as Kiftsgate and Bourton House. There are also some large private gardens which only open for us, such as Daylesford and Stowell Park.

There are however many more modest private gardens whose doors only open on the National Garden Scheme open day, such as Bowling Green Road in Cirencester with over 300 varieties of Hemerocallis. This tiny garden has now opened for over 40 years. The National collection of Rambling Roses is held at Moor Wood and that of Juglans and Pterocarya at Upton Wold.

Several very attractive Cotswold villages also open their gardens and a wonderful day can be had strolling from cottage to house marvelling at both the standard of the gardens and the beauty of the wonderful buildings, only to pause for the obligatory tea and cake!

Volunteers

County Organiser
Vanessa Berridge
01242 609535
vanessa.berridge@ngs.org.uk

County Treasurer
Pam Sissons
01242 573942
pam.sissons@ngs.org.uk

Social Media
Mandy Bradshaw
01242 512491
mandy.bradshaw@ngs.org.uk

Publicity
Ruth Chivers
01452 542493
ruth.chivers@ngs.org.uk

Booklet Coordinator
Vanessa Graham
07595 880261
vanessa.graham@ngs.org.uk

Assistant County Organisers
Valerie Kent
01993 823294
valerie.kent@ngs.org.uk

Colin & Verena Olle
01452 863750
colin.olle@ngs.org.uk

Rose Parrott
07853 164924
rosemary.parrott@ngs.org.uk

Heather Stopher
01453 882576
heather.stopher@ngs.org.uk

Gareth & Sarah Williams
01531 821654
dgwilliams84@hotmail.com

☐ @gloucestershirengs
☐ @Glosngs

Left: Mill Dene Garden

OPENING DATES

All entries subject to change. For latest information check www.ngs.org.uk

Map locator numbers are shown to the right of each garden name.

January

Sunday 31st
Home Farm 34

February

Snowdrop Festival

Sunday 7th
Trench Hill 68

Saturday 13th
Cotswold Farm 17

Sunday 14th
Cotswold Farm 17
Home Farm 34
Trench Hill 68

March

Sunday 7th
Home Farm 34

Sunday 21st
Trench Hill 68

Saturday 27th
South Lodge 62

Sunday 28th
◆ The Coach House
  Garden 15

April

Sunday 4th
Home Farm 34
Trench Hill 68

Monday 5th
Trench Hill 68

Sunday 11th
Highnam Court 32

Monday 12th
◆ Kiftsgate Court 36

Tuesday 13th
Wortley House 72

Wednesday 14th
Barnsley House 3
Lower Slaughter 42

Sunday 18th
Pear Tree Cottage 54
Upton Wold 69

Saturday 24th
The Gate 27

Sunday 25th
Blockley Gardens 7
Charlton Down House 13
Home Farm 34

Wednesday 28th
NEW Lords of the
  Manor Hotel 40

May

Saturday 1st
South Lodge 62

Sunday 2nd
Eastcombe and
  Bussage Gardens 22
Highnam Court 32
Ramblers 57
Trench Hill 68

Monday 3rd
Eastcombe and
  Bussage Gardens 22

Monday 10th
Brockworth Court 10

Wednesday 12th
◆ Lydney Park Spring
  Garden 43

Thursday 13th
Downton House 20
NEW Richmond Painswick
  Retirement Village 59

Friday 14th
The Old Vicarage 49

Saturday 15th
The Old Vicarage 49

Sunday 16th
20 Forsdene Walk 23
Green Bough 29
The Old Vicarage 49
◆ Stanway Fountain &
  Water Garden 64
Stowell Park 65

Friday 21st
◆ The Garden at
  Miserden 26

Saturday 22nd
Charingworth Court 12

Sunday 23rd
Charingworth Court 12
Forthampton Court 24

Wednesday 26th
Lower Farm House 41

Thursday 27th
NEW Richmond Painswick
  Retirement Village 59

Saturday 29th
Hookshouse Pottery 35

Sunday 30th
Hookshouse Pottery 35
Pasture Farm 52

Monday 31st
Hookshouse Pottery 35
Pasture Farm 52

June

Tuesday 1st
Hookshouse Pottery 35

Wednesday 2nd
Hookshouse Pottery 35
Rockcliffe 60

Thursday 3rd
Hookshouse Pottery 35
NEW Richmond Painswick
  Retirement Village 59

Friday 4th
Hookshouse Pottery 35

Saturday 5th
Hookshouse Pottery 35
Pasture Farm 52

Sunday 6th
Blockley Gardens 7
Highnam Court 32
Hodges Barn 33
Hookshouse Pottery 35
NEW Monks Spout
  Cottage 45
NEW Painswick Falcon
  Bowls Club 51

Monday 7th
Berkeley Castle 4
Hodges Barn 33

Wednesday 9th
Trench Hill 68

Saturday 12th
Cotswold Farm 17
NEW Little Orchard 39

NEW North Cerney Village
  Gardens 47

Sunday 13th
Charlton Down House 13
Cotswold Farm 17
The Gables 25
NEW Little Orchard 39
Stanton Village
  Gardens 63
Stowell Park 65
Weir Reach 70

Tuesday 15th
Wortley House 72

Wednesday 16th
Trench Hill 68
Weir Reach 70

Thursday 17th
NEW Richmond Painswick
  Retirement Village 59

Saturday 19th
Berrys Place Farm 5

Sunday 20th
Berrys Place Farm 5
Bisley Gardens 6
20 Forsdene Walk 23

Wednesday 23rd
Berrys Place Farm 5
Daylesford House 19
◆ Oxleaze Farm 50
Trench Hill 68

Thursday 24th
Berrys Place Farm 5

Saturday 26th
Perrywood House 55
South Lodge 62

Sunday 27th
NEW Great Oaks
  Hospice 28
The Manor 44
Moor Wood 46
Perrywood House 55

Monday 28th
NEW Ready Token
  House 58

Tuesday 29th
NEW Ready Token
  House 58

Wednesday 30th
NEW Ready Token
  House 58
Rockcliffe 60
Trench Hill 68

The Garden at Miserden

National Garden
Scheme gardens are
identified by their
yellow road signs
and posters. You
can expect a garden
of quality, character
and interest, a warm
welcome and plenty
of home-made
cakes!

THE GARDENS

1 AMPNEY BROOK HOUSE
School Lane, Ampney Crucis, Cirencester, GL7 5RT. Allan Hirst, 01285 851098, allan.hirst@clmail.co.uk. *From Cirencester go E on A417 toward Fairford. After passing the Crown of Crucis take 1st L and also immed L again onto School Lane and L again into the gated (open automatically) drive.* **Visits by arrangement Mar to Oct. Willing to discuss refreshments. Adm £5, chd free.** Striking Grade II Cotswold country house on 4.3 acres fronting Ampney Brook. The gardens are a haven for wildlife with fun and stimulating spaces year round. Inc woodland, kitchen garden, herbaceous borders, meadows, newly planted arbour. Ample areas and lawns for picnicking (encouraged). No wheelchair access to kitchen garden/greenhouse.

  🕭 🐐 ✿ 🚗 ☕ 🍷 🪑

2 AWKWARD HILL COTTAGE
Awkward Hill, Bibury, GL7 5NH. Mrs Victoria Summerley, 01285740289, v.summerley@hotmail.com, www.awkwardhill.co.uk. *Bibury, Gloucestershire. No parking at property, best to park in village and walk past Arlington Row up Awkward Hill, or up Hawkers Hill from Catherine Wheel pub.* **Sun 4 July, Sun 5 Sept (2-6). Adm £5. Home-made teas. Visits also by arrangement June to Sept for groups of 10 to 30. No children please (dog not safe with children).**
An ever-evolving country garden in one of the most beautiful villages in the Cotswolds, designed to reflect the local landscape and encourage wildlife. Planting is both formal and informal, contributing year round interest with lots of colour and texture. Pond and waterfall, beehives. Lots of seating areas. Wonderful views over neighbouring meadow and woodland, small pond with jetty, 2 sunny terraces and plenty of places to sit and relax, both in sun and shade.

  🐕 ✿ ☕ 🍷

3 BARNSLEY HOUSE
Barnsley, Cirencester, GL7 5EE. Calcot Health & Leisure Ltd, 01285 740000, reception@barnsleyhouse.com, www.barnsleyhouse.com. *4m NE of Cirencester. From Cirencester, take B4425 to Barnsley. House entrance on R as you enter village.* **Wed 14 Apr (10-3). Adm £5, chd free. Tea.**
The beautiful garden at Barnsley House, created by Rosemary Verey, is one of England's finest and most famous gardens inc knot garden, potager garden and mixed borders in her successional planting style. The house also has an extensive kitchen garden which will be open with plants and vegetables available for purchase. Narrow paths mean restricted wheelchair access but happy to provide assistance.

  🕭 🚗 🛏 ☕ 🍷

4 BERKELEY CASTLE
Berkeley, GL13 9PJ. Mrs RJG Berkeley, www.berkeley-castle.com. *Half-way between Bristol & Gloucester, 10mins from J13 &14 of M5. Follow signs to Berkeley from A38 & B4066. Visitors' entrance is on L of Canonbury St, just before town centre.* **Mon 7 June (11-4.30). Adm £6, chd £3. Light refreshments in The Courtyard, next to the Castle. Delicious home-made cakes, light lunches and locally-sourced items.**
Unique historic garden of a keen plantsman, with far-reaching views across the River Severn. Gardens contain many rare plants which thrive in the warm micro-climate against the stone walls of this medieval castle. Woodland, historic trees and stunning terraced borders. Difficult for wheelchairs due to terraced nature of gardens.

  🚗 ☕ 🍷

Warren Farmhouse, Stanton Village Gardens

5 BERRYS PLACE FARM

Bulley Lane, Churcham, Gloucester, GL2 8AS. Anne Thomas, 07950 808022, gary.j.thomas1953@gmail.com. *6m W of Gloucester. A40 towards Ross. Turning R into Bulley Lane at Birdwood.* **Sat 19, Sun 20, Wed 23, Thur 24 June (11-5). Adm £3.50, chd free. Home-made teas. Ploughman's lunches, cakes and cream tea.**

Country garden, approx 2 acres, surrounded by farmland and old orchards. Lawns and large sweeping mixed herbaceous borders with over 100 roses. Formal kitchen garden and beautiful rose arbour leading to lake and summerhouse with a variety of water lilies and carp. All shared with ducks.

GROUP OPENING

6 BISLEY GARDENS

Wells Road, Bisley, Stroud, GL6 7AG. *Gardens & car park well signed in Bisley village. Gardens on S edge of village at head of Toadsmoor Valley, N of A419 Stroud to Cirencester Road.* **Sun 20 June (2-6). Combined adm £5, chd free. Light refreshments at Paulmead.**

PAULMEAD
Judy & Philip Howard & Tom & Emma Howard.

PAX
Mr David Holdon & Mr Ramesh Mootoo.

WELLS COTTAGE
Mr & Mrs Michael Flint, 01452 770289, bisleyflints@gmail.com.

3 beautiful gardens with differing styles. Paulmead: 1 acre landscaped garden constructed in stages over last 25 years. Terraced in 3 main levels. Natural stream garden, herbaceous and shrub garden, formal vegetable garden, summerhouse overlooking pond. Unusual tree house. Development of new garden around hen house, inc ha ha. Pax: Small very well kept cottage garden, in a hidden away location, with box hedging and topiary. Great views. Wells Cottage: Just under 1 acre. Terraced on several levels with formal topiary and beautiful views over the valley. Much informal planting of trees and shrubs to give colour and texture. Lawns and

herbaceous borders. Collection of grasses. Formal pond area. Rambling roses on chain pergola. Vegetable garden with raised beds.

GROUP OPENING

7 BLOCKLEY GARDENS

Blockley, GL56 9DB. *3m NW of Moreton-in-Marsh. Just off the Moreton-in-Marsh to Evesham Rd A44.* **Sun 25 Apr, Sun 6 June (2-6). Combined adm £7, chd free. Home-made teas at St George's Hall on April 25, at Colebrook House & St George's Hall on June 6.**

CHURCH GATES
Mrs Brenda Salmon.
Open on all dates

COLEBROOK HOUSE
Mr & Mrs G Apsion.
Open on all dates

NEW ELM HOUSE
Chris & Val Scragg.
Open on all dates

◆ MILL DENE GARDEN
Mrs B S Dare, 01386 700457, info@milldenegarden.co.uk, www.milldenegarden.co.uk.
Open on all dates

MILL GARDEN HOUSE
Andrew & Celia Goodrick-Clarke.
Open on all dates

PEAR TREES
Cliff & Jo Murphy.
Open on Sun 6 June

PORCH HOUSE
Mr & Mrs Johnson.
Open on Sun 25 Apr

SNUGBOROUGH MILL
Rupert & Mandy Williams-Ellis, 01386 701310, rupert.williams-ellis@talk21.com.
Open on Sun 6 June

WOODRUFF
Paul & Maggie Adams.
Open on all dates

This popular historic hillside village has a great variety of high quality, well-stocked gardens - large and small, old and new. Blockley Brook, an attractive stream which flows right through the village, graces some of the gardens; these inc gardens of former water mills, with millponds attached. From some gardens there are wonderful rural views. Children welcome but close supervision is essential. Access to some gardens

quite steep and allowances should be made.

8 ◆ BOURTON HOUSE GARDEN

Bourton-on-the-Hill, GL56 9AE. Mr & Mrs R Quintus, 01386 700754, info@bourtonhouse.com, www.bourtonhouse.com. *2m W of Moreton-in-Marsh. On A44.* **For NGS: Sun 15 Aug (10-5). Adm £8, chd free. Light refreshments & home-made cakes in Grade I Listed C16 Tithe Barn. For other opening times and information, please phone, email or visit garden website.**

Award winning 3 acre garden featuring imaginative topiary, wide herbaceous borders with many rare, unusual and exotic plants, water features, unique shade house and many creatively planted pots. Fabulous at any time of year but magnificent in summer months and early autumn. Walk in 7 acre pasture with free printed guide to specimen trees available. 70% access for wheelchairs. Disabled toilet.

9 25 BOWLING GREEN ROAD

Cirencester, GL7 2HD. Mrs Sue Beck, 01285 653778, zen155198@zen.co.uk. *On NW edge of Cirencester. Take A4030 to Spitalgate/Whiteway T-lights, turn into The Whiteway (Chedworth turn), then 1st L into Bowling Green Rd, garden in bend in rd between Nos 23 & 27.* **Mon 5 July (11-4); Sun 11 July (2-5); Mon 12 July (11-4). Adm £3.50, chd free. Visits also by arrangement July to Sept for groups of up to 30. Tea/coffee/juice/biscuits can be provided for small groups by appointment.**

Welcome to this naturalistic garden, increasingly designed by plants themselves, where you can wander at will in a mini-jungle of curvaceous clematis, gorgeous grasses, romantic roses, heavenly hemerocallis and plentiful perennials, glimpsing a graceful giraffe, friendly frogs and a unicorn. Rated by visitors as an amazing hidden gem with a unique atmosphere. See: The Chatty Gardener, July 10 2019, https://wp.me/p6LDlk-2Dv - posting on Growing Hemerocallis and garden owner having daylily cultivar registered by UK Hybridizer to mark her 40th Anniversary of opening for NGS. Sadly, not suitable for wheelchair access.

10 ◆ BROCKWORTH COURT

Court Road, Brockworth, GL3 4QU. Tim & Bridget Wiltshire, 01452 862938, timwiltshire@hotmail.co.uk. *6m E of Gloucester. 6m W of Cheltenham. Adj St Georges Church on Court Rd. From A46 turn into Mill Lane, turn R, L, R at T junctions. From Ermin St, turn into Ermin Park, then R at r'about then L at next r'about.* **Mon 10 May, Mon 13 Sept (2-5.30). Adm £6, chd free. Home-made teas. Visits also by arrangement Apr to Oct for groups of 10+.**
This intense yet informal tapestry style garden beautifully complements the period manor house which it surrounds. Organic, naturalistic, with informal cottage-style planting areas that seamlessly blend together. Natural fish pond, with Monet bridge leading to small island with thatched Fiji house. Kitchen garden once cultivated by monks. Views to Crickley and Coopers Hill. Adj Norman Church (open). Historic tithe barn, historic manor house visited by Henry VIII and Anne Boleyn in 1535. Partial wheelchair access.

&♿ ✻ 🏡 ☕

11 ◆ CERNEY HOUSE GARDENS

North Cerney, Cirencester, GL7 7BX. Mr N W Angus & Dr J Angus, 01285 831300, janet@cerneygardens.com, www.cerneygardens.com. *4m NW of Cirencester. On A435 Cheltenham rd turn L opp Bathurst Arms, follow rd past church up hill, then go straight towards pillared gates on R (signed Cerney House).* **For NGS: Sun 4 July (10-7). Adm £5, chd £1. Home-made teas in The Bothy Tea room. For other opening times and information, please phone, email or visit garden website.**
A romantic English garden for all seasons. There is a secluded Victorian walled garden featuring herbaceous borders overflowing with colour. Early in the year we have a wonderful display of snowdrops, in spring we feature a magnificent display of tulips and during the summer the rambling romantic roses come to life. Koi carp pond. Enjoy our woodland walk, extended nature trail and new medicinal herb garden. Dogs welcome. Partial wheelchair access.

&♿ 🐕 ✻ 🏡 ☕

12 ◆ CHARINGWORTH COURT

Broadway Road, Winchcombe, GL54 5JN. Susan & Richard Wakeford, 07791 353779, susanwakeford@gmail.com, www.charingworthcourtcotswoldsgarden.com. *8m NE of Cheltenham. 400 metres N of Winchcombe town centre. Limited parking along Broadway Rd. Town car parks in Bull Lane (short stay) and all day parking (£1) in Back Lane. Map on our website.* **Sat 22, Sun 23 May (12-6). Adm £5, chd free. Home-made teas. Visits also by arrangement May & June for groups of 10 to 30.**
Artistically and lovingly created 1½ acre garden surrounding restored Georgian/Tudor house (not open). Relaxed country style with Japanese influences, large pond and walled vegetable/flower garden, created over 25 years from a blank canvas. Mature copper beech trees, Cedar of Lebanon and Wellingtonia; and younger trees replacing an earlier excess of Cupressus leylandii. Once again the garden will showcase a range of sculpture, most for sale with a percentage going to NGS charities. The garden website shows photographs of all the previous 8 exhibitions. Partial access due to gravel paths which can be challenging but several areas accessible without steps. Some disabled parking next to house.

&♿ 🐕 ☕

13 ◆ CHARLTON DOWN HOUSE

Charlton Down, Tetbury, GL8 8TZ. Neil & Julie Record. *2m SW of Tetbury, Glos. From Tetbury, take A433 towards Bath for 1½m; turn R (north) just before the Hare and Hounds, then R again after 200yds into Hookshouse Lane. Charlton Down House is 600yds on R.* **Sun 25 Apr, Sun 13 June (11-5). Home-made teas. Thur 1, Thur 8, Thur 29 July, Thur 5, Thur 19, Thur 26 Aug (1-5). Adm £6, chd free.**
Extensive country house gardens in 180 acre equestrian estate. Formal terraces, perennial borders, walled topiary garden, enclosed cut flower garden and large glasshouse. Newly planted copse. Rescue animals. Ample parking. Largely flat terrain; most garden areas accessible.

&♿ 🐕 🚗 ☕ 🌳

14 ◆ CLOUDS REST

Brockweir, Chepstow, NP16 7NW. Mrs Jan Basford. *In the Wye Valley, 6.7m N of Chepstow and 10.6m S of Monmouth, off A466, across Brockweir Bridge.* **Sun 12 Sept (12.30-5). Combined adm with The Patch £7, chd free. Home-made teas.**
The garden at Clouds Rest was started in 2012 from a south-westerly facing, stony paddock, with views across the Wye Valley. Its many gravel pathways meander through herbaceous beds with a mixture of roses, then a wide selection of Michaelmas daisies in September. Easy parking in our paddock. New additions inc woodland area. Partial wheelchair access.

✻ ☕

15 ◆ THE COACH HOUSE GARDEN

Church Lane, Ampney Crucis, Cirencester, GL7 5RY. Mr & Mrs Nicholas Tanner, 01285 850256, mel@thegenerousgardener.co.uk, www.thegenerousgardener.co.uk. *3m E of Cirencester. Turn into village from A417, immed before Crown of Crucis Inn. Over hump-back bridge, parking to R on cricket field (weather permitting) or signed nearby field. Disabled parking near house.* **For NGS: Sun 28 Mar (2-5). Adm £6, chd free. Coffee/tea home-made cakes and cream teas available for visiting groups and NGS opening. For other opening times and information, please phone, email or visit garden website.**
Approx 1½ acres, full of structure and design. Garden is divided into rooms inc rill garden, gravel garden, rose garden, herbaceous borders, green garden with pleached lime allee and potager. Created over last 30yrs by present owners and constantly evolving. New potting shed and greenhouse added in 2018. Visitors welcome during mid March - mid July (groups of 15+), please see above website. Rare plant sales (in aid of James Hopkins Trust), Garden Lecture Days and workshops. Partial wheelchair access. Ramp available to enable access to main body of garden, steps to other areas.

🐕 ✻ 🚗 ☕

Online booking is available for many of our gardens, at ngs.org.uk

GROUP OPENING

16 NEW COTSWOLD CHASE GARDENS

Spinners Road, Brockworth, Gloucester, GL3 4LR. *Follow signs off Ermin Street.* **Sun 4 July (2-5). Combined adm £5, chd free. Tea and cake at MidGlos Bowls Club.**
A group of small gardens all recently created on a new development. A wide range of designs, planting and imaginative ideas that show what can be achieved in a short space of time. The Chase is rich in open spaces which have been tastefully landscaped, with many foot/cycle paths, a sports field, children's play areas and three wildlife friendly balancing ponds. Lots of places to walk, take in the vistas (e.g. Coopers Hill) and view some new gardens. Parking, tickets, maps and toilets at MidGlos Bowls Club.

17 COTSWOLD FARM

Duntisbourne Abbots, Cirencester, GL7 7JS. John & Sarah Birchall, www.cotswoldfarmgardens.org.uk. *5m NW of Cirencester off old A417. From Cirencester L signed Duntisbourne Abbots Services, R and R underpass. Drive ahead. From Gloucester L signed Duntisbourne Abbots Services. Pass Services. Drive L.* **Sat 13, Sun 14 Feb (11-3). Light refreshments. Sat 12, Sun 13 June (2-5). Home-made teas. Adm £7.50, chd free. Donation to A Rocha.**
Arts and Crafts garden in lovely position overlooking quiet valley on descending levels with terrace designed by Norman Jewson in 1930s. Snowdrops named and naturalised, aconites in Feb. Winter garden. Bog garden best in May, white border overflowing with texture and scent. Shrubs, trees, shrub roses. 8 native orchids, hundreds of wildflowers and Roman snails. Family day out. Rare orchid walks. Picnics welcome. Wheelchair access to main terrace only (no wheelchair access to toilets).

18 DAGLINGWORTH HOUSE

Daglingworth, nr Cirencester, GL7 7AG. David & Henrietta Howard, 07970 122122, ettajhoward@gmail.com. *3m N of Cirencester off A417/419. House with blue gate beside church in centre of Daglingworth, at end of No Through Road.* **Visits by arrangement Apr to Sept for groups of 5 to 30. Adm £7, chd free. Light refreshments. Please discuss refreshments when phoning.**
Walled garden, temple, grotto, and pools. Classical garden of 2.5 acres, with humorous contemporary twist. Hedges, topiary shapes, herbaceous borders. Pergolas, grass garden, meadow, woodland, cascade and mirror canal. New sunken garden 2019. Pretty Cotswold village setting beside church. Visitor comment: 'It breaks every rule of gardening - but it's wonderful!'. Partial wheelchair access due to steps and level changes - narrow gates.

19 DAYLESFORD HOUSE

Daylesford, GL56 0YG. Lord & Lady Bamford. *5m W of Chipping Norton. Off A436. Between Stow-on-the-Wold & Chipping Norton.* **Wed 23 June (1-5). Adm £6, chd free. Light refreshments.**
Magnificent C18 landscape grounds created 1790 for Warren Hastings, greatly restored and enhanced by present owners under organic regime. Lakeside and woodland walks within natural wildflower meadows. Large formal walled garden, centred around orchid, peach and working glasshouses. Trellised rose garden. Collection of citrus within period orangery. Secret garden, pavilion formal pools. Very large garden with substantial distances. Partial wheelchair access.

Eastcombe and Bussage Gardens

20 DOWNTON HOUSE

Gloucester St, Painswick, GL6 6QN. **Ms Jane Kilpatrick.** *4m N of Stroud. Entry to garden via Hollyhock Lane only. Please note: No cars in Lane. Parking in Stamages Lane village car park below church or in Churchill Way (1st R off Gloucester Street – B4073).* **Thur 13 May (1.30-5). Adm £5, chd free. Home-made teas.**
Enthusiast's walled ⅓ acre garden in heart of historic Painswick. Planted for year round foliage colour and interest with many rare and unusual trees, shrubs and herbaceous plants. Particular focus on plants that thrive on thin limey soil in a changing climate. Lots of new planting since last open in 2018. Collection of tender plants in heated glasshouse.

21 EAST COURT

East End Road, Charlton Kings, Cheltenham, GL53 8QN. **Ben White.** *Parking will be signed.* **Sun 29, Mon 30 Aug (10-6). Adm £7, chd free. Home-made teas.**
A garden on 2 levels. The upper garden around the 1806 house (not open) is traditionally styled with formal beds and lawns and three majestic purple beech trees some 200 years old. The 2½ acre lower garden – now 4 years old – is in complete contrast. Winding brick paths lead through swathes of herbaceous plantings towered over by arching datisca, miscanthus and paulownia. Along a high, S-facing wall, we are experimenting successfully with less hardy plants, spiced up with a scattering of exotic annuals. The new pond has attracted a wide range of aquatic wildlife. The garden is always evolving, so visit us and share our excitement. Mostly accessible to wheelchairs – please ask for help where needed.

GROUP OPENING

22 EASTCOMBE AND BUSSAGE GARDENS

Eastcombe, Stroud, GL6 7EB. *3m E of Stroud. Tickets and maps available on the day from Eastcombe Village Hall, GL6 7EB and also from Redwood, nr Bussage Village Hall, GL6 8AZ. On street parking only. Tickets valid Sun and Mon.* **Sun 2, Mon 3 May (1.30-5.30). Combined adm £8, chd free. Cold drinks & ice creams available at some gardens.**

BREWERS COTTAGE
Jackie & Nick Topman.

CADSONBURY
Natalie & Glen Beswetherick.

NEW 17 FARMCOTE CLOSE
John & Sheila Coyle.

20 FARMCOTE CLOSE
Ian & Dawn Sim.

21 FARMCOTE CLOSE
Mr & Mrs Robert Bryant.

HAMPTON VIEW
Geraldine & Mike Carter.

HAWKLEY COTTAGE
Helen & Gerwin Westendorp.

12 HIDCOTE CLOSE
Mr K Walker.

HIGHLANDS
Helen & Bob Watkinson.

1 THE LAURELS
Andrew & Ruth Fraser.

MARYFIELD AND MARYFIELD COTTAGE
Mrs M Brown.

MOUNT PLEASANT
Mr & Mrs R Peyton.

REDWOOD
Heather Collins.

50 STONECOTE RIDGE
Julie & Robin Marsland.

VALLEY VIEW
Mrs Rebecca Benneyworth.

YEW TREE COTTAGE
Andy & Sue Green.

Medium and small gardens in a variety of styles and settings within

Ready Token House

this picturesque, hilltop village location with its spectacular views of the Toadsmoor Valley. In addition, there is one large garden located in the bottom of the valley, approachable only on foot as are some of the other gardens. NB there will not be a courtesy minibus this season. Full descriptions of each garden can be found on the NGS website. No dogs at 1The Laurels, 20 Farmcote Close, Yew Tree Cottage, please. Plants for sale at Eastcombe Village Hall and some gardens. Wheelchair access to some gardens - check which ones on NGS website.

23 20 FORSDENE WALK
Coalway, Coleford, GL16 7JZ.
Pamela Buckland, 01594 837179.
From Coleford take Lydney/ Chepstow Rd at T-lights. L after police station ½m up hill turn L at Xrds then 2nd R (Old Road) straight on at minor Xrds then L into Forsdene Walk. **Sun 16 May, Sun 20 June, Sun 11 July (2-6). Adm £3, chd free. Light refreshments. Visits also by arrangement May to Sept for groups of up to 20.**
Corner garden filled with interest and design ideas to maximise smaller spaces. A series of interlinking colour themed rooms, some on different levels. Packed with perennials, grasses and ferns. A pergola, small man-made stream, fruit and vegetables and pots in abundance on gravelled areas. Featured in Amateur Gardening.

24 FORTHAMPTON COURT
Forthampton, Tewkesbury,
GL19 4RD. Alan & Anabel Mackinnon. *W of Tewkesbury. From Tewkesbury A438 to Ledbury. After 2m turn L to Forthampton. At Xrds go L towards Chaceley. Go 1m turn L at Xrds.* **Sun 23 May (12.30-4.30). Adm £6, chd free. Home-made teas.**
Charming and varied garden surrounding north Gloucestershire medieval manor house (not open) within sight of Tewkesbury Abbey. Inc borders, lawns, roses and magnificent Victorian vegetable garden. Disabled drop off at entrance, some gravel paths and uneven areas.

25 THE GABLES
Riverside Lane, Broadoak,
Newnham on Severn,
GL14 1JE. Bryan & Christine Bamber, 01594 516323,
bryanbamber21@gmail.com. *1m NE of Newnham on Severn. Park in White Hart PH overspill car park, to R of PH when facing river. Please follow signs to car park. Walk, turning R along rd towards Gloucester for approx 250yds. Access through marked gate.* **Sun 13 June, Sun 15 Aug (11-5). Adm £4, chd free. Light refreshments. Visits also by arrangement May to Aug for groups of 10+.**
Large flat ¾ acre garden with formal lawns, colourful herbaceous borders from May - September, rose beds, shrubberies, hidden long border, mini stumpery with hostas, bamboos, grasses, wildflower meadow with soft fruits and fruit trees, wildlife pond created Autumn 2020, allotment size productive potager vegetable plot with herbaceous borders, greenhouse and composting bin area. Disabled parking information available at entrance. Partial wheelchair access but all areas of garden visible.

26 ◆ THE GARDEN AT MISERDEN
Miserden, nr Stroud, GL6 7JA.
Mr Nicholas Wills, 01285 821303,
estate.office@miserden.org,
www.miserden.org. *6m NW of Cirencester. Leave A417 at Birdlip, drive through Whiteway and follow signs for Miserden.* **For NGS: Fri 21 May (10-5). Adm £9, chd free. Light refreshments in The Garden Café in the Nursery. For other opening times and information, please phone, email or visit garden website.**
Winner of Historic Houses Garden of the Year 2018, the Garden at Miserden is a lovely, timeless walled garden with amazing views over a deer park and rolling Cotswold hills beyond. The garden was designed in C17 and still retains a wonderful sense of peace and tranquillity, especially with its spectacular 92m long mixed borders. Other magical features of this garden which has spanned generations include a unique sycamore tree that has grown through a Cotswold stone wall, grass steps, an ancient mulberry tree and a yew walk. Partial wheelchair access.

27 THE GATE
80 North Street, Winchcombe,
GL54 5PS. Vanessa Berridge & Chris Evans, 01242 609535,
vanessa.berridge@sky.com.
Winchcombe is on B4632 mid-way between Cheltenham and Broadway. Parking behind Library in Back Lane, 50 yds from The Gate. **Sat 24 Apr (2-5). Adm £3.50, chd free. Home-made teas. Visits also by arrangement Apr to Sept for groups of 5 to 20. Adm £7 to include home-made teas.**
Cottage-style garden planted with bulbs in spring, and with summer perennials, annuals, climbers and herbs in the walled courtyard of C17 former coaching Inn. Also a separate, productive, walled kitchen garden with espaliers and other fruit trees.

28 NEW GREAT OAKS HOSPICE
The Gorse, Coleford, GL16 8QF.
Ruth Keeble, 01594 811910. *On the outskirts of Coleford, next door to The Coombs nursing home. We are also known as Dean Forest Hospice. Onsite parking is available.* **Sun 27 June (10-5). Adm £5, chd free. Home-made teas on the top terrace. Visits also by arrangement in June for groups of up to 10.**
The Hospice is nestled in the heart of the Forest of Dean and has a large traditional all year round garden. At the entrance you will be greeted by lavender bushes giving an immediate feel of tranquility and as you follow the driveway down to the Hospice building you will be captivated by the stunning location which is hidden from street view. There is a sloping path into the garden allowing access for wheelchair users.

There are brilliant plant sales at many gardens. Look out for the symbol *in the garden description – and don't forget to bring a bag to carry your plants home in*

29 GREEN BOUGH
Market Lane, Greet,
Winchcombe, GL54 5BL. Mary
& Barry Roberts, 07966 528646,
barryandmary@gmail.com. 1¼m N
of Winchcombe. From Winchcombe
take B4078. After railway bridge,
R into Becketts Lane, immediately
L into Market Lane. Garden on R
at Mill Lane junction. Sun 16 May,
Sun 19 Sept (10-4). Adm £3, chd
free. Pre-booking essential,
please visit www.ngs.org.uk for
information & booking. Visits also
by arrangement Apr to Sept for
groups of up to 20. Home-made
teas/refreshments available for by
arrangement visits on request.
Small informal country garden
developed over 7 years, densely
planted for all seasons starting in early
spring with massed bulbs and flowers
surrounding the house, inc the grass
verge. Many of the plants are either
grown from seed or propagated
from cuttings by the owner, to give
generous drifts of colour.

30 GREENFIELDS, BROCKWEIR COMMON
Brockweir, NP16 7NU.
Jackie Healy, 07747 186302,
jackie@greenfields.garden,
www.greenfields.garden. Located
in Wye valley - midway between
Chepstow and Monmouth. A446:
from M'mouth: Thru Llandogo. L
to Brockweir, (from Chepstow, thru
Tintern. R to B'weir) over bridge,
pass PH up hill, 1st L, follow lane to
fork, L at fork. 1st property on R. No
coaches. Visits by arrangement
Apr to July for groups of up to
30. Joint opening with the Patch.
£7.00 for 2 gardens - or £4.50 for
1 garden. Home-made teas. All
visits and refreshments subject
to weather.
1½ acre plant person's gem of a
garden set in the beautiful Wye Valley.
Many mature trees and numerous
unusual plants and shrubs, all planted
as discrete gardens within a garden.
Greenfields is the passion and work
of head gardener Jackie who has a
long interest in the propagation of
plants. Featured in Garden Answers
magazine. Mostly wheelchair access.

31 GREENFIELDS, LITTLE RISSINGTON
Cheltenham, GL54 2NA. Mrs
Diana MacKenzie-Charrington. On
Rissington Road between Bourton-

on-the-Water and Little Rissington,
opp turn to Great Rissington (Leasow
Lane). SatNav using postcode does
not take you to house. Sun 11 July
(2-5). Adm £5, chd free. Home-
made teas.
The honey coloured Georgian
Cotswold stone house sits in 2
acres of garden, created by current
owners over last 20 years. Lawns
are edged with borders full of flowers
and flowering bulbs. A small pond
and stream overlook fields. Bantams
roam freely. Mature apple trees in
wild garden, greenhouse in working
vegetable garden. Sorry no dogs.
Partial wheelchair access.

32 HIGHNAM COURT
Highnam, Gloucester,
GL2 8DP. Mr & Mrs R J Head,
01684 292875 (Mike Bennett),
mike.highnamcourt@gmail.com,
www.HighnamCourt.co.uk. 2m
W of Gloucester. On A40/A48
junction from Gloucester to Ross or
Chepstow. At this r'about take exit at
3 o'clock if coming from Gloucester
direction. Do NOT go into Highnam
village. Sun 11 Apr, Sun 2 May, Sun
6 June, Sun 4 July, Sun 1 Aug,
Sun 5 Sept (11-6). Adm £5, chd
free. Light refreshments. Visits
also by arrangement Apr to Sept.
40 acres of Victorian landscaped
gardens surrounding magnificent
Grade I house (not open), set out by
artist Thomas Gambier Parry. Lakes,
shrubberies and listed Pulhamite
water gardens with grottos and
fernery. Exciting ornamental lakes,
and woodland areas. Extensive 1
acre rose garden and many features,
inc numerous wood carvings.
Contact: Mike Bennett, Events &
Visits Manager, 01684 292875, email
above. Some gravel paths and steps
into refreshment area. Disabled WC
outside.

33 HODGES BARN
Shipton Moyne, Tetbury,
GL8 8PR. Mr & Mrs N Hornby,
www.hodgesbarn.com. 3m S of
Tetbury. On Malmesbury side of
village. Sun 6, Mon 7 June (2-6).
Adm £6, chd free. Home-made
teas at the Pool House.
Very unusual C15 dovecote converted
into family home. Cotswold stone
walls host climbing and rambling
roses, clematis, vines, hydrangeas
and together with yew, rose and
tapestry hedges create formality

around house. Mixed shrub and
herbaceous borders, shrub roses,
water garden, woodland garden
planted with cherries and magnolias.

34 HOME FARM
Newent Lane, Huntley, GL19 3HQ.
Mrs T Freeman, 01452 830210,
torillfreeman@gmail.com. 4m S of
Newent. On B4216 ½m off A40 in
Huntley travelling towards Newent.
Sun 31 Jan, Sun 14 Feb, Sun 7
Mar, Sun 4, Sun 25 Apr (11-4).
Adm £3.50, chd free. 2022: Sun
30 Jan, Sun 13 Feb. Visits also by
arrangement Jan to Apr for groups
of up to 30.
Set in elevated position with
exceptional views. 1m walk through
woods and fields to show carpets of
spring flowers. Enclosed garden with
fern border, sundial and heather bed.
White and mixed shrub borders. Stout
footwear advisable in winter. Two
delightful cafés within a mile.

35 HOOKSHOUSE POTTERY
Hookshouse Lane,
Tetbury, GL8 8TZ.
Lise & Christopher White,
www.hookshousepottery.co.uk.
2½m SW of Tetbury. Follow signs
from A433 at Hare and Hounds
Hotel, Westonbirt. Alternatively
take A4135 out of Tetbury towards
Dursley and follow signs after ½m
on L. Daily Sat 29 May to Sun 6
June (11-5.30). Adm £4.50, chd
free. Home-made teas.
Garden offers a combination of
long perspectives and intimate
corners. Planting inc wide variety of
perennials, with emphasis on colour
interest throughout the seasons.
Herbaceous borders, new woodland
garden and flower meadow, water
garden containing treatment ponds
(unfenced) and flowform cascades.
Sculptural features. Kitchen garden
with raised beds, orchard. Run on
organic principles. Pottery showroom
with handthrown woodfired pots
inc frostproof garden pots.Art and
craft exhibition (May 29th to June
6th) includes garden furniture and
sculptures.Garden games and tree
house. Mostly wheelchair accessible.

36 ◆ KIFTSGATE COURT

Chipping Campden, GL55 6LN. Mr & Mrs J G Chambers, 01386 438777, info@kiftsgate.co.uk, www.kiftsgate.co.uk. *4m NE of Chipping Campden. Adj to Hidcote NT Garden.* For NGS: Mon 12 Apr, Mon 9 Aug (2-6). Adm £9.50, chd £3. Cream teas. For other opening times and information, please phone, email or visit garden website.

Magnificent situation and views, many unusual plants and shrubs, tree peonies, hydrangeas, abutilons, species and old-fashioned roses inc largest rose in England, Rosa filipes Kiftsgate. Regional Finalist, The English Garden's The Nation's Favourite Gardens 2019. Steep slopes and uneven surfaces.

37 NEW LANE'S COTTAGE

Winchcombe, Cheltenham, GL54 5BA. Norman & Zoe Carter, lanescottage@trelowen.com. *1 m S of entrance to Sudeley Castle. From Cheltenham on B4632, turn R after church down Vineyard Street, after 200yds bear R and follow road for 1m.* Visits by arrangement Apr to Oct for groups of up to 10. Parking limited to 4 vehicles. Adm £8, chd free. Tea or coffee and home-made cake inc in admission.

Set in a valley, this 2½ acre triangular garden was planted in 2018 after the renovation/extension of the listed cottage. Different landscaped areas, easy maintenance and wildlife habitat was key. The result is a garden with interest and colour through the year with orchard, woodland, spring bulbs, wildflower areas, cloud trees, formal borders and prairie planting. Although not really wheelchair friendly the garden can still be enjoyed by those with limited mobility.

38 78 LILLESFIELD AVENUE

Lilliesfield Avenue, Barnwood, Gloucester, GL3 3AH. Roger Le Couteur, Roger.lecouteur@hotmail.com. *3m E of Gloucester. Barnwood Road turn R into North Upton Lane then 1st L and follow signs or from Brockworth turn L into Brookfield Road by Lloyd's Pharmacy and follow signs.* Visits by arrangement Apr to Sept for groups of 10 to 20. Combined admission £5 with The Oaks.

Semi-detached house with a well stocked front garden and a small enclosed garden at the rear. The front has acers, palms and hydrangeas and other perennials. The rear garden is very well stocked with palms, acers, bamboo, hydrangeas and pines. Ornamental grasses, other perennials and fish pond. The close planting gives a jungle feel. Not suitable for wheelchairs, children or dogs.

Sudeley Castle & Gardens

39 NEW **LITTLE ORCHARD**
Slad, Stroud, GL6 7QD.
Mr & Mrs Terry & Rod
Clifford, 01452 813944,
terryclifford.tlc@gmail.com. *2m
from Stroud, 10 m from Cheltenham.
Last property on L in Slad village
before leaving 30mph speed limit
travelling from Stroud to Birdlip
on B4070. SatNav may not bring
you directly to property. Parking
on verge opposite.* **Sat 12, Sun
13 June (10.30-5). Adm £4, chd
free. Home-made teas. Cider
tasting available. Visits also by
arrangement May to Sept for
groups of 10 to 30. Home-made
teas and or cider tastings available
on request when booking.**
Steeply sloping, terraced acre plot
using many reclaimed materials,
stonework and statuary. Enhanced
with different styles of planting to
complement the natural surroundings,
varied areas inc apple orchard,
Mediterranean courtyard, vegetable
parterre and children's play area.
Stunning views of Slad Valley. Access
into adjoining Nature Reserve. Local
craft on sale. Children's Trail. Wheelchair
access possible but challenging due to
the severity of slopes and steps. Please
phone for further details.
ᗕ 🐴 ✤ ☕

40 NEW **LORDS OF THE MANOR
HOTEL**
Upper Slaughter, Cheltenham,
GL54 2JD. Mike Dron (Head
Gardener), 01451 820243,
reservations@lordsofthemanor.
com, *From Fosse way follow signs for the
Slaughters from close to Bourton
on the water (toward Stow). From
B4077 (Stanway Hill road) coming
from Tewkesbury direction, follow
signs 2mls after Ford village.* **Wed
28 Apr (10-4). Adm £5, chd £3.
Refreshments are available from
the hotel, from light snacks to
afternoon tea. Pre booking may
be necessary.**
Classic Cotswold country garden
with a very English blend of formal
and informal, merging beautifully with
the surrounding landscape. Beautiful
walled garden, established wildflower
meadow, the river Eye. The herb
garden and stunning bog garden
were designed by Julie Toll around
2012. Wildlife garden, croquet lawn,
courtyard and Victorian skating pond.
Mostly wheelchair access but steps
within walled garden and bog garden.
ᗕ 🐴 🛏 ☕

41 **LOWER FARM HOUSE**
Cliffords Mesne, Newent,
GL18 1JT. Gareth & Sarah
Williams. *2m S of Newent. From
Newent, follow signs to Cliffords
Mesne and Birds of Prey Centre
(1½ m). Approx ½ m beyond Centre,
turn L at Xrds (before church).* **Wed
26 May (1.30-5). Adm £5, chd free.
Light refreshments.**
2½ acre garden, inc woodland,
stream and large natural lily pond
with rockery and bog garden.
Herbaceous borders, pergola walk,
terrace with ornamental fishpond,
kitchen and herb garden; collections
of irises, hostas and paeonies. Many
interesting and unusual trees and
shrubs inc magnolias and cornus.
Some gravel paths.
ᗕ 🐴 ☕

42 **LOWER SLAUGHTER**
Cheltenham, GL54 2HP.
Jane Moore, George
Leonte, 01451 820456,
info@slaughtersmanor.co.uk,
www.slaughtersmanor.co.uk.
*Lower Slaughters Village is located
between Bourton-on-the-Water and
Stow-on-the-Wold. Use A429 road.
From A429, very close to petrol
station, take Copsehill Road, into The
Slaughters.* **Wed 14 Apr (11.30-
3.30). Adm £3.50, chd free. Light
refreshments.**
2 individual gardens filled with
spring bulbs, flowering cherries and
sparkling water features. Managed
by the same team of gardeners and
subject to ongoing development
and investment. The C17 Slaughters
Manor House is set within beautiful
grounds while across the road lies
The Slaughters Country Inn on the
banks of the River Eye with its wildlife
friendly feel. Gravel paths and lawns.
ᗕ 🐴 🛏 ☕

43 ◆ **LYDNEY PARK SPRING
GARDEN**
Lydney Park Estate, GL15 6BT.
The Viscount Bledisloe,
01594 842844, reception@
lydneyparkestate.co.uk,
www.lydneyparkestate.co.uk. *½ m
SW of Lydney. On A48 Gloucester
to Chepstow rd between Lydney &
Aylburton. Drive is directly off A48.*
**For NGS: Wed 12 May (10-5).
Adm £6.50, chd free. For other
opening times and information,
please phone, email or visit garden
website.**
Spring garden in 8 acre woodland
valley with lakes, profusion of
rhododendrons, azaleas and other
flowering shrubs. Formal garden;
magnolias and daffodils (April). Picnics
in deer park which has fine trees.
Important Roman temple site and
museum. Not suitable for wheelchairs
due to rough pathway through garden
and steps to WC.
🐴 🚐

44 **THE MANOR**
Little Compton, Moreton-
In-Marsh, GL56 0RZ. Reed
Foundation (Charity),
www.reedbusinessschool.co.uk.
*Next to church in Little Compton.
½ m from A44 or 2m from A3400.
Follow signs to Little Compton, then
yellow NGS signs.* **Sun 27 June,
Sun 22 Aug (2-5). Adm £8, chd
free. Home-made teas.**
Extensive Arts and Crafts garden with
meadow and arboreta. Footpaths
around our fields, playground
between car park and garden.
Croquet, and tennis free to play.
Children and dogs welcome! We
occasionally donate 15% of ticket
proceeds to a charity of our choice.
One ramp in main part of the
garden. Rock garden, tennis court
lawn and meadow not accessible
by wheelchair. Disabled drop-off at
entrance.
ᗕ 🐴 ☕

45 NEW **MONKS SPOUT
COTTAGE**
Glasshouse Hill, May Hill,
Longhope, GL17 0NN. Nigel &
Jane Jackson, 07767 858295,
monksspout@icloud.com. *1m
from National Trust May Hill, in the
hamlet of Glasshouse. Access via
lane (which is also a public footpath
called the Wysis Way) immediately
adjacent to Glasshouse Inn. Very
limited disabled parking at cottage.
Parking signed.* **Sun 6 June (12-5).
Adm £5, chd free. Home-made
teas. Visits also by arrangement
Feb to Oct for groups of up to 30.
No pets allowed in the garden.**
The ²⁄₃ acre garden is adjacent to
Castle Wood which is the backdrop
for a mix of herbaceous borders,
lawns and ponds, with greenhouse,
stream and large display of
insectivorous plants, mostly planted
outside but some in the greenhouse.
With several mature trees and recently
planted acers, there is a mix of shade
and sun creating both damp and
dry planting opportunities. Garden
sculptures. Adjacent public footpaths
through the wood (not part of the

garden) where visitors can see the remains of a ringwork castle dating from C12. Wheelchair access is not practical.

46 MOOR WOOD
Woodmancote, GL7 7EB. Mr & Mrs Henry Robinson, 01285 831397, henry@moorwoodhouse.co.uk, www.moorwoodroses.co.uk. *3½ m NW of Cirencester. Turn L off A435 to Cheltenham at North Cerney, signed Woodmancote 1¼ m; entrance in village on L beside lodge with white gates.* **Sun 27 June (2-6). Adm £5, chd free. Home-made teas. Visits also by arrangement in June for groups of up to 30.**
2 acres of shrub, orchard and wildflower gardens in beautiful isolated valley setting. Holder of National Collection of Rambler Roses. Not recommended for wheelchairs.

NPC

GROUP OPENING

47 NEW NORTH CERNEY VILLAGE GARDENS
North Cerney, Cirencester, GL7 7BZ. North Cerney Village. *Parking available off A435 (opp Bathurst Arms pub), which will be signed. Please access the village from Cirencester or Cheltenham using A435, not through the village itself.* **Sat 12 June (1-5). Combined adm £5, chd free. Light refreshments.**
Enjoy a range of gardens in this quintessential Cotswold village for our inaugural open gardens event.

48 NEW THE OAKS
4 Woodgate Close, Barnwood, Gloucester, GL4 3TN. Heather & David Neale, 01452 612696, Heather_neale@hotmail.co.uk. *3m E Gloucester. Turn into Newstead Road then 1st turning on R.* **Visits by arrangement May to Sept for groups of 10 to 20. Combined admission £5 with 78 Lillesfield Avenue.**
Well stocked garden with array of plants consisting of shrubs, trees, perennials and fishpond. Also a number of ornaments around the garden to complement the planting, and with a swing seat to sit back and enjoy the garden.

49 THE OLD VICARAGE
Murrells End, Hartpury, GL19 3DF. Mrs Carol Huckvale, 01452 700354, C.huckvale@btinternet.com. *5m NW of Gloucester. From Over r'about on A40 N Gloucester bypass, take A417 NW to Hartpury (Ledbury Road). After Maisemore, turn L at signs for Hartpury College. House 1m on R. Follow signs for parking.* **Fri 14, Sat 15, Sun 16 May (12.30-4.30). Adm £5, chd free. Home-made teas. Visits also by arrangement Feb to Apr for groups of 5 to 10.**
Tranquil garden of about 2 acres, with yew oval, mature trees, steps to croquet lawn, mixed borders around main lawn, potager, fruit trees. Work in progress developing wildflower meadow area and dry, shady woodland walk. Spring bulbs and wildflowers. Partial wheelchair access, disabled parking at house, gravel drive and path.

50 ♦ OXLEAZE FARM
Between Eastleach & Filkins, Lechlade, GL7 3RB. Mr & Mrs Charles Mann, 01367 850216, chipps@oxleaze.co.uk, www.oxleaze.co.uk. *5m S of Burford, 3m N of Lechlade off A361 to W (signed Barringtons). Take 2nd L then follow signs.* **For NGS: Wed 23 June (2-6). Adm £6, chd free. Home-made teas in Oxleaze Barn adjacent to garden. For other opening times and information, please phone, email or visit garden website.**
Set among beautiful traditional farm buildings, plantsperson's good size garden combining formality and informality. Year round interest; mixed borders, vegetable potager, new decorative fruit cages, pond/bog garden, bees, potting shed, meadow, and topiary for structure when flowers fade. Garden rooms off central lawn with corners in which to enjoy this Cotswold garden. Groups welcome by appt. Mostly wheelchair access.

51 NEW PAINSWICK FALCON BOWLS CLUB
New Street, Painswick, Stroud, GL6 6UN. Painswick Falcon Bowls Club, painswickfalconbowls.co.uk. *Behind the Falcon Inn Painswick. Up the ramp through the car park of the Falcon Inn Painswick.* **Sun 6 June (2-6). Adm £2, chd free. Tea/coffee/soft drinks, biscuits/cakes.**
A unique chance to visit the 2nd oldest bowls club in the world. Painswick Falcon Bowls Club has wrap around grounds, established borders, hanging baskets and ancient yews. In an idyllic setting, the Bowls Club dates back to 1554 when the green was laid by the lord of the manor to provide himself with recreation. Extensive old photos of the clubs history and a plaque mapping out the establishment of the green in 1554.

52 PASTURE FARM
Upper Oddington, Moreton-In-Marsh, GL56 0XG. Mr & Mrs John LLoyd. *3m W of Stow-on-the-Wold. Just off A436, mid-way between Upper & Lower Oddington.* **Sun 30, Mon 31 May (11-6); Sat 5 June (11-5). Adm £6, chd free. Home-made teas.**
Informal country garden developed over 30 years by current owners. Mixed borders, topiary, orchard and many species of trees. Gravel garden and rambling roses in 'the ruins'. A concrete garden and wildflower area leads to vegetable patch. Big spring-fed pond with ducks. Also bantams, chickens, black Welsh sheep and 1 Kunekune pig. Large plant sale 30/31 May with proceeds to Kate's Home Nursing. Public footpath across 2 small fields arrives at C11 church, St Nicholas, with doom paintings, set in ancient woodlands. Truly worth a visit. See Simon Jenkins' Book of Churches. Mostly wheelchair access.

53 THE PATCH

Hollywell Lane, Brockweir, Chepstow, NP16 7PJ. Mrs Immy Lee, 07801 816340, immylee1@hotmail.com. *In the Wye Valley, 6.7m N of Chepstow and 10.6m S of Monmouth, off A466, across Brockweir Bridge.* **Sun 12 Sept (12.30-5). Combined adm with Clouds Rest £7, chd free. Home-made teas. Visits also by arrangement Apr to July for groups of up to 30, combined with Greenfields, Brockweir Common. Teas subject to weather.**
Rural ¼ acre garden with far reaching views across the Wye Valley, containing 60+ repeat flowering roses and a variety of shrubs and perennials, providing interest around the year. The garden open day is in combination with Clouds Rest. The gardens are linked by an easy drive or a 15 min stony walk along the Offa's Dyke path. Ample parking at both gardens. Partial wheelchair access.
&. ✿ ☕

54 PEAR TREE COTTAGE

58 Malleson Road, Gotherington, nr Cheltenham, GL52 9EX. Mr & Mrs E Manders-Trett, 01242 674592, mmanderstrett@gmail.com. *4m N of Cheltenham. From A435, travelling N, turn R into Gotherington 1m after end of Bishop's Cleeve bypass at*
garage. *Garden on L approx 100yds past Shutter Inn.* **Sun 18 Apr (2-5). Adm £5, chd free. Tea/Coffee & cake available. Visits also by arrangement Mar to June for groups of up to 30.**
Mainly informal country garden of approx ½ acre with pond and gravel garden. Herbaceous borders, trees and shrubs surround lawns and seating areas. Wild garden and orchard lead to greenhouses, vegetable garden and beehives. Spring bulbs, early summer perennials and shrubs particularly colourful. Gravel drive and several shallow steps can be overcome for wheelchair users with prior notice.
&. 🐐 ✿ 🚗 ☕

55 PERRYWOOD HOUSE

Longney, Gloucester, GL2 3SN. Gill & Mike Farmer. *7m SW of Gloucester, 4m W of Quedgeley. From N: R off B4008 at Tesco r'about. R at next r'about then R at 2nd mini r'about, then signed. From S: L off A38 at Moreton Valence to Epney/Longney, over canal bridge, R at T junction then signed.* **Sat 26, Sun 27 June (11-5). Adm £4, chd free. Home-made teas.**
1 acre plant lover's garden in the Severn Vale surrounded by open farmland. Established over 20 years, an informal country garden with mature trees and shrubs, small pond, colourful herbaceous borders and containers. Plenty of places to sit and enjoy the garden and your tea. Lots of interesting plants for sale. All areas accessible with level lawns and gravel drives. Disabled parking available.
&. ✿ ☕

56 RADNORS

Wheatstone Lane, Lydbrook, GL17 9DP. Mrs Mary Wood, 01594 861690, mary.wood37@btinternet.com. *Lower Lydbrook in the Wye Valley. From Lydbrook, go through village. At T junction turn L into Stowfield Rd. Wheatstone Lane (300m) is 1st turning L, a small lane after the white cottages. Radnors is the last house.* **Visits by arrangement May to Oct for groups of up to 10. Adm £4, chd free. Light refreshments.**
5 acre hillside garden in AONB on bank above the River Wye. Focus on wildlife with naturalistic planting and weeds, some left for specific insects/birds. It has many paths, a wooded area, wildflower area, flower beds and borders, lawns, stumpery, fernery, vegetable beds and white garden. Of particular interest are the path along a disused railway line, and the summer dahlias. The garden has many narrow and uneven paths and steps and is not accessible to wheelchair users or those with mobility difficulties.
🐐 ✿ ☕ 🏕

East Court

57 RAMBLERS
Lower Common, Aylburton, Lydney, GL15 6DS. Jane & Leslie Hale. *1½m W of Lydney. Off A48 Gloucester to Chepstow Rd. From Lydney through Aylburton, out of de-limit turn R signed Aylburton Common, ¾m along lane.* **Sun 2 May (1.30-5). Adm £4, chd free. Home-made teas.**
Peaceful medium sized country garden with informal cottage planting, herbaceous borders and small pond looking through hedge windows onto wildflower meadow and mature apple orchard. Some shade loving plants and topiary. Large productive vegetable garden. Past winner of The English Garden magazine's Britain's Best Gardener's Garden competition.
✿ 🍵

58 NEW READY TOKEN HOUSE
Ready Token, Cirencester, GL7 5SX. Mark & Tabitha Mayall. *Hamlet between Bibury and Poulton. From Cirencester take Roman Road.* **Mon 28, Tue 29, Wed 30 June, Thur 1 July (10-5). Adm £7, chd free. Light refreshments.**
Beautiful, nature-friendly formal garden with herbaceous borders and mixed topiary set against backdrop of 300 year old wisteria clad country house and mature trees, set in 70 acres of rewilded parkland, with mown pathways through wildflower meadows for walks with stunning views to Vale of the White Horse; Roman sunken garden, well, pond, lake, organic vegetable garden with raised beds, greenhouse.
& Ⓓ 🍵

59 NEW RICHMOND PAINSWICK RETIREMENT VILLAGE
Stroud Road, Painswick, Stroud, GL6 6UL. *Just outside Painswick village.* **Thur 13, Thur 27 May, Thur 3, Thur 17 June (10-4). Adm £5, chd free. Home-made teas.**
Situated on the southern slopes of Painswick this 4 acre retirement village boasts formal lawns and borders planted for year round interest. A varied mix of herbaceous and perennial planting, with many areas of interest inc wildflower meadow and fruit trees that combine to attract an abundance of wildlife. The gardens are a blaze of colour throughout. Car parking, toilet facilities, cafe serving tea /coffee plus light bites, as well as a fully licensed restaurant. Make a day of it and

enjoy afternoon tea on the roof top garden. All this nestled in the beautiful cotswold countryside. Gentle slopes in wildflower meadow and around some areas of village.
& 🐕 🍵

60 ROCKCLIFFE
Upper Slaughter, Cheltenham, GL54 2JW. Mr & Mrs Simon Keswick, www.rockcliffegarden.co.uk. *2m from Stow-on-the-Wold. 1½m from Lower Swell on B4068 towards Cheltenham. Leave Stow on the Wold on B4068 through Lower Swell. Continue on B4068 for 1½m. Rockcliffe is well signed on R.* **Wed 2, Wed 30 June (10-5). Adm £7.50, chd free. Home-made teas. Donation to Kate's Home Nursing.**
Large traditional English garden of 8 acres inc pink garden, white and blue garden, herbaceous borders, rose terrace, large walled kitchen garden and greenhouses. Pathway of topiary birds leading up through orchard to stone dovecot. 2 wide stone steps through gate, otherwise good wheelchair access. Sorry no dogs.
& ✿ 🍵

61 ♦ SEZINCOTE
Moreton-in-Marsh, GL56 9AW. Mrs D Peake, 01386 700444, enquiries@sezincote.com, www.sezincote.co.uk. *3m SW of Moreton-in-Marsh. From Moreton-in-Marsh turn W along A44 towards Evesham; in 1½m (just before Bourton-on-the-Hill) turn L, by stone lodge with white gate.* **For NGS: Sun 11 July (2-5.30). Adm £7.50, chd £2.50. Home-made teas. Teas provided by and in aid of Longborough School. For other opening times and information, please phone, email or visit garden website.**
Exotic oriental water garden by Repton and Daniell with lake, pools and meandering stream, banked with massed perennials. Large semi-circular orangery, formal Indian garden, fountain, temple and unusual trees of vast size in lawn and wooded park setting. House in Indian manner designed by Samuel Pepys Cockerell. Garden on slope with gravel paths, so not all areas wheelchair accessible.
& 🚗 🍵

62 SOUTH LODGE
Church Road, Clearwell, Coleford, GL16 8LG. Andrew & Jane MacBean, 01594 837769, southlodgegarden@btinternet.com, www.southlodgegarden.co.uk. *2m S of Coleford. Off B4228. Follow signs to Clearwell. Garden on L of castle driveway. Please park on rd in front of church or in village. No parking on castle drive.* **Sat 27 Mar, Sat 1 May, Sat 26 June (1-5). Adm £4.50, chd free. Home-made teas. Visits also by arrangement Apr to June for groups of 20+.**
Peaceful country garden in 2 acres with stunning views of surrounding countryside. High walls provide a backdrop for rambling roses, clematis, and honeysuckles. Organic garden with large variety of perennials, annuals, shrubs and specimen trees with year round colour. Vegetable garden, wildlife and formal ponds. Rustic pergola planted with English climbing roses and willow arbour in gravel garden. Gravel paths and steep grassy slopes. Assistance dogs only.
& ✿ 🍵

GROUP OPENING

63 STANTON VILLAGE GARDENS
Stanton, nr Broadway, WR12 7NE. 01386 584659, susanhughes83@hotmail.co.uk. *3m S of Broadway. Off B4632, between Broadway (3m) and Winchcombe (6m).* **Sun 13 June (2-6). Combined adm £7.50, chd free. Home-made teas in several gardens around the village. Ice cream trike in village. Visits also by arrangement in June. Donation to Village charities.**
A selection of gardens open in this picturesque, unspoilt Cotswold village. Many houses border the street with long gardens hidden behind. Gardens vary, from houses with colourful herbaceous borders, established trees, shrubs and vegetable gardens to tiny cottage gardens. Some also have attractive, natural water features fed by the stream which runs through the village. Plants for sale and book stall. Free parking. An NGS visit not to be missed. A gem of Cotswold village. Church also open. The Mount Inn open for lunch. Gardens not suitable for wheelchair users due to gravel drives.

64 ◆ STANWAY FOUNTAIN & WATER GARDEN

Stanway, Cheltenham, GL54 5PQ. The Earl of Wemyss & March, 01386 584528, office@stanwayhouse.co.uk, www.stanwayfountain.co.uk. *9m NE of Cheltenham. 1m E of B4632 Cheltenham to Broadway rd on B4077 Toddington to Stow-on-the-Wold rd.* **For NGS: Sun 16 May, Sun 15 Aug (2-5). Adm £7, chd £2.50. Cream teas in Stanway Tea Room. For other opening times and information, please phone, email or visit garden website.**

20 acres of planted landscape in early C18 formal setting. The restored canal, upper pond and fountain have re-created one of the most interesting Baroque water gardens in Britain. Striking C16 manor with gatehouse, tithe barn and church. Britain's highest fountain at 300ft, the world's highest gravity fountain which runs at 2.45 and 4.00pm for 30 mins each time. Partial wheelchair access in garden, some flat areas, able to view fountain and some of garden. House is not wheelchair suitable.

65 STOWELL PARK

Northleach, Cheltenham, GL54 3LE. The Lord & Lady Vestey, www.stowellpark.co.uk. *8m NE of Cirencester. Off Fosseway A429 2m SW of Northleach.* **Sun 16 May, Sun 13 June (2-5). Adm £6, chd free. Home-made teas. Donation to another charity.**

Magnificent lawned terraces with stunning views over Coln Valley. Fine collection of old-fashioned roses and herbaceous plants, pleached lime approach to C14 house (not open). 2 large walled gardens containing vegetables, fruit, cut flowers and range of greenhouses inc vinery, peach and orchid houses. Long rose pergola and wide, plant filled borders. Fountain garden and water features. Open continuously for over 50yrs. Plants for sale May 16th opening only. Church open. Wheelchair access to walled gardens, terraces and teas. Partial access on some pathways.

66 NEW ◆ SUDELEY CASTLE & GARDENS

Winchcombe, GL54 5JD. Lady Ashcombe, 01242 604244, enquiries@sudeley.org.uk, www.sudeleycastle.co.uk. *8m NE Cheltenham, 10 m from M5 J9.* *SatNavs GL54 5LP. Free Parking.* **For NGS: Tue 6 July (10.30-3). Adm £10, chd £6. For other opening times and information, please phone, email or visit garden website.**

Sudeley Castle features 10 magnificent gardens, each with its own unique style and design. Surrounded by striking views of the Cotswold Hills, each garden reflects the fascinating 1000 year history of the Castle. This series of elegant gardens is set among the Castle and atmospheric ruins and inc a Knot garden, Queen's garden and Tudor physic garden. Sudeley Castle remains the only private castle in England to have a queen buried within the grounds - Queen Katherine Parr, the last and surviving wife of King Henry VIII – who lived and died in the castle. A circular route around gardens is wheelchair accessible although some visitors may require assistance from their companion.

67 TREE HILL

76 Gretton Road, Winchcombe, Cheltenham, GL54 5EL. Mark Caswell. *½ m N of Winchcombe. Leave Winchcombe via North St, straight ahead onto Gretton Rd for ½ m. From M5 J9, take A46 towards Evesham, at r'about take B4077 to Stow, then R to Gretton Rd.* **Sun 29 Aug (10-4). Adm £3, chd free.**

Stepping into this garden is akin to stepping into another world. As one visitor remarked 'this is unlike any English garden I have ever seen.' On this small and modest plot one finds huge leaves, tall exotic plants jostling for light with the small and the delicate. Created with a passion for the exotic landscapes of the Caribbean which was inspired by a visit to Barbados. Two stone steps down to the garden.

68 TRENCH HILL

Sheepscombe, GL6 6TZ. Celia & Dave Hargrave, 01452 814306, celia.hargrave@btconnect.com. *1½ m E of Painswick. From Cheltenham A46 take 1st turn signed Sheepscombe, follow for approx 1½ m. Or from the Butcher's Arms in Sheepscombe (with it on R) leave village and take lane signed for Cranham.* **Sun 7, Sun 14 Feb (11-4.30); Sun 21 Mar, Sun 4, Mon 5 Apr, Sun 2 May (11-6). Every Wed 9 June to 30 June (2-6). Sun 18 July, Sun 29 Aug, Sun 19 Sept (11-6). Adm £5, chd free.**

Home-made teas. Visits also by arrangement Feb to Sept. Any coach must be agreed by owners prior to visit.

Approx 3 acres set in small woodland with panoramic views. Variety of herbaceous and mixed borders, rose garden, extensive vegetable plots, wildflower areas, plantings of spring bulbs with thousands of snowdrops and hellebores, woodland walk, 2 small ponds, waterfall and larger conservation pond. Interesting wooden sculptures, many within the garden. Run on organic principles. Children's play area, wooden sculptures. Mostly wheelchair access but some steps and slopes.

69 UPTON WOLD

Moreton-in-Marsh, GL56 9TR. Mr & Mrs I R S Bond, www.uptonwold.co.uk. *4½ m W of Moreton-in-Marsh on A44. From Moreton/Stow ½ m past A424 turn R opp Deer Sign to road into fields then L at mini Xrds. From Evesham 1m past B4081 C/Campden Xrds turn L at end of stone wall to road into fields then as above.* **Sun 18 Apr (10-5). Adm £12, chd free. Home-made teas.**

One of the secret gardens of the Cotswolds, Upton Wold has commanding views; yew hedges; herbaceous walk; vegetable, pond and woodland gardens, labyrinth. An abundance of unusual and interesting plants, shrubs and trees. National Collections of juglans and pterocarya. A garden of interest to any garden and plant lover. 2 Star award from GGG.

70 WEIR REACH

The Rudge, Maisemore, Gloucester, GL2 8HY. Sheila & Mark Wardle. *3m NW of Gloucester. Turn into The Rudge by White Hart Pub. Parking 100m from garden.* **Sun 13, Wed 16 June (11-5). Adm £4, chd free. Home-made teas.**

Country garden by R Severn. Approx 2 acres, half cultivated with herbaceous beds and mixed borders plus fruit and vegetable cages. Clematis and acers, stone ornaments, small sculptures, bonsai collection. Planted rockery with waterfall and stream connect 2 ponds. Large specimen koi pond borders patio. Meadow with specimen and fruit trees and bamboo collection leading to river and country views.

71 ◆ WESTONBIRT SCHOOL GARDENS

Tetbury, GL8 8QG. Holfords of Westonbirt Trust, 01666 881373, jbaker@holfordtrust.com, www.holfordtrust.com. *3m SW of Tetbury. Please enter through main school gates on A433 - no access via side entrances.* For NGS: Sun 11 July (10-4). Adm £5, chd free. Tea, coffee & cake available to purchase. For other opening times and information, please phone, email or visit garden website.
28 acres. Former private garden of Robert Holford, founder of Westonbirt Arboretum. Formal Victorian gardens inc walled Italian garden now restored with early herbaceous borders and exotic border. Rustic walks, lake, statuary and grotto. Rare, exotic trees and shrubs. Beautiful views of Westonbirt House open with guided tours to see fascinating Victorian interior on designated days of the year. Afternoon tea with sandwiches and scones available for pre-booked private tours, groups of 20-60. Only some parts of garden accessible to wheelchairs. Ramps and lift allow access to house.

SPECIAL EVENT

72 WORTLEY HOUSE

Wortley, Wotton-Under-Edge, GL12 7QP. Simon & Jessica Dickinson, 01453 843174, jessica@wortleyhouse.co.uk. *1m from Wotton-under-Edge. Full directions will be provided with ticket.* Tue 13 Apr, Tue 15 June (2-5). Adm £15, chd free. Pre-booking essential, please visit www.ngs.org.uk/special-events for information & booking. Home-made teas included in ticket price. Visits also by arrangement Apr to June for groups of up to 30.
This diverse garden of over 20 acres has been created during the last 30 years by the current owners and inc walled garden, pleached lime avenues, nut walk, potager, ponds, Italian garden, shrubberies and wildflower meadows. Strategically placed follies, urns and statues enhance extraordinary vistas throughout, and the garden is filled with plants, arbours, roses through trees and up walls, and herbaceous borders. The stunning surrounding countryside is incorporated into the garden with views up the steep valley that are such a feature in this part of Gloucestershire. Wheelchair access to most areas, golf buggy also available.

Richmond Painswick Retirement Village

HAMPSHIRE

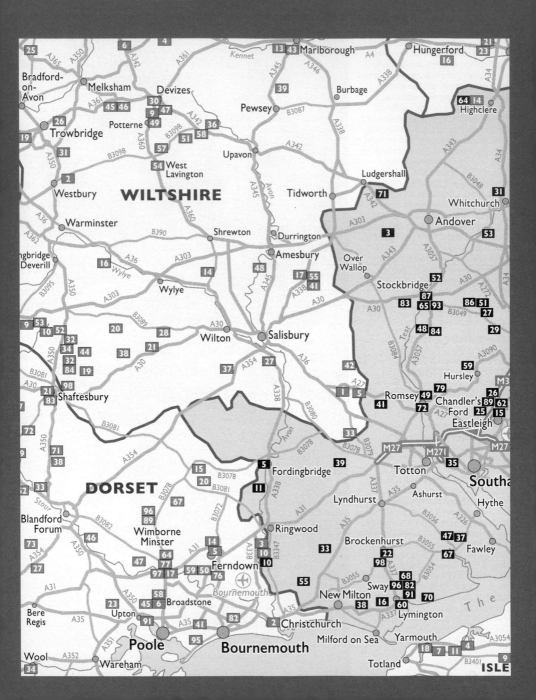

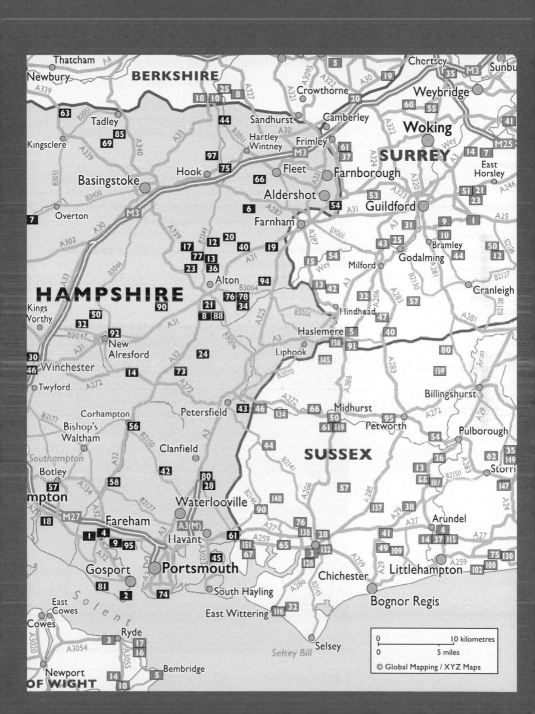

Volunteers

County Organiser
Mark Porter 01962 791054
markstephenporter@gmail.com

County Treasurer
Fred Fratter 01962 776243
fred@fratter.co.uk

Publicity
Pat Beagley 01256 764772
pat.beagley@ngs.org.uk

Social Media
Position Vacant
For details please contact
Mark Porter (as above)

Booklet Co-ordinator
Mark Porter (as above)

Assistant County Organisers

Central
Sue Cox 01962 732043
suealex13@gmail.com

Central West
Kate Cann 01794 389105
kategcann@gmail.com

East
Linda Smith 01329 833253
linda.ngs@btinternet.com

North
Cynthia Oldale 01420 520438
c.k.oldale@btinternet.com

North East
Lizzie Powell 01420 23185
lizziepowellbroadhatch@gmail.com

North West
Adam Vetere 01635 268267
adam.vetere@ngs.org.uk

South
Barbara Sykes 02380 254521
barandhugh@aol.com

South West
Elizabeth Walker 01590 677415
elizabethwalker13@gmail.com

West
Jane Wingate-Saul 01725 519414
jane.wingatesaul@ngs.org.uk

Hampshire is a large, diverse county. The landscape ranges from clay/gravel heath and woodland in the New Forest National Park in the south west, across famous trout rivers – the Test and Itchen – to chalk downland in the east, where you will find the South Downs National Park.

Our open gardens are spread right across the county and offer a very diverse range of interest for both the keen gardener and the casual visitor.

We have a large number of gardens with rivers running through them such as those in The Island, Bere Mill and Weir House; gardens with large vegetable kitchen gardens such as Bramdean House; and ten new gardens will open for the very first time.

You will be assured of a warm welcome by all our garden owners and we hope you enjoy your visits.

Below: Cherry Tree Barn, Searles Lane Gardens

@HampshireNGS   @HantsNGS   @hantsngs

OPENING DATES

All entries subject to change. For latest information check www.ngs.org.uk

Extended openings are shown at the beginning of the month.

Map locator numbers are shown to the right of each garden name.

February

Snowdrop Festival

Saturday 13th
The Down House 32

Sunday 14th
Little Court 51

Monday 15th
Little Court 51

Saturday 20th
The Down House 32

Sunday 21st
Bramdean House 14
◆ Chawton House 21
Little Court 51

Monday 22nd
Little Court 51

March

Sunday 14th
Bere Mill 7

Sunday 21st
Little Court 51

Saturday 27th
Pilley Hill Cottage 68

Sunday 28th
Pilley Hill Cottage 68

Wednesday 31st
Beechenwood Farm 6

April

Every Wednesday
Beechenwood Farm 6

Friday 2nd
Crawley Gardens 27

Sunday 4th
Pylewell Park 70
Twin Oaks 89

Monday 5th
Crawley Gardens 27
Twin Oaks 89

Sunday 11th
Bramdean House 14
Durmast House 33
Old Thatch & The
 Millennium Barn 66
28 St Ronan's Avenue 74
NEW Woodpeckers
 Care Home 98

Thursday 15th
Appleyards 5

Friday 16th
Appleyards 5

Saturday 17th
Appleyards 5
The Island 49
Pilley Hill Cottage 68

Sunday 18th
Appleyards 5
The Island 49
Pilley Hill Cottage 68
Terstan 87

Friday 23rd
Bluebell Wood 13

Saturday 24th
Bluebell Wood 13

Sunday 25th
◆ Spinners Garden 82

Wednesday 28th
Twin Oaks 89

May

Every Wednesday
Beechenwood Farm 6

Sunday 2nd
Bere Mill 7
Brick Kiln Cottage 17
The Cottage 25
Walhampton 91

Monday 3rd
Beechenwood Farm 6
The Cottage 25

Sunday 9th
The Cottage 25
The House in the
 Wood 47

Monday 10th
The Cottage 25

Thursday 13th
Appleyards 5
Bisterne Manor 10
Tanglefoot 86

Friday 14th
Appleyards 5

Saturday 15th
◆ Alverstoke Crescent
 Garden 2
Appleyards 5
146 Bridge Road 18
21 Chestnut Road 22
'Selborne' 76

Sunday 16th
Appleyards 5
146 Bridge Road 18
21 Chestnut Road 22
How Park Barn 48
'Selborne' 76
Tadley Place 85
Tanglefoot 86

Monday 17th
'Selborne' 76

Wednesday 19th
NEW Heckfield Place 44

Thursday 20th
Bisterne Manor 10

Friday 21st
Little Court 51

Sunday 23rd
Manor House 56
28 St Ronan's Avenue 74

Thursday 27th
Bisterne Manor 10
NEW Searles Lane
 Gardens 75

Friday 28th
NEW Searles Lane
 Gardens 75

Saturday 29th
NEW Searles Lane
 Gardens 75

Sunday 30th
Amport & Monxton
 Gardens 3
Pylewell Park 70
Romsey Gardens 72
NEW Searles Lane
 Gardens 75
The Thatched Cottage 88
Twin Oaks 89

Monday 31st
Amport & Monxton
 Gardens 3

Bere Mill 7
Romsey Gardens 72
The Thatched Cottage 88
Twin Oaks 89

June

**Every Wednesday
to Wednesday 9th**
Beechenwood Farm 6

Wednesday 2nd
NEW Heckfield Place 44

Thursday 3rd
Tanglefoot 86

Saturday 5th
21 Chestnut Road 22
Ferns Lodge 38
Froyle Gardens 40
1 Povey's Cottage 69
Spitfire House 83

Sunday 6th
Bramdean House 14
Brick Kiln Cottage 17
21 Chestnut Road 22
Ferns Lodge 38
Froyle Gardens 40
108 Heath Road 43
Manor House 56
1 Povey's Cottage 69
Shalden Park House 77
NEW South View
 House 80
Tanglefoot 86

Thursday 10th
Appleyards 5
Lake House 50

Friday 11th
Appleyards 5

Saturday 12th
Appleyards 5
The Thatched Cottage 88

Sunday 13th
Appleyards 5
Berry Cottage 8
Broadhatch House 19
Cranbury Park 26
Fritham Lodge 39
Lake House 50
Little Court 51
Old Thatch & The
 Millennium Barn 66
Terstan 87
The Thatched Cottage 88
Weir House 92

South View House

Wednesday 4th
West View 93

Friday 6th
Twin Oaks 89

Saturday 7th
Old Camps 63
Twin Oaks 89
Willows 96

Sunday 8th
The Homestead 45
Old Camps 63
NEW South View
House 80
Twin Oaks 89
Willows 96

Saturday 21st
Wheatley House 94

Sunday 22nd
Weir House 92
Wheatley House 94

Sunday 29th
Berry Cottage 8
The Deane House 29
Gilberts Dahlia Field 41
The Thatched Cottage 88

Monday 30th
Bleak Hill Nursery &
Garden 11
The Thatched Cottage 88

September

Sunday 5th
How Park Barn 48
Meon Orchard 58

Sunday 12th
Bere Mill 7
Bramdean House 14
108 Heath Road 43
NEW Woodpeckers
Care Home 98

Wednesday 15th
NEW Woodpeckers
Care Home 98

Thursday 16th
Redenham Park
House 71

Friday 17th
Redenham Park
House 71

Sunday 19th
Terstan 87

February 2022

Sunday 13th
Little Court 51

Monday 14th
Little Court 51

Sunday 20th
Little Court 51

Monday 21st
Little Court 51

By Arrangement

Arrange a personalised garden visit on a date to suit you. See individual garden entries for full details.

80 Abbey Road 1
Angels Folly 4
Beechenwood Farm 6
Bere Mill 7
Berry Cottage 8
8 Birdwood Grove 9
6 Breamore Close 15
NEW Briar Patch 16
Brick Kiln Cottage 17
Broadhatch House 19
Bumpers 20
Clover Farm 23
Colemore House
Gardens 24
The Cottage 25
Crookley Pool 28
The Deane House 29
The Dower House 30
Down Farm House 31

The Down House 32
Durmast House 33
Endhouse 35
Fairbank 36
Ferns Lodge 38
Hambledon House 42
108 Heath Road 43
The Homestead 45
Lake House 50
Little Court 51
Merdon Manor 59
Old Camps 63
The Old Rectory 64
Old Swan House 65
Pilley Hill Cottage 68
1 Povey's Cottage 69
Redenham Park
House 71
Sages 73
Silver Birches 78
Spindles 81
Spitfire House 83
Tanglefoot 86
Terstan 87
The Thatched Cottage 88
Trout Cottage, Stockbridge
Gardens 84
Twin Oaks 89
Walhampton 91
Weir House 92
Wheatley House 94
Willows 96

THE GARDENS

1 80 ABBEY ROAD
Fareham, PO15 5HW. Brian & Vivienne Garford, 01329 843939, vgarford@aol.com. *From M27 J9 take A27 E to Fareham for approx 2m. At top of hill, turn L at lights into Highlands Rd. Turn 4th R into Blackbrook Rd. Abbey Rd is 4th L.* **Visits by arrangement Apr to Aug for groups of up to 30. Light refreshments.**
A small garden designed to use all the available space. Many unusual plants, including a large collection of herbs and native wildflowers. Interesting use of containers and other ideas for the smaller garden. Two ponds and tiny meadow help attract a wide range of wildlife. Living willow seat, summerhouse and trained grapevine.
✿ ♿ ☕

2 ◆ ALVERSTOKE CRESCENT GARDEN
Crescent Road, Gosport, PO12 2DH. Gosport Borough Council, www.alverstokecrescentgarden.co.uk. *1m S of Gosport. From A32 & Gosport follow signs for Stokes Bay. Continue alongside bay to small r'about, turn L into Anglesey Rd. Crescent Garden signed 50yds on R.* **For NGS: Sat 15 May (10-4). Adm by donation. Home-made teas. For other opening times and information, please visit garden website.**
Restored Regency ornamental garden designed to enhance fine crescent (Thomas Ellis Owen 1828). Trees, walks and flowers lovingly maintained by community and council partnership. A garden of considerable local historic interest highlighted by impressive restoration and creative planting. Adjacent to St Mark's churchyard, worth seeing together. Heritage, history and horticulture, a fascinating package. Plant sale. Green Flag Award.
♿ 🐄 ✿ ☕

"The annual donation from the National Garden Scheme to Perennial is the cornerstone of our fundraising activities and encourages many of our donors". Perennial

GROUP OPENING

3 AMPORT & MONXTON GARDENS

Amport and Monxton, SP11 8AY. *3m SW of Andover. Turn off the A303 signed East Cholderton from the E or Thruxton village from the W. Follow signs to Amport. Car parking in field next to Amport village green. Please drive between the two villages.* Sun 30, Mon 31 May (2-5.30). Combined adm £6, chd free. Cream teas in village hall, Monxton.

BRIDGE COTTAGE
John & Jenny Van de Pette.

NEW CORNER COTTAGE, 15 SARSON
Ms Jill McAvoy.

WHITE GABLES
David & Coral Eaglesham.

Monxton and Amport are two pretty villages linked by Pill Hill Brook. Visitors have three gardens to enjoy. Bridge Cottage, a 2 acre haven for wildlife with the banks of the trout stream and lake planted informally with drifts of colour, a large vegetable garden, fruit cage, small mixed orchard and arboretum with specimen trees. A cottage style garden at White Gables with a collection of trees, along with old roses and herbaceous plants. Corner Cottage is a delightful cottage garden with a serpentine gravel path winding between gravel borders, clipped box hedging and old fashioned roses. Amport has a lovely village green, come early and bring a picnic. There is a very popular large plant sale at Bridge Cottage. No wheelchair access to White Gables and partial access to Corner Cottage.

4 ANGELS FOLLY

15 Bruce Close, Fareham, PO16 7QJ. Teresa & John Greenwood, 07545 242654, tgreenwood@ntlworld.com. *M27 W J10 under M27 bridge RH-lane, do U-turn. At r'about 3rd exit, across T-lights, lst R Miller Dr, 2nd R Somervell Dr, 1st R Bruce Close (2 disabled spaces). Free parking at Fareham Leisure Centre.* Sat 3, Sun 4, Sat 17, Sun 18 July (10-4.30). Adm £3.50, chd free. Home-made teas, cream teas & savoury scones. Visits also

by arrangement July to Sept for groups of 5 to 20.
The garden has a number of secluded areas each with their own character, including a Mediterranean garden, decking with raised beds and a seating area with a living wall. An arched folly, bench and fish pond leads to a raised planting bed and fireplace adjacent to a summerhouse. There is a wide range of colourful plants, hanging baskets and a lower secluded decked area with a planted gazebo and statue.

5 APPLEYARDS

Bowerwood Road, Fordingbridge, SP6 3BP. Bob & Jean Carr. *½m from Fordingbridge on B3078. After church & houses, 400yds on L as road climbs after bridge. Parking for 8 cars only. No parking on narrow road.* Thur 15, Fri 16, Sat 17, Sun 18 Apr, Thur 13, Fri 14, Sat 15, Sun 16 May, Thur 10, Fri 11, Sat 12, Sun 13 June (12-6). Adm £4, chd free. Pre-booking essential, please phone 01425 657631 or email bob.carr.rtd@gmail.com for information & booking. Home-made teas.
2 acre sloping s-facing garden, newly restored, overlooking pasture. 100+ trees, sloping lawns and paths though wooded sections with massed daffodils and bluebells in spring. Newly planted rhododendrons in wooded area. Herbaceous beds, two rose beds, shrubberies, two wildlife ponds, orchard, sloping rockery beds, soft fruit cages and greenhouse.

6 BEECHENWOOD FARM

Hillside, Odiham, Hook, RG29 1JA. Mr & Mrs M Heber-Percy, 01256 702300, beechenwood@totalise.co.uk. *5m SE of Hook. Turn S into King St from Odiham High St. Turn L after cricket ground for Hillside. Take 2nd R after 1½m, modern house ½m.* Every Wed 31 Mar to 9 June (2-5). Mon 3 May (2-5). Adm £4, chd free. Home-made teas. Visits also by arrangement.
2 acre garden in many parts. Lawn meandering through woodland with drifts of spring bulbs. Rose pergola with steps, pots with spring bulbs and later aeoniums. Fritillary and cowslip meadow. Walled herb garden with pool and exuberant planting. Orchard incl white garden and hot border. Greenhouse and vegetable

garden. Rock garden extending to grasses, ferns and bamboos. Shady walk to belvedere. 8 acre copse of native species with grassed rides. Assistance available with gravel drive and avoidable shallow steps.

7 BERE MILL

London Road, Whitchurch, RG28 7NH. Rupert & Elizabeth Nabarro, 07703 161074, rupertnab@gmail.com, www.beremillfarm.com. *9m E of Andover, 12m N of Winchester. In centre of Whitchurch, take London Rd at r'about. Uphill 1m, turn R 50yds beyond The Gables on R. Drop-off point for disabled at garden.* Sun 14 Mar, Sun 2, Mon 31 May, Sun 12 Sept (1.30-5). Adm £7.50, chd free. Home-made teas. Visits also by arrangement Feb to Sept. The garden is open to groups of any size at a fixed charge of £400. Donation to Smile Train.
On the Upper Test with water meadows and wooded valleys, this garden offers herbaceous borders, bog and Mediterranean plants, as well as a replanted orchard and two small arboretums. Features incl early bulbs, species tulips, Japanese prunus, peonies, wisteria, irises, roses, and semi-tropical planting. At heart it aims to complement the natural beauty of the site and to incorporate elements of oriental garden design and practice. The working mill was where Portals first made paper for the Bank of England in 1716. Unfenced and unguarded rivers and streams. Wheelchair access unless very wet.

8 BERRY COTTAGE

Church Road, Upper Farringdon, Alton, GU34 3EG. Mrs P Watts, 01420 588318. *3m S of Alton off A32. Turn L at Xrds, 1st L into Church Rd. Follow road past Massey's Folly, 2nd house on R, opp church.* Sun 13 June, Sun 11 July, Sun 29 Aug (1-5). Combined adm with The Thatched Cottage £10, chd free. Home-made teas. Visits also by arrangement May to Sept for groups of 10+.
Small organic cottage garden with yr-round interest, designed and maintained by owner, surrounding C16 house (not open). Spring bulbs, roses, clematis and herbaceous borders. The borders are colour themed and contain many unusual plants. Pond, bog garden and

shrubbery. Close to Massey's Folly built by the Victorian rector incl 80ft tower with unique handmade floral bricks, C11 church and some of the oldest yew trees in the county. Partial wheelchair access.

🟨 8 BIRDWOOD GROVE

Downend, Fareham, PO16 8AF. Jayne & Eddie McBride, 01329 280838, jayne.mcbride@ntlworld.com. *M27 J11, L lane slip to Delme r'about, L on A27 to Portchester over 2 T-lights, completely around small r'about, Birdwood Grove 1st L.* **Sat 17, Sun 18 July (12.30-5). Adm £3, chd free. Home-made teas. Visits also by arrangement July & Aug for groups of up to 10.**
The subtropics in Fareham! This small garden is influenced by the flora of Australia and New Zealand and incl many indigenous species and plants that are widely grown down under. The 4 climate zones; arid, temperate, lush fertile and a shady ternery, are all densely planted to make the most of dramatic foliage, from huge bananas to towering cordylines. Wheelchair access over short gravel path, not suitable for mobility scooters.

🔟 BISTERNE MANOR

Bisterne, Ringwood, BH24 3BN. Mr & Mrs Hallam Mills. *2½ m S of Ringwood on B3347 Christchurch Rd, 500yds past church on L. Entrance signed Stable Family Home Trust on L (blue sign), just past lodge. Disabled parking signed near the house.* **Thur 13, Thur 20, Thur 27 May (2-5). Adm £5, chd free.**
Glorious rhododendrons and azaleas form a backdrop for our C19 garden, first opened in the 1930s for the fledgling NGS. The C16 manor house (not open) overlooks a grand parterre with urns, box and yew hedges. Rare tree specimens grace fine lawns and wildflower planting, leading to a boundary woodland walk with glimpses of surrounding pastures. There is a small kitchen garden. Wheelchair access to a level garden with wide gravel paths.

🔢 BLEAK HILL NURSERY & GARDEN

Braemoor, Bleak Hill, Harbridge, Ringwood, BH24 3PX. Tracy & John Netherway,

www.bleakhillplants.co.uk. *2½ m S of Fordingbridge. Turn off A338 at Ibsley. Go through Harbridge village to T-junction at top of hill, turn R for ¼ m.* **Sun 4, Mon 5, Sun 18, Mon 19 July, Sun 1, Mon 2, Mon 30 Aug (2-5). Adm £4, chd free. Home-made teas. No refreshments on Mondays.**
Through the moongate and concealed from view are billowing borders contrasting against a seaside scene with painted beach huts and a boat on the gravel. Herbaceous borders fill the garden with colour wrapping around a pond and small stream. Greenhouses with cacti and Sarracenias. Vegetable patch and bantam chickens. Small adjacent nursery.

🔢 NEW BLOUNCE HOUSE

Blounce, South Warnborough, Hook, RG29 1RX. Tom Bartlam. *In hamlet of Blounce, 1m S of South Warnborough on B3349 from Odiham to Alton.* **Wed 28 July, Sun 1 Aug (1-4). Adm £4, chd free. Home-made teas.**
A 2 acre garden surrounding a classic Queen Anne house (not open). Mixed planting to give interest from spring to late autumn. Herbaceous borders with a variety of colour themes. In later summer an emphasis on dahlias, salvias and grasses.

Terstan

© Leigh Clapp

13 BLUEBELL WOOD

Stancombe Lane, Bavins, New Odiham Road, Alton, GU34 5SX. Mrs Jennifer Ospici, www.bavins.co.uk. *On the corner of Stancombe Lane & the B3349 2½ m N of Alton.* **Fri 23, Sat 24 Apr (11-4). Adm £5, chd free. Light refreshments.**
Unique 100 acre ancient bluebell woodland. If you are a keen walker you will have much to explore on the long meandering paths and rides dotted with secluded seats. Those who enjoy a more leisurely pace will experience the perfume of the carpet of blue, listen to the birdsong and watch the contrasting light through the trees nearer to the entrance of the woods. Refreshments will be served in an original rustic building and incl soups using natural woodland ingredients.

14 BRAMDEAN HOUSE

Bramdean, Alresford, SO24 0JU. Mr & Mrs E Wakefield, garden@bramdeanhouse.com. *4m S of Alresford; 9m E of Winchester; 9m W of Petersfield. In centre of village on A272. Entrance opp sign to the church. Parking is usually immed across the road from entrance.* **Suns 21 Feb, 11 Apr, 6 June, 4 July, 1 Aug, 12 Sept (2-4). Adm £7.50, chd free. Home-made teas. Donation to Bramdean Church.**
Beautiful 5 acre garden best known for its mirror image herbaceous borders. Also carpets of spring bulbs, especially snowdrops and a large and unusual collection of plants and shrubs giving yr-round interest. 1 acre walled garden featuring prize-winning vegetables, fruit and flowers. Small arboretum. Features incl a large collection of old fashioned sweet peas, an expansive collection of nerines and a boxwood castle. Home of the nation's tallest sunflower 'Giraffe'. Flowering cherries recently imported from Japan. Concessions for under 25s and/or on income support available in person on the day. Visits also by arrangement for groups of 5+ (non-NGS). Wheelchair access to walled garden and orchard areas may be difficult. Please no dogs, except guide dogs.

15 6 BREAMORE CLOSE

Boyatt Wood, Eastleigh, SO50 4QB. Mr & Mrs R Trenchard, 02380 611230, dawndavina6@yahoo.co.uk. *1m N of Eastleigh. M3 J12, follow signs to Eastleigh. Turn R at r'about into Woodside Ave, then 1st L into Broadlands Ave (park here). Breamore Close 3rd on L.* **Visits by arrangement Apr & May for groups of 10+. Adm £4, chd free. Home-made teas.**
Delightful plant lover's garden with coloured foliage and unusual plants, giving a tapestry effect of texture and colour. Over 60 hostas displayed in pots and many clematis scramble through roses. In April iris, acers. tulips epimediums and akebia. In May a wonderful wisteria scrambles over a pergola. The garden is laid out in distinctive planting themes with seating areas to sit and contemplate. Wheelchair access with small gravel areas.

16 NEW BRIAR PATCH

Northover Road, Pennington, Lymington, SO41 8GU. Annette Eales, nettyeales@gmail.com. *2m NW of Lymington. From N off A337, turn into Sway Rd, follow for 1½ m to Wheel Inn. Turn L into Ramley Rd & follow signs; Northover Rd is the 1st turning on the L.* **Sat 19, Sun 20, Sat 26, Sun 27 June (2-5). Adm £3, chd free. Cream teas & gluten free option. Visits also by arrangement Apr to Aug for groups of 10 to 30.**
Meander through this luscious cottage garden 140ft by 40ft with extended views over farmland. A glorious palette of pinks, blues, purples and burgundy excite the senses. A long central pergola divides the garden into numerous areas; woodland planting, a cutting garden, a shade and wildflower hedgerow. Herbaceous borders interspersed with salvaged vintage tables displaying plant collections.

17 BRICK KILN COTTAGE

The Avenue, Herriard, nr Alton, RG25 2PR. Barbara Jeremiah, 01256 381301, barbara@klca.co.uk. *4m NE of Alton. A339 Basingstoke to Alton, 7m out of Basingstoke turn L along The Avenue, past Lasham Gliding Club on R, then past Back Lane on L & take next track on L, one field later.* **Sun 2 May, Sun 6 June (11-4). Adm £4.50, chd free. Cream teas & home-made teas. Visits also by arrangement May & June.**
Bluebell woodland garden in 2 acres with a perimeter woodland path incl treehouse, pebble garden, billabong, stumpery, ferny hollow, bug palace, waterpool, shepherd's hut and a traditional cottage garden filled with herbs. The garden is maintained using eco-friendly methods as a haven for wild animals, butterflies, birds, bees and English bluebells. New feature children's reading area. Wildlife friendly garden in a former brick works. A haven in the trees. Wheelchair access limited to some parts of the garden.

18 146 BRIDGE ROAD

Sarisbury Green, Southampton, SO31 7EJ. Audrey & Jonathan Crutchfield. *4m W of Fareham. On A27 between Chapel Rd & Glen Rd. Free car parking by kind permission of the United Reformed Church on Chapel Rd.* **Sat 15, Sun 16 May (11.30-4.30). Adm £3.50, chd free. Home-made teas.**
A rambling, deceptively large, tranquil cottage garden that is accessed through a side gate and unexpectedly removed from the sometimes bustling A27. Creatively divided into rooms, richly filled borders, patios and lawns giving intense variety. These are interspersed with nooks, arbours, mirrors, statues and unexpected resting spots that offer promise, privacy and the chance to unwind and reflect.

19 BROADHATCH HOUSE

Bentley, Farnham, GU10 5JJ. Bruce & Lizzie Powell, 01420 23185, lizziepowellbroadhatch@gmail.com. *4m NE of Alton. Turn off A31 (Bentley bypass) through village, then L up School Lane. R to Perrylands, after 300yds drive on R.* **Sun 13 June (2-5.30); Mon 14, Tue 15 June (2-5). Adm £5, chd free. Home-made teas. Visits also by arrangement May to Sept.**
3½ acre garden set in lovely Hampshire countryside with views to Alice Holt. Divided into different areas by yew hedges and walled garden. Focussing on as long a season as possible on heavy clay. Two reflective pools help break up lawn areas; lots of flower borders and beds; mature trees. Working greenhouses and vegetable garden. Old sunken garden redesigned in 2020. Wheelchair access with gravel paths and steps in some areas.

20 BUMPERS

Sutton Common, Long Sutton, Hook, RG29 1SJ. Stella Wildsmith, 07766 754993, sfw@staxgroup.com. *From village of Long Sutton, turn up Copse Lane, immed opp duck pond. Follow lane for 1½m to top of steep hill, house on L.* **Sun 4 July (2-6). Adm £5, chd free. Home-made teas. Visits also by arrangement Apr to Sept for groups of 10+.**
Large country garden with beautiful views spread over 2 acres, mixed herbaceous and shrub borders and laid out in a series of individual areas. Some interesting sculptures and water features with informal paths through the grounds and a number of places to sit and enjoy the views. Wheelchair users, please park at front of house.
&♿ ❀ ☕

21 ◆ CHAWTON HOUSE

Chawton, Alton, GU34 1SJ. Julia Weaver, 01420 541010, info@chawtonhouse.org, www.chawtonhouse.org. *2m S of Alton. Take the road opp Jane Austen's House museum towards St. Nicholas Church. Property is at the end of this road on the L.* **For NGS: Sun 21 Feb (11-4.30). Adm £5, chd free. Light refreshments. For other opening times and information, please phone, email or visit garden website.**
Snowdrops are scattered through this 14 acre listed English landscape garden which is being restored. Sweeping lawns, ha-ha, wilderness, terraces, fernery and shrubbery walk surround the Elizabethan manor house. The walled garden designed by Edward Knight now incl rose garden, cutting beds, orchard and 'Elizabeth Blackwell' herb garden based on her book 'A Curious Herbal' of 1737-39. Refreshments are available in our tea room, in the old kitchen, serving hot and cold drinks, wine, light lunches, cream teas, home-made cakes and local ice creams. Due to slopes and gravel paths, we regret this garden is not suitable for wheelchairs.
🐂 🚐 ☕ 🪑

22 21 CHESTNUT ROAD

Brockenhurst, SO42 7RF. Iain & Mary Hayter, www.21-chestnut-rdgardens.co.uk. *New Forest, 4m S of Lyndhurst. At Brockenhurst turn R, B3055 Grigg Lane. Limited parking on Chestnut Rd, please use village car park nearby. Leave M27 J2, follow Heavy Lorry Route. Mainline station less than 10 mins walk.* **Sat 15, Sun 16 May, Sat 5, Sun 6 June, Sat 31 July, Sun 1 Aug (11.30-5.30). Adm £4, chd free. Light refreshments.**
Creative ideas aplenty, from formal to cottage style, fruit and veg, all planted with wildlife in mind. In the heart of the New Forest National Park we offer ponds, pergolas, arches and statues. Ever growing family of fairies for the children to visit. Art exhibition of owners paintings with a donation to NGS from any sales made on open days. Plants, bug houses and bird boxes usually available to purchase.
🐂 ❀ ☕ 🪑

23 CLOVER FARM

Shalden Lane, Shalden, Alton, GU34 4DU. Tom & Sarah Floyd, 01420 86294. *Approx 3m N of Alton in the village of Shalden. Take A339 out of Alton. After approx 2m turn R up Shalden Lane. At top, turn sharp R next to church sign.* **Visits by arrangement June to Sept for groups of 20+. Adm £10, chd free. Light refreshments.**
3 acre garden with far reaching views. Herbaceous borders and sloping lawns down to reflection pond, wildflower meadow, lime avenue, rose and kitchen garden and ornamental grass area.
&♿ 🐂 ❀ ☕

24 COLEMORE HOUSE GARDENS

Colemore, Alton, GU34 3RX. Mr & Mrs Simon de Zoete, 01420 588202, simondezoete@gmail.com. *4m S of Alton (off A32). Approach from N on A32, turn L (Shell Lane), ¼m S of East Tisted. Go under bridge, keep L until you see Colemore Church. Park on verge of church.* **Tue 22, Wed 23 June (2-6). Adm £6, chd free. Home-made teas. Visits also by arrangement May to July for groups of 5+.**
4 acres in lovely unspoilt countryside, featuring rooms containing many unusual plants and different aspects with a spectacular arched rose walk, water rill, mirror pond, herbaceous and shrub borders. Newly designed by David Austin Roses, an octagonal garden with 25 different varieties. Explore the interesting arboretum, grass gardens and thatched pavilion. Every year the owners seek improvement and the introduction of new, interesting and rare plants. We propagate and sell plants, many of which can be found in the garden. Some are unusual and not readily available elsewhere. For private visits, we endeavour to give a conducted tour and try to explain our future plans, rationale and objectives.
❀ ☕

25 THE COTTAGE

16 Lakewood Road, Chandler's Ford, Eastleigh, SO53 1ES. Hugh & Barbara Sykes, 02380 254521, barandhugh@aol.com. *Leave M3 J12, follow signs to Chandler's Ford. At King Rufus on Winchester Rd, turn R into Merdon Ave, then 3rd road on L.* **Sun 2, Mon 3, Sun 9, Mon 10 May (2-6). Adm £4, chd free. Home-made teas. Visits also by arrangement Apr & May.**
The house was built in 1905, but the ¾ acre garden has been designed, planted and cared for since 1950 by 2 keen garden loving families. Azaleas, camellias, trilliums and erythroniums under old oaks and pines. Herbaceous cottage style borders with many unusual plants for yr-round interest. Bog garden, ponds, kitchen garden. Bantams, bees and birdsong with over 30 bird species noted. Wildlife areas. NGS sundial for opening for 30yrs. Childrens' quiz. 'A lovely tranquil garden', Anne Swithinbank. Hampshire Wildlife Trust Wildlife Garden Award. Honey from our garden hives for sale.
❀ 🚐 ☕

26 CRANBURY PARK

Otterbourne, nr Winchester, SO21 2HL. Mrs Chamberlayne-Macdonald. *3m NW of Eastleigh. Main entrance on old A33 at top of Otterbourne Hill by bus stop. Entrances also in Hocombe Rd (opp Nichol Rd), Chandlers Ford & next to Otterbourne Church.* **Sun 13 June (2-6). Adm £5, chd free. Home-made teas. Donation to St John the Baptist.**
Extensive pleasure grounds laid out in late C18 and early C19 by Papworth; fountains, rose garden, specimen trees and pinetum, lakeside walk and fern walk. Family carriages and collection of prams will be on view, also photos of King George VI, Eisenhower and Montgomery reviewing Canadian troops at Cranbury before D-Day. Disabled WC. All dogs on leads please.
&♿ 🐂 ❀ ☕

GROUP OPENING

27 CRAWLEY GARDENS
Crawley, Winchester, SO21 2PR.
F J Fratter, 01962 776243,
fred@fratter.co.uk. *5m NW of
Winchester. Between B3049
(Winchester - Stockbridge) & A272
(Winchester - Andover). Parking
throughout village & in field at
Tanglefoot.* **Fri 2, Mon 5 Apr, Thur
24, Sun 27 June (2-5.30). Combined
adm £7.50, chd free. Home-made
teas in the village hall.**

LITTLE COURT
Mrs A R Elkington.
Open on Fri 2, Mon 5 Apr
(See separate entry)

PAIGE COTTAGE
Mr & Mrs T W Parker.
Open on all dates

TANGLEFOOT
Mr & Mrs F J Fratter.
Open on Thur 24, Sun 27 June
(See separate entry)

Crawley is an exceptionally pretty period village nestling in chalk downland with thatched houses, C14 church and village pond with ducks. The spring gardens are Little Court and Paige Cottage; the summer gardens are Paige Cottage and Tanglefoot; providing seasonal interest of varied character, and with traditional and contemporary approaches to landscape and planting. Most of the gardens have beautiful country views and there are other good gardens to be seen from the road. Little Court is a 3 acre traditional English country garden with carpets of spring bulbs and a large meadow. Paige Cottage is a 1 acre traditional English country garden surrounding a period thatched cottage (not open) with bulbs and wildflowers in spring, and old climbing roses in summer. Tanglefoot has colour themed borders, herb wheel, exceptional kitchen garden, traditional Victorian boundary wall supporting trained fruit including apricots; and a large wildflower meadow. Plants from the garden for sale at Little Court and Tanglefoot.

28 CROOKLEY POOL
Blendworth Lane, Horndean,
PO8 0AB. Mr & Mrs Simon
Privett, 02392 592662,
jennyprivett@icloud.com. *5m S of
Petersfield, 2m E of Waterlooville,
off the A3. From Horndean up
Blendworth Lane between bakery
& hairdresser. Entrance 200yds
before church on L with white
railings. Parking in field.* **Visits by
arrangement May to Sept.**
Here the plants decide where to grow. Californian tree poppies elbow valerian aside to crowd round the pool. Evening primroses obstruct the way to the door and the steps to wisteria shaded terraces. Hellebores bloom under the trees. Salvias, Pandorea jasminoides, Justicia, Pachystachys lutea and passion flowers riot quietly with tomatoes in the greenhouse. Not a garden for the

The Hospital of St Cross

neat or tidy minded, although this is a plantsman's garden full of unusual plants and a lot of tender perennials. Bantams stroll throughout. Oil and watercolour paintings of flowers found in the garden will be on display and for sale in the studio.

ở ✿ ☕ 🍵

29 THE DEANE HOUSE
Sparsholt, Winchester, SO21 2LR. Mr & Mrs Richard Morse, 07774 863004, chrissiemorse7@gmail.com. 3½m NW of Winchester. Turn off A3049 Stockbridge Rd, onto Woodman Lane, signed Sparsholt. Turn L at 1st cottage on L, white with blue gables, come to top of the drive. **Sun 29 Aug (2-5). Adm £5, chd free. Home-made teas. Visits also by arrangement Mar to Sept for groups of 10+.**
This beautiful 4 acre rural garden, nestling on a gentle slope, has been landscaped to draw the eye from one gentle terraced lawn to another with borders merging with the surrounding countryside. Featuring a good selection of specimen trees, a walled garden, prairie planting and herbaceous borders. Water features and sculptures. Children's play area. Sorry no dogs. Although the garden is on the side of a hill there is always a path to avoid steps. Plentiful parking.

ở ☕ 🍽 🍵

30 THE DOWER HOUSE
Springvale Road, Headbourne Worthy, Winchester, SO23 7LD. Mrs Judith Lywood, 01962 882848, hannahlomax@thedowerhousewinchester.co.uk, www.thedowerhousewinchester.co.uk. 2m N of Winchester. Entrance is directly opp watercress beds in Springvale Rd & near Cobbs Farm Shop & Kitchen. Parking at main entrance to house, following path to garden. **Sun 27 June (2-4.30). Adm £4.50, chd free. Light refreshments. Visits also by arrangement May to Sept for groups of 5 to 10.**
The Dower House is set within 5½ acres of gardens with meandering paths allowing easy access around the grounds. There are plenty of places to sit, relax and enjoy the surroundings. Areas of interest incl a scented garden, iris bed, geranium bed, shrubbery, a pond populated with fish and water lilies, bog garden, bluebell wood and secret courtyard garden.

ở 🐕 🍽 🍵

31 DOWN FARM HOUSE
Whitchurch, RG28 7FB. Pat & Steve Jones, 01256 892490, patthehound@gmail.com. 1½m from the centre of Whitchurch. From the centre of Whitchurch take the Newbury road up the hill, over railway bridge & after approx 1m turn L over the A34, after 300 metres turn L along bridleway. **Visits by arrangement Mar to July for groups of 5 to 30. Adm £12, chd free. Home-made teas included.**
Step back in time in this 2 acre garden, created from an old walled farmyard and the surrounding land. Many of the original features are used as hard landscaping, incl organic vegetables, succulents, alpine bed created from the old concrete capped well, informal and naturalistic planting, wooded area and orchard. The garden has been created slowly over the last 37 yrs. New small stumpery in 2019. Wheelchair access by gravel drive onto lawn.

ở 🐕 ✿ 🍵

32 THE DOWN HOUSE
Itchen Abbas, SO21 1AX. Jackie & Mark Porter, 01962 791054, markstephenporter@gmail.com, www.thedownhouse.co.uk. 5m E of Winchester on B3047. 5th house on R after the Itchen Abbas village sign if coming on B3047 from Kings Worthy. 400yds on L after Plough Pub if coming on B3047 from Alresford. **Sat 13, Sat 20 Feb (11-4). Adm £5, chd free. Pre-booking essential, please visit www.ngs.org.uk for information & booking. Home-made teas. Visits also by arrangement in Feb for groups of 20+. Pre-payment required.**
A 2 acre garden laid out in rooms overlooking the Itchen Valley, adjoining the Pilgrim's Way with walks to the river. In February come and see the snowdrops, winter aconites and crocus, plus borders of coloured dogwoods, willow stems and white birches. A garden of structure with pleached hornbeams, a rope-lined fountain garden and yew lined avenues, plus a pruned vineyard !

ở 🐕 🍵

33 DURMAST HOUSE
Bennetts Lane, Burley, BH24 4AT. Mr & Mrs P E G Daubeney, 01425 402132, philip@daubeney.co.uk, www.durmasthouse.co.uk. 5m SE of Ringwood. Off Burley to Lyndhurst Rd, near White Buck Hotel, C10

road. **Sun 11 Apr, Sun 27 June (2-5). Adm £5, chd free. Cream teas. Visits also by arrangement Mar to Sept. Donation to Delhi Commonwealth Women's Association Medical Clinic.**
Designed by Gertrude Jekyll, Durmast has contrasting hot and cool colour borders, formal rose garden edged with lavender and a long herbaceous border. Many old trees, Victorian rockery and orchard with beautiful spring bulbs. Rare azaleas; Fama, Princeps and Gloria Mundi from Ghent. Features incl rose bowers with rare French roses; Eleanor Berkeley, Psyche and Reine Olga de Wurtemberg. Jekyll borders with hot colours and another with cool colours, blue, yellow and white. Typical Jekyll long herbaceous border. Many old trees, including Cedar and Douglas firs. Wheelchair access on stone and gravel paths.

ở 🐕 ✿ 🚗 🍽 🍵

34 EAST WORLDHAM MANOR
Worldham Hill, East Worldham, Alton, GU34 3AX. Hermione Wood, www.worldham.org. 2m SE of Alton on B3004 in East Worldham. Coming from Alton, turn R by village hall, signed car park. **Sat 26, Sun 27, Mon 28 June (2-5). Combined adm with Selborne £6, chd free. Home-made teas.**
A rambling double-walled garden laid out in the 1870s with far-reaching views to the South Downs. Illustrates a substantial Victorian garden with fruit, vegetables and flower borders, rose garden, many shrubs, herbaceous plants, apple and pear orchard and large productive greenhouses. Gravel paths, some naturalised with white foxgloves and campanulas, wind through many parts of the garden.

ở 🐕 ✿ 🍵 🍽

The National Garden Scheme was set up in 1927 to raise funds for district nurses. We remain strongly committed to nursing and today our beneficiaries include Macmillan Cancer Support, Marie Curie, Hospice UK and The Queen's Nursing Institute

35 ENDHOUSE

6 Wimpson Gardens, Southampton, SO16 9ES. Kevin Liles, 02380 777590, k.liles@ntlworld.com, www.gardenatendhouse.com. *Exit M271 at J1 toward Lordshill. At 2nd r'about, turn R into Romsey Rd toward Shirley. After ½m turn R at Xrds into Wimpson Lane, 3rd on R Crabwood Rd (additional parking). Wimpson Gardens 4th on R.* **Visits by arrangement May to July for groups of 10 to 20. Adm £3.50, chd free. Home-made teas.**
Award-winning urban oasis of linked garden areas including small secret garden. Best in spring and summer months, but rich yr-round plant interest with tree ferns, palms, acers and wisteria. Deep herbaceous borders planted with hostas, grasses, agapanthus and alstroemerias. Significant exhibition of gallery quality sculpture and ceramics. Please visit www.ngs.org.uk for pop up openings.

36 FAIRBANK

Old Odiham Road, Alton, GU34 4BU. Jane & Robin Lees, 01420 86665, j.lees558@btinternet.com. *1½m N of Alton. From S, past Sixth Form College, then 1½m beyond road junction on R. From N, turn L at Golden Pot & then 50yds turn R. Garden 1m on L before road junction.* **Visits by arrangement May to Sept for groups of up to 30. Adm £4, chd free. Home-made teas.**
The planting in this large garden reflects our interest in trees, shrubs, fruit and vegetables. A wide variety of herbaceous plants provide colour and are placed in sweeping mixed borders that carry the eye down the long garden to the orchard and beyond. Near the house (not open), there are rose beds and herbaceous borders, as well as a small formal pond. There is a range of acers, ferns and unusual shrubs and 60 different varieties of fruit, together with a large vegetable garden. Wheelchair access with uneven ground in some areas.

37 FAIRWEATHER'S NURSERY

Hilltop, Beaulieu, SO42 7YR. Patrick Fairweather, 01590 612113, info@fairweathers.co.uk, www.fairweathers.co.uk. *1½m NE of Beaulieu village. Signed Hilltop Nursery on B3054 between Heath r'about (A326) & Beaulieu village.* **Sat 17, Sun 18 July (10-4). Adm £4, chd free. Cream teas in Aline Fairweather's garden.**
Fairweather's hold a specialist collection of over 400 agapanthus grown in pots and display beds, incl AGM award-winning agapanthus trialled by the RHS. Features incl guided tours of the nursery and demonstrations of how to get the best from agapanthus and companion planting. Agapanthus and a range of other traditional and new perennials for sale. Aline Fairweather's garden (adjacent to the nursery) will also be open, with mixed shrub and perennial borders containing many unusual plants. Also open Patrick's Patch at Fairweather's Garden Centre.

38 FERNS LODGE

Cottagers Lane, Hordle, Lymington, SO41 0FE. Sue Grant, 07860 521501, sue.grant@fernslodge.co.uk, www.fernslodge.co.uk. *Approx 5½m W of Lymington. From Silver St turn into Woodcock Lane, 100 metres to Cottagers Lane, parking in field ½m on L. From A337 turn into Everton Rd & drive approx 1½m, Cottagers Lane on R.* **Sat 5, Sun 6 June (2-5). Adm £3.50, chd free. Home-made teas. Visits also by arrangement Apr to Sept for groups of 5 to 20.**
Captivating ½ acre cottage garden, a happy jumble of colour, scent and interest, enveloping a Victorian lodge house with winding brick paths. Gazebo and terrace where azalea, hydrangea, fig, sweet pea, roses, agapanthus and many foxgloves jostle for position, joined by honeysuckle, clematis, passionflower and jasmine vie for your attention, as you venture into 3½ acres of Victorian garden in restoration. Wildlife abounds! We love our guests to enjoy the many peaceful seating areas around the garden and enjoy our amazing home-made cakes. Wheelchair access to some areas.

39 FRITHAM LODGE

Fritham, SO43 7HH. Sir Chris & Lady Powell. *6m N of Lyndhurst. 3m NW of M27 J1 (Cadnam). Follow signs to Fritham.* **Sun 13 June (2-4). Adm £4, chd free. Home-made teas.**
A walled garden of 1 acre in the heart of the New Forest, set within 18 acres surrounding a house that was originally a Charles 1 hunting lodge (not open). Herbaceous and blue and white mixed borders, pergolas and ponds. A box hedge enclosed parterre of roses, fruit and vegetables. Visitors will enjoy the ponies, donkeys, sheep and old breed hens on their meadow walk to the woodland and stream.

GROUP OPENING

40 FROYLE GARDENS

Lower Froyle, Froyle, GU34 4LG. www.froyle.com/ngs. *5m NE of Alton. Access to Lower Froyle from A31 between Alton & Farnham at Bentley, or via Upper Froyle at Hen & Chicken Pub, or via B3349 & Golden Pot Pub. Park at recreation ground, Lower Froyle. Map provided. Additional signed parking in Upper Froyle.* **Sat 5, Sun 6 June (2-6). Combined adm £7.50, chd free. Home-made teas at Froyle Village Hall, Lower Froyle.**

ALDERSEY HOUSE
Nigel & Julie Southern.

NEW **3 BURNHAM SQUARE**
Bernie & Vonny Wilks.

DAY COTTAGE
Nick & Corinna Whines, www.daycottage.co.uk.

OLD BREWERY HOUSE
Vivienne & John Sexton.

NEW **OLD COURT**
Sarah & Charlie Zorab.

WARREN COTTAGE
Gillian & Jonathan Pickering.

WELL LANE CORNER
Mark & Sue Lelliott.

You will certainly receive a warm welcome as Froyle Gardens open their gates this year, enabling visitors to enjoy a wide variety of gardens, which have undergone development since last year and will be looking splendid. Froyle is a beautiful village with many old and interesting buildings, our gardens harmonise well with the surrounding landscape and most have spectacular views. The gardens themselves are diverse with rich planting. You will see greenhouses, water features, vegetables, roses, clematis and wildflower meadows. Lots of ideas to take away with you, along with plants for sale. The delicious teas served in the village hall are famous and there is also a plant stall. Close by there is a children's playground with a zip wire where younger visitors can let off steam. Picnics welcome at the recreation ground. In conjunction with Froyle Open Gardens there is an exhibition of richly decorated historic vestments to be held in St Mary's Church, Upper Froyle GU34 4LB (separate donation). Parking by the church. No wheelchair access to Day Cottage and on request at Warren Cottage. Steep slope at 3 Burnham Square and long drive to Old Court.

41 GILBERTS DAHLIA FIELD
Gilberts Nursery, Dandysford Lane, Sherfield English, nr Romsey, SO51 6DT. Nick & Helen Gilbert, www.gilbertsdahlias.co.uk. *Midway between Romsey & Whiteparish on A27, in Sherfield English Village. From Romsey 4th turn on L, just before small petrol station on R, visible from main road.* **Sun 29 Aug (10-4). Adm £3, chd free. Light refreshments.**

This may not be a garden, but do come and be amazed by the sight of over 300 varieties of dahlias in our dedicated 1½ acre field. Prize-winning blooms are in all colours, shapes and sizes and can be closely inspected from wheelchair friendly hard grass paths. An inspiration for all gardeners.

42 HAMBLEDON HOUSE
East Street, Hambledon, PO7 4RX. Capt & Mrs David Hart Dyke, 02392 632380, dianahartdyke@gmail.com. *8m SW of Petersfield, 5m NW of Waterlooville. In village centre, driveway leading to house in East St. Do not go up Speltham Hill even if advised by SatNav.* **Visits by arrangement Apr to Oct for groups of 5+.**
3 acre partly walled plantsman's garden for all seasons. Large borders filled with a wide variety of unusual shrubs and perennials with imaginative plant combinations culminating in a profusion of colour in late summer. Hidden, secluded areas reveal surprise views of garden and village rooftops. Planting a large

central area, which started in 2011, has given the garden an exciting new dimension. Partial wheelchair access as garden is on several levels.

43 108 HEATH ROAD
Petersfield, GU31 4EL. Mrs Karen Llewelyn, 01730 269541, k.llewelyn@btinternet.com. *A3 N & S take A272 exit signed Midhurst. Take 1st exit from r'about, 1st R into Pullens Lane (B2199) & then 6th road on R into Heath Rd.* **Sun 6 June, Sun 12 Sept (2-5.30). Adm £3.50, chd free. Home-made teas. Visits also by arrangement June to Sept.**
⅔ acre garden close to town centre and Heath Pond. Greenhouse and succulent collection. Tropical plants, acers, small woodland walk. 30 metre long border with shade loving plants including many hostas and ferns. Patio garden, seasonal pots and late summer herbaceous border. Newly planted driveway borders. Wheelchair access after a 5 metre sloping gravel drive.

East Worldham Manor

SPECIAL EVENT

44 NEW HECKFIELD PLACE

Heckfield, RG27 0LD. Heckfield Management Ltd, 01189 326868, enquiries@heckfieldplace.com, www.heckfieldplace.com. *9m S of Reading. 4½m NW of Hartley Wintney on B3011. Two car parks, signed on the day.* **Wed 19 May, Wed 2 June (2-6). Adm £8, chd free. Pre-booking essential, please visit www.ngs.org.uk/ special-events for information & booking. Home-made teas in the glass house of the walled garden.**
Heckfield Place is a hotel on a 438 acre estate with an original 1927 NGS garden, now reopening with walled garden and pleasure grounds, including two lakes and arboretum. Both tamed and gently wild, the garden was created by Head Gardener William Walker Wildsmith in the C19 and has been lovingly restored through yrs of diligent work. Home-made teas including dairy, nut and gluten free options £4 (card payment on the day only). Wheelchair access to the upper walled garden on gravel pathway. No dogs please.
🚻 ❄ 🚍 ☕

45 THE HOMESTEAD

Northney Road, Hayling Island, PO11 0NF. Stan & Mary Pike, 02392 464888, jhomestead@aol.com, www.homesteadhayling.co.uk. *3m S of Havant. From A27 Havant & Hayling Island r'about, travel S over Langstone Bridge & turn immed L into Northney Rd. Car park entrance on R after Langstone Hotel.* **Sun 8 Aug (2-5.30). Adm £4, chd free. Home-made teas. Visits also by arrangement June to Sept for groups of 10+.**
1¼ acre garden surrounded by working farmland with views to Butser Hill and boats in Chichester Harbour. Trees, shrubs, colourful herbaceous borders and small walled garden with herbs, vegetables and trained fruit trees. Large pond and woodland walk with shade-loving plants. A quiet and peaceful atmosphere with plenty of seats to enjoy the vistas within the garden and beyond. Where old shrubs had become overgrown and tired, we have experimented with some new plant combinations. A recently constructed look-out platform provides some different views of the garden. Wheelchair access with some gravel paths.
🚻 🐑 ❄ ☕

46 ◆ THE HOSPITAL OF ST CROSS

St Cross Road, Winchester, SO23 9SD. The Hospital of St Cross & Almshouse of Noble Poverty, 01962 851375, porter@hospitalofstcross.co.uk, www.hospitalofstcross.co.uk. *½m S of Winchester. From city centre take B3335 (Southgate St & St Cross Rd) S. Turn L immed before The Bell Pub. If on foot follow riverside path S from Cathedral & College, approx 20 mins.* **For NGS: Sun 11 July (2-5). Adm £4, chd free. Home-made teas in the Hundred Men's Hall in the Outer Quadrangle. For other opening times and information, please phone, email or visit garden website.**
The Medieval Hospital of St Cross nestles in water meadows beside the River Itchen and is one of England's oldest almshouses. The tranquil, walled Master's Garden, created in the late C17 by Bishop Compton, now contains colourful herbaceous borders, old fashioned roses, interesting trees and a large fish pond. The Compton Garden has unusual plants of the type he imported when Bishop of London. Wheelchair access, but surfaces are uneven in places.
🚻 ❄ ☕

47 THE HOUSE IN THE WOOD

Beaulieu, SO42 7YN. Victoria Roberts. *New Forest. 8m NE of Lymington. Leaving the entrance to Beaulieu Motor Museum on R (B3056), take next R signed Ipley Cross. Take 2nd gravel drive on RH-bend, approx ½m.* **Sun 9 May (2-5.30). Adm £6, chd free. Cream teas.**
Peaceful 12 acre woodland garden with continuing progress and improvement. Very much a spring garden with tall, mature azaleas and rhododendrons interspersed with acers and other woodland wonders. A magical garden to get lost in with many twisting paths leading downhill to a pond and a more formal layout of lawns around the house (not open). Used in the war to train the Special Operations Executive. Partial wheelchair access.
🐑 🚗 ☕

48 HOW PARK BARN

Kings Somborne, Stockbridge, SO20 6QG. Kate & Chris Cann. *2m from Stockbridge, just outside Kings Somborne. A3057 from Stockbridge, turn R into Cow Drove Hill then follow NGS signs.* **Sun 16 May, Sun 5 Sept (2-5). Adm £5, chd free. Home-made teas.**
2 acre country garden in elevated position with uninterrupted panoramic views over the Test Valley. Set within 12 acres of chalk grassland. Large borders of naturalistic planting and shrubs with some slopes. Sweeping lawns connect different levels of garden interest. A tranquil setting within the landscape of C17 listed barn (not open). Walkers and cyclists welcome from nearby Test Way. Gravel drive and some slopes, but wheelchair access to most of the garden.
🚻 ❄ ☕

49 THE ISLAND

Greatbridge, Romsey, SO51 0HP. Mr & Mrs Christopher Saunders-Davies. *1m N of Romsey on A3057. Entrance alongside Greatbridge (1st bridge Xing the River Test), flanked by row of cottages on roadside. When arriving, please do not block narrow entrance or create a traffic jam on the main road.* **Sat 17, Sun 18 Apr, Sat 3, Sun 4 July (2-5). Adm £5, chd free. Home-made teas.**
6 acres either side of the River Test. Fine display of daffodils, tulips, spring flowering trees and summer bedding. Main garden has herbaceous and annual borders, fruit trees, rose pergola, lavender walk and extensive lawns. An arboretum planted in the 1930s by Sir Harold Hillier contains trees and shrubs providing interest throughout the yr. Please Note: No dogs allowed. Disabled WC available.
🚻 ☕ 🪑

50 LAKE HOUSE

Northington, SO24 9TG. Lord Ashburton, 07795 364539, lake. house.gardenvisits@gmail.com. *4m N of Alresford. Off B3046. Follow English Heritage signs to The Grange and then to Lake House.* **Thur 10, Sun 13 June (12.30-5). Adm £5, chd free. Home-made teas. Visits also by arrangement May to Oct for groups of 10+.**
Two large lakes in Candover Valley set off by mature woodland with waterfalls, abundant birdlife, long landscaped vistas and folly. 1½ acre walled garden with rose parterre, mixed borders, long herbaceous

border, rose pergola leading to moongate. Flowering pots, conservatory and greenhouses. Picnicking by lakes. Grass paths and slopes to some areas of the garden.

&. 🐾 ❀ 🚗 ☕ 💷

51 LITTLE COURT
Crawley, Winchester, SO21 2PU. Mrs A R Elkington, 01962 776365, elkslc@btinternet.com. *5m NW of Winchester. Between B3049 (Winchester - Stockbridge) & A272 (Winchester - Andover), 400yds from either pond or church.* **Sun 14, Mon 15, Sun 21, Mon 22 Feb (2-4.30); Sun 21 Mar, Fri 21 May, Sun 13, Mon 14 June, Sun 4 July, Sun 1 Aug (2-5.30). Adm £5, chd free. Home-made teas in Crawley Village Hall. 2022: Sun 13, Mon 14, Sun 20, Mon 21 Feb. Opening with Crawley Gardens on Fri 2, Mon 5 Apr. Visits also by arrangement Feb to July.**
This traditional walled sheltered garden is open in all seasons, specially exciting in spring. There are many flower beds and climbers, a kitchen garden, colourful bantams running free, and a south facing wildlife field with good butterflies. Many seats with distant views both within the garden and to the surrounding countryside. A garden for all seasons and ages. Regional Finalist, The English Garden's The Nation's Favourite Gardens 2019.

&. ❀ 🚗 💷 🌳

52 LONGSTOCK PARK
Leckford, Stockbridge, SO20 6EH. Leckford Estate Ltd, part of John Lewis Partnership, www.leckfordestate.co.uk. *4m S of Andover. From Leckford village on A3057 towards Andover, cross the river bridge & take 1st turning to the L signed Longstock.* **Sun 20 June (1-4). Adm £10, chd £2.**
Famous water garden with extensive collection of aquatic and bog plants set in 7 acres of woodland with rhododendrons and azaleas. A walk through the park leads to National Collections of *Buddleja* and *Clematis viticella*; arboretum and herbaceous border at Longstock Park Nursery. Refreshments at Longstock Park Farm Shop and Nursery (last orders at 3.30pm). Assistance dogs only.

&. ❀ 🚗 NPC 💷

53 NEW LOWER MILL
Mill Lane, Longparish, Andover, SP11 6PS. Mrs K-M Dinesen. *Off A303 from the W signed Longparish B3048 & from the E signed Barton Stacey, then Longparish.* **Sun 4 July (1-5.30). Adm £5, chd free. Home-made teas & soft drinks.**
Set in 15 acres of informally planted gardens, magnificent trees and water are but two features of this widely anticipated reopening. Stunning late summer and winter beds complement the existing array of grasses, perennials and shrubs leading to a captivating lake surrounded by wildflowers. A hidden sunken garden, tranquil riverside walks, a delightful water garden and much more awaits your discovery.

💷

54 26 LOWER NEWPORT ROAD
Aldershot, GU12 4QD. Mr & Mrs P Myles. *From the A331 coming off at the Aldershot junction, head towards Aldershot. Take the 1st R turn at the T-lights next to McDonalds into North Lane & then 1st L into Lower Newport Rd.* **Sat 26, Sun 27 June (11-4). Adm £3, chd free. Light refreshments**
A 'T' shaped small town garden full of ideas, split into four distinct sections, a semi-enclosed patio area with pots and water feature; a free-form lawn with a tree fern, perennials, bulbs and shrubs and 100 varieties of hosta; secret garden with a 20ft x 6ft raised pond, exotic planting backdrop and African carvings; and a potager garden with a selection of vegetable, roses and plant storage.

❀ 💷

55 ✦ MACPENNYS WOODLAND GARDEN & NURSERIES
Burley Road, Bransgore, Christchurch, BH23 8DB. Mr & Mrs T M Lowndes, 01425 672348, office@macpennys.co.uk, www.macpennys.co.uk. *6m SE of Ringwood, 5m NE of Christchurch. From Crown Pub Xrd in Bransgore take Burley Rd, following sign for Thorney Hill & Burley. Entrance ½m on R.* **For opening times and information, please phone, email or visit garden website.**
4 acre woodland garden originating from worked out gravel pits in the 1950s, offering interest yr-round, but particularly in spring and autumn. Attached to a large nursery that offers for sale a wide selection of homegrown trees, shrubs, conifers, perennials, hedging plants, fruit trees and bushes. Tearoom offering home-made cakes, afternoon tea (pre-booking required) and light lunches, using locally sourced produce wherever possible. Nursery closed Christmas through to the New Year. Partial wheelchair access on grass and gravel paths which can be bumpy with tree roots and muddy in winter.

&. 🐾 ❀ 🚗 ☕ 💷

56 MANOR HOUSE
Church Lane, Exton, SO32 3NU. Tina Blackmore, 01489 877529, manorhouseexton@gmail.com. *Off A32 just N of Corhampton. Pass The Shoe Inn on your L, go to the end of the road to a T-junction, turn R & Manor House is immed on the L, just below the church.* **Sun 23 May, Sun 6 June (2-5). Adm £5. Home-made teas.**
An enchanting 1 acre mature walled garden set in the Meon Valley. Yew hedges and flint walls divide the garden into rooms. The white garden planted with hydrangeas and roses. Herbaceous borders with colourful cottage garden favourites; delphiniums, roses, geraniums and salvias. A secluded, highly productive walled vegetable garden, a parterre with fountain, woodland area with wildflowers and pond.

&. ❀ 🚐 💷

57 MANOR LODGE
Brook Lane, Botley, Southampton, SO30 2ER. Gary & Janine Stone. *6m E of Southampton. From A334 to the W of Botley village centre, turn into Brook Lane. Manor Lodge is ½m on the R. Limited disabled parking. Continue past Manor Lodge to parking (signed).* **Sat 3, Sun 4 July (2-5). Adm £4, chd free. Home-made teas.**
Close to Manor Farm Country Park, this mid-Victorian house (not open), set in over 1½ acres is the garden of an enthusiastic plantswoman. A garden in evolution with established areas and new projects, a mixture of informal and formal planting, woodland and wildflower meadow areas. There are large established and new specimen trees, common and exotic perennials, planting combinations for extended seasonal interest. Fruit cages. Largely flat with hard paving, but some gravel and grass to access all areas.

&. 🐾 ❀ ☕ 💷

Froyle Gardens

58 MEON ORCHARD

Kingsmead, North of Wickham, PO17 5AU. Doug & Linda Smith, 01329 833253, meonorchard@btinternet.com. *5m N of Fareham. From Wickham take A32 N for 1½ m. Turn L at Roebuck Inn. Garden in ½ m. Park on verge or in field N of property.* **Sun 25 July, Sun 5 Sept (2-6). Adm £5, chd free. Home-made teas.**

2 acre garden designed and constructed by current owners. An exceptional range of rare, unusual and architectural plants incl National Collection of Eucalyptus. Dramatic foliage plants from around the world, see plants you have never seen before! Flowering shrubs in May and June; bananas, tree ferns, cannas, gingers, palms and perennials dominate July to Sep; streams and ponds, plus an extensive range of planters complete the display. Visitors are welcome to explore the 20 acre meadow and ½ m of Meon River frontage attached to the garden. Extra big plant sale of the exotic and rare on Sun 5 Sept. Garden fully accessible by wheelchair, reserved parking.

🚻 🐕 ✿ 🚗 NPC 🍵

59 MERDON MANOR

Merdon Castle Lane, Hursley, Winchester, SO21 2JJ. Mr & Mrs J C Smith, 01962 775215, vronk@fastmail.com. *5m SW of Winchester. From A3090 Winchester to Romsey road, turn R at Standon,* onto Merdon Castle Lane. Proceed for 1¼ m. Entrance on R between 2 curving brick walls. **Visits by arrangement May to Sept for groups of up to 20. Adm £6, chd free. Home-made teas.**

5 acre country garden surrounded by panoramic views; pond with ducks, damsel flies, dragonflies and water lilies; large wisteria, roses, fruit-bearing lemon trees, extensive lawns, impressive yew hedges and small secret walled garden with fountains. Black Hebridean sheep (St. Kildas). Very tranquil and quiet. Wheelchairs have to go down a drive to reach sunken garden.

🚻 ✿ 🚗 🍵

60 MOORE BARLOW

48 High Street, Lymington, SO41 9ZQ. Moore Barlow Solicitors. *Top end of Lymington High St, opp Poundland. Follow signs for Lymington town centre & use High St car parks.* **Sat 24 July (11-3); Sun 25 July (2-5). Adm £5, chd free. Home-made teas.**

Situated behind this elegant Georgian town house (not open) lies a surprising s-facing walled garden of 1 acre. From the raised terrace, enjoy the long vista across the croquet lawn to mature gardens beyond, with glimpses of the Isle of Wight. Amusing and varied topiary underplanted with mixed herbaceous borders and a new exciting water feature. Attractions close by incl the Lymington Saturday Market and the lively waterfront at the bottom of the attractive High St. Wheelchair access from rear carpark on gravel paths.

🚻 🐕 ✿ 🍵

61 23 NEW BRIGHTON ROAD

Emsworth, PO10 7PR. Lucy Watson & Mike Rogers. *7m W of Chichester, 2m E of Havant. From Emsworth main r'about head N, under the railway bridge, under the flyover & the garden is immed on the LH-side up a slope. Parking at Horndean Recreation Ground (approx 5 min walk).* **Sun 20 June (1.30-4.30). Adm £3.50, chd free. Home-made teas.**

Low maintenance, but verdantly planted long, narrow garden on gently sloping ground. The garden is divided into a series of sections in full sun to deep shade. Redesigned for easier maintenance, but no diminishment of horticultural love, the garden has changed enormously since last opening 3 yrs ago. Still many planted containers, well-stocked borders and a large collection of unusual plants. From Horndean Recreation Ground visitors will have the opportunity to walk through the small but delightfully planted Memorial Garden situated in the corner. No access for wheelchairs. Narrow, gravelled entrance may prove difficult for those with mobility issues.

🐕 ✿ 🍵

62 5 OAKFIELDS

Boyatt Wood, Eastleigh, SO50 4RP. Martin & Margaret Ward. M3 J12, follow signs to Eastleigh. 3rd exit at r'about into Woodside Ave, 2nd R into Bosville, 2nd R onto Boyatt Lane, 1st R to Porchester Rise & 1st L into Oakfields. Sat 31 July, Sun 1 Aug (2-5). Adm £3.50, chd free. Home-made teas.

A ⅓ acre garden full of interesting and unusual plants in predominately woodland beds with hydrangeas, lilies and highlighted with begonias. A pond with rockery, water cascade and flower beds formed from the intermittent winter streams, accommodate moisture loving plants. Colourful mixed herbaceous border and a terrace with architectural agapanthus and fuchsia. Wheelchair access over hard paths and some gravel.

&. ✿ ☕

63 OLD CAMPS

Newbury Road, Headley, Thatcham, RG19 8LG. Adam & Heidi Vetere, 07720 449702, gardens@oldcamps.co.uk, www.oldcamps.co.uk. Directions for Open Weekend only: Turn off the A339 into Galley Lane & after 100yds turn L into Plumtrees Farm. Follow concrete road to car park (signed). Walk 400yds to garden. Sat 7, Sun 8 Aug (10-5). Adm £7.50, chd free. Home-made teas. Visits also by arrangement June to Sept for groups of 20+.

As featured on Gardeners' World, a breathtaking garden set over an acre, which benefits from panoramic views of Watership Down. Surprises await, ranging from traditional herbaceous borders through desert/prairie planting, an enchanted knot garden, potager to exuberant subtropical schemes; featuring bananas, cannas, hedychiums and more. New additions incl the Ravine Garden and Orto. The garden is built on the site of a Roman Camp and Bath House. Picnics welcome in the orchard only (5 picnic pitches max). Partial wheelchair access. Non-disabled WC.

&. 🐴 ✿ Ⓓ ☕ ⌗

64 THE OLD RECTORY

East Woodhay, Newbury, RG20 0AL. David & Victoria Wormsley, 07801 418976, victoria@wormsley.net. 6m SW of Newbury. Turn off A343 between Newbury & Highclere to Woolton Hill. Turn L to East End, continue ¾m beyond East End. Turn R, garden opp St Martin's Church. Visits by arrangement Apr to Oct for groups of 20+. Adm £10, chd free. Home-made teas.

A classic English country garden of about 2 acres surrounding a Regency former rectory (not open). Formal lawns and terrace provide tranquil views over parkland. A large walled garden with grass paths, full of interesting herbaceous plants including topiary, roses and unusual perennials. A Mediterranean pool garden, wildflower meadow and fruit garden. Explore and enjoy.

🐴 ✿ ☕

65 OLD SWAN HOUSE

High Street, Stockbridge, SO20 6EU. Mr Herry Lawford, Herry@btinternet.com, www.facebook.com/oldswanhouse. 9m W of Winchester. The entrance to the garden is in Recreation Ground Lane, at the E end of the High St. Every Tue 6 July to 27 July (1.30-4.30). Adm £5, chd free. Home-made teas. Opening with Stockbridge Gardens on Thur 17, Sun 20 June. Visits also by arrangement June to Sept for groups of up to 10.

This town garden is designed around seven areas defined by the sun at different times of the day. Planting is modern perennial with euphorbias, rosemary and box used extensively. Particular interest is provided by a grass and gravel garden and a small wildflower meadow. There is an ancient hazel under a brick and flint wall, a loggia hung with creeper, an orchard and a pond.

&. 🐴 ☕

GROUP OPENING

66 OLD THATCH & THE MILLENNIUM BARN

Sprats Hatch Lane, Winchfield, Hook, RG27 8DD. www.old-thatch.co.uk. 3m W of Fleet. 1½m E of Winchfield Station, follow NGS signs. Sprats Hatch Lane is opp the Barley Mow Pub. Public car park at Barley Mow slipway is ½m from garden. Parking in field next to Old Thatch, if dry, disabled on-site. See garden website for park info. Sun 11 Apr, Sun 13 June (2-6). Combined adm £4, chd free.

Home-made teas. Pimms if hot & mulled wine if cool.

THE MILLENNIUM BARN
Mr & Mrs G Carter.

OLD THATCH
Jill Ede.

Who could resist visiting Old Thatch, a chocolate box thatched cottage (not open), featured on film and TV, a smallholding with a 5 acre garden and woodland alongside the Basingstoke Canal (unfenced). A succession of spring bulbs, a profusion of wildflowers, perennials and homegrown annuals pollinated by our own bees and fertilised by the donkeys, who await your visit. Over 30 named clematis and rose cultivars. Children enjoy our garden quiz, adults enjoy tea and home-made cakes. Arrive by narrow boat! Trips on 'John Pinkerton' may stop at Old Thatch on NGS days www.basingstoke-canal.org.uk. Also Accessible Boating shuttle available from Barley Mow wharf, approx every 45 mins. Signed parking for Blue Badge holders: please use entrance by the red telephone box. Paved paths and grass slopes give access to the whole garden.

&. 🐴 ✿ ☕

67 ✧ PATRICK'S PATCH

Fairweather's Garden Centre, High Street, Beaulieu, SO42 7YB. Patrick Fairweather, 01590 612307, info@fairweathers.co.uk, www.fairweathers.co.uk. SE of New Forest at head of Beaulieu River. Leave M27 at J2 & follow signs for Beaulieu Motor Museum. Go up High St & park in Fairweather's on LH side. For opening times and information, please phone, email or visit garden website.

Model kitchen garden with a full range of vegetables, trained top and soft fruit and herbs. Salads in succession used as an educational project for all ages. Maintained by volunteers, primary school children and a head gardener. We run a series of fun educational gardening sessions for children as well as informal workshops for adults. Open daily by donation from dawn to dusk. Wheelchair access on gravelled site.

&. ✿ ☕

68 PILLEY HILL COTTAGE
Pilley Hill, Lymington,
SO41 5QF. Steph & Sandy
Glen, 01590 677844,
stephglen@hotmail.co.uk. *New Forest. Pilley Hill Cottage is on the R as you come up Pilley Hill from Shallows Lane, before School Lane.* **Sat 27, Sun 28 Mar, Sat 17, Sun 18 Apr (2-5). Adm £3.50, chd free. Cream teas. Visits also by arrangement Mar & Apr.**
Pilley Hill Cottage garden changes constantly through the seasons. Entering through the creeper covered lych gate, the garden reveals itself via winding pathways with surprises around every corner. Dogwood, cornus and ghost bramble supply spring structure. Gnarled fruit trees provide shelter for bulbs, and wildflowers are making a welcome debut in our meadow strips. Please call to discuss access if you have a disability as there are some slopes and steps.

69 1 POVEY'S COTTAGE
Stoney Heath, Baughurst,
Tadley, RG26 5SN. Jonathan & Sheila Richards, 01256 850633,
smrjrichards@sky.com. *Between villages of Ramsdell & Baughurst, 10 mins drive from Basingstoke. Take A339 out of Basingstoke, direction Newbury. Turn R off A339 towards Ramsdell, then 4m to Stoneyheath. Pass under overhead power cables, take next L into unmade road & Povey's is 1st on the R.* **Sat 5, Sun 6 June, Sat 3, Sun 4 July (12-5). Adm £4, chd free. Home-made teas. Visits also by arrangement June to Sept for groups of 5 to 30.**
Herbaceous borders, trees and shrubs, a small orchard, greenhouses, fruit cage, vegetable garden and a wildflower meadow area. Beehives in one corner of the garden and chickens in another corner. A feature of the garden is an unusual natural swimming pond. Plant sale. See our Facebook page photos by searching One Poveys Cottage. Wheelchair access across flat grassed areas, but no hard pathways.

70 PYLEWELL PARK
South Baddesley, Lymington,
SO41 5SJ. Lord Teynham. *Coast road 2m E of Lymington. From Lymington follow signs for Car Ferry to Isle of Wight, continue for 2m to South Baddesley.* **Sun 4 Apr, Sun 30 May (2-5). Adm £5, chd free.**
A large parkland garden laid out in 1890. Enjoy a walk along the extensive informal grass and moss paths, bordered by fine rhododendron, magnolia and azalea. Wild daffodils bloom at Easter and bluebells in May. Large lakes are bordered by giant gunnera. Magnificent swans! Distant views of the Isle of Wight across the Solent. Lovely for families and dogs. Bring your own tea or picnic and wellingtons! Old glasshouses and other out buildings are not open to visitors. Wear suitable footwear for muddy areas.

71 REDENHAM PARK HOUSE
Redenham Park, Andover,
SP11 9AQ. Lady Olivia Clark, 01264 772511,
oliviaclark@redenhampark.co.uk. *Approx 1½m from Weyhill on the A342 Andover to Ludgershall road.* **Thur 16, Fri 17 Sept (2.30-5). Adm £6, chd £3. Home-made teas & cream teas in the thatched pool house. Visits also by arrangement June to Sept for groups of up to 30.**
Redenham Park built in 1784. The garden sits behind the house (not open). The formal rose garden is planted with white flowered roses.

Bere Mill

Steps lead up to the main herbaceous borders which peak in late summer. A calm green interlude, a gate opens into gardens with espaliered pears, apples, mass of scented roses, shrubs and perennial planting surrounds the swimming pool. A door opens onto a kitchen garden.

GROUP OPENING

72 ROMSEY GARDENS
Town Centre, Romsey, SO51 8LD. *All gardens are within walking distance of each other & are clearly signed. Use Lortemore Place public car park (SO51 8LD), free on Sundays & BH.* **Sun 30, Mon 31 May (10.30-4.30). Combined adm £6, chd free. Home-made teas.**

KING JOHN'S GARDEN
Friends of King John's Garden & Test Valley Borough. www.facebook.com/KingJohnsGarden/.
Open on Mon 31 May

4 MILL LANE
Miss J Flindall.
Open on all dates

THE NELSON COTTAGE
Margaret Prosser.
Open on all dates

Romsey is a small, unspoilt, historic market town with the majestic C12 Norman Abbey as a backdrop to 4 Mill Lane, a garden described by Joe Swift as 'the best solution for a long thin garden with a view'. King John's Garden with its fascinating listed C13 house (not open Sun), has all period plants that were available before 1700; it also has an award-winning Victorian garden with a courtyard (no dogs, please). The Nelson Cottage was formerly a pub; the ½ acre garden has a variety of perennial plants and shrubs with a wild grass meadow bringing the countryside into the town. No wheelchair access at 4 Mill Lane.

73 SAGES
Sages Lane, Privett, Alton, GU34 3NP. Joanne Edmonds, 07739 680052, sagesprivett@gmail.com. *Turn off A32 opp The Angel Hotel, then Sages is ½ m down the road on the L.* **Visits by arrangement in July**

for groups of 20 to 30. Adm £5, chd free.
A charming courtyard garden with a raised pond, pots of lilies and hydrangeas cascading over tiered flint walls, leads you up to the 2½ acre garden. Newly developed, designed and maintained by the owners in a setting of mature trees. Enjoy soft coloured planting, a glasshouse with unusual orchids, a tranquil Japanese style pond, a charming woodland shade garden and a modern kitchen garden. Wheelchair access via new sloping path to upper levels.

74 28 ST RONAN'S AVENUE
Southsea, Portsmouth, PO4 0QE. Ian & Liz Craig, www.28stronansavenue.co.uk. *St Ronan's Rd can be found off Albert Rd, Southsea. Follow signs from Albert Rd or Canoe Lake on seafront. Parking in Craneswater School.* **Sun 11 Apr, Sun 23 May (2-6). Adm £3.50, chd free. Home-made teas.** Town garden 145ft x 25ft, 700 metres from the sea. A mixture of tender, exotic and dry loving plants, along with more traditional incl king protea, bananas, ferns, agaves, echeverias, echium and puya. wildflower area and wildlife pond. Two different dry gardens showing what can be grown in sandy soil. Recycled items have been used to create sculptures.

GROUP OPENING

75 NEW SEARLES LANE GARDENS
Searles Lane, Hook, RG27 9EQ. *5 mins from M3 J5. Searles Lane is off A30 (London Rd) N of Hook, towards H Wintney. For disabled parking go to Searles Lane (RG27 9EQ). Other parking nearby, approx ¼ m, signed on the day.* **Thur 27, Fri 28, Sat 29, Sun 30 May (2-5). Combined adm £6.50, chd free. Home-made teas, gluten & dairy free options at Maple Cottage.**

NEW CHERRY TREE BARN
Lois & Patrick Bellew.
D

NEW CHERRY TREE BARN COTTAGE
Judy Orme-Bannister.

MAPLE COTTAGE
John & Pat Beagley.

Searles Lane is a long established area in Hook with Searles Farmhouse dating back to c1680. Visit 3 very different gardens bordering Whitewater Meadow. Maple Cottage is reopening after a 2 yr break, a ½ acre cottage style garden with herbaceous borders, veg plots and a tiny, but very active wildlife pond surrounded by hostas, astilbe, sarracenia and a quirky tree cave for kids. The neighbouring 1 acre contemporary garden at Cherry Tree Barn is just 2 yrs old and is the realisation of the owner's ideas working with garden designer Fiona Harrison. The area around the thatched barn has been planted with grasses, iris, salvia, euphorbia, phlomis and knautia. Drifts of oxeye daisies and other wildflowers on bunds line the perimeter. Hard landscaping incl terracing, decking, pathways and Japanese influenced Yatsuhashi Bridge through the main border. Small collection of sculptures completes the piece. The tiny garden at Cherry Tree Barn Cottage is full of lovingly attended, colourful pots. Plant sale at Maple Cottage, renowned for high quality! Wheelchair access with mainly grassed areas and avoidable steps

76 'SELBORNE'
Caker Lane, East Worldham, Alton, GU34 3AE. Mary Tigwell-Jones, www.worldham.org. *2m SE of Alton. On B3004 at Alton end of the village of East Worldham, near The Three Horseshoes Pub. Please note: 'Selborne' is the name of the house, it is not in the village of Selborne. Parking signed.* **Sat 15, Sun 16, Mon 17 May (2-5). Adm £4, chd free. Sat 26, Sun 27, Mon 28 June (2-5). Combined adm with East Worldham Manor £6, chd free. Sat 31 July, Sun 1, Mon 2 Aug (2-5). Adm £4, chd free. Home-made teas & picnics welcome.**
This much-loved ½ acre mature cottage-style garden provides visitors with surprises around every corner and views across farmland. Productive 60 yr old orchard of named varieties, densely-planted borders, shrubs and climbers, especially clematis. Metal and stone sculptures enhance the borders. Bug mansion. Enjoy tea in the shade of the orchard and the summerhouses and conservatory provide shelter. There will be a one-way system in place. Wheelchair access with some gravel paths.

77 SHALDEN PARK HOUSE

The Avenue, Shalden, Alton,
GU34 4DS. Mr & Mrs Michael
Campbell. *4½ m NW of Alton.
B3349 from Alton or M3 J5 onto
B3349. Turn W at Golden Pot Pub
marked Herriard, Lasham, Shalden.
Entrance ¼ m on L. Disabled parking
on entry.* **Sun 6 June (2-5). Adm £5,
chd free. Home-made teas.**
Large 4 acre garden to stroll around
with beautiful views. Herbaceous
borders incl kitchen walk and rose
garden, all with large-scale planting
and foliage interest. Pond, arboretum,
perfect kitchen garden and garden
statuary.

78 SILVER BIRCHES

Old House Gardens, East
Worldham, Nr Alton,
GU34 3AN. Jenny & Roger
Bateman, 07555 743029,
rogerjbateman@gmail.com. *2m
E of Alton off B3004. Turn L signed
Wyck & Binstead, then 1st R into
Old House Gardens.* **Visits by
arrangement May to Aug for
groups of 5 to 30. Adm £5, chd
free. Tea.**
½ acre garden completely
redesigned by the owners over
the last 11 yrs. Winding paths lead
through shrub and herbaceous
borders to fish pond with stream,
rockery and summerhouse. Rose
garden with arbour. Planting designed
for yr-round colour and interest using
foliage as well as flowers. Many sitting
areas and some unusual plants.
Partial wheelchair access.

79 ◆ SIR HAROLD HILLIER GARDENS

Jermyns Lane, Ampfield,
Romsey, SO51 0QA. Hampshire
County Council, 01794 369318,
info.hilliers@hants.gov.uk,
www.hants.gov.uk/hilliergardens.
*2m NE of Romsey. Follow brown
tourist signs off M3 J11, or off M27
J2, or A3057 Romsey to Andover.
Disabled parking available.* **For
opening times and information,
please phone, email or visit garden
website.**
Established by the plantsman Sir
Harold Hillier, this 180 acre garden
holds a unique collection of 12,000
different hardy plants from across
the world. It incl the famous Winter
Garden, Magnolia Avenue, Centenary
Border, Himalayan Valley, Gurkha
Memorial Garden, Magnolia Avenue,
spring woodlands, Hydrangea Walk,
fabulous autumn colour, 14 National
Collections and over 600 champion
trees. The Centenary Border is one
of the longest double mixed border
in the country, a feast from early
summer to autumn. Celebrated
Winter Garden is one of the largest
in Europe. Electric scooters and
wheelchairs are available for hire
(please pre-book). Accessible WC
and parking. Registered assistance
dogs only.

80 NEW SOUTH VIEW HOUSE

60 South Road, Horndean,
Waterlooville, PO8 0EP.
James & Victoria Greenshields.
*Between Horndean & Clanfield.
From N A3 towards Horndean. R
at T-junction, to r'about. From S A3
B2149 to Horndean. Continue on A3
N to r'about. 1st exit into Downwood
Way. 3rd L into South Rd. 3rd house
on R. Park in road.* **Sun 6 June, Sun
8 Aug (1-5). Adm £3.50, chd free.
Light refreshments.**
A fusion of traditional and
contemporary designs across a ½
acre site. The formal cottage garden
to the front incl a large herbaceous
border, small woodland garden and
formal topiary. The cleverly designed
garden to the rear features pond,
alpine garden, mini fruit orchard,
chickens and greenhouse and, for
entertaining, a patio with bar (not
open) and a large summerhouse. Well
worth a visit. Garden is accessed
along a gravel drive on a gentle slope.

81 SPINDLES

24 Wootton Road, Lee-on-
the-Solent, Portsmouth,
PO13 9HB. Peter & Angela
Arnold, 02393 115181,
angelliana62@gmail.com. *6m S
of Fareham. Exit A27, turn L onto
Gosport Rd A32. At r'about take 2nd
exit Newgate Lane B3385. Through
3 r'abouts, turn L into Marine Parade
B3333 onto Wootton Rd.* **Visits
by arrangement May to Aug for
groups of up to 30. Home-made
teas.**
Visit this small gold award-winning
garden where cottage style planting
merges with tropical and the exotic.
Over 50 roses vie for your attention as
you meander under clematis covered
arches towards a small wildlife pond,
tree ferns and bamboo. Interest to
new and experienced gardeners. Just
3 mins from the sea, why not make a
day of it, you won't be disappointed!
Wheelchair access very limited due to
narrow paths.

82 ◆ SPINNERS GARDEN

School Lane, Pilley, Lymington,
SO41 5QE. Andrew & Vicky
Roberts, 07545 432090,
info@spinnersgarden.co.uk,
www.spinnersgarden.co.uk.
*1½ m N of Lymington. Follow sign
to Boldre off the A337 between
Brockenhurst & Lymington. At top
of Pilley Hill turn R into School Lane.
Spinners Garden is at the end of a
row of houses on the R.* **For NGS:
Sun 25 Apr (2-5.30). Adm £6, chd
free. Cream teas on the patio.
For other opening times and
information, please phone, email or
visit garden website.**
Peaceful woodland garden
overlooking the Lymington valley with
many rare and unusual plants. The
garden continues to be developed
with new plants added to the
collections and the layout changed to
enhance the views. The house was
rebuilt in 2014 to reflect its garden
setting. Andy will take groups of 15 on
tours of the hillside with its woodland
wonders and draw attention to the
treats at their feet; trilliums, wood
anemones and erythroniums! Partial
wheelchair access.

83 SPITFIRE HOUSE

Chattis Hill, Stockbridge,
SO20 6JS. Tessa & Clive
Redshaw, 07711 547543,
tessa@redshaw.co.uk. *2m from
Stockbridge. Follow the A30 W
from Stockbridge for 2m. Go past
the Broughton/Chattis Hill Xrds, do
not follow SatNav into Spitfire Lane,
take next R to the Wallops & then
immed R again to Spitfire House.* **Sat
5 June (2-5). Adm £4, chd free.
Home-made teas. Visits also by
arrangement in June for groups of
10 to 30.**
A country garden situated high on
chalk downland. On the site of a
WW11 Spitfire assembly factory with
Spitfire tethering rings still visible.
This garden has wildlife at its heart
and incl fruit and vegetables, a small
orchard, wildlife pond, woodland
planting and large areas of wildflower
meadow. Wander across the downs
to be rewarded with extensive views.
Wheelchair access with areas of gravel
and a slope up to wildflower meadow.

GROUP OPENING

84 STOCKBRIDGE GARDENS

Stockbridge, SO20 6EX. *9m W of Winchester. On A30, at the junction of A3057 & B3049. All gardens & parking on High St.* **Thur 17, Sun 20 June (1.30-4.30). Combined adm £8, chd free. Home-made teas at St Peter's Church.**

LITTLE WYKE
Mrs Mary Matthews.

THE OLD RECTORY
Robin Colenso & Chrissie Quayle.

OLD SWAN HOUSE
Mr Herry Lawford.
(See separate entry)

SHEPHERDS HOUSE
Kim & Frances Candler.

TROUT COTTAGE
Mrs Sally Milligan,
sally@sallymilligan.co.uk.
Visits also by arrangement June to Sept for groups of up to 20.

Five gardens will open this year in Stockbridge, offering a variety of styles and character. Little Wyke, next to the Town Hall has a long mature town garden with mixed borders and fruit trees. Trout Cottage is a small walled garden, which will inspire those with small spaces and little time to achieve tranquillity and beauty. Full of approx 180 plants flowering for almost 10 mths of the yr, all set around a rectangular lawn. The Old Rectory has a partially walled garden with formal pond, fountain and planting near the house (not open) with a stream-side walk under trees, many climbing and shrub roses and a woodland area. Old Swan House, at the east end of the High St has modern mixed planting. There is a gravel grass garden and an orchard, as well as a pond. Shepherds House on Winton Hill with herbaceous borders, a new kitchen garden and a belvedere overlooking the pond. Wheelchair access to gardens, gravel path at Shepherds House.

 ♿ ☕

85 TADLEY PLACE

Church Lane, Baughurst, Tadley, RG26 5LA. Lyn & Ronald Duncan. *10 mins drive from Basingstoke, near Tadley.* **Sun 16 May (1-4). Adm £6, chd free. Light refreshments.**
Tadley Place is a Tudor manor house (not open) dating from the C15. The gardens surround the house and incl formal areas, a large kitchen garden and access to a fabulous bluebell wood with a pond. Wheelchair access to majority of garden on slightly uneven lawn.

♿ ☕

In our first year 609 private gardens opened their gates to all, for the modest sum of one shilling. Today the National Garden Scheme retains that combination of inclusivity and affordability

How Park Barn

86 TANGLEFOOT

Crawley, Winchester,
SO21 2QB. Mr & Mrs F J Fratter,
01962 776243, fred@fratter.co.uk.
*5m NW of Winchester. Between
B3049 (Winchester - Stockbridge) &
A272 (Winchester - Andover). Lane
beside Crawley Court (Arqiva). For
SatNav use SO21 2QA. Parking in
adjacent mown field.* **Thur 13, Sun
16 May, Thur 3, Sun 6 June, Thur
15, Sun 18 July (2-5.30). Adm
£5, chd free. Drinks & biscuits
included. Opening with Crawley
Gardens on Thur 24, Sun 27 June.
Visits also by arrangement May
to July.**
Developed by owners since 1976,
Tanglefoot's ½ acre garden is a
blend of influences, from Monet-
inspired rose arch and small wildlife
pond to Victorian boundary wall with
trained fruit trees. Highlights include
a raised lily pond, herbaceous bed
(a riot of colour later in the summer),
herb wheel, large productive kitchen
garden and unusual flowering
plants. In contrast to the garden,
a 2 acre field with views over the
Hampshire countryside has recently
been converted into spring and
summer wildflower meadows with
mostly native trees and shrubs;
it has delighted visitors in recent
summers. Plants from the garden for
sale. Visitors can picnic in the field,
but there are no facilities available.
Wheelchair access with narrow paths
in vegetable area.

87 TERSTAN

Longstock, Stockbridge,
SO20 6DW. Alexander & Penny
Burnfield, paburnfield@gmail.com,
www.pennyburnfield.wordpress.
com. *¾m N of Stockbridge.
From Stockbridge (A30) turn N to
Longstock at bridge. Garden ¾m
on R.* **Suns 18 Apr, 13 June, 18
July, 19 Sept (2-5). chd
free. Home-made teas. Visits also
by arrangement Apr to Sept for
groups of 10+.**
A garden for all seasons, developed
over 50 yrs into a profusely planted,
contemporary cottage garden in
peaceful surroundings. There is a
constantly changing display in pots,
starting with tulips and continuing with
many unusual plants. Features incl
gravel garden, water features, cutting
garden, showman's caravan and live
music. Wheelchair access with some
gravel paths and steps.

88 THE THATCHED COTTAGE

Church Road, Upper Farringdon,
Alton, GU34 3EG. Mr David &
Mrs Cally Horton, 01420 587922,
dwhorton@btinternet.com. *3m
S of Alton off A32. From A32, take
road to Upper Farringdon. At top of
the hill turn L into Church Rd, follow
round corner, past Masseys Folly
(large red brick building) & we are
the 1st house on the R.* **Sun 30,
Mon 31 May, Sat 12 June (1-5).
Adm £6, chd free. Sun 13 June
(1-5). Combined adm with Berry
Cottage £10, chd free. Sat 10 July
(1-5). Adm £6, chd free. Sun 11
July, Sun 29 Aug (1-5). Combined
adm with Berry Cottage £10, chd
free. Mon 30 Aug (1-5). Adm £6,
chd free. Home-made teas. Visits
also by arrangement for groups
of 10+. Donation to Jubilee Sailing
Trust.**
A 1½ acre garden hidden behind
a C16 thatched cottage (not open).
Borders burst with cottage garden
plants and a pond provides the
soothing sound of water. A pergola
of roses, clematis and honeysuckle
leads to chickens and ducks under
a walnut, one of several specimen
trees. A productive garden with fruit
trees, fruit cage and raised vegetable
beds. Enjoy the wildflower area and
gypsy caravan. Fully accessible by
wheelchair after a short gravel drive.

89 TWIN OAKS

13 Oakwood Road, Chandler's
Ford, Eastleigh, SO53 1LW. Syd
& Sue Hutchinson, 02380 907517,
syd@sydh.co.uk. *Leave M3 J12.
Follow signs to Chandlers Ford onto
Winchester Rd. After ½m turn R into
Hiltingbury Rd. After approx ½m
turn L into Oakwood Rd.* **Sun 4,
Mon 5, Wed 28 Apr, Sun 30, Mon
31 May, Fri 2, Sat 3, Sun 4 July,
Fri 6, Sat 7, Sun 8 Aug (1-5). Adm
£3.50, chd free. Home-made teas.
Visits also by arrangement Apr to
Aug for groups of 5 to 20.**
Continually evolving ⅓ acre suburban
woodland water garden, designed
and planted by owners, bordered by
mature oak beech and birch trees.
Enjoy spring colour from azaleas,
rhododendrons and bulbs. The lawn
meanders between informal beds and
ponds and bridges lead to a tranquil
pergola seating area overlooking a
wildlife pond. Rockery skirted by a
stream with a waterfall into a lily pond.
Aviary.

90 WALDEN

Common Hill, Medstead, Alton,
GU34 5LZ. Terri & Neil Burman.
*5m S of Alton. From N on A31 turn
R into Boyneswood Rd signed
Medstead. At small Xrds go R into
Roedowns Rd. At T-junction at village
green turn L. After church, Common
Hill is 1st turning on L.* **Fri 18, Sun
20 June (1-5). Adm £3.50, chd
free. Home-made teas.**
2 acre sloping garden on chalk,
designed by the owners with
panoramic views towards Winchester.
Restored 60ft rockery with many
alpine species. Early summer colour
with hardy perennials, peonies,
roses, mature shrubs and fruit trees.
Interesting sculptures enhance the
garden whilst recycled objects add
interest and humour. Parking for
disabled visitors in front of house on
paved driveway. The garden is mainly
grass with some slopes, assistance
required.

91 WALHAMPTON

Beaulieu Road, Walhampton,
Lymington, SO41 5ZG.
Walhampton School
Trust Ltd, 07928 385694,
d.hill@walhampton.com. *1m E of
Lymington. From Lymington follow
signs to Beaulieu (B3054) for 1m
& turn R into main entrance at 1st
school sign, 200yds after top of hill.*
**Sun 2 May (2-6). Adm £5, chd
free. Home-made teas in school
dining room (2-4). Visits also
by arrangement May to July for
groups of 5 to 20. Adm includes
refreshments. Donation to St
John's Church, Boldre.**
Glorious walks through large C18
landscape garden surrounding
magnificent mansion (not open).
Visitors will discover three lakes,
serpentine canal, climbable prospect
mount, period former banana house
and orangery, fascinating shell
grotto, plantsman's glade and Italian
terrace by Peto (c1907), drives and
colonnade by Mawson (c1914) with
magnificent views to the Isle of Wight.
Exedrae and sunken garden, rockery,
Roman arch, fountain and seating.
Guided garden history tours available
on the day. Wheelchair access with
gravel paths and some slopes.
Regret, no dogs allowed.

92 WEIR HOUSE

Abbotstone Road, Old Alresford, SO24 9DG. Mr & Mrs G Hollingbery, 07767 606729, jhollingbery@me.com. *½m N of Alresford. From New Alresford down Broad St (B3046), past Globe Pub, take 1st L signed Abbotstone. Weir House is 1st drive on L. Park in signed field.* **Sun 13 June, Sun 22 Aug (1-4). Adm £5, chd free. Visits also by arrangement for groups of 10+.** Spectacular riverside garden with sweeping lawn backed by old walls, yew buttresses and mixed perennial beds. Over 3 acres of garden including contemporary vegetable garden at its height in Aug/Sept, a contemporary garden around pool area, bog garden at its best in May/June and wilder walkways through wooded areas. Children and dogs welcome. Wheelchair access to most of the garden.

93 WEST VIEW

Old London Road, Stockbridge, SO20 6EL. Rebecca & Matthew Ferris. *Old London Rd is directly opp The White Hart Pub at the E end of Stockbridge High St. West View is 300yds on R, at the opp end of the road from the primary school.* **Sun 1 Aug (10.30-4); Wed 4 Aug (1-4). Adm £4, chd free. Home-made teas on 4th Aug only.** ½ acre garden designed and constructed by the current owners, built into the natural chalk cliff on levels. 60 steep steps take you up through a series of small garden rooms from pool area to sun deck, copper garden and white garden. The garden opens up as you get higher, culminating in a field with wildlife pond, shepherds hut, wildflower area, orchard and spectacular views of the Test Valley. On 1st Aug Food Festival In Stockbridge.

94 WHEATLEY HOUSE

Wheatley Lane, between Binsted & Kingsley, Bordon, GU35 9PA. Mr & Mrs Michael Adlington, 01420 23113, adlingtons36@gmail.com. *4m E of Alton, 5m SW of Farnham. Take A31 to Bentley, follow sign to Bordon. After 2m, R at Jolly Farmer Pub towards Binsted, 1m L & follow signs to Wheatley.* **Sat 21, Sun 22 Aug (1.30-5.30). Adm £5, chd free. Home-made teas. Visits also by arrangement May to Oct for groups of 10+.** Situated on a rural hilltop with panoramic views over Alice Holt Forest and the South Downs. The owner admits to being much more of an artist than a plantswoman, but has had great fun creating this 1½ acre garden full of interesting and unusual planting combinations. The sweeping mixed borders and shrubs are spectacular with colour throughout the season, particularly in late summer. The black and white border, now with bright red accents, is very popular with visitors. Wheelchair access with care on lawns, good views of garden and beyond from terrace.

95 WICOR PRIMARY SCHOOL COMMUNITY GARDEN

Portchester, Fareham, PO16 9DL. Louise Moreton, www.wicor.hants.sch.uk. *Halfway between Portsmouth & Fareham on A27. Turn S at Seagull Pub r'about into Cornaway Lane, 1st R into Hatherley Drive. Entrance to school is almost opp. Parking on-site, pay at main gate.* **Sun 27 June (12-4). Adm £3.50, chd free.** As shown on BBC Gardeners' World in 2017. Beautiful school gardens tended by pupils, staff and community gardeners. Wander along Darwin's path to see the coastal garden, Jurassic garden, orchard, tropical bed, stumpery, wildlife areas, allotments and apiary, plus one of the few camera obscuras in the south of England. Wheelchair access to all areas, flat ground.

96 WILLOWS

Pilley Hill, Boldre, Lymington, SO41 5QF. Elizabeth & Martin Walker, 01590 677415, elizabethwalker13@gmail.com, www.willowsgarden.co.uk. *New Forest. 2m N Lymington off A337. To avoid traffic in Lyndhurst, leave M27 at J2 & follow Heavy Lorry Route. Disabled parking at gate.* **Sat 7, Sun 8 Aug (2-5). Adm £4, chd free. Cream teas. Visits also by arrangement July to Sept for groups of 20+. Guided garden tour & Dahlia Demo on request.** Front borders overflow with colourful dahlias, cannas, crocosmias and swathes of heleniums. Exciting exotics contrast with a jungly mix of gunneras, ferns and giant hostas around the tranquil pond and lower bog garden, where ginger lilies may be in flower. Fascinating topiary greets you as you walk to the sunny upper borders of wonderful blue hydrangeas, dark leaved dahlias and billowing grasses. Elizabeth will give a Dahlia Demo at 3pm on all open days. Wheelchairs usually manage to access all parts of garden.

97 1 WOGSBARNE COTTAGES

Rotherwick, RG27 9BL. Miss S & Mr R Whistler. *2½m N of Hook. M3 J5, M4 J11, A30 or A33 via B3349.* **Sun 11, Mon 12 July (2-5). Adm £3, chd free. Home-made teas.** Small traditional cottage garden with a 'roses around the door' look, much photographed for calendars, jigsaws and magazines. Mixed flower beds and borders. Vegetables grown in abundance. Ornamental pond and alpine garden. Views over open countryside to be enjoyed whilst you take afternoon tea on the lawn. The garden has been open for the NGS for more than 30 yrs. Wheelchair access with some gravel paths.

98 NEW WOODPECKERS CARE HOME

Sway Road, Brockenhurst, SO42 7RX. Mr Charles Hubberstey. *New Forest. Sway Road from village centre. Past petrol station, then school, Woodpeckers on R.* **Sun 11 Apr, Sun 12, Wed 15 Sept (11-5). Adm £3.50, chd free. Home-made teas.** A vibrant and colourful garden surrounds our care home. We have active involvement from our residents who enjoy the wide paths, whether in a wheelchair or strolling on foot. The courtyard area, small orchard and allotments, all look particularly beautiful In spring and late summer with views through neighbouring fields. You may see deer to the west and ponies to the east and do spot the bug house!

The National Garden Scheme searches the length and breadth of England and Wales for the very best private gardens

HEREFORDSHIRE

Herefordshire is essentially an agricultural county, characterised by small market towns, black and white villages, fruit and hop orchards, meandering rivers, wonderful wildlife and spectacular, and often remote, countryside (a must for keen walkers).

As a major region in the Welsh Marches, Herefordshire has a long and diverse history, as indicated by the numerous prehistoric hill forts, medieval castles and ancient battle sites. Exploring the quiet country lanes can lead to many delightful surprises.

For garden enthusiasts the National Garden Scheme offers a range of charming and interesting gardens ranging from informal cottage plots to those of grand houses with parterres, terraces and parkland. Widely contrasting in design and plantings they offer inspiration and innovative ideas.

National collections of Asters and Siberian Iris can be found at The Picton Garden and Aulden Farm, respectively; and, for Galanthophiles, Ivycroft will not disappoint. The numerous specialist nurseries offer tempting collections of rare and unusual plants.

You can always be sure of a warm welcome at a National Garden Scheme open garden.

Volunteers

County Organiser
Lavinia Sole
07880 550235
lavinia.sole@ngs.org.uk

County Treasurer
Angela Mainwaring
01981 251331
angela.mainwaring@ngs.org.uk

Booklet Coordinator
Chris Meakins
01544 370215
christine.meakins@btinternet.com

Booklet Distribution
Graham Sole
01568 797522
grahamsole3@gmail.com

Social Media
Naomi Grove
naomi.grove@ngs.org.uk

Assistant County Organisers
David Hodgson
01531 640622
dhodgson363@btinternet.com

Sue Londesborough
01981 510148
slondesborough138@btinternet.com

Penny Usher
01568 611688
pennyusher@btinternet.com

 @NGSHerefordshire
 @HerefordNGS

Left: Moors Meadow Gardens

OPENING DATES

All entries subject to change. For latest information check www.ngs.org.uk

Extended openings are shown at the beginning of the month.

Map locator numbers are shown to the right of each garden name.

January

Thursday 28th
Ivy Croft   22

February

Snowdrop Festival

Every Thursday
Ivy Croft   22

Sunday 14th
The Old Corn Mill   33
◆ The Picton Garden   36

Friday 19th
Coddington Vineyard   10

Sunday 28th
The Old Corn Mill   33
◆ The Picton Garden   36

March

Thursday 4th
Ivy Croft   22

Sunday 7th
The Old Corn Mill   33

Saturday 13th
◆ The Picton Garden   36

Sunday 21st
The Old Corn Mill   33

Saturday 27th
◆ Ralph Court Gardens 38
NEW The Vern   45

Sunday 28th
◆ Ralph Court Gardens 38
NEW The Vern   45
Whitfield   49

Monday 29th
◆ Moors Meadow
Gardens   28

April

Thursday 1st
◆ Stockton Bury
Gardens   44

Saturday 3rd
NEW Wainfield   47

Sunday 4th
◆ The Picton Garden   36

Sunday 11th
Lower Hope   26
The Old Corn Mill   33

Saturday 17th
Revilo   39

Sunday 18th
Hill House Farm   19
Lower House Farm   27
Revilo   39
Woodview   50

Monday 19th
◆ The Picton Garden   36

Friday 23rd
Coddington Vineyard   10

Saturday 24th
Coddington Vineyard   10
NEW The Vern   45

Sunday 25th
◆ Caves Folly Nurseries 8
Longacre   24
The Old Corn Mill   33
NEW The Vern   45

Monday 26th
◆ Moors Meadow
Gardens   28

May

Saturday 1st
Aulden Farm   2
Ivy Croft   22

Sunday 2nd
Aulden Farm   2
Ivy Croft   22
◆ The Picton Garden   36

Monday 3rd
Aulden Farm   2
Ivy Croft   22

Saturday 8th
Dovecote Barn   11

Sunday 9th
Dovecote Barn   11
Lower House Farm   27
The Old Corn Mill   33

Saturday 15th
Southbourne & Pine
Lodge   43

Sunday 16th
Hill House Farm   19
Lower Hope   26
Southbourne & Pine
Lodge   43

Saturday 22nd
NEW The Vern   45

Sunday 23rd
Kentchurch Court   23
The Old Corn Mill   33
NEW The Vern   45

Monday 24th
◆ Moors Meadow
Gardens   28

Friday 28th
NEW Wainfield   47

Saturday 29th
NEW Wainfield   47

Sunday 30th
Coddington Vineyard   10
NEW Garway House   14
NEW The Hurst   21
Old Colwall House   32
Sheepcote   41

Monday 31st
NEW Garway House   14
NEW The Hurst   21
◆ The Picton Garden   36

June

Friday 4th
Castle Moat House   7

Saturday 5th
Castle Moat House   7

Sunday 6th
Brockhampton Cottage 3
Broxwood Court   4
Grendon Court   16
NEW The Nutshell   31

Saturday 12th
NEW The Forge   12
NEW Lower Farmhouse 25
Revilo   39

Sunday 13th
NEW The Forge   12
NEW Grange Court   15
NEW Lower Farmhouse 25
Lower House Farm   27
◆ The Picton Garden   36
Revilo   39

Saturday 19th
Aulden Arts and
Gardens   1
Byecroft   6
◆ Hereford Cathedral
Gardens   18

Sunday 20th
Aulden Arts and
Gardens   1
Byecroft   6
Hill House Farm   19
Whitfield   49

Tuesday 22nd
Hillcroft   20

Wednesday 23rd
Hillcroft   20

Thursday 24th
Hillcroft   20

Saturday 26th
◆ The Garden of the
Wind at Middle Hunt
House   13
NEW Wainfield   47

Sunday 27th
◆ The Garden of the
Wind at Middle Hunt
House   13
NEW Grange Court   15
Kentchurch Court   23
◆ The Picton Garden   36
The Vine   46

Monday 28th
◆ Moors Meadow
Gardens   28

July

Saturday 3rd
◆ Ralph Court Gardens
38

Sunday 4th
Lower Hope   26
◆ Ralph Court Gardens
38

Sunday 11th
◆ The Picton Garden   36
Woodview   50

Friday 16th
Herbfarmacy   17

Saturday 17th
Herbfarmacy   17

Sunday 18th
Hill House Farm   19
Woodview   50

Thursday 22nd
Herbfarmacy   17

Friday 23rd
Herbfarmacy   17

Sunday 25th
♦ The Picton Garden   36
Rhodds Farm   40

Monday 26th
♦ Moors Meadow
Gardens   28

August

Tuesday 10th
♦ The Picton Garden   36

Sunday 15th
Hill House Farm   19

Sunday 29th
Old Grove   34

Monday 30th
♦ The Picton Garden   36

September

Saturday 4th
Aulden Farm   2
Ivy Croft   22

Sunday 5th
Aulden Farm   2
Brockhampton Cottage   3
Broxwood Court   4
Grendon Court   16
Ivy Croft   22

Saturday 11th
Dovecote Barn   11

Sunday 12th
Dovecote Barn   11
Hill House Farm   19
Lower Hope   26

Monday 20th
♦ The Picton Garden   36

October

Saturday 9th
♦ Ralph Court Gardens
38

Sunday 10th
♦ Ralph Court Gardens
38

Wednesday 20th
♦ The Picton Garden   36

November

Saturday 13th
♦ Ralph Court
Gardens   38

Sunday 14th
♦ Ralph Court
Gardens   38

By Arrangement

Arrange a personalised
garden visit on a date to
suit you. See individual
garden entries for full
details.

Aulden Farm   2
Brighton House
Bury Court Farmhouse   5
Byecroft   6
Church Cottage   9
Coddington Vineyard   10
Hillcroft   20
Ivy Croft   22
Lower Hope   26
Lower House Farm   27
Mulberry House   29
Newton St Margarets
Gardens   30

Old Colwall House   32
The Old Corn Mill   33
Old Farm Cottage
The Old Rectory   35
Poole Cottage   37
Revilo   39
Shuttifield Cottage   42
NEW The Vern   45
Weston Hall   48
Whitfield   49
Woodview   50

National Garden
Scheme gardens are
identified by their
yellow road signs
and posters. You
can expect a garden
of quality, character
and interest, a
warm welcome and
plenty of home-
made cakes!

Ralph Court Gardens

THE GARDENS

GROUP OPENING

1 AULDEN ARTS AND GARDENS

Aulden, Leominster, HR6 0JT.
www.auldenfarm.co.uk/auldenarts.
4m SW of Leominster. From Leominster, take Ivington/Upper Hill rd, ³/₄m after Ivington church turn R signed Aulden. From A4110 signed Ivington, take 2nd R signed Aulden. **Sat 19, Sun 20 June (2-5). Combined adm £8, chd free. Home-made teas and ice cream.**
Lost in the back lanes of Herefordshire, 3 neighbours looking to celebrate art and our gardens, which vary from traditional to more zany – as does our art! Come and explore, sit awhile and enjoy yummy cakes! Aulden Farm art will be on display in our barn - canvases inspired by walks, stitchery by Nature. We are drawn to that fertile ground where realism meets abstraction. Honeylake Cottage; mature peaceful garden that encourages wildlife with a pond and numerous nest boxes. Cottage flowers predominate interspersed with fragrant English roses, lawns and a well-stocked traditional vegetable garden - the garden is an inspiration for artwork using a variety of media. Oak House: ²/₃ acre of evolving garden with amazing views towards Upper Hill. Currently has good bones but is fraying round the edges, divided into several areas including borders, ponds, chickens and a vegetable plot. Photography and textiles will be displayed throughout the garden.

2 AULDEN FARM

Aulden, Leominster, HR6 0JT. Alun & Jill Whitehead, 01568 720129, web@auldenfarm.co.uk, www.auldenfarm.co.uk. *4m SW of Leominster. From Leominster take Ivington/Upper Hill rd, ³/₄m after Ivington church turn R signed Aulden. From A4110 signed Ivington, take 2nd R signed Aulden.* **Sat 1, Sun 2, Mon 3 May, Sat 4, Sun 5 Sept (10-4). Combined adm with Ivy Croft £8, chd free. Home-made teas. Single garden opening £5. Visits also by arrangement Apr to Sept.**
Informal country garden, thankfully never at its Sunday best! 3 acres planted with wildlife in mind. Emphasis on structure and form, with a hint of quirkiness, a garden to explore with eclectic planting. Irises thrive around a natural pond, shady beds and open borders, seats abound, feels mature but ever evolving. Our own ice cream and home-burnt cakes, Lemon Chisel a speciality!

3 BROCKHAMPTON COTTAGE

Brockhampton, HR1 4TQ. Peter Clay. *8m SW of Hereford. 5m N of Ross-on-Wye on B4224. In Brockhampton take rd signed to B Crt nursing home, pass N Home after ³/₄m, go down hill and turn Left. Car park 500yds downhill on L in orchard.* **Sun 6 June, Sun 5 Sept (11-5). Adm £5, chd free. Also open Grendon Court. Picnic parties welcome by the lake.**
Created from scratch in 1999 by the owner and Tom Stuart-Smith, this beautiful hilltop garden looks S and W over miles of unspoilt countryside. On one side a woodland garden and 5 acre wildflower meadow, on the other side a Perry pear orchard and in valley below: lake, stream and arboretum. The extensive borders are planted with drifts of perennials in the modern romantic style. Allow 1hr 30 mins. Visit Grendon Court (11-4) after your visit to us.

4 BROXWOOD COURT

Broxwood, nr Pembridge, Leominster, HR6 9JJ. Richard Snead-Cox & Mike & Anne Allen. *From Leominster follow signs to Brecon A44/A4112. After approx 8m, just past Weobley turn off, go R to Broxwood/Pembridge. After 2m straight over Xrds to Lyonshall. 500yds on L over cattle grid.* **Sun 6 June, Sun 5 Sept (11-5.30). Adm £5, chd free. Home-made teas.**
Impressive 29 acre garden and arboretum, designed in 1859 by W. Nesfield for great-grandfather of present owner. Magnificent yew hedges and km long avenue of cedars and Scots pines. Spectacular view of Black Mountains, sweeping lawns, rhododendrons, gentle walks to summer house, chapel and ponds. Rose garden, mixed borders, rill, gazebo, sculpted benches and fountain. Peacocks and white doves. Some gravel, but mostly lawn. Gentle slopes. Disabled WC.

5 BURY COURT FARMHOUSE

Ford Street, Wigmore, Leominster, HR6 9UP. Margaret & Les Barclay, 01568 770618, l.barclay@zoho.com. *10m from Leominster, 10m from Knighton, 8m from Ludlow. On A4110 from Leominster, at Wigmore turn R just after shop & garage. Follow signs to parking and garden.* **Visits by arrangement Feb to Oct. Adm £4, chd free.**
³/₄ acre garden, 'rescued' since 1997, surrounds the 1820's stone farmhouse (not open). The courtyard contains a pond, mixed borders, fruit trees and shrubs, with steps up to a terrace which leads to lawn and vegetable plot. The main garden (semi-walled) is on two levels with mixed borders, greenhouse, pond, mini-orchard , many spring flowers, and wildlife areas. Year-round colour. Mostly accessible for wheelchairs by arrangement.

6 BYECROFT

Welshman's Lane, Bircher, Leominster, HR6 0BP. Sue & Peter Russell, 01568 780559, peterandsuerussell@btinternet. com, www.byecroft.weebly.com. *6m N of Leominster. From Leominster take B4361. Turn L at T-junction with B4362. ¼m beyond Bircher village turn R at war memorial into Welshman's Lane, signed Bircher Common.* **Sat 19, Sun 20 June (2-5.30). Adm £4, chd free. Home-made teas. Visits also by arrangement May to July for groups of 10 to 30.**
Developed almost from scratch over 14 yrs, Byecroft is a compact garden stuffed full of interesting plants, many grown from seed. Herbaceous borders, pergola with old roses, formal pond, lots of pots, vegetable garden, wildflower orchard, soft fruit area. Sue and Peter take particular pride in their plant sales table. Most areas accessible with assistance. Some small steps.

7 CASTLE MOAT HOUSE

Dilwyn, Hereford, HR4 8HZ. Mr & Mrs T Voogd. *6m W of Leominster. A44, after 4m take A4112 to Dilwyn. Garden by the village green.* **Fri 4, Sat 5 June (11-4). Adm £4, chd free. Light refreshments. and Light lunches.**
A 2 acre plot consisting of a more formal cottage garden that wraps

around the house. The remaining area is a tranquil wild garden which includes paths to a Medieval Castle Motte, part filled Moat and Medieval fish and fowl ponds. A haven for wildlife and people alike. The garden contains some steep banks and deep water, with limited access to Motte, Moat and ponds.

8 ◆ CAVES FOLLY NURSERIES
Evendine Lane, Colwall, WR13 6DX. Wil Leaper & Bridget Evans, 01684 540631, bridget@cavesfolly.com, www.cavesfolly.com. *1¼m NE of Ledbury. B4218. Between Malvern & Ledbury. Evendine Lane, off Colwall Green.* For NGS: Sun 25 Apr (2-5). Combined adm with Longacre £7, chd free. Home-made teas. Single garden adm £3.50. For other opening times and information, please phone, email or visit garden website.
Organic nursery and display gardens. Specialist growers of cottage garden plants herbs and alpines. All plants are grown in peat free organic compost. This is not a manicured garden! It is full of drifts of colour and wildflowers and a haven for wildlife.

9 CHURCH COTTAGE
Hentland, Ross-on-Wye, HR9 6LP. Sue Emms & Pete Weller, 01989 730222, sue.emms@mac.com, www.wyegardensbydesign.com. *6m from Ross-on-Wye. A49 from Ross. R turn Hentland/Kynaston. Sharp R to St Dubricius, narrow lane few passing places for ½m. Park at Church 150 metres to garden, can drop off at gate. Lane unsuitable for motor homes.* Visits by arrangement May to Aug for groups of up to 30. Adm £3.50, chd free.
Garden designer and plantswoman's ½ acre garden - feels much larger than it is! Series of garden rooms melting seamlessly into one another. Huge variety of plants, many of which are for sale, unusual varieties mixed with old favourites, providing interest over a long period. Wildlife pond, rose garden, potager, mixed borders, white terrace. Interesting plant combinations and design ideas to inspire.

✻

10 CODDINGTON VINEYARD
Coddington, HR8 1JJ. Sharon & Peter Maiden, 01531 641817, sgmaiden@yahoo.co.uk, www.coddingtonvineyard.co.uk. *4m NE of Ledbury. From Ledbury to Malvern A449, follow brown signs to Coddington Vineyard.* Fri 19 Feb (11-2); Fri 23, Sat 24 Apr (12-3); Sun 30 May (12-4). Adm £4, chd free. Tea. Wine. Visits also by arrangement Feb to Oct for groups of 10+.
5 acres incl 2 acre vineyard, listed farmhouse, threshing barn and cider mill. Garden with terraces, wildflower meadow, woodland with massed spring bulbs, large pond with wildlife, stream garden with masses of primula and hosta. Hellebores and snowdrops, hamamelis and parottia. Azaleas followed by roses and perennials. Lots to see all year. In spring, the gardens are a mass of bulbs. Wine, home-made ice cream and our own apple juice available.

11 DOVECOTE BARN
Stoke Lacy, Bromyard, HR7 4HJ. Gill Pinkerton & Adrian Yeeles. *4m S of Bromyard on A465. Turn into lane running alongside Stoke Lacy Church. Parking in 50 meters.* Sat 8, Sun 9 May, Sat 11, Sun 12 Sept (11-4). Adm £4, chd free.
Two-acre organic, wildlife-friendly, garden in the unspoilt Lodon Valley. Year-round interest: ornamental vegetable and fruit gardens, peaceful and romantic pond area, copse with spring and autumn colour, winter walk, wildflower meadow, and stunning planting on late summer prairie banks. The C17 barn, framed by cottage garden beds, has views over the garden towards the Malvern Hills. Gravel paths.

12 NEW THE FORGE
Byford, Hereford, HR4 7LD. Jane Meredith. *From Hereford, take A438 W direction Brecon for 7½m. Turn L for Byford. Proceed past Church down hill. At bottom follow road to L. After 100 yds sharp R into farmyard for parking.* Sat 12, Sun 13 June (11-5). Combined adm with Lower Farmhouse £7.50, chd free.
The Forge is a working garden. We grow as much of our own food as we have space for and are mostly self sufficient in vegetables and fruit. We also grow plants for dyeing as I run workshops teaching people how

to do this. We keep bees for honey. Over the past few years we have let areas of the garden rewild. Stunning views over the River Wye.

13 ◆ THE GARDEN OF THE WIND AT MIDDLE HUNT HOUSE
Walterstone, Hereford, HR2 0DY. Rupert & Antoinetta Otten, www.gardenofthewind.co.uk. *4m W of Pandy, 17m S of Hereford, 10m N of Abergavenny. A465 to Pandy, West towards Longtown, turn R at Clodock Church, 1m on R. Disabled parking available. Sat Nav may take you via a different route but indicates arrival at adjacent farm.* For NGS: Sat 26, Sun 27 June (11-5). Adm £5, chd free. Home-made teas. For other opening times and information, please visit garden website.
A modern garden using swathes of herbaceous plants and grasses, surrounding stone built farmhouse and barns with stunning views of the Black Mountains. Special features: rose border, hornbeam alley, formal parterre and water rill and fountains, William Pye water feature, architecturally designed greenhouse and RIBA bridge, vegetable gardens. Carved lettering and sculpture throughout, garden covering about 4 acres. Garden seating throughout the site on stone, wood and metal benches. Partial wheelchair access but WC facilities not easily accessible due to gravel.

14 NEW GARWAY HOUSE
Garway Hill, Hereford, HR2 8RT. Steve & Claire Owen. *12m SW of Hereford. A465 from Hereford, turn L at Pontrilas onto B4347, turn 1st R & immed L signed Orcop & Garway Hill, 3m to Bagwyllydiart then follow signs.* Sun 30, Mon 31 May (11-5). Adm £5, chd free. Light refreshments.
Peaceful location on Garway Hill with views across the valley to Orcop. Evolving 3 acre hillside garden with herbaceous beds, shrubs, lawns, orchard, woodland and wildflower meadow. Bulbs, rhododendrons, spring flowering shrubs and peonies bring spring and early summer colour with later season interest from perennials and grasses. Meadow full of orchids and insects from the end of May.

15 NEW GRANGE COURT
Pinsley Road, Leominster, HR6 8NL. Leominster Area Regeneration Company, www.grangecourt.org. *Grange Court in the centre of Leominster, accessible by car via Church St but not from Pinsley Rd via Etnam St. Short walk from Broad St & Etnam St car parks.* **Sun 13, Sun 27 June (10-4). Adm £5, chd free. Light refreshments.**
Set in the beautiful, tranquil Grange, Leominster's C17 market hall, Grange Court, has an ornate knot garden edged by the Saverne roses from our twin in France and the woodland garden to the front. Behind is the beautiful walled garden planted with a variety of roses, climbers and herbaceous flowers in the informal style of Marjory Fish. The Victorian Folly adds a little bit of kitsch! There will be a display of beautiful garden sculptures from the Metalsmiths, Claudia Petley and Paul Shepherd. Open for visitors to see the various displays and the extraordinary exhibition of the history of Leominster in embroidery by Leominster in Stitches. The gardens and building are fully accessible. We have accessible toilets, baby-changing facilities and a walk-in shower available.

16 GRENDON COURT
Upton Bishop, Ross-on-Wye, HR9 7QP. Mark & Kate Edwards. *3m NE of Ross-on-Wye. M50, J3 . Hereford B4224 Moody Cow PH, 1m open gate on R. From Ross. A40, B449, Xrds R Upton Bishop. 100yds on L by cream cottage.* **Sun 6 June, Sun 5 Sept (11-4). Adm £5, chd free. Also open Brockhampton Cottage. Picnics welcome in carpark field.**
A contemporary garden designed by Tom Stuart-Smith. Planted on 2 levels, a clever collection of mass-planted perennials and grasses of different heights, textures and colour give all-yr interest. The upper walled garden with a sea of flowering grasses makes a highlight. Views of the new pond and valley walk. Opening in conjuction with Brockhampton Cottage. Wheelchair access possible but some gravel.

17 HERBFARMACY
The Field, Eardisley, Hereford, HR3 6NB. Paul Richards, www.herbfarmacy.com. *11m NW of Hereford. Take A438 from Hereford, turn onto A4112 to Leominster, then A4111 to Eardisley. In Eardisley turn L off A4111 by Tram Inn & turn R 3 times. Farm is at end of No Through Road on L.* **Fri 16, Sat 17, Thur 22, Fri 23 July (10-4.30). Adm £5, chd free. Light refreshments.**
A 4 acre organic herb farm overlooking the Wye valley with views to the Black Mountains. Featured on BBC Countryfile, crops are grown for use in herbal skincare and medicinal products. Colourful plots of Echinacea, Marshmallow, Mullein (Verbascum) and Calendula will be on show as well as displays on making products. Refreshments will be available along with a shop selling Herbfarmacy products. Wheelchair Access possible when dry with assistance but some ground rough and some slopes.

18 ♦ HEREFORD CATHEDRAL GARDENS
Hereford, HR1 2NG. Dean of Hereford Cathedral, 01432 374202, Peter.Challenger@ herefordcathedral.org, www. herefordcathedral.org/garden-tours. *Centre of Hereford. Approach rds to the Cathedral are signed. Tours leave from information desk in the cathedral building or as directed.* **For NGS: Sat 19 June (10.30-3.30). Adm £5, chd £3. Light refreshments in the period College Hall by arrangement and the Cathedral Cafe. For other opening times and information, please phone, email or visit garden website.**
A guided tour of late 15th century historic award-winning gardens. Tours include colourful courtyard garden, an atmospheric cloisters garden enclosed by C15 buildings, the Vicars Choral garden, with plants with ecclesiastical connections and roses, the private Dean's Garden and the Bishop's Garden with fine trees and an outdoor chapel for meditation in ancient surroundings. Generally open every Wed & Sat May to September, guided tours 2.30. Please phone or see website for other dates and information on booked and guided tours. Partial wheelchair access. Tours can be adapted to suit individual needs. For more information

please visit www.herefordcathedral. org/accessibility.

19 HILL HOUSE FARM
Knighton, LD7 1NA. Amanda Gourlay & Gordon Franks, www.hillhousefarmgarden.com. *4m SE of Knighton. S of A4113 via Knighton (Llanshay Lane, 4m) or Bucknell (Reeves Lane, 3m).* **Sun 18 Apr, Sun 16 May, Sun 20 June, Sun 18 July, Sun 15 Aug, Sun 12 Sept (11-5). Adm £5, chd free. Self service teas, coffee and soft drinks.**
5 acre south facing garden developed over 50 years, set in magnificent hilly countryside. Some herbaceous and extensive lawns around the house. Mown paths lead through shrubs, roses and specimen trees down to the half acre Oak Pool 200ft below. There are 11 sculptures and as many sitting places scattered around this very peaceful garden. Good range of snack bars and biscuits. Suggested contributions into an honesty box. Surrounded by pastureland with distant views of the Black Mountains.

20 HILLCROFT
Coombes Moor, Presteigne, LD8 2HY. Liz O'Rourke & Michael Clarke, 01544 262795, lorconsulting@hotmail.co.uk. *North Herefordshire, 10m from Leominster. Coombes Moor is under Wapley Hill on the B4362, between Shobdon & Presteigne. Garden on R just beyond Byton Cross when heading west.* **Tue 22, Wed 23, Thur 24 June (11-5). Adm £4, chd free. Home-made teas. Visits also by arrangement June & July.**
The garden is part of a 5 acre site on the lower slopes of Wapley Hill in the beautiful Lugg Valley. The highlight in mid summer is the romantic Rose Walk, combining sixty roses with mixed herbaceous planting set in an old cider apple orchard. In addition to the garden area around the house, there is a secret garden, wildflower meadow and a vegetable and fruit area. Tea, coffee and cold drinks along with delicious home-made cakes can be enjoyed in the garden or the conservatory.

21 NEW THE HURST
Bosbury Road, Cradley, Malvern,
WR13 5LT. Clare Gogerty. *16m E
of Hereford, 11m W of Worcester.
'The Hurst' is situated directly on
the Bosbury Rd. The garden is to
the rear. From the junction of A4103
& B4220, garden is 0.7m towards
Bosbury on the L. On-road parking.*
Sun 30, Mon 31 May (1.30-5). Adm
£5, chd free.
A south-facing country garden/
smallholding of half an acre with
herbaceous perennials, grasses,
trees, a rockery and a cutting garden.
A productive veg plot and polytunnel
overlook a heritage orchard occupied
by a small flock of Herdwick sheep,
with a stream and views of the

Malvern Hills. Mostly level but gently
sloping gravel and grass track to
lower area. Orchard not suitable for
wheelchairs.

♿ ❋ ⛱

22 IVY CROFT
Ivington Green, Leominster,
HR6 0JN. Roger
Norman, 01568 720344,
ivycroft@homecall.co.uk,
www.ivycroftgarden.co.uk. *3m
SW of Leominster. From Leominster
take Ryelands Rd to Ivington. Turn
R at church, garden ¾m on R.
From A4110 signed Ivington, garden
1¾m on L.* Every Thur 28 Jan to
4 Mar (10-4). Adm £5, chd free.
Sat 1, Sun 2, Mon 3 May, Sat 4,

Sun 5 Sept (10-4). Combined adm
with Aulden Farm £8, chd free.
Home-made teas. Visits also by
arrangement.
A maturing rural garden with areas of
meadow, wood and orchard, blending
with countryside and providing habitat
for wildlife. The cottage is surrounded
by borders, raised beds, trained
pear trees and containers giving
all year interest. Paths lead to the
wider garden including herbaceous
borders, vegetable garden framed
with espalier apples and seasonal
pond with willows, ferns and grasses.
Snowdrops. Partial wheelchair
access.

♿ ❋ 🚐 ☕

Coddington Vineyard

23 KENTCHURCH COURT

Pontrilas, HR2 0DB. Mrs Jan Lucas-Scudamore, 01981 240228, jan@kentchurchcourt.co.uk, www.kentchurchcourt.co.uk. *12m SW of Hereford. From Hereford A465 towards Abergavanny, at Pontrilas turn L signed Kentchurch. After 2m fork L, after Bridge Inn. Garden opp church.* **Sun 23 May, Sun 27 June (11-5). Adm £5, chd free. Home-made teas.**

Kentchurch Court is sited close to the Welsh border. The large stately home dates to C11 and has been in the Scudamore family for over 1000yrs The deer-park surrounding the house dates back to the Knights Hospitallers of Dinmore and lies at the heart of an estate of over 5000 acres. Historical characters associated with the house incl Welsh hero Owain Glendower, whose daughter married Sir John Scudamore. The house was modernised by John Nash in 1795. First opened for NGS in 1927. Formal rose garden, traditional vegetable garden redesigned with colour, scent and easy access. Walled garden and herbaceous borders, rhododendrons and wildflower walk. Deer-park and ancient woodland. Extensive collection of mature trees and shrubs. Stream with habitat for spawning trout. Some slopes, shallow gravel.

24 LONGACRE

Evendine Lane, Colwall Green, WR13 6DT. Mr & Mrs C Hellowell. *3m S of Malvern. Off Colwall Green. Off B4218. No parking at the house - please park around Colwall Green, there is also parking at Caves Folly Nursery.* **Sun 25 Apr (2-5). Combined adm with Caves Folly Nurseries £7, chd free. Home-made teas. Single garden adm £4.50.**

2-acre garden-cum-arboretum developed since 1970. Island beds of trees and shrubs, some underplanted with bulbs and herbaceous perennials, present a sequence of contrasting pictures and views through the seasons. There are no 'rooms' - rather long vistas lead the eye and feet, while the feeling of spaciousness is enhanced by glimpses caught between trunks and through gaps in the planting. Over 50 types of conifer provide the background to maples, rhododendrons, azaleas, dogwoods, eucryphias etc. Gravel paths and lawns.

25 NEW LOWER FARMHOUSE

Byford, Hereford, HR4 7LD. Virginia & Martin Taylor. *7½m W of Hereford City. From Hereford take A438 W direction Brecon. After 7½m turn L for Byford. Go past*

Church to bottom of hill & follow road to L. After 100 yds sharp R into farmyard for parking & garden. **Sat 12, Sun 13 June (11-5). Combined adm with The Forge £7.50, chd free.**

Charming pastoral garden on banks of River Wye. Magnificent views. Informal but with hint of formality around black and white Herefordshire farmhouse. Attractive to wildlife. Fan shaped herbaceous border, Pemberton roses, beautiful Lavender hedge, white and hot borders, attractive trees, shrubs, raised beds, native hedge, wildflower area, vine-clad pergola, low stone walls, lovely unusual pots.

26 LOWER HOPE

Lower Hope Estate, Ullingswick, Hereford, HR1 3JF. Mr & Mrs Clive Richards, 01432 820557, cliverichards@crco.co.uk, www.lowerhopegardens.co.uk. *5m S of Bromyard. A465 N from Hereford, after 6m turn L at Burley Gate onto A417 towards Leominster. After approx 2m take 3rd R to Lower Hope. After ½m garden on L. Disabled parking available.* **Sun 11 Apr, Sun 16 May, Sun 4 July, Sun 12 Sept (2-5). Adm £7.50, chd £2. Visits also by arrangement Apr to Sept for groups of 20+.**

Garway House

Outstanding 5 acre garden with wonderful seasonal variations. Impeccable lawns with herbaceous borders, rose gardens, white garden, Mediterranean, Italian and Japanese gardens. Natural streams, man-made waterfalls, bog gardens. Woodland with azaleas and rhododendrons with lime avenue to lake with wildflowers and bulbs. Glasshouses with exotic plants and breeding butterflies. Prizewinning Hereford cattle and Suffolk sheep. Picnics welcome. Wheelchair access to most areas.

 ᚻ ❀ ☕ 🅿

27 LOWER HOUSE FARM
Vine Lane, Sutton, Tenbury Wells, WR15 8RL. Mrs Anne Durston Smith, 07891 928412, adskyre@outlook.com, www.kyre-equestrian.co.uk. *3m SE of Tenbury Wells; 8m NW of Bromyard. From Tenbury take A4214 to Bromyard. After approx 3m turn R into Vine Lane, then R fork to Lower House Farm.* Sun 18 Apr, Sun 9 May, Sun 13 June (10-4). Adm £5, chd free. Visits also by arrangement Apr to Sept. Picnics welcome.
Award-winning country garden surrounding C16 farm-house (not open) on working farm. Herbaceous borders, roses, box-parterre, productive kitchen and cutting garden, spring garden, ha-ha allowing wonderful views. Wildlife pond. Walkers and dogs can enjoy numerous footpaths across the farm land. Home to Kyre Equestrian Centre with access to safe rides and riding events.

 ᚻ 🐎 ☕ 🚌 🅿

28 ◆ MOORS MEADOW GARDENS
Collington, Bromyard, HR7 4LZ. Ros Bissell, 01885 410318/07942 636153, moorsmeadow@hotmail.co.uk, www.moorsmeadow.co.uk. *4m N of Bromyard, on B4214. ½ m up lane follow yellow arrows.* For NGS: Mon 29 Mar, Mon 26 Apr, Mon 24 May, Mon 28 June, Mon 26 July (10-5). Adm £7, chd £1.50. For other opening times and information, please phone, email or visit garden website.
7-acre organic hillside garden brimming with rarely seen plant species, an emphasis on working with nature to create a wildlife haven. With intriguing features and sculptures it is an inspiration to the garden novice as well as the serious plantsman. Wander through fernery, grass garden, extensive shrubberies, herbaceous beds, meadow, dingle, pools and kitchen garden. Resident Blacksmith. Huge range of unusual and rarely seen plants from around the world. Unique home-crafted sculptures.

29 MULBERRY HOUSE
Knapp Close, Goodrich, Ross-on-Wye, HR9 6JW. Tina & Adrian Barber, 01600 891372 & 07826 835300, tinaabarber@hotmail.co.uk. *Centre of Goodrich. 5m from Ross on Wye 7m from Monmouth. Close to Goodrich Castle in Wye Valley AONB. Goodrich signed from A40 or take B4234 from Ross on Wye. Park in village & follow signs to the garden.* Visits by arrangement May to Sept. Adm £5, chd free. Home-made teas. Tea and cake available.
A beautiful, peaceful and inspiring half acre garden created from scratch 14 years ago. Imaginatively planted and nurtured by a professional gardener. Winding paths lead you to themed herbaceous borders, areas of shrub planting including clipped box and ornamental grasses all providing a long season of interest with views out to beautiful listed buildings.

 ❀ ☕ 🅿

GROUP OPENING

30 NEWTON ST MARGARETS GARDENS
Newton St. Margarets, Hereford, HR2 0JU. Sue & Richard Londesborough, 01981 510148, slondesborough138@btinternet.com. *17m SW of Hereford, 9m SE of Hay-on-Wye. A465 S from Hereford, R onto B4348, L to Vowchurch & Michaelchurch Escley. After approx. 3m take either 1st or 3rd L turns. Further directions from owner. From Hay B4348, R to Vowchurch, & as above.* Visits by arrangement Apr to Sept. Directions for finding the garden from owner when booking. Combined adm £7, chd free.
Two country gardens with very different styles in the picturesque Herefordshire Golden valley. Brighton House: a plantswoman's garden of just over an acre, divided into several distinct areas crammed with many unusual and interesting plants. Old

Farm Cottage: Cottage garden with rill and bog garden. Brook side walk with wildflowers and mature oak trees. Views to the Black Mountains. Plants for sale at Brighton House.

 ❀ 🅿

31 NEW THE NUTSHELL
Goodrich, Ross-On-Wye, HR9 6HG. Louise Short. *The garden is half way between Ross on Wye & Monmouth, close to the A40 & the Cross Keys pub.* Sun 6 June (10-4). Adm £5, chd free. Light refreshments. Bacon rolls available from 9.30 until 12 noon.
The Nutshell is a cottage garden in approx half an acre, created from scratch over the last twenty years. The garden is made up of different areas separated by herbaceous borders and rose arches.There is a lovely selection of plants used incl many peonies and an extensive collection of hostas. The owner has a keen interest in propagation with two poly tunnels of plants available to purchase.

 ❀ ☕

32 OLD COLWALL HOUSE
Old Colwall, Malvern, WR13 6HF. Mr & Mrs Roland Trafford-Roberts, 07764 537181, garden1889@aol.com. *3m NE of Ledbury. From Ledbury, turn L off A449 to Malvern towards Coddington. Signed from 2½ m along lane. Signed from Colwall & Bosbury. Visitors will need to walk from car park to garden.* Sun 30 May (2-5). Adm £5, chd free. Light refreshments. Visits also by arrangement Apr to Oct for groups of up to 20.
Early C18 garden on a site owned by the Church till Henry VIII. Walled lawns and terraces on various levels. The heart is the yew walk, a rare survival from the 1700s: 100 yds long, 30ft high, cloud clipped, and with a church aisle-like quality inside. Later centuries have brought a summer house, water garden, and rock gardens. Fine trees, incl enormous veteran yew; fine views. Steep in places.

 🐎 ☕ 🅿

Online booking is available for many of our gardens, at ngs.org.uk

33 THE OLD CORN MILL

Aston Crews, Ross-on-Wye, HR9 7LW. Mrs Jill Hunter, 01989 750059, mjhunterross@outlook.com, www.theoldcornmillgarden.com. *5m E of Ross-on-Wye. A40 Ross to Gloucester. Turn L at T-lights at Lea Xrds onto B4222 signed Newent, Garden ½m on L. Parking for disabled down drive. DO NOT USE THE ABOVE POSTCODE IN YOUR SATNAV - try HR9 7LA.* **Sun 14, Sun 28 Feb, Sun 7, Sun 21 Mar, Sun 11, Sun 25 Apr, Sun 9, Sun 23 May (12-5). Adm £5, chd free. Light refreshments. Visits also by arrangement Feb to Oct. Admission price includes tea or coffee.**
Snowdrops in February, Daffodils in March, Tulips in April, Orchids in May. Do come and enjoy the calm of this very natural valley garden. Pretty good cakes too. Prize winning conversion of a ruined 18th century mill.

34 OLD GROVE

Llangrove, Ross-On-Wye, HR9 6HA. Ken & Lynette Knowles. *Between Ross & Monmouth. 2m off A40 at Whitchurch. Disabled parking at house otherwise follow signs for parking in nearby field.* **Sun 29 Aug (2-6). Adm £4, chd free.**
1½ acre garden plus two fields. SW facing with unspoilt views. Lots of mixed beds with plenty of late summer colour and many unusual plants. Large collection of dahlias and salvias; wildlife pond; informal herb garden; masses of pots; interesting trees. Seating throughout. Gentle slopes, some steps but accessible.

35 THE OLD RECTORY

Thruxton, Hereford, HR2 9AX. Mr & Mrs Andrew Hallett, 01981 570401 & 07774 129690, ar.hallett@gmail.com, www.thruxtonrectory.co.uk. *6m SW of Hereford. A465 to Allensmore. At Locks (Shell) garage take B4348 towards Hay-on-Wye. After 1½m turn L towards Abbey Dore & Cockyard. Car park 150yds on L.* **Visits by arrangement May to Sept for groups of 10+. Min adm £70 if less than 14 people. Adm £5, chd free.**
Constantly changing plantsman's garden stocked with unusual perennials and roses, woodland borders, gazebo, vegetable parterre and glasshouse. Mown paths meander through interesting collection of specimen trees and shrubs to the wildlife pond. This four acre garden, with breathtaking views over Herefordshire countryside has been created since 2007. Many places to sit. Most plants labelled. Mainly level with some gravel paths.

36 ♦ THE PICTON GARDEN

Old Court Nurseries, Walwyn Road, Colwall, WR13 6QE. Mr & Mrs Paul Picton, 01684 540416, oldcourtnurseries@btinternet.com, www.autumnasters.co.uk. *3m W of Malvern. On B4218 (Walwyn Rd) N of Colwall Stone. Turn off A449 from Ledbury or Malvern onto the B4218 for Colwall.* **For NGS: Sun 14, Sun 28 Feb, Sat 13 Mar (11-4); Sun 4, Mon 19 Apr, Sun 2, Mon 31 May, Sun 13, Sun 27 June, Sun 11, Sun 25 July, Tue 10, Mon 30 Aug, Mon 20 Sept, Wed 20 Oct (11-5). Adm £4, chd free. For other opening times and information, please phone, email or visit garden website. Donation to Plant Heritage.**
1½ acres W of Malvern Hills. Bulbs and a multitude of woodland plants in spring. Interesting perennials and shrubs in Aug. In late Sept and early Oct colourful borders display the National Plant Collection of Michaelmas daisies, backed by autumn colouring trees and shrubs. Many unusual plants to be seen, incl bamboos, more than 100 different ferns and acers. Features raised beds and silver garden. National Plant Collection of autumn-flowering asters and an extensive nursery that has been growing them since 1906. Wheelchair access gravel paths but all fairly level, no steps.

NPC

"Unfortunately, cancer was not in lockdown in 2020. The continued support of our long-standing and valued partner, the National Garden Scheme is more important than ever."
Macmillan

37 ♦ POOLE COTTAGE

Coppett Hill, Goodrich, Ross-on-Wye, HR9 6JH. Jo Ward-Ellison & Roy Smith, 07718 229813, jo@ward-ellison.com, www.herefordshiregarden.wordpress.com. *5m from Ross on Wye, 7m from Monmouth. Above Goodrich Castle in Wye Valley AONB. Goodrich signed from A40 or take B4234 from Ross.* **Visits by arrangement July to Sept for groups of up to 20. Adm £5, chd free. Home-made teas.**
A modern country garden in keeping with the local natural landscape. Home to designer Jo Ward-Ellison the 2-acre garden is predominantly naturalistic in style with a contemporary feel. A long season of interest with grasses and later flowering perennials. Some steep slopes, steps and uneven paths. Features include a small orchard, a pond loved by wildlife and kitchen garden with fabulous views.

38 ♦ RALPH COURT GARDENS

Edwyn Ralph, Bromyard, Hereford, HR7 4LU. Mr & Mrs Morgan, 01885 483225, ralphcourtgardens@aol.com, www.ralphcourtgardens.co.uk. *From Bromyard follow the Tenbury rd for approx 1m. On entering the village of Edwyn Ralph take 1st turning on R towards the church.* **For NGS: Sat 27, Sun 28 Mar, Sat 3, Sun 4 July, Sat 9, Sun 10 Oct, Sat 13, Sun 14 Nov (10-5). Adm £10, chd £7. Light refreshments. For other opening times and information, please phone, email or visit garden website.**
12 amazing gardens set in the grounds of a gothic rectory. A family orientated garden with a twist, incorporating an Italian Piazza, an African Jungle, Dragon Pool, Alice in Wonderland and the elves in their conifer forest and our new section 'The Monet Garden'. These are just a few of the themes within this stunning garden. Overlooking the Malvern Hills 120 seater Licenced Restaurant. Offering a good selection of daily specials, delicious Sunday roasts, Afternoon tea and our scrumptious homemade cakes. All areas ramped for wheelchair and pushchair access. Some grass areas, without help can be challenging during wet periods.

Sheepcote

39 REVILO

Wellington, Hereford, HR4 8AZ.
Mrs Shirley Edgar, 01432 830189,
Shirleyskinner@btinternet.com. *6m
N of Hereford. On A49 from Hereford
turn L into Wellington village, pass
church on R. Then just after barn
on L, turn L up driveway in front of
The Harbour, to furthest bungalow.*
**Sat 17, Sun 18 Apr, Sat 12, Sun
13 June (2-5). Adm £4, chd free.
Home-made teas. Visits also
by arrangement Apr to Sept for
groups of 5 to 30.**
Third of an acre garden surrounding
bungalow includes mixed borders,
meadow and woodland areas,
scented garden, late summer bed,
gravelled herb garden and vegetable/
fruit garden. Flower arranger's garden.
Featured on Hereford &Worcester
radio 2020. Wheelchair access to all
central areas of the garden from the
garage side.

& ⊶ ❋ ☕

40 RHODDS FARM

Lyonshall, HR5 3LW. Richard
& Cary Goode, 01544 340120,
cary.goode@russianaeros.com,
www.rhoddsfarm.co.uk. *1m E of
Kington. From A44 take small turning
S just E of Penrhos Farm, 1m E
of Kington. Continue 1m, garden
straight ahead.* **Sun 25 July (11-5).
Adm £5, chd free. Home-made
teas available for visitors to help
themselves.**
Created by the owner, a garden
designer, over the past 16 years, the
garden contains an extensive range of
interesting plants. Formal garden with
dovecote and 100 white doves, mixed
borders, double herbaceous borders
of hot colours, large gravel garden,
3 ponds, arboretum, perennial and
wildflower meadows and 13 acres of
woodland. A natural garden that fits
the setting with magnificent views.
Interesting and unusual trees, shrubs
and perennials. A natural garden
on a challenging site. A number of
sculptures by different artists. No
pesticides used. Not suitable for
wheelchairs.

❋ 🚌 ☕ ☕

41 SHEEPCOTE

Putley, Ledbury, HR8 2RD. Tim &
Julie Beaumont. *5m W of Ledbury
off the A438 Hereford to Ledbury
Rd. Passenger drop off; parking 200
yards.* **Sun 30 May (1-5.30). Adm
£5, chd free.**
⅓ acre garden taken in hand
from 2011 retaining many quality

plants, shrubs and trees from
earlier gardeners. Topiary holly, box,
hawthorn, privet and yew formalise
the varied plantings around the
croquet lawn and gravel garden;
beds with heathers, azaleas, lavender
surrounded by herbaceous perennials
and bulbs; small ponds; kitchen
garden with raised beds. Not suitable
for wheelchairs.

❋ ☕

42 SHUTTIFIELD COTTAGE

Birchwood, Storridge,
WR13 5HA. Mr & Mrs David
Judge, 01886 884243,
judge.shutti@btinternet.com.
*15m E of Hereford. Turn L off
A4103 at Storridg opp the Church
to Birchwood. After 1¼m L down
steep tarmac drive. Please park on
roadside at the top of the drive but
drive down if walking is difficult (150
yards).* **Visits by arrangement Apr
to Sept. Adm £5, chd free.**
Superb position and views.
Unexpected 3-acre plantsman's
garden, extensive herbaceous
borders, primula and stump bed,
many unusual trees, shrubs,
perennials, colour-themed for all-yr
interest. Anemones, bluebells,
rhododendrons, azaleas and camelias
are a particular spring feature.
Large old rose garden with many
spectacular climbers. Small deer
park, vegetable garden and 20 acre
wood. Wildlife ponds, wildflowers
and walks in 20 acres of ancient
woodland. Some sloping lawns and
steep paths and steps.

❋ ⇞

43 SOUTHBOURNE & PINE LODGE

Dinmore, Hereford, HR1 3JR.
Lavinia Sole & Frank Ryding. *8m
N of Hereford; 8m S of Leominster.
From Hereford on A49, turn R at
bottom of Dinmore Hill towards
Bodenham, gardens 1m on L. From
Leominster on A49, L onto A417,
2m turn R & through Bodenham,
following NGS signs to garden.* **Sat
15, Sun 16 May (11-5). Adm £6,
chd free.**
2 south facing linked gardens
totalling 4½ acres with panoramic
views over Bodenham Lakes to the
Black Mountains and Malvern Hills.
Southbourne: Steep access to 2
acres of terraced lawns, herbaceous
beds, shrubs, pond and ornamental
woodland. Pine Lodge: Goblin Wood
is 2½ acres featuring most of Britain's
native trees plus unusual oaks with

the emphasis on tree history and
folklore.

⊶ ⇞

44 ♦ STOCKTON BURY GARDENS

Kimbolton, HR6 0HA. Raymond
G Treasure, 07880 712649,
twstocktonbury@outlook.com,
www.stocktonbury.co.uk. *2m NE
of Leominster. From Leominster go
Ludlow on A49 turn R onto A4112.
Gardens 300yds on R.* **For NGS:
Thur 1 Apr (11-4.30). Adm £8.50,
chd £5. Light refreshments
in Tithe Barn Cafe. For other
opening times and information,
please phone, email or visit garden
website.**
Superb, sheltered 4-acre garden
with colour and interest from April
until the end of September. Extensive
collection of plants, many rare and
unusual set amongst medieval
buildings. Features pigeon house,
tithe barn, grotto, cider press, auricula
theatre, pools, secret garden, garden
museum and rill, all surrounded by
unspoilt countryside. Regional Finalist,
The English Garden's The Nation's
Favourite Gardens 2019. Garden
museum and Roman hoard. Home
made Teas. Partial wheelchair access.
Assistance may be required.

& ❋ 🚗 ☕

45 NEW THE VERN

Marden, Hereford,
HR1 3EX. Andrea Cowan-
Taylor, 07770 455759,
theverngardens@yahoo.com.
*Follow NGS signs from Marden
village through lanes to garden.* **Sat
27, Sun 28 Mar, Sat 24, Sun 25
Apr, Sat 22, Sun 23 May (11-4).
Adm £5, chd free. Visits also
by arrangement Mar to June for
groups of 10+.**
A 5 acre garden set on the banks
of the River Lugg in Herefordshire,
designed by Percy Cane for Richard
de Quincey. The garden is currently
undergoing renovation and replanting.
There is an extensive Walled Kitchen
Garden, a newly replanted Rose
Garden and a stunning collection
of rhododendron, azalea, acer and
magnolia.

& ❋ 🚗 ⇞

*All entries are subject to change.
For the latest information check
ngs.org.uk*

46 THE VINE

Tarrington, HR1 4EX. Richard & Tonya Price. *Between Hereford & Ledbury on A438. On School Lane, Tarrington village south of the A438. Park as directed. Disabled parking only at house.* **Sun 27 June (2-6). Adm £5, chd free. Cream teas.**
Mature, traditional garden in peaceful setting with stunning views of the surrounding countryside. Consisting of various rooms with mixed and herbaceous borders. Secret garden in blue/yellow/white, croquet lawn with C18 summer house, temple garden with ponds, herb and nosegay garden, vegetable/cutting/soft fruit garden around greenhouse on the paddock. Cornus avenues with obelisk and willow bower.

& ⚲ ☕ ⛱

47 NEW WAINFIELD

Peterstow, Ross-On-Wye, HR9 6LJ. Nick & Sue Helme. *From the A49 between Ross-on-Wye & Hereford. Take the B4521 to Skenfrith/Abergavenny. Wainfeild is 50yrds on R.* **Sat 3 Apr, Fri 28, Sat 29 May, Sat 26 June (10-4). Adm £5, chd free.**
3 acre informal, wildlife garden including rose garden, fruit trees, climbing roses and clematis. Delightful pond with waterfall. Climbing roses, many different clematis and honeysuckle. In spring, tulips, bluebells, crocuses and grasses followed by lush summer planting. Fruit walk with naturalised cowslips all set in an open area of interesting, unusual mature trees and sculptures. Wheelchair access on uneven grass.

&

48 WESTON HALL

Weston-under-Penyard, Ross-on-Wye, HR9 7NS. Mr P & Miss L Aldrich-Blake, 01989 562597, aldrichblake@btinternet.com. *1m E of Ross-on-Wye. on A40 towards Gloucester.* **Visits by arrangement Apr to Sept for groups of 5 to 30. Light refreshments by request at modest extra charge. Adm £5, chd free.**
6 acres surrounding Elizabethan house (not open). Large walled garden with herbaceous borders, vegetables and fruit, overlooked by Millennium folly. Lawns and mature and recently planted trees and shrubs, with many unusual varieties. Orchard, ornamental ponds and lake. 4 generations in the family, but still

Church Cottage

evolving year on year. Wheelchair access to walled garden only.

& ☕

49 WHITFIELD

Wormbridge, HR2 9BA. Mr & Mrs Edward Clive, 01981 570202, office@tamsinclive.co.uk, www.whitfield-hereford.com. *8m SW of Hereford. The entrance gates are off the A465 Hereford to Abergavenny rd, ½m N of Wormbridge.* **Sun 28 Mar, Sun 20 June (2-5). Adm £5, chd free. Home-made teas. Visits also by arrangement Apr to Sept for groups of 10 to 30.**
Parkland, wildflowers, ponds, walled garden, many flowering magnolias (species and hybrids), 1780 ginkgo tree, 1½ m woodland walk with 1851 grove of coastal redwood trees. Dogs on leads welcome. Delicious teas. Partial access to wheelchair users, some gravel paths and steep slopes.

& ⚲ ✳ 🚗 ☕

50 WOODVIEW

Great Doward, Whitchurch, Ross-on-Wye, HR9 6DZ. Janet & Clive Townsend, 01600 890477, clive.townsend5@homecall.co.uk. *6m SW of Ross-on-Wye, 4m NE of Monmouth. A40 Ross/Mon At Whitchurch follow signs to Symonds Yat west, then to Doward Park campsite. Take forestry rd 1st L garden 2nd L - follow NGS signs. (Don't rely on satnav).* **Sun 18 Apr, Sun 11, Sun 18 July (11-6). Adm £5, chd free. Light refreshments. Tea/coffee & cake. Visits also by arrangement Apr to Dec.**
Formal and informal gardens approx 4 acres in woodland setting. Herbaceous borders, hosta collection, mature trees, shrubs and seasonal bedding. Gently sloping lawns. Statuary and found sculpture, local limestone, rockwork and pools. Woodland garden, wildflower meadow and indigenous orchids. Collection of vintage tools and memorabilia. Croquet, clock golf and garden games.

& ⚲ ✳ ☕

HERTFORDSHIRE

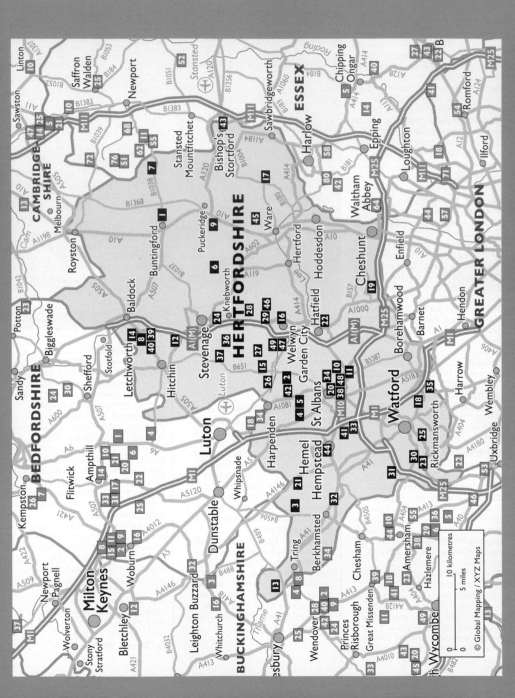

With its proximity to London, Hertfordshire became a breath of country air and a retreat for wealthy families wishing to escape the grime of the city – hence the county is peppered with large and small country estates, some of which open their garden gates for the National Garden Scheme.

Here in Hertfordshire we are immensely proud of our horticultural heritage. Our history of market gardening, which supplied vegetables, fruit, and cut flowers to London markets, dates back to the 1600s. We are home to historic houses whose gardens, parklands, and landscapes have associations with such celebrated names as Tradescant, Repton, Jekyll, and Lutyens. And our county is the birthplace of Ebenezer Howard's Garden City Movement.

Across sprawling urban areas and tiny rural hamlets, we have gardens that offer everything from contemporary garden design to rewilded spaces, from wildflower meadows to productive potagers, from ancient woodlands to tropical oases.

Many are open 'by arrangement' for private visits by groups of all sizes and we are happy to provide assistance in arranging group tours of multiple gardens.

After the challenges of 2020, we are all looking forward, more than ever, to opening our gardens in 2021 and hoping for some return to normality. Certainly, our garden owners are full of enthusiasm as they have shown in the numbers who are opening. However, we have to prepare for possible disruption and restrictions which might be imposed at short notice. If this happens our website will always carry the most up to date information about all gardens.

Below: Scudamore

f @HertfordshireNGS 𝕏 @HertfordshirNGS

Volunteers

County Organisers
Bella Stuart-Smith 01923 268414
bella.stuart-smith@ngs.org.uk

Kate Stuart-Smith 07551 923217
kate.stuart-smith@ngs.org.uk

County Treasurer
Peter Barrett
peter.barrett@ngs.org.uk

Publicity
Kerrie Lloyd-Dawson 07736 442883
kerrie.lloyddawson@ngs.org.uk

Photography
Barbara Goult 07712 131414
barbara.goult@ngs.org.uk

Social Media - Facebook
Anastasia Rezanova
anastasia.rezanova@ngs.org.uk

Social Media - Twitter
Charly Denton-Woods
charlotte.denton-woods@ngs.org.uk

Booklet Coordinator
Janie Nicholas 07973 802929
janie.nicholas@ngs.org.uk

New Gardens
Julie Wise 01438 821509
julie.wise@ngs.org.uk

Group Tours
Sarah Marsh 07813 083126
sarah.marsh@ngs.org.uk

Assistant County Organisers
Tessa Birch 07721682481
tessa.birch@ngs.org.uk

Kate de Boinville 07973 558838
kate.deboinville@ngs.org.uk

Christopher Melluish 01920 462500
c.melluish@btopenworld.com

Karen Smith 07850 406403
hertsgardeningangel@gmail.com

Lucy Swift 07808 737965
lucy.swift@ngs.org.uk

OPENING DATES

All entries subject to change. For latest information check www.ngs.org.uk

Extended openings are shown at the beginning of the month.

Map locator numbers are shown to the right of each garden name.

February

Snowdrop Festival

Friday 19th
1 Elia Cottage 17

Sunday 21st
1 Elia Cottage 17

March

Saturday 20th
◆ Hatfield House West Garden 22

April

Monday 5th
10 Cross Street 14

Sunday 11th
Alswick Hall 1
◆ St Paul's Walden Bury 37

Saturday 17th
The Mill House 29

Sunday 18th
The Mill House 29

Sunday 25th
Amwell Cottage 2
NEW The Old Rectory 31
Pie Corner 33
Serendi 40

May

Sunday 2nd
Patchwork 32

Sunday 9th
Brockholds Manor 9

◆ St Paul's Walden Bury 37

Sunday 16th
NEW 39 Firs Wood Close 19
The Manor House, Ayot St Lawrence 27

Friday 28th
Mackerye End House 26

Sunday 30th
NEW Beesonend House 5
15 Gade Valley Cottages 21

Monday 31st
43 Mardley Hill 28

June

Friday 4th
NEW 77 Warren Way 46

Saturday 5th
Brent Pelham Hall 7

Sunday 6th
Brent Pelham Hall 7
The Cherry Tree 12
9 Tannsfield Drive 44
Thundridge Hill House 45
NEW 77 Warren Way 46

Tuesday 8th
◆ Ashridge House 3

Wednesday 9th
◆ Ashridge House 3

Thursday 10th
◆ Ashridge House 3

Friday 11th
◆ Ashridge House 3

Saturday 12th
124 Highfield Way 23
Waterend House 47
The White Cottage 49

Sunday 13th
124 Highfield Way 23
Mackerye End House 26
◆ St Paul's Walden Bury 37

Sunday 20th
NEW 28 Dale Avenue 15

Friday 25th
28 Fishpool Street 20
Rustling End Cottage 36

Saturday 26th
42 Falconer Road 18
Rustling End Cottage 36

Sunday 27th
◆ Benington Lordship 6
Brockholds Manor 9
42 Falconer Road 18
28 Fishpool Street 20
St Stephens Avenue Gardens 38

July

Thursday 1st
NEW 71 Stansted Road 43

Friday 2nd
NEW 71 Stansted Road 43

Saturday 3rd
42 Falconer Road 18
NEW 71 Stansted Road 43

Sunday 4th
42 Falconer Road 18
Scudamore 39
Serge Hill Gardens 41
NEW 71 Stansted Road 43
9 Tannsfield Drive 44

Friday 9th
NEW 14 Watling Street 48

Saturday 10th
NEW 14 Watling Street 48

Sunday 11th
Beesonend Gardens 4

Friday 16th
102 Cambridge Road 10

Sunday 18th
102 Cambridge Road 10
15 Gade Valley Cottages 21
Morning Light 30

Sunday 25th
35 Digswell Road 16
NEW 12 Longmans Close 25
Morning Light 30

Friday 30th
44 Broadwater Avenue 8

August

Sunday 1st
44 Broadwater Avenue 8
NEW 12 Longmans Close 25

NEW Southdown Gardens 42

Friday 6th
Railway Cottage 34

Sunday 8th
Railway Cottage 34

Sunday 15th
Serendi 40

Sunday 22nd
Patchwork 32
Reveley Lodge 35

September

Daily from Saturday 4th to Sunday 12th
◆ The Celebration Garden 11

Sunday 5th
St Stephens Avenue Gardens 38

Sunday 12th
102 Cambridge Road 10
10 Cross Street 14

October

Sunday 17th
35 Digswell Road 16

November

Sunday 7th
42 Falconer Road 18

There are brilliant plant sales at many gardens. Look out for the symbol in the garden description – and don't forget to bring a bag to carry your plants home in

THE GARDENS

1 ALSWICK HALL

Hare Street Road, Buntingford, SG9 0AA. Mike & Annie Johnson, www.alswickhall.com/gardens. *1m from Buntingford on B1038. From the S take A10 to Buntingford, drive into town & take B1038 E towards Hare Street Village. Alswick Hall is 1m on R.* Sun 11 Apr (12-4.30). Adm £6, chd free. Pig roast.

Listed Tudor House with 5 acres of landscaped gardens set in unspoiled farmland. Two well established natural ponds with rockeries. Herbaceous borders, shrubs, woodland walk and wildflower meadow with a fantastic selection of daffodils, tulips, camassias and crown imperial. Spring blossom, formal beds, orchard and glasshouses. Licensed Bar, Hog Roast, Teas, delicious home-made cakes, plant stall and various other trade stands, children's entertainment. Good access for disabled with lawns and wood chip paths. Slight undulations.

2 AMWELL COTTAGE

Amwell Lane, Wheathampstead, AL4 8EA. Colin & Kate Birss. *½ m S of Wheathampstead. From St Helen's Church, Wheathampstead turn up Brewhouse Hill. At top L fork (Amwell Lane), 300yds down lane, park in field opp.* Sun 25 Apr (2-5). Adm £4.50, chd free. Home-made teas.

Informal garden of approx 2½ acres around C17 cottage. Large orchard of mature apples, plums and pear laid out with paths. Extensive lawns with borders, framed by tall yew hedges and old brick walls. A large variety of roses, stone seats with views, woodland pond, greenhouse, vegetable garden with raised beds and fire-pit area. Gravel drive.

3 ◆ ASHRIDGE HOUSE

Berkhamsted, HP4 1NS. Ashridge (Bonar Law Memorial) Trust, 01442 843491, events@ashridge.hult.edu, www.ashridgehouse.org.uk. *3m N of Berkhamsted. A4251, 1m S of Little Gaddesden.* For NGS: Tue 8, Wed 9, Thur 10, Fri 11 June (10-4). Adm £5, 16-18 yrs £2.50, chd free. Home-made teas. For other opening times and information, please phone, email or visit garden website.

The gardens cover 190 acres forming part of the Grade II Registered Landscape of Ashridge Park. Based on designs by Humphry Repton in 1813 modified by Jeffry Wyatville. Small secluded gardens, as well as a large lawn area leading to avenues of trees. 2013 marked the 200th anniversary of Repton presenting Ashridge with the Red Book, detailing his designs for the estate.

GROUP OPENING

4 BEESONEND GARDENS

Harpenden, AL5 2AN. *1m S of Harpenden on A1081, after 1m turn R into Beesonend Lane. Bear R into Burywick to T junction follow signs.* Sun 11 July (2-5.30). Combined adm £5, chd free. Home-made teas.

2 BARLINGS ROAD
Liz & Jim Machin.

7 BARLINGS ROAD
Chris Berendt.

Two gardens each reflecting their owner's individual interests and needs. 2 Barlings Road is packed with perennials, shrubs and climbers and a formal pond to provide yr-round structure and privacy. A courtyard suntrap and shade garden provide extra interest. 7 Barlings Road is a professionally landscaped garden

designed for minimal maintenance and maximum impact with mature trees, climbers, water feature, children's play area, pagoda and plant-filled borders. Wheelchair access to Barlings Road, level gardens accessed via side paths.

5 NEW BEESONEND HOUSE

Beeson Lane, Harpenden, AL5 2AD. John & Sarah Worth. *Turn off A1081(turn R from Harpenden & L from St Albans). Keep L up Beesonend Lane, past cottages & derelict stables on the L. Beesonend House is on L next to white stones.* Sun 30 May (11-5). Adm £4, chd free. Light refreshments.

Four garden areas. East garden laid lawn with herbaceous borders, pleached hedges, Japanese-themed border sitting under Eucalyptus tree. South garden laid lawn with herbaceous borders with semi-mature Lebanese cedar. South-West facing garden includes pond and orchard. North garden comprises circular lawn, pots, roses and herbaceous border. Vegetable garden with raised borders and a greenhouse.

The Old Rectory

6 ◆ BENINGTON LORDSHIP

Stevenage, SG2 7BS. Mr & Mrs R Bott, 01438 869668, garden@beningtonlordship.co.uk, www.beningtonlordship.co.uk. *4m E of Stevenage. In Benington Village, next to church. Signs off A602.* **For NGS: Sun 27 June (11-5). Adm £6, chd free. For other opening times and information, please phone, email or visit garden website.**
7 acre garden includes historic buildings, rose garden, walled kitchen garden, lakes, snowdrops, spring bulbs and herbaceous borders. Unspoilt panoramic views over surrounding parkland. As garden is on a steep slope there is only partial wheelchair access. Accessible WC available in parish hall.

7 BRENT PELHAM HALL

Brent Pelham, Buntingford, SG9 0HF. Alex & Mike Carrell. *From Buntingford take the B1038 E for 5m. From Clavering take the B1038 W for 3m.* **Sat 5, Sun 6 June (2-5). Adm £6, chd free. Home-made teas in the Estate Office.**
Surrounding a beautiful Grade 1 listed property, the gardens consist of 12 acres of formal gardens, redesigned in 2007 by the renowned landscaper Kim Wilkie. With two walled gardens, a potager, walled kitchen garden, greenhouses, orchard and a new double herbaceous border, there is lots to discover. The further 14 acres of parkland boast lakes and wildflower meadows. All gardened organically. Access by wheelchair to most areas of the garden, including paths of paving, gravel and grass.

8 44 BROADWATER AVENUE

Letchworth Garden City, SG6 3HJ. Karen & Ian Smith. *½ m SW Letchworth town centre. A1(M) J9 signed Letchworth. Straight on at 1st three r'abouts, 4th r'about take 4th exit then R into Broadwater Ave.* **Evening opening Fri 30 July (6-9). Wine. Sun 1 Aug (1-5). Home-made teas. Adm £5, chd free.**
Town garden in the Letchworth Garden City conservation area that successfully combines a family garden with a plantswoman's garden. Out of the ordinary, unusual herbaceous plants and shrubs. Constantly evolving to include lots of colour and texture. Topiary underpins the whole garden. Attractive front garden designed for year- round interest.

9 BROCKHOLDS MANOR

Old Hall Green, Ware, SG11 1HE. Richard & Juliet Penn Clark, 07931520152, events@pennclark.com, www.brockholdsmanor.com. *8m N of Ware 5 mins from A10. Once on Stockalls Lane, drive for approx. ½ m and look out for a tree-lined driveway/ track on R. Drive to the very bottom, past pond and Barn conversions.* **Sun 9 May, Sun 27 June (1-5). Adm £5, chd free. Pre-booking essential, please visit www.ngs.org.uk for information & booking. Home-made teas. Visits also by arrangement Feb to Oct for groups of 5 to 30.**
A new garden of 4 acres, surrounding a C15 farmhouse. The site of Brockholds dates from the C13. Small woodland walk in spring, an orchard planted 2 years ago with spring bulbs, remains of the original moat, New garden rooms with Yew and Beech hedges, 50 varieties of Old and English Roses, 20 varieties of peonies all planted since 2017. New Potager and Cuttings garden. Some wheelchair access.

10 102 CAMBRIDGE ROAD

St Albans, AL1 5LG. Anastasia & Keith, arezanova@gmail.com. *Nr Ashley Road end of Cambridge Road in The Camp neighbourhood on east side of city. S of the A1057 (Hatfield Rd). Take the A1057 from the A1(M) J3. Take the A1081 from M25 J22.* **Evening opening Fri 16 July (5.30-8.30). Wine. Sun 18 July, Sun 12 Sept (2-6). Home-made teas. Adm £4, chd free. Friday: Wines and authentic home-made samosas. Sunday: Selection of teas, barista/coffee-shop coffee, home-made cakes. Visits also by arrangement May to Oct for groups of up to 20. Donation to Alzheimer's Research UK.**
Contemporary space sympathetically redesigned in 2017 to keep as much of the existing plants, trees and shrubs in a 1930s semi's garden. Modern take on the classic garden in two halves: ornamental and vegetable. All-year interest gabion borders packed with perennials and annuals, central bed featuring a pond and a mature Japanese maple. All vegetables, annuals, and some perennials, grown from seed. 'Count the Frog' activity for children and young-at-heart. The garden features steps - there is no wheelchair access.

11 ◆ THE CELEBRATION GARDEN

North Orbital Road, St Albans, AL2 1DH. Aylett Nurseries Ltd, 01727 822255, info@aylettnurseries.co.uk, www.aylettnurseries.co.uk. *The Celebration Garden is adjacent to the Garden Centre. Aylett Nurseries is situated on the eastbound carriageway of the A414 S of St Albans, between the Park Street r'about & London Colney r'about. Drive through green gates at end of car park.* **For NGS: Daily Sat 4 Sept to Sun 12 Sept (10-5). Adm by donation. Light refreshments. For other opening times and information, please phone, email or visit garden website.**
The Celebration Garden is sited next to our famous Dahlia Field. Dahlias are planted amongst other herbaceous plants and shrubs. We also have a wildflower border complete with insect hotel. The garden is open all year to visit, during the garden centre opening hours, but it is especially spectacular from July to early autumn when the Dahlias are in flower. Annual Autumn Festival held in September. Refreshments available in the Dahlia Coffee House open all year round. Grass paths.

12 THE CHERRY TREE

Stevenage Road, Little Wymondley, Hitchin, SG4 7HY. Patrick Woollard & Jane Woollard, 07952655613, cherrywy@btinternet.com. *½ m W of J8 off A1M. Follow sign to Little Wymondley; under railway bridge & house is R at central island flower bed opp Bucks Head Pub. Parking in adjacent rds.* **Sun 6 June (1-5). Adm £4.50, chd free. Light refreshments. Visits also by arrangement June to Aug for groups of 5 to 10.**
The Cherry Tree is a small, secluded garden on several levels containing shrubs, trees and climbers, many of them perfumed. Much of the planting, including exotics, is in containers that are cycled in various positions throughout the seasons. A heated greenhouse and summerhouse maintain tender plants in winter. The garden has been designed to be a journey of discovery as you ascend. Not suitable for wheelchairs. Several steps throughout garden.

14 10 CROSS STREET

Letchworth Garden City, SG6 4UD. Renata & Colin Hume, www.cyclamengardens.com. *Nr town centre. From A1(M) J9 signed Letchworth, across 2 r'abouts, R at 3rd, across next 3 r'abouts L into Nevells Rd, 1st R into Cross St.* **Mon 5 Apr, Sun 12 Sept (2-5). Adm £4.50, chd free. Home-made teas.** A garden with mature fruit trees is planted for interest throughout the year. The structure of the garden evolved around three circles - two grass lawns and a wildlife pond. Borders connect the different levels of the garden with a large pond near the house. Not suitable for wheelchairs.

15 28 DALE AVENUE

Wheathampstead, St Albans, AL4 8LS. Judy Shardlow, www.heartwoodgardendesign.co.uk. *Up Lamer Lane (B651) through Lower Gustard Wood, past Mid Herts Golf Course. Turn R onto The Slype, continuing then L onto The Broadway & 1st R into Dale Avenue.* **Sun 20 June (2-5.30). Adm £5, chd £2. Light refreshments.** A large country garden designed by Heartwood Garden Design, with sweeping central lawn and deep mixed borders with evergreen and perennial plants and grasses. It includes a large island bed with a stepping stone path to a lawned path beneath a large Silver Birch. The border includes herbs, salvias, irises, agapanthus, lavender, eryngiums and many varieties of Harkness and David Austin roses.

16 35 DIGSWELL ROAD

Welwyn Garden City, AL8 7PB. Adrian & Clare de Baat, 01707 324074, adrian.debaat@ntlworld.com, www.adriansgarden.org. *½m N of Welwyn Garden City centre. From the Campus r'about in city centre take N exit just past the Public Library into Digswell Rd. Over the White Bridge, 200yds on L.* **Sun 25 July (2-5.30); Sun 17 Oct (1-4). Adm £4, chd free. Home-made teas. Visits also by arrangement July to Oct for groups of up to 20. Adm includes tea or coffee and cake.** Town garden of around a third of an acre with naturalistic planting inspired by the Dutch garden designer, Piet

Oudolf. The garden has perennial borders plus a small meadow packed with herbaceous plants and grasses. The contemporary planting gives way to the exotic, incl a succulent bed and under mature trees, a lush jungle garden incl bamboos, bananas, palms and tree ferns. Daisy Roots Nursery will be selling plants. Grass paths and gentle slopes to all areas of the garden.

17 1 ELIA COTTAGE

Nether Street, Widford, Ware, SG12 8TH. Margaret & Hugh O'Reilly, 01279 843324, hughoreilly56@yahoo.co.uk. *B1004 from Ware, Wareside to Widford past Green Man pub into dip at Xrd take R Nether St. 8m W of Bishop's Stortford on B1004 through Much Hadham at Widford sign turn L. B180 from Stanstead Abbots.* **Fri 19, Sun 21 Feb (12.30-4.30). Adm £4, chd free. Light refreshments. Visits also by arrangement Feb to Sept.** A third of an acre garden reflecting the seasons. Snowdrops are our first welcome visitors combined with crocus tommasinianus and hellebores. There is a stream with Monet-style bridge, pond and cascade plus water features. Plenty of seats and two summerhouses. Coffee/tea and cake available. Garden quiz. Steep nature of garden means we are sorry no wheelchair access.

18 42 FALCONER ROAD

Bushey, Watford, WD23 3AD. Mrs Suzette Fuller, 077142 94170. *M1 J5 follow signs for Bushey. From London A40 via Stanmore towards Watford. From Watford via Bushey Arches, through to Bushey High St, turn L into Falconer Rd, opp St James church.* **Sat 26, Sun 27 June, Sat 3, Sun 4 July (12-6). Home-made teas. Evening opening Sun 7 Nov (6-8). Adm £3.50, chd free. Mulled Wine on 7 Nov. Visits also by arrangement May to Sept for groups of up to 20.** Enchanting magical unusual Victorian style space. Bird cages and chimney pots feature, plus a walk through conservatory with orchids. Children are very welcome. Winter viewing for fairyland lighting, for all ages, bring a torch.

19 39 FIRS WOOD CLOSE

Potters Bar, EN6 4BY. Val & Peter Mackie. *Potters Bar High Street (A1000) fork R towards Cuffley, along the Causeway. Immed before the T-lights/Chequers pub, R down Coopers Lane Road. ½m on L Firs Wood Close.* **Sun 16 May (1-5). Adm £5, chd free. Home-made teas.** Situated within Northaw Park, a 3 acre garden made up of a woodland area, a field including a wildflower meadow and a more formal half acre around the house and lawn flanked by an eclectic mix of planted borders. Choice of seating areas include a breeze house, sunken fire pit area and colourful patio benches under the pergola covered with wisteria. Tranquil views up to Northaw village. Flat access from road to back garden and then on grass.

20 28 FISHPOOL STREET

St Albans, AL3 4RT. Jenny & Antony Jay. *28 Fishpool St. is at the Cathedral end near The Lower Red Lion pub. Free parking on Sunday in the Boys School car park otherwise town parking.* **Evening opening Fri 25 June (6-8). Wine. Sun 27 June (2-5). Home-made teas. Adm £5, chd free.** Sculpted box and yew hedging and a C17 Tripe House feature strongly in this tranquil oasis set in the vicinity of St Albans Cathedral. Gravel paths lead to a lawn surrounded by late flowering sustainable herbaceous perennial borders and a relaxed woodland retreat. Imaginative planting in all areas offer unique perspectives. Plants are on sale. Not suitable for wheelchairs due to differing levels.

The National Garden Scheme was set up in 1927 to raise funds for district nurses. We remain strongly committed to nursing and today our beneficiaries include Macmillan Cancer Support, Marie Curie, Hospice UK and The Queen's Nursing Institute

44 Broadwater Avenue

21 15 GADE VALLEY COTTAGES

Dagnall Road, Great Gaddesden, Hemel Hempstead, HP1 3BW. Bryan Trueman. *3m N of Hemel Hempstead. Follow B440 N from Hemel Hempstead. Past Water End. Go past turning for Great Gaddesden. Gade Valley Cottages on R. Park in village hall car park.* Sun 30 May, Sun 18 July (1.30-5). Adm £4, chd free. Home-made teas.

Medium sized sloping rural garden. Patio, lawn, borders and pond. Paths lead through a woodland area emerging by wildlife pond and sunny border. A choice of seating offers views across the beautiful Gade valley or quiet shady contemplation with sounds of rustling bamboos and bubbling water. Many Acers, hostas and ferns in shady areas. Hemerocallis, dahlias, iris, crocosmia and phlox found in sun.

22 ◆ HATFIELD HOUSE WEST GARDEN

Hatfield, AL9 5HX. The Marquess of Salisbury, 01707 287010, r.ravera@hatfield-house.co.uk, www.hatfield-house.co.uk. *Pedestrian Entrance to Hatfield House is opp Hatfield Railway Stn, from here you can obtain directions to the gardens. Free parking is available, please use AL9 5HX with a sat nav.* For NGS: Sat 20 Mar (11-3). Adm £11, chd £7. For other opening times and information, please phone, email or visit garden website. Donation to another charity.

Visitors can enjoy the spring bulbs in the lime walk, sundial garden and view the famous Old Palace garden, childhood home of Queen Elizabeth I. The adjoining woodland garden is at its best in spring with masses of naturalised daffodils and bluebells. Beautifully designed gifts, jewellery, toys and much more can be found in the Stable Yard shops. Visitors can also enjoy relaxing at the River Cottage Restaurant which serves a variety of delicious foods throughout the day. There is a good route for wheelchairs around the West garden and a plan can be picked up at the garden kiosk.

23 124 HIGHFIELD WAY

Rickmansworth, WD3 7PH. Mrs Barbara Grant. *1m E of J18 M25, 1m NW of Rickmansworth. Easily accessible from J17 or J18 M25.* Sat 12, Sun 13 June (2-5). Adm £4.50, chd free. Home-made teas.

This gently sloping garden with a plethora of plants and trees provides an ever-changing burst of colour and interest throughout the year. Mixed shrub and perennial borders, as well as 'cutting garden', woodland area, bog garden, fruit patch, greenhouse and an enormous pond with many fish including koi. The Open Day may also include sale of work of local artists. Plant sales, home-made cakes, tea and musical entertainment. Wide paths - although they are gently sloping in places.

24 ◆ KNEBWORTH HOUSE GARDENS

Knebworth, SG1 2AX. The Hon Henry Lytton Cobbold, 01438 812661, info@knebworthhouse.com, www.knebworthhouse.com. *nr Stevenage. Direct access from A1(M) J7 at Stevenage.* For opening times and information, please phone, email or visit garden website.

The present layout of Knebworth House's delightful Formal Gardens dates largely from the Edwardian era. Sir Edwin Lutyens' garden rooms and pollarded lime walks, Gertrude Jekyll's herb garden, the restored maze, yew hedges, roses and herbaceous borders are key features of the formal gardens with peaceful woodland walks beyond. The Garden Terrace Tea Room is available for visitors to the Park and Gardens, offering a selection of snacks and locally produced hot and cold lunches. The Gift Shop stocks a wide range of affordable gifts and souvenirs, seasonal plants and concert memorabilia, much of which is unique to Knebworth.

25 NEW 12 LONGMANS CLOSE

Byewaters, Watford, WD18 8WP. Mark Lammin, 07966 625559, markjango@msn.com, www.instagram.com/hertstinytropicalgarden. *Byewaters, Croxley Green. Leave M25 at J18 (A404) & follow signs for Rickmansworth/Croxley Green/Watford then follow A412 towards Watford & follow signs for Watford & Croxley Business Parks then follow*

the NGS signs. Sun 25 July, Sun 1 Aug (2-6). Adm £4, chd free. Home-made teas. Visits also by arrangement July to Sept for groups of up to 10.

Hertfordshire's Tiny Tropical Garden. See how dazzling colour, scent, lush tropical foliage, trickling water and clever use of pots in a densely planted small garden can transport you to the tropics. Stately bananas and canna rub shoulders with delicate lily and roses amongst a large variety of begonia, hibiscus, ferns and houseplants in a tropical theme more often associated with warmer climes.

26 MACKERYE END HOUSE

Mackerye End, Harpenden, AL5 5DR. Mr & Mrs G Penn. *3m E of Harpenden. A1 J4 follow signs Wheathampstead, turn R Marshalls Heath Lane. M1 J10 follow Lower Luton Road B653. Turn L Marshalls Heath Lane. Follow signs.* Evening opening Fri 28 May (6-9). Wine. Sun 13 June (12-5). Home-made teas. Adm £6, chd free. Friday evening wine with canapés.

C16 (Grade 1 listed) Manor House (not open) set in 15 acres of formal gardens, parkland and woodland, front garden set in framework of formal yew hedges. Victorian walled garden with extensive box hedging and box maze, cutting garden, kitchen garden and lily pond. Courtyard garden with extensive yew and box borders. West garden enclosed by pergola walk of old English roses. Walled garden access by gravel paths.

27 THE MANOR HOUSE, AYOT ST LAWRENCE

Welwyn, AL6 9BP. Rob & Sara Lucas. *4m W of Welwyn. 20 mins J4 A1M. Take B653 Wheathampstead. Turn into Codicote Rd follow signs to Shaws Corner. Parking in field, short walk to garden. A disabled drop-off point is available at the end of the drive.* Sun 16 May (11-5). Adm £5, chd free. Home-made teas.

A 6-acre garden set in mature landscape around Elizabethan Manor House (not open). 1-acre walled garden incl glasshouses, fruit and vegetables, double herbaceous borders, rose and herb beds. Herbaceous perennial island beds, topiary specimens. Parterre and temple pond garden surround the house. Gates and water features by

Arc Angel. Garden designed by Julie Toll. Home-made cakes and tea/coffee and produce & plants for sale.

✿ D ☕

28 43 MARDLEY HILL
Welwyn, AL6 0TT. Kerrie & Pete, www.agardenlessordinary.blogspot.co.uk. *5m N of Welwyn Garden City. On B197 between Welwyn & Woolmer Green, on crest of Mardley Hill by bus stop for Arriva 300/301.* Mon 31 May (1-5). Adm £4, chd free. Home-made teas. An unexpected garden created by plantaholics and packed with unusual plants. Focus on foliage and long season of interest. Constantly being developed and new plants sourced. Various areas: alpine bed; sunny border; deep shade; white-stemmed birches and woodland planting; naturalistic stream, pond and bog; chicken house and potted vegetables; potted exotics. Seating areas on different levels. Featured in Garden News and Garden Answers.

✿ ☕

29 THE MILL HOUSE
31 Mill Lane, Welwyn, AL6 9EU. Sarah & Ian. *Old Welwyn. J6 A1M approx ¾m to garden, follow yellow arrows to Welwyn Village.* Sat 17, Sun 18 Apr (2-5.30). Adm £5, chd free. Home-made teas. An abundant colourful display of tulips and spring flowers sets off this Listed house with semi-walled garden bordered by a bridged millstream and race. Ancient apple and box trees under planted with a display of perennials and herbaceous plants set off a garden full of promise, within which nestles a box parterre, vegetable planters magnolia tree and colourful terrace. Some gravel and uneven paths.

♿ ✿ ☕

30 MORNING LIGHT
7 Armitage Close, Loudwater, Rickmansworth, WD3 4HL. Roger & Patt Trigg, 01923 774293, roger@triggmail.org.uk. *From M25 J18 take A404 towards Rickmansworth, after ¾m turn L into Loudwater Lane, follow bends, then turn R at T-junction & R again into Armitage Close.* Sun 18, Sun 25 July (2-5.30). Adm £4, chd free. Pre-booking essential, please phone 01923 774293 or email roger@triggmail.org.uk for information & booking.

Home-made teas. Visits also by arrangement Apr to Sept for groups of up to 30. Compact, south-facing plantsman's garden, densely planted with mainly hardy and tender perennials and shrubs in a shady environment. Features include island beds, pond, chipped cedar paths and raised deck. Tall perennials can be viewed advantageously from the deck. Large conservatory (450 sq ft) stocked with sub-tropicals.

✿ 🚗 ☕

31 NEW THE OLD RECTORY
Church Lane, Sarratt, WD3 6HJ. Will & Kate Hobhouse. *From Sarratt Green the entrance & driveway are ½m on the R along Church Lane. From Chorleywood towards Sarratt, the entrance is 150 yards on L from the Church.* Sun 25 Apr (11-5.30). Adm £6, chd free. Home-made teas. A wild garden of 20 acres with specimen trees, a sculpture collection and semi ancient woodland with a stunning view over the Chess valley and the Chilterns. From the car park it is possible to follow paths in the grass but very little hard standing.

♿ 🐕 ✿ D ☕

32 PATCHWORK
22 Hall Park Gate, Berkhamsted, HP4 2NJ. Jean & Peter Block, 01442 864731, patchwork2@btinternet.com. *3m W of Hemel Hempstead. Entering E side of Berkhamsted on A4251, turn L 200yds after 40mph sign.* Sun 2 May, Sun 22 Aug (2-5). Adm £4, chd free. Light refreshments. Visits also by arrangement Apr to Oct for groups of 5 to 30. ¼-acre garden with lots of year-round colour, interest and perfume, particularly on opening days. Sloping site containing rockeries, 2 small ponds, herbaceous border, island beds with bulbs in Spring and dahlias in Summer, roses, fuchsias, hostas, begonias, patio pots and tubs galore - all set against a background of trees and shrubs of varying colours. Seating and cover from the elements. Traditional, labour intensive garden, many levels and views. Not suitable for wheelchairs, as side entrance is narrow, and there are many steps and levels.

🐕 ✿ ☕

33 PIE CORNER
Millhouse Lane, Bedmond, WD5 0SG. Bella & Jeremy Stuart-Smith, piebella1@gmail.com. *Between Watford & Hemel Hempstead. 1½m from J21 M25. 3m from J8 of M1. Go to the centre of Bedmond. Millhouse Lane is opp the shops. Entrance is 50m down Millhouse Lane.* Sun 25 Apr (2-5). Adm £5, chd free. Home-made teas. Visits also by arrangement Apr to Sept for groups of 10+. A garden designed to complement the modern classical house. Formal areas near the house, with views down the valley, include lawns and a formal pool. The garden becomes more informal towards the woodland edge. A dry garden leads through new meadow planting to the vegetable garden. Enjoy blossom, bulbs, wild garlic, bluebells and young rhododendron planting in late spring. Wheelchair access to all areas on grass or gravel paths except the formal pond where there are steps.

♿ ✿ ☕

34 RAILWAY COTTAGE
16 Sandpit Lane, St Albans, AL1 4HW. Siobhan & Barry Brindley. *½m N of St Albans city centre. J21a M25, follow St Albans A5183. Through town centre on A1081. Please use on-street parking in Battlefield, Lancaster, & Gurney Court Roads.* Evening opening Fri 6 Aug (5.30-8.30). Wine. Sun 8 Aug (2-5). Home-made teas. Adm £4.50, chd free. Sandwiched between a main road and railway line, this urban cottage garden is a hidden gem. Little paths are bordered by plants and flowers of all descriptions creating a mass of colour and a haven for insects and birds. There is seating throughout the garden with an eclectic collection of reclaimed items old and new. Raised bed vegetable plot with greenhouse, summerhouses and water features.

✿ ☕

The National Garden Scheme searches the length and breadth of England and Wales for the very best private gardens

35 REVELEY LODGE

88 Elstree Road, Bushey Heath, WD23 4GL. Reveley Lodge Trust, www.reveleylodge.org. *3½m E of Watford & 1½m E of Bushey Village. From A41 take A411 signed Bushey & Harrow. At mini-r'about 2nd exit into Elstree Rd. Garden ½m on L. Disabled parking only onsite.* Sun 22 Aug (2-6). Adm £5, chd free. Light refreshments.

2½-acre garden surrounding a Victorian house bequeathed to Bushey Museum in 2003. Planting includes a rose garden, medicinal bed, shrubs, perennials, and a woodland walk. A restored vegetable garden, a beautiful and productive space for vegetables and cut flowers. Plus, a conservatory, greenhouse, beehives and an Analemmatic (human) sundial constructed in stone believed unique to Hertfordshire.

Live music by Guitar n Brass: guitar/vocals and trombone / vocals performing a mix of happy jazz and smooth and mellow jazz. Partial wheelchair access.

♿ 🐖 ✺ ⛾

36 RUSTLING END COTTAGE

Rustling End, Codicote, SG4 8TD. Julie & Tim Wise, www.rustlingend.com. *1m N of Codicote. From B656 turn L into '3 Houses Lane' then R to Rustling End. House 2nd on L.* Evening opening Fri 25, Sat 26 June (5-8.30). Adm £5, chd free. Wine. Meander through our wildflower meadow to a cottage garden with contemporary planting. Behind lumpy hedges explore a garden. We've relaxed! A newly planted orchard with natural planting provides an environment for wild birds and small mammals. Our terrace features drought tolerant low maintenance simplistic planting. Our lawn around the pond is now a flowery mead and an abundant floral vegetable garden provides produce for the summer. Hens in residence.

✺ ⛾

37 ◆ ST PAUL'S WALDEN BURY

Whitwell, Hitchin, SG4 8BP. The Bowes Lyon family, stpaulswalden@gmail.com, , www.stpaulswaldenbury.co.uk. *5m S of Hitchin. On B651; ½m N of Whitwell village. From London leave A1(M) J6 for Welwyn (not Welwyn Garden City). Pick up signs to Codicote, then Whitwell.* For NGS: Sun 11 Apr, Sun 9 May, Sun 13 June (2-7). Adm £7.50, chd free. Home-made teas. Cakes, cream

71 Stansted Road

scones, sandwiches. **For other opening times and information, please email or visit garden website. Donation to St Paul's Walden Charity.**
Spectacular formal woodland garden, Grade 1 listed, laid out 1720, covering over 50 acres. Long rides lined with clipped beech hedges lead to temples, statues, lake and a terraced theatre. Seasonal displays of daffodils, cowslips, irises, magnolias, rhododendrons, lilies. wildflowers are encouraged. This was the childhood home of the late Queen Mother. Children welcome. 13 June, Open Garden combined with Open Farm Sunday with free tours of the farm. Wheelchair access to part of the garden. Steep grass slopes in places.

GROUP OPENING

38 ST STEPHENS AVENUE GARDENS
St Albans, AL3 4AD. *1m S of St Albans City Centre. From A414 take A5183 Watling St. At mini-r'about by St Stephens Church/King Harry Pub take B4030 Watford Rd. St Stephens Ave is 1st R.* Sun 27 June, Sun 5 Sept (2.30-5.30). Combined adm £5, chd free. Home-made teas. Gluten free and vegan cake available. WC.

20 ST STEPHENS AVENUE
Heather & Peter Osborne.

30 ST STEPHENS AVENUE
Carol & Roger Harlow.

Two gardens of similar size and the same aspect, developed in totally different ways. Dense planting at number 20 makes it impossible to see from one end to the other, enticing visitors to explore, and is designed to blur the boundaries of a long and narrow plot (53 x 12m). Paths meander through and behind the colour coordinated borders, giving access to all parts of the garden. Specimen trees, and fences clothed with climbers contribute to the peaceful seclusion. Varied plant habitats include cool shade, hot dry gravel and lush pondside displays. Seating is in both sun and shade, a large conservatory provides shelter. Number 30 has a southwest facing gravelled front garden that has a Mediterranean feel. Herbaceous plants, such as sea hollies and achilleas, thrive in the poor, dry soil.

Clipped box, beech and hornbeam in the back garden provide a cool backdrop for the strong colours of the herbaceous planting. A gate beneath a beech arch frames the view to the park beyond. Plants for sale at June opening only. Compost making demonstrations.

39 SCUDAMORE
1 Baldock Road, Letchworth Garden City, SG6 3LB. Michael & Sheryl Hann. *Opp Spring Rd, between Muddy Lane & Letchworth Lane. J9 A1M. Follow directions to Letchworth. Turn L to Hitchin A505. After 1m House on L after Muddy Lane. Parking in Muddy Lane & Spring Rd.* Sun 4 July (11-5). Adm £5, chd free. Home-made teas.
½ acre garden surrounding early C17 cottages that were converted and extended in 1920s to form current house. Garden of mature trees, many mixed herbaceous borders with shrubs, pond and stream, wet bed, wilder garden and orchard/vegetable area. Many sculptures add interest to this garden of both unusual and traditional planting.

40 SERENDI
22 Hitchin Road, Letchworth Garden City, SG6 3LT. Valerie, 01462 635386, valerie.aitken@ntlworld.com. *1m from city centre. A1(M) J9 signed Letchworth on A505. At 2nd r'about take 1st exit Hitchin A505. Straight over T-lights. Garden 1m on R.* Sun 25 Apr, Sun 15 Aug (11-5). Adm £5, chd free. Home-made teas.
Visits also by arrangement Apr to Sept for groups of 5 to 20.
A well designed plants woman's garden. A mass of tulips & other Spring bulbs, a silver birch grove, a 'dribble of stones', a rill, contemporary knot garden, and dry planted area with alliums. Later in the year an abundance of roses climbing 5 pillars, perennials, grasses and dahlias. A greenhouse for over wintering and a Griffin glass house with Brugmansia, Plumbago and Tibochina & other plants. Regional Finalist, The English Garden's The Nation's Favourite Gardens 2019. Gravel entrance driveway and paths, can be difficult. Plenty of lawns.

GROUP OPENING

41 SERGE HILL GARDENS
Serge Hill Lane, Bedmond, WD5 0RT. *½m E of Bedmond. Go to Bedmond & take Serge Hill Lane, where you will be directed past the lodge & down the drive.* Sun 4 July (2-5). Combined adm £8, chd free. Home-made teas at Serge Hill.

THE BARN
Sue & Tom Stuart-Smith.

SERGE HILL
Kate Stuart-Smith.

Two very diverse gardens. At its entrance the Barn has an enclosed courtyard, with tanks of water, herbaceous perennials and shrubs tolerant of generally dry conditions. To the N there are views over the 5-acre wildflower meadow, and the West Garden is a series of different gardens overflowing with bulbs, herbaceous perennials and shrubs. The house at Serge Hill results principally from the work of Charles Augustin Busby (1786-1817), an architect and engineer perhaps most renowned for a development west of Brighton christened Brunswick Town and for his 1808 publication: A series of designs for villas and country houses. In 1811 Busby exhibited his designs for Serge Hill House at the Royal Academy. It has wonderful views over the ha-ha to the park; a walled vegetable garden with a large greenhouse, roses, shrubs and perennials leading to a long mixed border. At the front of the house there is an outside stage used for family plays, and a ship.

GROUP OPENING

42 NEW **SOUTHDOWN GARDENS**
Harpenden, AL5 1EL. *Exit M1 J9 & turn R onto A5183 heading towards Redbourn; turn L at r'about towards Harpenden on B487; pass the White Horse Pub on the L; take 2nd exit at r'about onto Walkers Road.* **Sun 1 Aug (1-5). Combined adm £5, chd free.**

4 COLESWOOD ROAD
Mrs Linzi Claridge.

NEW **5 COLESWOOD ROAD**
Marilyn Couldridge.

Two gardens on opposite sides of Coleswood Road. Neither have a lawn; instead an abundance of mixed planting to make the most of each garden space. There is structure, height, loose planting, pots, water features. These gardens will show you what can be achieved within just a few years. No.4 is mainly gravel paths.

43 NEW **71 STANSTED ROAD**
Bishop's Stortford, CM23 2DT. Jill & Nigel Kerby. *E of town centre. 2½m W from J8 M11 - take A120 to 2nd r'about, then L onto Stansted Rd. Or 9m E from A10 - take A120 to 2nd r'about, R onto Stansted Rd.* **Thur 1, Fri 2, Sat 3, Sun 4 July (11-5). Adm £10, chd free. Pre-booking essential, please phone 07931255812 or email jill. kerby@icloud.com for information & booking. Home-made teas. Ticket includes guided tour and refreshments.**
An intriguing wildlife garden, set in ⅔ acre in the heart of town. Designed to encourage wildlife, with a wildflower meadow, three ponds and an extensive woodland border. Formal areas feature a romantic garden with many herbaceous plants chosen to encourage insects. Open to a limited number of visitors by pre-booking only.
❀ ☕

44 **9 TANNSFIELD DRIVE**
Hemel Hempstead, HP2 5LG. Peter & Gaynor Barrett, 01442 393508, peteslittlepatch@virginmedia. com, www.peteslittlepatch.co.uk. *Approx 1m NE of Hemel Hempstead*

town centre & 2m W of J8 on M1. *From M1 J8, cross r'about to A414 to Hemel Hempstead. Under ftbridge, cross r'about then 1st R across dual c'way to Leverstock Green Rd then on to High St Green. L into Ellingham Rd then follow signs.* **Sun 6 June, Sun 4 July (1.30-4.30). Adm £3.50, chd free. Home-made teas.**
Visits also by arrangement June to Aug for groups of 5 to 10. Limited refreshments available if requested when booking.
This small, town garden is decorated with over 450 plants which, together with the ever-present sound of water, create a welcoming oasis of calm for visitors to savour. Narrow paths divide, leading visitors on a voyage of discovery of the garden's many features. The owners regularly experiment with the garden planting scheme which ensures the look of the garden alters from year to year. Simple water features, metal sculptures, wall art and mirrors can be seen throughout the garden. As a time and cost saving experiment all hanging baskets are planted with hardy perennials most of which are normally used for ground cover.
❀ ☕

45 **THUNDRIDGE HILL HOUSE**
Cold Christmas Lane, Ware, SG12 0UE. Christopher & Susie Melluish, 01920 462500, c.melluish@btopenworld.com. *2m NE of Ware. ¾m from Maltons off the A10 down Cold Christmas Lane, crossing bypass.* **Sun 6 June (1-5.30). Adm £5. Cream teas. Visits also by arrangement Apr to Sept.**
Well-established garden of approx 2½ acres; good variety of plants, shrubs and roses, attractive hedges. Visitors often ask for the unusual yellow-only bed. Several delightful places to sit. Wonderful views in and out of the garden especially down to the Rib Valley to Youngsbury, visited briefly by Lancelot 'Capability' Brown. 'A most popular garden to visit'. Fine views down to Thundridge Old Church.
♿ ☕ ⛬

46 NEW **77 WARREN WAY**
Digswell, Welwyn, AL6 0DL. Caroline Goodchild. *Close to Welwyn North Station, Digswell. 2m N of Welwyn Garden City. 1m from J6 A1 (M). Approaching from the B1000 take exit at mini r'about into Station Rd. Turn L into Warren Way.*

Continue to top LH corner of Warren Way. **Evening opening Fri 4 June (6-8). Wine. Sun 6 June (2-5). Home-made teas. Adm £3.50, chd free.**
Family garden with access to surrounding woodland. Children welcome. Medium-sized multi-purpose family garden with an eclectic mix of planting. A variety of bulbs, perennial borders and mixed shrubs provide all year round interest and colour. Foxgloves, alliums and hellebores throughout the many beds.
♿ ❀ ☕

47 **WATEREND HOUSE**
Waterend Lane, Wheathampstead, St Albans, AL4 8EP. Mr & Mrs J Nall-Cain, 07736 880810, sj@nallcain.com. *2m E of Wheathampstead. Approx 10 mins from J4 of A1M. Take B653 to Wheathampstead, past Crooked Chimney Pub, after ½m turn R into Waterend Lane. Cross river, house is immed. on R.* **Sat 12 June (2-6). Adm £5, chd free. Home-made teas. Also open The White Cottage. Visits also by arrangement Feb to Oct for groups of 5 to 20.**
A hidden garden of 4 acres sets off an elegant Jacobean Manor House (not open). Steep grass slopes and fine views of glorious countryside. Formal flint-walled garden. Roses, peonies and irises in the summer. Formal beds and lots of colour throughout the year. Mature specimen trees, ponds, formal vegetable garden, bantams and Indian runner ducks. Hilly garden. Wheelchair access to lower gardens only.
♿ ☕

48 NEW **14 WATLING STREET**
St Albans, AL1 2PX. Phil & Becky Leach. *1m S of St Albans. Take A5183 up St Stephen's Hill & turn L at King Harry Pub. From A414 take A5183. No. 14 is off the main road, in cul-de-sac diagonally opposite St Stephen's Church (some parking in church car park).* **Evening opening Fri 9 July (6-9). Wine. Sat 10 July (2.30-5.30). Home-made teas. Adm £3.50, chd free.**
A vibrant garden replete with perennials grown from seed and striking foliage plants including Chinese rice-paper plants, foxglove trees, chusan palms, cardoons, echium and honey flowers. Gravel garden out front leads to a back garden divided into separate beds

(including bog garden, with giant rhubarb, ligularia and hosta) and which ends in a glade of bamboo shaded by maple trees. Some gravel paths and raised patio area.

& ✿ ☕

49 THE WHITE COTTAGE
Waterend Lane, Wheathampstead, St Albans, AL4 8EP. Sally Trendell, 01582 834617, sallytrendell@me.com. *2m E of Wheathampstead. Approx 10 mins from J5 A1M Take B653 to Wheathampstead. Soon after Crooked Chimney pub turn R into* Waterend Lane, garden 300yds on L. Parking in field opp. **Sat 12 June (2-6). Adm £5, chd free. Home-made teas. Also open Waterend House. Visits also by arrangement June to Aug for groups of 10 to 20.** Idyllic riverside retreat of over an acre in rural setting adjacent to a ford over the River Lea which widens to form the garden boundary. A wildlife haven cottage garden straight out of 'The Wind in the Willows'.

& 🛏 ☕ ⛱

In our first year 609 private gardens opened their gates to all, for the modest sum of one shilling. Today the National Garden Scheme retains that combination of inclusivity and affordability

Pie Corner

ISLE OF WIGHT

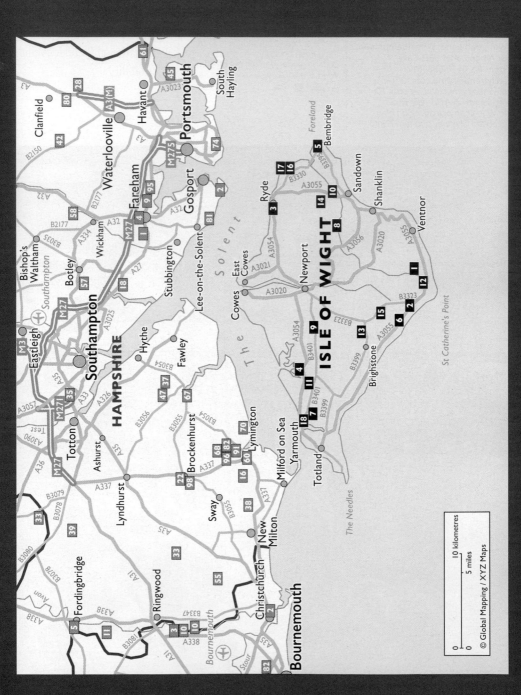

The island is a very special place to those who live and work here and to those who visit and keep returning. We have a range of natural features, from a dramatic coastline of cliffs and tiny coves to long sandy beaches.

Inland, the grasslands and rolling chalk downlands contrast with the shady forests and woodlands. Amongst all of this beauty nestle the picturesque villages and hamlets, many with gardens open for the National Garden Scheme. Most of our towns are on the coast and many of the gardens have wonderful sea views.

The one thing that makes our gardens so special is our climate. The moderating influence of the sea keeps hard frosts at bay, and the range of plants that can be grown is therefore greatly extended.

Conservatory plants are planted outdoors and flourish. Pictures taken of many island gardens fool people into thinking that they are holiday snaps of the Mediterranean and the Canaries.

Our gardens are very varied and our small enthusiastic group of garden owners are proud of their gardens, whether they are small town gardens or large manor gardens, and they love to share them for the National Garden Scheme.

Volunteers

County Organiser
Jennie Fradgley
01983 730805
Jennie.Fradgley@ngs.org.uk

County Treasurer
Position Vacant
For details please contact
Jennie Fradgley (as above)

Publicity
Position Vacant
For details please contact
Jennie Fradgley (as above)

Booklet Co-ordinator
Jennie Fradgley
(as above)

Assistant County Organisers
Mike Eastwood
01983 721060
mike@aristia.co.uk

Sally Parker
01983 612495
sallyparkeriow@btinternet.com

Below: Ningwood Manor

© Heather Edwards

OPENING DATES

All entries subject to change. For latest information check **www.ngs.org.uk**

Map locator numbers are shown to the right of each garden name.

May

Sunday 16th
Morton Manor 10

Sunday 23rd
Northcourt Manor Gardens 13

Sunday 30th
 Goldings 7
Meadowsweet 9
Thorley Manor 18

June

Saturday 5th
Knighton Farmhouse 8

Sunday 6th
Knighton Farmhouse 8
Red Cross Cottage 16
Salterns Cottage 17

Saturday 12th
Niton Gardens 12

Sunday 13th
Niton Gardens 12
The Old Rectory 15

Sunday 20th
◆ Nunwell House 14

Saturday 26th
Ashknowle House 1

Sunday 27th
Ashknowle House 1

July

Saturday 3rd
Blenheim House 3

Sunday 4th
Blenheim House 3

Sunday 11th
The Beeches 2

Saturday 17th
NEW Dove Cottage 5

Sunday 18th
NEW Dove Cottage 5

Saturday 31st
NEW East Dene 6

August

Sunday 1st
NEW East Dene 6

September

Sunday 5th
Morton Manor 10

By Arrangement

Arrange a personalised garden visit on a date to suit you. See individual garden entries for full details.

Blenheim House 3
Crab Cottage 4
Morton Manor 10
Ningwood Manor 11
Northcourt Manor Gardens 13
The Old Rectory 15

THE GARDENS

◻ ASHKNOWLE HOUSE

Ashknowle Lane, Whitwell, Ventnor, PO38 2PP. Mr & Mrs K Fradgley. *4m W of Ventnor. Take the Whitwell Rd from Ventnor or Godshill. Turn into unmade lane next to Old Rectory. Field parking. Disabled parking at house.* **Sat 26, Sun 27 June (12-4). Adm £4, chd free. Home-made teas.**
The mature garden of this Victorian house (not open) has great diversity including woodland walks, water features, colourful beds and borders. The large, well maintained kitchen garden is highly productive and boasts a wide range of fruit and vegetables grown in cages, tunnels, glasshouses and raised beds. Diversely planted and highly productive orchard incl protected cropping of strawberries, peaches and apricots.

◻ THE BEECHES

Chale Street, Chale, Ventnor, PO38 2HE. Mr Andrew Davidson, 01983 551876, andrewdavidson06@btinternet.com. *Turn off A3055 at Chale onto B3399. Entrance between Old Rectory & bus stop, in gap in stone wall.* **Sun 11 July (10-5). Adm £3, chd free. Home-made teas.**
The garden is laid out mainly to shrubs and border plants designed for colour and texture. A haven of peace with extensive 270 degree views of the countryside incl the south west coast of the island down to The Needles and Dorset beyond. Features incl a wildflower meadow and a deep pond home to fish and wildlife (children to be supervised because of deep water). Due to gravel driveway and some steps, wheelchair access is not easy, however we can accommodate with prior arrangement.

◻ BLENHEIM HOUSE

9 Spencer Road (use Market St entrance), Ryde, PO33 2NY. David Rosewarne & Magie Gray, 01983 614675, david65rosewarne@gmail.com. *Market St entrance behind Ryde Town Hall/Theatre off Lind St.* **Sat 3, Sun 4 July (11-4). Adm £4, chd free. Light refreshments. Visits also by arrangement May to Sept for groups of 5 to 30.**
A garden developed over 12 yrs, exploring the decorative qualities and long term effects of pattern making, colour and texture. This terraced 116ft x 30ft sloping site is centred on a twisting red brick path that both reveals and hides interesting and contrasting areas of planting, creating intimate and secluded spaces that belie its town centre location.

◻ CRAB COTTAGE

Mill Road, Shalfleet, PO30 4NE. Mr & Mrs Peter Scott, 07768 065756, mencia@btinternet.com. *4½ m E of Yarmouth. At New Inn, Shalfleet, turn into Mill Rd. Continue 400yds, drive through open NT gates. After 100yds pass Crab Cottage on L, turn L into gate. Limited parking in gravel drive.* **Visits by arrangement May to Sept. Adm £5, chd free. Home-made teas.**
1/4 acres on gravelly soil. Part glorious views across croquet lawn over Newtown Creek and Solent, leading through wildflower meadow to hidden water lily pond and woodland walk. Part walled garden protected from westerlies with mixed borders, leading to terraced sunken garden with ornamental pool and pavilion, planted with exotics, tender shrubs and herbaceous perennials. Croquet, plant sales and excellent tea, with cake on request. Wheelchair access over gravel and uneven grass paths.

5 NEW **DOVE COTTAGE**
Swains Lane, Bembridge,
PO35 5ST. Mr James & Mrs Alex
Hearn. *Head E on Lane End Rd,
passing Lane End Court Shops
on L. Take 3rd turning on L onto
Swains Lane. Dove Cottage is on
R.* **Sat 17, Sun 18 July (1.30-
5.30). Adm £3.50, chd free. Light
refreshments.**
An enclosed garden with lawn and
woodland area. A central path leads
through the woodland setting which
has mature variagated shrubs and
perennials, leading to the swimming
pool area and tennis court.

6 NEW **EAST DENE**
Atherfield Green, Ventnor,
PO38 2LF. Marc & Lisa Morgan-
Huws. *From A3055 Military Rd, turn
into Southdown. At end of road turn
L on to Atherfield Rd & the garden
is ahead on the R.* **Sat 31 July, Sun
1 Aug (11-5). Adm £4, chd free.
Light refreshments.**
Planted from 2 acres of farmland in
the 1970s, the garden is structured
around a variety of mature trees.
Taken over by brambles and nettles
prior to our arrival 5 yrs ago, the
garden is slowly being rediscovered
and restored, whilst still being a
haven for wildlife. Occupying a
windswept coastal location with
areas of woodland, fruit trees, mature

pond and informal planting. Alpacas.
Although the garden has level access,
the ground is unmade.

7 NEW **GOLDINGS**
Thorley Street, Thorley, Yarmouth,
PO41 0SN. John & Dee Sichel. *E
of Yarmouth. Follow directions for
Thorley from Yarmouth/Newport
road or from Wilmingham Lane. Then
follow NGS signs. Parking shared
with Thorley Manor.* **Sun 30 May
(2-5). Combined adm with Thorley
Manor £4, chd free. Home-made
teas at Thorley Manor.**
A country garden with many focuses
of interest. A newly planted orchard
already producing cider apples
in large amounts. A small, but
productive vegetable garden and
a well maintained lawn with shrub
borders and roses. A micro-climate
has been created by the adjustment
of levels to create a series of terraced
areas for planting.

8 **KNIGHTON FARMHOUSE**
Knighton Shute, Newchurch,
Sandown, PO36 0NT. Mr David
& Mrs Ali Cripps. *Approx 1m N of
Newchurch. Newchurch is accessed
from the A3056 Newport to
Sandown road, or Knighton from the
Downs road via Knighton Shute.* **Sat
5, Sun 6 June (2-5). Adm £4, chd
free. Home-made teas.**
This family garden has evolved over
the past 10 yrs and is continuing to
be developed. The C17 farmhouse
(not open) was surrounded only by
grass when we came here, and now
there are mixed beds and borders, a
productive organic vegetable garden,
cutting garden and fruit cage. There
are two small courtyard areas and
a new walled courtyard garden was
completed in 2020.

9 **MEADOWSWEET**
5 Great Park Cottages, off
Betty-Haunt Lane, Carisbrooke,
PO30 4HR. Gunda Cross. *4m SW
of Newport. From A3054 Newport/
Yarmouth road, turn L at Xrd
Porchfield/Calbourne, over bridge
into 1st lane on R. Parking along one
side, on grass verge & past house.*
**Sun 30 May (11.30-4.30). Adm £4,
chd free. Home-made teas.**
From windswept, barren 2 acre
cattle field to developing tranquil
country garden. Natural, mainly native
planting and wildflowers. Cottagey
front garden, herb garden, orchard,
fruit cage and large pond. The good
life and a haven for wildlife! Creative
planting, wildlife pond and woodland.
Very popular and unusual plants for
sale. Wheelchair access over flat level
garden with grass paths.

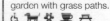

Salterns Cottage

© Heather Edwards

East Dene

ID MORTON MANOR

Morton Manor Road, Brading, Sandown, PO36 0EP. Mr & Mrs G Godliman, 07768 605900, patricia.godliman@yahoo.co.uk. *Off A3055 5m S of Ryde, just out of Brading. At Yarbridge T-lights turn into The Mall. Take next L into Morton Manor Rd.* **Sun 16 May, Sun 5 Sept (11-4). Adm £4, chd free. Home-made teas. Visits also by arrangement Apr to Oct.**

A colourful garden of great plant variety. Mature trees incl many acers with a wide variety of leaf colour. Early in the season a display of rhododendrons, azaleas and camellias and later hydrangeas and hibiscus. Ponds, sweeping lawns, roses set on a sunny terrace and much more to see in this extensive garden surrounding a picturesque C16 manor house (not open). Wheelchair access over gravel driveway.

II NINGWOOD MANOR

Station Road, Ningwood, nr Newport, PO30 4NJ. Nicholas & Claire Oulton, 07738 737482, claireoulton@gmail.com. *Nr Shalfleet. From Newport, turn L opp the Horse & Groom Pub. Ningwood Manor is 300-400yds on the L. Please use 2nd set of gates.* **Visits by arrangement May to Sept for groups of up to 30. Light refreshments.**

A 3 acre, landscape designed country garden divided into several rooms; a walled courtyard, croquet lawn, white garden and kitchen garden. They flow into each other, each with their own gentle colour schemes, the exception to this is the croquet lawn garden which is a riot of colour, mixing oranges, reds, yellows and pinks. Much new planting has taken place over the last few yrs. The owners have several new projects underway, so the garden is a work in progress. Features incl a vegetable garden with raised beds and a small summerhouse, part of which is alleged to be Georgian. Please inform us at time of booking if wheelchair access is required.

Online booking is available for many of our gardens, at ngs.org.uk

GROUP OPENING

12 NITON GARDENS
Niton, PO38 2AZ. *5m W of Ventnor. Parking at football ground (Blackgang Rd), in village & Allotment Rd car park. Tickets & maps from the library in the heart of the village on Niton Gardens open days only.* **Sat 12, Sun 13 June (11.30-4.30). Combined adm £5, chd free. Home-made teas.**

NEW 12 GREENLYDD CLOSE
Gaye Rolfe.

PUCKASTER CORNER
Mr & Mrs Ian McCallum.

SPRING COTTAGE
Mr & Mrs Neil White.

TALSA
Frances Pritchard.

TILLINGTON VILLA
Paul & Catherine Miller.

WINFRITH
Mrs Janet Tedman.

Niton is a delightful village with a busy community spirit, blessed with lovely churches, two pubs (one of which was renowned for smuggling), PO and shops, lovely walks and bridleways, school, football and cricket pitches and recreation ground. The gardens of this walk are very varied in both style and size and are full of colour, fragrance and interest; from cottage and country gardens to vegetable plot and havens for wildlife. The gardens are situated both in the heart of the village and the undercliff. We do hope you will enjoy them all.

13 NORTHCOURT MANOR GARDENS
Main Road, Shorwell, Newport, PO30 3JG. Mr & Mrs J Harrison, 01983 740415, john@northcourt.info, www.northcourt.info. *4m SW of Newport. On entering Shorwell from Newport, entrance at bottom of hill on R. If entering from other directions head through village in direction of Newport. Garden on the L, on bend after passing the church.* **Sun 23 May (12-5). Adm £5, chd free. Home-made teas. Visits also by arrangement Mar to Oct for groups of 5+.**
15 acre garden surrounding large C17 manor house (not open). Boardwalk along jungle garden. Stream and bog garden. A large variety of plants enjoying the different microclimates. Large collection of camellias and magnolias. Woodland walks. Tree collection. Roses and hardy geraniums in profusion. Picturesque wooded valley around the house. Bathhouse and snail mount leading to terraces. 1 acre walled garden. The house celebrated its 405th yr anniversary. A plantsman' garden. Wheelchair access on main paths only, some paths are uneven and the terraces are hilly.

14 ◆ NUNWELL HOUSE
Coach Lane, Brading, PO36 0JQ. Mr & Mrs S Bonsey, 01983 407240, info@nunwellhouse.co.uk, www.nunwellhouse.co.uk. *3m S of Ryde. Signed off A3055 as you arrive at Brading from Ryde & turn into Coach Lane.* **For NGS: Sun 20 June (1-4.30). Adm £5, chd free. Home-made teas. For other opening times and information, please phone, email or visit garden website.**
6 acres of tranquil and beautifully set formal and shrub gardens and old fashioned shrub roses prominent. Exceptional Solent views over historic parkland and Brading Haven from the terraces. Small arboretum and walled garden with herbaceous borders. House developed over 5 centuries and full of architectural interest.

15 THE OLD RECTORY
Kingston Road, Kingston, PO38 2JZ. Derek & Louise Ness, 01983 551701, louiseness@gmail.com, www.theoldrectorykingston.co.uk. *8m S of Newport. Entering Shorwell from Carisbrooke, take L turn at mini-r'about towards Chale (B3399). Follow road, house 2nd on L, after Kingston sign. Park in adjacent field.* **Sun 13 June (2-5). Adm £4.50, chd free. Home-made teas. Visits also by arrangement in June for groups of up to 30.**
Constantly evolving romantic country garden surrounding the late Georgian Rectory (not open). Areas of interest incl the walled kitchen garden, orchard, formal and wildlife ponds, a wonderfully scented collection of old and English roses and two perennial wildflower meadows. Partial wheelchair access, some gravel and grass paths.

16 RED CROSS COTTAGE
Salterns Road, Seaview, PO34 5AG. Katie Barnfather & Stephen Jones. *Enter Seaview from Springvale. Take signs for the Duver.* **Sun 6 June (11-5). Combined adm with Salterns Cottage £5, chd free.**
A great variety of plantings for colour and form. Much perennial colour enriched by seasonal plantings. Light refreshments at the Northbank Hotel and the Seaview Deli.

17 SALTERNS COTTAGE
Salterns Road, Isle of Wight, Seaview, PO34 5AH. Susan & Noël Dobbs. *Enter Seaview from W via Springvale, Salterns Rd links the Duver Rd with Bluett Ave.* **Sun 6 June (11-5). Combined adm with Red Cross Cottage £5, chd free.**
A glasshouse, a potager, exotic borders and fruit trees are some of the many attractions in this 40 metre x 10 metre plot. Salterns Cottage is a listed building built in 1640 and was bought in 1927 by Noel's grandmother Florence, married to Bram Stoker the author of Dracula. The garden was created by Susan in 2005 when she sold her school. Flooding and sandy soil poses a constant challenge to the planting. The greenhouse and potager all raised to cope with floods. Light refreshments at the Northbank Hotel and from the Seaview Deli. You can also picnic on the beach close by.

18 THORLEY MANOR
Thorley, Yarmouth, PO41 0SJ. Mr & Mrs Anthony Blest. *1m E of Yarmouth. From Bouldnor take Wilmingham Lane, house ½m on L.* **Sun 30 May (2-5). Combined adm with Goldings £4, chd free. Home-made teas.**
Mature informal gardens of over 3 acres surrounding manor house (not open). Garden set out in a number of walled rooms, perennial and colourful self-seeding borders, shrub borders, lawns, large old trees and an unusual island lawn, all seamlessly blending in to the surrounding farmland.

All entries are subject to change. For the latest information check ngs.org.uk

Volunteers

County Organiser
Jane Streatfeild 01342 850362
janestreatfeild@btinternet.com

County Treasurer
Andrew McClintock 01732 838605
andrew.mcclintock@ngs.org.uk

Publicity
Camilla Bidwell 07854 335557
camilla.bidwell@ngs.org.uk

Booklet Advertising
Nicola Denoon-Duncan
01233 758600
nicola.denoonduncan@ngs.org.uk

Booklet Co-ordinator
Ingrid Morgan Hitchcock
01892 528341
ingrid@morganhitchcock.co.uk

Booklet Distribution
Diana Morrish 01892 723905
diana.morrish@ngs.org.uk

Group Tours
Sue Robinson 01622 729568
sue.robinson@ngs.org.uk

Assistant County Organisers
Jacqueline Anthony 01892 518879
jacquelineanthony7@gmail.com

Marylyn Bacon 01797 270300
ngsbacon@ramsdenfarm.co.uk

Clare Barham 01580 241386
clarebarham@holepark.com

Mary Bruce 01795 531124
mary.bruce@ngs.org.uk

Liz Coulson 01233 813551
coulson.el@gmail.com

Kate Dymant 07766 201906
katedymant@hotmail.co.uk

Andy Garland
andy.garland@bbc.co.uk

Bridget Langstaff 01634 842721
bridget.langstaff@btinternet.com

Virginia Latham 01303 862881
lathamvj@gmail.com

Diana Morrish (as above)

Sue Robinson (as above)

Nicola Talbot 01342 850526
nicola@falconhurst.co.uk

@KentNGS

@NGSKent

@nationalgardenschemekent

Famously known as 'The Garden of England', Kent is a county full of natural beauty, special landscapes and historical interest.

Being England's oldest county, Kent unsurprisingly boasts an impressive collection of castles and historic sites, notably the spectacular Canterbury Cathedral, and the medieval Ightham Mote.

Twenty eight per cent of the county forms two Areas of Outstanding Natural Beauty: the Kent Downs and the High Weald. The landscapes of Kent are varied and breathtaking, and include haunting marshes, rolling downs, ancient woodlands and iconic white cliffs.

The gardens of Kent are well worth a visit too, ranging from the landscaped grounds of historic stately homes and castles, to romantic cottage gardens and interesting back gardens.

Never has a county been so close to London and yet feels so far away, so why not escape to the peace of a Kent garden? The variety of the gardens and the warmth of the garden owners will ensure a memorable and enjoyable day out.

Below: Bishopscourt

OPENING DATES

All entries subject to change. For latest information check www.ngs.org.uk

Map locator numbers are shown to the right of each garden name.

January

Saturday 16th
Spring Platt 99

Wednesday 20th
Spring Platt 99

Sunday 24th
Spring Platt 99

Sunday 31st
Spring Platt 99

February

Snowdrop Festival

By arrangement
The Old Rectory 77

Wednesday 3rd
Spring Platt 99

Saturday 6th
Knowle Hill Farm 59

Sunday 7th
Knowle Hill Farm 59

Monday 8th
Knowle Hill Farm 59

Tuesday 9th
Spring Platt 99

Saturday 13th
Copton Ash 27

Sunday 14th
◆ Goodnestone Park Gardens 47

Sunday 21st
Copton Ash 27
◆ Doddington Place 30
Frith Old Farmhouse 43

March

Wednesday 3rd
Hoath House 53

Sunday 14th
Copton Ash 27
Haven 51
Stonewall Park 101

Sunday 21st
Enchanted Gardens 35
◆ Godinton House & Gardens 45
◆ Mount Ephraim Gardens 71
The Old Barn 75

Sunday 28th
Copton Ash 27
Godmersham Park 46
◆ Great Comp Garden 49

April

Thursday 1st
◆ Ightham Mote 56

Sunday 4th
Copton Ash 27
Haven 51

Monday 5th
Haven 51

Saturday 10th
NEW The Knoll Farm 58
NEW The Silk House 95

Sunday 11th
NEW 52 Cobblers Bridge Road 25
Copton Ash 27
Frith Old Farmhouse 43
◆ Hole Park 54
NEW The Silk House 95

Saturday 17th
Bilting House 9
Bishopscourt 10

Sunday 18th
Bilting House 9
Goddards Green 44
Potmans Heath House 85

Wednesday 21st
◆ Hever Castle & Gardens 52

Friday 23rd
Oak Cottage and Swallowfields Nursery 74

Saturday 24th
Oak Cottage and Swallowfields Nursery 74

Sunday 25th
Balmoral Cottage 5
◆ Doddington Place 30
Frith Old Farmhouse 43

Wednesday 28th
◆ Riverhill Himalayan Gardens 90

May

Sunday 2nd
14 Anglesey Avenue 2
Copton Ash 27
May Cottage 67
Stonewall Park 101

Monday 3rd
14 Anglesey Avenue 2
Haven 51

Tuesday 4th
14 Anglesey Avenue 2

Saturday 8th
◆ Chilham Castle 22
Elgin House 33

Sunday 9th
◆ Boughton Monchelsea Place 12
NEW 52 Cobblers Bridge Road 25
Copton Ash 27
Elgin House 33
Frith Old Farmhouse 43
Ladham House 60
43 The Ridings 89
Sandown 92

Wednesday 12th
Balmoral Cottage 5
◆ Hole Park 54

Saturday 15th
Bilting House 9
Little Gables 62
43 The Ridings 89

Sunday 16th
Bilting House 9
◆ Godinton House & Gardens 45
Little Gables 62
The Orangery 79

Friday 21st
Oak Cottage and Swallowfields Nursery 74

Saturday 22nd
Oak Cottage and Swallowfields Nursery 74

18 Royal Chase 91

Sunday 23rd
Haven 51
NEW Linton Park 61
12 The Meadows 68
Old Bladbean Stud 76
18 Royal Chase 91
Sandown 92
Torry Hill 105
Whitstable Gardens 113

Sunday 30th
Copton Ash 27
Tram Hatch 106

Monday 31st
Falconhurst 38

June

Wednesday 2nd
◆ Doddington Place 30
NEW The Postern 84

Saturday 5th
Cherry Tree Cottage 20
Churchfield 23
Enchanted Gardens 35
Faversham Gardens 40
Little Gables 62
Manwood House 65
Tankerton Gardens 102
Topgallant 104
Vergers 108
West Court Lodge 111

Sunday 6th
NEW Arnold Yoke 3
Cherry Tree Cottage 20
Churchfield 23
Enchanted Gardens 35
Godmersham Park 46
Haven 51
◆ Hole Park 54
Little Gables 62
Nettlestead Place 72
Potmans Heath House 85
Sandown 92
Topgallant 104
Vergers 108
West Court Lodge 111
West Malling Early Summer Gardens 112

Monday 7th
Norton Court 73

Tuesday 8th
Norton Court 73

Wednesday 9th
NEW Cacketts
Farmhouse 17

Thursday 10th
Avalon 4
◆ Mount Ephraim
Gardens 71
◆ Riverhill Himalayan
Gardens 90

Friday 11th
Alderwood 1
Avalon 4

Saturday 12th
Alderwood 1
Avalon 4
NEW Beechmont Hall 6
Bishopscourt 10
Brompton Village
Gardens 16
NEW Cacketts
Farmhouse 17
The Farmhouse Garden
at Tyland Barn 39
NEW Sir John Hawkins
Hospital 96
Walmer Gardens 109

Sunday 13th
Avalon 4
NEW Beechmont Hall 6
Bishopscourt 10
◆ Boughton
Monchelsea Place 12
Brompton Village
Gardens 16
Chevening 21
Copton Ash 27
Downs Court 31
NEW Fisher Street Oast 41
◆ Goodnestone Park
Gardens 47
Lords 64
The Old Barn 75
NEW Sir John Hawkins
Hospital 96
Old Bladbean Stud 76
Torry Hill 105
Walmer Gardens 109
Wye Gardens 118

Wednesday 16th
Avalon 4
Upper Pryors 107

Thursday 17th
Avalon 4

Friday 18th
Avalon 4
◆ Godinton House &
Gardens 45

Saturday 19th
Avalon 4

Sunday 20th
Brewery Farmhouse 14
Copton Ash 27
Downs Court 31
Haven 51
Little Mockbeggar 63
Old Bladbean Stud 76
Sandown 92
◆ The World Garden at
Lullingstone Castle 117

Thursday 24th
Avalon 4

Friday 25th
Avalon 4
Pheasant Barn 82

Saturday 26th
NEW Arnold Yoke 3
Avalon 4
NEW 69 Capel Street 18
Eureka 37
Falconhurst 38
NEW The Old Vicarage 78
Womenswold
Gardens 114

Sunday 27th
Avalon 4
NEW 69 Capel Street 18
Eureka 37
Old Bladbean Stud 76
NEW The Old Vicarage 78
Smiths Hall 98
Womenswold
Gardens 114

Tuesday 29th
Pheasant Barn 82

Wednesday 30th
Pheasant Barn 82

July

Thursday 1st
Pheasant Barn 82

Saturday 3rd
◆ Belmont 7
Gravesend Garden for
Wildlife 48

Sunday 4th
◆ Belmont 7
Bidborough Gardens 8

Boldshaves 11
Deal Town Gardens 29
31 Forest Avenue 42
Goddards Green 44
Gravesend Garden for
Wildlife 48
Old Bladbean Stud 76
Sandown 92
Tram Hatch 106

Tuesday 6th
Pheasant Barn 82

Wednesday 7th
Pheasant Barn 82

Thursday 8th
Pheasant Barn 82

Saturday 10th
NEW Hammond Place 50
Pheasant Barn 82
NEW Tonbridge
School 103

Sunday 11th
3 Bramble Close 13
NEW Hammond Place 50
Pheasant Barn 82
43 The Ridings 89
NEW 23 Seafield Road 93

Saturday 17th
142 Cramptons Road 28
Eureka 37
Knowle Hill Farm 59
The Mount 70
Stable House,
Heppington 100

Sunday 18th
142 Cramptons Road 28
Eureka 37
Haven 51
Knowle Hill Farm 59
The Mount 70
Old Bladbean Stud 76
NEW Quaker's Hall
Allotments 86
◆ Quex Gardens 87
Sandown 92
Stable House,
Heppington 100
Torry Hill 105
Yoakley House 119

Monday 19th
Knowle Hill Farm 59

Saturday 24th
Cherry Tree Cottage 20
Hurst House 55
Woodlands Road
Allotments 115

Sunday 25th
Cherry Tree Cottage 20
Frith Old Farmhouse 43
Hurst House 55
Woodlands Road
Allotments 115

Saturday 31st
5 Montefiore Avenue 69
The Orangery 79
The Watch House 110

August

Sunday 1st
31 Forest Avenue 42
Haven 51
5 Montefiore Avenue 69
The Orangery 79
Sandown 92
Tram Hatch 106
The Watch House 110

Saturday 7th
Eureka 37
12 Woods Ley 116

Sunday 8th
Eureka 37
NEW Ramsgate
Gardens 88

Sunday 15th
◆ Cobham Hall 26
Sandown 92

Friday 20th
Avalon 4

Saturday 21st
Avalon 4
NEW The Silk House 95

Sunday 22nd
Avalon 4
NEW The Silk House 95

Saturday 28th
Eureka 37

Sunday 29th
Eureka 37

Monday 30th
Haven 51

September

Sunday 5th
31 Forest Avenue 42
Sandown 92

Sunday 12th
12 The Meadows 68

Goodnestone Park Gardens
© Leigh Clapp

THE GARDENS

1 ALDERWOOD

Penshurst Road, Penshurst, TN11 8HY. Jon Little. *From Leigh B2176 towards Penshurst. Approx 1.2m on R. Through stone pillars. See NGS yellow signs. From Penshurst B2176 towards Leigh, approx 0.9m turn L just after post box on L.* **Fri 11, Sat 12 June (10-4). Adm £7.50, chd free. Home-made teas.**

14 acres of recently restored formal gardens, parkland, woodland and large walled garden. Originally part of The Redleaf estate, the gardens were featured in JC Loudon's Garden Magazine in 1833. Sumptuous planting, incl delphiniums, nepeta, artichokes and roses, surrounds a formal pond. The 18th C walled garden incl a wildflower meadow, fruit and vegetables, pergolas and vibrant planting. We have tarmac and 'tar and chip' paths in much of the garden allowing access to the major areas - except the woodland and rockery.

&. ☕

2 14 ANGLESEY AVENUE

Maidstone, ME15 9SH. Mike & Hazel Brett, 01622 299932, mandh.brett@tiscali.co.uk. *2m S of Maidstone take A229 (bus routes 5 & 89) & after Swan pub 1st R into Anglesey Ave. Limited street parking.* **Sun 2, Mon 3, Tue 4 May (11-4). Adm £3, chd free. Tea. Visits also by arrangement Apr to June for groups of up to 20.**

Plantsman's 120ft x 30ft garden with many unusual plants. Raised beds, rockeries and troughs accommodating alpine/rock garden plants. Herbaceous & shrub borders plus a shady woodland area at the end of the garden with hellebores, erythroniums, trilliums, anemones etc.

&. ❀ ☕

3 NEW ARNOLD YOKE

Back Street, Leeds, Maidstone, ME17 1TF. Richard & Patricia Stileman. *5m E of Maidstone. Fm M20 J8 take A20 Lenham R to B2163 to Leeds. Thru Leeds R into Horseshoes La, 1st R into Back St. House ¾ m on L. Fm A274 follow B2163 to Langley L into Horseshoes La 1st R Back St.* **Sun 6 June (2-6).**

Adm £5, chd free. Home-made teas. Evening opening Sat 26 June (5-8). Adm £10. Light refreshments. Not suitable for chd.

Recently redesigned ½ acre garden, adjacent to 15th C Wealden Hall house with formal structure of box and yew embracing long mixed border, and centrally placed 'paradise garden' with water feature. NB for 26/06 Refreshments included in entry price. Prosecco/wine for a donation. Wheelchair access from the car park and around most parts of the garden.

&. ☕

4 AVALON

57 Stoney Road, Dunkirk, ME13 9TN. Mrs Croll, avalongarden8@gmail.com. *4m E of Faversham, 5m W of Canterbury, 2.5m E of J7 M2. M2 J7 or A2 E of Faversham take A299, first L, Staplestreet, then L, R past Mt Ephraim, turn L, R. From A2 Canterbury, turn off Dunkirk, bottom hill turn R, Staplestreet then R, R. Park in side roads.* **Thur 10 to Sun 13 June, Wed 16 to Sat 19 June, Thur 24 to Sun 27 June, Fri 20, Sat 21, Sun 22 Aug (11-3.30).**

Belmont

Adm £5, chd free. Pre-booking essential, please visit www.ngs. org.uk for information & booking. Visits also by arrangement in June for groups of 5 to 10.
½ acre sheltered woodland garden planted for all seasons on a NW slope with views of Thames estuary. Collections of roses, hostas and ferns plus rhododendrons, shrubs, trees, vegetables, fruit, unusual plants and cut flowers for local shows. It is planted by feeling, making it a reflective space and a plant lovers' garden. Plenty of seating for taking in the garden and resting from lots of steps.

5 BALMORAL COTTAGE
The Green, Benenden, Cranbrook, TN17 4DL. Charlotte Molesworth, thepottingshedholidaylet@gmail. com. *Few 100 yds down unmade track to W of St George's Church, Benenden.* **Sun 26 Apr, Wed 12 May (12-5). Adm £6, chd £2.50.**
An owner created and maintained garden now 33yrs mature. Varied, romantic and extensive topiary form the backbone for mixed borders. Vegetable garden, organically managed. Particular attention to the needs of nesting birds and small mammals lend this artistic plantswoman's garden a rare and unusual quality. No hot borders or dazzling dahlias here, where autumn exemplifies the 'season of mists and mellow fruitfulness'.

6 NEW BEECHMONT HALL
Gracious Lane, Sevenoaks, TN13 1TH. Lene Hansen, 01732461657, lene@netvigator.com. *There are four different 'Beechmonts' on this road. Please turn L after 'Little Beechmont' when coming from the Tonbridge Rd direction.* **Sat 12, Sun 13 June (10-5). Adm £15, chd free. Pre-booking essential, please visit www.ngs. org.uk for information & booking. Cream teas included in entry price. Visits also by arrangement in June for groups of 10 to 20.**
Panoramic views across the Weald towards Ashdown Forest create a spectacular borrowed landscape for the sweeping curves of this award-winning modern garden. Lovingly created over the last decade it celebrates a passion for supporting local wildlife, the elegant elements of Japanese aesthetics and unusual plants (Buddha's Hand, Pseudolarix,

Golden Elm and Wollemi pine amongst others). The garden is easily accessible from the carpark.

7 ♦ BELMONT
Belmont Park, Throwley, Faversham, ME13 0HH. Harris (Belmont) Charity, 01795 890202, administrator@belmont-house.org, www.belmont-house.org. *4½ m SW of Faversham. A251 Faversham-Ashford. At Badlesmere, brown tourist signs to Belmont.* **For NGS: Sat 3, Sun 4 July (12-5). Adm £5, chd free. Light refreshments. For other opening times and information, please phone, email or visit garden website.**
Belmont House is surrounded by large formal lawns that are landscaped with fine specimen trees, a pinetum and a walled garden containing long borders, wisteria and large rose border. Across the drive there is a second walled kitchen garden, restored in 2001 to a design by Arabella Lennox-Boyd. It includes vegetable and herbaceous borders, hop arbours and walls trained with a variety of fruit. The gardens are open daily all year round from 10 am - 6pm (dusk if earlier).

GROUP OPENING

8 BIDBOROUGH GARDENS
Bidborough, Tunbridge Wells, TN4 0XB. *3m N of Tunbridge Wells, between Tonbridge & Tunbridge Wells W off A26. Take B2176 Bidborough Ridge signed to Penshurst. Take 1st L into Darnley Drive, then 1st R into St Lawrence Ave, no. 2.* **Sun 4 July (1-5). Combined adm £5, chd free. Home-made teas at Boundes End. Gluten and dairy free cake available. Donation to Hospice in the Weald.**

21A BIDBOROUGH RIDGE
John & Wendy Johnson.

BOUNDES END
Carole & Mike Marks, 01892 542233, carole.marks@btinternet.com, www.boundesendgarden.co.uk. Visits also by arrangement June to Aug for groups of up to 20.

8 DOWER HOUSE CRESCENT
Judy & Bill Liddall.

SHEERDROP
Mr John Perry.

The Bidborough gardens (collect garden list from Boundes End, 2 St Lawrence Avenue) are in a small village at the heart of which are The Kentish Hare pub (book in advance), the church, village store and primary school. Boundes End - Carole & Mike Marks; An unusually shaped garden with both formal and informal areas. Sheerdrop - John Perry; hillside garden overlooking valley with view to Bidborough Church. Formal and woodland gardens. 21A Bidborough Ridge - John & Wendy Johnson; A plantsman's garden with borders full of shrubs and perennials and a small pond. 8 Dowerhouse Crescent - Judy & Bill Liddall; Shrubs and perennials planted for colour throughout the year. Productive vegetable garden. Partial wheelchair access, some gardens have steps.

9 BILTING HOUSE
nr Ashford, TN25 4HA. Mr John Erle-Drax, 07764 580011, erle-drax@marlboroughgallery.com. *5m NE of Ashford. A28, 5m E from Ashford, 9m S from Canterbury. Wye 1½ m.* **Sat 17, Sun 18 Apr, Sat 15, Sun 16 May (1.30-5.30). Adm £5, chd free. Cream teas. Visits also by arrangement Apr to Sept for groups of 10+.**
6 acre garden with ha-ha set in beautiful part of Stour Valley. Wide variety of rhododendrons, azaleas and ornamental shrubs. Woodland walk with spring bulbs. Mature arboretum with recent planting of specimen trees. Rose garden and herbaceous borders. Conservatory.

10 BISHOPSCOURT
24 St Margaret's Street, Rochester, ME1 1TS. Mrs Bridget Langstaff. *Central Rochester, nr castle & cathedral. On St Margaret's St at junction with Vines Lane.* **Sat 17 Apr (11-3); Sat 12, Sun 13 June (1-5). Adm £5, chd free. Home-made teas.**
The residence of the Bishop of Rochester, this is a mature one acre historic walled garden within the heart of Rochester. Mature trees, lawns, yew hedges, rose garden, fountain, mixed herbaceous borders and a greenhouse next to a small vegetable garden. Child friendly. WC incl disabled.

◫ BOLDSHAVES

Woodchurch, nr Ashford, TN26 3RA. Mr & Mrs Peregrine Massey, 01233 860283, masseypd@hotmail.co.uk, www.boldshaves.co.uk. *Between Woodchurch & High Halden off Redbrook St. From centre of Woodchurch, with church on L and Bonny Cravat/Six Bells PH on R, 2nd L down Susan's Hill, then 1st R after ½ mile before L after a few 100 yrds to Boldshaves. P as indicated.* **Sun 4 July (2-6). Adm £7.50, chd free. Home-made teas in the Cliff Tea House (weather permitting), otherwise in the Barn. Donation to Kent Minds.**

7-acre garden developed over past 25 years, partly terraced, S-facing, with wide range of ornamental trees and shrubs, walled garden, Italian garden, Diamond Jubilee garden, camellia dell, herbaceous borders (incl flame bed, red borders and rainbow border), vegetable garden, bluebell walks in April, woodland and ponds; wildlife haven renowned for nightingales and butterflies. **For details of other opening times see garden website www.boldshaves. co.uk** Home of the Wealden Literary Festival. Grass paths.

&. ❀ 🚗 ☕

◫ ◆ BOUGHTON MONCHELSEA PLACE

Church Hill, Boughton Monchelsea, Maidstone, ME17 4BU. Mr & Mrs Dominic Kendrick, 01622 743120, mk@boughtonplace.co.uk, www.boughtonplace.co.uk. *4m SE of Maidstone. From Maidstone follow A229 S for 3½m to major T-lights at Linton Xrds, turn L onto B2163, house 1m on R; or take J8 off M20 & follow Leeds Castle signs to B2163, house 5½m on L.* **For NGS: Sun 9 May, Sun 13 June (2-5.30). Adm £5, chd £1. Home-made teas. For other opening times and information, please phone, email or visit garden website.**

150 acre estate mainly park & woodland, spectacular views over own deer park & the Weald. Grade I manor house (not open). Courtyard herb garden, intimate walled gardens, box hedges, herbaceous borders, orchard. Planting is romantic rather than manicured. Terrace with panoramic views, bluebell woods, wisteria tunnel, David Austin roses, traditional greenhouse & kitchen garden. Visit St. Peter's Church next

door to see the huge stained glass Millennium Window designed by renowned local artist Graham Clark & the tranquil rose garden overlooking the deer park of Boughton Place. Regrettably, steep steps and narrow paths render the garden difficult for disabled visitors and unsuitable for wheelchairs.

❀ 🚗 ☕

◫ 3 BRAMBLE CLOSE

Wye, TN25 5QA. Dr M Copland. *two min walk from Wye Station car park. Bramble Close is off Bramble Lane, nearly opp Wye Motors and close to Wye Stn where parking is available.* **Sun 11 July (2-6). Adm £3, chd free. Opening with Wye Gardens on Sun 13 June.**

A very wild, experimental garden sown from seed 1987-89. wildflower meadow, pond and ditches, mown paths, native copse and hedges buzzing with wildlife - a unique experience. Demonstrating how plants manipulate diseases, insects and other animals to establish and maintain their natural population density. A completely wild meadow cut each year in September but supporting a large population of butterflies, moths and other insects, amphibians, reptiles, birds, mammals including bats. Some soft ground with uneven pathways.

◫ BREWERY FARMHOUSE

182 Mongeham Road, Great Mongeham, Deal, CT14 9LR. Mr David & Mrs Maureen Royston-Lee. *On the corner of Mongeham Rd & Northbourne Rd. Opp the village green, with a white picket fence at the front of the house, entrance to the gardens are through the Brewery yard at the back of the house entrance on Northbourne Rd.* **Sun 20 June (10.30-4.30). Adm £6, chd free. Light refreshments at the Village Hall.**

An established country walled garden divided into a series of 'rooms'. Roses, clematis, poppies abound in herbaceous borders and against walls and trellises. With water features, fruit tree pergola and a chinese pagoda. There is a gravel drive and one step up into the garden.

◫ 1 BRICKWALL COTTAGES

Frittenden, Cranbrook, TN17 2DH. Mrs Sue Martin, 01580 852425, sue.martin@talktalk.net, www.geumcollection.co.uk.

6m NW of Tenterden. E of A229 between Cranbrook & Staplehurst & W of A274 between Biddenden & Headcorn. Park in village & walk along footpath opp school. **Visits by arrangement Apr to June for groups of up to 20. Adm £5, chd free. Home-made teas.**

Although less than ¼ acre, the garden gives the impression of being much larger as it is made up of several rooms all intensively planted with a wide range of hardy perennials, bulbs and shrubs, with over 100 geums which comprise the National Collection planted throughout the garden. Pergolas provide supports for climbing plants and there is a small formal pond. Some paths are narrow and wheelchairs may not be able to reach far end of garden.

&. 🐎 ❀ 🚗 NPC ☕

GROUP OPENING

◫ BROMPTON VILLAGE GARDENS

Garden Street and Prospect Row, Brompton, Gillingham, ME7 5AL. Jennifer Jones. *Brompton is between Chatham & Gillingham A231 - Dock Rd next to Historic Dockyard. At r'about take A231 Wood St, opp. RSME Barracks enter Mansion Row, 1st L Garden St. Road parking in village.* **Sat 12, Sun 13 June (2-5). Combined adm £6, chd free. Light refreshments at Prospect Row or Mansion Row.**

NEW 6 GARDEN STREET
Mr John & Mrs Diane Brice.

26 GARDEN STREET
Mrs Lissie Larkin.

7 PROSPECT ROW
Ms Elaine Fowler.

14 PROSPECT ROW
Ms Audrey Iles.

16 PROSPECT ROW
Jennifer Jones.

20 PROSPECT ROW
Clive & Karen Perry.

In Brompton, the 'village in the Towns' on the Saxon Shore Way, and a stone's throw from the Historic Dockyard and other historic attractions, there are 6 town gardens in two adjacent streets showing different ideas for small plots. Designed by their owners, some in recent years, they form a combination of formal and informal designs giving alternative views about how to make interesting use of small spaces. These ideas incl a garden made entirely of containers, differing approaches to planting schemes and hard landscaping. The gardens range from low maintenance (20 Prospect Row), creative use of pots and containers (6 Garden St), Italianate style garden (20 Prospect Row),sunny courtyard with summer house (26 Garden St) and a more established garden with mature trees and all year colour attracting wildlife (14 Prospect Row), a plantswoman garden with unusual plants (16 Prospect Row) to a garden re-designed from a building site several years ago (7 Prospect Row).

✿ ☕

---

17 NEW CACKETTS FARMHOUSE

Haymans Hill, Horsmonden, TN12 8BX. Mr & Mrs Lance Morrish, diana.morrish@hotmail.co.uk. Take B2162 from Horsmonden towards Marden. 1st R into Haymans Hill, 200yds 1st L, drive immed to R of Little Cacketts. Wed 9, Sat 12 June (12-5). Adm £7.50, chd free. Home made teas. Visits also by arrangement for groups of up to 20. 1¼ acre garden surrounding C17 farmhouse (not open). Walled garden, bog garden and ponds, woodland garden with unusual plants, bug hotel. 4 acre hayfield with self planted wildflowers.

♿ ☕ ☕ ⌇

---

18 NEW 69 CAPEL STREET

Capel-Le-Ferne, Folkestone, CT18 7LY. John & Jenny Carter. Capel-le-Ferne. Take B2011 from Folkestone towards Dover. Past Battle of Britain Memorial on R and then take first L into Capel Street. 69 is 400yds on L. Sat 26, Sun 27 June (10.30-4.30). Adm £4, chd free. Light refreshments.
A contemporary urban cottage garden. A clever use of traditional and modern planting providing colour throughout the seasons. A rectangular garden where straight lines have been diffused by angles and planting. Space is provided for vegetables for self sufficiency. A quiet location occasionally amplified by a passing Spitfire. Walking distance to the famous Battle of Britain Memorial and pleasant walks along the White Cliffs of Dover. Garden access can be achieved via the garage.

♿ ☕ ☕

---

19 ◆ CHARTWELL

Mapleton Road, Westerham, TN16 1PS. National Trust, 01732 868381, chartwell@nationaltrust.org.uk, www.nationaltrust.org.uk/chartwell. 4m N of Edenbridge, 2m S of Westerham. Fork L off B2026 after 1½ m. For NGS: Wed 22 Sept (10-5). Adm £5, chd £2.50. For other opening times and information, please phone, email or visit garden website.
Informal gardens on hillside with glorious views over Weald of Kent. Water features and lakes together with red brick wall built by Sir Winston Churchill, former owner of Chartwell. Lady Churchill's rose garden. Avenue of golden roses runs down the centre of a must see productive kitchen garden. Hard paths to Lady Churchill's rose garden and the terrace. Some steep slopes and steps.

♿ ☕ ✿ 🚗 ☕ ⌇

---

20 CHERRY TREE COTTAGE

Brookestreet, Ash, Canterbury, CT3 2NP. Mr Neil & Mrs Kate Dymant. 8m from Canterbury. Turn off A257 into Hill's Court Rd (opp side of bypass to Ash village), continue ¾ m past Brookestreet Farmhouse, garden entrance on L after sharp L bend. Ample parking. Sat 5, Sun 6 June, Sat 24, Sun 25 July (11-4). Adm £6, chd free. Home-made teas.
A one-acre garden managed completely organically featuring a stream-side woodland garden, mixed borders, gravel garden, ponds, orchards, Japanese garden and formal vegetable beds. Home to a wide variety of wildlife. Most of the garden is wheelchair accessible, fairly flat with gravel paths except sloping woodland garden.

♿ ☕ ✿ ☕ ⌇

---

21 CHEVENING

nr Sevenoaks, TN14 6HG. The Board of Trustees of the Chevening Estate, www.cheveninghouse.com. 4m NW of Sevenoaks. Turn N off A25 at Sundridge T-lights on to B2211; at Chevening Xrds 1½ m turn L. Sun 13 June (2-5). Adm £7, chd £1. Home-made teas. Local ice cream available.
The pleasure grounds of the Earls Stanhope at Chevening House are today characterised by lawns and wooded walks around an ornamental lake. First laid out between 1690 and 1720 in the French formal style, in the 1770s a more informal English design was introduced. In the early C19 lawns, parterres and a maze were established and many specimen trees planted to shade woodland walks. A cascade, modelled on a 1718 predecessor, commemorates 250 years of the Stanhope family's ownership of Chevening and 50 years stewardship by the Board of Trustees. Group guided tours of park and gardens can sometimes be arranged with the Estate Office when the house is unoccupied. Gentle slopes, gravel paths throughout.

♿ ☕ ✿ ☕ ⌇

---

22 ◆ CHILHAM CASTLE

Canterbury, CT4 8DB. The Wheeler Family, 01227 733100, enquiries@chilham-castle.co.uk, www.chilham-castle.co.uk. 6m SW of Canterbury, 7m NE of Ashford, centre of Chilham Village. Follow NGS signs from A28 or A252 up to Chilham village square & through main gates of Chilham Castle. For NGS: Sat 8 May (10-4). Adm £5, chd free. Light refreshments. For other opening times and information, please phone, email or visit garden website.
The garden surrounds the Jacobean house 1616 (not open). C17 terraces with herbaceous borders. Topiary frames the magnificent views down with lake walk below. Extensive kitchen and cutting garden beyond spring bulb filled quiet garden. Established trees and ha-ha lead onto park. Check website for other events and attractions. Partial wheelchair access.

♿ ☕ ✿ 🚗 ☕ ⌇

"I was amazed to discover that the National Garden Scheme is Marie Curie's largest single funder and has given the charity nearly £10 million over 25 years. Their continued support makes such a difference to me and all Marie Curie Nurses on the frontline of the coronavirus crisis, as we continue to provide expert care and support to people at end of life." – Tracy McWilliams, Marie Curie Nurse

Open your garden with the National Garden Scheme and join a community of like-minded individuals, all passionate about gardens, and raise money for nursing and health charities. Big or small, if your garden has quality, character and interest we'd love to hear from you. Call us on 01483 211535 or email hello@ngs.org.uk

23 CHURCHFIELD

Pilgrims Way, Postling, Hythe, CT21 4EY. Chris & Nikki Clark. *2m NW of Hythe. From M20 J11 turn S onto A20. 1st L after ½ m on bend take rd signed Lyminge. 1st L into Postling.* **Sat 5, Sun 6 June (12-5). Combined adm with West Court Lodge £6, chd free. Home-made teas in Postling Village Hall.**
At the base of the Downs, springs rising in this garden form the source of the East Stour. Two large ponds are home to wildfowl and fish and the banks have been planted with drifts of primula, large leaved herbaceous bamboo and ferns. The rest of the 5 acre garden is a Kent cobnut platt and vegetable garden, large grass areas and naturally planted borders and woodland. Postling Church open for visitors. Areas around water may be slippery. Children must be carefully supervised.

24 THE COACH HOUSE

Kemsdale Road, Hernhill, Faversham, ME13 9JP. Alison & Philip West, 07801 824867, alison.west@kemsdale.plus.com. *3m E of Faversham. At J7 of M2 take A299, signed Margate. After 600 metres take 1st exit signed Hernhill, take 1st L over dual carriageway to T-junction, turn R & follow yellow NGS signs.* **Visits by arrangement June to Sept. Adm £5, chd free.**
The ¾ acre garden has views over surrounding fruit-producing farmland. Sloping terraced site and island beds with yr-round interest, a pond room, herbaceous borders containing bulbs, shrubs, perennials and a tropical bed. The different areas are connected by flowing curved paths.

Unusual planting on light sandy soil where wildlife is encouraged. Kent Wild for Wildlife gold award winner 2019. Most of garden accessible to wheelchairs but some slopes. Seating available in all areas.

25 NEW 52 COBBLERS BRIDGE ROAD

Herne Bay, CT6 8NT. Mercy Morris, 0786 0537664, mercy@home-plants.com, www.home-plants.com. *10 mins walk from centre of Herne Bay. Exit the A299 to Herne Bay. Cobblers Bridge Rd can be accessed from Sea St or Eddington Lane. Street parking, can be scarce at weekends.* **Sun 11 Apr, Sun 9 May, Sat 6 Nov (10.30-4). Adm £4. Pre-booking essential, please phone 0786 0537664 or email mercy@ home-plants.com for information & booking. Light refreshments. Visits also by arrangement Mar to Nov for groups of 4 to 6. Donation to Plant Heritage.**
As you walk through the house to the garden admire around 150 houseplants in a 1.5 bedroom house; from tiny airplants to philodendrons and monsteras. A range of cacti, succulents, tillandsia, tropical and half-hardy plants. A further selection of house plants are on view in the greenhouse outside. Collection of indoor plants suitable for most homes.

26 ◆ COBHAM HALL

Cobham, DA12 3BL. Mr D Standen (Bursar), 01474 823371, www.cobhamhall.com. *3m W of Rochester, 8m E of M25 J2. Ignore SatNav directions to Lodge Lane. Entrance drive is off Brewers Rd, 50 metres E from Cobham/Shorne A2 junction.* **For NGS: Sun 15 Aug (2-5). Adm £3, chd free. Cream teas in the Gilt Hall. For other opening times and information, please phone or visit garden website.**
1584 brick mansion (open for tours) and parkland of historical importance, now a boarding and day school for girls. Some herbaceous borders, formal parterres, drifts of daffodils, C17 garden walls, yew hedges and lime avenue. Humphry Repton designed 50 hectares of park. Most garden follies restored in 2009. Film location for BBC's Bleak House series and films by MGM and Universal. ITV serial The Great Fire. CBBC filmed

serial 1 & 2 of Hetty Feather. Gravel and slab paths through gardens. Land uneven, many slopes. Stairs and steps in Main Hall. Please call in advance to ensure assistance.

27 COPTON ASH

105 Ashford Road, Faversham, ME13 8XW. Drs Tim & Gillian Ingram, 01795 535919, coptonash@yahoo.co.uk, www.coptonash.plus.com. *½ m S of A2, Faversham. On A251 Faversham to Ashford rd. Opp E bound J6 with M2. Park in nearby laybys, single yellow lines are Mon to Fri restrictions.* **Sat 13 Feb (12-4). Sun 21 Feb (12-4), also open Doddington Place. Sun 14, Sun 28 Mar, Sun 4, Sun 11 Apr, Sun 2, Sun 9, Sun 30 May, Sun 13, Sun 20 June (12-5); Sun 10 Oct (12-4). Adm £4, chd free. Home-made teas. 2022: Sat 12, Sun 20 Feb. Visits also by arrangement Jan to June for groups of up to 30.**
Garden grown out of a love and fascination with plants. Contains very wide collection incl many rarities and newly introduced species raised from wild seed. Special interest in woodland flowers, snowdrops and hellebores with flowering trees and shrubs of spring. Refreshed Mediterranean plantings to adapt to a warming climate. Raised beds with choice alpines and bulbs. Small alpine nursery. Gravel drive, shallow step by house and some narrow grass paths.

28 142 CRAMPTONS ROAD

Sevenoaks, TN14 5DZ. Mr Bennet Smith. *3½ m from M25 J5. 1½ m N of Sevenoaks, off Otford Road (A225) between Bat & Ball T-lights & Otford. Access to garden via side/ rear passage. Limited parking in Cramptons Rd.* **Sat 17 July (10.30-6). Sun 18 July (10.30-6), also open Quaker's Hall Allotments. Adm £3.50, chd free. Home-made teas.**
A very small, lush & leafy plantsman's oasis. A tapestry of plants selected for texture, elegance, leaf shape, long-season interest or rarity: wildlife-friendly umbellifers; acers, sassafras, tetrapanax, pseudopanax & Aesculus wangii. Immerse yourself in plants & discover what can be created and combined in a tiny space! Kent Life Garden Awards Finalist (September 2019), ITV Love Your Garden.

18 Royal Chase

© Leigh Clapp

GROUP OPENING

29 DEAL TOWN GARDENS

Deal, CT14 6EB. *A258 to Deal. Signs from all town car parks, maps & tickets at all gardens.* **Sun 4 July (11-5). Combined adm £5, chd free.**

4 GEORGE ALLEY
Lyn Freeman & Barry Popple, 07889572676, lynandbarry10@yahoo.co.uk, www.gleaners.co.uk.

4 ROBERT STREET
Christine & Peter Hayes-Watkins.

8 ROBERT STREET
Mrs Gill Walshe.

16 ST ANDREW'S ROAD
Mr Martin Parkes & Mr Paul Green.

88 WEST STREET
Lyn & Peter Buller.

Start from any town car park (signs from here). 88 West Street: A non water cottage garden with perennials, shrubs, clematis and roses, shade garden. 4 George Alley: A pretty alley leads to a secret garden with courtyard, leading to a vibrant cottage garden with summer house. 4 Robert Street: A small walled garden divided into four areas. Bedding plants, perennials, shrubs, trees, mature bamboos and water feature. 8 Robert Street: A walled garden with an added bonus of a verge overflowing with colour. 16 St Andrew's Road: A tropical themed walled urban garden with large sun-drenched borders, lawn and family seating area. Refreshments in High Street, seafront and local public houses. Partial wheelchair access at 88 West Street 4 & 8 Robert Street. All other gardens no access.

❀

30 ◆ DODDINGTON PLACE

Church Lane, Doddington, Sittingbourne, ME9 0BB. Mr & Mrs Richard Oldfield, 01795 886101, enquiries@ doddingtonplacegardens.co.uk, www.doddingtonplacegardens. co.uk. *6m SE of Sittingbourne. From A20 turn N opp Lenham or from A2 turn S at Teynham or Ospringe (Faversham), all 4m.* **For NGS: Sun 21 Feb (11-5), also open Copton**

Ash. **Sun 25 Apr, Wed 2 June, Sun 19 Sept (11-5). Adm £9, chd £2.50. Home-made teas.** For other opening times and information, please phone, email or visit garden website.

10 acre garden,wide views; trees and cloud clipped yew hedges; woodland garden with azaleas and rhododendrons; Edwardian rock garden (not wheelchair accessible); formal garden with mixed borders. A flint and brick late 20th-century gothic folly; newly installed at the end of the Wellingtonia walk, a disused pinnacle from the southeast tower of Rochester Cathedral. Snowdrops in February. Chelsea Fringe Events. Wheelchair access possible to majority of gardens except rock garden.

31 DOWNS COURT

Church Lane, Boughton Aluph, Ashford, TN25 4EU. Mr Bay Green. *4m NE of Ashford. From A28 Ashford or Canterbury, after Wye Xrds take next turn NW to Boughton Aluph Church. Fork R at pillar box, garden only drive on R.* **Sun 13, Sun 20 June (2-5). Adm £5, chd free.**
3 acre downland garden on alkaline soil with fine trees, mature yew and box hedges, mixed borders. Shrub roses and rose arch pathway, small parterre. Sweeping lawns and lovely views over surrounding countryside.

32 EAGLESWOOD

Slade Road, Warren Street, Lenham, ME17 2EG. Mike & Edith Darvill, 01622 858702, mike.darvill@btinternet.com. *E on A20 nr Lenham, L into Hubbards Hill for approx 1m then 2nd L into Slade Rd. Garden 150yds on R. Coaches permitted by prior arrangement.* **Visits by arrangement Apr to Oct. Adm £5, chd free. Home-made teas. Donation to Demelza House Hospice.**
2 acre plantsman's garden. Wide range of trees and shrubs (many unusual), herbaceous material and woodland plants grown to give yr-round interest.

33 ELGIN HOUSE

Main Road, Knockholt, Sevenoaks, TN14 7LH. Mrs Avril Bromley. *Off A21 between Sevenoaks/Orpington at Pratts Bottom r'about, rd signed*

Knockholt (Rushmore Hill) 3m on R, follow yellow NGS signs. Main Rd is continuation of Rushmore Hill. **Sat 8, Sun 9 May (12-5). Adm £5, chd free.**
Victorian family house surrounded by a garden which has evolved over the last 50 years. Bluebells, rhododendrons, azaleas, wisteria, camellias, magnolias, mature trees, incl a magnificent cedar tree and spacious lawns. This garden is on the top of the North Downs.

❀

34 ◆ EMMETTS GARDEN

Ide Hill, Sevenoaks, TN14 6BA. National Trust, 01732 751507, emmetts@nationaltrust.org.uk, www.nationaltrust.org.uk/ emmetts-garden. *5m SW of Sevenoaks. 1½m S of A25 on Sundridge-Ide Hill Rd. 1½m N of Ide Hill off B2042.* **For NGS: Wed 22 Sept (10-5). Adm £5.50, chd £2.75.** For other opening times and information, please phone, email or visit garden website.
5 acre hillside garden, with the highest tree top in Kent, noted for its fine collection of rare trees and flowering shrubs. The garden is particularly fine in spring, while a rose garden, rock garden and extensive planting of acers for autumn colour extend the interest throughout the season. Hard paths to the Old Stables for light refreshments and WC. Some steep slopes. Volunteer driven buggy available for lifts up steepest hill.

35 ENCHANTED GARDENS

Sonoma House, Pilgrims Lane, Seasalter, Whitstable, CT5 3AP. Mrs Donna Richardson, 07967917161, donna@ enchantedgardenskent .co.uk, www.enchantedgardenskent.co.uk. *2m from Whitstable Town. Whitstable: At r'about, take exit onto A290 Canterbury. At r'about, take 3rd exit onto A299 ramp London/ Faversham Merge onto A299. Take slip rd offered on L, turn R for Pilgrims Lane.* **Sun 21 Mar, Sat 5, Sun 6 June, Sat 25, Sun 26 Sept (10-5). Adm £5, chd free. Home-made teas. Vegan and gluten free options available. Visits also by arrangement Apr to Oct.**
It has taken 25 years to create Enchanted Gardens from open farmland to the traditional cottage style garden it is today. I am

The World Garden at Lullingstone Castle

passionate about the loss of our pollinating insects and am organic. I have collections of roses, shrubs and perennials in herbaceous borders for all garden situations to provide colour and interest from February to December. Over 100 roses, 40+ peony, established natural habitats, pond, bog garden, mature trees.

ᯤ ✿ ☕ ⛉

37 EUREKA
Buckhurst Road, Westerham Hill, TN16 2HR. Gordon & Suzanne Wright. *Off A233, 1½m N of Westerham, 1m S from centre of Biggin Hill. 5m from J5 & J6 of M25 Parking at Westerham Heights Garden Centre at top of Westerham Hill on A233, 300yds from Eureka. Satnav: use TN16 2HW. Parking at house for those with walking difficulties.* **Sat 26, Sun 27 June, Sat 17, Sun 18 July, Sat 7, Sun 8, Sat 28, Sun 29 Aug (11-4). Adm £5, chd free. Home-made teas.**

Approx 1 acre garden with a blaze of colourful displays in perennial borders and the 8 cartwheel centre beds. Sculptures, garden art, chickens, lots of seating and stairs to a viewing platform. Many quirky surprises at every turn. 100s of annuals in tubs and baskets, a David Austin Shrub Rose border and a Free Treasure Trail with prizes for all children. Garden art incl 12ft dragon, a horse's head carved out of a 200yr old yew tree stump and a 10ft dragonfly on a reed. Wheelchair access to most of the garden.

ᯤ ⛉ ☕

38 FALCONHURST
Cowden Pound Road, Markbeech, Edenbridge, TN8 5NR. Mr & Mrs Charles Talbot, 01342 850526, nicola@falconhurst.co.uk, www.falconhurst.co.uk. *3m SE of Edenbridge. B2026 at Queens Arms pub turn E to Markbeech. 2nd drive on R before Markbeech village.* **Mon 31 May, Sat 26 June (11-4). Adm**

£7, chd free. Home-made teas. Visits also by arrangement Apr to Oct for groups of 20+.
4 acre garden with fabulous views devised and cared for by the same family for 170 yrs. Deep mixed borders with old roses, peonics, shrubs and a wide variety of herbaceous and annual plants; ruin garden; walled garden; interesting mature trees and shrubs; kitchen garden; wildflower meadows with woodland and pond walks. Market garden, pigs, orchard chickens; lambs in the paddocks.

ᯤ ✿ ⛉ ☕ ⛉

All entries are subject to change. For the latest information check ngs.org.uk

39 THE FARMHOUSE GARDEN AT TYLAND BARN
Chatham Road, Sandling, Maidstone, ME14 3BD. Kent Wildlife Trust, www. kentwildlifetrust.org.uk/nature-reserves/tyland-barn. *2½ m N of Maidstone. Fm M20 J6 take A229 to Chatham. 2nd L signposted Tyland Barn, R at Lower Bell Pub. Go under A229, R back onto A229 to Maidstone. L at Esso Garage & follow Tyland Barn sign & Yellow Signs.* **Sat 12 June (10-3). Adm £5, chd free. Light refreshments in the Visitor Centre Cafe.**
Encouraging wildlife into The Farmhouse Garden is the number one priority. No manicured lawns or use of chemicals but relaxed borders containing pollinator friendly flowers throughout the season. Wildflower meadows, bee hotels, log piles, small wildlife pond. Stepover apples, small fruit area. Plenty of ideas to make your garden wildlife friendly. Plants labelled. Wild Orchids. Look out for solitary bees, dragonflies & butterflies. Garden normally closed to the public. In Nature Park many native plants and wildflower meadows. Large pond. Tours with Head Gardener around the Nature park & gardeners on hand to answer questions. Parking for blue badge users. Wheelchair access to the Farmhouse Garden is possible but with care. Wheelchair accessible path around Nature park.

GROUP OPENING

40 FAVERSHAM GARDENS
Faversham, ME13 8QN. *On edge of town, short distance from A2 & train stn. From M2 J6 take A251, L into A2, R into The Mall. Combined tickets & maps from No. 58.* **Sat 5 June (10-5). Combined adm £5, chd free. Also open Tankerton Gardens. Teas widely available in Faversham.**

54 ATHELSTAN ROAD
Sarah Langton-Lockton OBE.
NEW 58 THE MALL
Jane Beedle.
19 NEWTON ROAD
Posy Gentles, www.posygentles.co.uk.

3 distinctive walled gardens in historic Faversham. Start at 58 The Mall. A contemporary garden packed with pollinators, created in 2018, with materials reused in wire gabions to create raised beds and a wildlife haven. On to 54 Athelstan Road, a maturing garden on a once-neglected site. Ornamental vegetable beds take centre stage. Climbing roses, clematis, thalictrums, Regale lilies, sibirica irises and unusual shrubs, sheltered by old walls. Next, 19 Newton Road, a long, thin town garden, where the plant loving owner has used billowing roses, shrubs, climbers and perennials to blur boundaries. The judicious planting of trees, and curving paths, veil rather than conceal the garden as you move through it. Teashops available in Faversham. Level access to paths and terrace 54 Athelstan Road.

41 NEW FISHER STREET OAST
Badlesmere, Faversham, ME13 0LB. Group Captain & Mrs Robert Perry. *4m from Faversham. Off A251 at Sheldwich Lees; follow Lees Court Rd & bear L into Fisher Street Rd.* **Sun 13 June (11-5). Adm £6, chd free. Home-made teas. Donation to St James' Church, Sheldwich.**
This 2 acre garden surrounds a pretty flint Oast & restored barn. Designed by the owners over 20 years, the garden has a distinctive structure with flint and brick walls, mature trees and topiary. Mixed planting in soft hues attracts wildlife in an Italian style garden room with fountain. Newly created pond; sculptures and vegetable/nursery garden. Extensive views over farmland and Perry Wood. There are a few single, shallow steps in various parts of the garden but these can be avoided.

42 31 FOREST AVENUE
Orchard Heights, Ashford, TN25 4GB. Tony & Wendy Green, 01233650710, tony.green41@yahoo.co.uk. *From Drovers Island Ashford take A20 Maidstone L @ 1st r'about Orchard Heights, R @ next r'about cont. to next r'about L into Forest Ave follow NGS signs.* **Sun 4 July, Sun 1 Aug, Sun 5 Sept (11-4). Adm £4, chd free. Home-made teas. Visits also by arrangement July to Sept for groups of 5 to 20.**
Small suburban garden evolved since May 2017 containing unusual and rare trees, shrubs and plants e.g. Multi-stemmed Ginkgo Biloba,

Japanese Redwood, Wolemi Pine and a collection of Salvias. There is a sunken garden with an exotic section and a Japanese style area.

43 FRITH OLD FARMHOUSE
Frith Road, Otterden, Faversham, ME13 0DD. Drs Gillian & Peter Regan, 01795 890556, peter.regan@cantab.net. *½ m off Lenham to Faversham rd. Fm A20 E of Lenham N up Hubbards Hill, follow signs Eastling; after 4m L into Frith Rd. Fm A2 in Faversham turn S (Brogdale Rd); continue 7m (thro' Eastling), R into Frith Rd.* **Sun 21 Feb (11-3). Light refreshments. Also open Doddington Place. Sun 11, Sun 25 Apr, Sun 9 May, Sun 25 July (11-5). Home-made teas. Adm £5, chd free. Visits also by arrangement Apr to Sept for groups of up to 30.**
A riot of plants growing together as if in the wild, developed over 40 yrs. No neat edges or formal beds, but several hundred interesting (& some very unusual) plants. Trees and shrubs chosen for year-round appeal. Special interest in bulbs and woodland plants. Visitor comments - 'one of the best we have seen, natural & full of treasures', 'a plethora of plants', 'inspirational', 'a hidden gem'. Altered habitat areas to increase the range of plants grown. Areas for wildlife.

44 GODDARDS GREEN
Angley Road, Cranbrook, TN17 3LR. John & Linde Wotton, 01580 715507, jpwotton@gmail.com, www.goddardsgreen.btck.co.uk. *½ m SW of Cranbrook. On W of Angley Rd. (A229) at junction with High St, opp War Memorial.* **Sun 18 Apr, Sun 4 July (12-4). Adm £5, chd free. Home-made teas. Visits also by arrangement May to Sept for groups of 10+. Coaches need to drop off and pick up in the road, outside the front gate.**
Gardens of about 5 acres, surrounding beautiful 500+yr old clothier's hall (not open), laid out in 1920s and redesigned since 1992 to combine traditional and modern planting schemes. Fountain and rill, water garden, fern garden, mixed borders of bulbs, perennials, shrubs, trees and exotics; birch grove, grass border, pond, kitchen garden, meadows, arboretum and mature orchard. Some slopes and steps, but

most areas (though not the toilets) are wheelchair accessible. Disabled parking is reserved near the house.

♿ 🐕 ♿ ☕ 💷 🍽

45 ◆ GODINTON HOUSE & GARDENS

Godinton Lane, Ashford, TN23 3BP. The Godinton House Preservation Trust, 01233 643854, info@godintonhouse.co.uk, www.godintonhouse.co.uk. *1½ m W of Ashford. M20 J9 to Ashford. Take A20 towards Charing & Lenham, then follow brown tourist signs.* **For NGS: Sun 21 Mar, Sun 16 May, Fri 18 June, Sun 26 Sept (1-5). Adm £7, chd free. Home-made teas. For other opening times and information, please phone, email or visit garden website.**
12 acres complement the magnificent Jacobean house. Terraced lawns lead through herbaceous borders, rose garden and formal lily pond to intimate Italian garden and large walled garden with delphiniums, potager, cut flowers and iris border. March/April the 3 acre wild garden is a mass of daffodils, fritillaries, primroses and other spring flowers. Delphinium Festival (12 June - 20 June). Garden workshops and courses throughout the yr. Partial wheelchair access to ground floor of house and most of gardens.

♿ 🐕 💷 🍽

46 GODMERSHAM PARK

Godmersham, CT4 7DT. Mrs Fiona Sunley, *5m NE of Ashford. Off A28, midway between Canterbury & Ashford.* **Sun 28 Mar, Sun 6 June (1-5). Adm £5, chd free. Home-made teas in the Orangery. Donation to Godmersham Church.**
24 acres of restored wilderness and formal gardens set around C18 mansion (not open). Topiary, rose garden, herbaceous borders, walled kitchen garden and recently restored Italian & swimming pool gardens. Superb daffodils in spring and roses in June. Historical association with Jane Austen. Also visit the Heritage Centre. Deep gravel paths.

♿ 🐕 🚗 💷

47 ◆ GOODNESTONE PARK GARDENS

Wingham, Canterbury, CT3 1PL. Francis Plumptre, 01304 840107, enquiries@ goodnestoneparkgardens.co.uk, www.goodnestoneparkgardens.

co.uk. *6m SE of Canterbury. Village lies S of B2046 from A2 to Wingham. Brown tourist signs off B2046.* **For NGS: Sun 14 Feb, Sun 13 June (11-5). Adm £7, chd £2. Light refreshments at The Old Dairy Cafe. For other opening times and information, please phone, email or visit garden website.**
One of Kent's outstanding gardens and the favourite of many visitors. 14 acres with views over parkland. Something special year-round from snowdrops and spring bulbs to the famous walled garden with old fashioned roses and kitchen garden. Outstanding trees and woodland garden with cornus collection and hydrangeas later. Two arboretums and a contemporary gravel garden.

♿ 🐕 ♿ ☕ 💷 🍽

48 GRAVESEND GARDEN FOR WILDLIFE

68 South Hill Road, Windmill Hill, Gravesend, DA12 1JZ. Judith Hathrill, 07810 550991, judith.hathrill@live.com. *On Windmill Hill 0.6m from Gravesend town centre, 1.8m from A2. From A2 take A227 towards Gravesend. At T-lights with Cross Lane turn R then L at next T-lights, following yellow NGS signs. Park in Sandy Bank Rd or Leith Park Rd.* **Sat 3, Sun 4 July (12-5). Adm £4, chd free. Cream teas. Gluten free oake also available. Visits also by arrangement May to Aug for groups of up to 20.**
Features and planting designed to attract and feed wildlife all year round with borders containing a colourful mix of trees, shrubs, perennials, annuals, herbs, grasses, wildflowers, fruit and vegetables. 3 wildlife ponds. Containers of fruit, flowers and vegetables on the terraces. Small lawn with seating, summerhouse and arbour to sit and enjoy the tranquillity. Information and leaflets about gardening for wildlife always available. Due to steep steps and uneven paths there is no wheelchair access to the garden.

♿ 💷

49 ◆ GREAT COMP GARDEN

Comp Lane, Platt, nr Borough Green, Sevenoaks, TN15 8QS. Great Comp Charitable Trust, 01732 885094, office@greatcompgarden.co.uk, www.greatcompgarden.co.uk. *7m E of Sevenoaks. 2m from Borough*

Green Station. Accessible from M20 & M26 motorways. A20 at Wrotham Heath, take Seven Mile Lane, B2016; at 1st Xrds turn R; garden on L *½ m.* **For NGS: Sun 28 Mar, Sun 31 Oct (10-5). Adm £8.50, chd free. For other opening times and information, please phone, email or visit garden website.**
Skilfully designed 7 acre garden of exceptional beauty. Spacious setting of well maintained lawns and paths lead visitors through plantsman's collection of trees, shrubs, heathers and herbaceous plants. Early C17 house (not open). Magnolias, hellebores and snowflakes (leucojum), hamamellis and winter flowering heathers are a great feature in the spring. A great variety of perennials in summer incl salvias, dahlias and crocosmias. Tearoom open daily for morning coffee, home-made lunches and afternoon teas. Most of garden accessible to wheelchair users. Disabled WC.

♿ ☕ 🚗 💷

50 NEW HAMMOND PLACE

High Street, Upnor, Rochester, ME2 4XG. Paul & Helle Dorrington. *3m NE of Strood or at A2. J1 take A289 twds Grain at r'about follow signs to Gillingham. After 2nd r'about take 1st L following signs to Upnor & Upnor Castle. Park in the free car park & continue by foot to the High St.* **Sat 10, Sun 11 July (10-5). Adm £3, chd free. Home-made teas.**
A small village garden around a Scandinavian style house growing an eclectic mix of fruit, vegetables and flowers. Greenhouse, Pond, Sauna Hut.

☕ 💷

National Garden Scheme gardens are identified by their yellow road signs and posters. You can expect a garden of quality, character and interest, a warm welcome and plenty of home-made cakes!

16 Northwood Road

51 HAVEN
22 Station Road, Minster,
Ramsgate, CT12 4BZ. Robin
Roose-Beresford, 01843 822594,
robin.roose@hotmail.co.uk. *Off A299
Ramsgate Rd, take Minster exit from
Manston r'bout, straight rd, R fork at
church is Station Rd.* **Sun 14 Mar,
Sun 4, Mon 5 Apr, Mon 3, Sun 23
May, Sun 6, Sun 20 June, Sun 18
July, Sun 1, Mon 30 Aug, Sun 19
Sept, Sun 17 Oct (10-4). Adm £5,
chd free. Visits also by arrangement
Mar to Oct for groups of up to 10.**
Award winning 300ft garden, designed
in the Glade style, similar to Forest
gardening but more open and with use
of exotic and unusual trees, shrubs
and perennials, with wildlife in mind,
devised and maintained by the owner,
densely planted in a natural style with
stepping stone paths. Two ponds (one
for wildlife, one for fish with water lilies),
gravel garden, rock garden, fernery,
Japanese garden, cactus garden,
hostas and many exotic, rare and
unusual trees, shrubs and plants incl
tree ferns and bamboos and yr-round
colour. Greenhouse cactus garden.
Collection of Bromeliads.

**52 ♦ HEVER CASTLE &
GARDENS**
Edenbridge, TN8 7NG. Hever
Castle Ltd, 01732 865224,
info@hevercastle.co.uk,
www.hevercastle.co.uk. *3m SE of
Edenbridge. Between Sevenoaks &
East Grinstead off B2026. Signed from
J5 & J6 of M25, A21, A264.* **For NGS:
Wed 21 Apr (10.30-6). Adm £15.55,
chd £9.75. Light refreshments in
grounds. For other opening times
and information, please phone,
email or visit garden website.**
Romantic double-moated castle, the
childhood home of Anne Boleyn, set
in 125 acres of formal and natural
landscape. Topiary, Tudor herb
garden, magnificent Italian garden
with classical statuary, sculpture and
fountains. 38-acre lake, yew and
water mazes. Walled rose garden
with over 4000 roses, 110 metre-long
herbaceous border. Fine Daffodil
displays in Mar and Tulip displays in
Apr. Partial wheelchair access.

53 HOATH HOUSE
Chiddingstone Hoath, Edenbridge,
TN8 7DB. Mr & Mrs Richard
Streatfeild, 07973842139,
richard@hoathhouse.com,
www.hoathhouse.com. *4m SE of
Edenbridge via B2026. At Queens
Arms PH turn E to Markbeech.
Approx 1m E of Markbeech.* **Wed 3
Mar (10-6.30). Adm £5, chd free.
Home-made teas. Light lunches
and teas available.**
Mediaeval/Tudor family house
surrounded by mature and unusual
young trees, knot garden, shaded
garden, herbaceous borders, yew
hedges. Situated in the Kentish
High Weald there are many stunning
viewpoints across the surrounding
countryside. This fine old garden
beginning to show the effect of new
ideas. Refreshments will be served
in the library where family will be on
hand to give a brief description of the
house and their long connection with
it. Steps, a gravel drive and uneven
paving stones may cause difficulty.

54 ♦ HOLE PARK
Benenden Road, Rolvenden,
Cranbrook, TN17 4JB. Mr
& Mrs Edward Barham,
01580 241344, info@holepark.com,
www.holepark.com. *4m SW of
Tenterden. Midway between Rolvenden
& Benenden on B2086. Follow brown
tourist signs from Rolvenden.* **For
NGS: Sun 11 Apr, Wed 12 May, Sun
6 June, Sun 10 Oct (11-6). Adm £9,
chd £1. Light refreshments. Picnics
only in picnic site and car park
please. For other opening times and
information, please phone, email or
visit garden website.**
Hole Park is proud to stand amongst
the group of gardens which first
opened in 1927 soon after it was
laid out by my great grandfather.
Our 15 acre garden is surrounded
by parkland with beautiful views
and contains fine yew hedges, large
lawns with specimen trees, walled
gardens, pools and mixed borders
combined with bulbs, rhododendrons
and azaleas. Massed bluebells in
woodland walk, standard wisterias,
orchids in flower meadow and
glorious autumn colours make this a
garden for all seasons. Wheelchairs
are available for loan and may be
reserved.

55 HURST HOUSE
Waltham Road, Hastingleigh,
Ashford, TN25 5JD. Mrs Lynn Smith,
01233 750120, btclynn@aol.com.
*On the top of the downs above
Wye Village, between Ashford and
Canterbury, 10 mins from A28.
From A28 go through Wye towards
Hastingleigh. Follow rd over Downs to
x-roads, L towards Waltham. Take 1st
R turn and 1st L. Follow parking signs.*
**Sat 24, Sun 25 July (11-5.30). Adm
£5, chd free. Home-made teas.
Visits also by arrangement June to
Aug for groups of 5+.**
3 acre idyllic garden surrounded by
bluebell woods. Winner Kent Life
garden of the year 2019. Mature
borders, winding beds, ponds, planting
designed for year round colour. Damp
shady borders/dry sunny borders,
unusual plants, huge pieris and unusual
trees. Stumpery planted with ferns/
shade loving plants – woodland walk/
painted forest with artwork on trees,
intriguing features and sculptures.
A massive owl mural, painted trees
and many sculptures in the woods,
with interest for all the family. Curving
beds with huge variety of plants,
flowing grasses and sculptural plants.
Gravel garden with winding borders
planted for colour throughout the year.
Colourful displays of flowering pots. Is
possible to use a wheelchair - gravel
drive and muddy paths. Disabled drop
off at Hurst House.

56 ♦ IGHTHAM MOTE
Mote Road, Ivy Hatch, Sevenoaks,
TN15 0NT. National Trust,
01732 810378, ighthammote@
nationaltrust.org.uk, www.
nationaltrust.org.uk/ightham-mote.
*6m E of Sevenoaks. Off A25, 2½m S
of Ightham. Buses from train stations:
Sevenoaks ('go' route 4 Mon-Fri to
Ightham Mote) or Borough Green
('Arriva' route 306/308 to Ightham or
Ightham Common).* **For NGS: Thur
1 Apr, Thur 7 Oct (10-5). Adm £18,
chd £9. Light refreshments on site
in the Mote Café. For other opening
times and information, please
phone, email or visit garden website.**
Lovely 14-acre garden surrounding
a picturesque medieval moated
manor house c1320, open for NGS
since 1927. Herbaceous borders,
lawns, C18 cascade, fountain pools,
courtyards and a cutting garden
provide formal interest; while the
informal north lake, stream, pleasure
grounds, stumpery, dell, fernery and
orchard contribute to the sense of
charm and tranquillity. Walks on
surrounding estate are open to dogs. A
range of meals, snacks, cakes, cream
teas, hot & cold drinks available. Please
check NT Ightham Mote website for
current access details, opening times,
last admission & ask for access guide
at visitor reception.

57 NEW KENFIELD HALL

Kenfield, Petham, Canterbury, CT4 5RN. Barnaby & Camilla Swire, kenfieldhallgarden@gmail.com. *Continue along Kenfield Road, down the hill and up the other side. Pass the farm and look out for the signs.* **Visits by arrangement Apr to July for groups of 5 to 30. Adm £10, chd free.**
The 8 acre gardens benefit from views within the AONB and its diverse wildlife, consisting of a sunken formal garden, lawns, mixed borders, herbaceous borders, spring bulbs, a Japanese water garden, orchards, an organic vegetable garden with glasshouses, a wildflower meadow, a rose garden and woodland garden.

58 NEW THE KNOLL FARM

Giggers Green Rd, Aldington, Ashford, TN25 7BY. Lord & Lady Aldington. *The Postcode leads to Goldenhurst, our drive entrance is opp & a little further down hill.* **Sat 10 Apr (12-5). Adm £8, chd free. Donation to Bonnington Church.**
10 acres of woodland garden with over 100 camellias, acer japonica, young specimen pines and oaks; bluebell wood; formal elements; Flock of Jacob Sheep around lake; long views across Romney Marsh. Picnics welcome by the lake. Paths throughout but the whole garden is on a slope.

59 KNOWLE HILL FARM

Ulcombe, Maidstone, ME17 1ES. The Hon Andrew & Mrs Cairns, 01622 850240, elizabeth@knowlehillfarm.co.uk, www.knowlehillfarmgarden.co.uk. *7m SE of Maidstone. From M20 J8 follow A20 towards Lenham for 2m. Turn R to Ulcombe. After 1½ m, L at Xrds, after ½ m 2nd R into Windmill Hill. Past Pepper Box PH, ½ m 1st L to Knowle Hill.* **Sat 6, Sun 7, Mon 8 Feb (11-3.30). Light refreshments. Sat 17, Sun 18, Mon 19 July (2-5). Home-made teas. Adm £6, chd free. Takeaway refreshments available. 2022: Sat 5, Sun 6, Mon 7 Feb. Visits also by arrangement Feb to Sept for groups of up to 30.**
2 acre garden created over 35yrs on S-facing slope of N Downs. Spectacular views. Snowdrops and hellebores, many tender plants, china roses, agapanthus, verbenas, salvias and grasses flourish on light soil. Topiary continues to evolve with birds

at last emerging. Lavender ribbons hum with bees. Pool enclosed in small walled white garden. A green garden completed in 2018. Access only for 35 seater coaches. Some steep slopes.

60 LADHAM HOUSE

Ladham Road, Goudhurst, TN17 1DB. Guy & Nicola Johnson. *8m E of Tunbridge Wells. On NE of village, off A262. Through village towards Cranbrook, turn L at The Goudhurst Inn. 2nd R into Ladham Rd, main gates approx 500yds on L.* **Sun 9 May (2-5). Adm £5, chd free. Home-made teas.**
Ten acres of garden with many interesting plants, trees and shrubs, incl rhododendrons, camellias, azaleas and magnolias. A beautiful rose garden, arboretum, an Edwardian sunken rockery, ponds, a vegetable garden & a woodland walk leading to bluebell woods. There is also a spectacular 60 meter twin border designed by Chelsea Flower Show Gold Medal winner, Jo Thompson. Small Classic Car Display.

61 NEW LINTON PARK

Heath Road, Linton, Maidstone, ME17 4AB. Linton Park Plc, www.camellia.plc.uk. *ME17 4AJ is the post code for North Lodge at the top of drive on the main rd. Visitors need to come down the drive for around ½ m to reach the main house & garden.* **Sun 23 May (10.30-4). Adm £10, chd free. Home-made teas.**
Overlooking the Weald of Kent this south facing, hillside garden is set in 450 acres of parkland. Re-imagined over the last 35 years, using original J C Loudon plans and historic maps, it boasts wonderful mature tree specimens of, amongst others, Copper Beech, Cedars, Limes and Oaks crowned by a magnificent avenue of Wellingtonia planted in the mid 19th century. Unsuitable for wheelchairs - gravel paths.

62 LITTLE GABLES

Holcombe Close, Westerham, TN16 1HA. Mrs Elizabeth James. *Centre of Westerham. Off E side of London Rd A233, 200yds from The Green. Please park in public car park. No parking available at house.* **Sat 15, Sun 16 May, Sat 5, Sun 6 June (2-5). Adm £4, chd free. Home-made teas.**

¾ acre plant lover's garden extensively planted with a wide range of trees, shrubs, perennials etc, incl many rare ones. Collection of climbing and bush roses. Large pond with fish, water lilies and bog garden. Fruit and vegetable garden. Large greenhouse.

63 LITTLE MOCKBEGGAR

Mockbeggar Lane, Biddenden, Ashford, TN27 8ES. Derek & Sue East. *Situated between Biddenden & Benenden. Opp Benenden Hospital, cont down Mockbeggar Lane for ½ m until you see the egg sign.* **Sun 20 June (2-5.30). Adm £5, chd free. Cream teas.**
Little Mockbeggar is a charming cottage garden crammed with herbaceous borders, shrubs, roses and climbers, incl a wildlife pond and vegetable garden. A large wildlife meadow full of oxeye daisies and many wildflowers of interest, with paths to meander through. No dogs please due to farm animals.

64 LORDS

Sheldwich, Faversham, ME13 0NJ. John Sell CBE & Barbara Rutter, 01795 536900, john@sellwade.co.uk. *On A251 4m S of Faversham & 3½ m N of Challock Xrds. From A2 or M2 take A251 towards Ashford. ½ m S of Sheldwich church find entrance lane on R adjacent to wood.* **Sun 13 June (2-5). Home-made teas. Also open Fisher Street Oast. Visits also by arrangement Apr to July for groups of 10 to 30.**
C18 walled garden. Mediterranean terrace and citrus standing. Flowery mead under apples, pears, quince, crab apple and medlar. Grass tennis court. A cherry orchard grazed by Jacob sheep. Pleached hornbeams, clipped yew hedges and topiary, lawns, ponds and wild area. Fine old sweet chestnuts, planes, copper beech and 120ft tulip tree. New this year: herb and salad bed, garden sculpture. Some gravel paths.

65 MANWOOD HOUSE

91 Strand Street, Sandwich, CT13 9HX. Mr Philip & Mrs Rebecca Croall. *Access to the garden is through white gates in Paradise Row, which is a small lane off Strand St.* **Sat 5 June**

(2-5). Adm £6, chd free. Light refreshments.
The abundantly planted gardens of Manwood House (not open) built in 1564 as the first home of the free school founded by Sir Roger Manwood, a favourite of Elizabeth 1. The gardens extend to the side and rear of the property and are bounded by flint walls. They feature a remarkable and ancient robinia tree, an ornamental pond and luxuriantly planted borders, extensively replanted in 2015. Sandstone paths into and through large parts of the garden, though narrow in places. A small number of shallow steps to access certain areas.

66 MARSHBOROUGH FARMHOUSE
Farm Lane, Marshborough, Sandwich, CT13 0PJ. David & Sarah Ash, 01304 813679. *1½ m W of Sandwich, ½ m S of Ash. From Ash take R fork to Woodnesborough. After 1m Marshborough sign. L into Farm Lane at white thatched cottage, garden 100yds on L. Coaches must phone for access information.* **Visits by arrangement in June for groups of 5+. Adm £6, chd free. Home-made teas.**
Interesting 2½ acre plantsman's garden, developed enthusiastically over 22 yrs by the owners. Paths and lawns lead to many unusual shrubs, trees and perennials in island beds, borders, rockery and raised dry garden creating yr-round colour and interest. Tender pot plants, succulents in glass house, pond and water features. Over 70 varieties of Salvia both hardy and tender.

67 MAY COTTAGE
52, St Botolphs Road, Sevenoaks, TN13 3AG. Graham & Maggie Moat. *Central Sevenoaks, close to the historic Vine cricket ground & Knole Park. Walking distance Sevenoaks station. 2m from J5 M25, on the B2020. Plenty of free parking on road.* **Sun 2 May (11-5). Adm £5, chd free.**
A ⅓ of an acre town family spring garden, comprising mature shrubs and specimen plants, with spring bulbs, cowslips, vegetable patch and local trees that shield the garden. Blue, white and pink bluebells spread amongst many shrubs featuring wire-netting bush (Corokia cotoneaster), Enkianthus campanulatus, Azalea April Showers, Japanese maples

together with blossom on a mature Bramley apple tree. One way system for all visitors. Wheelchair users exit via garage.

68 12 THE MEADOWS
Chelsfield, Orpington, BR6 6HS. Mr Roger & Mrs Jean Pemberton. *3m from J4 on M25. 10 mins walk from Chelsfield station. Exit M25 at J4. At r'about 1st exit for A224, next r'about 3rd exit - A224, ½ m, take 2nd L, Warren Rd. Bear L into Windsor Drive. 1st L The Meadway, follow signs to garden.* **Sun 23 May, Sun 12 Sept (11-5.30). Adm £5, chd free. Light refreshments.**
Front garden Mediterranean style gravel with sun loving plants. Rear ¾ acre garden in 2 parts. Semi-formal Japanese style area with tea house, two ponds, one Koi and one natural (lots of spring interest). Acers, grasses, huge bamboos etc and semi wooded area, children's path with 13ft high giraffe, Sumatran tigers and lots of points of interest. Children and well behaved dogs more than welcome. Designated children's area. Adults only admitted if accompanied by responsible child! Silver award for garden with the LGS. Winner of first prize for our back garden from Bromley in Bloom. Wheelchair access to all parts except small stepped area at very bottom of garden.

69 5 MONTEFIORE AVENUE
Ramsgate, CT11 8BD. Pauline & Mike Ashley. *Opp Thanet Bowls Club. A255 Hereson Rd from Ramsgate to Broadstairs. Turn R (L if from Broadstairs) at Garden Centre into Montefiore Av. Cross Dumpton Pk Dr into continuation of Montefiore Av. House on R.* **Sat 31 July, Sun 1 Aug (12-5). Adm £5, chd free. Light refreshments.**
An Edwardian walled garden densely planted with a variety of herbaceous and tropical plants. A courtyard with over 100 pots features acers, hostas, agaves, aeoniums and more. Through an arch, a lawned area with gazebo, borders, large koi pond, waterfall and two small wildlife ponds. A pergola leads to a vegetable garden and treehouse garden with a mix of roses, mediterranean and tropical plants. A highlight of the garden is the Koi pond, with a mix of large fish, and the wide variety of plants, both Mediterranean and tropical. The garden is situated in a cul-de-sac

with an entrance to King George VI park where the famous Italianate Greenhouse is located. Gravel paths and narrow paths in some areas.

70 THE MOUNT
Haven Street, Wainscott, Rochester, ME3 8BL. Marc Beney & Susie Challen. *3½ m N of Rochester. At M2 J1, take A289 twds Grain. At r'bout, R into Wainscott. Co-op ahead, turn R into Higham Rd. R into Islingham Farm Rd, parking in field on corner of Woodfield Way. House 7min walk up slight hill.* **Sat 17, Sun 18 July (10-4). Adm £5, chd free. Home-made teas.**
2 acres surround the house and include a renovated walled kitchen garden with fruit trees, roses, veg beds & a colourful herbaceous border; old grass tennis court with a mown labyrinth & a small nuttery around a wildlife pond; white & yellow terrace garden above a pleached lime path with iris and lavender & below, remains of Victorian glasshouses. Lovely countryside views. Mature specimen trees. Gravel drive, uneven paths and steps.

71 ◆ MOUNT EPHRAIM GARDENS
Hernhill, Faversham, ME13 9TX. Mr & Mrs E S Dawes & Mr W Dawes, 01227751496, info@mountephraimgardens.co.uk, www.mountephraimgardens.co.uk. *3m E of Faversham. From end of M2, then A299 take slip rd 1st L to Hernhill, signed to gardens.* **For NGS: Sun 21 Mar, Thur 10 June, Sun 26 Sept (11-5). Adm £7, chd £2.50. Cream teas in West Wing Tea Room serving home-made lunches and delicious afternoon teas. For other opening times and information, please phone, email or visit garden website.**
Mount Ephraim is a privately-owned family home set in ten acres of terraced Edwardian gardens with stunning views over the Kent countryside. Highlights incl a Japanese rock and water garden, arboretum, unusual topiary and a spectacular grass maze plus many mature trees, shrubs and spring bulbs. Partial wheelchair access; top part manageable, but steep slope. Disabled WC. Full access to tea room.

72 NETTLESTEAD PLACE
Nettlestead, ME18 5HA.
Mr & Mrs Roy Tucker,
www.nettlesteadplace.co.uk. *6m W/SW of Maidstone. Turn S off A26 onto B2015 then I'm on L, next to Nettlestead Church.* **Sun 6 June (12-4). Sun 3 Oct (2-4.30). Home-made teas. Adm £6, chd free.**
C13 manor house in 10 acre plantsman's garden. Large formal rose garden. Large herbaceous garden of island beds with rose and clematis walkway leading to a newly planned garden of succulents. Fine collection of trees and shrubs; sunken pond garden, a maze of Thuja, terraces, bamboos, glen garden, Acer lawn. Young pinetum adjacent to garden. Sculptures. Wonderful open country views. Maze. Gravel and grass paths. Most of garden accessible (but not sunken pond garden). Large steep bank and lower area accessible with some difficulty.
🚼 🐄 🚗 🍵 🥖

73 NORTON COURT
Teynham, Sittingbourne, ME9 9JU.
Tim & Sophia Steel, 07798804544, sophia@nortoncourt.net. *Off A2 between Teynham & Faversham. L off A2 at Esso garage in Norton Lane; next L into Provender Lane; L signed Church for car park.* **Mon 7, Tue 8 June (2-5). Adm £5, chd free. Home-made teas. Visits also by arrangement Apr to Sept for groups of 10 to 30.**
10 acre garden within parkland setting. Mature trees, topiary, wide lawns and clipped yew hedges. Orchard with mown paths through wildflowers. Walled garden with mixed borders and climbing roses. Pine tree walk. Formal box and lavender parterre. Tree house in the Sequoia. Church open, adjacent to garden. Flat ground except for 2 steps where ramp is provided.
🚼 🍵

74 OAK COTTAGE AND SWALLOWFIELDS NURSERY
Elmsted, Ashford, TN25 5JT.
Martin & Rachael Castle. *6m NW of Hythe. From Stone St (B2068) turn W opp the Stelling Minnis turning. Follow signs to Elmsted. Turn L at Elmsted village sign. Limited parking at house, further parking at Church (7mins walk).* **Fri 23, Sat 24 Apr, Fri 21, Sat 22 May (11-4). Adm £5, chd free. Home-made teas.**
Get off the beaten track and discover this beautiful ½ acre cottage garden in the heart of the Kent countryside. This plantsman's garden is filled with unusual and interesting perennials, including a wide range of salvias. There is a small specialist nursery packed with herbaceous perennials. Auriculas displayed in traditional theatres in April.
🌺 🍵 🎕

75 THE OLD BARN
Bells Farm Road, East Peckham, Tonbridge, TN12 5NA. John Greenslade. *Follow Bells Farm Rd, look out for brick walls with black sign saying 'The Old Barn'.* **Sun 21 Mar, Sun 13 June (11-3.30). Adm £4, chd free. Light refreshments.**
The Old Barn is a beautiful family garden surrounded by outbuildings and fields. In the last few years this garden has gone from neglected site to stunning gardens, with plenty more changes still to happen in the future. Large spring bulb drifts throughout the lawns. Japanese corner and various courtyard garden areas. This garden has a deep gravel drive, which may not be suitable for wheelchairs. The courtyard has a few steps within, so may not be accessible.
🐄 🍵 🎕

76 OLD BLADBEAN STUD
Bladbean, Canterbury, CT4 6NA. Carol Bruce, www.oldbladbeanstud.co.uk. *6m S of Canterbury. From B2068, follow signs into Stelling Minnis, turn R onto Bossingham Rd, then follow yellow NGS signs through single track lanes.* **Sun 23 May (10-6). Every Sun 13 June to 4 July (10-6). Sun 18 July (10-6). Pre-booking essential, please visit www.ngs. org.uk for information & booking. Adm £6, chd free.**
Five interlinked gardens all designed and created from scratch by the garden owner on three acres of rough grassland between 2003 and 2011. Romantic walled rose garden with over 90 old fashioned rose varieties, tranquil yellow and white garden, square garden with a tapestry of self sowing perennials and Victorian style greenhouse, 300ft long colour schemed symmetrical double borders and an organic fruit and vegetable garden. The gardens are maintained entirely by the owner and were designed to be managed as an ornamental ecosystem with a large number of perennial species encouraged to set seed, and with staking, irrigation, mulching and chemical use kept an absolute minimum. Please see the garden website at www.oldbladbeanstud. co.uk to read published articles, see photos of what's currently in bloom and for more visitor information.

77 THE OLD RECTORY
Valley Road, Fawkham, Longfield, DA3 8LX. Karin & Christopher Proudfoot, 01474 707513, keproudfoot@gmail.com. *1m S of Longfield. Midway between A2 & A20, on Valley Rd 1½m N of Fawkham Green, 0.3m S of Fawkham church, opp sign for Gay Dawn Farm/ Corinthian Sports Club. Parking on drive only. Not suitable for coaches.* **Visits by arrangement in Feb for groups of up to 20. Also open by arrangement Feb 2022. Adm £5, chd free. Home-made teas.**
1½ acres with impressive display of long-established naturalised snowdrops and winter aconites; over 100 named snowdrops added more recently. Garden developed around the snowdrops over 35yrs, incl hellebores, pulmonarias and other early bulbs and flowers, with foliage perennials, shrubs and trees, also natural woodland. Gentle slope, gravel drive, some narrow paths.
🚼 🌺 🍵

78 NEW THE OLD VICARAGE
Dully Road, Tonge, Sittingbourne, ME9 9NP. Sarah Varley. *Off A2 between Sittingbourne and Faversham. Please note location is Old Vicarage, and not Old Vicarage Bungalow. Parking available in field on corner of Dully road and A2.* **Sat 26, Sun 27 June (11-4.30). Adm £5, chd free. Home-made teas.**
A delightful mature garden of just under 3 acres, which incl an arboretum with some unusual trees, for example a tulip tree, a magnificent cedar, and ancient yews. Gravel paths meander through the wood and onto lawns where they pass formal hedging, mixed herbaceous borders, an active dovecote, a cut flower garden, a potager, a herb garden and more.
🚼 🌺 🍵

79 THE ORANGERY
Mystole, Chartham, Canterbury, CT4 7DB. Rex Stickland & Anne Prasse, 01227 738348, rex.mystole@btinternet.com. *5m SW of Canterbury. Turn off A28 through Shalmsford Street. In 1½m at Xrds turn R downhill. Cont, ignoring rds on L & R. Ignore*

drive on L (Mystole House only). At sharp RH bend in 600yds turn L into private drive. **Sun 16 May (1-6); Sat 31 July, Sun 1 Aug (1-5). Adm £5, chd free. Visits also by arrangement Mar to Sept for groups of 10+.**

1½ acre gardens around C18 orangery, now a house (not open). Magnificent extensive herbaceous border & impressive ancient wisteria. Large walled garden with a wide variety of shrubs, mixed borders & unusual specimen trees. Water features & intriguing collection of modern sculptures in natural surroundings. Refreshments on terrace with splendid views over ha-ha to the lovely Chartham Downs. Ramps to garden.

80 PARSONAGE OASTS
Hampstead Lane, Yalding, ME18 6HG. Edward & Jennifer Raikes, 01622 814272, jmraikes@parsonageoasts.plus.com. *6m SW of Maidstone. On B2162 between Yalding village & stn, turn off at Boathouse PH. over lifting bridge, cont 100 yds up lane. House & car park on L.* **Visits by arrangement for groups of up to 30. Coaches must park on road. Adm £4, chd free.**

Our garden has a lovely position on the bank of the R Medway. Typical Oast House (not open) often featured on calendars and picture books of Kent. 70yr old garden now looked after by grandchildren of its creator. ¾ acre garden with walls, daffodils, crown imperials, shrubs, clipped box and a spectacular magnolia. Small woodland on river bank. Best in spring, but always something to see. No refreshments but local Boathouse pub (5mins) & in the garden several nice places for a picnic. Unfenced river bank. Gravel paths. Bee hives.

81 ◆ PENSHURST PLACE & GARDENS
Penshurst, TN11 8DG. Lord & Lady De L'Isle, 01892 870307, contactus@penshurstplace.com, www.penshurstplace.com. *6m NW of Tunbridge Wells. SW of Tonbridge on B2176, signed from A26 N of Tunbridge Wells.* **For NGS: Tue 14 Sept (10.30-6). Adm £11, chd £6.50. For other opening times and information, please phone, email or visit garden website.**

11 acres of garden dating back to C14. The garden is divided into a series of rooms by over a mile of yew hedge. Profusion of spring bulbs, formal rose garden and famous peony border. Woodland trail and arboretum. Yr-round interest. Toy museum. Some paths not paved and uneven in places; own assistance will be required. 2 wheelchairs available for hire.

82 PHEASANT BARN
Church Road, Oare, ME13 0QB. Paul & Su Vaight, 07843 739301, suvaight46@gmail.com. *2m NW of Faversham. Entering Oare from Faversham, turn R at Three Mariners PH towards Harty Ferry. Garden 400yds on R, before church. Parking on roadside.* **Fri 25, Tue 29, Wed 30 June, Thur 1, Tue 6, Wed 7, Thur 8, Sat 10, Sun 11 July (11-4). Adm £6, chd free. Pre-booking essential, please visit www.ngs.org.uk for information & booking. Visits also by arrangement May to July for groups of up to 30. Contact owner directly.**

Series of smallish gardens around award-winning converted farm buildings in beautiful situation overlooking Oare Creek. Main area is nectar-rich planting in formal design with a contemporary twist inspired by local landscape. Also vegetable garden, dry garden, water features, wildflower meadow and labyrinth. July optimum for wildflowers. Kent Wildlife Trust Oare Marshes Bird Reserve within 1m. Two village inns serving lunches/dinners, booking recommended.

Gravesend Garden for Wildlife

83 PHEASANT FARM

Church Road, Oare, Faversham, ME13 0QB. Jonathan & Lucie Neame, 01795535366, pheasantfarm2019@gmail.com. *2m NW of Faversham. Enter Oare from Western Link Road. L at T-junction. R at Three Mariners PH into Church Rd. Garden 450yds on R, beyond Pheasant Barn, before church. Parking on roadside & as directed.* **Visits by arrangement Apr to June for groups of 10 to 30. Adm £10, chd free. Home-made teas. Entry price incl home-made refreshments.**

A walled garden surrounding C17 farmhouse with outstanding views over Oare marshes and creek. Main garden with shrubs and herbaceous plants. Infinity lawn overlooking Oare marshes and creek. Circular walk through orchard and adjoining churchyard. Two local public houses serving lunches. Some wheelchair access in main garden only.

84 NEW THE POSTERN

Postern Lane, Tonbridge, TN11 0QU. Mr & Mrs David Tennant, 07807229414. *Postern Lane runs E of Tonbridge off B2017, between Tonbridge & Tudeley.* **Wed 2 June (1.30-5). Adm £5, chd free. Visits also by arrangement. Donation to The Children's Society.**

4 acres with lawns, flowering shrubs, old and new shrub roses; apple and pear orchards. Georgian house (not open). Garden designed by Anthony du Gard Pasley. Wheelchair users could be dropped off in the main drive.

85 POTMANS HEATH HOUSE

Wittersham, TN30 7PU. Dr Alan & Dr Wilma Lloyd Smith. *1½m W of Wittersham. Between Wittersham & Rolvenden, 1m from junction with B2082. 200yds E of bridge over Potmans Heath Channel.* **Sun 18 Apr, Sun 6 June (2-6). Adm £5, chd free. Home-made teas.**

Large country garden divided into compartments each with a different style. Daffodils, tulips and many blossoming ornamental cherry and apple trees in Spring, often spectacular. Many and varied rose species; climbers a speciality. Early summer beds and borders. Orchards, some unusual trees, lawns, part walled vegetable garden, 2 greenhouses. NB House not open. Toilets at Wittersham Church. Some awkward slopes but generally accessible.

86 NEW QUAKER'S HALL ALLOTMENTS

Allotment Lane, Off Quaker's Hall Lane, Sevenoaks, TN13 3UZ. Sevenoaks Allotment Holders' Association on behalf of Sevenoaks Town Council, www.sevenoaksallotments.com. *Quaker's Hall Lane is off A225, St John's Hill. Site directly behind St John's Church.* **Sun 18 July (2-4). Adm by donation. Light refreshments.**

Our beautiful allotment gardens are 11½ acres right in the heart of Sevenoaks. A wide cross-section of allotment tenants grow a massive variety of flowers, fruit, veg and herbs using a number of different techniques. Gardeners cite healthy produce, exercise and relaxation in a beautiful open space as reasons to rent a plot. Stunning views of the North Downs make this a very special place to visit. Main paths are concrete with some steep slopes.

87 ◆ QUEX GARDENS

Quex Park, Birchington, CT7 0BH. Powell-Cotton Museum, 01843 842168, enquiries@powell-cotton.org, www.powell-cottonmuseum.org. *3m W of Margate. Follow signs for Quex Park on approach from A299 then A28 towards Margate, turn R into B2048 Park Lane. Quex Park is on L.* **For NGS: Sun 18 July (10-4). Adm £5, chd free. Light refreshments in Mama Feelgoods Café and Quex Barn. For other opening times and information, please phone, email or visit garden website.**

10 acres of woodland and gardens

Parsonage Oasts

with fine specimen trees unusual on Thanet, spring bulbs, wisteria, shrub borders, old figs and mulberries, herbaceous borders. Victorian walled garden with cucumber house, long glasshouses, cactus house, fruiting trees. Peacocks, dovecote, woodland walk, wildlife pond, children's maze, croquet lawn, picnic grove, lawns and fountains. Head Gardener will be available on NGS day to give tours & answer questions. Mama Feelgood's Boutique Café serving morning coffee, lunch or afternoon tea. Quex Barn farmers market selling local produce and serving breakfasts to evening meals. Picnic sites available. Garden almost entirely flat with tarmac paths. Sunken garden has sloping lawns to the central pond.

&. ✿ 🚗 🍺

GROUP OPENING

88 NEW **RAMSGATE GARDENS**
Ramsgate, CT11 9PX. Anne-Marie Nixey. *Enter Ramsgate on A299, continue on A255. At r'about take 2nd exit London Rd. Continue for less than 1m to the r'about and turn L onto Grange Rd.* **Sun 8 Aug (12-5). Combined adm £5, chd free. Light refreshments.**

> NEW **6 EDITH ROAD**
> Nicolette McKenzie.

> NEW **104 GRANGE ROAD**
> Anne-Marie Nixey.

> NEW **106 GRANGE ROAD**
> Mrs Sally Smart.

> NEW **12 WEST CLIFF ROAD**
> Brian Daubney.

Four evolving medium sized gardens in the beautiful, yet windy, coastal town of Ramsgate showing similar sized plots and how to make unique gardens out of them. Varied planting from traditional roses and bedding plants, to a range of vegetables and fruit trees, as well as use of recycled and sustainable materials and incorporating traditional family areas. Wheelchairs can access parts of 6 Edith Road and 104 Grange Road.

&. ✿ 🍺

---

89 **43 THE RIDINGS**
Chestfield, Whitstable, CT5 3QE. David & Sylvie Buat-Menard, 01227 500775, sylviebuat-menard@hotmail.com. *Nr Whitstable. From M2 heading E*

cont onto A299. In 3m take A2990. From r'about on A2990 at Chestfield, turn onto Chestfield Rd, 5th turning on L onto Polo Way which leads into The Ridings. **Sun 9, Sat 15 May (10-4). Sun 11 July (10-4), also open 23 Seafield Road. Adm £6, chd free. Home-made teas. Visits also by arrangement May to Sept for groups of 10 to 20.**
Delightful small garden brimming with interesting plants both in the front and behind the house. Many different areas. Dry gravel garden in front, raised beds with alpines and bulbs and borders with many unusual perennials and shrubs Kent Life First prize amateur garden for 2018. The garden ornaments are a source of interest for visitors. The water feature will be of interest for those with a tiny garden as planted with carnivorous plants. Many alpine troughs & raised beds as well as dry shade & mixed borders all in a small space.

🐑 ✿ 🚗 🍺

---

90 ❖ **RIVERHILL HIMALAYAN GARDENS**
Riverhill, Sevenoaks, TN15 0RR. The Rogers Family, 01732 459777, info@riverhillgardens.co.uk, www.riverhillgardens.co.uk. *2m S of Sevenoaks on A225. Leave A21 at A225 & follow signs for Riverhill Himalayan Gardens.* **For NGS: Wed 28 Apr, Thur 10 June (9-6). Adm £7, chd £7. Light refreshments in Otto's Cafe. Large cafe terrace or take away into the gardens. Pre-booking essential please visit www.riverhillgardens.co.uk. For other opening times and information, please phone, email or visit garden website.**
Beautiful hillside garden, privately owned by the Rogers family since 1840. Spectacular rhododendrons, azaleas and fine specimen trees. Edwardian Rock Garden with extensive fern collection, Rose Walk & Walled Garden with sculptural terracing. Bluebell walks. Extensive views across the Weald of Kent. Hedge maze, adventure playground, den building and Yeti spotting. Otto's Café serves speciality coffee, brunch, light lunches and cakes. Plant sales & quirky shed shop selling beautiful gifts & original garden ornaments. Disabled parking. Wheelchair access to Walled Garden, Rock Garden. Easy access to café, shop and tea terrace (no disabled WC).

&. 🐑 ✿ 🚗 🍺 🪑

---

91 **18 ROYAL CHASE**
Tunbridge Wells, TN4 8AY. Eithne Hudson, 07985305941, eithne.hudson@gmail.com. *Situated at the top of Mount Ephraim & The Common. Leave A21 at Southborough, follow signs for Tunbridge Wells. After St John's Rd take R turn at junction on Common & sharp R again. From South, follow A26 through town up over Common. Last turn on L.* **Sat 22, Sun 23 May (11.30-5). Adm £4, chd free. Visits also by arrangement May & June for groups of 10+.**
Town garden of a ¼ of acre on sandy soil which was originally Common and woodland. Lovely in late Spring with camellias, acers, rhododendrons and alliums. Large lawn with island beds planted with roses and herbaceous perennials and a small pond with lilies and goldfish. The owner is a florist and has planted shrubs and annuals specifically with flower arranging in mind. Spring bulbs in semi woodland setting in central Tunbridge Wells. Access through right hand side gate.

&. 🐑 🍺 🪑

---

92 **SANDOWN**
Plain Road, Smeeth, nr Ashford, TN26 6QX. Pamela Woodcock, 01303813478, pmwoodcock078@gmail.com. *4m SE of Ashford. Exit J10A onto A20, take 2nd L signed Smeeth, turn R Woolpack Hill, past garage on L, past nest I, garden on L. From A20 in Sellindge at Church, turn R carry on 1m. Park in layby on hill.* **Sun 9, Sun 23 May, Sun 6, Sun 20 June, Sun 4, Sun 18 July, Sun 1, Sun 15 Aug, Sun 5, Sun 19 Sept (11-4). Adm £5, chd free. Pre-booking essential, please visit www.ngs. org.uk for information & booking. Visits also by arrangement May to Sept for groups of up to 10. Telephone 01303 813478 or email pmwoodcock078@gmail.com.**
My small compact Japanese style garden and pond has visitor book comments such as: inspirational, just like Japan, a stunning hidden gem. There is a Japanese arbour, tea house/veranda, waterfall and stream. Acers, bamboos, ginkgo, fatsia japonica, clerodendrum trichotomum, pinus mugos, wisterias, hostas and mind your own business for ground cover. Wheelchair access to top section of garden only. 'Regret no small children owing to deep pond.

✿ 🪑

93 NEW **23 SEAFIELD ROAD**
Whitstable, CT5 2LW. Stephanie
Comins. *Tankerton/Chestfield. M2
heading east. Continue on to A299.
In 3m take A2990. After 3m at
Chestfield r'about turn L, then L on
to Herne Bay Rd. Then 3rd turning
on L.* **Sun 11 July (10-4). Adm
£4, chd free. Also open 43 The
Ridings.**
Small loved garden. Featuring a
tropical bed with exotics and cannas.
Mixed borders. Two circular beds
planted with perennials, shrubs and
bedding. Ornamental pond and
rockery. Large patio with selection of
pot plants.
🐾 ✲

94 NEW **45 SEYMOUR AVENUE**
Whitstable, CT5 1SA. Kevin
Tooher, sirplantalot@outlook.com.
*Near the centre of Whitstable town
& 400yrds from Whitstable station.
Take Thanet Way off A299 towards
Whitstable. 2nd r'about, L into
Millstrood Rd, bottom of hill R into
Old Bridge Rd, Station car park on
L & Seymour Ave on R.* **Sat 18,
Sun 19 Sept (11-4). Adm £5, chd
free. Home-made teas. Visits also
by arrangement Sept & Oct for
groups of 10 to 20.**
Larger than usual town centre garden
- about ¼ acre with wide range of
unusual plants grown on heavy wet
clay with lots of exotics growing in
containers, troughs and pots.
🐾 ✲ ☕

95 NEW **THE SILK HOUSE**
Lucks Lane, Rhoden Green, nr
Paddock Wood, Tunbridge Wells,
TN12 6PA. Adrian & Silke Barnwell,
01892836088, revinso@me.com.
*Rhoden Green. The Silk House is
about ½ m from Queen St end of
Lucks Lane, on a sharp bend. It's in
near Lucks Lane Fishery. Can also
be approached from Maidstone Rd,
entrance near Garden Centre.* **Sat
10, Sun 11 Apr (10-3.30); Sat 21,
Sun 22 Aug (10-4.30). Adm £7.50,
chd £2.50. Light refreshments.
Visits also by arrangement Mar to
Sept for groups of 5 to 10. Parking
limited to 12 cars, though some
parking on Lucks Lane is possible.**
Partly Japanese influenced stroll
garden, Koi & wildlife ponds, unusual
evergreen trees & bamboos, small
woodland walk, kitchen garden,
Mediterranean dry bank, new
sections being developed including
topiary, 2 acres overall, begun 4 years

ago. Designed, built & maintained
by owners. Tea, coffee, cakes and
snacks available to purchase with
profits to charities. Wheelchair users
will need to access via the oak framed
outbuildings area to avoid steps.
Some areas have steps and banks.
🐾 ✲ ☕

96 NEW **SIR JOHN HAWKINS
HOSPITAL**
High Street, Chatham,
ME4 4EW. Susan Fairlamb,
www.hawkinshospital.org.uk.
*On the N side of Chatham High St,
on the border between Rochester
& Chatham. Leave A2 at J1 &
follow signs to Rochester. Pass
Rochester Station & turn L at main
junction T-lights, travelling E towards
Chatham.* **Sat 12, Sun 13 June (11-
5). Adm £3, chd free. Cream teas.
Also open Bishopscourt.**
Built on the site of Kettle Hard - part
of Bishop Gundulph's Hospital of St
Bartholomew, the Almshouse is a
square of Georgian houses dating
from the 1790s. A delightful small
secluded garden overlooks the
River Medway, full of vibrant and
colourful planting. A lawn with cottage
style borders leads to the riverside
and a miniature gnome village
captivates small children and adults
alike. Disabled access via stairlift,
wheelchair to be carried separately.
🐾 🐾 ✲ ☕

97 ◆ **SISSINGHURST CASTLE
GARDEN**
Biddenden Road, Sissinghurst,
Cranbrook, TN17 2AB.
National Trust, 01580 710700,
sissinghurst@nationaltrust.org.
uk, www.nationaltrust.org.uk/
sissinghurst-castle-garden. *2m NE
of Cranbrook, 1m E of Sissinghurst
on Biddenden Rd (A262), see
our website for more information.*
**For NGS: Mon 27 Sept (10.30-
5.30). Adm £16, chd £8. Light
refreshments in Coffee Shop or
Restaurant at Sissinghurst Castle
Garden. For other opening times
and information, please phone,
email or visit garden website.**
Historic, poetic, iconic; a refuge
dedicated to beauty. Vita Sackville-
West and Harold Nicolson fell in love
with Sissinghurst Castle and created
a world renowned garden. More
than a garden, visitors can also find
Elizabethan and Tudor buildings, find
out about our history as a Prisoner
of War Camp and see changing
exhibitions. Free welcome talks and

estate walks leaflets. Café, restaurant,
gift, secondhand book and plant
shops are open from 10am-5.30pm.
Some areas unsuitable for wheelchair
access due to narrow paths and
steps.
🚗 ✲ 🐾 ☕

98 **SMITHS HALL**
Lower Road, West Farleigh,
ME15 0PE. Mr S Norman. *3m W of
Maidstone. A26 towards Tonbridge,
turn L into Teston Lane B2163. At
T-junction turn R onto Lower Rd
B2010. Opp Tickled Trout PH.* **Sun
27 June (11-5). Adm £5, chd free.
Home-made teas. Donation to
Heart of Kent Hospice.**
Delightful 3 acre gardens surrounding
a beautiful 1719 Queen Anne House
(not open). Lose yourself in numerous
themed rooms: sunken garden,
iris beds, scented old fashioned
rose walk, formal rose garden,
intense wildflowers, peonies, deep
herbaceous borders and specimen
trees. Walk 9 acres of park and
woodland with great variety of young
native and American trees and fine
views of the Medway valley. Cakes
available. Gravel paths
🚗 🐾 ☕

99 **SPRING PLATT**
Boyton Court Road, Sutton
Valence, Maidstone, ME17 3BY. Mr
& Mrs John Millen, 01622 843383,
carolyn.millen1@gmail.com,
www.kentsnowdrops.com. *5m SE
of Maidstone. From A274 nr Sutton
Valence follow yellow NGS signs.* **Sat
16, Wed 20, Sun 24, Sun 31 Jan,
Wed 3, Tue 9 Feb (10.30-3). Adm
£5, chd free. Light refreshments.
Home-made bread, soup, tea and
cakes. Limited parking. Please
ring to book an appointment for
the dates shown, 01622 843383.
Visits also by arrangement Jan
& Feb. Please phone to check
availability before visiting.**
One acre garden under continual
development with panoramic views
of the Weald. Over 700 varieties of
snowdrop grown in tiered display
beds with spring flowers in borders.
An extensive collection of alpine
plants in a large greenhouse.
Vegetable garden, natural spring fed
water feature and a croquet lawn.
Home-made soup, home-made
bread, tea/coffee and cake. Garden
on a steep slope and many steps.

100 STABLE HOUSE, HEPPINGTON

Street End, Canterbury, CT4 7AN. Charlie & Lucy Markes. *Stable House, Heppington, CT4 7AN. 2m S of Canterbury, off B2068. 1st R after Bridge Rd if coming from Canterbury, 400m after Granville pub on the L if coming towards Canterbury.* **Sat 17, Sun 18 July (1.30-5.30). Adm £4, chd free. Home-made teas.**

Approx 2 acres of relaxed, informal interlinked gardens surrounding converted Edwardian stable block in lovely setting with views towards the North Downs. Mixed borders with mature shrubs and perennials. Climbing roses, honeysuckle and clematis. Gravel garden, veg garden and walkthrough garden room beneath clocktower. Adjoining 5 acre wildflower meadow and 15 acre vineyard. Large wildflower meadow. Disabled parking by arrangement close to garden. Not all areas wheelchair accessible.

101 STONEWALL PARK

Chiddingstone Hoath, nr Edenbridge, TN8 7DG. Mr & Mrs Fleming. *4m SE of Edenbridge. Via B2026. ½ way between Markbeech & Penshurst.* **Sun 14 Mar, Sun 2 May (2-5). Adm £6, chd free. Home-made teas in the conservatory. Donation to Sarah Matheson Trust & St Mary's Church, Chiddingstone.**

Even from the driveway you can see a vast amount of self seeded daffodils, leading down to a romantic woodland garden in historic setting, featuring species such as rhododendrons, magnolias, azaleas and bluebells. You can see them in full flower on our May opening day. You can also discover a range of interesting trees, sandstone outcrops, wandering paths and lakes.

GROUP OPENING

102 TANKERTON GARDENS

Tankerton, CT5 2EP. *The gardens are in a triangle, either on Northwood Rd or directly off it, or parallel to it.* **Sat 5 June (10-5). Combined adm £6, chd free. Also open Faversham Gardens.** Local cafes are available in Tankerton High Street, Tower Parade and the Sea Front, within easy reach of all five gardens.

17A BADDLESMERE ROAD
Mr Derek Scoones.

NEW   14 NORTHWOOD ROAD
Philippa Langton.

16 NORTHWOOD ROAD
Mr Simon Courage.

12 STRANGFORD ROAD
Sarah Yallop.

NEW   18 STRANGFORD ROAD
Mia Young.

Five very different gardens within a mile of the sea, enjoying a mild climate. Baddlesmere Road featuring roses and densely planted for colour. The Strangford Road gardens are more traditional but with a quirky and wildlife elements. The Northwood Road gardens are more contemporary. Stepped access in three gardens (12 and 18 Strangford Road and 17a Baddlesmere Road) would make wheelchair access very difficult.

103 NEW   TONBRIDGE SCHOOL

High Street, Tonbridge, TN9 1JP. The Governors. *Various gardens around the school. Maps & guides for visitors. At N end of Tonbridge High St. Parking signed off London Rd (B245 Tonbridge-Sevenoaks).* **Sat 10 July (10-3). Adm £6, chd free. Light refreshments.**

In front of and behind Tonbridge School you will find the five main gardens that you can visit: Front of School Garden, The Garden of Remembrance, Smythe Library and Skinners Library Garden as well as the newly created Barton Science Centre Garden. Tea/coffee and cake available. Head Gardener available to chat. Toilets on site. All gardens can be accessed via wheelchair and gardeners will be on hand to help and guide.

The National Garden Scheme searches the length and breadth of England and Wales for the very best private gardens

104 TOPGALLANT

5 North Road, Hythe, CT21 5UF. Mary Sampson. *M20 exit 11, take A259 to Hythe, then going towards Folkestone at r'about, take 2nd L up narrow hill signed to Saltwood. L at junction with North Rd. House on L. See 'Vergers'.* **Sat 5, Sun 6 June (2-5). Combined adm with Vergers £6, chd free.**

Opening with Vergers, two very different hillside gardens, both terraced and developed to cope with the prevailing winds and the slope. Top Gallant is a secluded Sculptors' garden, with mature trees and shrubs, a wildlife pond. Decking and grass paths wind down through the garden giving glimpses of the sea. Relaxed planting for year round interest and to encourage wildlife. Ceramics studio open. Sloping hillside garden with many steps.

105 TORRY HILL

Frinsted/Milstead, Sittingbourne, ME9 0SP. Lady Kingsdown, 01795 830258, lady.kingsdown@btinternet.com. *5m S of Sittingbourne. From M20 J8 take A20 (Lenham). At r'about by Mercure Hotel turn L Hollingbourne (B2163). Turn R at Xrds at top of hill (Ringlestone Rd). Thereafter Frinsted-Doddington (not suitable for coaches), then Torry Hill/NGS signs. From M2 J5 take A249 towards Maidstone, then 1st L (Dredgar), Lagain (follow Dredgar signs), R at War Memorial, 1st L (Milstead), Torry Hill/NGS signs from Milstead. For disabled parking please follow the disabled signs.* **Sun 23 May, Sun 13 June, Sun 18 July (2-5). Adm £5, chd free. Home-made teas. Picnics welcome on cricket ground. Visits also by arrangement May to Aug for groups of 10 to 30. Donation to St. Dunstan's Church, Frinsted and Sounding Out, Saturday Music Centre c/o The King's School.**

8 acres; large lawns, specimen trees, flowering cherries, rhododendrons, azaleas and naturalised daffodils; walled gardens with lawns, shrubs, herbaceous borders, rose garden incl shrub roses, wildflower areas and vegetables. Extensive views to Medway and Thames estuaries. Some shallow steps. No wheelchair access to rose garden due to very uneven surface but can be viewed from pathway.

106 TRAM HATCH
Charing Heath, Ashford,
TN27 0BN. Mrs P
Scrivens, 07835 758388,
Info@tramhatch.com,
www.tramhatchgardens.co.uk.
*10m NW of Ashford. A20 towards
Pluckley, over motorway then 1st
R to Barnfield. At end, turn L past
Barnfield, Tram Hatch ahead.* **Sun 30
May, Sun 4 July, Sun 1 Aug (12-5).
Adm £5, chd free. Visits also by
arrangement May to Sept.**
Meander your way off the beaten
track to a mature, extensive garden
changing through the seasons. You
will enjoy a garden laid out in rooms -
what surprises are round the corner?
Large selection of trees, vegetable,
rose and gravel gardens, colourful
containers. The River Stour and the
Angel of the South enhance your visit.
Please come and enjoy, then relax in
our lovely garden room for tea. Water
features and statuary. The garden
is totally flat, apart from a very small
area which can be viewed from the
lane.

107 UPPER PRYORS
Butterwell Hill, Cowden, TN8 7HB.
Mr & Mrs S G Smith. *4½m
SE of Edenbridge. From B2026
Edenbridge-Hartfield, turn R at
Cowden Xrds & take 1st drive on R.*
**Wed 16 June (11-5). Adm £5, chd
free. Home-made teas.**
10 acres of English country garden
surrounding C16 house - a garden
of many parts; colourful profusion,
interesting planting arrangements,
immaculate lawns, mature woodland,
water and a terrace on which to
appreciate the view, and tea!

108 VERGERS
Church Road, Hythe, CT21 5DP.
Nettie & John Wren. *Hythe 6 mins
walk from Topgallant. M20 Exit 11,
take A259 to Hythe. Take directions
as for Topgallant. L at junc into
North Rd, then 1st L into Church Rd,
Vergers is next door to St Leonards
Church. Parking in Church car park.*
**Sat 5, Sun 6 June (2-5). Combined
adm with Topgallant £6, chd
free. Cream teas at St Leonard's
Church.**
South facing hillside garden which
has been reclaimed and developed
over the past five years by the present
owners. Steps from car park lead up
to the garden terrace, goldfish pond
and lawn. Winding paths slope up to

many seating areas with spectacular
views of the town and Channel.
Kitchen garden, wildlife area and
pond, bug hotels, mixed planting
for year round interest. Church and
Ossuary open Church Flower festival
planned.

GROUP OPENING

109 WALMER GARDENS
Walmer, Deal, CT14 7SQ. *Yellow
signs from A258. Dover Rd.* **Sat 12
June (11.30-4.30); Sun 13 June
(1.30-4.30). Combined adm £5,
chd free. Home-made teas at
Sunnybank & Pembroke Lodge.
Soft drinks at 31 Herschell Sq.**

NEW 31 HERSCHELL SQUARE
Kay Valentine & Johnny Morris.

**OLD CHURCH HOUSE, 26A
CHURCH STREET**
Christine Symons.

**NEW PEMBROKE LODGE, 140
DOVER ROAD**
Mr Steve & Mrs Jo Hammond,
PembrokeLodge140@gmail.com.
Visits also by arrangement June
to Aug for groups of 10 to 20.

**SUNNYBANK 12, HERSCHELL
ROAD EAST**
Mr James & Mrs Flora Cockburn.

Walking through Walmer you cannot
help noticing the proliferation of
seaside plants - giant echiums,
fuchsia hedges - which abound in
this quarter. In our selection of four
gardens you will find a rich vein of
horticultural activity. Give yourselves
and your friends a treat and come
and visit us.

110 THE WATCH HOUSE
7 Thanet Road, Broadstairs,
CT10 1LF. Dan Cooper
& John McKenna,
www.frustratedgardener.com.
*Town Centre Location. Off
Broadstairs High St on narrow side
rd. At Broadstairs station, cont along
High St (A255) towards sea front.
Turn L at Terence Painter Estate
Agent then immed turn R.* **Sat 31
July, Sun 1 Aug (12-4). Adm £5,
chd free. Light refreshments.
Also open 5 Montefiore Avenue.**
Adjoining an historic fishermen's
cottage, two small courtyard gardens

shelter an astonishing array of
unusual plants. Thanks to a unique
microclimate the east-facing garden
is home to a growing collection
of exotics, chosen principally for
exuberant, colourful, jungly foliage. In
the west-facing courtyard a garden
room leads onto a terrace where
flowering plants jostle for space.
Within a few mins walk of Viking Bay,
The Dickens Museum and Bleak
House.

111 WEST COURT LODGE
Postling Court, The Street,
Postling, nr Hythe, CT21 4EX.
Mr & Mrs John Pattrick. *2m NW
of Hythe. From M20 J11 turn S
onto A20. Immed 1st L. After ½m
on bend take rd signed Lyminge.
1st L into Postling.* **Sat 5, Sun 6
June (12-5). Combined adm with
Churchfield £6, chd free. Home-
made teas in Postling Village hall.**
S-facing one acre walled garden at
the foot of the N Downs, designed in
2 parts: main lawn with large sunny
borders and a romantic woodland
glade planted with shadow loving
plants and spring bulbs, small wildlife
pond. Lovely C11 church will be open
next to the gardens.

GROUP OPENING

**112 WEST MALLING EARLY
SUMMER GARDENS**
West Malling, ME19 6LW. *On A20,
nr J4 of M20. Park (Ryarsh Lane
& Station) in West Malling. Maps
& directions to 1&2 New Barns
Cottages (parking) are available from
gardens in town.* **Sun 6 June (12-
5). Combined adm £8, chd free.
Home-made teas at New Barns
Cottages only. Donation to St
Mary's Church, West Malling.**

ABBEY BREWERY COTTAGE
Dr & Mrs David & Lynda Nunn.

BROME HOUSE
John Pfeil & Shirley Briggs.

LUCKNOW, 119 HIGH STREET
Ms Jocelyn Granville.

NEW BARNS COTTAGES
Mr & Mrs Anthony Drake.

TOWN HILL COTTAGE
Mr & Mrs P Cosier.

WENT HOUSE
Alan & Mary Gibbins.

58 The Mall

West Malling is an attractive small market town with some fine buildings. Enjoy six lovely gardens that are entirely different from each other and cannot be seen from the road. Brome House and West House have large gardens with specimen trees, old roses, mixed borders, attractive kitchen gardens and garden features incl a coach house, Roman temple, fountain and parterre. Lucknow and Town Hill Cottage are walled town gardens with mature and interesting planting. Abbey Brewery Cottage is a recent jewel-like example of garden restoration and development. New Barns Cottages has serpentine paths leading through woodland to roomed gardens: tea and cakes in the courtyard garden of the cottages. Town Hill Cottage, Abbey Brewery Cottage and New Barns Cottages are more difficult to access but the other gardens have wheelchair access.

& 🐕 ☕

GROUP OPENING

113 WHITSTABLE GARDENS
Whitstable, CT5 4LT. *Off A299, or A290. Down Borstal Hill, L by garage into Joy Lane to collect map of participating gardens (also available at other gardens).* Parking at Joy Lane School and Gorrell Tank. **Sun 23 May (10-5). Combined adm £6, chd free. Home-made teas at Stream Walk Community Gardens and at the Umbrella Centre Cafe.**

87 ALBERT STREET
Paul Carey & Phil Gomm.

6 ALEXANDRA ROAD
Andrew Mawson & Sarah Rees.

NEW **8 ALEXANDRA ROAD**
Henry Kernighan.

NEW **21 ALEXANDRA ROAD**
Sarah Morgan.

56 ARGYLE ROAD
Emma Burnham & Mel Green.

NEW **42 CANTERBURY ROAD**
Elspeth Dougall.

NEW **76 CANTERBURY ROAD**
Deborah & Gary Parks.

THE GUINEA, 31 ISLAND WALL
Sheila Wyver.

NEW **15 JOY LANE**
Mr & Mrs Clare Godley.

19 JOY LANE
Francine Raymond,
www.kitchen-garden-hens.co.uk.

67A JOY LANE, JUPITER HOUSE
Ed Lamb, Shelagh O'Riordan.

96 JOY LANE
Vernon & Terrie Brown.

OCEAN COTTAGE
Katherine Pickering.

ST MARY'S TOWN GARDEN
Whitstable Umbrella Community Centre.

NEW **STARLINGS**
David & Pat Roberts.

STREAM WALK TRUST
Stream Walk Community Gardens.

NEW **10 WARWICK ROAD**
Lizzie Simpson.

NEW **WAYPOST HOUSE**
Zinnia Slade.

NEW **10 WEST CLIFF**
Lisa Feurtado,
www.fuchsiagreen.com.

Enjoy a day of eclectic gardens by the sea. 22 people are showing off their gardens, 10 of them for the first time, with 3 favourites returning, but others marked with yellow balloons, are there to admire from the street. From fishermen's yards to formal gardens, the residents of Whitstable are making the most of the mild climate. Choose a few from our leaflet, drop in and admire contemporary gardens, seaside gardens, rose gardens, gravel gardens, designers' gardens and wildlife friendly plots, both large and small and some with fabulous views. Stream Walk and the Umbrella Centre are at the heart of our community, ideal for those without gardening space of their own. We toil on heavy clay soils and are prone to northerly winds. By opening, we're hoping to encourage those new to gardening with our ingenuity and style, rather than rolling acres. To find out more see Whitstable Gardens on Facebook. Combined adm £6 per adult or £10 for 2. Plant stalls at 19 Joy Lane, Stream Walk and the Umbrella Centre.

❋ ☕

Ramsgate Gardens

GROUP OPENING

114 WOMENSWOLD GARDENS

Womenswold, Canterbury, CT4 6HE. Mrs Maggie McKenzie, 07941044767, maggiemckenzie@vfast.co.uk. *6m S of Canterbury, midway between Canterbury & Dover. Take B2046 for Wingham at Barham Xover. Turn 1st R, following signs.* **Sat 26, Sun 27 June (11-5). Combined adm £5, chd free. Home-made teas at Brambles, Womenswold. Visits also by arrangement June & July for groups of 10+.**

A diverse variety of cottage gardens in an idyllic situation in an unspoilt hamlet, mostly surrounding C13 Church. Cottage garden with variety of old climbing & shrub roses, clematis, vegetable bed & beehives; a garden in a setting of a traditional C17 thatched cottage; colourful garden with ponds, waterfalls, tropical area with many rare plants and a large collection of agapanthus; a garden with a large display of perennials, kniphofias & hemerocallis; a 2 acre plantsman's garden partially created in old chalk quarry-with vegetables, orchard, alpines, poly-tunnel with tender fruit . A very picturesque terraced cottage garden, with unusual plants, feature pond with lovely views of the church. Teas in lovely garden setting. Easy walking distance between gardens. North Downs Way runs through village. Many unusual plants for sale. Teas in lovely restful garden with home-made cakes. Produce stall; Church open. Additional parking in village with mini-bus running regularly for garden lying outside main village. Most gardens have good wheelchair access although some areas may be inaccessible. Minibus available.

115 WOODLANDS ROAD ALLOTMENTS

47 Tangmere Close, Gillingham, ME7 2TN. Medway Council, www.gillinghamhs.co.uk. *From A2 by Gillingham Golf Club onto Woodlands Rd, after railway bridge, 1st R onto Hazelmere Dr, 1st R onto Tangmere Cl.* **Sat 24, Sun 25 July (12-4). Adm £5, chd free. Home-made teas.**

162 individual allotments on the site include topiary, espalier fruits, many plots of unusual fruit and

vegetables, as well as floral and wildlife areas. Access is via a level concrete roadway. There will be ample opportunity to chat to many friendly allotmenteers, limited teas will be available with an opportunity to purchase plants or produce, making for a very worth-while visit. Beehives, local wildlife, and a plot imaginatively using many up-cycled materials. Concrete level path round much of the site.

  ♿ 🐄 ✿ ☕

116 12 WOODS LEY
Woods Ley, Ash, Canterbury, CT3 2HF. Philip Oostenbrink. *20 mins from Canterbury. From A257 turn into Chequer Lane. Go down Chequer Lane & turn L into Chilton Field. Turn the 2nd L into Woods Ley.* **Sat 7 Aug (1-5). Adm £3, chd free. Donation to Plant Heritage.**
A jungle style garden, showing what can be achieved in the smaller modern garden, filled with unusual foliage. The side garden has a selection of ferns and shade loving plants. The back garden is full of densely planted jungle plants. Indoors you will find the National Collection of Aspidistra elatior and sichuanensis. A plant lover's garden full of unusual plants. Private garden of the Head Gardener of Walmer Castle and Gardens. NGS garden owners' passes not accepted.

  ✿ NPC ☕

117 ◆ THE WORLD GARDEN AT LULLINGSTONE CASTLE
Eynsford, DA4 0JA. Mrs Guy Hart Dyke, 01322 862114, info@lullingstonecastle.co.uk, www.lullingstonecastle.co.uk. *1m from Eynsford. Over Ford Bridge in Eynsford Village. Follow signs to Roman Villa. Keep Roman Villa immed on R then follow Private Rd to Gatehouse.* **For NGS: Sun 20 June (12-5). Adm £9, chd £4.50. Light refreshments. For other opening times and information, please phone, email or visit garden website.**
The World Garden is located within a two-acre, 18th-century Walled Garden in the stunning grounds of Lullingstone Castle, where heritage meets cutting-edge horticulture. The garden is laid out in the shape of a miniature map of the world. Thousands of species are represented, all planted out in their respective beds. The World Garden Nursery offers a host of horticultural and homegrown delights, to reflect

the unusual and varied planting of the garden. Wheelchairs available upon request.

  ♿ ✿ 🚌 NPC ☕ 🪑

GROUP OPENING

118 WYE GARDENS
Ashford, TN25 5BP. *3m NE of Ashford. From A28 take turning signed Wye.* **Sun 13 June (2-6). Combined adm £5, chd free. Home-made teas at Wye Church.**

3 BRAMBLE CLOSE
Dr M Copland.
(See separate entry)

MIDDLEFIELD HOUSE
TN25 5EP. Kathy & Steve Bloom.

3 ORCHARD DRIVE
TN25 5AU. Liz Coulson.

32 OXENTURN ROAD
TN25 5BE. Rosemary Fitzpatrick.

SPRING GROVE FARM HOUSE
TN25 5EY. Heather Van den Bergh.

The gardens open in historic Wye village are all very different in character. 3 Bramble Close: a very wild experimental garden buzzing with wildlife demonstrates how plants maintain their natural population density. Middlefield House: a half-acre plot with sweeping views over open fields, woodland, borders, wildlife meadow, raised vegetable beds, sculpture and huge wildlife photographs peering through foliage. 3 Orchard Drive: a relatively new garden packed with plants and features (rose arches, benches, raised beds), a great example of what can be achieved in a modest plot. 32 Oxenturn Road: a wildlife garden featuring a wildflower meadow, a green-roofed shed and worm composting as well as shrubs, flowers and raised vegetable beds. Spring Grove Farm House: a large country garden full of colour and many interesting features including a lake, pond and a gravel garden. Wheelchair access to 32 Oxenturn Rd & Spring Grove Farm House only.

  ♿ ☕

119 YOAKLEY HOUSE
Drapers Close, Margate, CT9 4AH. Michael Yoakley's Charity, www.yoakleycare.co.uk. *Near Margate QEQM Hospital, Drapers*

Close is a cul de sac turning off St Peters Rd. Access to the Yoakley car park is at the end of Drapers Close, through the hedge. **Sun 18 July (2.30-4.30). Adm £5, chd £2.50. Light refreshments in marquee at Yoakley House.**
Set in 2½ acres of grounds, cultivated the old fashioned way to complement the ancient almshouses it serves. Well-manicured lawns with extensive borders and densely planted display beds: summer bedding, carpet bedding, specimen trees, shrubs and rockery plants, herbaceous planting, shrub rose beds with standard roses and magnificent hanging baskets. Accessible pathways from the main car park throughout the grounds.

  ♿ ☕

120 YOKES COURT
Coal Pit Lane, Frinsted, Sittingbourne, ME9 0ST. Mr & Mrs John Leigh Pemberton, leighpems@btinternet.com. *2.4m from Doddington. Turn off Old Lenham Rd to Hollingbourne. Take 1st R at Torry Hill Chestnut Fencing, then 1st L to Torry Hill. Take 1st L into Coal Pit Lane.* **Visits by arrangement June & July. Adm £4.50, chd free.**
3 acre garden surrounded by countryside. Hedges and herbaceous borders set in open lawns. Rose beds. New prairie planting. Serpentine walk way through wildflowers. Walled vegetable garden.

  ✿ ☕ 🪑

"I love the National Garden Scheme which has been the most brilliant supporter of Queen's Nurses like me. It was founded by the Queen's Nursing Institute which makes me very proud. As we battle Coronavirus on the front line in the community, knowing we have their support is a real comfort." – Liz Alderton, Queen's Nurse

OPENING DATES

All entries subject to change. For latest information check www.ngs.org.uk

Map locator numbers are shown to the right of each garden name.

February

Snowdrop Festival

Sunday 14th
Weeping Ash Garden 67

Sunday 21st
Weeping Ash Garden 67

April

Sunday 11th
NEW The Haven Garden 34

Saturday 17th
NEW Bold Hall Nursery 5
Dale House Gardens 18

Sunday 18th
Dale House Gardens 18

May

Sunday 2nd
Kington Cottage 40

Monday 3rd
◆ The Ridges 55

Saturday 8th
4 Brocklebank Road 8

Sunday 9th
Blundell Gardens 4
4 Brocklebank Road 8
NEW Derian House
Children's Hospice 20

Saturday 15th
NEW Willow Wood
Hospice Gardens 70

Sunday 16th
Sefton Park May
Gardens 60

Saturday 22nd
Halton Park House 33

Sunday 23rd

Halton Park House 33
◆ Hazelwood 35
Hillside Gardens 37
Parkers Lodge 52

Sunday 30th
Blundell Gardens 4
Bretherton Gardens 6
◆ Clearbeck House 13
NEW 194 Mottram Old
Road 49
14 Saxon Rd 59
NEW 45 Stourton
Road 64

Monday 31st
◆ Clearbeck House 13

June

Saturday 5th
136 Buckingham Road 9
Dent Hall 19

Sunday 6th
136 Buckingham Road 9
Dent Hall 19
Kington Cottage 40
Meresands Kennels &
Cattery 46
NEW St Thomas
Primary School 58

Saturday 12th
NEW Calder House Lane
Gardens 10
35 Ellesmere Road 24
Giles Farm 26
Grange Community
Garden 28
NEW Old Hollows Farm 50
The Old Vicarage 51
8 Water Head 66
11 Westminster Road 68

Sunday 13th
8 Andertons Mill 2
NEW Calder House Lane
Gardens 10
31 Cousins Lane 14
79 Crabtree Lane 15
Didsbury Village
Gardens 21
Giles Farm 26
Green Farm Cottage 29
45 Grey Heights View 30
◆ Hazelwood 35
6 Menivale Close 45
NEW Old Hollows Farm 50
The Old Vicarage 51
120 Roe Lane 57
8 Water Head 66

Saturday 19th
Dale House Gardens 18
NEW Willow Wood
Hospice Gardens 70

Sunday 20th
Dale House Gardens 18

Saturday 26th
Hale Village Gardens 32

Sunday 27th
Bridge Inn Community
Farm 7
◆ Clearbeck House 13
5 Crib Lane 16
Dutton Hall 23
Hale Village Gardens 32
NEW Mill Croft 48
NEW Raw Ridding
House 54

Monday 28th
5 Crib Lane 16

July

Saturday 3rd
Carr House Farm 11
Kington Cottage 40
NEW 4 Roe Green 56

Sunday 4th
Carr House Farm 11
NEW Freshfield
Gardens 25
◆ Hazelwood 35
Hutton Gardens 39
Kington Cottage 40

Sunday 11th
79 Crabtree Lane 15
Parkers Lodge 52
Woolton Village
Gardens 71

Saturday 17th
Ashton Walled
Community Gardens 3
Warton Gardens 65

Sunday 18th
NEW Gorse Hill Nature
Reserve 27
Hillside Gardens 37
72 Ludlow Drive 41
Lytham Hall 42
Maggie's, Oldham 44
Southlands 62
Warton Gardens 65

Sunday 25th
Bretherton Gardens 6
NEW Didsbury Village

Gardens The
Contemporary
Collection 22
Maggie's Manchester 43
14 Saxon Rd 59
NEW 45 Stourton
Road 64
Wigan & Leigh
Hospice 69

Saturday 31st
The Growth Project 31
NEW Higher Bridge
Clough House 36

August

Sunday 1st
NEW Higher Bridge
Clough House 36
Kington Cottage 40

Sunday 8th
NEW Derian House
Children's Hospice 20
Hillside Gardens 37
NEW Howick House 38
Plant World 53
Weeping Ash Garden 67

Saturday 21st
Stanhill Exotic Garden 63

Sunday 22nd
45 Grey Heights View 30
Stanhill Exotic Garden 63

Saturday 28th
8 Water Head 66

Sunday 29th
Bretherton Gardens 6
◆ Croxteth Hall Walled
Garden 17

Monday 30th
◆ The Ridges 55
8 Water Head 66

September

Saturday 4th
Grange Community
Garden 28

Sunday 5th
NEW Allerton & Mossley
Hill Gardens 1
Sefton Park September
Gardens 61

Sunday 12th
Weeping Ash Garden 67

Saturday 18th
Ashton Walled
  Community Gardens 3

October

Sunday 31st
Weeping Ash Garden 67

By Arrangement
Arrange a personalised
garden visit on a date to

suit you. See individual
garden entries for full
details.

Ashton Walled
Community Gardens 3
4 Brocklebank Road 8
NEW Calder House
  Lane Gardens 10
Carr House Farm 11
Casa Lago 12
79 Crabtree Lane 15
Dale House Gardens 18

Giles Farm 26
45 Grey Heights View 30
The Growth Project 31
Hazel Cottage,
  Bretherton Gardens 6
Kington Cottage 40
Maggie's Manchester 43
Maggie's, Oldham 44
Mill Barn 47
NEW Old Hollows Farm 50
Owl Barn, Bretherton
  Gardens 6
14 Saxon Rd 59

Southlands 62
8 Water Head 66
NEW 115 Waterloo Road,
  Hillside Gardens 37
NEW Willow Wood
  Hospice Gardens 70

Online booking is
available for many
of our gardens, at
ngs.org.uk

THE GARDENS

GROUP OPENING

1 NEW ALLERTON & MOSSLEY HILL GARDENS
Liverpool, L18 7JQ. *3m S of end of M62. From L lane at end of M62 follow Ring Road A5058 (S). Follow A5058 through 1st island. Keep to L lane through 2nd island then take B5180 for about ½ m to signage.* Sun 5 Sept (12-5). Combined adm £5, chd free. Light refreshments. at 146 Mather Avenue.

  NEW 33 GREENHILL ROAD
  L18 6JJ. Tony Rose.

  NEW 146 MATHER AVENUE
  L18 7HB. Barbara Peers.

  NEW 128 PITVILLE AVENUE
  L18 7JQ. Paul & Theresa Jevons.

At 33 Greenhill Road you will find a splendid collection of exotic and architectural plants, both containerised and ground-planted. The garden is mainly paved. Paths wend around a raised pond fed by a cobble-bedded stream, which provides a central focal point. The garden at 146 Mather Avenue has a very informal feel with an emphasis on providing for wildlife. There is a fishpond, wildlife pond and a woodpile for insects. The borders include perennials in mostly pinks and blues, a corner of hot colours and self-seeding wildflowers. The garden at 128 Pitville Avenue is an example of what can be achieved in very small space. It has well laid out paths, herbaceous borders, a pergola, arch and arbour seat. It also includes

a small pond, a water feature and stonework adornments.

 ✿ ☕ ▥

2 8 ANDERTONS MILL
Bentley Lane, Mawdesley, Ormskirk, L40 3TW. Mr & Mrs R Mercer. *Heading N from Mawdesley take 1st R into Dark Lane for 1m to T-junction turn L, garden 500yds on R.* Sun 13 June (11-5). Adm £3, chd free.
Cottage garden started in 2010 with colourful borders of perennials, shrubs and roses. Clematis and rose covered arches. Veg patch and large cut flower garden. Patio with raised beds and wonderful view of Harrock Hill.

& ▥ ✿

3 ASHTON WALLED COMMUNITY GARDENS
Pedders Lane, Ashton-On-Ribble, Preston, PR2 1HL. Annie Wynn, 07535 836364, letsgrowpreston@gmail.com, www.letsgrowpreston.org. *W of Preston. From M6 J30 head towards Preston turn R onto Blackpool Rd. Follow Blackpool Rd for 3.4m turn L onto Pedders Lane & next R onto the park. Entrance to walled garden is 50 metres on your L.* Sat 17 July, Sat 18 Sept (10-3). Adm £4, chd free. Home-made teas. Visits also by arrangement.
Formal raised beds within a walled garden, a peace garden and an edible garden. The formal part of the garden uses plants predominantly from just 3 families, rose, geranium and aster. It is punctuated by grasses and has been designed to demonstrate how diverse and varied plants can be from just the one family. Live musical entertainment Formal and informal

gardens workspace and plant sales edible garden.

& ▥ ✿ ☕

Mill Croft

GROUP OPENING

4 BLUNDELL GARDENS
Blundell Road, Hightown, Liverpool, L38 9EF. *11m N of Liverpool, 11m S of Southport. From M57/M58 Switch Island; join A5758 and A565 to Southport; L on B5193 to Hightown Village, over r'about, first L Village Way, R into Blundell Rd.* **Sun 9, Sun 30 May (1-5.30). Combined adm £5, chd free. Home-made teas at 11 Blundell Road. GF available.**

5 BLUNDELL AVENUE
Denise & Dave Ball.
Open on all dates

11 BLUNDELL AVENUE
Karen Rimmer.
Open on Sun 30 May

11 BLUNDELL ROAD
Joyce Batey.
Open on all dates

75 BLUNDELL ROAD
Shirley & Phil Roberts.
Open on all dates

4 gardens in Hightown Village, featuring: a quaint courtyard garden with colourful plant-filled pots, shrubs, raised borders and curiosities; a quirky small garden with delightful surprises (folly) and wide variety of plants; a wildlife-friendly organic garden with pond, pergola and fairy glen; and mature garden with large monkey puzzle tree, wooded azalea planted terrace, gravel path leading to Zen garden and pergola with a difference. Wheelchair access to lower part of 11 Blundell Rd.
✿ ☕

5 NEW BOLD HALL NURSERY
Hall Lane, Bold, St Helens, WA9 4SL. Elizabeth & Jonathan LLoyd. *3m from J8 M62. Hall Lane is off Gorsey Lane halfway down Hall Lane turn L at 1st cottage across from mirrors & L again into the driveway.* **Sat 17 Apr (11-3). Adm £5, chd free. Light refreshments.** The walled enclosure is of special interest as the principal component of an extensive, little altered and well-documented walled garden of 1844, It boasts beautiful mature gardens with expanses of lawns, shrub and herbaceous borders, wildlife and formal ponds. Featuring beautiful spring blossom having many mature fruit trees as well as azaleas and rhododendrons. Laid mainly to lawn with gravel pathway.
♿ ✿ ☕

"Unfortunately, cancer was not in lockdown in 2020. The continued support of our long-standing and valued partner, the National Garden Scheme is more important than ever."
Macmillan

Raw Ridding House

GROUP OPENING

6 BRETHERTON GARDENS

South Road, Bretherton, Leyland, PR26 9AS. *8m SW of Preston. Between Southport & Preston, from A59, take B5247 towards Chorley for 1m. Gardens signed from South Rd (B5247).* Sun 30 May, Sun 25 July, Sun 29 Aug (11-5.30). Combined adm £6, chd free. Home-made teas at Bretherton Congregational Church from 12 noon on all dates, also light lunches 25 July.

GLYNWOOD HOUSE
PR26 9AS. Terry & Sue Riding.

HAZEL COTTAGE
PR26 9AN. John & Kris Jolley, 01772 600896, jolley@johnjolley.plus.com. Visits also by arrangement May to Oct.

OWL BARN
PR26 9AD. Richard & Barbara Farbon, 01772 600750, barbaraandrichard.farbon@ngs.org.uk. Visits also by arrangement May to Aug for groups of up to 20.

PALATINE, 6 BAMFORDS FOLD
PR26 9AL. Alison Ryan.

PEAR TREE COTTAGE
PR26 9AS. John & Gwenifer Jackson.

Five contrasting gardens spaced across an attractive village with conservation area. Glynwood House finalist in 2019 Nation's Favourite Garden Competition has ¾ acre mixed borders, pond with drystone-wall feature, woodland walk, patio garden with pergola and spectacular open aspects. Hazel Cottage's former Victorian orchard plot has evolved into a series of themed spaces, while the adjoining land has a natural pond, meadow and developing native woodland. Owl Barn has herbaceous borders filled with cottage garden and hardy plants, productive kitchen garden with fruit, vegetables and cut flowers, two ponds with fountains and secluded seating area. NEW Palatine is a garden of 3 contrasting spaces started in 2019 around a modern bungalow to attract wildlife and give year-round interest. Pear Tree Cottage has informal beds featuring ornamental grasses, a solar-powered greenhouse watering system, a profusion of fruits, a pond, mature trees, and a fine backdrop of the West Pennine Moors. Full wheelchair access to Palatine, varying degrees of access at the other gardens due to narrow or uneven paths.

7 BRIDGE INN COMMUNITY FARM

Moss Side, Formby, Liverpool, L37 0AF. Bridge Inn Community Farm, www.bridgeinncommunityfarm.co.uk. *7m S of Southport. Formby by-pass A565, L onto Moss Side.* Sun 27 June (11-4). Adm £3.50, chd free. Light refreshments.
Bridge Inn Community Farm was established in 2010 in response to a community need. Our farm sits on a beautiful 4 acre small holding with views looking out over the countryside. We provide a quality service of training in a real life work environment and experience in horticulture, conservation and animal welfare.

8 4 BROCKLEBANK ROAD

Southport, PR9 9LP. Heather Sidebotham, 01704 543389, alansidebotham@yahoo.co.uk. *1¾ m N of Southport. Off A565 Southport to Preston Rd, opp North entrance to Hesketh Park.* Sat 8, Sun 9 May (11-5). Adm £4, chd free. Home-made teas. Visits also by arrangement May to Aug.
A walled garden landscaped with reclaimed materials from local historic sites. There are several water features, an extensive herbaceous border and areas of differing planting. Described by Matthew Wilson of Gardeners Question Time as a lovely garden with beautiful vistas at every turn. Designer and creator of the Large Gold Medal award winning NGS garden at Southport Flower Show 2019.

9 136 BUCKINGHAM ROAD

Maghull, L31 7DR. Debbie & Mark Jackson. *7m N of Liverpool. End M57/M58, take A59 towards Ormskirk. Turn L after car superstore onto Liverpool Rd Sth, cont' on past Meadows pub, 3rd R into Sandringham Rd, L into Buckingham Rd.* Sat 5 June (12-5.30). Adm £3, chd free. Sun 6 June (12-5.30).

Combined adm with St Thomas Primary School £3.50, chd free. Home-made teas. Donation to British Heart Foundation.
Small suburban cottage style garden herbaceous and hardy plants, roses, wisteria pergola and water features. Containers with perennials and annuals for added interest. Plenty of seating.

GROUP OPENING

10 NEW CALDER HOUSE LANE GARDENS

Calder House Lane, Bowgreave, Preston, PR3 1ZE. Mrs Margaret Richardson, 01995 604121, marg254@btinternet.com. *7m N of M6 J32. 7m S of M6 J33. J32, A6, Turn R on to B6340 1m. J33. Follow A6 for 7m, turn L onto Cock Robin Lane, Catterall turn L at the end, B6340. ½m turn R into Calder House Lane.* Sat 12 June (10.30-5); Sun 13 June (12-5). Combined adm £5, chd free. Home-made teas in the Friends Meeting House, adjacent to the gardens. Visits also by arrangement May to Aug.

NEW CALDER COACH HOUSE
Lila Thomas.

NEW CALDER COTTAGE
Lynn Jones.

NEW 1 CALDER HOUSE COTTAGE
Paul Hafren & Gaynor Gee.

NEW 2 CALDER HOUSE COTTAGE
Phil & Sarah Schofield.

NEW 3 CALDER HOUSE COTTAGE
Margaret & Mick Richardson.

The gardens range from a small satellite to a large (relatively new) garden featuring over 100 Hosta varieties. Mainly cottage gardens in style, each having its own features and interest. 1,2,3, Cottages have small front gardens with 3 very different designs. The Coach House a large cobbled yard with horse mounting steps and beds at the front, steps and herb terrace access the main lawn area at the rear. Calder Cottage a small cottage garden packed with interesting perennials, roses. No 1 the satellite garden at the top of the track, small but beautiful use of the space, making a tranquil haven. No 2 divided in to 3 rooms, cottage garden, a secluded seating area and finally a greenhouse, veg and fruit. No 3 the large garden planted with perennials, trees, roses, hostas and shrubs with a separate fruit and cut flower garden and greenhouse. All rear gardens apart from No1 have a step from the patio areas onto the main garden.

🐄 ❀ 🚗 ☕

🔟 CARR HOUSE FARM
Carr House Lane, Lancaster, LA1 1SW. Robin & Helen Loxam, 07801 663961. *SW of Lancaster City. From A6 Lancaster city centre turn at hospital, past B&Q & 1st R, straight under railway bridge into farm.* Sat 3, Sun 4 July (10-5). Adm £5, chd free. Home-made teas. Visits also by arrangement May to July. Donation to Fairfield Flora & Fauna Association.
A hidden gem in historic City of Lancaster. Farmhouse gardens incl Mediterranean, rustic and cottage flowers and trees intertwined beautifully with 2 ponds fed naturally by 'Lucy Brook' attracting all manner of wildlife. Apple, pear, plum, lemon and orange trees mix well within the scene. See rare breed cattle and enjoy nature walk in adjoining fields. Slope towards pond.

♿ 🐄 ❀ 🚗 ☕

🔟 CASA LAGO
1 Woodlands Park, Whalley, BB7 9UG. Carole Ann & Stephen Powers, 01254 824903, chows3@icloud.com. *2½m S of Clitheroe. From M6 J31, take A59 to Clitheroe. 9m take 2nd exit at r'about for Whalley. After 2m reach village & follow yellow signs. Parking in village car parks or nearby.* Visits by arrangement May to Sept for groups of up to 30. Adm £4.50, chd free. Light refreshments.
Climate Change / Lockdown impact has dictated new developments including a Trio of Water Features / Fountain Display, transformation of a drive into a creative oasis of Lakeland Stone, water, pocket planting areas lift and shift! Collections of Acers, Bonsai, Succulents, Hostas, Bamboos, grasses, black limestone wall, elevated glass areas. Koi Fish, Chow Chow pooches, guaranteed crowd pleasers!

♿ ❀ ☕

🔟 ◆ CLEARBECK HOUSE
Mewith Lane, Higher Tatham, Lancaster, LA2 8PJ. Peter & Bronwen Osborne, 01524 261029, bronwenmo@gmail.com, www.clearbeckgarden.org.uk. *13m E of Lancaster. Signed from Wray (M6 J34, A683, B6480) & Low Bentham.* For NGS: Sun 30, Mon 31 May (10-5). Sun 27 June (10-5), also open Mill Croft. Adm £4, chd free. Light refreshments. For other opening times and information, please phone, email or visit garden website.
It is our 31st. year with NGS 'A surprise round every corner' say visitors. They can enjoy streams, ponds, sculptures, boathouses and follies: Rapunzel's tower, temple, turf maze, giant fish made of CDs, walk-through pyramid. 2 acre wildlife lake attracts many species of insects and birds. Planting incl herbaceous borders, grasses, bog plants and many roses. Vegetable and fruit garden. Painting studio open. Children- friendly incl quiz. Artists and photographers welcome by arrangement. Wheelchair access -many grass paths, some sloped.

♿ 🐄 ❀ 🚗 ☕

🔟 31 COUSINS LANE
Rufford, Ormskirk, L40 1TN. Brenda & Roy Caslake. *From M6 J27, follow signs for Parbold then Rufford. Turn L onto the A59. Turn R at Hesketh Arms Pub. 4th turn.* Sun 13 June (10-4). Adm £3.50, chd free. Take time to re-energise with a glass of prosecco, a cream tea whilst enjoying the best of English pastimes.
The village cricket ground provides a backcloth to the garden with an open aspect to the north & west. Bordered by a stream flower beds run along three sides of the house and are populated in a cottage garden style. The subtle planting is enhanced by a design which provides discrete areas to come and enjoy.

❀ ☕

🔟 79 CRABTREE LANE
Burscough, L40 0RW. Sandra & Peter Curl, 01704 893713, peter.curl@btinternet.com, www.youtube.com/ watch?v=TqpxW7_8HT4. *3m NE of Ormskirk. A59 Preston - Liverpool Rd. From N before bridge R into Redcat Lane signed for Martin Mere. From S over 2nd bridge L into Redcat Lane after ¾m L into*

Crabtree Lane. Sun 13 June, Sun 11 July (11-4). Adm £4, chd free. Home-made teas. Visits also by arrangement June & July.
¾ acre all year round plants person's garden with many rare and unusual plants. Herbaceous borders and island beds leading to a pond and rockery, rose garden, spring area and autumn hot bed. Many stone features built with reclaimed materials. Shrubs and rhododendrons, Koi pond with waterfall, hosta and fern walk. Gravel garden with Mediterranean plants. Patio, surrounded by shrubs and raised alpine bed. Trees giving areas for shade loving plants. Flat grass and bark paths.

♿ 🐄 ❀ 🚗 ☕

🔟 5 CRIB LANE
Dobcross, Oldham, OL3 5AF. Helen Campbell. *5m E of Oldham. From Dobcross village-head towards Delph on Platt Lane, Crib Lane opp Dobcross Band Club - about 100 metres up, limited parking for disabled visitors only opp double green garage door, signed NGS.* Sun 27, Mon 28 June (1-4). Adm £3.50, chd free. Home-made teas.
A well loved and well used family garden - challenging as on a high terraced hillside and visited by deer, hares and the odd cow! Additional interests are wildlife ponds, wildflower areas, vegetable garden, a poly tunnel, four beehives with 40,000 bees and an art gallery and garden sculptures. Areas re thought annually, dug up and changed depending on time and aged bodies aches and pains! Local honey for sale.

❀ ☕

🔟 ◆ CROXTETH HALL WALLED GARDEN
Liverpool, L11 1EH. Liverpool City Council. Dina Younis, croxtethcountrypark@liverpool. gov.uk, , www.croxteth-hall.co.uk/. *6m NE of Liverpool. From M57 Exit J4 take A580 towards Liverpool. Look for brown tourist signs directing L. Main car park off Muirhead Ave East.* For NGS: Sun 29 Aug (11.30-5). Adm £3.50, chd free. Home-made teas. For other opening times and information, please email or visit garden website.
The two acre Victorian Walled Garden at Croxteth Hall was built around 1850. It produced a year round supply of fresh fruit, vegetable and cut flowers for the Hall until the last Earl died in 1972. Bedding display,

herbaceous and mixed borders, trained espalier and goblet fruit trees, herb garden, wildflower maze, peach house, rose beds, cut flower beds, soft fruits and a Fuchsia collection. The garden houses part of Liverpool's historic botanical collection under glass. The estate offers ample opportunities for walking. There is disabled permit parking available near to the Hall and Garden – use the service entrance from Croxteth Hall Lane (Satnav postcode L12 0HB).

 ♿ 🐄 ❀ 🚌 NPC 🍵

18 DALE HOUSE GARDENS
off Church Lane, Goosnargh, Preston, PR3 2BE. Caroline & Tom Luke, 01772 862464, tomlukebudgerigars@hotmail.com. *2½ m E of Broughton. M6 J32 signed Garstang Broughton, T-lights R at Whittingham Lane, 2½ m to Whittingham at PO turn L into Church Lane garden between nos 1 / & 19.* Sat 17, Sun 18 Apr, Sat 19, Sun 20 June (10-4). Adm £3.50, chd free. Home-made teas. Visits also by arrangement Apr to June. Donation to St Francis School, Goosnargh.
½ acre tastefully landscaped gardens comprising of limestone rockeries, well stocked herbaceous borders, raised alpine beds, well stocked koi pond, lawn areas, greenhouse and polytunnel, patio areas, specialising in alpines rare shrubs and trees, large collection unusual bulbs. All year round interest. Large indoor budgerigar aviary. 300+ budgies. New for 2021 a secret garden. Gravel path, lawn areas.

 ♿ ❀ 🚌 🍵

19 DENT HALL
Colne Road, Trawden, Colne, BB8 8NX. Mr Chris Whitaker-Webb & Miss Joanne Smith. *Turn L at end of M65. Follow A6068 for 2m; just after 3rd r'about turn R down B6250. After 1½ m, in front of church, turn R, signed Carry Bridge. Keep R, follow road up hill, garden on R after 300yds.* Sat 5, Sun 6 June (12-5). Adm £4, chd free. Home-made teas. Donation to MIND.
Nestled in the oldest part of Trawden villlage and rolling Lancashire countryside, this mature and evolving country garden surrounds a 400 year old grade II listed hall (not open); featuring a parterre, lawns, herbaceous borders, shrubbery, wildlife pond with bridge to seating area and a hidden summerhouse in

a woodland area. Plentiful seating throughout. Some uneven paths and gradients.

❀ 🍵

20 NEW DERIAN HOUSE CHILDREN'S HOSPICE
Chancery Road, Chorley, PR7 1DH. Gareth Elliot. *2m from Chorley town centre. From B5252 pass Chorley Hospital on L at r'about 1st exit to Chancery Lane Hospice on L after 0.4 m where parking around the building available & on the road. Sat Nav directions not always accurate.* Sun 9 May, Sun 8 Aug (9-5). Adm by donation. Home-made teas.
The gardens play a large part in creating an atmosphere of relaxation, tranquillity and joy. Distinct areas include the Smile Park, the Memorial, and Seaside Gardens. The gardens are an on going project, with two new areas recently created – an "enchanted" fairy garden and "Jurassic" dinosaur garden - both designed by the children in a shady corner offering a magical place to play.

 ♿ ❀ 🍵

GROUP OPENING

21 DIDSBURY VILLAGE GARDENS
Tickets available at all gardens. *5m S of Manchester. From M60 J5 follow signs to Northenden. Turn R at T-lights onto Barlow Moor Rd to Didsbury. From M56 follow A34 to Didsbury.* Sun 13 June (12-5). Combined adm £6, chd free. Home-made teas at Moor Cottage & 68 Brooklawn Drive.

68 BROOKLAWN DRIVE
M20 3GZ. Anne & Jim Britt, www.annebrittdesign.com.

3 THE DRIVE
M20 6HZ. Peter Clare & Sarah Keedy, www.theshadegarden.com.

MOOR COTTAGE
M20 6RW. William Godfrey.

NEW 1 OSBORNE STREET
M20 2QZ. Richard & Teresa Pearce-Regan.

NEW 40 PARRS WOOD AVENUE
M20 5ND. Tom Johnson.

38 WILLOUGHBY AVENUE
M20 6AS. Simon Hickey.

Didsbury is an attractive South Manchester suburb which retains its village atmosphere. There are interesting shops, cafes and restaurants, well worth a visit in themselves! This year we have 6 gardens demonstrating a variety of beautiful spaces- 2 new gardens for this year: a stylish contemporary garden with porcelain paving and two water features and , in contrast, a soft country garden with fabulous planting, We also have a large walled family garden surrounding a Georgian cottage, divided into several enchanting areas with towering Echiums and free range chickens. Another is an expertly planted shade garden with many choice rarities, whilst another reflects the charm of the cottage garden ethos with rose covered pergola, old fashioned perennials and tranquil raised pool. Our smaller gardens show beautifully how suburban plots, with limited space, can be packed full of interesting features and a range of planting styles. Dogs allowed at most gardens. Wheelchair access to some gardens.

 ♿ 🐕 ❀ 🍵

"The support of Hospice UK and the National Garden Scheme has been invaluable to hospice nurses across the country whilst we've been battling the coronavirus crisis, helping hospices such as Derian House to continue providing vital end of life and respite care to 400 children and young adults from across the North West. Thank you." – Katie Turner, Perinatal Nurse at Derian House Children's Hospice

GROUP OPENING

22 NEW **DIDSBURY VILLAGE GARDENS THE CONTEMPORARY COLLECTION**
Tickets available at all gardens. *Manchester. From M60 J5 follow signs to Northenden. Turn R at T-lights onto Barlow Moor Rd to Didsbury. From M56 follow A34 to Didsbury.* **Sun 25 July (12-5). Combined adm £6, chd free. Home-made teas and a gin bar.**

NEW **1 OSBORNE STREET**
M20 2QZ. Richard & Teresa Pearce-Regan.

NEW **FLAT 1, 5 PARKFIELD ROAD SOUTH**
M20 6DA. Kath & Rob Lowe.

NEW **11 WOLSELEY PLACE**
M20 3LR. Lesley-Ann Birley & Julian Curnuck.

A new collection of three stylish contemporary gardens in the suburban setting of Didsbury village. These gardens combine sleek design with the modern trend towards lush planting used to soften any harsh outlines. A combination of white rendered raised beds, pale porcelain paving, steel water cascades and modern Philippe Stark Bubble sofas are just some of the key notes that reference all that is best in the contemporary design field. Access is via a step in two gardens.

  ♿ ☕

23 **DUTTON HALL**
Gallows Lane, Ribchester, PR3 3XX. **Mr & Mrs A H Penny,** www.duttonhall.co.uk. *2m NE of Ribchester. Signed from B6243 & B6245 also directions on website.* **Sun 27 June (1-5). Adm £5, chd free. Home-made teas. Donation to Plant Heritage.**
An increasing range of interesting trees and shrubs have been added to the existing collection of old fashioned roses, including rare and unusual varieties and Plant Heritage National Collection of Pemberton Hybrid Musk roses. Formal garden at front with backdrop of C17 house (not open). Analemmatic Sundial, pond, meadow areas all with extensive views over Ribble Valley. Teas provided by St John's Church. Plant Heritage Plant Stall with unusual varieties for sale. Disabled access difficult due to different levels and steps.

❄ NPC ☕

24 **35 ELLESMERE ROAD**
Eccles, Salford, Manchester, M30 9FE. **Enid Noronha.** *3m W of Salford, 4m W of Manchester. From M60 exit at M602 for Salford. Take A576 for Trafford Park & Eccles, stay on A576. Turn L onto Half Edge Lane, keep L to Monton on Half Edge Lane. Turn R onto Stafford Rd & L onto Ellesmere Rd.* **Sat 12 June (12-5). Combined adm with 11 Westminster Road £5, chd free. Cream teas.**
Amidst the busy urban environment of Eccles in Salford, lies a hidden pocket

3 The Drive, Didsbury Village Gardens

© Fiona Lea

of grand houses with wide roads, and these two havens of tranquillity. 35 Ellesmere Rd is a peaceful country garden with deep herbaceous borders filled with shrubs, scented roses, and perennials. A climber covered pergola leads to a productive vegetable garden where raised beds and fruit trees add to the feeling of abundance.

& ✿ ☕

GROUP OPENING

25 NEW FRESHFIELD GARDENS
Freshfield Road, Formby, Freshfield, L37 3HW. *6m S of Southport. From Formby By-Pass turn L into Southport Rd (Esso at Junction). Go to mini r'bt & turn R into Green Ln (Grapes Pub on corner) Follow Rd to West Ln which is 2nd on R. Continue to Brewery Ln.* Sun 4 July (10-4). Combined adm £5, chd free. Home-made teas at 6 West Lane & 33 Brewery Lane.

33 BREWERY LANE
Sue & Dave Hughes.

NEW 5 WEST LANE
Kathy Taphouse.

6 WEST LANE
Laurie & Sue Lissett.

3 suburban gardens on sandy soil near to Formby sand dunes and NT nature reserve, home to the red squirrel. 5 West Lane is a W facing garden with herbaceous borders back and front with a fruit cage, herb parterre and small sink pond. 6 West Lane (opposite) has and a pergola and arches with mixed planting to rockeries and borders in sunny and shaded areas. Colourful rose and hydrangea displays with greenhouse and water features. Garden includes many baskets and containers. 33 Brewery Lane is North facing and has colour themed raised beds and a pond with a paved terrace and seating. Also a productive area with raised beds of vegetables from seeds grown in the greenhouse.

& ✿ ☕

26 GILES FARM
Four Acre Lane, Thornley, Preston, PR3 2TD. Kirsten & Phil Brown, 07925 603246, phil.brown32@aol.co.uk. *3m NE of Longridge. Pass through Longridge & follow signs for Chipping. After 2m turn R at the old school up Hope Lane. Turn L at the top and*

continue to the farm where there is ample parking. Sat 12, Sun 13 June (12-5). Adm £4, chd free. Light refreshments. A selection of sandwiches & home baked cakes. Visits also by arrangement May to July for groups of 10+.
Nestled high on the side of Longridge fell, with beautiful long-reaching views across the Ribble Valley, the gardens surround the old farmhouse and buildings. The gardens are ever evolving and include an acre of perennial wildflower meadows, wildlife pond, woodland areas and cottage gardens. There are plentiful areas to sit and take in the views. There are steps and uneven surfaces in the gardens. Disabled access difficult due to different levels, surface areas and steps.

☕

27 NEW GORSE HILL NATURE RESERVE
Holly Lane, Aughton, Ormskirk, L39 7HB. Jonathan Atkins (Reserve Manager), www.nwecotrust.org.uk. *1½ m S of Ormskirk. A59 from L'pool past Royal Oak pub take 1st L Gaw Hill Lane turn R Holly Lane. From Preston follow A59 across T-lights at A570 J & at r'about. After Xing lights turn R Gaw Hill Lane turn R Holly Lane.* Sun 18 July (11-4). Combined adm with 72 Ludlow Drive £5, chd free. Light refreshments.
Situated on a sandstone ridge offering spectacular views across the Lancashire Plain, our wildflower meadow in summer is brimming with a wide variety of wildflowers and grasses. The meadow wildlife pond is patrolled by dragonflies and damselflies and the air is full of butterflies and bees. Mown grassy paths take you through the meadow to enable close views of the flowers and insects. The woodland walk leading to the wild flower meadow has wheelchair accessible paths although access is limited in the meadow to mown grass paths.

🐕 ✿ ☕

28 GRANGE COMMUNITY GARDEN
opposite 79 Fir Trees Avenue, Ribbleton, Preston, PR2 6PQ. Annie Wynn, www.letsgrowpreston.org. *4m NE of Preston. From J30 of M6 head towards Preston. Turn R at r'about, take R at 2nd T-lights onto Ribbleton Lane B6243. After 2m turn R onto*

Grange Lane. R at Xrds. Turn R opposite 79. Sat 12 June, Sat 4 Sept (12-4). Adm £4, chd free. Light refreshments. Pizzas.
An oasis of a community garden with large polytunnels, vegetable and herbacious beds as well as an orchard, outdoor kitchen, pond and various willow structures. Willow chair, willow tepees and igloos. Pizza oven will be firing and you will have the chance to make your own pizza for a small donation Mud kitchen for children. There are disabled WCs on site and a main path and paved area to access visibility to all areas.

& ✿ ☕

29 GREEN FARM COTTAGE
42 Lower Green, Poulton-le-Fylde, FY6 7EJ. Eric & Sharon Rawcliffe. *500yds from Poulton-le-Fylde Village. M55 J3 follow A585 Fleetwood. T- lights turn L . Next lights bear L A586. Poulton 2nd set of lights turn R Lower Green. Cottage on L.* Sun 13 June (10-5). Adm £4, chd free. Home-made teas.
½ acre well established formal cottage garden. Feature koi pond, paths leading to different areas. Lots of climbers and rose beds. Packed with plants of all kinds. Many shrubs and trees. Well laid out lawns Collections of unusual plants. A surprise round every corner. Said by visitors to be a real hidden jewel.

☕

30 45 GREY HEIGHTS VIEW
Off Eaves Lane, Chorley, PR6 0TN. Barbara Ashworth, 07941 339702. *1m from Chorley Hospital. From Wigan/Coppull B5251. At (town centre) Xrds straight across . At r'about across to Lyons Ln and follow NGS signs. From M01, J8 follow signs AC Town Centre to Lyons Ln signed from here.* Sun 13 June, Sun 22 Aug (10-5). Adm £3, chd free. Cream teas. Visits also by arrangement June to Aug for groups of 10 to 30.
A small suburban garden with cottage garden style planting including fruit trees. Heavily planted with a profusion of perennials, roses and clematis. No repeat planting. Including a greenhouse, small vegetable and fruit area. An abundance of recycling and space saving ideas. Back drop of Healey Nab, and a stones throw from the Leeds - Liverpool Canal. Craft items for sale.

✿ ☕

31 THE GROWTH PROJECT
Kellett Street Allotments,
Rochdale, OL16 2JU. Karen
Hayday, 07464 546962,
k.hayday@hourglass.org.uk, www.
rochdalemind.org.uk/growth-
project. *From A627M. R A58 L
Entwistle Rd R Kellett St.* Sat 31
July (11.30-3.30). Adm £3.50, chd
free. Home-made teas. Home-
made lunches. Visits also by
arrangement June to Sept. Wed
and Thurs only from 11am-3pm.
Donation to The Growth Project.
The project is set on over an acre
and incl a huge variety of organic
veg, wildlife pond, insect hotels,
formal flower and wildflower borders,
potager and enchanted woodland
garden. See the mock Elizabethan
straw bale build and station, stroll
down the pergola walk to the
wildflower meadow and orchard.
Afternoon tea served in the Victorian
style ornate 'Woodland Green'
woodworking station. Jams and
cakes to buy. The new attraction
this year is the wildflower meadow
and orchard The Growth Project is a
partnership between Hourglass and
Rochdale and District Mind. Providing
cut flowers, veg, preserves and gifts
plus guides to give horticultural advice
and show you round. No disabled
WC, ground can be uneven.

GROUP OPENING

32 HALE VILLAGE GARDENS
Liverpool, L24 4BA. *6m S of M62
J6. Take A5300, A562 towards
L'pool, then A561, L for Hale opp the
old RSPCA. From S L'pool head for
the airport then L sign for Hale. The
82A bus from Widnes/Runcorn to
Liverpool stops in the village.* Sat 26,
Sun 27 June (1-5). Combined adm
£5, chd free. Light refreshments
at 66 Church Rd.

NEW 4 CHURCH ROAD
Mrs Chesters.

54 CHURCH ROAD
Norma & Ray Roe.

66 CHURCH ROAD
Liz Kelly-Hines & David Hines.

2 PHEASANT FIELD
Roger & Tania Craine.

The delightful village of Hale, is set
in rural South Merseyside between
Widnes and Liverpool Airport. It is
home to the cottage, sculpture and

grave of the famous giant known as
the Childe of Hale. Four gardens of
various sizes have been developed
by their present owners. There is
one mixed contemporary/country
style garden to the west of the village
which displays a delightful water
feature with lots of summer colour. In
Church Road to the east of the village
there are a further three country style
gardens. The first features a beautiful
swathe of herbaceous colour, the
second a wildlife pond and open
aspect with stunning views, and the
final one mixed planting for year round
interest plus an extensive allotment.

33 HALTON PARK HOUSE
Halton Park, Halton, Lancaster,
LA2 6PD. Mr & Mrs Duncan
Bowring. *7min drive from both J34
& 35 M6. On Park Lane, approx.
1½ m from Halton or Caton. Park
Lane accessed either from Low Rd
or High Rd out of Halton. From Low
Rd turn into Park Lane through pillars
over cattle grid.* Sat 22, Sun 23 May
(10-6). Adm £4, chd free. Cream
teas.
Approximately 6 acres of garden, with
gravel paths leading through large
mixed herbaceous borders, terraces,
orchard and terraced vegetable
beds. Wildlife pond and woodland
walk in dell area, extensive lawns,
greenhouse and herb garden. Plenty
of places to sit and rest! Cream teas
and home made cakes. Gravel paths
(some sloping) access viewing points
over the majority of the garden. Hard
standing around the house.

Open your garden with the
National Garden Scheme
and join a community of
like-minded individuals, all
passionate about gardens, and
raise money for nursing and
health charities. Big or small,
if your garden has quality,
character and interest we'd
love to hear from you. Call
us on 01483 211535 or email
hello@ngs.org.uk

34 NEW THE HAVEN GARDEN
Snow Hill Lane, Scorton, Preston,
PR3 1BA. Mrs Nikki Cookson,
www.woodcroftcrafts.co.uk.
*Between Preston & Lancaster. Leave
A6 follow signs to Scorton. From
village centre, go up Snowhill Lane &
cross the motorway bridge. Entrance
to Haven Garden is via Woodcroft
Crafts.* Sun 11 Apr (11-4). Adm £5,
chd free. Pre-booking essential,
please visit www.ngs.org.uk for
information & booking.
Situated in the beautiful Forest of
Bowland, The Haven Garden is set
in four acres of private woodland.
There is magic around every corner
and something for all the family
including stickmen, fairy houses, a
wishing well, 30ft. high train bridge,
waterwheel and a labyrinth prayer
walk. The garden was originally
landscaped over 100 years ago and
has been developing ever since. At
the entrance to the Haven Garden is
Woodcroft Crafts, a little shop-in-
a-shed selling handmade gifts and
Christian crafts made from natural or
recycled materials.

35 ♦ HAZELWOOD
North Road, Bretherton, Leyland,
PR26 9AY. Jacqueline Iddon &
Thompson Dagnall, 01772 601433,
jacquelineiddon@gmail.com,
www.jacquelineiddon.co.uk. *8m
SW of Preston. Between Southport
& Preston, from A59, take B5247
for 1m then L onto (B5248) Garden
signed from North Rd.* For NGS:
Sun 23 May, Sun 13 June, Sun 4
July (12-5). Adm £3.50, chd free.
Cream teas. For other opening
times and information, please
phone, email or visit garden
website.
New for 2021 Alpine House, cutting
and vegetable garden. 1½ acre
garden and hardy plant nursery,
gravel garden with pots and seating
area, shrubs, herbaceous borders,
stream-fed pond with woodland
walk. Oak-framed, summerhouse,
log cabin Sculpture gallery fronted
by cottage garden beds. Sculpture
demonstration at 2 pm. Beach
area. Teas in Coach house in aid of
Queenscourt Hospice. Extensive
sculpture collection, the work of
Thompson Dagnall Printmaking
Studio With prints and cards By Tilly
Dagnall. Majority of the garden is
accessible to wheelchairs.

36 NEW HIGHER BRIDGE CLOUGH HOUSE

Coal Pit Lane, Rossendale, BB4 9SB. Karen Clough. *4m from Rawtenstall. Take A681 into Waterfoot. Turn L onto the B6238. Approx ½ m turn R onto Shawclough Rd. Follow the road up until you reach the yellow signs.* Sat 31 July, Sun 1 Aug (12-6). Adm £3.50, chd free. Cream teas.

Nestled within Rossendale farm land in an exposed site the garden is split by a meandering stream. There are raised mixed borders with shrubs and perennials enclosed by a stone wall, with loose planting of shrubs and water loving plants around the stream. A local farmer once told me 'you won't grow owt up 'ere... so the challenge is on. Plant sales and refreshments The Rossendale area provides some excellent walking and rambling sites. Rawtenstall has the famous East Lancashire railway, which is worth a visit.

GROUP OPENING

37 HILLSIDE GARDENS

Clovelly Drive, Southport, PR8 3AJ. *3m S of Southport. Gardens signed from A565 Waterloo Rd & A5267 Liverpool Rd.* Sun 23 May (11-5). Combined adm £3.50, chd free. Sun 18 July (11-5). Combined adm £6, chd free. Sun 8 Aug (11-5). Combined adm £3.50, chd free. Light refreshments at 23 Ashton Rd. 115 Waterloo Rd,339 Liverpool Rd, wine at 18 Clovelly Drive.

23 ASHTON ROAD
PR8 4QE. John & Jennifer Mawdsley.
Open on Sun 18 July, Sun 8 Aug

33 CLOVELLY DRIVE
PR8 3AJ. Bob & Eunice Drummond.
Open on Sun 23 May, Sun 18 July

LINKS VIEW, 18 CLOVELLY DRIVE
PR8 3AJ. Christine & Dave McGarry.
Open on Sun 18 July

339 LIVERPOOL ROAD
PR8 3DE. Ian & Sue Dexter.
Open on Sun 18 July

NEW 115 WATERLOO ROAD
PR8 4QN. Antony & Rebecca Eden, antony@alary.co.uk.
Open on all dates
Visits also by arrangement May to Aug.

The gardens of Hillside are full of variety and interest, each having ponds or water features which create a relaxed atmosphere. 23 Ashton Road is separated into three rooms with interesting shrubs, herbaceous plants, greenhouses and a vegetable plot. 18 Clovelly Drive, curved lawn with a patio, glass features around the fencing add a different dimension, with chain saw carvings interspersed amongst plants, individual gate. WC available. 33 Clovelly Drive, mature trees and shrubs set off the lawn with sweeping colour-themed herbaceous borders. Rhododendrons and azaleas give colour in spring. 339 Liverpool Road full of beautiful and unusual trees and shrubs. It is now being brought back to life and developed with herbaceous planting and grasses. Excellent spring colour. WC available. 115 Waterloo Road, new large family garden, full of colour with shrubs and annuals, having many seating areas. Partial wheelchair access to 3 gardens, fully accessible to 2.

38 NEW HOWICK HOUSE

Howick Park Avenue, Penwortham, Preston, PR1 0LS. Galloways Society for Blind. *Follow A59 towards Southport using the Sat Nav PR1 0LS until you come to Howick Park Ave, turn R, you will see the Galloway sign - enter the main car park.* Sun 8 Aug (1-4.30). Adm £4, chd free. Light refreshments.

The Gardens include 2 poly tunnels, greenhouses and sheds. We have built raised beds to accommodate vegetables and flowers. We have various fruit trees and a variety of herbs grown in recycled tyres. All waste is recycled and have a large compost area. Last year we started a wildlife garden and it is still an ongoing project. Most of our plants and equipment have been donated. solid paths all round.

GROUP OPENING

39 HUTTON GARDENS

Tolsey Drive, Hutton, Preston, PR4 5SH. Heather & John Lund. *2m SW of Preston. Take A59 towards Southport, at the r'about head towards Longton on Liverpool Rd. Tolsey Dr is 100 yds on the L. Signed from A59 r'about.* Sun 4 July (11.30-4.30). Combined adm £4, chd free. Home-made teas

at 10 Tolsey Drive. At 2 Tolsey Drive home-made sausage rolls, cheese and onion rolls and home-made rhubarb gin.

2 TOLSEY DRIVE
Vicki & Alex Cullen.

5 TOLSEY DRIVE
Marilyn & James Woods.

10 TOLSEY DRIVE
Heather & John Lund.

Hutton is a small village on the outskirts of Preston. There are 3 long, fairly narrow gardens which vary in style. 2 Tolsey Drive has an open aspect with sweeping lawns tapering to a point, mature trees, shrubs and developing flower beds. 5 Tolsey Drive is an informal garden with cottage planting, a wildlife pond, hedgerow and trees, with many peaceful places to sit and contemplate. 10 Tolsey Drive is an eclectic garden with three distinct areas. A formal lawn with large herbaceous borders with some unusual plants, a working section with greenhouses growing vegetables and exotic plants and a calm shady garden. As featured in Lancashire Life as the Recycled Garden with Wonky Wall, water feature and planters made from recycled materials from the house renovations. Wheelchair access to patio area only in all gardens.

40 KINGTON COTTAGE

Kirkham Road, Treales, Preston, PR4 3SD. Mrs Linda Kidd, 01772 683005. *M55 J3. Take A585 to Kirkham, exit Preston St, L into Carr Lane to Treales Village. Cottage on L in front of Derby Arms parking here by kind permission of owner.* Sun 2 May, Sun 6 June, Sat 3, Sun 4 July, Sun 1 Aug (10-5). Adm £3.50, chd free. Home-made teas. Visits also by arrangement May to Aug for groups of 10 to 30.

Nestling in the beautiful village of Treales this generously sized Japanese garden has many authentic and unique Japanese features, along side its 2 ponds linked by a river. The stroll garden leads down to the tea house garden. The planting and the meandering pathway blend together to create a tranquil meditative garden in which to relax. Runner up in Daily Mail Best Kept Garden Competition. Wheelchair access to some areas, uneven paths.

41 72 LUDLOW DRIVE

Ormskirk, L39 1LF. Marian & Brian Jones. *½ m W of Ormskirk on A570. From M58 J3 follow A570 to Ormskirk town centre. Continue on A570 towards Southport. At A570 junction with A59 cross T-lights after ½ m turn R at Spar garage onto Heskin Lane then 1st R Ludlow Drive.* Sun 18 July (11-4). Combined adm with Gorse Hill Nature Reserve £5, chd free. Home-made teas. Gluten free and vegan cake also provided.

A beautiful town garden overflowing with a wide variety of bee friendly planting. Developed over 15 years it includes a gravel garden and a newly developed Jewel garden to the front. The back garden has raised shade and sunny borders with roses and clematis. There is new Victorian greenhouse with a collection of Pelargoniums and succulents also a raised pond, and alpine troughs. Wheelchair access to front and rear garden.

& ❋ ☕

42 LYTHAM HALL

Ballam Rd, Lytham, Lytham St Annes, FY8 4JX. Paul Lomax. *Follow the brown tourist signs for Lytham Hall Sat Nav users use postcode FY8 4TQ.* Sun 18 July (10-4.30). Adm £4.50, chd free. Light refreshments. Picnic tables on the East Lawn.

Lytham Hall is a C18 Georgian house set in 78 acres of historic woodlands, with a parterre, herbaceous border, south aspect garden, two wildlife ponds. A mount, that can be climbed with views over the parkland and 4km of paths. There is a RHS Gold award winning vegetable garden and potager. Farmland animals, apiary and garden nursery. 1m drive from the main gates to the Hall. There is a separate designated path for pedestrians past fields and through woodland. An outside catering vehicle will be available for hot drinks and snacks in addition to the cafe. Wheelchair access available to most areas. Gravel paths in woodland. Coaches by appointment only.

& 🐕 ❋ 🚌 ☕ 🪑

43 MAGGIE'S MANCHESTER

Kinnaird Road, Manchester, M20 4QL. Jemma Halman, 0161 641 4857, jessica.ruth@maggiescentres.org, maggies.org. *At the end of Kinnaird Rd which is off Wilmslow Rd opposite the Christie Hospital.* Sun 25 July (12-3.30). Adm by donation. Light refreshments. Visits also by arrangement.

The architecture of Maggie's Manchester, designed by world-renowned architect Lord Foster, is complemented by gardens designed by Dan Pearson, Best in Show winner at Chelsea Flower Show. Combining a rich mix of spaces, including the working glass house and vegetable garden, the garden provides a place for both activity and contemplation. The colours and sensory experience of nature becomes part of the Centre through micro gardens and internal courtyards, which relate to the different spaces within the building. Wheelchair access to most of the garden from the front entrance.

& 🐕 ☕

44 MAGGIE'S, OLDHAM

The Royal Oldham Hospital, Rochdale Road, Oldham, OL1 2JH. Maggie's Centres, 0161 989 0550, oldham@maggies.org, www.maggiescentres.org/oldham. *Maggie's in the grounds of the Royal Oldham Hospital, next door to A&E. It's the wooden building on stilts & the garden lies underneath the building.* Sun 18 July (11-3). Adm £3, chd free. Light refreshments. Visits also by arrangement Apr to Oct.

The garden is framed by enclosing walls. The building 'floating' aloft is like a drop curtain to the scene, creating a picture window effect. The trees soar upwards filling the volume of space. A woodland understorey weaves between the structure of the numerous white birch and crispy bark of the pine trunks. The garden could be described as an ornamental woodland. A member of the team can guide you to the wheelchair entrance to the garden if required.

& 🐕 🚌 ☕

45 6 MENIVALE CLOSE

Southport, PR9 9RY. Ann-Marie Hutson. *4m N of Southport. A565 from Preston. 2nd exit at Plough r'bout. At BP garage turn R. Then 3rd R. At T junction turn R, then 1st L. A565 from Southport at BP garage turn L. Then 3rd R. At T junction turn R. then 1st L.* Sun 13 June (10-5.30). Combined adm with 120 Roe Lane £3.50, chd free. Light refreshments.

The house having been built in the 1970's, the garden follows a cottage garden style with brick edged island beds, each bed having a seasonal focus. Surrounded by neighbouring trees, the garden is sheltered from the NW winds. Partial access for wheelchairs.

❋ ☕

46 MERESANDS KENNELS & CATTERY

Holmeswood Rd, Rufford, Ormskirk, L40 1TG. Mrs Bridget Street. *6m N of Ormskirk. From the A59 Rufford, turn L on to Holmeswood Rd, past the Hesketh Arms on R. 1m on L Signed to Meresands Kennels & Cattery.* Sun 6 June (10.30-4.30). Adm £5, chd free. Light refreshments.

An enchanting woodland garden of approx 4 acres with formal gardens and colourful koi pond. Walking through the laburnum arch you can explore the woodland area. Winding paths lead you through rhododendrons and mature trees, with areas for sitting, a stumpery, a hobbit house and wildflower area. There is a natural pond overlooked by a pavilion, with lovely views and an abundance of wildlife. Woodland walks, natural pond, hidden follies. Not all the garden is accessible by wheelchair.

& ☕

47 MILL BARN

Goosefoot Close, Samlesbury, Preston, PR5 0SS. Chris Mortimer, 07742924124, chris@millbarn.net, www.millbarn.net. *6m E of Preston. From M6 J31 2½m on A59/A677 B/burn. Turn S. Nabs Head Lane, then Goosefoot Lane.* Visits by arrangement May to July fr groups of up to 20. Book by text or email with 24 hrs notice. Open from 11am to 5pm. Adm £5, chd free. Cups of hot drinks only unless other arrangements have been specifically made.

The unique and quirky garden at Mill Barn is a delight: or rather a series of delights. Along the R Darwin, through the tiny secret grotto, past the suspension bridge and view of the fairytale tower, visitors can a stroll past folly, sculptures, lily pond, and lawns, enjoy the naturally planted flowerbeds, then enter the secret garden and through it the pathways of the wooded hillside beyond. A garden developed on the site of old mills gives a fascinating layout which evolves at many levels. The garden jungle provides a smorgasbord of

111 Main Street, Warton Gardens

flowers to attract insects throughout the season. Children enjoy the garden very much. Partial wheelchair access, visitors have not been disappointed in the past.

48 NEW MILL CROFT

Wennington, Lancaster, LA2 8NX. **Linda Ashworth.** *From J34 of M6 take A600 towards Kirkby Lonsdale. Turn R on B6480 towards Bentham.* **Sun 27 June (10-5). Combined adm with Raw Ridding House £6, chd free. Light refreshments. Also open Clearbeck House.**

A rural garden extending to about 3.25 acres with a summer flowering meadow, stream, trees, shrubs, herbaceous planting, vegetable plot and stumpery. Various seating areas to rest and enjoy local wildlife and the surrounding views. An ongoing experiment to discover what will thrive in heavy clay with occasional flooding, exposure to wind and in a frost pocket. New projects always underway. Most areas are accessed over grass; the ground may be soft when wet. Areas of sloping land. Flower beds, stumpery and views seen from main drive.

49 NEW 194 MOTTRAM OLD ROAD

Gee Cross, Hyde, SK14 3BA. **Mrs Anne Dickinson.** *10m SE of Manchester on M60, M67. Leave M67 at r'about (motorway terminates) take exit for Gee Cross.* **Sun 30 May (11-4). Adm £5, chd free. Home-made teas.**

A terraced garden on several levels with far reaching views over Manchester and the Pennines. Created by the owner's late husband Harry. Acers and Rhododendrons feature alongside many species of perennials. A meandering naturalistic water feature constructed from locally sourced stone cascades down to the pond. Woodland garden leads to a vegetable garden and work area. Many seating areas to enjoy.

50 NEW OLD HOLLOWS FARM

Old Hollow Lane, Banks, Southport, PR9 8DU. **Janet Baxter,** 07067 274387, **j.baxter728@btinternet.com.** *5m N of Southport. A565 r'about take turning for Banks. Straight through Banks. New Lane Pace Rd ends on bend turn L on farm rd into Old Hollow Lane parking in yard at lane end.* **Sat 12, Sun 13 June (11-5). Adm £3.50, chd free. Home-made teas. Visits also by arrangement Mar to June for groups of 20+.**

Old Hollow Cottage Garden open in memory of my late husband Alec Baxter. Whose straw hat was always popping up somewhere. Waste ground and pit 16yrs ago - Now a peaceful tranquil garden on the edge of the Ribble estuary. Patio area garden room - large lawned areas - trees - large herbaceous border - arboretum. The wildlife pond has a large reed bed - Walks along the estuary embankment. Display of Vintage Bicycles - Driftwood Display - Bee demonstration Local Crafts and Artisans. No disabled WC facilities. Level in main area only.

51 THE OLD VICARAGE
Church Road, Astley, Manchester, M29 7FS. Susan & Neil Kinsella. *½ way between Leigh & Worsley off A580 East Lancs Rd. Opp the site of the old St Stephens Astley CE Church & next to Dam House, 250 yds from the mini r'about junction at Church Rd (A5082) & Manchester Rd(A572). Close to the Bull's Head.* **Sat 12, Sun 13 June (10-4). Adm £4, chd free. Home-made teas.**
Recently restored medium sized gardens of Grade 2* vernacular Georgian vicarage comprising both formal and informal planting and statuary as the garden narrows to follow the woodland stream which runs through it. There are six separate areas/rooms which gradually merge into the surrounding trees ending several hundred yards from the frontage to the property. Most of the garden is flat with some gravel paths.

  ♿ 🐕 ✽ 🍵

52 PARKERS LODGE
28 Lodge Side, Bury, BL8 2SW. Keith Talbot. *2m W of M66 J2. From A58 then B6196 Ainsworth Rd turn R onto Elton Vale Rd, drive past the sports club onto a small estate, & follow the road round to car park.* **Sun 23 May, Sun 11 July (10-4.30). Adm £5, chd free. Light refreshments.**
Parkers Lodge is set on the site of a demolished Victorian Mill that was used for bleaching. Since the houses were completed in 2014 a small group of volunteers have been working to transform what was a jungle into a manicured wild space that still allows the local wildlife to flourish. Set in 12 acres with 2 lakes with 50 percent of them open for a leisurely stroll round.

🍵

53 PLANT WORLD
Myerscough College, St Michaels Road, Bilsborrow, Preston, PR3 0RY. Myerscough college, 01995 642264, tmelia@myerscough.ac.uk, www.myerscough.ac.uk/commercial-services-equine-services/plant-world-gardens/. *5m N J32 M6. From M6 take J32 and head North up the A6. Turn L onto St Michaels Rd take 2nd entrance into the college signed Plant World.* **Sun 8 Aug (10.30-4). Adm £3.50, chd free. Cream teas.**
A gardener's paradise with an acre of RHS gold award winning gardens

and glasshouses, including tropical, temperate and desert zones. Themed and herbaceous borders, pinetum, fruit garden, bog garden, pond and woodland garden. With many plants labelled for identification purposes. Plant sales with many rare and unusual specimens. Myerscough Tea rooms overlook the stunning gardens serving a wide range of delicious locally produced cakes and sandwiches. Indoor and outdoor seating. Guided tours throughout the day. Dogs welcome in the gardens, sales area and the café. Partial access for wheelchairs. Paths are grass and it can be wet in some areas.

🐕 ✽ 🚌 🚐 🍵 🪑

54 NEW RAW RIDDING HOUSE
Monk's Gate, Tatham, Lancaster, LA2 8NH. Rebecca & Richard Sanderson. *10m from J34 M6. Off the B6480 between Wennington & Wray. 1m E of Wray, turn L onto Monk's Gate, signed for Tatham Church. Park at church unless disabled.* **Sun 27 June (10-5). Combined adm with Mill Croft £6, chd free. Home-made teas. Also open Clearbeck House. Gluten Free cake available.**
A blend of contemporary design with relaxed planting with decks and terraces to accommodate the sloping site without compromising the amazing views. Large herbaceous beds, lawns, a small wildlife pond and wildflower meadow. Hidden seating areas. Parking at church with a 5min walk to garden. No steps but gently sloping gravelled or bark chip paths to all areas may require assistance to navigate.

  ♿ 🐕 🍵

55 ◆ THE RIDGES
Weavers Brow (cont. of Cowling Rd), Limbrick, Chorley, PR6 9EB. Mr & Mrs J M Barlow, 01257 279981, barbara@barlowridges.co.uk, www.bedbreakfast-gardenvisits.com. *2m SE of Chorley town centre. From M6 J27. From M61 J8. Follow signs for Chorley A6 then signs for Cowling & Rivington. Passing Morrison's up Brook St, mini r'about 2nd exit, Cowling Brow. Pass Spinners Arms on L. Garden on R.* **For NGS: Mon 3 May, Mon 30 Aug (11-5). Adm £5, chd free. Light refreshments. by ladies of St James Church. For other opening times and information,**

please phone, email or visit garden website.
3 acres, incl old walled orchard garden, cottage-style herbaceous borders, with perfumed rambling roses and clematis thru fruit trees. Arch leads to formal lawn, surrounded by natural woodland, shrub borders and specimen trees with contrasting foliage. Woodland walks and dell. Natural looking stream, wildlife ponds. Walled water feature with Italian influence, and walled herb garden. Classical music played. Wheelchair access some gravel paths and woodland walks not accessible.

  ♿ 🐕 ✽ 🚐 🍵

56 NEW 4 ROE GREEN
Worsley, Manchester, M28 2JB. Geoff & Pauline Ogden. *5m W of Manchester. Roe Green is adjacent to J14 of M60 motorway. Please park on Old Clough Lane.* **Sat 3 July (12-5). Adm £4, chd £1. Home-made teas. Cordial and biscuits for children.**
A 300 year old cottage located on the edge of the village green in the picturesque conservation area of Roe Green. This is a plantswoman's quintessential Cottage Garden, packed with many unusual perennials and mixed wildflowers that thrive in this sheltered walled space. In addition to a woodland area, there is an old bothy which adds to the delightful atmosphere of this garden.

🐕 ✽ 🍵

57 120 ROE LANE
Southport, PR9 7PJ. Mrs Mavis Standing. *2m N Southport. A565 N on Lord St at r'about 3rd exit to Manchester Rd continue to Roe Lane.* **Sun 13 June (10-5.30). Combined adm with 6 Menivale Close £3.50, chd free. Home-made teas.**
Mature garden encouraged to develop over 25 years and includes trees, shrubs, succulents and perennials. Diverse corners and small wildlife pond.

🐕 ✽ 🍵

58 NEW ST THOMAS PRIMARY SCHOOL
Kenyons Lane, Lydiate, Liverpool, L31 0BP. Mr Mark Ward. *8m N of Liverpool. From end of M57/58 take A59 towards Ormskirk after 2.3m turn L into Lydiate Lane.* **Sun 6 June (12-5.30). Combined adm with 136 Buckingham Road £3.50, chd**

free. **Home-made teas.**
A collection of gardens planted and cared for by the children and school gardener, which features our 'Bookshelf Beach', blissful prayer garden, boy garden and tranquil wildlife garden with colourful bee friendly planting. The gardens provide a place for both activity and contemplation and year-round colour and interest. There are picnic tables and an adventure playground at the rear.

59 14 SAXON RD
Birkdale, Southport, PR8 2AX. Margaret & Geoff Fletcher, 01704 567742, margaret.fletcher@ngs.org.uk. *1m S of Southport. Garden signed from A565 Southport to Liverpool road.* **Sun 30 May, Sun 25 July (11-5). Combined with 45 Stourton Road £4, chd free. Home-made teas. Visits also by arrangement May to Sept for groups of 10+.**
A walled garden surrounding a Victorian house with formal and informal planting mainly cottage garden style, accessed by bark and gravel paths. New planting for 2021

GROUP OPENING

60 SEFTON PARK MAY GARDENS
Sefton Drive, Sefton Park, Liverpool, L8 3SD. *1m S of Liverpool city centre. From end of M62 take A5058 Queens Drive ring rd S through Allerton to Sefton Park. Parking roadside and in the park.* **Sun 16 May (12-5). Combined adm £5, chd free. Light refreshments.**

THE COMMUNITY ORCHARD AND WILDLIFE GARDEN
L17 2AT. The Society of Friends, www.tann.org.uk.

6 CROXTETH GROVE
L8 3SA. Stuart Speeden.

PARKMOUNT
L17 3BP. Jeremy Nicholls.

THAT BLOOMIN' GREEN TRIANGLE
L8 2XA. Mrs Helen Hebden.

This is a fascinatingly varied group of Liverpool gardens. In That Bloomin' Green Triangle, the most recent project has been the creation of a delightful indoor Winter Garden in what were two derelict houses. It was the guerrilla gardening by residents which led to the area's regeneration. The Community Orchard in Arundel Avenue is a quiet haven in the former Quaker Burial Ground hidden away behind high walls. Parkmount opens for the first time in spring - a chance to see its glorious borders at a different time of year. And Croxteth Grove garden is a great illustration of what can be achieved in a small space. Tours of the Bloomin' Green Triangle at 1 and 3 pm starting from the Winter Garden in Cairns Street.

GROUP OPENING

61 SEFTON PARK SEPTEMBER GARDENS
Sefton Drive, Sefton Park, Liverpool, L8 3SD. *From end of M62, take A5058 Queens Drive S through Allerton to Sefton Park & follow yellow signs. Parking roadside and in Sefton Park.* **Sun 5 Sept (12-5). Combined adm £5, chd free. Light refreshments.**

FERN GROVE COMMUNITY GARDEN
L8 0RY. Liverpool City Council.

37 PRINCE ALFRED ROAD
L15 8HH. Jane Hammett.

SEFTON PARK ALLOTMENTS
L17 1AS. Sefton Park Allotments Society.

SEFTON VILLA
L8 3SD. Patricia Williams.

17 SYDENHAM AVENUE
L17 3AU. Fatima Aabbar-Marshall.

Three beautiful private gardens planted for late summer colour and interest, nearly 100 allotments, and children's activities and beekeeping demonstration at Fern Grove Community. The secret garden in Prince Alfred Road is a hidden sandstone walled space with glorious planting, a lovely greenhouse, summerhouse and bothy. Gorgeous planting in Sefton Villa and Sydenham Ave. Vegetable produce and lovely dahlia displays at the allotments. No wheelchair access at 37 Prince Alfred Road. Disabled WC at Sefton Park allotments.

62 SOUTHLANDS
12 Sandy Lane, Stretford, M32 9DA. Maureen Sawyer & Duncan Watmough, 0161 283 9425, moe@southlands12.com, www.southlands12.com. *3m S of Manchester. Sandy Lane (B5213) is situated off A5181 (A56) ¼m from M60 J7.* **Sun 18 July (12-5.30). Adm £5, chd free. Home-made teas. Cake-away service (take a slice of your favourite cake home) Home-made Ice-cream. Visits also by arrangement June to Aug for groups of 5+.**
Described by visitors as 'totally inspirational', this artists' multi-award winning garden unfolds into a series of beautiful spaces including Mediterranean, Ornamental and Woodland gardens. Organic kitchen garden with large glasshouse containing vines. Recently redesigned herbaceous borders, hanging baskets and stunning container plantings throughout the garden, 2 ponds and a water feature. 21st year opening celebrations. Artist's work on display.

63 STANHILL EXOTIC GARDEN
10 Stanhill Street, Oswaldtwistle, Accrington, BB5 4QE. Tez Donnolly, www.facebook.com/stanhillgarden. *From M65 J5 take A6077 towards Shadsworth, first R to B623. 2.3m, L at Black Dog Pub, 0.2m R into Thwaites Rd for parking. Then cross Rd to Stanhill St on foot. Garden at rear.* **Sat 21, Sun 22 Aug (12.30-4). Adm £4, chd free. Light refreshments.**
Behind a row of terraced houses lies a hidden gem. A lush exotic style garden containing palms, tree ferns, bananas, and many other unusual plants. A mixture of paving and bark chipping paths takes you on a journey round the tropical garden, encountering a pond and an aviary along the way.

The National Garden Scheme searches the length and breadth of England and Wales for the very best private gardens

64 NEW **45 STOURTON ROAD**
Southport, PR8 3PL. Pat & Bill
Armstrong. *4m S of Southport.*
Take A565 Southport to Liverpool
rd. At Ainsdale village r'about
proceed along main road to T-lights
at Kenilworth Road. Turn R & then
immediate R into Stourton Road.
Sun 30 May, Sun 25 July (11-5).
Combined adm with 14 Saxon Rd
£4, chd free. Home-made teas.
A garden which has evolved over
several years and features a wide
selection of trees, shrubs and
perennials to take you through the
year. Differing points of interest -
summerhouse, topiary, areas of
cottage style planting punctuated
with floral creations, groupings of
succulents and varied seating areas
to relax. A surprise round every
corner.

GROUP OPENING

65 **WARTON GARDENS**
Warton, LA5 9PJ. 01524 727770,
claire@lavenderandlime.co.uk.
1½ m N of Carnforth. From M6 J35
take A601M NW for 1m, then N on
A6 for 0.7m turn L signed Warton
Old Rectory. Warton Village 1m
down Borwick Lane. From Carnforth
pass train stn & follow signs Warton
& Silverdale. **Sat 17, Sun 18 July**
(11-4.30). Combined adm £5, chd
free. Home-made teas at 109/111
Main Street.

BRIAR COTTAGE
LA5 9PT. Mr Bendall.

2 CHURCH HILL AVENUE
LA5 9NU. Mr & Mrs J Street.

NEW **107 MAIN STREET**
LA5 9PL. Becky Hindley,
www.pickingposies.co.uk.

111 MAIN STREET
LA5 9PJ. Mr & Mrs J
Spendlove, 01524 727 770,
claire@lavenderandlime.co.uk.

NEW **WARTON ALLOTMENT**
HOLDERS
LA5 9QU. Mrs Jill Slaughter.

The gardens and allotments are spread
across the village and offer a wide
variety of planting and design ideas.
Our group offers a contemporary
garden on limestone pavement, flower
picking garden, unusual herbaceous
and vegetable garden, 21 allotments
and a large more formal garden.
Parking at the bottom of village (LA5
9NU) or the public car park, next to
Old School Brewery (LA5 9PL). Warton
has 2 Pubs . It is the birthplace of
the medieval ancestors of George
Washington, whose family coat of
arms can be seen in St Oswald's
Church. The ruins of the Old Rectory

Meresands Kennels & Cattery

(English Heritage) is the oldest surviving building in the village. Ascent of Warton Crag (AONB), provides panoramic views across Morecambe Bay to the Lakeland hills beyond. All gardens have steps and uneven surfaces unsuitable for wheelchair access.

66 8 WATER HEAD

Fulwood, Preston, PR2 3TU. Phil Parkinson, philinterfaith@gmail.com. *3m N of Preston. 2m from M6 J32. A6 into Preston, R at Black Bull Lane. At 2nd r'about, R, Cadley Causeway. R at T junction r'about. Signage after the bend. Parking in Hollins Grove please.* **Sat 12, Sun 13 June, Sat 28, Mon 30 Aug (10.30-5.30). Adm £3.50, chd free. Light refreshments. Visits also by arrangement Mar to Oct for groups of up to 10.**

A medium sized, rigorously structured, suburban garden in 5 rooms with a canal bank for extras. Over 70 roses, entirely crimson and white in the first room, a large fernery with rare specimens, Germanica iris, rudbeckia, hosta, hellebore, canna, and tree collections, Langdale slate paving, and an imaginative use of levels and height, the whole being suffused with a contemplative, centred energy.

67 WEEPING ASH GARDEN

Bents Garden & Home, Warrington Road, Glazebury, WA3 5NS. John Bent, www.bents.co.uk. *15m W of Manchester. Located next to Bents Garden & Home, just off the A580 East Lancs Rd at Greyhound r'about near Leigh. Follow brown 'Garden Centre' signs.* **Sun 14, Sun 21 Feb (11-3); Sun 8 Aug, Sun 12 Sept (10-4.30); Sun 31 Oct (11-3). Adm by donation.**

Created by retired nurseryman and photographer John Bent, Weeping Ash is a garden of all-year interest with a beautiful display of early snowdrops. Broad sweeps of colour lend elegance to this stunning garden which is much larger than it initially seems with hidden paths and wooded areas creating a sense of natural growth. Bents offers a choice of dining destinations from including The Fresh Approach Restaurant, Caffe nel Verde and its Mediterranean style Tapas Bar. Partial wheelchair access and weather dependent.

68 11 WESTMINSTER ROAD

Eccles, Manchester, M30 9HF. George & Lynne Meakin. *3m W of Salford, 4m W of Manchester. 1st exit M602 Manchester direction 1st exit (r'about) 2nd T-lights turn L. After Xing turn R, Victoria Rd 2nd R (Westminster Rd) no.11 on L.* **Sat 12 June (12.30-5.30). Combined adm with 35 Ellesmere Road £5, chd free. Home-made teas at Ellesmere Rd. Wine & nibbles at 11 Westminster Rd.**

There is pretty front garden with topiary chickens and well stocked with perennials, mature trees, and box hedging. The garden is divided by a trellis and rose arch which separates the flower beds and lawn from the fruit growing area, and there are a large number of fuchsias grown in pots. A coach house to the rear of the garden where wine/nibbles will be served. The garden is on the flat except for a small area in the front garden, care will need to be taken in case the path is slippery.

69 WIGAN & LEIGH HOSPICE

Kildare Street, Hindley, Wigan, WN2 3HZ. Wigan & Leigh Hospice, thehospiceogardener.com. *1½m SW of Wigan. From Wigan on A577 Leigh/Manchester Rd. In Hindley turn R at St Peter's Church onto Liverpool Rd A58. After 280 metres turn R into Kildare St.* **Sun 25 July (11-4). Adm £4, chd free. Light refreshments.**

Large attractive gardens surround the Hospice creating a place of tranquillity. At the front are beautiful raised beds and a new courtyard garden, at the rear 3 large ponds. Outside patients' rooms are colourful tubs and flower beds. A memorial daisy garden. A wildflower garden has been created - 'The Amberswood Garden'. The gardens are a haven for wildlife. Garden awarded 'Gold' in NW in Bloom. Fully accessible, including WC.

70 NEW WILLOW WOOD HOSPICE GARDENS

Willow Wood Close, Mellor Road, Ashton-Under-Lyne, OL6 6SL. Willow Wood Hospice, 07501 017703, smcordingley@hotmail.com, www.willowwood.info. *3mins from J23 M60. Exit onto A6140 towards Ashton-u-Lyne turn R onto Manchester Rd/A635 then slight R onto Park Parade/A635 keep R,*

stay on A635 turn L onto Mellor Rd L onto Willow Wood Close. **Sat 15 May, Sat 19 June (11-4). Adm £4, chd free. Light refreshments. Sats, cream teas, cakes, light snacks, hot and cold beverages, Pimms, wine and cocktails. Visits also by arrangement May to Oct for groups of up to 10. Visit the hospice website 'About Us' section for weekday openings details.**

The sensory gardens, redesigned and maintained by two former National Trust gardeners and a wonderful team of volunteers, comprise a series of formal and informal gardens. These spaces contain lush, romantic planting design, filled with scent, texture, studies in colour combinations and sculptural form, water features, a kitchen garden, wildlife friendly planting and two woodland areas. The Sensory and Tranquil gardens are wheelchair accessible.

GROUP OPENING

71 WOOLTON VILLAGE GARDENS

Woolton, Liverpool, L25 0QF. *7m S of Liverpool. Woolton Rd B5171 or Menlove Ave A562 follow signs for Woolton.* **Sun 11 July (12-5). Combined adm £5, chd free. Light refreshments.** at Hillside Dr, Speke Rd & Oakwood Rd.

NEW GREEN RIDGES, RUNNYMEDE CLOSE
L25 5JU. Sarah & Michael Beresford.

23 HILLSIDE DRIVE
L25 5NR. Bruce & Fiona Pennie.

71 MANOR ROAD
L25 8QF. John & Maureen Davies.

NEW 89 OAKWOOD ROAD
L26 1XD. Sean Gargan.

231 SPEKE ROAD
L25 0LA. Paul & Helen Ekoku, 07765 379967, iekoku@yahoo.co.uk.

A group of gardens surrounding the NW and Britain in Bloom award winning Woolton village. The gardens are all different reflecting their owners individuality and style. All within a short walk or drive from each other. Wheelchair access to some gardens.

LEICESTERSHIRE & RUTLAND

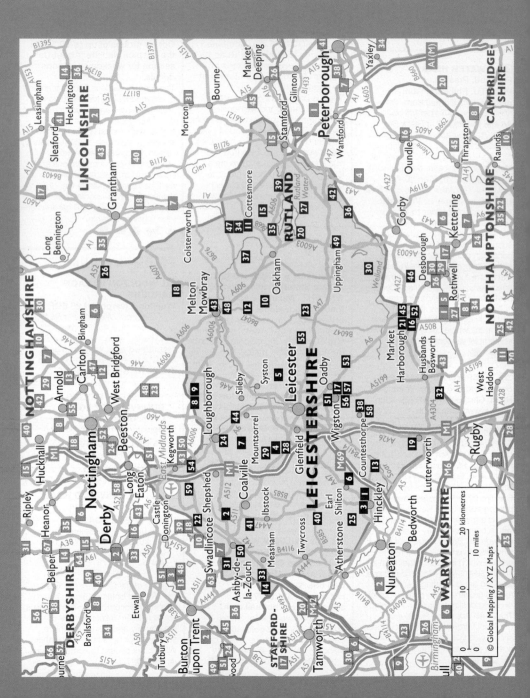

Leicestershire is a landlocked county in the Midlands with a diverse landscape and fascinating heritage providing a range of inspiring city, market town and village gardens.

Our gardens include Victorian terraces that make the most of small spaces and large country houses with historic vistas. We have an arboretum with four champion trees and a city allotment with over 100 plots. We offer something for everyone, from the serious plants person to the casual visitor.

You'll receive a warm welcome at every garden gate. Visit and get ideas for your own garden or to simply enjoy spending time in a beautiful garden. Most gardens sell plants and offer tea and cake, many are happy to take group bookings. We look forward to seeing you soon!

'Much in Little' is Rutland's motto. They say small is beautiful and never were truer words said.

Rutland is rural England at its best. Honey-coloured stone cottages make up pretty villages nestling amongst rolling hills; the passion for horticulture is everywhere you look, from stunning gardens to the hanging baskets and patio boxes showing off seasonal blooms in our two attractive market towns of Oakham and Uppingham.

There's so much to see in and around Rutland whatever the time of year, including many wonderful National Garden Scheme gardens.

Below: Stoke Albany House

@NGSLeicestershire
@leicestershire_ngs
@rutlandngs

@LeicsNGS
@rutlandngs

Volunteers

Leicestershire

County Organiser
Pamela Shave 01858 575481
pamela.shave@ngs.org.uk

County Treasurer
Martin Shave 01455 556633
martin.shave@ngs.org.uk

Publicity
Carol Bartlett 01616 261053
carol.bartlett@ngs.org.uk

**Booklet Co-ordinator
(Leicestershire & Rutland)**
Sharon Maher 01162 711680
sharon.maher@ngs.org.uk

Social Media
Zoe Lewin 07810 800 007
zoe.lewin@ngs.org.uk

Twitter
Donna Smith 07905 297766
donna.smith@ngs.org.uk

Talks
Pat Beeson 07940 771185
pat.beeson@ngs.org.uk

Karen Gimson 07930 246974
k.gimson@btinternet.com

**Local Groups
Communication Officer**
Rosemary Collict 01164 296742
rosemary.collict@ngs.org.uk

Assistant County Organisers
Janet Rowe 01162 597339
janetnandrew@btinternet.com

Gill Hadland 01162 592170
gillhadland1@gmail.com

Roger Whitmore and
Shirley Jackson 01162 787179
whitmorerog@hotmail.co.uk

Rutland

County Organisers
Sally Killick 07799 064565
sally.killick@ngs.org.uk

Lucy Hurst 07958 534778
lucy.hurst@ngs.org.uk

County Treasurer
Sandra Blaza 01572 770588
sandra.blaza@ngs.org.uk

Publicity
Lucy Hurst (as above)

Social Media
Nicola Oakey 07516 663358
nicola.oakey@ngs.org.uk

OPENING DATES

All entries subject to change. For latest information check www.ngs.org.uk

Extended openings are shown at the beginning of the month.

Map locator numbers are shown to the right of each garden name.

February

Snowdrop Festival

Saturday 20th
Hedgehog Hall 23
Westview 53

Sunday 21st
The Acers 1
Hedgehog Hall 23
Westview 53

March

Saturday 6th
NEW 8 Hinckley Road 25

Sunday 7th
NEW 8 Hinckley Road 25

Sunday 14th
Gunthorpe Hall 20

Saturday 27th
Oak Cottage 31

Sunday 28th
Oak Cottage 31

April

Sunday 25th
The Old Hall 34
Tresillian House 48
Westbrooke House 52

May

Sunday 2nd
Hedgehog Hall 23

Monday 3rd
Hedgehog Hall 23

Sunday 9th
Burrough Hall 10

Sunday 16th
1 The Dairy, Hurst Court 14
The Old Vicarage, Burley 35
Westview 53

Saturday 29th
Goadby Marwood Hall 18

Sunday 30th
10 Brook Road 7
Mountain Ash 29
Nevill Holt Hall 30
The Old Vicarage, Whissendine 37

Monday 31st
10 Brook Road 7
Mountain Ash 29

June

Every Wednesday
Stoke Albany House 46

Friday 4th
Redhill Lodge 42

Saturday 5th
28 Gladstone Street 17

Sunday 6th
28 Gladstone Street 17
4 Packman Green 38
NEW Silver Birches 44
Uppingham Gardens 49
◆ Whatton House 54

Saturday 12th
88 Brook Street 8
109 Brook Street 9

Sunday 13th
88 Brook Street 8
109 Brook Street 9
NEW Chapel End 11
Crossfell House 12
44 Fairfield Road 16
Manton Gardens 27
NEW Snowdrop Ridge 45

Saturday 19th
NEW Wigston Framework Knitters Museum 56

Sunday 20th
NEW Barkby Hall 5
Dairy Cottage 13
Tresillian House 48
Westbrooke House 52

NEW Wigston Framework Knitters Museum 56

Wednesday 23rd
The Old Vicarage, Burley 35

Saturday 26th
Oak Tree House 32
The Old Barn 33

Sunday 27th
Oak Tree House 32
The Old Barn 33
4 Packman Green 38
15 The Woodcroft 59

July

Every Wednesday
Stoke Albany House 46

Saturday 3rd
NEW 8 Hinckley Road 25
NEW Warwick Glen 50
Wigston Gardens 57

Sunday 4th
Exton Hall 15
NEW 8 Hinckley Road 25
NEW Warwick Glen 50
Wigston Gardens 57

Sunday 11th
Green Wicket Farm 19
Prebendal House 39
Willoughby Gardens 58

Wednesday 14th
Green Wicket Farm 19

Saturday 17th
NEW Wigston Framework Knitters Museum 56

Sunday 18th
59 Thistleton Road 47
NEW Wigston Framework Knitters Museum 56

Sunday 25th
NEW 12 Hastings Close 22
119 Scalford Road 43

Saturday 31st
221 Markfield Road 28

August

Every Sunday from Sunday 8th
Honeytrees Tropical Garden 26

Every day to Sunday 8th
221 Markfield Road 28

Sunday 1st
28 Ashby Road 3
44 Fairfield Road 16
NEW Snowdrop Ridge 45

Sunday 8th
NEW The Old Vicarage, Harringworth 36

Sunday 15th
NEW The Old Vicarage, Harringworth 36

Sunday 29th
Tresillian House 48

September

Saturday 4th
Oak Tree House 32

Sunday 5th
Oak Tree House 32
Washbrook Allotments 51

Sunday 12th
Westview 53

October

Sunday 10th
Hammond Arboretum 21

Saturday 30th
Tresillian House 48

By Arrangement

Arrange a personalised garden visit on a date to suit you. See individual garden entries for full details.

The Acers 1
Aqueduct Cottage 2
Bank Cottage 4
Barracca 6
88 Brook Street 8
109 Brook Street 9
NEW Chapel End 11
Dairy Cottage 13
1 The Dairy, Hurst Court 14
Farmway, Willoughby Gardens 58
Goadby Marwood Hall 18

National Garden Scheme gardens are identified by their yellow road signs and posters. You can expect a garden of quality, character and interest, a warm welcome and plenty of home-made cakes!

THE GARDENS

1 THE ACERS
10 The Rills, Hinckley, LE10 1NA. Mr Dave Baggott, 07983639683, davebaggott18@hotmail.com. *Off B4668 out of Hinckley. Turn into Dean Rd then 1st on the R.* Sun 21 Feb (10.30-4). Adm £5, chd free. Home-made teas. **Visits also by arrangement Feb to Oct.**
Medium sized garden, with a Japanese theme incl a zen garden, Japanese tea house, koi pond, more than 20 different varieties of Acers, many choice alpines, Trilliums, Cyclamen, Erythroniums, Cornus, Hamamelis and dwarf conifers. Over 150 different varieties of Snowdrops in spring. Large greenhouse.

2 AQUEDUCT COTTAGE
Gelsmoor Road, Coleorton, Coalville, LE67 8JF. Jayne Wright, 07713 624595, jaynewright38@yahoo.co.uk. *Nr Ashby de la Zouch. Corner of Gelsmoor Rd and Aqueduct Rd. Access is via gate on Aqueduct Rd.* **Visits by arrangement. In June and September only.** Adm £7, chd free. Home-made teas. Adm incl refreshments
Mature, classic English garden, in excess of 3 acres. It is flanked by a disused (1836) railway line which is wooded and boasts a large variety of specimen trees. There are formal perennial beds and specimen rose beds with a lot of roses! A small classic fish pond in the formal part of the garden and a 30m open pond in a wildlife friendly setting. Please advise if dietary requirements are needed at time of booking an appointment.

3 28 ASHBY ROAD
Hinckley, LE10 1SL. Stan & Carol Crow. *Hinckley is SW of Leicester with easy access off the M69 or A5. Ashby Rd is on the A447 running N from Hinckley to Ibstock. Easy parking in rd.* Sun 1 Aug (11-5). Adm £3.50, chd free. Home-made teas.
Behind the post war house lies a garden packed full of colour and interesting planting combinations. Approximately 150 feet long. Natural sculptures and pots abound. Wildlife area with pond and wildflowers to attract bees and butterflies into the garden. Plenty of hideaway seating to view the garden at your leisure. Vegetable garden with cut flower area. Eco friendly as much as possible.

4 BANK COTTAGE
90 Main Street, Newtown Linford, Leicester, LE6 0AF. Jan Croft, 01530 244865, gardening91@icloud.com. *6m NW Leicester. 2.5m from M1 jct 22. From Leicester on LHS after Markfield Lane. Just before a public footpath sign if coming via Markfield Lane or Anstey or on the R just after public footpath sign if coming via Warren Hill.* **Visits by arrangement Feb to Oct for groups of up to 20.** Adm £3, chd free.
Traditional cottage garden set on different levels leading down to the River Lin. Providing colour all year round but at its prettiest in the Spring. Aconites, snowdrops, blue and white bells, primroses, alliums, aquilegia, poppies, geraniums, roses, honeysuckle, Philadelphia, acers, wisteria, autumn crocus, perennial sweet pea, perennial sunflowers, some fruit trees, small pond. Full of wildlife.

5 NEW BARKBY HALL
Barkby, LE7 3QB. Mr & Mrs A J Pochin. *Entrance is by the lodge on the Queniborough Road, postcode LE7 3QJ. From Barkby village follow long brick wall until a white cottage, turn in and proceed down drive.* Sun 20 June (11-6). Adm £5, chd free. Light refreshments.
The current garden has been through an extensive restoration and renovation, with formal gardens of circa 8 acres. Areas of interest include the Rose garden, fragrant garden, white garden, Japanese garden, Camellia houses, 1 acre walled kitchen garden under production. Mixture of herbaceous borders, Azalea areas, woodland walk planted with Rhododendrons, orchard, formal lawns, parkland. Large areas are accessible by wheelchair along gravel paths. Some areas have inclines, and other areas have raised viewing points up steps.

6 BARRACCA
Ivydene Close, Earl Shilton, LE9 7NR. Mr John & Mrs Sue Osborn, 01455 842609, susan.osborn1@btinternet.com, www.barraccagardens.com. *10m W of Leicester. From A47 after entering Earl Shilton, Ivydene Close is 4th on L from Leicester side of A47.* **Visits by arrangement Feb to July for groups of 10+. Visit includes the price of home made teas as well as entry fee.** Adm £8, chd free. Home-made teas. Other refreshments available on request
1 acre garden with lots of different areas, silver birch walk, wildlife pond with seating, apple tree garden, Mediterranean planted area and lawns surrounded with herbaceous plants and shrubs. Patio area with climbing roses and wisteria. There is also a utility garden with greenhouse, vegetables in beds, herbs and perennial flower beds, lawn and fruit cage. Part of the old gardens owned by the Cotton family who used to open approx 9 acres to the public in the 1920's. Partial wheelchair access.

7 10 BROOK ROAD
Woodhouse Eaves, carol_fowle@
yahoo.co.uk, Loughborough,
LE12 8RS. Geoff & Carol Fowle.
*4m south of Loughborough, opp
Bulls Head PH. Parking at pub.* Sun
30, Mon 31 May (11-5.30). Adm
£4, chd free. Light refreshments.
1 acre mature country garden created
into different and interesting areas
incl a stream & pond, lawn areas
with island beds containing roses,
rhododendrons, azaleas & Japanese
maples, created by the owners over
20 yrs from a neglected Victorian
garden. Many recycled & quirky
materials made into garden features,
over 200 pots & containers and a
traditional summer house to enjoy
your refreshments. Partial wheelchair
access.

8 88 BROOK STREET
Wymeswold, LE12 6TU. Adrian
& Ita Cooke, 01509 880155,
itacooke@btinternet.com. *4m
NE of Loughborough. From A6006
Wymeswold turn S by church onto
Stockwell, then E along Brook
St. Roadside parking on Brook
St.* Sat 12, Sun 13 June (2-5).
Combined adm with 109 Brook
Street £5, chd free. Visits also
by arrangement May & June for
groups of 10 to 30. Admission to
88 Brook St only will be £3, or £5
incl tea and cakes.
The ½ acre garden is set on a
hillside, which provides lovely views
across the village, and comprises
3 distinct areas: firstly, a cottage
style garden; then a series of
water features incl a stream and a
'champagne' pond; and finally at the
top there is a vegetable plot, small
orchard and wildflower meadow.
The ponds attract great crested and
common newts, frogs, toads and
grass snakes.

9 109 BROOK STREET
Wymeswold, LE12 6TT. Maggie
& Steve Johnson, 07973692931,
steve@brookend.org,
www.brookend.org. *4m NE of
Loughborough. From A6006
Wymeswold turn S onto Stockwell,
then E along Brook St. Roadside
parking along Brook St. Steep drive
with limited disabled parking at
house.* Sat 12, Sun 13 June (2-5).
Combined adm with 88 Brook
Street £5, chd free. Home-made
teas. Gluten free option available.

Visits also by arrangement May &
June for groups of 10 to 30.
South facing, ¾ acre, gently sloping
garden with views to open country.
Modern garden with mature features.
Patio with roses and clematis,
wildlife and fish ponds, mixed
borders, vegetable garden, orchard,
hot garden and woodland garden.
Something for everyone! Optional tour
of rain water harvesting. Some gravel
paths.

10 BURROUGH HALL
Burrough on the Hill, LE14 2QZ.
Richard & Alice Cunningham.
*Somerby Rd, Burrough on the Hill.
Close to B6047. 10 mins from A606.
20 mins from Melton Mowbray.* Sun
9 May (2-5). Adm £4, chd free.
Home-made teas.
Burrough Hall was built in 1867
as a classic Leicestershire hunting
lodge. The garden, framed by mature
trees and shrubs, was extensively
redesigned by garden designer
George Carter in 2007. The garden
continues to develop. This family
garden designed for all generations
to enjoy is surrounded by magnificent
views across High Leicestershire.
In addition to the garden there will
be a small collection of vintage and
classic cars on display. Gravel paths
and lawn.

11 NEW CHAPEL END
Main Street, Barrow, Oakham,
LE15 7PE. Mrs Soo Spector,
soo@soospectorgardens.co.uk.
*At the right hand end of the village
behind the large village green. Take R
fork when road divides. Down slope
on road, with main village green in
front of you. Park off road on the
green. Chapel End is down the track
just beyond green under weeping
willow tree.* Sun 13 June (1-5). Adm
£5, chd free. Home-made teas.
Visitors are welcome to picnic
in the parkland paddock but
please no barbecues. Visits also
by arrangement May to Sept for
groups of up to 10.
A rural garden with different areas
of interest: topiary driveway, rose
garden, herbaceous border, kitchen
garden, rhododendron border,
country side views across Vale of
Catmose, wild flower meadows, pond
with accessible island for wildlife,
open meadow with tree varieties.

12 CROSSFELL HOUSE
4d Nether End, Great Dalby,
Melton Mowbray, LE14 2EY.
Jane & Ian West. *3m S of Melton
Mowbray on B6047. On entering
Great Dalby from Melton Mowbray
remain on B6047. Crossfell House
is on L approx 300 yrds from village
30mph sign. From the N or E,
continue towards Melton Crossfell
House on R.* Sun 13 June (11-5).
Adm £5, chd free. Home-made
teas.
A formal garden consisting of a
terraced border of informal cottage
style planting and a rockery, flanked
by a border of shrubs, two small
areas of lawn and a sweeping path
leading to a two acre meadow
with wild grasses, flowers and a
wildlife pond. Paths criss-cross the
meadows, culminating in spectacular
countryside views from our
Shepherd's Hut and picnic area. Far
reaching views. Wheelchair access
to patio and garden area, only partial
wheelchair access to meadow.

13 DAIRY COTTAGE
15 Sharnford Road, Sapcote,
LE9 4JN. Mrs Norah Robinson-
Smith, 01455 272398,
nrobinsons@yahoo.co.uk. *9m
SW of Leicester. Sharnford Rd joins
Leicester Rd in Sapcote to B4114
Coventry Rd. Follow NGS signs at
both ends.* Sun 20 June (1-5). Adm
£3.50, chd free. Home-made teas.
Tea & cakes £2.00. Visits also
by arrangement May to July for
groups of 10 to 30.
From a walled garden with colourful
mixed borders to a potager
approached along a woodland path,
this mature cottage garden combines
extensive perennial planting with
many unusual shrubs and specimen
trees. More than 90 clematis and
30 climbing roses are trained up
pergolas, arches and into trees
– so don't forget to look up! Also
a Hornbeam hedge covered with
varieties of viticella clematis.

**14 1 THE DAIRY, HURST
COURT**
Netherseal Road,
Chilcote, Swadlincote,
DE12 8DU. Alison Dockray,
alisondockray56@gmail.com. *3m
from J11 of the M42. 10m from
Tamworth & Ashby de la Zouch.
Hurst Court is situated on Netherseal
Rd but sat navs show it as Church*

Rd. Due to limited parking it is advisable to park on Netherseal Rd. **Sun 16 May (1.30-5). Adm £4, chd free. Light refreshments. Visits also by arrangement May to Aug. Weekdays only.**
Developed over the last 11 years from a muddy ¼ acre patch attached to a barn conversion, the garden now contains a large Japanese Koi Carp pond, goldfish pond and a stream with planting having a Japanese connection viewed from winding paths. With many rhododendrons, camellias and azaleas combined with cloud pruning and a variety of other plants, an air of peace pervades the garden.

15 EXTON HALL
Cottesmore Road, Exton, LE15 8AN. Viscount & Viscountess Campden, www.extonpark.co.uk.
Exton, Rutland. 5m E of Oakham. 8m from Stamford off A1 (A606 turning). **Sun 4 July (12-4). Adm £5, chd free. Home-made teas.**
Extensive park, lawns, specimen trees and shrubs, lake, private chapel and C19 house (not open). Pinetum, woodland walks, lakes, ruins, dovecote and formal herbaceous garden. Coffee trailer. Whilst there is wheelchair access, areas of the garden are accessible along grass or gravel paths which, weather dependent, may make access difficult.

16 44 FAIRFIELD ROAD
Market Harborough, LE16 9QJ. Steve Althorpe and Judith Rout.
Opp primary school. **Sun 13 June, Sun 1 Aug (12-5). Adm £3.50, chd free. Home-made teas.**
This wildlife friendly, ⅓ acre plot in the heart of Market Harborough, is a garden in which you can relax. There are large borders informally planted with a good variety of perennials and shrubs, a substantial pond full of amphibians and invertebrates and a copse There's also plenty of seating at which you can enjoy all the container planting and homemade refreshments. Hot drinks (incl hot chocolate) & home made cakes available. Block paved drive and patio with gentle slope down on to lawn.

17 28 GLADSTONE STREET
Wigston Magna, LE18 1AE. Chris & Janet Huscroft. *4m S of Leicester. Off Wigston by-pass (A5199) follow signs off McDonalds r'about.* **Sat 5, Sun 6 June (11-5). Adm £3, chd free. Home-made teas. Opening with Wigston Gardens on Sat 3, Sun 4 July.**
Our mature 70'x15' town garden is divided into rooms and bisected by a pond with a bridge. It is brimming with unusual hardy perennials, incl collections of ferns and hostas. David Austin roses chosen for their scent feature throughout, incl a 30' rose arch. A shade house with unusual hardy plants and a Hosta Theatre. Regular changes to planting. Wigston Framework Knitters Museum and garden nearby - open Sunday afternoons.

The Old Barn

18 GOADBY MARWOOD HALL

Goadby Marwood, LE14 4LN. Mr & Mrs Westropp, 01664 464202, vwestropp@gmail.com. *4m NW of Melton Mowbray. Between Waltham-on-the-Wolds & Eastwell, 8m S of Grantham. Plenty of parking space available.* Sat 29 May (11-5). Adm £5, chd £3. Home-made teas in Village Hall. Visits also by arrangement Jan to Nov.
Redesigned in 2000 by the owner based on C18 plans. A chain of 5 lakes (covering 10 acres) and several ironstone walled gardens all interconnected. Lakeside woodland walk. Planting for yr-round interest. Landscaper trained under plantswoman Rosemary Verey at Barnsley House. Beautiful C13 church open. Water, Walled gardens. Gravel paths and lawns.

19 GREEN WICKET FARM

Ullesthorpe Road, Bitteswell, Lutterworth, LE17 4LR. Mrs Anna Smith, 01455 552646, greenfarmbitt@hotmail.com. *2m NW of Lutterworth J20 M1. From Lutterworth follow signs through Bitteswell towards Ullesthorpe. Garden situated behind Bitteswell Cricket Club. Use this as a landmark rather than relying totally on satnav.* Sun 11, Wed 14 July (2-5.30). Adm £4, chd free. Cream teas. Visits also by arrangement June to Sept for groups of up to 30.
Created in 2008 on a working farm with clay soil and a very exposed site. Many unusual hardy plants along with a lot of old favourites have been used to provide a long season of colour and interest. Anemone nemorosa, Pacific coast iris and Salvias are of particular interest. Formal pond and water features. Some gravel paths.

20 GUNTHORPE HALL

Gunthorpe, nr Oakham, LE15 8BE. Tim Haywood, 01572 737514 ask for lettings. *A6003 between Oakham & Uppingham; 1m from Oakham, up drive between lodges.* Sun 14 Mar (2-5). Adm £5, chd free. Light refreshments.
Large garden in a country setting with extensive views across the Rutland landscape with the carpets of daffodils being a key seasonal feature. The Stable Yard has been transformed into a series of parterres. The kitchen garden and borders have all been rejuvenated over the last five years. Gravel path allows steps to be avoided.

21 HAMMOND ARBORETUM

Burnmill Road, Market Harborough, LE16 7JG. The Robert Smyth Academy, www.hammondarboretum.org.uk. *15m S of Leicester on A6. From High St, follow signs to The Robert Smyth Academy via Bowden Lane to Burnmill Rd. Park in 1st entrance on L.* Sun 10 Oct (2-4.30). Adm £5, chd free. Home-made teas provided by the Academy Parents Association.
A site of just under 2½ acres containing an unusual collection of trees and shrubs, many from Francis Hammond's original planting dating from 1913 to 1936 whilst headmaster of the school. Species from America, China and Japan with malus and philadelphus walks and a moat. Proud owners of 3 champion trees identified by national specialist. Walk plans available. Some steep slopes.

22 NEW 12 HASTINGS CLOSE

Breedon-On-The-Hill, Derby, DE73 8BN. Mr and Mrs P Winship. *5m N from Ashby de la Zouch. Follow NGS signs in the Village. Parking around the Village Green. Please do not park in the close due to limited parking.* Sun 25 July (1-5). Adm £3, chd free.
A medium sized prairie garden, organically managed and planted in the Piet Oudolf style. Also many roses and a wide range of perennials. The back garden has a small 'white' garden with box hedging and more colourful style perennial borders. The garden has narrow paths and steps so is not suitable for wheelchairs. The garden is also open for Breedon on the Hill village open garden day.

23 HEDGEHOG HALL

Loddington Road, Tilton on the Hill, LE7 9DE. Janet & Andrew Rowe. *8m W of Oakham. 2m N of A47 on B6047 between Melton & Market Harborough. Follow yellow NGS signs in Tilton towards Loddington.* Sat 20, Sun 21 Feb (11-4). Also open Westview. Sun 2, Mon 3 May (11-4). Home-made teas. Adm £4, chd free.
½ acre organically managed plant lover's garden. Steps leading to three stone walled terraced borders filled with shrubs, perennials, bulbs and a patio over looking the valley. Lavender walk, herb border, beautiful spring garden, colour themed herbaceous borders. Courtyard with collection of hostas and acers and terrace planted for yr-round interest with topiary and perennials. Snowdrop collection. Cake, tea or coffee at May opening. Regret, no wheelchair access to terraced borders. Disabled parking in the road outside the White House.

24 134 HERRICK ROAD

Loughborough, LE11 2BU. Janet Currie, janet.currie@me.com, , www.thesecateur.com. *1m SW of Loughborough centre. From M1 J23 take A512 Ashby Rd to L'boro. At r'about R onto A6004 Epinal Way. At Beacon Rd r'about L, Herrick Rd 1st on R.* Visits by arrangement June to Aug for groups of 5 to 10. Adm £3, chd £1.50. Tea. Dietary requirements catered for, please phone to discuss
A small garden brimming with texture, colour and creative flair. Trees, shrubs and climbers give structure. A sitting area surrounded by lilies, raised staging for herbs and alpines. A lawn flanked with deeply curving and gracefully planted beds of perennials growing through willow structures made by Janet. A shaded area under the Bramley apple tree, small greenhouse and potting area. Janet makes attractive willow plant supports to her own designs, and to order for equally enthusiastic gardeners!

25 NEW 8 HINCKLEY ROAD

Stoke Golding, Nuneaton, CV13 6DU. Mr John & Mrs Stephanie Fraser. *3m NW of Hinckley. Approach from any direction into village then follow NGS signs. Please park roadside with due consideration to other residents properties.* Sat 6 Mar (12-4); Sun 7 Mar (12-3); Sat 3 July (12-4); Sun 4 July (12-3). Adm £3.50, chd free. Light refreshments.
A small SSW garden with pond/ water features to add interest to the 30+ hellebores in March and packed with colourful and some unusual perennials including climbers to supplement the trees and shrubs in July. Garden established and recently re-established by current owner to provide seating for a variety of views of the garden. Many perennials in the

garden are represented in the plants for sale. Wheelchair access to garden room patio which provides a view of the lower part of the garden.

26 HONEYTREES TROPICAL GARDEN

85 Grantham Road, Bottesford, NG13 0EG. Julia Madgwick & Mike Ford. *7m E of Bingham on A52. Turn into village. Garden is on L on slip road behind hedge going out of village towards Grantham. Parking on grass opp property.* **Every Sun 8 Aug to 29 Aug (11-4). Adm £4, chd free. Home-made teas.** Tropical and exotic with a hint of jungle! Raised borders with different themes from lush foliage to arid cacti. Exotic planting as you enter the garden gives way on a gentle incline to surprises to incl glass houses dedicated to various climatic zones interspersed with more exotic planting, ponds and a stream. A representation of in excess of 20 years plant hunting. There are steps and some gravel but access by wheelchair to most parts of the garden.

GROUP OPENING

27 MANTON GARDENS

Oakham, LE15 8SR. *3m N of Uppingham and 3m S of Oakham. Manton is on S shore of Rutland Water ¼ m off A6003. Please park carefully in village.* **Sun 13 June (1-5). Combined adm £5, chd free.**

22 LYNDON ROAD
Chris & Val Carroll.

MANTON GRANGE
Anne & Mark Taylor.

MANTON LODGE
Caroline Burnaby-Atkins, 01572737269, info@mantonlodge.co.uk, www.mantonlodge.co.uk.

3 gardens in small village on S shore of Rutland Water. Manton Grange - 2½ acre garden with interesting trees, shrubs and herbaceous borders, incl a rose garden, water features, a lime tree walk and clematis pergola. 22 Lyndon Rd - a beautiful combination of cottage garden and unusual plants in overflowing borders, hanging baskets and decorative

pots. Manton Lodge - steeply sloping garden with wonderful views and colourful beds of shrubs, roses and perennials, with ornamental pond and terrace. Wheelchair access to Manton Grange only.

28 221 MARKFIELD ROAD

Groby, Leicester, LE6 0FT. Jackie Manship, 01530 249363, jmanship@btinternet.com. *From M1 J22 take A50 towards Leicester. In approx 3m at the T-lights junction with Lena Drive turn L. Parking available along this road, no parking on A50.* **Daily Sat 31 July to Sun 8 Aug (10-4). Adm £4, chd free. Light refreshments. Light lunches, ploughmans available (please order in advance). Cream teas, sandwiches/rolls, hot/cold drinks available daily (no prior order needed). Visits also by arrangement for groups of 20 to 30. Private evening bookings available during 31st Jul - 8th Aug 6.00 - 10.00pm.** A S facing plot of land nestled between the village of Groby and Markfield approx 1 acre in size. The hidden treasures are deceptive from the front of the property which sits on one of the main trunk roads out of Leicester. Packed with interest and

created over the last 20 years from a dishevelled overgrown plot you will be presented with a garden full of delight.

29 MOUNTAIN ASH

140 Ulverscroft Lane, Newtown Linford, LE6 0AJ. Mike & Liz Newcombe, 01530 242178, mjnew12@gmail.com. *7m SW of Loughborough, 7m NW of Leicester, 1m NW of Newtown Linford. Head ½ m N along Main St towards Sharpley Hill, fork L into Ulverscroft Ln & Mountain Ash, is about ½ m along on the L. Parking is along the opp verge.* **Sun 30, Mon 31 May (11-5). Adm £5, chd free. Home-made teas. Visits also by arrangement May to Sept for groups of 20+.** 2 acre garden with stunning views across Charnwood countryside. Nr the house are patios, lawns, water feature, flower & shrub beds, fruit trees, soft fruit cage, greenhouses & vegetable plots. Lawns slope down to a gravel garden, large wildlife pond and small areas of woodland with walks through many species of trees. Many statues and ornaments. Several places to sit and relax around the garden. Only the top part of the garden around the house is reasonably accessible by wheelchair.

Manton Lodge

30 NEVILL HOLT HALL

Drayton Road, Nevill Holt, Market Harborough, LE16 8EG. Mr David Ross. *5m NE of Market Haborough. Signed off B664 at Medbourne.* **Sun 30 May (12-4). Adm £5, chd free. Light refreshments.**
The gardens are at their peak in late May/early June. Whilst the walled kitchen garden shows signs of a harvest to come, our two other walled gardens and adjoining cedar lawn garden are full of colour and texture, all working together to create a summer scene that feels warm and summery, even on the dullest days. Please do join us for tea and homemade cakes – we would love to share our garden and sculpture collection with you. Nevill Holt Opera Event partner with the RHS.

31 OAK COTTAGE

Well Lane, Blackfordby, Swadlincote, DE11 8AG. Colin and Jenny Carr. *Blackfordby, just over 1m from (and between) Ashby-de-la-Zouch or Swadlincote. From Ashby-de-la-Zouch take Moira Rd, turn R on Blackfordby Lane. As you enter Blackfordby, turn L to Butt Lane and quickly R to Strawberry Lane. Park then it's a 2 min walk to Well Lane entrance.* **Sat 27, Sun 28 Mar (10-4). Adm £4, chd free. Home-made teas.**

½ acre garden set around Blackfordby's "hidden" listed thatched cottage, which itself is more than 300 years old. In total there are 3.4 acres of paddocks, front and rear gardens to explore with extensive displays of Hellebores, Snakes Heads and mature Magnolias in Spring. The lower paddock has been planted with 425 native trees as part of the National Forest Freewoods scheme, with a large pond created at its base. The central swathe is being developed with wildflowers. At the top of the rear garden there is a chicken run, old and new orchards and a peach house. Garden is heavily sloped in parts.

32 OAK TREE HOUSE

North Road, South Kilworth, LE17 6DU. Pam & Martin Shave. *15m S of Leicester. From M1 J20, take A4304 towards Market Harborough. At North Kilworth turn R, signed South Kilworth. Garden on L after approx 1m.* **Sat 26 June (11-4); Sun 27 June (11-2); Sat 4 Sept (11-4); Sun 5 Sept (11-2). Adm £4, chd free. Light refreshments.**
⅔ acre beautiful country garden full of colour, formal design, softened by cottage style planting. Modern sculptures. Large herbaceous borders, vegetable plots, pond, greenhouse, shady area, colour-themed borders. Extensive collections in pots, home to everything from alpines to trees. Trees with attractive bark. Many clematis and roses. Dramatic arched pergola. Constantly changing garden. Access to patio and greenhouse via steps.

33 THE OLD BARN

Rectory Lane, Stretton-En-Le-Field, Swadlincote, DE12 8AF. Gregg and Claire Mayles, 07870160318, greggmayles@gmail.com. *On A444, 1½m from M42/A42 J11. Rectory Lane is a concealed turn off the A444, surrounded by trees. Look out for brown 'church' signs. Postcode in Sat Nav usually helps.* **Sat 26, Sun 27 June (11-5). Adm £4, chd free. Home-made teas. Visits also by arrangement May to Sept.**
2 acres in leafy hamlet, lots of interest. Main garden has lawns, many colourful shrubs and tree-lined cobbled paths. Walled garden with fishpond, pergola with climbers and cottage garden planting. Orchard, meadow with paths and open views. Lots of wildlife, incl our Peafowl. Plenty of drinks, cakes and seats to enjoy them! Redundant medieval church close, open to visitors. Main garden fully wheelchair accessible. Walled garden partial access due to gravel. Orchard and meadow is accessible, but uneven.

Tresillian House

34 THE OLD HALL

Main Street, Market Overton,
LE15 7PL. Mr & Mrs Timothy Hart.
6m N of Oakham. 6m N of Oakham; 5m from A1via Thistleton. 10m E from Melton Mowbray. **Sun 25 Apr (2-6). Adm £5, chd free. Light refreshments incl Hambleton Bakery cakes.**
Set on a southerly ridge. Stone walls and yew hedges divide the garden into enclosed areas with herbaceous borders, shrubs, and young and mature trees. There are interesting plants flowering most of the time. In 2020 a Japanese Tea House was added at the bottom of the garden. Partial wheelchair access. Gravel and mown paths. Return to house is steep. It is, however, possible to just sit on the terrace.

& 🐕 ✳ 🚍 ☕

35 THE OLD VICARAGE, BURLEY

Church Road, Burley, nr
Oakham, LE15 7SU. Jonathan
& Sandra Blaza, 01572 770588,
sandra.blaza@googlemail.com,
www.theoldvicarageburley.com.
1m NE of Oakham. In Burley just off B668 between Oakham & Cottesmore. Church Rd is opp village green. **Sun 16 May (11-5). Home-made teas. Evening opening Wed 23 June (6-9). Wine. Adm £5, chd free. Visits also by arrangement May & June for groups of 10+.**
The Old Vicarage is a relaxed country garden, planted for year round interest and colour. There are lawns and borders, a lime walk, rose gardens and a sunken rill garden with an avenue of standard wisteria. The walled garden produces fruit, herbs, vegetables and cut flowers. There are two orchards, an acer garden and areas planted for wildlife including woodland, a meadow and a pond. Regional Finalist, The English Garden's The Nation's Favourite Gardens 2019. Some gravel and steps between terraces.

& 🐕 ✳ 🚍 ☕

36 NEW THE OLD VICARAGE, HARRINGWORTH

Seaton Road, Harringworth,
Corby, NN17 3AF. Mr & Mrs Alan
Wordie. *5m SE of Uppingham, 9m SW of Stamford & 10m west of A1/ A47 junction near Peterborough. Next to Harringworth church.* **Sun 8, Sun 15 Aug (1-5). Adm £5, chd free. Home-made teas.**
An established and romantic

country garden of rooms, which include a 'quiet garden' leading into the adjacent church, bee garden, mixed long border, nuttery, orchard, kitchen garden, paddock and paths stretching to the River Welland. Working bee hives and mischievous guinea fowl. Home-made teas, plant stall and honey. Bring a picnic if you wish and watch the kingfishers. Partial wheelchair access.

& 🐕 ✳ 🚍 ☕ 🏕

37 THE OLD VICARAGE, WHISSENDINE

2 Station Road, Whissendine,
LE15 7HG. Prof Peter
& Dr Sarah Furness,
www.pathology.plus.com/Garden.
Garden situated up hill from St Andrew's church in Whissendine. **Sun 30 May (2-5). Adm £5, chd free. Home-made teas in St Andrew's Church, Whissendine.**
⅔ acre packed with variety. Terrace with topiary, a formal fountain courtyard and raised beds backed by gothic orangery. Herbaceous borders surround main lawn. Wisteria tunnel leads to raised vegetable beds and large ornate greenhouse, four beehives, Gothic hen house plus rare breed hens. Hidden white walk, unusual plants. New Victorian style garden room. Featured on BBC Gardeners' World in 2019.

🐕 ✳ ☕

38 4 PACKMAN GREEN

Countesthorpe, Leicester,
LE8 5WS. Roger Whitmore &
Shirley Jackson. *Countesthorpe. 5m S of Leicester. Garden is close to village centre pass bank of shops off Scotland way.* **Sun 6, Sun 27 June (10-4). Adm £2, chd free. Tea.**
A small town house garden packed with hardy perennials surrounded by climbing roses and clematis. A pond and rose arch complete the picture to make it an enclosed peaceful haven filled with colour and scent. Lots of inexpensive plants for sale besides a refreshing cuppa and slice of cake.

✳ ☕

39 PREBENDAL HOUSE

Crocket Lane, Empingham,
LE15 8PW. Matthew & Rebecca
Eatough. *5m E of Oakham. Facing church, large gates on R.* **Sun 11 July (1.30-5). Adm £6, chd free. Home-made teas.**
There has been a house on the site since the 11th century. The present

house was built in 1688 owned by the diocese of Lincoln until the mid 19th century and then absorbed into the Normanton Park Estate. The present owners have been in residence since 2016. The garden is a combination of open parklands, yew tree walks, herbaceous borders and a large formal 18th century walled garden. The majority of the garden is accessible by wheelchair.

& 🐕 ✳ ☕

40 4 PRIORY ROAD

Market Bosworth, Nuneaton,
CV13 0PB. Mrs Margaret
Barrett, 01455 290112,
info@margaretbarrett.co.uk. *½m W of centre of Market Bosworth, off Heath Rd.* **Visits by arrangement May to Sept for groups of 5 to 20. Adm by donation.**
Mixed borders, shrubs and small trees. Island beds, lawn, shaded border, small pond. Fruit, cool greenhouse. Approx 0.25 acres. Artist's Studio open in conservatory. Sorry no wheelchairs as too many steps in garden.

41 RAVENSTONE HALL

Ashby Road, Ravenstone,
Coalville, LE67 2AA. Jemima
Wade, 07976 302260,
jemimawade@hotmail.com.
Ravenstone village is situated off the A511 between Ashby de la Zouch & Coalville. The house is 1st on the L if approached from Ashby. **Visits by arrangement Mar to Sept for groups of up to 20. Adm £6, chd free. Home-made teas. Tea and Cake £4**
The garden was transformed in 2009 when the new owners moved to the house and planting and development has been on-going since that time. Azaleas and rhododendrons are planted on the bank of the drive. Beech trees and a beech hedge line the main front lawn. There is a woodland walk planted with bluebells. The main garden comprises a sunken garden, a rose garden, gravel path with herbaceous planting one side and iris and tulip on the other, a vegetable garden, an orchard, and a koi pond situated in a courtyard with white flowering plants and mixed foliage. Bluebell walk for groups from mid April, please call for details.

& 🐕 🚍 ☕

42 REDHILL LODGE

Seaton Road, Barrowden, Oakham, LE15 8EN. Richard & Susan Moffitt, www.m360design.co.uk. *Redhill Lodge is 1m from village of Barrowden along Seaton Rd.* **Evening opening Fri 4 June (6-9). Adm £8, chd free. Wine.** Bold contemporary design with formal lawns, grass amphitheatre and turf viewing mound, herbaceous borders & rose garden. Prairie style planting showing vibrant colour in late summer. Also natural swimming pond surrounded by Japanese style planting, bog garden and fernery.

&. 🐎 🚗 ☕ 🍷

43 119 SCALFORD ROAD

Melton Mowbray, LE13 1JZ. Richard & Hilary Lawrence, 01664 562821, randh1954@me.com. *½m N of Melton Mowbray. Take Scalford Rd from town centre past Cattle Market. Garden 100yds after 1st turning on L (The Crescent). Some parking available on the drive but The Crescent is an easy walk.* **Sun 25 July (11-5). Adm £4, chd free. Home-made teas incl gluten free cakes. Visits also by arrangement June to Aug for groups of 10 to 30.** Larger than average town garden which has evolved over the last 30 yrs. Mixed borders with traditional and exotic plants, enhanced by container planting particularly begonias. Vegetable parterre and greenhouse. Various seating areas for viewing different aspects of the garden. Water features incl ponds. New additions to the garden are a White Border and a Succulents bed. Partial wheelchair access.

🌼 ☕

44 NEW SILVER BIRCHES

82A Leicester Road, Quorn, Loughborough, LE12 8BB. Ann & Andrew Brown. *400m from Quorn Village centre on Leicester Rd. The property is down a gravel drive between 80 & 82 Leicester Rd. Park on Leicester Rd. Please do not bring your car down the drive.* **Sun 6 June (11-5). Adm £4, chd free. Home-made teas.** A tree framed garden of nearly half an acre leading down to the River Soar. The formal area with an attractive summerhouse has well stocked borders, specifically planted for year round interest of both people and wildlife. The rest of the garden incl a

vegetable plot, fruit trees, chickens, a wild riverside area and fernery. Plenty of nectar rich flowers and food plants for bees and birds. Wheelchair access not easy on main gravel driveway. Some paths too narrow for wheelchairs.

&. 🌼 ☕

45 NEW SNOWDROP RIDGE

35 The Ridgeway, Market Harborough, LE16 7HG. Donna and Steve Smith. *15m S of Leicester on A6. From High St follow signs to Robert Smyth Academy via Bowden Lane to Burnmill Rd & take R turn to Ridgeway West which becomes The Ridgeway.* **Sun 13 June, Sun 1 Aug (10-3). Adm £3, chd free. Light refreshments.** A garden reclaimed over the last four years and still a work in progress, constantly changing and evolving. The 320 sqm garden is on three levels and predominantly a cottage style garden where self seeding is encouraged to give a natural effect. The planting incl many traditional cottage plants and is interspersed with gravel areas, barrel ponds, a greenhouse and lawns with colourful borders. Second hand books for sale, and hand made crafts.

🐎 🌼 ☕

46 STOKE ALBANY HOUSE

Desborough Road, Stoke Albany, Market Harborough, LE16 8PT. Mr & Mrs A M Vinton, www.stokealbanyhouse.co.uk. *4m E of Market Harborough. Via A427 to Corby, turn to Stoke Albany, R at the White Horse (B669) garden ½m on the L.* **Every Wed 2 June to 28 July (2-4.30). Adm £5, chd free. Donation to Marie Curie Cancer Care.** 4 acre country house garden; fine trees and shrubs with wide herbaceous borders and sweeping striped lawn. Good display of bulbs in spring, roses June and July. Walled grey garden; nepeta walk arched with roses, parterre with box and roses. Mediterranean garden. Heated greenhouse, potager with topiary, water feature garden and sculptures.

&. 🐎 🌼 🚗

47 59 THISTLETON ROAD

Market Overton, Oakham, LE15 7PP. Wg Cdr Andrew Stewart JP, 01572 767662, stewartaj59@gmail.com. *6m NE of Oakham.* **Sun 18 July (1-5). Adm £5, chd free. Visits also by**

arrangement May to Sept. Over the last 15 years, the owners of No. 59 have transformed 1.8 acres of bare meadow into a wildlife friendly garden full of colour and variety. Against a backdrop of mature trees, the garden includes a small kitchen garden, rose pergola, large pond, shrubbery with 'Onion Day Bed', orchard, wild flower meadow, a small arboretum, a woodland walk and large perennial borders a riot of colour. Small area of shingle to access main path, woodland walk inaccessible.

&. 🌼

48 TRESILLIAN HOUSE

67 Dalby Road, Melton Mowbray, LE13 0BQ. Mrs Alison Blythe, 01664 481997, alisonblythe@tresillianhouse. com, www.tresillianhouse.com. *Situated on B6047 Dalby Rd, S of Melton town centre. (Melton to Gt Dalby/Market Harborough rd). Parking on site.* **Sun 25 Apr, Sun 20 June, Sun 29 Aug (11-4); Sat 30 Oct (10.30-3.30). Adm £4.50, chd free. Light refreshments. Ploughman's lunches and cream teas in April, June and August. Stew and Dumplings and Cream Teas in October. Visits also by arrangement Mar to Oct for groups of up to 30.** ¾ acre garden re-established by current owner. Beautiful blue cedar trees, specimen tulip tree. Variety of trees, plants and bushes reinstated. Original bog garden and natural pond. Koi pond maturing well; glass garden room holds exhibitions & recitals. Vegetable plot. Cowslips and bulbs in Springtime. Unusual plants, trees and shrubs. Contemporary area added 2020. Quiet and tranquil oasis. Ploughmans lunches, cream teas, home-made cakes always available. Small Art Exhibition hosted by local artists. Relax in August with cream tea whilst listening to traditional jazz from Springfield Ensemble. Keep warm in October with stew & dumplings or soup Slate paths, steep in places but manageable.

&. 🐎 🌼 🚗 🏠 ☕ 🪑

GROUP OPENING

49 UPPINGHAM GARDENS

7 Stockerston Road, Uppingham, Oakham, LE15 9UD. Lawrence & Jennifer Fenelon. *Uppingham town centre. Close to Oakham, Corby & Rutland Water.* **Sun 6 June (1-5).**

Snowdrop Ridge

Combined adm £5, chd free. Home-made teas.

HILLSIDE
Mr & Mrs Lawrence Fenelon.

THE ORCHARD
Doug & Margaret Stacey.

 ROBIN HILL
Kathy & Andrew Robinson.

Three contrasting gardens opening in the historic market town of Uppingham. Start in the Market Square where teas, tickets and route maps are available in the Church Hall, accessed through passage from front of church. The Orchard and Robin Hill are side by side on R. Their south facing gardens have views over a stream to sheep pastures beyond and are full of interesting trees, shrubs, bulbs and perennials. Robin Hill has recently had a major makeover and is in the course of re-development. Hillside is a 1 acre S facing garden with terraces, patio, new rose planting, orchard, vegetable garden, woodland walk and spring fed pond. Wheelchair access is available at Hillside but garden is on steep slope in parts.

♿ 🐔 ✱ ☕

50 NEW ▶ **WARWICK GLEN**
29 Willesley Road, Ashby-De-La-Zouch, LE65 2QA. Alison & Alan Cross. *1m SE of Ashby de la Zouch. At J12 of A42 take Ashby/Willesley rd towards Ashby de la Zouch. After 1m turn L into Willesley Rd past golf club. Warwick Glen 200yrds on L.* **Sat 3, Sun 4 July (10-4). Adm £3.50, chd free. Cream teas.**
⅓ acre packed with a diverse plant variety. Large frontage incl trees, shrubs, perennials and fruit trees. Also 22 square metres of raised vegetables and a Hosta collection. Rear garden consists of patio, gravel and grass areas, rockery, small tropical area, perennial area, octagonal greenhouse and children's woodland walk and fernery. Also fruit, herbs and other plants. Squash and biscuits will be available.

✱ ☕

51 ▶ **WASHBROOK ALLOTMENTS**
Welford Road, Leicester, LE2 6FP. Sharon Maher. *Approx 2½m S of Leicester, 1½ m N of Wigston. No onsite parking. Welford Rd difficult to park on. Please use nearby side rds & Pendlebury Drive (LE2 6GY).* **Sun 5 Sept (11-3). Adm £3.50, chd free. Home-made teas.**

A hidden oasis. There are over 100 whole, half and quarter plots growing a wide variety of fruit, vegetables and flowers. Meadows, anderson shelters and a composting toilet! Circular route around the site is uneven in places but is suitable for wheelchairs.

♿ 🐔 ✱ ☕

52 ▶ **WESTBROOKE HOUSE**
52 Scotland Road, Little Bowden, Market Harborough, LE16 8AX. Bryan & Joanne Drew, 07872 316153, Jwsd1980@hotmail.co.uk. *½m S Market Harborough. From Northampton Rd follow NGS arrows.* **Sun 25 Apr, Sun 20 June (10.30-5). Adm £5, chd free. Cream teas. Visits also by arrangement May & June for groups of 20+.**
Westbrooke House is a late Victorian property built in 1887. The gardens comprise 6 acres in total and are approached through a tree lined driveway of mature limes and giant redwoods. Key features are walled flower garden, walled kitchen garden, lower garden, pond area, spring garden, lawns, woodland paths and a meadow with a wild flower area, ha-ha and hornbeam avenue.

✱ 🏠 ☕

"I love the National Garden Scheme which has been the most brilliant supporter of Queen's Nurses like me. It was founded by the Queen's Nursing Institute which makes me very proud. As we battle Coronavirus on the front line in the community, knowing we have their support is a real comfort." – Liz Alderton, Queen's Nurse

53 WESTVIEW
1 St Thomas's Road, Great Glen, Leicester, LE8 9EH. Gill & John Hadland, 01162592170, gillhadland1@gmail.com. *7m S of Leicester. Take either r'about from A6 into village centre then follow NGS signs. Please park in Oaks Rd.* Sat 20, Sun 21 Feb, Sun 16 May, Sun 12 Sept (11-4). Adm £3, chd free. Home-made teas. Hot soup and home-made bread rolls also available in February. **Visits also by arrangement Feb to Sept for groups of 5 to 20.**
Organically managed small walled cottage garden with year-round interest. Rare and unusual plants, many grown from seed. Formal box parterre herb garden, courtyard garden, herbaceous borders, woodland garden, small wildlife pond, greenhouse, vegetable and fruit garden. Display of alpines. Collection of Galanthus (Snowdrops.) Recycled materials used to make quirky garden ornaments. Restored Victorian outhouse functions as a garden office and houses a collection of old garden tools and ephemera.

❄ ☕

54 ◆ WHATTON HOUSE
Long Whatton, Loughborough, LE12 5BG. The Crawshaw Family, 01509 431193, hello@whattonhouse.co.uk, www.whattonhouse.co.uk. *4m NE of Loughborough. On A6 between Hathern & Kegworth; 2½m SE of M1J24.* For NGS: Sun 6 June (11.30-6). Adm £5, chd free. Light refreshments. For other opening times and information, please phone, email or visit garden website.

Redhill Lodge

Come explore our tranquil gardens. Often described by visitors as a hidden gem, this 15 acre C19 Country House garden is a relaxing experience for all the family. Listen to the birds, and enjoy walking through the many fine trees, spring bulbs and shrubs, large herbaceous border, traditional rose garden, ornamental ponds and lawns. Refreshments available. Gravel paths.

55 THE WHITE HOUSE FARM

Ingarsby, nr Houghton-on-the-Hill, LE7 9JD. Pam & Richard Smith, 0116 259 5448, Pamsmithtwhf@aol.com. *7m E of Leicester. 12m W of Uppingham. Take A47 from Leicester through Houghton-on-the-Hill towards Uppingham. 1m after Houghton, turn L (signed Tilton). After 1m turn L (signed Ingarsby), garden is 1m further on.* **Visits by arrangement May to Sept for groups of 20+. Adm £5, chd free. Home-made teas.**

Former Georgian farm in 2 acres of country garden. Beautiful views. Box, yew & beech hedges divide a cottage garden of gaily coloured perennials and roses; a formal herb garden; a pergola draped with climbing plants; an old courtyard with roses, shrubs & trees. Herbaceous borders lead to pools with water lilies & informal cascade. Orchard, wild garden and lake. Home for lots of wildlife.

56 [NEW] WIGSTON FRAMEWORK KNITTERS MUSEUM

42-44 Bushloe End, Wigston, LE18 2BA. Wigston Framework Knitters Museum, 07814042889, chris.huscroft@tiscali.co.uk, www.wigstonframeworkknitters.org.uk. *4m S of Leicester. On A5199 Wigston bypass, follow yellow signs from Esso petrol station onto D502. Museum about ¼m on R, local car park on Paddock St.* **Sat 19, Sun 20 June, Sat 17, Sun 18 July (11-5). Adm £3, chd free. Light refreshments. Visits also by arrangement June to Aug for groups of 10 to 20. Museum buildings open for an extra charge.**

Victorian walled garden approx 70'x80' with traditional cottage garden planting. Managed by a group of volunteers and is undergoing a 2 year restoration, some replanting

started in autumn 2020. Garden located in the grounds of a historic museum, (an extra charge applies). A unique garden in the centre of Wigston which still retains an air of tranquillity.

GROUP OPENING

57 WIGSTON GARDENS

Wigston, LE18 3LF. Zoe Lewin. *Just south of Leicester off A5199.* **Sat 3, Sun 4 July (11-5). Combined adm £6, chd free. Home-made teas at Little Dale Wildlife Garden, 40 Rolleston Road and 28 Gladstone Street.**

28 GLADSTONE STREET
Chris & Janet Huscroft.
(See separate entry)

[NEW] **2A HOMESTEAD DRIVE**
Mrs Sheila Bolton.

LITTLE DALE WILDLIFE GARDEN
Zoe Lewin & Neil Garner, www.facebook.com/zoesopengarden.

40 ROLLESTON ROAD
Jenni & Glen Proudman.

[NEW] **'VALLENVINA' 6 ABINGTON CLOSE**
Mr Steve Hunt.

Wigston Gardens consists of 5 relatively small gardens all within a 2 mile radius of each other. There is something different to see at each garden from traditional and formal to a taste of the exotic via wildflowers, upcycling and interesting artefacts and new this year some prairie style planting! A couple of the gardens are within walking distance of one another but you will need to travel by car to visit all of the gardens in the group or it will make for quite a long walk and you'll need your comfy shoes.

GROUP OPENING

58 WILLOUGHBY GARDENS

Willoughby Waterleys, LE8 6UD. *9m S of Leicester. From A426 heading N turn R at Dunton Bassett lights. Follow signs to Willoughby. From Blaby follow signs to Countesthorpe. 2m S to Willoughby.* **Sun 11 July (11-5). Combined adm £5, chd free.**

2 CHURCH FARM LANE
Valerie & Peter Connelly.

FARMWAY
Eileen Spencer, 01162 478321, eileenfarmway9@msn.com.
Visits also by arrangement July & Aug for groups of up to 30.

HIGH MEADOW
Phil & Eva Day.

JOHN'S WOOD
John & Jill Harris.

KAPALUA
Richard & Linda Love.

3 ORCHARD ROAD
Diane Brearley.

[NEW] **3 YEW TREE CLOSE**
Emma Clanfield.

Willoughby Waterleys lies in the South Leicestershire countryside. The Norman Church will be open, hosting a film of the local bird population filmed by a local resident. 7 gardens will be open. John's Wood is a 1½ acre nature reserve planted to encourage wildlife. 2 Church Farm Lane has been professionally designed with many interesting features. Farmway is a plant lovers garden with many unusual plants in colour themed borders. High Meadow has been evolving over 10yrs. Incl mixed planting and ornamental vegetable garden. 3 Orchard Road is a small garden packed with interest. Kapalua has an interesting planting design incorporating views of open countryside. 3 Yew Tree Close is a wrap around garden that naturally creates a series of rooms with cottage garden style borders.

59 15 THE WOODCROFT

Diseworth, Derby, DE74 2QT. Nick & Sue Hollick. *The Woodcroft is off The Green, parking on The Woodcroft.* **Sun 27 June (11-5). Adm £4, chd free. Home-made teas.**

⅓ acre garden developed over 40 years with mature choice trees and shrubs, old and modern shrub roses, fern garden, wildlife area, alpine troughs, seasonal containers and mixed herbaceous borders, planted with a garden designer's eye and a plantsman's passion.

LINCOLNSHIRE

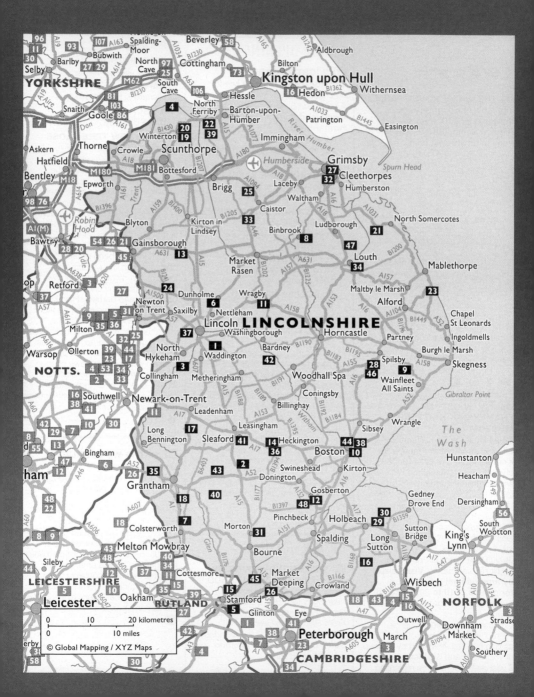

© Global Mapping / XYZ Maps

Lincolnshire is a county shaped by a rich tapestry of fascinating heritage, passionate people and intriguing traditions; a mix of city, coast and countryside.

The city of Lincoln is dominated by the iconic towers of Lincoln Cathedral. The eastern seaboard contains windswept golden sands and lonely nature reserves. The Lincolnshire Wolds is a nationally important landscape of rolling chalk hills and areas of sandstone and clay, which underlie this attractive landscape.

To the south is the historic, religious and architectural heritage of The Vales, with river walks, the fine Georgian buildings of Stamford and historic Burghley House. In the east the unqiue Fens landscape thrives on an endless network of waterways inhabited by an abundance of wildlife.

Beautiful gardens of all types, sizes and designs are cared for and shared by their welcoming owners. Often located in delightful villages, a visit to them will entail driving through quiet roads often bordered by verges of wild flowers.

Lincolnshire is rural England at its very best. Local heritage, beautiful countryside walks, aviation history and it is the home of the Red Arrows.

Volunteers

County Organisers

Helen Boothman
01652 628424
helen.boothman@ngs.org.uk

Lesley Wykes
01673 860356
lesley.wykes@ngs.org.uk

County Treasurer

Helen Boothman
(as above)

Publicity

Margaret Mann
01476 585905
marg_mann2000@yahoo.com

Erica McGarrigle
01476 585909
ericamcg@hotmail.co.uk

Assistant County Organisers

Tricia Elliot 01427 788517
t.elliott575@gmail.com

Sally Grant 01205 750486
sallygrant50@btinternet.com

Stephanie Lee 01507 442151
marigoldloo@btinternet.com

Jenny Leslie 01529 497317
jenny@collegefarmbraceby.com

Rita Morgan 01472 597529
rita.morgan1@sky.com

Sylvia Ravenhall 01507 526014
sylvan@btinternet.com

Jo Rouston 01673 858656
jo@rouston-gardens.co.uk

Joanna Townell 07446 239367
joannatownell@icloud.com

f @LincolnshireNGS
y @LincsNGS

Left: Marigold Cottage

OPENING DATES

All entries subject to change. For latest information check **www.ngs.org.uk**

Map locator numbers are shown to the right of each garden name.

April

Friday 2nd
◆ Easton Walled
   Gardens   7

Sunday 4th
Ashfield House   1
Woodlands   47

Monday 5th
Firsby Manor   9

Saturday 10th
◆ Burghley House Private
   South Gardens   5

Sunday 11th
◆ Burghley House Private
   South Gardens   5

Sunday 18th
◆ Goltho House   11
The Old Rectory   28

May

Sunday 2nd
Dunholme Lodge   6
Woodlands   47

Monday 3rd
Firsby Manor   9

Sunday 9th
66 Spilsby Road   38

Sunday 16th
NEW Fydell House   10
Ludney House Farm   21
The Old Rectory   28

Saturday 22nd
NEW Kings Hill Lodge   17
Oasis Garden - Your
   Place   27
Willoughby Road
   Allotments   44

Sunday 23rd
Oasis Garden - Your
   Place   27

Sunday 30th
◆ Mill Farm   25

June

Sunday 6th
Ludney House Farm   21
West Syke   41
NEW White House
   Farm   42
Woodlands   47

Saturday 12th
NEW St Andrews
   Hospice   32

Sunday 13th
Little Ponton Hall   18
Manor Farm   22
The Old Vicarage   29
Old White House   30
Skellingthorpe Hall   37
Springfield   39

Saturday 19th
NEW 3 Stone Lane   40

Sunday 20th
Home Farm   15
Little Ponton Hall   18
Shangrila   36
NEW 3 Stone Lane   40
Windrush   46

Sunday 27th
Gosberton Gardens   12
Ludney House Farm   21
◆ Mill Farm   25
Wildwood   43

July

Sunday 4th
Dunholme Lodge   6
NEW The Fern Nursery   8
Woodlands   47

Sunday 11th
Ludney House Farm   21

Sunday 18th
Ballygarth   4
Yew Tree Farm   48

Sunday 25th
Ludney House Farm   21

August

Sunday 1st
NEW The Fern Nursery   8
Woodlands   47

Saturday 7th
School House   33

Sunday 15th
NEW 19 Low Street   19
NEW 21 Low Street   20

Sunday 22nd
Willoughby Road
   Allotments   44

September

Sunday 5th
◆ Hall Farm   13
Woodlands   47

Saturday 25th
Inley Drove Farm   16

Sunday 26th
◆ Goltho House   11
Inley Drove Farm   16

October

Sunday 3rd
Woodlands   47

By Arrangement

Arrange a personalised garden visit on a date to suit you. See individual garden entries for full details.

Ashfield House   1
Aswarby Park   2
Aubourn Hall   3
Firsby Manor   9
23 Handley Street   14
Inley Drove Farm   16
Little Ponton Hall   18
Ludney House Farm   21
Marigold Cottage   23
Mere House   24
NEW Mill House   26
The Old Rectory   28
The Old Vicarage   29
NEW The Plant Lover's
   Garden   31
NEW St Andrews
   Hospice   32
School House   33
The Secret Garden of
   Louth   34
Sedgebrook Manor   35
NEW Willow Croft   45
Woodlands   47

Hall Farm

THE GARDENS

1 ASHFIELD HOUSE
Lincoln Road, Branston, Lincoln, LN4 1NS. John & Judi Tinsley, 07977 505682, john@tinsleyfarms.co.uk. *3m S of Lincoln on B1188. Northern outskirts of Branston on the B1188 Lincoln Rd. Signed 'Tinsley Farms - Ashfield'. Near bus stop, follow signs down drive.* **Sun 4 Apr (11-4). Adm £4, chd free. Light refreshments. Visits also by arrangement Apr to Oct for groups of 20 to 30. Refreshments incl in adm.**
See 140 Flowering Cherries, 30 Magnolias. Many thousands of Spring bulbs, Sweeping lawns, a lake and a fascinating arboretum. Grass paths - not ideal for wheelchairs. Fairly level garden.
&. 🐐 ☕

2 ASWARBY PARK
Aswarby, Sleaford, NG34 8SD. Mr & Mrs George Playne, 01529 455222/07770 721646, cplayne1@gmail.com, www.aswarbyestate.co.uk. *5m S of Sleaford on A15. Take signs to Aswarby. Entrance is straight ahead by Church through black gates.* **Visits by arrangement Apr to Sept for groups of 10 to 30. Adm £7, chd free. Light refreshments at Aswarby Park.**
Formal and woodland garden in a parkland setting of approx 20 acres. Yew Trees form a backdrop to borders and lawns surrounding the house which is a converted stable block. Walled garden contains a greenhouse with a Muscat vine, which is over 300 years old. Large display of daffodils, snowdrops and climbing roses in season. Partial wheelchair access on gravel paths and drives.
&. ❀ ☕

3 AUBOURN HALL
Harmston Road, Aubourn, Lincoln, LN5 9DZ. Mr & Mrs Christopher Nevile, 01522 788224, becky@aubournhall.co.uk, www.aubourngardens.com. *7m SW of Lincoln. Signed off A607 at Harmston & off A46 at Thorpe on the Hill.* **Visits by arrangement Apr to Sept. Adm £8, chd free. Light refreshments.**
Approx 9 acres. Lawns, mature trees, shrubs, roses, mixed borders,

rose garden, large prairie and topiary garden, spring bulbs, woodland walk and ponds. C11 church adjoining. Access to garden is fairly flat and smooth. Depending on weather some areas may be inaccessible to wheelchairs. Parking in field not on tarmac.
&. 🚗 ☕

4 BALLYGARTH
Post Office Lane, Whitton, Scunthorpe, DN15 9LF. Joanne & Adrian Davey. *From Scunthorpe on A1077 follow signs to West Halton. Through West Halton approx 3m to Whitton. Follow signs for parking at Village Hall.* **Sun 18 July (11-4.30). Adm £3, chd free. Home-made teas in Whitton Village Hall.**
Set in the rural village of Whitton our end terraced house has approx ⅓ acre garden with large herbaceous and grass borders and two water features. Seating areas overlooking the garden, countryside. Many home made garden artifacts using recycled materials incl a small folly. Everything in wood, brick and metal has been made by us. Drop off for those with limited mobility but parking is at village hall.
&. ❀ ✿ ☕

5 ◆ BURGHLEY HOUSE PRIVATE SOUTH GARDENS
Stamford, PE9 3JY. Burghley House Preservation Trust, 01780 752451, burghley@burghley.co.uk, www.burghley.co.uk. *1m E of Stamford. From Stamford follow signs to Burghley via B1443.* **For NGS: Sat 10, Sun 11 Apr (11-3.30). Adm £5, chd £3. For other opening times and information, please phone, email or visit garden website.**
On 10 and 11 April the Private South Gardens at Burghley House will open for the NGS with spectacular spring bulbs in park like setting with magnificent trees and the opportunity to enjoy Capability Brown's famous lake and summerhouse. Entry to the Private South Gardens via Orangery. The Garden of Surprises, Sculpture Garden and House are open as normal. (Regular adm prices apply). Fine Food Market. Gravel paths.
&. ✿ 🚗 ☕

6 DUNHOLME LODGE
Dunholme, Lincoln, LN2 3QA. Hugh & Lesley Wykes. *4m NE of Lincoln. Turn off A46 towards Welton at the r'about. After ½m turn L up long private road. Garden at top.* **Sun 2 May, Sun 4 July (11-5). Adm £4, chd free. Light refreshments.**
3 acre garden. Spring bulb area, shrub borders, fern garden, natural pond, wildflower area, orchard and vegetable garden. Developing 2 acre arboretum. RAF Dunholme Lodge Museum and War Memorial in the grounds. Lincoln Ukulele Band performances on both open days. Vintage vehicles on both days. Most areas wheelchair accessible but some loose stone and gravel.
&. 🐐 ✿ 🚗 ☕

7 ◆ EASTON WALLED GARDENS
Easton, NG33 5AP. Sir Fred & Lady Cholmeley, 01476 530063, info@eastonwalledgardens.co.uk, www.visiteaston.co.uk. *7m S of Grantham. 1m from A1, off B6403.* **For NGS: Fri 2 Apr (11-4). Adm £7.95, chd £3.75. Light refreshments & lunches. For other opening times and information, please phone, email or visit garden website.**
A 400-year-old, restored, 12 acre garden set in the heart of Lincolnshire. Home to snowdrops, sweet peas, roses and meadows. The tearoom serves delicious light lunches and cream teas and there is a well-stocked gift shop and plants and garden accoutrements for sale. Other highlights include a turf maze, yew tunnel and bird hide. Regret no wheelchair access to lower gardens but tearoom, shop and upper gardens all accessible.
&. ✿ 🚗 🚎 ☕ ⛱

The National Garden Scheme was set up in 1927 to raise funds for district nurses. We remain strongly committed to nursing and today our beneficiaries include Macmillan Cancer Support, Marie Curie, Hospice UK and The Queen's Nursing Institute

West Syke

8 NEW **THE FERN NURSERY**
Grimsby Road, Binbrook, Market
Rasen, LN8 6DH. Neil Timm,
www.fernnursery.co.uk. *On
B1203 from Market Rasen. On
the Grimsby road from Binbrook
Square 400 Metres.* **Sun 4 July,
Sun 1 Aug (10-4). Adm £3, chd
free. Refreshments Covid 19
Dependant.**
The garden has been designed as
a wildlife garden with a number of
features of interest to both visitors and
wildlife, helped by having a natural
stream running through the garden,
which supplies water to a pond and
water features. Visitors can also
enjoy rock features, acid beds, and a
sheltered winter garden with a sundial
at its centre. A path leads to a small
wood with the main fern collection. In
addition there is a semi formal garden
and bowling green, where they will
often see a game being played, large
shrubs, a bank of drought tolerant
plants and herbaceous perennials,
while steps, seats, a gazebo, bridge
and many other features add interest
to the garden. Partial wheelchair
access, gravel paths. WC.
&. 🐖 ❀ ☕

9 FIRSBY MANOR
Firsby, Spilsby, PE23 5QJ. David
& Gill Boldy, 01754 830386,
gillboldy@gmail.com. *5m E of
Spilsby. From Spilsby take B1195
to Wainfleet all Saints. In Firsby, turn
R into Fendyke Rd. Firsby Manor is*

0.8m along lane on L. **Mon 5 Apr,
Mon 3 May (1-4). Adm £3, chd
free. Home-made teas. Visits also
by arrangement Feb to Sept for
groups of 10 to 20.**
Firsby Manor is a lovely garden which
has been developed to provide peace
and pleasure for humans as well as a
restful haven for wildlife. Snowdrops
appear in February, followed by over
a hundred daffodil cultivars in April.
By June the garden is full of cottage
garden perennials. During 2020 we
created a larger terrace in the secret
garden and have started to plant the
banks of the pond. Partial wheelchair
access due to large areas of shingle
and uneven ground and no toilet
access.
🐖 ❀ ☕

10 NEW **FYDELL HOUSE**
South Square, Boston, PE21 6HU.
Boston Preservation Trust, www.
bostonpreservationtrust.com/
fydell-garden.html. *Central Boston
down South Street. Through the
Market Square, past Boots the
Chemist. One way street. By
Guildhall. There are three car parks
within 200 yards of the house.
Disabled parking in council car park
opposite the house.* **Sun 16 May
(10-4). Adm £4, chd free. Cream
teas.**
Within three original red brick walls a
formal garden was created in 1995.
Yew buttresses, arbours and four
parterres use dutch themes. The

borders contain herbaceous plants
and shrubs. The north facing border
holds shade loving plants. There
is a mulberry and walnut tree The
astrolabe was installed in 1997. A
Victorian rockery is built from slag
from ironworks in Boston. Wheelchair
access is along the south alleyway
from the front to the back garden.
&. ❀ ☕ 🏓

11 ◆ **GOLTHO HOUSE**
Lincoln Road, Goltho, Wragby,
Market Rasen, LN8 5NF. Mr & Mrs
S Hollingworth, 01673 857768,
bookings@golthogardens.com,
www.golthogardens.com. *10m
E of Lincoln. On A158, 1m before
Wragby. Garden on L (not in Goltho
Village).* **For NGS: Sun 18 Apr, Sun
26 Sept (12-4). Adm £7.50, chd
free. For other opening times and
information, please phone, email or
visit garden website.**
4½ acre garden started in 1998 but
looking established with long grass
walk flanked by abundantly planted
herbaceous borders forming a focal
point. Paths and walkway span
out to other features incl nut walk,
prairie border, wildflower meadow,
rose garden and large pond area.
Snowdrops, hellebores and shrubs for
winter interest.
&. ❀ 🛏 🏓 ☕

GROUP OPENING

12 GOSBERTON GARDENS
Gosberton, Spalding, PE11 4NQ.
Entering Gosberton on A152, from Spalding , Salem St on L & Mill Lane on R opp the War Memorial. **Sun 27 June (12-5). Combined adm £5, chd free. Cream Teas and home-made cakes. Tea, Coffee and cold drinks also available.**

MILLSTONE HOUSE
Mrs J Chatterton.

4 SALEM STREET
Roley & Tricia Hogben.

The village of Gosberton welcomes visitors to 2 private houses to view their gardens. We hope that everyone will find interesting features during their tour and enjoy the 2 locations. Millstone House. Colourful herbaceous borders are hidden by a privet hedge. Dappled shade creates a feeling of relaxation at the rear of the house. 4 Salem Street. Delightful secluded garden. Mixed borders including a small water feature lead to a productive vegetable plot. Partial wheelchair access.

&. ✿ ☕

13 ◆ HALL FARM
Harpswell, Gainsborough, DN21 5UU. Pam & Mark Tatam, 01427 668412, pam.tatam@gmail.com, www.hall-farm.co.uk. *7m E of Gainsborough. On A631, 1½ m W of Caenby Corner.* **For NGS: Sun 5 Sept (1-5). Adm £4, chd free. Light refreshments. For other opening times and information, please phone, email or visit garden website.**
The 3 acre garden encompasses formal and informal areas, a sunken garden, courtyard with rill, walled Mediterranean garden, double herbaceous borders for late summer, lawns, bog garden, giant chess set, flower and grass meadow. It is a short walk to the medieval moat, which surrounds over an acre of wild semi-woodland garden with picnic table, benches and 'beach'. Free seed collecting on Sun 5 Sept. Most of garden suitable for wheelchairs.

&. 🐕 ✿ 🚗 ☕

14 23 HANDLEY STREET
Heckington, nr Sleaford, NG34 9RZ. Stephen & Hazel Donnison, 01529 460097, donno5260@gmail.com. *A17 from Sleaford, turn R into Heckington. Follow rd to the Green. L, follow rd past Church, R into Cameron St. At end of this rd L into Handley St.* **Visits by arrangement June to Sept for groups of up to 20. Adm £3, chd free. Home-made teas. Home-made teas and cakes.**
Compact, quirky garden, large fish pond. Further 5 small wildlife ponds. Small wooded area and Jurassic style garden with Tree ferns. Densely planted flower borders featuring Penstemons. Large patio with seating.

✿ ☕

15 HOME FARM
Little Casterton Road, Ryhall, Stamford, PE9 4HA. Steve & Karen Bourne. *1½ m N of Stamford. Off A6121 at mini r'about, towards Little Casterton.* **Sun 20 June (10-5). Adm £5, chd free. Home-made teas.**
Nine acres including formal garden with herbaceous, rose and shrub borders. Recent additions include Mediterranean border and Rugosa rose hedge interspersed with striking Tibetan cherry trees. Raised vegetable beds, asparagus and rhubarb beds in front of the ha-ha. Fruit cage with soft fruit, orchard of old local varieties of top fruit. Woodland walk.

&. ☕

16 INLEY DROVE FARM
Inley Drove, Sutton St James, Spalding, PE12 0LX. Francis & Maisie Pryor, 01406 540088, maisietaylor7@gmail.com, www.pryorfrancis.wordpress.com/. *Just off rd from Sutton St James to Sutton St Edmund. 2m S of Sutton St James. Look for yellow NGS signs on double bend.* **Sat 25, Sun 26 Sept (11-4). Adm £4.50, chd free. Home-made teas. Visits also by arrangement Apr to July for groups of 10 to 30. Restricted parking.**
Over 3 acres of Fenland garden and meadow plus 0½ acre wood developed over 20yrs. Garden planted for colour, scent and wildlife. Double mixed borders and less formal flower gardens all framed by hornbeam hedges. Unusual shrubs and trees, incl fine stand of Black Poplars, vegetable garden, woodland walks and orchard. Disabled WC

outside. Some gravel and a few steps but mostly flat grass.

&. ✿ ☕ 🏕

17 NEW KINGS HILL LODGE
Gorse Hill Lane, Caythorpe, Grantham, NG32 3DY. Tim & Carol Almond. *Off the A607 approx 10m from Grantham towards Lincoln. From S turn L into Church Lane (R from N) at Xrds. Pass church on R, after 100m turn R onto Waterloo Road. Then sharp L, drive 150m then L into Gorse Hill Lane with Lodge on R.* **Sat 22 May (1-5). Adm £4, chd free. Home-made teas.**
A new garden of about 1000 sq metres developed over 8 years from 800 sq metres of neglected lawn and 200 sq metres overgrown, boggy shrubbery. Now consists of 250 sq metres of managed lawn with the remainder put to over 20 mixed herbaceous beds, 7 large raised vegetable beds, cedar greenhouse, summerhouse, 4 sitting out areas, water feature with 15 different climbing roses around the house. There is a circular path around the house which is wide enough for a wheelchair with care. Wide-wheeled chairs would be able to use the lawn.

&. ✿ ☕

18 LITTLE PONTON HALL
Grantham, NG33 5BS. Bianca & George McCorquodale, www. littlepontonhallgardens.org. uk. *2m S of Grantham. ½ m E of A1 at S end of Grantham bypass. Disabled parking.* **Sun 13, Sun 20 June (11-4). Adm £5, chd free. Light refreshments. Visits also by arrangement Feb to Aug.**
3 to 4 acre garden. River walk. Spacious lawns with cedar tree over 200 years old. Formal walled kitchen garden and listed dovecote, with herb garden. Victorian greenhouses with many plants from exotic locations. Wheelchair access on hard surfaces, unsuitable on grass. Disabled WC.

&. 🐕 ✿ 🚗 ☕

"The annual donation from the National Garden Scheme to Perennial is the cornerstone of our fundraising activities and encourages many of our donors". Perennial

19 NEW **19 LOW STREET**
Winterton, DN15 9RT. Jane &
Allan Scorer. *On A1077 4m N of
Scunthorpe & 7m S of the Humber
Bridge. Garden signed from A1077.
Parking on Low Street & nearby
Market Place (2 mins walk from
Low Street).* **Sun 15 Aug (10-4).
Combined adm with 21 Low
Street £5, chd free.**
A half acre garden with an emphasis
on dense sub tropical planting,
although there are other distinct
areas too, including an arid bed,
greenhouses, an extensive variety of
pots and planters, ornamental pond,
wildlife pond set in informal area, soft
fruit, cut flower garden and seating
areas. Wheelchair is difficult but not
impossible as areas of the garden
have bark or gravel paths.

⅃ 🐕 ✳ 🍵 🌲

20 NEW **21 LOW STREET**
Winterton, DN15 9RT. Brian Dale
& Nigel Bradford. *Garden is in the
centre of the village & parking is
limited. A short walk from the parking
area near the church & Coop.* **Sun
15 Aug (10-4). Combined adm
with 19 Low Street £5, chd free.
Home-made teas.**
The garden is approx ½ acre and
when we moved into the house, five
yrs ago, it mainly consisted of shrubs
that had been allowed free rein. We
have enlarged the existing beds
and created new beds and borders
which are now mostly planted with
herbaceous perennials. There is a
large area given over to vegetables
and soft fruits which is now producing
successfully. Wheelchair access to
the first section of the garden only,
approximately 150sq metres.

🐕 🍵

21 **LUDNEY HOUSE FARM**
Ludney, Louth, LN11 7JU.
Jayne Bullas, 07733 018710,
jayne@theoldgatehouse.com.
Between Grainthorpe & Conisholme.
**Sun 16 May, Sun 6, Sun 27 June,
Sun 11, Sun 25 July (1.30-4). Adm
£6.50, chd free. Refreshments
incl in adm. Visits also by
arrangement May to July for
groups of 5 to 30.**
A beautiful landscaped garden of
several defined spaces containing
formal and informal areas. There is
an excellent mix of trees, shrubs,
perennials, rose garden and
wildflower area which is home to the
beehives. In spring there is a nice

selection of bulbs and spring flowers.
There is also a new pond area , There
are plenty of seats positioned around
to sit and enjoy a cuppa and piece
of cake! Wheelchair access to most
parts.

⅃ 🍵

22 **MANOR FARM**
Horkstow Road, South Ferriby,
Barton-upon-Humber, DN18 6HS.
Geoff & Angela Wells. *3m from
Barton-upon-Humber on A1077,
turn L onto B1204, opp Village Hall.*
**Sun 13 June (11-5). Combined
adm with Springfield £5, chd free.
Home-made teas.**
A garden which is much praised by
visitors. Set within approx 1 acre
with mature shrubberies, herbaceous
borders, gravel garden and pergola
walk. Rose bed, white garden
and fernery. Many old trees with
preservation orders. Wildlife pond set
within a paddock.

⅃ 🐕 🚗 🍵

23 **MARIGOLD COTTAGE**
Hotchin Road, Sutton-on-Sea,
LN12 2NP. Stephanie Lee &
John Raby, 01507 442151,
marigoldlee@btinternet.com,
www.rabylee.uk/marigold/. *16m N
of Skegness on A52. 7m E of Alford
on A1111. 3m S of Mablethorpe
on A52. Turn off A52 on High St at
Cornerhouse Cafe. Follow rd past
playing field on R. Rd turns away
from the dunes. House 2nd on L.*
**Visits by arrangement Feb to Nov
for groups of up to 30. Adm £3,
chd free.**
Slide open the Japanese gate to
find secret paths, lanterns, a circular
window in a curved wall, water
lilies in pots and a gravel garden,
vegetable garden and propagation
area. Take the long drive to see the
sea. Back in the garden, find a seat,
enjoy the birds and bees. We face
the challenges of heavy clay and salt
ladened winds but look for unusual
plants not the humdrum for these
conditions. Most of garden accessible
to wheelchairs along flat, paved
paths.

⅃ 🐕 ✳ 🚗 🚌 🌲

24 **MERE HOUSE**
Stow Road, Sturton by Stow,
Lincoln, LN1 2BZ. Nigel &
Alice Gray, 07932 442349,
alicemerehouse@icloud.com. *10m
NW of Lincoln between Sturton &
Stow. 1m from centre of Sturton*

*village heading to Stow, house on L.
NB: Postcode will not bring you far
enough out of Sturton village.* **Visits
by arrangement Apr to Sept for
groups of up to 30. Adm incl tea
and cakes. Adm £6, chd free.**
Approx 1½ acres of established
garden planted for the first time in
1975, redesigned in 1996. Renovated
over the last 8yrs to incl new beds
with drift planting but still incl the
formal parterre. Spring bulbs and
late summer colour are highlights.
There is also a cutting garden,
pleached hedges, vegetable garden
and orchard. Work in progress incl
a new garden project with a long
herbaceous border. There is the
highly acclaimed Cross Keys pub in
Stow Village that does a very good
lunch and also the Tilbridge Tastery
in Sturton Village. The garden is
wheelchair accessible on grass.

⅃ 🍵

25 ◆ **MILL FARM**
Caistor Road, Grasby, Caistor,
DN38 6AQ. Mike & Helen
Boothman, 01652 628424,
boothmanhelen@gmail.com,
www.millfarmgarden.co.uk. *3m
NW of Caistor on A1084. Between
Brigg & Caistor. From Cross Keys
pub towards Caistor for approx
200yds. Do not go into Grasby
village.* **For NGS: Sun 30 May,
Sun 27 June (11-4). Adm £4, chd
free. Home-made teas. For other
opening times and information,
please phone, email or visit garden
website.**
A much loved garden by visitors,
which continues to be developed
and maintained to a high standard by
the owners. Over 3 acres of garden
with many diverse areas. Formal
frontage with shrubs and trees. The
rear is a plantsman haven with a
peony and rose garden, specimen
trees, vegetable area, old windmill
adapted into a fernery, alpine house
and shade house with a variety of
shade loving plants. Herbaceous
beds with different grasses and hardy
perennials. Small nursery on site with
home grown plants available. Open
by arrangement for groups. Mainly
grass, but with some gravelled areas.

⅃ 🐕 ✳ 🚗 🍵 🍵

*Online booking is available
for many of our gardens, at
ngs.org.uk*

26 NEW MILL HOUSE

Stamford Road, Market Deeping, Peterborough, PE6 8AB. Mr & Mrs James Wherry, 01778 342343, thewherrys@hotmail.com. *8m E of Stamford. From Stamford take A16 towards Market Deeping. At A16/A15 r'about take B1525 into Market Deeping. First house on RHS by 30mph signs. Parking for up to 20 cars.* **Visits by arrangement May & June for groups of 10 to 30. Mid week visits only. Admission incl refreshments. Adm £7.50, chd free.**

4 acre informal garden with the river Welland running through, pond, stream and bog garden. Separate and different areas with many interesting and unusual plants and trees, yew house, tree house, pet cemetery and other quirky features. Parking for up to 20 cars only on site over narrow bridge.

🚶 🐕

27 OASIS GARDEN - YOUR PLACE

Rear of Your Place, 236 Wellington Street, Grimsby, DN32 7JP. Grimsby Neighbourhood Church, www.yourplacegrimsby.com. *Enter Grimsby (M180) over flyover, along Cleethorpes Rd. Turn R into Victor St, Turn L into Wellington St. Your Place is on the R on junction of* Wellington St & Weelsby St. **Sat 22, Sun 23 May (11-3). Adm £3, chd free. Light refreshments.**

The multi award winning Oasis Garden, Your Place, recently described by the RHS as the 'Most inspirational garden in the six counties of the East Midlands', is approximately 1½ acres and nestles in the heart of Great Grimsby's East Marsh Community. A working garden producing 15k plants per year, grown by local volunteers of all ages and abilities. Lawns, fruit, vegetable, perennial and annual beds.

🚶 ✿ ☕

28 THE OLD RECTORY

Church Lane, East Keal, Spilsby, PE23 4AT. Mrs Ruth Ward, 01790 752477, rfjward@btinternet.com. *2m SW of Spilsby. Off A16. Turn into Church Lane by PO.* **Sun 18 Apr, Sun 16 May (1.30-4). Adm £4, chd free. Home-made teas. Visits also by arrangement Feb to Nov. Refreshments by arrangement.**

Beautifully situated, with fine views, rambling cottage garden on different levels falling naturally into separate areas, with changing effects and atmosphere. Steps, paths and vistas to lead you on, seats well placed for appreciating special views or relaxing and enjoying the peace. Dry border, vegetable garden, orchard, woodland walk, wildflower meadow. Yr-round interest. Welcoming to wildlife. Partial wheelchair access.

🐕 🚗 ☕

29 THE OLD VICARAGE

Low Road, Holbeach Hurn, PE12 8JN. Mrs Liz Dixon Spain, 01406 424148, lizdixonspain@gmail.com. *2m NE of Holbeach. Turn off A17 N to Holbeach Hurn, past post box in middle of village, 1st R at war memorial into Low Rd. Old Vicarage is on R approx 400yds Parking in grass paddock.* **Sun 13 June (1-5). Combined adm with Old White House £5, chd free. Home-made teas at Old White House. Visits also by arrangement Mar to Sept for groups of up to 30.**

2 acres of garden with 150yr old tulip, plane and beech trees: borders of shrubs, roses, herbaceous plants. Shrub roses and herb garden in old paddock area, surrounded by informal areas with pond and bog garden, wildflowers, grasses and bulbs. Small fruit and vegetable gardens. Kids love exploring winding paths through the wilder areas. Garden is managed environmentally. Fun for kids! Gravel drive, some paths, mostly grass access.

🚶 🐕 ✿ 🚗 ☕

St Andrews Hospice

30 OLD WHITE HOUSE
Baileys Lane, Holbeach Hurn,
PE12 8JP. Mrs A Worth. *2m N
of Holbeach. Turn off A17 N to
Holbeach Hurn, follow signs to
village, cont through, turn R after
Rose & Crown pub at Baileys Lane.*
**Sun 13 June (1-5). Combined adm
with The Old Vicarage £5, chd
free. Home-made teas.**
1½ acres of mature garden, featuring
herbaceous borders, roses, patterned
garden, herb garden and walled
kitchen garden. Large catalpa, tulip
tree that flowers, ginko and other
specimen trees. Flat surfaces, some
steps, wheelchair access to all areas
without using steps.

**31 NEW THE PLANT LOVER'S
GARDEN**
Morton, Bourne, PE10 0XF.
Danny & Sophie, 07850 239393,
plantloversgarden@outlook.com.
*Situated on the edge of beautiful S
Lincolnshire close to the borders of
Cambridgeshire & Rutland in the East
of England.* **Visits by arrangement
Mar to Aug for groups of up to
20. No WC facilities available at
present. Adm £3.50, chd free.
Light refreshments.**
An inspirational garden, to see what
can be achieved with a new garden.
In 6 years with its current owners,
now a verdant space of colour, form
and style, with brimming raised beds
and packed herbaceous borders,
peppered throughout with clipped
topiary balls. With an ever changing
interest and a variety of blooms
throughout the months of March to
August. A plant lover's delight. Most
areas accessible by wheelchair.

32 NEW ST ANDREWS HOSPICE
Peaks Lane, Grimsby, DN32 9RP.
Rob Baty, 01472 350908. *Follow
Signs for St Andrews Hospice.*
**Sat 12 June (12-4.30). Adm £5,
chd free. Home-made teas
on site catering. Visits also by
arrangement May to Dec for
groups of 5 to 10.**
The St Andrews gardens are beautifully
landscaped spaces which focus on the
lovely nature that surrounds us. They
are suitable for all ages providing all
with a tranquil and calm environment.
The gardens are fully accessible with
plenty of room for all to enjoy the
outdoors together. Herbaceous beds
and ample seating areas.

33 SCHOOL HOUSE
Market Rasen Road, Holton-
le-Moor, Market Rasen,
LN7 6AE. Chris & Rosemary
Brown, 07778 464068,
67chris.47holton@gmail.com.
*15m N of Lincoln. A46 towards
Caistor, take B1434 to Holton le
Moor. School House on R next
to village (Moot) hall.* **Sat 7 Aug
(12.30-5). Adm £5, chd free. Light
refreshments. Adm incl tea/
coffee and biscuits. Visits also
by arrangement May to Aug for
groups of 5 to 20.**
An all around the house garden,
ranging from shaded early area
to summer and autumn flowering
areas. Central gravel garden with
alliums followed by agapanthus
and supplemented with grasses.
Designed and built by Chris Brown a
now retired garden designer.

**34 THE SECRET GARDEN OF
LOUTH**
68 Watts Lane, Louth,
LN11 9DG. Jenny & Rodger
Grasham, 07977 318145,
sallysing@hotmail.co.uk,
www.facebook.com/
thesecretgardenoflouth. *½m S of
Louth town centre. For Sat Nav and
to avoid opening/closing gate on
Watts Lane, use postcode LN119DJ
this is Mount Pleasant Ave, leads
straight to our house front.* **Visits
by arrangement July & Aug for
groups of up to 20. Children
under 14yrs free. Adm £3. Chd
under 14yrs free. Tea/coffee
available.**
Blank canvas of ⅕ acre in early 90s.
Developed into lush, colourful, exotic
plant packed haven. A whole new
world on entering from street. Exotic
borders, raised exotic island, long hot
border, ponds, stumpery . Intimate
seating areas along garden's journey.
Facebook page - The Secret Garden
of Louth. Children, find where the
frogs are hiding! Butterflies and bees
but how many different types? Feed
the fish, find Cedric the spider, Simon
the snake, Colin the Crocodile and
more. Grass pathways, main garden
area accessible. Wheelchairs not
permitted on the bridge.

35 SEDGEBROOK MANOR
Church Lane, Sedgebrook,
Grantham, NG32 2EU. Hon
James & Lady Caroline Ogilvy,
01949 842337. *2m W of Grantham*

on A52. *In Sedgebrook village by
church.* **Visits by arrangement.
Adm by donation.**
Yew and box topiary surround this
charming Manor House (not open).
Croquet lawn, herbaceous border
and summer house. Bridge over small
pond and two larger ponds. Ancient
mulberry tree. Tennis court, vegetable
garden. Swimming pool in enclosed
garden. Wheelchair access to most
areas Dogs on leads.

36 SHANGRILA
Little Hale Road, Great Hale,
Sleaford, NG34 9LH. Marilyn
Cooke & John Knight. *On B1394
between Heckington & Helpringham.*
**Sun 20 June (11-5). Adm £4.50,
chd free. Home-made teas.**
Approx 3 acre garden with sweeping
lawns long herbaceous borders,
colour themed island beds, hosta
collection, lavender bed with seating
area, topiary, acers, small raised
vegetable area, 3 ponds and new
exotic borders new Japanese zen
garden. Wheelchair access to all
areas.

37 SKELLINGTHORPE HALL
Lincoln Road, Skellingthorpe,
Lincoln, LN6 5UU. Charlie & Anne
Coltman. *4m W of Lincoln, 500 yds
from A46 Lincoln relief rd. At the
r'about (mid way between A57 &
B1190) signed Skellingthorpe take
NW exit. Skellingthorpe Hall entrance
is circa. 500yds on R opposite
Waterloo Lane.* **Sun 13 June (10.30-
4). Adm £3.50, chd free. Home-
made teas. Some allergies are
not catered for please ask.**
A 3½ acre landscaped garden with
long views across the ha-ha to the
park and beyond. Extensive lawns
with mature trees. Shrubs and
perennial borders created over the
last 25yrs. Spring bulbs naturalised
in the grass and tulips in the borders.
Pond and paved area by the small
conservatory and a larger one in
the main garden, A large vegetable
garden with green houses. Some
paths can be an effort to navigate.

38 66 SPILSBY ROAD
Boston, PE21 9NS. Rosemary &
Adrian Isaac. *From Boston town
take A16 towards Spilsby. On L after
Trinity Church. Parking on Spilsby
Rd.* **Sun 9 May (11-4). Adm £4,**

chd free. Home-made teas.
1⅓ acre with mature trees, moat,
Venetian Folly, summer house and
orangery, lawns and herbaceous
borders. Children's Tudor garden
house, gatehouse and courtyard.
Wide paths most paths suitable for
wheelchairs.

&. ✿ ☕

39 SPRINGFIELD
Main Street, Horkstow, Barton-
Upon-Humber, DN18 6BL. Mr &
Mrs G Allison. 4m from Barton on
Humber. Take the A1077 towards
Scunthorpe & in South Ferriby bear L
onto the B1204, after 2m Springfield
is on the hillside on L. Sun 13
June (11-5). Combined adm with
Manor Farm £5, chd free.
This beautiful hillside garden on the
edge of the Wolds was renovated
and redesigned in 2011 from an
overgrown state. It features many
shrubs and perennials with a rose
pergola and stunning views over the
Ancholme Valley.

🐈

40 NEW 3 STONE LANE
Little Humby, Grantham,
NG33 4HX. Paul & Marijka Hance.
Leave the A52 follow signs to
Ropsley, through Ropsley. Proceed
along the Humby Road. After1m,
turn R & proceed uphill into Little
Humby. Direction signs are on the
village green. Garden is located
at rear of property, through L
access gate. Sat 19, Sun 20 June
(10-4). Adm £5, chd free. Light
refreshments.
Our family garden is approx. two
thirds of an acre and slopes gently
downwards from the terrace towards
the valley. It has been created over
the last 12 years from a simple
lawn to encompass a wide range
of Herbaceous Plants, Shrub and
Climbing Roses, Orchard and
Wildlife pond. All areas of the garden
are accessible. The gravel drive is
however a little bumpy.

&. 🐕 ✿ ☕

41 WEST SYKE
38 Electric Station Road, Sleaford,
NG34 7QJ. Ada Trethewey. From
A17, A15 & A153 take bypass exit
at r'about to Sleaford town centre.
At HSBC & Nationwide turn R into
Westgate. Turn R, on to Electric
Station Road 38 at end of rd. Ample
parking. Sun 6 June (12-5). Adm
£4, chd free. Light refreshments.

Over 1 acre, comprising bog gardens,
rockeries, 3 large ponds, wildflower
meadow, lawns and cottage garden
planting. Rambling roses a feature.
Designed for wildlife habitats;
sustainable principles and a wealth of
native species. Large well established
garden evolved of 30yrs. Lawned
paths. WC wheelchair accessible.

&. ✿ ☕ 🛏 🚡

42 NEW WHITE HOUSE FARM
Metheringham Fen, Lincoln,
LN4 3AW. Bruce & Lisa Spencer-
Knott. From Metheringham turn on to
Fen Lane opposite the Fire Station &
proceed for approx 4m to our property
on the L. Sun 6 June (11-4). Adm £4,
chd free. Home-made teas.
White House Farm is a stunning
Georgian manor and idyllic focal point
for the Spencer-Knott's beautiful
English garden. A sensitive fusion of
French renaissance and quintessential
English naturalistic garden design
inspired this elegant, serene and
restorative space. David Austin Rose
collection. Posies of cut flowers
Mixture of hard paving and lawn.

&. ✿ 🚗 ☕

43 WILDWOOD
Aisby, Grantham, NG32 3NE. Paul &
Joy King. Halfway between Grantham
& Sleaford. Off A52 for Dembelby or
Oasby. Follow signs to Aisby. In village
take track with a footpath sign on the
west side of the green opp the village
hall. Plenty of parking on site. Sun
27 June (11-5). Adm £5, chd free.
Home-made teas.
Created over 9 yrs from fields
and existing hard landscaping.
Windbreaks protect the garden which
is heavy clay, wet in winter, dry in
summer. There are many borders
(some sloping) planted with unusual
and rare trees, shrubs, climbers,
perennials, bulbs and alpines. This is
a wildlife-friendly garden with a bee
orchid patch and fruit and vegetable
areas. Grass, gravel and paving
provide access. A modern glass Huf
Haus, the only one in Lincolnshire,
and one of the few in the country
open as part of the NGS scheme.
There are over 250 Huf Haus in the
UK, most in the south. Most paths are
grass or gravel.

&. ✿ ☕

Yew Tree Farm

Firsby Manor

44 WILLOUGHBY ROAD ALLOTMENTS

Willoughby Road, Boston, PE21 9HN. Willoughby Road Allotments Association, willoughbyroadallotments.org.uk. *Entrance is adjacent to 109 Willoughby Road. Street Parking only.* **Sat 22 May, Sun 22 Aug (10.30-4). Adm £4, chd free. Light refreshments.** Set in 5 acres the allotments comprise 60 plots growing fine vegetables, fruit, flowers and herbs. There is a small orchard and wildflower area and a community space adjacent. Grass paths run along the site. Several plots will be open to walk round. There will be a seed and plant stall. Community area with kitchen and disabled WC. Large Polytunnel with raised beds inside and out. Accessible for all abilities.

&. 🐐 ❀ ☕

45 NEW WILLOW CROFT

Wisthorpe, Stamford, PE9 4PE. Diana Holden, 01778 560746, diholden4@hotmail.com. *In centre of Wilsthorpe village one of a pair of bungalows. A15 to Waterside garden centre between Bourne & Market Deeping onto King Street take 1st R & then 1st R again.* **Visits by arrangement May to Sept for groups of up to 20. Home-made teas. Refreshments by arrangement.** Medium sized garden packed with plants to provide colour and interest over an extended period including Roses, Japanese Acers, Clematis, over 100 Hostas,, Salvias, Fuschia, Cannas,, Dahlias and many other varieties. Raised vegetable beds, Gravel garden with grasses, and succulents. Lots of nooks and crannies and seating.

❀ ☕

46 WINDRUSH

Main Road, East Keal, Spilsby, PE23 4BB. Ian & Suzie MacDonald. *On A16, 4m S of Spilsby, opp A155 turning signed West Keal.* **Sun 20 June (11-4). Adm £3.50, chd free. Home-made teas.** Country garden of approx 4 acres with herbaceous borders, shrub and climbing roses, clematis and grasses. Woodland walk and ponds, and vegetable garden. Meadow planted in 2018 with orchard of Lincolnshire apples. Shallow steps and some uneven ground and paths Gravel paths.

&. ❀ ☕

47 WOODLANDS

Peppin Lane, Fotherby, Louth, LN11 0UW. Ann & Bob Armstrong, 01507 603586, annbobarmstrong@btinternet.com, www.woodlandsplants.co.uk. *2m N of Louth off A16 signed Fotherby. Please park on R verge opp allotments & walk approx 350 yds to garden. If full please park considerately elsewhere in the village. No parking at garden. Please do not drive beyond designated area.* **Sun 4 Apr, Sun 2 May, Sun 6 June, Sun 4 July, Sun 1 Aug, Sun 5 Sept (11-5); Sun 3 Oct (11-4). Adm £4, chd free. Home-made teas. Visits also by arrangement Mar to Nov.** A lovely mature woodland garden where a multitude of unusual plants are the stars, many of which are available from the well stocked RHS listed nursery. During the summer of 2020 several small areas have been redesigned and a great deal of fresh planting has taken place. Award winning professional artist's studio/gallery open to visitors. Specialist collection of Codonopsis for which Plant Heritage status has been granted. Wheelchair users able to access some areas with care. Parking at the home for those with limited mobility.

&. 🐐 ❀ NPC ☕

48 YEW TREE FARM

Westhorpe Road, Gosberton, Spalding, PE11 4EP. Robert & Claire Bailey-Scott. *Nr Spalding. Enter the village of Gosberton. Turn into Westhorpe Rd, opp The Bell Inn, cont for approx. 1.5 miles. Property is 3rd on R after bridge.* **Sun 18 July (11-5). Adm £4, chd free. Home-made teas.** A lovely country garden, 1½ acres. Large herbaceous and mixed borders surround well kept lawns. Wildlife pond with two bog gardens, woodland garden and shaded borders containing many unusual plants. A Mulberry tree forms the centre piece of one lawn. Orchard, wildflower meadow and large vegetable plot. Gravel driveway, some gravel paths.

&. 🐐 ❀ 🚌 ☕

Mill House

LONDON

ESSEX

Id EN1 EN3

21 N9 Chingford
Edmonton E4
od N18 Woodford IG7 RM4
en Green IG8 RM5 RM1 RM3
N17 Tottenham E17 E18 IG6 RM6 Romford RM11
Walthamstow IG5 RM7 RM2 Upminster
N15 IG4 IG RM12 RM14
N16 E10 E11 IG3 RM8 RM11
N5 E5 Ilford IG1 RM8
NI E8 E9 E7 E12 E RM10
Stratford E15 Barking IG11
E2 E3 E13 E6 RM9 Rainham
EC1 E1 RM13
EC2 London
E4 EC3 City
DON E14 E16 SE28 Thamesmead
SE1 DA18
SE16 SE10 SE2 DA17
SE17 SE8 SE7 DA8
SE15 SE14 Greenwich SE18 DA7
Peckham SE3 DA16 Bexleyheath
SE5 SE4 Lewisham Eltham DA6 DA1
ich SE24 SE22 SE13 SE9 DA5 DA
SE21 SE23 SE12 DA15 Bexley
SE27 SE6 DA14 Sidcup
SE26 Chislehurst
SE19 BR1 BR7
7 SE20 Bromley BR5
SE25 BR3 BR2 BR
ydon CR0 BR
Addington BR4 Orpington
CR BR6
CR2 BR2

River Thames

KENT

TN14
Biggin Hill TN16 TN

0   5   10 kilometres
0   5 miles
© Global Mapping / XYZ Maps

Volunteers

County Organiser
Penny Snell
01932 864532
pennysnellflowers@btinternet.com

County Treasurer
Maria Rowden
07970981999
maria.rowden@ngs.org.uk

Publicity
Penny Snell (as above)

Booklet Co-ordinator
Sue Phipps
07771 767196
sue@suephipps.com

Booklet Distributor
Joey Clover
020 8870 8740
joey.clover@ngs.org.uk

Social Media
Sue Phipps (as above)

Alex Redfern 07976849344
alex.redfern@ngs.org.uk

Assistant County Organisers

Central London
Eveline Carn
07831 136069
evelinecbcarn@icloud.com

Clapham & surrounding area
Sue Phipps (as above)

Croydon & outer South London
Janine Wookey
07711 279636
j.wookey@btinternet.com

Dulwich & surrounding area
Clive Pankhurst
07941 536934
alternative.ramblings@gmail.com

E London
Teresa Farnham
07761 476651
farnhamz@yahoo.co.uk

Finchley & Barnet
Debra Craighead 07415 166617
dcraighead@icloud.com

Hackney
Philip Lightowlers
020 8533 0052
plighto@gmail.com

Hampstead
Joan Arnold
020 8444 8752
joan.arnold40@gmail.com

Islington
Penelope Darby Brown
020 7226 6880
penelope.darbybrown@ngs.org.uk

Gill Evansky
020 7359 2484
gevansky@gmail.com

**Northwood, Pinner,
Ruislip & Harefield**
Hasruty Patel
07815 110050
hasruty@gmail.com

NW London
Susan Bennett & Earl Hyde
020 8883 8540
suebearlh@yahoo.co.uk

Outer NW London
James Duncan Mattoon
020 8830 7410
jamesmattoon@msn.com

Outer W London
Julia Hickman
020 8339 0931
julia.hickman@virgin.net

SE London
Janine Wookey
07711 279636
j.wookey@btinternet.com

SW London
Joey Clover (as above)

Tower Hamlets
Vivien Taylor 07903 933881
vivien.taylor@ngs.org.uk

W London, Barnes & Chiswick
Siobhan McCammon
07952 889866
siobhan.mccammon@gmail.com

f @LondonNGS

@LondonNGS

@londonngs

From the tiniest to the largest, London gardens offer exceptional diversity. Hidden behind historic houses in Spitalfields are exquisite tiny gardens, while on Kingston Hill there are 9 acres of landscaped Japanese gardens.

The oldest private garden in London boasts 5 acres, while the many other historic gardens within these pages are smaller – some so tiny there is only room for a few visitors at a time – but nonetheless full of innovation, colour and horticultural excellence.

London allotments have attracted television cameras to film their productive acres, where exotic Cape gooseberries, figs, prizewinning roses and even bees all thrive thanks to the skill and enthusiasm of city gardeners.

The traditional sit comfortably with the contemporary in London – offering a feast of elegant borders, pleached hedges, topiary, gravel gardens and the cooling sound of water – while to excite the adventurous there are gardens on barges and green roofs to explore.

The season stretches from April to October, so there is nearly always a garden to visit somewhere in London. Our gardens opening this year are the beating heart of the capital just waiting to be visited and enjoyed.

LONDON GARDENS LISTED BY POSTCODE

Inner London Postcodes

E and EC London

Spitalfields Gardens E1
26 College Gardens E4
Lower Clapton Gardens E5
London Fields Gardens E8
333 Victoria Park Road, Flat 2 E9
17 Greenstone Mews E11
37 Harold Road E11
Aldersbrook Gardens L12
12 Western Road E13
87 St Johns Road E17
83 Cowslip Road E18
25 Mulberry Way E18
Amwell Gardens EC1
The Inner and Middle Temple Gardens EC4

N and NW London

26 Arlington Avenue N1
25 Arlington Square N1
Arlington Square Gardens N1
Barnsbury Group N1
4 Canonbury Place N1
De Beauvoir Gardens N1
41 Ecclesbourne Road N1
91 Englefield Road N1
58 Halliford Street N1
57 Huntingdon Street N1
King Henry's Walk Garden N1
2 Lonsdale Square N1
Malvern Terrace Gardens N1
5 Northampton Park N1
19 St Peter's Street N1
131 Southgate Road N1
36 Thornhill Square N1
66 Abbots Gardens N2
7 Deansway N2
12 Lauradale Road N2
24 Twyford Avenue N2
31 Hendon Avenue N3
18 Park Crescent N3
32 Highbury Place N5
19 Cholmeley Park N6
Southwood Lodge N6
9 View Road N6
5 Blackthorn Av, Apartment 5 N7
9 Furlong Road N7

10 Furlong Road N7
20 Furlong Road N7
33 Huddleston Road N7
1a Hungerford Road N7
23 & 24b Penn Road N7
19 Coolhurst Road N8
12 Fairfield Road N8
11 Park Avenue North N8
12 Warner Road N8
35 Weston Park N8
55 Dukes Avenue N10
Princes Avenue Gardens N10
5 St Regis Close N10
25 Springfield Avenue N10
33 Wood Vale N10
92 Brownlow Road N11
94 Brownlow Road N11
9 Churston Gardens N11
Golf Course Allotments N11
9 Shortgate N12
11 Shortgate N12
2 Conway Road N14
70 Farleigh Road N16
53 Manor Road N16
15 Norcott Road N16
36 Ashley Road N19
24 Langton Avenue N20
21 Oakleigh Park South N20
20 Hillcrest N21
1 Wades Grove N21
10 York Road N21
Railway Cottages N22
Wolves Lane Horticultural Centre N22
The Gable End Gardens NW1
Garden of Medicinal Plants, Royal College of Physicians NW1
69 Gloucester Crescent NW1
70 Gloucester Crescent NW1
The Holme NW1
36 Park Village East NW1
93 Tanfield Avenue NW2
Willesden Green Gardens NW2
Marie Curie Hospice, Hampstead NW3
The Mysteries of Light Rosary Garden NW5
Highwood Ash NW7
Oakwood Gardens NW11
86 Willifield Way NW11
74 Willifield Way NW11

SE and SW London

Garden Barge Square at Tower Bridge Moorings SE1
The Garden Museum SE1
Lambeth Palace SE1
49 Lee Road SE3
33 Newman Road SE3
24 Grove Park SE5
Lee Group SE12
41 Southbrook Road SE12
28 Granville Park SE13

101 Pepys Road SE14
Choumert Square SE15
Lyndhurst Square Garden Group SE15
4 Becondale Road SE19
103 and 105 Dulwich Village SE21
Gardens of Court Lane SE21
38 Lovelace Road SE21
Peckham Rye Group SE22
58 Cranston Road SE23
Forest Hill Garden Group SE23
39 Wood Vale SE23
5 Burbage Road SE24
28 Ferndene Road SE24
2 Shardcroft Avenue SE24
South London Botanical Institute SE24
Stoney Hill House SE26
27 Thorpewood Avenue SE26
7 Norwood Park Road SE27
Cadogan Place South Garden SW1
Eccleston Square SW1
Spencer House SW1
Brixton Water Lane Gardens SW2
Chelsea Physic Garden SW3
51 The Chase SW4
4 Franconia Road SW4
1 Maple Close SW4
Royal Trinity Hospice SW4
35 Turret Grove SW4
152a Victoria Rise, SW4
The Hurlingham Club SW6
10 Streatham Common South SW16
36 Melrose Road SW18
61 Arthur Road SW19
97 Arthur Road SW19
123 South Park Road SW19
11 Ernle Road SW20
Paddock Allotments & Leisure Gardens SW20

W and WC London

Rooftopvegplot W1
Hyde Park Estate Gardens W2
34 Buxton Gardens W3
41 Mill Hill Road W3
65 Mill Hill Road W3
Zen Garden at Japanese Buddhist Centre W3
Chiswick Mall Gardens W4
The Orchard W4
36 Park Road W4
56 Park Road W4
38 York Road W5
10 Loris Road W6
Maggie's West London W6
27 St Peters Square W6

1 York Close W7
Edwardes Square W8
57 St Quintin Avenue W10
Arundel & Elgin Gardens W11
Arundel & Ladbroke Gardens W11
12 Lansdowne Road W11
49 Loftus Road W12
14 Doughty Street WC1
57 Tonbridge House WC1

Outer London postcodes

45 Cotswold Way EN2
Oak Farm/Homestead EN2
West Lodge Park EN4
190 Barnet Road EN5
45 Great North Road EN5
55 College Road HA3
42 Risingholme Road HA3
1 Manningtree Road HA4
470 Pinner Road HA5
4 Ormonde Road HA6
Hornbeams HA7
Perth Road Garden IG2
74 Glengall Road IG8
7 Woodbines Avenue KT1
The Watergardens KT2
The Circle Garden KT3
Berrylands Gardens KT5
15 Catherine Road KT6
3 Elmbridge Lodge KT7
Hampton Court Palace KT8
5 Pemberton Road KT8
61 Wolsey Road KT8
239a Hook Road KT9
40 Ember Lane KT10
9 Imber Park Road KT10
40 The Crescent SM2
Maggie's SM2
7 St George's Road TW1
20 Beechwood Avenue TW9
Kew Green Gardens TW9
Marksbury Avenue Gardens TW9
28 Taylor Avenue TW9
Trumpeters' House & Sarah's Garden TW9
31 West Park Road TW9
Ormeley Lodge TW10
Petersham House TW10
Stokes House TW10
16 Links View Road TW12
9 Warwick Close TW12
Wensleydale Road Gardens, Hampton TW12
Dragon's Dream UB8
Church Gardens UB9
Swakeleys Cottage, 2 The Avenue UB10
Silverwood WD3

OPENING DATES

All entries subject to change. For latest information check www.ngs.org.uk

February

Snowdrop Festival
Sunday 7th
NEW Hornbeams HA7

March

Saturday 27th
4 Canonbury Place, N1

Sunday 28th
74 Glengall Road, IG8

April

Thursday 1st
♦ Chelsea Physic Garden, SW3

Sunday 11th
NEW 39 Wood Vale, SE23

Wednesday 14th
NEW 39 Wood Vale, SE23

Saturday 17th
Hyde Park Estate Gardens, W2

Sunday 18th
19 Coolhurst Road, N8
Edwardes Square, W8
Malvern Terrace Gardens, N1
Petersham House, TW10
33 Wood Vale, N10
7 Woodbines Avenue, KT1

Wednesday 21st
36 Thornhill Square, N1

Thursday 22nd
♦ Hampton Court Palace, KT8

Saturday 24th
11 Ernle Road, SW20
Maggie's West London, W6

Sunday 25th
Arundel & Ladbroke Gardens, W11

51 The Chase, SW4
11 Ernle Road, SW20
101 Pepys Road, SE14
South London Botanical Institute, SE24

Tuesday 27th
51 The Chase, SW4

May

Sunday 2nd
NEW 7 Deansway, N2
5 St Regis Close, N10
Southwood Lodge, N6

Monday 3rd
King Henry's Walk Garden, N1

Wednesday 5th
12 Lansdowne Road, W11

Saturday 8th
NEW Maggie's, SM2

Sunday 9th
NEW 19 Cholmeley Park, N6
19 Coolhurst Road, N8
Eccleston Square, SW1
11 Ernle Road, SW20
Highwood Ash, NW7
27 St Peters Square, W6
86 Underhill Road, SE22
The Watergardens, KT2
West Lodge Park, EN4
38 York Road, W5

Wednesday 12th
57 Huntingdon Street, N1

Saturday 15th
The Circle Garden, KT3
NEW 9 View Road, N6

Sunday 16th
5 Burbage Road, SE24
58 Halliford Street, N1
Princes Avenue Gardens, N10
NEW 42 Risingholme Road HA3
Stoney Hill House, SE26

Monday 17th
Lambeth Palace, SE1

Saturday 22nd
Cadogan Place South Garden
The Hurlingham Club, SW6

Sunday 23rd
61 Arthur Road, SW19
Arundel & Elgin Gardens, W11
190 Barnet Road, EN5
Chiswick Mall Gardens, W4
The Circle Garden, KT3
Dragon's Dream, UB8
55 Dukes Avenue, N10
Forest Hill Garden Group, SE23
9 Furlong Road, N7
NEW 10 Furlong Road
NEW 20 Furlong Road, N7
Garden Barge Square at Tower Bridge Moorings, SE1
Kew Green Gardens, TW9
London Fields Gardens, E8
10 Loris Road, W6
53 Manor Road, N16
36 Melrose Road, SW18
15 Norcott Road, N16
21 Oakleigh Park South, N20
7 St George's Road, TW1
19 St Peter's Street, N1
2 Shardcroft Avenue, SE24
12 Western Road, E13

Saturday 29th
NEW Lee Group, SE12
16 Links View Road, TW12
NEW The Mysteries of Light Rosary Garden, NW5

Sunday 30th
36 Ashley Road, N19
40 Ember Lane, KT10
91 Englefield Road, N1
12 Fairfield Road, N8
70 Farleigh Road, N16
31 Hendon Avenue, N3
9 Imber Park Road, KT10
Kew Green Gardens, TW9
16 Links View Road, TW12
41 Southbrook Road, SE12
27 Thorpewood Avenue, SE26

Monday 31st
36 Ashley Road, N19
Church Gardens, UB9

June

Saturday 5th
NEW Wensleydale Road Gardens, Hampton, TW12
Zen Garden at Japanese Buddhist Centre, W3

Sunday 6th
Amwell Gardens, EC1R
103 and 105 Dulwich Village, SE21
Barnsbury Group, N1
Berrylands Gardens, KT5
Brixton Water Lane Gardens, SW2
Choumert Square, SE15
NEW 26 College Gardens, E4
De Beauvoir Gardens, N1
69 Gloucester Crescent, NW1
70 Gloucester Crescent, NW1
20 Hillcrest, N21
1a Hungerford Road, N7
NEW 24 Langton Avenue, N20
2 Lonsdale Square, N1
Marksbury Avenue Gardens, TW9
Oak Farm/Homestead, EN2
4 Ormonde Road, HA6
36 Park Road, W4
NEW 36 Park Village East, NW1
Peckham Rye Group, SE22
23 & 24b Penn Road, N7
123 South Park Road, SW19
Stokes House, TW10
1 Wades Grove, N21
61 Wolsey Road, KT8
Zen Garden at Japanese Buddhist Centre, W3

Wednesday 9th
26 Arlington Avenue, N1
25 Arlington Square, N1

Friday 11th
239a Hook Road, KT9

Saturday 12th
Marie Curie Hospice, Hampstead, NW3
5 Northampton Park, N1
Spitalfields Gardens, E1

Sunday 13th
66 Abbots Gardens, N2
97 Arthur Road, SW19
NEW 55 College Road,
    HA3
40 The Crescent, SM2
28 Granville Park, SE13
37 Harold Road, E11
49 Lee Road, SE3
49 Loftus Road, W12
Lower Clapton Gardens,
    E5
Lyndhurst Square Garden
    Group, SE15
5 Pemberton Road, KT8
NEW 333 Victoria Park
    Road, Flat 2, E9
12 Warner Road, N8

Tuesday 15th
The Inner and Middle
    Temple Gardens, EC4

Saturday 19th
4 Franconia Road, SW4
Zen Garden at Japanese
    Buddhist Centre, W3

Sunday 20th
Ormeley Lodge, TW10
18 Park Crescent, N3
5 St Regis Close, N10
Trumpeters' House &
    Sarah's Garden, TW9
74 Willifield Way, NW11
10 York Road, N21
Zen Garden at Japanese
    Buddhist Centre, W3

Saturday 26th
The Holme, NW1
Paddock Allotments
    & Leisure Gardens,
    SW20

Sunday 27th
5 Blackthorn Av,
    Apartment 5, N7
NEW 15 Catherine Road,
    KT6
NEW The Gable End
    Gardens
Gardens of Court Lane,
    SE21
32 Highbury Place, N5
The Holme, NW1
38 Lovelace Road, SE21
25 Mulberry Way, E18
11 Park Avenue North,
    N8
33 Weyman Road, SE3
7 Woodbines Avenue,
    KT1

July

Saturday 3rd
33 Huddleston Road, N7
Rooftopvegplot, W1

Sunday 4th
NEW Aldersbrook
    Gardens, E12
Arlington Square
    Gardens, N1
51 The Chase, SW4
14 Doughty Street,
    WC1N
NEW 3 Elmbridge Lodge,
    KT7
NEW 28 Ferndene Road,
    SE24
NEW 1 Maple Close, SW4
NEW Oakwood Gardens,
    NW11
Railway Cottages, N22
Rooftopvegplot, W1
57 St Quintin Avenue,
    W10
NEW 9 Shortgate, N12
NEW 11 Shortgate
131 Southgate Road, N1
25 Springfield Avenue,
    N10
Swakeleys Cottage, 2
    The Avenue, UB10
57 Tonbridge House,
    WC1H
NEW 152a Victoria Rise,
    SW4

Sunday 11th
83 Cowslip Road, E18
41 Ecclesbourne Road,
    N1
17 Greenstone Mews,
    E11
NEW 7 Norwood Park
    Road, SE27
28 Taylor Avenue, TW9
24 Twyford Avenue, N2
31 West Park Road, TW9

Thursday 15th
♦ Hampton Court
    Palace, KT8

Sunday 18th
45 Cotswold Way, EN2
18 Park Crescent, N3
56 Park Road, W4
57 St Quintin Avenue,
    W10
35 Turret Grove, SW4
35 Weston Park, N8
38 York Road, W5

Saturday 24th
NEW 34 Buxton Gardens,
    W3

Sunday 25th
NEW 92 Brownlow Road
94 Brownlow Road, N11
NEW 34 Buxton Gardens,
    W3
87 St Johns Road, E17
5 St Regis Close, N10
93 Tanfield Avenue, NW2
86 Willifield Way, NW11

Monday 26th
87 St Johns Road, E17

Saturday 31st
NEW Perth Road Garden
    IG2

August

Sunday 1st
45 Great North Road, EN5
4 Manningtree Road,
    HA4
NEW 9 Warwick Close,
    TW12

Saturday 7th
4 Becondale Road, SE19
The Holme, NW1

Sunday 8th
4 Becondale Road, SE19
20 Beechwood Avenue,
    TW9
51 The Chase, SW4
NEW 58 Cranston Road,
    SE23
69 Gloucester Crescent,
    NW1
70 Gloucester Crescent,
    NW1
The Holme, NW1

Wednesday 11th
Garden of Medicinal
    Plants, Royal College
    of Physicians, NW1

Saturday 21st
1 York Close, W7

Sunday 22nd
41 Mill Hill Road, W3
65 Mill Hill Road, W3
NEW Silverwood WD3
1 York Close, W7

Sunday 29th
NEW 10 Streatham
    Common South, SW16

*In our first year
609 private gardens
opened their gates
to all, for the modest
sum of one shilling.
Today the National
Garden Scheme retains
that combination
of inclusivity and
affordability*

September

Sunday 5th
9 Churston Gardens, N11
2 Conway Road, N14
Golf Course Allotments,
    N11
24 Grove Park, SE5
NEW Hornbeams HA7
12 Lauradale Road, N2
NEW 42 Risingholme
    Road HA3
Royal Trinity Hospice,
    SW4

Saturday 11th
♦ The Garden Museum,
    SE1

Sunday 12th
Oak Farm/Homestead,
    EN2
The Orchard, W4
4 Ormonde Road, HA6

Sunday 19th
53 Manor Road, N16
NEW Willesden Green
    Gardens, NW2
NEW Wolves Lane
    Horticultural Centre,
    N22

Sunday 26th
470 Pinner Road, HA5

October

Sunday 3rd
51 The Chase, SW4

Sunday 17th
The Watergardens, KT2

Sunday 24th
West Lodge Park, EN4

By Arrangement

Arrange a personalised garden visit on a date to suit you. See individual garden entries for full details.

36 Ashley Road, N19

190 Barnet Road, EN5

4 Becondale Road, SE19

5 Burbage Road, SE24

51 The Chase, SW4

NEW 7 Deansway, N2

NEW 3 Elmbridge Lodge, KT7

69 Gloucester Crescent, NW1

70 Gloucester Crescent, NW1

45 Great North Road, EN5

27 Horniman Drive, Forest Hill Garden Group, SE23

1a Hungerford Road, N7

69 Kew Green, Kew Green Gardens, TW9

71 Kew Green, Kew Green Gardens, TW9

84 Lavender Grove, E8, London Fields Gardens, E8

49 Loftus Road, W12

53 Manor Road, N16

53 Mapledene Road, London Fields Gardens, E8

41 Mill Hill Road, W3

65 Mill Hill Road, W3

NEW The Mysteries of Light Rosary Garden, NW5

15 Norcott Road, N16

21 Northchurch Terrace, De Beauvoir Gardens, N1

NEW Oakwood Gardens, NW11

7 St George's Road, TW1

27 St Peters Square, W6

57 St Quintin Avenue, W10

5 St Regis Close, N10

41 Southbrook Road, SE12

Southwood Lodge, N6

Stokes House, TW10

93 Tanfield Avenue, NW2

58A Teignmouth Road, NW2 4DX, Willesden Green Gardens, NW2

24 Twyford Avenue, N2

NEW 333 Victoria Park Road, Flat 2, E9

West Lodge Park, EN4

33 Weyman Road, SE3

86 Willifield Way, NW11

74 Willifield Way, NW11

33 Wood Vale, N10

THE GARDENS

66 ABBOTS GARDENS, N2

East Finchley, N2 0JH. Stephen & Ruth Kersley. *8 mins walk from rear exit East Finchley tube on Causeway to East End Rd. 2nd L into Abbots Gardens. 143 stops at Abbots Gardens on East End Rd. 102, 263 & 234 all go to East Finchley High Rd.* **Sun 13 June (2-6). Adm £4, chd free. Light refreshments.** Combination of grass & glass: designed for year round interest with calm yet dramatic environment created through plant form, texture, asymmetrical geometry, water features & restricted colour palette. Stephen studied Garden design at Capel, Ruth is a glass artist. Glass amphorae and mosaics catch the eye amongst grasses, ornamental shrubs & perennials; rose bedecked archway to quiet space with vegetable plot, silver birches, slate pebble fountain.

GROUP OPENING

NEW **ALDERSBROOK GARDENS, E12**

Wanstead, E12 5ES. *Empress Ave is a turning off Aldersbrook Rd. 101 bus from Manor Park or Wanstead Stns. From Manor Park Stn take the 3rd R off Aldersbrook Rd, from Wanstead, drive past St Gabriel's Church, take the 6th turning on L.* **Sun 4 July (12-5). Combined adm £7, chd free. Home-made teas at Empress Avenue, Clavering Rd and Park Rd.**

> **NEW** **1 CLAVERING ROAD**
> Theresa Harrison.
>
> **NEW** **4 EMPRESS AVENUE**
> Ruth Martin.
>
> **NEW** **21 PARK ROAD**
> Theresa O'Driscoll & Barry Reeves.
>
> **NEW** **47 ST MARGARETS ROAD**
> Jane Karavasili.

Four different gardens situated on the Aldersbrook Estate - between Wanstead Park and Wanstead Flats. Park Rd is a colour themed garden with evergreen shrubs for year round structure and a vine covered pergola leading to a vegetable area. The St Margaret's Road garden has a small

front south facing garden where as well as off street parking, tomatoes and chillies grow, the back garden is designed with a theme of circles and curves with closely planted borders. 1 Clavering Rd is an end of terrace garden where incremental space to the side has been adapted to create a kitchen garden and chicken coop - excess produce is eagerly received by 5 resident hens. Borders and beds contain variety of planting. At 4 Empress Avenue a largish garden is divided in two with a vegetable growing area, areas to attract more wildlife incl pond, 2 mixed borders; one with hot colours and one with white planting.

GROUP OPENING

AMWELL GARDENS, EC1R

EC1R 1YE. *South Islington. Tube: Angel, 5 mins walk. Buses: 19, 38 to Rosebery Ave; 30, 73 to Pentonville Rd.* **Sun 6 June (2-5). Combined adm £8, chd free. Home-made teas at 11 Chadwell St.**

> **11 CHADWELL STREET**
> Mary Aylmer & Andrew Post.
>
> **LLOYD SQUARE**
> Lloyd Square Garden Committee.
>
> **27 MYDDELTON SQUARE**
> Sally & Rob Hull.
>
> **NEW RIVER HEAD**
> NRH Residents.
>
> **NEW** **4 WHARTON STREET**
> Barbara Holliman.
>
> **49 WHARTON STREET**
> David Sulkin & Geoffrey Milton.

The Amwell Gardens Group is in a secluded corner of Georgian Clerkenwell. Contrasting gardens incl Lloyd Square, a mature space with drifting borders in the centre of the Lloyd Baker Estate and the nearby gardens surrounding the historic New River Head, where a stylish fountain & pergola have replaced the outer pond, which distributed fresh water to London. 27 Myddelton Square is a courtyard garden with lush cottage garden planting, complementing this elegant terraced setting & winning Best Back Garden award for Islington in Bloom competition 2020. New 4 Wharton St demonstrates imaginative planting in a N-facing garden while opposite, 49 Wharton St boasts a fruiting apricot tree, wisteria & bandstand.11 Chadwell St is a

164 Court Lane

tranquil setting for tea & delicious cakes amongst roses & honeysuckle. Live music on the bandstand at 49 Wharton Street.

26 ARLINGTON AVENUE, N1
N1 7AX. Thomas Blaikie. *South Islington.* **Evening opening Wed 9 June (6-8). Adm £4, chd free. Wine. Also open 25 Arlington Square. Opening with Arlington Square Gardens on Sun 4 July.**
A friend once said 'Your garden is like a tiny corner of some much grander horticultural vision'. I assume it was a compliment. Somehow, over 150 plants (approx) are crammed into the minute space. Visitors always like my slender willow tree (Salix Exigua). In June, the highlights should be a good selection of old roses, climbing & shrub, big dramatic alliums, lupins

& tall spires of verbascum. A roses evening.

25 ARLINGTON SQUARE, N1
N1 7DP. Michael Foley. *South Islington.* **Evening opening Wed 9 June (6-8). Adm £4, chd free. Wine. Also open 26 Arlington Avenue. Opening with Arlington Square Gardens on Sun 4 July.**
Plants are my passion; they pack the central bed & spill out onto the other borders. Camellias, hydrangeas, clematis & roses dominate with other unusual plants. The exuberance of the roses, like Compassion, Ferdinand Pichard, Mutabilis, Rosa Mundi, Gentle Hermione, Sir John Mills & the City of York on the arch, hits you like a painter's palette with its flush of the year. A roses evening.

"Unfortunately, cancer was not in lockdown in 2020. The continued support of our long-standing and valued partner, the National Garden Scheme is more important than ever."
Macmillan

41 Southbrook Road

GROUP OPENING

ARLINGTON SQUARE GARDENS, N1

N1 7DP.
www.arlingtonassociation.org.uk.
South Islington. Off New North Rd via Arlington Ave or Linton St. Buses: 21, 76, 141, 271. **Sun 4 July (2-5.30). Combined adm £8, chd free. Home-made teas at St James' Vicarage, 1A Arlington Square.**

26 ARLINGTON AVENUE, N1
Thomas Blaikie.
(See separate entry)

21 ARLINGTON SQUARE
Alison Rice.

25 ARLINGTON SQUARE, N1
Michael Foley.
(See separate entry)

27 ARLINGTON SQUARE
Geoffrey Wheat & Rev Justin Gau.

30 ARLINGTON SQUARE
James & Maria Hewson.

5 REES STREET
Gordon McArthur & Paul Thompson-McArthur.

ST JAMES' VICARAGE, 1A ARLINGTON SQUARE
John & Maria Burniston.

Behind the early Victorian facades of Arlington Square and Arlington Avenue are 7 contrasting town gardens; 6 plantsmen's gardens & a delightful spacious garden at the Vicarage with an impressive herbaceous border created over the last few years, & mature London Plane trees. The group reflects the diverse tastes and interests of each garden owner, who know each other through the community gardening of Arlington Square. It is hard to believe you are minutes from the bustle of the City of London. Live music at St James' Vicarage, 1A Arlington Square.

🍵

61 ARTHUR ROAD, SW19
Wimbledon, SW19 7DN. Daniela McBride. *Tube: Wimbledon Park, then 8 mins walk. Mainline: Wimbledon, 18 mins walk.* **Sun 23 May (2-6). Adm £5, chd free. Home-made teas.**
This steeply sloping garden comprises woodland walks, filled with flowering shrubs and ferns. Azaleas, Rhododendrons and Acers bring early season colour; in early summer the focus is the many roses grown around the garden. Partial wheelchair access to top lawn and terrace only, steep slopes elsewhere.

♿ 🍵

97 ARTHUR ROAD, SW19
Wimbledon, SW19 7DP. Tony & Bella Covill. *Wimbledon Park tube, then 200yds up hill on R.* **Sun 13 June (2-6). Adm £5, chd free. Light refreshments.**
½ acre garden of an Edwardian house. Garden est. for more than 25yrs, constantly evolving with a large

variety of plants and shrubs. It has grown up around several lawns with ponds and fountains, encouraging an abundance of wildlife and a bird haven. A beautiful place with much colour, foliage and texture. New gravel garden, planting to attract butterflies. Wild meadow.

ARUNDEL & ELGIN GARDENS, W11

Kensington Park Road, Notting Hill, W11 2JD. Residents of Arundel Gardens & Elgin Crescent, www.arundelandelgingardens.org. *Entrance opp 174 Kensington Park Rd, Nearest tube within walking distance: Ladbroke Grove (3mins) or Notting Hill (8mins). Buses: 52, 452, 23, 228 all stop opp garden entrance.* **Sun 23 May (2-6). Adm £4, chd free. Home-made teas.**
A friendly and informal garden square with mature and rare trees, plants and shrubs laid out according to the original Victorian design of 1862, one of the best preserved gardens of the Ladbroke estate. The central hedged garden area is an oasis of tranquillity with extensive and colourful herbaceous borders. The garden incl several topiary hedges, a rare Mulberry tree, a pergola and benches from which the vistas can be enjoyed. Gardeners Chris Jelston & Anna Park. Play areas for young children.

ARUNDEL & LADBROKE GARDENS, W11

Kensington Park Road, Notting Hill, W11 2PT. Arundel & Ladbroke Gardens Committee, www.arundelladbrokegardens.co.uk. *Entrance on Kensington Park Rd, between Ladbroke & Arundel Gardens. Tube: Notting Hill Gate or Ladbroke Grove. Buses: 23, 52, 228, 452. Alight at stop for Portobello Market/Arundel Gardens.* **Sun 25 Apr (2-6). Adm £4, chd free. Home-made teas.**
This private communal garden is one of the few that retains its attractive mid Victorian design of lawns and winding paths. A woodland garden at its peak in spring, with rhododendrons, flowering dogwoods, early roses, bulbs, ferns and rare exotica. Live music on the lawn during tea. Playground for small children. A few steps and gravel paths to negotiate.

36 ASHLEY ROAD, N19

N19 3AF. Alan Swann & Ahmed Farooqui, swann.alan@googlemail.com. *Between Stroud Green & Crouch End. Underground: Archway or Finsbury Park. Overground: Crouch Hill. Buses: 210 or 41 from Archway to Hornsey Rise. W7 from Finsbury Park to Heathville Road. Car: Free parking in Ashley Road at weekends.* **Sun 30, Mon 31 May (2-6). Adm £3.50, chd free. Home-made teas. incl vegan and gluten free options. Visits also by arrangement May to Sept for groups of up to 30.**
A lush town garden rich in textures, colour and forms. At its best in late spring as Japanese maple cultivars display great variety of shape and colour whilst ferns unfurl fresh, vibrant fronds over a tumbling stream and alpines and clematis burst into flower on the rockeries and pergola. A number of micro habitats incl ferneries, bog garden, stream and ponds, rockeries, alpines and shade plantings. Young ferns and plants propagated from specimens in the garden for sale. Pop-up tearoom with selection of cakes and home-brewed ginger beer with indoor seating and views over the garden. Entrance to the garden is down a flight of 7 steps.

190 BARNET ROAD, EN5

Arkley, Barnet, EN5 3LF. Hilde & Lionel Wainstein, 07949764007, hildewainstein@hotmail.co.uk. *1m S of A1, 2m N of High Barnet tube. Garden located on corner of A411 Barnet Rd & Meadowbanks cul-de-sac. Nearest tube: High Barnet, then 107 bus, Glebe Lane stop. Ample unrestricted roadside parking.* **Sun 23 May (2-6). Adm £4, chd free. Home-made teas. Visits also by arrangement May to Sept for groups of up to 30.**
The Upcycled garden. Garden designer's walled garden, 90ft x 36ft. Modern, idiosyncratic design, year round interest. Flowing herbaceous drifts around a central pond. Upcycled containers, recycled objects and home-made sculptures. Copper trellis divides space into contrasting areas. Garden continues to evolve as planted areas are expanded. Rusty tin can 'Derek Jarman' style garden. Selection of home-made cakes worthy of Mary Berry! The favourite at our last opening was raspberry and white chocolate layer cake. Gluten

free cakes also available. Homemade jams and a range of interesting plants for sale, all propagated from the garden. Single steps within garden.

GROUP OPENING

BARNSBURY GROUP, N1

N1 1DB. *Barnsbury, London N1. Tube: King's Cross, Caledonian Rd or Angel. Overground: Caledonian Rd & Barnsbury. Buses: 17, 91, 259 to Caledonian Rd.* **Sun 6 June (2-6). Combined adm £8, chd free. Home-made teas at 57 Huntingdon Street N1 1BX. Also open 2 Lonsdale Square.**

◆ BARNSBURY WOOD
London Borough of Islington, ecologycentre@islington.gov.uk.

44 HEMINGFORD ROAD
Peter Willis & Haremi Kudo.

57 HUNTINGDON STREET, N1
Julian Williams.
(See separate entry)

36 THORNHILL SQUARE, N1
Anna & Christopher McKane.
(See separate entry)

Within Barnsbury's historic Georgian squares & terraces, discover these four contrasting spaces. Barnsbury Wood is London's smallest nature reserve, a hidden secret & Islington's only site of mature woodland, a tranquil oasis of wild flowers & massive trees just minutes from Caledonian Road. Wildlife info available. The gardens have extensive collections of unusual plants; 57 Huntingdon St is a secluded garden room - an understorey of silver birch & hazel, ferns, native perennials & grasses and two container ponds to encourage wildlife. 44 Hemingford Road is a small, dense composition of trees (some unusual), shrubs, perennials & lawns & a small pond. 36 Thornhill Square, a 120 ft garden with a country atmosphere, filled with old & new roses & many herbaceous perennials. A bonsai collection will astound! These gardens have evolved over many years & show what can be achieved in differing spaces with the right plants growing in the right conditions, surmounting the difficulties of dry walls & shade. Plants for sale at 36 Thornhill Square.

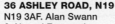

4 BECONDALE ROAD, SE19

Gipsy Hill, Norwood, SE19 1QJ. Christopher & Wendy Spink, cjkitdesign@yahoo.co.uk. *Off Gipsy Hill. Nearest stn Gipsy Hill. Buses 3 & 322. Some parking on Becondale rd.* **Evening opening Sat 7 Aug (7.30-10.30). Adm £7. Wine. Sun 8 Aug (1-6). Adm £4.50, chd free. Home-made teas. Also open 58 Cranston Road. Visits also by arrangement June to Sept for groups of up to 10.**

This garden takes vertical planting to new heights - but is not for those who are afraid to walk a gangplank. It is packed with a mass of exotic and rich planting with a Mediterranean feel from Bougainvillea to bananas and palms to plumbago. Steeply sloping, it maximises every bit of height with plants cascading over high rise balconies and dropping down to a theatrically styled well of a garden. Mediterranean planting on four levels.

Regret, with narrow paths/steep steps, garden not suitable for small children or mobility challenged.

20 BEECHWOOD AVENUE, TW9

Kew, Richmond, TW9 4DE. Dr Laura de Beden, www.lauradebeden.co.uk. *Within walking distance of Kew Gardens Tube Station on E side exit.* **Evening opening Sun 8 Aug (5.30-7). Adm £5.50, chd free. Wine.**

Delightful town garden minutes away from Royal Botanic Gardens and Kew Retail Park. Minimalist layout by the designer owner offsets exquisite favourite plant combinations. A writing shed holds pride of place as safe refuge & main idea production centre. Topiary, pots, sculpture, surprises (the latest in the new small fernery) and good humour are all on offer for an inspiring innovative visit.

GROUP OPENING

BERRYLANDS GARDENS, KT5

Berrylands, Surbiton, KT5 9AF. *2m S of Kingston-upon-Thames. From A3 take A240 joining Ewell Rd. Take Hollyfield Rd on R, cross King Charles Rd into Alexandra Dr. Map to other gardens from here.* **Sun 6 June (2-5). Combined adm £5, chd free. Home-made teas at 1 The Crest.**

68 ALEXANDRA DRIVE
Andy Hutchings.

1 THE CREST
Robert & Julia Humphries.

64 PINE GARDENS
Barbara Hutchings.

A selection of three gardens all within 10 mins walk of each other, all owned by different members of the same family. The gardens are all very varied ; country cottage, hidden garden, giants head, even a shed made out of reclaimed materials. Have an enjoyable afternoon being nosey, buying plants and sampling my sister's fantastic cakes. We look forward to seeing you.

5 BLACKTHORN AV, APARTMENT 5, N7

N7 8AQ. Juan Carlos Cure Hazzi. *Barnsbury. 5 min walk from Highbury & Islington Stn. Building is on S side of Arundel Sq.* **Sun 27 June (11-5). Adm £5, chd free. Pre-booking essential, please visit www.ngs. org.uk for information & booking. Light refreshments. Also open 32 Highbury Place.**

Small patio garden with beautiful connection with the house. Plenty of colour, texture and year-round interest with lush tropical, sub-tropical and temperate plants. Featured in BBC Gardeners' World & several gardening magazines. A finalist in the BBC's Small Space category competition.

The Mysteries of Light Rosary Garden

All entries are subject to change. For the latest information check ngs.org.uk

GROUP OPENING

BRIXTON WATER LANE GARDENS, SW2

Brixton, SW2 1QB. *Tube: Brixton. Mainline: Herne Hill, both 10 mins. Buses: 3, 37, 196 or 2, 415, 432 along Tulse Hill.* **Sun 6 June (2-5.30). Combined adm £5, chd free. Tea.**

60 BRIXTON WATER LANE
Caddy & Chris Sitwell.

62 BRIXTON WATER LANE
Daisy Garnett & Nicholas Pearson.

Two 90ft gardens backing onto Brockwell Park with original apple trees from the old orchard. No 60 has a large garden with floral borders and a mature wisteria covering the house. Strong colour comes from laburnum and lilac under which teas will be served. Number 62 is a country garden with exuberant borders of soft colours, a productive greenhouse and a mass of pots on the terrace. Plenty of colour from old fashioned roses, peonies and other perennials.

NEW 92 BROWNLOW ROAD
Bounds Green, N11 2BS. Donna Armstrong & Toby Barrance. *Back of the property. Entrance down side of house. Gate will be open for visitors on the day. Look for arrows.* **Sun 25 July (2-6). Combined adm with 94 Brownlow Road £5, chd free. Light refreshments. Also open 5 St Regis Close.**
Contemporary garden for entertaining and relaxation. Structured Yew hedging frames a raised circular lawn edged in Corten steel, surrounded by specimen trees and ornamental grasses, offering many points of interest throughout the garden. The design features a mixture of materials and playful use of color including sandstone patio and hardwood deck surrounded by a wall of Jasmine.

94 BROWNLOW ROAD, N11
Bounds Green, N11 2BS. Spencer Viner, www.northeleven.co.uk. *Close to N Circular. Tube: Bounds Green then 5 mins walk, direction N Circular. Corner of Elvendon Rd & Brownlow Rd.* **Sun 25 July (2-6). Combined adm with 92 Brownlow Road £5,** chd free. Light refreshments. Also open 5 St Regis Close.
A small courtyard for meditation created by designer owner transports the visitor to foreign places of imagination and tranquility. Features incl. reclaimed materials, trees, water, pergola, pleached limes. Pared back simplicity contrasts with newly created prairie planting in front of house on busy main road. Design and horticultural advice available.

5 BURBAGE ROAD, SE24
Herne Hill, SE24 9HJ. Crawford & Rosemary Lindsay, 020 7274 5610, rl@rosemarylindsay.com, www.rosemarylindsay.com. *Nr junction with Half Moon Lane. Herne Hill & N Dulwich mainline stns, 5 mins walk. Buses: 3, 37, 40, 68, 196, 468.* **Sun 16 May (11-5). Adm £4, chd free. Visits also by arrangement Mar to June.**
The garden of a member of The Society of Botanical Artists and regular writer for Hortus magazine. 150ft x 40ft with large and varied range of plants- many unusual. Herb garden, packed herbaceous borders for sun and shade, climbing plants, pots, terraces, lawns. Immaculate topiary. Gravel areas to reduce watering. All the box has been removed because of attack by blight and moth, and replaced with suitable alternatives to give a similar look. A garden that delights from spring through summer. See our website for what the papers say. Incl in The London Garden Book A-Z. Plants for sale.

NEW 34 BUXTON GARDENS, W3
Acton, W3 9LQ. Alexey Kuzmin. *10-15 min walk from: - Acton Town Station (Piccadilly Line) - Acton Main (Overground) - Acton Central (Overground) - West Acton (Central Line).* **Sat 24, Sun 25 July (12-5). Adm £5. Home-made teas.**
A 1,000 ft² of a bold take on the conventional, the garden that celebrates the black, purple and copper over the green. As if the chocolate has been 'spilled, split or sprayed' to frame the blue slate with black: the cottage garden, the meadow and the veg patch. Hear the black buzz on the banks of its tropical ponds with the waterfall. Meet the Black & Bold garden, a wee wild yet practical. Cascading ponds with goldfish, frogs and water lilies. 'Hanging Rock' - observation platform over the pond. 'Secret Patio'.

CADOGAN PLACE SOUTH GARDEN
Sloane Street, Chelsea, SW1X 9PE. The Cadogan Estate, www.cadogan.co.uk. *Entrance to garden opp 97 Sloane St.* **Sat 22 May (10-4). Adm £5, chd free. Home-made teas.**
Many surprises, unusual trees and shrubs are hidden behind the railings of this large London square. The first square to be developed by architect Henry Holland for Lord Cadogan at the end of C18, it was then called the London Botanic Garden. Mulberry trees planted for silk production at end of C17. Cherry trees, magnolias and bulbs are outstanding in spring. Beautiful 300 year old Black Mulberry Tree (originally planted to produce silk, but incorrect variety!). This was once the home of the Royal Botanic Garden. Now featuring a Bug Hotel & Children's Playground.

4 CANONBURY PLACE, N1
N1 2NQ. Mr & Mrs Jeffrey Tobias. *Highbury & Islington Tube & Overground. Buses: 271 to Canonbury Square. Located in old part of Canonbury Place, off Alwyne Villas, in a cul de sac.* **Sat 27 Mar (2.30-5.30). Adm £3.50, chd free. Home-made teas.**
Join us on the very first day of British Summer Time to enjoy the romance of early Spring flowering in our historic, secluded 100ft garden with its architectural stonework & ancient hidden fountain all echoing the 1780 house. Plenty of distanced seating and contactless payment available. Spectacular mature trees. Mostly pots, interesting shrubs & climbers. Artisan pastries and sourdough bread from the legendary Dusty Knuckle Bakery on sale (as supplied to Fortnum & Masons, Ottolenghi etc.). Wonderful home-made cakes.

NEW 15 CATHERINE ROAD, KT6
Surbiton, KT6 4HA. Malcolm Simpson & Stefan Gross. *5-10 min walk from Surbiton Station. Public transport is recommended. Note that, while car parking is not charged in the immed area at weekends, is also very limited.* **Sun 27 June (12-5). Combined adm with 7 Woodbines Avenue £7.50, chd free.**
A well loved town garden approx 40 ft by 70 ft with deep borders of shrubs and perennial planting and with sculptures, a small folly and a walled area.

51 THE CHASE, SW4

SW4 0NP. Mr Charles Rutherfoord & Mr Rupert Tyler, 02076270182, Charles@charlesrutherfoord.net, www.charlesrutherfoord.net. *Off Clapham Common Northside. Tube: Clapham Common. Buses: 137, 452.* **Sun 25 Apr (12-5). Light refreshments. Evening opening Tue 27 Apr (5-8). Light refreshments. Sun 4 July, Sun 8 Aug, Sun 3 Oct (12-5). Adm £4.50, chd free. Visits also by arrangement Mar to Oct for groups of 5 to 30.** Member of the Society of Garden Designers, Charles has created the garden over 35yrs. In 2015 the main garden was remodelled, to much acclaim. Spectacular in spring, when 2000 tulips bloom among Camellias, irises and tree peonies. Scented front garden. Rupert's geodetic dome shelters seedlings, succulents and subtropicals. Roses, Brugmansia, Hibiscus, & Dahlias later in the season.

◆ CHELSEA PHYSIC GARDEN, SW3

Swan Walk, SW3 4HS. Chelsea Physic Garden Company, 020 7352 5646, enquiries@ chelseaphysicgarden.co.uk, www.chelseaphysicgarden.co.uk. *Tube: Sloane Square (15 mins). Bus: 170. Parking: Battersea Park (charged). Entrance in Swan Walk off Royal Hospital Rd.* **For NGS: Thur 1 Apr (11-5). Adm £13.50, chd £9.50. Brunches, lunch and afternoon tea at The Physic Garden Café. For other opening times and information, please phone, email or visit garden website.** Explore London's oldest botanic garden situated in the heart of Chelsea for 350 years. With a unique living collection of around 5000 plants, this walled garden is a celebration of the importance of plants and their beauty. Highlights incl Europe's oldest pond rockery, the Garden of Edible and Useful Plants, the Garden of Medicinal Plants and the World Woodland Garden. Free tours available, led by knowledgeable guides. Walks, Talks and Workshops. Café, Shop, Toilets. Wheelchair access is via 66 Royal Hospital Rd.

GROUP OPENING

CHISWICK MALL GARDENS, W4

Chiswick Mall, Chiswick, W4 2PS. *Car: Towards Hogarth r'about, A4 (W) turn Eyot Grds S. Tube: Stamford Brook or Turnham Green. Buses: 27, 190, 267 & 391 to Young's Corner. From Chiswick High Rd or Kings St S under A4 to river.* **Sun 23 May (2-5). Combined adm £6, chd free. Tea at Eyot Gardens.**

16 EYOT GARDENS
Dianne Farris.

NEW MILLER'S COURT
Miller's Court Tenants Ltd.

ST PETERS WHARF
Barbara Brown.

SWAN HOUSE
Mr & Mrs George Nissen.

Gardens on or near the River Thames: a riverside garden in an artists' complex, a large walled garden with the emphasis on foliage and shade-loving plants, a small walled garden demonstrating the clever use of restricted space and a communal garden on the river bank with well planted borders.

NEW 19 CHOLMELEY PARK, N6

Highgate, N6 5EL. Rhian and Andrew Bliss. *Accessed from Causton Rd. From Highgate High St L down Cholmeley Park at Channing School. From Highgate Tube Stn either Highgate Ave to Peacock Walk or Archway Rd to Causton Rd.* **Sun 9 May (2-5). Adm £4, chd free. Home-made teas.** A lived-in and evolving family garden. Situated on a corner site, the plot is very varied as are the uses to which space has been put over time - roses replaced a climbing frame and the goal became a productive border. Recently updated to squeeze in a green house, potting shed and raised beds. Planting influenced by family's experiences (e.g. of living in Japan) as well as horticultural merit.

CHOUMERT SQUARE, SE15

Peckham, SE15 4RE. The Residents. *Off Choumert Grove. Trains from London Victoria, London Bridge, London Blackfriars, Clapham Junction to Peckham Rye; buses (12,* 36, 37, 63, 78, 171, 312, 345). Free Car park (1 min) in Choumert Grove. **Sun 6 June (1-6). Adm £4, chd free. Afternoon teas, a variety of home-made cakes & Pimms. Donation to St Christopher's Hospice.** About 46 mini gardens with maxi planting in Shangri-la situation that the media has described as a Floral Canyon, which leads to small communal secret garden. The day is primarily about gardens and sharing with others our residents' love of this little corner of the inner city; but it is also renowned for its demonstrable community spirit. Stalls in the style of a village fete and Live Music. The popular open gardens will combine this year with our own take on a village fete with home produce stalls, arts, crafts and music. No steps within the Square just a tiny step to a raised paved space in the communal garden area.

CHURCH GARDENS, UB9

Church Hill, Harefield, Uxbridge, UB9 6DU. Patrick & Kay McHugh, www.churchgardens.co.uk/. *From Harefield Village, continue for ¼ m down Church Hill. From A40 Uxbridge junction, follow signs to Harefield. Turn off Church Hill towards St Mary's Church.* **Mon 31 May (11-5). Adm £5, chd £2. Home-made teas.** Harefield's own 'secret garden'. C17 Renaissance walled gardens on the outskirts of Harefield incl a geometrically designed organic kitchen garden, 60m long herbaceous borders, trained fruit trees, fruit cage, alpines, herb garden, vine mount, paradise garden and an orchard with rare arcaded wall, dating back to the early 1600's. Unique opportunity to view ongoing restoration project.

9 CHURSTON GARDENS, N11

Bounds Green, N11 2NJ. Pauline Hamilton. *Tube Bounds Green or Overground Bowes Park both 10 mins walk. Buses 102, 184, 299 and 221 look for NGS arrows.* **Sun 5 Sept (2-6). Adm £4, chd free. Also open Golf Course Allotments.** A charming garden designed by the Capel Manor trained owner. Elegant hard landscaping combines with informal pretty planting. Specimen trees, evergreens and grasses are interspersed with bulbs, roses, clematis and herbaceous perennials in strong pinks, plums, purples and orange. A small cutting garden is full of dahlias Refreshments available at Golf Course Allotments.

THE CIRCLE GARDEN, KT3
33 Cambridge Avenue, New Malden, KT3 4LD. Vincent & Heidi Johnson-Paul-McDonnell, www.thecirclegarden.com. *1¼ m N of A3 Malden junction. Bus: 213. 10 mins walk from New Malden train stn; A3 signposted for Kingston; 213 bus stop located a short distance from end of rd; our house is pink!* **Sat 15, Sun 23 May (2-6). Adm £4, chd free. Home-made teas.**
You are welcomed to a traditional cottage front garden and a back garden that offers visionary and sensory rewards.

 26 COLLEGE GARDENS, E4
CHINGFORD, E4 7LG. Lynnette Parvez. *2m from Walthamstow. 15 mins walk Chingford mainline stn. 97 bus from Walthamstow Central tube*

stn. *Alight at College Gardens then short walk down hill.* **Sun 6 June (2-5). Adm £4, chd free. Home-made teas.**
Large suburban garden, approx ⅔ acre. Sun terrace leads to established borders and variety of climbing roses. Beyond this, wildlife pond and lawn, small woodland walk with spring plants and orchard. A further garden area was recently uncovered and will be restored and planted over time.

 55 COLLEGE ROAD, HA3
Harrow Weald, Harrow, HA3 6EF. Jenny Phillips. *Buses 258,340,182,140 Near Waitrose, Harrow Lawn Tennis Club Tube/ train Harrow & Wealdstone Station (10 mins walk or bus). Garden is few steps from tennis club.* **Sun 13 June (2-5). Adm £3.50, chd free. Tea.**
A delightful garden with mature shrubs and small trees. The current owner has made very clever use of textures, shapes and colour palates to make the garden very charming. It can be enjoyed from many different

aspects and the full length of the gardens delights. There are lots of places to sit and relax with lots of quirky features. Two lauquat trees which are unusual to see in London.

2 CONWAY ROAD, N14
Southgate, N14 7BA. Eileen Hulse. *Buses: 121 & W6 from Palmers Green or Southgate to Broomfield Park stop. Walk up Aldermans Hill R into Ulleswater Road 1st L into Conway Rd.* **Sun 5 Sept (2-6). Adm £4.50, chd free. Home-made teas.**
A passion, nurtured from childhood, for growing unusual plants has culminated in two contrasting gardens. The original, calming with lawn, pond and greenhouse, compliments the adjoining, Mediterranean terraced rooms with pergolas clothed in exotic climbers, vegetable beds and cordon fruit, Tumbling Achoqcha, figs, datura and Rosa banksiae mingle creating a horticultural adventure.

4 Franconia Road

19 COOLHURST ROAD, N8

Hornsey, N8 8EP. Jane Muirhead. *Exit tube at Highgate onto Priory Gardens, L on Shepherd's Hill, R on Stanhope Rd, L on Hurst Ave, R on Coolhurst Rd (15mins) W7 from Finsbury Park to Crouch End & 5min walk 41 & 91 busses nearby.* **Sun 18 Apr, Sun 9 May (2-6). Adm £4, chd free.**
Evolving, organic and wildlife garden with dappled sunlight, bees, birds and butterflies. Informal woodland planting under magnificent deciduous trees with interesting shrubs, box shapes and perennials. Small vegetable patch and wild flower garden where knapweed, campion, honesty, wild carrot and sweet rocket have naturalised. Large lawn with seating. Some steps and uneven surfaces. Recently created Bee and Bug Hotels. Gravel Garden and Wildlife pond.

45 COTSWOLD WAY, EN2

Oakwood, Enfield, EN2 7HD. Ian Brownhill & Michael Hirschl. *Short bus ride from either Oakwood tube stn or Enfield Chase train stn. Use buses 121 or 307 and alight at Cotswold Way.* **Sun 18 July (2-6). Adm £5, chd free. Home-made teas.**
Featured in the August 2020 issue of Country Living magazine this contemporary, sunny, medium sized garden looks out over London's Green Belt. The garden contains a number of features incl an imposing outdoor fireplace with dining area, raised Koi pond and large deck. These provide a backdrop to the overflowing, densely planted borders featuring a wide range of grasses, perennials and annuals.

83 COWSLIP ROAD, E18

South Woodford, E18 1JN. Fiona Grant. *5 mins walk from Central Line tube. Close to exit for A406.* **Sun 11 July (2-5). Adm £5, chd free. Home-made teas. Also open 17 Greenstone Mews.**
80 ft long wildlife-friendly garden. On two levels at rear of Victorian semi. Patio has a selection of containers with a step down to the lawn past a pond full of wildlife. Flowerbeds stuffed with an eclectic mix of perennials. Ample seating on patio and under an ancient pear tree. Wheelchair access through side of house to patio. Steps down to main garden and into kitchen for homemade teas.

NEW ▸ 58 CRANSTON ROAD, SE23

Forest Hill, SE23 2HB. Mr Sam Jarvis and Mr Andres Sampedro. *12mins walk from nearest Overground stns Forest Hill or Honor Oak. Nearest bus stops: Stanstead Rd/Colfe Rd (185, 122), Kilmorie Rd (185, 171) or Brockley Rise/Cranston Rd (122, 171).* **Sun 8 Aug (12-6). Adm £3.50, chd free. Teas, cakes and tortilla.**
This exotic-style plant-lover's garden features a modern landscaped path and carefully curated subtropical planting. Vivid evergreens – incl palms, cordylines, loquat and cycad – provide structure and year-round interest. Tree ferns, bananas and tetrapanax add to the striking foliage, while cannas, dahlias and agapanthus provide vibrant pops of colour against the black-painted boundaries.

40 THE CRESCENT, SM2

Belmont, Sutton, SM2 6BJ. Mrs Barbara Welch. *Off B2230 Brighton Rd, Belmont. Train: 5 min walk from Belmont Stn, from Sutton Stn take Bus 280 to Belmont. Over bridge into Station Rd, 1st L into The Crescent. No 40 ½ way up on L. Street parking.* **Sun 13 June (1-5). Adm £4, chd free. Home-made teas.**
A rectangular suburban garden, 80' x 50' on chalk, with formal border and island beds surrounded by clipped box hedges, box and yew topiary, with many different shrubs incl philadelphus, deutzias, weigela, lilacs, roses, interwoven with cottage-style perennials. Small paths dissect the borders from central lawn leading to various enclosed areas, some with seating, incl a circular tree seat, a water feature and a rose-covered pergola. Terrace with second water feature, many pots for ericaceous plants, fuchsias and seasonal planting.

GROUP OPENING

DE BEAUVOIR GARDENS, N1

N1 4HU. *Highbury & Islington tube then 30 bus; Angel tube then 38, 56 or 73 bus; Bank tube then 21, 76 or 141 bus. 10 mins walk from Dalston Overground Stns. Street parking available.* **Sun 6 June (11-3). Combined adm £8, chd free. Home-made teas at 158 Culford Rd.**

158 CULFORD ROAD

Gillian Blachford.

NEW ▸ ♦ 100 DOWNHAM ROAD

Ms Cecilia Darker.

64 LAWFORD ROAD

21 NORTHCHURCH TERRACE

Nancy Korman, nancylkorman@hotmail.co.uk. **Visits also by arrangement May & June for groups of up to 20.**

Four gardens to explore in De Beauvoir, a leafy enclave of Victorian villas near to Islington and Dalston. The area boasts some of Hackney's keenest gardeners and a thriving garden club. 100 Downham Rd features garden sculpture, giant echiums, two green roofs. The walled garden at 21 Northchurch Terrace has a formal feel, with deep herbaceous borders, pond, fruit trees, pergola, patio pots and herb beds. 64 Lawford Road is a small cottage style garden with old fashioned roses, espaliered apples and scented plants. 158 Culford Road is a long narrow garden with a path winding through full borders with shrubs, small trees, perennials and many unusual plants.

NEW ▸ 7 DEANSWAY, N2

East Finchley, N2 0NF. Joan Arnold & Tom Heinersdorff, 07850 764543, joan.arnold40@gmail.com. *From East Finchley Tube Stn exit along the Causeway to East End Rd then L down Deansway. From Bishops Ave, head N up Deansway towards East End Rd close to the top.* **Sun 2 May (12.30-6). Adm £5, chd free. Home-made teas. Gluten free cakes available. Visits also by arrangement Apr to July for groups of 10 to 20.**
A garden of stories, statues, shapes and structures surrounded by trees and hedges. Bird friendly, cottage style with scented roses, clematis, mature shrubs, a weeping mulberry and abundant planting. Containers, Spring bulbs and grape vine provide all-year colour. Developing secret shady woodland area with ferns and hostas. Easy access through the side passage to the patio but there is one shallow step on to the lawn and main garden which would need assistance.

14 DOUGHTY STREET, WC1N

WC1N 2PL. Gillian Darley & Michael Horowitz QC. *Off Guilford St or Theobalds Rd. Tube: Chancery Lane or Russell Square. Bus: 19,38,55. Garden S of Guilford St* on W side of Doughty St with brown LCC plaque to Sidney Smith. **Sun 4 July (2-5.30). Adm £3.50, chd free. Also open 57 Tonbridge House.** Small paved rear garden with a wilderness feel, surprisingly since the house opens directly off the street. Medlar tree, vine, plants jostling for space and pots add extra interest. Pleasing disorder best describes it.

DRAGON'S DREAM, UB8
Grove Lane, Uxbridge, UB8 3RG. Chris & Meng Pocock. *Garden is in a small lane that is very close to Hillingdon Hospital. Parking is available in nearby Royal Lane. Buses from Uxbridge Tube Station: U1,U3,U4,U5,U7. Buses from West Drayton : U1, U3.* **Sun 23 May (2-5). Adm £4, chd free. Home-made teas. Malaysian curry puffs also available.**
This is an unusual and secluded garden with two contrasting sections divided by a brick shed that has been completely covered by a rampant wisteria and a climbing hydrangea and rose. Other highlights include rare dawn redwood tree, large yucca and a huge gunnera manicata . More features incl a romneya poppy, ferns, rose and herb beds, tree peonies, acers and a pond.

55 DUKES AVENUE, N10
Muswell Hill, N10 2PY. Jo de Banzie & Duncan Lampard, www.jodebanzie.com. *W3 Bus (Alexandra Palace Garden Centre stop) or W7 Bus (Muswell Hill stop). Free on-street parking available on Dukes Av.* **Sun 23 May (2-6). Adm £4, chd free. Light refreshments.**
A photographer's small town garden uses curves and spheres to add shape and interest to a pretty, shady space. Gravel, paving, decking and a planting platform in an old apple tree create structure, whilst black bamboo, ferns and box provide the backdrop for a gentle palette of white and purple planting. Exhibition of Botanica Photographs. Unfortunately not suitable for wheelchairs due to access via stairs through house.

GROUP OPENING

103 AND 105 DULWICH VILLAGE, SE21
SE21 7BJ. *Rail: N Dulwich or W Dulwich then 10 -15 mins walk. Tube: Brixton then P4 bus, alight*

Dulwich Picture Gallery stop. *Street parking.* **Sun 6 June (2-5). Combined adm £8, chd free. Home-made teas at 103 Dulwich Village. Donation to Macmillan Cancer Care.**

103 DULWICH VILLAGE
Mr & Mrs N Annesley.

105 DULWICH VILLAGE
Mr & Mrs A Rutherford.

2 Georgian houses with large gardens, 3 mins walk from Dulwich Picture Gallery and Dulwich Park. 103 Dulwich Village is a country garden in London with a long herbaceous border, lawn, pond, roses and fruit and vegetable gardens. 105 Dulwich Village is a very pretty garden with many unusual plants, lots of old fashioned roses, fish pond and water garden. Amazing collection of plants for sale from both gardens. Please bring your own bags for plants. Music provided by the Colomb Street Ensemble wind band.

41 ECCLESBOURNE ROAD, N1
N1 3AF. Steve Bell & Sandie Macrae. *Canonbury. Tube: Highbury & Islington 12 mins walk or 271 bus to Ecclesbourne Rd. Rail: Essex Road, 6 mins walk or bus 73, 38, 56, 341, 476 to Northchurch Rd. down Halliford St, R onto Ecclesbourne Rd.* **Sun 11 July (12-6). Adm £4, chd free. Home-made teas.**
A delightful, interesting and surprising artists' minimalist Mediterranean garden complemented by new contemporary architecture. This garden has a hoggin surface and flourishing agapanthus, trees and pots on two levels. There is a birch, a handkerchief tree, herbs, trailing clematis, roses and, unbelievably a bay tree hand-gilded in gold leaf.

ECCLESTON SQUARE, SW1
SW1V 1NP. Roger Phillips & the Residents. *Off Belgrave Rd nr Victoria Stn, parking allowed on Suns.* **Sun 9 May (2-5). Adm £5, chd free. Home-made teas.**
Planned by Cubitt in 1828, the 3 acre square is subdivided into mini gardens with camellias, iris, ferns and containers. Dramatic collection of tender climbing roses and 20 different forms of tree peonies. National Collection of ceanothus incl

more than 70 species and cultivars. Notable important additions of tender plants being grown and tested. World collection of ceanothus, tea roses and tree peonies.

EDWARDES SQUARE, W8
South Edwardes Square, Kensington, W8 6HL. Edwardes Square Garden Committee, www. edwardes-square-garden.co.uk. *South Edwardes Square. Tube: Kensington High St & Earls Court. Buses: 9, 10, 27, 28, 31, 49 & 74 to Odeon Cinema. Entrance in South Edwardes Square.* **Sun 18 Apr (10.30-4). Adm £5, chd free. Tea, coffee and cakes.**
One of London's prettiest secluded garden squares. 3½ acres laid out differently from other squares, with serpentine paths by Agostino Agliothe, Italian artist and decorator who lived at No.15 from 1814-1820, and a beautiful Grecian temple which is traditionally the home of the head gardener. Romantic rose tunnel winds through the middle of the garden. Good displays of bulbs and blossom. Pimms available if sunny. Tea and Cakes. Children's play area. WC. Wheelchair access through Main Gate, South Edwardes Square.

NEW 3 ELMBRIDGE LODGE, KT7
Weston Green Road, Thames Ditton, KT7 0HY. Mrs Julia Hickman, 020 8339 0031, julia@hickmanweb.co.uk. *House opp Esher College on Weston Green Rd & 5 mins walk from Thames Ditton station.* **Sun 4 July (2-5). Adm £4.50, chd free. Light refreshments. Visits also by arrangement May to July for groups of 10 to 20.**
Mature garden 70' x 35' designed by Cleve West five years ago. From a sunny terrace steps lead through two oak pergolas festooned with climbing plants. Gravelled areas feature unusual and drought tolerant plants. Decorative greenhouse. Generous perennial border brims with a kaleidoscope of plants. Beyond is a productive vegetable area opposite a catalpa with wild flowers beneath.

15 Waite Davies Road

40 EMBER LANE, KT10

Esher, KT10 8EP. Sarah & Franck Corvi. *½ m from centre of Esher. From the A307, turn into Station Rd which becomes Ember Lane.* **Sun 30 May (1-5). Combined adm with 9 Imber Park Road £5, chd free. Home-made teas.**

A contemporary family garden designed and maintained by the owners with distinct areas for outdoor living. A 70 ft East facing plot with some unusual planting and several ornamental trees including a mature cercis canadensis. Home made teas can be enjoyed outside in the garden or inside the large airy kitchen.

91 ENGLEFIELD ROAD, N1

N1 3LJ. Antoinette and Michael. *East Canonbury. Highbury & Islington tube or Canonbury Overground stn; 30 bus to Southgate Rd stop, 5 min walk. Angel tube; 30, 56 or 73 bus to Ockendon Rd, 5 min walk.* **Sun 30 May (2-5). Adm £3.50, chd free. Home-made teas.**

This garden surprises. South facing yet shaded by a 40ft magnolia & a mature apple tree, a hammock slung between, making a sublime place to relax & read. The philosophy is 'if you like a plant you can find a space', so dense & varied planting, more at home in a country garden, creates an eclectic mix & it works! A patio crammed with pots & hanging baskets seeks to entertain. Garden is approached via steep steps at side of house.

11 ERNLE ROAD, SW20

Wimbledon, SW20 0HH. Theresa-Mary Morton. *¼ m from Wimbledon Village, 200yds from Crooked Billet pub. Exit A3 at A238 to Wimbledon, turning L at Copse Hill. Mainline: Wimbledon or Raynes Park. Tube: Wimbledon; Bus: 200 to High Cedar Drive; 200 yds walk.* **Sat 24, Sun 25 Apr (2.30-6). Adm £5, chd free. Home-made teas. Evening opening Sun 9 May (6-8). Adm £8, chd free. Wine.**

Established suburban garden of ⅓ acre on sandy acid soil, spatially organised into separate sections: oak pergola framing the main vista, woodland garden, pool, yew circle, flower garden and summerhouse. Beaten gravel paths, one step up to main garden.

♿ ☕

12 FAIRFIELD ROAD, N8

N8 9HG. Christine Lane. *Tube: Finsbury Park & then W3 (Weston Park stop) or W7 (Crouch End Broadway), alternatively Archway & then 41 bus (Crouch End Broadway) it's then a short walk.* **Sun 30 May (2-5.30). Adm £4, chd free. Also open 36 Ashley Road.**

A tranquil garden created on two levels with a lawn, a cobbled zen garden and a patio on the lower level and a secluded woodland garden with sculptures on the higher level. There is a variety of trees, shrubs and flowers, as well as succulents, palms and bamboos all of which creates different atmospheres and plants in pots give height and interest. Seating provides places to contemplate and enjoy.

✿

70 FARLEIGH ROAD, N16

Stoke Newington, N16 7TQ. Mr Graham Hollick. *Short walk from junction of Stoke Newington High St & Amhurst Rd.* **Sun 30 May (10.30-6). Adm £3.50, chd free. Home-made teas.**

A diverse garden in a Victorian terrace with an eclectic mix of plants, many in vintage pots reflecting the owner's interests. A small courtyard leads onto a patio surrounded by pots followed by a lawn flanked by curving borders. At the rear is a paved area with raised beds containing vegetables.

NEW 28 FERNDENE ROAD, SE24

Herne Hill, SE24 0AB. Mr David & Mrs Lynn Whyte. *Overlooking Ruskin Park. Buses 68, 468, 42. A 5 min walk from Denmark Hill. Train stations: Herne Hill, Denmark Hill, Loughborough Junction. All 15 min walk. House overlooks Ruskin Park. Free parking.* **Sun 4 July (1-5). Adm £4, chd free. Home-made teas. Home-made scones with garden produce jam.**

It's all about structure and careful planting in this dramatically sloping S.S.E. facing garden, 30m wide x 18m. A lively blend of perennials and shrubs show a definite Kiwi influences. The kitchen garden with raised beds and soft fruits is wonderfully secluded. Lower level planting has a coastal feel. Upper level has a 'hot colour' border. Borrowed views of mature trees and big skies set it off. The garden office/Summer house, constructed from sustainably sourced materials, has a rubble roof to attenuate water runoff. Recycling and re-use of materials.

✿ ☕

FOREST HILL GARDEN GROUP, SE23

SE23 3BJ. *Off S Circular (A205) behind Horniman Museum & Gardens. Station: Forest Hill, 10 mins walk. Buses: 176, 185, 197, 356, P4 - alight for this bus' Horniman Drive'. Buses: 176, 185, 197, 356 & P4 stop at Horniman Museum. P4 also stops Horniman Dr.* **Sun 23 May (1-6). Combined adm £8. Home-made teas at 53 Ringmore Rise. Donation to St Christopher's Hospice and Marsha Phoenix Trust.**

7 CANONBIE ROAD
June Wismayer.

THE COACH HOUSE, 3 THE HERMITAGE
Pat Rae.

HILLTOP, 28 HORNIMAN DRIVE
Frankie Locke.

27 HORNIMAN DRIVE
Rose Agnew, 020 8699 7710, roseandgraham@talktalk.net. **Visits also by arrangement Apr to Oct for groups of 10 to 20.**

53 RINGMORE RISE
Valerie Ward.

25 WESTWOOD PARK
Beth & Steph Falkingham-Blackwell.

Six lively gardens in eclectic and differing styles on the highest hill in SE London, nr Horniman Museum, with fine views over the Downs. All within a short walk of each other. For small donation, lifts available for those with mobility problems. An intricately planted sloping woodland garden. A tame robin still demands cheese in the walled courtyard of an artist's 18C coach house. New design with water features, sculptures and unusual wisteria. A country-style garden with vegetable and cutting beds, meadow, assorted chickens and a new activity for children. Relax among vibrant colours and enjoy breathtaking views. Admire the tiered, bee and butterfly-enticing organic garden. Unwind in spacious and fragrant surroundings with delicious cakes, while looking out over the London skyline. Plants for sale: 27 and 28 Horniman Dr, 7 Canonbie Rd. Ceramics and sculptures for sale at the Coach House. Teas 53 Ringmore Rise. Gardens have slopes/steps.

4 FRANCONIA ROAD, SW4

Abbeville Village, Clapham, SW4 9ND. Paul Harris. *Abbeville Village. From Clapham Common tube, walk 600 metres S to Elms Rd (opp The Windmill hotel) walk along Elms Rd 400 metres to Abbeville Rd, then turn L. Franconia Rd is next on R after 100 metres.* **Sat 19 June (10-5). Adm £5, chd free. Light refreshments.**

A private residential garden with a SW aspect. Soil type heavy clay. The garden is a tranquil spot despite the close proximity of neighbours. Receiving sunshine throughout the day, the design offers a shady terrace as well as a sun terrace, separated by a sunken lawn. An exotic planting scheme helps one to escape, whilst being practical for the growing conditions found in a London climate. Floating pergola offering shade to the wildlife hotel and concealing the working area.

9 FURLONG ROAD, N7

Islington, N7 8LS. Nigel Watts & Tanuja Pandit. *Close to Highbury & Islington Tube/Overground. Tube & Overground: Highbury & Islington, 3 mins walk along Furlong Rd, 2nd L. Furlong Rd joins Holloway Rd & Liverpool Rd. Buses: 43, 271, 393.* **Sun 23 May (12-6). Adm £3, chd free.**

Award winning small garden designed by Karen Fitzsimon, making clever use of an awkwardly shaped plot. Curved lines are used to complement a modern extension. Raised beds contain a mix of tender & hardy plants to give an exotic feel & incl loquat, banana, palm, cycad & tree fern. Contrasting traditional front garden. Featured in Small Family Gardens and Modern Family Gardens by Caroline Tilston. Refreshments at 10 & 20 Furlong Road.

NEW 10 FURLONG ROAD

N7 8LS. Gavin & Nicola Ralston. *Close to Highbury & Islington Tube/Overground. Tube & Overground: Highbury & Islington, 3 mins walk along Holloway Rd, 2nd L. Furlong Rd runs between Holloway Rd & Liverpool Rd. Buses on Holloway Rd: 43, 271, 393; other buses 4, 19, 30,.* **Sun 23 May (2-5.30). Adm £5, chd free. Home-made teas. Also open 20 Furlong Road.**

A green oasis in the heart of a densely populated area, 10 Furlong Road is an open, sunny garden of considerable size for its urban location. Its new owners, previously at Canonbury House, extensively remodelled the garden in 2019, building on a foundation of trees, shrubs & roses, adding herbaceous planting, seating, winding brick path, pergola, raised vegetable bed & wildflower circle. Variety of teas & soft drinks, with a range of cakes & biscuits, all home made; plenty of seating dotted around the garden.

NEW 20 FURLONG ROAD, N7

N7 8LS. Mr Simon Toms. *Close to Highbury & Islington Tube/Overground. Tube & Overground: Highbury & Islington, 3 mins walk along Holloway Rd, 2nd L. Furlong Rd joins Holloway Rd & Liverpool Rd. Buses: 43, 271, 393.* **Sun 23 May (12-5). Adm £5, chd free. Light refreshments. Also open 9 Furlong Road.**

A mature garden hidden away from the busy streets of Islington. The garden draws from planting to be found in Cornish spring gardens, incl rhododendrons, magnolias, cherry & cercis trees. There is a degree of formality with topiary, pleached magnolias & lawn mixed with a touch of fun in the form of banana trees, palms & cannas. The rhododendrons will be in bloom.

NEW THE GABLE END GARDENS

52 Hawley Road, Camden Town, NW1 8RG. Magda Segal. *Set back off Chalk Farm Rd opp the Stables Market. On the L approx. 600yds down from Chalk Farm tube station. The nearest bus stop is served by the 31, 24, 168 & 88 Buses.* **Sun 27 June (2-6). Adm £4, chd free. Home-made teas. Also open The Holme.**

The Gable End Gardens consist of a series of planted areas that together create a wildlife haven in the heart of Camden Town, within them can be found a colony of sparrows, beehives and ponds. The planting, based around rescued and donated plants, is dominated by a mature fig tree at the front of the property and a willow in the rear garden, providing forage for the bees and cover for the birds. A must-see example of what can be achieved to help wildlife in urban environments. Plus prize winning tree pits. Scottish home baking incl. real

Scottish tea cake and shortbread. The gardens are not wheelchair accessible, the rest can be enjoyed with comparative ease.

GARDEN BARGE SQUARE AT TOWER BRIDGE MOORINGS, SE1

31 Mill Street, SE1 2AX. Mr Nick Lacey, towerbridgemoorings.org. *5 mins walk from Tower Bridge. Mill St off Jamaica Rd, between London Bridge & Bermondsey Stns, Tower Hill also nearby. Buses: 47, 188, 381, RV1.* **Sun 23 May (2-5). Adm £4, chd free. Home-made teas on the 'ArtsArk,' a floating platform with tables and chairs. Donation to RNLI.**

Series of seven floating barge gardens connected by walkways and bridges. Gardens have an eclectic range of plants for yr-round seasonal interest. Marine environment: suitable shoes and care needed. Small children must be closely supervised.

◆ THE GARDEN MUSEUM, SE1

5 Lambeth Palace Road, SE1 7LB. The Garden Museum, www.gardenmuseum.org.uk. *Lambeth side of Lambeth Bridge. Tube: Lambeth North, Vauxhall, Waterloo. Buses: 507 Red Arrow from Victoria or Waterloo mainline & tube stns, also 3, 77, 344.* **For NGS: Sat 11 Sept (10.30-4). Adm £6.50, chd free. Light refreshments in the Garden Cafe. For other opening times and information, please visit garden website.**

At the heart of the Garden Museum is the Sackler Garden. Designed by Dan Pearson as an 'Eden' of rare plants, the garden is inspired by John Tradescant's journeys as a plant collector. Taking advantage of the sheltered, warm space, Dan has created a green retreat in response to the bronze and glass architecture, conjuring up a calm, reflective atmosphere. Visitors will also see a permanent display of paintings, tools, ephemera and historic artefacts: a glimpse into the uniquely British love affair with gardens. The Garden Cafe is an award-winning lunch venue that is considered one of the best Museum restaurants in the country. The Museum is accessible for wheelchair users via ramps and access lift.

GARDEN OF MEDICINAL PLANTS, ROYAL COLLEGE OF PHYSICIANS, NW1

11 St Andrews Place, Regents Park, NW1 4LE. Royal College of Physicians of London, garden.rcplondon.ac.uk/. *Opp the SE corner of Regents Park. Tubes: Great Portland St & Regent's Park. Garden is one block N of station exits, on Outer Circle opp SE corner of Regent's Park.* **Wed 11 Aug (11-4). Adm £5, chd free.**

We have 1,100 different plants connected with the history of plants in medicine: plants named after physicians; plants which make modern medicines; those used for millennia; plants which cause epidemics; plants used in herbal medicine in all the continents of the world and plants from the College's Pharmacopoeia of 1618 on which Nicholas Culpeper based his Herbal. Guided tours by physicians explaining the uses of the plants, their histories and stories about them. Books about the plants in the medicinal garden will be on sale. The entry to the garden is at far end of St Andrews Place. Wheelchair ramps at steps. Wheelchair lift for lavatories. No parking on site.

&

GROUP OPENING

GARDENS OF COURT LANE, SE21

Dulwich Village, SE21 7EA. *Court Lane, Dulwich. Buses P4, 12, 40, 176, 185 (to Dulwich Library) 37. Mainline; North Dulwich then 12 mins walk. Ample free parking.* **Sun 27 June (2.30-5.30). Combined adm £8, chd free. Home-made teas.**

122 COURT LANE, SE21
Jean & Charles Cary-Elwes.

NEW 148 COURT LANE
Mr Anthony & Mrs Sue Wadsworth.

164 COURT LANE
James & Katie Dawes.

164 was recently redesigned to create a more personal and intimate space with several specific zones. A modern terrace and seating area leads onto a lawn with abundant borders and a beautiful mature oak. A rose arch leads to the vegetable beds and greenhouse. 122 has a countryside feel, backing onto Dulwich Park with colourful herbaceous borders, live Jazz on the terrace, a children's trail, plant sales and a wormery demonstration. 148, a spacious garden, developed over 20 years backs on to Dulwich Park. From the wide sunny terrace surrounded by tall, golden bamboos and fan palms, step down into a garden designed with an artist's eye for colour, form, texture and flow, creating intimate spaces beneath mature trees and shrubs, with colourful perennials. Jazz, children's trail, wormery demonstration, tea and cakes at 164. Refreshments and plant sale at 148. 122 has wheelchair access to the terrace which has a good view of the garden.

---

74 GLENGALL ROAD, IG8

Woodford Green, IG8 0DL. Mr & Mrs J Woolliams. *0208 504 1709, email jgmwoolliams@outlook.com, 5 mins walk from Woodford Central line, off Snakes Lane West. Buses nearby incl 275,179,W13 & 20. No parking restrictions on Suns.* **Sun 28 Mar (2-5). Adm £4, chd free. Tea.**
A secluded S-facing cottage style garden, developed over 25 years for yr-round interest. Areas incl 2 lawns, a rock garden, small wildlife pond with bog garden, shade borders and a gravel garden, linked by several paths. The garden is at its best in Springtime Throughout are mixtures of trees shrubs perennials bulbs bamboos grasses and climbers.

69 GLOUCESTER CRESCENT, NW1

Camden, NW1 7EG. Sandra Clapham, 020 7485 5764, set69@gloscres.com. *Between Regent's Park & Camden Town tube station. Tube: Camden Town 2 mins, Mornington Crescent 10 mins. Metered parking in Oval Rd.* **Sun 6 June (2-5.30), also open 36 Park Village East. Sun 8 Aug (2-5.30), also open The Holme. Combined adm with 70 Gloucester Crescent £6, chd free. Visits also by arrangement Apr to Oct for groups of up to 30.**
Delightful little cottage front garden, opening jointly with No. 70 next door. It shows what can be done with a front garden as an attractive alternative to a paved parking space. Ursula Vaughan Williams lived here and the very old iceberg rose at the front, the yellow roses, the border of London pride and the Crinum powellii Roseа lily in a pot are all inherited from her. Many plants have been added since, incl a bed of tomatoes and a delicious 24yr-old grape vine, trained up and along the balcony, that produces generous amounts of grape jelly!

&

The National Garden Scheme searches the length and breadth of England and Wales for the very best private gardens

33 Weyman Road

70 GLOUCESTER CRESCENT, NW1

70 Gloucester Crescent, NW1 7EG. Lucy Gent, 07531 828752 (texts please), gent.lucy@gmail.com. *Between Regent's Park & Camden Town tube station. Tube: Camden Town 2 mins, Mornington Crescent 10 mins. Metered parking in Oval Rd.* **Sun 6 June (2-5.30), also open 36 Park Village East. Sun 8 Aug (2-5.30), also open The Holme. Combined adm with 69 Gloucester Crescent £6, chd free. Visits also by arrangement Apr to Oct for groups of up to 30.**

Here is an oasis in Camden's urban density, where resourceful planting outflanks challenges of space and shade and Mrs Dickens, who once lived here, is an amiable ghost. Open days occur alongside another distinctive local garden, an August opening shows how wonderful the month can be in a town garden. September is also a great show, colour highlights amidst beautiful foliage. Many cafes in nearby Parkway.

GOLF COURSE ALLOTMENTS, N11

Winton Avenue, N11 2AR. GCAA Haringey, www.golfcourseallotments.co.uk. *Junction of Winton Av & Blake Rd. Tube: Bounds Green. Buses: 102, 184, 299 to Sunshine Garden Centre, Durnsford Rd. Through park to Bidwell Gdns. Straight on up Winton Ave. No cars on site.* **Sun 5 Sept (1-4.30). Adm £4, chd free. Home-made teas. Also open 9 Churston Gardens. Light lunches available.**

Large, long established allotment with over 200 plots, some organic. Maintained by culturally diverse community growing wide variety of fruit, vegetables and flowers enjoyed by the bees. Picturesque corners and quirky sheds - a visit feels like being in the countryside. Autumn Flower and Produce Show features prize winning horticultural and domestic exhibits and beehives. Tours of best plots. Fresh allotment produce, chutneys, jams, honey, cakes and light refreshments for sale. Wheelchair access to main paths only. Gravel and some uneven surfaces. WC incl disabled.

28 GRANVILLE PARK, SE13

Blackheath, SE13 7EA. Joanna Herald, www.joannaherald.com. *Lewisham Station 10 mins walk uphill, buses 53 & 386 plus busses to Lewisham & Blackheath. Free parking on Sunday.* **Sun 13 June (2-5). Combined adm with 49 Lee Road £8, chd free. Light refreshments.**

Garden designer's relaxing and softly contoured hillside family garden has recently had a gentle facelift with a 'useful' room added to the bottom of the garden and a wildlife-friendly pool resized and re-sited. Mixed herbaceous and shrubs encompass a lawn and gravel garden. The sunken terrace packed with pots provides a suntrap seating area. 100ft x 35ft. Plants will be on offer. wildlife pond. Narrow but level side access to the garden.

45 GREAT NORTH ROAD, EN5

Barnet, EN5 1EJ. Ron & Miriam Raymond, 07880 500617, ron.raymond91@yahoo.co.uk. *1m S of Barnet High St, 1m N of Whetstone. Tube: Midway between High Barnet & Totteridge & Whetstone stns. Buses 34, 234, 263, 326, alight junction Great N Rd & Lyonsdown Rd. 45 Great North Rd is on the corner of Cherry Hill.* **Sun 1 Aug (2-5.30). Adm £3, chd free. Light refreshments. Visits also by arrangement July & Aug for groups of 5 to 20.**

45 Great North Road is designed to give a riot of colour in late summer. The 90ft x 90ft cottage style front garden is packed with interesting and perennials. Tiered stands line the side entrance with over 64 pots displaying a variety of flowering and foliage plants.The rear garden incl nearly 100 tubs and hanging baskets. Small pond surrounded by tiered beds. Magnificent named tuberous begonias. Children's fun trail for 3-6yr olds and adult garden quiz with prizes. Partial wheelchair access.

103 Dulwich Village

17 GREENSTONE MEWS, E11

Wanstead, E11 2RS. Mrs T Farnham. *Wanstead. Tube: Snaresbrook or Wanstead, 5 mins walk. Bus: 101, 308, W12, W14 to Wanstead High St. Greenstone Mews is accessed via Voluntary Place which is off Spratt Hall Rd.* **Sun 11 July (12.30-4). Adm £5. Light refreshments. Also open 83 Cowslip Road. Tea or coffee incl in adm price.**

Tiny Slate paved garden. Approx 15 feet square. Sunken reused bath now a fishpond surrounded by climbers clothing fences underplanted with hardy perennials, herbs, vegetables, shrubs and perennials grown from cuttings. Height provided by established Palm. Ideas aplenty for small space gardening. Regret garden unsuitable for children. Wheelchair access through garage. Limited turning space.

24 GROVE PARK, SE5

Camberwell, SE5 8LH. Clive Pankhurst, www.alternative-planting.blogspot.com. *Chadwick Rd end of Grove Park. Stns: Peckham Rye or Denmark Hill, both 10 mins walk. Good street parking.* **Sun 5 Sept (11-4.30). Adm £5, chd free. Home-made teas.**

An inspiring exotic jungle of lush big leafed plants and Southeast Asian influences. Towering exotica transport you to the tropics. Huge hidden garden created from the bottom halves of two neighbouring gardens gives the 'wow' factor and unexpected size. Lawn and lots of hidden corners give spaces to sit and enjoy. Renowned for delicious home-made cake.

58 HALLIFORD STREET, N1

N1 3EQ. Jennifer Tripp Black. *Canonbury. Tube: Highbury & Islington or Essex Rd. From Essex Rd, house numbers consecutive on LH side by 3rd speed bump. Overground: Canonbury.* **Sun 16 May (2-6). Adm £4, chd free. Home-made teas. Vegan & full fat cakes. Donation to The Royal Marsden Cancer Charity (Katie's Lymphoedema Fund).**

An English country garden in the heart of Islington. Lush planting: cytisus battandieri 'Yellow Tail' tree, two apple trees, and a quince. Roses: bush & climbing, clematis, rhododendrons, exotic palms, cannas, abutilons,

heucheras, salvias, cordelines. Small greenhouse, antique pergola, pots containing hosta collection & window boxes. The front garden welcomes with special roses & agapanthus.

SPECIAL EVENT

◆ HAMPTON COURT PALACE, KT8

East Molesey, KT8 9AU. Historic Royal Palaces, www.hrp.org.uk. *Follow brown tourist signs on all major routes. Junction of A308 with A309 at foot of Hampton Court Bridge.* **For NGS: Evening opening Thur 22 Apr, Thur 15 July (6-8). Adm £12, chd free. Pre-booking essential, please visit www.ngs.org.uk/special-events for information & booking. pre booking essential. For other opening times and information, please visit garden website.**

Take the opportunity to join 2 special NGS private tours, after the wonderful historic gardens have closed to the public. Spring Walk in April and Mid-Summer abundance in July in the wonderful gardens of Hampton Court Palace. Some un-bound gravel paths.

37 HAROLD ROAD, E11

Leytonstone, E11 4QX. Dr Matthew Jones Chesters. *Tube: Leytonstone exit L subway 5 mins walk. Overground: Leytonstone High Rd 5 mins walk. Buses: 257 & W14. Parking at station or limited on street.* **Sun 13 June (1-5). Adm £4, chd free. Home-made teas. Home-made preserves sold.**

50ft x 60ft pretty corner garden arranged around 7 fruit trees. Fragrant climbers, woodland plants and shade-tolerant fruit along north wall. Fastigiate trees protect raised vegetable beds and herb rockery. Long lawn bordered by roses and perennials on one side; prairie plants on the other. Patio with raised pond, palms and rhubarb. Planting designed to produce fruit, fragrance and lovely memories. Garden map with plant names and planting plans included.

31 HENDON AVENUE, N3

Finchley, N3 1UJ. Sandra Tomaszewska Day. *Finchley Central. 15 mins walk from Finchley Central Tube. Buses: 326 & 143.*

Car: 5 mins from A1 via Hendon Ln. No parking restrictions on Sun. **Sun 30 May (2-6). Adm £4.50, chd free. Light refreshments.**

An extensive garden with mature trees, shrubs and perennials divided into areas. Herbaceous beds in semi shade, a raised triangular bed with lavender, agapanthus and roses. Two arches, draped with grapevines, wisteria, kiwi and clematis, guide you into a tranquil, white garden and a wildlife pond, lead to tropical and Mediterranean beds with olive, bean Albizia julibrissin Rosa trees and palms. Refreshments served from the pool house; relax by the pool in the tropical garden. Partial wheelchair access.

32 HIGHBURY PLACE, N5

N5 1QP. Michael & Caroline Kuhn. *Highbury Fields. Highbury & Islington Tube; Overground & National Rail. Buses: 4, 19, 30, 43, 271, 393 to Highbury Corner. 3 mins walk up Highbury Place which is opp stn.* **Sun 27 June (2-5.30). Adm £4, chd free. Home-made teas.**

An 80ft garden behind a C18 terrace house. An upper York stone terrace leads to a larger terrace surrounded by overfilled beds of cottage garden style planting. Further steps lead to a lawn by a rill and a lower terrace. A willow tree dominates the garden: amelanchiers, fruit trees & dwarf acers, winter flowering cherry, lemon trees and magnolia.

HIGHWOOD ASH, NW7

Highwood Hill, Mill Hill, NW7 4EX. Mrs P Gluckstein. *Totteridge & Whetstone on Northern line, then bus 251 stops outside - Rising Sun/ Mill Hill stop. By car: A5109 from Apex Corner to Whetstone. Garden located opp The Rising Sun PH.* **Sun 9 May (2-5). Adm £5, chd free.**

Created over the last 56 years, this 3¼ acre garden features rolling lawns, two large interconnecting ponds with koi, herbaceous and shrub borders and a modern gravel garden. A garden for all seasons with many interesting plants and sculptures. May should be perfect for the camellias, rhododendrons and azaleas. A country garden in London. Partial access for wheelchairs, lowest parts too steep.

20 HILLCREST, N21
Winchmore Hill, N21 1AT. Gwyneth & Ian Williams. *Tube: Southgate then W9 to Winchmore Hill Green followed by a short walk via Wades Hill. Train: Winchmore Hill, turn R towards the Green, then a short walk via Wades Hill.* Sun 6 June (1.30-5.30). Adm £4, chd free. Home-made teas.
A beautiful NW facing hillside garden offers a panoramic horizon and afternoon sun. A terrace has seating in an iron-work gazebo. Wide stone steps descend to a pretty, evergreen pergola, water feature and oasis of dappled sunlight. Beyond lie rose arches, a lawn, with shrubs and trees bordering a meandering path towards alpine rockeries, a summer house, a miniature wildlife pond and secret garden. Steep Steps.

THE HOLME, NW1
Inner Circle, Regents Park, NW1 4NT. Lessee of The Crown Commission. *In centre of Regents Park on The Inner Circle. Within 15 mins walk from Great Portland St or Baker St Underground Stations, opp Regents Park Rose Garden Cafe.* Sat 26, Sun 27 June, Sat 7, Sun 8 Aug (2.30-5.30). Adm £5, chd free.
4 acre garden filled with interesting and unusual plants. Sweeping lakeside lawns intersected by islands of herbaceous beds. Extensive rock garden with waterfall, stream and pool. Formal flower garden with unusual annual and half hardy plants, sunken lawn, fountain pool and arbour. Gravel paths and some steps which gardeners will help wheelchair users to negotiate. Gravel paths and some steps which gardeners will help wheelchair users to negotiate.

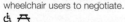

239A HOOK ROAD, KT9
Chessington, KT9 1EQ. Mr & Mrs D St Romaine, www.gardenphotolibrary.com. *4m S of Kingston. A3 from London, turn L at Hook underpass onto A243. Gdn 300yds on L. Parking opp in Park or on rd, no restrictions at night. Buses K4, 71, 465 from Kingston & Surbiton to North Star PH.* Evening opening Fri 11 June (8-10). Adm £6, chd free. Wine. Light refreshments.
Completely redesigned. From the patio with raised vegetable bed and garden room a central path with standard hollies, urns and box balls,

leads to a water feature at the end of the garden. Themed rectangular beds with obelisks form a grid either side of the central axis, surrounded on three sides by sun, shade and wildflower borders. All cleverly lit to enhance the atmosphere in the evenings.

NEW HORNBEAMS HA7
Priory Drive, Stanmore, HA7 3HN. Mrs B Stalbow. *5m SE of Watford. Tube: Stanmore. Priory Drive private rd off Stanmore Hill (A4140 Stanmore-Bushey Heath Rd).* Sun 7 Feb, Sun 5 Sept (2-5). Adm £5, chd free.
Carpets of bulbs and cyclamen, witch hazels, cornus kousa Satomi of special interest in February. A formal walkway with wisteria-covered arches alternate with squares of box surrounding flower-filled pots. Roses and lavender replace the small plot of vegetables which are now grown in containers. Where there was a tennis court there is now a memorial garden for my late husband & daughter.

33 HUDDLESTON ROAD, N7
N7 0AD. Gilly Hatch & Tom Gretton. *5 mins from Tufnell Park Tube. Tube: Tufnell Park. Buses: 4, 134, 390 to Tufnell Park. Follow Tufnell Park Rd to 3rd rd on R.* Sat 3 July (2-6). Adm £4, chd free. Home-made teas.
The rambunctious front garden weaves together perennials, grasses and ferns, while the back garden makes a big impression in a small space. After 40+yrs, the lawn is now a wide curving path, a deep sunny bed on one side; mixing shrubs and perennials in an ever changing blaze of colour, on the other; a screen of varied greens and textures. This flowery passage leads to a secluded area.

1A HUNGERFORD ROAD, N7
N7 9LA. David Matzdorf, davidmatzdorf@blueyonder.co.uk, , www.growingontheedge.net. *Between Camden Town & Holloway. Tube: Caledonian Rd. Buses: 17, 29, 91, 253, 259, 274, 390 & 393. Parking free on Sundays.* Sun 6 June (1-6). Adm £3, chd free. Also open 23 & 24b Penn Road. Visits also by arrangement Apr to Oct for groups of up to 10.
Unique eco house with walled, lush

front garden in modern exotic style, densely planted with palms, Acacia, bamboo, ginger lilies, bananas, ferns, Yuccas, Abutilons & unusual understorey plants. Floriferous & ambitious green roof resembling Mediterranean or Mexican hillside, planted with Yuccas, Dasylirions, Nolinas, Agaves, Cacti, Aloes, Cistus, Euphorbias, grasses, alpines, Sedums & aromatic herbs. Sole access to roof is via built in vertical ladder. Garden and roof each 50ft x 18ft (15m x 6m).

57 HUNTINGDON STREET, N1
Islington, N1 1BX. Julian Williams. *Barnsbury, Islington. Overground: Caledonian Road & Barnsbury. Tube: Kings Cross or Highbury & Islington. Buses: Caledonian Road:17, 91, 259, 274. Hemingford Rd: 153 from Angel.* Evening opening Wed 12 May (6-8). Adm £4, chd free. Wine & soft drinks. Opening with Barnsbury Group on Sun 6 June.
A secluded woodland 'garden room' below an ash canopy, and framed by timber palisade supporting roses, hydrangea and ivy. An understorey of silver birch and hazel provides the setting for shade loving ferns, mainly native perennials, and grasses. Oak pathways offset the informal planting and lead to a tranquil central space with bench seating and two container ponds to encourage wildlife.

THE HURLINGHAM CLUB, SW6
Ranelagh Gardens, SW6 3PR. The Members of the Hurlingham Club, www.hurlinghamclub.org.uk. *Main gate at E end of Ranelagh Gardens. Tube: Putney Bridge (110yds). NB: No onsite parking. Meter parking on local streets & restricted parking on Sats (9-5).* Sat 22 May (10-4.30). Adm £5, chd free.
Rare opportunity to visit this 42 acre jewel with many mature trees, 2 acre lake with water fowl, expansive lawns and a river walk. Capability Brown and Humphry Repton were involved with landscaping. The gardens are renowned for their roses, herbaceous and lakeside borders, shrubberies and stunning bedding displays. The riverbank is a haven for wildlife with native trees, shrubs and wild flowers.

GROUP OPENING

HYDE PARK ESTATE GARDENS, W2

Kendal Street, W2 2AN. Church Commissioners for England, www.hydeparkestate.com. *The Hyde Park Estate is bordered by Sussex Gardens, Bayswater Rd and Edgware Rd. Nearest tube stations incl. Marble Arch, Paddington and Edgware Rd.* Sat 17 Apr (10-4). Combined adm £5, chd free. Pre-booking essential, please visit www.ngs.org.uk for information & booking.

CONISTON COURT
Church Commissioners for England, www.hydeparkestate.com.

DEVONPORT
Church Commissioners for England.

THE QUADRANGLE
Church Commissioners for England.

NEW ◆ **REFLECTIONS 2020**
Mr Mark McKeown.

THE WATER GARDENS
Church Commissioners for England.

Five gardens only open to the public through the National Garden Scheme. Each garden planted sympathetically to reflect the surroundings and featuring a new garden created during 2020. The gardens on the Hyde Park Estate are owned and managed by the Church Commissioners for England and play a key part in the environmental and ecological strategy on the Hyde Park Estate. The Estate overs 90 acres of which 12.5% is 'green' - not only with the garden spaces but by installing planters on unused paved areas, green roofs on new developments and olive trees throughout Connaught Village. Most of the gardens can be accessed by wheelchairs. There are some steps at The Water Gardens for the upper levels.

&. 🐕 ☕

9 IMBER PARK ROAD, KT10

Esher, KT10 8JB. Jane & John McNicholas. *½ m from centre of Esher. From the A307, turn into Station Rd which becomes Ember Lane. Go past Esher train station on R. Take 3rd rd on R into Imber Park Rd.* Sun 30 May (1-5). Combined adm with 40 Ember Lane £5, chd free. Home-made teas.
An established cottage style garden, always evolving, and designed and maintained by the owners who are passionate about gardening and collecting plants. The garden is S-facing, with well stocked, large, colourful herbaceous borders containing a wide variety of perennials, evergreen and deciduous shrubs, a winding lawn area and a small garden retreat. There is a short gravel path at the side of the house.

🌾 ☕

SPECIAL EVENT

THE INNER AND MIDDLE TEMPLE GARDENS, EC4

Crown Office Row, Inner Temple, EC4Y 7HL. The Honourable Societies of the Inner and Middle Temples, www.innertemple.org.uk/www.middletemple.org.uk. *Entrance: Main Garden Gate on Crown Office Row, access via Tudor Street gate or Middle Temple Lane gate.* Tue 15 June (10.30-3). Adm £55. Pre-booking essential, please visit www.ngs.org.uk/special-events for information & booking. Light refreshments.
Inner Temple Garden is a haven of tranquillity and beauty with sweeping lawns, unusual trees and charming woodland areas. The well known herbaceous border shows off inspiring plant combinations from early spring through to autumn. The award winning gardens of Middle Temple are comprised of a series of courtyards. Adm incl conducted tour of the gardens by Head Gardeners. Light lunch in Middle Temple Hall, subject to confirmation. Please advise in advance if wheelchair access is required.

&. ☕

GROUP OPENING

KEW GREEN GARDENS, TW9

Kew, TW9 3AH. *NW side of Kew Green. Tube: Kew Gardens. Mainline Stn: Kew Bridge. Buses: 65, 391. Entrance via riverside.* Sun 23 May (2-5). Combined adm £8, chd free. Evening opening Sun 30 May (6-8). Combined adm £10, chd free. Wine.

65 KEW GREEN
Giles & Angela Dixon.

67 KEW GREEN
Lynne & Patrick Lynch.

69 KEW GREEN
John & Virginia Godfrey, virginiagodfrey69@gmail.com. **Visits also by arrangement in June.**

71 KEW GREEN
Mr & Mrs Jan Pethick, linda@bpethick.co.uk. **Visits also by arrangement in June.**

73 KEW GREEN
Sir Donald & Lady Elizabeth Insall.

These five adjacent long gardens run for 100 yds from the back of historic houses on Kew Green down to the Thames towpath. Together they cover nearly 1½ acres, and in addition to the style and structures of the individual gardens they can be seen as one large space, exceptional in London. The borders between the gardens are mostly relatively low and the trees and large shrubs in each contribute to viewing the whole, while roses and clematis climb between gardens giving colour to two adjacent gardens at the same time. We try to have music as the sound carries through the five gardens. South East Regional Winner, The English Garden's The Nation's Favourite Gardens 2019.

🌾 ☕

KING HENRY'S WALK GARDEN, N1

11c King Henry's Walk, N1 4NX. Friends of King Henry's Walk Garden, www.khwgarden.org.uk. *Buses incl: 21, 30, 38, 56, 141, 277. Behind adventure playground on KHW, off Balls Pond Rd.* Mon 3 May (2-4.30). Adm £3.50, chd free. Home-made teas. Donation to Friends of KHW Garden.
Vibrant ornamental planting welcomes the visitor to this hidden oasis leading into a verdant community garden with secluded woodland area, bee hives, wildlife pond, wall trained fruit trees, & plots used by local residents to grow their own fruit and vegetables. Disabled access WC.

&. 🌾 ☕

All entries are subject to change. For the latest information check ngs.org.uk

LAMBETH PALACE, SE1
Lambeth Palace Rd, SE1 7JU.
The Church Commissioners,
www.archbishopofcanterbury.org.
*Entrance via Main Gatehouse facing
Lambeth Bridge. Station: Waterloo.
Tube: Westminster, Vauxhall all
10 mins walk. Buses: 3, C10, 77,
344, 507.* **Evening opening Mon
17 May (5-9). Adm £6, chd free.
Wine.**
Lambeth Palace has one of the
oldest and largest private gardens
in London. It has been occupied by
Archbishops of Canterbury since
1197. Formal courtyard boasts
historic White Marseilles fig planted
in 1556. Parkland style garden
features mature trees, woodland
and native planting. There is a
formal rose terrace, summer gravel
border, scented chapel garden and
active beehives. Garden Tours will
be available. Ramped path to rose
terrace, disabled WC.

NEW 24 LANGTON AVENUE, N20
Whetstone, N20 9DA. **Quentin &
Xihomara Zentner.** *Tube: Totteridge:
Buses 263;125;234. 10 mins walk
from Whetstone High St. M&S on R.
Turn R at Buckingham Ave. House
behind conifers corner Langton &
Buckingham. Parking in side road.*
**Sun 6 June (2-6). Adm £4.50, chd
free. Light refreshments.**
Contemporary front and rear garden
with a big heart. Designed by Chelsea
winner Jilayne Rickards as part of
house renovation. With naturalistic
planting, it reflects the lives of its
owners incorporating Corten steel
screens (Arabic motifs), water feature,
fire pit and ample covered seating
area. The garden is transformed by
lighting into an intimate space at
night. Corten steel hoop sculpture.
Winner of a prestigious BALI award.

There are brilliant plant
sales at many gardens.
Look out for the symbol in
the garden description –
and don't forget to bring
a bag to carry your plants
home in

12 LANSDOWNE ROAD, W11
W11 3LW. **The Lady Amabel Lindsay.**
*Tube: Holland Park. Buses: 12, 88, 94,
148, GL711, 715 to Holland Park, 4
mins walk up Lansdowne Rd.* **Wed 5
May (2.30-6.30). Adm £5, chd free.
Light refreshments.**
A country garden in the heart of
London. An old mulberry tree, billowing
borders, rambling Rosa banksiae,
and a greenhouse of climbing
pelargoniums. Partial wheelchair
access to level paved surfaces.

12 LAURADALE ROAD, N2
Fortis Green, N2 9LU. **David
Gilbert and Mary Medyckyj,
www.sites.google.com/site/
davidgilbertportfolio/garden.**
*300 metres from 102 & 234 bus
stops. 500 metres from 43 & 134
bus stops. 10 min walk from East
Finchley Underground Stn Look for
arrows.* **Sun 5 Sept (12-5.30). Adm
£4, chd £1. Home-made teas.**
Exotic, huge, recently-designed
garden, featuring tropical/
Mediterranean-zone plants. Dramatic,
architectural planting incl bananas,
large tree ferns and rare palms, weave
along curving stone paths, culminating
in a paradise garden. A modern
take on the rockery embeds glacial
boulders amid dry zone plants, incl
many succulents. New developments
this year. Sculptures by artist owner.

GROUP OPENING

NEW LEE GROUP, SE12
Lee, SE12 8LJ. *Walkable distance
between these Lee gardens.
Southbrook rd is off S. circular,
nr Burnt Ash rd. tns: Lee & Hither
Green. Buses: P273, 202 Waite
Davies rd, stns Lee (10min), Hither
Green & Grove Pk (20min). Buses:
202, 261 &160 stop near.* **Sat 29
May (2-5.30). Combined adm £7,
chd free. Home-made teas.**

**41 SOUTHBROOK ROAD,
SE12**
Barbara & Marek Polanski.
(See separate entry)

**NEW 13 WAITE DAVIES ROAD,
SE12 0NE**
Janet Pugh.

**NEW 15 WAITE DAVIES ROAD,
SE12 0NE**
Will Jennings.

One large garden and two small
make up this diverse group in south
east London. The two small ones,
next door to each other, make the
very best of their spaces with roses
winding up and down and in and
out, a mass of perennial colour and
delights in every corner. The third is
an elegant and spacious garden with
wonderful structures and ponds and
yet even more wonderful rambling
and scrambling roses!

49 LEE ROAD, SE3
Blackheath, SE3 9RT. **Jane Glynn
& Colin Kingsnorth.** *5 mins walk
from Blackheath mainline stn. Turn R
up hill to mini r'about. Take lst rd past
Blackheath Halls, 5 mins downhill on
Lee Rd, on. L. Buses: 202, 89, 54 or
108 to Blackheath.* **Sun 13 June (2-5).
Combined adm with 28 Granville
Park £8, chd free. Home-made teas.**
Escape all signs of city life and unwind
for an hour or two in this sprawling
acre of calm. Take a stroll up formal
lawns and chose your route through
the hornbeam hedges into a diverse
mix of wild and formal planting.
Discover rambling roses, woodland
walk, and a well stocked vegetable
plot. Don't forget to stop and soak
up the beautiful surroundings at one
of the many benches along the way.
Prolific vegetable garden, shady
seating, many roses. Side access to
garden for wheelchairs.

16 LINKS VIEW ROAD, TW12
Hampton Hill, TW12 1LA. **Guy &
Virginia Lewis.** *5 mins walk from
Fulwell station. On 281,267,285
and R70 bus routes.* **Sat 29, Sun
30 May (2-6). Adm £4, chd free.
Home-made teas. incl gluten
free. Cream teas and herbal teas.**
A surprising garden featuring acers,
hostas and fern collection and
other unusual shade loving plants.
Many climbing roses, clematis and
herbaceous border. Rockery and folly
with waterfall, bog garden and small
pond, with grotto. A formal pond. A
mini meadow with chickens. Summer
house and Greenhouse with tender
pelargonium collection. A verandah
with pots, further seating area with
exotic plants. Backyard chickens Folly
with shell grotto Greenhouse. Many
seating areas for teas. Children's
quiz. 2 very friendly dogs. Wheelchair
access with help.

49 LOFTUS ROAD, W12

W12 7EH. Emma Plunket,
emma@plunketgardens.com,
www.plunketgardens.com.
*Shepherds Bush or Shepherds Bush
Market tube, train or bus to Uxbridge
Rd. Free street parking.* **Sun 13
June (2.30-6.30). Adm £5, chd
free. Light refreshments. Visits
also by arrangement June to Aug
for groups of 5 to 10.**
Professional garden designer Emma
Plunket opens her acclaimed walled
garden. Richly planted, it is the ultimate
hard working city garden with all year
structure and colour incorporating fruit
and herbs. Set against a backdrop of
trees, this city haven is unexpectedly
open and peaceful. Garden plan, plant
list and advice available.

GROUP OPENING

LONDON FIELDS GARDENS, E8

Hackney, E8 3JW. *7 mins walk from
67, 149, 242, 243 bus stop Middleton
Rd, 10 mins from 30, 38, 55 stops on
Dalston Lane. 7 mins from Haggerston
Overground or 10 mins walk through
London Fields from Mare St buses*
**Sun 23 May (2-5.30). Combined
adm £8, chd free. Home-made teas
at 61 Mapledene Road.**

84 LAVENDER GROVE, E8
Anne Pauleau,
a.pauleau@hotmail.co.uk.
**Visits also by arrangement Mar
to Oct for groups of up to 20.**

53 MAPLEDENE ROAD
Tigger Cullinan, 020 7249 3754,
tiggerine8@blueyonder.co.uk.
**Visits also by arrangement May
to Aug for groups of up to 10.**

55 MAPLEDENE ROAD
Amanda & Tony Mott.

61 MAPLEDENE ROAD
Ned & Katja Staple.

NEW 92 MIDDLETON ROAD
Mr Richard & Mrs Louise Jarrett.

A fascinating and diverse collection of
gardens in London Fields within easy
walking distance of each other. South
facing are twin gardens in Lavender
Grove, one a courtyard with tropical
backdrop, and its other half a highly
scented, romantic cottage garden. In
Middleton Road, an elegant garden
with a circular theme incl roses, acers
and examples of stone lettering. The
Mapledene Road gardens are north
facing and have much the same

space but totally different styles. 53 is
an established plantaholic's garden in
4 sections, with not a spare unplanted
inch. 55 has a Moorish-inspired
terrace leading to a wildlife garden
with plants chosen to attract birds,
bees and butterflies. 61 is a family
garden with a wide English lawn
where teas are served surrounded by
roses and a wild flower meadow.

2 LONSDALE SQUARE, N1

N1 1EN. Jenny Kingsley,
www.artisticmiscellany.com.
*Barnsbury. Tube: Highbury &
Islington or Angel. Along Liverpool
Rd, walk up Richmond Ave. 1st R,
entrance via passageway on R.* **Sun
6 June (2-6). Adm £3, chd free.
Also open Barnsbury Group.**
One could describe our garden as
a person: small and unpretentious.
Attractively formed by yew & box
hedges and beds with hellebores,
fatsia, hydrangeas, arum lily &
lively climbing roses, star jasmine
& solanum. Plantors with olive
trees, lavender & pansies are fine
companions, mauve, white & emerald
favoured colours. She walks delicately
on cobblestones, a fountain calms
her soul.

10 LORIS ROAD, W6

Hammersmith, W6 7QA. Mrs
Cordelia Fraser Trueger. *Loris Rd
is a cul-de-sac off Lena Gardens,
located behind Shepherds Bush
Rd, midway between Hammersmith
Underground & Shepherds Bush
Underground stations.* **Evening
opening Sun 23 May (5-7). Adm
£5, chd free. Light refreshments.
Home-made pizza.**
Long narrow garden designed
by award winning Jo Thompson.
This garden is separated into four
distinct rooms by clever use of
3D hardscaping, so as to obscure
surrounding houses. Espaliered apple
trees, steel-framed pergolas, seating
areas & a mix of ground textures
together with a working pizza oven.
Host to more than 20 different roses
and 15 clematis as well as many
perennials.

38 LOVELACE ROAD, SE21

Dulwich, SE21 8JX. José & Deepti
Ramos Turnes. *Midway between
West Dulwich & Tulse Hill stations.
Buses: 2, 3 & 68.* **Sun 27 June
(12.30-4.30). Adm £4, chd free.**

Home-made teas.
This gem of a garden has an all-white
front and a family friendly back. The
garden slopes gently upwards with
curving borders, packed with an
informal mix of roses, perennials and
annuals. The garden is designed to
be an easy to maintain oasis of calm
at the end of a busy day. There are
several smile-inducing features e.g.
a dragon, the Cheshire cat and a
stream. Lots of seating and delicious
cake. Here you'll find: a brook with
stepping stones and wild-life ponds,
wide curving borders packed with
colourful perennials - plants for shade,
some that love full sun and several
that thrive on neglect, two magnificent
40 year old acers, raised vegetable
beds & fruit trees and a children's
play area.

GROUP OPENING

LOWER CLAPTON GARDENS, E5

8 Almack Road, Lower Clapton,
E5 0RL. *Lower Clapton. 12 mins
walk from Hackney Central, Hackney
Downs or Homerton stns. Buses 38,
55, 106, 242, 253, 254 or 425, alight
Lower Clapton Rd or Powerscroft
Rd.* **Sun 13 June (2-6). Combined
adm £6, chd free. Home-made
teas at 77 Rushmore Rd.**

8 ALMACK ROAD
Philip Lightowlers.

NEW 51 GLENARM ROAD
Ms Alice Chadwick.

77 RUSHMORE ROAD
Penny Edwards.

Lower Clapton is an area of mid
Victorian terraces sloping down to the
River Lea. These gardens reflect their
owner's tastes and interests. One
new garden this year: 15 Glenarm
Rd features lush, English gardening
with Mediterranean overtones plus
a miniature allotment. 77 Rushmore
Rd has a fruit and vegetable garden
and wildlife pond. 8 Almack Rd is
a long thin garden with two rooms,
one cool and peaceful the other with
hot colours, succulents and tropical
foliage.

GROUP OPENING

LYNDHURST SQUARE GARDEN GROUP, SE15
Lyndhurst Square, SE15 5AR. Lyndhurst Square Garden Group. *Rail & overground services to Peckham Rye Station or by bus.* **Sun 13 June (1-5.30). Combined adm £6, chd free. Home-made teas in the square. Teas, coffee and soft drinks with delicious home-made cakes. Donation to MIND, Mental Health Charity.**

5 LYNDHURST SQUARE
Martin Lawlor & Paul Ward.

6 LYNDHURST SQUARE
Iain Henderson & Amanda Grygelis.

7 LYNDHURST SQUARE
Pernille Ahlström & Barry Joseph.

Three very attractive gardens open in this small, elegant square of 1840s listed villas located in Peckham SE London. Each approx 90ft x 50ft has its own shape and style as the Square curves in a U shape. Number 5, the design combines Italianate and Gothic themes with unusual herbaceous plant and towering Echiums. Plants for sale. Number 6 the design features architectural plants, pergola, vegetable garden and espalier apple trees. Number 7 Simplicity, Swedish style, is key with roses and raised beds, framed by yew hedges. Unfortunately there isn't wheelchair access due to narrow paths and uneven surfaces.

🐾 ❄ ☕

NEW **MAGGIE'S, SM2**
17 Cotswold Road, Sutton, SM2 5NG. *Maggie's at The Royal Marsden is located on the corner of Cotswold Road via the staff entrance to The Royal Marsden NHS Foundation Trust.* **Sat 8 May (11-3.30). Adm £4. Light refreshments.**
The garden surrounding the centre designed by world-famous Dutch Landscape Architect Piet Oudolf, has many different zones, some enjoying full sun and others in dappled shade, whilst the pathway from the hospital meanders under mature trees. Plant communities are carefully chosen, incl 14 different grasses, a palette of 6 hardy ferns and more than 50 different perennials.

☕ 🏕

MAGGIE'S WEST LONDON, W6
Charing Cross Hospital, Fulham Palace Road, Hammersmith, W6 8RF. *Follow Fulham Palace Rd from Hammersmith station towards Charing Cross Hospital. The centre is on the corner of St Dunstan's Rd & is painted tomato-orange so is very visible.* **Sat 24 Apr (10-2). Adm £4, chd free. Tea.**
The garden at Maggie's West London was designed by Dan Pearson in 2008. It is now a well-established space offering therapy and peace to those affected by cancer each year. The gardens surround the vivid orange walls of the centre. The path leading to the centre meanders through scented beds and mature trees. Visitors have access to various courtyards with a wonderful array of flora incl fig trees, grape vines and even a mature pink silk mimosa. Wheelchair access ground floor gardens and courtyards are accessible. Roof gardens not accessible.

♿ ☕

GROUP OPENING

MALVERN TERRACE GARDENS, N1
Malvern Terrace, N1 1HR. *Barnsbury, Islington. Malvern Terrace is off Thornhill Rd (nr The Albion PH) between Hemingford Rd & Liverpool Rd.* **Sun 18 Apr (2.30-5.30). Combined adm £4, chd free. Home-made teas.**
Front gardens in this pretty cobbled cul-de-sac of terraced houses, overlooking Thornhill Road Gardens in the heart of Barnsbury. Wisteria & roses contribute to the picturesque street. Delicious home-made teas & music add a sense of festivity enjoyed by neighbours & visitors from further afield. One of Islington's hidden gems - an oasis of peace in the city. Live music. Access for wheelchairs but cobbles are hard going.

♿ 🐾 ☕

---

4 MANNINGTREE ROAD, HA4
Ruislip, HA4 0ES. Costas Lambropoulos & Roberto Haddon. *Manningtree Rd is just off Victoria Rd, 10-15 mins walk from South Ruislip tube station.* **Sun 1 Aug (2-6). Adm £5, chd free. Home-made teas. Cakes and savouries. Home-made jams and biscuits also for sale.**

Compact garden with an exotic feel that combines hardy architectural plants with more tender ones. A feeling of a small oasis incl plants like Musa Basjoo, Ensette Montbelliardii, tree ferns, black bamboo etc. Potted mediterranean plants on the patio incl a fig tree and two olive trees.

❄ ☕

---

53 MANOR ROAD, N16
Stoke Newington, N16 5BH. Jonathan Trustram, 07808 870800, jonathantrustram@live.co.uk. *Nr Stoke Newington station & Heathland Rd 106 bus stop.* **Sun 23 May (3-6.30); Sun 19 Sept (1-5). Adm £5, chd free. Home-made teas. Jams and chutneys for sale. Visits also by arrangement Mar to Nov for groups of up to 10.**
Big garden for London, thickly enclosed by ivy, roses and jasmine, crowded with plants, many unusual: eryngiums, thalictrums, salvias, pelargoniums, eucomis, inulas, lilies, indigofera, azara, myrtle. Small sculptural rock garden. Soft fruit. Lots of poorly policed self-seeders. Organic credentials finally lost in 30 years war against slugs. Popular plant sale.

🐾 ❄ ☕

NEW **1 MAPLE CLOSE, SW4**
Clapham, SW4 8LL. Brian Hannath. *Turn into Clarence Ave from Kings Ave. Maple Close 1st on L.* **Sun 4 July (2-5). Adm £3.50, chd free. Also open 152a Victoria Rise.**
A 'dell' garden inspired by Arthur Rackham. Green oasis of ferns, ivy, hostas and fish pond. Rustic vine-covered arch leads to formal garden with box hedging containing flower beds, gazebo and lion head wall fountain. The very small plot is inspirational for anyone with a tiny town garden.

🐾

MARIE CURIE HOSPICE, HAMPSTEAD, NW3
7 Deansway, NW3 5NS. Tracy Annunziato. *In the heart of Hampstead. Nearest tube: Belsize Park. Buses: 46, 268 & C11 all stop nr the Hospice.* **Sat 12 June (2.30-5). Adm £3.50, chd free. Cream teas. and a range of other refreshments available.**
This peaceful and secluded two part garden surrounds the Marie Curie Hospice, Hampstead. A garden, tended by dedicated volunteers, makes for a wonderful space

for patients to enjoy the shrubs and seasonal colourful flowers. The garden has seating areas for relaxation either in the shade or in the sunshine with great views of the garden and in company with squirrels running through the trees. Step free access to garden and WC.

GROUP OPENING

MARKSBURY AVENUE GARDENS, TW9

Richmond, TW9 4JE. *Approx 10 mins walk from Kew Gardens tube. Exit westbound platform to North Rd. Take 3rd L into Atwood Ave. Marksbury Ave is 1st R. Buses 190, 419 or R68.* **Sun 6 June (2.30-5.30). Combined adm £7, chd free. Home-made teas at 60 Marksbury Avenue.**

26 MARKSBURY AVENUE
Sue Frisby.

59 MARKSBURY AVENUE
Clarissa and Michael Fletcher.

60 MARKSBURY AVENUE
Gay Lyle.

61 MARKSBURY AVENUE
Siobhan McCammon.

NEW 62 MARKSBURY AVENUE
Sarah Halaka.

Although of similar size, these five neighbouring gardens in Kew are all very different. The first includes many New Zealand natives, a camomile lawn and several fruit trees incl fig, apricot and vines. On the other side of the road are two more gardens. One leads through several concealed areas, using shade from a variety of ornamental trees, to a small vegetable garden. Next door is an English cottage garden with a touch of wildness comprising different 'rooms' containing a variety of plants incl hostas, geum and astrantia. There is a pond with frogs and newts and a parrotia persica provides dappled shade for alliums and ferns. Crossing the road visitors will find a new family garden, created from scratch in 2018, while next door delicious homemade cakes and tea will be served on the lawn while visitors can enjoy a fine display of patio roses and, perhaps, identify over twenty varieties of clematis. All five gardens are within two minutes' walk. Wheelchair access at all except No. 59.

36 MELROSE ROAD, SW18

36 Melrose Road Wandsworth, SW18 1NE. John Tyrwhitt. *¼ m E of A3 Wandsworth. Entrance behind wooden gates is on Viewfield Rd (on corner of Melrose Rd).* **Sun 23 May (2-6). Adm £4.50, chd free. Tea. Cake and wine also available.**
Unusual walled patio garden, providing colour from April through October and year-round interest. Densely planted with roses, architectural plants, shrubs and climbers. Private and not overlooked. In May roses, geraniums and geums should be in full flow, with acers at their freshest. As summer closes, agapanthus, repeat flowering roses and geraniums are accompanied by exotic cannas and dahlias.

41 MILL HILL ROAD, W3

Acton, W3 8JE. Marcia Hurst, 020 8992 2632/07989 581940, marcia.hurst@sudbury-house.co.uk. *Tube: Acton Town, cross zebra crossing, turn R, Mill Hill Rd 2nd R off Gunnersbury Lane. Many local Buses & Overground.* **Sun 22 Aug (2-6). Combined adm with 65 Mill Hill Road £6, chd free. Home-made teas. Visits also by arrangement June to Aug for groups of 5 to 20.**
Large S-facing garden terrace with pots, lawn bordered by lavender hedge & box topiary. Border incl. salvias, clematis & annuals looking best in July & Aug. In 2019 the owner acquired a large adjoining plot doubling the size of the garden, creating a meadow, gravel garden & pond, mature trees & native hedges for a wildlife garden. Lots of space to sit & enjoy the garden. Good selection of the plants growing in the garden are for sale in pots with planting and growing advice from the knowledgeable owner.

65 MILL HILL ROAD, W3

W3 8JF. Anna Dargavel, 07802 241965, annadargavel@mac.com. *Tube: Acton Town, turn R, Mill Hill Rd 2nd R off Gunnersbury Lane.* **Sun 22 Aug (2-6). Combined adm with 41 Mill Hill Road £6, chd free. Visits also by arrangement May to Sept for groups of 5 to 20.**
Garden designer's own garden. A secluded and tranquil space, paved, with changes of level and borders. Sunny areas, topiary, a greenhouse

and interesting planting combine to provide a wildlife haven. A pond and organic principles are used to promote a green environment and give a stylish walk to a studio at the end of the garden.

25 MULBERRY WAY, E18

South Woodford, E18 1EB. Mrs Laura Piercy-Farley. *100 metres from South Woodford tube station. ¼ m M11 JW. Public transport central line to South Woodford use westbound exit. Cross the pedestrian crossing & turn L. Garden is 100 meters on the R opp public car park.* **Sun 27 June (1-5). Adm £3, chd free. Light refreshments.**
A pretty Victorian terraced London house with a dog friendly Italian patio style garden. The garden has a tranquil white theme with a preference for white hydrangeas. The garden has all year interest with box hedges, bay trees evergreen shrubs and climbers. There are places to sit, lounge, eat and relax.

NEW THE MYSTERIES OF LIGHT ROSARY GARDEN, NW5

St Dominic's Priory (the Rosary Shrine), Southampton Road, Kentish Town, NW5 4LB. Raffaella Morini on behalf of the Church and Priory, 07778 526434, garden@raffaellamorini.com. *Entrance to the garden is from Alan Cheales Way on the RHS of the church, next to the school.* **Sat 29 May (1.30-5.30). Adm £4.50, chd free. Home-made teas. Visits also by arrangement Apr to Sept for groups of up to 20.**
A small walled garden behind the Priory Church of Our Lady of the Rosary and St Dominic, commissioned by the Dominican Friars as a spiritual and meditative space representing the "Mysteries of Light" of the Holy Rosary. The sandstone path marks out a Rosary with black granite beads, surrounded by flowers traditionally associated with the Virgin Mary: roses, lilies, iris, periwinkle, columbine. The garden is fully accessible by wheelchair users, having a stone path and a wheelchair friendly gravel path.

15 NORCOTT ROAD, N16
Stoke Newington,
N16 7BJ. **Amanda & John
Welch,** 0208 806 5723,
amandashetlandwelch@gmail.com.
*Buses: 67, 73, 76, 106, 149, 243,
393, 476, 488. Clapton & Rectory Rd
overground stns. One way system:
by car approach from Brooke Rd
which crosses Norcott Rd, garden
in S half of Norcott Rd.* **Sun 23
May (2-6). Adm £3.50, chd free.
Home-made teas. Visits also by
arrangement for groups of up to
20. Telephone 0208 806 5723.**
A large walled garden developed by
the present owners over 40 years
with pond, aged fruit trees and an
abundance of herbaceous plants,
many available in our plant sale. We
enjoy opening at different times of
the year. This year we will be opening
in Spring. We have plenty of room
for people to sit about in the garden
enjoying their tea.

5 NORTHAMPTON PARK, N1
N1 2PP. **Andrew Bernhardt & Anne
Brogan.** *Backing on to St Paul's
Shrubbery, Islington. 5 mins walk
from Canonbury stn, 10 mins from
Highbury & Islington Tube (Victoria
Line) Bus: 30, 277, 341, 476.* **Sat
12 June (1-6). Adm £4, chd free.
Light refreshments. Prosecco &
strawberries.**
Early Victorian S-facing walled
garden, (1840's) saved from neglect
and developed over the last 25yrs.
Cool North European blues, whites
and greys moving to splashes of red/
orange Mediterranean influence. The
contrast of the cool garden shielded
by a small park creates a sense
of seclusion from its inner London
setting.

NEW **7 NORWOOD PARK ROAD,
SE27**
West Norwood, SE27 9UB.
**Miss Victoria Twyman & Mr M
McKown.** *Off Elder Rd. Nearest
stn West Norwood. Bus 432 or 10
min walk from West Norwood train
station. Some parking on Norwood
Park Rd.* **Sun 11 July (1.30-5.30).
Adm £4.50, chd free. Light
refreshments.**
The skills of two artists created
this pretty cottage style garden in
a suburban setting. Victoria does
the planting and Mark created the
structures to this 100ft garden divided
into three areas. The plants are

vibrant, and the star of the summer
show is an eye-catching plum tree
which in July hangs heavy with
luscious fruits. Artists studio doubles
up as a relaxing spot for evening
sundowners.

OAK FARM/HOMESTEAD, EN2
Cattlegate Road, Crews Hill,
Enfield, EN2 9DS. **Genine & Martin
Newport.** *5 mins. from M25 J24
& J25. Follow yellow signs. Few
mins walk from Crews Hill station.
Opp Warmadams. Entrance in
Homestead.* **Sun 6 June, Sun 12
Sept (12-4). Adm £5, chd free.
Home-made teas. Tea served in
Barn.**
In the heart of Crews Hill is our 3
acre garden & meadow. It has been
reclaimed over 30 years from pig
farm to relaxed planting, a haven
for wildlife. Walled garden leads to
veg plot, greenhouse, chickens and
orchard. Romantic woodland walk,
lawns and stone ornaments Martin
built the house, the brick walls and
metal work. Abundant cyclamen in
woodland September. Arboretum and
lovely country views.

**21 OAKLEIGH PARK SOUTH,
N20**
N20 9JS. **Carol and Robin Tullo.**
*Totteridge & Whetstone tube on
N. Line (15 min walk or 251 bus) &
Oakleigh Park station (10 mins). Also
34 & 125 from High Rd. Plenty of
street parking.* **Sun 23 May (2-6).
Adm £4, chd free. Home-made
teas.**
A Spring opening after 2020
cancellation. A mature 200ft garden
framed by a magnificent 100 year
old ash tree. Path leads to a pond
area fed by a natural spring within
landscaped terraced paving. Beyond
is a herb and vegetable area, orchard
with bulbs and wild flowers and the
working part of the garden. A mix of
sunny borders, pond marginals and
woodland shade areas with seating.
Level access to terrace and lawn.
Path up to pond area but raised levels
beyond.

GROUP OPENING

NEW **OAKWOOD GARDENS,
NW11**
NW11 6RN. **Michael
Franklin,** 07836 541383,

mikefrank@onetel.com. *Nearest
tube Golders Green. H2 bus to
Northway.* **Sun 4 July (2-5.30).
Combined adm £9.50, chd free.
Home-made teas at 92 Oakwood
Road. Visits also by arrangement
May to Aug for groups of 5 to 20.**

85 NORTHWAY
Susan Fischgrund.

NEW **92 OAKWOOD ROAD**
Tracy Harris.

94 OAKWOOD ROAD
Michael & Adrienne Franklin.

Three contrasting gardens. At
94 Oakwood Road, a very pretty
woodland garden in a tranquil setting
with herbaceous beds and many
unusual plants to buy. Next door
at 92, where teas will be served,
new owners are restoring a once
beautiful garden to its former glory. At
85 Northway, a sophisticated town
garden offers a haven of peace on
a busy suburban corner. Access to
patio giving an overall view of the
garden.

THE ORCHARD, W4
40A Hazledene Road, Chiswick,
W4 3JB. **Vivien Cantor.** *10 mins
walk from Chiswick mainline &
Gunnersbury tube. Off Fauconberg
Rd. Close to junction of A4 & Sutton
Court Rd.* **Sun 12 Sept (2-5.30).
Adm £5, chd free. Home-made
teas.**
Informal, romantic ¼ acre garden
with mature flowering trees, shrubs
and imaginative planting in flowing
herbaceous borders. Climbers, fern
planting and water features with
ponds, a bridge and waterfall in this
ever evolving garden.

ORMELEY LODGE, TW10
Ham Gate Avenue, Richmond,
TW10 5HB. **Lady Annabel
Goldsmith.** *From Richmond Park
exit at Ham Gate into Ham Gate Ave,
1st house on R. From Richmond
A307, after 1½ m, past New Inn on
R. At T-lights turn L into Ham Gate
Ave.* **Sun 20 June (3-6). Adm £5,
chd free. Home-made teas.**
Large walled garden in delightful
rural setting on Ham Common.
Wide herbaceous borders and box
hedges. Walk through to orchard with
wild flowers. Vegetable garden, knot
garden, aviary and chickens. Trellised

tennis court with roses and climbers. A number of historic stone family dog memorials. Dogs not permitted.

 ♿ ☕ 🌱

4 ORMONDE ROAD, HA6
Moor Park, Northwood, HA6 2EL. **Hasruty & Yogesh Patel.** *Approx 5m from J17 & 18, M25; 6½ m from J5, M1. From Batchworth Lane take Wolsey Rd exit at mini r'about. Ormonde Rd is 2nd turning on L. Ample parking on Ormonde Rd & surrounding rds.* **Sun 6 June, Sun 12 Sept (2-5). Adm £6, chd free. Home-made teas.**
Beautifully planted frontage entices visitors to a large rear garden. A calm oasis enclosed by mature hedging. A rare variegated flowering tulip tree provides dappled shade alongside rhododendrons, peonies, magnolias and diverse acers. Lavender hues of phlox foam along the raised patio. Much interest throughout the whole garden due to attention paid to successional planting. New lily pond. There is an area in the garden that we are trying to leave alone so it can be rewilded.

 ♿ ☕

PADDOCK ALLOTMENTS & LEISURE GARDENS, SW20
51 Heath Drive, Raynes Park, SW20 9BE. **Paddock Horticultural Society.** *Bus: 57, 131, 200 to Raynes Pk station then 10 min walk or bus 163. 152 to Bushey Rd 7 min walk; 413, 5 min walk from Cannon Hill Lane. Street parking.* **Sat 26 June (12-5). Adm £4, chd free. Light refreshments.**
An allotment site not to be missed, over 150 plots set in 5½ acres. Our tenants come from diverse communities growing a wide range of flowers, fruits and vegetables, some plots are purely organic others resemble English country gardens. Winner of London in Bloom Best Allotment on four occasions. Plants and produce for sale. Ploughman's Lunch available. Paved and grass paths, mainly level.

 ♿ ❀ ☕

11 PARK AVENUE NORTH, N8
Crouch End, N8 7RU. **Mr Steven Buckley & Ms Liz Roberts.** *Bus: 144, W3, W7. Tube: Finsbury Park or Turnpike Lane. Rail: Hornsey or Alexandra Palace.* **Sun 27 June (11.30-5.30). Adm £4, chd free. Light refreshments.**

The Orchard

An exotic 250ft garden. Dramatic, mainly spiky, foliage dominates, with the focus on palms, agaves, dasylirions, aeoniums, tree ferns, nolinas, cycads, bamboos, yuccas, bananas, cacti, puyas and many succulents. Aloes are a highlight. Trees incl peach, Cussonia spicata and Szechuan pepper. Rocks and terracotta pots lend a Mediterranean accent. Vegetables grow in oak raised beds and a glasshouse.

 ☕

18 PARK CRESCENT, N3
Finchley, N3 2NJ. **Rosie Daniels.** *Tube: Finchley Central. Buses: 13 to Victoria Park, also 125, 460, 626, 683. Walk from Ballards Lane into Etchingham Pk Rd, 2nd L Park Crescent.* **Sun 20 June, Sun 18 July (2-6). Adm £4.50, chd free. Home-made teas.**
This charming constantly evolving "Secret Garden" is designed and densely planted by the owner. Tumbling roses & clematis in June, lots of salvias and more clematis in July. Two very small ponds, tub water features, bird haven. Stepped

terrace with lots of pots. New glass installations and sculptures by owner. Have tea for two in newly created & planted secluded hideaway. Children's treasure hunt. Extensive collection of clematis and salvias. London Gardens Society: Silver Gilt Award 2019.

 ❀ ☕

36 PARK ROAD, W4
Chiswick, W4 3HH. **Meyrick & Louise Chapman.** *Adjacent to Chiswick House Gardens. Chiswick BR: 6 mins walk up Park Rd. District Line: Turnham Green 15 mins walk. Buses: E3 & 272 alight Chesterfield Rd then 4 mins walk. Free Street parking.* **Sun 6 June (2.30-6.30). Adm £4, chd free. Wine will be served for a donation.**
City garden with distinct structure and formality based on a series of rooms within hedging. Designed to create a flavour to each room using perennials, roses, hostas and ferns against a repeated background of yew, azalea and camellia. Green wall and reflective pool. Partial wheelchair access.

 ♿ ☕

56 PARK ROAD, W4
Chiswick, W4 3HH. Richard Treganowan. *Adjacent to Chiswick House. Chiswick BR: 6 mins walk up Park Rd. District Line: Turnham Green 15 mins walk. Buses: E3 & 272 alight Chesterfield Rd then 4 mins walk. Free street parking.* Sun 18 July (2.30-6.30). Adm £4, chd free.
Screaming exotica! Be prepared to be astounded by many 'different' plants which, for some, no amount of head scratching will provide an answer. Created over 25 years by Richard Treganowan and his late wife Diane. Bold and large leaved plants have been introduced which have thrived owing to an effective microclimate and attention to soil conditions. Mature stumpers Featured in The Times.

 36 PARK VILLAGE EAST, NW1
Camden Town, NW1 7PZ. Christy Rogers. *Tube: Mornington Crescent or Camden Town 7 mins. Opp railway just S of Mornington Street Bridge. Free parking all weekend.* Sun 6 June (2-6). Adm £5, chd free. Home-made teas. Also open

70 Gloucester Crescent.
A large peaceful garden behind a sympathetically modernised John Nash house. Re-landscaped in 2014, retaining the original mature sycamores and adding hornbeam hedges dividing a woodland area and orchard from a central large lawn, mixed herbaceous border and rose bank. Children enjoy an artificial grass slide. Musical entertainment is provided by young local musicians. Grass ramp down from driveway to main garden, however, much steeper than stipulated by wheelchair regulations.

GROUP OPENING

PECKHAM RYE GROUP, SE22
East Dulwich, SE22 0LP. *All gardens can be walked along the length of Peckham Rye. Stns Peckham Rye; Honor Oak. Buses: 12, 37, 63, 197 and 363.* Sun 6 June (2.30-5.30). Combined adm £8, chd free. Home-made teas.

4 CORNFLOWER TERRACE, SE22
Clare Dryhurst.

174 PECKHAM RYE, SE22
Mr & Mrs Ian Bland.

4 PIERMONT GREEN, SE22
Janine Wookey.

Small, medium and large: three picturesque gardens dotted along the length of Peckham Rye Park. One making its early transformation into a woodland space, another home to an array of unusual and eclectic plants, while the third may be tiny in size but is packed with big personality. An abundance of roses link the three.

5 PEMBERTON ROAD, KT8
East Molesey, KT8 9LG. Armi Maddison, no5workshops.com. *Please enter the garden down the side path to the RHS of the house.* Sun 13 June (2-5). Adm £4, chd free. Light refreshments.
An artist's sheltered and secluded gravel garden, designed alongside our new build in 2015. Many grasses, pink blue and white planting with occasional 'pops' of bright colour, a galvanised drinking trough with bullrushes and water lilies, a large mature central acer tree combine with several sitting areas to extend our

92 Middleton Road

living space into this fabulous outdoor room. Modern house and artist studio boarder the garden on 3 sides with large sliding doors making the garden our sheltered outside room.

23 & 24B PENN ROAD, N7
Lower Holloway, N7 9RD. Pierre Delarue & Mark Atkinson. *Between Camden Town & Holloway. Bus: 29 or 253 to Hillmarton Rd or 91 to Camden Rd stops. Tube: Caledonian Road on Piccadilly line, 6 mins walk.* **Sun 6 June (1.30-5.30). Adm £5, chd free. Home-made teas. Also open 1a Hungerford Road.**
2 neighbouring gardens famous for their planting as well as delicious teas & home-made cakes. Access through a leafy passage at No.23, fronted with prairie wilderness. The main 25x70ft back garden presents a mixture of old-fashioned roses & Mediterranean plants to achieve an exotic feel. Specimen plants incl a Red Barked Arbutus, a Santa Cruz Ironwood, an Orange Barked Myrtle & a tall 'stripped' Trachycarpus. A neoclassic studio with patio and newly installed water feature act as focal point. The 25x46ft 'pleasure garden' at 24b is now accessible through a gate & features three contrasting borders: dry, woodland & roses. A selection of refreshments served inside the Garden Studio alongside a wide range of delicious homemade Bundt cakes in a variety of interesting flavours, now integral to the NGS experience at 23 Penn Road.

101 PEPYS ROAD, SE14
New Cross, SE14 5SE. Mrs Helen Le Fevre. *Entrance via side gate. Situated at top of Telegraph Hill, off the A2. Overground New Cross Gate 15mins walk. Buses to New Cross Bus garage: 21,36,53,136,171,172,453. Street parking available.* **Sun 25 Apr (2-5.30). Adm £4.50, chd free. Light refreshments.**
This garden will surprise you with its unexpected length (160ft) and beautiful individual 'rooms' on different levels. Features incl informal planting with splendid mature trees, shrubs and an emphasis on colourful foliage, many raised organic veg beds, a hidden greenhouse and nature pond. Bee and pollinator friendly borders, with ingenious snail defences create a charming haven for wildlife.

NEW PERTH ROAD GARDEN IG2
110 Perth Road, Gant's Hill, Ilford, IG2 6AS. Mark Kenny. *5-10 mins from the M11 & the North Circular. Gant's Hill tube & Ilford train station are within walking distance - 15 mins each.* **Sat 31 July (1-5). Adm £3.50, chd free. Light refreshments.**
A suburban garden of a typical Ilford terrace house. This long, thin garden is populated by a range of plants which form a rich tapestry combing the Tropical, the Mediterranean, Japan and the UK. Cannas, bananas, buddleias and acers fight it out for space alongside a multiplicity of other plant types. A garden whose journey can be best described as Quaker Meeting Hall to Baroque Cathedral. The garden is long and and the pathways are small. Be prepared to disappear in its jungle like structure.

PETERSHAM HOUSE, TW10
Petersham Road, Petersham, Richmond, TW10 7AA. Francesco & Gael Boglione, www.petershamnurseries.com. *Stn: Richmond, Bus 65 to Dysart. Entry to garden off Petersham Rd, through Petersham Nurseries, Parking very limited on Church Lane.* **Sun 18 Apr (11-5). Adm £5, chd free. Light refreshments in the Nurseries.**
Broad lawn with large topiary, generously planted double borders. Productive vegetable garden with chickens. Adjoins Petersham Nurseries with extensive plant sales, shop and café serving lunch, tea and cake.

"I love the National Garden Scheme which has been the most brilliant supporter of Queen's Nurses like me. It was founded by the Queen's Nursing Institute which makes me very proud. As we battle Coronavirus on the front line in the community, knowing we have their support is a real comfort." – Liz Alderton, Queen's Nurse

470 PINNER ROAD, HA5
Pinner, HA5 5RR. Nitty Chamcheon. *N Harrow Station, L to T-lights, L at next T-lights, cross to be on Pinner Rd. L - 3rd house from T-lights. Parking: Pinner Rd & George V Ave - yellow lines stop after 15 yds.* **Sun 26 Sept (2-5). Adm £5, chd free. Home-made teas.**
Once (23 yrs ago) a back yard with just a lawn in the first half and the second half a jungle with a very mature apple and pear tree; now a beautiful garden. A path passing through fruit and vegetable garden to the secret log cabin after a bridge over the pond with waterfall in front of a tree house in the pear tree. An attempt has been made to extend the season as far as possible.

GROUP OPENING

PRINCES AVENUE GARDENS, N10
Muswell Hill, N10 3LS. *Buses: 43 & 134 from Highgate tube, also W7, 102, 144, 234, 299. Princes Ave opp M&S in Muswell Hill Broadway, or John Baird PH in Fortis Green.* **Sun 16 May (2-6). Combined adm £5, chd free. Home-made teas.**

17 PRINCES AVENUE
Patsy Bailey & John Rance.

28 PRINCES AVENUE
Ian & Viv Roberts.

In a beautiful Edwardian avenue in the heart of Muswell Hill Conservation Area, two very different gardens reflect the diverse life styles of their owners. The charming garden at No 17 is designed for relaxing and entertaining. Although south facing it is shaded by large surrounding trees - among which is a ginko. The garden features a superb hosta and fern display. No 28 is a well established traditional garden reflecting the charm typical of the era. Mature trees, shrubs, mixed borders and woodland garden creating an oasis of calm just off the bustling Broadway. Live music at No 17 from the Secret Life Sax Quartet, 4:30 - 5:30 pm.

GROUP OPENING

RAILWAY COTTAGES, N22
2 Dorset Road, Alexandra Palace, N22 7SL. *Tube: Wood Green, 10 mins walk. Overground: Alexandra Palace, 3 mins. Buses W3, 184. 3 mins. Free parking in local streets on Suns.* **Sun 4 July (2-5.30). Combined adm £5, chd free. Home-made teas.**

2 DORSET ROAD
Jane Stevens.

4 DORSET ROAD
Mark Longworth.

14 DORSET ROAD
Cathy Brogan.

22 DORSET ROAD
Mike & Noreen Ainger.

24A DORSET ROAD
Eddie & Jane Wessman.

A row of historical railway cottages, tucked away from the bustle of Wood Green nr Alexandra Palace, takes the visitor back in time. The tranquil country style garden at 2 Dorset Rd flanks three sides of the house. Clipped hedges contrast with climbing roses, clematis, honeysuckle, abutilon, grasses and ferns. Trees incl mulberry, quince, fig, apple and a mature willow creating an interesting shady corner with a pond. There is an emphasis on scented flowers that attract bees and butterflies and the traditional medicinal plants found in cottage gardens. No 4 is a pretty secluded garden (accessed through the rear of no 2) and sets off the sculptor owners figurative and abstract work. There are three front gardens open for view. No 14 is an informal, organic, bee friendly garden, planted with fragrant and useful herbs, flowers and shrubs. No 22 is nurtured by the grandson of the original railway worker occupant. A lovely place to sit and relax and enjoy the varied planting. No.24a reverts to the potager style cottage garden with raised beds overflowing with vegetables and flowers. Popular plant sale.

NEW 42 RISINGHOLME ROAD HA3
Harrow, HA3 7ER. Brenda White. *Buses 258, 340,182,140 Salvatorian College/St. Joseph's Catholic Church, Wealdstone. Tube/train*

Harrow & Wealdstone Station (10 mins walk or bus). Road opp the Salvatorian College. **Sun 16 May, Sun 5 Sept (11-3.30). Adm £4, chd free. Light refreshments.**
35 metre long paved suburban garden. It is divided into different areas incl a Mediterranean themed area, a raised bed vegetable garden, an aviary and summer house. Feature plants incl acers, azaleas, camellias, ferns, hydrangeas, rhododendrons, cannas and grasses as well as box hedging and topiary.

ROOFTOPVEGPLOT, W1
122 Gt Titchfield Street, W1W 6ST. Miss Wendy Shillam, 020 7637 0057, coffeeinthesquare@me.com, www.rooftopvegplot.com. *Fitzrovia. Located on the 5th floor, flat roof of a private house. Ring the doorbell marked Shillam & Smith to be let into the building.* **Sat 3, Sun 4 July (11-5). Adm £5, chd free. Pre-booking essential, please visit www.ngs. org.uk for information & booking. Light refreshments.**
A nutritional garden, where fruit & veg grow amongst complementary flowers in six inches of soil, in raised beds on a flat roof. This is a tiny garden, so tours are restricted to six persons. Home-made cakes & growing & nutritional tips from Wendy Shillam, a keen environmentalist with an extensive knowledge of green nutrition. Tomatoes & cucumber growing in a greenhouse. Grapevine, Apple, Elder, Jasmine, Japanese wineberry. Strawberries, Potatoes. Salads, garlic, annuals, roses, marigolds, nasturtiums, sweet & garden peas, climbing beans & courgettes. 2021 will see us trying Indian (climbing) spinach. No wheelchair access. 5 flights of stairs to reach the garden. Resting places on landings & in the studio at the top.

ROYAL TRINITY HOSPICE, SW4
30 Clapham Common North Side, SW4 0RN. Royal Trinity Hospice, www.royaltrinityhospice.org.uk. *Tube: Clapham Common. Buses: 35, 37, 345,137 (37 + 137 stop outside).* **Sun 5 Sept (10.30-4.30). Adm £3, chd free. Light refreshments.**
Royal Trinity's beautiful, award winning gardens play an important therapeutic role in the life and function of Royal Trinity Hospice. Over the years, many people have enjoyed our gardens and today they continue to be enjoyed by

patients, families and visitors alike. Set over nearly 2 acres, they offer space for quiet contemplation, family fun and make a great backdrop for events. Ramps and pathways.

7 ST GEORGE'S ROAD, TW1
St Margarets, Twickenham, TW1 1QS. Richard & Jenny Raworth, 020 8892 3713, jraworth@gmail.com. *1½ m SW of Richmond. Off A316 between Twickenham Bridge & St Margarets r'about.* **Evening opening Sun 23 May (6-8). Adm £7, chd free. Wine. Visits also by arrangement May to July for groups of 10 to 30.**
Exuberant displays of Old English roses and vigorous climbers with unusual herbaceous perennials. Massed scented crambe cordifolia. Pond with bridge converted into lush bog garden and waterfall. Large N-facing luxuriant conservatory with rare plants and climbers. Pelargoniums a speciality. Sunken garden Pergola covered with climbing roses and clematis. New white garden. Water feature and fernery. Reading Garden.

87 ST JOHNS ROAD, E17
Walthamstow, Walthamstow, E17 4JH. Andrew Bliss. *15 mins walk from W'stow tube/overground or 212/275 bus. Alight at St Johns Rd stop.10 mins walk from Wood St overground. Very close to N Circular.* **Sun 25, Mon 26 July (12.30-5). Adm £4, chd £0.50. Home-made teas.**
My garden epitomizes what can be achieved with imagination, design and colour consideration in a small typical terraced outdoor area. Its themes are diverse and incl a fernery, Jardin Majorelle, and 3 individual seating areas. All enhanced with circles, mirrors and over planting to create an atmosphere of tranquility within an urban environment.

27 ST PETERS SQUARE, W6
W6 9NW. Oliver & Gabrielle Leigh Wood, oliverleighwood@hotmail.com. *Tube to Stamford Brook exit station & turn S down Goldhawk Rd. At T-lights cont ahead into British Grove. Entrance to garden at 50 British Grove 100 yds on L.* **Sun 9 May (2-6). Adm £5, chd free.**

152a Victoria Rise

Home-made teas. **Visits also by arrangement Apr to July.**
This long, secret space, is a plantsman's eclectic semi-tamed wilderness. Created over the last 10yrs it contains lots of camellias, magnolias and fruit trees. Much of the hard landscaping is from skips and the whole garden is full of other people's unconsidered trifles of fancy incl a folly and summer house.

19 ST PETER'S STREET, N1
N1 8JD. Adrian Gunning. *Angel, Islington. Tube: Angel. Bus: Islington Green.* Sun 23 May (2.30-5). Adm £3.50, chd free.
Charming secluded town garden with climbing roses, trees, shrubs, climbers, pond, patio with containers, and a gazebo with a trompe l'oeil mural.

57 ST QUINTIN AVENUE, W10
W10 6NZ. Mr H Groffman, 020 8969 8292. *Less than 1m from Ladbroke Grove or White City tube. Buses: 7, 70, 220 all to North Pole Rd.* Sun 4, Sun 18 July (2-5.30). Adm £5, chd free. **Home-made teas. Visits also by arrangement July & Aug.**

Award winning 30 x 40 ft garden with a diverse selection of plants incl shrubs for foliage effects. Patio with colour themed bedding material. Focal points throughout. Clever use of mirrors and plant associations. New look front garden, now rear patio layout, new plantings for 2021 with a good selection of climbers and wall shrubs. This year's special theme celebrates the 150th anniversary of The Royal Albert Hall.

5 ST REGIS CLOSE, N10
Alexandra Park Road, Muswell Hill, N10 2DE. Ms S Bennett & Mr E Hyde, 020 8883 8540, suebearlh@yahoo.co.uk. *Tube: Bounds Green then 102 or 299 bus, or E. Finchley take 102. Alight St Andrews Church. 134 or 43 bus stop at end of Alexandra Pk Rd, follow arrows.* Sun 2 May, Sun 20 June (2-6.30). Sun 25 July (2-6.30), also open 92 Brownlow Road. Adm £4, chd free. **Home-made teas. Gluten free available. Herbal teas. Visits also by arrangement Apr to Oct for groups of 10+. Short talk on history of the garden. Cost according to catering.**
Cornucopia of sensual delights. Artist's garden famous for

architectural features and delicious cakes. New Oriental Tea House. Baroque temple, pagodas, Raku tiled mirrored wall conceals plant nursery. American Gothic shed overlooks Liberace Terrace and stairway to heaven. Maureen Lipman's favourite garden, combines colour, humour, trompe l'oeil with wildlife friendly ponds, waterfalls, weeping willow, lawns, abundant planting. A unique experience awaits! Unusual architectural features including Oriental Tea House overlooking carp pond. Mega plant sale. Open Studio with ceramics and cards. Wheelchair access to all parts of garden unless waterlogged.

2 SHARDCROFT AVENUE, SE24
Herne Hill, SE24 0DT. Catriona Andrews. *Short walk from Herne Hill rail station & bus stops. Buses: 3, 68, 196, 201, 468 to Herne Hill. Closest tube Brixton. Parking in local streets.* Sun 23 May (2-6). Adm £4.50, chd free. **Home-made teas.**
A designer's garden with a loose, naturalistic feel, planted ecologically to benefit wildlife. A pond, nesting boxes and log piles.

9 SHORTGATE, N12

Woodside Park, N12 7JP. John & Jane Owen. *Bus: 326, alight at the green on Southover, follow signs. Tube: Woodside Park, exit from northbound platform, follow signs. Parking: surrounding roads, not in Shortgate.* **Sun 4 July (1.30-5.30). Combined adm with 11 Shortgate £7, chd free. Home-made teas.**
A large secluded garden at the end of a quiet cul-de-sac. Trees screen surrounding houses, providing a wooded walkway with hidden surprises. There is a large lawn with herbaceous borders, a rockery and two small ponds, a vegetable garden and a fruit cage. 19 water butts help reduce the use of tap water, and create a naturally fed irrigation system created by owner as do drought tolerant plants, both exotic and traditional. Several steps and uneven surfaces.

❄ ☕

11 SHORTGATE

North Finchley, N12 7JP. Jennifer O'Donovan. *Bus: 326, alight at the green on Southover, follow signs. Tube: Woodside Park, exit from northbound platform, follow signs. Parking: surrounding roads, not in Shortgate.* **Sun 4 July (1.30-5.30). Combined adm with 9 Shortgate £7, chd free. Home-made teas at No.9 Shortgate.**
This corner plot is a spacious and elegant garden. It has sunny herbaceous beds, a neat vegetable patch, shade bearing trees with tranquil seating and a greenhouse busy with plants. Developed over 35 years - there wasn't even one tree when the owner arrived.

🐕 ❄ ☕

SILVERWOOD WD3

London Road, Rickmansworth, WD3 1JR. Ian & Kumud Gandhi. *3m from the M25 motorway at J18. The property is situated directly opp the Batchworth Park Golf club on London Rd. Please park at the golf club & walk across the rd to the house.* **Sun 22 Aug (2-5). Adm £5, chd free. Home-made teas.**
A well designed wrap around garden of 1-acre backing on to Moor Park Golf Club with beautiful mature oaks dating back hundreds of years, and generous borders stocked with a variety of interesting plants. It has been developed over the last 10 years with the addition of many interesting trees and shrubs as well as a herb garden used for medicinal and culinary purposes. The grounds were formerly part of the crown estate used by King Henry VIII for hunting deer at his hunting lodge Moor Park which is now the part of the Moor Park estate golf club. For wheelchair access: Drive into the garden and park closest to the side entrance where you can access the garden as the initial driveway is gravel.

♿ ☕

SOUTH LONDON BOTANICAL INSTITUTE, SE24

323 Norwood Road, SE24 9AQ. South London Botanical Institute, www.slbi.org.uk. *Mainline stn: Tulse Hill. Buses: 68, 196, 322 & 468 stop at junction of Norwood & Romola Rds.* **Sun 25 Apr (2-5). Adm £3.50, chd free. Home-made teas. Donation to South London Botanical Institute.**
London's smallest botanical garden, densely planted with 500 labelled species grown in themed borders. Spring highlights incl unusual bulbs, ferns and flowering trees. Wildflowers flourish beside medicinal herbs. Carnivorous, scented, native and woodland plants are featured, growing among rare trees and shrubs. The fascinating SLBI building is also open. Our small cafe serves home-made teas.

☕

123 SOUTH PARK ROAD, SW19

123 South Park Road, SW19 8RX. Susan Adcock. *Mainline & tube: Wimbledon, 10 mins; S Wimbledon tube 5 mins. Buses: 57, 93, 131, 219 along High St. Entrance in Bridges Rd (next to church hall) off South Park Rd.* **Sun 6 June (2-6). Adm £4, chd free. Light refreshments.**
This small, romantic L-shaped garden has a high treetop deck overlooking a woodland area, with a second deck below and small hut, paving from the garden room with pots and seating, several small water containers, a fish pond and a secluded courtyard with raised beds for flowers and herbs, as well as a discreet hot tub. Lots of ideas for giving a small space atmosphere and interest.

☕

41 SOUTHBROOK ROAD, SE12

Lee, SE12 8LJ. Barbara & Marek Polanski, 07818022983, polanski101@yahoo.co.uk. *enter through sideway to back of house. Southbrook Rd is situated off S* Circular, off Burnt Ash Rd. *Train: Lee & Hither Green, both 10 mins walk. Bus: P273, 202.* **Sun 30 May (2-5.30). Adm £3.50, chd free. Home-made teas. Opening with Lee Group on Sat 29 May. Visits also by arrangement May to July for groups of 10 to 20. Group must be accompanied by representatives or Carers.**
Developed over 14yrs, this large garden has a formal layout, with wide mixed herbaceous borders full of colour, surrounded by mature trees, framing sunny lawns, a central box parterre and an Indian pergola. Ancient pear trees festooned in June with clouds of white Kiftsgate and Rambling Rector roses. Discover fish and damselflies in 2 lily ponds. Many sheltered places to sit and relax. Enjoy refreshments in a small classical garden building with interior wall paintings, almost hidden by roses climbing way up into the trees. Orangery. Parterre and Gazebo, wall fountain, lily pond, with fish. Side access available for standard wheelchairs. Gravel driveway, a few steps in sidepath.

♿ ☕

131 SOUTHGATE ROAD, N1

N1 3JZ. John Le Huquet and Vicki Primm-Sexton. *East Canonbury. Bank or Old St tube then 21 or 141 bus to Englefield Rd stop (outside house). Highbury & Islington tube, 30 bus to Southgate Rd stop, 5 min wk. Angel tube, 38, 56 or 73 bus to Ockendon Rd, 5 min walk.* **Sun 4 July (12-6). Adm £3.50, chd free. Light refreshments.**
Open for the second time, this vivacious little walled town garden is densely planted with over 50 species of sun-loving perennials, creating an intense visual experience. The lush, naturalistic planting showcases a jamboree of jewel-like blooms weaving through softly waving grasses & delicate umbellifers. Specially commissioned Corten steel wall screens & a charming idiosyncratic shed. Garden reached by a staircase.

☕

SOUTHWOOD LODGE, N6

33 Kingsley Place, Highgate, N6 5EA. Mrs S Whittington, 020 8348 2785, suewhittington@hotmail.co.uk. *Tube: Highgate then 6 mins uphill walk along Southwood Lane. 4 min walk from Highgate Village along Southwood Lane. Buses: 143, 210,*

214, 271. Sun 2 May (2-5.30). Adm £4, chd free. Home-made teas. Visits also by arrangement Apr to Aug for groups of up to 30. Lunch for groups of 10+ or teas (any number) by arrangement.
Densely planted garden hidden behind C18 house (not open). Many unusual plants, some propagated for sale. Ponds, waterfall, frogs, toads, newts. Topiary shapes formed from self sown yew trees. Sculpture carved from three trunks of a massive conifer which became unstable in a storm. Hard working greenhouse! Only one open day this year so visits by appointment especially welcome. Secret Life Sax Quartet will perform in the garden from 2.30pm.

◆ SPENCER HOUSE, SW1
27 St James' Place, Westminster, SW1A 1NR. RIT Capital Partners, www.spencerhouse.co.uk. *From Green Park tube: Exit station on south side, walk down Queen's Walk, turn L through narrow alleyway. Turn R & Spencer House will be in front of you.* For opening times and information, please visit garden website.
Originally designed in the eighteenth century by Henry Holland (son-in-law to Lancelot 'Capability' Brown), the garden was among the grandest in the West End. Restored since 1990 under the Chairmanship of Lord Rothschild, the garden, with a delightful view of the adjacent Royal Park, now evokes its original layout with planting suggested by early nineteenth-century nursery lists.

GROUP OPENING

SPITALFIELDS GARDENS, E1
E1 6QE. *Nr Spitalfields Market, 10 mins walk from Aldgate E Tube & 5 mins walk from Liverpool St stn. Overground: Shoreditch High St - 3 mins walk.* Sat 12 June (11-4). Combined adm £15, chd free. Home-made teas at Town House (5 Fournier St) 29 & 31 Fournier St.

26 ELDER STREET
The Future Laboratory.

20 FOURNIER STREET
Ms Charlie de Wet.

29 FOURNIER STREET
Juliette Larthe.

31 FOURNIER STREET
Tom Holmes.

21 PRINCELET STREET
Marianne & Nicholas Morse.

37 SPITAL SQUARE
Society for the Protection of Ancient Buildings, www.spab.org.uk.

21 WILKES STREET
Rupert Wheeler.

Discover a selection of courtyard gardens, some very small, behind fine C17 French Huguenots merchants' & weavers' houses in Spitalfields. Visit the courtyard of 37 Spital Square, on the site of the C12 priory of St Mary's Spital, now the Society for the Protection of Ancient Buildings, founded by William Morris in 1877. Experience a 'vertical' garden & roof terrace in nearby Elder Street, an architect-designed garden in Wilkes Street, three small courtyards in Fournier Street, & a larger imaginatively created garden in Princelet Street. Each garden owner has adapted their particular urban space to complement an historic house: vegetables, herbs, vertical & horizontal beds, ornamental pots, statuary & architectural artefacts abound . Plants for sale at SPAB.

25 SPRINGFIELD AVENUE, N10
Muswell Hill, N10 3SU. Heather Hampson & Nigel Ragg. *Buses 102 299 W7 134 43 From main r'about at Muswell Hill descend towards Crouch End. Springfield Av 1st L.* Sun 4 July (2-6.30). Adm £4, chd free. Home-made teas.
A mystical and secluded garden packed with home ideas. Spread over 3 atmospheric terraces up to a backdrop of the trees of Alexandra Palace. Travel through perennial planted beds of hydrangeas and climbing roses shaded by the old apple tree. The middle terrace is lawn surrounded by shrubs and climbers. The south facing paved terrace has sunny flower beds, tub water features and pots. This is a garden full of surprises from the secluded setting to the sense of journey from a medium sized city garden. Three main terraces rising to a south facing summer house. It has its challenges as the bottom is north facing and heavy clay to the top which is south facing and hot. Uneven steps.

STOKES HOUSE, TW10
Ham Street, Ham, Richmond, TW10 7HR. Peter & Rachel Lipscomb, 020 8940 2403, rlipscomb@virginmedia.com. *2m S of Richmond off A307. Trains & tube to Richmond & train to Kingston which link with 65 bus stopping at Ham Common every 6 mins.* Sun 6 June (2-5). Adm £4, chd free. Home-made teas. Visits also by arrangement May to Sept for groups of 20 to 30.
Originally an orchard, this ½ acre walled country garden surrounding Georgian house (not open) is abundant with roses, clematis and perennials. There are mature trees incl ancient mulberries and wisteria. The yew hedging, pergola and box hedges allow for different planting schemes throughout the year. Supervised children are welcome to play on the slide and swing. Herbaceous borders, brick garden, wild garden, large compost area and interesting trees. Teas, garden tour, history of house and garden for group visits. Wheelchair access via double doors from street with 2 wide steps. Unfortunately no access for larger motorised chairs.

STONEY HILL HOUSE, SE26
Rock Hill, SE26 6SW. Cinzia & Adam Greaves. *Off Sydenham Hill. Train: Sydenham, Gipsy Hill or Sydenham Hill (closest) stations. Buses: To Crystal Palace, 202 or 363 along Sydenham Hill. House at end of cul-de-sac on L coming from Sydenham Hill.* Sun 16 May (2-6). Adm £6, chd free. Home-made teas. Chilled Prosecco will also be available to enjoy whilst listening to the saxophone quartet.
Garden and woodland of approx 1 acre providing a secluded secret green oasis in the city. Paths meander through mature rhododendron, oak, yew and holly trees, offset by pieces of contemporary sculpture. The garden is on a slope and a number of viewpoints set at different heights provide varied perspectives. The planting in the top part of the garden is fluid and flows seamlessly into the woodland. Fresalca, a wonderful saxophone quartet, will be playing for the afternoon. Swings and woodland tree-house. Please be aware that access to the main part of the garden is via shallow steps or a grassy slope alongside the steps, so assistance will be required.

Chelsea Physic Garden

10 STREATHAM COMMON SOUTH, SW16
SW16 3BT. Lindy and Mark Cunniffe. *By Streatham Common. Two nearby train stations, Streatham & Streatham Common both about 10 mins walk. Buses from Brixton incl 250,109,159, 118 & 133. Free Parking in the area.* **Sun 29 Aug (2-6). Adm £5, chd free. Light refreshments.**
This large south facing garden is divided into naturally separated spaces with many seating areas. There are many paths, a green house, woodland area, traditional lawn, small wildlife area and many mature trees such as ash, willow and mulberry. The Chinese inspired garden gives a modern twist. There is a small tranquil garden with koi pond and seating as well as an award winning front garden.

SWAKELEYS COTTAGE, 2 THE AVENUE, UB10
Ickenham, Uxbridge, UB10 8NP. Lady Singleton Booth. *Take the B466 to Ickenham from the A40 at Hillingdon Circus. Go 1m into Ickenham village. Coach and Horses PH on R, turn L into Swakeley's Rd.*

After the shops, The Avenue is on L. **Sun 4 July (2-5). Adm £4, chd free. Home-made teas.**
Classic English cottage garden. The garden has been designed by Lady Booth and her late husband Sir Christopher Booth. The garden is charming and wraps around a 600 year old cottage. It consists of herbaceous borders which are dotted with vegetables, garden herbs & fruit trees. There is an abundance of colour and some very interesting plants incorporating different styles.

93 TANFIELD AVENUE, NW2
Dudden Hill, NW2 7SB. Mr James Duncan Mattoon, 020 8830 7410. *Dudden Hill - Neasden. Nearest station: Neasden - Jubilee line then 10 mins walk; or various bus routes to Neasden Parade or Tanfield Ave.* **Sun 25 July (2-6). Adm £5, chd free. Tea. Visits also by arrangement May to Sept for groups of up to 10.**
New Chamomile lawn and newly finished Arabic style watercourse, complete a 10 year development of this Plantsman's hillside paradise garden! Raised Deco deck with panoramic views, plunges down

steps into Mediterranean and subtropical oasis, overflowing with many rare and exotic plants e.g. Caesalpinia, Hedychium, Puya,! To rear, jungle shade terrace offers cool views of paradise on sunny days. Previous garden was Tropical Kensal Rise (Doyle Gardens), featured on BBC2 Open Gardens and in Sunday Telegraph.

28 TAYLOR AVENUE, TW9
Kew, Richmond, TW9 4ED. Inma Lapena. *10 mins walk from Kew Gardens Station. Follow North Rd until Atwood Ave. Turn L & then follow to end where it becomes Taylor Ave. If driving turn R off the South Circular.* **Sun 11 July (2-6). Combined adm with 31 West Park Road £5, chd free. Home-made teas.**
The garden is about 30 metres long and about 10 metres wide. It is an urban garden typical of any semi-detached house in London. There are borders on both sides with many varieties of plants.

36 THORNHILL SQUARE, N1

N1 1BE. Anna & Christopher McKane. *Barnsbury. Tube: King's Cross, Caledonian Rd or Angel. Overground: Caledonian Rd & Barnsbury. Buses: 17, 91, 259 to Caledonian Rd.* **Evening opening Wed 21 Apr (6-8). Adm £4, chd free. Wine. Opening with Barnsbury Group on Sun 6 June.** Old and new roses incl. a huge Banksian rose, hardy geraniums, clematis. Several specimen trees in curved beds give a country garden atmosphere in this 120ft long garden, a Judas tree as a focus. Dramatic hosta collection in the front basement area. Small bonsai collection and many unusual plants propagated for sale. Unsuitable for wheelchairs.

27 THORPEWOOD AVENUE, SE26

Sydenham, SE26 4BU. Barbara Nella. *½ m from Forest Hill Stn. Off the S Circular (A205) turning up Sydenham Hill nr Horniman Gdns, or Dartmouth Rd from Forest Hill stn. Buses: 122, 176, 312 to Forest Hill library.* **Sun 30 May (1.30-5.30). Adm £4.50, chd free. Light refreshments. Also open 41 Southbrook Road.**
This ½ acre sloping garden blends a cool English woodland look with hotter Mediterranean styles and, with its bamboos and bananas, travels even further afield, all the way to the Antipodes. A particular joy lies in the garden's heart where woodland opens out to reveal a sunny glade packed with alpine delights. Barbara is a compulsive plants person - so come prepared to buy some interesting plants.

57 TONBRIDGE HOUSE, WC1H

57 Tonbridge House, Tonbridge Street, WC1H 9PG. Sue Heiser. *S of Euston Rd. Tube: King's Cross & St Pancras Stn & Russell Sq. Behind Camden Town Hall off Judd St. Turn into Bidborough St which becomes Tonbridge St. Side entrance to garden.* **Sun 4 July (2-5.30). Adm £3.50, chd free. Home-made teas. Also open 14 Doughty Street.**
Unexpected oasis off the Euston Road, overlooked on all sides by tall buildings. Mixed informal planting, with seating, rockery, pergola and shady areas. Mature magnolia, sycamore, holly and some long established shrubs and perennials incl

ferns, hostas and heuchera. Small vegetable beds and herbs. Evolved over 40 yrs on a low budget with plenty of help from friends. Wheelchair access from street and throughout garden.

TRUMPETERS' HOUSE & SARAH'S GARDEN, TW9

Trumpeters' House, Old Palace Yard, Richmond, TW9 1PD. Baroness Van Dedem. *Richmond riverside. 5 mins walk from Richmond Station via Richmond Green in Old Palace Yard. No Parking in OPY - parking on Richmond Green & nearby Old Deer Park car park.* **Sun 20 June (2-5). Adm £5, chd free. Home-made teas.**
The 2 acre garden is on the original site of Richmond Palace. Long lawns stretch from the house to banks of the River Thames. There are clipped yews, a box parterre and many unusual shrubs and trees, a rose garden and oval pond with carp. The ancient Tudor walls are covered with roses and climbers. Discover Sarah's secret gravel garden, orchard, dovecote and mid C18 summerhouse. Wheelchair access on grass and gravel.

35 TURRET GROVE, SW4

Clapham Old Town, SW4 0ES. Wayne Amiel, www.turretgrove.com. *Off Rectory Grove. 10 mins walk from Clapham Common Tube & Wandsworth Rd Mainline. Buses: 87, 137.* **Sun 18 July (10-5). Adm £4.50, chd free. Home-made teas.**
As featured on BBC Two - Gardeners' World, 2018, this north facing garden shows what can be achieved in a small space (8m x 20m). The owner, who makes no secret of disregarding the rule book, describes this visual feast of intoxicating colours as Clapham meets Jamaica. This is gardening at its most exuberant, where bananas, bamboos, tree ferns and fire bright plants flourish beside the traditional. Children very welcome.

24 TWYFORD AVENUE, N2

East Finchley, N2 9NJ. Rachel Lindsay and Jeremy Pratt, 07930 632902, jeremypr@blueyonder.co.uk. *Twyford Ave runs parallel to Fortis Green, between East Finchley &*

Muswell Hill. Tube: Northern line to East Finchley. Buses 102, 143, 234, 263 to East Finchley. Buses 43, 134, 144, 234 to Muswell Hill. Buses 102 & 234 stop at end of rd. Garden signposted from Fortis Green. **Sun 11 July (2-6). Adm £4, chd free. Home-made teas. Visits also by arrangement.**
Sunny, 120 foot S facing garden, planted for colour. Brick-edged borders and over-flowing containers packed with masses of traditional herbaceous and perennial cottage garden plants and shrubs. Shady area at rear evolving as much by happy accident as design. Some uneven ground. Water feature. Greenhouse bursting with cuttings. Many places to sit and think, chat or doze. Sale of honey & Bee products.

86 UNDERHILL ROAD, SE22

East Dulwich, SE22 0QU. Claire & Rob Goldie. *Between Langton Rise & Melford Rd. Stn: Forest Hill. Buses: P13, 363, 63, 176, 185 & P4.* **Sun 9 May (2-6). Adm £3.50, chd free. Home-made teas.**
A generous family space bursting with tulips and spring colour. Step down from the elegant slate terrace to the bustling gravel garden and then wind your way through to a green embrace. Feast on delicious cakes and relax in secluded seating areas.

NEW 333 VICTORIA PARK ROAD, FLAT 2, E9

Flat 2, Homerton, E9 5DX. Mrs E Cole, 07810 641463, elsacole01@gmail.com. *Overground - 10 mins walk from Homerton Station. Buses 30, 26, 388 alight Wick Road. Tube - Bethnal Green station + 388 bus.* **Sun 13 June (1-5.30). Adm £3, chd free. Light refreshments. Visits also by arrangement June to Sept for groups of up to 10.**
N facing urban garden in which oak sleepers manage a difference in levels equivalent to 1½ floors. Steps lead from a patio to a gravelled landing and then up to a lawn, several borders, a rose arch and a sunny seating area. The geometry is softened by fruit trees, roses, hydrangeas, camellias, ferns and geraniums. Beware: not a garden for people with mobility issues as the steps have no handrails.

152A VICTORIA RISE, SW4
Clapham, SW4 0NW. Benn Storey.
Clapham. Entry via basement flat.
Sun 4 July (12-5). Adm £4.50, chd free. Also open 1 Maple Close.
In its 5th year this terraced garden is 21m long and 8m wide. Planting ranges from the lush greens of the courtyard to the frothy, bee-friendly plants of the main level to the espalier fruit trees and vegetables of the productive levels. A copper beech hedge hides a cozy arbour seat and fire pit at the top of the plot out of view from surrounding neighbours.

 9 VIEW ROAD, N6
Highgate, N6 4DJ. Paul and Sophia Davison. *A 9 mins walk (½ m) from Highgate Tube. Buses: 134, 43, 263 to Highgate Tube, 143 to North Hill. View Rd is a turning off North Hill. Please enter the garden via the path to the LHS of house.* **Sat 15 May (2-5). Adm £4, chd free. Home-made teas.**
A well stocked, informally planted front garden leads to a generous rear garden which has been lovingly coaxed back from its wild, overgrown state by the current owners. Large lawn, beautiful hornbeam hedge spanning the garden, mature trees (incl handkerchief trees) bulbs and perennials. Fruit trees, grasses, and a greenhouse plus an enchanting tree house. A garden which continues to evolve. Front garden fully wheelchair accessible, access to the terrace part of the rear garden step free. Steps down to the lawn. Some uneven paths.

1 WADES GROVE, N21
Winchmore Hill, N21 1BH. C & K Madhvani. *Tube: Southgate then W9 to Winchmore Hill Green then short walk. Train: Winchmore Hill then short walk via Wades Hill.* **Sun 6 June (2-6). Adm £3.50, chd free. Home-made teas.**
Tiny secluded space in charming peaceful cul-de-sac as featured on the TV series presented by Dee Hart Dyke and Miranda Hart, 'All Gardens Great and Small'. Views divided by selection of mature and young trees. Planting loose and naturalistic. Focus on scented plants, edibles, ground and wall coverings. Designed to encourage wildlife. Interesting use of recycled materials. Visitors welcome to explore tranquil adjoining Quaker gardens (free) via secret entrance.

12 WARNER ROAD, N8
N8 7HD. Linnette Ralph. *Near Alexandra Palace, between Crouch End & Muswell Hill. Turning off Priory Rd. Tube to Finsbury Park then W3 bus to Hornsey Fire Station or bus 144 from Turnpike Lane. Bus W7 to Priory Road.* **Sun 13 June (12-5.30). Adm £4, chd free. Home-made teas.**
A secluded courtyard area leads through to a circular lawn which is surrounded by borders with mixed planting. The curved path leads to a circular pond with seating area. A picket fence divides this area from the kitchen garden with its raised beds, potting shed and another seating area. Used in articles to illustrate what can be done with a long, narrow, London garden.

 9 WARWICK CLOSE, TW12
Hampton, TW12 2TY. Chris Churchman. *2m W of Twickenham, 2m N of Hampton Court, overlooking Bushy Park. 100 metres from Hampton Open Air Swimming Pool.* **Sun 1 Aug (9-6). Adm £3.50, chd £1. Home-made teas.**
A small suburban garden in south west London. Front garden with espaliered roses featuring roses, lavender and stipa. Shade garden with rare ferns and herbaceous. Rear garden has a simple rectangular lawn with prairie style plantings crossed with sub tropical species.

THE WATERGARDENS, KT2
Warren Road, Kingston-upon-Thames, KT2 7LF. The Residents' Association. *1m E of Kingston. From Kingston take A308 (Kingston Hill) towards London; after approx ½ m turn R into Warren Rd. No. 57 bus along Coombe Lane West, alight at Warren Rd.* **Sun 9 May, Sun 17 Oct (2-4). Adm £5, chd free.**
Japanese landscaped garden originally part of Coombe Wood Nursery, planted by the Veitch family in the 1860s. Approx 9 acres with ponds, streams and waterfalls. Many rare trees which, in spring and autumn, provide stunning colour. For the tree lover this is a must see garden. Gardens attractive to wildlife.

GROUP OPENING

 WENSLEYDALE ROAD GARDENS, HAMPTON, TW12
Wensleydale Road, Hampton, TW12 2LX. *Hampton. Post code TW12 2LX.* **Sat 5 June (12-5). Combined adm £5, chd free. Tea at Parke House.**

> **PARKE HOUSE**
> Mr & Mrs Mike & Andrea Harris.
> **68 WENSLEYDALE ROAD**
> Julie Melotte.
> **70 WENSLEYDALE ROAD**
> Mr Steve Pickering.

Three suburban gardens, all different. 68 is a charming garden with abundant and colourful planting and interesting 'found' objects. A round lawn with roses and evergreen borders, leading to wildlife pond and bog garden. 70 is a traditional garden with classic layout. Greenhouse, summerhouse, pergolas and a rockery. Parke House at No. 74 is a large traditional garden laid to lawn. Mixed flower borders and shrubs. Feature ginkgo tree. No 68, access via narrow side entrance not accessible to wheelchairs. No 70 and 74 wheelchair accessible.

WEST LODGE PARK, EN4
Cockfosters Road, Hadley Wood, EN4 0PY. Beales Hotels. 020 8216 3904, headoffice@bealeshotels.co.uk, www.bealeshotels.co.uk/westlodgepark/. *1m S of Potters Bar. On A111. J24 from M25 signed Cockfosters.* **Sun 9 May (2-5); Sun 24 Oct (1-4). Adm £6, chd free. Light refreshments. Visits also by arrangement Apr to Oct.**
Open for the NGS for over 30yrs, the 35 acre Beale Arboretum consists of over 800 varieties of trees and shrubs, incl National Collection of Hornbeam cultivars (Carpinus betulus) and National collection of Swamp Cypress (Taxodium). Network of paths through good selection of conifers, oaks, maples and mountain ash - all specimens labelled. Beehives and 2 ponds. Stunning collection within the M25. Guided tours available. Breakfasts, morning coffee/biscuits, restaurant lunches, light lunches, dinner all served in the hotel. Please see website.

31 WEST PARK ROAD, TW9

Kew, Richmond, TW9 4DA. Anna Anderson. *Close to the E side of Kew Gardens station. From Richmond bound exit from Kew Gardens station, West Park Rd is straight ahead & No.31 is the 2nd house on the LHS.* Sun 11 July (2-6). Combined adm with 28 Taylor Avenue £5, chd free. Teas and cakes available at 28 Taylor Ave. Modern botanical garden with an oriental twist. Emphasis on foliage and an eclectic mix of unusual plants, a reflecting pool and willow screens. Dry bed, shady beds, mature trees and a private paved dining area with dappled light and shade.

12 WESTERN ROAD, E13

Plaistow, E13 9JF. Elaine Fieldhouse. *Stn: Upton Park, 3mins walk. Buses: 58, 104, 330, 376. No parking restrictions on Suns.* Sun 23 May (1-4). Adm £4, chd free. Home-made teas. Gluten free/vegan options available.
Urban oasis, 85ft garden designed and planted by owners. Relying heavily on evergreen, ferns, foliage and herbaceous planting. Rear of garden leads directly onto a 110ft allotment and a half allotment adjoining it - part allotment, part extension of the garden - featuring topiary, medlar tree, mulberry tree, 2 ponds, small fruit trees, raised beds and small iris collection.

35 WESTON PARK, N8

35 Weston Park, N8 9SY. Mrs Theresa & Mr Keith Rutter. *Tube Finsbury Park then W3 bus (Weston Park stop) or W7 (Crouch End Broadway.) Or tube to Archway then 41 bus (Crouch End Broadway).* Short walk from each one. Sun 18 July (2-6). Adm £4, chd free.
SE facing large garden with a wide range of plants, shrubs and trees suited to varying conditions incl a bog garden. Summer colour in the densely planted beds and pots incl dahlias, cannas and salvias. Elements such as golden bamboo, phormiums, grasses and sculptures provide structure. A curving path leads up to an artists's studio.

✿

33 WEYMAN ROAD, SE3

Blackheath, SE3 8RY. Kevin & Cosetta Lawlor, 07977670111, kevlawlor2000@yahoo.com. *Weyman Rd is off Shooters Hill Rd.*

Stns: Kidbrooke or Blackheath. Bus: 178 from Kidbrooke stn, 89 from Blackheath stn, 386 from Greenwich. Free parking in Weyman Rd. Sun 27 June (12-6). Adm £5, chd free. Light refreshments. Visits also by arrangement May to July for groups of 5 to 20. Please call in advance to arrange a visit.
A small but immaculate garden, that is as pretty as a picture in a wonderfully colour co-ordinated melange of pinks, whites, purples and blues. Divided into rooms to entice the visitor on, this is a couples' joint effort with hard and soft landscaping that offers unexpected delights round every corner. The garden finishes with a stunning display of succulents and an alpine rockery and waterfall.

GROUP OPENING

NEW WILLESDEN GREEN GARDENS, NW2

NW2. *Mapesbury, nr Willesden Green Tube Station & Kilburn Tube Station. Less than 10 mins walk from either tube station, plus regular buses pass along Walm Ln.* Sun 19 Sept (1-5). Combined adm £12, chd free. Home-made teas at Teignmouth Road & 106 Dartmouth Road.

106 DARTMOUTH ROAD, NW2 4HB
Ms Hester Coley.

NEW 131C DARTMOUTH ROAD, NW2 4ES
John Thorogood.

58A TEIGNMOUTH ROAD, NW2 4DX
Drs Elayne & Jim Coakes, 020 8208 0082, elayne.coakes@btinternet.com, www.facebook.com/gardening4bees.
Visits also by arrangement Apr to Sept for groups of up to 20. Small extra charge for tea and cake. All money goes to charity.

A small group of large urban gardens set in the green mature London Plane tree lined roads of a North London Arts and Crafts Conservation Area. These gardens demonstrate the richness of planting that London clay affords and the warmth provided by an urban setting. The area is rich in wildlife with abundant birdlife including locally nesting sparrow hawks; butterflies, bees and bats and a green

wildlife corridor running alongside the several rail lines that border the area. Examples of patios, planting on levels, ponds and pergolas within garden owner designed gardens are shown plus the Conservation Area incl community planted triangles and an active garden club. Water features, wildlife friendly planting, terracing; teas and cakes. Teignmouth Road garden is largely wheelchair accessible.

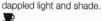

86 WILLIFIELD WAY, NW11

Golders Green, NW11 6YJ. Diane Berger, 020 8455 0455, dianeberger@hotmail.co.uk. *Hampstead Garden Suburb. Golders Green Tube Northern Line Turn R out of station - Then either 82, 102 or 460 north to Temple Fortune (3 stops).* Sun 25 July (1.30-5.30). Adm £4, chd free. Home-made teas. Visits also by arrangement June to Sept for groups of 5 to 20.
Wildlife pond surrounded by bamboo & rodgersia, opens out to spectacularly colourful deep herbaceous beds numerous examples of rudbeckia, echinacea, crocosmia, kniphofia and ornamental grasses. The clematis and wisteria laden pergola leads to an arbour seat and then on to a secret decked area surrounded by mature shrubs and trees. Winner of London Gardens Society Best Large Back Garden 2016, 2017 and 2018. Wheelchair access to front lawn only.

74 WILLIFIELD WAY, NW11

NW11 6YJ. David Weinberg, 07956 579205, davidwayne@hotmail.co.uk. *Hampstead Garden Suburb. H2 bus from Golders Green will stop outside or take the 102,13 or 460 to Hampstead Way & walk up Asmuns Hill and turn Right.* Sun 20 June (1.30-5.30). Adm £5, chd free. Cream teas. Visits also by arrangement May to Aug for groups of 5 to 20.
A very peaceful traditional English country cottage garden. Borders full of herbaceous perennials and hydrangeas with a central rose bed surrounded by a box parterre with plenty of space to sit and take it all in. Winner of Suburb In Bloom 2017 and awarded silver gilt by London Garden Society 2019 and Best Lockdown garden in Hampstead and Highgate 2020. Wheelchair access to patio area only.

♿ 🐐 ☕

"I was amazed to discover that the National Garden Scheme is Marie Curie's largest single funder and has given the charity nearly £10 million over 25 years. Their continued support makes such a difference to me and all Marie Curie Nurses on the frontline of the coronavirus crisis, as we continue to provide expert care and support to people at end of life."
– Tracy McWilliams, Marie Curie Nurse

61 WOLSEY ROAD, KT8
East Molesey, KT8 9EW. Jan & Ken Heath. *Less than 10 mins walk from Hampton Court Palace & station - very easy to find.* **Sun 6 June (2-6). Adm £5, chd free. Home-made teas.**
Romantic, secluded and peaceful garden of two halves designed and maintained by the owners. Part is shaded by two large copper beech trees with woodland planting. The second reached through a beech arch has cottage garden planting, pond and wooden obelisks covered with roses. Beautiful octagonal gazebo overlooks pond plus an oak framed summerhouse designed and built by the owners. Extensive seating throughout the garden to sit quietly and enjoy your tea and cake, either in the cool shade of the gazebo under the copper beech trees, relaxing in the summerhouse or enjoying the full sunshine elsewhere in the garden.

NEW **WOLVES LANE HORTICULTURAL CENTRE, N22**
Wolves Lane, N22 5JD. **Wolves Lane Flower Company, www.wolveslaneflowercompany.com.** *Wood Green. Situated nearest the corner of Woodside Rd & Wolves Ln. Please use the name of the centre to search on Google Maps rather than the postcode.* **Sun 19 Sept (11-3). Adm £5, chd free.**
WLFC is a micro urban flower farm with an organic & sustainable approach. It incl a 40 metre glasshouse and external growing plots housed within an old horticultural site in Wood Green. Marianne Mogendorff and Camila Klich, two flower obsessives, are committed to the belief that beauty in floristry needn't be compromised at the expense of the environment. In September the cutting garden is a joy. The centre as a whole has a palm house and cacti house. The flower farm includes a large glasshouse and smaller propagation house plus external, allotment style external plots. There is level access within Wolves Lane Horticultural Centre and wheelchair accessible toilets. Some of the external site is on a slope/has steps.

NEW **39 WOOD VALE, SE23**
Forest Hill, SE23 3DS. Nigel Crawley. *Entrance through Thistle Gates 48 Melford Rd. Train & Overground stns: Forest Hill & Honor Oak. Victoria stn to W. Dulwich, then P4. Buses: 363 Elephant & Castle to Wood Vale/Melford rd; 176, 185 & 197 to Lordship La/Wood Vale* **Sun 11 Apr (1-5). Adm £4.50, chd free. Home-made teas. Evening opening Wed 14 Apr (5-7). Adm £6, chd free. Wine.**
Diverse garden dominated by a gigantic perry pear forming part of one of the East Dulwich orchards. The emphasis in the garden is on its inhabitants; white comfrey and pear blossom keeps the bees busy in for the spring. Home to a variety of birds incl. green woodpecker, dunnock and redwing. The clumps of narcissi around the old apple tree are a sight to see. Surprising green oasis in Forest Hill. Close to Sydenham Woods, Horniman gdns and Camberwell Old Cemetery. Level but rough terrain in the lane.

33 WOOD VALE, N10
Highgate, N10 3DJ. Mona Abboud, 020 8883 4955, monaabboud@hotmail.com, www.monasgarden.co.uk. *Tube: Highgate, 10 mins walk. Buses: W3, W7 to top of Park Rd.* **Sun 18 Apr (1.30-5.30). Adm £4, chd free. Light refreshments. Visits also by arrangement May to Sept for groups of 5+.**
This 100m-long unique and award winning garden is home to the Corokia National Collection along with a great number of other unusual Australasian, Mediterranean and exotic plants complemented by perennials and grasses which thrive thanks to 250 tons of topsoil, gravel and compost brought in by wheelbarrow. Emphasis on structure, texture, foliage and shapes brought alive by distinctive pruning. Visit www.monasgarden.co.uk for more information.

7 WOODBINES AVENUE, KT1
Kingston-upon-Thames, KT1 2AZ. Mr Tony Sharples & Mr Paul Cuthbert. *Take K2, K3, 71 or 281 bus. From Surbiton, bus stop outside Waitrose & exit bus Kingston University Stop. From Kingston, walk or get the bus from Eden Street (opp Heals).* **Sun 18 Apr (12-5). Adm £4.50, chd free. Sun 27 June (12-5). Combined adm with 15 Catherine Road £7.50, chd free. Light refreshments.**
We have created a winding path through our 70ft garden with trees, evergreen structure, perennial flowers and grasses. We are also opening early in 2021 for our enormous tulip display of many types and also, later in midsummer. We like to create a garden party, so feel welcome to stay as long as you like.

1 YORK CLOSE, W7
Hanwell, W7 3JB. Tony Hulme & Eddy Fergusson. *By road only, entrance to York Close via Church Rd. Nearest station Hanwell mainline. Buses E3, 195, 207.* **Sat 21 Aug (2-6). Adm £6, chd free. Sun 22 Aug (2-6). Adm £5, chd free. Wine.**
Tiny quirky, prize winning garden extensively planted with eclectic mix incl hosta collection, many unusual and tropical plants. Plantaholics paradise. Many surprises in this unique and very personal garden. Pimms available on Saturday, drinks available on Sunday

10 YORK ROAD, N21
N21 2JL. Androulla & Harry Tsappas. *Winchmore Hill. Buses: 329 and W8 bus routes.* **Sun 20 June (2-6). Adm £4, chd free. Home-made teas.**

This suburban garden is full of country perennials canopied with beautiful trees such as Indian bean, olive and acer trees, and has a pretty, dainty look inspired by country cottages. There is a large pond with koi fish which is surrounded by luscious grasses and a rockery. The garden incl a wood choppers' enclave, and an array of wildlife such as frogs, butterflies and bees are prevalent.

38 YORK ROAD, W5
Ealing, W5 4SG.
Nick & Elena Gough,
www.thedistinctivegardener.com.
Northfields & South Ealing Tube - 5 mins. Buses: E3, 65. 5 mins. Free parking in local streets. Off Northfield Av & South Ealing Rd. **Sun 9 May, Sun 18 July (10-6). Adm £4.50, chd free. Tea.**

A hidden oasis full of surprises and built on several different levels. This walled corner garden was restored and expanded by its garden designer owner, having been acquired in 2014. There are a number of beautiful and diverse areas within it, all of which add to its special atmosphere, incl a woodland dell path, circular sun terrace, large pond with waterfalls and flower filled parterre.

ZEN GARDEN AT JAPANESE BUDDHIST CENTRE, W3
Three Wheels, 55 Carbery Avenue, Acton, W3 9AB. Reverend Prof K T Sato, www.threewheels.org.uk. *Tube: Acton Town 5 mins walk, 200yds off A406.* **Sat 5, Sun 6, Sat 19, Sun 20 June (2-5). Adm £3.50, chd free. Matcha tea ceremony £3.**

Pure Japanese Zen garden (so no flowers) with 12 large and small rocks of various colours and textures set in islands of moss and surrounded by a sea of grey granite gravel raked in a stylised wave pattern. Garden surrounded by trees and bushes outside a cob wall. Oak framed wattle and daub shelter with Norfolk reed thatched roof. Talk on the Zen garden between 3-4pm. Buddha Room open to public.

7 Deansway

NORFOLK

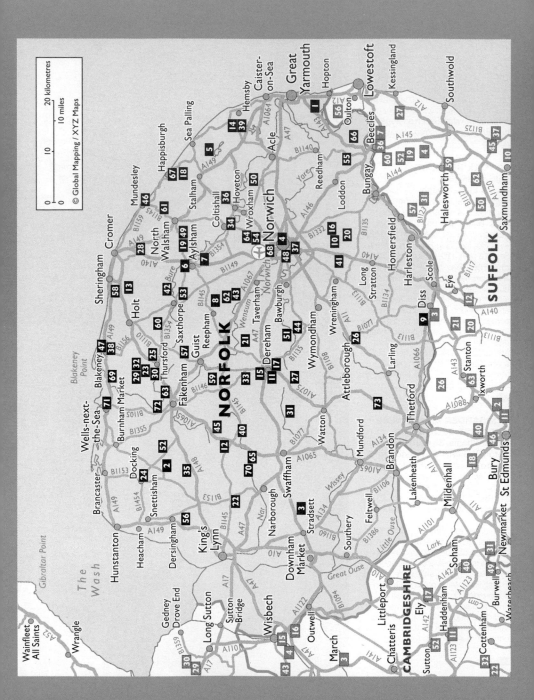

Norfolk is a lovely low-lying county, predominantly agricultural with an abundance of wildlife and a beautiful coastline.

Visitors come here because they are attracted not only to the peaceful and spacious countryside, but also the medieval churches and historical houses. There is an extensive coastal footpath and a large network of rivers, together with the waterways of the Norfolk Broads. Norwich the capital is a fine city.

We are fortunate to have a loyal group of garden owners; Sandringham was one of the original gardens to open for the scheme in 1927 and has been supporting us continuously ever since.

Whilst many of our gardens have opened their gates for over half a century, others will be opening for the very first time. Located throughout the county, they range from those of the large estates and manor houses, to the smaller cottages, courtyards and town gardens, accommodating different styles of old and traditional, newly constructed and contemporary, designed and naturalistic.

So why not come and experience for yourself the county's rich tapestry of big skies, open countryside, attractive architecture and delightful gardens.

Below: Mannington Estate

Volunteers

County Organiser
Julia Stafford Allen
01760 755334
julia.staffordallen@ngs.org.uk

County Treasurer
Andrew Stephens OBE
07595 939769
andrew.stephens@ngs.org.uk

Publicity
Carol Allen 07368 238528
carol.allen@ngs.org.uk

Social Media
Claire Reinhold 07884 435484
reinhold.claire@gmail.com

Photographer
Simon Smith 01362 860530
simon.smith@ngs.org.uk

Booklet Co-ordinator
Juliet Collier 07986 607170
juliet.collier@ngs.org.uk

New Gardens Organiser
Fiona Black 01692 650247
fiona.black@ngs.org.uk

Assistant County Organisers
Jenny Clarke 01508 550261
jenny.clarke@ngs.org.uk

Nick Collier 07733 108443
nick.collier@ngs.org.uk

Sue Guest 01362 850317
guest63@btinternet.com

Sue Roe 01603 455917
sueroe8@icloud.com

Retty Wace 01328 820028
retty.wace@ngs.org.uk

Graham Watts 01362 690065
graham.watts@ngs.org.uk

f @ngsnorfolk   @Norfolkngs   @norfolkngs

OPENING DATES

All entries subject to change. For latest information check www.ngs.org.uk

Map locator numbers are shown to the right of each garden name.

February

Snowdrop Festival

Sunday 7th
NEW ◆ The Bressingham Gardens 9

Saturday 13th
Horstead House 34

Saturday 20th
◆ Hindringham Hall 29

Sunday 21st
Bagthorpe Hall 2

March

Monday 1st
◆ Raveningham Hall 55

Saturday 13th
◆ East Ruston Old Vicarage 18

Sunday 28th
Gayton Hall 22

April

Saturday 10th
Manor Farm, Coston 44

Sunday 11th
◆ Mannington Estate 42
The Old House 50

May

Sunday 2nd
Wretham Lodge 73

Monday 3rd
◆ Hoveton Hall Gardens 36
Wretham Lodge 73

Sunday 9th
NEW Quaker Farm 54

Tuesday 18th
◆ Stody Lodge 60

Saturday 22nd
The Merchants House 47

Sunday 23rd
NEW East Wing 19
Lexham Hall 40
NEW Oaks Farm Cottage 49
Warborough House 69

Sunday 30th
Bolwick Hall 7
Warborough House 69

June

Thursday 3rd
NEW Gunthorpe Hall 25

Sunday 6th
Barton Bendish Hall 3
Grove House 24
NEW Home Farmhouse 32
The Old Rectory, Syderston 52
Oulton Hall 53
Wells-Next-The-Sea Gardens 71

Saturday 12th
Blickling Lodge 6
NEW Church Cottage 14
NEW La Foray 39

Sunday 13th
NEW Booton Hall 8
NEW Church Cottage 14
Elsing Hall Gardens 21
High House Gardens 27
NEW La Foray 39

Wednesday 16th
High House Gardens 27
Thorpland Hall 63

Saturday 19th
Kettle Hill 38

Sunday 20th
Broadway Farm 11
Manor House Farm, Wellingham 45
Walcott House 67

Thursday 24th
NEW ◆ The Bressingham Gardens 9

Saturday 26th
Elm House 20

Sunday 27th
NEW Highview House 28
Holme Hale Hall 31
NEW Warren House 70

July

Sunday 4th
Black Horse Cottage 5
Manor Farm House, Swannington 43
Swannington Manor 62
Tyger Barn 66

Sunday 11th
Bishop's House 4

Sunday 18th
Dunbheagan 17
Honeysuckle Walk 33
NEW 26 Ipswich Road 37
West Barsham Hall 72

Wednesday 21st
Lexham Hall 40

Sunday 25th
Dale Farm 15
30 Hargham Road 26
NEW 26 Ipswich Road 37
NEW 3 Meadow Close 46
North Lodge 48

August

Sunday 1st
The Long Barn 41
North Lodge 48
NEW 61 Trafford Way 64
Tudor Lodgings 65
33 Waldemar Avenue 68

Thursday 5th
Sheringham Hall 58

Saturday 7th
NEW Swafield Hall 61

Sunday 8th
Brick Kiln House 10
◆ Severals Grange 57
NEW Swafield Hall 61

September

Sunday 5th
Holme Hale Hall 31
33 Waldemar Avenue 68

Sunday 12th
Chapel Cottage 12
High House Gardens 27

Wednesday 15th
High House Gardens 27

Sunday 19th
Manor Farm, Coston 44
Silverstone Farm 59

October

Saturday 16th
◆ East Ruston Old Vicarage 18

By Arrangement

Arrange a personalised garden visit on a date to suit you. See individual garden entries for full details.

Acre Meadow 1
Blickling Lodge 6
Brick Kiln House 10
Broadway Farm 11
Chapel Cottage 12
Chestnut Farm 13
Dale Farm 15
NEW Dove Cottage 16
Dunbheagan 17
Gayton Hall 22
NEW Holly House 30
Holme Hale Hall 31
NEW Home Farmhouse 32
Honeysuckle Walk 33
Horstead House 34
The Old Rectory, Brandon Parva 51
Silverstone Farm 59
Tudor Lodgings 65
Tyger Barn 66
Walcott House 67
33 Waldemar Avenue 68
West Barsham Hall 72
Wretham Lodge 73

THE GARDENS

1 ACRE MEADOW

New Road, Bradwell, Great Yarmouth, NR31 9DU. Mr Keith Knights, 07476 197568, kk.acremeadow@gmail.com, www.acremeadow.co.uk. *Between Bradwell & Belton in Arable surroundings. Take Belton & Burgh Castle turn (New Rd) from r'about at Bradwell on A143 Great Yarmouth to Beccles rd. Then 400 yards on R drive in very wide gateway. For sat nav use NR31 9JW. Visits by arrangement July to Sept. Sundays and Mondays, although groups 20+ can arrange visits for any day. Adm £4, chd free. Light refreshments.*
Exotic and insect attracting plants. A riot of late summer colour with lots of dark foliage, in deep, densely planted borders in an immersive garden on former greenhouses site of currently 1/10th of an acre. Separate areas of interest include Pond area and 80 foot border for pollinators. Wheelchair access to main garden limited to viewing area at both ends due to narrow gravel paths.

2 BAGTHORPE HALL

Bagthorpe, Bircham, King's Lynn, PE31 6QY. Mr & Mrs D Morton, 07979 746591, dgmorton@hotmail.com. *3½m N of East Rudham, off A148. Take turning opposite The Crown in East Rudham. Look for white gates in trees, slightly set back from the road. Sun 21 Feb (11-4). Adm £5, chd free.*
A delightful circular walk which meanders through a stunning display of snowdrops naturally carpeting a woodland floor, and then returning through a walled garden. No refreshments.

3 BARTON BENDISH HALL

Fincham Road, Barton Bendish, King's Lynn, PE33 9DL. The Barton Bendish Gardening Team. *5m E of Downham Market off A1122. On entering Barton Bendish follow yellow signs to field parking. Sun 6 June (11-5). Adm £6, chd free. Home-made teas in Hall court yard.*
Traditional country estate garden of 10-acres. Woodland drive, orchard, kitchen garden with soft fruits, espaliered fruit trees, vegetables, glasshouse full of scented pelargoniums. Walled herb and cut flower garden. Herbaceous borders, south facing terrace. Informal area around pond. Sculptures. Some wheelchair access.

4 BISHOP'S HOUSE

Bishopgate, Norwich, NR3 1SB. The Bishop of Norwich, www.dioceseofnorwich.org/gardens. *Located in the city centre near the Law Courts & The Adam & Eve Pub Parking available at Town Centre car parks including one by the Adam and Eve pub. Sun 11 July (11-4). Adm £5, chd free. Home-made teas.*
4 acre walled garden dating back to the C12. Extensive lawns with specimen trees. Borders with many rare and unusual shrubs. Spectacular herbaceous borders flanked by yew hedges. Rose beds, meadow labyrinth, kitchen garden, woodland walk and long border with hostas and bamboo walk. Popular plant sales. Gravel paths and some slopes.

Online booking is available for many of our gardens, at ngs.org.uk

5 BLACK HORSE COTTAGE

The Green, Hickling, Norwich, NR12 0YA. Yvonne Pugh. *3m E of Stalham. Turn E off A149 at Catfield, turn L onto Heath Rd, 1½m to centre of Hickling village. Next to The Greyhound Inn (Good food!). Sun 4 July (10.30-5). Adm £5, chd free. Home-made teas.*
Thatched house close to Hickling Broad and NWT Nature Reserve. Plantsman's garden 2 acres professionally redesigned. Spacious borders and islands with diverse range of characterful planting. Particular emphasis on achieving full year round interest. Wide range of managed mature specimen trees. Many long two-way vistas. Wide mown walkways through large meadow. Various sitting opportunities! Sorry, no dogs allowed.

Swannington Manor

6 BLICKLING LODGE

Blickling, Norwich, NR11 6PS.
Michael & Henrietta Lindsell,
nicky@lindsell.co.uk. *½m N of
Aylsham. Leave Aylsham on Old
Cromer rd towards Ingworth. Over
hump back bridge & house is on R.*
Sat 12 June (11-5). Adm £5, chd
free. Light refreshments. **Visits
also by arrangement.**
Georgian house (not open) set in 17
acres of parkland including cricket
pitch, mixed border, walled kitchen
garden, yew garden, woodland/water
garden.

7 BOLWICK HALL

Marsham, NR10 5PU. Mr & Mrs G
C Fisher. *8m N of Norwich off A140.
From Norwich, heading N on A140,
just past Marsham take 1st R after
Plough Pub, signed 'By Road' then
next R onto private drive to front of
Hall.* Sun 30 May (1-5). Adm £5,
chd free. Home-made teas.
Landscaped gardens and park
surrounding a late Georgian hall. The
original garden design is attributed to
Humphry Repton. The current owners
have rejuvenated the borders, planted
gravel and formal gardens and clad the
walls of the house in old roses. Enjoy
a woodland walk around the lake as
well as a stroll through the working
vegetable and fruit garden with its
double herbaceous border. Please ask
at gate for wheelchair directions.

8 NEW BOOTON HALL

Church Road, Booton, Norwich,
NR10 4NZ. Piers & Cecilia Willis.
*1m E of Reepham. Drive out of
Reepham east on Norwich Rd, after
1m turn L into Church Rd. After 500
metres turn L through white wooden
gates just after bendy road sign on L.*
Sun 13 June (11-5). Adm £4, chd
free. Home-made teas.
Walled garden with formal layout and
tiered lawns re-designed 6 years ago
and attached to C17/18 hall. Lawns
between the house and parkland
meadows, shrub beds, pond planting,
small orchard and short woodland
walk. Parking in one of the meadows
close to drive entrance, prior drop-
off by house possible in extremis.
Wheelchair access, but some gravel
quite deep, and some steps, so may
require some help.

9 NEW ◆ THE BRESSINGHAM GARDENS

Low Road, Bressingham,
Diss, IP22 2AA. Adrian Bloom,
www.thebressinghamgardens.com.
*Bressingham Gardens entrance lies
to the W of Bressingham Garden
Centre & Bressingham Steam
Museum, S of the A1066 rd, 2½m
W of Diss, 14m E of Thetford.* For
NGS: Sun 7 Feb (11-4). Adm
£5, chd free. Evening opening
Thur 24 June (6-9). Adm £7, chd
free. **For other opening times and
information, please visit garden
website.**
17 acres of gardens which visitors
describe as 'the best they have seen'.
Follow the trail around the island beds
of the late Alan Bloom's Dell Garden,
through the Fragrant Garden, and
Adrian's Wood with Giant Redwoods,
to famous Foggy Bottom garden, with
dramatic rivers of planting. Admire the
garden's beauty in winter or enjoy a
rare opportunity for an evening visit.

Manor House Farm, Wellingham

10 BRICK KILN HOUSE

Priory Lane, Shotesham, Norwich, NR15 1UJ. Jim & Jenny Clarke, jennyclarke@uwclub.net. *6m S of Norwich. From Shotesham All Saints church Priory Lane is 200m on R on Saxlingham Rd.* **Sun 8 Aug (10-4). Adm £5, chd free. Home-made teas. Visits also by arrangement June to Aug.**

2-acre country garden with a large terrace, lawns and colourful herbaceous borders. There are garden sculptures and a stream running through a diversely planted wood. Parking in field but easy access to brick path.

🚶 ☕ 🪑

11 BROADWAY FARM

The Broadway, Scarning, Dereham, NR19 2LQ. Michael & Corinne Steward, 01362 693286, corinneasteward@gmail.com. *16m W of Norwich. 12m E of Swaffham From A47 W take a R into Fen Rd, opposite Drayton Hall Lane. From A47 E take L into Fen Rd, then immed L at T-junction, immed R into The Broadway.* **Sun 20 June (11-5). Adm £4, chd free. Visits also by arrangement June & July for groups of 10 to 30.**

Half acre cottage garden surrounding a C14 clapboard farmhouse. Colourful herbaceous borders with a wide range of perennial and woody plants and a well planted pond, providing habitat for wildlife. A plantswoman's garden !

🚶 ☕ 🪑

12 CHAPEL COTTAGE

Rougham, King's Lynn, PE32 2SE. Sarah Butler, 01328 838347, sarahbutler4@gmail.com. *15m E from King's Lynn, 8m SW from Fakenham. Off B1145. Parking in the centre of the village.* **Sun 12 Sept (10.30-4.30). Adm £4, chd free. Home-made teas. Visits also by arrangement for groups of 10 to 20.**

A naturalistic cottage garden that was created by the owner who is interested in the relationship of wildlife to plants, as well as gardens for wellbeing. Divided into charming peaceful areas that include a pond, kitchen garden, herb area and beehives. Plenty of seating. Situated in an attractive rural village with Church open.

🐑 ☕

13 CHESTNUT FARM

Church Road, West Beckham, Holt, NR25 6NX. Mr & Mrs John McNeil Wilson, 01263 822241, judywilson100@gmail.com. *2½m S of Sheringham. From A148 opp Sheringham Park entrance. Take the rd signed BY WAY TO WEST BECKHAM, about ¾m to the garden. NB Satnav will take you to pub, garden at village sign.* **Visits by arrangement Feb to Nov for groups of up to 20. Adm £5, chd free.**

Mature 3-acre garden with collections of many rare and unusual plants and trees. 100+ varieties of snowdrops, drifts of crocus with seasonal flowering shrubs. Later, wood anemones, fritillary meadow, wildflower walk, pond, small arboretum, croquet lawn and colourful herbaceous borders. Wheelchair access tricky if wet.

🚶 🐑 ☕

14 NEW CHURCH COTTAGE

57 White Street, Martham, Great Yarmouth, NR29 4PQ. Jo & Nigel Craske. *Approx 9m N of Gt Yarmouth off A149. Take B1152 signed to Martham & Winterton. On White Street pass Village Hall & Bell Meadow. Garden on L down gravel drive before Church & sharp RH bend.* **Sat 12, Sun 13 June (12-5). Combined adm with La Foray £5, chd free.**

Delightful small cottage garden attached to 200 year old house with views of St Mary's Church. Divided into different areas with lots of plants, many typical but others more unusual. Areas including cool, shady border, hot sunny herbaceous border, small gravel garden, walled garden, sunken area with collection of hostas and greenhouse with pelargoniums and succulents. Unsuitable for wheelchairs.

🐑

15 DALE FARM

Sandy Lane, Dereham, NR19 2EA. Graham & Sally Watts, 01362 690065, grahamwatts@dsl.pipex.com. *16m W of Norwich. 12m E of Swaffham. From A47 take B1146 signed to Fakenham, turn R at T-junction, ¼m turn L into Sandy Lane (before pelican crossing).* **Sun 25 July (11-5). Adm £5, chd free. Home-made teas. Visits also by arrangement June to Sept for groups of 10+.**

2-acre plant lovers' garden with a large spring-fed pond. Over 1000 plant varieties in exuberantly planted borders with sculptures. Also, gravel, vegetable, nature and waterside gardens. Collection of 130 hydrangeas! Music during the afternoon and remote-controlled model boats on pond for children, some grass paths and gravel drive. Wide choice of plants for sale.

🐑 ☕ 🪑

16 NEW DOVE COTTAGE

Wolferd Green, Shotesham All Saints, Norwich, NR15 1YU. Sarah Cushion, sarah.cushion815@btinternet.com. *From Norwich travel to far end of Poringland, turn R at Poringland rd/Shotesham rd junction by the church, continue along rd for approx 2½m, house pink cottage on L.* **Visits by arrangement June to Aug for groups of up to 30. Adm £4, chd free.**

⅓ acre densely planted colourful cottage garden including a small pond, summer house, and extensive views across the countryside.

☕

17 DUNBHEAGAN

Dereham Road, Westfield, NR19 1QF. Jean & John Walton, 01362 696163, jandjwalton@btinternet.com. *2m S of Dereham. From Dereham turn L off A1075 into Westfield Rd by the Vauxhall garage/Premier food store. Straight ahead at Xrds into lane which becomes Dereham Rd. Garden on L.* **Sun 18 July (1-5). Adm £5, chd free. Home-made teas. Visits also by arrangement June & July for groups of 10+.**

Relax and enjoy walking among extensive borders and island beds - a riot of colour all summer. Vast collection of rare, unusual and more recognisable plants in this ever changing plantsman's garden. If you love flowers, you'll love it here. We aim for the WOW factor. Lots of changes for 2021. Featured in Nick Bailey's book 365 Days of Colour. Music during the afternoon. Gravel driveway.

🐑 ☕

All entries are subject to change. For the latest information check ngs.org.uk

18 ♦ EAST RUSTON OLD VICARAGE
East Ruston, Norwich, NR12 9HN. Alan Gray & Graham Robeson, 01692 650432, erovoffice@btconnect.com, www.eastrustonoldvicarage.co.uk. *3m N of Stalham. Turn off A149 onto B1159 signed Bacton, Happisburgh. After 2m turn R 200yds N of East Ruston Church (ignore sign to East Ruston).* For NGS: Sat 13 Mar, Sat 16 Oct (12-5). Adm £11, chd £3. Light refreshments. For other opening times and information, please phone, email or visit garden website.
Large garden with traditional borders and modern landscapes inc. Walled Gardens, Rose Garden, Exotic Garden, Topiary and Box Parterres, Water Features, Mediterranean Garden, a monumental Fruit Cage, Containers to die for in spring and summer. Cornfield and Meadow Gardens, Vegetable and Cutting Gardens, Parkland and Heritage Orchard, in all 32 acres. Rare and unusual plants abound. Regional Finalist, The English Garden's The Nation's Favourite Gardens 2019.
& ❀ 🚌 NPC 🍵

19 NEW EAST WING
Suffield Road, Felmingham, North Walsham, NR28 0JZ. Stephanie Kershaw. *16m north of Norwich (towards Cromer), off the B1145 between Felmingham and Banningham.* Sun 23 May (12-5). Combined adm with Oaks Farm Cottage £5, chd free. Light refreshments.
A driveway flanked by structural phormiums and ornamental grasses leads to East Wing's open lawns, walled beds and various seating areas including a glass 'pod', as well as beehives, fruit trees and raised vegetable beds. Impressive interlinking gardens of a total of 2 acres, with sumptuous planting, ponds and trees. Wheelchair access between gardens is via the road - please ask at gate for directions.
& ❀ 🍵

20 ELM HOUSE
The Green, Saxlingham Nethergate, Norwich, NR15 1TH. Mrs Linda Woodwark. *8m S of Norwich. From Norwich take A140 to Ipswich, at Newton Flotman turn L at sign to Saxlingham Nethergate, at end of the rd, turn R into village, follow NGS signs.* Sat 26 June (11-

5). Adm £4.50, chd free. Home-made teas.
A garden for plant lovers. Set in the quiet countryside this garden hosts many trees, shrubs and an abundance of herbaceous perennials. A large natural pond that is home to many wild ducks, a vegetable garden, and alpines. In the paddock there are two pygmy goats and chickens. There is a certain amount of wheelchair access.
& ❀ 🍵

21 ELSING HALL GARDENS
Elsing Hall, Hall Road, Elsing, NR20 3DX. Patrick Lines & Han Yang Yap, www.elsinghall.com. *6km NW of Dereham. From A47 take the N Tuddenham exit. From A1067 take the turning to Elsing opp the Bawdeswell Garden Centre. Location details can be found on www. elsinghall.com.* Sun 13 June (10-4). Adm £8, chd free.
C15 fortified manor house (not open) with working moat. 10 acre gardens and 10 acre park surrounding the house. Significant collection of old roses, walled garden, formal garden, marginal planting, Gingko avenue, viewing mound, moongate, interesting pinetum and terraced garden. Please note that there are no WC facilities available.

22 GAYTON HALL
Gayton, King's Lynn, PE32 1PL. Viscount & Viscountess Marsham, 01485 528432, ciciromney@icloud.com. *6m E of King's Lynn. Off the B1145. At village sign take 2nd exit off Back Street to entrance.* Sun 28 Mar (1-5). Adm £5, chd free. Home-made teas. Visits also by arrangement Mar to Oct for groups of 10 to 30.
This rambling semi-wild 20-acre water garden, has over 2 miles of paths, and contains lawns, lakes, streams, bridges and woodland. Primulas, astilbes, hostas, lysichiton and gunnera. Spring bulbs. A variety of unusual trees and shrubs, many labelled, have been planted over the years. Wheelchair access to most areas, gravel and grass paths.
& 🐐 🍵

24 GROVE HOUSE
High Street, Docking, King's Lynn, PE31 8NH. Mr & Mrs Charles Polito. *Limited parking at house.* Sun 6 June (10-5). Adm £4.50, chd free. Light refreshments.

Also open The Old Rectory, Syderstone. Teas available all day.
Village garden just over an acre, created during the last four years by the owners. Includes fine lawn surrounded by flower borders, vegetable and cutting garden with extensive sweet peas and small orchard. Partial access due to gravelled paths.
❀ 🍵

25 NEW GUNTHORPE HALL
Hall Lane, Gunthorpe, Melton Constable, NR24 2PA. Jeremy & Marie Denholm, 01263-861-373, enquiries@gunthorpehall.co.uk, www.gunthorpehall.co.uk. *At Gunthorpe Village Green turn up Heath Lane. Straight ahead at the end, will be the gates to Gunthorpe Hall. Enter here, keep R & park in the paddock to the R just past the lake.* Thur 3 June (2-6). Adm £5, chd free.
Gunthorpe Hall and Garden (grade 11 listed). Private home and 40 year project restoring the Hall and Garden including - parterre, topiaries, mature and very old trees and shrubs, espalier fruit trees, variety of mixed borders, bath pond, lake and working, productive walled veg garden. Paths are gravel and there are some slopes and steps to the parterre. Most of the garden is accessible by wheelchair.
& 🐎 🚂 🏖

26 30 HARGHAM ROAD
Attleborough, NR17 2ES. Darren & Karen Spencer. *Turn off A11 at Breckland Lodge, continue 2m into Attleborough, turn R opp Sainsbury's into Hargham Road and we're 200 yds on the R.* Sun 25 July (11-5). Adm £4, chd free. Home-made teas.
Step into a vibrant, colourful haven in just under a ⅓ acre. Created from scratch with self built structures and home-made water features. Borders filled with perennials and annuals with a variety of exotics. Follow the pathway round to a modern allotment area.
& 🍵

27 HIGH HOUSE GARDENS
Blackmoor Row, Shipdham, Thetford, IP25 7PU. Sue & Fred Nickerson. *6m SW of Dereham. Take the airfield or Cranworth Rd off A1075 in Shipdham. Blackmoor Row is signed.* Sun 13, Wed 16 June,

Sun 12, Wed 15 Sept (2-5.30). Adm £5, chd free.

3 acre plantsman's garden developed and maintained by the current owners, over the last 40 years. Garden consists of colour themed herbaceous borders with an extensive range of perennials, box edged rose and shrub borders, woodland garden, pond and bog area, orchard and small arboretum. Plus large vegetable garden. Gravel paths.

28 NEW HIGHVIEW HOUSE

Norwich Road, Roughton, Norwich, NR11 8NA. Graham & Sarah Last. *3m S of Cromer. From mini r'about in centre of Roughton head N towards Cromer. Just after lay by on R. Large black gates set in red brick wall.* Sun 27 June (1-5). Adm £5, chd free. Home-made teas.

Two acre garden featuring a large range of herbaceous borders. Long season of interest with a large range of perennial plants maintained by the owners. Many Salvia varieties feature through the planting. Interesting garden structures including a Japanese pagoda garden and large decked area. Pathways take you through each area of the garden. Suitable in most parts for wheelchair access. Football goals may be used by visitors at their own risk. For wheelchair access park on the main house driveway with access to the brickweave pathway. There are some slopes that will need to be accomodated.

29 ♦ HINDRINGHAM HALL

Blacksmiths Lane, Hindringham, NR21 0QA. Mr & Mrs Charles Tucker, 01328 878226, hindhall@btinternet.com, www.hindringhamhall.org. *7m from Holt/Fakenham/Wells. Turn L. off A148 at Crawfish Pub towards Hindringham. L into Blacksmiths Lane after village hall on R.* For NGS: Sat 20 Feb (10-4). Adm £8, chd free. For other opening times and information, please phone, email or visit garden website.

A garden surrounding a Grade 2* Tudor Manor House (not open) enveloped by a complete medieval moat plus 3-acres of fishponds. Working walled vegetable garden, Victorian nut walk, formal beds, bog and stream gardens. Something of interest throughout the year continuing well into autumn. Spring wild garden. Some access for wheelchairs able to cope with gravel paths.

Chapel Cottage

30 NEW **HOLLY HOUSE**
9, Balls Lane, Thursford, Fakenham, NR21 0BX. Tim Doncaster, 01328 878393, tmjdoncaster@gmail.com. *6miles from Holt/Fakenham. thru village past the Thursford collection Balls lane last lefthand turning . Holly House 500yds on L.* Visits by arrangement Feb to Oct for groups of up to 10. Closed Weds and Sats. Adm £4, chd free. Mature 3-acre year round wildlife garden, with a mix of trees, shrubs and perennials including natives. Woodland walk. In constant flux due to honey fungus, other wildlife and gardeners change of whim.

31 **HOLME HALE HALL**
Holme Hale, Swaffham, Thetford, IP25 7ED. Mr & Mrs Simon Broke, 01760 440328, simon.broke@hotmail.co.uk. *2m S of Necton off A47. 1m E of Holme Hale village.* Sun 27 June, Sun 5 Sept (12-4). Adm £6, chd free. Light refreshments. Visits also by arrangement May to Sept.
Walled kitchen garden designed by Arne Maynard and replanted in 2016/17. A soft palette of herbaceous plants include some unusual varieties. Topiaries, vegetables, trained fruits and roses. 130 year old wisteria. Wildflower meadow. Long season of interest. Wildlife friendly. Historic buildings. Wheelchair access available to some areas.
♿ 🐕 ✿ 🚗 ☕

32 NEW **HOME FARMHOUSE**
91 The Street, Hindringham, NR21 0PS. John & Rachel Hannyngton, 01328 878571, rhannyngton@btinternet.com. *Hindringham is between the villages of Binham & Thursford. From A148 between Fakenham & Holt, turn North at The Crawfish pub.* Sun 6 June (10-5). Adm £5, chd free. Home-made teas. Visits also by arrangement June & July for groups of up to 20. Opening dates 14,15 & 22 June and 14 &15 July 1-5.
An imaginative, informal 2-acre garden, developed over the past 25 years with orchard, working potager, herbaceous borders, themed beds, 'white' courtyard, Piet Oudolf-style perennial & grass border, wildflower meadow and woodland walk. Art and ceramics exhibition in the old barn.
✿ ☕

33 **HONEYSUCKLE WALK**
Litcham Road, Gressenhall, Dereham, NR20 4AR. Simon & Joan Smith, 01362860530, simon.smith@ngs.org.uk. *17m w of Norwich. Follow B1146 & signs to Gressenhall Rural Life Museum then follow Litcham Road into Gressenhall village.* Sun 18 July (11-5). Adm £5, chd free. Home-made teas. Visits also by arrangement July & Aug for groups of 10 to 30.
2 acre woodland garden featuring two ponds, river and stream boundaries, 80 ornamental trees and a collection of 400 different hydrangeas and 100 hostas. Woodland access is by ¼ m grass track. Many rare hydrangea and hosta plants for sale.
🐖 ✿ ☕

34 **HORSTEAD HOUSE**
Mill Road, Horstead, Norwich, NR12 7AU. Mr & Mrs Matthew Fleming, 07771 655 637, caro.fleming@sky.com. *6m NE of Norwich on North Walsham Road, B1150. Down Mill Rd opp the Recruiting Sergeant pub.* Sat 13 Feb (11-4). Adm £5, chd free. Home-made teas. We will be serving drinks and cakes. Visits also by arrangement in Feb for groups of up to 20.
Stunning display of beautiful snowdrops carpet the woodland setting with winter flowering shrubs. Another beautiful feature is the dogwoods growing on a small island in the R. Bure, which flows through the garden. Small walled garden. Wheelchair access to main snowdrop area.
♿ 🐕 🛏 ☕

35 ✦ **HOUGHTON HALL WALLED GARDEN**
Bircham Road, New Houghton, King's Lynn, PE31 6TY. The Cholmondeley Gardens Trust, 01485 528569, info@houghtonhall.com, www.houghtonhall.com. *11m W of Fakenham. 13m E of King's Lynn. Signed from A148.* For opening times and information, please phone, email or visit garden website.
The award-winning, 5-acre walled garden includes a spectacular double-sided herbaceous border, rose parterre, wisteria pergola and glasshouses. Mediterranean garden and kitchen garden with arches and espaliers of fruit trees. Antique statues, fountains and contemporary sculptures. Gravel and grass paths.

Electric buggies available in the walled garden.
♿ 🚗 ☕ 🪑

36 ✦ **HOVETON HALL GARDENS**
Hoveton Hall Estate, Hoveton, Norwich, NR12 8RJ. Mr & Mrs Harry Buxton, 01603 784297, office@hovetonhallestate.co.uk, www.hovetonhallestate.co.uk. *8m N of Norwich. 1m N of Wroxham Bridge. Off A1151 Stalham Rd - follow brown tourist signs.* For NGS: Mon 3 May (10.30-5). Adm £8, chd £4. Light refreshments on site independent cafe - Garden Kitchen. Picnics are allowed for garden patrons only. For other opening times and information, please phone, email or visit garden website.
15 acre gardens and woodlands taking you through the seasons. Mature walled herbaceous and kitchen gardens. Informal woodlands and lakeside walks. Nature Spy activity trail for our younger visitors. A varied events programme runs throughout the season. Please visit our website for more details. Light lunches and afternoon tea from our on-site Garden Kitchen Cafe. The gardens are approx 75% accessible to wheelchair users. We offer a reduced entry price for wheelchair users and their carers.
♿ ✿ 🚗 🛏 ☕ 🪑

37 NEW **26 IPSWICH ROAD**
Norwich, NR2 2LZ. Pat Adcock. *Parking available opposite the garden at Harford Manor School.* Sun 18, Sun 25 July (11-5). Adm £3.50, chd free. Light refreshments. Tea, coffee, home-made cakes.
This is no ordinary garden! Half a mile from Norwich City centre this half acre garden is dedicated to sustainability and biodiversity. Organic and wildlife friendly there are resident chickens, bees, a large wildlife pond, fish pond, vegetables and ornamental plants. The owner is a Master Composter, composting advice available. There is a home-made system of water conservation and storage. Access to rear garden via a gravel area. and may be difficult for some wheelchairs.
✿ ☕

38 **KETTLE HILL**
The Downs, Langham Road, Blakeney, NR25 7PN. Mrs R Winch. *½m from Blakeney off B1156 towards Langham Double*

white gates on R opp Scargills Electric Company & campsite which is on the L. Up the long drive double white gates at entrance. **Sat 19 June (11-5). Adm £5, chd free. Home-made teas. Tea, coffee, Elderflower and cake.**
A garden with herbaceous borders, wildflower meadows secret garden. Stunning rose garden Beautiful woods.Excellent views across Morston to sea, framed by lavender, roses and sky. Gardens have been developed by owner with design elements from George Carter /Tamara Bridge. gravel driveway leading to paths and flat lawned areas. Partial access to woods. No disabled WC. Nearest is at Blakeney village hall.

39 NEW LA FORAY
White Street, Martham, Great Yarmouth, NR29 4PQ. Ameesha & Alan Williams. *10m N of Gt Yarmouth approx 2m off A149 follow yellow signs at Martham junction. Parking in village.* **Sat 12, Sun 13 June (12-5). Combined adm with Church Cottage £5, chd free. Home-made teas.**
A lovely garden with many interesting and unusual features created and maintained by the owners. Various Mixed borders with an array of vintage and unusual items incorporated to add interest. A Wildlife Retreat, Stumpery, collection of Hostas and Succulents and a Vintage Summerhouse to relax in. Child's Gypsy Caravan, Treehouse Birdhide and Friendly Pet Goats. Teas in aid of Donor Family Network.

40 LEXHAM HALL
nr Litcham, PE32 2QJ. Mr & Mrs Neil Foster, www.lexhamestate.co.uk. *6m N of Swaffham off B1145. 2m west of Litcham.* **Sun 23 May, Wed 21 July (11-5). Adm £7.50, chd free. Home-made teas.**
Parkland with lake and river walks surround C17/18 Hall (not open). Formal garden with terraces, roses and mixed borders. Traditional working kitchen garden with crinkle-crankle well. Year round interest with rhododendrons, azaleas, camellias and magnolias in the 3-acre woodland garden in May. July sees the walled garden borders at their peak.

41 THE LONG BARN
Flordon Road, Newton Flotman, Norwich, NR15 1QX. Mr & Mrs Mark Bedini. *6m S of Norwich along A140. Leave A140 in Newton Flotman towards Flordon. Exit Newton Flotman & approx 150 yards beyond 'passing place' on L, turn L into drive. Note that SatNav does not bring you to destination.* **Sun 1 Aug (11-4.30). Adm £5, chd free. Home-made teas.**
Beautiful informal garden with new haha creating infinity views across ancient parkland. Strong Mediterranean influence with plenty of seating in large paved courtyard under pollarded Plane trees. Unusual wall-sheltered group of 4 venerably gnarled olive trees, figs and vines, then herbaceous borders, tiered lawns merging northwards into woodland garden. Walks through Parkland the tiny River Tas - serene! Wheelchair drop off at front of house. Gradual lawn slope accesses upper tier of garden. WC requires steps.

42 ◆ MANNINGTON ESTATE
Mannington, Norwich, NR11 7BB. The Lord & Lady Walpole, 01263 584175, admin@walpoleestate.co.uk, www.manningtonestate.co.uk. *18m NW of Norwich. 2m N of Saxthorpe via B1149 towards Holt. At Saxthorpe/Corpusty follow signs to Mannington.* **For NGS: Sun 11 Apr (11-5). Adm £6, chd free. Light refreshments. For other opening times and information, please phone, email or visit garden website.**
20-acres feature shrubs, lake and trees. Period gardens. Borders. Sensory garden. Extensive countryside walks and trails. Moated manor house and Saxon church with C19 follies. wildflowers and birds. The tearooms offer home-made locally sourced food with home-made teas. Gravel paths, one steep slope.

43 MANOR FARM HOUSE, SWANNINGTON
Manor Drive, Swannington, NR9 5NR. Mr & Mrs John Powles. *7m NW of Norwich. Almost halfway between A1067 (to Fakenham) & B1149 (to Holt). In the middle of Swannington village. All parking in next door Romantic Garden Nursery.* **Sun 4 July (10-5). Combined adm with Swannington Manor £5,**
chd free. Light refreshments at Swannington Manor.
Small garden including knot garden with central lead fountain, topiary and lavender garden enclosed by hornbeam hedging. Terrace with large pots of lavender and agapanthus. Access to the Romantic Garden Nursery, which adjoins the garden.

44 MANOR FARM, COSTON
Coston Lane, Coston, Barnham Broom, NR9 4DT. Mr & Mrs J D Hambro. *10m W of Norwich. Off B1108 Norwich - Watton Rd. Take B1135 to Dereham at Kimberley. After approx 300yds sharp L bend, go straight over down Coston Lane, house & garden on L.* **Sat 10 Apr, Sun 19 Sept (11-5). Adm £5, chd free. Home-made teas.**
Wonderful 3 acre country garden set in larger estate. Several small garden rooms with both formal and informal planting. Climbing roses, walled kitchen garden, white, grass and late summer themes, classic herbaceous and shrub borders, box parterres and large areas of wildflowers. Various sculptures dotted round the garden. Dogs and picnics most welcome. 1m field walk. Something for everyone. Some gravel paths and steps.

45 MANOR HOUSE FARM, WELLINGHAM
Fakenham, Kings Lynn, PE32 2TH. Robin & Elisabeth Ellis, 01328 838227, libbyelliswellingham@gmail.com, www.manor-house-farm.co.uk. *7m W from Fakenham. 8m E from Swaffham, ½m off A1065. By the Church.* **Sun 20 June (11-5). Adm £6, chd free. Home-made teas.**
Charming 4 acre country garden surrounds an attractive farmhouse. Formal quadrants, 'hot spot' of grasses and gravel, small arboretum, pleached lime walk, vegetable parterre and rose tunnel. Unusual walled 'Taj' garden with old-fashioned roses, tree peonies, lilies and formal pond. A variety of herbaceous plants. Small herd of Formosan Sika deer. Some wheelchair access.

46 NEW **3 MEADOW CLOSE**
Mundesley, Norwich, NR11 8LW.
Martin & Karen Day. *Unadopted rd off High St in Mundesley. (Beckmeadow Way) Follow rd bearing L, 1st turning on L onto short tarmac rd. No 3 directly in front of you at top of small hill.* Sun 25 July (10-4). Adm £4, chd free. Home-made cakes, scones, jam tea and coffee.
A garden for all seasons and the senses. Created over the last 4 years from the desolation that was here previously! Plants have been chosen for fragrance, tactility and to extend colour throughout the year. Also functional as fruit and vegetables are also grown throughout the garden. Many of the plants have been grown by seed or cutting by Martin in his greenhouse.

47 **THE MERCHANTS HOUSE**
Blakeney, Holt, NR25 7NT. Mr & Mrs David Marris. *Centre of Blakeney Village. Garden located N of A149 (New Road) up Little Lane in Blakeney.* Sat 22 May (11.30-4.30). Adm £5, chd free. Home-made teas.
2 acres of walled 'secret' garden with terrace, woodland walk, parterre, shrub borders, orchard, kitchen garden, herbaceous border and ice house. Dogs on leads. Most of the garden is suitable for wheelchairs.

48 **NORTH LODGE**
51 Bowthorpe Road, Norwich, NR2 3TN. Bruce Bentley & Peter Wilson. *1½ m W of Norwich City Centre. Turn into Bowthorpe Rd off Dereham Rd, garden 150 metres on L. By bus: 21, 22, 23, 23A/B, 24 & 24A from City centre, Old Catton, Heartsease, Thorpe & most of W Norwich. Parking available outside.* Sun 25 July, Sun 1 Aug (11-5). Adm £4, chd free. Home-made teas.
Delightful town garden surrounding Victorian Gothic Cemetery Lodge. Full of follies created by current owners, including a classical temple, oriental water garden and formal ponds. Original 80ft-deep well! Predominantly herbaceous planting. House extension won architectural award. Slide show of house and garden history. Wheelchair access possible but difficult. Sloping gravel drive followed by short, steep, narrow,

brickweave ramp. WC not easily wheelchair accessible.

49 NEW **OAKS FARM COTTAGE**
Felmingham, North Walsham, NR28 0JZ. Stephane Lustig. Sun 23 May (12-5). Combined adm with East Wing £5, chd free. Light refreshments.
Oaks Farm Cottage owned by local garden designer (@stephanelustig), which transitions from beautiful pond with delightful summerhouse including well stocked herbaceous borders, an orchard, terraces and a lawn bordered by mature trees. Look out for the free-range chickens, and stylish fire pit along the way. Both gardens have been created within the last 15 years. Wheelchair access between gardens is via the road - please ask at gate for directions.

50 **THE OLD HOUSE**
Ranworth, NR13 6HS. The Hon Mrs Jacquetta Cator. *9m NE of Norwich. Nr South Walsham, below historic Ranworth Church.* Sun 11 Apr (11-4). Adm £5, chd free. Home-made teas.
Attractive linked and walled gardens alongside Ranworth Inner Broad. Bulbs, shrubs and Villandry inspired potager with interesting sculptures throughout. Mown rides through arboretum with spectacular views of the church and the broad. Some rough grass and gravel. Dogs allowed in arboretum but not in the garden itself. Wheelchair access some rough grass and gravel.

51 **THE OLD RECTORY, BRANDON PARVA**
Stone Lane, Brandon Parva, NR9 4DL. Mr & Mrs S Guest, 07867 840149, guest63@btinternet.com. *9m W of Norwich. Leave Norwich on B1108 towards Watton, turn R at sign for Barnham Broom. L at T-junction, stay on rd approx 3m until L next turn to Yaxham. L at Xrds.* Visits by arrangement May to July for groups of up to 20. Picnics allowed. Adm £5, chd free.
4 acre, mature garden with large collection (70) specimen trees, huge variety of shrubs and herbaceous plants combined to make beautiful mixed borders. The garden comprises several formal lawns and borders, woodland garden incl rhododendrons,

pond garden, walled garden and pergolas covered in wisteria, roses and clematis which create long shady walkways. Croquet lawn open for visitors to play.

52 **THE OLD RECTORY, SYDERSTONE**
Creake Road, Syderstone, King's Lynn, PE31 8SF. Mr & Mrs Tom White. *Access from Creake Road, or side gate opposite Village Hall.* Sun 6 June (11-5). Adm £4.50, chd free. Home-made teas in Village Hall nearby. Also open Grove House, Docking.
Charming Old Rectory garden designed in 1999 by Arne Maynard; with lawns, box, hornbeam, yew and beach hedging, pleached crab apple, wisteria and climbing roses, parterre of English shrub roses, herbaceous beds, shrubbery and orchard.

53 **OULTON HALL**
Oulton, Aylsham, NR11 6NU. Bolton Agnew. *4m NW of Aylsham. From Aylsham take B1354. After 4m turn L for Oulton Chapel, Hall ½ m on R. From B1149 (Norwich/Holt rd) take B1354, next R, Hall ½ m on R.* Sun 6 June (1-5). Adm £5, chd free. Home-made teas.
C18 manor house (not open) and clocktower set in 6-acre garden with lake and woodland walks. Chelsea designer's own garden - herbaceous, Italian, bog, water, wild, verdant, sunken and parterre gardens all flowing from one tempting vista to another. Developed over 25 yrs with emphasis on structure, height and texture, with a lot of recent replanting in the contemporary manner.

54 NEW **QUAKER FARM**
Quaker Lane, Spixworth, Norwich, NR10 3FL. Mr & Mrs Peter Cook. *3 miles north of Norwich. From Norwich take Buxton Road towards Spixworth, through Old Catton. Turning signposted Quaker Lane just over the NDR (A1270).* Sun 9 May (10.30-5). Adm £5, chd free. Light refreshments.
An acre of garden surrounding a traditional Norfolk farmhouse which has evolved over the last 40 plus years. Comprises garden rooms of herbaceous borders, gravel garden, an avenue of shrubs and a more recently created woodland garden. Garden games weather permitting!

Suitable for wheelchair users - a small area of gravel to negotiate.

SPECIAL EVENT

55 ◆ RAVENINGHAM HALL

Raveningham, Norwich,
NR14 6NS. Sir Nicholas &
Lady Bacon, 01508 548480,
sonya@raveningham.com,
www.raveningham.com. *14m SE
of Norwich. 4m from Beccles off
B1136.* **For NGS: Mon 1 Mar (2-4).
Adm £20, chd free. Pre-booking
essential, please visit www.
ngs.org.uk/special-events for
information & booking. For other
opening times and information,
please phone, email or visit garden
website.**
Join Raveningham Hall owner
and former President of the Royal
Horticultural Society, Sir Nicholas
Bacon for a guided walk around his
gardens. Traditional country house
garden in glorious parkland setting.
Restored Victorian conservatory,
walled kitchen garden, herbaceous
borders, newly planted stumpery.
Arboretum established after the 1987
gale, Millennium lake and sculpture by
Susan Bacon.

56 ◆ SANDRINGHAM GARDENS

Sandringham, PE35 6EH. Her
Majesty The Queen, 01485 545408,
visits@sandringhamestate.co.uk,
www.sandringhamestate.co.uk.
*6m NW of King's Lynn. Please see
website for detailed directions.* **For
opening times and information,
please phone, email or visit garden
website.**
60-acres of glorious gardens and
woodland with lakes, rare plants
and trees. Colour and all year round
interest; spring-flowering bulbs,
rhododendrons and azaleas, lavender
and roses. Dazzling autumn colour.
Gravel paths are not deep, some long
distances. Please tel or visit website
for our Accessibility Guide.

57 ◆ SEVERALS GRANGE

Holt Road B1110, Wood
Norton, NR20 5BL. Jane
Lister, 01362 684206,
hoecroft@hotmail.co.uk. *8m S
of Holt, 6m E of Fakenham. 2m N
of Guist on L of B1110. Guist is
situated 5m SE of Fakenham on
A1067 Norwich rd.* **For NGS: Sun
8 Aug (1-5). Adm £5, chd free.
Home-made teas. For other
opening times and information,
please phone or email.**
The gardens surrounding Severals

Grange are a perfect example of
how colour, shape and form can be
created by the use of foliage plants,
from large shrubs to small alpines.
Movement and lightness are achieved
by interspersing these plants with a
wide range of ornamental grasses,
which are at their best in late summer.
Splashes of additional colour are
provided by a variety of herbaceous
plants. Some gravel paths but help
can be provided.

58 SHERINGHAM HALL

Upper Sheringham, Sheringham,
NR26 8TB. *5m W of Cromer, 6m
E of Holt.* **Thur 5 Aug (2-5). Adm
£10. Pre-booking essential,
please visit www.ngs.org.uk for
information & booking. Tea.**
A unique opportunity to see this
wonderful garden. Designed by Repton
the park is open regularly, the hall and
walled garden are private. Kitchen
garden recreated by Arabella Lennox-
Boyd with restored glasshouses and
cold frames. Replanted orchards,
hornbeam temple, new herbaceous
borders, parterres and a white garden.
Wildflower meadow between hot and
cool borders in the east and restored
Repton pleasure grounds in the west.
Repton's walks have been reopened.

Highview House

59 SILVERSTONE FARM

North Elmham, Dereham,
NR20 5EX. George Carter,
grcarter@easynet.co.uk,
georgecartergardens.co.uk. *Nearer
to Gateley than North Elmham. From
North Elmham church head N to
Guist. Take 1st L onto Great Heath
Rd. L at T-junction. Take 1st R signed
Gateley, Silverstone Farm is 1st drive
on L by a wood.* Sun 19 Sept (2-5).
Adm £5, chd free. Tea. **Visits also
by arrangement Apr to Sept for
groups of 20 to 30.**
Garden belonging to George Carter
described by the Sunday Times as
'one of the 10 best garden designers
in Britain'. 1830s farmyard and formal
gardens in 2 acres. Inspired by C17
formal gardens, the site consists of a
series of interconnecting rooms with
framed views and vistas designed in
a simple palette of evergreens and
deciduous trees and shrubs such
as available in that period. Books by
the owner for sale. Level site mostly
wheelchair accessible.

60 ◆ STODY LODGE

Melton Constable,
NR24 2ER. Mr & Mrs Charles
MacNicol, 01263 860572,
enquiries@stodyestate.co.uk,
www.stodylodgegardens.co.uk.
*16m NW of Norwich, 3m S of Holt.
Off B1354. Signed from Melton
Constable on Holt Rd. For SatNav
NR24 2ER. Gardens signed as you
approach.* For NGS: Tue 18 May (1-
5). Adm £8, chd free. Home-made
teas. **For other opening times and
information, please phone, email or
visit garden website.**
Spectacular gardens with one of the
largest concentrations of rhododendrons
and azaleas in East Anglia. Created in
the 1920s, the gardens also feature
magnolias, camellias, a variety of
ornamental and specimen trees, late
daffodils, tulips and bluebells. Expansive
lawns and magnificent yew hedges.
Woodland walks and 4-acre Water
Garden filled with over 2,000 vividly-
coloured azalea mollis. Home-made teas
provided by selected local and national
charities. Access to most areas of the
garden. Gravel paths to Azalea Water
Gardens with some uneven ground.

61 NEW SWAFIELD HALL

Knapton Road, Swafield, North
Walsham, NR28 0RP. Tim
Payne & Boris Konoshenko,
swafieldhall.co.uk. *Swafield Hall is*

*approx ½m along Knapton Rd (also
called Mundesley Rd) from its start in
village of Swafield. You need to add
about 300 yards to the location given
by most Sat Navs.* Sat 7, Sun 8 Aug
(10-5). Adm £5, chd free. Home-
made teas.
C16 Manor House with Georgian
additions (not open) set within 4 acres
of gardens including a parterre and
various rooms including a summer
garden, orchard, cutting garden, pear
tunnel, secret oriental garden (with
nine flower beds based on a Persian
carpet), the Apollo Promenade of
theatrical serpentine hedging, a duck
pond and woodland walk. The whole
garden is accessible by wheelchair.

62 SWANNINGTON MANOR

Norwich, NR9 5NR. Gregory & Sue
Darling. *7m NW of Norwich. Almost
halfway between A1067 (to Fakenham)
& B1149 (to Holt). Parking at Romantic
Garden Nursery.* Sun 4 July (10-5).
Combined adm with Manor Farm
House, Swannington £5, chd free.
Light refreshments.
C17 manor house (not open) creates a
stunning backdrop to this garden which
is framed by 300yr old hedges, thought
to be unique in this country. 7 acres
includes mixed shrub and herbaceous
borders, a water garden, sunken rose
garden, a recently installed potager
style vegetable garden, specimen trees
and sloping lawns make this garden
both delightful and unusual. Some
access for wheelchairs able to cope
with gravel paths.

63 THORPLAND HALL

Thorpland, Fakenham, NR21 0HD.
Mr & Mrs N R Savory. *1m N of
Fakenham bypass A148. Turn off
r'about on A148 north of Morrisons
superstore, signed Thorpland Rd.
Additional parking for wheelchair users
only.* Wed 16 June (11-5). Adm £6,
chd free. Home-made teas.
Mature English garden in a superb
setting, containing ruins of an old
chapel, and surrounding a fine Tudor
house (not open). 6 acres of garden
with some recent re-planting, incl
peony beds, herbaceous borders and
a rose garden. The walled garden is in
full working order and kept to a high
standard. Small lake, and restored
shepherd's hut, are surrounded by
interesting trees and shrubs. Gravel
paths, paved areas and lawns. Some
grass paths maybe a little uneven.

64 NEW 61 TRAFFORD WAY

Spixworth, Norwich, NR10 3QL. Mr
& Mrs Colin Ryall. *Park in Spixworth
Community car park in Crosswick
Lane.* Sun 1 Aug (10-5). Adm £4,
chd free. Light refreshments.
Small garden showing what can be
achieved with careful planting and
the use of pleached hornbeam trees.
Gravel garden with flower beds and
pot plants. Colourful herbaceous
borders with roses.

65 TUDOR LODGINGS

Castle Acre, King's Lynn,
PE32 2AN. Gus & Julia
Stafford Allen, 01760 755334,
jstaffordallen@btinternet.com.
Parking in field below the house. Sun
1 Aug (11-5). Adm £5, chd free.
Light refreshments. **Visits also by
arrangement July to Sept.**
The 2- acre garden fronts a C15 flint
house (not open), and incorporates
part of the Norman earthworks.
C18 dovecote, abstract topiary,
lawn and a 'Mondrian' knot garden.
Ornamental grasses and hot border.
Productive fruit cage and cutting
garden. A natural wild area includes
a shepherd's hut and informal
pond. Bantams and Ducks. Partial
wheelchair access, please ask for
assistance beforehand. Disabled WC.

66 TYGER BARN

Wood Lane, Aldeby, Beccles,
NR34 0DA. Julianne Fernandez,
juliannefernandez@btinternet.
com, www.chasing-arcadia.com.
*Approx 1m from Toft Monks. From
A143 towards Great Yarmouth at
Toft Monks turn R into Post Office
Lane opp White Lion Pub, after ¼m
turn L into Wood Lane. After ½m
Tyger Barn is 2nd house on L.* Sun
4 July (12-4). Adm £5, chd free.
Light refreshments. **Visits also
by arrangement June & July for
groups of 10 to 30.**
Featured in The English Garden, EDP
Norfolk and Suffolk magazines, Tyger
Barn is a modern country garden
started in 2007. It includes extensive
borders with hot, exotic and 'ghost'
themes, a secret cottage garden,
wildflower swathes and colonies of bee
orchids. A traditional hay meadow and
ancient woodland provide a beautiful
setting. Garden is mainly level, but
is divided by a shingle drive which
wheelchairs may find difficult to cross.

67 WALCOTT HOUSE
Walcott Green, Walcott,
Norwich, NR12 0NU. Nick &
Juliet Collier, 07986 607170,
julietcollier1@gmail.com. *3m N of
Stalham. Off the Stalham to Walcott rd
(B1159).* Sun 20 June (11-5). Adm
£5, chd free. Light refreshments.
Visits also by arrangement in July for
groups of 5 to 30.
A 12 acre site with over 1 acre of
formal gardens based on model C19
Norfolk farm buildings. Woodland and
damp gardens; arboretum; vistas with
tree lined avenues; woodland walks.
Small single steps to negotiate when
moving between gardens in the yards.
& 🐕 ☕ 👜

68 33 WALDEMAR AVENUE
Hellesdon, Norwich,
NR6 6TB. Sonja Gaffer &
Alan Beal, 07798 522380,
sonja.gaffer@ntlworld.com, www.
facebook.com/Hellesdontropicalgarden
*Waldemar Ave is situated approx
400 yards off Norwich ring road
towards Cromer on A140.* Sun 1
Aug, Sun 5 Sept (10-5). Adm £4,
chd free. Home-made teas. Visits
also by arrangement Aug & Sept
for groups of 10 to 30.
A surprising and large suburban
garden of many parts with an exciting
mix of exotic and tropical plants
combined with unusual perennials,
many grown from seed. A quirky
palm-thatched Tiki hut is an eye
catching feature. Teas will be served
and you can sit by the pond which is
brimming with wildlife and rare plants.
A large collection of succulents will be
on show and there will be plants to
buy. Wheelchair access surfaces are
mostly of lawn and concrete and are
on one level. Two entrance gates are
33 and 39 inches wide respectively.
& ✳ 👜

69 WARBOROUGH HOUSE
2 Wells Road, Stiffkey, NR23 1QH.
Mr & Mrs J Morgan. *13m N of
Fakenham, 4m E of Wells-Next-The-
Sea on A149 in the centre of Stiffkey
village. Parking is available & signed
at garden entrance. Coasthopper
bus stop outside garden. Please DO
NOT park on the main road as this
causes congestion.* Sun 23, Sun
30 May (11-4). Adm £6, chd free.
Home-made teas.
7-acre garden on a steep chalk slope,
surrounding C19 house (not open)
with views across the Stiffkey valley
and to the coast. Woodland walks,

formal terraces, shrub borders, lawns
and walled garden create a garden of
contrasts. Garden slopes steeply in
parts. Paths are gravel, bark chip or
grass. Disabled parking allows access
to garden nearest the house and teas.
& ✳ 👜

70 NEW WARREN HOUSE
Narford Road, West Acre, King's
Lynn, PE32 1UG. Henry Birkbeck.
*Between West Acre & Castle Acre,
1m E of West Acre at the Narford
Road Xrds.* Sun 27 June (11-
5). Adm £4.50, chd free. Light
refreshments.
Farmhouse garden 3-acres,
overlooking the river Nar and its
meadows. Established and new
areas with lawns and a variety of
hedges, paths and borders. Featuring
a pergola, fruit trees, lily pond,
small ornamental rose garden, a
greenhouse, a vegetable patch and a
long south-facing wall. Limited access
on mainly grass and gravel paths.
& ✳ 👜 🌲

GROUP OPENING

**71 WELLS-NEXT-THE-SEA
GARDENS**
Wells-Next -The-Sea, NR23 1DP.
*10m N of Fakenham. All gardens
near Coasthopper 'Burnt Street'
or 'The Buttlands' bus stops. Car-
parking for all gardens in Market
Lane area.* Sun 6 June (11-4.30).
Combined adm £5, chd free.
Light refreshments at Poacher
Cottage, 15 Burnt Street, Wells,
NR23 1HS.

CAPRICE
Clubbs Lane, Wells-next-the-Sea.
David & Joolz Saunders.

HIRAETH
11 Burnt Street, Wells-next-the
Sea. Jen Davies.

7 MARKET LANE
Wells-next-the-Sea. Hazel Ashley.

NORFOLK HOUSE
17 Burnt Street, Wells-next-the-
Sea. Katrina & Alan Jackson.

POACHER COTTAGE
15 Burnt Street. Roger & Barbara
Oliver.

Wells-next-the-Sea is a small, friendly
coastal town on the North Norfolk
Coast. Popular with families, walkers
and bird watchers. The harbour has
shops, cafes, fish and chips. Beach

served by a narrow gauge railway.
Fine parish church. The five town
gardens, though small, demonstrate a
variety of design and a wide selection
of planting. Wheelchair access all
gardens except Norfolk House which
has limited access.
& ☕ 👜

72 WEST BARSHAM HALL
Fakenham, NR21 9NP. Mr & Mrs
Jeremy Soames, 01328 863519,
susannasoames@gmail.com. *3m N
of Fakenham. From Fakenham take
A148 to Cromer then L on B1105
to Wells. After ½m L again to Wells.
After 1½m R The Barshams & West
Barsham Hall.* Sun 18 July (10.30-
5.30). Adm £5, chd free. Morning
coffee and light refreshments,
afternoon cream teas. Visits also
by arrangement May to Sept.
Price not specified as depends on
refreshment requirements.
Large garden with lake, approx
10-acres. Mature yew hedging and
sunken garden originally laid out
by Gertrude Jeykll. Swimming pool
garden, shrub borders, kitchen
garden with herbaceous borders, fruit
cage, cutting garden and bog garden.
Separate old fashioned cottage
garden also open. Some slopes and
gravel paths.
& 🐴 ✳ ☕ 👜

73 WRETHAM LODGE
East Wretham, IP24 1RL.
Mr Gordon Alexander & Mr
Ian Salter, 01953 498997,
grdalexander@btinternet.com.
*6m NE of Thetford. A11 E from
Thetford, L up A1075, L by village
sign, R at Xrds then bear L.* Sun 2,
Mon 3 May (11-5). Adm £5, chd
free. Tea at Church. Visits also by
arrangement Apr to Sept.
10 acre garden surrounding former
Georgian rectory (not open). In spring
masses of species tulips, hellebores,
fritillaries, daffodils and narcissi;
bluebell walk and small woodland
walk. Topiary pyramids and yew
hedging lead to double herbaceous
borders. Shrub borders and rose
beds (home of the Wretham Rose).
Traditionally maintained walled garden
with fruit, vegetables and perennials.
& 🐕 👜 🌲

NORTH EAST

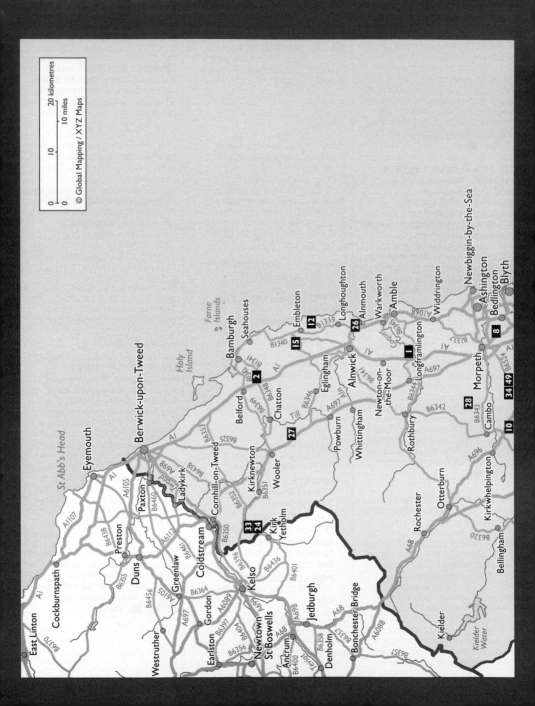

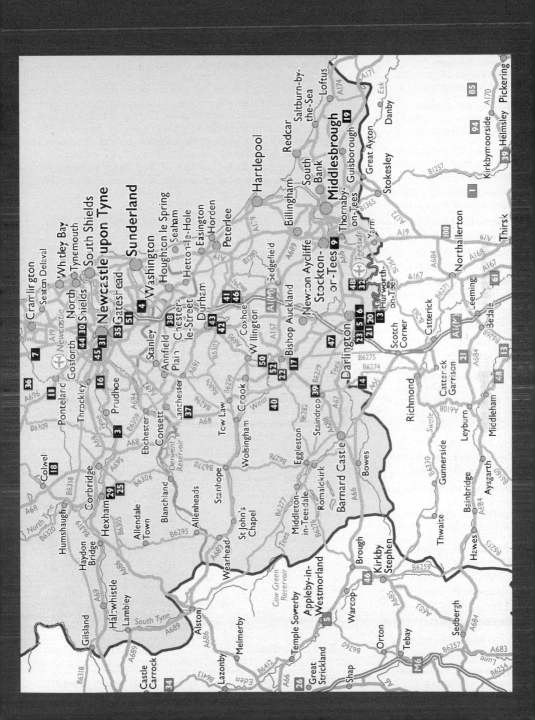

Volunteers

County Durham
County Organiser
Iain Anderson
01325 778446
iain.anderson@ngs.org.uk

County Treasurer
Sue Douglas 07712 461002
sue.douglas@ngs.org.uk

Booklet Co-ordinator
Sheila Walke 07837 764057
sheila.walke@ngs.org.uk

Assistant County Organisers
Gill Knights 01325 483210
gillianknights55@gmail.com

Helen Jackson 07985 699960
helen.jackson@ngs.org.uk

Dorothy Matthews 01325 354434
matthews.dorothy@googlemail.com

Monica Spencer 01325 286215
monica.spencer@ngs.org.uk

Margaret Stamper 01325 488911
margaretstamper@tiscali.co.uk

Gill Naisby 01325 381324
gillnaisby@gmail.com

Sue Walker 01325 481881
walker.sdl@gmail.com

Northumberland & Tyne
and Wear County Organiser
& Booklet Coordinator
Maureen Kesteven 01914 135937
maureen.kesteven@ngs.org.uk

County Treasurer
David Oakley 07941 077594
david.oakley@ngs.org.uk

Publicity, Talks Co-ordinator &
Social Media
Liz Reid 01914 165981
liz.reid@ngs.org.uk

Assistant County Organisers
Maxine Eaton 077154 60038
maxine.eaton@ngs.org.uk

Natasha McEwen 07917 754155
natashamcewengd@aol.co.uk

Liz Reid (as above)

Susie White 07941 077594
susie@susie-white.co.uk

David Young 01434 600699
david.young@ngs.org.uk

County Durham: an unsung county.

County Durham lies between the River Wear and the River Tees and is varied and beautiful.

Our National Garden Scheme gardens are to be found in the city, high up in the dales, in the attractive villages of South Durham and outskirts of industrial Teesside. Something different every week.

Choose your old favourites but also check out our wonderful new gardens. Country gardens include an exciting newly developed woodland and garden at The Stables, traditional village gardens at High Bank Farm and Homelands and a wildlife garden at East Middleton. Our town gardens can surprise you too. The quirky new "Garden and Studio at 67" should also be on your visiting list. This year we have the "Hidden Gardens of Croft Road" back again, a lovely collection of different gardens with many changes since their last opening.

All our garden owners are looking forward to welcoming you.

Northumberland is a county of ancient castles and wild coastline, of expansive views and big skies.

This beautiful landscape with its picturesque valleys provides the backdrop for a delightful range of gardens.

There's a great variety of styles from the walled vegetable gardens and sumptuous borders of large country houses to exciting contemporary planting. An historic manor house has a Gertrude Jekyll connection whilst a recently made garden is experimental and sustainable and full of ideas for organic growing.

Another unusual garden has been made on an exceptionally steep slope, showing that you can transform even the most difficult of sites. There are city gardens too: a tiny courtyard packed with colour or the landscaped garden rooms around a hospice. From a series of cottage plots in a fishing village to meadows to arboreta and dramatic water features, there is so much to inspire.

Right: Capheaton Hall Cabbage

f @gardensopenforcharity   🐦 @NGSNorthumberl1

📷 @ngsnorthumber   🐦 @Durham_TeesNGS

OPENING DATES

All entries subject to change. For latest information check **www.ngs.org.uk**

Map locator numbers are shown to the right of each garden name.

April

Sunday 18th
Adderstone House	2
◆ Whalton Manor Gardens	49

May

Sunday 2nd
45 Blackwell	5
46 Blackwell	6

Sunday 23rd
24 Bede Crescent	4
Blagdon	7
Walworth Gardens	47

Sunday 30th
Croft Hall	13
NEW The Moore House	34

June

Sunday 6th
The Beacon	3
Lilburn Tower	27

Saturday 12th
Acton House	1
Maggie's	30

Sunday 13th
NEW High Bank Farm	21
NEW Holmelands	23
The Shilling House Garden	45

Saturday 19th
NEW Lesbury Gardens	26
◆ Whalton Manor Gardens	49

Sunday 20th
NEW Lesbury Gardens	26
Loughbrow House	29
◆ Mindrum Garden	33

Saturday 26th
24 Bede Crescent	4
Craster's Hidden Gardens	12
Fallodon Hall	15
Woodlands	51

Sunday 27th
NEW Longwitton Hall	28
Oliver Ford Garden	37
NEW Rose Cottage	41
NEW The Stables	46

July

Friday 2nd
Hidden Gardens of Croft Road	20

Saturday 3rd
Capheaton Hall	10
Kirky Cottage	24

Sunday 4th
NEW 2 Briarlea	8
Capheaton Hall	10
Hidden Gardens of Croft Road	20
Marie Curie Hospice	31

Saturday 10th
NEW East Middleton	14
Woodbine House	50

Sunday 11th
Ferndene House	16
St Margaret's Allotments	43

Saturday 17th
NEW St Oswald's Hospice	44

Sunday 18th
NEW The Moore House	34

Saturday 24th
Middleton Hall Retirement Village	32

Sunday 25th
The Beacon	3
24 Bede Crescent	4

August

Sunday 1st
Heather Holm	19
Lambshield	25
St Cuthbert's Hospice	42

Sunday 8th
NEW Garden and studio at 67	17
Gardener's Cottage Plants	18
Ravensford Farm	40

Saturday 14th
Coldcotes Moor Garden	11

Sunday 15th
Coldcotes Moor Garden	11
Hillside Cottages	22

September

Saturday 4th
Ogle Garden	36

By Arrangement

Arrange a personalised garden visit on a date to suit you. See individual garden entries for full details.

Acton House	1
Adderstone House	2
The Beacon	3
24 Bede Crescent	4
NEW 2 Briarlea	8
NEW 4 Briarwalk	9
Coldcotes Moor Garden	11
Fallodon Hall	15
Ferndene House	16
Hillside Cottages	22
Kirky Cottage	24
Lambshield	25
Lilburn Tower	27
Loughbrow House	29
Ogle Garden	36
25 Park Road South	38
Ravensford Farm	40
NEW 28 Washington Avenue	48
Woodlands	51
Woodside House	52

© Carole Drake

Acton House

THE GARDENS

1 ACTON HOUSE

Felton, Morpeth, NE65 9NU. Mr Alan & Mrs Eileen Ferguson, Head Gardener 07779 860217, jane@actonhouseuk.com. *A1 N of Morpeth. From dual carriageway A1 north of Felton, use junction for Amble & Felton. If coming through Felton follow road north out of village for 1m & in both cases follow signs from Acton turning.* **Sat 12 June (1-4). Adm £5, chd free. Light refreshments. Visits also by arrangement May to Aug for groups of 5 to 30.**

This walled garden has structure, colour and variety of planting, with abundant herbaceous perennials and different grasses. Planted in the spring of 2011, it has sections devoted to fruit and vegetables, David Austin rose borders, standard trees and climbers spreading over the brick walls. There are additional mixed borders, a ha-ha, and developing woodland planting, in total extending over 5 acres. Herbaceous perennial plantings include species and varieties favoured by butterflies and bees.

 ♿ 🐕 ❀ 🚗 ☕

2 ADDERSTONE HOUSE

Adderstone Mains, Belford, NE70 7HS. John & Pauline Clough, 01668 219171, john.clough@me.com. *1m S of Belford off A1. Best approached from the N off A1 Adderstone Mains is 1m S of Belford, turn L off A1. From S, 0.7m N of Purdy Lodge on R but turn dangerous, continue to Belford, turn & approach from N. Ample parking.* **Sun 18 Apr (11-3). Adm £6, chd free. Pre-booking essential, please visit www.ngs.org.uk/events for information & booking. Light refreshments in the Barn beside the Japanese Garden. Visits also by arrangement Apr to Sept for groups of 10+.**

A Victorian house, gardens, millpond with folly and grounds of 10½ acres. Established and 'new projects' in development with Sean Murray (RHS Chelsea Challenge winner). New Japanese Garden with cherry blossom, Walled garden; an orchard leads to the sunken garden; rose walk and formal garden, then on to the mill pond, fruit/vegetable plots and vineyard. Plant stall, teas and grape juice from vineyard.

🐕 ❀ 🚗 ☕ 🪑

3 THE BEACON

10 Crabtree Road, Stocksfield, NE43 7NX. Derek & Patricia Hodgson, 01661 842518, patandderek@btinternet.com. *12m W of Newcastle. From A69 follow signs into village. Station & cricket ground on L. Turn R into Cadehill Rd then 1st R into Crabtree Rd (cul de sac) Park on Cadehill.* **Sun 6 June, Sun 25 July (1.30-5). Adm £5, chd free. Cream teas. Visits also by arrangement Apr to Sept for groups of 10+.**

This garden illustrates how to make a cottage garden on a steep site with loads of interest at different levels. Planted with acers, roses and a variety of cottage garden and formal plants. Water runs gently through it and there are tranquil places to sit and talk or just reflect. Stunning colour and plant combinations. Wildlife friendly - numerous birds, frogs, newts, hedgehogs. Haven for butterflies and bees. Owner available for entertaining group talks. Steep, so not wheelchair friendly but wheelchair users have negotiated the drive and enjoyed the view of the main garden.

❀ 🚗 ☕

4 24 BEDE CRESCENT

Washington, NE38 7JA. Sheila Brookes, 07596619035, sheilab2424@gmail.com. *From A1231 follow directions to Washington Old Hall, turn L onto Abbey Rd, then L onto Village Lane. Turn L into Bede Cres just before the Black Bush, garden is the end house at the grassed oval.* **Sun 23 May, Sat 26 June, Sun 25 July (12.30-4.30). Adm £3, chd free. Cold drinks, biscuits & crisps available if desired. Visits also by arrangement May to Aug for groups of 5 to 20.**

A lesson in how to make a small shady place colourful and interesting. Small 'courtyard style' garden, with central paved area, surrounded by borders containing shrubs and box balls for year round structure, and packed with colourful Astilbes, lilies and clematis for summer impact. Small patio front garden, with gravel border planted with box balls and containerised shrubs. Light refreshments available from several nearby cafes and pubs around our pretty Village Green, (Multi Gold medal winner in Northumbria in Bloom). Access into rear paved 'courtyard garden' via a side gate, (which should accommodate a small wheelchair, although not a wide entrance).

5 45 BLACKWELL

Darlington, DL3 8QT. Cath & Peter Proud. *SW Darlington, next to R.Tees. ½ way along Blackwell in Darlington, which links Bridge Rd (on A66 just past Blackwell Bridge) & Carmel Rd South. Alternatively, turn into Blackwell from Post Office on Carmel Rd South.* **Sun 2 May (1-5). Combined adm with 46 Blackwell £5, chd free. Teas, coffees and soft drinks and light refreshments will be available.**

No. 45 Blackwell rises from the River Tees up to a 1.5 acre garden with many mature trees, wildlife meadow, pond with waterfall, herb and Mediterranean garden, vegetable bed, greenhouse, camellia border, lawns and colourful Spring borders and containers full of Spring bulbs. A one-way route will be in operation. Refreshments, preserves and plants may be on sale. Well behaved dogs on leads welcome. Wheelchair access to patios and to lower lawn with assistance.

 ♿ 🐕 ❀ ☕ 🪑

6 46 BLACKWELL

Darlington, DL3 8QT. Christopher & Yvonne Auton. *(Please refer to directions to 45 Blackwell, Darlington - joint garden opening).* **Sun 2 May (1-5). Combined adm with 45 Blackwell £5, chd free. Teas at 45 Blackwell.**

46 Blackwell is a plantsman's garden - compact, but full of unusual specimens and collections of plants and trees, with a pond and summerhouse in the back garden. Wheelchair access to front garden and partial access to back garden.

 ♿ 🐕 ❀ ☕

7 BLAGDON

Seaton Burn, NE13 6DE. Viscount Ridley, www.blagdonestate.co.uk. *5m S of Morpeth on A1. 8m N of Newcastle on A1, N on B1318, L at r'about (Holiday Inn) & follow signs to Blagdon. Entrance to parking area signed.* **Sun 23 May (1-4). Adm £5, chd free. Home-made teas.**

Unique 27 acre garden encompassing formal garden with Lutyens designed 'canal', Lutyens structures and walled kitchen garden. Valley with stream and various follies, quarry garden and woodland walks. Large numbers of ornamental trees and shrubs planted over many generations. National Collections of Acer, Alnus and Sorbus. Trailer rides around the estate (small additional charge). Partial wheelchair access.

8 NEW 2 BRIARLEA

Hepscott, Morpeth, NE61 6PA. Richard & Carolyn Torr, 07470 391936, richardtorr6@gmail.com. *Entering Hepscott from A192, Briarlea is 2nd rd on R. Please park considerately within the village. Please do not park on Briarlea - 2 disabled spaces only available on Briarlea.* **Sun 4 July (11-5). Adm £4, chd free. Light refreshments at 7 Thornlea, approx 100m away, with further cactus collection and Plant Stall. Visits also by arrangement June & July for groups of up to 20.**
A plant lover's garden of just under ½ acre, rejuvenated and redeveloped over the last 4 years, featuring gravel areas, stream side planting and many herbaceous borders. Planting features hostas, grasses, roses, penstemons, ferns and a wide range of shrubs and other perennials. A vegetable area with raised beds includes greenhouses housing an extensive collection of cacti and other succulents. Good access to most of garden, however, some steps and gravel areas.

9 NEW 4 BRIARWALK

Briar Walk, Hartburn, Stockton-On-Tees, TS18 5BQ. Mr & Mrs Bob & Joan Cornwell, 07985 606898, bobcornwell@ntlworld.com. *Briarwalk is a small private Rd opposite the Stockton Arms on Darlington Rd Hartburn.*

Visits by arrangement Mar to Oct for groups of up to 10. Adm £4, chd free. Light refreshments.
Briarwalk is a large informal garden with a selection of trees and shrubs and a diverse collection of plants to provide year round interest. The first of the new season colour starts in late Jan with the hellebores and continues through Feb, Mar and April with the spring bulbs, alliums, foxgloves. The colour progresses throughout the year with a final flourish of autumn colour from the acers.

10 CAPHEATON HALL

Capheaton, Newcastle Upon Tyne, NE19 2AB. William & Eliza Browne-Swinburne, 01913 758152, capheatonhall@gmail.com, www.capheatonhall.co.uk/ accomodation. *24m N of Newcastle off A696. From S turn L onto Silver Hill rd signed Capheaton. From N, past Wallington/Kirkharle junction, turn R.* **Sat 3, Sun 4 July (10-4). Adm £7.50, chd free. Pre-booking essential, please visit www.ngs. org.uk/events for information & booking. Home-made teas.**
Set in parkland, Capheaton Hall has magnificent views over the Northumberland countryside. Formal ponds sit south of the house, which has C19 conservatory, and a walk to a Georgian folly of a chapel. The outstanding feature is the very productive walled kitchen garden, at its height in late summer, mixing colourful vegetables, espaliered fruit with annual and perennial flowering

borders. Victorian glasshouse and conservatory. Wheelchair access to the walled garden is limited by steps and gravel paths, but the lawns are closely mown and generally flat.

11 COLDCOTES MOOR GARDEN

Ponteland, Newcastle Upon Tyne, NE20 0DF. Ron & Louise Bowey, info@theboweys.co.uk. *Off A696 N of Ponteland. From S, leave Ponteland on A696 towards Jedburgh, after 1m take L turn marked 'Milbourne 2m'. After 400yds turn L into drive.* **Sat 14, Sun 15 Aug (11-4). Adm £6, chd free. Pre-booking essential, please visit www.ngs. org.uk for information & booking. Light refreshments. Visits also by arrangement June to Sept for groups of 20+. Arrive by car or 'small' coach.**
The garden, landscaped grounds and woods cover around 15 acres. The wooded approach opens out to lawned areas surrounded by ornamental and woodland shrubs and trees. A courtyard garden leads to an ornamental walled garden, beyond which is an orchard, vegetable garden, flower garden and rose arbour. To the south the garden looks out over a lake and field walks, with woodland walks to the west. Small children's play area. Most areas can be accessed though sometimes by circuitous routes or an occasional stop. WC access involves three steps.

Ferndene House

GROUP OPENING

12 CRASTER'S HIDDEN GARDENS

1, 2 and 5 Chapel Row, Craster, Alnwick, NE66 3TU. Sue Chapman, Gill Starkey, June Drage. *Park in the pay & display parking at entrance to village. Walk to the harbour & turn L along Dunstanburgh Rd. Look for NGS sign on L before you reach the castle field. No parking in the village.* **Sat 26 June (11-4). Combined adm £5, chd free. Pre-booking essential, please visit www.ngs. org.uk/events for information & booking.**

In sight of Dunstanburgh Castle, in the picturesque village of Craster, sit 3 very different cottage gardens showing what can be achieved in a windy seaside setting. Formerly allotments attached to fishermen's cottages - No.1 is newly developed, has two ponds, colour-themed borders and prairie style planting. No. 2 has sea views, many perennial bee friendly plants, roses galore and a lovely herbaceous border. No. 5 is a ¼ acre, divided into several separate areas: woodland garden, pond, fruit and vegetables plus borders featuring many unusual perennials, shrubs, grasses and ferns. The garden is managed to encourage wildlife. Teas at RNLI Unsuitable for wheelchairs - steep and uneven ground.

✽

13 CROFT HALL

Croft-on-Tees, DL2 2TB. Mr & Mrs Trevor Chaytor Norris. *3m S of Darlington. On A167 to Northallerton, 6m from Scotch Corner. Croft Hall is 1st house on R as you enter village from Scotch Corner.* **Sun 30 May (2-5.30). Adm £5, chd free. Home-made teas.**

A lovely lavender walk leads to a Queen Anne-fronted house (not open) surrounded by a 5 acre garden, comprising a stunning herbaceous border, large fruit and vegetable plot, two ponds and wonderful topiary arched wall. Pretty rose garden and mature box Italianate parterre are beautifully set in this garden offering peaceful, tranquil views of open countryside. No dogs. Wheelchair access, some gravel paths.

& ✽ ☕

14 NEW EAST MIDDLETON

Caldwell, Richmond, DL11 7QL. Deny & Dave Holden. *12m W of Darlington. 9m N of Richmond. 7m SE of Barnard Castle. Turn off B6274 in Caldwell village towards Hutton Magna for 1m. ½m up track with passing places. Field Parking. Hard standing parking for wheelchair users.* **Sat 10 July (1-6). Adm £4, chd free. Light refreshments.**

Wildlife friendly half acre includes pond, mixed planting of perennials, trees and shrubs, vegetables and courtyard garden with greenhouse. Additional land with orchard and small arboretum started in 2000. Wheelchair friendly. Plants, art and craft work for sale. Creature trail. Small exhibit of classic motorbikes. Dogs on leads welcome.

& 🐕 ✽ ☕

15 FALLODON HALL

Alnwick, NE66 3HF. Mr & Mrs Mark Bridgeman, 01665 576252, luciabridgeman@gmail.com, www.bruntoncottages.co.uk. *5m N of Alnwick, 2m off A1. From the A1 turn onto the B6347 signed Christon Bank & Seahouses. Turn into the Fallodon gates after exactly 2m, at Xrds. Follow drive for 1m.* **Sat 26 June (2-5). Adm £5, chd free. Home-made teas. Visits also by arrangement May to Sept for groups of 10 to 30.**

Extensive, well established garden, with a hot greenhouse beside the bog garden. The late C17 walls of the kitchen garden surround cutting and vegetable borders and the fruit greenhouse. Natasha McEwen replanted the sunken garden from 1898, and the redesigned 30m border newly planted in 2017. Woodlands, pond and arboretum over 10 acres to explore. Grave of Sir Edward Grey, Foreign Secretary during WW1, famous ornithologist and fly fisherman, is in the woods near the pond and arboretum. C17 walls of Kitchen garden contain a fireplace, to heat the fruit trees of the Salkeld family, renowned for their gardening expertise. Partial wheelchair access.

& 🐕 ✽ 🛏 ☕

16 FERNDENE HOUSE

2 Holburn Lane Court, Holburn Lane, Ryton, NE40 3PN. Maureen Kesteven, 0191 413 5937, maureen.kesteven@ngs. org.uk, www.facebook.com/ northeastgardenopenforcharity/. *In Ryton Old Village, 8m W of Gateshead. Off B6317, on Holburn Lane. Park on street or in Co-op carpark on High St, cross rd through Ferndene Park following yellow signs.* **Sun 11 July (12.30-4). Adm £5, chd free. Home-made teas. Light lunches, pizza and prosecco available. Visits also**

High Bank Farm

by arrangement Apr to Aug for groups of 10+.

³⁄₄ acre garden surrounded by trees. Informal areas of herbaceous perennials, more formal box bordered area, sedum roof, wildlife pond, gravel and bog gardens. Willow work. Early interest - hellebores, snowdrops, daffodils, bluebells and tulips. Summer interest from wide range of flowering perennials, including a meadow planted in 2020. 1½ acre mixed broadleaf wood with beck running through. Driveway, from which main borders can be seen, is wheelchair accessible but 'phone for assistance.

&. ✿ ☕

17 NEW **GARDEN AND STUDIO AT 67**

Manor Road, St. Helen Auckland, Bishop Auckland, DL14 9ER. Mitsi B Kral, www.instagram.com/gardenat67/. *From A68 West Auckland, turn on to Station Rd towards St Helen Auckland avoiding the by-pass follow the rd over the bridge and on to Manor Rd, garden on R by pedestrian crossing.* **Sun 8 Aug (11-4). Adm £4, chd free. Home-made teas.**

The garden at 67 is at the rear of a Georgian house which was formerly a pub. Artist Mitsi has landscaped the walled garden over the last 2 years. The garden is a magical eclectic mix of unusual plants, packed with colour. There are wildlife ponds, raised vegetable garden, greenhouse, rose garden, herb garden and salvaged items. Wheelchair access is to the main feature of the garden, other paths are gravel.

&. 🐃 ✿ 🚗 ☕

18 **GARDENER'S COTTAGE PLANTS**

Gardener's Cottage, Bingfield, Newcastle Upon Tyne, NE19 2LE, Andrew Davenport, www.gcplants.co.uk. *6m N of Corbridge. From N turn L off A68 signed Bingfield, after approx 0.6m turn L at T junction, garden on R approx 0.9m. From S turn R off A68 signed Ringfield, garden on R approx 1.8m.* **Sun 8 Aug (11-4). Adm £5, chd free. Tea. Variety of home-made cakes and biscuits.**

This compact (¼ acre) experimental garden and nursery provides an education in organic and sustainable gardening. Organic vegetable, fruit, herb and floral gardens show the use

of mulching, composting, growing systems and other ideas from the garden's inventive creator. Wildflowers thrive amongst cultivated varieties in a range of attractive and wildlife friendly ornamental borders. Collection of unusual, historic and rare wildflowers of the mid-Tyne valley. Designated composting education area featuring composting systems, innovations and facilities.

✿ ☕

19 **HEATHER HOLM**

Stanghow Road, Stanghow, Saltburn-By-The-Sea, TS12 3JU. Arthur & June Murray. *Stanghow is 5 km E of Guisborough on A171. Turn L at Lockwood Beck (signed Stanghow) Heather Holm is on the R past the Xrds.* **Sun 1 Aug (12-4). Adm £4, chd free. Home-made teas.**

At 700 feet above sea level on the edge of the North Yorkshire Moors, Stanghow has won numerous RHS Gold awards for Best Small Village, winning again in 2017, and was a Champion among Champions in the 2013 Britain in Bloom competition. Heather Holm has extensive topiary which adds form and structure to this ¼ acre garden. No dogs. Wheelchair access to viewing deck only.

✿ ☕

GROUP OPENING

20 **HIDDEN GARDENS OF CROFT ROAD**

Darlington, DL2 2SD. Carol Pratt. *2m S of Darlington on A167. ³⁄₄ m S from A167/A66 r'about between Darlington & Croft.* **Fri 2, Sun 4 July (1-5). Combined adm £6, chd free. Home-made teas at Oxney Flatts & Orchard Gardens.**

NAGS HEAD FARM
Jo & Ian Fearnley.

NEW COTTAGE
Jane & John Brown.

ORCHARD GARDENS
Gill & Neil Segger.

OXNEY COTTAGE
Gypsy Nichol.

OXNEY FLATTS FARM
Carol & Chris Pratt.

5 very different and interesting gardens, well named as 'Hidden Gardens'. All are behind tall hedges. Oxney Cottage is a very

pretty cottage garden with lawns, herbaceous borders and roses, colourful and varied unusual plants. Nags Head Farm has a wonderful rill running alongside a sloping garden with a variety of plants leading to a quiet, peaceful courtyard. There is a large vegetable garden in which stands a magnificent glass-house with prolific vines and chilli plants. A woodland walk enhances the tranquility of this garden. Orchard Gardens is a large interesting garden of mixed planting, stump sculptures, colourful themed beds, fruit trees and different imaginative ornaments. Oxney Flatts has well-stocked herbaceous borders and a wildlife pond. New Cottage a recently acquired garden has a new fruit and vegetable area, original herbaceous border, pond and small alpine garden. Partial wheelchair access, grassy and mostly flat. Some small areas of gravel. Dogs only welcome in some gardens on a lead.

&. 🐃 ✿ ☕

21 NEW **HIGH BANK FARM**

Cleasby Road, Stapleton, Darlington, DL2 2QE. Lesley Thompson. *We are located on rd leading to Cleasby & Manfield from the Stapleton Rd. Drive 300 yards, you will see a sign saying 'Nursery on a Farm', take turning & park in car park.* **Sun 13 June (1-5). Combined adm with Holmelands £4, chd free. Home-made teas.**

The front garden has an abundance of roses, leading onto the Orchard. The back garden has a lovely cottage garden, with a secluded summer house to relax in, together with paved area with beautiful pots and you can sit and listen to the trickling water from our lovely fishpond. No dogs.

&. ☕

National Garden Scheme gardens are identified by their yellow road signs and posters. You can expect a garden of quality, character and interest, a warm welcome and plenty of home-made cakes!

GROUP OPENING

22 HILLSIDE COTTAGES

Low Etherley, Bishop Auckland, DL14 0EZ. Mary Smith, Eric & Delia Ayres, 01388 832727, mary@maryruth.plus.com. *Off the B6282 in Low Etherley, nr Bishop Auckland. To reach the gardens walk down the track opp number 63 Low Etherley. Please park on main rd. Limited disabled parking at the cottages.* **Sun 15 Aug (1.30-4.30). Combined adm £5, chd free. Home-made teas. Visits also by arrangement Feb to Nov for groups of up to 30.**

1 HILLSIDE COTTAGE
Eric & Delia Ayres.

2 HILLSIDE COTTAGE
Mrs M Smith.

The mature gardens of these two C19 cottages offer contrasting styles. At Number 1, grass paths lead you through a layout of trees and shrubs including many interesting specimens. Number 2 is based on island beds and has a cottage garden feel with a variety of perennials among the trees and shrubs and also incl a wild area, vegetables and fruit and a greenhouse with a collection of succulents and cacti. Both gardens have ponds and water features. There are a wide variety of specimen trees in both gardens including a Wollomi pine and a beautiful Monkey Puzzle tree. Children will be able to see pond life close up. At number 2 there are two stone railway sleepers from the nearby 1825 Stockton and Darlington railway line. There are steps in both gardens.

23 NEW HOLMELANDS

Cleasby, Darlington, DL2 2QY. Nicky & Clare Vigors. *1st open yard on R approaching village from Stapleton.* **Sun 13 June (1-5). Combined adm with High Bank Farm £4, chd free. Tea.**
Semi formal garden with orchard and large vegetable patch. Herbaceous beds and climbing roses as well as standard roses. Laid out in a semi formal design to create a harmonious effect. Attractive mixed borders, formal topiary and lavender at the front of the house.

24 KIRKY COTTAGE

12 Mindrum Farm Cottages, Mindrum, TD12 4QN. Mrs Ginny Fairfax, 01890 850246, ginny@mindrumgarden.co.uk, mindrumestate.com/mindrum-garden/kirky-cottage-garden/. *6m SW of Coldstream. 9m NW of Wooler on B6352. 4m N of Yetholm village.* **Sat 3 July (11-5). Adm £5, chd free. Home-made teas. Visits also by arrangement for groups of 10+.**
It is 7 years since Ginny Fairfax created Kirky Cottage Garden in the beautiful Bowmont Valley surrounded, and protected by, the Border Hills. A gravel garden in cottage garden style, old roses, violas and others jostle with favourites from Mindrum. A lovely, abundant garden and, with Ginny's new and creative ideas, ever evolving.

25 LAMBSHIELD

Hexham, NE46 1SF. David Young, 01434 600699, david.young@ngs.org.uk. *2m S of Hexham. Take the B6306 from Hexham. After 1.6m turn R at chevron sign. Lambshield drive is 2nd on L after 0.6m.* **Sun 1 Aug (12.30-4.30). Adm £6, chd free. Home-made teas. Visits also by arrangement May to July for groups of 10+.**
3 acre country garden begun in 2010 with strong structure and exciting plant combinations. Distinct areas and styles with formal herbaceous, grasses, contemporary planting, cottage garden, pool and orchard. Cloud hedging, pleached trees, and topiary combine with colourful and exuberant planting. Modern sculpture. Oak building and fencing by local craftsmen. New woodland garden being developed. Level ground but gravel paths not suitable for wheelchairs.

GROUP OPENING

26 NEW LESBURY GARDENS

Lesbury Village Hall, Lesbury, Alnwick, NE66 3PP. lesburyvillagehall.co.uk/. *From A1068 Alnwick to Ashington, take first left at the first roundabout before the river after entering the 30 mph limit. Village hall is on the left, past the pub after 300 metres.* **Sat 19, Sun 20 June (10-4). Combined adm £6, chd free. Home-made teas and tickets in the village hall.**
The gardens are in the Parish of Lesbury, a delightful conservation area of historic interest, having a number of listed buildings, with the Village Hall at its heart. The village has an active gardening club and the open gardens range from a large riverside garden with formal area, small 'woodland' and mixed borders; medium sized gardens with a wide range of perennial plants; and smaller cottage style gardens. Teas in aid of Village Hall activities.

27 LILBURN TOWER

Alnwick, NE66 4PQ. Mr & Mrs D Davidson, 01668 217291, lilburntower@outlook.com. *3m S of Wooler. On A697.* **Sun 6 June (2-5). Adm £5, chd free. Home-made teas. Visits also by arrangement May to Sept for groups of 10+.**
10 acres of magnificent walled and formal gardens set above river; rose parterre, topiary, scented garden, Victorian conservatory, wildflower meadow. Extensive fruit and vegetable garden, large glasshouse with vines. 30 acres of woodland with walks. Giant lilies, meconopsis around pond garden. Rhododendrons and azaleas. Also ruins of Pele Tower, and C12 church. Partial wheelchair access.

28 NEW LONGWITTON HALL

Longwitton, Morpeth, NE61 4JJ. Michael & Louise Spriggs. *2m N of Hartburn off B6343. Entrance at east end of Longwitton village.* **Sun 27 June (12-4). Adm £5, chd free. Home-made teas.**
6 acre historic site with glorious views to the south. Sheltered, mature garden, with specimen trees and acers, redeveloped with new borders. Circular rose garden, crescent shaped pool surrounded by foliage plants, 'standing stone' feature and a rhododendron and azalea glade leading to newly planted laburnum tunnel. Tree peonies and yew walk. Good wheelchair access.

Online booking is available for many of our gardens, at ngs.org.uk

29 LOUGHBROW HOUSE

Hexham, NE46 1RS. Mrs
K A Clark, 01434 603351,
patriciaclark351@hotmail.com. *1m
S of Hexham on B6306. Dipton Mill
Rd. Rd signed Blanchland, ¼m take
R fork; then ¼m at fork, lodge gates &
driveway at intersection.* **Sun 20 June
(1-4). Adm £5, chd free. Home-
made teas. Light lunches. Visits
also by arrangement. Driveway too
narrow for large coaches.**
A real country house garden with
sweeping, colour themed herbaceous
borders set around large lawns. Unique
Lutyens inspired rill with grass topped
bridges and climbing rose arches.
Part walled kitchen garden and paved
courtyard. Bog garden with pond.
New rose bed. Wild flower meadow
with specimen trees. Woodland quarry
garden with rhododendrons, azaleas,
hostas and rare trees. Home-made
jams and chutneys for sale.

30 MAGGIE'S

Melville Grove, Newcastle
Upon Tyne, NE7 7NU. www.
maggiescentres.org/newcastle.
*Driving into the grounds of the
Freeman Hospital, Maggie's opp
entrance to Northern Centre for
Cancer Care. Nearest parking the
Freeman Hospital multi-storey.* **Sat
12 June (12-3). Adm by donation.**

Home-made teas.
The garden, by Chelsea medal winner,
Sarah Price, is a sheltered sun trap.
Banked wildflower beds and multiple
planters, with seasonal displays, at
ground level, plus two roof gardens.
This gives a choice of outside spaces
for visitors to enjoy. Copper beech,
cherry blossom, crocus, bulbs,
wildflowers and herbs give a colourful
seasonal planting palette. A tranquil
oasis. 2021 extension planned.
Regional Finalist, The English Garden's
The Nation's Favourite Gardens 2019.
Partial wheelchair access, gravel in the
garden and roof garden, but the main
section can be accessed.

31 MARIE CURIE HOSPICE

Marie Curie Drive, Newcastle
Upon Tyne, NE4 6SS. www.
mariecurie.org.uk/help/hospice-
care/hospices/newcastle/about. *In
West Newcastle just off Elswick rd.
At the bottom of a housing estate.
Turning is between MA Brothers &
Dallas Carpets.* **Sun 4 July (2-4).
Adm by donation. Cream teas in
our Garden Café.**
The landscaped gardens of the
purpose-built Marie Curie Hospice
overlook the Tyne and Gateshead
and offer a beautiful, tranquil place for
patients and visitors to sit and chat.
Rooms open onto a patio garden

with gazebo and fountain. There
are climbing roses, evergreens and
herbaceous perennials. The garden
is well maintained by volunteers.
Come and see the work NGS funding
helps make possible. Plant sale
and refreshments available. The
Hospice and Gardens are wheelchair
accessible.

32 MIDDLETON HALL RETIREMENT VILLAGE

Middleton St. George,
Darlington, DL2 1HA. www.
middletonhallretirementvillage.
co.uk. *From A67 D'ton/Yarm, turn
at 2nd r'about signed to Middleton
St George. Turn L at the mini r'about
& immediately R after the railway
bridge, signed Low Middleton. Main
entrance is ¼m on L.* **Sat 24 July
(10-4). Adm £5, chd £2. Cream
teas in The Orangery, on-site.**
Like those in our retirement
community, the extensive grounds
are gloriously mature, endlessly
interesting and with many hidden
depths. 45 acres of features beckon;
natural woodland and parkland,
Japanese, Mediterranean and Butterfly
Gardens, croquet lawn, putting green,
allotments, ponds, wetland and bird
hide. All wheelchair accessible and
linked by a series of Woodland Walks.

45 Blackwell

33 ◆ MINDRUM GARDEN
Mindrum, TD12 4QN. Mr & Mrs T Fairfax, 01890 850634, tpfairfax@gmail.com, www.mindrumestate.com. *6m SW of Coldstream, 9m NW of Wooler. Off B6352. Disabled parking close to house.* **For NGS: Sun 20 June (2-5). Adm £5, chd free. Home-made teas. For other opening times and information, please phone, email or visit garden website.**
7 acres of romantic planting with old fashioned roses, violas, hardy perennials, lilies, herbs, scented shrubs, and intimate garden areas flanked by woodland and river walks. Glasshouses, newly created pond area leading to hillside rock garden with water course down to a natural pond, delightful stream, woodland and wonderful views across Bowmont valley. Large plant sale, mostly home grown. Partial wheelchair access due to landscape. Wheelchair accessible WC available.

34 NEW THE MOORE HOUSE
Whalton, Morpeth, NE61 3UX. **Phillip & Filiz Rodger.** *5m W of Morpeth & 4m N of Belsay on B6524. House in middle of village. Park in village. Follow NGS signs.* **Sun 30 May, Sun 18 July (2-5). Adm £5, chd free. Light refreshments in Community Hall.**
2 ½ acre mature garden extensively

but sympathetically redesigned by Sean Murray, BBC /RHS Great Chelsea Garden Challenge winner. Garden divided into sections. Wide use and mix of flowering perennials and grasses, also gravel garden and stone rill. Decorative stonework around beds in patio area. Emphasis on scent, colour, texture and form with all year round interest. Uneven surfaces. Gravel paths and some areas unsuitable for the less able and wheelchairs.

35 ◆ NGS BUZZING GARDEN
East Park Road, Gateshead, NE9 5AX. Gateshead Council, maureen.kesteven@ngs.org.uk, www.gateshead.gov.uk/article/3958/Saltwell-Park. *Between Pets' Corner & Saltwell Towers. Pedestrian entrance from East Park Rd or car park in Joicey Road. Open all year as part of the 55 acre Saltwell Park, known as The People's Park.* **For opening times and information, please visit garden website.**
The Buzzing Garden is a unique collaboration between the National Garden Scheme North East, Trädgårdsresan, Region Västra Götaland and Gateshead Council. It was funded by sponsorship. The garden is a tribute to the importance of international friendship. The Swedish design reflects the landscape of West Sweden, with coast, meadow and

woodland areas. Many of the plant species grow wild in Sweden, providing a welcoming vision for visitors and a feast for pollinators. Planted in 2019 the garden is maturing. Visitors can make a Donation to NGS. Dogs on leads. Wide tarmac path around the garden and mown grass paths through the meadow, but much of the garden is loose gravel.

36 OGLE GARDEN
Bonas Hill Farmhouse, Ogle, Ponteland, Newcastle upon Tyne, NE20 0AS. Sara Hunter, 07506 155323, oglegarden@gmail.com, www.oglegarden.com. *In Ogle Village turn R at T-junction. Follow rd for 1m. Turn R up dead-end lane to reach Ogle Garden.* **Sat 4 Sept (11-5). Adm £5, chd free. Home-made teas. Visits also by arrangement Feb to Oct for groups of up to 20. Guided tours available as well as garden experiences such as workshops.**
Ogle Garden has been under development by a flower-grower/florist since July 2018. Transformed from a sheep paddock, it is now developing into an interesting country garden. Featuring a rose walk , duck pond, cutting garden, a centuries-old well, island beds, a shade walk, dahlia garden and herbaceous borders in a rural setting.

The Moore House

37 OLIVER FORD GARDEN

Longedge Lane, Rowley, Consett, DH8 9HG. Bob & Bev Tridgett, www.gardensanctuaries.co.uk. *5m NW of Lanchester. Signed from A68 in Rowley. From Lanchester take rd towards Sately. Garden will be signed as you pass Woodlea Manor.* **Sun 27 June (1-5). Adm £4, chd free. Home-made teas.**

A peaceful, contemplative 3 acre garden developed and planted by the owner and BBC Gardener of the Year as a space for quiet reflection. Arboretum specialising in bark, stream, wildlife pond and bog garden. Semi-shaded Japanese maple and dwarf rhododendron garden. Rock garden and scree bed. Insect nectar area, orchard and 1½ acre meadow. Annual wildflower area. Terrace and ornamental herb garden,. Has a number of sculptures around the garden. Managed to maximise wildlife. Unfortunately not suitable for wheelchairs.

38 25 PARK ROAD SOUTH

Chester le Street, DH3 3LS. Mrs A Middleton, 0191 388 3225, midnol2@gmail.com. *4m N of Durham. Located at S end of A167 Chester-le-St bypass rd. Precise directions provided when booking visit.* **Visits by arrangement May to Dec. Adm £3.50, chd free. Light refreshments.**

A stunning town garden with all year round interest. Herbaceous borders with unusual perennials, grasses, shrubs surrounding lawn and paved area. Courtyard planted with foliage and small front gravel garden. The garden owner is a very knowledgeable plantswoman who enjoys showing visitors around her inspiring garden. No minimum size of group. Plants for sale.

39 ◆ RABY CASTLE

Staindrop, Darlington, DL2 3AH. Lord Barnard, 01833 660202, admin@raby.co.uk, www.raby.co.uk/raby-castle. *12m NW of Darlington, 1m N of Staindrop. On A688, 8m NE of Barnard Castle.* **For opening times and information, please phone, email or visit garden website.**

Raby Castle is one of the founding gardens of the NGS and has been opening for charity since 1927. The 18th century Walled Gardens are set within the grounds of Raby Castle.

Designers such as Thomas White and James Paine have worked to establish the Gardens, which now extend to 5 acres, displaying herbaceous borders, old yew hedges, formal rose gardens and informal heather and conifer gardens. The Stables Cafe and Shop are located in the Coach Yard at the entrance of the Walled Gardens. Assistance will be needed for wheelchairs.

40 RAVENSFORD FARM

Hamsterley, DL13 3NH. Jonathan & Caroline Peacock, 01388 488305, caroline@ravensfordfarm.co.uk. *7m W of Bishop Auckland. From A68 at Witton-le-Wear turn off W to Hamsterley. Go through village & turn L just before tennis courts at west end.* **Sun 8 Aug (2-5). Adm £5, chd free. Home-made teas. Visits also by arrangement Apr to Oct for groups of up to 20. Single track lane, not suitable for coaches.**

The virus closure in 2020 allowed us to seize the moment and undertake some new projects, as well as do some overdue repairs, so it has been a welcome chance to give this very varied, 3-acre country garden a general 'refresh'. We look forward to sharing the results with visitors in August, when we shall be selling plants, serving home-made teas, and enjoying music in the background as usual. Some gravel, so assistance will be needed for wheelchairs. Assistance dogs only.

41 NEW ROSE COTTAGE

Old Quarrington, Durham, DH6 5NN. Mr & Mrs Richard & Ann Cowen. *1m from J61 A1(M). All access is via Crow Trees Lane Bowburn. Satnav may be misleading!* **Sun 27 June (11-4). Combined adm with The Stables £5, chd free. Home-made teas at The Stables.**

There are 3 areas, a main garden to the front and side of the house, a smaller 'Mediterranean' type garden to the rear and a woodland garden also to the rear. The main garden is planted with wildlife friendly flowers, has slate paths and a pond that attracts 2 species of newts. The woodland garden is still 'work in progress' and a pond has been dug with water pumped to provide a stream.

42 ST CUTHBERT'S HOSPICE

Park House Road, Durham, DH1 3QF. Paul Marriott, CEO, www.stcuthbertshospice.com. *1m SW of Durham City on A167. Turn into Park House Rd, the Hospice is on the L after bowling green car park. Parking available.* **Sun 1 Aug (11-4). Adm £4, chd free. Light refreshments. Teas, coffees, juice and cakes.**

5 acres of mature gardens surround this CQC outstanding-rated Hospice. In development since 1988, the gardens are cared for by volunteers. Incl a Victorian-style greenhouse and large vegetable, fruit and cut flower area. Lawns surround smaller scale specialist planting, and areas for patients and visitors to relax. Woodland area with walks, sensory garden, and an 'In Memory' garden with stream. Plants and produce for sale. We are active participants in Northumbria in Bloom and Britain in Bloom, with several awards in recent years, including overall winner in 2015, 2018 & 2019 for the Care/ Residential /Convalescent Homes / Day Centre / Hospices category. Almost all areas are accessible for wheelchairs. Hot and cold drinks, home-made cakes and snacks served in the coffee shop.

GROUP OPENING

43 ST MARGARET'S ALLOTMENTS

Margery Lane, Durham, DH1 4QU. *From A1 take A690 to City Centre/ Crook. Straight ahead at T-lights after 4th r'about. 10mins walk from bus or rail station.* **Sun 11 July (2-5). Combined adm £5, chd free.**

5 acres of over 100 allotments against the spectacular backdrop of Durham Cathedral. This site has been cultivated since the Middle Ages. Enthusiastic gardeners, many using organic methods, cultivate plots which display a great variety of fruit, vegetables and flowers. Guided tour may be available. Many unusual vegetables. Display of creative and fun competition for plot holders. The site has some steep and narrow paths.

44 NEW **ST OSWALD'S HOSPICE**
Regent Avenue, Newcastle Upon Tyne, NE3 1EE. Katy Taberham, www.stoswaldsuk.org/who-we-are/. *Parking available on site & on street (max 2 hrs stay).* **Sat 17 July (10-2). Adm by donation. Cream teas.**
The recently redesigned in-patients' garden is a delightful, tranquil breakout space for patients and their visitors. Adjacent to the new 'family room' and designed around an existing pond, offering plenty of opportunities to sit and reflect on the mix of colours, textures, scents and sounds within it. Planted and maintained by a wonderful team of dedicated volunteers. Wheelchair accessible.

45 **THE SHILLING HOUSE GARDEN**
342 West Road, Fenham, Newcastle Upon Tyne, NE4 9JU. David Wallace. *On West Rd. Nearest side street, Grange Rd. Park on West Rd (when no restrictions in force) or in Grange Rd.* **Sun 13 June (3-6). Adm £6, chd free. Pre-booking essential, please visit www.ngs.org.uk for information & booking. Light refreshments.**
Unusual terraced garden begun in 1996 and still evolving. A calm green space characterised by structural form strongly related to the house's distinctive architecture. Hedges create a verdant framework - clipped yew, privet, beech, hornbeam and bay. Flowers are mainly white, with a mature Kiftsgate rose. Recent removal of all box has resulted in a new, less formal planting scheme. Opening to a limited number of visitors by pre-booking only. There will be a display about the history of the unique house and gardens, with original garden plans and photographs.

46 NEW **THE STABLES**
Old Quarrington, Bowburn, Durham, DH6 5NN. John & Claire Little. *1m from J61 of A1(M). All vehicle access via Crow Trees Lane, Bowburn. Satnav may be misleading!* **Sun 27 June (11-4). Combined adm with Rose Cottage £5, chd free. Home-made teas.**
A large family garden full of hidden surprises and extensive views. The main garden is about an acre including gravel garden, vegetable patch, orchard, play area, lawn and woodland gardens. There's a further four acres to explore which include wildlife ponds, woodlands, walks, hens, ducks, bees and alpacas. Large maze. Play area available for young children. Proceeds from refreshments going to local mountain rescue team. The main grass paths can be soft and bumpy in places. Some of the woodland paths are not accessible to wheelchairs.

GROUP OPENING

47 **WALWORTH GARDENS**
Walworth, Darlington, DL2 2LY. Iain & Margaret. *Approx 5m W of Darlington on A68 or ½m E of Piercebridge on A67. Follow brown signs to Walworth Castle Hotel. Just up the hill from the Castle entrance, follow NGS yellow signs down private track. Tickets & Teas at Quarry End.* **Sun 23 May (1.30-5). Combined adm £6, chd free. Home-made teas at Quarry End.**

THE ARCHES
Stephen & Becky Street-Howard.

CASTLE BARN
Joe & Sheila Storey.

THE DOVECOTE
Tony & Ruth Lamb.

QUARRY END
Iain & Margaret Anderson.

Coldcotes Moor Garden

Even if you have visited before you will see our 4 gardens in a new setting as we open for the first time in the Spring. Quarry End is a 1.5 acre woodland garden developed over 20 years on the site of an ancient quarry. The woodland has hellebores, woodland geraniums, comfrey, cow parsley, celandine,and maybe bluebells by then. In the main garden see over 20 different aquilegias as well as flowering shrubs, a C18 Ice House and extensive views over south Durham. In contrast Castle Barn, The Dovecote and The Arches are less than 10 years old and offer a wonderful mix of garden treasures. As well as spring flowers in borders, you will see extensive vegetable and herb gardens, grape vines, a Japanese garden with a Koi pond and a collection of specimen acers in their spring glory, fruit blossom in the orchards, a wildlife pond. There are beehives and chickens and a large play area so bring the family. Some areas of Quarry End are not accessible for wheelchairs users but most is grassy and flat.

♿ 🐔 ❊ ☕

48 NEW ▶ 28 WASHINGTON AVENUE
Middleton St. George, Darlington, DL2 1HE. Mrs Vanessa Hart, van2695@msn.com. *Situated between Middleton St George & Oak Tree.* Visits by arrangement May to Sept for groups of up to 20. Adm by donation. Light refreshments.
A garden created for the love of colour, plants and wildlife, filled with English roses, clematis, jasmines and passion flowers. Borders are packed with as many bee, butterfly and insect loving plants as I can fit in. An eclectic mix of water features and garden ornaments add to the quirkiness of this garden and there is something to stimulate every sense. A beautiful place to sit in the sun.

♿ ☕

49 ♦ WHALTON MANOR GARDENS
Whalton, Morpeth, NE61 3UT. Mr T R P S Norton, gardens@whaltonmanor.co.uk, , www.whaltonmanor.co.uk. *5m W of Morpeth. On the B6524, the house is at E end of the village & will be signed.* For NGS: Sun 18 Apr (11-3); Sat 19 June (11-4). Adm £5, chd free. Home-made teas in The

Game Larder. For other opening times and information, please email or visit garden website.
The historic Whalton Manor, altered by Sir Edwin Lutyens in 1908, is surrounded by 3 acres of magnificent walled gardens, designed by Lutyens with the help of Gertrude Jekyll. The gardens, developed by the Norton family since the 1920s incl extensive herbaceous borders, spring bulbs, 30yd peony border, rose garden, listed summerhouses, pergolas and walls, festooned with rambling roses and clematis. Partial wheelchair access to main area but otherwise stone steps and gravel paths.

♿ 🐔 ❊ 🚑 🚐 ☕

50 WOODBINE HOUSE
22 South View, Hunwick, Crook, DL15 0JW. Stewart Irwin & Colin Purvis. *On main rd through the village opp village green. B6286 off A689 Bishop Auckland - Crook or A690 Durham - Crook. On street parking, entrance to the rear of the property RHS of house.* Sat 10 July (1-5). Adm £5, chd free. Home-made teas.
The garden is approx a quarter of an acre, divided into two, one half used as a vegetable garden with large greenhouse. The other half of the garden is lawn with well stocked (and some unusual planting) herbaceous borders and small pond. Bees are kept in the vegetable garden. All refreshments are home-made. No dogs. Wheelchair access to refreshment area but paths in garden are not wide enough for wheelchairs.

♿ ❊ ☕

51 WOODLANDS
Pear-th Hall Road, Springwell Village, Gateshead, NE9 7NT. Liz Reid, 07719 875750, liz.reid@ngs.org.uk. *3½m N Washington Galleries. 4m S Gateshead town centre. On B1288 turn opp Guide Post pub (NE9 7RR) onto Peareth Hall Rd. Continue for ½m passing 2 bus stops on L. 3rd drive on L past Highbury Ave.* Sat 26 June (1.30-4.30). Adm £4, chd free. Home-made teas. Beer and wine available. Visits also by arrangement June & July for groups of 10 to 30.
Mature garden on a site of approx one seventh acre- quirky, with tropical themed planting and Caribbean inspired bar. Also an area of cottage garden planting. A fun garden with colour throughout the year, interesting

plants, informal beds and borders and pond area. On 26 June, Springwell Village plans to hold a 'Forties Weekend' with many attractions e.g. World War II battle re-enactments and a military camp on the nearby Bowes Railway (SAM) site. There are also plans for craft and other stalls and live music throughout the village.

❊ ☕

52 WOODSIDE HOUSE
Witton Park, Bishop Auckland, DL14 0DU. Charles & Jean Crompton, 01388 609973, ctjcrompton@gmail.com, www.woodsidehousewittonpark .com. *2m N of Bishop Auckland. From Bishop Auckland take A68 to Witton Park. In village DO NOT follow SatNav. park on main st, walk down track next to St Pauls Church.* Visits by arrangement Apr to Sept. Adm £5, chd free. Light refreshments.
Stunning 2 acre, mature, undulating garden full of interesting trees, shrubs and plants. Superbly landscaped (by the owners) with island beds, flowing herbaceous borders, an old walled garden, rhododendron beds, fernery, 3 ponds, grass bed and vegetable garden. Delightful garden full of interesting and unusual features: much to fire the imagination. Featured in the Telegraph, Amateur Gardening. Winner of Bishop Auckland in Bloom. Partial wheelchair access.

♿ ❊ 🚐 ☕

"The support of Hospice UK and the National Garden Scheme has been invaluable to hospice nurses across the country whilst we've been battling the coronavirus crisis, helping hospices such as Derian House to continue providing vital end of life and respite care to 400 children and young adults from across the North West. Thank you." – Katie Turner, Perinatal Nurse at Derian House Children's Hospice

NORTHAMPTONSHIRE

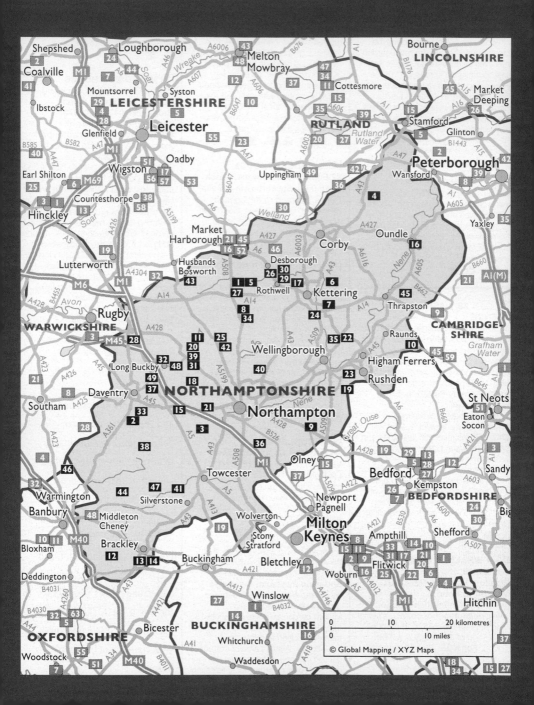

Shepshed
Loughborough
Coalville
M1
24
44
Soar
Wreake
A6006
43
48
Melton
Mowbray
A676
B676
Bourne
LINCOLNSHIRE
A1
B1176

41
7
A6
A46
A607
A606
12
37
47
34
11
Cottesmore
45
A15
Market
Deeping

Mountsorrel
Syston
29
LEICESTERSHIRE
28
B6047
10
15
35
A606
RUTLAND
39
15
A1
5
Stamford
A16
Glinton
42
8
39
A605

Ibstock
Glenfield
Leicester
55
B6047
23
A47
Uppingham
20
27
A6003
Rutland
Water
49
A47
Peterborough
B1443
A15
2

Earl Shilton
B582
A47
M1
25
M69
6
Wigston
51
17
56
57
53
B6047
A6
36
42
4
A43
Wansford
A605
Yaxley
35

Hinckley
B585
40
B582
3
1
13
Soar
A426
Countesthorpe
38
58
A5199
Market
Harborough
Welland
30
A427
A6003
Corby
Oundle
16
Nene
A605
B660
21
A1(M)

Lutterworth
19
A4304
32
Husbands
Bosworth
43
A508
21
45
16
52
46
A6
A6116
427
45
B662
Thrapston
A14
CAMBRIDGE-
SHIRE
Grafham
Water

Rugby
M6
M1
A14
1
5
27
26
30
29
17
Rothwell
A14
6
Kettering
7
24
A509
35
22
Raunds
10
45
59

WARWICKSHIRE
A428
3
M45
28
8
34
A43
35
22
Higham Ferrers
A45
St Neots

Southam
21
A426
A45
Long Buckby
32
48
11
20
39
31
25
42
Wellingborough
40
23
Rushden
19
B645
Eaton
Socon

Daventry
8
49
37
18
NORTHAMPTONSHIRE
19
51
3

33
2
15
21
Northampton
Nene
9
A509
Great Ouse
19
29
5
13
28
27
12
3
Sandy

Warmington
28
38
36
B526
Olney
37
15
A428
26
7
BEDFORDSHIRE
Kempston
Bedford
A603

4
46
A5
A508
M1
Towcester
Newport
Pagnell
A421
24
30
Big

Banbury
32
44
47
41
Silverstone
Wolverton
19
Milton
Keynes
8
Ampthill
A600
33
14
10
Shefford
A507

Bloxham
M40
10
11
48
Middleton
Cheney
A43
Stony
Stratford
Bletchley
15
11
2
1
31
17
20
6
Flitwick
22
Hitchin

Deddington
Brackley
12
13
14
Buckingham
A421
12
Woburn
25
A6
A5
A4012
M1
A507

B4031
A4260
32
63
5
27
Winslow
1
14
16
BUCKINGHAMSHIRE
A6146
A4012
18
34
15
27

OXFORDSHIRE
Woodstock
55
51
7
A34
M40
Bicester
A41
Whitchurch
Waddesdon
A418
B4011

0   10   20 kilometres
0   10 miles
© Global Mapping / XYZ Maps

The county of Northamptonshire is famously known as the 'Rose of the Shires', but is also referred to as the 'Shire of Spires and Squires', and lies in the East Midlands area of the country bordered by eight other counties.

Take a gentle stroll around charming villages with thatch and stone cottages and welcoming inns. Wander around stately homes, discovering art treasures and glorious gardens open for the National Garden Scheme at Kelmarsh Hall, Holdenby House, Castle Ashby, Cottesbrooke Hall and Boughton House. In contrast visit some village groups, which include small imaginatively designed gardens. Explore historic market towns such as Oundle and Brackley in search of fine footwear, antiques and curiosities.

The serenity of our waterways with Woodcote Villa at Long Buckby Wharf will delight, and our winding country lanes and footpaths will guide you around a rural oasis, far from the pressures of modern living, where you can walk knee deep in bluebells at Doughton House, view hellebores and spring flowers at 67/69 High Street, Finedon or The Old Vicarage at Norton through the seasons, to the late autumn colours of Briarwood.

Our first garden opens in February and the final opening occurs in October, giving a glimpse of gardens throughout the seasons.

Volunteers

County Organisers
David Abbott
01933 680363
david.abbott@ngs.org.uk

Gay Webster
01604 740203
gay.webster@ngs.org.uk

County Treasurer
David Abbott (as above)

Publicity
David Abbott (as above)

Photographer
Snowy Ellson
07508 218320
snowyellson@googlemail.com

Booklet Coordinator
William Portch
01536 522169
william.portch@ngs.org.uk

Talks
Elaine & William Portch
01536 522169
elaine.portch@yahoo.com

Assistant County Organisers
Amanda Bell
01327 860651
amanda.bell@ngs.org.uk

Lindsey Cartwright
01327 860056
lindsey.cartwright@ngs.org.uk

Jo Glissmann-Hill
07725 258755
joanna.glissmann-hill@ngs.org.uk

Philippa Heumann
01327 860142
pmheumann@gmail.com

Elaine & William Portch
(as above)

⬛ @Northants Ngs
🟦 @NorthantsNGS
⬜ @northantsngs

Left: 5 Larkhall Lane,
Harpole Gardens

OPENING DATES

All entries subject to change. For latest information check **www.ngs.org.uk**
Map locator numbers are shown to the right of each garden name.

February

Snowdrop Festival

Sunday 28th
◆ Boughton House 6
67-69 High Street 22

March

Sunday 28th
Woodcote Villa 49

April

Sunday 11th
Briarwood 7
Flore Gardens 15
Titchmarsh House 45

Sunday 18th
◆ Kelmarsh Hall & Gardens 27
The Old Vicarage 37

Sunday 25th
The Bungalow 8
◆ Cottesbrooke Hall Gardens
Limetrees 31
Nightingale Cottage 34

May

Sunday 2nd
Great Brington Gardens 18

Sunday 9th
Guilsborough Gardens 20

Sunday 16th
Greywalls 19

Sunday 23rd
Badby Gardens 2
NEW The Cottage 12
NEW Nonsuch 35
Old Rectory, Quinton 36
Walnut House 46

Sunday 30th
Newnham Gardens 33

Monday 31st
Titchmarsh House 45

June

Sunday 6th
Evenley Gardens 13
Foxtail Lilly 16
67-69 High Street 22
1 Hinwick Close 24
◆ Kelmarsh Hall & Gardens 27
Wisteria House 48

Saturday 12th
Titchmarsh House 45

Sunday 13th
3 Baptists Close 3
NEW Cobblers Cottage 10
Harpole Gardens 21
Hostellarie 26
NEW 14 Leys Avenue 29
16 Leys Avenue 30
Spratton Gardens 42

Weedon Lois & Weston Gardens 47

Saturday 19th
Flore Gardens 15

Sunday 20th
Flore Gardens 15
Kilsby Gardens 28
Rosearie-de-la-Nymph 40
Slapton Gardens 41
Sulgrave Gardens 44

Sunday 27th
Arthingworth Open Gardens 1
67-69 High Street 22
Rosearie-de-la-Nymph 40

July

Sunday 4th
◆ Castle Ashby Gardens 9
Ravensthorpe Gardens 39

Sunday 11th
◆ Holdenby House & Gardens 25

Saturday 17th
◆ Evenley Wood Garden 14

Sunday 18th
The Bungalow 8
Long Buckby Gardens 32
Nightingale Cottage 34

Sunday 25th
Blatherwycke Estate 4

August

Sunday 8th
136 High Street 23

September

Sunday 5th
16 Leys Avenue 30
Old Rectory, Quinton 36

Sunday 12th
Briarwood 7
◆ Coton Manor Garden 11

Sunday 19th
Wisteria House 48
Woodcote Villa 49

October

Sunday 17th
◆ Boughton House 6

February 2022

Sunday 27th
67-69 High Street 22

By Arrangement

Arrange a personalised garden visit on a date to suit you. See individual garden entries for full details.

3 Baptists Close 3
Bosworth House 5
Briarwood 7
The Bungalow 8
The Close, Harpole Gardens 21
Dripwell House, Guilsborough Gardens 20
Foxtail Lilly 16
Glendon Hall 17
Greywalls 19
67-69 High Street 22
136 High Street 23
Hostellarie 26
16 Leys Avenue 30
19 Manor Close, Harpole Gardens 21
Old West Farm 38
Ravensthorpe Nursery, Ravensthorpe Gardens 39
Titchmarsh House 45
Wisteria House 48
Woodcote Villa 49

Blatherwycke Estate

Kelmarsh Hall & Gardens

THE GARDENS

GROUP OPENING

◼ ARTHINGWORTH OPEN GARDENS

Arthingworth, nr Market Harborough, LE16 8LA. *6m S of Market Harborough. From Market Harborough via A508, after 4m take L to Arthingworth. From Northampton, A508 turn R just after Kelmarsh. Park cars in Arthingworth village & tickets for sale in the village hall.* **Sun 27 June (1-5). Combined adm £6, chd free. Home-made teas at Bosworth House & village hall.**

Arthingworth has been welcoming NGS visitors for a decade. It is a village affair with 8 to 9 gardens opening and 2 pop-up tearooms with home baked cakes. We now have some regulars who keep us on our toes and we love it. Come and enjoy the diversity, we aim to give visitors an afternoon of discovery. Our gardens have been chosen because they are all different in spirit and tended by young and weathered gardeners. We have gardens with stunning views, traditional with herbaceous borders and vegetables, walled, and artisan. The village is looking forward to welcoming you. St Andrew's Church, Grade II* listed will be open and the village is next to the national cycle path. Wheelchair access to some gardens.

GROUP OPENING

◩ BADBY GARDENS

Badby, Daventry, NN11 3AR. *3m S of Daventry on E-side of A361.* **Sun 23 May (2-6). Combined adm £5, chd free. Home-made teas in St Mary's Church.**

CHAPEL HOUSE
Moira & Peter Cooper.

THE OLD HOUSE
Mr & Mrs Robert Cain.

SHAKESPEARES COTTAGE
Jocelyn Hartland-Swann & Pen Keyte.

SOUTHVIEW COTTAGE
Alan & Karen Brown.

Delightful hilly village with attractive old houses of golden coloured Hornton stone, set around a C14 church and two village greens (no through traffic). There are four gardens of differing styles; a wisteria-clad thatched cottage (not open) with a sloping garden and modern sculptures; a traditional garden featuring a spectacular view across fields to Badby Wood; an elevated garden with views over the village; and a secluded garden with five distinct areas, pond and fernery, patio, lawn, small orchard and formal vegetable garden. We look forward to welcoming you to our lovely village!

◧ 3 BAPTISTS CLOSE

Bugbrooke, Northampton, NN7 3RJ. Claire Smith, 07798 905563, smithsgarden16@gmail.com. *2 mins from A5, 5m from J16 M1. From A5 signs to Bugbrooke, follow road to village that becomes Church St. Park close to Five Bells Pub on L. Baptists Close is on L (no parking in close).* **Sun 13 June (1.30-5). Adm £4, chd free. Home-made teas. Visits also by arrangement Mar to Sept for groups of 5 to 20.**

A medium sized plant lover's garden with tranquil pastoral farmland behind. Designed and planted about 10 yrs ago from a previous farmyard site, the garden has fully mature lime trees down one side and is a mix of shrubs with foliage interest and colourful herbaceous borders. Spring garden under the walnut tree, n-facing potager, traditional borders and short woodland walk. Wheelchair access on pathway around house (not open) and lawned area. Large gravel paths and woodland walk not easily accessible.

"The annual donation from the National Garden Scheme to Perennial is the cornerstone of our fundraising activities and encourages many of our donors".
Perennial

4 BLATHERWYCKE ESTATE

Blatherwycke, Peterborough, PE8 6YW. Mr George, Owner & S Bonney, Head Gardener. *Blatherwycke is signed off the A43 between Stamford & Corby. Follow road through village & the gardens entrance is next to the large river bridge.* **Sun 25 July (11-4). Adm £4, chd free. Home-made teas.** Blatherwycke Hall was demolished in the 1940s and its gardens lost. In 2011 the renovation of the derelict 4 acre walled gardens started. So far a large kitchen garden, wall trained fruit trees, extensive herbaceous borders, parterre, wildflower meadows, tropical bed, shrub borders and large arboretum have been planted. Restoration of the crinkle crankle wall has also begun. Wheelchair access over grass and gravel paths, some slopes and steps.

&⚘ 🐕 ✿ 🚗 ☕ 🍷

5 BOSWORTH HOUSE

Oxendon Road, Arthingworth, Nr Market Harborough, LE16 8LA. Mr & Mrs C E Irving-Swift, 01858 525202, irvingswift@btinternet.com. *From the phone box, when in Oxendon Rd, take the little lane with no name, 2nd to the R.* **Visits by arrangement May to July for groups of 10 to 20. For groups of 15+ guided tour by Cecile Irving-Swift for 1½ hours.** Approx 3 acres, almost completely organic garden and paddock with fabulous panoramic views and magnificent Wellingtonia. The garden also incl herbaceous borders, orchard, cottage garden with greenhouse, vegetable garden, herbs and strawberries and spinney. Early in the season a pleasing display of snowdrops, daffodils, bluebells and tulips, please visit www.ngs.org.uk for pop up openings. Partial wheelchair access.

&⚘ ☕ 🍷

6 ♦ BOUGHTON HOUSE

Geddington, Kettering, NN14 1BJ. Duke of Buccleuch & Queensberry, KT, 01536 515731, info@boughtonhouse.co.uk, www.boughtonhouse.org.uk. *3m NE of Kettering. From A14, 2m along A43 Kettering to Stamford, turn R into Geddington, house entrance 1½m on R.* **For NGS: Sun 28 Feb, Sun 17 Oct (1-5). Adm £6, chd £3. Light refreshments. Pre-booking essential, please visit garden website. For other opening times and information, please phone, email or visit garden website.**

The Northamptonshire home of the Duke and Duchess of Buccleuch. The garden opening incl opportunities to see the historic walled garden and herbaceous border, and the sensory and wildlife gardens. The wilderness woodland will open for visitors to view the spring flowers or the autumn colours. As a special treat the garden originally created by Sir David Scott (cousin of the Duke of Buccleuch) will also be open. Designated disabled parking. Gravel around house, please see our accessibility document for further information.

&⚘ ✿ 🚗 ☕ 🍷

7 BRIARWOOD

4 Poplars Farm Road, Barton Seagrave, Kettering, NN15 5AF. William & Elaine Portch, 01536 522169, elaine.portch@yahoo.com, www.elaineportch-gardendesign.co.uk. *1½ m SE of Kettering Town Centre. J10 off A14 turn onto Barton Rd (A6) towards Wicksteed Park. R into Warkton Lane, after 200 metres R into Poplars Farm Rd.* **Sun 11 Apr, Sun 12 Sept (10-4). Adm £4.50, chd free. Light refreshments. Visits also by arrangement Apr to Oct for groups of 10 to 30. Adm includes refreshments.** A garden for all seasons with quirky original sculptures and many faces. Firstly, a south aspect lawn and borders containing bulbs, shrubs, roses and rare trees with yr-round interest; hedging, palms, climbers, a wildlife, fish and lily pond, terrace with potted bulbs and unusual plants in odd containers. Secondly, a secret garden with garden room, small orchard, raised bed potager and greenhouse. Good use of recycled and repurposed materials throughout the garden, including a unique self-build garden room, sculpture and planters.

&⚘ ✿ ☕

8 THE BUNGALOW

Harborough Road, Maidwell, Northampton, NN6 9JA. David & Ann Sharman, 01604 686243. *Approx 10m N of Northampton. On A508 between Northampton & Market Harborough, opp Westaways Garage.* **Sun 25 Apr, Sun 18 July (10-5). Combined adm with Nightingale Cottage £4, chd free. Home-made teas in the village hall. Visits also by arrangement Apr to Sept for groups of 5 to 20.** The garden is on a steep, terraced site, which tumbles down to a stream.

Bridges lead to a recently acquired wooded area, equally steep. Ann has artistically incorporated a quirky collection of car boot finds, which adds interest to the exuberant planting. Self-seeding is encouraged, giving a natural effect. There are several secluded seating areas for relaxation. Plants for sale. A steep site and many steps, sadly not suitable for the less mobile.

✿ 🍷

9 ♦ CASTLE ASHBY GARDENS

Castle Ashby, Northampton, NN7 1LQ. Earl Compton, 01604 422180, petercox@ castleashbygardens.co.uk, www.castleashbygardens.co.uk. *6m E of Northampton. 1½ m N of A428, turn off between Denton & Yardley Hastings. Follow brown tourist signs (SatNav will take you to the village, look for brown signs).* **For NGS: Sun 4 July (10-5.30). Adm £9.60, chd £3.60. For other opening times and information, please phone, email or visit garden website.** 35 acres within a 10,000 acre estate of both formal and informal gardens, incl Italian gardens with orangery and arboretum with lakes, all dating back to the 1860s, as well as a menagerie which incl meerkats and marmosets. Play area, tearooms and gift shop. Tickets may need to be pre-booked, please check garden website. Wheelchair access on gravel paths.

&⚘ 🐕 🚗 🍷 🧺

10 NEW COBBLERS COTTAGE

Church Street, Hargrave, Wellingborough, NN9 6BW. Neil & Ros Sheppard. *Follow B645 from either Kimbolton or Higham Ferrers until you see signs for Hargrave. Turn into village onto Church Rd & follow signs (opp north gate of church).* **Sun 13 June (11-3). Adm £4, chd free. Light refreshments.** Since 1998 we have created a garden with distinctive large borders bursting with flowers, shrubs and trees, linked by gravel paths. There is a pond which is full of life, metal and stone sculptures abound and there are several seats for relaxation. Vegetables and fruit grow both outside and inside the greenhouse. The garden inspires our grandchildren to play, discover and imagine. There will be two performances by the Hargrave Singers and a display of artworks by local artist. Wheelchair access to majority of garden by gravel paths and a slope which bypasses the entrance steps.

&⚘ 🐕 🍷

1 ◆ **COTON MANOR GARDEN**
Coton, Northampton,
NN6 8RQ. Mr & Mrs Ian
Pasley-Tyler, 01604 740219,
www.cotonmanor.co.uk. *10m N
of Northampton, 11m SE of Rugby.
From A428 & A5199 follow tourist
signs.* **For NGS: Sun 12 Sept
(12-5.30). Adm £8, chd £3. Light
refreshments at Stableyard
Cafe. For other opening times and
information, please phone or visit
garden website.**
Winner of The English Garden's The
Nation's Favourite Gardens 2019,
this 10 acre garden set in peaceful
countryside with old yew and holly
hedges and extensive herbaceous
borders, containing many unusual
plants. One of Britain's finest
throughout the season, the garden is
at its most magnificent in Sept and
is an inspiration as to what can be
achieved in late summer. Adjacent
specialist nursery with over 1000
plant varieties propagated from the
garden. Partial wheelchair access as
some paths are narrow and the site is
on a slope.

2 NEW **THE COTTAGE**
Main Street, Charlton, Banbury,
OX17 3DR. Richard Dland. *Located
opp the Rose & Crown Pub, behind
tall green gates & a number of small
white posts in front of house.* **Sun 23
May (2-5.30). Combined adm with
Walnut House £5, chd free.**
First laid out in the early C20 by FE
Smith, the gardens have recently
been updated and replanted. The
gardens incl a pond, streams and two
lakes, alongside a courtyard garden,
a kitchen garden and a number of
borders. A keen tennis fan, FE Smith
also laid two grass tennis courts,
which still form part of the garden.

◆ **COTTESBROOKE HALL
GARDENS**
Cottesbrooke,
Northampton, NN6 8PF.
Mr & Mrs A R Macdonald-
Buchanan, 01604 505808,
welcome@cottesbrooke.co.uk,
www.cottesbrooke.co.uk. *10m
N of Northampton. Signed from
J1 on A14. Off A5199 at Creaton,
A508 at Brixworth.* **For NGS: Sun
25 Apr (2-5.30). Adm £7, chd
£4. Home-made teas. For other
opening times and information,
please phone, email or visit garden
website.**

Award-winning gardens by Geoffrey
Jellicoe, Dame Sylvia Crowe, James
Alexander-Sinclair and more recently
Arne Maynard. Formal gardens and
terraces surround Queen Anne house,
with extensive vistas onto the lake
and C18 parkland containing many
mature trees. Wild and woodland
gardens, a short distance from the
formal areas, are exceptional in
spring. Partial wheelchair access as
paths are grass, stone and gravel.
Access map identifies best route.

Boughton House

GROUP OPENING

13 EVENLEY GARDENS
Evenley, Brackley, NN13 5SG. *1m S of Brackley off the A43. Gardens situated around the village green & Church Lane. Follow signs around the village. Tickets available at each garden, to cover entry to all gardens.* **Sun 6 June (2-6). Combined adm £5, chd free. Home-made teas in St George's Church (2.30-5).**

15 CHURCH LANE
Carrie & Kevin O'Regan.

FINCH COTTAGE
Cathy & Chris Ellis.

14 THE GREEN
Nic Hamblin.

NEW **22 THE GREEN**
Giles & Alison Kendall.

38 THE GREEN
Anna & Matt Brown.

Evenley is a charming village with a central village green surrounded by many period houses (not open), an excellent village shop and The Red Lion pub which offers first class food and a warm welcome. Evenley gardens are a mix of established gardens and those being developed over the past 5 yrs. They all have mixed borders with established shrubs and trees. There are also orchards and vegetable gardens in some. Partial wheelchair access to most gardens across gravel drives, narrow paths and some steps.

14 ◆ EVENLEY WOOD GARDEN
Evenley, Brackley, NN13 5SH. Whiteley Family, 07788 207428, alison@evenleywoodgarden.co.uk, www.evenleywoodgarden.co.uk. *¾m S of Brackley. Turn off at Evenley r'about on A43 & follow signs within the village to the garden which is situated off the Evenley & Mixbury road.* **For NGS: Sat 17 July (10-4). Adm £6, chd free. Light refreshments. For other opening times and information, please phone, email or visit garden website.**
Please come and celebrate summer in the woods when the roses and lillies will have taken over from the azaleas and rhododendrons. A wonderful opportunity to see all that has been developed in the woods since Timothy Whiteley acquired them in 1980 with

the continuation of his legacy. Morning tea or coffee, lunch with a glass of wine and home-made cakes will be available in the café. Please take care as all paths are grass.

GROUP OPENING

15 FLORE GARDENS
Flore, Northampton, NN7 4LQ. *Off the A45 2m W of M1 J16. Avoid the by-pass. SatNav NN7 4LS for car park. Coaches welcome, please phone 01327 341225 for coach parking information.* **Sun 11 Apr (2-6); Sat 19, Sun 20 June (11-6). Combined adm £6, chd free. Home-made teas in Chapel School Room (Apr). Morning coffee & teas in Church & light lunches & teas in Chapel School Room (June). Donation to All Saints Church & United Reform Church, Flore (June).**

24 BLISS LANE
John Miller.
Open on Sun 11 Apr

THE CROFT
John & Dorothy Boast.
Open on all dates

THE GARDEN HOUSE
Edward & Penny Aubrey-Fletcher.
Open on Sat 19, Sun 20 June

17 THE GREEN
Mrs Wendy Amos.
Open on Sat 19, Sun 20 June

THE OLD BAKERY
John Amos & Karl Jones,
www.johnnieamos.co.uk.
Open on all dates

PRIVATE GARDEN OF BLISS LANE NURSERY
Christine & Geoffrey Littlewood.
Open on all dates

ROCK SPRINGS
Tom Higginson & David Foster.
Open on all dates

RUSSELL HOUSE
Peter Pickering & Stephen George, 01327 341734, peterandstephen@btinternet.com, www.RussellHouseFlore.com.
Open on all dates

NEW **64 SUTTON STREET**
Heather & Andy Anderson.
Open on Sat 19, Sun 20 June

NEW **6 THORNTON CLOSE**
William & Lesley Craghill.
Open on Sat 19, Sun 20 June

THE WHITE COTTAGE
Tony & Gill Lomax.
Open on all dates

Flore gardens have been open since 1963 as part of the Flore Flower Festival. The partnership with the NGS started in 1992 with openings every year since, incl 2020 following COVID-19 restrictions. Flore is an attractive village with views over the Upper Nene Valley. We have a varied mix of gardens, large and small, incl new gardens this yr. They have all been developed by friendly, enthusiastic and welcoming owners. Our gardens range from the traditional to the eccentric providing yr-round interest. There are greenhouses, gazebos, summerhouses and seating opportunities to rest while enjoying the gardens. Partial wheelchair access to most gardens, some assistance may be required.

16 FOXTAIL LILLY
41 South Road, Oundle, PE8 4BP. Tracey Mathieson, 01832 274593, foxtaillilly41@gmail.com, www.foxtail-lilly.co.uk. *1m from Oundle town centre. From A605 at Barnwell Xrds take Barnwell Rd, 1st R to South Rd.* **Sun 6 June (11-5). Adm £4.50, chd free. Home-made teas. Visits also by arrangement June to Sept.**
A cottage garden where perennials and grasses are grouped creatively together amongst gravel paths, complementing one another to create a natural look. Some unusual plants and quirky oddities create a different and colourful informal garden. Lots of flowers for cutting and a shop in the barn. New meadow pasture turned into new cutting garden. Plants, flowers and gift shop.

17 GLENDON HALL
Kettering, NN14 1QE. Rosie Bose, 01536 711732, rosiebose@googlemail.com. *1½m E of Rothwell. A6003 to Corby (A14 J7) W of Kettering, turn L onto Glendon Rd signed Rothwell, Desborough, Rushton. Entrance 1½m on L past turn for Rushton.* **Visits by arrangement for groups of up to 30. Adm £4, chd free.**
Mature specimen trees, topiary, box hedges and herbaceous borders stocked with many unusual plants. Large walled kitchen gardens with glasshouse and a shaded area, well

stocked with ferns. Some gravel and slopes, but wheelchair access via longer route.

GROUP OPENING

18 GREAT BRINGTON GARDENS

Northampton, NN7 4JJ. *7m NW of Northampton. Off A428 Rugby Rd. From Northampton, turn 1st L past main gates of Althorp. Programmes & maps available at free car park.* **Sun 2 May (11-5). Combined adm £8, chd free. Home-made teas & light lunches.**

NEW 2 HAMILTON LANE
Ruth Hawker.

15 HAMILTON LANE
Mr & Mrs Robin Matthews.

NEW THE RECTORY
Rev Andrea Watkins.

NEW RIDGEWAY HOUSE
Janet & Keith White.

ROSE COTTAGE
David Green & Elaine MacKenzie.

THE STABLES
Mrs J George.

SUNDERLAND HOUSE
Mrs Margaret Rubython.

THE WICK
Ray & Sandy Crossan.

YEW TREE HOUSE
Mrs Joan Heaps.

Great Brington is proud of over 25 yrs association with the NGS and arguably one of the most successful one day scheme events in the county. This yr 9 gardens will open, they are situated in a circular and virtually flat walk around the village. Our gardens provide inspiration and variety; continuing to evolve each yr and designed, planted and maintained by their owners. The village on the Althorp Estate is particularly picturesque and well worth a visit, predominantly local stone with thatched houses, a church of historic interest, and walkers can extend their walk to view Althorp House. Visitors will enjoy a warm welcome with plants for sale, tea and coffee at St Mary's Church (11-1), light lunches in the Reading Room (12-2) and tea, coffee and cake in the church (2-5). There are also secluded spots for artists to draw, sketch and paint. Small coaches for groups welcome by prior arrangement, please email

friends_of_stmarys@btinternet.com. Wheelchair access to some gardens.

19 GREYWALLS

Farndish, NN29 7HJ. Mrs P M Anderson, 01933 353495, greywalls@dbshoes.co.uk. *2½m SE of Wellingborough. A609 from Wellingborough, B570 to Irchester, turn to Farndish by cenotaph. House adjacent to church.* **Sun 16 May (2-5). Adm £3.50, chd free. Light refreshments. Visits also by arrangement for groups of 10 to 30.** A 2 acre mature garden surrounding the old vicarage (not open). The garden features an alpine house with raised alpine beds, stunning water features and natural ponds with views over open countryside. Historic church next door is open.

GROUP OPENING

20 GUILSBOROUGH GARDENS

Guilsborough, NN6 8RA. *10m NW of Northampton. 10m E of Rugby. Between A5199 & A428. J1 off A14. Car parking at Guilsborough Surgery, on West Haddon Rd, NN6 8QE. Information from car park or village hall next to primary school.* **Sun 9 May (1-5). Combined adm £7, chd free. Home-made teas in village hall.**

DRIPWELL HOUSE
Mr J W Langfield & Dr C Moss, 01604 740140, cattimoss@gmail.com.
Visits also by arrangement May & June for groups of 10 to 30. Combined visits with Gower House next door.

FOUR ACRES
Mark & Gay Webster.

THE GATE HOUSE
Mike & Sarah Edwards.

GOWER HOUSE
Ann Moss.

THE OLD HOUSE
Richard & Libby Seaton Evans.

THE OLD VICARAGE
John & Christine Benbow.

PEACE GARDEN
Guilsborough Church of England Primary School.

Enjoy a warm welcome in this village with its very attractive rural setting of rolling hills and reservoirs. This year

we have 7 contrasting gardens for you to visit. Several of us are passionate about growing fruit and vegetables, and walled kitchen gardens and a potager are an important part of our gardening. Plants both rare and unusual from our plantsmen's gardens are for sale, a true highlight here. No wheelchair access at Dripwell House and The Gate House.

GROUP OPENING

21 HARPOLE GARDENS

Harpole, NN7 4BX. *On A45 4m W of Northampton towards Weedon. Turn R at The Turnpike Hotel into Harpole. Village maps given to all visitors.* **Sun 13 June (1-6). Combined adm £5, chd free. Home-made teas at The Close.**

BRYTTEN-COLLIER HOUSE
James & Lucy Strickland.

CEDAR COTTAGE
Spencer & Joanne Hannam.

THE CLOSE
Michael Orton-Jones, 07714 896500, michael@orton-jones.com.
Visits also by arrangement.

14 HALL CLOSE
Marion & Charley Oliver.

NEW 5 LARKHALL LANE
Greg Hearne.

19 MANOR CLOSE
Caroline & Eamonn Kemshed, 01604 830512, carolinekemshed@live.co.uk.
Visits also by arrangement in June for groups of up to 20.

THE MANOR HOUSE
Mrs Katy Smith.

THE OLD DAIRY
David & Di Ballard.

Harpole is an attractive village nestling at the foot of Harpole Hills with many houses built of local sandstone. Visit us and delight in a wide variety of gardens of all shapes, sizes and content. You will see luxuriant lawns, mixed borders with plants for sun and shade, mature trees, shrubs, herbs, alpines, water features and tropical planting. We have interesting and quirky artifacts dotted around, garden structures and plenty of seating for the weary. Wheelchair access at Brytten-Collier House, The Close and The Old Dairy only.

22 67-69 HIGH STREET
Finedon, NN9 5JN. Mary &
Stuart Hendry, 01933 680414,
sh_archt@hotmail.com. *6m SE
Kettering. Garden signed from A6 &
A510 junction.* **Sun 28 Feb (10.30-
3.30); Sun 6, Sun 27 June (2-6).
Adm £3.50, chd free. Soup & roll
in Feb (incl in adm). Cream teas
in June. 2022: Sun 27 Feb.
Visits also by arrangement Feb
to Sept.**
1/3 acre rear garden of C17 cottage
(not open). Early spring garden with
snowdrops and hellebores, summer
and autumn mixed borders, many
obelisks and containers, kitchen
garden, herb bed, rambling roses and
at least 60 different hostas. All giving
varied interest from Feb through to
Oct. Large selection of home-raised
plants for sale (all proceeds to NGS).
🐕 ✿ 🚗 🍵

23 136 HIGH STREET
Irchester, Wellingborough,
NN29 7AB. Ade & Jane Parker,
jane692@btinternet.com. *200yds
past the church on the bend as
you leave the village going towards
the A45. Please park on High St.
Disabled parking only in driveway.*
**Sun 8 Aug (11-4). Adm £3.50, chd
free. Light refreshments. Visits
also by arrangement.**
1/2 acre garden with various different
borders including those planted
for shade, sun and bee friendly
situations. Alpine houses, raised
beds and planted stone sinks.
Wildlife pond. Seasonal planted tubs.
Wheelchair access over mainly grass,
with some gravel pathways.
♿ ✿ 🍵

24 1 HINWICK CLOSE
Kettering, NN15 6GB. Mrs Pat
Cole-Ashton. *J9 A14 A509
Kettering. At Park House r'about take
4th exit to Holdenby. Hinwick Close
3rd exit on R. From Kettering A509,
at Park House r'about take the 1st
exit to Holdenby, Hinwick Close 3rd
exit on R.* **Sun 6 June (12-5). Adm
£3.50, chd free. Home-made teas
& savouries.**
A garden reclaimed from rubble
surrounding a new build house (not
open). In the past 8 yrs Pat and
Snowy have transformed this space
into a wildlife haven. The garden
has numerous influences; seaside,
woodland and English country
garden. Ponds and waterfalls add to
the delights. Vintage signs, numerous

figures and seating areas at different
vantage points are dotted throughout
the garden.
♿ ✿ 🍵

**25 ◆ HOLDENBY HOUSE &
GARDENS**
Holdenby House, Holdenby,
Northampton, NN6 8DJ. Mr & Mrs
James Lowther, 01604 770074,
office@holdenby.com,
www.holdenby.com. *7m NW of
Northampton. Off A5199 or A428
between East Haddon & Spratton.*
**For NGS: Sun 11 July (11-4).
Adm £5.50, chd £4. Adm subject
to change. Cream teas & light
refreshments. For other opening
times and information, please
phone, email or visit garden
website.**
Holdenby has a historic Grade I
listed garden. The inner garden
including Rosemary Verey's renowned
Elizabethan Garden and Rupert
Golby's Pond Garden and long
borders. There is also a delightful
walled kitchen garden. Away from
the formal gardens, the terraces of
the original Elizabethan Garden are
still visible, one of the best preserved
examples of their kind. Accessible,
but contact garden for further details.
♿ 🍵

26 HOSTELLARIE
78 Breakleys Road,
Desborough, NN14 2PT. Stella
Freeman, 01536 760124,
stelstan78@aol.com. *6m N
of Kettering. 5m S of Market
Harborough. From church & war
memorial turn R into Dunkirk Ave,
then 3rd R. From cemetery L into
Dunkirk Ave, then 4th L.* **Sun 13
June (1.30-5). Combined adm
with 14 Leys Avenue & 16 Leys
Avenue £5, chd free. Home-made
teas & a gluten free option. Visits
also by arrangement June & July
for groups of 10 to 30.**
Town garden that once was an
allotment plot. The length has been
divided into different rooms; a
courtyard garden with a sculptural
clematis providing shade, colour
themed flower beds, ponds and water
features, cottage garden and gravel
borders, clematis and roses, all linked
by lawns and grass paths. Collection
of over 50 different hostas.
✿ 🚗 🍵

**27 ◆ KELMARSH HALL &
GARDENS**
Main Road, Kelmarsh,
Northampton, NN6 9LY. The
Kelmarsh Trust, 01604 686543,
enquiries@kelmarsh.com,
www.kelmarsh.com. *Kelmarsh
is 5m S of Market Harborough &
11m N of Northampton. From A14,
exit J2 & head N towards Market
Harborough on the A508.* **For NGS:
Sun 18 Apr, Sun 6 June (11-5).
Adm £4, chd free. The tearoom
offers light lunches, cream teas
& cakes. For other opening times
and information, please phone,
email or visit garden website.**
Kelmarsh Hall is an elegant
Palladian house set in glorious
Northamptonshire countryside with
highly regarded gardens, which are
the work of Nancy Lancaster, Norah
Lindsay and Geoffrey Jellicoe. Hidden
gems incl an orangery, sunken garden,
long border, rose gardens and, at the
heart of it all, a historic walled garden.
Highlights throughout the seasons incl
fritillaries, tulips, roses and dahlias.
Beautiful interiors brought together
by Nancy Lancaster in the 1930s,
in a palladian style hall designed by
James Gibbs. The recently restored
laundry and servant's quarters in the
Hall are open to the public, providing
visitors the incredible opportunity to
experience life 'below stairs'. Blue
badge disabled parking close to the
Visitor Centre entrance. Paths are
loose gravel, wheelchair users advised
to bring a companion.
♿ 🐕 ✿ 🚗 🍵 🍴

GROUP OPENING

28 KILSBY GARDENS
Middle Street, Kilsby, Rugby,
CV23 8XT. *5m SE of Rugby. 6m N
of Daventry on A361.* **Sun 20 June
(1-5.30). Combined adm £6, chd
free. Light refreshments at Kilsby
Village Hall (1-5).**
Kilsby's name has long been
associated with Stephenson's famous
railway tunnel and an early skirmish
in the Civil War. The houses and
gardens of the village offer a mixture
of sizes and styles, which reflect
its development through time. We
welcome you to test the friendliness
for which we are renowned. Please
visit www.ngs.org.uk closer to the
opening date to see which gardens
will be opening.
♿ 🍵

29 NEW **14 LEYS AVENUE**
Desborough, Kettering, NN14 2PY.
Dave & Linda Pascan. *6m N
of Kettering, 5m S of Market
Harborough. From St Giles church
(with spire) in Desborough & War
Memorial turn into Dunkirk Ave & 5th
R into Leys Ave.* **Sun 13 June (1.30-
5). Combined adm with 16 Leys
Avenue & Hostellarie £5, chd free.**
Town garden with patio, two lawns
bordered by curved pathway and a
rockery with 'lion's head' waterfall
feature. Mature trees give structure
to herbaceous borders planted with
flowering shrubs, lupins, clematis,
primulas and more. Brick pillars and
climbing plants provide the entrance
through to a small orchard with
vegetable patch and greenhouse.

30 **16 LEYS AVENUE**
Desborough, Kettering, NN14 2PY.
Keith & Beryl Norman,
01536 760950,
bcn@stainer16.plus.com. *6m
N of Kettering, 5m S of Market
Harborough. From church & War
Memorial turn R into Dunkirk Ave &
5th R into Leys Ave.* **Sun 13 June
(1.30-5). Combined adm with
Hostellarie & 14 Leys Avenue £5,
chd free. Sun 5 Sept (2-5). Adm
£3, chd free. Bottled water, pre-
wrapped cake & biscuits. Visits
also by arrangement June to Sept
for groups of 20 to 30.**
A town garden with two water
features, plus a stream and a pond
flanked by a 12ft clinker built boat.
There are six raised beds which are
planted with vegetables and dahlias.
A patio lined with acers has two steps
down to a gravel garden with paved
paths. Mature trees and acers give
the garden yr-round structure and
interest. Wheelchair access by two
steps from patio to main garden.

31 **LIMETREES**
1 Priestwell Court, East Haddon,
Northampton, NN6 8BT. Barry &
Sally Hennessey. *15 mins from J16
or J18 on M1, or J1 on A14 just off
A428. Garden is at the junction of
Tilbury Rd, Main St & Ravensthorpe
Rd. Strictly no parking allowed in
Priestwell Court.* **Sun 25 Apr (2-6).
Adm £4, chd free.**
Limetrees is a medium sized garden,
only 4 yrs in the making by the
current owners, but already looking
mature and well established thanks

to the existing mature trees around
the mostly ironstone and cob-
walled boundaries. Several hundred
herbaceous plants and shrubs from
previous gardens have been planted
and added to. Plants propagated from
those in the garden and raised from
seed will be on sale. Some paths may
be narrow for wheelchairs, but you can
view most of the mainly level garden.

GROUP OPENING

32 **LONG BUCKBY GARDENS**
Northampton, NN6 7RE. *8m NW
of Northampton, midway between
A428 & A5. Long Buckby is signed
from A428 & A5. 10 mins from J18
M1. Long Buckby train station is
½m from centre of the village.* **Sun
18 July (1-6). Combined adm £6,
chd free. Home-made teas, ice
creams, cold drinks & Pimm's.**

3 COTTON END
Roland & Georgina Wells.

4 COTTON END
Sue & Giles Baker.

THE GROTTO
Andy & Chrissy Gamble.

3A KNUTSFORD LANE
Tim & Jan Hunt.

**LAWN COTTAGE, 36 EAST
STREET**
Michael & Denise Nichols.

10 LIME AVENUE
June Ford.

NEW **THE OLD BOAT**
Dawn & Steven Chilvers.

4 SKINYARD LANE
William & Susie Mitchell.

WISTERIA HOUSE
David & Clare Croston.
(See separate entry)

WOODCOTE VILLA
Sue & Geoff Woodward.
(See separate entry)

Ten gardens in the historic villages
of Long Buckby and Long Buckby
Wharf, including a new garden, which
has been developed over 2 yrs into a
family garden. During our rest yr, the
gardeners have been busy creating
new borders, features and extensively
replanting. We welcome visitors old
and new to see our wonderful variety
of gardens. Something for everyone.
The gardens in the group vary in
size and style, from courtyard and
canal side to cottage garden, some

are established and others evolving.
They incl water features, pergolas,
garden structures and chickens, but
the stars are definitely the plants.
Bursting with colour, visitors will find
old favourites and the unusual, used
in a variety of ways; trees, shrubs,
perennials, climbers, annuals, fruit
and vegetables. Of course there
will be teas and plants for sale to
complete the visit. Come and see
us for a friendly welcome and a
good afternoon out. Full or partial
wheelchair access to all gardens,
except 4 Skinyard Lane.

GROUP OPENING

33 **NEWNHAM GARDENS**
Newnham, Daventry, NN11 3HF.
*2m S of Daventry on B4037 between
the A361 & A45. Continue to the
centre of the village & follow signs
for the car park, just off the main
village green.* **Sun 30 May (11-5).
Combined adm £5, chd free. Light
refreshments in village hall.**

THE BANKS
Sue & Geoff Chester,
www.suestyles.co.uk.

THE COTTAGE
Jacqueline Minor.

HILLTOP
David & Mercy Messenger

NEW **STONE HOUSE**
Pat & David Bannerman.

WREN COTTAGE
Mr & Mrs Judith Dorkins.

Five lovely gardens set in a beautiful
old village cradled by the gentle hills
of south Northamptonshire. The
varied gardens, set around traditional
village houses, look enchanting at this
special time of year and, for 2021, we
are especially pleased to have a new
garden opening. Spend the day with
us enjoying the gardens, buying at
our large plant sale, strolling around
the village lanes and visiting our
C14 church. Why not treat yourself
to a tasty light lunch, scrumptious
cakes and refreshments in the village
hall. Please note that the village
and gardens are hilly in parts and
while most gardens are accessible
to wheelchairs, others are more
restricted.

34 NIGHTINGALE COTTAGE
Draughton Road, Maidwell, Northampton, NN6 9JF. Ken & Angela Palmer. *Approx 10m N of Northampton. Approach Maidwell on A508. Turn L (from Market Harborough) or R (from Northampton). Follow NGS signs.* **Sun 25 Apr, Sun 18 July (10-5). Combined adm with The Bungalow £4, chd free. Home-made teas in the village hall.** A garden designed by the owners over the last 10 yrs, surrounding an old barn. There are many pots and cottage type planting incl some vegetables. There are also some surprises of wooden animals made from prunings, steps and hidden seating areas, but no grass. It is a garden which is always evolving as different species self-seed and are often left to grow. There is a small water feature over rocks into a stones pool where the many garden creatures drink. Plants sale at The Bungalow only.

35 NEW NONSUCH
11 Mackworth Drive, Finedon, Wellingborough, NN9 5NL. *David & Carrie Whitworth. Off Wellingborough Rd (A510) onto Bell Hill, then to Church Hill, 2nd L after church, entrance on the L as you enter Mackworth Drive.* **Sun 23 May (1-5). Adm £3.50, chd free. Drinks, cakes & biscuits.** 1/3 acre country garden within a conservation boundary stone wall. Mature trees, enhanced by many rare and unusual shrubs and perennial plants. A garden for all seasons with several seating areas. Wheelchair access on a level site with paved and gravel paths.

SPECIAL EVENT

36 OLD RECTORY, QUINTON
Preston Deanery Road, Quinton, Northampton, NN7 2ED. **Alan Kennedy & Emma Wise, www.garden4good.co.uk.** *M1 J15, 1m from Wootton towards Salcey Forest. House is next to the church.* **Sun 23 May, Sun 5 Sept (10-4). Adm £10, chd free. Pre-booking essential, please visit www. ngs.org.uk/special-events for information & booking. Home-made teas. Lunches 11am-2pm to pre-order via garden website.**

A beautiful contemporary 3 acre rectory garden designed by multi-award-winning designer, Anoushka Feiler. Taking the Old Rectory's C18 history and its religious setting as a key starting point, the main garden at the back of the house has been divided into six parts; a kitchen garden, glasshouse and flower garden, a woodland menagerie, a pleasure garden, a park and an orchard. Elements of C18 design such as formal structures, parterres, topiary, long walks, occasional seating areas and traditional craft work have been introduced, however with a distinctly C21 twist through the inclusion of living walls, modern materials and features, new planting methods and abstract installations. Pop up shop selling plants, garden produce, local honey, home-made bread and bath products. Wheelchair access with gravel paths.

37 THE OLD VICARAGE
Daventry Road, Norton, Daventry, NN11 2ND. Barry & Andrea Coleman.

Norton is about 2m E of Daventry, 11m W of Northampton. From Daventry follow signs to Norton for 1m. On A5 N from Weedon follow road for 3m, take L turn signed Norton. On A5 S take R at Xrds signed Norton, 6m from Kilsby. Garden is R of All Saints Church. **Sun 18 Apr (1-5). Adm £5, chd free. Home-made teas in orangery.** The vicarage days bequeathed dramatic and stately trees to the modern garden. The last 40 yrs of evolution and the happy accidents of soil-type, and a striking location with lovely vistas have shaped the garden around all the things that make April so thrilling, including prodigious sweeps of primulas of many kinds and the trees in blossom. The interesting and beautiful C14 church of All Saints will be open to visitors.

38 OLD WEST FARM
Little Preston, Daventry, NN11 3TF. Mr & Mrs G Hoare, caghoare@gmail.com. *7m SW Daventry, 8m W Towcester, 13m NE Banbury. 3/4 m E of Preston Capes on road to Maidford. Last house*

Woodcote Villa

on R in Little Preston with white flagpole. Beware, the postcode applies to all houses in Little Preston. **Visits by arrangement May & June for groups of 10 to 30. Light refreshments.**
Large rural garden developed over the past 40 yrs on a very exposed site, planted with hedges and shelter. Roses, shrubs and borders aiming for yr-round interest. Partial wheelchair access over grass.

GROUP OPENING

39 RAVENSTHORPE GARDENS
Ravensthorpe, NN6 8ES. *7m NW of Northampton. Signed from A428. Please start & purchase tickets at Ravensthorpe Nursery.* **Sun 4 July (1.30-5.30). Combined adm £5, chd free. Home-made teas in village hall.**

CORNERSTONE
Lorna Jones.

QUIETWAYS
Russ Barringer.

RAVENSTHORPE NURSERY
Mr & Mrs Richard Wiseman, 01604 770548, ravensthorpenursery@ hotmail.com.
Visits also by arrangement Apr to Oct.

TREETOPS
Ros & Gordon Smith.

Attractive village in Northamptonshire uplands near to Ravensthorpe reservoir and Top Ardles Wood Woodland Trust, which have bird watching and picnic opportunities. Established and developing gardens set in beautiful countryside displaying a wide range of plants, many of which are available from the nursery. Offering inspirational planting, quiet contemplation, beautiful views, water features and gardens encouraging wildlife. Disabled WC in village hall.

40 ROSEARIE-DE-LA-NYMPH
55 The Grove, Moulton, Northampton, NN3 7UE. Peter Hughes, Mary Morris, Irene Kay, Steven Hughes & Jeremy Stanton. *N of Northampton town. Turn off A43 at small r'about to Overstone Rd. Follow NGS signs in village. The garden is on the Holcot Rd out of*

Moulton. **Sun 20, Sun 27 June (11-5). Adm £4.50, chd free. Home-made teas & light refreshments.**
We have been developing this romantic garden for over 10 yrs and now have over 1800 roses, incl English, French and Italian varieties. Many unusual water features and specimen trees. Roses, scramblers and ramblers climb into trees, over arbours and arches. Collection of 140 Japanese maples. Mostly flat wheelchair access, but there is a standard width doorway to negotiate.

GROUP OPENING

41 SLAPTON GARDENS
Slapton, Towcester, NN12 8PE. *A hamlet 4m W of Towcester. ¼m N of the Towcester to Wappenham road.* **Sun 20 June (2-5.30). Combined adm £6, chd free. Home-made teas at Home Farm Cottage.**

BOXES FARM
James & Mary Miller.

BRADDEN COTTAGE
Mrs Philippa Heumann.

1 CHAPEL LANE
Kathryn McLaughlin.

CORNER HOUSE
Amanda & Stuart Bell.

NEW **THE OLD RECTORY**
Mr & Mrs Kit James, Kitjames7@gmail.com.

SOWBROOK HOUSE
Mrs Caroline Coke.

Slapton is a very pretty Northamptonshire hamlet with only 30 houses'. Six gardens will be opening for the NGS this year including one new, The Old Rectory a garden originally designed by James Alexander-Sinclair and enhanced by the current owners. Corner House is an established garden with charming rooms and Sowbrook House has shrub and herbaceous borders, as well as a more formal small walled garden. Bradden Cottage and 1 Chapel Lane are cottage gardens and Boxes Farm is a ¼ acre garden of rooms round a restored Grafton farmhouse (not open), created over the last 30 yrs. There is the lovely C12 St Botolph's Chuch with rare Medieval wall paintings which will be open to visitors. Partial wheelchair access with narrow and gravel paths.

GROUP OPENING

42 SPRATTON GARDENS
Smith Street, Spratton, NN6 8HP. *6½m NNW of Northampton. On A5199 between Northampton & Welford. S from J1, A14. Car Park at Spratton Hall School with close access to gardens.* **Sun 13 June (11-5). Combined adm £7, chd free. Home-made teas.**

THE COTTAGE
Mrs Judith Elliott.

DALE HOUSE
Fiona & Chris Cox.

FORGE COTTAGE
Daniel & Jo Bailey.

28 GORSE ROAD
Lee Miller.

11 HIGH STREET
Philip & Frances Roseblade.

MULBERRY COTTAGE
Kerry Herd.

NORTHBANK HOUSE
Helen Millichamp.

NEW **OLD HOUSE FARM**
Dr Susie Marchant.

STONE HOUSE
John Forbear.

VALE VIEW
John Hunt.

WALTHAM COTTAGE
Norma & Allan Simons.

NEW **11 YEW TREE LANE**
Niall & Becks O'Brien.

As well as attractive cottage gardens alongside old Northampton stone houses, Spratton also has unusual gardens, including those showing good use of a small area; one dedicated to encouraging wildlife with views of the surrounding countryside; newly renovated gardens and those with new planting; courtyard garden; gravel garden with sculpture; mature gardens with fruit trees and herbaceous borders. Tea, cakes and rolls will be available in the Norman St. Andrew's Church, Church Road. The King's Head Pub will be open, lunch reservations recommended.

Online booking is available for many of our gardens, at ngs.org.uk

Mulberry Cottage, Spratton Gardens

© Leigh Clapp

43 SULBY GARDENS

Sulby, Northampton, NN6 6EZ. **Mrs Alison Lowe.** *16m NW of Northampton, 2m NE of Welford off A5199. Past Wharf House Hotel, take 1st R signed Sulby. After R & L bends, turn R at sign for Sulby Hall Farm. Turn R at junction, garden is 1st L. Parking limited, no vans or buses please.* **For opening times and information, please visit www.ngs.org.uk**

Interesting and unusual property, on the Leicestershire border between Welford and Husbands Bosworth, covering 12 acres comprising working Victorian kitchen garden, orchard, and late C18 icehouse, plus species-rich nature reserve incl woodland, feeder stream to River Avon, a variety of ponds and established wildflower meadows.

GROUP OPENING

44 SULGRAVE GARDENS

Banbury, OX17 2RP. *8m NE of Banbury. Just off B4525 Banbury to Northampton road, 7m from J11 off M40. Car parking at church hall.* **Sun 20 June (2-6). Combined adm £6, chd free. Home-made teas.**

MILL HOLLOW BARN
David & Judith Thompson.

RECTORY FARM
Charles & Joanna Smyth-Osbourne, 01295 760261, sosbournejm@gmail.com.

THREEWAYS
Alison & Digby Lewis.

VINECROFT
Claire & Jon Sadler.

THE WATERMILL
Mr & Mrs T Frost.

NEW **WOOTTON HOUSE**
Zoe & Richard McCrow.

Sulgrave is a small historic village having recently celebrated its strong American connections as part of the 150 yrs of the signing of the Treaty of Ghent. Six gardens opening; Threeways, a small walled cottage garden packed with interest. Rectory Farm has lovely views, a rill, well and planted arbours. Mill Hollow Barn, a large garden with lakes, streams, ponds and many rare and interesting trees, shrubs and perennials. The Watermill, a contemporary garden designed by James Alexander-Sinclair, set around a C16 watermill and mill pond. Vinecroft, a contemporary garden by Alexander John Design with roses, climbers, perennials and shrubs. New for 2021 this large quirky garden at Wootton House with a good selection of rare and interesting plants. An award-winning community owned and run village shop will be open.

45 TITCHMARSH HOUSE
Chapel Street, Titchmarsh, NN14 3DA. Sir Ewan & Lady Harper, 01832 732439, ewan@ewanh.co.uk, www.titchmarsh-house.co.uk. 2m N of Thrapston. 6m S of Oundle. Exit A14 at junction signed A605, Titchmarsh signed as turning E towards Oundle & Peterborough. **Sun 11 Apr, Mon 31 May (2-6); Sat 12 June (12.30-5). Adm £5, chd free. Home-made teas at parish church (Apr & May). BBQ lunch & teas at village fete (June). Visits also by arrangement Apr to June for groups of 5+.**
4½ acres extended and laid out since 1972. Special collections of magnolias, spring bulbs, iris, peonies and roses with many rare trees and shrubs. Walled ornamental vegetable garden and ancient yew hedge. Some newly planted areas; please refer to the website. Collections of flowering trees and other unusual plants such as rare Buddleias, Philadelphus, Deutzias and Abelias. Wheelchair access to most of the garden without using steps. No dogs.

46 WALNUT HOUSE
Main Street, Charlton, Banbury, OX17 3DR. Sir Paul & Lady Hayter. In Main St, Charlton between Banbury & Brackley. **Sun 23 May (2-5.30). Combined adm with The Cottage £5, chd free.**
Large garden behind C17 farmhouse (not open). Colour themed borders and separate small gardens with beech and yew hedges, each with their own character. Orchard with wildflowers and an old-fashioned vegetable garden. Wilderness (in C18 sense) and archery lawn. Garden started in 1992 with new hot and gravel garden created in 2011. Wheelchair access to garden with gravel paths.

GROUP OPENING

47 WEEDON LOIS & WESTON GARDENS
Weedon Lois, Towcester, NN12 8PJ. 7m W of Towcester. 7m N of Brackley. Turn off A43 at Towcester towards Abthorpe & Wappenham & turn R for Weedon Lois. Or turn off A43 at Brackley, follow signs to Helmdon & Weston. **Sun 13 June (1-6). Combined adm £5, chd free. Home-made teas in Baptist Chapel, Weston.**

NEW **4 HELMDON ROAD**
Mrs S Wilde.

NEW **8A HIGH STREET**
Mr & Mrs J Archard-Jones.

NEW **MIDDLETON HOUSE**
Mark & Donna Cooper.

OLD BARN
Mr & Mrs John Gregory.

RIDGEWAY COTTAGE
Jonathan & Elizabeth Carpenter.

4 VICARAGE RISE
Ashley & Lindsey Cartwright.

Two adjacent villages set in the rolling south Northamptonshire countryside, with a handsome medieval church in Weedon Lois. The extension churchyard contains the grave of the poet, Dame Edith Sitwell who lived in Weston Hall (not open), marked with a gravestone by Henry Moore. This year we have three new gardens, including one which shows that even the steepest site can be transformed into a colourful, user friendly garden, another designed by James Alexander-Sinclair, which continues to be developed by the enthusiastic owner, and the third, an interesting contemporary design set around a traditional period house (not open). There is also a long established plantsman's garden, herbaceous borders, vegetables, roses, woodland planting, a large wildlife pond and much more. So we hope you will join us for our open day, enjoy looking round our gardens, visit the plant stalls and tuck into our famous home-made teas. Not all gardens suitable for wheelchairs.

48 WISTERIA HOUSE
19a East Sreet, Long Buckby, Northampton, NN6 7RB. David & Clare Croston, 07771 911892, dad.croston@gmail.com. 8m NW of Northampton, midway between A428 & A5. Long Buckby is signed from A428 & A5. 10 mins from J18 M1. East St is on the N side of B5385, between centre of village & the A428. **Sun 6 June (11-6). Adm £3.50, chd free. Light refreshments. Sun 19 Sept (11-5). Combined adm with Woodcote Villa £5, chd free. Home-made teas. Opening with Long Buckby Gardens on Sun 18 July. Visits also by arrangement May to Sept for groups of 10+. Adm incl combined visit with Woodcote Villa & refreshments.**
Artist and plantsman's garden framed by wisteria on Georgian house (not open) and pergolas. Constantly evolving themed areas around the garden offer various seating and views. Ferns and bamboos surround a wild area that leads to lawn, long herbaceous border, pond rill and circular mixed borders. Gravel area with grasses and shrubs. Vegetable plot, rockery and greenhouses. Partial wheelchair access, gravelled areas may be difficult.

49 WOODCOTE VILLA
Old Watling Street (A5), Long Buckby Wharf, Long Buckby, Northampton, NN6 7FW. Sue & Geoff Woodward, geoff.and.sue@btinternet.com. 2m NE of Daventry. From M1 J16, take Flore by pass, turn R at A5 r'about for approx 3m. From Daventry follow Long Buckby signs, but turn L at A5 Xrds. From M1 J18 signed Kilsby, follow A5 S for approx 6m. **Sun 28 Mar (11-5). Adm £3.50, chd free. Sun 19 Sept (11-5). Combined adm with Wisteria House £5, chd free. Home-made teas. Opening with Long Buckby Gardens on Sun 18 July. Visits also by arrangement Mar to Sept for groups of 10+. Adm incl combined visit with Wisteria House & refreshments.**
In a much admired location, this delightful canalside garden has a large variety of plants, styles, structures and unusual bygones. Bulbs will feature in March, with colourful planting in July and Sept, all set against a backdrop of trees and shrubs. Lovely places to sit, relax and watch the boats and wildlife. Plants for sale. Sorry no WC. Wheelchair access via ramp at entrance to garden.

NOTTINGHAMSHIRE

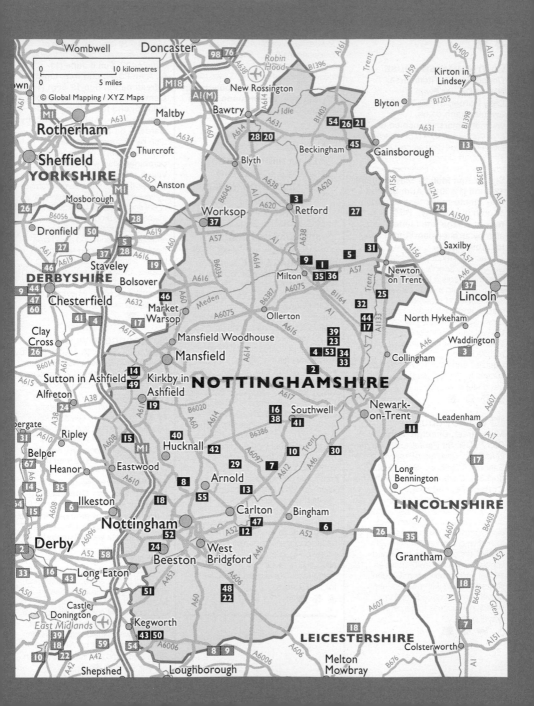

Nottinghamshire is best known as Robin Hood country. His legend persists and his haunt of Sherwood Forest, now a nature reserve, contains some of the oldest oaks in Europe. The Major Oak, thought to be 800 years old, still produces acorns.

Civil War battles raged throughout Nottinghamshire, and Newark's historic castle bears the scars. King Charles I surrendered to the Scots in nearby Southwell after a night at The Saracen's Head, which is still an inn today. Southwell is also home to the famous Bramley apple, whose descendants may be found in many Nottinghamshire gardens.

We have groups of cottage gardens, we have beautiful farmhouse gardens, and we have gardens with special woodland walks. We have gardens with particular concern for environmental issues, we have gardens full of rare exotic imports, and we have artists' gardens of inspiration. All will offer a warm welcome, and most will offer a brilliant tea.

Volunteers

County Organiser
Georgina Denison
01636 821385
georgina.denison@ngs.org.uk

County Treasurer
Nicola Cressey
01159 655132
nicola.cressey@gmail.com

Publicity
Julie Davison
01302 719668
julie.davison@ngs.org.uk

Social Media
Malcolm Turner
01159 222831
malcolm.turnor14@btinternet.com

Booklet Co-ordinators
Malcolm and Wendy Fisher
0115 966 4322
wendy.fisher111@btinternet.com.

Assistant County Organisers
Judy Geldart
01636 823832
judygeldart@gmail.com

Beverley Perks
01636 812181
perks.family@talk21.com

Mary Thomas
01509 672056
nursery@piecemealplants.co.uk

Andrew Young
01623 863327
andrew.young@ngs.org.uk

 @National Garden Scheme Nottinghamshire

@nottsngs

Left: Beesthorpe Hall Farm

OPENING DATES

All entries subject to change. For latest information check www.ngs.org.uk

Map locator numbers are shown to the right of each garden name.

February

Snowdrop Festival

Saturday 13th
NEW 1 Highfield Road 18

Sunday 14th
Church Farm 9

Sunday 21st
The Poplars 44

April

Saturday 17th
NEW Oasis Community
Gardens 37

Sunday 18th
◆ Felley Priory 15

Saturday 24th
Capability Barn 7

Sunday 25th
Capability Barn 7
NEW 1 Highfield Road 18
Normanton Hall 32

May

Sunday 2nd
NEW Jasmine Cottage 26
NEW Westmoor House 54

Sunday 9th
The Old Vicarage 38

Sunday 16th
Ivy Bank Cottage 25
38 Main Street 29
◆ Norwell Nurseries 34

Sunday 23rd
Bridge Farm 4
6 Hope Street 24
Upper Grove Farm 53

Saturday 29th
NEW Sutton Bonington
Gardens 50

Sunday 30th
Papplewick Hall 40
NEW Sutton Bonington
Gardens 50

Monday 31st
Halam Gardens and
Wildflower Meadow 16

June

Sunday 6th
Primrose Cottage 45

Sunday 13th
Askham Gardens 1
The Chimes 8
NEW Hollinside 19
Home Farm House,
17 Main Street 22

Patchings Art Centre 42
Rose Cottage 48

Sunday 20th
Church House 10
NEW Manor Farm 30
Park Farm 41
The Poplars 44

Sunday 27th
Norwell Gardens 33

Monday 28th
5 Burton Lane 6

Wednesday 30th
Norwell Gardens 33

July

Saturday 3rd
Lodge Mount 27

Sunday 4th
East Meets West 14
Spring Bank House 49
6 Weston Close 55

Saturday 10th
The Old Vicarage 38

Sunday 11th
Hopbine Farmhouse,
Ossington 23
Ossington House 39

Sunday 18th
Beesthorpe Hall Farm 2
Cornerstones 12
Thrumpton Hall 51

Sunday 25th
The Coach House 11

August

Sunday 1st
5a High Street 17

Saturday 7th
Oak Barn Exotic
Garden 36

Sunday 8th
Church House 10
The Old Vicarage 38

Sunday 15th
Meadow Farm 31
Oasis Community
Gardens 37

Sunday 22nd
University Park
Gardens 52

Monday 30th
5 Burton Lane 6
Piecemeal 43

September

Sunday 12th
Broadlea 5
Oak Barn Exotic
Garden 36

Sunday 19th
Riseholme,
125 Shelford Road 47

October

Sunday 3rd
◆ Norwell Nurseries 34

By Arrangement

Arrange a personalised garden visit on a date to suit you. See individual garden entries for full details.

Bolham Manor 3
5 Burton Lane 6
Capability Barn 7
Church Farm 9
The Coach House 11
Cornerstones 12
Dumbleside 13
NEW Forge Cottage,
Sutton Bonington
Gardens 50
5a High Street 17
NEW 2 Holmefield
Croft 20
Holmes Villa 21
Home Farm House, 17
Main Street 22
NEW Lyndhurst 28
Meadow Farm 31
Norwood Cottage 35
NEW Oasis Community
Gardens 37
The Old Vicarage 38
Park Farm 41
Piecemeal 43
Rhubarb Farm 46
Riseholme,
125 Shelford Road 47
Rose Cottage 48
6 Weston Close 55

Manor Farm

THE GARDENS

GROUP OPENING

1 ASKHAM GARDENS
Markham Moor, Retford, NG22 0RP. *6m S of Retford. On A638, in Rockley village turn E to Askham or on A57 at East Markham turn N to Askham.* **Sun 13 June (1-5). Combined adm £5, chd free.**

DOVECOTE COTTAGE
Mrs C Slack.

NURSERY HOUSE
Mr & Mrs D Bird.

ORCHARD HOUSE
David Garner & Jane Ball.

NEW THE STABLES
Daniel & Ros Barnes.

Variety of pleasant English village gardens, with a flower festival in the church. Nursery House is a plantsman's garden, secluded and private; featuring an attractive water feature. Dovecote Cottage is an enchanting terraced cottage garden with amazing roses on the walls, perennial beds and attractive raised fish pond. Orchard House's back garden contains flower beds, shrubs, fruit trees, lawns, fish pond, raised beds with vegetables. The Stables incorporates a fantastic hedged kitchen garden and large koi pond, herbaceous bed in the making and new colourful beds surround the house. Deep gravel at Nursery House. Steep slopes at Dovecote Cottage.

2 BEESTHORPE HALL FARM
Caunton, Newark, NG23 6AT. Pamela & Peter Littlewood. *On the Maplebeck rd. 1/2 way between Caunton & Maplebeck. From A616 at Caunton take the Maplebeck/Eakring rd. 1m farm on R. 2 fields off the rd.* **Sun 18 July (1-5). Adm £3.50, chd free. Home-made teas.**
Large country garden full of exciting planting including an extensive ecologically designed, unique gravel garden.

3 BOLHAM MANOR
Bolham Way, Bolham, Retford, DN22 9JG. Pam & Butch Barnsdale, 01777 703528, pamandbutch@hotmail.co.uk. *1m from Retford. A620 Gainsborough Rd from Retford, turn L onto Tiln Lane, signed 'A620 avoiding low bridge'. At sharp R bend take rd ahead to Tiln then L Bolham Way.* **Visits by arrangement Feb to Sept for groups of 5 to 30. Refreshments to be discussed when booking. Adm £3.50, chd free.**
This 3 acre mature garden provides year round interest . In February, 'Dancing willow Ladies' greet you amongst swathes of snowdrops, narcissus and early bulbs. Topiary features and sculptures, guide you through the different areas of the garden with its well-planted terraced and herbaceous borders, ponds and orchard. Partial wheelchair access to parts of garden.

4 BRIDGE FARM
Norwell Woodhouse, Newark, NG23 6NG. Rachel Cook. *If entering the village from Norwell, the property is on the R just after the dairy farm.* **Sun 23 May (1-4). Combined adm with Upper Grove Farm £4, chd free.**
A large country garden in a quiet village with a wide variety of plants providing flower and foliage colour all year round. A contemporary swimming pond with a tranquil decking area to sit and ponder, views of open fields to the rear. Patio and courtyard, along with raised flower and vegetable beds.

5 BROADLEA
North Green, East Drayton, Retford, DN22 0LF. David & Jean Stone. *Broadlea. From A1 take A57 E towards Lincoln. East Drayton is signed L off A57, approx 2m from A1. North Green runs N from church. Garden last gate on R.* **Sun 12 Sept (2-5). Adm £3, chd free. Light refreshments. Tea and cakes £2.50.**
Our aim in this 1 acre garden is to have interest throughout the yr and attract wildlife. There is plenty to see, woodland walk, many perennials, shrubs and spring bulbs. Large pond is a haven for wildlife and a kitchen garden together with wild bank and dyke add attraction to the formal vistas. Partial wheelchair access.

6 5 BURTON LANE
Whatton in the Vale, NG13 9EQ. Ms Faulconbridge, 01949 850942, jpfaulconbridge@hotmail.co.uk, www.ayearinthegardenblog.wordpress.com. *3m E of Bingham. Follow signs to Whatton from A52 between Bingham & Elton. Garden nr Church in old part of village. Follow yellow NGS signs.* **Mon 28 June, Mon 30 Aug (11.30-4.30). Adm £4, chd free. Home-made teas. Visits also by arrangement June to Sept for groups of up to 20.**
Modern cottage garden which is productive and highly decorative. We garden organically and for wildlife. The garden is full of colour and scent from spring to autumn. Several distinct areas, incl fruit and vegetables. Large beds are filled with over 500 varieties of plants with paths through so you can wander and get close. Also features seating, gravel garden, pond, shade planting and sedum roof. Attractive village with walks.

7 CAPABILITY BARN
Gonalston Lane, Hoveringham, NG14 7JH. Malcolm & Wendy Fisher, 01159 664322, wendy.fisher111@btinternet.com, www.capabilitybarn.com. *8m NE of Nottingham. A612 from Nottingham through Lowdham. Take 1st R into Gonalston Lane. Garden is 1m on L.* **Sat 24, Sun 25 Apr (11-4.30). Adm £4, chd free. Home-made teas. Visits also by arrangement May & June for groups of 20+. Admission price includes refreshments.**
Imaginatively planted large country garden with something new each year. April brings displays of Daffodils, Hyacinths and Tulips along with erythroniums, brunneras and primulas. Wisteria, Magnolia, Rhodos and apple blossom greet May. A backdrop of established trees, shrubs and shady paths give a charming country setting. Large vegetable/fruit gardens with orchard/meadow completes the picture.

The National Garden Scheme searches the length and breadth of England and Wales for the very best private gardens

THE CHIMES

37 Glenorchy Crescent, Heronridge, NG5 9LG. Stan & Ellen Maddock. *4m N of Nottingham. A611 towards Hucknall on to Bulwell Common. Turn R at Tesco Top Valley up to island. Turn L 100 yds. 1st L then 2nd L on to Glenorchy Crescent to bottom.* **Sun 13 June (12-5). Adm £3, chd free. Home-made teas.**

We would like to invite you to pass through our archway and into our own little oasis on the edge of a busy city. Come and share our well-stocked small garden, full of roses, peonies, lilies and much more. Visit us and be surprised. We look forward to seeing you.

CHURCH FARM

Church Lane, West Drayton, Retford, DN22 8EB. Robert & Isabel Adam, 01777838250, robertadam139@btinternet.com. *A1 exit Markham Moor. A638 Retford 500 yds signed West Drayton. ³⁄₄ m, turn R, into Church Lane,* *1st R past church. Ample parking in farm yard.* **Sun 14 Feb (10.30-4). Adm £3.50, chd free. Light refreshments. Refreshments served from 11.30am. Visits also by arrangement in Feb for groups of 5 to 20.**

The garden is essentially a spring garden with a small woodland area which is carpeted with many snowdrops, aconites and cyclamen which have seeded into the adjoining churchyard, with approx. 180 named snowdrops growing in island beds, along with hellebores and daffodils. Limited amount of snowdrops are for sale.

CHURCH HOUSE

Hoveringham, NG14 7JH. Alex & Sue Allan. *6m E of Nottingham. To the R of the Church & Church Hall in the centre of Hoveringham village.* **Sun 20 June (12-5), also open Manor Farm. Sun 8 Aug (12-5). Adm £4, chd free.**

Stunning small gem of a garden, beautifully planted with immaculate hostas which greet you along the gravelled path leading past a mini pseudo-roof garden at eye level. Relax in an oasis of cottage garden style planting where the chickens, small pond, vegetable beds, charming auricula theatre and espalier fruit trees combine to create a delightfully peaceful atmosphere. Deep gravel on driveway and path into the garden so, we regret, no wheelchair access.

THE COACH HOUSE

Fosse Road, Farndon, Newark, NG24 3SF. Sir Graeme & Lady Svava Davies, 07740829495, graeme.davies@london.ac.uk. *On old A46 W of Farndon approx 250yds on R past new overbridge turn off to Hawton. Entrance driveway marked Private Rd.* **Sun 25 July (1-5). Adm £5, chd free. Home-made teas. Visits also by arrangement July & Aug for groups of 5 to 20.**

The Coach House has two major garden areas set in approximately 0.9 acres. Both have several distinct

Ossington House

sections and the planting throughout is designed to give colour and interest through the seasons. There are many rare and unusual species among the mainly perennial plants, shrubs, grasses and trees. The planting in each area of the garden is fully detailed in a brochure for visitors.

12 CORNERSTONES

15 Lamcote Gardens, Radcliffe-on-Trent, Nottingham, NG12 2BS. Judith & Jeff Coombes, 0115 8458055, judithcoombes@gmail.com, www.cornerstonesgarden.co.uk. *4m E of Nottingham. From A52 take Radcliffe exit at RSPCA junction, then 2nd L just before hairpin bend.* **Sun 18 July (1.30-5). Adm £4, chd free. Home-made teas. Visits also by arrangement July & Aug for groups of 10+.**
Plant lovers' garden, approaching ½ acre. Flowing colour themed and specie borders with rare, exotic and unusual plants, provide a wealth of colour and interest, whilst the unique fruit and vegetable garden generates an abundance of produce. Bananas, palms, fernery, fish pond, bog garden, lovely summerhouse area and greenhouse. Enjoy tea and delicious home-made cake in a beautiful setting. Wheelchair access but some bark paths and unfenced ponds.

13 DUMBLESIDE

17 Bridle Road, Burton Joyce, NG14 5FT. Mr P Bates, 01159 313725, cpbates2015@gmail.com. *5m NE of Nottingham. The Bridle Rd is an unsurfaced, single track, R hand fork off Lambley Ln. Leave passengers at our gate and park 50 yds beyond where the rd branches 3 ways.* **Visits by arrangement Jan to Oct. Adm £6.50, chd free. Home-made teas. incl in adm fee.**
Gorgeous 2 acres of varied habitat. Stream with primulas, iris, tree ferns and the like; 50yds of mixed herbaceous borders; gardening in grass with wild flowers, Spring and Autumn bulbs; woodland walks of massed cyclamen, snowdrops & anemones and a nice sunny raised gravel bed for alpines and small plants. Plant lovers' delight! Access to main garden and borders but steep slopes towards stream therefore partial access only for wheelchairs.

14 EAST MEETS WEST

85 Cowpes Close, Sutton-In-Ashfield, NG17 2BU. Kate & Mel Calladine. *Close to Quarrydale School entrance to Carsic Housing Estate. The Cl has limited parking for those with limited mobility. Parking on Stoneyford Rd, NG17 2DU would relieve congestion. Cross on zebra & walk down jitty into Cowpes Cl following NGS signposting.* **Sun 4 July (12-5). Combined adm with Spring Bank House £5, chd free.**
East: We have a number of sizable acers, bamboos and Japanese lanterns. A stream flows past a cloud tree into a pond with goldfish and water lilies. West: A trompe d'oeil arch creates a magical garden illusion with an arch shaped 'rainbow' flower bed providing a flamboyant colour arrangement. The front drive has pink/white borders, and 'green' camouflage for bins. No steps from pavement to garden. Cobbled path allows wheelchair access to full length of garden.

15 ◆ FELLEY PRIORY

Underwood, NG16 5FJ. Ms Michelle Upchurch for the Brudenell Family, 01773 810230, michelle@felleypriory.co.uk, www.felleypriory.co.uk. *8m SW of Mansfield. Off A608 ½m W M1 J27.* **For NGS: Sun 18 Apr (10-4). Adm £5, chd free. Light refreshments. For other opening times and information, please phone, email or visit garden website.**
Garden for all seasons with yew hedges and topiary, snowdrops, hellebores, herbaceous borders and rose garden. There are pergolas, a white garden, small arboretum and borders filled with unusual trees, shrubs, plants and bulbs. The grass edged pond is planted with primulas, bamboo, iris, roses and eucomis. Bluebell woodland walk. Orchard with extremely rare daffodils. Regional Finalist, The English Garden's The Nation's Favourite Gardens 2019.

Oasis Community Gardens

GROUP OPENING

16 HALAM GARDENS AND WILDFLOWER MEADOW

nr Southwell, NG22 8AX. *Village gardens within walking distance of one another. Hill's Farm wildflower meadow is a short drive of ½ m towards Edingley village, turn R at brow of hill as signed.* **Mon 31 May** (12.30-4.30). **Combined adm £6.50, chd free. Home-made teas at The Old Vicarage, Halam.**

HILL FARM HOUSE
Victoria Starkey.

HILL'S FARM
John & Margaret Hill.

THE OLD VICARAGE
Mrs Beverley Perks.
(See separate entry)

Lovely mix of a very popular beautiful, well-known, organic, rural plant lovers' garden with clematis, striking viburnums amongst herbaceous borders and sweeping lawns, lots of pond life; contrasting with small wrap-around cottage garden and 6 acre wildflower meadow - part of an organic farm - visitors can be assured of an inspiring discussion with the farmer who is passionate about the benefits of this method of farming for our environment and beef quality. 12th Century Church open - surrounded by freely planted, attractive churchyard - all welcome to enjoy this peaceful haven in English rural village setting. Old Vicarage - gravel entrance to lower flat garden. Further up is on gentle hill. Hill Farm House - gravel entrance. Hills Farm flat field.

"I love the National Garden Scheme which has been the most brilliant supporter of Queen's Nurses like me. It was founded by the Queen's Nursing Institute which makes me very proud. As we battle Coronavirus on the front line in the community, knowing we have their support is a real comfort." – Liz Alderton, Queen's Nurse

17 5A HIGH STREET

Sutton-on-Trent, NG23 6QA. Kathryn & Ian Saunders, 07827920236, kathrynsaunders.optom@gmail.com. *6m N of Newark. Leave A1 at Sutton on Trent , follow Sutton signs. L at Xrds. 1st R turn (approx 1m) onto Main St. 2nd L onto High St. Garden 50 yds on R. Park on rd.* **Sun 1 Aug** (1-4). **Adm £4, chd free. Light refreshments. Visits also by arrangement June & July for groups of 10+. Open mid June to early August.**

Manicured lawns are the foil for this plantsman's garden. Vistas lead past succulents to tropical areas and vibrant herbaceous beds. Ponds run through the plot, leading to woodland walks and a dedicated fernery (180+ varieties, with magnificent tree ferns). Topiary links all the different planting areas to great effect. Around 1000 named varieties of plants with interesting plant combinations.

18 NEW 1 HIGHFIELD ROAD

Nuthall, Nottingham, NG16 1BQ. Richard & Sue Bold. *4mins from J26 of the M1- 4m NW of Nottingham City. From J26 of the M1 take the A610 towards Notts, R-hand lane at r'about, take turn-off to Horsendale. Follow rd to Woodland Dr, then 2nd R is Highfield Rd.* **Sat 13 Feb, Sun 25 Apr** (10-4). **Adm £3, chd free. Light refreshments.**

Offers a large variety of plants, shrubs, rhodos, azaleas, bulbs and alpines. Large heuchera collection gives colour and texture all year. Espaliered archway, with apples, plums & pears. Soft fruit collection. Patio with many pots and troughs. Greenhouses with tomatoes, cucumbers and peppers grown hydroponically. Vegetable plot. February snowdrops 500 varieties in garden and pots, sales available. Wheelchair access to the side garden and patio area only.

19 NEW HOLLINSIDE

252 Diamond Avenue, Kirkby-In-Ashfield, Nottingham, NG17 7NA. Sue & Bob Chalkley. *1m E of Kirkby in Ashfield at the Xrds of the A611 & B6020. Please use the signed parking 200m on the L towards Kirkby away from the busy junction.* **Sun 13 June** (1-4.30). **Adm £3, chd**

free. Light refreshments.

A formal front garden with terraced lawns and borders lead to a shaded area with ferns, camellias, roses, hydrangeas and other flowering shrubs. The rear garden has a wildlife pond, a wild flower meadow, a summerhouse and a victorian style greenhouse. There are topiary box and yews and a box parterre planted with roses. Majority of garden is suitable for wheelchairs. Disabled parking near house by prior arrangement.

20 NEW 2 HOLMEFIELD CROFT

Low Road, Scrooby, DN10 6BS. Mr Phil Walton & Mrs Julie Davison, 01302 719668, julie.davison@ngs.org.uk. *1m from Bawtry on Road towards Retford., turn into Scrooby village at Pilgrim Fathers pub, take 2nd L at church, follow Low Rd around 90 degree bend. Signage in Beech hedge on L.* **Visits by arrangement in May for groups of up to 20. Weeks beginning 3rd May and 24th May only. Adm £5, chd free. Home-made teas. Tea, coffee and cake included in admission price of £5.**

Wrap around small garden in village setting with colour and interest squeezed into every area. Joint opening with Lyndhurst, Chapel Lane, Scrooby. Topiary shapes of Yew and box frames an unusual shaped front garden with tree and circular spring border, leading to L shaped rear garden featuring unusual shrubs and trees, 6 distinct small gardens and tulips, tulips everywhere.

21 HOLMES VILLA

Holmes Lane, Walkeringham, Gainsborough, DN10 4JP. Peter & Sheila Clark, 01427 890233, clarkshaulage@aol.com. *4m NW of Gainsborough. A620 from Retford or A631 from Bawtry/Gainsborough & A161 to Walkeringham then towards Misterton. Follow NGS signs for 1m. Plenty of parking. Reserved disabled parking.* **Visits by arrangement Feb to June. Tea and biscuits can be arranged. Adm £3, chd free.**

1¾ acre plantsman's garden offering yr-round interest and inspiration starting with carpets of snowdrops, mini daffodils, hellebores and spring bulbs. Unusual collection of plants and shrubs for winter. Come and be surprised at the different fragrant and interesting plants in early spring. Places to sit and ponder, gazebos, arbours, wildlife pond, hosta garden, old tools on display and scarecrows. A flower arranger's artistic garden. Special parking for those requiring wheelchair access in yard.

22 HOME FARM HOUSE, 17 MAIN STREET

Keyworth, Nottingham, NG12 5AA. Graham & Pippa Tinsley, 0115 9377122, Graham_Tinsley@yahoo.co.uk, www.homefarmgarden.wordpress. com. *7m S of Nottingham. Follow signs for Keyworth from A60 or A606 & head for church. Garden about 50yds down Main St. Parking on the street or at village hall or Bunny Lane car parks.* **Sun 13 June (12-5). Combined adm with Rose Cottage £6, chd free. Home-made teas. Visits also by arrangement May to Sept for groups of 5 to 30.**
A large garden behind old farmhouse in the centre of the village. The old orchard, herbaceous and vegetable gardens are preserved and the cart shed has been rebuilt as a pergola. Large areas of unmown grass planted with perennials, ponds, turf mound and many trees. High hedges create hidden areas incl. rose and winter gardens. All combine in an intriguing blend of wildness and formality. Grade 2 listed barn open with display by Keyworth & District History Society. Interesting and unusual perennials for sale by Piecemeal Plants (www. piecemealplants.co.uk). Access via gravel yard. Some steps and slopes.

23 HOPBINE FARMHOUSE, OSSINGTON

Hopbine Farmhouse, Main Street, Ossington, NG23 6LJ. Mr & Mrs Geldart. *From A1 N take exit marked Carlton, Sutton-on-Trent, Weston etc. At T-Junction turn L to Kneesall. Drive 2m to Ossington. In village turn R to Moorhouse & park in field.* **Sun 11 July (2-5). Combined adm with Ossington House £5, chd free. Home-made teas.**
A small garden in separate halves; the southern half has a long herbaceous border with an arbour over which climb honeysuckle, clematis and r. Chinensis mutabilis. A full central bed has many favourite salvias, euphorbia and veronicastrum. The highlight of the intimate walled garden is a "waterfall" of clematis Summer Snow

and Wisley, rambling through roses Iceberg and Ghislaine de Feligonde. Some narrow paths.

24 6 HOPE STREET

Beeston, Nottingham, NG9 1DR. Elaine Liquorish. *From M1 J25, A52 for Nottm. After 2 r'abouts, turn R for Beeston at The Nurseryman (B6006). Beyond hill, turn R into Bramcote Dr. 3rd turn on L into Bramcote Rd, then immediately R into Hope St.* **Sun 23 May (1.30-5.30). Adm £4, chd free. Cream teas.**
A small garden packed with a wide variety of plants providing flower and foliage colour year round. Collections of alpines, bulbs, mini, small and medium size hostas (60+), ferns, grasses, carnivorous plants, succulents, perennials, shrubs and trees. A pond and a greenhouse with subtropical plants. Troughs and pots. Home-made crafts. Shallow step into garden, into greenhouse and at rear. No wheelchair access to plant sales area.

25 IVY BANK COTTAGE

The Green, South Clifton, Newark, NG23 7AG. David & Ruth Hollands. *12m N of Newark. From S, exit A46 N of Newark onto A1133 towards Gainsborough. From N, exit A57 at Newton-on-Trent onto A1133 towards Newark.* **Sun 16 May (1-5). Adm £3.50, chd free. Home-made teas.**
A traditional cottage garden, with herbaceous borders, fruit trees incl a Nottinghamshire Medlar, vegetable plots and many surprises incl a stumpery, a troughery, dinosaur footprints and even fairies! Many original features: pigsties, double privy and a wash house. Children can search for animal models and explore inside the shepherd's van. Seats around and a covered refreshment area.

26 NEW JASMINE COTTAGE

High Street, Walkeringham, Doncaster, DN10 4LJ. Alan Jackson. *Jasmine Cottage High Street Walkeringham DN10 4LJ. Jasmine Cottage is situated at the junction of High St & South Moor Rd.* **Sun 2 May (12-3). Combined adm with Westmoor House £4, chd free.**
Jasmine Cottage is a prizewinning, small, personal, traditional Cottage garden, in the centre of the village.

A wide variety of herbaceous plants, flowers and vegetables can be seen, in a relaxed setting. A delight for the enthusiastic gardener. This garden is open in conjunction with West Moor House, West Moor Road.

27 LODGE MOUNT

Town Street, South Leverton, Retford, DN22 0BT. Mr A Wootton-Jones. *4m E of Retford. Opp Miles Garage on Town St.* **Sat 3 July (1-5). Adm £4, chd free. Home-made teas.**
Lodge Mount Garden is the vision founded and established by Helen just before she died of cancer in 2012. The garden has been open to the public ever since her death. Helen's original plans can be seen through the flourishing growth and development of her plants and trees which now grow strong displaying the vibrant colours which she hoped they would. Access is via a long sloping driveway. There is one step into the main garden which is flat.

28 NEW LYNDHURST

Chapel Lane, Scrooby, Doncaster, DN10 6AE. Jean & Mike Rush, 01302 719668, jandp2@icloud.com. *Turn into Scrooby from Bawtry at 1st L Chapel Lane. Lyndhurst is 1st bungalow on R.* **Visits by arrangement in May for groups of up to 20. Week beginning 3rd May & week beginning 24th May (with 2 Holmefield Croft). Adm £5, chd free. Home-made teas at 2 Holmefield Croft, Scrooby. Refreshments included in entry fee**
Developed over 6 years from a blank canvas to have all year round interest, but especially wonderful in spring. Small garden packed with colour and unusual plants.

Online booking is available for many of our gardens, at ngs.org.uk

38 MAIN STREET
Woodborough, Nottingham,
NG14 6EA. Martin Taylor
& Deborah Bliss. *Turn off
Mapperley Plains Rd at sign for
Woodborough. Alternatively, follow
signs to Woodborough off A6097
(Epperstone bypass). Property is
between Park Av & Bank Hill.* **Sun
16 May (1-5). Adm £4, chd free.
Home-made teas.**
Varied ⅓ acre. Bamboo fenced Asian
species area with traditional outdoor
wood fired Ofuro bath, herbaceous
border, raised rhododendron bed,
vegetables, greenhouse, pond area
and art studio and terrace.

30 NEW **MANOR FARM**
Moor Lane, East Stoke, Newark,
NG23 5QD. Mr Greg & Mrs Pam
Stevens. *Travelling towards Newark
on the A46 exit at East Stoke/Elston
slip road follow signs to East Stoke.
Travelling towards Nottingham on the
A46, exit at Farndon r'about towards
East Stoke.* **Sun 20 June (12-6).
Adm £4, chd free. Home-made
teas. Also open Park Farm.**
Stunning ½ acre country garden
planted for colour and perfume -
food for the soul. Begun in 2015
still evolving with new secret garden
to compliment the packed borders

and shrubberies, gravel garden,
rose pergola, box parterre and
mature trees. Level site. Paved paths
throughout the garden.

31 MEADOW FARM
Broadings Lane, Laneham,
Retford, DN22 0NF. Maureen
Hayward, 01777228284,
maureen.hayward@zen.co.uk.
*10m E of Retford. From A1, take
A57 towards Lincoln 5m, L towards
Laneham. Village of Laneham, from
Main St, L onto Broading Lane, 2nd
on R, parking available past gateway.*
**Sun 15 Aug (11-4). Adm £4.50,
chd free. Light refreshments.
Visits also by arrangement July
to Sept.**
A 1 acre garden that takes you
through a hosta garden into a jungle
with bananas, tree ferns, cannas.
Winding path takes you to a parterre
with topiary and sequoiadendron
giganteum pendula trees and water
features. Also incl a woodland walk,
sculptures, wisteria circular walkway,
seating throughout, multiple areas
of herbaceous planting and unusual
specimen trees and plants with Arctic
Cabin. Wheelchair access through
separate gate.

32 NORMANTON HALL
South Street, Normanton-on-Trent,
NG23 6RQ. His Honour John & Mrs
Machin. *3m SE of Tuxford. Leave
A1 at Sutton Carlton/Normanton-on-
Trent junction. Turn L onto B1164
in Carlton. In Sutton-on-Trent turn
R at Normanton sign. Go through
Grassthorpe, turn L at Normanton
sign.* **Sun 25 Apr (1-5). Adm £4,
chd free. Light refreshments.**
3 acres with mature oak, lime beech
and yew and recently planted trees.
Vegetable area. New plantings of
bulbs, rhododendrons and a camellia
walk. Arboretum planted with
unusual, mainly hardwood trees which
are between three and twelve years
old. Also specimen oaks and beech.
New planting of wood anemones.
Recently established parkland. All
surfaces level from car park.

GROUP OPENING

33 NORWELL GARDENS
Newark, NG23 6JX. *6m N of
Newark. Halfway between Newark &
Southwell. Off A1 at Cromwell turning,
take Norwell Rd at bus shelter. Or
off A616 take Caunton turn.* **Sun 27
June (1-5). Evening opening Wed
30 June (6.30-9). Combined adm**

Lodge Mount

£5, chd free. Home-made teas in Village Hall (27 June) and Norwell Nurseries (30 June).

FAUNA FOLLIES
Lorraine & Roy Pilgrim.

JUXTA MILL
Janet McFerran.

NORWELL ALLOTMENTS / PARISH GARDENS
Norwell Parish Council.

◆ NORWELL NURSERIES
Andrew & Helen Ward.
(See separate entry)

THE OLD MILL HOUSE, NORWELL
Mr & Mrs M Burgess.

NEW ROSE COTTAGE
Mr Iain & Mrs Ann Gibson.

This is the 25th yr that Norwell has opened a range of different, very appealing gardens all making superb use of the beautiful backdrop of a quintessentially English countryside village. Incl a garden and nursery of national renown and the rare opportunity to walk around vibrant allotments with a wealth of gardeners from seasoned competition growers to plots that are substitute house gardens, bursting with both flower colour and vegetables in great variety. To top it all there are a plethora of breathtaking village gardens showing the diversity that is achieved under the umbrella of a cottage garden description! The beautiful medieval church and its peaceful churchyard with grass labyrinth will be open for quiet contemplation.

 ❀ ♿ ☕

The National Garden Scheme was set up in 1927 to raise funds for district nurses. We remain strongly committed to nursing and today our beneficiaries include Macmillan Cancer Support, Marie Curie, Hospice UK and The Queen's Nursing Institute

---

34 ◆ NORWELL NURSERIES
Woodhouse Road, Norwell, NG23 6JX. Andrew & Helen Ward, 01636 636337, wardha@aol.com, www.norwellnurseries.co.uk. *6m N of Newark halfway between Newark & Southwell. Off A1 at Cromwell turning, take rd to Norwell at bus stop. Or from A616 take Caunton turn.* **For NGS: Sun 16 May, Sun 3 Oct (2-5). Adm £3.50, chd free. Home-made teas. Opening with Norwell Gardens on Sun 27, Wed 30 June. For other opening times and information, please phone, email or visit garden website.** Jewel box of over 2,500 different, beautiful and unusual plants sumptuously set out in a one acre plantsman's garden incl shady garden with orchids, woodland gems, cottage garden borders, alpine and scree areas. Pond with opulently planted margins. Extensive herbaceous borders and effervescent colour themed beds. Sand beds showcase Mediterranean, North American and alpine plants. Nationally renowned nursery open with over 1,000 different rare plants for sale. Autumn opening features UK's largest collection of hardy chrysanthemums for sale and the National Collection of Hardy Chrysanthemums. New borders incl the National Collection Of Astrantias. Innovative sand beds. Grass paths, no wheelchair access to woodland paths.

 ♿ ❀ ♿ NPC ☕

35 NORWOOD COTTAGE
Church Street, East Markham, Newark, NG22 0SA. Anne Beeby, 01777872799, anne.beeby@sky.com. *7m S of Retford. Take A1 Markham Moor r'about exit for A57 Lincoln. Turn at East Markham junction & follow signs for the church. Property adjacent the church.* **Visits by arrangement in July for groups of up to 30. Adm by donation. Light refreshments.** Norwood Cottage is an outstanding example of a plantswoman's take on a cottage garden, combining plants to create a visual feast of texture, colour and form. Partial wheelchair access to parts of garden only. Steps leading to some areas

 ♿ 🐄 ♿ ☕

36 OAK BARN EXOTIC GARDEN
Oak Barn Church Street, East Markham, Newark, NG22 0SA. Simon Bennett & Laura Holmes, www.facebook.com/OakBarn1.

From A1 Markham Moor junction take A57 to Lincoln. Turn R at Xrds into E Markham. L onto High St & R onto Plantation Rd. Enter farm gates at T-Junction, garden located on L. **Evening opening Sat 7 Aug (6-9). Light refreshments. Sun 12 Sept (1-5). Home-made teas. Adm £3.50, chd free.** On entering the oak lych-style gate you will be met with the unexpected dense canopy of greenery and tropical foliage. The gravel paths wind under towering palms and bananas which are underplanted with cannas and gingers. On the lowest levels houseplants are bedded out from the large greenhouse to join the summer displays. They surround the Jungle Hut and new raised walkway.

 ❀ ♿ ☕

37 NEW OASIS COMMUNITY GARDENS
2a Longfellow Drive, Kilton Estate, Worksop, S81 0DE. Steve Williams, 07795 194957, Stevemark126@hotmail.com, www.oasis-centre.org.uk. *From Kilton Hill (leading to the Worksop hospital), take 1st exit to R (up hill) onto Kilton Cres, then 1st exit on R Longfellow Dr. Car Park off Dickens Rd (1st R).* **Sat 17 Apr (10-6). Cream teas in Oasis Community Centre - 'Food for Life'. Sun 15 Aug (10-6). Light refreshments in Oasis Community Centre - 'Food for Life'. Adm £4, chd free. Meals can be arranged for group bookings by arrangement. Visits also by arrangement Mar to Oct.** Oasis Gardens is a community project transformed from abandoned field to an award winning garden. Managed by volunteers the gardens boast over 30 project areas, several garden enterprises and hosts many community events. Take a look in the Cactus Kingdom, the Children's pre-school play village, Wildlife Wonderland or check out a wonderful variety of trees, plants, seasonal flowers and shrubs. The Oasis Gardens hosts the first Liquorice Garden in Worksop for 100 years. The site hosts the 'Flowers for Life' project which is a therapeutic gardening project growing and selling cut flowers and floristry. There is disabled access from Longfellow Drive. From the town end there is a driveway after the first fence on the right next to house number 2.

 ♿ ❀ ☕ 🪑

38 THE OLD VICARAGE

Halam Hill, Halam, NG22 8AX. Mrs Beverley Perks, 01636 812181, perks.family@talk21.com. *1m W of Southwell. Please park diagonally into beech hedge on verge with speed interactive sign or in village - a busy road so no parking on roadside.* **Sun 9 May, Sat 10 July (12.30-4.30). Sun 8 Aug (1-4.30), also open Church House. Adm £4, chd free. Home-made teas at The Old Vicarage. Opening with Halam Gardens and Wildflower Meadow on Mon 31 May. Visits also by arrangement May to Aug for groups of 10+.**

Single handed labour of love has grown out of 2 acre hillside pony paddocks into much admired landscape gardens. One time playground for children, a source of pleasure for village openings and 19 happy years for NGS. An artful eye for design/texture/colour/love of unusual plants/trees makes this a welcoming gem to visit/share. Possible last opening as building with new garden planting at bottom. Beautiful C12 Church open only a short walk into the village or across field through attractively planted churchyard - rare C14 stained glass window. Gravel drive - undulating levels as on a hillside - plenty of cheerful help available.

🐄 ❋ 🚌 ☕

39 OSSINGTON HOUSE

Moorhouse Road, Ossington, Newark, NG23 6LD. Georgina Denison. *10m N of Newark, 2m off A1. From A1 N take exit marked Carlton, Sutton-on-Trent, Weston etc. At T-junction turn L to Kneesall. Drive 2m to Ossington. In village turn R to Moorhouse & park in field next to Hopbine Farmhouse.* **Sun 11 July (2-5). Combined adm with Hopbine Farmhouse, Ossington £5, chd free. Home-made teas in The Hut, Ossington.**

Vicarage garden redesigned in 1960 and again in 2014. Chestnuts, lawns, formal beds, woodland walk, poolside planting, orchard. Terraces, yews, grasses. Ferns, herbaceous perennials, roses and a new kitchen garden. Disabled parking available in drive to Ossington House.

♿ ❋ 🚌 Ⓓ ☕

40 PAPPLEWICK HALL

Hall Lane, Papplewick, Nottingham, NG15 8FE. Mr & Mrs J R Godwin-Austen, www.papplewickhall.co.uk. *7m N of Nottingham. 300yds out N end of Papplewick village, on B683 (follow signs to Papplewick from A60 & B6011). Free parking at Hall.* **Sun 30 May (2-5). Adm £4, chd free. Donation to St James' Church.**

This historic, mature, 8 acre garden, mostly shaded woodland, abounds with rhododendrons, hostas, ferns, and spring bulbs. Suitable for wheelchair users, but sections of the paths are gravel.

♿

41 PARK FARM

Crink Lane, Southwell, NG25 0TJ. Ian & Vanessa Johnston, 01636 812195, v.johnston100@gmail.com. *1m SE of Southwell. From Southwell town centre go down Church St, turn R on Fiskerton Rd & 200yds up hill turn R into Crink Lane. Park Farm is on 2nd bend.* **Sun 20 June (1-5). Adm £4, chd free. Home-made teas. Also open Church House. Visits also by arrangement Apr to Aug for groups of up to 30. Guided visits by arrangement 50p per person extra.**

3 acre garden remarkable for its extensive variety of trees, shrubs and perennials, many rare or unusual. Long colourful herbaceous borders, roses, woodland garden, alpine/scree garden and a large wildlife pond. New for 2021- developing area for acid loving plants, and new woodland planting and paths. Spectacular views of the Minster across a wildflower meadow and ha-ha.

♿ 🐄 ❋ ☕ ⛲

42 PATCHINGS ART CENTRE

Oxton Road, Calverton, Nottingham, NG14 6NU. Chas & Pat Wood, www.patchingsartcentre.co.uk. *N of Nottingham city take A614 towards Ollerton. Turn R on to B6386 towards Calverton & Oxton. Patchings is on L before turning to Calverton. Brown tourist directional signs.* **Sun 13 June (11-3). Adm £3, chd free. Light refreshments at Patchings Café.**

Promoting the enjoyment of art. Established in 1988, a 50 acre site with visitor centre in converted farm buildings. Each year a 4 day art festival has a national reputation. The grounds and gardens developed to inspire and encourage artists to paint in the open, whilst providing an enjoyable and tranquil setting for visitors. New for 2021 - The Artists' Trail. Four galleries with changing exhibitions. Card gallery, gift shop, café and art materials. New for 2021 Patchings Artists' Trail - a walk through art history, famous paintings within glass take visitors through the centuries, meeting well known artists from the past along the way. Grass and compacted gravel paths with some undulations and uphill sections accessible to wheelchairs with help. Please enquire for assistance.

♿ 🐄 🚌 ☕

43 PIECEMEAL

123 Main Street, Sutton Bonington, Loughborough, LE12 5PE. Mary Thomas, 01509 672056, nursery@piecemealplants.co.uk. *2m SE of Kegworth (M1 J24). 6m NW of Loughborough. Almost opp St Michael's Church at the N end of the village.* **Mon 30 Aug (2-6). Adm £4, chd free. Opening with Sutton Bonington Gardens on Sat 29, Sun 30 May. Visits also by arrangement June to Aug for groups of 5 to 10. For 10+ please contact to discuss.**

Tucked behind early 19th century cottages, a tiny walled garden featuring a wide range of unusual shrubs, many flowering and most displayed in around 400 terracotta pots bordering narrow paths. Also climbers, perennials and even a few trees! Focus is on distinctive form, foliage shape and colour combination. Collection of ferns around well. Half-hardy and tender plants fill the conservatory. Plant list available on request.

🐄 ❋ ☕

44 THE POPLARS

60 High Street, Sutton-on-Trent, Newark, NG23 6QA. Sue & Graham Goodwin-King. *7m N of Newark. Leave A1 at Sutton/Carlton/Normanton-on-Trent junction. In Carlton turn L onto B1164. Turn R into Hemplands Ln then R into High St. 1st house on R. Limited parking.* **Sun 21 Feb (11-4). Light refreshments. Sun 20 June (1-5.30). Cream teas. Adm £4, chd free. Home-made soup will be offered on snowdrop open day in February.**

Mature ½ acre garden on the site of a Victorian flower nursery, now a series of well planted areas each with its own character. Exotics courtyard, pond and oriental style gravel garden, 'jungle' with thatched shack, black and white garden, woodland area, walled potager and fernery. Lawns, borders, charming sitting places and over 350 snowdrop varieties in early spring for the galanthophiles. Some gravel paths and shallow steps, but most areas accessible.

🐕 ✿ ☕

45 PRIMROSE COTTAGE

Bar Road North, Beckingham, Doncaster, DN10 4NN. Terry & Brenda Wilson. *8m N of Retford. A631 to Beckingham r'about, enter village, L to village green, L to Bar Rd.* **Sun 6 June (2-5). Adm £3, chd free. Home-made teas.**
Old fashioned cottage garden. Walled herbaceous border, well stocked shrubbery, many old roses, kitchen garden, summerhouse and greenhouse. Secret fernery and courtyard, herb garden, looking onto bespoke archway hosting many clematis.

✿ ☕

46 RHUBARB FARM

Hardwick Street, Langwith, nr Mansfield, NG20 9DR. Rhubarb Farm, enquiries@rhubarbfarm.co.uk, , www.rhubarbfarm.co.uk. *On NW border of Nottinghamshire in village of Nether Langwith. From A632 in Langwith, by bridge (single file traffic) turn up steep Devonshire Drive. N.B. Then turn off Satnav. Take 2nd L into Hardwick St. Rhubarb Farm at end. Parking to R of gates.* **Visits by arrangement June to Sept for groups of 10 to 20. Appointments will be taken for Wednesdays only. Adm £2.50, chd free. Cream teas. Cream teas made by Rhubarb Farm volunteers available in our on-site café.**
52 varieties of fruit and vegetables organically grown not only for sale but for therapeutic benefit. This 2 acre social enterprise provides training and volunteering opportunities to 90 ex offenders, drug and alcohol misusers, and people with mental and physical ill health and learning disability. 8 polytunnels, 2 composting toilets and 100 hens and pigs. 3x65ft polytunnels, forest school barn, willow dome & arch, 100 hens, flower borders, small shop, raised beds, comfrey bed and comfrey fertiliser factory, composting toilets. Chance to meet and chat with volunteers with a variety of needs, who come to gain skills, confidence and training. Main path suitable for wheelchairs but bumpy. Not all site accessible. Cafe & composting toilet wheelchair-accessible. Mobility scooter available.

♿ ✿ 🚌 ☕

The Coach House

47 RISEHOLME, 125 SHELFORD ROAD

Radcliffe on Trent, NG12 1AZ. John & Elaine Walker, 01159 119867. *4m E of Nottingham. From A52 follow signs to Radcliffe. In village centre take turning for Shelford (by Co-op). Approx ¾m on L.* **Sun 19 Sept (1.30-4.30). Adm £4, chd free. Home-made teas. Visits also by arrangement June to Sept for groups of 20+.**
Imaginative and inspirational is how the garden has been described by visitors. A huge variety of perennials, grasses, shrubs and trees combined with an eye for colour and design. Jungle area with exotic lush planting contrasts with tender perennials particularly salvias thriving in raised beds and in gravel garden with stream. Unique and interesting objects complement planting. Gravel drive and paths.

🐾 ✿ 🚌 ☕

48 ROSE COTTAGE

81 Nottingham Road, Keyworth, Nottingham, NG12 5GS. Richard & Julie Fowkes, 0115 9376489, richardfowkes@yahoo.co.uk. *7m S of Nottingham. Follow signs for Keyworth from A606. Garden (white cottage) on R 100yds after Sainsburys. From A60, follow Keyworth signs & turn L at church, garden is 400yds on L.* **Sun 13 June (12-5). Combined adm with Home Farm House, 17 Main Street £6, chd free. Home-made teas. Visits also by arrangement May to Sept for groups of 5 to 20.**
Small cottage garden (300sqm) informally designed and packed full of butterfly and bee friendly plants. Sedum roof, mosaics and water features add unique interest. There is a decked seating area and summerhouse. A wildlife stream meanders down between two ponds and bog gardens. Some narrow paths and steps. A woodland area leads to fruit, flowers and a herb spiral. Art studio will be open. Paintings and art cards designed by Julie will be on sale.

🐾 ✿ ☕

49 SPRING BANK HOUSE

84 Kirkby Road, Sutton-in-Ashfield, NG17 1GH. Mr Peter Robinson. *From Mansfield take A38 to Sutton-in-Ashfield. Turn R on to Station Rd, turn L on to High Pavement & continue on to Kirkby*

Rd. **Sun 4 July (12-5). Combined adm with East Meets West £5, chd free.**
Mediterranean planting, a bog garden, a summerhouse and a white garden, a water table from India and lion statues from Nepal, this garden offers much variety. Dug from wasteland in 2012 its clear design displays hundreds of rare plants. Two terraces punctuate its slope and a woodland area with hardy exotics and a wildlife pond offer cobbled paths and mown grass for a choice of garden circuits. Coaches may be possible but will need to park elsewhere.

🐾 ✿ 🚌 ☕

GROUP OPENING

50 NEW SUTTON BONINGTON GARDENS

Main Street, Sutton Bonington, Loughborough, LE12 5PE. Mary Thomas. *2m SE of Kegworth (M1 J24). 6m NW of Loughborough. From A6 Kegworth, past University of Nottingham Sutton Bonington campus. From Loughborough, A6 then A6006. From Nottingham, A60 then A6006.* **Sat 29, Sun 30 May (2-6). Combined adm £6, chd free. Home-made teas at Forge Cottage and 118 Main Street.**

NEW FORGE COTTAGE
Judith & David Franklin, 01509673578, drj.franklin@yahoo.com.
Visits also by arrangement June to Aug for groups of 10 to 20.

NEW 118 MAIN STREET
Alistair Cameron & Shelley Nicholls.

PIECEMEAL
Mary Thomas.
(See separate entry)

All 3 village gardens are located at the north end of the village, just a few minutes walk from each other. Piecemeal has a tiny walled garden and small conservatory, both filled with a wide variety of unusual plants. Forge Cottage garden, reclaimed from a blacksmith's yard and a little larger, has vibrant curved herbaceous borders. 118 Main Street is large with well-established trees and varied planting as well as an orchard/wild flower meadow. Varied features including Japanese-inspired seating area, redesigned patio with water feature, as well as more traditional

established borders surrounding lawn and pond. Attractive greenhouse. Wheelchair access to some parts of the larger two of the gardens.

🐾 ✿ ☕ 🚌

51 THRUMPTON HALL

Thrumpton, NG11 0AX. Miranda Seymour, www.thrumptonhall.com. *7m S of Nottingham. M1 J24 take A453 towards Nottingham. Turn L to Thrumpton village & cont to Thrumpton Hall.* **Sun 18 July (1.30-4.30). Adm £5, chd free. Home-made teas.**
2 acres incl. lawns, rare trees, lakeside walks, flower borders, rose garden, new pagoda, and box-bordered sunken herb garden, all enclosed by C18 ha-ha and encircling a Jacobean house. Garden is surrounded by C18 landscaped park and is bordered by a river. Rare opportunity to visit Thrumpton Hall (separate ticket). Jacobean mansion, unique carved staircase, Great Saloon, State Bedroom, Priest's Hole.

♿ 🐾 ✿ 🚌 ☕ 🎪

52 UNIVERSITY PARK GARDENS

Nottingham, NG7 2RD. University of Nottingham, www.nottingham.ac.uk/estates/grounds/. *Approx 4m SW of Nottingham city centre & opp Queens Medical Centre. NGS visitors: Please purchase admission tickets in the Millennium Garden (in centre of campus), signed from N & W entrances to University Park & within internal road network.* **Sun 22 Aug (12-3). Adm £4, chd free. Light refreshments in the Lakeside Pavilion Cafe for limited catering & light refreshments adjacent to the Millennium Garden.**
University Park has many beautiful gardens incl the award-winning Millennium garden with its dazzling flower garden, timed fountains and turf maze. Also the huge Lenton Firs rock garden, and the Jekyll garden. During summer, the Walled Garden is alive with exotic plantings. In total, 300 acres of landscape and gardens. Picnic area, cafe, walking tours, information desk, workshop, accessible minibus to feature gardens within campus. Plants for sale in Millennium garden. Some gravel paths and steep slopes.

♿ 🐾 ✿ 🚌 ☕ 🎪

Spring Bank House

53 UPPER GROVE FARM
Norwell Woodhouse, Newark,
NG23 6NG. Kathryn Wiltshire.
*A616 either from Newark (10m)
or Ollerton (7m) at Xroads turn
to Laxton R to Norwell 1st white
farmhouse on L, look out for the big
chimney pots!* **Sun 23 May (1-4).
Combined adm with Bridge Farm
£4, chd free. Home-made teas at
Upper Grove Farm.**
We are a country garden around a
farmhouse. We have a small orchard
and informal beds of spring/summer
flowers. Most of garden accessible to
wheelchairs, some gravel paths.

54 NEW WESTMOOR HOUSE
West Moor Road, Walkeringham,
Doncaster, DN10 4LR. Frances
Loates. *Between Bawtry &
Gainsborough. At Beckingham*
take A161N to Walkeringham. L on
Sidsaph Hill/High St, then L on West
Moor Rd. From Epworth take A161S
to Walkeringham. R on Sidsaph Hill/
High St, then L on West Moor Rd.
**Sun 2 May (12-3). Combined adm
with Jasmine Cottage £4, chd
free. Home-made teas at the
Village Hall, Stockwith Rd.**
Welcome to our garden, a view from
every aspect. A small garden, full of
colour and interesting plants. Mature
fruit and blossom trees in Spring. Bog
and pond area and an aviary with a
variety of beautiful birds.

55 6 WESTON CLOSE
Woodthorpe, Nottingham,
NG5 4FS. Diane & Steve
Harrington, 0115 9857506,
mrsdiharrington@gmail.com. *3m
N of Nottingham. A60 Mansfield Rd.*
Turn R at T-lights into Woodthorpe
Dr. 2nd L Grange Rd. R into The
Crescent. R into Weston Close.
Please park on The Crescent. **Sun
4 July (1-5). Adm £3, chd free.
Home-made teas including
gluten free options. Visits also
by arrangement June to Aug for
groups of 10 to 30.**
Set on a substantial slope with 3
separate areas surrounding the
house. Dense planting creates a
full, varied yet relaxed display incl
many scented roses, clematis and a
collection of over 80 named mature
hostas in the impressive colourful
rear garden. Walls covered by many
climbers. Home propagated plants
for sale.

OXFORDSHIRE

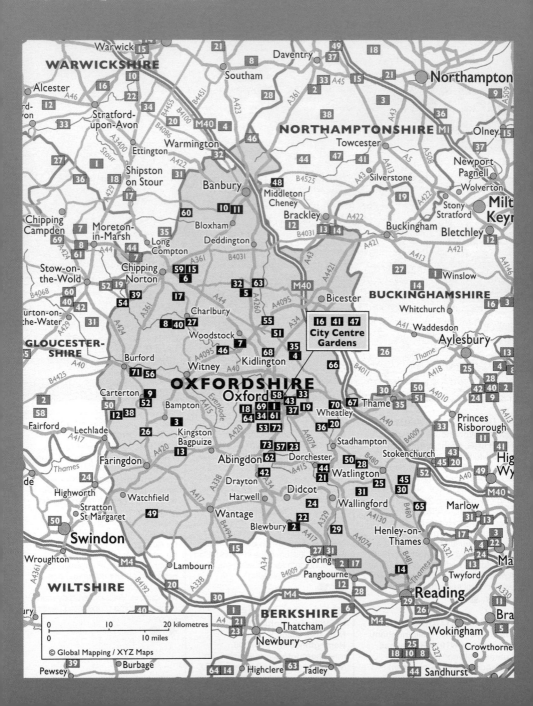

In Oxfordshire we tend to think of ourselves as one of the most landlocked counties, right in the centre of England and furthest from the sea.

We are surrounded by Warwickshire, Northamptonshire, Buckinghamshire, Berkshire, Wiltshire and Gloucestershire, and, like these counties, we benefit from that perfect British climate which helps us create some of the most beautiful and famous gardens in the world.

Many gardens open in Oxfordshire for the National Garden Scheme between spring and late-autumn. Amongst these are the perfectly groomed college gardens of Oxford University, and the grounds of stately homes and palaces designed by a variety of the famous garden designers such as William Kent, Capability Brown, Rosemary Verey, Tom Stuart-Smith and the Bannermans of more recent fame.

But we are also a popular tourist destination for our honey-coloured mellow Cotswold stone villages, and for the Thames which has its spring near Lechlade. More villages open as 'groups' for the National Garden Scheme in Oxfordshire than in any other county, and offer tea, hospitality, advice and delight with their infinite variety of gardens.

All this enjoyment benefits the excellent causes that the National Garden Scheme supports.

Left: Stonor Park

Volunteers

County Organiser
Marina Hamilton-Baillie
01367 710486
marina_hamilton_baillie@hotmail.com

County Treasurer
David White 01295 812679
davidwhite679@btinternet.com

Publicity
Priscilla Frost 01608 811818
info@oxconf.co.uk

Social Media
Lara Cowan
laracowan@icloud.com

Dr Jill Edwards 07971 201352
jill.edwards@ngs.org.uk

Booklet Co-ordinator
Position Vacant
For details please contact
Marina Hamilton-Baillie (as above)

Assistant County Organisers
Lynn Baldwin 01608 642754
elynnbaldwin@gmail.com

Ann Conibear 01805 373494
annconibear494@gmail.com

Lara Cowan (as above)

Dr David Edwards
07973 129473
david.edwards@ngs.org.uk

Dr Jill Edwards (as above)

Penny Guy 01865 862000
penny.theavon@virginmedia.com

Michael Hougham 01865 890020
mike@gmec.co.uk

Pat Hougham 01865 890020
pat@gmec.co.uk

Lyn Sanders
01865 739486
sandersc4@hotmail.com

 @NGSOxfordshire

 @ngs_oxfordshire

 @ngs_oxfordshire

OPENING DATES

All entries subject to change. For latest information check www.ngs.org.uk

Map locator numbers are shown to the right of each garden name.

February

Snowdrop Festival

Sunday 7th
NEW 23 Hid's Copse
Road   34
Stonehaven   64

March

Sunday 21st
◆ Waterperry Gardens 67

April

Sunday 4th
Ashbrook House   2

Monday 5th
Kencot Gardens   38

Sunday 11th
Buckland Lakes   13
NEW 38 Leckford
Road   41
Lime Close   42
Magdalen College   43

Saturday 17th
50 Plantation Road   58

Sunday 18th
50 Plantation Road   58

Saturday 24th
50 Plantation Road   58

Sunday 25th
Claridges Barn   17
Hollyhocks   35
The Old Vicarage,
Bledington   54
50 Plantation Road   58

May

Sunday 2nd
◆ Broughton Grange   11
Old Boars Hill Gardens 53

Wednesday 5th
Kingham Lodge   39

Sunday 16th
The Grove   32
NEW 38 Leckford Road   41
Meadow Cottage   45

Sunday 23rd
◆ Blenheim Palace   7

Thursday 27th
NEW Black Knap House   6

Sunday 30th
Barton Abbey   5
Bolters Farm   8
Broughton Poggs &
Filkins Gardens   12
NEW Kings Cottage   40
Lime Close   42

Monday 31st
Bolters Farm   8
Friars Court   26
NEW Kings Cottage   40

June

Sunday 6th
Failford   23
116 Oxford Road   57
Whitehill Farm   71

Tuesday 8th
NEW ◆ Stonor Park   65

Wednesday 9th
Claridges Barn   17

Thursday 10th
Wootton Gardens   73

Sunday 13th
Brize Norton Gardens   9
Claridges Barn   17
East Hagbourne
Gardens   22
Iffley Gardens   37

Sunday 20th
◆ Broughton Grange   11
Gothic House   27
Green and Gorgeous   29
Orchard House   56
The Priory Garden   59
Steeple Aston
Gardens   63
Wheatley Gardens   70

Sunday 27th
◆ Broughton Castle   10
Chalkhouse Green
Farm   14
Sibford Gardens   60
NEW West Oxford
Gardens   69

Wednesday 30th
Midsummer House   49

July

Sunday 4th
Cumnor Village
Gardens   18
Dorchester Gardens   21
Middleton Cheney
Gardens   48
The Old Vicarage,
Bledington   54
South Hayes   61

Wednesday 7th
Midsummer House   49

Saturday 10th
Headington Gardens   33

Sunday 18th
The Filberts   24

Sunday 25th
◆ Broughton Grange   11
Merton College Oxford
Fellows' Garden   47

August

Saturday 7th
16 Oakfield Road   52

Sunday 8th
16 Oakfield Road   52

Saturday 21st
Aston Pottery   3

Sunday 22nd
Aston Pottery   3

September

Saturday 4th
Bolters Farm   8
Christ Church Masters',
Pocock & Cathedral
Gardens   16
NEW Kings Cottage   40
Meadowmist   46

Sunday 5th
Ashbrook House   2

Bolters Farm   8
NEW Kings Cottage   40
Meadowmist   46

Thursday 9th
The Grange   28

Sunday 12th
◆ Broughton Grange   11

Sunday 19th
Old Whitehill Barn   55
◆ Waterperry Gardens   67

February 2022

Sunday 6th
NEW 23 Hid's Copse
Road   34
Stonehaven   64

Sunday 13th
Hollyhocks   35

By Arrangement

Arrange a personalised garden visit on a date to suit you. See individual garden entries for full details.

17 Abberbury Road,
Iffley Gardens   37
Appleton Dene   1
Ashbrook House   2
Bannisters   4
Bolters Farm   8
Carter's Yard, Sibford
Gardens   60
Chivel Farm   15
Claridges Barn   17
103 Dene Road   19
Denton House   20
Failford   23
Foxington   25
The Grange   28
Greenfield Farm   30
NEW Greyhound
House   31
Hollyhocks   35
Home Close   36
86 Hurst Rise Road,
West Oxford
Gardens   69
NEW 38 Leckford
Road   41
Lime Close   42
Manor House   44
Meadow Cottage   45
Mill Barn   50
Monks Head   51

"Unfortunately, cancer was not in lockdown in 2020. The continued support of our long-standing and valued partner, the National Garden Scheme is more important than ever." Macmillan

THE GARDENS

Gothic House

© Andrew Lawson

❶ APPLETON DENE

Yarnells Hill, Botley, Oxford, OX2 9BG. Mr & Mrs A Dawson, 07701 000977, annrobe@aol.com. *3m W of Oxford. Take W road out of Oxford, through Botley Rd, pass under A34, turn L into Westminster Way, Yarnells Hill 2nd on R, park at top of hill. Walk 200 metres. Disabled parking at the house.* Visits by arrangement May to Sept for groups of 5 to 30. Adm £4, chd free. Home-made teas. Wine.
Beautiful secluded garden set in a hidden valley. The ¼ acre garden on a steeply sloping site surrounds a mature tulip tree. There is a skilfully incorporated level lawn area overlooked by deep colour themed borders incl a wide variety of plants for long seasonal interest. Not suitable for wheelchairs as there is a steep slope and steps.

❷ ASHBROOK HOUSE

Blewbury, OX11 9QA. Mr & Mrs S A Barrett, 01235 850810, janembarrett@me.com. *4m SE of Didcot. Turn off A417 in Blewbury into Westbrook St. 1st house on R. Follow yellow signs for parking in Boham's Rd.* Sun 4 Apr, Sun 5 Sept (2-5.30). Adm £5, chd free. Home-made teas. Visits also by arrangement Apr to Sept.
The garden where Kenneth Grahame read Wind in the Willows to local children and where he took inspiration for his description of the oak doors to Badger's House. Come and see, you may catch a glimpse of Toad and friends in this 3½ acre chalk and water garden, in a beautiful spring line village. In spring the banks are a mass of daffodils and in late summer the borders are full of unusual plants.

❸ ASTON POTTERY

Aston, Bampton, OX18 2BT. Mr Stephen Baughan, www.astonpottery.co.uk. *4m S of Witney. On the B4449 between Bampton & Standlake.* Sat 21, Sun 22 Aug (12-5). Adm £5, chd free. Light refreshments in the café.
6 stunning borders set around Aston Pottery. 72 metre double hornbeam border full of riotous perennials. 80 metre long hot bank of Alestromeria, Salvias, Echinacea and Knifophia. Quadruple dahlia border with over 600 dahlias, grasses and asters. Tropical garden with bananas, cannas and Ricinus. Finally, 80 metres of 120 different annuals planted in 4 giant successive waves of over 6000 plants. Our Hornbeam Walk is planted to give a new experience to the knowing eye every month. If bees and butterflies are your thing, come between July and August for a chance to see our rare Black Bumble Bee 'Bombus ruderatus'.

❹ BANNISTERS

Middle Street, Islip, Kidlington, OX5 2SF. Wendy Price, 01865 375418. *2m E of Kidlington & approx 5m N of Oxford. From A34, exit Bletchingdon & Islip. B4027 direction Islip, turn L into Middle St, beside Great Barn.* Visits by arrangement July to mid-Sept for groups of up to 20. Visits can be combined with Hollyhocks
A hidden garden of perennials and grasses, naturalistic planting contrasted with trained fruit trees and shrubs. A contemporary interpretation of an old garden. Access over gravel paths and some shallow steps.

5 BARTON ABBEY

Steeple Barton, OX25 4QS. Mr & Mrs P Fleming. *8m E of Chipping Norton. On B4030, ½m from junction of A4260 & B4030.* **Sun 30 May (2-5). Adm £5, chd free. Home-made teas.**

15 acre garden with views from house (not open) across sweeping lawns and picturesque lake. Walled garden with colourful herbaceous borders, separated by established yew hedges and espalier fruit, contrasts with more informal woodland garden paths with vistas of specimen trees and meadows. Working glasshouses and fine display of fruit and vegetables.

6 NEW BLACK KNAP HOUSE

Priory Road, Heythrop, Chipping Norton, OX7 5TP. Mr Karl Devine. *Follow signs for Heythrop from the A44. We are 1m down the lane on the RH-side. Parking at Black Knap House.* **Evening opening Thur 27 May (6-8). Adm £20. Pre-booking essential, please visit www.ngs. org.uk for information & booking. Wine.**

A relatively new garden, commenced in 2010, on the site of an ancient quarry. When the garden began, it was no more than a collection of falling down kennels, barns and rough woodland sat on very uneven ground formed of spoil heaps from old stone excavations. The garden now comprises a walled kitchen garden, extensive prairie garden and formal lawn with rill.

7 ◆ BLENHEIM PALACE

Woodstock, OX20 1PX. His Grace the Duke of Marlborough, 01993 810530, customerservice@ blenheimpalace.com, www.blenheimpalace.com. *8m N of Oxford. The S3 bus runs every 30 mins from Oxford train station & Oxford's Gloucester Green bus station to Blenheim. Oxford Bus Company's 500 leaves from Oxford Parkway & stops at Blenheim.* **For NGS: Sun 23 May (10-4). Adm £5, chd free. Light refreshments. For other opening times and information, please phone, email or visit garden website.**

Blenheim Gardens, originally laid out by Henry Wise, incl the formal Water Terraces and Italian Garden by Achille Duchêne, Rose Garden, Arboretum, and Cascade. The Secret Garden offers a stunning garden paradise in all seasons. Blenheim Lake, created by Capability Brown and spanned by Vanburgh's Grand Bridge, is the focal point of over 2,000 acres of landscaped parkland. The Pleasure Gardens complex incl the Herb and Lavender Garden and Butterfly House. Other activities incl the Marlborough Maze, adventure play area, giant chess and draughts. Wheelchair access with some gravel paths, uneven terrain and steep slopes. Dogs allowed in park only.

8 BOLTERS FARM

Chilson, Pudlicote Lane, Chipping Norton, OX7 3HU. Robert & Amanda Cooper, art@amandacooper.co.uk. *Centre of Chilson village. On arrival in the hamlet of Chilson, heading N, we are the last in an old row of cottages on R. Please drive past & park considerately on the L in the lane.* **Sun 30, Mon 31 May, Sat 4, Sun 5 Sept (2-6). Combined adm with Kings Cottage £6, chd free. Light refreshments & gluten free options. Visits also by arrangement May to Sept for groups of 5 to 20. Donation to Hands Up Foundation.**

A cherished old cottage garden restored over the last 14 yrs. Tumbly moss covered walls and sloping lawns down to a stream with natural planting and character. Wheelchairs have to negotiate sloping deep gravel and numerous steps.

The Grange

GROUP OPENING

🞂 BRIZE NORTON GARDENS
Brize Norton, OX18 3LY.
www.bncommunity.org/ngs. *3m SW of Witney. Brize Norton Village, S of A40, between Witney & Burford. Parking at Elderbank Hall. Coaches welcome with plenty of parking nearby. Tickets & maps available at Elderbank Hall & at each garden.* Sun 13 June (1-6). Combined adm £5, chd free. Home-made teas in Elderbank Village Hall.

BARNSTABLE HOUSE
Mr & Mrs P Butcher, www.ourgarden.org.uk.

THE CHAPEL
Chris & Jayne Woodward.

CHURCH FARM HOUSE
Philip & Mary Holmes.

CLUMBER
Mr & Mrs S Hawkins.

MIJESHE
Mr & Mrs M Harper.

MILLSTONE
Bev & Phil Tyrell.

PAINSWICK HOUSE
Mr & Mrs T Gush.

ROSE COTTAGE
Brenda & Brian Trott.

95 STATION ROAD
Mr & Mrs P A Timms.

STONE COTTAGE
Mr & Mrs K Humphris.

Doomsday village on the edge of the Cotswold's offering a number of gardens open for your enjoyment. You can see a wide variety of planting incl ornamental trees and grasses, herbaceous borders, traditional fruit and vegetable gardens. Features incl a Mediterranean style patio, courtyard garden, terraced roof garden, water features; plus gardens where you can just sit, relax and enjoy the day. Plants for sale at individual gardens. A Flower Festival will take place in the Brize Norton St Britius Church. Partial wheelchair access to some gardens.
& 🐄 ✻ 🚗 ☕

🞂 BROUGHTON CASTLE
Banbury, OX15 5EB. Martin Fiennes, 01295 276070, info@broughtoncastle.com, www.broughtoncastle.com. *2½m SW of Banbury. On Shipston-on-Stour road (B4035).* For NGS: Sun 27 June (2-4.30). Adm £6, chd free. Home-made teas. For other opening times and information, please phone, email or visit garden website.
1 acre; shrubs, herbaceous borders, walled garden, roses, climbers seen against background of C14-C16 castle surrounded by moat in open parkland. House also open (additional charge).
& 🐄 🚗 ☕

🞂 BROUGHTON GRANGE
Wykham Lane, Broughton, Banbury, OX15 5DS. S Hester, www.broughtongrange.com. *¼m out of village. From Banbury take B4035 to Broughton. Turn L at Saye & Sele Arms Pub up Wykham Lane (one way). Follow road out of village for ¼m. Entrance on R.* For NGS: Suns 2 May; 20 June; 3 July; 12 Sept (10-5). Adm £9, chd free. Cream teas. For other opening times and information, please visit garden website.
An impressive 25 acres of gardens and light woodland in an attractive Oxfordshire setting. The centrepiece is a large terraced walled garden created by Tom Stuart-Smith in 2001. Vision has been used to blend the gardens into the countryside. Good early displays of bulbs followed by outstanding herbaceous planting in summer. Formal and informal areas combine to make this a special site incl newly laid arboretum with many ongoing projects. Terraces are accessible from side doors, but garden is on a slope with steps and gravel paths.
✻ 🚗 Ⓓ ☕

GROUP OPENING

🞂 BROUGHTON POGGS & FILKINS GARDENS
Lechlade, GL7 3JH. www.filkins.org.uk. *3m N of Lechlade. 5m S of Burford. Just off A361 between Burford & Lechlade on the B4477. Map of the gardens available.* Sun 30 May (2-6). Combined adm £6.50, chd free. Home-made teas in Filkins Village Hall. Ice creams at village shop.

BROUGHTON POGGS MILL
Charlie & Avril Payne.

3 THE COACH HOUSE
Peter & Brenda Berners-Price.

THE CORN BARN
Ms Alexis Thompson.

THE FIELD HOUSE
Peter & Sheila Gray.

FILKINS ALLOTMENTS
Filkins Allotments.

FILKINS HALL
Filkins Hall Residents.

LITTLE PEACOCKS
Colvin & Moggridge.

PEACOCK FARMHOUSE
Pauline & Peter Care.

PIGEON COTTAGE
Lynne Savege.

PIP COTTAGE
G B Woodin.

THE TALLOT
Ms M Swann & Mr D Stowell.

TAYLOR COTTAGE
Ronnie & Ian Bailey.

WELL COTTAGE
Christiaan Richards & Michelle Woodworth.

WILLOW COTTAGE
Emma Sparks.

14 gardens incl flourishing allotments in these beautiful and vibrant Cotswold stone twin villages. Scale and character vary from the grand landscape setting of Filkins Hall, to the small but action packed Pigeon Cottage, Taylor Cottage and The Tallot. Droughton Poggs Mill has a rushing mill stream with an exciting bridge. Pip Cottage combines topiary, box hedges and a fine rural view. In these and the other equally exciting and varied gardens horticultural interest abounds. Features incl plant stall by professional nursery, home-made teas, Swinford Museum of Cotswolds tools and artefacts, and Cotswold Woollen Weavers. Many gardens have gravel driveways, but most are suitable for wheelchair access. Most gardens welcome dogs on leads.
& 🐄 ✻ 🚗 ☕

The National Garden Scheme searches the length and breadth of England and Wales for the very best private gardens

🟦 BUCKLAND LAKES

nr Faringdon, SN7 8QW. The Wellesley Family. *3m NE of Faringdon. Buckland is midway between Oxford (14m) & Swindon (15m), just off the A420. Faringdon 3m, Witney 8m. Follow yellow NGS signs which will lead you to driveway & car park by St Mary's Church.* **Sun 11 Apr (2-5). Adm £5, chd free. Home-made teas at Memorial Hall.** Donation to RWMT (community bus).

Descend down wooded path to two large secluded lakes with views over undulating historic parkland, designed by Georgian landscape architect Richard Woods. Picturesque mid C18 rustic icehouse, cascade with iron footbridge, thatched boathouse and round house, and exedra. Many fine mature trees, drifts of spring bulbs and daffodils amongst shrubs. Norman church adjoins. Cotswold village. Children must be supervised due to large expanse of unfenced open water.

🐕 🚗 ☕ 💷

🟦 CHALKHOUSE GREEN FARM

Chalkhouse Green, Kidmore End, Reading, RG4 9AL. Mr & Mrs J Hall, www.chgfarm.com. *2m N of Reading, 5m SW of Henley-on-Thames. Situated between A4074 & B481. From Kidmore End take Chalkhouse Green Rd. Follow NGS yellow signs.* **Sun 27 June (2-6). Adm £4, chd free. Home-made teas.**

1 acre garden and open traditional farmstead. Herbaceous borders, herb garden, shrubs, old fashioned roses, trees incl medlar, quince and mulberries, walled ornamental kitchen garden and cherry orchard. Rare breed farm animals incl British White cattle, Suffolk Punch horses, donkeys, pigs, geese, chickens, ducks and turkeys. Plant and jam stall, donkey and pony rides, swimming in covered pool, grass tennis court, trailer rides, farm trail, WWII bomb shelter, heavy horse and bee display and live music. Partial wheelchair access.

♿ 🐕 ❀ ☕

🟦 CHIVEL FARM

Heythrop, OX7 5TR. Mr & Mrs J D Sword, 01608 683227, rosalind.sword@btinternet.com. *4m E of Chipping Norton. Off A361 or A44. Parking at Chivel Farm.* **Visits by arrangement Mar to Sept. Refreshments by request.**

Beautifully designed country garden with extensive views, designed for continuous interest that is always evolving. Colour schemed borders with many unusual trees, shrubs and herbaceous plants. Small formal white garden and a conservatory.

♿ 🚗 💷

🟦 CHRIST CHURCH MASTERS', POCOCK & CATHEDRAL GARDENS

St Aldate's, Oxford, OX1 1DP. Christ Church, www.chch.ox.ac.uk/gardens-and-meadows. *5 mins walk from Oxford city centre. Entry from St Aldate's through the Memorial Gardens, into Christ Church Meadow, then turn L into Masters' Garden gate after main visitor entrance. No parking available.* **Sat 4 Sept (10-4). Adm £5, chd free.**

Three walled gardens, not normally open to visitors, with herbaceous, shrub, Mediterranean and tropical borders. Incl the magnificent 'Jabberwocky' Tree, an Oriental Plane planted in the mid 1600s, as well as other links to Alice in Wonderland, St Frideswide and Harry Potter. Wheelchair access over gravel paths.

♿

🟦 CLARIDGES BARN

Charlbury Road, Chipping Norton, OX7 5XG. Drs David & Jill Edwards, 07973 129473, drdavidedwards@hotmail.co.uk. *3m SE of Chipping Norton. Take B4026 from Chipping Norton to Charlbury after 3m turn R to Dean, we are 200 metres on the R. Please park on the verge.* **Sun 25 Apr (11-4); Wed 9, Sun 13 June (11-5). Adm £4, chd free. Light refreshments. Visits also by arrangement Apr to Sept.**

3½ acres of family garden, wood and meadow hewn from a barley field on limestone brash. Situated on top of the Cotswolds, it is open to all weathers, but rewarding views and dog walking opportunities on hand. Large vegetable, fruit and cutting garden, wildlife pond and 5 cedar greenhouses, all loved by rabbits, deer and squirrel. Herbaceous borders and woodland gardens with gravel and flagged paths, divided by stone walls. Claridges Barn dates back to the 1600s, the cottage 1860s, converted about 30 yrs ago. Plants for sale. Mainly level site with flagstone and gravel paths, flat lawns and some uneven steps. The gravel driveway can be hard work for wheelchair users.

🐕 ❀ 🚗 ☕ 💷

GROUP OPENING

🟦 CUMNOR VILLAGE GARDENS

Leys Road, Cumnor, Oxford, OX2 9QF. *4m W of central Oxford. From A420, exit for Cumnor & follow B4017 into the village. Parking on road & side roads, behind PO, or behind village hall in Leys Rd.* **Sun 4 July (2-6). Combined adm £5, chd free. Home-made teas in United Reformed Church Hall, Leys Road.**

NEW 19 HIGH STREET
Janet Cross.

10 LEYS ROAD
Penny & Nick Bingham.

41 LEYS ROAD
Philip & Jennie Powell.

NEW 43 LEYS ROAD
Anna Stevens.

STONEHAVEN
Dr Dianne & Prof Keith Gull. (See separate entry)

The gardens feature a wide variety of plants, shrubs, trees, vegetables, fruit and wildflowers. Two of the gardens are cottage style with unusual perennials and many shrubs and trees; another has a Japanese influence exhibiting unusual plants many with black or bronze foliage and gravel areas; and two new gardens for 2021, one, a courtyard garden with extremes of light and shade and planting that cleverly reflects the conditions and is easy to maintain, the other garden has many different rooms and a seating area around a pond. Wheelchair access to 19 High Street and 41 Leys Road. WC facilities in United Reformed Church Hall.

♿ ❀ 💷

🟦 103 DENE ROAD

Headington, Oxford, OX3 7EQ. Steve & Mary Woolliams, 01865 764153, stevewoolliams@gmail.com. *S Headington, near Nuffield. Dene Rd accessed from The Slade from the N, or from Hollow Way from the S. Both access roads are B4495. Garden on sharp bend.* **Visits by arrangement Apr to Sept for groups of up to 10. Adm £3.50, chd free. Home-made teas & light refreshments.**

A surprising eco-friendly garden with

borrowed view over the Lye Valley Nature Reserve. Lawns, a wildflower meadow, pond and large kitchen garden are incl in a suburban 60ft x 120ft sloping garden. Fruit trees, soft fruit and mixed borders of shrubs, hardy perennials, grasses and bulbs, designed for seasonal colour. This garden has been noted for its wealth of wildlife incl a variety of birds, butterflies and other insects, incl the rare Brown Hairstreak butterfly, the rare Currant Clearwing moth and the Grizzled Skipper.

20 DENTON HOUSE
Denton, Oxford, OX44 9JF. Mr & Mrs Luke, 01865 874440, **waveney@jandwluke.com.** *In a valley between Garsington & Cuddesdon.* Visits by arrangement Apr to Sept for groups of up to 20. Adm £5, chd free. Home-made teas.
Large walled garden surrounds a Georgian mansion (not open) with shaded areas, walks, topiary and many interesting mature trees, large lawns, herbaceous borders and

rose beds. The windows in the wall were taken in 1864 from Brasenose College Chapel and Library. Wild garden and a further walled fruit garden.

GROUP OPENING

21 DORCHESTER GARDENS
Dorchester-On-Thames, Wallingford, OX10 7HZ. *Off A4074 or A415 signed to Dorchester. Parking at Old Bridge Meadow, at SE end of Dorchester Bridge. Disabled parking at 26 Manor Farm Road (OX10 7HZ).* **Sun 4 July (2-5). Combined adm £5, chd free. Home-made teas in downstairs tearoom at Abbey Guesthouse. Last orders at 4.30pm.**

26 MANOR FARM ROAD
David & Judy Parker.

6 MONKS CLOSE
Leif & Petronella Rasmussen.

7 ROTTEN ROW
Michael & Veronica Evans.

Three contrasting gardens in a historic village surrounding the medieval Abbey and the scene of many Midsomer Murders. 26 Manor Farm Road (OX10 7HZ) was part of an old larger garden, which now has a formal lawn and planting, vegetable garden and greenhouse. From the yew hedge down towards the River Thame, which often floods in winter, is an apple orchard underplanted with spring bulbs. 6 Monks Close (OX10 7JA) is idyllic and surprising. A small spring-fed stream and sloping lawn surrounded by naturalistic planting runs down to a monastic fish pond. Bridges over this deep pond lead to the River Thame with steep banks. 7 Rotten Row (OX10 7LJ) is Dorchester's lawless garden, a terrace with borders leads to a lovely geometric garden supervised by a statue of Hebe. Access is from the allotments. Due to deep water at 26 Manor Farm Road and 6 Monks Close children should be accompanied at all times. Wheelchair access to 26 Manor Farm Road, partial access to other gardens. No dogs, please.

Upper Green

GROUP OPENING

22 EAST HAGBOURNE GARDENS

East Hagbourne, OX11 9LN. *½ m S of Didcot. Enter village via B4016, or from A417 through West Hagbourne & Coscote. Cycle path 44.* Sun 13 June (2-5). Combined adm £5, chd free. Home-made teas.

BOTTOM BARN
Mr Colin & Dr Jean Millar.

BUCKELS
Felicity Topping.

THE GABLES
Sally & Bill Barksfield.

5 HIGGS CLOSE
Mrs Jenny Smith.

KINGFISHER HOUSE
Robert & Nicola Ainger.

LIME TREE COTTAGE
Jane & Robin Bell.

LITTLE THATCH
Mark Granger & Chris Beverley.

NEW **2 LOWER CROSS COTTAGES**
Allison Huckle.

YEW TREE COTTAGE
Sarah & Jonathan Beynon.

Pretty Domesday village, timber frame and local clunch stone houses. Open every 3 yrs. Follow main road or explore pretty alleyways to visit 9 gardens; new for 2021, 2 Lower Cross Cottages with gravel planting, ferns, shepherds hut and rockery; Bottom Barn, in the shadow of the C12 church with topiary and perennial planting around a converted C18 barn; Lime Tree Cottage with water plants, mature trees and secret garden; Buckels, a sculptured walled garden with cypresses, box, heritage fruit trees and herbs; 5 Higgs Close, a colourful garden with herbaceous borders, small pond and vegetable patch; Kingfisher House, a family garden with mixed herbaceous planting. Little Thatch, a cottage garden with a stream, pond planting, perennials, lavender beds and kitchen garden; The Gables, a garden full of colour with roses, perennials, lavender, box and trees; Yew Tree Cottage, a secluded garden to the front of a thatched cottage (not open) with roses, clematis and honeysuckle. Features incl access to wildflower meadow, plant sales and refreshments, along with a beautiful C12 church. The Fleur de Lys Pub is also open for food and drinks.

23 FAILFORD

118 Oxford Road, Abingdon, OX14 2AG. Miss R Aylward, 01235 523925, aylwardsdooz@hotmail.co.uk. *118 is on the LH-side of Oxford Rd when coming from Abingdon Town, or on the RH-side when approaching from the N. Entrance to this garden is via 116 Oxford Rd.* Sun 6 June (11-4). Combined adm with 116 Oxford Road £5, chd free. Home-made teas. **Visits also by arrangement June to Sept for groups of 10 to 30.**
This town garden is an extension of the home divided into rooms both formal and informal. Features incl: walkways through shaded areas, arches, a beach, grasses, fernery, roses, topiaries, acers, hostas and heucheras. Be inspired by the wide variety of planting, many unusual and quirky features, all within an area 570 sq ft. Partial wheelchair access due to gravel areas, narrow pathways and uneven surfaces.

24 THE FILBERTS

High Street, North Moreton, Didcot, OX11 9AT. Mr & Mrs Prescott. *About 2m from Didcot, signed from A4130 Didcot-Wallingford road. The Filberts is near The Bear at Home pub, on opp side of High St. Parking is on The Croft (Recreation Ground) accessed via Bear Lane.* Sun 18 July (2-5). Adm £5, chd free. Home-made teas in village hall.
1 acre garden used for teaching RHS courses demonstrating many different styles: formal colour themed garden with lily and fish ponds; island beds for old roses, grasses; large informal pond; colourful mixed borders; secluded Japanese area; over 100 varieties of clematis; formal parterre with roses; vegetable garden; fruit cage; orchard. Picnics possible on the Recreation Ground before or after visiting the garden. Access via ramp or grass slope through side gate to garden. Many narrow and gravel paths.

25 FOXINGTON

Britwell Salome, Watlington, OX49 5LG. Ms Mary Roadnight, 01491 612418, mary@foxington.co.uk. *At Red Lion Pub take turning to Britwell Hill. After 350yds turn into drive on L.* Visits by arrangement Apr to Sept for

groups of 10 to 20. Adm £6, chd free. Home-made teas. Special dietary options by prior request. Stunning views to the Chiltern Hills provide a wonderful setting for this impressive garden, remodelled in 2009. Patio, heather and gravel gardens enjoy this view, whilst the back and vegetable gardens are more enclosed. The relatively new planting is maturing well and the area around the house (not open) is full of colour. There is an orchard and a flock of white doves. Well behaved dogs are welcome, but they must be kept on a lead at all times as there are many wild animals in the garden, wildflower meadow, wood and neighbouring fields. Wheelchair access throughout the garden on level paths with no steps.

&. 🐐 🅳 ☕ 🎪

26 FRIARS COURT

Clanfield, OX18 2SU. Charles Willmer, www.friarscourt.com. *4m N of Faringdon. On A4095 Faringdon to Witney road. ½ m S of Clanfield.* Mon 31 May (2-5). Adm £4, chd free. Home-made teas.
Over 3 acres of formal and informal gardens with flower beds, borders and specimen trees, lie within the remaining arms of a C16 moat which partially surrounds the large C17 Cotswold stone house. Bridges span the moat with water lily filled ponds to the front whilst beyond the gardens is a woodland walk. A museum about Friars Court is located in the old Coach House. A level path goes around part of the gardens and the museum is accessed over gravel.

&. ☕

27 GOTHIC HOUSE

Charlbury, Oxford, OX7 3PP. Mr & Mrs Andrew Lawson. *In the centre of Charlbury, between church & The Bell Hotel.* Sun 20 June (2-5). Combined adm with The Priory Garden £5, chd free. Home-made teas.
⅓ acre walled garden, designed for sculpture display and colour association. Area of 14 planted squares replaces lawn. False perspective, pleached lime walk, trellis, terracotta containers. New wildlife pond. Special interest, a colony of 16 nest sites for swifts. Home-made teas in the Charlbury Community Centre (limited parking nearby). Partial wheelchair access.

28 THE GRANGE
1 Berrick Road, Chalgrove, OX44 7RQ. Mrs Vicky Farren, 01865 400883, vickyfarren@mac.com. *12m E of Oxford & 4m from Watlington, off B480. The entrance to The Grange is at the grass triangle between Berrick Rd & Monument Rd, by the pedestrian crossing. GPS is not reliable in the final 200yds.* Evening opening Thur 9 Sept (5.30-7.30). Adm £15. Pre-booking essential, please visit www.ngs.org.uk for information & booking. Prosecco & canapes. Visits also by arrangement June to Oct for groups of up to 30.
11 acres of garden including herbaceous borders and a field turned into a prairie with many grasses and a wildflower meadow. There is a lake with bridges and a planted island, a dried riverbed and a labyrinth. A brook runs through the garden with a further pond, arboretum, an old orchard and partly walled vegetable garden. There is deep water, steps and bridges, which may be slippery when wet. Partial wheelchair access on many grass paths.

♿ 🐄 😊 💤

29 GREEN AND GORGEOUS
Little Stoke, Wallingford, OX10 6AX. Rachel Siegfried, www.greenandgorgeousflowers.co.uk. *3m S of Wallingford. Off B4009 between N & S Stoke, follow single track road down to farm.* Sun 20 June (1-5). Adm £5, chd free. Cream teas.
6 acre working flower farm next to River Thames. Cut flowers (many unusual varieties) in large plots and polytunnels, planted with combination of annuals, bulbs, perennials, roses and shrubs, plus some herbs, vegetables and fruit to feed the workers! Flowers selected for scent, novelty, nostalgia and naturalistic style. Floristry demonstrations, PYO sweet peas, plant and craft stalls. Wheelchair access on short grass paths, with large concrete areas.

♿ 🐄 ✿ ☕ 💤

30 GREENFIELD FARM
Christmas Common, nr Watlington, OX49 5HG. Andrew & Jane Ingram, 01491 612434, andrew@andrewbingram.com. *4m from J5 M40, 7m from Henley. J5 M40, A40 towards Oxford for ½ m, turn L signed Christmas Common. ¾ m past Fox & Hounds Pub, turn L at Tree Barn sign.* Visits by arrangement June to Sept for groups of up to 30. Adm £4, chd free.
10 acre wildflower meadow surrounded by woodland, established 22 yrs ago under the Countryside Stewardship Scheme. Traditional Chiltern chalkland meadow in beautiful peaceful setting with 100 species of perennial wildflowers, grasses and 5 species of orchids. ½ m walk from parking area to meadow. Opportunity to return via typical Chiltern beechwood.

🐄 😊

31 NEW GREYHOUND HOUSE
The Street, Ewelme, Wallingford, OX10 6HU. Mrs Wendy Robertson, flowerfrond@gmail.com. *Greyhound House is on the main road at the school end of Ewelme.* Visits by arrangement June to Sept for groups of 10 to 30. Adm £5, chd free.
2 acre garden set in historic Chiltern village of Ewelme. Mixed herbaceous borders, ornamental grass walk, walled courtyard with dahlias and late summer flowering perennials, houseplant theatre, fernery, cutting garden, orchard and woodland. Partial wheelchair access by a long gravel drive and narrow paths leading to the walled area.

✿ ☕ 🪑

32 THE GROVE
North Street, Middle Barton, Chipping Norton, OX7 7BZ. Ivor & Barbara Hill. *7m E Chipping Norton. On B4030, 2m from junction A4260 & B4030, opp Cinnamon Stick restaurant. Parking in street.* Sun 16 May (1.30-5). Adm £4, chd free. Home-made teas.
Mature informal plantsman's ⅓ acre garden, planted for yr-round interest around C19 Cotswold stone cottage (not open). Numerous borders with wide variety of unusual shrubs, trees and hardy plants; several species weigela, syringe, viburnum and philadelphus. Pond area, well stocked greenhouse. Plant list and garden history available. Home-made preserves for sale. Wheelchair access to most of garden.

♿ ✿ ☕

GROUP OPENING

33 HEADINGTON GARDENS
Old Headington, Oxford, OX3 9BT. *2m E from centre of Oxford. After T-lights in the centre of Headington,* heading towards Oxford, take the 2nd turn on R into Osler Rd. Further gardens at end of road in Old Headington. Sat 10 July (2-6). Combined adm £6, chd free. Home-made teas at Ruskin College.

THE COACH HOUSE
David & Bryony Rowe.

MONCKTON COTTAGE
Julie Harrod & Peter McCarter, 01865 751471, petermccarter@msn.com. 🪑

40 OSLER ROAD
Nicholas & Pam Coote.

RUSKIN COLLEGE
Ruskin College, www.ruskincrinklecrankle.org/.

9 STOKE PLACE
Clive Hurst.

Headington Gardens represent an eclectic group of gardens extending in an old village with little lanes, high stone walls and mature trees. The gardens reflect many different styles and eras from the Grade II listed Crinkle Crankle wall in Ruskin College grounds to the exotic and Mediterranean, to high hedged formality with courtyards, cottages and modern gardens. Partial wheelchair access to most gardens due to gravel paths and steps.

♿ ✿ ☕ 🪑

34 NEW 23 HID'S COPSE ROAD
Oxford, OX2 9JJ. Kathy Eldridge. *W of central Oxford, halfway up Cumnor Hill. Turn into Hids Copse Rd from Cumnor Hill. We are the last house on the R before T-junction.* Sun 7 Feb (2-5). Combined adm with Stonehaven £4, chd free. 2022: Sun 6 Feb. Opening with West Oxford Gardens on Sun 27 June.
The garden surrounds a house built in the early 30s (not open) on a ½ acre plot. Designed to be wildlife friendly, it has a kitchen garden, two ponds with frogs and newts, a wildflower meadow and woodland areas under many trees. Small winding paths lead to areas to sit and contemplate. The planting is dependant on the area within the garden, but an especial treat are the snowdrops in early spring.

♿ 🐄 ☕ 💤

35 HOLLYHOCKS
North Street, Islip, Kidlington, OX5 2SQ. Avril Hughes, 01865 377104, ahollyhocks@btinternet.com. *3m NE of Kidlington. From A34, exit Bletchingdon & Islip. B4027 direction Islip, turn L into North St.* Sun 25 Apr (1.30-5.30). Adm £3.50, chd free. Home-made teas. Combined adm with Monks Head, Bletchingdon or Bannisters, Islip. 2022: Sun 13 Feb. Visits also by arrangement Feb to Sept for groups of up to 10.
Plantswoman's small Edwardian garden brimming with yr-round interest, especially planted to provide winter colour, scent and snowdrops. Divided into areas with bulbs, May tulips, herbaceous borders, roses, clematis, shade and woodland planting especially Trillium, Podophyllum and Arisaema, late summer salvias and annuals give colour. Large pots and troughs add seasonal interest and colour. Some steps to access the garden.

36 HOME CLOSE
29 Southend, Garsington, OX44 9DH. Mrs M Waud & Dr P Giangrande, 01865 361394, m.waud@btinternet.com. *3m SE of Oxford. N of B480, opp Garsington Manor.* Visits by arrangement Apr to Sept. Adm £4.50, chd free. Refreshments by prior request.
2 acre garden with listed house (not open), listed granary and 1 acre mixed tree plantation with fine views. Unusual trees and shrubs planted for yr-round effect. Terraces, stone walls and hedges divide the garden and the planting reflects a Mediterranean interest. Vegetable garden and orchard.

GROUP OPENING

37 IFFLEY GARDENS
Iffley, Oxford, OX4 4EF. *2m S of Oxford. Within Oxford's ring road, off A4158 Iffley road, from Magdalen Bridge to Littlemore r'about, to Iffley village. Map provided at each garden.* Sun 13 June (2-6). Combined adm £5, chd free. Home-made teas in village hall.

17 ABBERBURY ROAD
Mrs Julie Steele, 01865 712039, info@juliesummers.co.uk. Visits also by arrangement May to July.

25 ABBERBURY ROAD
Rob & Bridget Farrands.

86 CHURCH WAY
Helen Beinart & Alex Coren.

122 CHURCH WAY
Sir John & Lady Elliott.

THE MALT HOUSE
Helen Potts.

NEW 431 MEADOW LANE
Evelyn Sanderson.

NEW 4A TREE LANE
Pemma & Nick Spencer-Chapman.

Secluded old village with renowned Norman church, featured on cover of Pevsner's Oxon Guide. Visit 7 gardens ranging in variety and style from the large Malt House garden to mixed family gardens with shady borders and vegetables. Varied planting throughout the gardens including herbaceous borders, shade loving plants, roses, fine specimen trees and plants in terracing. Features incl water features, formal gardens, small lake, Thames riverbank and a new permaculture vegetable garden. Plant Sale at The Malt House. Wheelchair access to some gardens.

GROUP OPENING

38 KENCOT GARDENS
Kencot, Lechlade, GL7 3QT. *5m NE of Lechlade. E of A361 between Burford & Lechlade.* Mon 5 Apr (2-6). Combined adm £5, chd free. Light refreshments at village hall.

THE ALLOTMENTS
Amelia Carter Charity.

BELHAM HAYES
Mr Joseph Jones.

THE GARDENS
Mark & Jayne Hodds.

HILLVIEW HOUSE
Andrea Moss.

IVY NOOK
Gill & Wally Cox.

KENCOT HOUSE
Tim & Katie Gardner.

MANOR FARM
Henry & Kate Fyson.

WELL HOUSE
Janet & Richard Wheeler.

The 2021 Kencot Gardens group will consist of 7 gardens and the allotments. The Allotments, tended by 8 people, growing a range of vegetables, flowers and fruit. Ivy Nook, with spring flowers, shrubs, rockery, small pond, magnolia and fruit trees. Belham Hayes, mature cottage garden, mixed herbaceous borders, two old fruit trees. Emphasis on scent and colour coordination. The Gardens, 1/3 acre garden with two old apple trees and newly planted trees, old well and herbaceous beds. Plentiful spring bulbs. Manor Farm has a 2 acre walled garden, bulbs, wood anemones, fritillaries in mature orchards, pleached lime walk, 130yr old yew ball and Black Hamburg vine. Chickens and occasionally pigs and lambs. Kencot House, 2 acre garden, a haven for wildlife, gingko tree, shrubs, spring bulbs, clockhouse, summerhouse and a carved C13 archway. Well House, 1/3 acre garden with mature trees, hedges, wildlife pond, waterfall, small bog area, spring bulbs and rockeries. Hill View, 2 acres, lime tree drive, established trees, shrubs, perennial borders, daffodils and aconites. No wheelchair access to The Allotments, other gardens maybe difficult due to gravel and uneven paths.

---

39 KINGHAM LODGE
West End, Kingham, Chipping Norton, OX7 6YL. Christopher Stockwell, 01608 658226, info@sculptureatkinghamlodge.com, www.sculptureatkinghamlodge.com. *From Kingham Village, West St turns into West End at tree in middle of road, bear R & you will see black gates for Kingham Lodge immed on L. Follow signs for parking.* Wed 5 May (11-5.30). Adm by donation. Home-made lunches & teas.
Many ericaceous plants not normally seen in the Cotswolds grow on 5 acres of garden, planted over 2 decades. Big display of rhododendron, laburnum arch and azaleas. Formal 150 metre border, backed with trellis, shaded walks with multi-layered planting, an informal quarry pond, formal mirror pond, massive pergola, parterre and unique Islamic garden. Sculpture show from 1-10 May 2021. Islamic pavilion with fountains and rills. Disabled parking on gravel at entrance and level access to all areas of the garden.

40 NEW **KINGS COTTAGE**
Chilson, Pudlicote Lane, Chipping Norton, OX7 3HU. Mr Michael Anderson. *S end of Chilson village. Entering Chilson from the S, off the B4437 Charlbury to Burford road, we are the 1st house on the L. Please drive past & park considerately in centre of village.* Sun 30, Mon 31 May, Sat 4, Sun 5 Sept (2-6). Combined adm with Bolters Farm £6, chd free. Light refreshments. An old row of cottages with mature trees and yew hedging. Over the last 5 yrs a new design and planting scheme has been started to reduce the areas of lawn, bring in planting to complement the house (not open) and setting, introduce new borders and encourage wildlife. A work in progress. Partial wheelchair access via gravel drive. Moderate slopes, grass and some sections of garden only accessible via steps.

41 NEW **38 LECKFORD ROAD**
Oxford, OX2 6HY. Dinah Adams, 01865 511996, dinah_zwanenberg@fastmail.com. *Central Oxford. North on Woodstock Rd take 3rd L. Coming into Oxford on Woodstock Rd, 1st R after Farndon Rd. Some 2hr parking nearby.* Sun 11 Apr, Sun 16 May (12-4). Adm £4, chd free. Light refreshments in garden layg(u). **Visits also by arrangement Mar to Oct for groups of 5 to 10.**
Behind the rather severe facade of a Victorian town house (not open) is a very protected long walled garden with mature trees. The planting reflects the varied levels of shade and incl rare and unusual plants. The trees in the front garden deserve attention. The back garden is divided into three distinct parts, each with a very different character. Amongst other things there is a hornbeam roof.

42 **LIME CLOSE**
35 Henleys Lane, Drayton, Abingdon, OX14 4HU. M C de Laubarede, mail@mclgardendesign.com, www.mclgardendesign.com. *2m S of Abingdon. Henleys Lane is off main road through Drayton. Please Note: When visiting Lime Close, please respect local residents & park considerately.* Sun 11 Apr, Sun 30 May (2-5.30). Adm £5, chd free. Cream teas. **Visits also by**

arrangement Feb to end of June & Sept to Nov only for groups of 10+. Donation to CLIC Sargent Care for Children.
4 acre mature plantsman's garden with rare trees, shrubs, roses and bulbs. Mixed borders, raised beds, pergola, topiary and shade borders. Herb garden by Rosemary Verey. Listed C16 house (not open). Cottage garden by MCL Garden Design, planted for colour, an iris garden with 100 varieties of tall bearded irises. Winter bulbs. New arboretum with rare and exotic trees from Asia and USA.

43 **MAGDALEN COLLEGE**
Oxford, OX1 4AU. Magdalen College, www.magd.ox.ac.uk. *Entrance In High St.* Sun 11 Apr (1-6). Adm £7, chd £6 (chd under 7 yrs free). Light refreshments in the Old Kitchen.
60 acres incl deer park, college lawns, numerous trees 150-200 yrs old; notable herbaceous and shrub plantings. Magdalen meadow where purple and white snake's head fritillaries can be found is surrounded by Addison's Walk, a tree lined circuit by the River Cherwell developed since the late C18. Ancient herd of 60 deer. Press bell at the lodge for porter to provide wheelchair access.

44 **MANOR HOUSE**
Manor Farm Road, Dorchester-on-Thames, OX10 7HZ. Simon & Margaret Broadbent, 01865 340101, manor@dotoxon.uk. *8m SSE of Oxford. Disabled parking at house.* Visits by arrangement July & Sept only for groups of 10+.
2 acre garden in beautiful setting around Georgian house (not open) and medieval abbey. Spacious lawn leading to riverside copse of towering poplars with fine views of Dorchester Abbey. Terrace with rose and vine covered pergola around lily pond. Colourful herbaceous borders, small orchard and vegetable garden. Wheelchair access on gravel paths.

45 **MEADOW COTTAGE**
Christmas Common, Watlington, OX49 5HR. Mrs Zelda Kent-Lemon, 01491 613779, zelda_kl@hotmail.com. *1m from Watlington. Coming from*

Oxford M40 to J6. Turn R & go to Watlington. Turn L up Hill Rd to top. Turn R after 50yds, turn L into field. Sun 16 May (11.30-5.30). Adm £5, chd free. Home-made teas. **Visits also by arrangement Feb to Sept.**
1¾ acre garden adjoining ancient bluebell woods, created by the owner from 1995 onwards. Many areas to explore including a professionally designed vegetable garden, large composting areas, wildflower garden and pond, old and new fruit trees, many shrubs, much varied hedging and a tall treehouse which children can climb under supervision. Indigenous trees and C17 barn (not open). For visits by arrangement come and see the wonderful snowdrops in Feb and during the month of May visit the bluebell woodland. Partial wheelchair access over gravel driveway and lawns.

46 **MEADOWMIST**
46 Millwood End, Long Hanborough, Witney, OX29 8BY. Andrea McDowell. *4m NE of Witney. The garden is on a no through road, with some parking at the side of the house.* Sat 4, Sun 5 Sept (10-5). Adm £5, chd £1. Light refreshments.
A cottage garden with an emphasis on family entertaining. We have raised beds for cut flowers in the front garden and a hot tub, sun deck and trampoline in the back garden, all surrounded by cottage garden plants and over 80 dahlias. Please bring cash for a cup of fresh lemonade, sorry, no change available.

47 **MERTON COLLEGE OXFORD FELLOWS' GARDEN**
Merton Street, Oxford, OX1 4JD. Merton College, 01865 276310. *Merton St runs parallel to High St about halfway down.* Sun 25 July (10-5). Adm £6, chd £6.
Ancient mulberry, said to have associations with James I. Specimen trees, long mixed border, established herbaceous bed. View of Christ Church meadow.

GROUP OPENING

48 MIDDLETON CHENEY GARDENS

Middleton Cheney, Banbury, OX17 2ST. *3m E of Banbury. From M40 J11 follow A422 signed Middleton Cheney. Parking at nursery school car park, next to 19 Glovers Lane, OX17 2NU. Map available at all gardens.* Sun 4 July (1-6). Combined adm £5, chd free. Home-made teas at Peartree House. Picnics welcome at Springfield House.

CHURCH COTTAGE
David & Sue Thompson.

CROFT HOUSE
Richard & Sandy Walmsley.

THE GABLES
Adam & Wanda Teeuw.

19 GLOVERS LANE
Michael Donohoe & Jane Rixon.

38 MIDWAY
Margaret & David Finch.

PEARTREE HOUSE
Roger & Barbara Charlesworth.

14 QUEEN STREET
Brian & Kathy Goodey.

SPRINGFIELD HOUSE
Lynn & Paul Taylor.

Large village with C13 church with renowned William Morris stained glass. 8 open gardens with a variety of sizes, styles and maturity. Of the smaller gardens, one contemporary garden contrasts formal features with colour-filled beds, borders and exotic plants. Another modern garden has flowing curves that create an elegant, serene feeling. One features cottage and Mediterranean planting with summerhouse and statuary. A mature small front and back garden is planted profusely with a feel of an intimate haven. A garden that has evolved through family use, features rooms and dense planting. A newly planted garden features a mixture of courtyards, pergola, borders and a stunning view of the church. One of the larger gardens continues with the renovation of a long lost garden, along with restoring areas of orchard, beds and borders. Another has an air of mystery with hidden corners and an extensive water feature weaving its way throughout the garden. Wheelchair access to 5 of the 8 gardens.

 ♿ ✿ 🐾 ☕ 🪑

49 MIDSUMMER HOUSE

Woolstone, Faringdon, SN7 7QL. Penny Spink. *7m W & 7m S of Faringdon. Woolstone is a small village off B4507, below Uffington White Horse Hill. Take road towards Uffington from the White Horse Pub.* Wed 30 June, Wed 7 July (2-6). Adm £4, chd free. Home-made teas.
On moving to Midsummer House 6 yrs ago, the owners created the garden using herbaceous plants brought with them from their previous home at Woolstone Mill House (not open this yr). Herbaceous border, parterre with new topiary, and espaliered Malus Everest. The garden designed by owner's son Justin Spink, a renowned garden designer and landscape architect. Picnics welcome in the field opposite. Wheelchair access over short gravel drive at entrance.

 ♿ 🚗 ☕

50 MILL BARN

25 Mill Lane, Chalgrove, OX44 7SL. Pat Hougham, 01865 890020, pat@gmec.co.uk. *12m E of Oxford. Chalgrove is 4m from Watlington off B480. Mill Barn is in Mill Lane, W of Chalgrove, 300yds S of Lamb Pub.* Visits by arrangement May to Sept for groups of 5+. Adm £4, chd free. Combined visit with The Manor garden £6. Home-made teas. Wine & canapés for evening visits.
Mill Barn has an informal cottage garden with a variety of flowers throughout the seasons. Rose arches and a pergola lead to a vegetable plot surrounded by a cordon of fruit trees, all set in a mill stream landscape. The Manor garden has a lake and wildlife areas, mixed shrubs and herbaceous beds that surround the C15 Grade 1 listed Manor House (not open).

 ♿ 🐾 ☕

51 MONKS HEAD

Weston Road, Bletchingdon, OX5 3DH. Sue Bedwell, 01869 350155, bedwell615@btinternet.com. *Approx 4m N of Kidlington. From A34 take B4027 to Bletchingdon, turn R at Xrds into Weston Rd.* Visits by arrangement Jan to Nov. Home-made teas.
Plantaholics' garden for all yr interest. Bulb frame and alpine area, greenhouse. Changes evolving all the time.

 ✿ 🚗 🐾 ☕

52 16 OAKFIELD ROAD

Carterton, OX18 3QN. Karen & Jason. *Off A40 Oxford W bound Carterton & Brize Norton junction. Go past RAF Brize Norton main gate, head for town center. Through T-lights, take 3rd L Foxcroft Dr. Follow road around to T-junction, then turn R into Oakfield Rd.* Sat 7, Sun 8 Aug (12.30-5). Adm £4, chd free. Cold refreshments.
A small tropical inspired garden that incorporates good use of the small space. Overflowing with palms, bamboo, cannas, tree ferns, hedychiums, tetrapanax papyrifera 'rex' and colocasia. Small walkway leading to a decked area with hot tub and thatched gazebo with its own Tiki bar. Also small pond, lawn, patio area and greenhouse, in an area of 10 x 12 metres. Partial wheelchair access on patio area only.

 ♿ ☕

GROUP OPENING

53 OLD BOARS HILL GARDENS

Jarn Way, Boars Hill, Oxford, OX1 5JF. *3m S of Oxford. From S ring road towards A34 at r'about follow signs to Wootton & Boars Hill. Up Hinksey Hill take R fork. 1m R into Berkley Rd to Old Boars Hill.* Sun 2 May (1.30-5.30). Combined adm £6, chd free. Home-made teas at Uplands.

BLACKTHORN
Louise Edwards.

HEDDERLY HOUSE
Mrs Julia Bennett.

HOLLY TREE HOUSE
Jillian Morrow.

TALL TREES
David Clark.

UPLANDS
Lyn Sanders, 01865 739486, sandersc4@hotmail.com.
Visits also by arrangement Mar to Oct for groups of 5+.

Five delightful gardens in a conservation area. Hedderly House has a 3 acre terraced hillside garden with wooded walks and ponds with extensive views over the Vale of the White Horse. Blackthorn is a large and sloping garden with paths wandering through trees and lovely borders. Uplands is a southerly facing cottage garden with an extensive range of plants for colour for every season and a new formal area. Tall

Monks Head

Trees is providing the plant sale surrounded by pots, rhododendrons and gravel gardens. Holly Tree House is a glorious garden with so many surprises and wonderful planting.

🚶 🐐 ✳ 🚗 ☕

54 THE OLD VICARAGE, BLEDINGTON

Main Road, Bledington, Chipping Norton, OX7 6UX. Sue & Tony Windsor, 01608 658525, tony.g.windsor@gmail.com. *6m SW of Chipping Norton. 4m SE of Stow-on-the-Wold. On the main street B4450 through Bledington. Not next to church.* **Sun 25 Apr (11-5); Sun 4 July (2-6). Adm £4, chd free. Visits also by arrangement May to July.** 1½ acre garden around an early Victorian vicarage (1843) not open. Borders and beds filled with spring bulbs, hardy perennials, shrubs and trees. Informal rose garden with over 300 David Austin roses. Small pond and vegetable garden. Paddock with trees, shrubs and herbaceous border. Planted for yr-round interest. Gravel driveway and gently sloped garden can be hard work for wheelchair users.

🚶 🐐 ✳ 🚗 ☕

55 OLD WHITEHILL BARN

Old Whitehill, Tackley, Kidlington, OX5 3AB. Gill & Paul Withers. *10m N of Oxford. 3m from Woodstock. Hamlet ¾m S of Tackley. Accessed from either A4260 & A4095.* **Sun 19 Sept (2-5). Adm £4, chd free. Home-made teas.** 1 acre country garden on a sloping site around a stone barn conversion. Created by the owners from a farmyard and surrounding field over last 20 yrs. Sunny walled courtyard. Colour themed borders. Field of formal and informal areas, mature hedging, orchard, meadow grass and enclosed vegetable garden.

✳ ☕

56 ORCHARD HOUSE

Asthall, Burford, OX18 4HH. Dr Elizabeth Maitreyi, 07939 111605, oneconsciousbreath@gmail.com. *3m E of Burford. 1st house on R as you come downhill into Asthall from the A40.* **Sun 20 June (2-6). Adm £5, chd free. Home-made teas. Visits also by arrangement Apr to June for groups of 5 to 30. Refreshments by request. Donation to MS Society.** A 5 acre garden still in the making. Formal borders and courtyard.

New paths and hedges created in 2018 are developing adding more character and planting opportunities. The whole design is now visible. We are creating an English garden finely balanced between the formal and the wild. Beautiful sculptures add to the planting and there is woodland to walk in beyond. Rare breed piglets also to be seen. Plants for sale and picnics welcome. Dogs on a lead only. Sadly, no WC. Wheelchair access with small step onto drive and one step down to garden.

🚶 🐐 ✳ 🛏 ☕

There are brilliant plant sales at many gardens. Look out for the symbol in the garden description – and don't forget to bring a bag to carry your plants home in

57 116 OXFORD ROAD
Abingdon, OX14 2AG. Mr &
Mrs P Aylward, 01235 523925,
aylwardsdooz@hotmail.co.uk.
*116 Oxford Rd is on the RH-side
if coming from A34 N exit, or on
the LH-side after Picklers Hill turn
if approaching from Abingdon
town centre.* **Sun 6 June (11-4).
Combined adm with Failford £5,
chd free. Home-made teas. Visits
also by arrangement June to Sept
for groups of up to 20.**
This young town garden brings
alive the imagination of its creators.
It will inspire both the enthusiast
and the beginner. The use of
recycled materials, lots of colour
and architectural plants. A folly/
greenhouse is just one of many quirky
features. Raised beds, rockeries,
specimen plants, beds of roses
and hostas. Pushes the boundaries
of conventional gardening. There
is something for everyone! Partial
wheelchair access by gravel driveway.
Uneven surfaces and paths.
✿ ☕ 🍴

58 50 PLANTATION ROAD
Oxford, OX2 6JE. Philippa
Scoones. *Central Oxford. N on
Woodstock Rd take 2nd L. Coming
into Oxford on Woodstock Rd turn
R after Leckford Rd. Best to park on
Leckford Rd. No disabled parking
near house.* **Sat 17, Sun 18, Sat 24,
Sun 25 Apr (2-5). Adm £3.50, chd
free. Home-made teas.**
Surprisingly spacious city garden
designed in specific sections.
N-facing front garden, side alley filled
with shade loving climbers. S-facing
rear garden with hundreds of tulips
in spring, unusual trees incl Mount
Etna Broom, conservatory, terraced
area and secluded water garden with
water feature, woodland plants and
alpines. Good design ideas for small
town garden and 100s of pots that
add to the overall atmosphere.
♿ ✿ ☕

59 THE PRIORY GARDEN
Church Lane, Charlbury, OX7 3PX.
Dr D El Kabir & Colleagues. *6m SE
of Chipping Norton. Large Cotswold*

*village on B4022 Witney-Enstone Rd,
near St Mary's Church.* **Sun 20 June
(2-5). Combined adm with Gothic
House £5, chd free. Home-made
teas.**
1½ acre of formal terraced topiary
gardens with Italianate features.
Foliage colour schemes, shrubs,
parterres with fragrant plants, old
roses, water features, sculpture and
inscriptions aim to produce a poetic,
wistful atmosphere. Formal vegetable
and herb garden. Arboretum of over
3 acres borders the River Evenlode
and incl wildlife garden and pond.
Home-made teas in the Charlbury
Community Centre (limited parking
nearby). Partial wheelchair access.
♿ 🐕 ☕

GROUP OPENING

60 SIBFORD GARDENS
Sibford Ferris, OX15 5RE. *7m W
of Banbury. Near the Warwickshire
border, S of B4035, in centre of
Sibford Ferris village at T-junction
& additional gardens near Wykham*

Chivel Farm

Arms Pub in Sibford Gower. **Sun 27 June (2-6). Combined adm £7, chd free. Home-made teas at Sibford Gower Village Hall (opp the church).**

BUTTSLADE HOUSE
James & Sarah Garstin.

CARTER'S YARD
Sue & Malcolm Bannister, 01295 780365, sebannister@gmail.com. **Visits also by arrangement May to Sept for groups of 10 to 30.**

NEW **HOLMBY HOUSE**
John & Sally Wass.

HOME CLOSE
Graham & Carolyn White.

NEW **THE LONG HOUSE**
Jan & Diana Thompson.

In two charming small villages of Sibford Gower and Sibford Ferris, off the beaten track with thatched stone cottages, five contrasting gardens ranging from an early C20 Arts and Crafts house (not open) and garden, to varied cottage gardens bursting with bloom, interesting planting and some unusual plants. No wheelchair access to Carter's Yard, partial access to other gardens.

61 SOUTH HAYES
Yarnells Hill, Oxford, OX2 9BG. **Mark & Louise Golding.** *2m W of Oxford. Take Botley Rd heading W out of Oxford, pass under A34, turn L onto Westminster Way, Yarnells Hill is 2nd road on R. Park at top of hill & walk 100 metres down the lane.* **Sun 4 July (2-6). Adm £4, chd free. Home-made teas.**

One acre garden transformed to make the most of a steeply sloping site. Formal planting close to the house (not open) and a raised decking area which overlooks the garden. Gravel pathways are retained by oak sleepers and pass through swathes of hardy perennials and grasses, leading to the lower garden with natural wildlife pond and fruit trees. Garden once owned by the late Primrose Warburg, an Oxford galanthophile. Not suitable for wheelchairs due to steep slope and gravel filled steps.

62 64 SPRING ROAD
Abingdon, OX14 1AN. **Janet Boulton, 01235 524514, j.boulton89@btinternet.com, www.janetboulton.co.uk.** *S Abingdon from A34 take L turn after police station into Spring Rd. Minute's drive to number 64 on L.* **Visits by arrangement May to Sept for groups of up to 10. Adm £5, chd free.**
A very special small but unique artist's garden (4½ x 30½ metres) behind a Victorian terrace house (not open). Predominantly green with numerous inscribed sculptures relating to art, history and the human spirit. Inspired by gardens the owner has painted, especially Little Sparta in Scotland. Visitors are invited to watch a film about this celebrated garden before walking around the garden itself.

GROUP OPENING

63 STEEPLE ASTON GARDENS
Steeple Aston, OX25 4SP. *11m N of Oxford, 9m S of Banbury. ½m E of A4260.* **Sun 20 June (2-6). Combined adm £6, chd free. Home-made teas in village hall.**

ACACIA COTTAGE
Jane & David Stewart.

CANTERBURY HOUSE
Peter & Harriet Higgins.

CEDAR COTTAGE
Robert Scott.

COMBE PYNE
Chris & Sally Cooper.

KRALINGEN
Mr & Mrs Roderick Nicholson.

THE LONGBYRE
Mr & Mrs V Billings.

PRIMROSE GARDENS
Richard & Daphne Preston, 01869 340512, richard.preston5@btopenworld.com. **Visits also by arrangement Apr to July for groups of 10+. Can provide brief history of garden dating back to 1890s.**

Steeple Aston, often considered the most easterly of the Cotswold villages, is a beautiful stone built village with gardens that provide a wide range of interest. A stream meanders down the hill as the landscape changes from sand to clay. The 7 open gardens incl small floriferous cottage gardens, large landscaped gardens, natural woodland

areas, ponds and bog gardens, and themed borders. Primrose Gardens, Longbyre and Canterbury House have gravel and Acacia Cottage also has some steps. Kralingen has slopes to stream and bog garden.

64 STONEHAVEN
6 High Street, Cumnor, Oxford, OX2 9PE. **Dr Dianne & Prof Keith Gull.** *4m from central Oxford. Exit to Cumnor from the A420. In centre of village opp PO. Parking at back of PO.* **Sun 7 Feb (2-5). Combined adm with 23 Hid's Copse Road £4, chd free. 2022: Sun 6 Feb. Opening with Cumnor Village Gardens on Sun 4 July.**
Front, side and rear garden of a thatched cottage (not open). Front is partly gravelled and side courtyard has many pots. Rear garden overlooks meadows with old apple trees underplanted with ferns, wildlife pond, unusual plants, many with black or bronze foliage, planted in drifts and repeated throughout the garden. Planting has mild Japanese influence; rounded, clipped shapes interspersed with verticals. Snowdrops feature in Feb. There are two pubs in the village serving food; The Bear & Ragged Staff and The Vine. Wheelchair access to garden via gravel drive.

65 NEW ✦ **STONOR PARK**
Stonor, Henley-On-Thames, RG9 6HF. **Lady Ailsa Stonor, 01491 638587, administrator@stonor.com, www.stonor.com.** *Stonor is located between the M4 (J8/J9) & the M40 (J6) on the B480 Henley-on-Thames to Watlington road. If you are approaching Stonor on the M40 from the E, please exit at J6 only.* **For NGS: Tue 8 June (10-5). Adm £6.50, chd free. Light refreshments. For other opening times and information, please phone, email or visit garden website.**
Surrounded by dramatic sweeping valleys and nestled within an ancient deer park, you will find the gardens at Stonor, which date back to Medieval times. Visitors love the serenity of the our C17 walled, Italianate Pleasure Garden and herbaceous perennial borders beyond.

66 UPPER GREEN
Brill Road, Horton cum Studley, Oxford, OX33 1BU. Susan & Peter Burge, 01865 351310, sue.burge@ndm.ox.ac.uk, www.uppergreengarden.co.uk. *6½ m NE of Oxford. Enter village, turn R up Horton Hill. At T-junction turn L into Brill Rd. Upper Green 250yds on R, 2 gates before pillar box. Roadside parking.* Visits by arrangement Feb to Sept for groups of up to 30.
Mature ½ acre wildlife friendly garden packed with interest and colour throughout the yr. Over 1500 plants in garden database. Wildlife friendly. Herbaceous borders, gravel bed, rock bed, ferns, hot border, potager, bog and pond. Snowdrop collection. Alpines. Apple trees support rambling roses. Spectacular views. Great compost! Sorry, no dogs. Gravel drive limits wheelchair access.

67 ◆ WATERPERRY GARDENS
Waterperry, Wheatley, OX33 1JZ. School of Economic Science, 01844 339226, office@waterperrygardens.co.uk, www.waterperrygardens.co.uk. *7½ m from Oxford city centre. From E M40 J8, from N M40 J8a. Follow brown tourist signs. For SatNav please use OX33 1LA.* For NGS: Sun 21 Mar (10-5); Sun 19 Sept (10-5.30). Adm £8.95, chd free. Light refreshments in the teashop (10-5). For other opening times and information, please phone, email or visit garden website.
Waterperry Gardens are extensive, well-maintained and full of interesting plants. From The Virgin's Walk with its shade-loving plants to the long classical herbaceous border, brilliantly colourful from late May until October. The Mary Rose Garden illustrates modern and older roses, and the formal garden is neatly designed and colourful with a small knot garden, herb border and wisteria tunnel. Newly redesigned walled garden, river walk, statues and pear orchard. Riverside walk may be inaccessible to wheelchair users if very wet.
ᵭ ✿ ⊕ NPC ☕

68 WAYSIDE
82 Banbury Road, Kidlington, OX5 2BX. Margaret & Alistair Urquhart, 01865 460180, alistairurquhart@ntlworld.com. *5m N of Oxford. On R of A4260 travelling N through Kidlington.* Visits by arrangement May to Sept for groups of up to 20. Adm £3, chd free. Tea.
¼ acre garden shaded by mature trees. Mixed border with some rare and unusual plants and shrubs. A climber clothed pergola leads past a dry gravel garden to the woodland garden with an extensive collection of hardy ferns. Conservatory and large fern house with a collection of unusual species of tree ferns and tender exotics. Important collection of hardy garden ferns. Partial wheelchair access.
ᵭ ✿ ☕

GROUP OPENING

69 NEW WEST OXFORD GARDENS
Cumnor Hill, Oxford, OX2 9HH. *Take Botley interchange off A34 from N or S. Follow signs for Oxford & then turn R at Botley T-lights, opp MacDonalds & follow NGS yellow signs. Street parking.* Sun 27 June (2-6). Combined adm £5, chd free. Home-made teas at 10 Eynsham Road.

NEW **10 EYNSHAM ROAD**
Jon Harker.

NEW **23 HID'S COPSE ROAD**
Kathy Eldridge.
(See separate entry)

86 HURST RISE ROAD
Ms P Guy & Mr L Harris, penny.theavon@virginmedia.com. Visits also by arrangement May to July for groups of 10 to 20. Refreshments by prior request.

NEW **6 SCHOLAR PLACE**
Helen Ward.

Four gardens opening in this new group for 2021. 86 Hurst Rise Road (OX2 9HH), a small garden abounding in perennials, roses, clematis and shrubs displayed in layers and at different levels around a circular lawn and path with sculptural features. Delightful colour. 10 Eynsham Road Botley (OX2 9BP), a New Zealander's take on an English-style garden including many rose varieties, white garden, herbaceous borders and pond. 23 Hid's Copse Road (OX2 9JJ), a ½ acre, wildlife friendly plot under many trees with two ponds, a wildflower meadow, kitchen garden and contemplative areas. 6 Scholar Place (OX2 9RD), an urban garden on a gently upward slope with a selection of small trees, shrubs and perennials on the top. Steps up lead to a small pond. The side bordering a public footpath has a wilder feel. Gravel steps at 6 Scholar Place and pebble path at 86 Hurst Rise Road.
ᵭ ☕

GROUP OPENING

70 WHEATLEY GARDENS
High Street, Wheatley, OX33 1XX. 07813 339480, echess@hotmail.co.uk. *5m E of Oxford. Leave A40 at Wheatley, turn into High St. Gardens at W end of High St, S side.* Sun 20 June (2-6). Combined adm £4, chd free. Cream teas at The Manor House. Visits also by arrangement Mar to Sept for groups of up to 30.

BREACH HOUSE GARDEN
Liz Parry.

THE MANOR HOUSE
Mrs Elizabeth Hess.

THE STUDIO
Ann Buckingham.

Three adjoining gardens in historic Wheatley are: Breach House Garden with many shrubs, perennials, a contemporary reflective space and a wild meadow with ponds; The Studio with walled garden and climbing roses, clematis, herbaceous borders, vegetables and fruit trees; The Elizabethan Manor House (not open), a romantic oasis with formal box walk, herb garden, rose arches and old rose shrubbery. Various musical events. Wheelchair access with assistance due to gravel paths, two shallow steps and grass.
ᵭ 🐐 ✿ ☕

71 WHITEHILL FARM
Widford, Burford, OX18 4DT. Mr & Mrs Paul Youngson, 01993 822894, anneyoungson@btinternet.com. *1m E of Burford. From A40 take road signed Widford. Turn R at bottom of hill, 1st house on R with ample car parking.* Sun 6 June (2-6). Adm £4, chd free. Home-made teas. Visits also by arrangement May to Aug. Refreshments for groups of 10+.
2 acres of hillside gardens and woodland with spectacular views overlooking Burford and Windrush valley. Informal plantsman's garden built up by the owners over 25 yrs. Herbaceous and shrub borders,

Broughton Poggs & Filkins Gardens

ponds and bog area, old fashioned roses, ground cover, ornamental grasses, bamboos and hardy geraniums. Large cascade water feature, pretty tea patio and wonderful Cotswold views.

🐕 ✿ ☕ 🪑

72 WHITSUN MEADOWS
Berkeley Road, Boars Hill, Oxford, OX1 5ET. Jane & Nigel Jones, 07500 722722, mail@jonesoxford.co.uk. *3m SW of Oxford. From S ring road towards A34 at r'about follow signs to Wootton & Boars Hill. Up Hinksey Hill take R fork. 1m R into Berkley Rd to Old Boars Hill.* Visits by arrangement May to Sept for groups of 10 to 30.
The garden at Whitsun Meadows has a magical backdrop with a mixture of mature Scots pine, English oak, acer and cherry. Against this the owners have developed an interesting mixture of herbaceous borders, hosta beds, gravel gardens and a wildflower meadow defined by a curving cleft chestnut post and rail fencing. All on a pleasingly level site of 2 acres. The site is level and pathways have been designed to be wheelchair friendly.

♿

GROUP OPENING

73 WOOTTON GARDENS
Wootton, OX13 6DP. *Wootton is 3m SW of Oxford. From Oxford ring road S, take turning to Wootton. Parking for all gardens at the Bystander Pub & on the road.* Thur 10 June (1-5). Combined adm £4, chd free. Home-made teas.

13 AMEY CRESCENT
Sylv & Liz Gleed.

60 BESSELSLEIGH ROAD
Freda East.

67 BESSELSLEIGH ROAD
Jean Beedell.

14 HOME CLOSE
Kev & Sue Empson.

35 SANDLEIGH ROAD
Hilal Baylav Inkersole.

5 inspirational small gardens, all with very different ways of providing a personal joy. 13 Amey Crescent is a gravel garden with grasses and prairie plants. The garden has a small wildlife pond, alpine house and troughs. 14 Home Close is a garden containing mainly shrubs with an unusually shaped lawn. There is also a pond, raised bed and a vegetable garden. 60 Besselsleigh Road is 'The Deadwood Stage' with ornamental wood, sticks, stones, toadstools, many creatures to find in the secret garden with wild flower greenery. 67 Besselsleigh Road is a compact and vibrant garden. Its deep borders are brimming with a variety of colourful plants shaped around a pathway. 35 Sandleigh Road is a mature garden, laid to lawn on two levels. Grown mostly from cuttings the garden is brimming with vibrant flowers, pondside planting and glass art. Partial wheelchair access.

♿ ✿ ☕

SHROPSHIRE

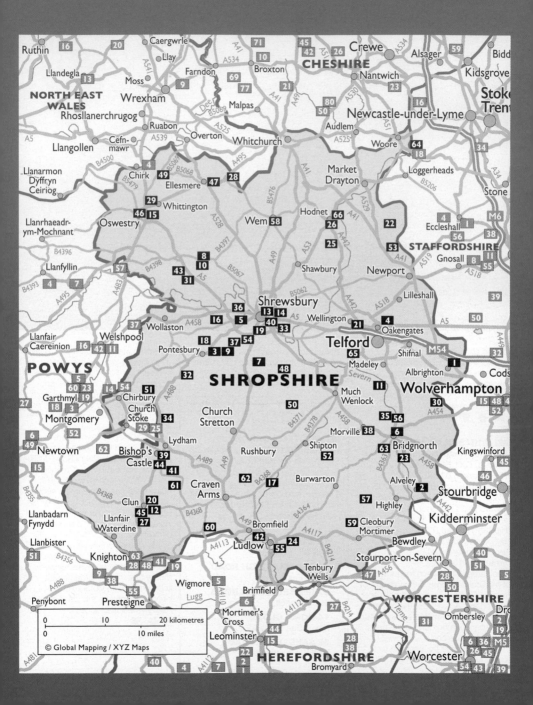

Welcome to Shropshire and our beautiful gardens where our lovely, historical towns (we have no cities within the county) dwell in harmony with the countryside and our farming communities.

Shropshire was also the birthplace of modern industry at Ironbridge near Telford, named after the famous architect Thomas Telford, where many modern businesses flourish.

Our gardens range from small, exquisite cottage and urban gardens to large country estates, several of which opened at the start of the National Garden Scheme in 1927. Our garden history is also unique: Shropshire being the birthplace of the first "celebrity" gardener, Percy Thrower, and the ultimate horticulturist, Charles Darwin.

In 2021, we have many new gardens opening for the first time or returning to the Scheme after many years break. We also have several new gardens that could not open in 2020 due to the Covid 19 Pandemic, so there is much to look forward to. In the event of changes having to be made to this year's programme, we strongly advise visitors to check the NGS website first before planning a visit to our gardens: www.ngs.org.uk. Many of our gardens will be also be offering the facility to pre-book tickets on-line before visiting, so do please check ahead.

Thank you for your continued support, it is much appreciated by us and our beneficiaries.

Volunteers

County Organiser
Allison Walter
01743 627900
allison.walter@ngs.org.uk

County Treasurer
Elaine Jones
01588 650323
elaine.jones@ngs.org.uk

Publicity
Ruth Dinsdale
01948 710924
ruth.dinsdale@ngs.org.uk

Twitter/Facebook
Vicky Kirk
01743 821429
victoria.kirk@ngs.org.uk

Instagram
Logan Blackburn
logan.blackburn.work@gmail.com

Booklet Co-ordinator
Fiona Chancellor
01952 507675
fiona.chancellor@ngs.org.uk

Talks Organiser
Chris Neil
01743 821651
billneil@me.com

Assistant County Organisers
Sue Griffiths
Sheila Jones
Sue & Ron Herepath
And all the other volunteers in Shropshire who help to make the Scheme work in the county

f TheNationalGarden SchemeShropshire

@shropshireNGS

Left: Oswestry Gatacre Allotments & Gardens Association

OPENING DATES

All entries subject to change. For latest information check www.ngs.org.uk

Extended openings are shown at the beginning of the month.

Map locator numbers are shown to the right of each garden name.

April

Sunday 11th
Edge Villa 18

Sunday 18th
Ruthall Manor 52

Sunday 25th
NEW Westwood House 63

Tuesday 27th
Brownhill House 8

May

Sunday 2nd
Longner Hall 33
Lyndale House 35

Monday 3rd
Ruthall Manor 52

Wednesday 5th
Goldstone Hall Gardens 22

Friday 7th
Avocet 3
Cheriton 9

Saturday 8th
Kinton Grove 31

Sunday 9th
Henley Hall 24
Kinton Grove 31
Oteley 47
Sunningdale 58

Wednesday 12th
NEW The Bramleys 7

Saturday 15th
NEW Rorrington Lodge 51
Ruthall Manor 52

Sunday 16th
Ancoireán 2
NEW 3 Haye Court 23

Tuesday 18th
Pooh Corner 49

Saturday 22nd
Upper Shelderton House 60

Sunday 23rd
5 Church Street 10
Preen Manor 50
Stanley Hall Gardens 56
Upper Shelderton House 60

Tuesday 25th
Brownhill House 8

Wednesday 26th
Goldstone Hall Gardens 22

Saturday 29th
Beaufort 4

Sunday 30th
The Gardeners Lodge 21
The Mount 39
The Old Vicarage,
    Bishops Castle 44
Walcot Hall 61

Monday 31st
Walcot Hall 61

June

Saturday 5th
Windy Ridge 65

Sunday 6th
Edge Villa 18
Windy Ridge 65

Wednesday 9th
Goldstone Hall Gardens 22
NEW Little Hogstow 32

Saturday 12th
Ruthall Manor 52

Sunday 13th
Morville Hall Gardens 38
Ruthall Manor 52

Saturday 19th
NEW The Secret Gardens
    at Steventon Terrace 55

Sunday 20th
◆ Delbury Hall Walled
    Garden 17
NEW 3 Haye Court 23
2 School Cottages 54

Friday 25th
NEW The Albrighton Trust
    Moat & Gardens 1

Saturday 26th
NEW The Albrighton Trust
    Moat & Gardens 1

Brownhill House 8
NEW Westhope College 62

Sunday 27th
NEW Clove Hitch 11
◆ Hodnet Hall Gardens 26
The Mount 39
Stottesdon Village Open
    Gardens 57

Wednesday 30th
Goldstone Hall
    Gardens 22

July

**Daily from
Tuesday 20th to
Saturday 24th**
Offcot 43

Friday 2nd
Avocet 3
Cheriton 9

Saturday 3rd
48 Bramble Ridge 6
Cruckfield House 16

Sunday 4th
NEW Clun Group
    Gardens Open 12
Upper Marshes 59

Wednesday 7th
3 Oakeley Mynd 41

Saturday 10th
Ruthall Manor 52

Sunday 11th
NEW High Hatton Hall 25
Ruthall Manor 52

Wednesday 14th
Goldstone Hall
    Gardens 22
Merton 36

Friday 16th
NEW 12 Colley Close 13
NEW 74 Conway Drive 14

Saturday 17th
NEW 153
    Willoughbridge 64

Sunday 18th
◆ Delbury Hall Walled
    Garden 17
Lower Brookshill 34
Sambrook Manor 53
NEW 153
    Willoughbridge 64

Monday 19th
Lower Brookshill 34

Saturday 24th
NEW Oswestry Gatacre
    Allotments & Gardens
    Association 46

Sunday 25th
NEW Esme's Garden 19
NEW Nancy's Garden 40
NEW Oswestry Gatacre
    Allotments & Gardens
    Association 46

Friday 30th
NEW The Albrighton Trust
    Moat & Gardens 1

Saturday 31st
NEW The Albrighton Trust
    Moat & Gardens 1
27 Croxon Rise 15

August

Sunday 1st
27 Croxon Rise 15
Merton 36
2 School Cottages 54

Saturday 7th
Cruckfield House 16

Sunday 8th
Bowbrook Allotment
    Community 5
NEW The Ferns 20

Wednesday 11th
Merton 36

Sunday 15th
NEW Moat Hall 37

Sunday 22nd
NEW Pitchford Hall 48

Sunday 29th
Sambrook Manor 53

September

Saturday 4th
NEW Oakly Park 42
Windy Ridge 65

Sunday 5th
Windy Ridge 65

Wednesday 8th
◆ Wollerton Old Hall 66

Sunday 12th
◆ Delbury Hall Walled
    Garden 17

Wednesday 15th
Goldstone Hall Gardens 22

Sunday 19th
NEW Horatio's Garden 29

By Arrangement
Arrange a personalised garden visit on a date to suit you. See individual garden entries for full details.

All entries are subject to change. For the latest information check ngs.org.uk

---

THE GARDENS

1 NEW THE ALBRIGHTON TRUST MOAT & GARDENS
Blue House Lane, Albrighton, Wolverhampton, WV7 3FL.
Stephen Jimson, 01902 372441, moat@albrightontrust.org.uk, albrightontrust.org.uk. *Off A41 & adjacent to the motorway network. We are located very close to Albrighton train station & close to the RAF Cosford base.* **Fri 25, Sat 26 June, Fri 30, Sat 31 July (11-3). Adm £5, chd free. Home-made teas. Visits also by arrangement June & July.**
The Albrighton Trust gardens are designed around the remains of a C13 fortified manor house and an ancient moat, offering excellent recreational and educational opportunities for anyone wanting to visit and enjoy this award winning outdoor space. Our gardens have won an outstanding award in the Heart of England Bloom competition for 8 years running. Our gardens are wheelchair accessible. Disabled WC, and personal care area with hoist. Assistance dogs only.

2 ANCOIREÁN
24 Romsley View, Alveley, WV15 6PJ. Judy & Peter Creed, 01746 780504, pdjc@me.com. *6m S Bridgnorth off A442 Bridgnorth to Kidderminster rd. N from Kidderminster turn L just after Royal Oak PH. S from Bridgnorth turn R after Squirrel pub. Take 3rd turning on R & follow NGS signs.* **Sun 16 May (1-5). Adm £5, chd free. Home-made teas. Visits also by arrangement May to July for**

groups of 20 to 30.
Natural garden layout on several levels developed over 30yrs, with a large variety of herbaceous plants and shrubs, water features, wooded area with bog garden containing numerous varieties of ferns and hostas, and colourful alpine scree. Features, wooded area, stumpery, ornamental grass border and Spring bulb collection, clematis collection, acer and azalea beds. Selection of plants for sale. Close to Severn Valley Railway and Country Park and Dudmaston Hall NT.

3 AVOCET
3 Main Road, Plealey, SY5 0UZ. Malc & Jude Mollart, 01743 791743, malcandjude@btinternet.com. *6m SW of Shrewsbury. From A5 take A488 signed Bishops Castle, approx ½m past Lea Cross Tandoori turn L signed Plealey. In ¾m turn L, garden on R. SatNav unreliable!* **Fri 7 May, Fri 2 July (10-4.30). Combined adm with Cheriton £6, chd free. Pre-booking essential, please visit www.ngs.org.uk for information & booking. Home-made teas. Visits also by arrangement Apr to July for groups of up to 30.**
Cottage style garden with modern twists owned by plantaholics, shared with wildlife. Designed around a series of garden rooms for year round interest: wildlife pool, mixed borders, seaside and gravel gardens, succulents, trained fruit, sculpture. Constantly evolving. Opportunity to see 2 neighbouring gardens of same size, with similar organic principles and for wildlife but different interpretations. Collection of vintage garden tools.

4 BEAUFORT
Coppice Drive, Moss Road Wrockwardine Wood, Telford, TF2 7BP. Mike King, www.carnivorousplants.uk.com. *Approx 2m N from Telford town centre. From Asda Donnington, turn L at lights on Moss rd, ⅓m, turn L into Coppice Drive. 4th Bungalow on L with solar panels.* **Sat 29 May (10-5). Adm £5, chd free. Light refreshments.**
If carnivorous plants are your thing, then come and visit our National Collection of Sarracenia (pitcher plants); also over 100 different Venus flytrap clones (Dionaea muscipula), Sundews (Drosera) and Butterworts (Pinguicula.) - over 6000 plants in total. Large greenhouses at Telford's first carbon negative house, a great place to visit - kids will love it! Not wheelchair accessible into greenhouses.

5 BOWBROOK ALLOTMENT COMMUNITY

Mytton Oak Road, Shrewsbury, SY3 5BT. Bowbrook Allotment Community, bowbrookallotments. wordpress.com. *On western edge of Shrewsbury between A5 Bypass & Royal Shrewsbury Hospital. From A5 Shrewsbury Bypass take B4386 (Mytton Oak Road) towards Shrewsbury, following signs to Royal Shrewsbury Hospital. The allotment entrance is ½ m from A5 r'about, on R, opp Oak Lane.* **Sun 8 Aug (2-6). Adm £6, chd free. Light refreshments.**
Recipient of RHS National Certificate of Distinction, this 6.5 acre site, comprising 93 plots, displays wide ranging cultivation methods. Members cultivate organically with nature in mind using companion planting to attract natural predators. Green spaces flourish throughout and include Gardens of the 4 Seasons, orchards and many wildlife features including wildflower meadows and pond. Children are encouraged to be part of the community and have their own special places: story telling willow dome, willow tunnel, sensory garden and turf spiral. See the now well-developed Contemplation Garden & Prairie Garden. New all-ability plot developed in 2020. Wheelchair access possible with care. Generally flat wide grass paths allow access to all the main areas of the site, although paths may be bumpy!
&♿ ☕

6 48 BRAMBLE RIDGE

Bridgnorth, WV16 4SQ. Heather, 07572 706706, heatherfran48@gmail.com. *From Bridgnorth N on B4373 signed Broseley. 1st on R Stanley Lane, 1st R Bramble Ridge. From Broseley S on B4373, nr Bridgnorth turn L into Stanley Lane, 1st R Bramble Ridge.* **Sat 3 July (12-5). Adm £5, chd free. Light refreshments. Visits also by arrangement Apr to Sept for groups of 10 to 20.**
Steep cottage-style garden with many steps; part wild, part cultivated, terraced in places and overlooking the Severn valley with views to High Rock and Queens Parlour. The garden features shrubs, perennials, wildlife pond, summerhouse and, in the wilderness area, wildflowers, fruit trees and a sandstone outcrop. My garden means a lot to me and I am looking forward to sharing it with others.
✿ ☕

7 NEW THE BRAMLEYS

Condover, Shrewsbury, SY5 7BH. Toby & Julie Shaw. *Through the village of Condover towards Dorrington. Pass junction by village hall follow rd round, cross the bridge, in approx 100 metres there is a drive on L. Look for NGS signs.* **Wed 12 May (11-5). Adm £5, chd free. Home-made teas.**
A large country garden extending to 2 acres with a variety of trees and shrubs, herbaceous borders and a woodland with the Cound Brook flowing through. Through the gate a courtyard garden with a greenhouse, herb garden and flower beds for cutting flowers. The aspect looks out on to open countryside. Pathways close to the house suitable for wheelchairs, but steps to woodland area and pathways would make access restrictive in places.
♿ ☕

8 BROWNHILL HOUSE

Ruyton XI Towns, SY4 1LR. Roger & Yoland Brown, 01939 261121, brownhill@eleventowns.co.uk, www.eleventowns.co.uk. *9m NW of Shrewsbury on B4397. On the B4397 in the village of Ruyton XI Towns.* **Tue 27 Apr, Tue 25 May, Sat 26 June (10-5). Adm £5, chd free. Pre-booking essential, please visit www.ngs.org.uk for information & booking. Home-made teas. Visits also by arrangement Apr to July.**
A unique 2 acre hillside garden with many steps and levels bordering R Perry. Visitors can enjoy a wide variety of plants and styles from formal terraces to woodland paths. The lower areas are for the sure-footed and mobile while the upper levels have many places to sit and enjoy the views. Kit cars on show.
✿ 🚐 ☕

9 CHERITON

Plealey, SY5 0UY. Vicky Wood. *6m SW of Shrewsbury. From A5, take A488 signed Bishops Castle; approx ½ m past Lea Cross Tandoori, turn L signed to Plealey; in ¾ m turn L, garden on R. SatNav unreliable.* **Fri 7 May, Fri 2 July (10-4.30). Combined adm with Avocet £6, chd free. Pre-booking essential, please visit www.ngs.org.uk for information & booking. Home-made teas.**
Family garden where every space has a function, with annual and perennial cut flower borders, fruit and vegetables, polytunnel, chickens, wildlife pond, outdoor cooking and green woodworking areas. Full of colour with stunning borrowed views to the fields and hill beyond. Opportunity to see 2 neighbouring gardens of same size with similar organic principles and for wildlife with different interpretations.
✿ ☕

10 5 CHURCH STREET

Ruyton X1 Towns, SY4 1LA. Steve & Jill Owen. *From A5 take B4397 to Ruyton XI Towns. Enter the village, carry straight on passing The Talbot pub on R. House on the L. 50m further on, opposite school. Parking opposite.* **Sun 23 May (12-5). Adm £5, chd free. Home-made teas.**
Long village garden divided into sections affording different styles of planting: herbaceous borders with a wide variety of plants, shrubs and trees giving year-round interest through texture and colour; wildflower area and wildlife pond. Summer house, vegetable area and greenhouse. Not suitable for wheelchairs.
✿ ☕

11 NEW CLOVE HITCH

Coalport Road, Broseley, TF12 5AN. Nick & Mia Harrington. *½ m E of Broseley on Coalport Road. From E (Woodbridge Inn) (6'6" height limit) drive to top of hill - parking on R, 300m after farm. From Broseley (W) parking on L, 500m beyond postcode.* **Sun 27 June (11-4). Adm £5, chd free. Home-made teas.**
Award winning Paul Richards designed garden set in a spectacular location with ranging views over the Severn Valley. The 0.5 acres comprises 5 areas - small mixed orchard set to wildflower meadow; circular main lawn with haha, deep full borders and sculptures; wildlife pond with deck that allows visitors to walk out over the water; circular elevated patio; working area with raised beds.
🐕 ✿ 🅳 ☕

All entries are subject to change. For the latest information check ngs.org.uk

GROUP OPENING

12 NEW **CLUN GROUP GARDENS OPEN**

Bridge Street, Clun, SY7 8JP.
Clun & Environs. Clun is situated on B4368, 9m W of Craven Arms on A488, 30m S of Shrewsbury.
Sun 4 July (1-5). Combined adm £7.50, chd free. Home-made teas at Hightown Community Room, Vicarage Road, Clun.

Clun and the Clun Valley nestle in the Shropshire Hills Area of Outstanding Natural Beauty. A minimum of 8 gardens are open, with a wide range of planting styles and innovative landscaping. Cottage gardens, terracing, wildflower areas, specialist plants and vegetable production. One garden is outside of Clun with beautiful vistas and space for large scale planting. Partial wheelchair access to some gardens, grass and gravel access in others.

13 NEW **12 COLLEY CLOSE**

Shrewsbury, SY2 5YN. Kevin & Karen Scurry. *Off Telford Way nr Shrewsbury police stn. From r'about by police stn head down Telford Way at next r'about take 2nd exit onto Oswell Rd then 1st R into Colley Close. Please use shuttle bus between gardens as parking limited in Colley Close.* **Evening opening Fri 16 July (5-9). Combined adm with 74 Conway Drive £6, chd free. Light refreshments.**

This garden has 3 ponds including a large koi pond and waterfall, wildlife pond and goldfish pond all surrounded by interesting planting. A large collection of specimen bonsai trees accent the koi pond giving an oriental feel to the garden. The front garden has a large collection of grasses. The garden has good wheelchair access and many seating areas to view the garden from many different angles.

14 NEW **74 CONWAY DRIVE**

Shrewsbury, SY2 5UY. Andrew & Deborah Abel. *Off Monkmoor Rd nr Shrewsbury Police Stn. At Lord Hill towards town turn R at T-lights towards police stn. At r'about 2nd exit, then 1st R into Conway Dr. Please use free shuttle bus between the gardens as parking is limited in Colley Close.* **Evening opening Fri 16 July (5-9). Combined adm with 12 Colley Close £6, chd free. Light refreshments.**

A delightful urban garden with extensive use of pots to accentuate various garden rooms. Garden is well planted with many specimen trees and shrubs in a stylish, contemporary setting. Borders contain herbs, herbaceous and tropical planting mingled throughout the garden. Featuring a rockery, waterfall and fish pond. Various seating areas make it a lovely place to spend a summer's evening

Albrighton Trust Moat & Gardens

15 27 CROXON RISE
Oswestry, SY11 2YQ. Natalie & Tony Bainbridge. *Eastern Oswestry. ½m from A5 Oswestry Bypass. Take B4580 to Oswestry. Then 1st L Harlech Rd, 1st exit r'about Cabin lane. Take 5th L Aston Way, 1st L & 1st L again Croxon Rise. Please park courteously in surrounding rds.* **Sat 31 July, Sun 1 Aug (12-5). Adm £5, chd free. Home-made teas.**
Packing a lot into a small space, this secluded, beautifully planted town garden offers interest at every turn. Mixed borders provide colour through to late summer. A raised bed vegetable garden, greenhouse and cut flower garden, soft fruit cage, seating areas, and water features. A rose garden surrounds a mature cherry tree and south facing wall protects step-over and cordon fruits.

16 CRUCKFIELD HOUSE
Shoothill, Ford, SY5 9NR. Geoffrey Cobley, 01743 850222. *5m W of Shrewsbury. A458 from Shrewsbury, turn L towards Shoothill.* **Sat 3 July, Sat 7 Aug (2-5). Adm £6, chd free. Home-made teas. Visits also by arrangement June & July for groups of 20+.**

An artist's romantic 3 acre garden, formally designed, informally and intensively planted with a great variety of unusual herbaceous plants. Nick's garden, with many species trees, shrubs and wildflower meadow, surrounds a large pond with bog and moisture-loving plants. Ornamental kitchen garden. Rose and peony walk. Courtyard fountain garden, large shrubbery and extensive clematis collection Extensive topiary, and lily pond.

17 ◆ DELBURY HALL WALLED GARDEN
Delbury Hall Estate, Mill Lane, Diddlebury, SY7 9DH. Mr & Mrs Richard Rallings, 01584 841 222, info@myndhardyplants.co.uk, www.myndhardyplants.co.uk. *8m W of Craven Arms. Follow the B4368 to Diddlebury. Drive into village; the entrance to the Walled Garden is on R 300 metres past St Peter's church.* **For NGS: Sun 20 June, Sun 18 July, Sun 12 Sept (10-5). Adm £5, chd free. Home-made teas. English wine from our own vineyard. For other opening times and information, please phone,**

email or visit garden website.
A two acre early Victorian Walled Garden with large herbaceous borders, a vegetable and a herb garden, shrubbery, vines and very old fruit trees. In addition there is a large Penstemon and Hemerocallis collection. A signposted walk guides you around the front of Delbury Hall and the lake. Extensive range of perennials and English wine from the vineyard for sale. Gravel and grass paths. Dogs are welcome in the Walled Garden but not in the Hall Gardens.

18 EDGE VILLA
Edge, nr Yockleton, SY5 9PY. Mr & Mrs W F Neil, 01743 821651, billfneil@me.com. *6m SW of Shrewsbury. From A5 take either A488 signed to Bishops Castle or B4386 to Montgomery for approx 6m then follow NGS signs.* **Sun 11 Apr, Sun 6 June (2-5). Adm £5, chd free. Home-made teas. Visits also by arrangement Apr to Aug for groups of 10+.**
Two acres nestling in South Shropshire hills. Self-sufficient vegetable plot. Chickens in orchard,

Offcot

foxes permitting. Large herbaceous borders. Dewpond surrounded by purple elder, irises, candelabra primulas and dieramas. Large selection of fragrant roses. Wendy House for children. Some German and French spoken. Some gravel paths.

19 NEW ESME'S GARDEN
Church Row, Meole Village, Shrewsbury, SY3 9EX. Nancy Estrey & Peter Alltree. *Through wicker gate by No 3 Church Row & follow path to the garden at the top. For Church Row, take next turning on the L; on-street parking on Church Row, Church Road & Upper Road or take shuttle bus from Nancy's garden.* **Sun 25 July (1-5). Combined adm with Nancy's Garden £6, chd free. Light refreshments in Peace Memorial Hall, Meole Village.**
Esme's Garden is situated in a secluded spot in the centre of Meole Village, with the beautiful Meole Church as a backdrop. Designed purely for its exuberant planting, gravel paths take you on a romantic journey through trees, shrubs and perennials, eventually leading to a central oasis by a tranquil pond. There are also many unexpected features. Both gardens are nurtured by Nancy. On the level, but a lot of gravel paths which might prove difficult for wheelchairs.

20 NEW THE FERNS
Newport Street, Clun, SY7 8JZ. Andrew Dobbin, 01588 640064, andrew.clun@outlook.com. *Enter Clun from Craven Arms. Take 2nd R signed into Ford St. At T junction turn R for parking in the Memorial Hall car park (100 yds). Retrace steps to T junction. The Ferns is on L.* **Sun 8 Aug (12-6). Adm £5, chd free. Visits also by arrangement in Aug. Teas are available in the village at the River Cafe and the Malthouse.**
A formal village garden of 3/4 acre, approached via a drive lined with crab apple and pear trees. On the right is the autumn garden, giving fine views of the surrounding hills. From the front courtyard garden a path leads through double herbaceous borders full of late summer colour, to further rooms, of yew, beech and box. There is also a rear courtyard with tender exotics. Lily pond and statuary.

Wheelchairs difficult, but possible for some of the garden.

21 THE GARDENERS LODGE
2 Roseway, Wellington, TF1 1JA. Amanda Goode, www. lovegrowshereweb.wordpress. com/. *1½ m (4 mins) from J7 (M54). B5061 Holyhead Rd 2nd L after NT 'Sunnycroft'. (New Church Rd) We are the cream house on corner of NCR & Roseway.* **Sun 30 May (9.30-4.30). Adm £5, chd free. Light refreshments.**
There was just one tree in the garden when the current owner purchased The Gardener's Lodge, the ground having been cleared, ready to sell as a building plot. However the owner had other plans for it: the garden is now eclectically divided, arranged and planted into areas: Mediterranean, Cottage, Indian etc but, seamlessly, each section merges and leads into the next developing idea. A small urban garden with seating, water features and shade.

22 GOLDSTONE HALL GARDENS
Goldstone, Market Drayton, TF9 2NA. John Cushing, 01630 661202, enquiries@goldstonehall.com, www.goldstonehall.com. *5m N of Newport on A41. Follow brown & white signs from Hinstock. From Shrewsbury A53, R for A41 Hinstock & follow brown & white signs & NGS signs.* **Wed 5, Wed 26 May, Wed 9, Wed 30 June, Wed 14 July, Wed 15 Sept (12-5). Adm £6, chd free. Home-made teas in award winning oak framed pavilion in the midst of the garden with cakes created by our talented chef. Visits also by arrangement Apr to Sept for groups of 10+.**
5 acres with highly productive beautiful kitchen garden. Unusual vegetables and fruits - alpine strawberries; heritage tomatoes, salad, chillies, celeriac. Roses in Walled Garden from May; Double herbaceous in front of old English garden wall at its best July and August; Sedums and Roses stunning in September. Lawn aficionados will enjoy the stripes. Majority of garden can be accessed on gravel and lawns.

23 NEW 3 HAYE COURT
Lower Forge, Eardington, Bridgnorth, WV16 5LQ. Eileen Paradise, 01746 764884, eileenparadise22@gmail.com. *2½ m S of Bridgnorth on B4555. Continue for 1m past Eardington over railway bridge; do not take next L but carry on for further 200 metres then take L - follow yellow NGS signs.* **Sun 16 May, Sun 20 June (11-5). Adm £5, chd free. Visits also by arrangement May to July for groups of up to 20.**
Enchanting garden with year round interest; innovative design by plantswoman/floral designer. Started 5 years ago but well established; a stylish, peaceful sanctuary. Contemporary sculpture, elegant statuary; Glorious views across Severn Valley, paved paths, clipped box, cloud pruning, unusual planting schemes in delicate hues; teas served in sheltered courtyard. Not suitable for wheelchairs. Quiche & salad at lunchtime; tea/coffee/cold drinks/cakes all day.

24 HENLEY HALL
Henley, Ludlow, SY8 3HD. Helen & Sebastian Phillips, www.henleyhallludlow.com. *2m E of Ludlow. 1½ m from A49. Take A4117 signed to Clee Hill & Cleobury Mortimer. On reaching Henley, the gates to Henley Hall are on R when travelling in the direction towards Clee Hill.* **Sun 9 May (10.30-5). Adm £6, chd free. Home-made teas.**
Henley Hall offers a mixture of formal and informal gardens. The historic elements include a formal lawn, stone staircase with balustrades and a ha-ha, and a beautiful walled garden and 'pulmonary' water feature created in 1874. Highlights include walk ways along the banks of the river Ledwyche with decorative weirs, a stone arched bridge, and woodland paths. Some parts of the garden are accessible by wheelchair.

"The annual donation from the National Garden Scheme to Perennial is the cornerstone of our fundraising activities and encourages many of our donors." Perennial

25 NEW HIGH HATTON HALL

High Hatton, nr Shawbury, Shrewsbury, SY4 4EY. Sarah & Tim Leslie. *From Shrewsbury (A49), at r'about, take 2nd exit onto A53. After approx 8m, turn R where signed to continue to village of High Hatton.* **Sun 11 July (12-6). Adm £6, chd free. Light refreshments.** Beautifully located gardens surrounding a stunning grade II listed hall. A large garden with expansive herbaceous borders surrounding formal lawns. A large walled kitchen garden, rose borders and pergola. A developing orchard and a spectacular courtyard with large display of hanging baskets and summer pots adding further colour.

&. ◖

26 ♦ HODNET HALL GARDENS

Hodnet, Market Drayton, TF9 3NN. Sir Algernon & The Hon Lady Heber-Percy, 01630 685786, secretary@hodnethall.com, www.hodnethallgardens.org. *5½m SW of Market Drayton. 12m NE Shrewsbury. At junction of A53 & A442.* **For NGS: Sun 27 June (11-5). Adm £8.50, chd £1. Light refreshments in the 17th century tea room. For other opening times and information, please phone, email or visit garden website.** The 60+ acres of Hodnet Hall Gardens are amongst the finest in the country. There has been a park and gardens at Hodnet for many hundreds of years. Magnificent forest trees, ornamental shrubs and flowers planted to give interest and colour from early Spring to late Autumn. Woodland walks alongside pools and lakes, home to abundant wildlife. Productive walled kitchen garden and historic dovecot. For details please see website and Facebook page. Maps are available to show access for our less mobile visitors.

&. 🐕 ◖ NPC ◖

27 THE HOLLIES

Rockhill, Clun, SY7 8LR. Pat & Terry Badham, 01588 640805, patbadham@btinternet.com. *10m W of Craven Arms. 8m S of Bishops Castle. From A49 Craven Arms take B4368 to Clun. Turn L onto A488 continue for 1½m. Bear R signed Treverward. After 50 yards turn R at Xrds, property is 1st on L.* **Visits by arrangement June to Sept. Adm £5, chd free.**

A garden of approx 2 acres at 1200ft which was started in 2009. Features include a kitchen garden with raised beds and fruit cage. Large island beds and borders with perennials, shrubs and grasses, specimen bamboos and trees. Birch grove and wildlife dingle with stream, rain permitting! Refreshments can be arranged at a local cafe in Clun (daytime hours). Wheelchair access is available to the majority of the garden over gravel and grass.

&. 🚐 ◖ 🛆

28 HOLLY HOUSE

Welshampton, SY12 0QA. Mike & Ruth Dinsdale, 01948 710924, ruth.dinsdale@btinternet.com. *On A495 ½m E of Welshampton towards Whitchurch. From Welshampton head towards Whitchurch for ½m; Follow yellow NGS signs.* **Visits by arrangement May to Sept for groups of up to 30. Adm £5, chd free. Home-made teas.** A densely planted ⅓ acre plantswoman's country garden featuring some rare and unusual plants, herbaceous borders and ornamental pond. Leading to a 1 acre field with both mature and newly planted specimen trees, a natural wildlife pond, wildflower "project", vegetable garden and rural views. The garden is not suitable for wheelchairs .

🐎 ❀ ◖ 🛆

29 NEW HORATIO'S GARDEN

The Robert Jones & Agnes Hunt Orthopaedic Hospital, Gobowen, Oswestry, SY10 7AG. Imogen Jackson, www.horatiosgarden.org.uk. *At the spinal unit. From A5 follow signs to Orthopaedic Hospital; parking in pay & display car park by hospital.* **Sun 19 Sept (2-5). Adm £5, chd free. Home-made teas.** Beautifully designed by Bunny Guinness and delightfully planted, this garden opened to great acclaim in September 2019. Part-funded by the National Garden Scheme, the garden offers a place of peace and therapy for patients at the spinal unit. Raised beds, creative use of space, beautiful specimen trees, potting shed, sculpture, all in soothing hues. A pretty rill runs the length of the garden. Very good access throughout.

&. ❀ ◖

31 KINTON GROVE

Kinton, SY4 1AZ. Tim & Judy Creyke, 01743 741263, judycreyke@icloud.com. *Off A5, between Shrewsbury & Oswestry, approx 1m from Nesscliffe. Follow the A5 to the r'about at the Oswestry end, ie NW end, of the Nesscliffe by-pass. Follow the signs to Kinton, approx 1m.* **Sat 8, Sun 9 May (1-5). Adm £5, chd free. Cream teas. Visits also by arrangement Apr to Sept for groups of up to 30.** A garden of ¾ acre surrounding a Georgian house. The garden features well-filled herbaceous borders, roses, a gravel area, raised vegetable beds, and a wide range of interesting trees and shrubs. Hedges, including some as old as the house, divide up the garden. Lovely views across the Breidden hills and plenty of pleasant places to sit, whilst enjoying cream tea and cakes.

❀ ◖

32 NEW LITTLE HOGSTOW

Mill Lane, Plox Green, Minsterley, Shrewsbury, SY5 0HU. Jon & Anne Yeeles. *Leave Shrewsbury, on A488. 1m beyond Minsterley, continue straight on at Plox Green Xrds. After 500 yds. turn R up Mill Lane at NGS sign.* **Wed 9 June (11-5). Adm £5, chd free. Home-made teas.** A delightful garden of just under 2 acres, surrounding C17 cottage. Situated within an AONB with views to Long Mountain and Stiperstones. Featuring a wildlife pond, hay meadows, mature apple orchard complete with hens, small greenhouse, soft fruit and raised vegetable beds, With herbaceous perennial borders, roses and woodland, all aiming to be a haven for wildlife and human visitors. Wheelchair access to most areas.

&. 🐕 ◖

33 LONGNER HALL

Atcham, Shrewsbury, SY4 4TG. Mr & Mrs R L Burton. *4m SE of Shrewsbury. From M54 follow A5 to Shrewsbury, then B4380 to Atcham. From Atcham take Uffington Road, entrance ¼m on L.* **Sun 2 May (2-5). Adm £5, chd free. Home-made teas in aid of Atcham Church.** A long drive approach through parkland designed by Humphry Repton. Walks lined with golden yew through extensive lawns, with views over Severn Valley. Borders containing roses, herbaceous and shrubs, also ancient yew wood. Enclosed one acre

High Hatton Hall

walled garden open to NGS visitors. Mixed planting, garden buildings, tower and game larder. Short woodland walk around old moat pond which is not suitable for wheelchairs.

34 LOWER BROOKSHILL

Nind, SY5 0JW. Patricia & Robin Oldfield, 01588650137, robin.oldfield@live.com. *3m N of Lydham on A488. Take signed turn to Nind & after ½ m sharp L & follow narrow rd for another ½ m. Drive v.slowly & use temporary passing places.* **Sun 18, Mon 19 July (1-5). Adm £5, chd free. Cream teas.**
10 acres of hillside garden and woods at 950ft within the AONB. Begun in 2010 from a derelict and overgrown site, cultivated areas now rub shoulders with the natural landscape using fine borrowed views over and down a valley. Includes brookside walks, a 'pocket' park, four ponds (including a Monet lily pond), mixed borders and lawns, cottage garden (with cottage) and wildflowers. Lots

of lovely picnic spots. Sadly not wheelchair friendly

35 LYNDALE HOUSE

Astley Abbotts, Bridgnorth, WV16 4SW. Bob & Mary Saunders. *2m out of Bridnorth off B4373. From High Town Bridgnorth take B4373 Broseley Rd for 1½ m, then take lane signed Astley Abbotts & Colemore Green.* **Sun 2 May (1-5). Adm £5, chd free. Home-made teas.**
1½ acre garden which has been lovingly tended for 25yrs. Rose terrace under planted with tulips and alliums. Large lawns interspersed with well planted flower beds. New Japanese themed pool with Koi. Courtyard with colourfully planted pots and topiary. Masses of tulips in the spring, and beautiful roses in the summer. Plenty of seating. Water features. Tea on the terrace over looking countryside. Topiary garden, waterfall to pool, many unusual trees. Please ask owner about wheelchair friendly access.

36 MERTON

Shepherds Lane, Dicton, Shrewsbury, SY3 8BT. David & Jessica Pannett, 01743 850773, jessicapannett@hotmail.co.uk. *3m W of Shrewsbury. Follow B4380 from Shrewsbury past Shelton for 1m Shepherd's Lane turn L garden signed on R or from A5 by pass at Churncote r'about turn towards Shrewsbury 2nd turn L Shepherds Lane.* **Wed 14 July (1-5). Sun 1 Aug (1-5). Tea. Wed 11 Aug (1-5). Adm £5, chd free. Visits also by arrangement May to Sept.**
Mature ½ acre botanical garden with a rich collection of trees and shrubs including unusual conifers from around the world. Hardy perennial borders with seasonal flowers and grasses plus an award winning collection of hosta varieties in a woodland setting. Outstanding gunneras in a waterside setting with moisture loving plants. Level paths and lawns.

37 NEW MOAT HALL
Annscroft, Shrewsbury, SY5 8AZ.
Martin & Helen Davies. *Take the Longden road from Shrewsbury to Hook a Gate. Our lane is 2nd on R after Hook a Gate & before Annscroft. Single track lane for ½m.* **Sun 15 Aug (12-6). Adm £5, chd free. Home-made teas.**
1 acre garden around an old farmhouse within a dry moat which can be walked around. Well organised and extensive kitchen garden, fruit garden and orchard for self-sufficiency. Colourful herbaceous borders; stumpery; many interesting stone items including troughs, cheese weights, staddle stones some uncovered in the garden. Plenty of seating areas on the lawns for enjoying the garden and a homemade tea. Wheelchair access: mostly lawn with 1 grass ramp and 1 concrete ramp; kitchen garden has 2' wide paved paths.

GROUP OPENING

38 MORVILLE HALL GARDENS
Morville, Bridgnorth, WV16 5NB.
3m W of Bridgnorth. On A458 at junction with B4368. **Sun 13 June (2-5). Combined adm £5, chd free. Home-made teas in Morville Church.**

THE COTTAGE
Ms A Nichol-Smith.

1 THE GATE HOUSE
Mr & Mrs Rowe.

2 THE GATE HOUSE
Mrs G Medland.

MORVILLE HALL
Mr & Mrs A Lewis & The National Trust.

SOUTH PAVILION
Mrs Joy Jenkinson.

An interesting group of gardens that surround a beautiful Grade I listed mansion (house not open). The Cottage has a pretty walled garden with plenty of colour, while 1 and 2 The Gate House are cottage-style gardens with colourful borders, formal areas, lawns and wooded glades. The three acre Morville Hall Garden has a parterre, medieval stew pond, shrub borders, and large lawns, with lovely views across the Mor Valley. South Pavilion features new thoughts and new designs in a small courtyard garden. Please note that National Trust membership does not give free admission to this charity opening, and sorry, no dogs are allowed in the gardens. Also, please note that The Dower House Garden will not be open. Mostly level ground, but plenty of gravel paths and lawns to negotiate.

39 THE MOUNT
Bull Lane, Bishops Castle, SY9 5DA. Heather Willis, 01588 638288, adamheather@btopenworld.com. *Off A488 Shrewsbury to Knighton Rd. at top of the town, 130 metres up Bull Lane on R. No parking at property, parking free in Bishops Castle. Sun 30 May opening same date as neighbouring Walcot Hall & Old Vicarage.* **Sun 30 May, Sun 27 June (12.30-5.30). Adm £5, chd free. Light refreshments. Visits also by arrangement Apr to Sept for groups of up to 30.**
An acre of garden that has evolved over 24 years, with 4 lawns, a rosebed in the middle of the drive with pink and white English roses, and herbaceous and mixed shrub borders. There are roses planted throughout the garden and in the spring daffodils and tulips abound. Two large beech trees frame the garden with a view that sweeps down the valley over fields and then up to the Long Mynd. Access easy to parts of the garden itself, but not to WC which is in the house up steps.

40 NEW NANCY'S GARDEN
11 Elmfield Road, Shrewsbury, SY2 5PB. Nancy Estrey & Peter Alltree. *Off Belvedere Avenue. Parking is available at Shirehall Council car park off Belvedere Avenue. A free shuttle bus is available between Nancy's Garden & Esme's Garden where there is limited on-street parking.* **Sun 25 July (1-5). Combined adm with Esme's Garden £6, chd free. Light refreshments at Peace Memorial Hall, Meole Village.**
Nancy's garden is a small suburban garden with lots of form, foliage and colour. Winding paths lead you to a circular lawn, patio and pond. A secluded patio at the top of the garden is surrounded by lush planting that gives a sub tropical feel. A summer house used throughout the summer is a true outside room.

A plantswomen's garden featuring a wealth of plants. Beautifully designed to give colour and form all year round. Tranquil seating and pond. Both gardens nurtured by Nancy.

41 3 OAKELEY MYND
Stank Lane, Bishops Castle, SY9 5EX. Derek & Eileen Mattey. *Approx 2m E of Bishop's Castle, off B4385. Follow B4385 for approx ½m, turn L at NGS sign. Garden located on R, 1m up the hill. Limited parking at garden. On open day, please use free shuttle bus from Bishop's Castle rugby club.* **Wed 7 July (11-5). Adm £5, chd free. Home-made teas.**
Just under 1 acre, south facing, wildlife-friendly country garden. Informal planting, with wildflowers, herbaceous perennials, shrubs, small wildlife pond, produce garden, greenhouse and fruit trees. Hillside setting at 935'/285m with lovely views over the Shropshire Hills. Not suitable for wheelchairs.

SPECIAL EVENT

42 NEW OAKLY PARK
Bromfield, Ludlow, SY8 2JW. Lord & Lady Plymouth. *Off A49 (Shrewsbury to Hereford Rd) opp Ludlow Farm Shop. Take immediate L turn towards Bromfield Church, follow rd over bridge & take L hand fork. Watch for yellow NGS signs.* **Sat 4 Sept (1-5). Adm £7.50, chd free. Pre-booking essential, please visit www.ngs.org.uk/special-events for information & booking. Light refreshments at The Clive Arms and the Ludlow Kitchen, part of the estate.**
One of Shropshire's magnificent estates, rarely open to the public, named after its centuries-old oak trees. At the confluence of the River Onny and Teme, with extensive views to Ludlow and the south Shropshire Hills. Well-planted borders, rose garden, fern grotto, lake, interesting historical buildings and features. 2 acres of productive walled garden supplying produce to Ludlow Farm site. Accommodation available at The Clive Arms.

43 OFFCOT

Kynaston, Kinnerley, SY10 8EF.
Tom Pountney. *Just off A5 on
Nesscliffe r'about towards Knockin.
Then take the 1st L towards
Kinnerley. Follow NGS signs from
this road.* **Daily Tue 20 July to Sat
24 July (10-4). Adm £5, chd free.
Light refreshments.**
A cottage garden with lots of winding
pathways leading to different focal
points. The garden is packed with
a wide range of evergreen and
deciduous trees and shrubs and
underplanted with herbaceous
perennials. There is a natural looking
pond with a running stream feeding
into it. A haven for wildlife. So many
different areas to see and enjoy
including the garden bar.

44 THE OLD VICARAGE, BISHOPS CASTLE

Church Lane, Bishops Castle,
SY9 5AF. **Helen & Jerry Robinson.**
*Near the junction of Church Street &
Kerry Lane, directly behind St John's
Church. Pedestrian access through
the main gate on Church Lane or
through the garden gate to the rear
of St John's churchyard.* **Sun 30
May (12-6). Adm £5, chd free.
Light refreshments at The Mount
and Walcot Hall.**
Extending to just over one and a half
acres, the gardens include lawns
surrounded by perennial beds and
mature shrubs, a pond, an orchard
and a romantic ruin, fragments of the
lost thirteenth century church. Tending
toward the wild, in early summer the
garden is full of colour; foxgloves,
wisteria, rhododendron, alliums and
ornamental trees. Garden themed arts
and crafts for sale. Level access over
gravel drive and lawns.

45 THE OLD VICARAGE, CLUN

Vicarage Road, Clun, SY7 8JG.
**Peter & Jay Upton, 01588 640775,
jay@salopia.plus.com.** *16m NW of
Ludlow. Over bridge at Clun towards
Knighton (parking in public car park
by bridge); walk up to church, turn L
into Vicarage Rd; house on R next to
church. Refreshments in Clun village.*
**Visits by arrangement May to Sept
for groups of up to 30. Adm £5,
chd free.**
A revived, old vicarage garden:
a slow retrieval and recovery
revealing a wealth of features and
plants chosen by plantsmen vicars:
buddleia globosa fascinates bees

153 Willoughbridge

and butterflies; glorious oriental
poppies fascinate visitors. 'The Tree',
an enormous Leyland cypress (5th
biggest girth in the world); the most
dramatic feature is a formal wisteria
allee with alliums - a symphony of
mauve and purple.

46 NEW OSWESTRY GATACRE ALLOTMENTS & GARDENS ASSOCIATION

Lloyd Street, Oswestry, SY11 1NL.
**Graham Mitchell, 01691 654961,
gsfmitchell@gmail.com,
gatacre.wordpress.com.** *2 parts to
the allotments either side of Liverpool
Road. From Whittington r'about on
A5 turn towards Oswestry B4580. In
½ m go across staggered junction
(L then R). Continue 100y then bear
sharp L. After 200y straight over
r'about, then 3rd on L.* **Sat 24, Sun
25 July (10.30-3.30). Adm £5, chd
free. Home-made teas. Visits also
by arrangement July & Aug for
groups of 5 to 10.**
Two adjacent, large allotment
communities in the heart of Oswestry;
well supported and well-loved by local
residents. Vast array of fruit, flowers,
vegetables in every shape and form
grown on the allotments. Allotment

holders will be on-hand to talk about
their produce and give advice.
Entry ticket covers both sides of the
allotments. The site was featured on
BBC's Countryfile on Sept 1st 2019.
The main pathways are wheelchair
friendly, but the small paths tend to
be rather steep or narrow. There are
disabled WC on one side of the site.

"I was amazed to discover that
the National Garden Scheme
is Marie Curie's largest single
funder and has given the charity
nearly £10 million over 25
years. Their continued support
makes such a difference to me
and all Marie Curie Nurses on
the frontline of the coronavirus
crisis, as we continue to provide
expert care and support to
people at end of life." – Tracy
McWilliams, Marie Curie Nurse

47 OTELEY

Ellesmere, SY12 0PB. Mr RK Mainwaring, 01691 622514, office@oteley.com. *1m SE of Ellesmere. Entrance out of Ellesmere past Mere, opp Convent nr to A528/495 junction.* **Sun 9 May (2-5). Adm £5, chd free. Tea. Visits also by arrangement May to Sept for groups of 10+.**

10 acres running down to The Mere. Walled kitchen garden, architectural features, many old interesting trees. Rhododendrons, azaleas, wild woodland walk and views across Mere to Ellesmere. First opened in 1927 when the National Garden Scheme started. Beautiful setting on the Mere. Wheelchair access if dry.

SPECIAL EVENT

48 NEW PITCHFORD HALL

Pitchford, Condover, Shrewsbury, SY5 7DN. James Nason & Rowena Colthurst, www.pitchfordestate.com. *At Pitchford between Condover & Acton Burnell, S of Shrewsbury.* **Sun 22 Aug (12-5). Adm £7.50, chd free. Pre-booking essential, please visit www.ngs.org.uk/special-events for information & booking. Light refreshments in the restored Orangery.**

Viewed by Country Life magazine as one of the most beautiful historic houses in the country, the Grade 1 listed Pitchford Hall and its gardens are returning to their former glory. Featuring the world's oldest treehouse, dating from the late 1600s, the gardens are undergoing sympathetic restoration revealing a wealth of features and interest. Come and see one of Shropshire's real treasures! One of the original NGS gardens to open in 1927. Partial wheelchair access in some areas.

49 POOH CORNER

6 Laburnum Close, St Martins, SY11 3HU. Sue Napper. *5m NE of Oswestry in St Martin's village. Follow NGS signs from A5. Please park courteously outside property and in surrounding roads.* **Tue 18 May (11-4). Adm £5, chd free. Light refreshments. Light lunches will be available in addition to tea/coffee and cakes.**

A plants-woman's garden giving particular emphasis to shade loving perennials and unusual shrubs and climbers, some of which are rarely grown outdoors. Relatively compact in size and divided into 4 distinct areas providing diverse growing conditions for a wide variety of plants including ferns, primulas and alpines.

Partial wheelchair access only but mainly level throughout.

50 PREEN MANOR

Church Preen, SY6 7LQ. Mr & Mrs J Tanner. *6m W of Much Wenlock; For sat nav, please use postcode SY5 6LF. From A458 Shrewsbury to Bridgnorth rd turn off at Harley follow signs to Kenley & Church Preen. From B4371 Much Wenlock to Church Stretton road go via Hughley to Church Preen.* **Sun 23 May (2-6). Adm £6, chd free. Home-made teas.**

Fine historical and architectural 6-acre garden on site of Cluniac priory and former Norman Shaw mansion. Compartmentalised with large variety of garden rooms, including kitchen parterre and fernery. Formal terraces with fine yew and hornbeam hedges have panoramic views over parkland to Wenlock Edge. Dell, woodland walks and specimen trees. Gardens also open by appointment (not NGS).

51 NEW RORRINGTON LODGE

Rorrington, Chirbury, near Montgomery, SY15 6BX. Adrian & Samantha Boyes. *Between Chirbury & Marton. Signed from Marton & Wotherton.* **Sat 15 May (2-5.30). Adm £5, chd free. Home-made teas.**

Pooh Corner

Hillside garden with spectacular 360 views of the surrounding countryside and Welsh Hills. Interesting collection of trees and shrubs in their spring beauty lovely places to wander with some very steep banks but accessible around most of the garden. Attractive parterre-style planting around the house and pool. Plenty of places to sit, have a cup of tea and admire the views. Wheelchair access around most of the garden except steep banks.

52 RUTHALL MANOR
Ditton Priors, WV16 6TN. Mr & Mrs G T Clarke, 01746 712608, clrk608@btinternet.com. *7m SW of Bridgnorth. At Ditton Priors Church take road signed Bridgnorth. then 2nd L. Garden 1m.* **Sun 18 Apr, Mon 3, Sat 15 May, Sat 12, Sun 13 June, Sat 10, Sun 11 July (11-5.30). Adm £5, chd free. Tea. Visits also by arrangement Apr to Sept.**
Offset by a mature collection of specimen trees, the garden is divided into intimate sections, carefully linked by winding paths. The front lawn flanked by striking borders, extends to a gravel, art garden and ha-ha. Clematis and roses scramble through an eclectic collection of wrought-iron work, unique pottery and secluded seating. A stunning horse pond features primulas, iris and bog plants. Lots of lovely shrubs to see. Jigsaws for sale bring or buy. Wheelchair access to most parts.

53 SAMBROOK MANOR
Sambrook, TF10 8AL. Mrs E Mitchell, 01952 550256, eileengran@hotmail.com . *Between Newport & Ternhill, 1m off A41. In the village of Sambrook.* **Sun 18 July, Sun 29 Aug (12.30-5). Adm £5, chd free. Home-made teas. Visits also by arrangement May to Sept for groups of 10+.**
Deep, colourful, well-planted borders offset by sweeping lawns surrounding an early C18 manor house (not open). Wide ranging herbaceous planting with plenty of roses to enjoy; the arboretum below the garden, with views across the river, has been further extended with new trees. The waterfall and Japanese garden are now linked by a pretty rill. Lovely garden to visit for all the family. Woodland area difficult for wheelchairs.

54 2 SCHOOL COTTAGES
Hook-a-Gate, Shrewsbury, SY5 8BQ. Andrew Roberts & Dru Yarwood, andyrobo@hotmail.co.uk. *From B4380 Roman Rd in Shrewsbury by cemetery take the Longden Rd island. 1.8m past schools continue to the village of Hook-a-Gate. Garden is on L past Hill Side Nursery. Parking at Nursery.* **Sun 20 June, Sun 1 Aug (12-4.30). Adm £5, chd free. Home-made teas. Visits also by arrangement May to Aug for groups of 5+.**
Delights and surprises at every turn and a little Oriental mystique. Interest as soon as you enter the property, from little nooks and crannies to fabulous views towards Wales. Combination of sunken garden, borders galore and numerous water features. Woodland walk and specimen trees. Self sufficient veg, fruit, pigs and chickens all in a 3 acre site on an incline. Alterations since 2019.

GROUP OPENING

55 NEW THE SECRET GARDENS AT STEVENTON TERRACE
Steventon Terrace, Steventon New Road, Ludlow, SY8 1JZ. 01584 876037, carolynwood2152@yahoo.co.uk. *Gardens are located behind row of terraced cottages. Easily accessible from A49; on-street parking; Park & Ride stops outside the garden.* **Sat 19 June (1-5.30). Combined adm £5, chd free. Home-made teas. cakes, coffee, ice cream. Visits also by arrangement June to Sept for groups of up to 30.**
Very secret gardens behind a row of Victorian terraced cottages in Ludlow: some large, some small, but a great variety of planting and styles. All very different and a fascinating place to see how individuals have interpreted and use their green space. Includes the well-known 'Secret Garden' at No 21 which has won many awards for its design, planting and ingenuity. Come and see our secret spaces and enjoy an afternoon with us!

56 STANLEY HALL GARDENS
Bridgnorth, WV16 4SP. Mr & Mrs M J Thompson. *½ m N of Bridgnorth. Leave Bridgnorth by N gate B4373; turn R at Stanley Lane. Pass Golf Course Club House on L & turn L at Lodge.* **Sun 23 May (2-5.30). Adm £5, chd free. Home-made teas.**
Georgian landscaped drive with rhododendrons, fine trees in parkland setting, woodland walks and fish ponds. Restored ice house. Dower House (Mr and Mrs C Wells): 4 acres of specimen trees, contemporary sculpture, walled vegetable garden and potager; South Lodge (Mr Tim Warren) Hillside cottage garden. Wheelchair access to the main gardens.

GROUP OPENING

57 STOTTESDON VILLAGE OPEN GARDENS
Stottesdon, DY14 8TZ. *In glorious S Shropshire near Cleobury Mortimer (A4117/B4363). 30m from Birmingham (M5/42), 15m E of Ludlow (A49/4117) and 10m south of Bridgnorth (A458/442) Stottesdon is between Clee Hill and the Severn Valley. NGS Signed from B4363. SatNav DY14 8TZ.* **Sun 27 June (2-6). Combined adm £5, chd free. Light refreshments in the Parish Church.**
Located in unspoilt countryside near the Clee Hills, up to 10 gardens and the heritage church in Stottesdon village are open to visitors. Several places have stunning views. Some gardens feature spaces for outdoor living. Many are traditional or more modern 'cottage gardens', containing fruit, vegetables and livestock. There are contrasting vegetable gardens including one devoted to permaculture principles and one to growing championship winners. Take teas and refreshments in the Norman church and join a unique guided tour of the historic Tower, Bells and Turret Clock. A garden-related competition to be held and be judged by garden visitors. Dogs on leads please. Heritage Church open - tower tours. Lunches will be available at The Fighting Cocks pub before the 2pm NGS opening - call pub on 01746 718270 to pre-book (essential). Most gardens have some wheelchair access. Those gardens not suitable for wheelchair access will be listed.

58 SUNNINGDALE
9 Mill Street, Wem, SY4 5ED. Mrs Susan Griffiths, 01939 236733, sue.griffiths@btinternet.com. *Town centre. Wem is on B5476. Parking in public car park Barnard St. The property is opp the purple house below the church. Some on street free parking on the High St.* **Sun 9 May (11-4). Adm £4, chd free. Home-made teas. Visits also by arrangement Feb to Nov for groups of 10+.**
A good half acre town garden. A wildlife haven for a huge variety of birds including nesting gold crests. A profusion of excellent nectar rich plants means that butterflies and other pollinators are in abundance. Interesting plantings with carefully collected rare plants and unusual annuals means there is always something new to see, in a garden created for all year round viewing. Koi pond and natural stone waterfall rockery. Antique and modern sculpture. Sound break yew walkway. Large perennial borders, with rare plants, unusual annuals, exotic climbers, designed by owner as an all year round garden. Why not include a visit by having lunch at the floral Castle Hotel in Wem. Although the garden is on the level there are a number of steps mostly around the pond area; paths are mainly gravel or flags; there is a flat lawn.

59 UPPER MARSHES
Catherton Common, Hopton Wafers, nr Kidderminster, DY14 0JJ. Jo & Chris Bargman. *3m NW of Cleobury Mortimer. From A4117 follow signs to Catherton. Property is on Common land 100yds at end of track.* **Sun 4 July (12-5). Adm £5, chd free. Home-made teas.**
Commoner's stone cottage and 3 acre small holding. 800' high. Garden has been developed to complement its unique location on edge of Catherton common with herbaceous borders, vegetable plot, herb garden. Short walk down to a spring fed wildlife pond. Plenty of seats to stop and take in the tranquillity. Optional circular walk across Wildlife Trust common to SSI field. Various animals and poultry.

60 UPPER SHELDERTON HOUSE
Shelderton, Clungunford, SY7 0PE. Andrew Benton & Tricia McHaffie. *Between Ludlow & Craven Arms. Heading from Shrewsbury to Ludlow on A49, take 1st R after Onibury railway crossing. Take 3rd R signed Shelderton. After approx 2½ m the house is on L. Look out for yellow NGS signs.* **Sat 22, Sun 23 May (1.30-5). Adm £5, chd free. Home-made teas.**
Set in a stunning tranquil position, our naturalistic and evolving 6½ acre garden was originally landscaped in 1962. Most of the trees, azaleas and rhododendrons were planted then. There is a wonderful new kitchen garden designed and planted by Jayne and Norman Grove. Ponds and woodland walk encourage wildlife. A large sweeping lawn leads in various directions revealing a multitude of colourful rhododendron and azalea beds, ponds a varied collection of trees and a very productive kitchen garden. There are plenty of tranquil seating areas from which to enjoy a moment in our garden. Wheelchairs: unfortunately our garden isn't flat and there are gravel paths, however if you contact us beforehand we may be able to offer a solution.

61 WALCOT HALL
Lydbury North, SY7 8AZ. Mr & Mrs C R W Parish, www.walcothall.com. *4m SE of Bishop's Castle. B4385 Craven Arms to Bishop's Castle, turn L by Powis Arms, in Lydbury North, Sun 30 opening date same as neighbouring The Mount & Old Vicarage in Bishops Castle (4m).* **Sun 30, Mon 31 May (1.30-5.30). Adm £5, chd free. Light refreshments.**
Arboretum planted by Lord Clive of India's son, Edward. Cascades of rhododendrons, azaleas amongst specimen trees and pools. Fine views of Sir William Chambers' Clock Towers, with lake and hills beyond. Walled kitchen garden; dovecote; meat safe; ice house and mile-long lakes. Outstanding ballroom where excellent teas are served. Russian wooden church, grotto and fountain now complete and working; tin chapel. Relaxed borders and rare shrubs. Lakeside replanted, and water garden at western end re-established. The garden adjacent to the ballroom is accessible via a sloping bank, as is the walled garden and arboretum.

62 NEW WESTHOPE COLLEGE
Westhope, Craven Arms, SY7 9JL. Anne Dyer, www.westhope.org.uk. *At Westhope off B4368 Craven Arms/Bridgnorth rd. From A49 at Craven Arms turn L (from Shrewsbury) or R (from Ludlow). Follow B4368 then signs on L to Westhope College (On R from Bridgnorth direction). Watch for yellow NGS signs.* **Sat 26 June (10-5). Adm £5, chd free. Tea.**
More than 3,500 common orchids live in our meadow: a delight to see. There are wildflowers, wildlife and a lovely woodland walk with stream/pond. The now restored walled garden is full of produce and flowers. The pretty front garden welcomes visitors to the college, part of which opened for the National Garden Scheme in the 1980s. The large main garden has been rewilded for over 30 years. Good wheelchair access throughout.

63 NEW WESTWOOD HOUSE
Oldbury, Bridgnorth, WV16 5LP. Hugh & Carolyn Trevor-Jones. *Take the Ludlow Road B4364 out of Bridgnorth. Past the Punch Bowl Inn, turn 1st L, Westwood House signed on R.* **Sun 25 Apr (2-5). Adm £5, chd free. Home-made teas.**
A country garden, well designed and planted around the house, particularly known for its tulips. Sweeping lawns offset by deeply planted mixed borders; newly planted pool garden. Extensive productive kitchen and cutting garden, with everything designed to attract wildlife for organic growth. Far reaching views of this delightful corner of the county and woodland walks to enjoy. Reasonable access around the house, but gravel paths and some steps.

64 NEW 153 WILLOUGHBRIDGE
Market Drayton, TF9 4JQ. John Butcher & Sarah Berry, 07817443837, butchinoz@hotmail.com, www.instagram.com/jobobutchy/. *Approx ½ m from the Dorothy Clive Garden, turn onto Minn Bank from A51, 3rd drive on L. Parking will be signed.* **Sat 17, Sun 18 July (1-5). Adm £5, chd free. Pre-booking essential, please visit www.ngs. org.uk for information & booking. Home-made teas. Visits also by arrangement May to Sept for**

groups of up to 20.

Small and beautiful garden packed full of scented insect friendly plants. The garden includes a long cottage-prairie garden full of climbers, English roses, grasses and herbaceous perennials, a woodland-jungle garden consisting of huge scented tree lilies, tree ferns, acers, bamboo and bananas, 2 green roofs, 3 bug hotels and a large Victorian greenhouse full of traditional and exotic plants. As seen on Gardeners' World Sept 2020.

65 WINDY RIDGE

Church Lane, Little Wenlock, TF6 5BB. George & Fiona Chancellor, 01952 507675, fiona.chancellor@ngs.org.uk. *2m S of Wellington. Follow signs for Little Wenlock from N (J7, M54) or E (off A5223 at Horsehay). Parking signed. Do not rely on SatNav.* **Sat 5, Sun 6 June, Sat 4, Sun 5 Sept (12-5).**

Adm £6, chd free. Home-made teas. Visits also by arrangement May to Sept for groups of 10+.

Universally admired for its structure, inspirational planting and balance of texture, form and all-season colour, the garden more than lives up to its award-winning record. Developed over 30 years, 'open plan' garden rooms display over 1000 species (mostly labelled) in a range of colour-themed planting styles, beautifully set off by well-tended lawns, plenty of water and fascinating sculpture. Some gravel paths but help available.

66 ◆ WOLLERTON OLD HALL

Wollerton, Market Drayton, TF9 3NA. Lesley & John Jenkins, 01630 685760, info@wollertonoldhallgarden.com, www.wollertonoldhallgarden.com. *4m SW of Market Drayton. On A53 between Hodnet & A53-A41*

junction. Follow brown signs *Tickets must be pre-booked through www.wollertonoldhallgarden.com.* For NGS: Wed 8 Sept (11-5). Adm £8.50, chd £1. Light refreshments in our excellent cafe. For other opening times and information, please phone, email or visit garden website.

4-acre garden created around C16 house (not open). Formal structure creates variety of gardens each with own colour theme and character. Planting is mainly of perennials many in their late summer/early Autumn hues, particularly the Asters. Winner of many awards and nationally acclaimed. Ongoing lectures by Gardening Celebrities including Chris Beardshaw, and other garden designers and personalities. Partial wheelchair access.

Clove Hitch

© Joe Wainwright

SOMERSET, BRISTOL AREA & SOUTH GLOUCESTERSHIRE incl BATH

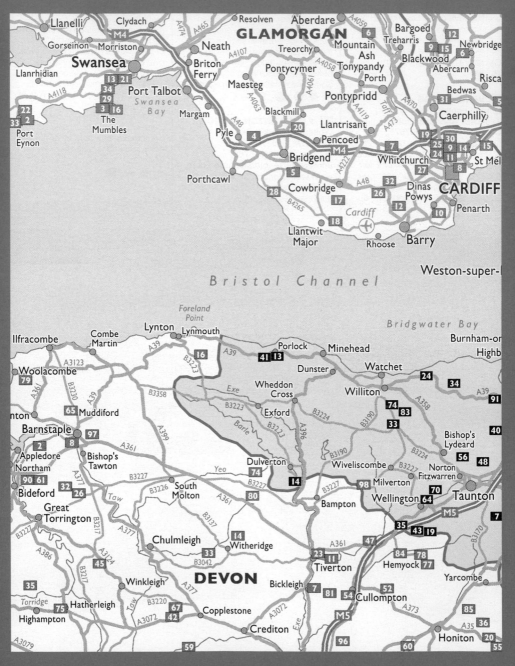

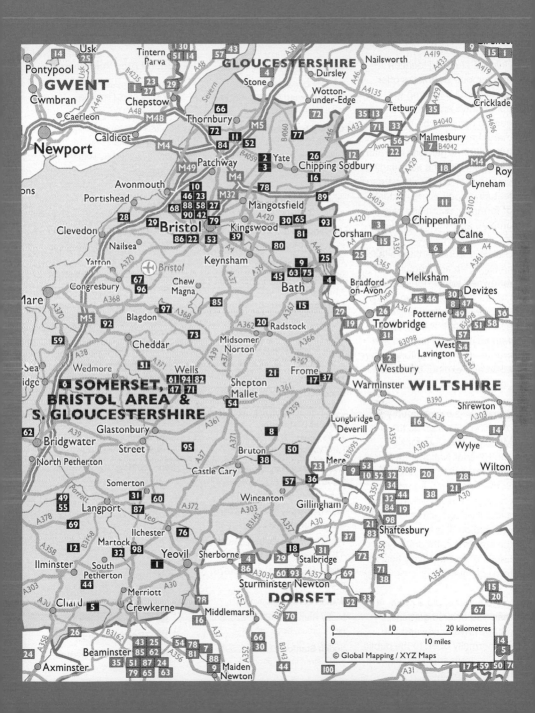

Somerset, Bristol, Bath and South Gloucestershire make up a National Garden Scheme 'county' of captivating contrasts, with castles and countryside and wildlife and wetlands, from amazing cities to bustling market towns, coastal resorts and picturesque villages.

Bristol's stunning location and famous landmarks offer a wonderful backdrop to our creative and inspiring garden owners who have made tranquil havens and tropical back gardens in urban surroundings. The surrounding countryside is home to gardens featuring contrasting mixtures of formality, woodland, water, orchard and kitchen gardens.

Bath is a world heritage site for its Georgian architecture and renowned for its Roman Baths. Our garden visitors can enjoy the quintessentially English garden of Bath Priory Hotel with its billowing borders and croquet lawn, or venture further afield and explore the hidden gems in nearby villages.

Somerset is a rural county of rolling hills such as the Mendips, the Quantocks and Exmoor National Park contrasted with the low-lying Somerset Levels. Famous for cheddar cheese, strawberries and cider; agriculture is a major occupation. It is home to Wells, the smallest cathedral city in England, and the lively county town of Taunton.

Visitors can explore more than 150 diverse gardens, mostly privately owned and not normally open to the public ranging from small urban plots to country estates.

Somerset Volunteers

County Organiser
Laura Howard 01460 282911
laura.howard@ngs.org.uk

County Treasurer
Jill Wardle 07702 274492
jill.wardle@ngs.org.uk

Publicity
Roger Peacock
roger.peacock@ngs.org.uk

Social Media
Janet Jones 01749 850509
janet.jones@ngs.org.uk

Rae Hick 07972 280083
raehick@gmail.com

Photographer
Sue Sayer 07773 181891
suesayer58@hotmail.com

Presentations
Dave & Prue Moon 01373 473381
davidmoon202@btinternet.com

Booklet Co-ordinator
John Simmons 07855 944049
john.acheta@btinternet.com

Booklet Distributor
Katie Lewis 01761 221477
katie.lewis@ngs.org.uk

Assistant County Organisers
Liz Anderson 07871 103257
lizandersoncello@gmail.com

Marsha Casely 07854 882616
marsha.casely@ngs.org.uk

Patricia Davies-Gilbert 01823
412187 pdaviesgilbert@gmail.com

Alison Highnam 01258 821576
allies1@btinternet.com

Janet Jones (as above)

Nicky Ramsay 01643 862078
nicky.ramsay@ngs.org.uk

Judith Stanford 01761 233045
judith.stanford@ngs.org.uk

Bristol Area Volunteers

County Organiser
Su Mills 01454 615438
su.mills@ngs.org.uk

County Treasurer
Harsha Parmar 07889 201185
harsha.parmar@ngs.org.uk

Publicity
Myra Ginns 01454 415396
myra.ginns@ngs.org.uk

Booklet Co-ordinator
John Simmons 07855 944049
john.acheta@btinternet.com

Booklet Distributor
John Simmons 07855 944049
john.acheta@btinternet.com

Assistant County Organisers
Tracey Halladay 07956 784838
thallada@icloud.com

Christine Healey 01454 612795
christine.healey@uwclub.net

Margaret Jones 01225 891229
ian@weircott.plus.com

Jeanette Parker 01454 299699
jeanette_parker@hotmail.co.uk

Jane Perkins 01454 414570
janekperkins@gmail.com

Irene Randow 01275 857208
irene.randow@sky.com

f @visitsomersetngs

@SomersetNGS

@ngs_bristol_s_glos_ somerset

OPENING DATES

All entries subject to change. For latest information check www.ngs.org.uk

Map locator numbers are shown to the right of each garden name.

Extended openings are shown at the beginning of the month.

January

Sunday 31st
Rock House · 72

February

Snowdrop Festival

Tuesday 2nd
◆ Elworthy Cottage · 33

Sunday 7th
◆ Elworthy Cottage · 33
Greystones · 42
Rock House · 72

Tuesday 9th
◆ Elworthy Cottage · 33

Sunday 14th
◆ East Lambrook Manor Gardens · 32

Tuesday 23rd
◆ Elworthy Cottage · 33

Sunday 28th
Algars Manor · 2
Algars Mill · 3

March

Wednesday 3rd
◆ Hestercombe Gardens · 48

Sunday 14th
Midney Gardens · 60

Monday 15th
Midney Gardens · 60

Saturday 20th
Lower Shalford Farm · 57

Sunday 21st
Rock House · 72

Sunday 28th
Rock House · 72

Tuesday 30th
◆ Elworthy Cottage · 33

April

Every Tuesday to Tuesday 20th
◆ Elworthy Cottage · 33

Saturday 3rd
Midney Gardens · 60

Sunday 4th
Midney Gardens · 60

Monday 5th
◆ Elworthy Cottage · 33

Friday 9th
NEW The Downs Preparatory School · 29

Saturday 10th
Brindham Bungalow
NEW The Downs Preparatory School · 29

Sunday 11th
Brindham Bungalow
Fairfield · 34
Rose Cottage · 73

Wednesday 14th
◆ Greencombe Gardens · 41

Thursday 15th
Bath Priory Hotel · 9

Saturday 17th
Westbrook House · 95

Sunday 18th
NEW Calsson House · 15
NEW Claylands · 19
◆ The Yeo Valley Organic Garden at Holt Farm · 97

Saturday 24th
◆ The Walled Gardens of Cannington · 91

Sunday 25th
Algars Manor · 2
Algars Mill · 3
Greystones · 42
4 Haytor Park · 46
Lucombe House · 58
◆ The Walled Gardens of Cannington · 91
Watcombe · 92

May

Sunday 2nd
NEW Claylands · 19
Gants Mill & Garden · 38

Thursday 6th
◆ Kilver Court Gardens · 54

Friday 7th
Midney Gardens · 60

Sunday 9th
Court House · 24
Holland Farm · 50
Midney Gardens · 60
◆ Milton Lodge · 61

Saturday 15th
◆ East Lambrook Manor Gardens · 32

Sunday 16th
The Red Post House · 69
Watcombe · 92

Saturday 22nd
Lower Shalford Farm · 57

Sunday 23rd
NEW Somerset Street Display Gardens · 79
◆ Stoberry Garden · 82

Saturday 29th
NEW The Hayes · 45
John's Corner · 53

Sunday 30th
NEW Claylands · 19
NEW Coombe Cottage · 22
NEW 81 Coombe Lane · 23
NEW The Hayes · 45
John's Corner · 53
Rendy Farm · 70
NEW Sunnymead · 86

Monday 31st
◆ Elworthy Cottage · 33

June

Tuesday 1st
Brindham Bungalow

Wednesday 2nd
◆ Hestercombe Gardens · 48

Thursday 3rd
Midney Gardens · 60

Saturday 5th
NEW 13 Glenarm Walk · 39
Stoneleigh Down · 84

Sunday 6th
NEW 13 Glenarm Walk · 39
Midney Gardens · 60
◆ Milton Lodge · 61
Model Farm · 62
Stoneleigh Down · 84

Wednesday 9th
Wrington Gardens · 96

Thursday 10th
Watcombe · 92

Saturday 12th
The Old Rectory, Doynton · 65

Sunday 13th
NEW 11 Brushford · 14
9 Catherston Close · 17
NEW Claylands · 19
Coleford House · 21
Wrington Gardens · 96

Monday 14th
NEW 11 Brushford · 14

Tuesday 15th
◆ Elworthy Cottage · 33

Thursday 17th
9 Catherston Close · 17
◆ Special Plants · 81

Saturday 19th
Batcombe House · 8
Lympsham Gardens · 59
NEW Oldbury on Severn Gardens · 66
Westbrook House · 95

Sunday 20th
Frome Gardens · 37
Lympsham Gardens · 59
NEW Oldbury on Severn Gardens · 66
Stogumber Gardens · 83

Saturday 26th
Babbs Farm · 6
Midney Gardens · 60
The School Yard · 77

Sunday 27th
Babbs Farm · 6
Crete Hill House · 27
4 Haytor Park · 46
Midney Gardens · 60
Nynehead Court · 64
Penny Brohn UK · 68
NEW Rowdon · 74
Yews Farm · 98

July

Saturday 3rd
165 Newbridge Hill   63
The Rib   71

Sunday 4th
[NEW] Coombe Cottage   22
Holland Farm   50
Honeyhurst Farm   51
◆ Milton Lodge   61
165 Newbridge Hill   63
[NEW] Sunnymead   86
Swift House   88

Monday 5th
Honeyhurst Farm   51
[NEW] The Royal Crescent
Hotel & Spa   75

Tuesday 6th
◆ Elworthy Cottage   33
[NEW] The Royal Crescent
Hotel & Spa   75

Wednesday 7th
9 Catherston Close   17

Thursday 8th
[NEW] ◆ Berwick Lodge   10

Friday 9th
[NEW] ◆ Berwick Lodge   10

Saturday 10th
◆ East Lambrook
Manor Gardens   32

Sunday 11th
[NEW] Badgers Holt   7
Cox's Hill House   26
The Hay Barn   44
Nynehead Court   64
[NEW] Somerset Street
Display Gardens   79
◆ University of Bristol
Botanic Garden   90

Wednesday 14th
◆ Greencombe
Gardens   41

Thursday 15th
◆ Special Plants   81

Saturday 17th
American Museum &
Gardens   4
[NEW] Goathurst
Gardens   40

Sunday 18th
Fernhill   35
[NEW] Goathurst
Gardens   40
Hangeridge Farmhouse   43
Stowey Gardens   85

Saturday 24th
[NEW] Court View   25

Sunday 25th
Court House   24
[NEW] Court View   25
◆ Elworthy Cottage   33
Park Cottage   67
Sutton Hosey Manor   87

Tuesday 27th
◆ Elworthy Cottage   33

August

Sunday 1st
◆ Elworthy Cottage   33
Fernhill   35
[NEW] The Hayes   45
Park Cottage   67

Monday 2nd
[NEW] The Hayes   45

Friday 6th
[NEW] The Downs
Preparatory School   29

Saturday 7th
[NEW] The Downs
Preparatory School   29
◆ Jekka's Herbetum   52

Thursday 19th
◆ Special Plants   81

Saturday 28th
John's Corner   53

Sunday 29th
John's Corner   53

Monday 30th
◆ Elworthy Cottage   33

September

Saturday 4th
Stoneleigh Down   84

Sunday 5th
Stoneleigh Down   84

Thursday 9th
◆ Kilver Court Gardens 54

Saturday 11th
Batcombe House   8

Sunday 12th
The Hay Barn   44
Midney Gardens   60
Yews Farm   98

Thursday 16th
◆ Special Plants   81

Saturday 18th
◆ The Walled Gardens of
Cannington   91

Sunday 19th
◆ The Walled Gardens of
Cannington   91

Sunday 26th
Midney Gardens   60

October

Thursday 14th
◆ Special Plants   81

January 2022

Sunday 30th
Rock House   72

February 2022

Sunday 6th
Rock House   72

Thursday 10th
◆ East Lambrook Manor
Gardens   32

By Arrangement

Arrange a personalised
garden visit on a date to
suit you. See individual
garden entries for full
details.

Rugg Farm

THE GARDENS

1 ABBEY FARM

Montacute, TA15 6UA. Elizabeth McFarlane, 01935 823556, abbey.farm64@gmail.com. *4m from Yeovil. Follow A3088, take slip rd to Montacute, turn L at T-junction into village. Turn R between Church & King's Arms (no through rd).* Visits by arrangement May & June for groups of 10 to 20. Evening tours with a glass of wine are recommended. Adm £6, chd free. If light refreshments are required please ask for details.
2½ acres of mainly walled gardens on sloping site provide the setting for Cluniac Medieval Priory gatehouse. Interesting plants incl roses, shrubs, grasses, clematis. Herbaceous borders, white garden, gravel garden. Small arboretum. Pond for wildlife - frogs, newts, dragonflies. Fine mulberry, walnut and monkey puzzle trees. Seats for resting. Restored Grade 2 listed dovecote. Gravel area and one steep slope.

2 ALGARS MANOR

Station Rd, Iron Acton, BS37 9TB. Mrs B Naish. *9m N of Bristol, 3m W of Yate/Chipping Sodbury. Turn S off Iron Acton bypass B4059, past village green and past White Hart PH, 200yds, then over level Xing. No access from Frampton Cotterell via lane; ignore SatNav.* Parking at Algars Manor. Sun 28 Feb (1-4); Sun 25 Apr (2-5). Combined adm with Algars Mill £7, chd free. Tea in April.
2 acres of woodland garden beside River Frome, mill stream, native plants mixed with collections of 60 magnolias and 70 camellias, rhododendrons, azaleas, eucalyptus and other unusual trees and shrubs. Daffodils, snowdrops and other early spring flowers. Teas at Algars Manor in April. Partial wheelchair access only, gravel paths, some steep and uneven slopes.

3 ALGARS MILL

Frampton End Rd, Iron Acton, Bristol, BS37 9TD. Mr & Mrs John Wright. *9m N of Bristol, 3m W of Yate/Chipping Sodbury. (For directions see Algars Manor).* Sun 28 Feb (1-4). Sun 25 Apr (2-5). Home-made teas in April. Combined

adm with Algars Manor £7, chd free.
2 acre woodland garden bisected by River Frome; spring bulbs, shrubs; very early spring feature (Feb-Mar) of wild Newent daffodils. 300-400yr-old mill house (not open) through which millrace still runs.

4 AMERICAN MUSEUM & GARDENS

Claverton Manor, Bath, BA2 7BD. American Museum & Gardens, www.americanmuseum.org. *Signposted from Bath city centre and from the A36 Warminster road.* Sat 17 July (10-5). Adm £8, chd £5.50. Admission amount for garden only. The American Museum & Gardens takes you on a journey through the history of America, from its early settlers to the C20 with its remarkable collection of folk and decorative arts. The New American Garden follows a free form planting style featuring many native American plants. The Winding Way encircles the lawn, threads through the American Rose Collection, and skirts the natural amphitheatre. We have a full access statement available on our website.

5 NEW AVALON

Higher Chillington, Ilminster, TA19 0PT. Dee & Tony Brook, 07506 688191, dee1jones@hotmail.com. *From A30 take turning signed to Chillington opp Swandown Lodges. Take 2nd L down Coley Lane and 1st L Moor Lane. Avalon is the large pink house. Parking limited, so car sharing advised if possible.* Visits by arrangement May to Aug for groups of up to 20. Max 6 cars or 2 minibuses. Adm £5, chd £2. Home-made teas.
Secluded hillside garden with wonderful views as far as Wales. The lower garden has large herbaceous borders, a sizeable wildlife pond and 2 greenhouses filled with RSA succulents. The middle garden has mixed borders, wild spotted orchids on the lawn, allotment area and a small orchard. The upper garden has a spring fed water course with ponds, plus many terraces with different planting schemes. Partial wheelchair access across lower lawns & side paths. Steep slope & gravel paths. Wheelchairs will require to be pushed & attended at all times.

Batcombe House

© Eva Nemeth

6 BABBS FARM
Westhill Lane, Bason Bridge, Highbridge, TA9 4RF.
Sue & Richard O'Brien, www.babbsfarm.co.uk. *1½m E of Highbridge, 1½m SSE of M5 exit 22. Turn into Westhill Lane off B3141 (Church Rd), 100yds S of where it joins B3139 (Wells-Highbridge rd).* **Sat 26, Sun 27 June (12-5). Adm £6, chd free. Pre-booking essential, please visit www.ngs. org.uk/events for information & booking. Home-made teas. and plants if circumstances allow.**
1 acre plantsman's garden on Somerset Levels, gradually created out of fields surrounding old farmhouse over last 30 yrs and still being developed. Trees, shrubs and herbaceous perennials planted with an eye for form and shape in big flowing borders.

🐕 ❀ 🍵

7 NEW BADGERS HOLT
Frost Street, Thurlbear, Taunton, TA3 5BA. Mr Neil Jones Ms Sharon Bradford. *Following SatNav will bring you to parking. House approx 100 yds further along Frost Street.* **Sun 11 July (10-4). Adm £4, chd free. Light refreshments.**
Newly created quintessential English Cottage garden, boasting beautiful views over the Blackdown Hills. Wheelchair side access to main garden. Some narrow stone steps and paths.

♿ ❀ 🍵

8 BATCOMBE HOUSE
Gold Hill, Batcombe, Shepton Mallet, BA4 6HF.
Libby Russell, libby@ mazzullorusselllandscapedesign. com, www. mazzullorusselllandscapedesign. com. *In centre of Batcombe, 3m from Bruton. Parking will be between Batcombe House and church at centre of village and clearly marked.* **Sat 19 June, Sat 11 Sept (2-6). Adm £6.50, chd free. Cream teas. Visits also by arrangement May to Sept for groups of 10 to 30.**
Plantswoman's and designer's garden of two parts – one a riot of colour through kitchen terraces, potager leading to wildflower orchard; the other a calm contemporary amphitheatre with large herbaceous borders and interesting trees and shrubs. Always changing. Dogs are allowed but on a lead. Wheelchairs welcome but garden is steep with

steps - difficult access to parts of garden for wheelchairs.

🐕 ❀ 🅓 🍵

9 BATH PRIORY HOTEL
Weston Rd, Bath, BA1 2XT.
Jane Moore, Head Gardener, 01225 331922, info@thebathpriory.co.uk, www.thebathpriory.co.uk. *Close to centre of Bath. Metered parking in Royal Victoria Park. No 4, 14, 39 and 37 buses from City centre. Please note: Disabled parking only in Hotel grounds.* **Thur 15 Apr (2-5). Adm £3.50, chd free. Home-made teas.**
Discover 3 acres of mature walled gardens. Quintessentially English, the garden has billowing borders, croquet lawn, wildflower meadow and ancient specimen trees. Spring is bright with tulips and flowering cherries. Perennials and tender plants provide summer highlights while the kitchen garden supplies herbs, fruit and vegetables to the restaurant. Gravel paths and some steps.

♿ ❀ 🛏 🍵

10 NEW ◆ BERWICK LODGE
Berwick Drive, Bristol, BS10 7TD.
Sarah Arikan, 0117 958 1590, info@berwicklodge.co.uk, www.berwicklodge.co.uk. *Leave M5 at J17. A4018 towards Bristol West. 2nd exit on r'about, straight on at mini r'about. At next r'about by Old Crow pub go 360° turn back down A4018, follow sign saying BL off to L.* **For NGS: Thur 8, Fri 9 July (12-5). Adm £5. Cream teas. For other opening times and information, please phone, email or visit garden website.**
Berwick Lodge, named Bristol's hidden gem by its customers, is an independent hotel with beautiful gardens on the outskirts of Bristol. Built in 1890, this Victorian Arts and Crafts property is set within 18 acres of gardens and woodland, 4 of which offer a peaceful garden for use by its visitors. The garden has been extensively developed in recent years and enjoys pretty views across to Wales. 2 elegant water fountains. Original Victorian summerhouse with veranda sits near lavender border, restored to house wedding ceremonies. Head Gardener Robert Dunster is an ex-Royal gardener, working for Prince Charles at Highgrove.

♿ 🛏 🍵

11 NEW 1 BIRCH DRIVE
Alveston, Bristol, BS35 3RQ.
Myra Ginns, 01454 415396, m.ginns1@btinternet.com. *14m N of Bristol. Alveston on A38 Bristol to Gloucester. At Alveston House Hotel turn into David's Lane. At end, L then R onto Wolfridge Ride. Birch Drive 2nd on R.* **Visits by arrangement Jan to Mar for groups of 5 to 10. Pre booking with garden owner essential. Adm £5, chd free. Light refreshments.**
A newly planted garden with particular interest in the spring. From early January a wide range of bulbs will be in flower, along with unusual named varieties of Anemone, Hellebore, Hepatica and Crocus. The main feature of the garden is the Snowdrops, collected since 2008. About 100 named varieties, flowering between November and March, will interest Snowdrop collectors. Ramp to decked area gives a view of the garden.

♿ ❀ 🍵

12 BRADON FARM
Isle Abbotts, Taunton, TA3 6RX.
Mr & Mrs Thomas Jones, deborahjstanley@hotmail.com. *Take turning to Ilton off A358. Bradon Farm is 1½m out of Ilton on Bradon Lane.* **Visits by arrangement June to Aug for groups of 10+. Adm £6, chd free. Home-made teas.**
Classic formal garden demonstrating the effective use of structure. Much to see incl parterre, knot garden, pleached lime walk, formal pond, herbaceous border, orchard and wildflower planting.

♿ 🛏 🍵

"I love the National Garden Scheme which has been the most brilliant supporter of Queen's Nurses like me. It was founded by the Queen's Nursing Institute which makes me very proud. As we battle Coronavirus on the front line in the community, knowing we have their support is a real comfort." – Liz Alderton, Queen's Nurse

Badgers Holt

Tormarton Court

BRINDHAM BUNGALOW
Wick lane, Wick, Glastonbury, BA6 8JR. Ms Elizabeth Anderson. *Foot of Glastonbury Tor. From A39 take Wells Road, 1st L after West Mendip Hospital sign signed Brindham Lane bungalow on R.* **Sat 10, Sun 11 Apr, Tues 1 June (2-4). Adm £4, chd free. Home-made teas.**
Very peaceful garden with stunning views of the Glastonbury Tor and surrounding countryside. Flower borders well stocked with roses, ceanothus, buddleja, and flowering cherries to name but a few. Wild area with bulbs, hydrangeas and cornus.

🐐 ☕

13 BROOMCLOSE
off the road to Porlock Weir, Porlock, TA24 8NU. David & Nicky Ramsay, 01643 862078, nickyjramsay@googlemail.com. *Off A39 on Porlock Weir Rd, between Porlock and West Porlock. From Porlock take rd signed to Porlock Weir. Leave houses of Porlock behind & after about 500 yards we are 1st drive on L marked Broomclose. NB Some SatNavs direct wrongly from Porlock - so beware.* **Visits by arrangement Mar to Oct for groups of up to 30. Adm £4, chd free. Home-made teas.**
Gluten-free cake available.
Large, varied garden set around early 1900s Arts and Crafts house overlooking the sea. Original stone terraces, mediterranean garden, pond, extensive borders, copse, camellia walk, large orchard with beehives. Maritime climate favours unusual trees, shrubs and herbaceous plants. We are increasingly looking to plant drought tolerant species. Mixed orchard including 45 apple varieties - many recently grafted from local heritage collections. Wildlife actively encouraged.

🐐 ☕ 🪑

14 NEW 11 BRUSHFORD
Brushford, Dulverton, TA22 9AP. Nicky & David Stewart-Smith. *Brushford is on B3222 between Dulverton and Exbridge. NGS signs from road will direct you to 11 Brushford at far end of village beyond the church.* **Sun 13, Mon 14 June (10-5). Adm £4, chd free. Home-made teas.**
Small but delightful organic cottage garden, designed in a relaxed and informal style with wildlife in mind. The garden is made up of rooms.

Wander through the flower garden with abundant rambling roses a riot of colour and scent to the secret meadow and pond with resident runner ducks and small river. Back through the quirky potager and cut flower beds where you will meet the bantams. Finish with a cup of tea and cake and browse the exhibition of local crafts in The Piggery.

🐐 ❀ ☕ 🪑

15 NEW CAISSON HOUSE
Combe Hay, Bath, BA2 7EF. Amanda Honey. *Near Wheatsheaf pub in Combe Hay. Take road to Combe Hay from Odd Down Park and Ride.* **Sun 18 Apr (11-4). Adm £7, chd free. Home-made teas.**
This is a wonderfully eclectic garden set in the most beautiful English countryside around a gorgeous Georgian house built in 1815. It is a mix of herbaceous borders, topiaries, ponds and rills, a walled garden with fruit trees, greenhouses, flower and vegetable beds. There are wildflower meadows surrounding the garden and the disused Somerset Coal Canal running through the property.

☕

16 NEW CAMERS
Badminton Road, Old Sodbury, Bristol, BS37 6RG. Mr & Mrs Michael Denman, 01454 327929, jodenman@btinternet.com, www.camers.org. *2m E of Chipping Sodbury. Entrance in Chapel Lane off A432 at Dog Inn.* **Visits by arrangement May to Sept for groups of 10+. POA, chd free. Refreshments by arrangement.**
Elizabethan farmhouse (not open) set in 4 acres of constantly developing garden and woodland with spectacular views over Severn Vale. Garden full of surprises, formal and informal areas planted with wide range of species to provide year-round interest. Parterre, topiary, Japanese garden, bog and prairie areas, white and hot gardens, woodland walks. Some steep slopes.

♿ 🐐 ❀ ☕

17 9 CATHERSTON CLOSE
Frome, BA11 4HR. Dave & Prue Moon. *15m S of Bath. Town centre W towards Shepton Mallet (A361). R at Sainsbury's r'about, follow lane for ½m. L into Critchill Rd. Over Xrds, 1st L Catherston Close.* **Sun 13, Thur 17 June, Wed 7 July (12-5). Adm £4, chd free. Opening with**

Frome Gardens on Sun 20 June. Wander around the corner to see the unexpected, a small town garden which has grown to ⅓ acre! Colour-themed shrub and herbaceous borders, patio, pergolas, pond and evolving wild meadow areas lead to wonderful far reaching views. Productive 'no dig' vegetable and fruit garden with greenhouse. Exhibition of garden photography from near and far, by the garden owner, displayed in summerhouse. Gold winner Frome-in-Bloom. Several shallow steps, gravel paths.

♿ ❀ ☕

18 CHERRY BOLBERRY FARM
Furge Lane, Henstridge, BA8 0RN. Mrs Jenny Raymond, 01963 362177, cherrybolberryfarm@tiscali.co.uk. *6m E of Sherborne. In centre of Henstridge, R at small Xrds signed Furge Lane. Continue straight up lane, over 2 cattle grids, garden at top of lane on R.* **Visits by arrangement in June for groups of 6+. Adm £6, chd free. Home-made teas.**
45 yr-old award winning, owner designed and maintained, 1 acre garden planted for yr-round interest with wildlife in mind. Colour themed Island beds, shrub and herbaceous borders, unusual perennials and shrubs, old roses and an area of specimen trees. Lots of hidden areas, brilliant for hide and seek! Vegetable and flower cutting garden, greenhouses, nature ponds. Wonderful extensive views. Garden surrounded by our dairy farm which has been in the family for over 100 years.

♿ 🐐 ❀ 🚗 ☕

19 NEW **CLAYLANDS**
Wrangway, Wellington, TA21 9QG.
John & Ruth Little, 07840 716317,
ruth.little21@yahoo.com.
*Wrangway Rd. From Wellington on
A38 turn L at Build Base towards
Pleamore Cross. L at the Xrds
towards Hemyock. Cross over the
motorway and turn R at Warwick
Farm. Follow signs.* **Sun 18 Apr (10-
5). Sun 2, Sun 30 May (10-5). Tea
in conservatory. Sun 13 June (10-
5). Adm £5, chd free. Pre-booking
essential, please visit www.ngs.
org.uk for information & booking.
Visits also by arrangement Apr
to June for groups of up to 20.
Parking is limited, especially if
wet. Please telephone.**
1.5 acres of woodland and 1.2 acres
of garden. 3 ponds, bog garden
connected by cascades. Herbaceous,
2 pergolas, hot garden and
summerhouse. Variety of specimen
trees and shrubs, particularly
magnolias. Several seating places.
Refreshments may be available.
Conservatory and most of garden are
accessible to wheelchairs, with the
exception of the woodland.

&. 🐕 ☕ 🪑

20 **COLDHARBOUR COTTAGE**
Radford Hill, Radford, Radstock,
BA3 2XU. Ms Amanda
Cranston, 01761 470600,
amanda.cranston@yahoo.co.uk.
*Bath & NE Somerset. South of
Bath, please ask for directions when
booking.* **Visits by arrangement**
May to Sept for groups of up to
10. Adm £5.50, chd £2.50. Tea.
Acre of rural garden set in peaceful
countryside. Very old garden with
fruit trees, lily ponds and colourful
herbaceous borders. White garden,
summer path planted with peonies,
roses and lavender, wide variety
of hydrangeas. Kitchen garden,
greenhouse with grape vine, herb
garden and cutting flowers plot,
romantic arches with climbing roses
and clematis. 2 summerhouses. A
garden attractive to birds and other
wildlife. Wheelchair access to grassed
areas weather-dependant. Some
narrow and uneven paths.

&. ✽ ☕ 🪑

21 **COLEFORD HOUSE**
Underhill, Coleford, Radstock,
BA3 5LU. Mr James Alexandroff.
*Coleford House is opp Kings Head
Pub in Lower Coleford with black
wrought iron gates just before bridge
over river. Parking in field 400 metres
away.* **Sun 13 June (10-4.30). Adm
£5, chd free. Home-made teas.**
The river Mells flows through this
picturesque garden with large lawns,
wildflower planting, ornamental pond,
woodland, substantial herbaceous
borders, walled garden, arboretum/
orchard, kitchen garden, vegetable
garden, bat house and orangery.
Some art work will be on sale. Most
of garden is wheelchair friendly.

&. 🐕 ☕ 🪑

22 NEW **COOMBE COTTAGE**
161 Long Ashton Road, Long
Ashton, Bristol, BS41 9JQ. Peter &
Sheena Clark. *3m SW of Bristol city
centre. Exit off A370 (Long Ashton
by-pass) onto B3128. L at junction
with Long Ashton Road (sign post
Long Ashton and opp gateway to
Ashton Court). Garden 100 yds on
R opp field.* **Sun 30 May, Sun 4
July (2-5). Combined adm with
Sunnymead £5, chd free. Home-
made teas.**
Georgian cottage overlooking field;
fronted with flagstone paved pathway
behind hedge and containing
ericaceous beds; linking with small
walled side garden laid to lawn with
herbaceous borders; arch leads to
rear paved patio with old water cistern;
beyond is hedged lawn bordered by
rockery, small pond and beds stocked
with flowers and shrubs.

🐕 ☕

23 NEW **81 COOMBE LANE**
Stoke Bishop, Bristol, BS9 2AT. Karl
Suchy. *4m from Bristol city centre. J17
of M5, then follow A4018, direction
Bristol for approx 2m. Turn R onto
Canford Rd A4162. After 0.8m turn L
onto Coombe Lane. Destination on R.*
**Sun 30 May (1-5). Adm £5, chd free.
Home-made teas.**
Hidden Victorian walled garden.
You'll encounter substantial mixed
borders containing traditional and
contemporary planting. Large
lawns, numerous seating areas,
summerhouse and French inspired

Court View

patio with coppiced lime trees. Parterre and large raised Koi pond surrounded by bananas and tree ferns. Access to parterre and Koi pond might be difficult for wheelchair due to narrow gravel path, main part of garden is accessible.

24 COURT HOUSE

East Quantoxhead, TA5 1EJ. Mr & Mrs Hugh Luttrell. *12m W of Bridgwater. Off A39, house at end of village past duck pond. Enter by Frog Street (Bridgwater/Kilve side from A39). Car park £1 in aid of church.* **Sun 9 May, Sun 25 July (2-5). Adm £5, chd free. Home-made teas. For other opening times and information, please phone or email.** Lovely 5 acre garden, trees, shrubs (many rare and tender), herbaceous and 3 acre woodland garden with spring interest and late summer borders. Traditional kitchen garden (chemical free). Views to sea and Quantocks. Gravel, stone and some mown grass paths.

25 NEW COURT VIEW

Solsbury Lane, Batheaston, Bath, BA1 7HB. Maria & Jeremy Heffer, www.thebathgreenhouse.com. *3m E of Bath. In Batheaston High St take turning on L signed Northend, St Catherine. At top of rise L into Solsbury Lane. Court View is 2nd driveway on L. Parking in village.* **Sat 24, Sun 25 July (11-5). Adm £5, chd free. Home-made teas. All refreshment money to Bath Men's Shed.** 2 acre S-facing gardens with 1/3 acre devoted to cut flowers and foliage. Colourful mix of annuals, biennials and perennials. Spectacular views from terraced lawns, box parterre, small orchard and meadow area. A floral experience for garden lovers, artisan florists, flower arrangers and anyone interested in the revival of beautiful, diverse and locally grown British cut flowers.

26 COX'S HILL HOUSE

Horton, Bristol, BS37 6QT. Charles Harman. *Approx 1m W of A46, 4m N of M4 J18. What3words app - giggle.listen.nicer.* **Sun 11 July (2-5.30). Adm £5, chd free. Teas in aid of the Nelson Trust.** 1 acre garden created by current owners over past 12 yrs and laid out

over 2 levels, each backed by a high stone wall, with panoramic views to south and west. Small vegetable/ cutting garden and formal lawn flanked by pleached limes on lower level; upper level has herbaceous, shrub and yellow themed borders and orchard with meadow grass and wildflowers.

27 CRETE HILL HOUSE

Cote House Lane, Durdham Down, Bristol, BS9 3UW. John Burgess. *2m N of Bristol city centre, 3m S J16 M5. A4018 Westbury Rd from city centre, L at White Tree r'about, R into Cote Rd, continue into Cote House Lane across the Downs. 2nd house on L. Parking on street.* **Sun 27 June (1-5). Adm £4, chd free. Home-made teas.** C18 house in hidden corner of Bristol. Mainly SW facing garden, 80'x40', with shaped lawn, heavily planted traditional mixed borders - shrub, rose, clematis and herbaceous. Pergola with climbers, terrace with pond, several seating areas. Shady walled garden. Roof terrace (44 steps) with extensive views. A couple of seating areas not accessible by wheelchair as accessed though borders via stepping stones. Whole garden can be viewed.

28 THE DAIRY, CHURCH LANE

Clevedon Road, Weston-in-Gordano, Bristol, BS20 8PZ. Mrs Chris Lewis, 01275 849214, chris@dairy.me.uk. *Weston-in-Gordano is on B3124 Portishead to Clevedon rd. Find Parish Church on main rd and take lane down side of churchyard for 200m. Access by coach involves a walk of 400m.* **Visits by arrangement May to Sept for groups of 10 to 30. Payment on booking. Adm £7, chd free. Home-made teas.** The garden surrounds a barn conversion and has been developed from concrete milking yards and derelict land. Once the site of Weston in Gordano Manor House, the ambience owes much to the use of medieval stone which had lain undiscovered in the land for over 2 centuries. Changes of level, with steps and gravel paths, sadly make wheelchair access difficult.

29 NEW THE DOWNS PREPARATORY SCHOOL

Charlton Drive, Wraxall, Bristol,

BS48 1PF. thedownsschool.co.uk. *4.3m from J19 of the M5 or 8.3m from the centre of Bristol. Follow signs for Noahs Ark Zoo Farm from motorway or centre of Bristol.* **Fri 9, Sat 10 Apr, Fri 6, Sat 7 Aug (10-4). Adm £6, chd free. Tea, coffee, home-made cakes and biscuits.** 65 acres wrap around the Grade II listed Charlton House, once part of a wider estate incl the well-known Tyntesfield National Trust property. Come and discover historic garden features, stumpery, pond and greenhouse. In addition, an edible and medicinal bed and well-presented annual bedding displays framed by beautiful views across open parkland with specimen trees dotted around the estate. Gravel paths and some steps. Majority of garden is wheelchair accessible.

30 DOYNTON HOUSE

Bury Lane, Doynton, Bristol, BS30 5SR. Frances & Matthew Lindsey-Clark, Franceslc11@gmail.com. *5m S of M4 J18, 6m N of Bath, 8m E of Bristol. Doynton is NE of Wick (turn off A420 opp Bath Rd) and SW of Dyrham (signed from A46). Doynton House is at S end of Doynton village, opp Culloysgate/ Horsepool Lane. Park in signed field.* **Visits by arrangement Apr to Sept. Please contact us regarding refreshments. Adm £7.50,** A variety of garden areas separated by old walls and hedges. Mixed borders, lawns, wall planting, parterre, rill garden, walled vegetable garden, cottage beds, pool garden, dry garden, peach house and greenhouse. Bees. Chickens. Maturing meadow area. Paths are of hoggin, stone and gravel. The grade of the gravel makes it a hard push in places but all areas are wheelchair accessible.

31 EAST END FARM

Pitney, Langport, TA10 9AL. Mrs A M Wray, 01458 250598. *2m E of Langport. Please telephone for directions.* **Visits by arrangement in June. Adm £5, chd free. Light refreshments.** Approx 1/3 acre. Timeless small garden of many old-fashioned roses in beautiful herbaceous borders set amongst ancient listed farm buildings. Mostly wheelchair access - some narrow paths.

Cox's Hill House

32 ◆ **EAST LAMBROOK MANOR GARDENS**
Silver Street, East Lambrook, TA13 5HH. Mike & Gail Werkmeister, 01460 240328, enquiries@eastlambrook.com, www.eastlambrook.com. *2m N of South Petherton. Follow brown tourist signs from A303 South Petherton r'about or B3165 Xrds with lights N of Martock.* **For NGS: Sun 14 Feb pre-booking essential, please see eastlambrook.com for information & booking, Sat 15 May, Sat 10 July (10-5). Adm £6, chd free. Discount vouchers not valid. 2022: Thur 10 Feb. For other information, please phone, email or visit garden website.**
The quintessential English cottage garden created by C20 gardening legend Margery Fish. Plantsman's paradise with contemporary and old-fashioned plants grown in a relaxed and informal manner to create a remarkable garden of great beauty and charm. With noted collections of snowdrops, hellebores and geraniums

and the excellent specialist Margery Fish Plant Nursery. Also open 1 Feb - 31 Oct, Tues to Sat plus Suns in Feb; (10-5). Partial wheelchair access.

33 ◆ **ELWORTHY COTTAGE**
Elworthy, Taunton, TA4 3PX. Mike & Jenny Spiller, 01984 656427, mike@elworthy-cottage.co.uk, www.elworthy-cottage.co.uk. *12m NW of Taunton. On B3188 between Wiveliscombe and Watchet.* **For NGS: Tue 2, Sun 7, Tue 9, Tue 23 Feb (11-3); Tue 30 Mar, Mon 5 Apr (11-4). Every Tue 6 Apr to 20 Apr (11-4). Mon 31 May, Tue 15 June, Tue 6, Sun 25, Tue 27 July, Sun 1, Mon 30 Aug (11-4). Adm £4, chd free. Home-made teas. For other opening times and information, please phone, email or visit garden website.**
1 acre plantsman's garden in tranquil setting. Island beds, scented plants, clematis, unusual perennials and ornamental trees and shrubs to provide yr-round interest. In spring,

pulmonarias, hellebores and more than 350 varieties of snowdrops. Planted to encourage birds, bees and butterflies, lots of birdsong. wildflower areas and developing wildflower meadow, decorative vegetable garden, living willow screen. Seats for visitors to enjoy views of the surrounding countryside. Garden attached to plantsman's nursery, open at the same time.

34 **FAIRFIELD**
Stogursey, Bridgwater, TA5 1PU. Lady Acland Hood Gass. *7m E of Williton. 11m W of Bridgwater. From A39 Bridgwater to Minehead rd turn N. Garden 1½ m W of Stogursey on Stringston rd. No coaches.* **Sun 11 Apr (2-5). Adm £5, chd free. Home-made teas.**
Woodland garden with many interesting bulbs incl naturalised anemones, fritillaria with roses, shrubs and fine trees. Paved maze. Views of Quantocks and sea.

35 FERNHILL
Whiteball, Wellington,
TA21 0LU. Peter Bowler, www.
sampfordarundel.org.uk/fernhill.
*3m W of Wellington. At top of
Whiteball hill on A38 on L going
West, just before dual carriageway,
parking on site, Blue Badge parking
only in front of house please.* **Sun
18 July, Sun 1 Aug (2-5). Adm £4,
chd free. Home-made teas.**
In approx 2 acres, a delightful garden
to stir your senses, with a myriad
of unusual plants and features.
Intriguing almost hidden paths leading
through English roses and banks of
hydrangeas. Scenic views stretching
up to the Blackdowns and its famous
monument. Truly a Hide and Seek
garden ..... for all ages. Well stocked
herbaceous borders, octagonal
pergola and water garden with slightly
wild boggy area and a specimen
Dawn Redwood dating from 1960.
Wheelchair access to tea terrace is
from drive and front of house only.
&

36 FOREST LODGE
Pen Selwood, BA9 8LL.
Mr & Mrs James & Lucy
Nelson, 07974 701427,
lucillanelson@gmail.com. *1½m
N of A303, 3m E of Wincanton.
Leave A303 at B3081 (Wincanton
to Gillingham rd), up hill to Pen
Selwood, L towards church. ½m,
garden on L - low curved wall and
sign saying Forest Lodge Stud.*
**Visits by arrangement Apr to Oct
for groups of 5 to 30. Adm £7, chd
free. Donation to Heads Up Wells,
Balsam Centre Wincanton.**
3 acre mature garden with many
camellias and rhododendrons in May.
Lovely views towards Blackmore
Vale. Part formal with pleached
hornbeam allée and rill, part water
garden with lake. Wonderful roses in
June. Unusual spring flowering trees
such as Davidia involucrata, many
beautiful cornus. Interesting garden
sculpture. Wheelchair access to front
garden only, however much of garden
viewable from there.
&

GROUP OPENING

37 FROME GARDENS
Frome, BA11 4HR. *15m S of Bath.
9 Catherston Cl signed r'about
A361 W of town. Elmfield Hse, New
Buildings Lane, A362 Portway T-light,
Locks Hill, 5th L signed. Please park
considerately on nearby roads.*

**Sun 20 June (12-5). Combined
adm £6, chd free.**

9 CATHERSTON CLOSE
Dave & Prue Moon.
(See separate entry)

NEW ELMFIELD HOUSE
Nigel & Frances Day.

A warm welcome awaits you at
both contrasting town gardens
developed to share the lives and
character of the current owners. 9
Catherston Close will bowl you over
on seeing how this town garden
grew into ⅓ acre! Colour themed
borders, pergolas, pond, evolving wild
meadow area with fine far reaching
views. No dig veg and fruit plots.
Exhibition of owner's photography
on display. Elmfield House, entering
via a wicket gate engenders a feeling
of secrecy, surprise and welcome.
Surrounded by high stone walls lies a
60' by 100' secret garden with formal
lawn, mature trees and overflowing
flowerbeds. An elaborately decorated
folly completes the picture.
Catherston Cl few gravel paths,
slopes, shallow steps. Elmfield Hse
few shallow steps.
&

38 GANTS MILL & GARDEN
Gants Mill Lane, Bruton,
BA10 0DB. Elaine & Greg Beedle,
gantsmill.co.uk. *½m SW of Bruton.
From Bruton centre take Yeovil rd,
A359, under railway bridge, 100yds
uphill, fork R down Gants Mill Lane.
Parking for wheelchair users.* **Sun
2 May (2-5). Adm £6, chd free.
Home-made teas.**
¾ acre flower garden with tulips and
wallflowers. Vegetable garden. The
gardens are overlooked by the historic
watermill, open on NGS day. Firm
wide paths round garden. Narrow
entrance to mill not accessible to
wheelchairs. WC.
&

39 NEW 13 GLENARM WALK
Brislington, Bristol, BS4 4LS.
Martin Fitton. *A4 Bristol to Bath.
A4 Brislington, at Texaco Garage at
bottom of Bristol Hill turn into School
Rd and immediate R into Church
parade. Car Park 1st turn on R or
take 2nd R into Glenarm Rd, and
2nd R again for car park.* **Sat 5, Sun
6 June (1-4). Adm £4, chd free.
Home-made teas.**
As you walk through the gate you

will be welcomed by Japanese Koi.
Then take a step to another level to
the relaxing Japanese garden rooms
surrounded by Acers and cloud
trees. Past the fire pit area where
you can embrace the true sound of
Japan, continue through a gate to
a peaceful Zen garden. There you
will find seating to enjoy the serene
atmosphere.

GROUP OPENING

40 NEW GOATHURST GARDENS
Goathurst, Bridgwater, TA5 2DF.
*4m SW of Bridgwater, 2½m W of
North Petherton. Close to church in
village. Park in field at N end village
(signed) 500 yds from gardens.* **Sat
17, Sun 18 July (2-5). Combined
adm £5, chd free. Home-made
teas at The Lodge.**

NEW THE LODGE
Sharon & Richard Piron.

OLD ORCHARD
Mr Peter Evered.

Only 30 yds apart, 2 beautiful
examples of quintessentially English
cottage gardens in a rural village
setting. Old Orchard: ¼ acre garden
planted to complement the cottage
with over 100 clematis viticella
interplanted with a range of shrubs,
herbaceous perennials, annuals
and summer bulbs. The Lodge: ⅓
acre garden surrounding a thatched
cottage comprising flower borders
packed with shrubs and perennials.
Fruit and vegetable beds are also
a feature. Both gardens have
wheelchair access to most areas.
&

41 ◆ GREENCOMBE GARDENS
Porlock, Minehead,
TA24 8NU. Greencombe
Garden Trust, 01643 862363,
info@greencombe.org,
www.greencombe.org. *Just W
of Porlock on wooded slopes of
Exmoor. Follow A39 to west end
of Porlock and turn onto B3225 to
Porlock Weir. Drive ½m and turn
L at Greencombe Gardens sign.
Go up drive; parking signed.* **For
NGS: Wed 14 Apr, Wed 14 July
(2-6). Adm £7, chd £1. Cream
teas. For other opening times and
information, please phone, email or
visit garden website. Donation to
Plant Heritage.**

Organic woodland garden of international renown, Greencombe stretches along a sheltered hillside and offers outstanding views over Porlock Bay. Moss-covered paths meander through a collection of ornamental plants that flourish beneath a canopy of oaks, hollies, conifers and chestnuts. Camellias, rhododendrons, azaleas, lilies, roses, clematis, and hydrangeas blossom among 4 National Collections. Champion English Holly tree (Ilex aquifolium), the largest and oldest in the UK. A millennium chapel hides in the wood. An ecologically constructed Green Room holds garden records, collection information, and paintings by Exmoor artist Jon Hurford.

42 GREYSTONES

Hollybush Lane, Bristol, BS9 1JB. Mrs P Townsend. *2m N of Bristol city centre, close to Durdham Down in Bristol, backing onto the Botanic Garden. A4018 Westbury Rd, L at White Tree r'about, L into Saville Rd, Hollybush Lane 2nd on R. Narrow lane, parking limited, recommended to park in Saville Rd.* **Sun 7 Feb (12-3). Sun 25 Apr (2-5). Home-made teas. Adm £3.50, chd free. No teas on 7 Feb.**
Peaceful garden with places to sit and enjoy a quiet corner of Bristol. Interesting courtyard, raised beds, large variety of conifers and shrubs leads to secluded garden of contrasts - sun drenched beds with olive tree and brightly coloured flowers to shady spots, with acers, hostas and ferns. Snowdrops, hellebores and spring bulbs. Rambling roses, small orchard, espaliered pears, koi pond. Paved footpath provides level access to all areas.

43 HANGERIDGE FARMHOUSE

Wrangway, Wellington, TA21 9QG. Mrs J M Chave, 07812 648876, hangeridge@hotmail.co.uk. *2m S of Wellington. Off A38 Wellington bypass signed Wrangway. 1st L towards Wellington monument, over motorway bridge 1st R.* **Sun 18 July (2-5). Adm £3, chd free. Home-made teas. Visits also by arrangement June to Aug for groups of 5 to 10.**
Rural fields and mature trees surround this 1 acre informal garden offering views of the Blackdown and Quantock Hills. Magnificent hostas and heathers, colourful flower beds,

cascading wisteria and roses and a trickling stream. Relax with home-made refreshments on sunny or shaded seating admiring the views and birdsong.

44 THE HAY BARN

Kingstone, Ilminster, TA19 0NS. Philippa Sage. *Kingstone is 1m out of Ilminster on Crewkerne road. Coming from Ilminster take Crewkerne rd, at Kingstone Church turn L, road turns to gravel follow round to L. 1st house on R.* **Sun 11 July, Sun 12 Sept (1.30-5.30). Adm £5, chd free. Home-made cakes and tea/coffee/cold drinks. Gluten free options available.**
Delightful garden created over past 9 yrs, wrapping around an attractive Moolham stone barn conversion creating a wonderful sense of peace and tranquility. Paths invite you around the garden to view the collection of unusual plants, shrubs and trees providing yr-round colour and interest. Gravel areas provide ideal planting for drought loving plants. Trees link the garden into the countryside. Front of house and driveway gravel, making pushing a wheelchair hard. Please ask for assisted parking.

45 NEW THE HAYES

Newton St. Loe, Bath, BA2 9BU. Jane Giddins. *From the Globe Pub on A4 take Pennyquick which is signed to Odd Down. Take 1st turning R which is just after the bend. At top of hill turn R then L.* **Sat 29, Sun 30 May, Sun 1, Mon 2 Aug (11-4). Adm £6, chd free. Home-made teas.**
Stunning in all seasons. 1 acre garden on edge of Duchy of Cornwall village. Herbaceous borders and formal lawns and terraces; informal garden of trees and long grass, bulbs and meadow flowers; formal potager and greenhouse; small orchard with espalier apple trees. Tulips, wisteria, alliums, foxgloves, gladioli, dahlias, asters. Wonderful views. All of garden can be accessed in a wheelchair but some grassy inclines.

46 4 HAYTOR PARK

Bristol, BS9 2LR. Mr & Mrs C J Prior, 07779 203626, p.l.prior@gmail.com. *3m NW of Bristol city centre. From A4162 Inner Ring Rd take turning into Coombe Bridge Ave, Haytor Park is 1st on L. Please no parking in Haytor Park.* **Sun 25 Apr, Sun 27 June (1.30-5). Adm £3.50, chd free. Visits also by arrangement Apr to Aug for groups of 10 to 20.**
Continually changing over 34 yrs, a peaceful and very personal sanctuary in a leafy suburb. Packed with plants along meandering paths, in quirky spaces, around a wildlife pond and on a green roof. Plants cascade at every level, from many arches and novel screens. Children will enjoy the dragon hunt competition. Celebrate 21 yrs opening for the NGS with a plantaholic!

47 HENLEY MILL

Henley Lane, Wookey, Wells, BA5 1AW. Peter & Sally Gregson, 01749 676966, millcottageplants@gmail.com, www.millcottageplants.co.uk. *2m W of Wells, off A371 towards Cheddar. From A371 L signed Wookey into Henley Lane. Immed L between stone pillars, drive 100yds down drive. Henley Mill is white house with black doors. Parking for 10 cars. Coach drop-off at end of drive.* **Visits by arrangement Apr to Sept. Adm £5.50, chd free. Home-made teas.**
Developed around River Axe, with a zigzag boardwalk at river level, lies 2½ acres of scented garden with roses, hydrangea borders, shady folly garden and late summer borders with grasses and perennials. Kitchen and cutting garden. Deck overhangs the mill leat and looks down onto gunneras, Siberian iris and miscanthus. Due to the river children must be accompanied by an adult at all times please. Wild area of native English daffodils, fritillaries and cowslips in spring. Rare Japanese hydrangeas and new Chinese epimediums in shade. Collections of Itoh peonies and Benton bearded irises that should be in flower in May/June. Paths may get muddy after heavy rain, please bring suitable footwear.

48 ◆ HESTERCOMBE GARDENS

Cheddon Fitzpaine, Taunton, TA2 8LG. Hestercombe Gardens Trust, 01823 413923, info@hestercombe.com, www.hestercombe.com. *3m N of Taunton, less than 6m from J25 of M5. Follow brown daisy signs. SatNav postcode TA2 8LQ.* **For NGS: Wed 3 Mar, Wed 2 June (10-4). Adm £13.30, chd £6.65. Discount/prepaid vouchers not valid on NGS charity days kindly donated by Hestercombe. For other opening times and information, please phone, email or visit garden website.**

Magnificent Georgian landscape garden designed by artist Coplestone Warre Bampfylde, Victorian terrace and shrubbery and an exquisite example of a Lutyens/Jeykll designed formal garden; C17 water garden opening 2021. Enjoy 50 acres of woodland walks, temples, terraces, pergolas, lakes and cascades. Restaurant and café, restored watermill and barn, historic house and family garden trails. Gravel paths, steep slopes, steps. All abilities access route marked. Tramper mobility scooter available, booking recommended in advance.

49 HILLCREST

Curload, Stoke St Gregory, Taunton, TA3 6JA. Charles & Charlotte Sundquist, 01823 490852, chazfix@gmail.com. *At top of Curload. From A358 turn L along A378, then branch L to North Curry and Stoke St. Gregory. L ½ m after Willows & Wetlands centre. Hillcrest is 1st on R with parking directions.* **Visits by arrangement Apr to Sept for groups of up to 30. Admission price includes tea/coffee and home-made cakes. Adm £8, chd free.**

Garden boasts stunning views of the Somerset Levels, Burrow Mump and Glastonbury Tor and even on a hazy day this 6 acre garden offers plenty of interest. Woodland walks, varied borders, flowering meadow and several ponds; kitchen garden, greenhouses, orchards and unique standing stone as focal point. Most of garden is level. Long gently sloping path through flower meadow to lower pond and wood.

50 HOLLAND FARM

South Brewham, Bruton, BA10 0JZ. Mrs Nickie Gething. *3m N of Wincanton. From Wincanton: B3081for 3m toward Bruton. R to S Brewham. Follow NGS signs. From Mere: B3092 toward Frome. Under A303, past Stourhead Hse. L to Alfred's Tower/Kilmington. Follow NGS signs.* **Sun 9 May, Sun 4 July (2-5). Adm £6, chd free. Home-made teas in Garden Room.**

The house and garden were created from a derelict farmyard 15 yrs ago. Garden now boasts a number of exquisite rooms, divided by hornbeam and yew hedging. Some rooms are tranquil, with a simple water feature or trees, others burst with a variety of planting. House frames a stunning French-style courtyard where the sound of water echoes. Hornbeam avenue leads the visitor to a swimming lake with lakeside planting. Partial wheelchair access.

Rowdon

Claylands

51 HONEYHURST FARM
Honeyhurst Lane, Rodney Stoke,
Cheddar, BS27 3UJ. Don &
Kathy Longhurst, 01749 870322,
donlonghurst@btinternet.com,
www.ciderbarrelcottage.co.uk. *4m
E of Cheddar. From Wells (A371) turn
into Rodney Stoke signed Wedmore.
Pass church on L and continue
for almost 1m. Car park signed.
From Cheddar (A371) turn R signed
Wedmore, through Draycott to car
park.* **Sun 4, Mon 5 July (2-5). Adm
£4, chd free. Home-made teas.
Visits also by arrangement May to
Aug for groups of 10 to 30.**
²/₃ acre part walled rural garden with
babbling brook and 4 acre traditional
cider orchard, with views. Specimen
hollies, copper beech, paulownia,
yew and poplar. Pergolas, arbour
and numerous seats. Mixed informal
shrub and perennial beds with many
unusual plants. Many pots planted
with shrubs, hardy and half-hardy
perennials. Level, grass and some
shingle.
&♿ 🐾 ❀ 🚗 🛏 ☕ 🍵

52 ◆ JEKKA'S HERBETUM
Shellards Lane, Alveston, Bristol,
BS35 3SY. Mrs Jekka McVicar,
01454 418878, sales@jekkas.com,
www.jekkas.com. *7m N of M5 J16
or 6m S from J14 of M5. 1m off A38
signed Itchington. From M5 J16,
A38 to Alveston, past church turn
R at junction signed Itchington. M5
J14 on A38 turn L after T-lights to
Itchington.* **For NGS: Sat 7 Aug**

(10-4). Adm £5. Pre-booking
essential, please visit www.
jekkas.com for information &
booking. Light refreshments. Our
herb inspired café will be open
offering seasonal herb-based
treats, home-made cakes and
coffee as well as Jekka's herbal
infusions. For other opening times
and information, please phone,
email or visit garden website.
Jekka's Herbetum is a living
encyclopaedia of herbs, displaying the
largest collection of culinary herbs in
Europe. A wonderful resource for plant
identification for the gardener and a
gastronomic experience for chefs and
cooks. Wheelchair access possible
however terrain is rough from car park
to Herbetum.
&♿ ❀ ☕

53 JOHN'S CORNER
2 Fitzgerald Road, Bedminster,
Bristol, BS3 5DD. John Hodge.
*3m from city centre. S of Bristol, off
St.John's Lane, Totterdown end.
1st house on R, entrance at side of
house. On number 91 bus route.
Parking in residential street.* **Sat 29,
Sun 30 May, Sat 28, Sun 29 Aug
(1-5). Adm £4, chd free.**
Unusual and interesting city garden in
Bedminster with a mixture of exciting
plants and features. Ponds, ferns
and much more. Eden project style
greenhouse with collection of cacti.
Not all areas accessible by wheelchair.
&♿ 🐾 ☕

54 ◆ KILVER COURT GARDENS
Kilver Street, Shepton
Mallet, BA4 5NF. Roger &
Monty Saul, 01749 340410,
info@kilvercourt.com,
www.kilvercourt.com. *Directly off
A37 rd to Bath, opp Showerings
factory in Shepton Mallet. Disabled
parking in lower car park.* **For
NGS: Thur 6 May, Thur 9 Sept
(10-4). Adm £7.50, chd free. Light
refreshments in Sharpham Pantry
Restaurant & Harlequin Café, pl
check closing time on the day.
Discount/prepaid vouchers are
not valid on the 2 NGS charity
days kindly donated by Kilver
Court. Last entry into the garden
is 3.15pm. For other opening times
and information, please phone,
email or visit garden website.**
Visitors can wander by the millpond,
explore the formal and informal
gardens and enjoy a replica of the
splendid Chelsea Flower Show
Gold Medal winning rockery where
a gushing recirculated stream flows
from pool to pool and waterfalls
into the lake. All this set against
the stunning backdrop of Charlton
Viaduct with the recently planted
100m border beyond. Some slopes,
rockery not accessible for wheelchairs
but can be viewed.
&♿ ❀ 🚗 ☕

*All entries are subject to change.
For the latest information check
ngs.org.uk*

55 LANE END HOUSE

Curload, Stoke St Gregory, Taunton, TA3 6JA. Eric & Veronica Martin, 01823 491261, va.martin@gmail.com. *Taunton A358 turn L on A378, Fork L to North Curry & Stoke St Gregory. L ½ m after Willows & Wetlands Centre. 1st house on l. Limited parking - up to 6 cars.* **Visits by arrangement May to Aug for groups of up to 20. Adm £5, chd free.**
Mature Somerset Levels garden set on heavy clay with mixed borders, orchard, veg patch, greenhouses and free range chickens. The recent introduction of a ½ acre field into the garden sees mature trees and shrubs sharing the garden with newly planted specimen trees, a pond and wildflower areas that support our own honey bees and a variety of wildlife. Unique sculptures enhance the 1 acre plot. Level gardens with gravel courtyard.

&

56 LITTLE YARFORD FARMHOUSE

Kingston St Mary, Taunton, TA2 8AN. Brian Bradley, 01823 461350, dilly.bradley@gmail.com. *1½ m W of Hestercombe, 3½ m N of Taunton. From Taunton on Kingston St Mary rd. At 30mph sign turn L at Parsonage Lane. Continue 1¼ m W, to Yarford sign. Continue 400yds. Turn R up concrete rd.* **Visits by arrangement Apr to Sept. Easy reach M5 (J25) and Hestercombe for coffee/lunch etc. Adm £5, chd free. Cream teas/light refreshments by prior arrangement.**
Unusual 5 acre garden embracing C17 house (not open) overgrown with climbing plants. Natural pond and 90ft waterlily pond. Its special interest are its 300+ rare and unusual tree cultivars: the best collection of broad leaf and conifer cultivars in Somerset West (listed on NGS website); those trees not available to Bampfylde Warre at Hestercombe in C18. An exercise in landscaping; contrast planting and creating views both within the garden and without to the vale and the Quantock Hills. Featured in Somerset Living 2019. On Google map. Mostly wheelchair access.

& 🐎 ✿ 🚗 ☕

57 LOWER SHALFORD FARM

Shalford Lane, Charlton Musgrove, Wincanton, BA9 8HE. Mr & Mrs David Posnett. *Lower Shalford is 2m NE of Wincanton. Leave A303 at Wincanton go N on B3081 towards Bruton. Just beyond Otter Garden Centre turn R Shalford Lane, garden is ½ m on L. Parking opp house.* **Sat 20 Mar (10-3); Sat 22 May (10-4). Adm £5, chd free. Light refreshments.**
Fairly large open garden with extensive lawns and wooded surroundings with drifts of daffodils in spring. Small winterbourne stream running through with several stone bridges. Walled rose/parterre garden, hedged herbaceous garden and several ornamental ponds. Partial wheelchair access.

58 LUCOMBE HOUSE

12 Druid Stoke Ave, Stoke Bishop, Bristol, BS9 1DD. Malcolm Ravenscroft, 01179 682494, famrave@gmail.com. *4m NW of Bristol centre. At top of Druid Hill. Garden on R 200m from junction.* **Sun 25 Apr (2-5). Adm £3.50, chd free. Home-made teas. Also open 4 Haytor Park. Visits also by arrangement Apr to Sept.**
For tree lovers of all ages!! In addition to the 255 yr old Lucombe Oak - registered as one of the most significant trees in the UK - there are over 30 mature English trees planted together with ferns and bluebells to create an urban woodland - plus a newly designed Arts & Craft front garden. A new path through the woodland will be completed in time for the NGS opening. It is hoped that a musical group featuring flautists will be present. Rough paths in woodland area, 2 steps to patio.

& ✿ ☕

GROUP OPENING

59 LYMPSHAM GARDENS

Church Road, Lympsham, Weston-super-Mare, BS24 0DT. *5m S of Weston-super-Mare and 5m N of Burnham on Sea. 2m M5 J22. Entrance to all 3 gardens initially from main gates of Manor at junction of Church Rd and Lympsham Rd.* **Sat 19, Sun 20 June (2-5). Combined adm £6, chd free. Cream teas at The Manor.**

CHURCH FARM
Andy & Rosemary Carr.

NEW **HAWTHORNS**
Mrs Victoria Daintree.

LYMPSHAM MANOR
James & Lisa Counsell.

NEW **SOMERDOWN FARM**
Claire & Martin Sleight.

At the heart of the stunning village of Lympsham next to C15 church are the gardens of 3 of the village's most historic homes. The Manor, a 200 yr old gothic rectory manor house with 2 octagonal towers, is set in 10 acres of formal/semi-formal garden, surrounded by paddocks and farmland. Fully working Victorian kitchen garden and greenhouse, arboretum of trees from all parts of the world, a large pond and beautiful old rose garden. Church Farm has a ¾ acre informal English country garden with herbaceous borders, shrub-lined paths, raised beds and a small courtyard garden. Somerdown Farm is a Gothic Victorian village house built by Stephenson with a meandering garden encircling the house, a long rockery border, curving herbaceous borders with an interesting variety of shrubs, gravelled gardens, rose arches and very unusual swagged roses along the driveway. Hawthorns, a short drive from the village, is an informal cottage garden with colourful borders and an extensive vegetable garden.

& 🐎 ✿ ☕

---

60 MIDNEY GARDENS

Mill Lane, Midney, Somerton, TA11 7HR. David Chase & Alison Hoghton, 01458 274250, davidandalison@midneygardens.co.uk. *1m SE of Somerton. 100yds off B3151. From Podimore r'about on A303 take A372. After 1m R on B3151 towards Street. After 2m L on bend into Mill Lane.* **Sun 14, Mon 15 Mar, Sat 3, Sun 4 Apr, Fri 7, Sun 9 May, Thur 3, Sun 6, Sat 26, Sun 27 June, Sun 12, Sun 26 Sept (11-5). Adm £5, chd £1.50. Home-made teas. Visits also by arrangement Apr to Sept for groups of 10+.**
1.4 acre garden with beautiful and varied planting and quirky features. It showcases many unusual plants and collections of daffodils, alliums and buddleias. Winding paths, wildlife friendly planting and places to sit create an atmosphere of calm. Plant nursery offers wide selection of herbaceous perennials, alpines, herbs and grasses. Featured in The Secret Gardens of Somerset by Abigail Willis.

✿ 🚗 ☕

61 ◆ MILTON LODGE

Old Bristol Road, Wells, BA5 3AQ. Simon Tudway Quilter, 01749 679341, www.miltonlodgegardens.co.uk. ½m N of Wells. From A39 Bristol-Wells, turn N up Old Bristol Rd; car park 1st gate on L signed. **For NGS: Sun 9 May, Sun 6 June, Sun 4 July (2-5). Adm £5, chd free. Home-made teas. Discount/prepaid vouchers are not valid on the NGS charity days kindly donated by Milton Lodge. For other opening times and information, please phone or visit garden website.**
Architectural terraces transformed from sloping land with profusion of plants capitalising on views of Wells Cathedral and Vale of Avalon. Grade II terraced garden was restored to its former glory by current owner's parents who moved here in 1960, orchard replaced with a collection of ornamental trees. Herbaceous borders and blooming roses, incl Gertrude Jekyll, flourish next to well tended lawns. Now a peaceful, serene and relaxing atmosphere within the garden following the ravages of two World Wars. Cross Old Bristol Rd to 7 acre woodland garden, the Combe, a natural peaceful contrast to formal garden of Milton Lodge. Unsuitable for wheelchairs/pushchairs or those with limited mobility due to slopes and differing levels.

62 MODEL FARM

Perry Green, Wembdon, Bridgwater, TA5 2BA. Mr & Mrs Dave & Roz Young, 01278 429953, daveandrozontour@hotmail.com, www.modelfarm.com. 4m from J23 of M5. Follow Brown signs from r'about on A39 2m W of Bridgwater. **Sun 6 June (2-5.30). Adm £5, chd free. Tea, coffee, squash and home-made cakes.**
4 acres of flat gardens to S of Victorian country house. Created from a field in last 11 yrs and still being developed. A dozen large mixed flower beds planted in cottage garden style with wildlife in mind. Wooded areas, mixed orchard, lawns, wildflower meadows and wildlife pond. Plenty of seating throughout the gardens with various garden sculptures. Lawn games incl croquet and various garden sculptures from Somerset artists.

63 165 NEWBRIDGE HILL

Bath, BA1 3PX. Helen Hughesdon, www.thehiddengardensofbath.co.uk. On the western fringes of Bath (A431), 100m on L after Apsley Road. Several bus routes go to Newbridge Hill. **Sat 3, Sun 4 July (10-5). Adm £4, chd free. Home-made teas.**
Incorporating 'Sculpture to Enhance a Garden' three sculptors showcase their work in this Bath in Bloom award winning garden which offers herbaceous borders, unusual plants, vegetable garden, greenhouse, shade garden, treehouse and swing, and a sunny terrace overlooking the garden where light lunches and home-made cakes are served. Pieces of sculpture for sale. Regret no wheelchair access.

64 NYNEHEAD COURT

Nynehead, Wellington, TA21 0BN. Nynehead Care Ltd, www.nyneheadcare@aol.com. 1½m N of Wellington. M5 J26 B3187 towards Wellington. R on r'about marked Nynehead & Poole, follow lane for 1m, take Milverton turning at fork, turning into Chipley Road. **Sun 27 June, Sun 11 July (12-4). Adm £4.50, chd free. Light refreshments in Orangery.**
Nynehead Court is a private residential care home for the elderly. We are subject to Government regulations re covid-19 in terms of our opening, before travelling to Nynehead please call 01823 662481 or email Nyneheadcare@aol.com or check our facebook page for up to date details. We require all visitors to wear a face mask and maybe have temperature taken on arrival, thank you. Garden tour with Head Gardener at 12.30 & 2.30. Partial wheelchair access: cobbled yards, gentle slopes, chipped paths, liable to puddle during or after rain. Please wear suitable footwear.

65 THE OLD RECTORY, DOYNTON

18 Toghill Lane, Doynton, Bristol, BS30 5SY. Edwina & Clive Humby, www.doyntongardens.tumblr.com. At heart of village of Doynton, between Bath and Bristol. Parking in Bury Lane, on L just after junction with Horsepool Lane. Car parking is signed and charges may apply as there is another local event being held that day. **Sat 12 June (11-4). Adm £5, chd free. Home-made**

teas by WI.
Doynton's Grade II-listed Georgian Rectory's walled garden and extended 15 acre estate. Renovated over 12 yrs, it sits within AONB. Garden has diversity of modern and traditional elements, fused to create an atmospheric series of garden rooms. Large landscaped kitchen garden featuring canal, vegetable plots, fruit cages and tree house. Partial wheelchair access, some narrow gates and uneven surfaces.

GROUP OPENING

66 NEW OLDBURY ON SEVERN GARDENS

Oldbury-On-Severn, Bristol, BS35 1QA. 3m W of Thornbury. A38 N from Bristol to Thornbury. Signed Oldbury on Severn. Turn L into village. Follow yellow signs. **Sat 19, Sun 20 June (1-5). Combined adm £6, chd free. Home-made teas at Vindolanda.**

NEW CHAPEL COTTAGE
Ann Martin & Alan Taylor.

NEW CHERRY TREE COTTAGE
Jenny & Keith Miller.

NEW CHRISTMAS COTTAGE
Angela Conibere & Doug Mills.

NEW VINDOLANDA
Jan & Rob Willcox.

Four developing country gardens of varying size and design. All have water features with a variety of interesting plants and borders. One has intriguing old agricultural artifacts and is situated on an historical site, another has an area with a developing Japanese feel, next a parterre garden and a traditional cottage garden of delight. All are reasonably level with some gravel paths.

67 PARK COTTAGE

Wrington Hill, Wrington, Bristol, BS40 5PL. Mr & Mrs J Shepherd. Halfway between Bristol & Weston S Mare. 10m S of Bristol on A370 at Cleeve turn L onto Cleeve Hill Rd (opp sports field). Continue on Cleeve Hill Road for 1½ miles. Car park in paddock on R 50m from garden on L. **Sun 25 July, Sun 1 Aug (11-5). Adm £5, chd free.**
Take a colourful journey through 1¼ acres of this 'Alice in Wonderland'

garden. Divided by high hedges is an established perennial flower garden developed over 28 yrs. Explore the labyrinth of different areas incl potager, jungle garden, rainbow border, white garden, green gallery, dahlia border and 90 ft of double herbaceous borders. Large Victorian-style greenhouse displays tender plants. Countryside views and plenty of seating. No WC available. Mostly good wheelchair access, some narrow bark chip paths. Narrow flagstone bridge with steps.

68 PENNY BROHN UK
Chapel Pill Lane, Pill, BS20 0HH. Penny Brohn UK, 01275 370073, fundraising@pennybrohn.org.uk, www.pennybrohn.org.uk. *4m W of Bristol. Off A369 Clifton Suspension Bridge to M5 (J19 Gordano Services). Follow signs to Penny Brohn UK and to Pill/Ham Green (5 mins).* **Sun 27 June (11-4). Adm £4, chd free. Light refreshments. Hot and cold drinks, cakes and light lunches are available.**
3½ acre tranquil garden surrounds Georgian mansion with many mature

trees, wildflower meadow, flower garden and cedar summerhouse. Fine views from historic gazebo overlooking R Avon. Courtyard gardens with water features. Garden is maintained by volunteers and plays an active role in the Charity's Living Well with Cancer approach. Plants, teas, music and plenty of space to enjoy a picnic. Gift shop. Tours of centre to find out more about the work of Penny Brohn UK. Some gravel and grass paths.

69 THE RED POST HOUSE
Fivehead, Taunton, TA3 6PX. The Rev Mervyn & Mrs Margaret Wilson. *10m E of Taunton. On the corner of A378 and Butcher's Hill, opp garage. From M5 J25, take A358 towards Langport, turn L at T-lights at top of hill onto A378. Garden is at Langport end of Fivehead.* **Sun 16 May (2-5). Adm £4, chd free. Home-made teas.**
⅓ acre walled garden with shrubs, borders, trees, circular potager, topiary. We combine beauty and utility. Further 1½ acres, lawn,

orchard and vineyard. Plums, 40 apple and 20 pear, walnut, quince, medlar, mulberry, fig. Fruit trees may be in blossom. Mown paths, longer grass. Views aligned on Ham Hill. Summerhouse with sedum roof, belvedere. Garden in its present form developed over last 15 yrs. Paths are gravel and grass, belvedere is not wheelchair accessible. Dogs on leads.

70 RENDY FARM
Oake, Taunton, TA4 1BB. Mr & Mrs N Popplewell. *1m from Oake on rd from Oake to Nynehead.* **Sun 30 May (2-5.30). Adm £5, chd free. Home-made teas.**
3 acre garden with formal walled front garden, raised vegetable beds and greenhouse enclosed by hornbeam, box and yew hedging. Decorative fruit and cut flower beds, meadow with large wild pond and orchard with stream running through, marked by pollarded willows and small garden surrounding shepherd's hut. Regret no wheelchair access.

9 Catherston Close

The Royal Crescent Hotel and Spa

71 THE RIB

St. Andrews Street, Wells,
BA5 2UR. Paul Dickinson & David
Morgan-Hewitt. *Wells City Centre,
adjacent to East end of Wells
Cathedral and opp Vicars Close.
There is absolutely no parking at or
very near this city garden. Visitors
should use one of the 5 public car
parks and enjoy the 10-15 minutes
stroll through the city to The Rib.* **Sat
3 July (10-5). Adm £5, chd free.**
The Rib is one of the few houses in
England that can boast a cathedral
and a sacred well in its garden. Whilst
the garden is compact, it delivers a
unique architectural and historical
punch. Long established trees,
interesting shrubs and more recently
planted mixed borders frame the view
in the main garden. Orchard. Cottage
garden. Lunch, tea and WC facilities
available in Wells marketplace - 5
minutes walk. Slightly bumpy but
short gravel drive and uneven path
to main rear garden. 2-3 steps up to
orchard and cottage gardens.

72 ROCK HOUSE

Elberton, BS35 4AQ. Mr & Mrs
John Gunnery, 01454 413225. *10m
N of Bristol. 3½m SW Thornbury.
From Old Severn Bridge on M48 take
B4461 to Alveston. In Elberton, take
1st turning L to Littleton-on Severn
and turn immed R.* **Sun 31 Jan,
Sun 7 Feb, Sun 21, Sun 28 Mar
(11-4). Adm £4, chd free. 2022:
Sun 30 Jan, Sun 6 Feb. Visits also
by arrangement for groups of up
to 10.**
2 acre garden. Pretty woodland vistas
with many snowdrops and daffodils,
some unusual. Spring flowers, cottage
garden plants and roses. Old yew tree
and pond. Partial wheelchair access.

73 ROSE COTTAGE

Smithams Hill, East Harptree,
Bristol, BS40 6BY. Bev &
Jenny Cruse, 01761 221627,
bandjcruse@gmail.com. *5m N of
Wells, 15m S of Bristol. From B3114
turn into High St in EH. L at Clock
Tower and immed R into Middle St,
up hill for 1m. From B3134 take EH
rd opp Castle of Comfort, continue
1½m. Car parking in field opp
cottage.* **Sun 11 Apr (2-5). Adm £5,
chd free. Home-made teas. Visits
also by arrangement Apr to June
for groups of up to 30. Please
confirm your visit 2 weeks prior to
your date.**

This 1-acre hillside cottage garden
welcomes spring, carpeted with
seasonal bulbs, primroses and
hellebores, bordered by stream
and established mixed hedges. The
garden is evolving with new planting.
Plenty of seating areas to enjoy
panoramic views over Chew Valley,
home-made teas and the music of
the Congresbury Brass Band. Wildlife
area with pond in corner of car pk
field develops with interest. Hillside
cottage garden full of spring and
summer colour of roses and hardy
geraniums. Organically gardened and
planted to encourage wildlife. Partial
wheelchair access, hillside setting.

74 NEW ROWDON

Monksilver, Taunton, TA4 4JD.
Mr & Mrs David Gliddon. *From
Stogumber take rd to Monksilver.
After 1m, take the drive on R down
to farm.* **Sun 27 June (10-5). Adm
£4.50, chd free. Home-made teas.**
Formal and informal areas in a garden
still under construction. Herbaceous
borders, shrubbery and rose arbour
lead to small walled kitchen garden.
Rockery and small bog garden are
recent additions. 5 acre lake provides
wildlife interest. Steps provide access
to some areas. Teas will be available.

75 NEW THE ROYAL CRESCENT HOTEL & SPA

16 Royal Crescent, Bath,
BA1 2LS. 01225 823 333,
info@royalcrescent.co.uk,
www.royalcrescent.co.uk/. *Nearest
car park is in Charlotte Street. Walk
through front door of hotel, carry
straight on, exit via back door to
gardens.* **Mon 5, Tue 6 July (11-
3.30). Adm £5, chd free. Tea &
coffee available to purchase from
onsite bar. Booking is essential
for lunch or afternoon tea.**
One acre of secluded gardens sits
waiting for you behind this iconic
hotel, lovingly curated by our
gardeners. Gently winding lavender
paths take you across the beautiful
lawns, with various stunning floral
displays along the route. Century-
old trees, rescue hedgehogs, rose
bushes, and interesting statues; this
garden has it all. Groups wishing to
take refreshments must call ahead.
Garden map available on website.
Wheelchairs can access most garden
pathways.

76 RUGG FARM

Church Street, Limington, nr
Yeovil, BA22 8EQ. Morene
Griggs, Peter Thomas &
Christine Sullivan, 01935 840503,
griggsandthomas@btinternet.com.
*2m E of Ilchester. From A303 exit
on A37 to Yeovil/ Ilchester. At 1st
r'about L to Ilchester/Limington,
2nd R to Limington, continue 1½m.*
**Visits by arrangement May to July.
Adm £5, chd free. Adm with light
refreshments £7.**
2 acre garden created since 2007
around former farmhouse and farm
buildings. Diverse areas of interest.
Ornamental, kitchen and cottage
gardens, courtyard container planting,
orchard, wildlife meadows and pond,
developing shrubberies, woodland
plantings and walk (unsuitable for
wheelchairs). Exuberant annuals and
perennials throughout. Open again
after 4 yrs. Garden best late May
to mid July. Metalwork designs by
Andy Stevenson Garden Sculptures.
Compost champion in residence.
Some gravel paths.

77 THE SCHOOL YARD

2 High Street, Wickwar, Wotton-
Under-Edge, GL12 8NE. Jeanette
& Tony Parker. *12m N of Bristol.
Between T-lights at N end of High
Street. If using SatNav please note
High Street Wickwar not High Street
Wotton under Edge. Parking in village
- please park respectfully and consider
our neighbours.* **Sat 26 June (1.30-
5.30). Adm £5, chd free. Home-
made teas at village hall.**
Garden arranged around a former
Victorian school. Vegetable plot with
sunken garden and large greenhouse.
Raised beds edged by espalier
and step-over fruit trees. Terraced
flower garden with variety of trees
and shrubs. Mediterranean
courtyard with ancient yew tree and
olive tree. Pond and deep shade
garden. Variety of English apples and
pear trees. Partial wheelchair access,
gravel paths and steps.

*In our first year 609 private gardens
opened their gates to all, for the
modest sum of one shilling. Today
the National Garden Scheme retains
that combination of inclusivity
and affordability*

78 SERRIDGE HOUSE
Henfield Rd, Coalpit Heath,
BS36 2UY. Mrs J Manning,
01454 773188,
janserigehouse@gmail.com. *9m
N of Bristol. On A432 at Coalpit
Heath T-lights (opp church), turn
into Henfield Rd. R at PH, ½ m small
Xrds, garden on corner with Ruffet
Rd, park on Henfield Rd.* **Visits by
arrangement July & Aug for groups
of 10+. Adm £7, chd free. Cream
teas. Home-made teas/glass of
wine (evening).**
2½ acre garden with mature trees,
heather and conifer beds, island beds
mostly of perennials, woodland area
with pond. Colourful courtyard with
old farm implements. Lake views and
lakeside walks. Unique tree carvings.
Mostly flat grass and concrete
driveway. Wheelchair access to lake
difficult.

&♿ 🚗 ☕🍷

**79 NEW SOMERSET STREET
DISPLAY GARDENS**
23-25 Somerset Street,
Kingsdown, Bristol, BS2 8LZ.
John & Heather Frenkel & others.
*At top of Cotham Hill, 2nd exit at
mini-r'about into Cotham Rd. At end
turn L, then 2nd R into Fremantle
Road (not Lane). At end, turn R into
Somerset St. Parking limited, find
nearest space.* **Sun 23 May, Sun 11
July (2-5). Adm £3.50, chd free.**
A linked row of walled gardens dating
from the C18, separated from their
houses by a narrow setted street.
Designed as display gardens for the
enjoyment of passers-by, they form
a welcome green space of shrubs
and perennials between terraces of
Georgian houses. Features incl box
parterre, pleached hornbeam hedge,
fan-trained apple trees, pond, mature
trees and mixed borders. Steps,
paths, no wheelchair access.

80 SOUTH KELDING
Brewery Hill, Upton Cheyney,
Bristol, BS30 6LY. Barry &
Wendy Smale, 0117 9325145,
wendy.smale@yahoo.com. *Halfway
between Bristol and Bath. Upton
Cheyney lies ½ m up Brewery Hill off
A431 just outside Bitton. Detailed
directions and parking arrangements
given when appt made. Restricted
access means pre-booking essential.*
**Visits by arrangement May to Sept
for groups of 5 to 30. Adm £5.50,
chd free. Home-made cakes and
tea/coffee are an additional £3.50
pp.**

7 acre hillside garden offering
panoramic views from its upper levels,
with herbaceous and shrub beds,
prairie-style scree beds, orchard,
native copses and small arboretum
grouped by continents. Large wildlife
pond, boundary stream and wooded
area featuring shade and moisture-
loving plants. In view of slopes
and uneven terrain this garden is
unsuitable for disabled access.

🐕 ☕🍷

81 ◆ SPECIAL PLANTS
Greenway Lane, Cold
Ashton, SN14 8LA. Derry
Watkins, 01225 891686,
derry@specialplants.net,
www.specialplants.net. *6m N of
Bath. From Bath on A46, turn L into
Greenways Lane just before r'about
with A420.* **For NGS: Thur 17
June, Thur 15 July, Thur 19 Aug,
Thur 16 Sept, Thur 14 Oct (11-5).
Adm £5, chd free. Home-made
teas. For other opening times and
information, please phone, email or
visit garden website.**
Architect-designed ¾ acre hillside
garden with stunning views. Started
autumn 1996. Exotic plants. Gravel
gardens for borderline hardy plants.
Black and white (purple and silver)
garden. Vegetable garden and
orchard. Hot border. Lemon and lime
bank. Annual, biennial and tender
plants for late summer colour. Spring
fed ponds. Bog garden. Woodland
walk. Allium alley. Free list of plants
in garden.

🐕 ✳ ☕🍷

82 ◆ STOBERRY GARDEN
Stoberry Park, Wells,
BA5 3LD. Frances & Tim
Young, 01749 672906,
stay@stoberry-park.co.uk,
www.stoberryhouse.co.uk. *½ m
N of Wells. From Bristol - Wells on
A39, L into College Rd and immed
L through Stoberry Park, signed.*
**For NGS: Sun 23 May (11-5.30).
Adm £5, chd free. Pre-booking
essential, please visit www.ngs.
org.uk for information & booking.
Discount/prepaid vouchers not
valid on NGS charity day kindly
donated by Stoberry Garden.
For other opening times and
information, please phone, email or
visit garden website.**
With breathtaking views over Wells
Cathedral, this 6 acre family garden
planted sympathetically within
its landscape provides stunning
combinations of vistas accented with

wildlife ponds, water features, 1½
acre walled garden, gazebo, and
lime walk. Colour and interest every
season; spring bulbs, irises, salvias,
wildflower circles, new wildflower
meadow walk and fernery. Interesting
sculpture artistically integrated. Do
visit the maturing TV garden created
in 6 weeks for Alan Titchmarsh's TV
programme. Regret no wheelchair
access. No dogs.

✳ 🛏 ☕🍷

GROUP OPENING

83 STOGUMBER GARDENS
Station Road, Stogumber,
TA4 3TQ. *11m NW of Taunton. 3m
W of A358. Signed to Stogumber, W
of Crowcombe. Village maps given
to all visitors.* **Sun 20 June (2-6).
Combined adm £6, chd free.
Home-made teas in Village Hall.**

BRAGLANDS BARN
Simon & Sue Youell,
www.braglandsbarn.com.

KNOLL COTTAGE
Elaine & John Leech,
01984 656689,
john@Leech45.com,
www.knoll-cottage.co.uk.
Visits also by arrangement June
to Sept for groups of 5 to 30.

ORCHARD DEANE
Brenda & Peter Wilson.

POUND HOUSE
Barry & Jenny Hibbert.

WICK BARTON
Sara & Russ Coward.

5 delightful and very varied gardens
in a picturesque village on the edge
of the Quantocks. 3 surprisingly large
gardens near village centre, plus 2
very large gardens on outskirts of
village, with many rare and unusual
plants. Conditions range from
waterlogged clay to well-drained
sand. Features incl courtyard, ponds,
bog gardens, rockery, extensive
mixed beds, vegetable and fruit
gardens, and a collection of over
80 different roses. Fine views of
surrounding countryside from some
gardens. Wheelchair access to main
features of all gardens.

&♿ 🐕 ✳ 🚗 ☕🍷

84 STONELEIGH DOWN

Upper Tockington Road, Tockington, Bristol, BS32 4LQ. Su & John Mills, 01454 615438, susanlmills@gmail.com. *12m N of Bristol. On LH side of Upper Tockington Road when travelling from Tockington towards Olveston. Set back from road up gravel drive. Parking in village.* Sat 5, Sun 6 June, Sat 4, Sun 5 Sept (2-5). Adm £5, chd free. Home-made teas. Visits also by arrangement Apr to Oct for groups of 10 to 30. Approaching ⅔ acre, the south-facing garden has curved gravel pathways around an S-shaped lawn that connects themed areas: sub-tropical; summer walk; acers; oriental pond; winter garden; spring garden. On a level site, it has been densely planted with trees, shrubs, perennials and bulbs for yr-round interest. Plenty of places to sit. Steps into courtyard. Look at the website for pop up openings. Featured in the Gardeners' World Calendar 2021. Finalist in BBC Gardeners' World Magazine 'Gardens of the Year 2010', and filmed for BBC Gardeners' World and ITV's 'Love Your Garden'. Featured in 2019 March issue of Gardeners' World magazine and 2020 'Chelsea Special' issue of Garden Answers.

GROUP OPENING

85 STOWEY GARDENS

Stowey, Bishop Sutton, Bristol, BS39 5TL. *10m W of Bath. Stowey Village on A368 between Bishop Sutton and Chelwood. From Chelwood r'about take A368 to Weston-s-Mare. At Stowey Xrds turn R to car park, 150yds down lane, ample off road parking opp Dormers. Limited disabled parking at each garden which will be signed.* Sun 18 July (2-6). Combined adm £5, chd free. Home-made teas at Stowey Mead, pedestrian entrance for teas via lower gate at Stowey Xrds on A368, entrance to R of seat, diagonally opp lane to Dormers.

DORMERS
Mr & Mrs G Nicol.

◆ MANOR FARM
Richard Baines & Alison Fawcett, 01275 332297.

STOWEY MEAD
Mr Victor Pritchard.

A broad spectrum of interest and styles developing year on year. But there is far more than this in these gardens; the visitor's senses will be aroused by the sights, scents and diversity of these 3 gardens in the tiny, ancient village of Stowey. Flower-packed beds, borders and pots, roses, topiary, hydrangeas, exotic garden, many unusual trees and shrubs, orchards, vegetables, ponds, specialist sweet peas, lawns, Stowey Henge, and a ha ha. An abundant collection of mature trees and shrubs. Ample seating areas, wonderful views from each garden. Not to be missed, something of interest for everyone, all within a few minutes walk of car park at Dormers. Plant sales at Dormers. Well behaved dogs on short leads welcome. Wheelchair access restricted in places, many grassed areas in each garden.

86 NEW SUNNYMEAD

153 Long Ashton Road, Long Ashton, Bristol, BS41 0JQ. Anna-Liisa & Ian Blanks-Walden (Colonel Rtd, MBE). *Located opp The Angel Inn, with a LALHS historic green plaque; sat on the main Long Ashton Road opp Church Lane junction.* Sun 30 May, Sun 4 July (2-5). Combined adm with Coombe Cottage £5, chd free. Home-made teas. Sunnymead is an old medieval village property, dating back to 1403, with a wrap around family garden (0.4 acre), 1700s pond and rockery, 200yr old box hedge, various period features, ongoing projects and Finnish theme. Opened in memory of Marja-Liisa Walden. Cream teas, cakes and other treats for sale.

87 SUTTON HOSEY MANOR

Long Sutton, TA10 9NA. Roger Bramble, rbramble@bdbltd.co.uk. *2m E of Langport, on A372. Gates N of A372 at E end of Long Sutton.* Sun 25 July (2.30-6). Adm £5, chd £2. Home-made teas. Visits also by arrangement Aug to Dec for groups of 10 to 30. 3 acres, of which 2 walled. Lily canal through pleached limes leading to amelanchier walk past duck pond; rose and juniper walk from Italian terrace; judas tree avenue; pteica walk. Ornamental potager. Drive-side shrubbery. Music by players of Young Musicians Symphony Orchestra.

Coombe Cottage

88 SWIFT HOUSE
9 Lyndale Avenue, Stoke Bishop, Bristol, BS9 1BS. Mark & Jane Glanville, www.bristolgarden.weebly.com. *NW Bristol - 4m from city centre. 2m from M5 J18 Portway (A4). Turn into Sylvan Way. At lights take R onto Shirehampton Rd. After ¾m turn R into Sea Mills Lane. Lyndale Ave is 2nd rd on L - we're half way up.* **Sun 4 July (1-4.30). Adm £5, chd free.** We've designed our city garden to look beautiful, provide fruit and vegetables and to be a haven for wildlife. Plants are grown to provide food and shelter throughout the year with the emphasis on flowers that are nectar rich. We have the largest breeding colony of swifts in Bristol www.bristolswifts.co.uk. Plants for sale from 50p.

89 TORMARTON COURT
Church Road, Tormarton, GL9 1HT. Noreen & Bruce Finnamore, 01454 218236, home@thefinnamores.com. *3m E of Chipping Sodbury, off A46 at J18 M4. Follow signs to Tormarton from A46 then follow signs for car parking.* **Visits by arrangement Mar to June for groups of 10 to 30. Adm £5, chd free. Light refreshments. Please contact for refreshment arrangements.** 11 acres of formal and natural gardens in stunning Cotswold setting. Features incl roses, herbaceous, kitchen garden, Mediterranean garden, mound and natural pond. Extensive walled garden, spring glade and meadows with young and mature trees.

90 ◆ UNIVERSITY OF BRISTOL BOTANIC GARDEN
Stoke Park Road, Stoke Bishop, Bristol, BS9 1JG. University of Bristol Botanic Garden, 0117 4282041, botanic-gardens@bristol.ac.uk, botanic-garden.bristol.ac.uk/. *¼m W of Durdham Downs. Located in Stoke Bishop next to Durdham Downs 1m from city centre. After crossing the Downs to Stoke Hill, Stoke Park Rd is 1st on R.* **For NGS: Sun 11 July (10-4.30). Adm £8, chd free. Light refreshments. Hot and cold drinks, sandwiches, salads, cakes and ice cream provided by local delicatessen Chandos Deli. For other opening times and information, please phone, email or visit garden website. Donation to University of Bristol Botanic Garden.** Exciting and contemporary award winning Botanic Garden with dramatic displays illustrating collections of Mediterranean flora, rare native, useful plants (incl European & Chinese herbs) and those that illustrate plant evolution. Large floral displays illustrating pollination in flowering plant and evolution. Glasshouses, home to giant Amazon waterlily, tropical fruit, medicinal plants, orchids, cacti and unique sacred lotus collection. For other opening times see website. Regional Finalist, The English Garden's The Nation's Favourite Gardens 2019. Upon request wheelchair available to borrow from Welcome Lodge. Wheelchair friendly primary route through garden incl. glasshouses, accessible WCs.

91 ◆ THE WALLED GARDENS OF CANNINGTON
Church Street, Cannington, TA5 2HA. Bridgwater College, 01278 655042, walledgardens@btc.ac.uk, www.canningtonwalledgardens.co.uk. *Part of Bridgwater & Taunton College Cannington Campus, 3m NW of Bridgwater. On A39 Bridgwater-Minehead rd - at 1st r'about in Cannington 2nd exit, through village. War memorial, 1st L into Church Street then 1st L.* **For NGS: Sat 24, Sun 25 Apr, Sat 18, Sun 19 Sept (10-4). Adm £5.95, chd free. Light refreshments. For other opening times and information, please phone, email or visit garden website.** Within the grounds of a medieval priory, the Walled Gardens of Cannington are a gem waiting to be discovered! Classic and contemporary features incl hot herbaceous border, blue garden, sub-tropical walk and Victorian style fernery, amongst others. Botanical glasshouse where arid, sub-tropical and tropical plants can be seen. Tea room, plant nursery, gift shop, also events throughout the year. So plenty to see and do for all the family! Gravel paths. Motorised scooter can be borrowed free of charge (only one available).

Vindolanda, Oldbury on Severn Gardens

92 WATCOMBE

92 Church Road, Winscombe, BS25 1BP. Peter & Ann Owen, 01934 842666, peterowen449@btinternet.com. *12m SW of Bristol, 3m N of Axbridge. 100 yds after yellow signs on A38 turn L, (from S), or R. (from N) into Winscombe Hill. After 1m reach The Square. Watcombe on L after 150yds.* **Sun 25 Apr, Sun 16 May, Thur 10 June (2-5). Adm £4, chd free. Home-made cakes and cream teas, gluten free available. Visits also by arrangement Apr to July.** ¾-acre mature Edwardian garden with colour-themed, informally planted herbaceous borders. Strong framework separating several different areas; pergola with varied wisteria, unusual topiary, box hedging, lime walk, pleached hornbeams, cordon fruit trees, 2 small formal ponds and growing collection of clematis. Many unusual trees and shrubs. Small vegetable plot. Some steps but most areas accessible by wheelchair with minimal assistance.

94 WELLFIELD BARN

Walcombe Lane, Wells, BA5 3AG. Virginia Nasmyth, 01749 675129. *½ m N of Wells. From A39 Bristol to Wells rd turn R at 30 mph sign into the narrow Walcombe Lane. Entrance at 1st cottage on R, parking signed.* **Visits by arrangement June to Aug for groups of 10 to 30. Adm £5.50, chd free. Home-made teas.** Refreshments are not included in admission price, please confirm group numbers 2 weeks prior to visit. 24 yrs ago this concrete farmyard began to grow into a tranquil ½-acre garden which is, today, still evolving. Structured design integrates house, lawn and garden with the landscape. Wonderful views, ha-ha, pond/bog garden, mixed borders, hydrangea bed, hardy geraniums, formal sunken garden, grass walks with interesting young and semi-mature trees. A well planned garden created by the owners to provide colour and stature throughout the property. Featured on BBC Gardeners World by Carol Klein who presented our collection of hardy geraniums. Moderate slopes in places, some gravel paths.

95 WESTBROOK HOUSE

West Bradley, BA6 8LS. Keith Anderson & David Mendel, andersonmendel@aol.com. *4m E of Glastonbury. From A361 at W Pennard follow signs to W Bradley (2m). From A37 at Wraxall Hill follow signs to W Bradley (2m).* **Sat 17 Apr, Sat 19 June (11-5). Adm £5, chd free. Visits also by arrangement Apr to Aug for groups of 10+. Donation to West Bradley Church.** 4 acres comprising 3 distinct gardens around house with mixed herbaceous and shrub borders leading to meadow and orchard with spring bulbs, species roses and lilacs. Planting and layout began 2004 and continues to the present. Instagram: keithbfanderson and david_paint.

GROUP OPENING

96 WRINGTON GARDENS

School Road, Wrington, Bristol, BS40 5NB. *At top of School Rd, by junction with Ropers Lane, Long Lane and Old Hill. 10m SW of Bristol, midway between A38 at Redhill or Lower Langford and A370 at Congresbury. Beware SatNav may take you to Orchard Close. Warfords is opp The Hanging Tree. Car parking in nearby field.* **Wed 9, Sun 13 June (2-5.30). Combined adm £5, chd free. Home-made teas at Mathlin Cottage.**

MATHLIN COTTAGE
Tony & Sally Harden.

WARFORDS
Roger & Susan Vincent.

2 contrasting properties on the edge of a delightful and historic village. Mathlin Cottage: a cottage garden accessed by shallow steps with wonderful views of the Mendip Hills. Border, pergola and walkway in front garden; more cottage-style back garden, full of bee/butterfly plants. Small pond, greenhouse and salad/soft fruit plot. Large specimen of Paul's Himalayan Musk tumbles over beech hedge dividing flower garden from fruit and vegetable areas. Plenty of seating and places to relax. Warfords: pretty garden remodelled since 2015, designed to encompass the wide and wonderful backdrop of the Mendip Hills. Plants of interest for all seasons, incl alpines and sub-tropical species such as echiums. S-facing terrace in back garden, shaded by tulip tree. Grapevine. Beyond the greenhouse, with orchids, is a sunken gravel vegetable and fruit garden with raised beds. Plants for sale. Mathlin Cottage: wheelchair access for tea only. Warfords: wheelchair access in most of garden.

97 ◆ THE YEO VALLEY ORGANIC GARDEN AT HOLT FARM

Bath Road, Blagdon, BS40 7SQ. Mr & Mrs Tim Mead, 01761 258155, visit@yeovalleyfarms.co.uk, www.yeovalley.co.uk. *12m S of Bristol. Off A368. Entrance approx ½ m outside Blagdon towards Bath, on L, then follow garden signs past dairy.* **For NGS: Sun 18 Apr (11-5). Adm £6, chd £2. Light refreshments. For other opening times and information, please phone, email or visit garden website.** One of only a handful of ornamental gardens that is Soil Association accredited, 6½ acres of contemporary planting, quirky sculptures, bulbs in their thousands, purple plaza, glorious meadow and posh vegetable patch. Great views, green ideas. Events, workshops and exhibitions held throughout the year - see website for further details. Level access to cafe, around garden some grass paths, some uneven bark and gravel paths. Accessibility map available at ticket office.

98 YEWS FARM

East Street, Martock, TA12 6NF. Louise & Fergus Dowding, 01935 822202, fergus.dowding@btinternet.com, instagram.dowdinglouise. *Turn off main road through village at Market House, onto East Street, past White Hart and Post Office on R, Yews Farm 150 yds on R, opp Foldhill Lane. Turn around if you get to Nag's Head.* **Sun 27 June, Sun 12 Sept (2-5). Adm £8, chd free. Visits also by arrangement May to Sept for groups of 20+.** Theatrical planting in large south facing walled garden. Sculptural planting for height, shape, leaf and texture. Box topiary. High maintenance pots. Self seeding hugely encouraged. Prolific cracked concrete garden in farmyard with hens and pigs. Working organic kitchen garden. Greenhouses bursting with summer vegetables. Organic orchard and active cider barn - taste the difference!

STAFFORDSHIRE
Birmingham & West Midlands

Staffordshire, Birmingham and part of the West Midlands is a landlocked 'county', one of the furthest from the sea in England and Wales.

It is a National Garden Scheme 'county' of surprising contrasts, from the 'Moorlands' in the North East, the 'Woodland Quarter' in the North West, the 'Staffordshire Potteries' and England's 'Second City' in the South East, with much of the rest of the land devoted to agriculture, both dairy and arable.

The garden owners enthusiastically embraced the National Garden Scheme from the very beginning, with seven gardens opening in the inaugural year of 1927, and a further thirteen the following year.

The county is the home of the National Memorial Arboretum, the Cannock Chase Area of Outstanding Natural Beauty and part of the new National Forest.

There are many large country houses and gardens throughout the county with a long history of garden-making and with the input of many of the well known landscape architects.

Today, the majority of National Garden Scheme gardens are privately owned and of modest size. However, a few of the large country house gardens still open their gates for National Garden Scheme visitors.

Volunteers

County Organiser
John & Susan Weston
01785 850448
john.weston@ngs.org.uk

County Treasurer
Brian Bailey
01902 424867
brian.bailey@ngs.org.uk

Publicity
Graham & Judy White
01889 563930
graham&judy.white@ngs.org.uk

Booklet Co-ordinator
Peter Longstaff
01785 282582
peter.longstaff@ngs.org.uk

Assistant County Organisers
Jane Cerone
01827 873205
janecerone@btinternet.com

Ken & Joy Sutton
01889 590031
kenandjoysutton@ngs.org.uk

Alison & Peter Jordan
01785 660819
alisonandpeterjordan@ngs.org.uk

@StaffsNGS

Left: Wild Thyme Cottage

OPENING DATES

All entries subject to change. For latest information check www.ngs.org.uk

Map locator numbers are shown to the right of each garden name.

February

Snowdrop Festival

Sunday 14th
5 East View Cottages 20

March

Sunday 14th
◆ Castle Bromwich
Hall Gardens 9

Sunday 21st
The Beeches 3
Millennium Garden 31
23 St Johns Road 42

April

Sunday 11th
The Beeches 3

Sunday 18th
23 St Johns Road 42

Sunday 25th
Millennium Garden 31

May

Sunday 2nd
Hall Green Gardens 26
NEW 12 Meres Road 29
Yew Tree Cottage 55

Friday 14th
23 St Johns Road 42

Sunday 16th
Cats Whiskers 10
NEW Monarchs Way 32
10 Paget Rise 36

Sunday 23rd
The Beeches 3
Courtwood House 16
NEW 33 Gorway Road 23

Wednesday 26th
The Secret Garden 43

Sunday 30th
NEW Butt Lane Farm 8
The Old Dairy House 34

Monday 31st
Bridge House 6
The Old Dairy House 34

June

Wednesday 2nd
◆ Middleton Hall &
Gardens 30

Thursday 3rd
◆ Middleton Hall &
Gardens 30

Sunday 6th
NEW ◆ Compton Care 15
The Garth 22
The Pintles 38
12 Waterdale 47
19 Waterdale 48

Thursday 10th
Bankcroft Farm 2

Friday 11th
23 St Johns Road 42

Saturday 12th
Hall Green Gardens 26

Sunday 13th
Ashcroft and Claremont 1
Hall Green Gardens 26
3 Marlows Cottages 28
Priory Farm 39
8 Rectory Road 40
91 Tower Road 46

Wednesday 16th
The Secret Garden 43

Thursday 17th
Bankcroft Farm 2

Friday 18th
Yarlet House 54

Saturday 19th
Colour Mill 14

Sunday 20th
The Beeches 3
Chapel House 11
2 Woodland Crescent 52

Wednesday 23rd
5 East View Cottages 20

Thursday 24th
Bankcroft Farm 2

Sunday 27th
Brooklyn 7
NEW Cheadle
Allotments 12
5 East View Cottages 20
The Garth 22
Grafton Cottage 24
Marie Curie Hospice
Garden 27
15 New Church Road 33

July

Thursday 1st
Yew Tree Cottage 55

Friday 2nd
22 Greenfield Road 25

Saturday 3rd
NEW Cross Roads,
Huddlesford Lane 17
22 Greenfield Road 25

Sunday 4th
Bournville Village 5
NEW ◆ Compton Care 15
NEW Cross Roads,
Huddlesford Lane 17
12 Waterdale 47
19 Waterdale 48

Thursday 8th
Yew Tree Cottage 55

Friday 9th
Grafton Cottage 24

Saturday 10th
NEW Fifty Shades of
Green 21
The Old Vicarage 35

Sunday 11th
NEW Fifty Shades of
Green 21
Grafton Cottage 24
3 Marlows Cottages 28
2 Woodland Crescent 52

Wednesday 14th
The Secret Garden 43

Thursday 15th
Yew Tree Cottage 55

Sunday 18th
The Beeches 3
56 St Agnes Road 41
Tanglewood Cottage 44

Sunday 25th
198 Eachelhurst Road 19
Grafton Cottage 24
The Wickets 50

Yew Tree Cottage 55

Thursday 29th
The Wickets 50

August

Sunday 1st
NEW ◆ Compton Care 15
Grafton Cottage 24

Wednesday 4th
The Secret Garden 43

Sunday 8th
Grafton Cottage 24

Thursday 12th
Church Cottage 13
Colour Mill 14

Sunday 15th
The Beeches 3

Saturday 28th
NEW Fifty Shades of
Green 21

Sunday 29th
The Wickets 50

September

Sunday 5th
The Beeches 3
8 Rectory Road 40

Sunday 12th
Bridge House 6
NEW 12 Meres Road 29

October

Saturday 23rd
◆ Dorothy Clive
Garden 18

Sunday 24th
◆ Dorothy Clive
Garden 18

Saturday 30th
Wild Thyme Cottage 51

Sunday 31st
Wild Thyme Cottage 51

February 2022

Sunday 13th
5 East View Cottages 20

By Arrangement

Arrange a personalised garden visit on a date to suit you. See individual garden entries for full details.

THE GARDENS

GROUP OPENING

1 ASHCROFT AND CLAREMONT

Stafford Road, Eccleshall, ST21 6JP. Mrs G Bertram. *7m W of Stafford. J14 M6. At Eccleshall end of A5013 the garden is 100 metres before junction with A519. On street parking nearby. Note: Some Satnavs give wrong directions.* Sun 13 June (2-5). Combined adm £4.50, chd free. Home-made teas at Ashcroft.

ASHCROFT

Peter & Gillian Bertram.

26 CLAREMONT ROAD

Maria Edwards.

Weeping limes hide Ashcroft, a 1 acre garden of green tranquillity, rooms flow seamlessly around the Edwardian house. Covered courtyard with lizard water feature, sunken herb bed, kitchen garden, greenhouse, wildlife boundaries home to five hedgehogs increasing yearly. Deep shade border, woodland area and ruin with stone carvings and stained glass sculpture. Claremont is a master class in clipped perfection. An artist with an artist's eye has blurred the boundaries of this small Italianate influenced garden. Overlooking the aviary, a stone lion surveys the large pots and borders of vibrant planting, completing the Feng-Shui design of this beautiful all seasons garden. Tickets, teas and plants available at Ashcroft. Partial wheelchair access at Claremont.

♿ ✿ ☕ 🌿

2 BANKCROFT FARM

Tatenhill, Burton-on-Trent, DE13 9SA. Mrs Penelope Adkins. *2m NW of Burton-on-Trent. 1m NW of Burton upon Trent take Tatenhill Rd off A38 on Burton/Branson flyover, 1m 1st house on L after village sign. Parking on farm.* Thur 10, Thur 17, Thur 24 June (1.30-4.30). Adm £3.50, chd free. Lose yourself for an afternoon in our 1½-acre organic country garden. Arbour, gazebo and many other seating areas to view ponds and herbaceous borders, backed with shrubs and trees with emphasis on structure, foliage and colour. Productive fruit and vegetable gardens, wildlife areas and adjoining 12 acre native woodland walk. Picnics welcome. Gravel paths.

♿ 🚗

3 THE BEECHES

Mill Street, Rocester, ST14 5JX. Ken & Joy Sutton. *5m N of Uttoxeter. On B5030 from Uttoxeter turn R at 2nd r'about into village by JCB factory. At Red Lion Pub & mini r'about take rd signed Mill Street. Garden 250yds on R. Parking at JCB academy Sundays only.* Sun 21 Mar, Sun 11 Apr (1-4.30). Tea. Sun 23 May, Sun 20 June, Sun 18 July, Sun 15 Aug, Sun 5 Sept (1.30-5). Home-made teas. Adm £5, chd free.
A stunning plant lover's large garden with countryside views, box garden, shrubs, rhododendrons and azaleas, vibrant colour-themed herbaceous borders, scented roses, clematis and vegetable garden, prairie style island beds planted with grasses, annuals and unusual perennials, plants cut back and shrubs pruned to reveal an under planting of spring jewels from bulbs, hellebores and early flowering plants. (March & April). Partial wheelchair access.

♿ ✿ ☕ 🌿

4 BIRCH TREES

Copmere End, Eccleshall, ST21 6HH. Susan & John Weston, 01785 850448, john.weston@ngs.org.uk. *1½m W of Eccleshall. On B5026, turn at junction signed Copmere End. After ½m straight across Xrds by Star Inn.* Visits by arrangement in July for groups of 5 to 20. Adm £5, chd free. Tea & biscuits included
Surprising ½ acre SW-facing sun trap hidden from the road which takes advantage of the 'borrowed landscape' of the surrounding countryside. Take time to explore the pathways between the island beds which contain many rare & unusual herbaceous plants and shrubs; also vegetable patch, stump bed, alpine house, orchard and water features.

♿ ☕

GROUP OPENING

5 BOURNVILLE VILLAGE

Birmingham, B30 1QY. Bournville Village Trust, www.bvt.org.uk. *Gardens spread across 1,000 acre estate. Walks of up to 30 mins between some. Map supplied on day. Additional parking: Wyevale Garden Centre, (B30 2AE) & Rowheath Pavilion, (B30 1HH).* **Sun 4 July (11-5). Combined adm £6, chd free. Home-made teas at various locations. Light meals available.**

103 BOURNVILLE LANE
B30 1LH. Mrs Jennifer Duffy.

52 ELM ROAD
Mr & Mrs Jackie Twigg.

82 HAY GREEN LANE
B30 1UP. Mr Tony Walpole & Mrs Elsie Wheeler.

11 KESTREL GROVE
Mr Julian Stanton.

32 KNIGHTON ROAD
B31 2EH. Mrs Anne Ellis & Mr Lawrence Newman.

NEW MASEFIELD COMMUNITY GARDEN
Mrs Sally Gopsill.

NEW 8 NEWENT ROAD
Mr & Mrs Edward and Mary Rutledge.

Bournville Village is showcasing 7 gardens - 2 of which are new to the scheme. Bournville is famous for its large gardens, outstanding open spaces and of course its chocolate factory in a garden! Free information sheet/map available on the day. Gardens spread across the 1,000 acre estate, with walks of up to 30 minutes between sites. For those with a disability, full details of access are available on the NGS website. Visitors with particular concerns with regards to access are welcome to call Bournville Village Trust on 0300 333 6540 or email: CommunityAdmin@bvt.org.uk. Music and singing available across a number of sites. Please check on the day.

6 BRIDGE HOUSE

Dog Lane, Bodymoor Heath, B76 9JD. Mr & Mrs J Cerone, 01827 873205, janecerone@btinternet.com. *5m S of Tamworth. From A446 at Belfry Island take A4091 to Tamworth, after 1m turn R onto Bodymoor Heath Lane & continue 1m into village, parking in field opp garden.* **Mon 31 May, Sun 12 Sept (2-5). Adm £4, chd free. Home-made teas. Visits also by arrangement May to Sept for groups of 5 to 30.**

1 acre garden surrounding converted public house. Divided into smaller areas with a mix of shrub borders, azalea and fuchsia, herbaceous and bedding, orchard, kitchen garden with large greenhouse. Pergola walk, formal fish pool, pond, bog garden and lawns. Kingsbury Water Park and RSPB Middleton Lakes Reserve located within a mile.

7 BROOKLYN

Gratton Lane, Endon, Stoke-on-Trent, ST9 9AA. Janet & Steve Howell. *4m W of Leek. 6m from Stoke-on-Trent on A53 turn at Black Horse Pub into centre of village, R into Gratton Lane 1st house on R. Parking signed in village.* **Sun 27 June (12-5). Adm £3.50, chd free. Cream teas.**

A cottage garden in the picturesque old village of Endon. Pretty front garden overflows with roses, geraniums and astrantia, box surrounds a central sundial. Rear garden features shady area with hostas and ferns, small waterfall and pond. Steps lead to rear lawn, large well stocked borders and summerhouse. Several seating areas with village and rural views. Enjoy tea and cake in the potting shed. Traditional cottage garden in rural village location.

8 NEW BUTT LANE FARM

Butt Lane, Ranton, Stafford, ST18 9JZ. Pete Gough, 07975928968, claire-pickering@hotmail.co.uk. *Off Butt lane. 6m W of Stafford. Take A518 to Haughton, turn R Station Rd (signed Ranton) 2m turn L at Butt Ln. Or from Great Bridgeford head W B5405 for 3m. Turn L Moorend Ln, 2nd L Butt Lane.* **Sun 30 May (10-4). Adm £3, chd free. Light refreshments. Home-made Pizza and BBQ available. Visits also by arrangement May to Aug for groups of 10+.**

Developing cottage style garden entering from unspoilt farmland through small wooded area onto lawns surrounded by floral beds and productive fruit and vegetable areas. With unusual areas of interest, greenhouse area, well and outside kitchen. Several craft stalls to be in attendance. Lamb petting. Small nature walk (not suitable for wheelchairs or mobility impaired).

9 ◆ CASTLE BROMWICH HALL GARDENS

Chester Road, Castle Bromwich, Birmingham, B36 9BT. Castle Bromwich Hall & Gardens Trust, 0121 749 4100, admin@cbhgt.org.uk, www.castlebromwichhallgardens.org.uk. *4m East of Birmingham centre. 1m J5 M6 (exit N only).* **For NGS: Sun 14 Mar (10.30-3.30). Adm £4, chd £1. Light refreshments. For other opening times and information, please phone, email or visit garden website.**

10 acres of restored C17/18 walled gardens attached to a Jacobean manor (now a hotel) just minutes from J5 of M6. Formal yew parterres, wilderness walks, summerhouses, holly maze, espaliered fruit and wild areas. Paths are either lawn or rough hoggin - sometimes on a slope. Most areas generally accessible, rough areas outside the walls difficult when wet.

10 CATS WHISKERS

42 Amesbury Rd, Moseley, Birmingham, B13 8LE. Dr Alfred & Mrs Michele White. *Opp back of Moseley Hall Hospital. Past Edgbaston Cricket ground straight on at r'about & up Salisbury Rd. Amesbury Rd, 1st on R.* **Sun 16 May (1-5). Adm £5, chd free. Light refreshments.**

A plantsman's garden developed over the last 38 years but which has kept its 1923 landscape. The front garden whilst not particularly large is full of interesting trees and shrubs; the rear garden is on 3 levels with steps leading to a small terrace and further steps to the main space. At the end of the garden is a pergola leading to the vegetable garden and greenhouse. Two flights of steps in back garden.

Online booking is available for many of our gardens, at ngs.org.uk

Church Cottage

CHAPEL HOUSE
Coton Clanford, Stafford,
ST18 0PE. Mr Richard Clamp, 2m
*W of J14 M6. From M6 J14 take
A5013 to Eccleshall, in 1½ m L onto
B5405 to Woodseaves, in ¼ m L
to Seighford. After 1m 3rd L onto
Bunns Bank/Clanford Lane. Continue
for 1m.* Sun 20 June (11-4). Adm
£4, chd free. Home-made teas.
Mature country garden in rural
location, with 3 distinct areas in 1½
acres. Multiple borders bring colour
and foliage in late Spring in the
formal garden. The vegetable and
fruit garden provides ample produce
and feature a small herb bed; while
a woodland area creates a natural
habitat for wildlife, incl a pond. A
number of seating areas around the
garden provide respite to enjoy the
tranquillity.

NEW CHEADLE
ALLOTMENTS
Delphouse Road, Cheadle, Stoke-
on-Trent, ST10 2NN. Cheadle
Allotment Association. *On the
A521 1m to the W of Cheadle town
centre.* Sun 27 June (1-5). Adm

£5, chd free. Home-made teas.
Refreshments are incl in the adm
price.
The allotments, which were opened
in 2015, are located on the western
edge of Cheadle (Staffs). There are 29
plots growing a variety of vegetables,
fruits and flowers. A new addition in
2019 was a community area, with
an adjacent wildlife area. A small
community orchard is currently being
developed. Wheelchair access on all
main paths.

CHURCH COTTAGE
Aston, Stone, ST15 0BJ. Andrew
& Anne Worrall, 01785 815239,
acworrall@aol.com. *1m S of Stone,
Staffordshire. N side of St Saviour's
Church, Aston. Go S 150 metres on
A34 after junction with A51. Turn L at
Aston Village Hall. 200 metres down
lane into churchyard for parking.*
Thur 12 Aug (12-4.30). Adm £3.50,
chd free. Home-made teas. Visits
also by arrangement July & Aug
for groups of 10 to 20.
An acre of cottage garden with large
pond, waterfall and stream. Trees,
sculptures, small orchard and wild

flowers. Dahlias by award-winning
grower Dave Bond. Views across the
R Trent, St Saviour's Church and Trent
and Mersey Canal nearby.

COLOUR MILL
Winkhill, Leek, ST13 7PR.
Jackie Pakes, 01538 308680,
jackie.pakes@icloud.com,
www.colourmillbandb.co.uk.
*7m E of Leek. Follow A523 from
either Leek or Ashbourne, look for
NGS signs on the side of the main
rd which will direct you down to
Colour Mill.* Sat 19 June, Thur 12
Aug (1.30-5). Adm £4, chd free.
Home-made teas. Visits also by
arrangement June to Aug.
1½ acre S-facing garden, created in
the shadow of a former iron foundry,
set beside the delightful R Hamps.
Informal planting in a variety of rooms
surrounded by beautiful 7ft beech
hedges. Large organic vegetable
patch complete with greenhouse and
polytunnel. Maturing trees provide
shade for the interesting seating
areas. River walk through woodland
and willows. Herbaceous borders.

15 NEW ◆ **COMPTON CARE**
4 Compton Road West,
Wolverhampton, WV3 9DH.
Mr Martin Guyler,
www.comptoncare.org.uk. *2m W
of Wolverhampton city centre. From
Wolverhampton ring rd, take A454 to
Bridgenorth for 2m.* **For NGS: Sun 6
June, Sun 4 July (10.30-4.30), also
open 12 & 19 Waterdale. Sun 1 Aug
(10.30-4.30). Adm £3.50, chd free.
Light refreshments. Coffee Shop
menu also available. For other
opening times and information,
please visit garden website.**
5 acre Victotrian estate, home to a
well-supported hospice. Steeped
in history and nestled in the leafy
suburbs of Wolverhampton. Formal
& informal borders set against a
backdrop of mature specimen
trees, full of wildlife, ample seating &
wheelchair friendly paths. Recently
developed Garden of Reflection
designed by an award-winning
garden designer. Coffee shop, toilets
& on-site parking.

16 **COURTWOOD HOUSE**
3 Court Walk, Betley, CW3 9DP.
Mike Reeves. *6m S of Crewe. On
A531 toward Keele & Newcastle
under Lyme or from J16 off M6,
pickup A531 off A500 on Nantwich
rd, into village by Betley Court.* **Sun
23 May (12.30-5). Adm £3.50, chd
£2. Light refreshments.**
Small L-shaped, walled garden,
which is designed as a walk-through
sculpture. Mainly shrubs with
structures and water features, hidden
spaces and seating areas, with strong
shapes and effects utilising a wide
range of materials, incl. a synthetic
lawn. Small art gallery with acrylic
paintings by owner for sale.

17 NEW **CROSS ROADS,
HUDDLESFORD LANE**
Back Lane, Whittington, Lichfield,
WS14 9NL. Mike Kinghan & Julia
Spencer, 01543 432238. *3m east
of Lichfield, Staffs. 1m E of the A38
and 1m N of the A51. Postcode/
satnav takes you to Back Lane. Park
on Back Lane or Chapel Lane. Walk
30 yds up Huddlesford Lane (which
is halfway along Back Lane) to house
entrance.* **Sat 3, Sun 4 July (12-5).
Adm £3.50, chd free. Pre-booking
essential, please visit www.
ngs.org.uk for information &
booking. Cream teas. Visits also
by arrangement June to Aug for
groups of up to 10.**
1.3 acres of stunning wildlife-friendly
perennial meadows, (usually full of
bees and butterflies), grass paths,
large pond, with orchard of 50
heritage fruit trees, bee hives, wild
areas, interesting shrubs, polytunnel,
and small copse. Not your average

Grafton Cottage

garden! Not really suitable for wheelchairs unfortunately - bumpy grass paths.

🐕 ☕

18 ◆ DOROTHY CLIVE GARDEN
Willoughbridge, Market Drayton, TF9 4EU. Willoughbridge Garden Trust, 01630 647237, info@dorothyclivegarden.co.uk, www.dorothyclivegarden.co.uk. *3m SE of Bridgemere Garden World. From M6 J15 take A53 W bound, then A51 N bound midway between Nantwich & Stone, near Woore. For* NGS: Sat 23, Sun 24 Oct (10-4). Adm £5, chd £2. Cream teas in the Tearooms. Vintage Afternoon Tea can be pre-booked. **For other opening times and information, please phone, email or visit garden website.**
12 informal acres, incl superb woodland garden, alpine scree, gravel garden, fine collection of trees and spectacular flower borders. Renowned in May when woodland quarry is brilliant with rhododendrons. Waterfall and woodland planting. Laburnum Arch in June. Creative planting has produced stunning summer borders. Large Glasshouse. Spectacular autumn colour. Much to see, whatever the season. The Dorothy Clive Tea Rooms will be open throughout the weekend during winter for refreshments, lunch and afternoon tea. Open all week in summer. Plant sales, Gift room Picnic area and children's play area for a wide age range. Wheelchairs (inc. electric) are available to book through the tea rooms . Disabled parking is available. Toilets on both upper and lower car parks.

♿ 🐕 ✿ 🚌 ☕

19 198 EACHELHURST ROAD
Walmley, Sutton Coldfield, B76 1EW. Jacqui & Jamie Whitmore. *5mins N of Birmingham. M6 J6, A38 Tyburn Rd to Lichfield, continue to T-lights at Lidl & continue on Tyburn Rd, at island take 2nd exit to destination rd.* Sun 25 July (12.30-4.30). Adm £3, chd free. Home-made teas.
A long garden approx 210ft x 30ft divided by arches and pathways. Plenty to explore incl wildlife pond, corner arbour, cottage garden and hanging baskets leading to formal garden with box-lined pathways, well, stocked garden, gazebo and hen coop,however they mostly roam the garden, then through to raised seating area, overlooking Pype Hayes golf course, with summer house bar and courtyard garden.

✿ ☕

20 5 EAST VIEW COTTAGES
School Lane, Shuttington, nr Tamworth, B79 0DX. Cathy Lyon-Green, 01827 892244, cathyatcorrabhan@hotmail.com, www.ramblinginthegarden. wordpress.com. *2m NE of Tamworth. From Tamworth, Amington Rd or Ashby Rd to Shuttington. From M42 J11, B5493 for Seckington & Tamworth, 3 mls L turn to Shuttington. Pink house nr top of School Lane, disabled at house.* Sun 14 Feb (11-3); Wed 23 June (1-4); Sun 27 June (1-5). Adm £4, chd free. Home-made teas. 2022: Sun 13 Feb. **Visits also by arrangement June & July for groups of 5 to 30. Groups also welcome in Feb.**
Deceptive & quirky plantlover's garden, full of surprises & always something new. Informally planted themed borders, cutting beds, woodland & woodland edge, stream, water features, sitooterie, folly, greenhouses & many artefacts. Roses, clematis, perennials, potted hostas. Snowdrops & witch hazels in Feb. Benches & seating areas for contemplation & enjoying home-made cake. 'Wonderful hour's wander'.

🐕 ✿ ☕

21 NEW FIFTY SHADES OF GREEN
20 Bevan Close, Shelfield, Walsall, WS4 1AB. Annmarie and Andrew Swift, 07963041402, annmarie.1963@hotmail.co.uk. *Walsall. M6 J10 take A454 to Walsall for 1.6m turn L at Lichfield St for ⅓m, turn L onto A461 Lichfield Rd for 2m, at Co-op T-lights turn L onto Mill rd then follow yellow signs.* Sat 10, Sun 11 July (10-6). Adm £3.50, chd free. Home-made teas. Sat 28 Aug (2-10). Adm £3.50. Light refreshments. **Visits also by arrangement May to Sept.**
A small garden 15m x11m, taking several years to create & landscape one area at a time. We have many distinctive areas of interest inc 2 ponds linked by a stream with a stone waterfall, 5 unique water features, places to sit, watch & relax. We encourage & welcome wildlife. Our planting style is varied inc Ferns Hostas, Palms Bananas Bamboos architectural plants for

foliage & well over 40 trees. The garden has a bridge over a stream & deep water. Steps to and from some areas. Gravel pathways. Please note: we are located in a cul-de-sac & parking is limited. Please park with consideration. Easier parking is available in Broad Lane. Find us on Facebook.

✿ ☕

22 THE GARTH
2 Broc Hill Way, Milford, Stafford, ST17 0UB. Anita & David Wright, 01785 661182, anitawright1@yahoo.co.uk, www.anitawright.co.uk. *4½m SE of Stafford. A513 Stafford to Rugeley rd; at Barley Mow turn R (S) to Brocton; L after ½m.* Sun 6, Sun 27 June (2-6). Adm £4, chd free. Cream teas. **Visits also by arrangement May to Sept for groups of 20+.**
½ acre garden of many levels on Cannock Chase AONB. Acid soil loving plants. Series of small gardens, water features, raised beds. Rare trees, island beds of unusual shrubs and perennials, many varieties of hosta and ferns. Varied and colourful foliage, summerhouses, arbours and quiet seating to enjoy the garden. Ancient sandstone caves.

🐕 ✿ ☕

23 NEW 33 GORWAY ROAD
Walsall, WS1 3BE. Gillian Brooks. *M6 J9 turn N onto Bescot Rd, at r'about take Wallows L (A4148), in 2 m at r'about, 1st exit Birmingham Rd, L to Jesson Rd, L to Gorway Rd.* Sun 23 May (10.30-3.30). Adm £3.50, chd free. Home-made teas. Cottage style Edwardian house garden. Late spring bulbs and roses. Willow tunnel, pond & rockery. Garden viewing is over flat grass and paths. Disabled access through garage.

♿ ☕

24 GRAFTON COTTAGE
Barton-under-Needwood,
DE13 8AL. Margaret & Peter
Hargreaves, 01283 713639,
marpeter1@btinternet.com. *6m N
of Lichfield. Leave A38 for Catholme
S of Barton, follow sign to Barton
Green, L at Royal Oak, ¼m.* **Sun 27
June (11.30-5); Fri 9 July (1.30-5);
Sun 11, Sun 25 July, Sun 1, Sun
8 Aug (11.30-5). Adm £4.50, chd
free. Home-made teas. Visits
also by arrangement June to
Aug. Minimum admission £90 if
less than 20 people. Donation to
Alzheimer's Research UK.**
A visitors comment, such a wealth
of colour and clever planting.
Unusual herbaceous plants with
new additions introduced, perfume
from old fashioned Roses, Sweet
peas, dianthus, phlox and lilies,
viticella clematis, salvias and violas.
Cottage garden annuals and use
of foliage plants play a part in the
garden. Coloured themed borders,
amphitheatre, brook, parterre to
celebrate 25 years of opening.
✿ 🚗 ☕

25 22 GREENFIELD ROAD
Stafford, ST17 0PU. Alison &
Peter Jordan, 01785 660819,
alison.jordan2@btinternet.com.
*3m S of Stafford. Follow the A34 out
of Stafford towards Cannock. 2nd L
onto Overhill Rd.1st R into Greenfield
Rd.* **Evening opening Fri 2 July
(7-9). Adm £5, chd free. Wine.
Sat 3 July (11.30-4.30). Adm £3,
chd free. Cream teas. Visits also
by arrangement May to Aug for
groups of 5 to 20.**
Suburban garden,working towards
all round interest. In spring bulbs and
stunning azaleas and rhododendrons.
June onwards perennials and grasses.
A garden that shows being diagnosed
with Parkinson's needn't stop you
creating a peaceful place to sit and
enjoy. You might be able to catch
a glimpse of Pete's model railway
running. Come in the evening and enjoy
a glass of wine and live music. Flat
garden but with some gravelled areas.
♿ 🐕 ✿ ☕

GROUP OPENING

26 HALL GREEN GARDENS
Hall Green, Birmingham, B28 8SQ.
*Off A34, 3m city centre, 6m from
M42 J4. From City Centre start
at 120 Russell Rd B28 8SQ .
Alternatively, from M42 start at 638*
Shirley Rd B28 9LB. **Sun 2 May,
Sat 12, Sun 13 June (1.30-5.30).
Combined adm £5, chd free.
Home-made teas on May 2nd at
120 Russell Road, on June 12th
and 13th at 111 Southam Road.**

42 BODEN ROAD
Mrs Helen Lycett.
Open on Sat 12, Sun 13 June

36 FERNDALE ROAD
Mrs E A Nicholson, 0121 777
4921.
Open on all dates
Visits also by arrangement May
to Aug for groups of up to 20.

120 RUSSELL ROAD
Mr David Worthington,
07552 993911,
hildave@hotmail.com.
Open on all dates
Visits also by arrangement May
to Sept for groups of up to 30.

638 SHIRLEY ROAD
Dr. and Mrs. M. Leigh.
Open on Sat 12, Sun 13 June

111 SOUTHAM ROAD
Ms Val Townend & Mr Ian Bate.
Open on Sat 12, Sun 13 June

A group of diverse suburban gardens.
42 Boden Rd: Large restful garden,
mature trees, cottage borders,
seating areas and small vegetable
area. 36 Ferndale Rd: Florist's
large suburban garden, ponds and
waterfalls, and fruit garden. 120
Russell Rd: Plantsman's garden,
formal raised pond and hosta
collection and unusual perennials,
container planting. 111 Southam
Rd: Mature garden with well defined
areas incl ponds, white garden,
rescue hens and a majestic cedar.
638 Shirley Rd: Large young garden
with herbaceous borders, vegetables
and greenhouses. Partial wheelchair
access only at some gardens.
♿ ✿ ☕

**27 MARIE CURIE HOSPICE
GARDEN**
Marsh Lane, Solihull, B91 2PQ. Mrs
Do Connolly, www.mariecurie.org.
uk/westmidlands. *Close to J5 M42
to E of Solihull Town Centre. M42 J5,
travel towards Solihull on A41. Take
slip toward Solihull to join B4025 &
after island take 1st R onto Marsh
Lane. Hospice is on R. Limited onsite
parking - available for blue badge
holders.* **Sun 27 June (11-4). Adm
£4, chd free. Light refreshments at
the hospice bistro.**

The gardens contain two large,
formally laid out patients' gardens
incl a ball fountain water feature and
large rose/clematis arches, indoor
courtyards, a long border adjoining the
car park and a beautiful wildlife and
pond area. The volunteer gardening
team hope that the gardens provide
a peaceful and comforting place for
patients, their visitors and staff.
♿ 🐕 ✿ ☕

28 3 MARLOWS COTTAGES
Little Hay Lane, Little Hay,
WS14 0QD. Phyllis Davies. *4m S of
Lichfield. Take A5127, Birmingham
Rd. Turn L at Park Lane (opp Tesco
Express) then R at T junction into
Little Hay Lane, ½m on L.* **Sun 13
June, Sun 11 July (11-4). Adm
£3.50, chd free. Home-made teas.**
Long, narrow, gently sloping cottage
style garden with borders and beds
containing abundant herbaceous
perennials and shrubs leading to
vegetable patch.
✿ ☕

29 NEW 12 MERES ROAD
Halesowen, B63 2EH. Nigel &
Samantha Hopes, 01384 413070,
contact@hopesgardenplants.co.uk,
www.hopesgardenplants.co.uk. *4m
from M5 J3. Follow the A458 out of
Halesowen heading W for 2m, L turn
onto Two Gates Lane just after Round
of Beef pub, this becomes Meres Rd.
House is on R.* **Sun 2 May, Sun 12
Sept (11-4). Adm £3.50, chd free.
Light refreshments. Visits also
by arrangement June to Sept for
groups of 10 to 30.**
A little oasis in the heart of the Black
Country, with stunning views across
to Shropshire. The borders are filled
with interest and colour throughout
the year. A small family garden for
plant hunters and collectors, we hold a
National Collection of Border Auriculas,
with a passion for Snowdrops,
Hellebores, Epimedium, Bearded Iris,
Roscoea, Cyclamen and Kniphofia.
✿ NPC ☕

*The National Garden Scheme
searches the length and
breadth of England and Wales
for the very best private
gardens*

Butt Lane Farm

30 ◆ MIDDLETON HALL & GARDENS

Middleton, Tamworth, B78 2AE. Middleton Hall Trust, 01827 283095, enquiries@middleton-hall.co.uk, www.middleton-hall.co.uk. *4m S of Tamworth, 2m N of J9 M42. On A4091 between The Belfry & Drayton Manor.* **For NGS: Wed 2, Thur 3 June (11-4). Adm £7, chd £2. Light refreshments at The Courtyard at Middleton Hall. For other opening times and information, please phone, email or visit garden website.**

Our formal gardens form part of the 42 estate of the Grade II* Middleton Hall, the seventeenth-century home of naturalists Francis Willoughby and John Ray. The formal gardens are made up of a Walled Garden, lawns and an orchard. The Walled Garden contain a variety of herbaceous and seasonal planting with specimen plants that have a botanical and/ or historical significance to our site. Bake 180 Coffee Shop will also be open serving lunches and other light refreshments. Wheelchair access - Walled Garden paths are gravel, Glade and Orchard paths are grass. No access to 1st floor of the Hall and Nature Trail.

  ♿ 🐕 ❀ ☕ ⛱

31 MILLENNIUM GARDEN

London Road, Lichfield, WS14 9RB. Carol Cooper. *1m S of Lichfield. Off A38 along A5206 towards Lichfield ¼ m past A38 island towards Lichfield. Park in field on L. Yellow signs on field post.* **Sun 21 Mar, Sun 25 Apr (1-5). Adm £3.50, chd free. Home-made teas.** 2-acre garden with mixed spring bulbs in the woodland garden and host of golden daffodils fade slowly into the summer borders in this English country garden. Designed with a naturalistic edge and with the environment in mind. A relaxed approach creates a garden of quiet sanctuary with the millennium bridge sitting comfortably, with its surroundings of lush planting and mature trees. Well stocked borders give shots of colour to lift the spirit and the air fills with the scent of wisterias and climbing roses. A stress free environment awaits you at the Millennium Garden. Park in field then footpath round garden. Some uneven surfaces.

  ♿ ☕

33 Gorway Road

32 NEW MONARCHS WAY

Park Lane, Coven,
Wolverhampton, WV9 5BQ.
Eileen and Bill Johnson,
www.monarchsway-garden.co.uk.
From Port Lane (main road between Codsall & Brewood) turn E onto Park Lane. Monarchs Way is on the sharp bend on Park Lane. **Sun 16 May (11-5). Adm £6, chd free. Pre-booking essential, please visit www.ngs.org.uk for information & booking. Light refreshments. Visits also by arrangement Feb to Oct for groups of up to 20. We have very limited parking, six cars max.**
We bought a 1¾ acre bare, treeless blank canvas in 2010 with grass around three feet high! Since then we have designed a Tudor folly and jungle hut, built a pergola, excavated a lily pond with bog garden created a cottage garden, orchard, and vegetable garden. We have planted hundreds of conifer, evergreen, fruit and flowering trees, roses, perennials and designed numerous flower beds. Access is easy for most of the garden.

33 15 NEW CHURCH ROAD

Sutton Coldfield, B73 5RT. Owen & Lloyd Watkins. *4½m NW of Birmingham city centre. Take A38M then A5127 to 6 Ways r'about, take 2nd exit then bear L Summer Rd then Gravelly Lane over Chester Rd to Boldmere Rd, 5th R New Church Rd.* **Sun 27 June (1-4). Adm £4, chd free. Light refreshments.**
A garden created and recreated, evolving over 10 years. From the new patio an extended pond takes centre stage. A new small stumpery and dry shade area have recently been added. Herbaceous borders, tree ferns and bamboo hedging, a rockery and woodland area. Rear of the garden has raised vegetable beds, fruit trees and chickens. Wheelchair access possible although some areas may not be accessible. Several steps, ramps available.

34 THE OLD DAIRY HOUSE

Trentham Park, Stoke-on-Trent, ST4 8AE. Philip & Michelle Moore. *S edge of Stoke-on-Trent. Next to Trentham Gardens. Off Whitmore Rd. Please follow NGS signs or signs for Trentham Park Golf Club. Parking in church car park.* **Sun 30, Mon 31**

May (1.30-5). Adm £4, chd free. Home-made teas.
Grade 2 listed house which originally formed part of the Trentham Estate forms backdrop to this 2-acre garden in parkland setting. Shaded area for rhododendrons, azaleas plus expanding hosta and fern collection. Mature trees, 'cottage garden', long borders and stumpery. Narrow brick paths in vegetable plot. Large courtyard area for teas. Wheelchair access - some gravel paths but lawns are an option.

35 THE OLD VICARAGE

Fulford, nr Stone, ST11 9QS. Mike & Cherry Dodson. *4m N of Stone. From Stone A520 (Leek). 1m R turn to Spot Acre & Fulford, turn L down Post Office Terrace, past village green/Pub towards church. Parking signed on L.* **Sat 10 July (11-4.30). Adm £4, chd free. Home-made teas.**
1½ acres of formal sloping garden around Victorian house. Sit on the terrace or in the summerhouse to enjoy home-made cakes and tea amongst mature trees, relaxed herbaceous borders, roses and a small pond. Move to the organic vegetable garden with raised beds, fruit cage and very big compost heaps! In complete contrast, easy walk around the natural setting of a two-acre reclaimed lake planted with native species designed to attract wildlife. Waterfall, jetty, fishing hut, acer and fern glade plus arboretum provide more interest. Children will enjoy meeting the chickens and horses. A garden of contrasts, easy formality around the Victorian house and the accent on very natural planting around the lake, managed for wildlife and sustainability. Wheelchair access to most areas.

36 10 PAGET RISE

Paget Rise, Abbots Bromley, Rugeley, WS15 3EF. Mr Arthur Tindle. *4m W of Rugeley 6m S of Uttoxeter and 12m N of Lichfield. From Rugeley: B5013 E. At T junc turn R on B5014. From Uttoxeter take the B5013 S then B5014. From Lichfield take A515 N then turn L on B5234. In Abbots Bromley follow NGS yellow signs.* **Sun 16 May (11-5). Adm £3, chd free. Light refreshments.**
This small 2 level garden has a strong Japanese influence. Rhododendrons

and a wide range of flowering shrubs. Many bonsai-style Acer trees in shallow bowls occupy a central gravel area with stepping stones. The rear of the garden has a woodland feel with a fairy dell under the pine tree. A little gem of a garden! Arthur hopes visitors will be inspired to take away ideas to use in their own garden. Arthur is a watercolour artist & will be displaying a selection of his paintings for sale.

37 PAUL'S OASIS OF CALM

18 Kings Close, Kings Heath, Birmingham, B14 6TP. Mr Paul Doogan, 0121 444 6943, gardengreen18@hotmail.co.uk. *4m from city centre. 5m from M42 J4. Take A345 to Kings Heath High St then B4122 Vicarage Rd. Turn L onto Kings Rd then R to Kings Close.* **Visits by arrangement May to Aug for groups of up to 20. Adm £2.50, chd free. Cream teas.**
Garden cultivated from nothing into a little oasis. Measuring 18ft x 70ft. It's small but packed with interesting and unusual plants, water features and 7 seating areas. It's my piece of heaven.

38 THE PINTLES

18 Newport Road, Great Bridgeford, Stafford, ST18 9PR. Peter & Leslie Longstaff, 01785 282002, peter.longstaff@ngs.org.uk. *From J14 M6 take A5013 towards Eccleshall, in Great Bridgeford turn L onto B5405 after 600 metres turn L onto Great Bridgeford Village Hall car park The Pintles is opp the hall main doors.* **Sun 6 June (1.30-5). Adm £3.50, chd free. Home-made teas. Visits also by arrangement June & July for groups of 10 to 30.**
Located in the village of Great Bridgeford this traditional semi-detached house has a medium sized wildlife friendly garden designed to appeal to many interests. There are two greenhouses, 100s of cacti and succulents, vegetable and fruit plot, wildlife pond, weather station and hidden woodland shady garden. Plenty of outside seating to enjoy the home-made cakes and refreshments. Steps or small ramp into main garden.

Cross Roads, Huddlesford Lane

39 PRIORY FARM
Mitton Road, Bradley, Stafford, ST18 9ED. Debbie Farmer. *3½m W Penkridge. At Texaco island on A449 in Penkridge take Bungham Ln. Continue for 2½m past Swan & Whiston Hall to Mitton. Turn R to Bradley, continue 1m to Priory Farm on L.* **Sun 13 June (10.30-4). Adm £5, chd free. Home-made teas. BBQ subject to availability.**
Priory Farm looks forward to welcoming you all in 2021. The gardens have matured and we have made some changes to enhance this truly hidden gem. If you are looking for a tranquil hideaway to wander around the dells and lake, then to enjoy delicious homemade cakes and weather permitting BBQ (using only local produce) then look no further. Partial wheelchair access.

40 8 RECTORY ROAD
Solihull, B91 3RP. Nigel & Daphne Carter. *Town centre. Located in the town centre: off Church Hill Rd, turn into Rectory Rd, bear L down the rd, house on the R 100 metres down.*

Sun 13 June (11-6). Adm £4, chd free. Home-made teas. Evening opening Sun 5 Sept (7-10.30). Adm £4. Wine.
Stunning town garden divided into areas with different features. A garden with unusual trees and intensively planted borders. Walk to the end of the garden and step into a newly created Japanese garden complete with pond, fish and traditional style bridge. Sit on the shaded decking area. A garden to attract bees and butterflies with plenty of places to relax.

41 56 ST AGNES ROAD
Moseley, Birmingham, B13 9PN. Michael & Alison Cullen. *3m from city centre. From Moseley T-lights take St Mary's Row which becomes Wake Green Rd. After ½m turn R into St Agnes Rd and L at the church. Number 56 is approx 200 metres on L.* **Sun 18 July (1-5.30). Adm £3, chd free. Home-made teas.**
Immaculately maintained, medium-sized, urban garden with curving borders surrounding a formal lawn punctuated with delicate acers and contemporary sculpture. Seating by a Victorian-style fish pond with a fountain and waterfall offers a peaceful setting to enjoy the tranquillity of this elegant garden.

42 23 ST JOHNS ROAD
Rowley Park, Stafford, ST17 9AS. Fiona Horwath, 07908 918181, fiona_horwath@yahoo.co.uk. *½m S of Stafford Town Centre. Just a few mins from J13 M6, towards Stafford. After approx 2m on the A449, turn L into St. John's Rd after bus-stop.* **Sun 21 Mar, Sun 18 Apr (2-5). Adm £4, chd free. Home-made teas. Evening opening Fri 14 May (6.30-8.30). Adm £6, chd free. Wine. Fri 11 June (2-5). Adm £4, chd free. Home-made teas. Refreshments incl in adm Friday 14th May. Visits also by arrangement Mar to July for groups of 20+.**
Pass through the black and white gate of this Victorian house into a

part-walled gardener's haven. Bulbs and shady woodlanders in Spring and masses of herbaceous plants and climbers. Sit and enjoy home-made cakes by the pond or Victorian-style greenhouse. Gardener is keen Hardy Plant Society member and sows far too many seeds, so always something good for sale! Our new outdoor kitchen is great for refreshments! The waterlily wildlife pond remains - with the greenhouses - the beating heart of the garden. A growing interest in alpines is leading to a proliferation of troughs. Whilst ferns, the quiet groen stars of shady areas are also increasing in number!

43 THE SECRET GARDEN
3 Banktop Cottages, Little Haywood, ST18 0UL. Derek Higgott & David Aston. *5m SE of Stafford. A51 from Rugeley or Weston signed Little Haywood A513 Stafford Coley Ln, Back Ln R into Coley Gr. Entrance 50 metres on L.* **Wed 26 May, Wed 16 June, Wed 14 July, Wed 4 Aug (11-4). Adm £4, chd free. Home-made teas.** Wander past the other cottage gardens and through the evergreen arch and there before you is a fantasy for the eyes and soul. Stunning garden approx ½ acre, created over the last 30yrs. Strong colour theme of trees and shrubs, underplanted with perennials, 1000 bulbs and laced with clematis; other features incl water, laburnum and rose tunnel and unique buildings. Is this the jewel in the crown? New water feature and a warm Bothy for inclement days. Most areas have had a complete make-over due to Covid lockdown. Wheelchair access - some slopes.

44 TANGLEWOOD COTTAGE
Crossheads, Colwich, Stafford, ST18 0UG. Dennis & Helen Wood, 01889 882857, shuvitdog@gmail.com. *5m SE of Stafford. A51 Rugeley/Weston R into Colwich. Church on L school on R, under bridge R into Crossheads Ln follow railway approx ¼ m (it does lead somewhere). Parking signed on grass opp Brick Kiln Cottage.* **Sun 18 July (10.30-3). Adm £3, chd free. Light refreshments. Visits also by arrangement June to Aug for groups of 20+.** A country cottage garden incl koi carp pool, vegetables, fruit and an array of wonderful perennials. Meander

through the garden rooms, enjoying sights, sounds and fragrances. Many seats to absorb the atmosphere incl the courtyard to enjoy Helen's home-made fayre - cakes and meals. Year on year people spend many hours relaxing with us and don't forget Charlie the parrot and 6 chickens. Art/jewellery/crafts/book sales. Teas/coffees. Cakes and light breakfasts and lunches. Lots of gravel paths, some steps.

46 91 TOWER ROAD
Four Oaks, Sutton Coldfield, B75 5EQ. Heather & Gary Hawkins. *3m N Sutton Coldfield. From A5127 at Mere Green island, turn onto Mere Green Rd towards Sainsburys, L at St James Church, L again onto Tower Rd.* **Sun 13 June (1.30-5.30). Adm £3, chd free. Cream teas.** 163ft S-facing garden with sweeping borders and island beds planted with an eclectic mix of shrubs and perennials. We have made a few changes this year: removing a tired rookery has given us more room for another seating to enjoy the sunshine. A vast array of home-made cakes will tempt you during your visit. An ideal setting for sunbathing, children's hide and seek and lively garden parties. Large selection of home-made cream teas to eat in the garden or take away. Plant sale on front drive for garden visitors and passers by. More than just an Open Garden, we like to think of it as a garden party! There are 2 steps up into the garage and 2 steps down to the garden, which is then completely flat.

47 12 WATERDALE
Compton, Wolverhampton, WV3 9DY. Mr & Mrs Colin Bennett. *1½m W of Wolverhampton city centre. From Wolverhampton Ring Rd take A454 towards Bridgnorth for 1m. Waterdale is on the L off A454 Compton Rd West.* **Sun 6 June, Sun 4 July (11.30-5.30). Combined adm with 19 Waterdale £5, chd free. Also open Compton Care.** A riot of colour welcomes visitors to this quintessentially English garden. The wide central circular bed and side borders overflow with classic summer flowers, incl the tall spires of delphiniums, lupins, irises, campanula, poppies and roses. Clematis tumble over the edge of the decked terrace, where visitors can sit among pots of begonias and

geraniums to admire the view over the garden.

48 19 WATERDALE
Compton, Wolverhampton, WV3 9DY. Anne & Brian Bailey, 01902 424867, m.bailey1234@btinternet.com. *1½m W of Wolverhampton city centre. From Wolverhampton Ring Rd take A454 towards Bridgnorth for 1m. Waterdale is on L off A454 Compton Rd West.* **Sun 6 June, Sun 4 July (11.30-5.30). Combined adm with 12 Waterdale £5, chd free. Home-made teas. Also open Compton Care. Visits also by arrangement June to Aug for groups of 10 to 30.** A romantic garden of surprises, which gradually reveals itself on a journey through deep, lush planting, full of unusual plants. From the sunny, flower filled terrace, a ruined folly emerges from a luxuriant fernery and leads into an oriental garden, complete with tea house. Towering bamboos hide the way to the gothic summerhouse and mysterious shell grotto. Find us on Facebook at 'Garden of Surprises'.

50 THE WICKETS
47 Long Street, Wheaton Aston, ST19 9NF. Tony & Kate Bennett, 01785 840233, ajtonyb@talktalk.net. *8m W of Cannock, 10m N of Wolverhampton, 10m E of Telford. M6 J12 W towards Telford on A5; 3m R signed Stretton; 150yds L signed Wheaton Aston; 2m L; over canal, garden on R or at Bradford Arms on A5 follow signs.* **Sun 25, Thur 29 July, Sun 29 Aug (1.30-5). Adm £3.50, chd free. Home-made teas. Visits also by arrangement July & Aug.** There's a delight around every corner and lots of quirky features in this most innovative garden. Its themed areas incl a fernery, grasses bed, hidden gothic garden, succulent theatre, cottage garden beds and even a cricket match! It will certainly give you ideas for your own garden as you sit and have tea and cake. Afterwards, walk by the canal and enjoy the beautiful surrounding countryside. Wheelchair access - two single steps in garden and 2 gravel paths.

51 WILD THYME COTTAGE

Woodhouses, Barton Under Needwood, Burton-On-Trent, DE13 8BS. Ray & Michele Blundell. *B5016 Midway between villages of Barton & Yoxall. From A38 through Barton Village B5016 towards Yoxall. Or from A513 at Yoxall centre take Town Hill sign for Barton. Garden is at Woodhouses, approx 2m from either village.* Sat 30, Sun 31 Oct (11-4.30). Adm £4, chd free. Light refreshments.
The garden surrounds a self built Oak frame house just 20 years old but looks 400 and extends to ⅓ acre. Rare trees incl white bark birch and a collection of over 120 Japanese Maples. With Prairie planting of ornamental grasses and herbaceous perennials successional shrub. Growers of Zantedeschia (Arum Lily's). Plants sales. Emphasis is on successive interest throughout the year. Great Dixter inspired porch and garden entrance Oak Lychgate. Ornamental grasses. Rare trees and shrubs. Partial wheelchair access on gravel and grass.
& ✿ ☕

52 2 WOODLAND CRESCENT

Finchfield, Wolverhampton, WV3 8AS. Mr & Mrs Parker, 01902 332392, alisonparker1960@hotmail.co.uk. *2m SW of Wolverhampton City centre. From Inner Ring Rd take A41 W to Tettenhall. At junction with A459 turn L onto Merridale Rd. Straight over Bradwell Xrds then R onto Trysull Rd. 3rd R onto Coppice Rd & 1st R onto Woodland Crescent.* Sun 20 June, Sun 11 July (11.30-4.30). Adm £3.50, chd free. Home-made teas. Visits also by arrangement June & July for groups of 5 to 30.
A semi-detached townhouse garden 120 x 25 ft, every inch packed with perennials, trees, acers, shrubs, roses, hostas and clematis. Wildlife pond and nesting boxes to attract birds. Productive vegetable and fruit garden. Informal style but clipped box and topiary animals add some formality.
✿ ☕

53 WOODLEIGHTON GROVE KARIBU GARDEN

9 Woodleighton Grove, Uttoxeter, ST14 8BX. Graham & Judy White, 01889 563930 or 0788 178 1547, graham&judy.white@ngs.org.uk. *Take B5017 Marchington Rd from Uttoxeter Town Centre. Go over Railway Bridge, then take 1st exit at r'about, then 3rd exit at next r'about into Highwood Rd, After ¼ m turn R then 1st L.* Visits by arrangement June & July. Adm £8, chd free. Home-made teas. Admission price includes the refreshments.
A tranquil & intriguing garden informally planted on two levels. Unique landscaping & design features incl a Giant Insect Hotel; Bell Tower; Folly; Dovecote; Boardwalk; Stumpery; Archways; Natural Stream; Bridges & Waterfall. Various collections of Plants & Artefacts incl 400 cacti, succulents & sempervivum plus over 250 vintage and antique horticultural & agricultural hand tools. W.C available.
🚗 ☕

54 YARLET HOUSE

Yarlet, Stafford, ST18 9SD. Mr & Mrs Nikolas Tarling. *2m S of Stone. Take A34 from Stone towards Stafford, turn L into Yarlet School & L again into car park.* Fri 18 June (10-1). Adm £4, chd free. Light refreshments. Donation to Staffordshire Wildlife Trust.
4 acre garden with extensive lawns, walks, lengthy herbaceous borders and traditional Victorian box hedge. Water gardens with fountain and rare lilies. Sweeping views across Trent Valley to Sandon. New peaceful Japanese garden for 2021. Victorian School Chapel. 9 hole putting course. Boules pitch. Yarlet School Art Display. Gravel paths.
& 🐴 🚗 ☕

55 YEW TREE COTTAGE

Podmores Corner, Long Lane, White Cross, Haughton, ST18 9JR. Clive & Ruth Plant, 07591 886925, pottyplantz@aol.com. *4m W of Stafford. Take A518 W Haughton, turn R Station Rd (signed Ranton) 1m, then turn R at Xrds ¼ m on R.* Sun 2 May (2-5); Thur 1, Thur 8, Thur 15 July (11-4); Sun 25 July (2-5). Adm £4, chd free. Home-made teas. Visits also by arrangement May to July for groups of 10+.
Hardy Plant Society member's garden brimming with unusual plants. All-yr-round interest incl Meconopsis, Trillium. ½ -acre incl gravel, borders, vegetables and plant sales. National Collection Dierama featured on BBC Gardeners World flowering first half July. Facebook 'Dierama Species in Staffordshire'. Covered vinery for tea if weather is unkind, and seats in the garden for lingering on sunny days. Partial wheelchair access, grass and paved paths, some narrow and some gravel.
& ✿ 🚗 NPC ☕ 🪑

56 YEW TREES

Whitley Eaves, Eccleshall, Stafford, ST21 6HR. Mrs Teresa Hancock, 07973432077, hancockteresa@gmail.com. *7m from J14 M6. Situated on A519 between Eccleshall (2.2m) and Woodseaves, (1m). Traffic cones & signs will highlight the entrance.* Visits by arrangement July to Sept. Adm £4, chd free. Light refreshments. Donation to Plant Heritage.
1 acre garden divided into rooms by mature hedging, shrubs & trees enjoying views over the surrounding countryside. Large patio area with containers & seating. Other features incl pond, topiary, vegetable plot, hen run & wildlife area. If the garden is visited during July there are 8 acres of natural wild flower meadow to enjoy before it is cut for hay.
& ☕ 🪑

"I was amazed to discover that the National Garden Scheme is Marie Curie's largest single funder and has given the charity nearly £10 million over 25 years. Their continued support makes such a difference to me and all Marie Curie Nurses on the frontline of the coronavirus crisis, as we continue to provide expert care and support to people at end of life."
– Tracy McWilliams, Marie Curie Nurse

Monarchs Way

SUFFOLK

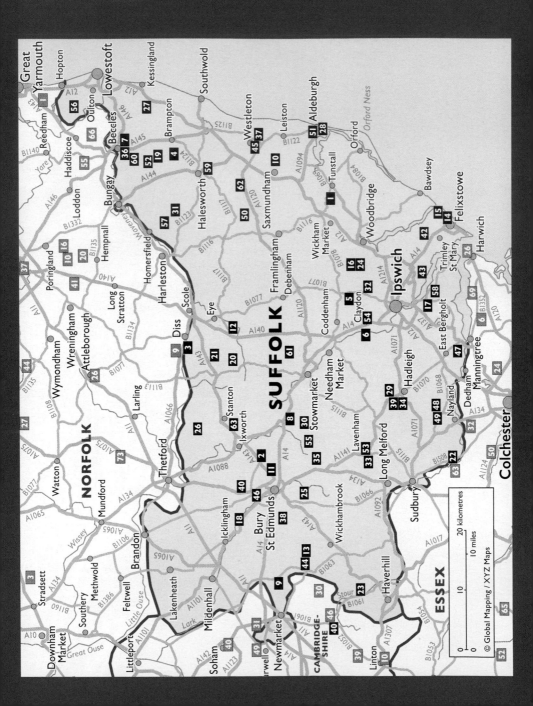

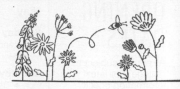

Suffolk has so much to offer – from charming coastal villages, ancient woodlands and picturesque valleys – there is a landscape to suit all tastes.

Keen walkers and cyclists will enjoy Suffolk's low-lying, gentle countryside, where fields of farm animals and crops reflect the county's agricultural roots.

Stretching north from Felixstowe, the county has miles of Heritage Coast set in an Area of Outstanding Natural Beauty. The Suffolk coast was the inspiration for composer Benjamin Britten's celebrated work, and it is easy to see why.

To the west and north of the county are The Brecks, a striking canvas of pine forest and open heathland, famous for its chalky and sandy soils – and one of the most important wildlife areas in Britain.

A variety of gardens to please everyone open in Suffolk, so come along on an open day and enjoy the double benefit of a beautiful setting and supporting wonderful charities.

Volunteers

County Organiser
Jenny Reeve
01638 715289
jenny.reeve@ngs.org.uk

County Treasurer
David Reeve
01638 715289
dreeve43@gmail.com

Publicity
Jenny Reeve
(as above)

Social Media
Barbara Segall
01787 312046
barbara.segall@ngs.org.uk

Booklet Co-ordinator
Michael Cole
01473 272920
michael.m.d.cole@btinternet.com

Assistant County Organisers
Jane Bastow 01986 782413
jane.bastow@btinternet.com

Michael Cole 01473 272920
michael.m.d.cole@btinternet.com

Gillian Garnham 01394 448122
gill.garnham@btinternet.com

Yvonne Leonard 01638 712742
yj.leonard@btinternet.com

Marie-Anne Mackenzie
01728 831155
marieanne_mackenzie@yahoo.co.uk

Wendy Parkes 01473 785504
wendy.parkes@ngs.org.uk

Barbara Segall
(as above)

Peter Simpson 01787 249845
bergholt2002@btopenworld.com

@SuffolkNGS
@SuffolkNGS

Left: Berghersh Place

OPENING DATES

All entries subject to change. For latest information check **www.ngs.org.uk**

Map locator numbers are shown to the right of each garden name.

February

Snowdrop Festival

Sunday 14th
◆ Blakenham Woodland Garden 6
Gable House 19
The Old Rectory, Brinkley 41

Sunday 21st
The Laburnums 31

March

Sunday 28th
Great Thurlow Hall 23

April

Sunday 18th
◆ The Place for Plants, East Bergholt Place Garden 47

Sunday 25th
◆ Blakenham Woodland Garden 6
NEW The Old Rectory 40

National Garden Scheme gardens are identified by their yellow road signs and posters. You can expect a garden of quality, character and interest, a warm welcome and plenty of home-made cakes!

May

Sunday 2nd
Hawstead Place 25
Moat House 38

Monday 3rd
NEW ◆ The Red House 51

Sunday 9th
◆ Fullers Mill Garden 18
◆ The Place for Plants, East Bergholt Place Garden 47

Sunday 16th
NEW Finndale House 16
NEW Grundisburgh House 24
The Priory 49

Saturday 22nd
NEW Granary Barn 21

Sunday 23rd
Great Thurlow Hall 23

Saturday 29th
◆ Wyken Hall 63

Sunday 30th
Berghersh Place 5
Bridges 8
Freston House 17
Lavenham Hall 33
◆ Wyken Hall 63

Monday 31st
11 Brookside 9
Cattishall Farmhouse 11
Holm House 30
NEW ◆ The Red House 51

June

Sunday 6th
NEW Ashe Park 1
NEW Barton Mere 2
◆ Somerleyton Hall Gardens 56

Saturday 12th
The Old Rectory, Nacton 43
NEW Smallwood Farmhouse 55

Sunday 13th
Great Bevills 22
Heron House 28
The Old Rectory, Nacton 43
5 Parklands Green 46
Wood Farm, Gipping 61

Saturday 19th
Brambly Hedge 7
The Rooks 54

Sunday 20th
Hillside 29
The Old Rectory, Kirton 42
Wenhaston Grange 59
Wood Farm, Sibton 62

Friday 25th
Church Cottage 12

Sunday 27th
Becks End Farm 4
Church Cottage 12
246 Ferry Road 14
NEW Field Cottage 15
Ousden House 44

July

Sunday 4th
Batteleys Cottage 3

Sunday 11th
Redisham Hall 52
NEW Squires Barn 57

Saturday 17th
The Old Rectory, Brinkley 41

Sunday 18th
Loreto 36
NEW Manor House 37
The Old Rectory, Brinkley 41
Paget House 45

Sunday 25th
NEW The Old Rectory 40

August

Saturday 7th
Gislingham Gardens 20

Sunday 8th
Gislingham Gardens 20
River Cottage 53

Sunday 22nd
Henstead Exotic Garden 27

Sunday 29th
Bridges 8
NEW The Priory, Laxfield Rd 50

September

Sunday 5th
NEW ◆ The Red House 51

October

Sunday 3rd
◆ Fullers Mill Garden 18

Sunday 10th
◆ The Place for Plants, East Bergholt Place Garden 47

February 2022

Sunday 20th
The Laburnums 31

By Arrangement

Arrange a personalised garden visit on a date to suit you. See individual garden entries for full details.

Berghersh Place 5
Brambly Hedge 7
By the Crossways 10
12 Chapel Farm Close, Gislingham Gardens 20
Dip-on-the-Hill 13
Gable House 19
Gislingham Gardens 20
Helyg 26
Heron House 28
Hillside 29
Holm House 30
Ivy Chimneys, Gislingham Gardens 20
Larks' Hill 32
Lillesley Barn 34
NEW The Lodge 35
Moat House 38
Old Gardens 39
The Old Rectory, Nacton 43
Paget House 45
5 Parklands Green 46
Polstead Mill 48
NEW Stone Cottage 58
White House Farm 60
Wood Farm, Gipping 61

THE GARDENS

1 NEW ASHE PARK
Ivy Lodge Road, Campsea
Ashe, Woodbridge, IP13 0QB.
Mr Richard Keeling. *Using the postcode in Sat Nav will bring you to the entrance to Ashe Park on Ivy Lodge Rd. Drive through entrance signed Ashe Park, past the gate cottage on L & follow signs to car park.* **Sun 6 June (10.30-5). Adm £6, chd free.**
In the 350 years of this 12 acre garden's existence it has gone through many changes. Gertrude Jekyll said of it in 1905 that it was interesting but it had no coherence of design. Since then the main house has been destroyed and the gardens have been re-aligned, however the bones of the gardens have been retained namely canals, massive cedar trees, yew hedges, walled garden and many more features. Partial wheelchair access, some gravel paths and steps.
&

2 NEW BARTON MERE
Thurston Road, Great Barton,
IP31 2PR. Mr & Mrs C O Stordcrup. *2m E of Bury St Edmunds. From Bury St Edmunds on A143 through Gt Barton turn R at Dunbury Arms PH. Entrance ½m on L. From Thurston take Gt Barton Rd from railway bridge. Entrance 1½ m on R.* **Sun 6 June (1-5). Adm £5, chd free. Home-made teas.**
C16 house (not open) with later Georgian façade, set in 50 acres of parkland. Extensive lawns with views over the Mere. Rose garden and herbaceous borders mostly surrounded by C16 walls, two courtyards and large conservatory. Productive vegetable garden, old orchard and beautiful grass tennis court. Gravel paths.
& 🐎 ☕ 🍽

3 BATTELEYS COTTAGE
The Ling, Wortham, Diss, IP22 1ST.
Mr & Mrs Andy & Linda Simpson.
3m W of Diss. Turn signed from A143 Diss/Bury Rd at Wortham. By Church turn R at T-junction. At top of hill turn L. Go down hill & round sharp L corner. **Sun 4 July (11.30-4.30). Adm £5, chd free. Home-made teas and light refreshments.**
A varied one acre garden planted for abundance in all seasons. Formality

and informality, a mix of winding bark paths, light and shade, secluded spots to sit, new vistas at every turn. Fitting into its rural setting, it supports a wealth of bird life. There is a diversity of planting in densely planted borders as well as pots, sculptures, meadow, ponds, stream and vegetable areas to inspire you. Wheelchair access to most parts of the garden, gravel, grass and bark paths.
& ❋ 🚗 ☕ 🍽

4 BECKS END FARM
School Road, Westhall,
Halesworth, IP19 8QZ. Mr & Mrs John & Marjorie Milbank. *E of Halesworth on A144 between Halesworth & Bungay. From Halesworth turn R at the Spexhall Xrds. Continue along & take 3rd turning L. Becks End Farm is the first house on L.* **Sun 27 June (11-5.30). Adm £5, chd free. Home-made teas.**
Set in one and a half acres comprising lawns, trees, hard landscaping, wide herbaceous flower borders, orchard, natural pond, vegetable garden and central pole fruit cage. Pleached Hornbeam hedges screen the paddock and stable block. The garden surrounds an old Victorian Schoolhouse on the edge of the village and has been designed and developed since 2007 when acquired by the present owners. No wheelchair access because of steps.

5 BERGHERSH PLACE
Ashbocking Road, Witnesham,
Ipswich, IP6 9EZ. Mr & Mrs T C Parkes, 01473 785504,
wendyparkes@live.com. *N of Witnesham village, B1077 double bends. Farm entrance, concrete drive, approx 1m S of Ashbocking Xrds. Entrance on sharp bend so please drive slowly. Turn in between North Lodge & Berghersh House.* **Sun 30 May (12-5). Adm £5, chd free. Home-made teas. Gluten-free options. Visits also by arrangement in June for groups of 10 to 30. Guided visits available.**
Peaceful walled and hedged gardens surround elegant Regency house (not open) among fields above the Fynn Valley. Circular walk from the farm buildings, around house with lawns and mature trees to a pretty view of the valley. Mound, ponds, bog area and orchard paddock. Informal family garden with shrub and perennial beds. Garden created over last 20 years by

current owner. Parking for elderly and visitors with limited mobility at end of farmyard close to family garden. Most areas are accessible to wheelchair users.
& 🐎 🍽

6 ◆ BLAKENHAM WOODLAND GARDEN
Little Blakenham,
Ipswich, IP8 4LZ. M Blakenham, 07917612355,
blakenham@btinternet.com, www.
blakenhamwoodlandgarden.org.
uk. *4m NW of Ipswich. Follow signs at Little Blakenham, 1m off B1113 or go to Blakenham Woodland Garden web-site.* **For NGS: Sun 14 Feb, Sun 25 Apr (10-4). Adm £5, chd £2.50. Home-made teas (NGS days only). For other opening times and information, please phone, email or visit garden website.**
Beautiful 6 acre woodland garden with variety of rare trees and shrubs, Chinese rocks and a landscape spiral form. Lovely in spring with snowdrops (Feb opening), daffodils and camellias followed by magnolias and bluebells (Apr opening). Woodland Garden open 1 Feb to 28 June.
& ❋ 🚗 ☕

7 BRAMBLY HEDGE
Lowestoft Road, Beccles,
NR04 7DC. Lynton & Teresa Cooper, 01502 715678,
Happy-chickens@hotmail.co.uk.
From town centre head towards Lowestoft, along Ingate - on R after lights. From Lowestoft direction head through Worlingham on L before lights. Roadside parking Ellough Rd. NR34 7AA. **Sat 19 June (11-4). Adm £3.50, chd free. Home-made teas. Visits also by arrangement June to Sept for groups of 10 to 20.**
A garden of contrasts and surprises. A modern introduction to a quirky woodland finish, the visitor can explore a huge variety of shrubs and trees with many unusual specimens. Add in a beautiful raised-bed potager and just when you think you have seen it all, and relax with tea and cake on one of the many seats dotted around, tucked away is a gem you missed the first time. Partial wheelchair access.
❋ ☕

8 BRIDGES

The Street, Woolpit, Bury St Edmunds, IP30 9SA. Mr Stanley Bates & Mr Michael Elles. *Through green coach gates marked Deliveries. From A14 take slip rd to Woolpit, follow signs to centre of village, road curves to R. Bridges is on L & covered in Wisteria & opp Co-op.* **Sun 30 May, Sun 29 Aug (11-5). Adm £5, chd free. Home-made teas.**

C15 Grade 2 terraced house in the centre of a C12 Suffolk village with walled garden to the rear of the property. Additional land was acquired 20 years ago, and this garden was developed into formal and informal planting. The main formal feature is the Shakespeare Garden featuring the bust of Shakespeare, and the 'Umbrello' a recently constructed pavillion in an Italianate design. Various surprises for children of all ages. Usually a Wind Quintet playing in the main garden. Two public houses The Swan and The Bull in village open for lunch. Not suitable for wheelchairs.

9 11 BROOKSIDE

Moulton, Newmarket, CB8 8SG. Elizabeth Goodrich & Peter Mavroghenis. *Near the Packhorse Bridge & Pub. 3m due E of Newmarket on B1085.* **Mon 31 May (2-5). Adm £5, chd free.**

Pre-booking essential, please phone 07973 622136 or email goodrichelizabeth1@gmail.com for information & booking. Light refreshments, tea, coffee and home-made cakes.

1½ acres over 4 levels. Traditional hedges, mature trees and roses at the front while rear garden landscaped in contemporary style. Lower terrace with water feature and fig trees. Terraced beds with ornamental grasses, knot garden, hosta courtyard. Metal retaining wall to former paddock, many specimen trees, apple espalier bordered greenhouse with kitchen garden, apples and vines.

10 BY THE CROSSWAYS

Kelsale, Saxmundham, IP17 2PL. Mr & Mrs William Kendall, miranda@bythecrossways.co.uk. *2m NE of Saxmundham, just off Clayhills Rd. ½m N of town centre, turn R to Theberton on Clayhills Rd. After 1½m, 1st L to Kelsale, then turn L immed after white cottage.* **Visits by arrangement May to Sept for groups of 5+. Adm by donation.**

Three acre wildlife garden designed as a garden within a working organic farm where wilderness areas lie next to productive beds. Large semi-walled vegetable and cutting garden and a spectacular crinkle-crankle

wall. Extensive perennial planting, grasses and wild areas. This garden is not highly manicured which more traditionally minded gardeners may find alarming! The garden is mostly flat, with paved or gravel pathways around the main house, a few low steps and extensive grass paths and lawns.

11 CATTISHALL FARMHOUSE

Cattishall, Great Barton, Bury St Edmunds, IP31 2QT. Mrs J Mayer, joannamayer42@googlemail.com. *3m NE of Bury St Edmunds. Approach Great Barton from Bury on A143 take 1st R turn to church. If travelling towards Bury take last L turn to church as you leave the village. At church bear R & follow lane to Farmhouse on R.* **Mon 31 May (1-5). Adm £4, chd free. Home-made teas.**

Approx 2 acre farmhouse garden enclosed by a flint wall and mature beech hedge laid mainly to lawns with both formal and informal planting and large herbaceous border. There is an abundance of roses, small wildlife pond and recently developed kitchen garden incl a wildflower area and fruit cages. Chickens, bees and a boisterous Labrador also live here. Generally flat with some gravel paths. The occasional small step.

The Old Rectory

12 CHURCH COTTAGE
Braiseworth Lane, Braiseworth, IP23 7DT. Mr Rajat Jindal. *Enter from A140 on a 1m narrow track road or through Eye.* **Fri 25, Sun 27 June (11-4). Adm £5, chd free. Pre-booking essential, please visit www.ngs.org.uk for information & booking. Light refreshments.**
A naturalistic garden set in 2.5 acres and developed from scratch over the last 10 years. A combination of formal and informal features including a parterre, cottage garden surrounding an C18 thatched cottage, ponds, large prairie style beds, wildflower meadow, Hornbeam cubes, two copses, pleached limes, raised vegetable beds, willow dome and specimen trees. Wheelchair accessible with care, mainly on grass.

& ♞ ⚘ ⚍

13 DIP-ON-THE-HILL
Ousden, Newmarket, CB8 8TW. Geoffrey & Christine Ingham, 01638 500329, gki1000@cam.ac.uk. *5m E of Newmarket; 7m W of Bury St Edmunds. From Newmarket: 1m from junction of B1063 & B1085. From Bury St Edmunds follow signs for Hargrave. Parking at village hall. Follow NGS sign at the end of the lane.* **Visits by arrangement June to Sept for groups of up to 20. Adm £5, chd free. Tea.**
Approx one acre in a dip on a S-facing hill based on a wide range of architectural/sculptural evergreen trees, shrubs and groundcover; pines; grove of Phillyrea latifolia; 'cloud pruned' hedges; palms; large bamboo; ferns; range of kniphofia and Croscosmia. Winner: 'Britain's Best Garden' 2018, Telegraph/ Yorkshire Tea. Visitors may wish to arrange a visit when visiting gardens nearby. No wheelchair access. No dogs.

⚍

14 246 FERRY ROAD
Felixstowe, IP11 9RU. Mrs Sally Gallant. *At Felixstowe Golf Club you will find Ferry Rd opposite, travel up Ferry Rd & house is 4th on L.* **Sun 27 June (11-5). Combined adm with Field Cottage £10, chd free. Home-made teas.**
Established garden of approximately ¼ acre, giving year round colour and interest. With places to sit and ponder and various walkways leading you around this secluded coastal garden, moments from the sea.

Some mature trees, and attractive hedging, shady walkway and wooden boardwalk, herbaceous borders and Japanese themed feature. Sunny sheltered sitting areas. Not suitable for wheelchair users due to steps. Well behaved dogs on a lead.

♞ ⚘ ⚍

15 NEW FIELD COTTAGE
Marsh Lane, Felixstowe, IP11 9RN. Mr & Mrs Nigel Papworth. *From Colneis Road, Felixstowe, go to Ferry Road & turn L, take R hand fork along Ferry Road. Next turn L into Marsh Lane. You will see signs a white fence which is Field Cottage.* **Sun 27 June (11-5). Combined adm with 246 Ferry Road £10, chd free. Light refreshments.**
Created over thirty years ago by a Suffolk lady garden designer. Features include a row of Norman style cloisters with climbing roses including Albertine [pink] and a large pond with Gunnera. From part of the rear garden there are distant views of the River Deben. Most of the garden is fairly level, the ground rising behind the pond.

& ⚍

16 NEW FINNDALE HOUSE
Woodbridge Road, Grundisburgh, Woodbridge, IP13 6UD. Bryan & Catherine Laxton. *On B1079, 2m NW of Woodbridge. Parking in field (postcode IP13 6PU) blue badge parking by the house. Short walk to Grundisburgh House.* **Sun 16 May (11-5). Combined adm with Grundisburgh House £5, chd free.**
A Georgian house surrounded by 10 acres of garden and meadows which are bisected by a 'Monet' style bridge over the River Lark. The garden was designed 30 years ago and has been refreshed by recent additions. Many mature trees, colourful herbaceous borders, thousands of daffodils followed by tulips and alliums, roses in the summer and dahlias to round the year off. Productive kitchen garden. Lunches available at The Dog PH in village. Gravel drive.

& ♞ ⚘ ⚍

17 FRESTON HOUSE
The Street, Freston, Ipswich, IP9 1AF. Mr & Mrs Andrew & Judith Whittle, www.frestonhouse.co.uk. *Go under the Orwell Bridge from Ipswich towards Holbrook. At the junction for Holbrook & Woolverstone, turn sharp R signed Freston. After 300m, turn L into*

a no through road. **Sun 30 May (12-5). Adm £5.50, chd free. Light refreshments.**
20 acre garden and large Georgian rectory set in parkland, planted from 2006 onwards by the current owners. Individual colour-themed, roomed gardens, cottage garden and formal long borders with mass plantings of hundreds of shrubs and perennials. A one acre kitchen garden, wildlife pond, large winter garden (originally inspired by Anglesey Abbey) and a gravel garden. Woodlands with over 1,000 varieties of hosta (one of the largest collections in the country) and other shade loving plants. Over 200 varieties of bearded iris. Some gravel paths.

& ♞ ⚘ ⚗ ⚍

18 ♦ FULLERS MILL GARDEN
West Stow, IP28 6HD. Perennial, 01284 728888, fullersmillgarden@perennial.org.uk, www.fullersmillgarden.org.uk. *6m NW of Bury St Edmunds. Turn off A1101 Bury to Mildenhall Rd, signed West Stow Country Park, go past Country Park continue for ¼ m, garden entrance on R. Sign at entrance.* **For NGS: Sun 9 May, Sun 3 Oct (11-5). Adm £5, chd free. Home-made teas. For other opening times and information, please phone, email or visit garden website.**
An enchanting 7 acre garden on the banks of the river Lark. A beautiful site with light dappled woodland and a plantsman's paradise of rare and unusual shrubs, perennials and marginals planted with great natural charm. Euphorbias and lilies are a particular feature with the late flowering colchicums, including many rare varieties, being of great interest in Autumn. Tea, coffee and soft drinks. Home-made cakes. Partial wheelchair access around garden.

& ⚘ ⚗ ⚍

"The annual donation from the National Garden Scheme to Perennial is the cornerstone of our fundraising activities and encourages many of our donors." Perennial

19 GABLE HOUSE

Halesworth Road, Redisham, Beccles, NR34 8NE. John & Brenda Foster, 01502 575298, gablehouse@btinternet.com. *5m S of Beccles. Signed from A12 at Blythburgh and A144 Bungay/ Halesworth Rd.* **Sun 14 Feb (11-4). Adm £4.50, chd free. Light refreshments. incl warming soups available in February. Visits also by arrangement Apr to Sept for groups of 10+.**

We have a large collection of snowdrops, cyclamen, hellebores and other flowering plants for the Snowdrop Day in February. Many bulbs and plants will be for sale. Hot soup and home-made teas available. Greenhouses contain rare bulbs and tender plants. A one acre garden with lawns and scree with water feature. We have a wide range of unusual trees, shrubs, perennials and bulbs collected over the last fifty years.

GROUP OPENING

20 GISLINGHAM GARDENS

Mill Street, Gislingham, IP23 8JT. 01379 788737, alanstanley22@gmail.com. *4m W of Eye. Gislingham 2½m W of A140. 9m N of Stowmarket, 8m S of Diss. Disabled parking at Ivy Chimneys. Parking limited to disabled parking at Chapel Farm Close.* **Sat 7, Sun 8 Aug (11-4.30). Combined adm £5, chd free. Light refreshments at Ivy Chimneys. Tea, coffee squash savouries cake vegetarian option. Visits also by arrangement June to Sept for groups of up to 30. Adm incl refreshments or £4 if no teas.**

12 CHAPEL FARM CLOSE
Ross Lee, 01379788737, alanstanley22@gmail.com.

IVY CHIMNEYS
Iris & Alan Stanley, 01379 788737, alanstanley22@btinternet.com.

2 varied gardens in a picturesque village with a number of Suffolk timbered houses. Ivy Chimneys is planted for yr round interest with ornamental trees, some topiary, exotic borders and fishpond set in an area of Japanese style. Wisteria draped pergola supports a productive vine. Also a separate ornamental vegetable garden. Small orchard on front lawn. 12 Chapel Farm Close is a tiny garden, exquisitely planted and

an absolute riot of colour. Despite the garden's size, the owner has planted a Catalpa, a Cornus Florida Rubra and many unusual plants. It is a fine example of what can be achieved in a small space. Wheelchair access to Ivy Chimneys. Access for smaller wheelchairs only at 12 Chapel Farm Close.

21 NEW GRANARY BARN

Little Green, Burgate, Diss, IP22 1QQ. Wendy Keeble. *23m NE of Bury St Edmunds; 6m SW of Diss; 1m S of A143. On A143 between Wortham & Botesdale, take turning signed Burgate Little Green (Buggs Ln) & follow NGS signs to car park 200 yd walk from garden. No parking on verges or common please.* **Sat 22 May (11-4.30). Adm £4, chd free.**

If re-wilding is the latest trend, we must be high fashion! From 2 acres of closely mown grass 10 yrs ago, the garden now has formal planting near the house, box parterre and pergola, herbaceous borders, a productive orchard, vegetable and soft fruit gardens and a large perennial wildflower meadow with 120 native trees and pond. Our aim is to attract and sustain wildlife during every season. No refreshments available, but plants, artwork and crafts are offered for sale. Partial wheelchair access can be organised on arrival.

22 GREAT BEVILLS

Sudbury Road, Bures, CO8 5JW. Mr & Mrs G T C Probert. *4m S of Sudbury. Just N of Bures on the Sudbury rd B1508.* **Sun 13 June (2-5.30). Adm £4, chd free. Home-made teas.**

Overlooking the Stour Valley the gardens surrounding an Elizabethan manor house are formal and Italianate in style with Irish yews and mature specimen trees. Terraces, borders, ponds and woodland walks. A short drive away from Great Bevills visitors may wish to also see the C13 St Stephen's Chapel with wonderful views of the Old Bures Dragon recently re-created by the owner. Woodland walks give lovely views over the Stour Valley. Gravel paths.

23 GREAT THURLOW HALL

Great Thurlow, Haverhill, CB9 7LF. Mr George Vestey. *12m S of Bury St Edmunds, 4m N of Haverhill. Great Thurlow village on B1061 from Newmarket; 3½ N of junction with A143 Haverhill/Bury St Edmunds rd.* **Sun 28 Mar, Sun 23 May (2-5). Adm £5, chd free. Home-made teas in the Church.**

13 acres of beautiful gardens set around the River Stour, the banks of which are adorned with stunning displays of daffodil and narcissi together with blossoming trees in spring. Herbaceous borders, rose garden and extensive shrub garden come alive with colour from late spring onwards, there is also a large walled kitchen garden and arboretum.

24 NEW GRUNDISBURGH HOUSE

Woodbridge Road, Grundisburgh, Woodbridge, IP13 6UD. Mrs Linden Hibbert. *Not, as the postcode claims, at the junction but just off B1079. Search for The Grundisburgh House Gallery (google maps) for precise location. Last house in village on L. Disabled parking at house. Parking at IP13 6PU. Main car park in nearby field.* **Sun 16 May (11-5). Combined adm with Finndale House £5, chd free.**

3-acre garden wrapping around classic Georgian house, highlights include natural swimming pond, formal garden, spring bulbs and fruit blossom, roses, irises and hydrangeas. New projects are on-going but include planting more hedging, ornamental trees and pleaching. There is a pop-up art gallery in the old coach house which is open intermittently throughout the year. The Grundisburgh Dog serves hot food. 3m from Woodbridge with numerous restaurants and cafes. Wheelchair access paved terrace and path in the formal garden, decking with two steps on swimming pond terrace. Remainder is gravel path.

25 HAWSTEAD PLACE

Bull Lane, Pinford End, Bury St Edmunds, IP29 4AB. Mr & Mrs R Brown, www.suffolkbarn.co.uk. *On A143 take a L down Sharpes Lane in Horringer. At the end of Sharpes Lane the Hawstead Place drive is directly in front of you. Be careful. Go down the drive & follow the signs for*

Cattishall Farmhouse

the Suffolk Barn. **Sun 2 May (2-5). Adm £5, chd free. Home-made teas.**

Rear C15 walled garden with perennial and shrub borders. Late spring colour provided by bulbs, tulips, narcissi, camassias and alliums planted en masse. Small walled kitchen garden growing vegetables, fruit and cut flowers for the house. Three small mixed borders on the front lawn with views to parkland beyond. Walks around the estate. No steps so wheelchairs are possible and we have a disabled toilet.

26 HELYG

Thetford Road, Coney Weston, Bury St Edmunds, IP31 1DN. Jackie & Briant Smith, 01359 220106, briant. broadsspirituality@gmail.com. *From Barningham Xrds/shop turn off the B1111 towards Coney Weston & Knettishall Country Park. After approx 1m Helyg will be found on L behind some large willow trees.* **Visits by arrangement May to Sept for groups of 10 to 20. Refreshments by arrangement with the Garden Owner. Adm £4, chd free.**

A garden that combines many seating areas providing tranquillity and relaxation with novel ideas and planting to stimulate interest. There

are different areas with their own feature plants including naturalised spring bulbs, hundreds of hostas, roses, rhododendrons, camellias, buddleias, fuchsias, azaleas, herbs, vegetables and exotics. Many varied, colourful and inspirational areas. Most of the garden is wheelchair accessible.

&

27 HENSTEAD EXOTIC GARDEN

Church Road, Henstead, Beccles, NR34 7LD. Andrew Brogan, www.hensteadexoticgarden.co.uk. *Equal distance between Beccles, Southwold & Lowestoft approx 5m. 1m from A12 turning after Wrentham (signed Henstead) very close to B1127.* **Sun 22 Aug (11-5). Adm £5, chd free. Home-made teas. Various teas, caefetiere coffees, choice of top quality local delicatessen made cakes.**

2 acre exotic garden featuring 100 large palms, 20+ bananas and giant bamboo, some of biggest in the UK. Streams, 20ft tiered walkway leading to Thai style wooden covered pavilion. Mediterranean and jungle plants around 3 large ponds with fish. Winner Britain's Best Garden 2015 on ITV as voted by Alan Titchmarsh. Unique garden buildings, streams, waterfalls, rock walkways, different levels, Victorian grotto, giant compost

toilet etc. Partial disabled access to the garden.

28 HERON HOUSE

Priors Hill Road, Aldeburgh, IP15 5EP. Mr & Mrs Jonathan Hale, 01728 452200, jonathanrhhale@aol.com. *At the southeastern junction of Priors Hill Rd & Park Rd. Last house on Priors Hill Rd on south side, at the junction where it rejoins Park Rd.* **Sun 13 June (2-5). Adm £5, chd free. Tea and cakes for sale in conservatory attached to house. Visits also by arrangement Apr to Oct.**

2 acres with superb views over the North Sea, River Alde and marshes. Unusual trees, herbaceous beds, shrubs and ponds with a waterfall in large rock garden, and a stream and bog garden. Some half hardy plants in the coastal micro-climate. Partial wheelchair access.

The National Garden Scheme searches the length and breadth of England and Wales for the very best private gardens

29 HILLSIDE

Union Hill, Semer, Ipswich, IP7 6HN. Mr & Mrs Neil Mordey, 07753 332022, Mordeysue@gmail.com. *Car park through field gate off A1141.* **Sun 20 June (10.30-4.30). Adm £5, chd free. Tea, coffee and soft drinks with home-made cakes and sandwiches. Visits also by arrangement May to Sept for groups of 10 to 30.**
This garden in its historic setting of 10.5 acres has sweeping lawns running down to a spring fed carp pond. The formal garden has island beds of mixed planting for a long season of interest. The wild area of meadow has been landscaped with extensive tree planting to complement the existing woodland. There is also a small walled kitchen garden. Most areas are accessible although the fruit and vegetable garden is accessed over a deep gravel drive.

30 HOLM HOUSE

Garden House Lane, Drinkstone, Bury St Edmunds, IP30 9FJ. Mrs Rebecca Shelley, Rebecca.shelley@hotmail.co.uk. *7m SE of Bury St Edmunds. Coming from the E exit A14 at J47, from W J46. Follow signs to Drinkstone, then Drinkstone Green. Turn into Rattlesden Road & look for Garden House Lane on L. Then 1st house on L.* **Mon 31 May (10.30-5). Adm £6, chd free. Home-made teas. Visits also by arrangement May to Sept for groups of 10+. Adm includes tea and cake.**
Approx 10 acres including: orchard and lawns with mature trees and clipped Holm Oaks; formal garden with topiary, parterre and mixed borders; rose garden; woodland walk with hellebores, camellias, rhododendrons and bulbs; lake, woodland planting and wildflower meadow; cut flower garden with greenhouse; large kitchen garden with impressive greenhouse; Mediterranean courtyard with mature olive tree. Much of the garden is wheelchair accessible, but not the kitchen garden.

31 THE LABURNUMS

The Street, St James South Elmham, Halesworth, IP19 0HN. Mrs Jane Bastow. *6m W of Halesworth, 7m E of Harleston & 6m S of Bungay. Parking at nearby village hall. For disabled parking please phone to arrange. Yellow* signs from 8m out in all directions. **Sun 21 Feb (11-4). Adm £4.50, chd free. Home-made teas. Hot/ Cold drinks,variety of home-made cakes and hot soup with crusty bread. Gluten free will be available. 2022: Sun 20 Feb.**
Again open for Snowdrops, but also the beautiful Hellebores and other flowering plants and shrubs. More snowdrops will be added to the 20,000 odd Snowdrops already planted in the last few years. The herbaceous beds and borders are always being added to by the 'plantaholic' owner. A garden for wildlife with a large variety of birds and creatures. A haven to relax in and enjoy nature in all forms. Plant stall with a variety of plants and bulbs. Newly restored pond and sunken garden. Conservatory packed with tender plants. Large glasshouse. Gravel drive. Partial wheelchair access to front garden. Steps to sunken garden. Concrete path in back garden.

32 LARKS' HILL

Clopton Road, Tuddenham St Martin, IP6 9BY. Mr John Lambert, 01473 785248, lordtuddenham@gmail.com. *3m NE of Ipswich. From Ipswich take B1077, go through village, take the Clopton Rd to the L, after 300 metres at the brow of the hill you will see the house. Follow the car parking signs.* **Visits by arrangement Apr to Aug for groups of 20+. Adm £5, chd free. Home-made teas.**
The gardens of eight acres comprise woodland, a newly-planted conifer garden, and formal areas, and fall away from the house to the valley floor. A hill within a garden and in Suffolk at that! Hilly garden with a modern castle keep with an interesting and beautiful site overlooking the gentle Fynn valley and the village beyond. A fossil of a limb bone from a Pliosaur that lived at least sixty million years ago was found in the garden in 2013. The discovery was reported in the national press but its importance has been recognised world-wide. A booklet is available to purchase giving all the details. Our Big Shed Café can comfortably seat thirty or so and there is additional seating outside. Sit and talk and plan what to do next.

33 LAVENHAM HALL

Hall Road, Lavenham, Sudbury, CO10 9QX. Mr & Mrs Anthony Faulkner, www.katedenton.com. *Next to Lavenham's iconic church & close to High St. From church turn off the main road down the side of church (Potland Rd). Go down hill. Car Park on right after 100 metres.* **Sun 30 May (10.30-5). Adm £5, chd free.**
5 acre garden built around the ruins of the original ecclesiastical buildings on the site and the village's 1 acre fishpond. The garden incl deep borders of herbaceous planting with sweeping vistas and provides the perfect setting for the sculptures which Kate makes in her studio at the Hall and exhibits both nationally and internationally under her maiden name of Kate Denton. 40 garden sculptures on display. There is a gallery in the grounds which displays a similar number of indoor sculptures and working drawings. Mostly wheelchair accessible. Note gravel paths / slopes may limit access to certain areas. Follow signs for separate disabled parking.

34 LILLESLEY BARN

The Street, Kersey, Ipswich, IP7 6ED. Mr Karl & Mrs Bridget Allen, 07939 866873, bridgetinkerseybarn@gmail.com. *In village of Kersey, 2m NW Hadleigh. Driveway is 200 metres above 'The Bell' pub. Lillesley Barn is situated behind 'The Ancient Houses'.* **Visits by arrangement May to Aug. Adm £3.50, chd free. Home-made teas. Opening jointly with Old Gardens. Refreshments may be subject to change.**
Dry gravel garden (inspired by the Beth Chatto Garden) including variety of mediterranean plants, ornamental grasses, herbs and collection of succulents. Large herbaceous border, rose arbours and small orchard with poultry. Golden willow hedge and fruit trees. The garden contains various species of birch, elder, amelanchier and willow. All within an acre of garden bordered on two sides by fields. Meals available at The Bell Inn. Not suitable for Wheelchair access due to gravel, uneven ground and a slope.

35 NEW THE LODGE

Bury Road, Bradfield St. Clare, Bury St Edmunds, IP30 0ED. Christian & Alice Ward-Thomas, 07768 347595, alice.baring@btinternet.com. *4m S of Bury St Eds. From N turn L up*

Water Lane off A134, at Xrds, turn R & go exactly 1m on R. From S turn R up Ixer Lane, R at T-junction, ½m on R. White iron gate. Before Post Box. **Visits by arrangement May to Sept for groups of up to 30. Weekdays only. July and August limited (school holidays). Depending on group size, refreshments may be available.**
Family garden, surrounded by old parkland (mostly oaks), established from blank canvass over 20yrs but recently remodelled. Rose courtyard (mostly English roses) gravel garden including alpines, Herbaceous border, tulip and wildlife meadow, Veg garden, extensive grassland with mown paths. Life is busy and help is minimal so please don't expect immaculacy! Parking on gravel but wheelchair access is possible.

♿ ☕

36 LORETO
5 Ringsfield Road, Beccles, NR34 9PQ. Mark & Jan Oakley. *Next to Sir John Leman High School, Beccles. Signed from Beccles town centre & the Bungay Rd/London Rd T-lights.* **Sun 18 July (11-4). Adm £4, chd free. Home-made teas. Tea and cakes including gluten free will be available.**
Stepping around the side of this Arts and Crafts home, you'll discover a gorgeous colour themed garden all set in about 1 acre. The herbaceous beds and borders harmonise beautifully, with sumptuous purples, cool pastels and hot scarlet collections. As you wander through the garden you'll find fun topiary and some fine specimen trees. Please park next door at the Sir John Leman High School NR34 9PG. Disabled parking and level access to the rear of the property in Ashmans Road.

♿ ✿ 🚗 ☕

37 NEW MANOR HOUSE
Leiston Road, Middleton, Saxmundham, IP17 3NS. Mandy Beaumont & Steve Thorpe. *See Paget House entry for directions to parking. Once there, Manor House is an 8-minute signed walk.* **Sun 18 July (11-5). Combined adm with Paget House £6, chd free. Home-made teas at Paget House.**
Just over an acre of garden on a triangular plot which was once the northern tip of a medieval green. Started from a wilderness in 2015 - now an ambulatory garden with seating to enjoy many new tree plantings, borders, meadows and

a vegetable garden all created from scratch. Biodiversity is encouraged and 'right plant, right place' ethos. Also includes a 'Silent Space' area.

✿ ✿ ☕

38 MOAT HOUSE
Little Saxham, Bury St Edmunds, IP29 5LE. Mr & Mrs Richard Mason, 01284 810941, suzanne@ countryflowerssuffolk.com. *2m SW of Bury St Edmunds. Leave A14 at J42 – leave r'about towards Westley. Through Westley Village, at Xrds R towards Barrow/Saxham. After 1.3m turn L down track. (follow signs).* **Sun 2 May (1-5.30). Adm £5, chd free. Home-made teas. Visits also by arrangement Apr to Sept for groups of 20+.**
Set in a 2 acre historic and partially moated site. This tranquil mature garden has been developed over 20yrs. Bordered by mature trees the garden is in various sections incl a sunken garden, rose and clematis arbours, herbaceous borders with hydrangeas and alliums, small arboretum. A Hartley Botanic greenhouse erected and partere have been created. Each year the owners enjoy new garden projects. Secluded and peaceful setting, each year new additions and wonderful fencing.

♿ ✿ ✿ 🚗 ☕

39 OLD GARDENS
The Street, Kersey, Ipswich, IP7 6ED. Mr & Mrs David Anderson, 01473 828044, davidmander15@gmail.com. *10m W of Ipswich. Up hill, approximately 100 metres from The Bell Inn Public House. On the same side of the road.* **Visits by arrangement May to Aug. Opening jointly with Lillesley Barn. Adm £3.50, chd free. Home-made teas. Refreshments available at Lillesley Barn.**
Entered from The Street, a natural garden with wildflowers under a copper beech tree. To the rear, a formal garden designed by Cherry Sandford with a sculpture by David Harbour. Meals available at The Bell Inn.

✿ ☕

40 NEW THE OLD RECTORY
Ingham, Bury St Edmunds, IP31 1NQ. Mr J & Mrs E Hargreaves. *5m N of Bury St Edmunds & A14. The garden is on R immed as you come into Ingham, coming N on A134 from Bury St*

Edmunds, just before the church. **Sun 25 Apr, Sun 25 July (11-5). Adm £4, chd free. Cream teas.**
A large garden in a lovely setting surrounding an early Victorian Rectory. We are developing a vibrant colourful flower display including dahlias, geraniums, insect-friendly borders and lawns. Lavender walk and parterre. Spring bulbs and large wisteria (April). Wide variety of mature trees including weeping willows. Children friendly. Wheelchair access through 1m wide gate.

♿ ☕

41 THE OLD RECTORY, BRINKLEY
Hall Lane, Brinkley, CB8 0SB. Mr & Mrs Mark Coley. *Hall Lane is a turning off Brinkley High St. At the end of Hall Lane, white gates on the R.* **Sun 14 Feb (11.30-4.30). Adm £4, chd free. Light refreshments in Brinkley Village Hall. Sat 17, Sun 18 July (2-6). Adm £5, chd free. Home-made teas in Brinkley Village Hall.**
Two acre garden started in 1973 when there was nothing there except a few large trees and snowdrops. Interesting trees planted since include a large liriodendron (tulip tree) planted in 1975. In the spring, many snowdrops, helebores and winter flowering shrubs. Mixed herbaceous borders with roses, interesting perennials and pockets of annuals grown from seed. Traditional potager with box hedges. Partial wheelchair access.

✿ ☕

42 THE OLD RECTORY, KIRTON
Church Lane, Kirton, Ipswich, IP10 0PT. Mr & Mrs N Garnham. *Adjacent to Kirton parish church on corner of Church Lane & Burnt House Lane. Car parking available in Church Hall car park next door.* **Sun 20 June (12-6). Adm £5, chd free. Home-made teas.**
This traditional English garden contains mixed flower borders for both sun and shade providing year-round colour, fragrance and succour for bees and butterflies. The 3 acre garden contains many mature trees, including varieties of oak, which give it a park-like feel. Many plants are labelled. Delicious teas served by rose and lavender border. Plant stall. Not suitable for disabled.

✿ ☕

43 THE OLD RECTORY, NACTON

Nacton, IP10 0HY. Mrs Elizabeth & Mr James Wellesley Wesley, 01473 659673, tizyww@gmail.com. *The garden is down the road to Nacton from the first A14 turn off after Orwell bridge going N/E. Parking on Church Rd opp Old Rectory driveway. This will be signed.* **Sat 12, Sun 13 June (10.30-4.30). Adm £5, chd free. Home-made teas. Visits also by arrangement Apr to Oct for groups of up to 30.** Just under 2 acres of garden divided into areas for different seasons: mature trees and herbaceous borders, herb/picking garden, rose garden. Light soil so many self sown flowers. Damp area with emphasis on foliage (Rheum, Darmera, Rodgersia, Hellebores). Most recently a terrace to West side of the house, still a work in progress after 30 years, looking at how to bring in more butterflies. The Ship Inn is 2m up road in Levington and does good pub lunches. Lovely walks on the Orwell Estuary with extensive bird life especially at low tide. Nacton is very accessible from Ipswich and surrounding areas. Most areas are accessible for disabled visitors, however several grassy slopes and various different levels so some energy needed to get everywhere!

44 OUSDEN HOUSE

Ousden, Newmarket, CB8 8TN. Mr & Mrs Alastair Robinson. *Newmarket 6m, Bury St Edmunds 8m. Ousden House stands at the west end of the village next to the Church.* **Sun 27 June (2-5.30). Adm £6, chd free. Home-made teas.** A large spectacular garden with fine views over the surrounding country. Herbaceous borders, rose garden and lawns leading to Spring woodland, and lake. Additional special features include a long double crinkle-crankle yew hedge leading from the clock tower and a moat garden densely planted with hellebores, flowering shrubs and moisture loving plants. Tea is served in the house or the courtyard and in aid of St. Peter's Church Ousden. Extensive garden on various levels not suitable for wheelchairs or people who find difficulty in walking.

45 PAGET HOUSE

Back Road, Middleton, Saxmundham, IP17 3NY. Julian & Fiona Cusack, 01728 649060, julian.cusack@btinternet.com. *From A12 at Yoxford take B1122 towards Leiston. Turn L after 1.2m at Middleton Moor. After 1m enter Middleton & drive straight ahead into Back Road. Turn 1st R on Fletchers Lane for car park.* **Sun 18 July (11-5). Combined adm with Manor House £6, chd free. Home-made teas. Gluten free options. Visits also by arrangement Apr to Sept. Guided visits can be arranged.** The garden is designed to be wildlife friendly with wild areas meeting formal planting. There is an orchard and a vegetable plot. There are areas of woodland, laid hedges, a pond supporting amphibians and dragonflies, a wildflower meadow and an abstract garden sculpture by local artist Paul Richardson. We record over 40 bird species each year and a good showing of butterflies, dragonflies and wildflowers including orchids. Gravel drive and mown paths.

46 5 PARKLANDS GREEN

Fornham St Genevieve, Bury St Edmunds, IP28 6UH. Mrs Jane Newton, newton.jane@talktalk.net. *2m Northwest of Bury St Edmunds off B1106. Plenty of parking on the green.* **Sun 13 June (11-4). Adm £5, chd free. Home-made teas. Visits also by arrangement May to Sept.** 1½ acres of gardens developed since the 1980s for all year interest. There are mature and unusual trees and shrubs and riotous herbaceous borders. Explore the maze of paths to find 4 informal ponds, a tree house, the sunken garden, greenhouses and woodland walks. Partial wheelchair access only.

47 ◆ THE PLACE FOR PLANTS, EAST BERGHOLT PLACE GARDEN

East Bergholt, CO7 6UP. Mr & Mrs Rupert Eley, 01206 299224, sales@placeforplants.co.uk, www.placeforplants.co.uk. *2m E of A12, 7m S of Ipswich. On B1070 towards Manningtree, 2m E of A12. Situated on the edge of East Bergholt.* **For NGS: Sun 18 Apr, Sun 9 May (2-5). Adm £7, chd free. Sun 10 Oct (1-4). Adm**

£6, chd free. Home-made teas. For other opening times and information, please phone, email or visit garden website. 20-acre woodland garden originally laid out at the turn of the last century by the present owner's great grandfather. Full of many fine trees and shrubs, many seldom seen in East Anglia. A fine collection of camellias, magnolias and rhododendrons, topiary, and the National Collection of deciduous Euonymus. Partial Wheelchair access in dry conditions - it is advisable to telephone before visiting.

48 POLSTEAD MILL

Mill Lane, Polstead, Colchester, CO6 5AB. Mrs Lucinda Bartlett, 07711 720418, lucyofleisure@hotmail.com. *Between Stoke by Nayland & Polstead on the R Box. From Stoke by Nayland take rd to Polstead - Mill Lane is 1st on L & Polstead Mill is 1st house on R.* **Visits by arrangement May to Sept for groups of 10+. Adm £6, chd free. Home-made teas. A range of refreshments available from coffee and biscuits to full cream teas and even light lunches.** The garden has been developed since 2002, it has formal and informal areas, a wildflower meadow and a large productive kitchen garden. The R Box runs through the garden and there is a mill pond, which gives opportunity for damp gardening, while much of the rest of the garden is arid and is planted to minimise the need for watering. Featured in Secret Gardens of East Anglia. Partial wheelchair access.

49 THE PRIORY

Stoke by Nayland, Colchester, CO6 4RL. Mrs H F A Engleheart. *5m SW of Hadleigh. Entrance on B1068 to Sudbury (NW of Stoke by Nayland).* **Sun 16 May (2-5). Adm £5, chd free. Home-made teas.** Interesting 9 acre garden with fine views over Constable countryside; lawns sloping down to small lakes and water garden; fine trees, rhododendrons and azaleas; walled garden; mixed borders and ornamental greenhouse. Wide variety of plants. Wheelchair access over most of garden, some steps.

50 NEW ▶ THE PRIORY, LAXFIELD RD

Badingham, Woodbridge, IP13 8LS. Mr Nick Smith. *A1120 to Badingham. At White Horse pub, turn in Low Street. 0.25m R into Mill Road. Uphill for 0.75m. L at post box onto Laxfield Road. 0.5m past Priory Cottage, follow NGS signs.* **Sun 29 Aug (11-4). Adm £5, chd free. Home-made teas.**

Designed by Frederic Whyte - a Chelsea Gold Medal winner - we wanted to create a garden with a contemporary feel, that had clear structure and focused on a refined palette of plants. The challenge was to clearly delineate between Priory Barn's guest areas, the kitchen garden, more formal borders and informal planting leading to the established orchard. Wheelchair access slightly restricted by gravel paths and steps.

&. ♨ 🛋

51 NEW ◆ THE RED HOUSE

Golf Lane, Aldeburgh, IP15 5PZ. Britten Pears Arts, **brittenpears.org/.** *Top of Aldeburgh, approx. 1m from the sea. From A12, take A1094 to Aldeburgh. Follow the brown sign directing you towards the r'about, take 1st exit: B1122 Leiston Road. Golf Lane is 2nd L, follow sign to 'The Red House'.* **For NGS: Mon 3, Mon 31 May, Sun 5 Sept (11-4). Adm £5, chd free. Home-made teas. For other opening times and information, please visit garden website.**

The former home of the renowned British composer Benjamin Britten and partner, the tenor Peter Pears. The five acre gardens provide an atmospheric setting for the house they shared and contain many plants loved by the couple. Mixed herbaceous borders, kitchen garden, contemporary planting and a new summer tropical border. The house has recently become part of the new Britten Pears Arts charity. The Red House is lovingly preserved (for Museum opening times, see website). It cares for the collections left by the two men, including their Archive which holds an extraordinary wealth of material documenting their lives. The garden offers a peaceful setting to this beautiful corner of Suffolk. There are brick, concrete, gravel paths and grass. Some areas are uneven and may require effort to navigate. A wheelchair is available upon request.

&. 🐕 ♨ 🛋

The Old Rectory, Nacton

52 REDISHAM HALL

Redisham Rd, Redisham, nr Beccles, NR34 0LZ. Philip & Lucy Everington, **www.redishamhallnurseries.co.uk.** *5m S of Beccles. From A145, turn W on to Ringsfield-Bungay rd. Beccles, Halesworth or Bungay, all within 5m.* **Sun 11 July (2-6). Adm £5, chd free. Home-made teas.**

C18 Georgian house (not open). 6.5 acre garden set in 400 acres parkland and woods including a traditional 2-acre Walled kitchen garden with Peach house, Vinery and Glasshouses. Large fruit cage, Espalier fruit trees, many varieties of vegetables, Herbs and Flower beds. Lawns, Herbaceous borders, Shrubberies, Woodland garden, Ha-Ha wall, Ponds, Arboretums and Mature trees. SORRY NO DOGS. On site Plant Nursery & Tearoom. The Walled Kitchen Garden is open to the public during the summer months. Wheelchair access is possible with assistance, gravel paths, steps, some slopes and uneven surfaces. Parking is on uneven parkland.

&. ❄ ♨

53 RIVER COTTAGE

Lower Road, Lavenham, Sudbury, CO10 9QJ. Clare & Michael Moore. *Please use Village Car Parks. Walk down Prentice Street, off Market Place. Look for The Angel & The Great House. Proceed to the bottom, then turn R on Lower Rd.* **Sun 8 Aug (11-5). Adm £5, chd free. Home-made teas.**

The previous owners created a tranquil garden which runs for 400ft alongside the River Brett. The current owners have been at River Cottage since June 2020 and have embarked on the process of not only maintaining Sue and Geoff's beautiful garden but adding one or two of their own touches. The garden includes hostas, hyrdangeas, dahlias, lilies, roses, peonies, clematis and grasses.

🐕 ❄ ♨

Online booking is available for many of our gardens, at ngs.org.uk

54 THE ROOKS

Old Paper Mill Lane, Claydon, Ipswich, IP6 0AL. Mrs Marilyn Gillard. *Off A14 at J52. From r'about turn in Claydon & take 1st available R, into Old Ipswich Rd where parking is available. The Rooks is 200m, 1st house in Old Paper Mill Lane with thatched roof.* **Sat 19 June (11-4). Adm £5, chd free. Home-made teas and cakes.**

The garden surrounds a thatched cottage built c1600. The garden has been divided into rooms to give different feelings of interest, with seating in most areas. There are two ponds, mixed borders, trees, statues and features. Not suitable for wheelchairs because of slopes and steps.

All entries are subject to change. For the latest information check ngs.org.uk

55 NEW SMALLWOOD FARMHOUSE

Smallwood Green, Bradfield St George, Bury St Edmunds, IP30 0AJ. Mr & Mrs P Doe. *On Hessett/Felsham road, S of A14, E of A134 on Bradfield St George to Felsham rd On main rd between Hessett & Felsham, ignore sign to Smallwood Green and follow NGS signage* **Sat 12 June (11-5). Adm £5, chd free.**

The garden is a combination of traditional cottage planting and contemporary styles. At its heart, a C16 farmhouse provides the backdrop to a variety of old English roses, a profusion of clematis and honeysuckle, a herb garden and potager. There are two natural ponds and a more recently planted gravel garden, whilst paths wind through an ancient meadow and orchard Partially wheelchair accessible, although not suitable in damp or wet weather

56 ◆ SOMERLEYTON HALL GARDENS

Somerleyton, NR32 5QQ. Lord Somerleyton, 01502 734901, info@somerleyton.co.uk, www.somerleyton.co.uk. *5m NW of Lowestoft. From Norwich (30mins) - on the B1074, 7m SE of Great Yarmouth (A143). Coaches should follow signs to the rear west gate entrance.* **For NGS: Sun 6 June (11-5). Adm £7.95, chd £5.95. Light refreshments in Cafe. For other opening times and information, please phone, email or visit garden website.**

12 acres of beautiful gardens contain a wide variety of magnificent specimen trees, shrubs, borders and plants providing colour and interest throughout the yr. Sweeping lawns and formal gardens combine with majestic statuary and original Victorian ornamentation. Highlights incl the Paxton glasshouses, pergola, walled garden and famous yew hedge maze. House and gardens remodelled

Moat House

in 1840s by Sir Morton Peto. House created in Anglo-Italian style with lavish architectural features and fine state rooms. All areas of the gardens are accessible, path surfaces are gravel and can be a little difficult after heavy rain. Wheelchairs available on request.

&. ✿ ⌂ ☕ 🪑

57 NEW SQUIRES BARN
St. Cross South Elmham, Harleston, IP20 0PA. Stephen & Ann Mulligan. *6m W of Halesworth, 6m E of Harleston & 7m S of Bungay. On New Road between St Cross & St James. Parking in field opposite. Yellow signs from 8m out in all directions.* Sun 11 July (10.30-4). Adm £4, chd free. Light refreshments. Hot/cold drinks, variety of home-made cakes, some gluten free, on lawn.
A young and evolving garden of 3 acres, including an orchard, cutting and kitchen garden with greenhouse and fruit cage, large ornamental pond with water lilies, fish and waterfall, wildflower area, willow spiral, island beds of mixed planting and a growing range of trees .Views over surrounding countryside. Gravel drive with pool. Plant stall. The garden is largely grass with some slight slopes. Seating is available across the garden. One single flight of steps can be bypassed.

&. ✿ ☕

58 NEW STONE COTTAGE
34 Main Road, Woolverstone, Ipswich, IP9 1BA. Mrs Jen Young, 01473 780156, j_s_young@icloud.com. *Take B1456 towards Shotley. When leaving Woolverstone Village the yellow signs will direct you to Stone Cottage.* Visits by arrangement Apr to July for groups of up to 20. Adm £3, chd free. Home-made teas.
An idyllic Suffolk Country cottage, surrounded by a garden created and maintained by the owner from a derelict space into a beautiful, calming garden. Over a hundred roses, unusual delphiniums, irises, spring bulbs and many more are planted together in interesting colour combinations that provide year round interest and perfume. A lovely space to sit and relax. Areas of gravel paths.

&. 🐾 ✿ ☕

59 WENHASTON GRANGE
Wenhaston, Halesworth, IP19 9HJ. Mr & Mrs Bill Barlow. *Turn SW from A144 between Bramfield & Halesworth. Take the single track rd (signed Walpole 2) Wenhaston Grange is approx ½ m, at the bottom of the hill on L.* Sun 20 June (11-4). Adm £5, chd free. Home-made teas.
Over 3 acres of varied gardens on a long established site which has been extensively landscaped and enhanced over the last 15 yrs. Long herbaceous borders, old established trees and a series of garden rooms created by beech hedges. Levels and sight lines have been carefully planned. The veg garden is now coming on nicely and there is also a wildflower meadow and woodland garden. The garden is on a number of levels, with steps so wheelchair access would be difficult.

🐾 ☕

60 WHITE HOUSE FARM
Ringsfield, Beccles, NR34 8JU. Jan Barlow, Justin 07780 901233, coppertops707@aol.com. *2m SW of Beccles. From Beccles take B1062 to Bungay, after 1¼ turn L signed Ringsfield. Continue for approx 1m. Parking opp church. Garden 300yds on L.* Visits by arrangement Apr to Sept for groups of 10 to 30. Adm £6, chd free. Tea and coffee (incl adm).
Tranquil park-type garden approx 30 acres, bordered by farmland and with fine views. Comprising formal areas, copses, natural pond, woodland walk, vegetable garden and orchard. Picnickers welcome. NB The pond and beck are unfenced. Uneven paving around house. Partial wheelchair access to the areas around the house.

&. 🐾 ✿ ☕

61 WOOD FARM, GIPPING
Back Lane, Gipping, Stowmarket, IP14 4RN. Mr & Mrs R Shelley, 07809 503019, els@maritimecargo.com. *From A14 take A1120 to Stowupland, Turn L opp Petrol Station, Turn R at T- junction, follow for approx 1m turn L at Walnut Tree Farm, then imm R & follow for 1m along country lane. Wood Farm is on L.* Sun 10 June (1-4.30). Adm £5, chd free. Home-made teas. Large Party Barn with Facilities. Visits also by arrangement May & June for groups of 10+.
Wood Farm is an old farm with ponds, orchards and a magnificent

8 acre wildflower meadow (with mown paths) bordered with traditional hedging, trees and woodland. The large cottage garden was created in 2011 with a number of beds planted with flowers, vegetables and topiary. Wildlife is very much encouraged in all parts of the garden (particularly bees and butterflies). Partial wheelchair access.

&. 🐾 ☕

62 WOOD FARM, SIBTON
Halesworth Road, Sibton, Saxmundham, IP17 2JL. Andrew & Amelia Singleton. *4m S of Halesworth. Turn off A12 at Yoxford onto A1120. Turn R after Sibton Nursery towards Halesworth. Take the 2nd drive on R after the White Horse PH.* Sun 20 June (2-6). Adm £5, chd free. Home-made teas.
Country garden surrounding old farmhouse, divided into colour themed areas, incl white garden, hot courtyard and blue and yellow border. Large (unfenced) ponds, vegetable garden, wild white flowering shrub area and mown walks. Garden designer owner. Teas in aid of The Friends of St Peter's Church, Sibton. Some gravel paths which are not suitable for wheelchairs.

&. 🐾 ☕

63 ◆ WYKEN HALL
Stanton, IP31 2DW. Sir Kenneth & Lady Carlisle, 01359 250262, kenneth.carlisle@wykenvineyards. co.uk, www.wykenvineyards.co.uk. *9m NE of Bury St Edmunds. Along A143. Follow signs to Wyken Vineyards on A143 between Ixworth & Stanton.* For NGS: Sat 29, Sun 30 May (10-5). Adm £5, chd free. Light refreshments in Restaurant and Cafe. For other opening times and information, please phone, email or visit garden website.
4-acres around the old manor, and an RHS Partner Garden. The gardens include knot and herb gardens, old-fashioned rose garden, kitchen and wild garden, nuttery, pond, gazebo and maze; herbaceous borders and old orchard. Woodland walk and vineyard nearby. Restaurant (booking 01359 250287.) and shop. Vineyard. Farmers' Market Sat 9 - 1.

&. ✿ ☕

SURREY

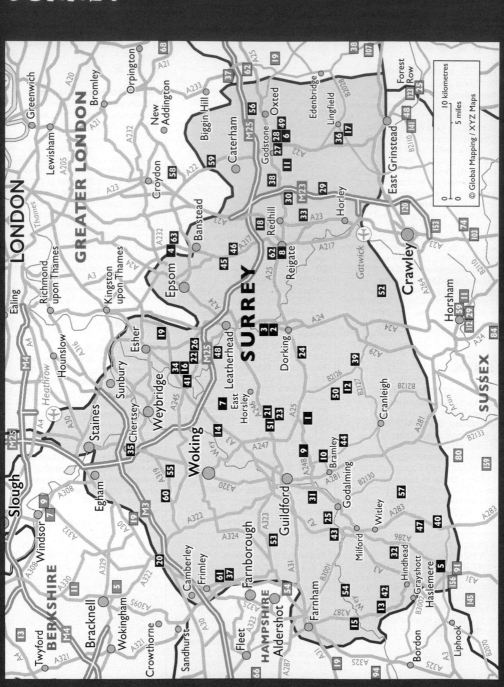

© Global Mapping / XYZ Maps

As a designated Area of Outstanding Natural Beauty, it's no surprise that Surrey has a wealth of gardens on offer.

With its historic market towns, lush meadows and scenic rivers, Surrey provides the ideal escape from the bustle of nearby London.

Set against the rolling chalk uplands of the unspoilt North Downs, the county prides itself on extensive country estates with historic houses and ancient manors. Visitors are inspired by the breathtaking panorama from Polesden Lacey, lakeside views at The Old Croft or timeless terraces at Albury Park.

Surrey is the heartland of the National Garden Scheme at Hatchlands Park and the RHS at Wisley, both promoting a precious interest in horticulture. Surrey celebrates a landscape coaxed into wonderful vistas by great gardeners such as John Evelyn, Capability Brown and Gertrude Jekyll.

With many eclectic gardens to visit, there's certainly plenty to treasure in Surrey.

Below: Little Court House

Volunteers

County Organiser
Margaret Arnott
01372 842459
margaret.arnott@ngs.org.uk

County Treasurer
Nigel Brandon
020 8643 8686
nbrandon@ngs.org.uk

Booklet Co-ordinator
Annabel Alford-Warren
01483 203330
annabel.alford-warren@ngs.org.uk

Social Media
Annette Warren
07790 045354
annette.warren@ngs.org.uk

Assistant County Organisers
Anne Barnes
01306 730196
spurfold@btinternet.com

Jan Brandon
020 8643 8686
janmbrandon@outlook.com

Penny Drew
01252 792909
penelopedrew@yahoo.co.uk

Di Grose
01883 742983
di.grose@ngs.org.uk

Annie Keighley
01252 838660
annie.keighley12@btinternet.com

Caroline Shuldham
01932 596960
c.m.shuldham@btinternet.com

Jean Thompson
01483 425633
norney.wood@btinternet.com

f @surrey.ngs

@SurreyNGS

@surreyngs

OPENING DATES

All entries subject to change. For latest information check **www.ngs.org.uk**
Extended openings are shown at the beginning of the month.
Map locator numbers are shown to the right of each garden name.

January

Every Wednesday
Timber Hill   55

February

Snowdrop Festival

Every Wednesday to Wednesday 10th
Timber Hill   55

Sunday 14th
◆ Gatton Park   18

Monday 15th
Timber Hill   55

Thursday 18th
Timber Hill   55

Sunday 21st
Shieling   46

March

Every Wednesday from Wednesday 17th
Timber Hill   55

Tuesday 9th
◆ The Sculpture Park   42

Sunday 21st
Albury Park   1

Monday 29th
◆ Vann   57

Tuesday 30th
◆ Vann   57

Wednesday 31st
◆ Vann   57

April

Thursday 1st
◆ Vann   57

Sunday 4th
Caxton House   8
Shieling   46

Monday 5th
Timber Hill   55

Thursday 8th
◆ Dunsborough Park   14

Saturday 10th
11 West Hill   58

Sunday 11th
Coverwood Lakes   12
11 West Hill   58

Sunday 18th
Coverwood Lakes   12
NEW Little Orchards   29

Thursday 22nd
Lower House   32

Sunday 25th
◆ Hatchlands Park   21

May

Every day from Monday 3rd to Sunday 9th
◆ Vann   57

Sunday 2nd
NEW ◆ Crosswater Farm   13
The Garth Pleasure Grounds   17

Monday 3rd
Coverwood Lakes   12
NEW ◆ Crosswater Farm   13
Shieling   46

Friday 7th
◆ Ramster   40

Sunday 9th
Coverwood Lakes   12
The Garth Pleasure Grounds   17
NEW Little Orchards   29
Westways Farm   60

Thursday 13th
Lower House   32

Tuesday 18th
Woodland   63

Thursday 20th
Woodland   63

Saturday 22nd
Hall Grove School   20

Sunday 23rd
Chilworth Manor   9
NEW Little Orchards   29
The Manor House
◆ Titsey Place Gardens   56

Wednesday 26th
NEW Little Orchards   29

Saturday 29th
15 The Avenue   4

Sunday 30th
High Clandon Estate Vineyard   23
Monks Lantern   35
The Therapy Garden   53
57 Westhall Road   59

Monday 31st
15 The Avenue   4
57 Westhall Road   59

June

Every day from Monday 7th to Sunday 13th
◆ Vann   57

Saturday 5th
NEW 2 Knott Park House   26

Sunday 6th
Fairmile Lea   16
NEW Little Orchards   29
Moleshill House   34
Oakleigh   37
The Old Rectory   38
Shackleford Garden Safari   43
Wildwood   61

Monday 7th
Timber Hill   55

Saturday 12th
NEW Leigh Place   27
Little Priory   30

Sunday 13th
NEW Leigh Place   27
Little Priory   30
◆ Loseley Park   31
Macmillan Cancer Support Centre   33
57 Westhall Road   59

Monday 14th
Timber Hill   55
57 Westhall Road   59

Tuesday 15th
◆ The Sculpture Park   42

Thursday 17th
Lower House   32

Friday 18th
Shieling   46

Saturday 19th
Shieling   46
11 West Hill   58

Sunday 20th
◆ Titsey Place Gardens   56
11 West Hill   58

Friday 25th
Ashcombe   2
Ashleigh Grange   3

Saturday 26th
Ashcombe   2

Sunday 27th
Ashcombe   2
Ashleigh Grange   3
NEW Little Orchards   29

Wednesday 30th
Ashleigh Grange   3

July

Sunday 4th
Pratsham Grange   39

Thursday 8th
Lower House   32

Saturday 10th
Southlands Lodge   49
Woodbury Cottage   62

Sunday 11th
41 Shelvers Way   45
Southlands Lodge   49
Woodbury Cottage   62

Saturday 17th
Earleywood   15
16 Hurtmore Chase   25

Sunday 18th
Earleywood   15
16 Hurtmore Chase   25
NEW Tanhouse Farm   52
◆ Titsey Place Gardens   56

Sunday 25th
Heathside   22

August

Sunday 1st
Moleshill House 34
Randalls Allotments 41

Sunday 15th
◆ Titsey Place
Gardens 56

Sunday 22nd
Pratsham Grange 39
41 Shelvers Way 45

September

Saturday 11th
Woodbury Cottage 62

Sunday 12th
[NEW] 21 Glenavon
Close 19
Hill Farm 24
[NEW] Little Orchards 29
Sleepy Hollow 47
Woodbury Cottage 62

Sunday 19th
The Therapy Garden 53

October

Saturday 2nd
Hall Grove School 20

Sunday 3rd
Albury Park 1

Sunday 10th
Coverwood Lakes 12

January 2022

Wednesday 12th
Timber Hill 55

Wednesday 19th
Timber Hill 55

Wednesday 26th
Timber Hill 55

February 2022

Monday 7th
Timber Hill 55

Monday 14th
Timber Hill 55

Monday 21st
Timber Hill 55

By Arrangement

Arrange a personalised
garden visit on a date to
suit you. See individual
garden entries for full
details.

Ashleigh Grange 3
15 The Avenue 4
Bardsey 5
The Bothy 6
Bridge End Cottage 7
Caxton House 8
2 Chinthurst Lodge 10
Coldharbour House 11
Coverwood Lakes 12
The Garth Pleasure
Grounds 17
Hall Grove School 20
Heathside 22
Little Court House 28
[NEW] Little Orchards 29
Moleshill House 34

Monks Lantern 35
[NEW] Oaklands 36
Oakleigh 37
The Old Rectory 38
Shamley Wood Estate 44
41 Shelvers Way 45
Shieling 46
2 Slyfield Farm
Cottages 48
Southlands Lodge 49
Spurfold 50
Stuart Cottage 51
[NEW] Tanhouse Farm 52
Tilford Cottage 54
Timber Hill 55
Westways Farm 60
Wildwood 61
Woodbury Cottage 62
Woodland 63

THE GARDENS

1 ALBURY PARK
**Albury, GU5 9BH. Trustees of
Albury Estate.** *5m SE of Guildford.
From A25 take A248 towards
Albury for ¼ mile, then up New Rd,
entrance to Albury Park immediately
on L.* **Sun 21 Mar, Sun 3 Oct (2-5).
Adm £5, chd free. Home-made
teas.**
14 acre pleasure grounds laid out
in 1670s by John Evelyn for Henry
Howard, later 6th Duke of Norfolk.
¼ m terraces, fine collection of trees,
lake and river. Gravel path and slight
slope.

2 ASHCOMBE
**Chapel Lane, Westhumble,
Dorking, RH5 6AY. Vivienne &
David Murch.** *From A24 at Boxhill/
Burford Bridge follow signs to
Westhumble. Through village &
L up drive by ruined chapel (1m
from A24).* **Evening opening Fri
25 June (6-8.30). Combined
adm with Ashleigh Grange £8,
chd free. Wine. Sat 26 June**
(1-5). Adm £4.50, chd free. Light
refreshments. Sun 27 June (1-5).
Combined adm with Ashleigh
Grange £6, chd free. Wine.
Plantaholics 1½ acre wildlife
friendly sloping garden on chalk and
flint. Enclosed 3rd of acre cottage
garden with large borders of roses,
delphiniums & clematis. Amphibian
pond. Secluded decking and Patio
area with colourful acers and views
over garden & Boxhill. Gravel bed
of salvia and day lilies. House
surrounded by banked flower beds
and lawn leading to bee and butterfly
garden. Wine and Pimms available.
Gravel paths, steps and a sloping site.

In our first year 609 private
gardens opened their gates to
all, for the modest sum of one
shilling. Today the National
Garden Scheme retains that
combination of inclusivity
and affordability

3 ASHLEIGH GRANGE
**off Chapel Lane, Westhumble,
RH5 6AY. Clive & Angela
Gilchrist, 01306 884613,
ar.gilchrist@btinternet.com.** *2m
N of Dorking. From A24 at Boxhill/
Burford Bridge follow signs to
Westhumble. Through village & L
up drive by ruined chapel (1m from
A24).* **Evening opening Fri 25 June
(6-8.30). Combined adm with
Ashcombe £8, chd free. Wine.
Sun 27 June (1-5). Combined
adm with Ashcombe £6, chd
free. Home-made teas. Wed 30
June (2-5.30). Adm £4.50, chd
free. Home-made teas. Visits also
by arrangement May to July for
groups of 5+. Sorry no access for
coaches.**
Plant lover's chalk garden on 3½
acre sloping site in charming rural
setting with delightful views. Many
areas of interest incl rockery and
water feature, raised ericaceous bed,
prairie style bank, foliage plants,
woodland walk, fernery and folly.
Large mixed herbaceous and shrub
borders planted for dry alkaline soil
and widespread interest.

◪ 15 THE AVENUE

Cheam, Sutton, SM2 7QA. Jan & Nigel Brandon, 020 8643 8686, janmbrandon@outlook.com. *1m SW of Sutton. By car; exit A217 onto Northey Av, 2nd R into The Avenue. By train; 10 mins walk from Cheam station. By bus; use 470.* **Evening opening Sat 29 May (5-9). Adm £8, chd £3. Wine. Mon 31 May (1-5). Adm £5, chd free. Home-made teas. Visits also by arrangement May to July for groups of 10+.**

A contemporary garden designed by RHS Chelsea Gold Medal Winner, Marcus Barnett. Four levels divided into rooms by beech hedging and columns; formal entertaining area, contemporary outdoor room, lawn and wildflower meadow. Over 100 hostas hug the house. Silver birch, cloud pruned box, ferns, grasses, tall bearded irises, contemporary sculptures. Partial wheelchair access, terraced with steps; sloping path provides view of whole garden but not all accessible.

◫ BARDSEY

11 Derby Road, Haslemere, GU27 1BS. Maggie & David Boyd, 01428 652283, maggie.boyd@live.co.uk, www.bardseygarden.co.uk. *¼ m N of Haslemere station. Turn off B2131 (which links A287 to A286 through town) 400yds W of station into Weydown Rd, 3rd R into Derby Rd, garden 400yds on R.* **Visits by arrangement June & July for groups of 10+. Adm £10, chd free. Home-made teas.**

Unexpected 2 acre garden in the heart of Haslemere. Several distinct areas containing scent, colour, texture and movement. Stunning pictorial meadow within an ilex crenata parterre. Prairie planted border provides a modern twist. Large productive fruit and vegetable garden. Natural ponds and bog gardens. Several unusual sculptures. Beehives and bug hotel. Ducks and chickens supply the eggs for cakes. Classic MGs. First third of garden level, other two thirds sloping.

◬ THE BOTHY

Tandridge Court, Tandridge Lane, Oxted, RH8 9NJ. Diane & John Hammond, 07785612478, dhammo@hotmail.co.uk. *5 mins from J6 of M25, towards Oxted. From N, at r'about on A25 between A22 & Oxted go S to Tandridge after*

¼ m follow signs. From S, N up Tandridge Ln from Ray Ln, on exiting Tandridge village follow signs. Do not use Jackass Ln from A25. **Visits by arrangement June & July for groups of 10 to 20. Adm £5, chd free.**

A 1½ acre hillside garden on 5 levels. Accessed by steps and slopes, creating varying views, with specimen trees, incl a Sequoia, Acers and Banana. Explore wild banks, perennial beds, woodland and relaxing areas along with a productive kitchen garden with raised beds, fruit trees, fruit cage and large greenhouse. Each level contains either water features, garden artwork, chickens and more. Check NGS website for pop-up openings. The garden entrance is down a number of steps and on different levels and therefore it is unsuitable for people with mobility issues.

◰ BRIDGE END COTTAGE

Ockham Lane, Ockham, GU23 6NR. Clare & Peter Bevan, 01483 479963, c.fowler@ucl.ac.uk. *Nr RHS Gardens, Wisley. At Wisley r'about turn L onto B2039 to Ockham/Horsley. After ½ m turn L into Ockham Lane. House ½ m on R. From Cobham go to Blackswan Xrds.* **Visits by arrangement May to July for groups of 10 to 30. Adm £5.50, chd free. Light refreshments in the garden room.**

A two acre country garden with different areas of interest, incl perennial borders, mature trees, pond and streams, small herb parterre, fruit trees and a vegetable patch. An adjacent two acre field was sown with perennial wildflower seed in May 2013 and has flowered well each summer. Large perennial wildflower meadow. Partial wheelchair access.

◱ CAXTON HOUSE

67 West Street, Reigate, RH2 9DA. Bob Bushby, 01737 243158 / 07836201740, bob.bushby@sky.com. *On A25 towards Dorking, approx ¼ m West of Reigate. Parking on Rd or past Black Horse On Flanchford Rd.* **Sun 4 Apr (2-5). Adm £5, chd free. Cream teas. Visits also by arrangement Apr to Sept for groups of 10+.**

Lovely large spring garden with Arboretum, 2 well stocked ponds, large collection of hellebores and spring flowers. Pots planted with colourful displays. Interesting plants. Small Gothic folly built by owner. Herbaceous

borders with grasses, perennials, spring bulbs, and parterre. New bed with wild daffodils and prairie style planting in summer. new wildflower garden in arboretum. Wheelchair access to most parts of the garden.

◲ CHILWORTH MANOR

Halfpenny Lane, Chilworth, Guildford, GU4 8NN. Mia & Graham Wrigley, www.chilworthmanorsurrey.com. *3½ m SE of Guildford. From centre of Chilworth village turn into Blacksmith Lane. 1st drive on R on Halfpenny Lane.* **Sun 23 May (11-5). Adm £6, chd free. Home-made teas.**

The grounds of the C17 Chilworth Manor, create a wonderful tapestry: a jewel of an C18 terraced walled garden, topiary, herbaceous borders, sculptures, mature trees and stew ponds that date back a 1000 years. A fabulous, peaceful garden for all the family to wander and explore or just to relax and enjoy! Perhaps our many visitors describe it best: "Magical", "a sheer delight', 'elegant and tranquil", "a little piece of heaven", "spiffing!".

◳ 2 CHINTHURST LODGE

Wonersh Common, Wonersh, Guildford, GU5 0PR. Mr & Mrs M R Goodridge, 01483 535108, michaelgoodridge@ymail.com. *4m S of Guildford. From A281 at Shalford turn E onto B2128 towards Wonersh. Just after Waverley sign, before village, garden on R, stable entrance opp Little Tangley.* **Visits by arrangement May to July for groups of 10+. Adm £6, chd free. Home-made teas.**

One acre yr-round enthusiast's atmospheric garden, divided into rooms. Herbaceous borders, dramatic white garden, specimen trees and shrubs, gravel garden with water feature, small kitchen garden, fruit cage, two wells, ornamental ponds, herb parterre and millennium parterre garden. Some gravel paths, which can be avoided.

◴ COLDHARBOUR HOUSE

Coldharbour Lane, Bletchingley, Redhill, RH1 4NA. Mr Tony Elias, 01883 742685, eliastony@hotmail.com. *Coldharbour Lane off Rabies Heath Rd ½ m from A25 at Bletchingley &*

0.9m from Tilburstow Hill Rd. Park in field & walk down to house. **Visits by arrangement Apr to Oct. Adm £8, chd free.**
This 1½ acre garden offers breathtaking views to the South Downs. Originally planted in the 1920's, it has since been adapted and enhanced. Several mature trees and shrubs incl a copper beech, a Canadian maple, magnolias, azaleas, rhododendrons, camellias, wisterias, berberis georgeii, vitex agnus-castus, fuschias, hibiscus, potentillas, mahonias, a fig tree and a walnut tree.

⓬ COVERWOOD LAKES
Peaslake Road, Ewhurst, GU6 7NT. The Metson Family, 01306 731101, farm@coverwoodlakes.co.uk, www.coverwoodlakes.co.uk. 7m SW of Dorking. From A25 follow signs for Peaslake; garden ½ m beyond Peaslake on Ewhurst rd. **Sun 11, Sun 18 Apr, Mon 3, Sun 9 May, Sun 10 Oct (11-5). Adm £6, chd free. Light refreshments. Visits also by arrangement Apr to Oct for groups of 20+.**
14 acre landscaped garden in stunning position high in the Surrey Hills with 4 lakes and bog garden. Extensive rhododendrons, azaleas and fine trees. 3½ acre lakeside arboretum. Marked trail through the 180 acre working farm with Hereford cows and calves, sheep and horses, extensive views of the surrounding hills. Light refreshments, incl home produced beef burgers, gourmet coffee and home-made cakes.

⓭ NEW ◆ CROSSWATER FARM
Crosswater Lane, Churt, Farnham, GU10 2JN. David & Susanna Millais, 01252 792698, sales@rhododendrons.co.uk, www.rhododendrons.co.uk. 0m S of Farnham, 6m NW of Haslemere. From A287 turn E into Jumps Rd ½ m N of Churt village centre. After ¼ m turn acute L into Crosswater Lane & follow signs for Millais Nurseries. **For NGS: Sun 2, Mon 3 May (10-5). Adm £5, chd free. Home-made teas. For other opening times and information, please phone, email or visit garden website.**
Idyllic 5 acre woodland garden. Plantsman's collection of rhododendrons and azaleas, including rare species collected in the Himalayas, and hybrids raised by the family. Everything from alpine dwarfs to architectural large leaved trees. Ponds, stream and companion plantings including sorbus, magnolias and Japanese acers. Re-opening after 5 years with new developments and wildflower meadows. Woodland garden and specialist rhododendron, azalea and magnolia plant centre. Grass paths may be difficult for wheelchairs after rain.

⓮ ◆ DUNSBOROUGH PARK
Ripley, GU23 6AL. Baron & Baroness Sweerts de Landas Wyborgh, 01483 225366, office@sweerts.com, www.dunsboroughpark.com. 6m NE of Guildford. Entrance via Newark Lane, Ripley through Tudor-style gatehouses and courtyard up drive. Car park signposted. Sat Nav ref GU23 6BZ. **For NGS: Thur 8 Apr (9.30-4). Adm £8, chd free.** Pre-booking essential, please phone 01483 225366 or email office@sweerts.com for information & booking. For other opening times and information, please phone, email or visit garden website.
6-acre garden redesigned by Penelope Hobhouse & Rupert Golby. April Tulip Festival: formal planting in borders/colourful informal display in meadow under oaks. May: Wisteria. June: Roses & Peonies. Sept: Dahlias 50+ varieties. Garden rooms, lush herbaceous borders, standard wisteria, 70ft Ginkgo hedge, potager & 300yr-old mulberry tree. Rose Walk, Italian Garden & Water Garden with folly bridge. April: Spectacular tulip displays in meadow/Penelope Hobhouse borders/'hot' border/Peacock Gate. May: Wisteria. June: Peonies and Roses. 56 Dahlia varieties in Sept. See www.dunsboroughpark.com to book other garden open dates during the year. Produce and flowers for sale when available. Gravel paths and grass, cobbled over folly bridge.

Crosswater Farm

15 EARLEYWOOD

Hamlash Lane, Frensham, Farnham, GU10 3AT. Mrs Penny Drew. *3m S of Farnham just off A287. From A31 Farnham take A287 to Frensham. R lst turn passed Edgeborough School. From A3, N on A287 to Frensham. Parking available, More House School Lower Car Park top of Hamlash Lane.* **Sat 17, Sun 18 July (11-5). Adm £5, chd free. Home-made teas. Also open 16 Hurtmore Chase.**

Award winning (2019), ½ acre garden with colourful shrubberies, unusual trees and mixed borders throughout. Summer flowering shrubs, particularly hydrangeas and hardy hibiscus, together with herbaceous perennials in colour themed borders and shade loving plants under the trees. Unusual tender perennials in containers. Productive greenhouse and small pond. New features to be enjoyed in 2021. John Negus, well known horticulturalist, gardening writer and broadcaster will be in the garden to answer gardening questions during the weekend. Also demonstration of lavender weaving and craft items for sale. Disabled drop off at front gate, short gravel drive then level lawns throughout.

&. ✿ ☕

16 FAIRMILE LEA

Portsmouth Road, Cobham, KT11 1BG. Steven Kay. *2m NE of Cobham. On Cobham to Esher rd. Access by lane adjacent to Moleshill House & car park for Fairmile Common woods.* **Sun 6 June (2-5). Combined adm with Moleshill House £6.50, chd free. Home-made teas.**

Victorian sunken garden fringed by rose beds and lavender with a pond in the centre. An old acacia tree stands in the midst of the lawn. Interesting planting on a large mound camouflages an old underground air raid shelter. Caged vegetable garden. Formality adjacent to wilderness.

&. 🐕 ☕

17 THE GARTH PLEASURE GROUNDS

Newchapel Road, Lingfield, RH7 6BJ. Mr Sherlock & Mrs Stanley, ab_post@yahoo.com, , www.oldworkhouse.webs.com. *From A22 take B2028 by the Mormon Temple to Lingfield. The Garth is on the L after 1½m, opp Barge Tiles. Parking: Gun Pit Rd in Lingfield & limited space for disabled at Barge Tiles.* **Sun 2, Sun 9 May (2-6). Adm £7, chd free.**

Home-made teas. Visits also by arrangement May to Aug for groups of 5 to 30.

Mature 9 acre Pleasure Grounds created by Walter Godfrey in 1919 present an idyllic setting surrounding the former parish workhouse refurbished in Edwardian style. The formal gardens, enchanting nuttery, a spinney with many mature trees and a pond attract wildlife. Wonderful bluebells in spring. The woodland gardens and beautiful borders full of colour and fragrance for year round pleasure. Many areas of interest incl pond, woodland garden, formal gardens, spinney w/ large specimen plants incl 500yr old oak, many architectural features designed by Walter H Godfrey. Partial wheelchair access in woodland, iris and secret gardens.

&. 🐕 ✿ 🚐 🚌 ☕ 🏕

18 ◆ GATTON PARK

Reigate, RH2 0TW. Royal Alexandra & Albert School, 01737 649068, events@gatton-park.org.uk, www.gattonpark.co.uk. *3m NE of Reigate. 5 mins from M25 J8 (A217) or from top of Reigate*

Oaklands

Hill, over M25 then follow sign to Merstham. Entrance off Rocky Lane accessible from Gatton Bottom or A23 Mersthan. **For NGS: Sun 14 Feb (12-5). Adm £5, chd free. Pre-booking essential, please visit www.ngs.org.uk for information & booking. For other opening times and information, please phone, email or visit garden website. Donation to another charity.**
Historic 260-acre estate in the Surrey Hills AONB. 'Capability' Brown parkland with ancient oaks. Discover the Japanese garden, Victorian parterre and breathtaking views over the lake. Seasonal highlights include displays of snowdrops and aconites in February. Ongoing restoration projects by the Gatton Trust. Bird hide open to see herons nesting. Free guided tours and activities for children. A selection of hot and cold drinks, cakes and snacks available.

🐾 ✿ 🏠 ☕

19 **NEW** **21 GLENAVON CLOSE**
Claygate, Esher, KT10 0HP.
Selina & Simon Botham,
www.designsforallseasons.co.uk.
2m SE of Esher. From A3 S exit Esher, R at T-lights. Cont straight through Claygate village bear R at two mini r'bouts. Past church & rec. Turn L at bollards into Causeway. At end straight over to Glenavon Cl. **Sun 12 Sept (1.30-5). Adm £4, chd free. Home-made teas.**
A relaxing 66ft x 92ft secluded garden created by garden designer Selina Botham and her husband Simon. An awkward shaped suburban plot with swathes of grasses and perennials around a spacious lawn and curving paths invite exploration. River inspired barefoot walk through soft blue planting leads to the garden studio and pond. Sculpted bug hotels and birdhouses and wildlife-friendly planting. Guided tours by RHS Gold medal winning designer, exhibit of private garden design work and RHS show garden concepts.

☕

20 **HALL GROVE SCHOOL**
London Road (A30), Bagshot,
GU19 5HZ. Mr & Mrs A R Graham,
www.hallgrove.co.uk. *6m SW of Egham. M3 J3, follow A322 1m until sign for Sunningdale A30, 1m E of Bagshot, opp Longacres garden centre, entrance at footbridge. Ample car parking.* **Sat 22 May, Sat 2 Oct (2-5). Adm £5, chd free. Home-made teas. Visits also by**

arrangement May to Sept.
Formerly a small Georgian country estate, now a co-educational preparatory school. Grade II listed house (not open). Mature parkland with specimen trees. Historical features include ice house, recently restored walled garden, lake, woodland walks, rhododendrons and azaleas. Live music at 3pm.

♿ ✿ ☕

21 ♦ **HATCHLANDS PARK**
East Clandon, Guildford, GU4 7RT.
National Trust, 01483 222482,
hatchlands@nationaltrust.org.uk,
www.nationaltrust.org.uk/
hatchlands-park. *4m E of Guildford. Follow brown signs to Hatchlands Park (NT).* **For NGS: Sun 25 Apr (10-5). Adm £9.40, chd £4.80. Light refreshments. For other opening times and information, please phone, email or visit garden website.**
Garden and park designed by Repton in 1800. Follow one of the park walks to the stunning bluebell wood in spring (2.5km/1.7m round walk over rough and sometimes muddy ground). In autumn enjoy the changing colours on the long walk. Partial wheelchair access to parkland, rough, undulating terrain, grass and gravel tracks, dirt tracks, cobbled courtyard. Tramper booking essential.

♿ 🐾 🏠 ☕

22 **HEATHSIDE**
10 Links Green Way,
Cobham, KT11 2QH. Miss
Margaret Arnott & Mr Terry
Bartholomew, 01372 842459,
margaret.arnott@ngs.org.uk. *1½m E of Cobham. Through Cobham A245, 4th L after Esso garage into Fairmile Lane. Straight on into Water Lane. Links Green Way 3rd turning on L.* **Sun 25 July (11-5). Adm £5, chd free. Home-made teas. Visits also by arrangement May to Aug for groups of 5+. Wine & canapés available for evening visits.**
⅓ acre terraced, plants persons garden, designed for yr-round interest. Gorgeous plants all set off by harmonious landscaping. Many urns & pots give seasonal displays. Several water features add tranquil sound. Topiary shapes provide formality. Stunning colour combinations excite. Dahlias & begonias a favourite. Beautiful Griffin Glasshouse housing the exotic. Many inspirational ideas. Situated 5m from RHS Wisley.

✿ 🏠 ☕

23 **HIGH CLANDON ESTATE VINEYARD**
High Clandon, Off Blakes
Lane, East Clandon,
GU4 7RP. Mrs Sibylla Tindale,
www.highclandon.co.uk. *A3 Wisley junction, L for Ockham/Horsley for 2m to A246. R for Guildford for 2m, then 100yds past landmark Hatchlands NT, turn L into Blakes Lane straight up hill through gates High Clandon to vineyard entrance. Extensive parking in our woodland area.* **Sun 30 May (11-4). Adm £6, chd free. Cream teas. Gold awarded English Sparkling wine available by the glass.**
Vistas, gardens, sculptures, multi-Gold vineyard in beautiful Surrey Hills AONB. On 12 acres. Panoramic views to London, water features, Japanese garden, wildflower meadow, truffière, apiary, vineyard, English sparkling wine High Clandon Cuvée sold from atmospheric Glass Barn. Twice winner Cellar Door of Year. Sculptures in Vineyard exhibition. Percentage of sales to Cherry Trees charity. Atmospheric Glass Barn used for wine tastings and art exhibitions. Over 150 works of art on show in gardens and Vineyard. High Clandon Cuvée multi-gold awarded Sparkling wine available at £6 per glass. Available by the bottle. Homemade teas & cakes. One acre Wildflower Meadow with rare butterflies. There is a short stretch of 4 metres of gravel to cover in one direction. Otherwise access is on lawned paths all on firm ground.

♿ 🐾 ☕

24 **HILL FARM**
Logmore Lane, Westcott, Dorking,
RH4 3JY. Helen Thomas. *1m W of Dorking. Parking on Westcott Heath just past Church. Entry to garden just opp.* **Sun 12 Sept (11-4). Adm £5, chd free. Home-made teas.**
1¾ acre set in the magnificent Surrey Hills landscape. The garden has a wealth of different natural habitats to encourage wildlife, and planting areas which come alive through the different seasons. A wildlife pond, woodland walk, a tapestry of heathers and glorious late summer grasses and perennials. A garden to be enjoyed by all. Everyone welcome. Vegetable and cut flower garden and wildflower meadow and lime kiln in wildings area. Sloping garden, most areas are accessible to wheelchair users. Most paths are grass so care needed if very wet.

♿ 🐾 D ☕

25 16 HURTMORE CHASE
Hurtmore, Godalming, GU7 2RT.
Mrs Ann Bellamy. *4m SW of Guildford. From Godalming follow signs to Charterhouse & cont about ¼m beyond Charterhouse School. From A3 take Norney, Shackleford & Hurtmore turn off & proceed E for ½m.* **Sat 17 July (11-5); Sun 18 July (1-5). Adm £5, chd free. Also open Earleywood.**
A secluded medium sized (approx ¼ acre) garden comprised mainly of a large lawn divided into discrete areas by shrubs, trees and colourful flowerbeds. On the bungalow side of the lawn there is a patio area with hanging baskets, troughs and planted pots with fuchsias being a speciality. In the opposite corner there is a shaded arbour bordered by hostas. Award winner - Godalming in Bloom 2019. Cash only for Refreshments please. Kerb height step up from patio to lawn so minor assistance may be required for wheelchairs.

&. ✿ ❦

26 NEW 2 KNOTT PARK HOUSE
Wrens Hill, Oxshott, Leatherhead, KT22 0HW. Joanna Nixon. *On the A244, which is off the A3, or S from Leatherhead. Turn onto Wrens Hill, the road next to The Bear pub, continue up c. 200m, take R fork, part of the first big house on L.* **Sat 5 June (11-5). Adm £4, chd free. Home-made teas.**
A south-facing terrace with far-reaching views is planted with herbs & lavender. Steps lead down to a lower area extending to ¼ acre. It is wildlife friendly and maintained without pesticides. On 4 levels, it features pollinator loving plants, trees & shrubs. These include irises, alliums, sedum, geums, acers, flowering currant and succulents. Scented flowering shrubs/climbers attract bees. No access to wheelchairs due to steps.

✿ ❦

27 NEW LEIGH PLACE
Leigh Place Lane, Godstone, RH9 8BN. Mike & Liz McGhee. *Head along B2236 Eastbourne Rd from Godstone village. Turn 2nd L onto Church Lane. Follow parking directions.* **Sat 12, Sun 13 June (10-4). Adm £5, chd free. Tea.**
Leigh Place garden has 25 acres, on greensand. Part of the Godstones Pond's with SSSI and lakeside paths, walled garden; including cutting garden, orchard and vegetable

quadrant with greenhouses. Wildflower meadow, beehives and large rock garden. WC, Teas, children and dog friendly. Gravel paths with slopes. The walled garden, has breedon gravel paths suitable for wheelchairs and pushchairs.

🐾 ❦

28 LITTLE COURT HOUSE
Jackass Lane, Tandridge, Oxted, RH8 9NH. Jane & Graham King, 07798 626767, littlecourtfarm@hotmail.com. *2m from J6 on M25. Head S on A22 from J6, L onto A25 towards Oxted, at mini-r'about R onto Tandridge Ln, 1st R onto Jackass Ln, 1st drive on RHS after 500 metres. Do not enter Jackass Ln from A25.* **Visits by arrangement Apr to Sept for groups of up to 30. Adm £5, chd free. Home-made teas.**
¾ acre garden created from scratch in beautiful country setting. Small courtyard style front garden, gate through to charming terraced rear garden with lawn, herbaceous borders, wisteria and rose covered pergola, archway through to wider area of garden with topiary yew hedge and further to kitchen garden with greenhouse, potting shed, fruit trees, raised cutting and veg beds.

🐾 ❦

29 NEW LITTLE ORCHARDS
Prince Of Wales Road, Outwood, Redhill, RH1 5QU. Nic Howard, 01883744020, info@we-love-plants.co.uk. *A few hundred metres N of the Dog and Duck Pub.* **Sun 18 Apr, Sun 9, Sun 23, Wed 26 May, Sun 6, Sun 27 June, Sun 12 Sept (11-5). Adm £5, chd free. Home-made teas. Visits also by arrangement Apr to Sept for groups of 10 to 30.**
Garden Designer's contemporary cottage garden that has been planted for all year round interest using a tapestry of foliage textures as well as flower interest. The garden is arranged as a series of connected garden areas that flow between the old gardeners' cottage and the old stables. The small spaces are full of character, with paved areas, brick walls and vintage garden paraphernalia.

✿ 🚗 ⅅ ☕ ❦

30 LITTLE PRIORY
Sandy Lane, South Nutfield, RH1 4EJ. Liz & Richard Ramsay.

1½m E of Redhill. From Nutfield, on A25, turn into Mid St, following sign for South Nutfield. 1st R into Sandy Lane. Follow signs to parking on R, approx ½m. **Evening opening Sat 12 June (5-8). Adm £6.50, chd free. Wine. Sun 13 June (11-5). Adm £5, chd free. Light refreshments. Also open Macmillan Cancer Support Centre.**
Explore a 5 acre country garden where old blends with new. Wander the flower garden; lose yourself in the old orchard meadow; discover cherries and kiwi fruit in the greenhouse in the Victorian kitchen garden; relax beside the large pond, and be inspired by the views. Partial wheelchair access.

&. ✿ ❦

31 ◆ LOSELEY PARK
Guildford, GU3 1HS. Mr & Mrs A G More-Molyneux, 01483 304440/405112, pa@loseleypark.co.uk, www.loseleypark.co.uk. *4m SW of Guildford. For Sat Nav please use GU3 1HS, Stakescorner Lane.* **For NGS: Sun 13 June (11-5). Adm by donation. Light refreshments. For other opening times and information, please phone, email or visit garden website.**
Delightful 2½ acre walled garden. Award winning rose garden (over 1,000 bushes, mainly old fashioned varieties), extensive herb garden, fruit/flower garden, white garden with fountain, and spectacular organic vegetable garden. Magnificent vine walk, herbaceous borders, moat walk, ancient wisteria and mulberry trees. Refreshments available.

&. ✿ 🚗 ❦

32 LOWER HOUSE
Bowlhead Green, Godalming, GU8 6NW. Georgina Harvey. *1m from A3 leaving at Thursley/Bowlhead Green junction. Follow Bowlhead Green signs. Using A286 leave at Brook (6m from Haslemere & 3m from Milford). Follow NGS signs for approx 2m. Field Parking.* **Thur 22 Apr, Thur 13 May, Thur 17 June, Thur 8 July (11.30-4.30). Adm £6, chd free. Home-made teas.**
The original Gertrude Jekyll plans dated 1916 still give rhythm to the way you travel around the garden although the planting is now constantly changing as so many plants are finishing their lifecycle

and need replacing to balance their surroundings. Alternative routes avoiding steps for wheelchairs users, although some paths could be narrow.

THE MANOR HOUSE
Three Gates Lane, Haslemere, GU27 2ES. Mr & Mrs Gerard Ralfe.
NE of Haslemere. From Haslemere centre take A286 towards Milford. Turn R after Museum into Three Gates Lane. At T-junction turn R into Holdfast Lane. Car park on R. **Sun 23 May (12-5). Adm £5, chd free Home-made teas.**
Described by Country Life as 'The Hanging Gardens of Haslemere', the well-established Manor House gardens are tucked away in a valley of the Surrey Hills. Set in 6 acres, it was one of Surrey's inaugural NGS gardens, with fine views, an impressive show of azaleas, wisteria, beautiful trees underplanted with bulbs and enchanting water gardens.

33 MACMILLAN CANCER SUPPORT CENTRE
East Surrey Hospital, Canada Avenue, Redhill, RH1 5RH. East Surrey Macmillan Cancer Support Centre. *Follow signs for East Surrey Hospital. Drive past main hospital entrance; Centre is opp E Entrance. Follow parking signs.* **Sun 13 June (11-4). Adm £3, chd free. Light refreshments. Also open Little Priory.**

The new Macmillan Cancer Support Centre offers holistic support to people affected by cancer. The two small gardens surrounding the state of art Centre provide a peaceful haven to visitors. The thoughtful planting ensures there is interest in the garden throughout the year and offers a therapeutic retreat away from the hospital environment. Also open Little Priory, 10 minutes away.

34 MOLESHILL HOUSE
The Fairmile, Cobham, KT11 1BG. Penny Snell, pennysnellflowers@ btinternet.com, , www.pennysnellflowers.co.uk. *2m NE of Cobham. On A307 Esher to Cobham Rd next to free car park by A3 bridge, at entrance to Waterford Close.* **Sun 6 June (2-5). Combined adm with Fairmile Lea £6.50, chd free. Sun 1 Aug (2-5). Adm £5, chd free. Home-made teas at Fairmile Lea 6 June and at Moleshill House 1 August. Visits also by arrangement May to Sept for groups of 10+.**
Romantic garden. Short woodland path leads from dovecote to beehives. Informal planting contrasts with formal topiary box and garlanded cisterns. Colourful courtyard and pots, conservatory, pond with fountain, pleached avenue, circular gravel garden replacing most of the lawn. Gipsy caravan garden, green wall and stumpery. Espaliered crab apples. Chickens and Bees. Garden 5 mins from Claremont Landscape Garden,

Painshill Park and Wisley, also adjacent to excellent dog walking woods.

35 MONKS LANTERN
Ruxbury Road, Chertsey, KT16 9NH. Mr & Mrs J Granell, 01932 569578, janicegranoll@hotmail.com. *1m NW from Chertsey. M25 J11, signed A320/Woking. R'about 2nd exit A320/Staines, straight over next r'about. L onto Holloway Hill, R Hardwick Lane. ½m, R over motorway bridge, on Almners then Ruxbury Rd.* **Sun 30 May (1.30-5.30). Adm £5, chd free. Light refreshments. Visits also by arrangement June to Aug for groups of 5 to 20.**
A delightful garden with borders arranged with colour in mind: silvers and white, olive trees, nicotiana and senecio blend together. Large rockery and an informal pond. A weeping silver birch leads to the oranges and yellows of a tropical bed, with large bottle brush, hardy palms and fatsia japonica. There is a display of hostas, cytisus battandieri and a selection of grasses in an island bed. Aviary with small finches. Workshop with handmade guitars,and paintings. Pond with ornamental ducks and fish. Music and wine. Wheelchairs are welcome and we reserve a couple of parking spaces at the entrance to the garden, as the gravel drive is not easy for wheelchairs.

Oakleigh

36 NEW OAKLANDS

Eastbourne Road, Blindley Heath, Lingfield, RH7 6LG. Joy & Justin Greasley, 01342 837369, joy@allaero.com. *S of Blindley Heath Village. From M25 take A22 to East Grinstead. At Blindley Heath T-lights carry straight on, after 800yds take sign on L to Nestledown Boarding Kennels. Oaklands is down the lane on the L.* Visits by arrangement June to Sept for groups of 5 to 30. Adm £5, chd free. Light refreshments.

A contemporary garden of approximately half an acre designed and developed by the owners from new 23 years ago and surrounded by SSSI. Starting from scratch, and with very little gardening experience the garden, like the owners it is maturing nicely. Vibrant colour, large pots, lawn, herbaceous planting with plenty of seating areas complimented by some unusual features. Level access on paths to main features.

37 OAKLEIGH

22 The Hatches, Frimley Green, GU16 6HE. Angela O'Connell, 01252 668645, angela.oconnell@icloud.com. *Frimley Green. 100 metres from village green. On street parking.* Sun 6 June (1.30-5). Combined adm with Wildwood £4.50, chd free. Home-made teas at Wildwood. Visits also by arrangement in June for groups of 10 to 20. Joint visits with Wildwood. Option of evening visits with wine.

A magical long garden with a few surprises. There are plenty of colours and textures with a great variety of different plants. Wander past the long borders and under a rose arch and you will find the garden opens up to two large colour themed mixed beds. The style is naturalistic with just a hint of elegance. Fruit and vegetables grow by the summerhouse and pots adorn the top patio. The garden is mostly level but the side alley is narrow and there are two steps up onto the lawn. There are no paths only an uneven lawn.

38 THE OLD RECTORY

Sandy Lane, Brewer Street, Bletchingley, RH1 4QW. Mr & Mrs A Procter, 01883 743388 or 07515 394506, trudie.y.procter@googlemail.com. *Top of village nr Red Lion PH, turn R into Little Common Lane then R Cross Rd into Sandy Lane. Parking nr house, disabled parking in courtyard.* Sun 6 June (11-4). Adm £5, chd free. Home-made teas. Visits also by arrangement Mar to Aug for groups of 10 to 30.

Georgian Manor House (not open). Quintessential Italianate topiary garden, statuary, box parterres, courtyard with columns, water features, antique terracotta pots. Much of the 4 acre garden is the subject of ongoing reclamation. This incl the ancient moat, woodland with fine specimen trees, one of the largest Tulip trees in the country, a rill, sunken and exotic garden. New water garden. Gravel paths.

39 PRATSHAM GRANGE

Tanhurst Lane, Holmbury St Mary, RH5 6LZ. Alan & Felicity Comber. *12m SE of Guildford, 8m SW of Dorking. From A25 take B2126, after 4m turn L into Tanhurst Lane. From A29 take B2126, before Forest Green turn R on B2126 then 1st R to Tanhurst Lane.* Sun 4 July, Sun 22 Aug (12.30-4.30). Adm £5, chd free. Home-made teas.

5 acre garden overlooked by Holmbury Hill and Leith Hill. Features incl 2 ponds joined by cascading stream, extensive scented rose and blue hydrangea beds. Also herbaceous borders, cutting flower garden, 2 white beds. Some steps, steep slopes, slippery when wet, gravel paths. Deep ponds and a drop from terrace.

40 ◆ RAMSTER

Chiddingfold, GU8 4SN. Mrs R Glaister, 01428 654167, office@ramsterhall.com, www.ramsterevents.com. *Ramster is on A283 1½m S of Chiddingfold, large iron gates on R, the entrance is signed from the road.* For NGS: Fri 7 May (10-4). Adm £8, chd free. Light refreshments in the teahouse. For other opening times and information, please phone, email or visit garden website.

A stunning, mature woodland garden set in over 20 acres, famous for its rhododendron and azalea collection and its carpets of bluebells in Spring. Enjoy a peaceful wander down the grass paths and woodland walk, explore the bog garden with its stepping stones, or relax in the tranquil enclosed tennis court garden. The tea house, found by the entrance to the garden, serves sandwiches, cakes and drinks, and is open every day while the garden is open. The teahouse is wheelchair accessible, some paths in the garden are suitable for wheelchairs.

41 RANDALLS ALLOTMENTS

Old Nursery Lane, Cobham, KT11 1JN. Cobham Garden Club, www.cobhamgardenclub.com. *1.2m from Moleshill House going towards Cobham, off the A307 Esher to Cobham Rd, on the RHS opp Sheargold Pianos & behind Premier Service Station. Entrance is ½ way down Old Nurseries Lane. (Postcode is a guide only, not for the site.)* Sun 1 Aug (2-5). Adm £3.50, chd free. Home-made teas. Also open Moleshill House.

Well presented allotments in Cobham, growing unusual vegetables, fruit & salad items from different countries. Guided tours are available by the knowledgeable owners of the plots. Green Fingers Garden Shop on site. If you have ever thought about 'Growing Your Own' this will inspire you. The Greenfingers Shop will be open for visitors as a special opportunity, offering plants, garden sundries, composts etc. All at highly competitive prices. We will also have local honey for sale. Tea/coffee and homemade cakes will also be available to enjoy, and the opportunity to join our club. There is wheelchair access, but some areas are uneven and will need careful attention.

42 ◆ THE SCULPTURE PARK

Tilford Road, Churt, Farnham, GU10 2LH. Eddie Powell, 01428 605453, sian@thesculpturepark.com, www.thesculpturepark.com. *Corner of Jumps and Tilford Rd, Churt. Look out for Bell & The Dragon pub, we are directly opp. You can use our car park or the pub where refreshments are available.* For NGS: Tue 9 Mar, Tue 15 June (10-5). Adm £5, chd £3. For other opening times and information, please phone, email or visit garden website.

This garden sculpture exhibition is set within an enchanting arboretum and wildlife inhabited water garden. You should set aside between two and four hours for your visit as there are two miles of trail within 10 acres. Our displays evolve and diversify as the seasons pass, with vivid and lush colours of the rhododendrons in May and June to the enchanting frost in the depths of winter. Around one third of The Sculpture Park is accessible to wheelchairs. Disabled toilets are available.

&. 🐎

GROUP OPENING

43 SHACKLEFORD GARDEN SAFARI

Godalming, GU8 6AY. 5m SW of Guildford. Go to centre of Shackleford Village which is ½m towards Elstead from A3 Hurtmore/ Shackleford junction. Follow signs for parking. **Sun 6 June (10.30-5). Combined adm £12, chd free. Home-made teas at Norney Wood.**

DOLPHIN HOUSE
Mr & Mrs C Bell.

HEADLANDS
David & Jackie Soworbutts.

NEW HUNTSMORE
Mrs Fiona Vanstone.

NORNEY WOOD
Mr & Mrs R Thompson, www.norneywood.co.uk.
Ⓓ

Enjoy a leisurely Sunday visiting four inspirational gardens in the village of Shackleford in the Surrey Hills, finishing with delicious tea and cake at Norney Wood. Three of the gardens have opened previously and one garden is opening for the first time. Meet the garden owners and hear what inspired them, challenged them and gives them great pleasure. Meander through the naturalistic planting, meadow and woodland of Headlands. See the historical walls and the sophisticated and richly planted borders of Dolphin House. For the first time, enjoy the tranquility of the cottage garden at Huntsmore with its majestic Cedar of Lebanon at its centre And finally, relax in the gardens of Norney Wood amidst the heavenly scented Gertrude Jekyll roses to enjoy delicious home-made teas. Follow signs for parking and start your visit at either Headlands or Dolphin House. Surrey Hills village, historic buildings and structures, country gardens, cottage garden, contemporary landscaping,

woodland, ancient trees, formal and informal water features, Society of Garden Designer garden. Partial wheelchair access available at most gardens. Gravel paths, steps and slopes.

&. ❀ 🍵

44 SHAMLEY WOOD ESTATE

Woodhill Lane, Shamley Green, Guildford, GU5 0SP. Mrs Claire Merriman, 07595 693132, claire@merriman.co.uk. 5m (15 mins) S of Guildford in village of Shamley Green. Entrance is approx ¼m up Woodhill Lane from centre of Shamley Green. **Visits by arrangement Feb to Nov for groups of 10+. Adm £8, chd free. Home-made teas. incl gluten free options.**

A relative newcomer, this garden is worth visiting just for the setting! Sitting high on the North Downs, the garden enjoys beautiful views of the South Downs and is approached through a 10 acre deer park. Set within approximately 3 acres, there is a large pond and established rose garden. More recent additions incl a stream, fire pits, dry garden, heather, vegetable garden and woodland walk. Most of garden accessible by wheelchair. Large ground level WC but step up to access area.

&. ❀ Ⓓ 🍵

45 41 SHELVERS WAY

Tadworth, KT20 5QJ. Keith & Elizabeth Lewis, 01737 210707, kandelewis@ntlworld.com. 6m S of Sutton off A217. 1st turning on R after Burgh Heath T-lights heading S on A217. 400yds down Shelvers Way on L. **Sun 11 July, Sun 22 Aug (2-5.30). Adm £5, chd free. Home-made teas. Visits also by arrangement Apr to Sept for groups of 10+.**

Visitors say 'one of the most colourful gardens in Surrey'. In spring, a myriad of small bulbs with specialist daffodils and many pots of colourful tulips. Choice perennials follow. Cobbles and shingle support grasses and self sown plants with a bubble fountain. Annuals and herbaceous plants ensure colour well into September. A garden for all seasons.

❀ 🚐 🍵

46 SHIELING

The Warren, Kingswood, Tadworth, KT20 6PQ. Drs Sarah & Robin Wilson, 01737 833370, sarahwilson@doctors.org.uk. Kingswood Warren Estate. Off A217, gated entrance just before church on S-bound side of dual carriageway after Tadworth r'about . ¾m walk from Station. Parking on The Warren or by church on A217. **Sun 21 Feb, Sun 4 Apr (2-4); Mon 3 May (11-4). Adm £5, chd free. Home-made teas. Evening opening Fri 18, Sat 19 June (5.30-9). Adm £8, chd free. Wine. Visits also by arrangement May to July for groups of 10+.**

One acre garden restored to its original 1920s design. Formal front garden with island beds and shrub borders. Unusual large rock garden and mixed borders with collection of beautiful slug free hostas and uncommon woodland perennials. The rest is an interesting woodland garden with acid loving plants, a new shrub border and a stumpery Play area for children. Plant list provided for visitors. Lots for children to do with swing, balance bars and Wendy house in woodland glade. Some narrow paths in back garden. Otherwise resin drive, grass and paths easy for wheelchairs.

&. 🐎 Ⓓ 🍵 🪑

47 SLEEPY HOLLOW

Pook Hill, Chiddingfold, Godalming, GU8 4XR. Mrs Susannah Money. Exit A3 at Milford, follow A283 towards Petworth for 3m, R into Combe Ln. Proceed 1½m then R into Prestwick Ln, bear L into Pook Hill. **Sun 12 Sept (1-5). Adm £5, chd free. Home-made teas.**

Wander around the formal gardens which incl a knot garden with box hedges, a vegetable garden and greenhouse and a stunning cutting garden. Take a stroll through light woodland along a meandering stream. Incl three paddocks, the garden is just over 15 acres. Enjoy tea on our relaxing terrace looking out over the extensive lawn and summer beds.

🐎 🍵

Online booking is available for many of our gardens, at ngs.org.uk

48 2 SLYFIELD FARM COTTAGES
Cobham Road, Stoke D'Abernon, Cobham, KT11 3QH.
Mr Clive & Mrs Chrissie Shaw, 01932 863460,
chrissieanneshaw@yahoo.co.uk.
2½ M SE of Cobham. Take A245 from Cobham towards Fetcham go over M25 & we are 0.2m (325 metres) on L up a small farm drive & immediately R. (We are not at Slyfield House). **Visits by arrangement May & June for groups of 10 to 20. Adm £4, chd free. Home-made teas.**
A calm and relaxing cottage garden planned and planted for all year round interest with the 'right plant, right place' mantra, nature and the environment in mind using foliage colour and texture as well as a variety of different plants and scents (from chocolate and candy floss to cherry pie and lemon sorbet).

49 SOUTHLANDS LODGE
Southlands Lane, Tandridge, Oxted, RH8 9PH. Colin David & John Barker, 07500009603, colin3d@gmail.com. *3min from J6 of M25. From Tandridge Village head S down hill for ¼ m to T-junction with Southlands Lane. Turn L along Southlands Lane, approx ¼ m to Southlands Lodge entrance gate.* **Sat 10, Sun 11 July (11-4). Adm £5, chd free. Pre-booking essential, please visit www.ngs. org.uk for information & booking. Home-made teas. Visits also by arrangement July & Aug for groups of 5 to 20. Note beehives: visitors with allergies need to carry an EPI Pen.**
A surprising garden, on the edge of a woodland, enjoying open views of adjacent fields. Formal and informal areas are scaled to compliment a late Georgian gatehouse. Mass planting offsets selected specimens, mostly chosen for their pollen and nectar that helps sustain the resident honey bees. An enclosed vegetable garden with greenhouse, a flock of Shetland sheep and chickens add interest. The garden contains a working apiary of 10 beehives. Cayuga ducks and Guinea Fowl roam free through the property. An 'insect hotel' features. Gluten free cakes will be available, all cakes will contain nuts and dairy.

50 SPURFOLD
Radnor Road, Peaslake, Guildford, GU5 9SZ. Mr & Mrs A Barnes, 01306 730196, spurfold@btinternet.com. *8m SE of Guildford. A25 to Shere then through to Peaslake. Pass Village stores & L up Radnor Rd.* **Visits by arrangement May to July for groups of 10 to 30. Adm £6, chd free. Home-made teas.**
2½ acres, large herbaceous and shrub borders, formal pond with Cambodian Buddha head, sunken gravel garden with topiary box and water feature, terraces, beautiful

Dunsborough Park

lawns, mature rhododendrons and azaleas, woodland paths, and gazebos. Garden contains a collection of Indian elephants and other objets d'art. Topiary garden created 2010 and new formal lawn area created in 2012.

✳ ☕

51 STUART COTTAGE
Ripley Road, East Clandon, GU4 7SF. John & Gayle Leader, 01483 222689, gayle@stuartcottage.com, www.stuartcottage.com. *4m E of Guildford. Off A246 or from A3 through Ripley until r'about, turn L & cont through West Clandon until T-lights, then L onto A246. East Clandon 1st L.* **Visits by arrangement June to Sept for groups of 10+. Refreshments for groups by arrangement.**
Walk in to this tranquil partly walled ½ acre garden to find an oasis of calm. Beds grouped around the central fountain offer floral continuity through the seasons with soft harmonious planting supported by good structure with topiary, a rose/clematis walk and wisteria walk. Outside the wall is the late border with its vibrant colours and fun planting. The decorative kitchen garden with raised beds edged with tiles is always of interest to visitors, many say it is one of the best gardens that they visit and just a lovely place to be.

♿ 🐕 ✳ 🚗 ☕

52 NEW TANHOUSE FARM
Rusper Road, Newdigate, RH5 5BX. Mrs N Fries, 01306 631334. *8m S of Dorking. On A24 turn L at r'about at Beare Green. R at T-junction in Newdigate 1st farm on R approx ⅔m. Signposted Tanhouse Farm Shop.* **Sun 18 July (1-5). Adm £5, chd free. Tea at Tanhouse Farm Shop, next to car park. Lunches and teas served til 5pm. Visits also by arrangement June to Aug for groups of 10 to 30.**
Country garden created by owners since 1987. 1 acre of charming rambling gardens surrounding a C16 house (not open). Herbaceous borders; small lake and stream with ducks and geese, and an orchard with wild garden and meadow walk, with plentiful seats and benches to stop for contemplation.

♿ 🐕 ☕

53 THE THERAPY GARDEN
Manor Fruit Farm, Glaziers Lane, Normandy, Guildford, GU3 2DT. The Centre Manager, www.thetherapygarden.org. *SW of Guildford. Take A323 travelling from Guildford towards Aldershot, turn L into Glaziers Ln in centre of Normandy village opp War Memorial. The Therapy Garden is 200yds on L.* **Sun 30 May, Sun 19 Sept (10-4). Adm £6, chd free. Light refreshments.**
The Therapy Garden is a horticulture and education charity that uses gardening to generate positive change for those living with mental health challenges. Like so many, we faced unprecedented challenges in 2020 due to the pandemic and had to shut our gates to clients for a period but we came out the other side like a snowdrop re-emerging in the spring and we would be delighted to welcome you back! The Therapy Garden is a registered charity that provides social and therapeutic horticulture to different groups in the local community. We are a working garden full of innovation with an onsite shop selling plants and produce. Barbecue, salads and sandwiches, teas, coffees and cakes all available. Paved pathways throughout most of the garden, many with substantial handrails.

♿ 🐕 ✳ 🚗 ☕

54 TILFORD COTTAGE
Tilford Road, Tilford, GU10 2BX. Mr & Mrs R Burn, 01252 795423 or 07712 142728, rodneyburn@outlook.com, www.tilfordcottagegarden.co.uk. *3m SE of Farnham. From Farnham station along Tilford Rd. Tilford Cottage opp Tilford House. Parking by village green.* **Visits by arrangement Apr to Sept for groups of 5+. Min group size 6, max group size 40. Adm £10, chd free.**
Artist's garden designed to surprise, delight and amuse. Formal planting, herb and knot garden. Numerous examples of topiary combine beautifully with the wildflower river walk. Japanese and water gardens, hosta beds, rose, apple and willow arches, treehouse and fairy grotto all continue the playful quality especially enjoyed by children. Local pub within walking distance. Partial wheelchair access, top half of the garden only.

♿ ✳

55 TIMBER HILL
Chertsey Road, Chobham, GU24 8JF. Nick & Lavinia Sealy, 01932 873875, lavinia@chobham.net, www.timberhillgarden.com. *4m N of Woking. 2½m E of Chobham & ⅓m E of Fairoaks aerodrome on A319 (N side). 1¼m W of Ottershaw, J11 M25. (If approaching from Ottershaw the A319 is the Chobham Road). See website for more detail.* T: 01932 873875. **Every Wed 6 Jan to 10 Feb (12-2). Tea. Mon 15, Thur 18 Feb (12-2). Tea. Every Wed 17 Mar to 31 Mar (12-2). Tea. Mon 5 Apr (1.30-4.30). Home-made teas. Mon 7, Mon 14 June (2-4.30). Tea. Adm £7, chd free. DIY Tea/coffee available in Jan, Feb & Mar. Home-made teas in Barn on Easter Monday. Please book teas in advance for June openings. 2022: Wed 12, Wed 19, Wed 26 Jan, Mon 7, Mon 14, Mon 21 Feb. Visits also by arrangement in Oct for groups of 10+.**
16 acres of garden, park & woodland giving an undulating walk with views to North Downs. Stunning winter garden available to enjoy winter walks on most Wednesdays from January - March. Lonicera & witch hazel walk, abundant snowdrops, aconites & a sea of crocuses, spectacular camellias (featured in Surrey Life) followed by magnolias & wild cherries. May brings bluebells, azaleas, rhododendrons. See our website for more information. Nature & wildlife trails for adults & children. For other information, pop up openings & events, see Timber Hill website or please telephone.

♿ 🐕 ☕

"I love the National Garden Scheme which has been the most brilliant supporter of Queen's Nurses like me. It was founded by the Queen's Nursing Institute which makes me very proud. As we battle Coronavirus on the front line in the community, knowing we have their support is a real comfort." – Liz Alderton, Queen's Nurse

Huntsmore

56 ◆ TITSEY PLACE GARDENS
Titsey, Oxted, RH8 0SA. The
Trustees of the Titsey Foundation,
07889052461, office@titsey.org,
www.titsey.org. *3m N of Oxted.
A25 between Oxted & Westerham.
Follow brown heritage signs to Titsey
Estate from A25 at Limpsfield or see
website for directions.* **For NGS:
Sun 23 May, Sun 20 June, Sun 18
July, Sun 15 Aug (1-4.30). Adm
£6, chd £2. Light refreshments.
For other opening times and
information, please phone, email or
visit garden website.**
One of the largest surviving historic
estates in Surrey. Magnificent
ancestral home and gardens of
the Gresham family since 1534.
Walled kitchen garden restored early
1990s. Golden Jubilee rose garden.
Etruscan summer house adjoining
picturesque lakes and fountains.
15 acres of formal and informal
gardens in an idyllic setting. Highly
Commended in the 2019 Horticulture
Week Custodian Awards. Tea room
serving delicious homemade cakes
and selling local produce is open
from 12.30-5.00 pm on open days.
Walks through the estate woodland
are open all year round. Pedigree herd
of Sussex Cattle roam the park. Last
admissions to gardens at 4pm. Dogs
on leads allowed in picnic area, car
park and woodland walks. Disabled
car park alongside tearooms.

  ♿ 🐕 🚐 ☕ ♨

57 ◆ VANN
Hambledon, Godalming,
GU8 4EF. Caroe Family,
01428 683413, vann@caroe.com,
www.vanngarden.co.uk. *6m S
of Godalming. A283 to Lane End,
Hambledon. On NGS days only,
follow yellow Vann signs for 2m.
Please park in the field as signposted, not in the road. At other
times follow the website instructions.*
**For NGS: Mon 29, Tue 30, Wed 31
Mar, Thur 1 Apr (10-4). Daily Mon
3 May to Sun 9 May (10-4). Daily
Mon 7 June to Sun 13 June (10-
4). Adm £8, chd free. Home-made
teas. Home-made tea, cakes and
biscuits available 'DIY style' on
some weekdays and weekends
- call to confirm. Picnics are
allowed in the field. For other
opening times and information,
please phone, email or visit garden
website.**
Five-acre, 2* English Heritage
registered garden surrounding house
dating back to 1542 with Arts and

Crafts additions by W D Caröe incl a Bargate stone pergola. At the front, brick paved original cottage garden; to the rear a lake, yew walk with rill and Gertrude Jekyll water garden. Snowdrops and hellebores, spring bulbs, spectacular Fritillaria in Feb/March. Island beds, crinkle crankle wall, orchard with wildflowers. Vegetable garden. Gertrude Jekyll water garden and spring bulbs. Also open by appointment for individuals or groups. Some paths not suitable due to uneven stone walkways. Please ring prior to visit to request disabled parking.

58 11 WEST HILL

Sanderstead, CR2 0SB. **Rachel & Edward Parsons.** *M25, J6, A22, 2.9m r'about 4th exit to Succombs Hill, R to Westhall Rd, at r'about 2nd exit to Limpsfield Rd, r'about 2nd exit on Sanderstead Hill 0.9m a sharp R to West Hill, plse park on West Hill.* **Sat 10, Sun 11 Apr, Sat 19, Sun 20 June (2-5). Adm £5, chd free. Home-made teas. Donation to British Hen Welfare Trust.**
A hidden gem tucked away... a beautiful country cottage style garden set in ½ acre, designed by Sam Aldridge of Eden Restored. The garden flows through pathways, lawn, vegetable and play areas. Flower beds showcase outstanding tulips. Informal planting areas throughout the garden allows you to absorb the wonderful garden, whilst observing our rescued chickens and rabbits!

59 57 WESTHALL ROAD

Warlingham, CR6 9BG. **Robert & Wendy Baston.** *3m N of M25. M25, J6, A22 London, at Whyteleafe r'about, take 3rd R, under railway bridge, turn immed R into Westhall Rd.* **Sun 30, Mon 31 May, Sun 13, Mon 14 June (2-5). Adm £5, chd free. Home-made teas. Donation to Warlingham Methodist Church.**
Reward for the sure footed – many steps to 3 levels! Swathes of tulips. Mature kiwi and grape vines. Mixed borders. Raised vegetable beds. Box, bay, cork oak and yew topiaries. Amphitheatre of potted plants on lower steps. Stunning views of Caterham and Whyteleafe from top garden. Olive tree floating on a circular 'pond' of white and pink flowers. Flint walls, water spilling onto pebbles in secluded lush setting, vegetable borders, summerhouse,

apple tree with child swing, gravel garden. Not suitable for wheelchairs.

60 WESTWAYS FARM

Gracious Pond Road, Chobham, GU24 8HH. **Paul & Nicky Biddle, 01276 856163, nicolabiddle@rocketmail.com.** *4m N of Woking. From Chobham Church proceed over r'about towards Sunningdale, 1st Xrds R into Red Lion Rd to junction with Mincing Lane.* **Sun 9 May (11-5). Adm £5, chd free. Home-made teas. Visits also by arrangement Apr to June for groups of 10+.**
6 acre garden surrounded by woodlands planted in 1930s with mature and some rare rhododendrons, azaleas, camellias and magnolias, underplanted with bluebells, lilies and dogwood; extensive lawns and sunken pond garden. Working stables and sandschool. Lovely Queen Anne House (not open) covered with listed *Magnolia grandiflora*. Victorian design glasshouse. New planting round garden room.

61 WILDWOOD

34 The Hatches, Frimley Green, Camberley, GU16 6HE. **Annie Keighley, 01252 838660, annie.keighley1@btinternet.com.** *3m S of Camberley. M3 J4 follow A325 to Frimley Centre, towards Frimley Green for 1m. Turn R by the green, R into The Hatches for on street parking.* **Sun 6 June (1.30-5). Combined adm with Oakleigh £4.50, chd free. Home-made teas. Visits also by arrangement in June for groups of 5 to 20. Please email owner to customise your visit and catering requirements.**
Visitors love the hidden surprises in this romantic cottage garden with tumbling roses, topiary and towering magnolia grandiflora. Holly hedges hide a secret haven with wildlife pond, dell, fernery and shaded loggia. Quirky organic potager with raised beds, fruit trees, tadpole nursery, greenhouses and composting areas. Surrey Wildlife Trust winner - Large Private Garden category. Wildlife gardening introduction by owner. Gravel drive and paths. Care needed by pond.

62 WOODBURY COTTAGE

Colley Lane, Reigate, RH2 9JJ. **Shirley & Bob Stoneley, 01737 244235, woodburycottage@gmail.com.** *1m W of Reigate. M25 J8, A217 (Reigate). Immed before level crossing turn R into Somers Rd, cont as Manor Rd. At end turn R into Coppice Ln & follow signs to car park. Do not come up Colley Ln from A25.* **Sat 10, Sun 11 July, Sat 11, Sun 12 Sept (1-5). Adm £5, chd free. Home-made teas. Visits also by arrangement July to Oct for groups of 10 to 20.**
Cottage garden just under ¼ acre. It is stepped on a slope, enhanced by its setting under Colley Hill and the North Downs. We grow a colourful diversity of plants incl perennials, annuals and tender ones. A particular feature throughout the garden is the use of groups of pots containing unusual and interesting plants. The garden is colour themed and is still rich and vibrant in September.

63 WOODLAND

67 York Rd, Cheam, Sutton, SM2 6HN. **Sean Hilton, 07879498654, shilton.shx@gmail.com.** *York Rd runs parallel to Belmont Rise, main A217 from Sutton to Banstead. From A217 turn L (from N) or R (from S) at T lights into Dorset Rd. York Rd is 1st L.* **Evening opening Tue 18, Thur 20 May (5.30-8). Adm £6, chd free. Wine. Visits also by arrangement May to Aug for groups of 5 to 30.**
Woodland is a suburban walled garden of approx ⅕ acre with a range of plants shrubs and young trees. Hard landscaping has created terraces, two lawns and four levels from the upward incline of the garden. Good selection of shrubs, perennials, hostas, bamboos and ferns. Central features of the garden are a pond with pergola, a weeping copper beech and a Victorian Plant House. All but the top level accessible by wheelchair.

All entries are subject to change. For the latest information check ngs.org.uk

SUSSEX

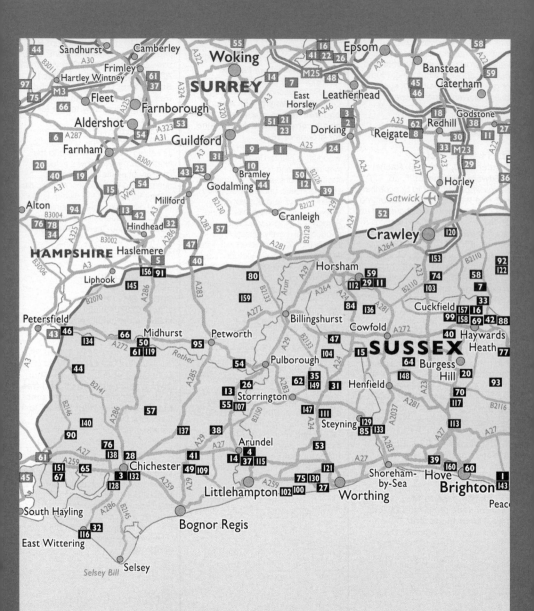

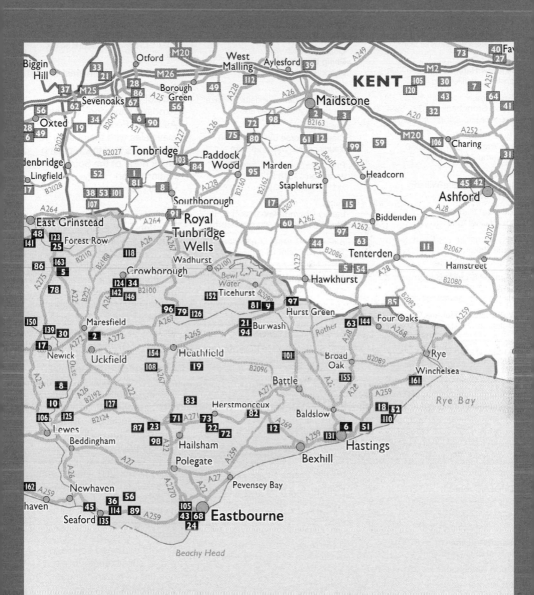

Biggin Hill

Otford

West Malling

Aylesford 39

M20

M26

KENT

73

40 Fav 27

33 21

37

M25

62

28 86 67

Borough Green

49

A228

112

105 120

30

7

86

Sevenoaks

A25

56

Maidstone

43

64

41

56

Oxted

19

34

B2042

6 90

A26

72 98

2

B2163

32

28 6

49

A21

Tonbridge

103

84

Paddock Wood

75

80

61 12

3

A20

106

M20

Charing

A252

31

denbridge

52

B2027

95

Marden

B2162

B2160

Staplehurst

A229

Headcorn

45 42

Lingfield

17

B2028

81

8

A228

17

Southborough

B2079

15

Ashford

A28

38 53 101

107

91

Royal Tunbridge Wells

A264

60

44

97

63

Biddenden

A262

11

B2067

A207C

48 141

123 25

Forest Row

Wadhurst

A267

B2086

5 54

Tenterden

Hamstreet

East Grinstead

86

163 5

B2110

B2100

A26

118

Bewl Water

A229

Hawkhurst

A28

B2080

78

A22

124 34

142 146

B2100

Crowborough

152

Ticehurst

B2099

81 9

97

85

B2082

A259

150

96 79 126

A267

21 94

Burwash

Rother

63 144

Four Oaks

A268

139 30

Maresfield

A265

101

Broad Oak

B2089

Rye

17

2

A272

Heathfield

19

B2096

155

Winchelsea

Newick

A26

154

108

A267

Battle

A21

161

Uckfield

B2192

A22

83

Herstmonceux

82

A271

12

A269

18 52

110

8

127

71 73

22 72

Baldslow

A259

131

6 51

10

87 23

98

Hailsham

A22

A259

12

Hastings

106 125

B2124

Rye Bay

Lewes

Beddingham

Polegate

Bexhill

162

A259

Newhaven

56

A27

A270

Pevensey Bay

45 36 114

89

A259

haven

135

Seaford

105

43 68

24

Eastbourne

Beachy Head

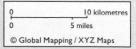

East & Mid Sussex Volunteers

County Organiser, Booklet & Advertising Co-ordinator
Irene Eltringham-Willson
01323 833770
irene.willson@btinternet.com

County Treasurer
Andrew Ratcliffe 01435 873310
andrew.ratcliffe@ngs.org.uk

Publicity
Geoff Stonebanks 01323 899296
sussexeastpublicity@ngs.org.uk

Twitter
Liz Warner 01273 586050
liz.warner@ngs.org.uk

Talks
Irene Eltringham-Willson
(as above)

Booklet Distributor
Liz Warner (as above)

Assistant County Organisers
Jane Baker 01273 842805
jane.baker@ngs.org.uk

Michael & Linda Belton
01797 252984
belton.northiam@gmail.com

Victoria Brocklebank 01825 890348
victoria.brocklebank@ngs.org.uk

Shirley Carman-Martin
01444 473520
shirleycarmanmartin@gmail.com

Isabella Cass 07908 123524
oaktreebarn@hotmail.co.uk

Linda Field 01323 720179
lindafield3@gmail.com

Diane Gould 01825 750300
lavenderdgould@gmail.com

Aideen Jones 01323 899452
sweetpeasa52@gmail.com

Dr Denis Jones 01323 899452
sweetpeasd49@gmail.com

Susan Laing 01444 892500
splaing@btinternet.com

Sarah Ratcliffe 01435 873310
sallyrat@btinternet.com

Geoff Stonebanks (as above)

Liz Warner (as above)

David Wright 01435 883149
david.wright@ngs.org.uk

f @SussexNGSEast
🐦 @SussexNGS
📷 @ngseastsussex

West Sussex Volunteers

County Organisers
Meryl Walters 07766 761926
meryl.walters@ngs.org.uk

Maggi Hooper 07793 159304
maggi.hooper@ngs.org.uk

County Treasurer
Philip Duly 01428 661089
philipduly@tiscali.co.uk

Publicity
Position Vacant, for details please
contact Philip Duly (as above)

Social Media
Claudia Pearce 07985 648216
claudiapearce17@gmail.com

Photographer
Judi Lion 07810 317057
judi.lion@ngs.org.uk

Talks
Lesley Chamberlain 07950 105966
chamberlain_lesley@hotmail.com

Assistant County Organisers
Teresa Barttelot 01798 865690
tbarttelot@gmail.com

Sanda Belcher 01428 723259
sandambelcher@gmail.com

Lesley Chamberlain (as above)

Patty Christie 01730 813323
sussexwestngs@gmail.com

Elizabeth Gregory 01903 892433
elizabethgregory1@btinternet.com

Peter & Terri Lefevre 01403 256002
teresalefevre@outlook.com

Judi Lion (as above)

Carrie McArdle 01403 820272
carrie.mcardle@btinternet.com

Ann Moss 01243 370048
ann.moss@ngs.org.uk

Claudia Pearce (as above)

Fiona Phillips 07884 398704
fiona.h.phillips@btinternet.com

Susan Pinder 01403 820430
nasus.rednip@gmail.com

Diane Rose 07789 565094
dirose8@me.com

f @Sussexwestngs
🐦 @SussexWestNGS
📷 @sussexwestngs

Sussex is a vast county with two county teams, one covering East and Mid Sussex and the other covering West Sussex.

Over 80 miles from west to east, Sussex spans the southern side of the Weald from the exposed sandstone heights of Ashdown Forest, past the broad clay vales with their heavy yet fertile soils and the imposing chalk ridge of the South Downs National Park, to the equable if windy coastal strip.

Away from the chalk, Sussex is a county with a largely wooded landscape with imposing oaks, narrow hedged lanes and picturesque villages. The county offers much variety and our gardens reflect this. There is something for absolutely everyone and we feel sure that you will enjoy your garden visiting experience - from rolling acres of parkland, country and town gardens, to small courtyards and village trails. See the results of the owner's attempts to cope with the various conditions, discover new plants and talk with the owners about their successes.

Many of our gardens are open by arrangement, so do not be afraid to book a visit or organise a visit with your local gardening or U3A group.

Should you need advice, please e-mail sussexeastpublicity@ngs.org.uk for anything relating to East and Mid Sussex or sussexwestngs@gmail.com for anything in West Sussex.

OPENING DATES

All entries subject to change. For latest information check www.ngs.org.uk

Extended openings are shown at the beginning of the month.

Map locator numbers are shown to the right of each garden name.

February

Snowdrop Festival

By Arrangement
Pombury House 117

Every Thursday from Thursday 11th
The Old Vicarage 111

Saturday 6th
5 Whitemans Close 158

Monday 8th
5 Whitemans Close 158

Thursday 11th
Manor of Dean 95
5 Whitemans Close 158

Saturday 13th
◆ Denmans Garden 41

Monday 15th
5 Whitemans Close 158

Tuesday 16th
5 Whitemans Close 158

Wednesday 17th
5 Whitemans Close 158

March

Every Thursday
The Old Vicarage 111

Tuesday 9th
Manor of Dean 95

Sunday 21st
The Hamblin Centre 65
◆ King John's Lodge 81
Manor of Dean 95

Saturday 27th
Butlers Farmhouse 22
Down Place 44

Sunday 28th
Butlers Farmhouse 22
Down Place 44

April

Every Wednesday from Wednesday 21st
Fittleworth House 54

Every Thursday
The Old Vicarage 111

Sunday 4th
NEW Cloud Cottage & Hollyoaks 31
Hammerwood House 66

Monday 5th
The Old Vicarage 111

Saturday 10th
The Garden House 60

Sunday 11th
Findon Place 53
The Garden House 60

Thursday 15th
Manor of Dean 95

Saturday 17th
Banks Farm 8
Rymans 128
Winchelsea's Secret Gardens 161

Sunday 18th
Banks Farm 8
Penns in the Rocks 118
Rymans 128

Tuesday 20th
◆ Borde Hill Garden 16
◆ Great Dixter House, Gardens & Nurseries 63
Peelers Retreat 115

Wednesday 21st
Cupani Garden 36

Saturday 24th
◆ Denmans Garden 41

Sunday 25th
◆ Clinton Lodge 30
Cupani Garden 36
Peelers Retreat 115

Tuesday 27th
Peelers Retreat 115

May

Every Wednesday to Wednesday 12th
Fittleworth House 54

Every Thursday
The Old Vicarage 111

Saturday 1st
◆ King John's Lodge 81

Sunday 2nd
◆ King John's Lodge 81
Stanley Farm 145

Monday 3rd
Copyhold Hollow 33
The Old Vicarage 111

Wednesday 5th
◆ Sheffield Park and Garden 139

Saturday 8th
Cookscroft 32
NEW Forest Ridge 58
Holly House 78
NEW Redriff 124
NEW Silver Springs 142
Stone Cross House 146

Sunday 9th
Champs Hill 26
NEW Forest Ridge 58
Hammerwood House 66
Holly House 78
Mountfield Court 101
NEW Redriff 124
NEW Silver Springs 142

Tuesday 11th
Manor of Dean 95
Peelers Retreat 115

Wednesday 12th
The Walled Garden at Tilgate Park 153

Thursday 13th
The Walled Garden at Tilgate Park 153
3 Whitemans Close 157
5 Whitemans Close 158

Friday 14th
96 Ashford Road 6

Saturday 15th
96 Ashford Road 6
Harlands Gardens 69
Holford Manor 77

Sunday 16th
Harlands Gardens 69
NEW 28 Larkspur Way 84
Legsheath Farm 86
Penns in the Rocks 118

Tuesday 18th
Bignor Park 13

Wednesday 19th
Balcombe Gardens 7
Cookscroft 32

Thursday 20th
South Grange 144

Friday 21st
Caxton Manor 25
2 Quarry Cottages 123

Saturday 22nd
96 Ashford Road 6
Caxton Manor 25
NEW Clarence Road Gardens 29
54 Elmleigh 50
2 Quarry Cottages 123
NEW St Mary's Hospital 132

Sunday 23rd
NEW Clarence Road Gardens 29
54 Elmleigh 50
The Hamblin Centre 65
Holford Manor 77
Sienna Wood 141

Tuesday 25th
Peelers Retreat 115

Saturday 29th
51 Carlisle Road 24
Grovelands 64
◆ The Priest House 122

Sunday 30th
51 Carlisle Road 24
Dittons End 43
Foxglove Cottage 59
Grovelands 64
Hardwycke 68
Peelers Retreat 115
Seaford Gardens 135

Monday 31st
Copyhold Hollow 33
Oaklands Farm 104
The Old Vicarage 111
NEW Pangdean Farm 113

Online booking is available for many of our gardens, at ngs.org.uk

June

Every Wednesday and Thursday from Wednesday 23rd
◆ The Apuldram Centre 3

Every Thursday
The Old Vicarage 111

Tuesday 1st
NEW Pangdean Farm 113

Wednesday 2nd
◆ Highdown Gardens 75

Thursday 3rd
Kitchenham Farm 82

Saturday 5th
12 Ainsworth Avenue 1
Durford Abbey Barn 46
NEW Eastergate House
Gardens 49
Holford Manor 77
Limekiln Farm 87
Lordington House 90
NEW Pine Tree
Cottage 120
Skyscape 143
Waterworks &
Friends 155

Sunday 6th
12 Ainsworth Avenue 1
Durford Abbey Barn 46

NEW Eastergate House
Gardens 49
Hassocks & Ditchling
Garden Trail 70
◆ High Beeches
Woodland and
Water Garden 74
Limekiln Farm 87
Lordington House 90
Offham House 106
Old Stonelynk Edge 110
NEW Pine Tree Cottage 120
Selhurst Park 137
Skyscape 143

Monday 7th
◆ Clinton Lodge 30

Tuesday 8th
Peelers Retreat 115

Wednesday 9th
Fittleworth House 54
4 Hillside Cottages 76
Rolfs Farm 126

Thursday 10th
NEW Bramley 17
3 Whitemans Close 157
5 Whitemans Close 158

Friday 11th
NEW Bramley 17
Cupani Garden 36
Holly House 78
1 Pest Cottage 119

Saturday 12th
54 Elmleigh 50
Mayfield Gardens 96
NEW 28 Oliver Road 112

Sunday 13th
Ashdown Park Hotel 5
Breanross 18
NEW Deaks Lane
Gardens 40
East Grinstead
Gardens 48
54 Elmleigh 50
Fairlight End 52
Findon Place 53
Mayfield Gardens 96
1 Pest Cottage 119
Seaford Gardens 135

Wednesday 16th
Cupani Garden 36
Fittleworth House 54
Rolfs Farm 126

Thursday 17th
NEW Crowborough
Gardens 34
NEW 3 Normandy
Drive 102

Friday 18th
The Garden House 60
Town Place 150

Saturday 19th
Balcombe Gardens 7
Cookscroft 32
◆ King John's Lodge 81
Rymans 128
Town Place 150
Winchelsea's Secret
Gardens 161

Sunday 20th
Cupani Garden 36
Down Place 44
◆ King John's Lodge 81
NEW 28 Larkspur Way 84
Peelers Retreat 115
33 Peerley Road 116
Ringmer Park 125
Rymans 128
St Barnabas House 130
Town Place 150

Monday 21st
◆ Clinton Lodge 30
Down Place 44

Tuesday 22nd
Butlers Farmhouse 22
Kitchenham Farm 82
Peelers Retreat 115

Wednesday 23rd
Butlers Farmhouse 22

Saturday 26th
Burwash Hidden
Gardens 21
Luctons 92
Mandalay 94
◆ The Priest House 122
NEW St Mary's
Hospital 132

Sunday 27th
Burwash Hidden
Gardens 21
NEW Gorselands 62
Herstmonceux Parish
Trail 73
Luctons 92
Mandalay 94
NEW 28 Oliver Road 112
NEW St Leonards-on-Sea
Group 131

Tuesday 29th
Bignor Park 13
Driftwood 45
Luctons 92

July

Every Wednesday and Thursday to Thursday 15th
◆ The Apuldram Centre 3

Every Thursday
The Old Vicarage 111

Friday 2nd
Cupani Garden 36

Saturday 3rd
70 Dale View 39
54 Elmleigh 50
Follers Manor 56

Sunday 4th
Bexhill-on-Sea Trail 12
70 Dale View 39
54 Elmleigh 50
Follers Manor 56
NEW Gorselands 62
Whithurst Park 159

Wednesday 7th
Foxglove Cottage 59
NEW Wadhurst Park 152

Friday 9th
◆ St Mary's House
Gardens 133

Camberlot Hall

© Leigh Clapp

"I was amazed to discover that the National Garden Scheme is Marie Curie's largest single funder and has given the charity nearly £10 million over 25 years. Their continued support makes such a difference to me and all Marie Curie Nurses on the frontline of the coronavirus crisis, as we continue to provide expert care and support to people at end of life."
– Tracy McWilliams, Marie Curie Nurse

By Arrangement

Arrange a personalised garden visit on a date to suit you. See individual garden entries for full details.

D & S Haus

THE GARDENS

1 12 AINSWORTH AVENUE

Ovingdean, Brighton, BN2 7BG.
Jane & Chris Curtis. *From Brighton
take A259 coast road E, passing
Roedean School on L. Take 1st L at
r'about into Greenways & 2nd R into
Ainsworth Ave. No 12 on R.* **Sat 5, Sun
6 June (1-5). Combined adm with
Skyscape £5, chd free. Tea & cake
at Skyscape.**
A small (36ft x 16ft at back and 36ft
x 36ft at front) coastal garden that
began in 2014, and is now becoming
established. Pergola, arches and arbour
strategically positioned to provide
privacy and a choice of places to sit
with different aspects, even a glimpse
of the sea. Mixed planting, mostly new,
but several mature trees incl a large
walnut that creates a focal point.

2 NEW ALPINES

High Street, Maresfield, Uckfield,
TN22 2EG. Ian & Cathy Shaw,
07887 825032, Info@shaw.buzz.
*1½ m N of Uckfield. Garden approx
150 metres N of Budletts r'about
towards Maresfield. Blue Badge
parking at garden, other parking in
village.* **Sat 17 July (11-4). Adm £5,
chd free. Light refreshments. Visits
also by arrangement June to Aug.**
This is a young (7 yrs), 1 acre garden
with large and rampant mixed
borders, each loosely following a
limited colour palette of unusual
combinations. Numerous new trees
incl orchard with beehives and
meadow grasses. Wildlife pond with
bog garden, developing stumpery
and fernery. Veggie patch with raised
beds, fruit cage, pretty greenhouse
and bee garden. Largely on one level
with a few steps round greenhouse.

3 ◆ THE APULDRAM CENTRE

Appledram Lane South,
Apuldram, Chichester, PO20 7PE.
01243 783370, info@apuldram.org,
www.apuldram.org. *1m S of A27.
Follow A259 Fishbourne & turn into
Appledram Lane South. For NGS:*
**Every Wed and Thur 23 June to 15
July (10-3). Adm £4, chd free. Light
refreshments. For other opening
times and information, please
phone, email or visit garden website.**
Award-winning wildlife and sensory
garden. Winding pathway sensitively

planted with bee loving plants, all
grown from seed, which leads to
a tranquil pond and garden. Full of
surprises, this garden encompasses
many interesting features incl
sculptures and other art works
created by our adult trainees with
learning difficulties. Teas, coffees,
light lunches and home-made cakes
served in the café until 4pm. Local
produce incl home-made apple juice,
local honey and a large selection of
bedding plants and perennials for sale.
Wheelchair access throughout sensory
garden. Restricted access to working
greenhouses and vegetable plots.

4 ◆ ARUNDEL CASTLE & GARDENS

Arundel, BN18 9AB. Arundel
Castle Trustees Ltd, 01903 882173,
visits@arundelcastle.org,
www.arundelcastle.org. *In the centre
of Arundel, N of A27.* **For opening
times and information, please phone,
email or visit garden website.**
Ancient castle, family home of the
Duke of Norfolk. 40 acres of grounds
and gardens which incl hot subtropical
borders, English herbaceous borders,
stumpery, wildflower garden, two
glasshouses with exotic fruit and
vegetables, walled flower and organic
kitchen gardens. C14 Fitzalan Chapel
white garden.

5 ASHDOWN PARK HOTEL

Wych Cross, East Grinstead,
RH18 5JR. Mr Kevin Sweet,
01342 824988,
reservations@ashdownpark.co.uk,
www.elitehotels.co.uk. *6m S of East
Grinstead. Turn off A22 at Wych Cross
T-lights.* **Sun 13 June (1-5). Adm £5,
chd free. Tea.**
186 acres of parkland, grounds and
gardens surrounding Ashdown Park
Hotel. Our Secret Garden is well worth
a visit with many new plantings. Large
number of deer roam the estate and
can often be seen during the day. Enjoy
and explore the woodland paths, quiet
areas and views. Access over gravel
paths and uneven ground with steps.

6 96 ASHFORD ROAD

Hastings, TN34 2HZ. Lynda
& Andrew Hayler. *From A21
(Sedlescombe Rd N) towards
Hastings, take 1st exit on r'about
A2101, then 3rd on L (approx 1m).*
Fri 14, Sat 15, Sat 22 May (1-5).

Adm £3, chd free.
Small (100ft x 52ft) Japanese
inspired front and back garden. Full
of interesting planting with many
acers, azaleas and bamboos. Over
100 different hostas, many miniature.
Lower garden with greenhouse
and raised beds. Also an attractive
Japanese Tea House.

GROUP OPENING

7 BALCOMBE GARDENS

*Follow B2036 N from Cuckfield for
3m. ¼ m N of Balcombe Station,
turn L immed before Balcombe
Primary School. From N, take J10A
from M23 & follow S for 2½ m.
Gardens signed within village.*
**Wed 19 May, Sat 19 June (12-5).
Combined adm £7.50, chd free.
Home-made teas at Stumlet.**

STUMLET
Oldlands Avenue, RH17 6LW.
Max & Nicola Preston Bell.

46 WESTUP FARM COTTAGES
London Road, RH17 6JJ.
Chris Cornwell, 01444 811891,
chris.westup@btinternet.com.
**Visits also by arrangement Apr
to Sept.**

WINTERFIELD
Oldlands Avenue, RH17 6LP. Sue
& Sarah Howe, 01444 811380,
sarahjhowe_uk@yahoo.co.uk.
**Visits also by arrangement Apr
to June.**

Within the Balcombe AONB there are
three quite different gardens that are
full of variety and interest, which will
appeal to plant lovers. Set amidst the
countryside of the High Weald, No 46
is a classic cottage garden with unique
and traditional features linked by
intimate paths through lush and subtle
planting. There is also a new herb
and butterfly garden. In the village,
Winterfield is a country garden packed
with uncommon shrubs and trees,
herbaceous borders, a summerhouse,
pond and wildlife area. At nearby
Stumlet the garden is restful, there are
places to sit and enjoy a little peace,
scent and colour. Re-designed to incl
interesting plants for lasting enjoyment
by different generations. A garden to
watch develop in the future. Wheelchair
access at Winterfield and Stumlet.

8 BANKS FARM
Boast Lane, Barcombe, Lewes, BN8 5DY. Nick & Lucy Addyman. *From Barcombe Cross follow signs to Spithurst & Newick. 1st road on R into Boast Lane towards the Anchor Pub. At sharp bend carry on into Banks Farm.* **Sat 17, Sun 18 Apr (11-4). Adm £5, chd free.**
9 acre garden set in rural countryside. Extensive lawns and shrub beds merge with the more naturalistic woodland garden set around the lake. An orchard, vegetable garden, ponds and a wide variety of plant species add to an interesting and very tranquil garden. Refreshments served outside, so may be limited during bad weather. Wheelchair access to lower part of garden, sloping grass paths may be difficult.

&. 🐄 🚗 ☕

9 NEW THE BARN
Burgham Farm, Sheepstreet Lane, Etchingham, TN19 7AZ. Eleanor Knowles & Mary Barnes. *Approx 1½m along Etchingham to Ticehurst road. A farm road leaves Sheepstreet Lane heading E downhill towards the River Limden. The carpark is 400 metres on the R. Partial one-way system operates.* **Sat 10, Sun 11 July (10-5). Adm £5, chd free. Light refreshments in the courtyard garden.**
Set in approx 1 acre, this former hop garden has evolved using a variety of influences from famous gardens and the interests of the owners. The planting is of semi mature trees and shrubs, underplanted with a variety of herbaceous perennials and large borders of mixed herbaceous plants. The variety of roses and dahlias are of particular interest. Wheelchair access to majority of the garden with uneven grass in places.

&. ✿ 🚗 ☕

10 THE BEECHES
Church Road, Barcombe, Lewes, BN8 5TS. Sandy Coppen, 01273 401339, sand@thebeechesbarcombe.com, www.thebeechesbarcombe.com. *From Lewes, A26 towards Uckfield for 3m, turn L signed Barcombe. Follow road for 1½m, turn L signed Hamsey & Church. Follow road for approx ½m & parking in field on R.* **Sat 17, Sun 18, Wed 21 July (2-5). Adm £6, chd free. Home-made teas. Visits also by arrangement June to Aug for groups of 10 to 20.**
C18 walled garden with cut flowers, vegetables, salads and fruit. Separate orchard and rose garden. Herbaceous borders, a hot border and extensive lawns. A hazel walk is being developed and a short woodland walk. An old ditch has been made into a flowing stream with gunnera, ferns, tree ferns, hostas and a few flowers going into a pond. Wheelchair access without steps, but some ground is a little bumpy.

&. 🐄 ✿ 🚑 ☕

11 4 BEN'S ACRE
Horsham, RH13 6LW. Pauline Clark, 01403 266912, brian.clark8850@yahoo.co.uk. *E of Horsham. A281 via Cowfold, after Hilliers Garden Centre, turn R by Tesco on to St Leonards Rd, straight over r'about to Comptons Lane, next R Heron Way, 2nd L Grebe Cres, 1st L Ben's Acre.* **Visits by arrangement June to Aug for groups of 10 to 30. Adm £9, chd free. Home-made teas included.**
Described as inspirational, a visual delight on different levels with ponds, rockery, summerhouse and arbours, all interspersed with colourful containers and statuettes. The compartmentalised layout of the garden lies at the heart of the design. Small themed sections nest within borders full of harmonising perennials, climbers, roses and more. See what a diverse space can be created on a small scale. Seating throughout the garden, 5 mins from Hilliers Garden Centre, 15 mins from Leonardslee and NT Nymans. Visit us on YouTube Pauline & Brian's Sussex Garden.

🐄 ✿ 🚑 ☕

GROUP OPENING

12 BEXHILL-ON-SEA TRAIL
Bexhill & Little Common. Follow individual NGS signs to gardens from main roads. Tickets & maps at all gardens. **Sun 4 July (12-5). Combined adm £6, chd free. Home-made teas & light lunch at Westlands.**

NEW 64 COLLINGTON AVENUE
TN39 3RA. Dr Roger & Ruth Elias.

NEW SMALL HOUSE
Sandhurst Lane, TN39 4RG. Veronika & Terry Rogers.

WESTLANDS
36 Collington Avenue, TN39 3NE. Madeleine Gilbart & David Harding.

Westlands with mature shrubs in the front and a large walled rear garden with shrubs, trees, lawn, fruit and vegetables. Small House is a beautiful garden surrounded by trees and extensive flower beds, planted in tiers with a viewing deck over the garden. 64 Collington Avenue is a bit of a surprise as it incorporates the garden next door too. 34 yrs in the making, it has a beautiful mix of herbaceous planting with structural shrubs, vegetables, planting to encourage wildlife and stunning roses. Plant sale at Westlands. Partial wheelchair access to some gardens.

&. ✿ ☕

13 BIGNOR PARK
Pulborough, RH20 1HG. The Mersey Family, www.bignorpark.co.uk. *5m S of Petworth & Pulborough. Well signed from B2138. Nearest villages Sutton, Bignor & West Burton. Approach from the E, directions & map available on website.* **Tue 18 May, Tue 29 June, Tue 14 Sept (2-5). Adm £5, chd free. Home-made teas.**
11 acres of peaceful garden to explore with magnificent views of the South Downs. Interesting trees, shrubs, wildflower areas, with swathes of daffodils in spring. The walled flower garden has been replanted with herbaceous borders. Temple, Greek loggia, Zen pond and unusual sculptures. Former home of romantic poet Charlotte Smith, whose sonnets were inspired by Bignor Park. Spectacular Cedars of Lebanon and rare Lucombe Oak. Wheelchair access to shrubbery and croquet lawn, gravel paths in rest of garden and steps in stables quadrangle.

&. 🐄 ☕

14 4 BIRCH CLOSE
Arundel, BN18 9HN. Elizabeth & Mike Gammon, 01903 882722, mgammon321@gmail.com. *1m S of Arundel. From A27 & A284 r'about at W end of Arundel take Ford Rd. Immed turn R & follow Torton Hill Rd to Dalloway Rd (straight on), Birch Close on L after bend. About 1m from r'about. Parking on roads.* **Visits by arrangement Apr & May for groups of 10 to 30. Adm £4, chd free. Light refreshments on request.**
⅓ acre of woodland garden on edge of Arundel. Wide range of mature trees and shrubs with many hardy

perennials. Emphasis on extensive selection of spring flowers and clematis (incl many montanas). All in a tranquil setting with secluded corners, meandering paths and plenty of seating.

🍵 🎍

15 NEW BLACK BARN

Steyning Road, West Grinstead, Horsham, RH13 8LR. Jane Gates, 07774 980819, jane.er.gates@gmail.com. *Between Partridge Green & A24 on Steyning Rd at West Grinstead. From A24 follow Steyning road signed West Grinstead for 1m, passing Park Lane & Catholic Church on the L, continue round two bends, Black Barn is on the L down a gravel drive.* Visits by arrangement Apr to Sept for groups of 5 to 20. Adm £5, chd free. Home-made teas.
Black Barn Garden was designed and planted in Jan 2018. The site was cleared and only the mature trees remained. A hedgerow separating the garden from the field was removed and a large pond has been created in the southwest corner. A large gravel garden was created to the south of the barn and a terrace laid. The existing terrace at the back was increased and a large bed was created. Wheelchair access over grass only.

♿ 🐕 🍵 🎍

16 ◆ BORDE HILL GARDEN

Borde Hill Lane, Haywards Heath, RH16 1XP. Borde Hill Garden Ltd, 01444 450326, info@bordehill.co.uk, www.bordehill.co.uk. *1½m N of Haywards Heath. 20 mins N of Brighton, or S of Gatwick on A23 taking exit 10a via Balcombe.* For NGS: Tue 20 Apr (10-5). Adm £9.95, chd £6.70. Cream teas. For other opening times and information, please phone, email or visit garden website.
Rare plants and stunning landscapes make Borde Hill Garden the perfect day out for horticultural enthusiasts, families and those who love beautiful countryside. Enjoy tranquil outdoor rooms, woodland walks, playgrounds, picnic areas, home-cooked food and events throughout the season. Wheelchair access to 17 acres of formal garden.

♿ 🐕 ✿ 🚗 🍵 🎍

17 NEW BRAMLEY

Lane End Common, North Chailey, Lewes, BN8 4JH. Marcel & Lee Duyvesteyn. *From A272 North Chailey mini-r'abouts take NE turn to A275 signed Bluebell Railway, after 1m turn R signed Fletching. 'Bramley' is 300yds on R. Free car park opp & roadside parking.* Thur 10, Fri 11 June (1-5.30). Adm £5, chd free. Home-made teas.
Rural 1 acre garden. Planting began in 2006 and incl an orchard avenue with wildflowers and long grassed areas and beehives. Vegetable garden, numerous formal and informal ornamental flowerbeds and borders. Planted for strong seasonal effect and structure. Interesting planting and design. SSSI nearby.

✿ 🍵

18 BREANROSS

Pett Road, Pett, Hastings, TN35 4HA. Tim & Libby Rothwell. *4m E of Hastings. From Hastings, take A259 towards Rye. At White Hart/Beefeater turn R into Friars Hill. Descend into Pett village. White boarded house on R. Parking & WC at village hall.* Sun 13 June (11-4). Adm £3, chd free. Also open Fairlight End.
The steeply sloping garden offered the new owners a fairly blank canvas when they bought the property in Oct 2014. There were two old espalier apple trees, a mature perimeter hedge and a quirky stone wall, but no borders. Now there is a wealth of planting, a cedar greenhouse with cold frames, a woodland border and an informal water feature. There is a display of before and after photos, and lots of ideas for making the most of a steeply sloping site.

✿

19 BRIGHTLING DOWN FARM

Observatory Road, Dallington, TN21 9LN. Val & Pete Stephens, 07770 807060, valstephens@icloud.com. *1m from Woods Corner. At Swan Pub, Woods Corner, take road opp to Brightling. Take 1st L to Burwash & almost immed, turn into 1st driveway on L.* Visits by arrangement on Thurs 3, Fri 4 June, Fri 24 Sept & Fri 1 Oct only for groups of 10 to 30. Adm £10, chd free. Home-made teas included.
The garden has several different areas incl a Zen garden, water garden, walled vegetable garden with two large greenhouses, herb garden, herbaceous borders and

a new woodland walk. The garden makes clever use of grasses and is set amongst woodland with stunning countryside views. Winner of the Society of Garden Designers award. Most areas can be accessed with the use of temporary ramps.

♿ 🅳 🍵

GROUP OPENING

20 BURGESS HILL NGS GARDENS

10m N of Brighton, off B2113. Keen walkers can reach 20 The Ridings first by walking from train station. Bus stops in area. Sun 25, Mon 26 July (1-5). Combined adm £6, chd free. Home-made teas at 14 Barnside.

14 BARNSIDE AVENUE
RH15 0JU. Brian & Sue Knight.

20 THE RIDINGS
RH15 0LW. Rachelle & Malcolm Russell.

30 SYCAMORE DRIVE
RH15 0GH. John Smith & Kieran O'Regan.

59 SYCAMORE DRIVE
RH15 0GG. Steve & Debby Gill.

This diverse group of four gardens is a mixture of established and small new gardens. Two of the group are a great example of what can be achieved over an 11 yr period from a blank canvas in a new development (Sycamore Drive), while 20 The Ridings is a mature garden with productive outside grapevine (pruning advice given). 14 Barnside Avenue has a splendid wisteria. Many useful ideas for people living in new build properties with small gardens and heavy clay soil. Partial wheelchair access to some gardens.

♿ 🐕 🍵

"Unfortunately, cancer was not in lockdown in 2020. The continued support of our long-standing and valued partner, the National Garden Scheme is more important than ever."
Macmillan

GROUP OPENING

21 BURWASH HIDDEN GARDENS
Burwash, TN19 7EN.
www.mandalaygarden.co.uk. *In Burwash village on A265. Burwash is 3m W of junction with A21 at Hurst Green; 6m E of Heathfield. Follow signs for parking from High St.* **Sat 26, Sun 27 June (1-5). Combined adm with Mandalay £7, chd free. Home-made teas in Swan Meadow Sports Pavilion, Ham Lane.**

BRAMDEAN
Fiona & Paul Barkley.

IVY HOUSE
Susan & Dom Beddard.

LINDEN COTTAGE
Philip & Anne Cutler,
01435 883149,
david.wright@ngs.org.uk.
Visits also by arrangement June to Sept for groups of 10 to 20. Combined adm with Mandalay.

LONGSTAFFES
Dorothy Bysouth.

MOUNT HOUSE
Richard & Lynda Maude-Roxby.

Six distinctive, characterful village gardens within easy walking distance, varying in design and style mostly behind centuries old period houses here in the heart of the High Weald AONB. All located in and around the picturesque High St in Burwash, home for so many yrs of Rudyard Kipling who lived at nearby Bateman's. Restricted wheelchair access to rear gardens as mainly terraced properties.

22 BUTLERS FARMHOUSE
Butlers Lane, Herstmonceux, BN27 1QH. Irene Eltringham-Willson, 01323 833770, irene.willson@btinternet.com, www.butlersfarmhouse.co.uk. *3m E of Hailsham. Take A271 from Hailsham, go through village of Herstmonceux, turn R signed Church Rd, then approx 1m turn R. Do not use SatNav!* **Sat 27, Sun 28 Mar, Tue 22, Wed 23 June (2-5). Adm £5, chd free. Sat 21, Sun 22 Aug (2-5). Adm £6, chd free. Home-made teas. Visits also by arrangement Mar to Oct. Refreshments on request.**

Lovely rural setting for 1 acre garden surrounding C16 farmhouse with views of South Downs. Pretty in spring with daffodils, hellebores and primroses. Come and see our meadow in June and perhaps spot an orchid or two. Quite a quirky garden with surprises round every corner including a rainbow border, small pond, Cornish inspired beach corners, a poison garden and secret jungle garden. Plants for sale. Picnics welcome. Live jazz in August. Most of garden accessible by wheelchair.

23 CAMBERLOT HALL
Camberlot Road, Lower Dicker, Hailsham, BN27 3RH. Nicky Kinghorn, 07710 566453, nickykinghorn@hotmail.com. *500yds S of A22 at Lower Dicker, 4½m N of A27 Drusillas r'about. From A27 Drusilla's r'about through Berwick Station to Upper Dicker & L into Camberlot Rd after The Plough pub, we are 1m on L. From A22 we are 500yds down Camberlot Rd on R.* **Sat 14, Sun 15 Aug (2-5). Adm £6, chd free. Home-made teas. Visits also by arrangement June to Sept for groups of 5 to 20.**
A 3 acre country garden with a lovely view across fields and hills to the South Downs. Created from scratch over the last 8 yrs with all design, planting and maintenance by the owner. Lavender lined carriage driveway, naturalistic border, vegetable garden, shady garden, 30 metre white border and exotic garden. New part-walled garden and summerhouse with new planting. Wheelchair access over gravel drive and some uneven ground.

24 51 CARLISLE ROAD
Eastbourne, BN21 4JR. E & N Fraser-Gausden. *200yds inland from seafront (Wish Tower), close to Congress Theatre.* **Sat 29, Sun 30 May (2-5). Adm £4, chd free. Home-made teas.**
Small walled, s-facing garden (82ft x 80ft) with mixed beds intersected by stone paths and incl small pool. Profuse and diverse planting. Wide selection of shrubs, old roses, herbaceous plants and perennials mingle with specimen trees and climbers. Constantly revised planting to maintain the magical and secluded atmosphere.

25 CAXTON MANOR
Wall Hill, Forest Row, RH18 5EG. Adele & Jules Speelman. *1m N of Forest Row, 2m S of East Grinstead. From A22 take turning to Ashurstwood, entrance on L after ⅓m, or 1m on R from N.* **Fri 21, Sat 22 May (2-5). Adm £5, chd free. Home-made teas. Donation to St Catherine's Hospice, Crawley.**
Delightful 5 acre Japanese inspired gardens planted with mature rhododendrons, azaleas and acers surrounding large pond with boathouse, massive rockery and waterfall, beneath the home of the late Sir Archibald McIndoe (house not open). Japanese tea house and Japanese style courtyard. **Also open 2 Quarry Cottages (separate admission).**

26 CHAMPS HILL
Waltham Park Road, Coldwaltham, Pulborough, RH20 1LY. Mrs Mary Bowerman, info@thebct.org.uk, www.thebct.org.uk. *3m S of Pulborough. On A29 turn R to Fittleworth into Waltham Park Rd, garden 400 metres on R.* **Sun 9 May, Sun 15 Aug (2-5). Adm £5, chd free. Tea. Visits also by arrangement Apr to Sept for groups of 10+.**
A natural landscape, the garden has been developed around three disused sand quarries, with far-reaching views across the Amberley Wildbrooks to the South Downs. A woodland walk in spring leads you past beautiful sculptures, against a backdrop of colourful rhododendrons and azaleas. In summer the garden is a colourful tapestry of heathers, which are renowned for their abundance and variety.

27 CHANNEL VIEW
52 Brook Barn Way, Goring-by-Sea, Worthing, BN12 4DW. Jennie & Trevor Rollings, 01903 242431, tjrollings@gmail.com. *1m W of Worthing, nr seafront. Turn S off A259 into Parklands Ave, L at T-junction into Alinora Cres. Brook Barn Way is immed on L.* **Visits by arrangement May to Sept for groups of 5 to 30. Adm £5, chd free. Home-made teas.**
A seaside Tudor cottage garden, cleverly blending the traditional, antipodean and subtropical with dense planting, secret rooms and intriguing sight-lines. Sunny patios,

insect friendly flowers and unusual structures supporting numerous roses, clematis and other climbers, as well as brick and flint paths radiating from a wildlife pond. Countless planted hanging baskets and containers, sinuous beds packed with flowers and foliage, with underplanting to ensure a 3D experience. Please visit www.ngs.org.uk for pop up openings. Partial wheelchair access.

28 CHANTERELLE

The Lane, Chichester, PO19 5PY. Mrs Julia Farwell, 07552 219243, jafarwell@icloud.com. *Turn off A286 into The Avenue, then turn R into The Lane. House on corner of The Avenue & The Lane.* **Visits by arrangement Apr to Oct for groups of 10 to 20. Adm £4.50, chd free. Light refreshments.**
Mature, well maintained town garden cleverly designed in different areas for yr-round interest. Developed by the present owner over a period of 16 yrs. Wrap around garden with ornamental shrubs and a wide variety of interesting plants. Perennial planting, rhododendrons, azaleas, actinidia, crinodendron, camellias, large magnolia tree and a water feature.

GROUP OPENING

29 NEW CLARENCE ROAD GARDENS

Horsham, RH13 5SJ. *From Cowfold A281 turn R onto B2180 Clarence Rd. Street parking free after 10am. 10 mins from station & 15 mins walk from town centre carparks.* **Sat 22, Sun 23 May (10-4). Combined adm £7.50, chd free. Home-made teas at 23 & 98 Clarence Road.**

NEW 10 CLARENCE ROAD
David Hide.

NEW 23 CLARENCE ROAD
Caron Gillet.

NEW 27 CLARENCE ROAD
Mrs Jacqueline Ovenden.

NEW 29 CLARENCE ROAD
Ellen & Andy Bateson.

NEW 41 CLARENCE ROAD
Alison Caldwell.

NEW 81 CLARENCE ROAD
Mrs Sarah Roberts.

NEW 98 CLARENCE ROAD
Dee & James Child.

7 small gardens hidden behind a street of Victorian houses (not open), all are very different in character. A wildlife garden, rose garden, plants person garden, formal garden as well as perennial cottage gardens. There are wildlife ponds a mulberry tree and specialist plants from New Zealand and the Mediterranean. Each is a unique treasure waiting to be discovered. Plant sales at 10 & 81 Clarence Road.

30 ◆ CLINTON LODGE

Fletching, TN22 3ST. Lady Collum, 01825 722952, garden@clintonlodge.com, www.clintonlodgegardens.co.uk. *4m NW of Uckfield. Clinton Lodge is situated in Fletching High St, N of Rose & Crown Pub. Off road parking provided. It is important visitors do not park in street. Parking available from 1pm. Gardens open 2pm.* **For NGS: Sun 25 Apr, Mon 7, Mon 21 June, Mon 2 Aug (2-5.30). Adm £6, chd free. Home-made teas. For other opening times and information, please phone, email or visit garden website. Donation to local charities.**
6 acre formal and romantic garden overlooking parkland with old roses, William Pye water feature, double white and blue herbaceous borders, yew hedges, pleached lime walks, Medieval style potager, vine and rose allée, wildflower garden, small knot garden and orchard. Caroline and Georgian house (not open).

Brightling Down Farm

© Suzie Gibbons

31 NEW CLOUD COTTAGE & HOLLYOAKS

Ivy Close, Ashington, Pulborough, RH20 3LW. Ian & Elizabeth Gregory, Bill & Pam Whittaker. *1st turning L on London Rd, off B2133 r'about. Cloud Cottage is at bottom of close on L.* **Sun 4 Apr, Sun 22 Aug (2-5). Adm £4, chd free. Home-made teas at Cloud Cottage.**

Cloud Cottage - At the front is a sycamore with shade loving plants. The rear garden is divided into a variety of areas. Hollyoaks - A small cottage style garden with specimen trees, heavily planted borders with shrubs, perennials, climbing roses, clematis and in late summer dahlias and fuchsias. A pond and a variety of containers surround the patio. At the front is a scree bed.

32 COOKSCROFT

Bookers Lane, Earnley, Chichester, PO20 7JG. Mr & Mrs J Williams, 01243 513671, williams. cookscroft330@btinternet.com, www.cookscroft.co.uk. *6m S of Chichester. At end of Birdham Straight A286 from Chichester, take L fork to East Wittering B2198. 1m before sharp bend, turn L into* *Bookers Lane, 2nd house on L. Parking available.* **Sat 8, Wed 19 May (11-4). Light refreshments. Evening opening Sat 19 June (5-9). Wine. Adm £5, chd free. Visits also by arrangement Apr to July for groups of up to 30.**

A garden for all seasons which delights the visitor. Started in 1988, it features cottage, woodland and Japanese style gardens, water features and borders of perennials with a particular emphasis on southern hemisphere plants. Unusual plants for the plantsman to enjoy, many grown from seed. The differing styles of the garden flow together making it easy to wander anywhere. Wheelchair access over grass paths and unfenced ponds.

33 COPYHOLD HOLLOW

Copyhold Lane, Borde Hill, Haywards Heath, RH16 1XU. Frances Druce, 01444 413265, frances.druce@yahoo.com. *2m N of Haywards Heath. Follow signs for Borde Hill Gardens. With Borde Hill Gardens on L over brow of hill, take 1st R signed Ardingly. Garden ½m. If the drive is full, please park in lane.* **Mon 3, Mon 31 May (11-4). Adm £4, chd free. Home-made teas.**

Visits also by arrangement Mar to June.

A different NGS experience in 2 north-facing acres. The cottage garden surrounding C16 house (not open) gives way to steep slopes up to woodland garden, a challenge to both visitor and gardener. Species primulas a particular interest of the owner. Stumpery. Not a manicured plot, but with a relaxed attitude to gardening, an inspiration to visitors.

GROUP OPENING

34 NEW CROWBOROUGH GARDENS

Myrtle Road, Crowborough, TN6 1EY. *Crowborough is 10m S of Tunbridge Wells on A26. Four gardens close to town centre, please park in public car parks. Bellemarie, Fermor Way is approx 1m drive from town centre. Leave town via Whitehill Rd & follow NGS signs.* **Thur 17 June (1-5). Combined adm £6, chd free. Home-made teas.**

NEW **BELLEMARIE**
Glenn & Allison Ford.

NEW **HOATH COTTAGE**
Frances Arrowsmith.

NEW **8 MILL LANE**
Brenda Smart.

NEW **MOLE END**
Pauline Bastick.

NEW **SWEET SPRINGS**
Janet Gamba.

Five gardens are opening for the first time, four of which are within easy walking distance of the town centre. Please use central car parks as parking near these gardens is difficult. Hoath Cottage is an informal garden dedicated to supporting wildlife. 8 Mill Lane is a compact garden with clematis walk and well stocked borders. Mole End is a garden for all seasons, with deep borders, vegetable and fruit garden and a beehive. Sweet Springs has an extensively planted large garden wrapped around the cottage (not open), with vastly different soil conditions throughout. Bellemarie is a small garden providing a colourful taste of the Caribbean, located 1m south of town centre; car required, on-street parking.

Sennicotts

© Judi Lion

35 NEW CUMBERLAND HOUSE

Cray's Lane, Thakeham, Pulborough, RH20 3ER. George & Jane Blunden. *At junction of Cray's Lane & The Street, nr the church.* **Wed 21, Sun 25 July (2-5). Combined adm with Thakeham Place Farm £7, chd free. Home-made teas at Thakeham Place Farm.**

A Georgian village house (not open), next to the C12 church with a beautiful, mature, ¾ acre English country garden comprising a walled garden laid out as a series of rooms with well-stocked flower beds, two rare Ginkgo trees and yew topiary, leading to an informal garden with vegetable, herb and fruit areas, pleached limes and a lawn shaded by a copper beech tree. Wheelchair access through gate at right-hand side of house.

&. ☕

36 CUPANI GARDEN

8 Sandgate Close, Seaford, BN25 3LL. Dr D Jones & Ms A Jones OBE, 01323 899452, sweetpeasa52@gmail.com, www.cupanigarden.com. *From A259 follow signs to Alfriston, E of Seaford. R off the Alfriston Rd onto Hillside Ave, 2nd L, 1st R & 1st R. Park in adjoining streets. Bus 12A Brighton/Eastbourne, get off Millberg Rd stop & walk down alley to the garden.* **Wed 21, Sun 25 Apr (1-4); Fri 11, Wed 16, Sun 20 June, Fri 2, Mon 19 July (1-5). Adm £5, chd free. Visits also by arrangement June to Aug for groups of 5 to 20. Various food options, call to discuss.**

Cupani is a small tranquil haven with a delightful mix of trees, shrubs and perennial borders in different themed beds. Courtyard garden, gazebo, summerhouse, water features, sweet pea obelisks, and a huge range of plants. See TripAdvisor reviews. The Garden has undergone a major renovation in 2020 and now incl a dry/gravel planting and some more tropical planting. Delicious afternoon tea and a good range of lunches. Vegetarian, vegan and gluten free options. See menu and to pre-book lunches on our website. Plants, jams, and china for sale. Not suitable for wheelchairs, steps to courtyard and steep steps to WC.

🐖 ✳ ☕

37 NEW D & S HAUS

41 Torton Hill Road, Arundel, BN18 9HF. Darrell Gale & Simon Rose. *1m SW of Arundel town square. From A27 Ford Rd/Chichester Rd r'about, take exit to Ford & immed turn R into Torton Hill Rd. Continue uphill & at large oak tree, keep L & we are on L going down the hill.* **Sun 25 July (12-5). Adm £4.50, chd free. Home-made teas.**

A lush suburban garden, 25ft x 200ft, which the owners have transformed over the yrs. Both front and rear gardens contain a mass of palms, bananas, bamboos and all manner of spiky and large luxuriant foliage. Rules are not followed, as the delights of colour, shape and texture have driven its design, with desert plants next to bog plants, a pond, stream and raised flint bed. Tropical planting with lots of unusual plants, quirky sculptures and features. No wheelchair access as many steps, narrow areas and gravel.

✳ ☕

38 DALE PARK HOUSE

Madehurst, Arundel, BN18 0NP. Robert & Jane Green, 01243 814260, robertgreenfarming@gmail.com. *4m W of Arundel. Take A27 E from Chichester or W from Arundel, then A29 (London) for 2m, turn L to Madehurst & follow red arrows.* **Visits by arrangement June & July for groups of 10+. Adm £4.50, chd free. Home-made teas.**

Set in parkland, enjoying magnificent views to the sea. Come and relax in the large walled garden which features an impressive 200ft herbaceous border. There is also a sunken gravel garden, mixed borders, a small rose garden, dreamy rose and clematis arches, an interesting collection of hostas, foliage plants and shrubs, and an orchard and kitchen garden. Wheelchair access not easy.

&. ✳ 🚗 ☕

The National Garden Scheme searches the length and breadth of England and Wales for the very best private gardens

39 70 DALE VIEW

Hove, BN3 8LB. Chris & Tony Ashby-Steed. *1m from Hove junction of A27. From Hove junction of A27, at r'about take A2038 towards Hove. Continue down to mini-r'about by Grenadier Pub & take 2nd exit into West Way. Dale View is 1st L after zebra crossing.* **Sat 3, Sun 4 July (11-5). Adm £5, chd free. Home-made teas.**

Star of C4's Gogglebox for 5 yrs, Chris Ashby-Steed and husband Tony have completely relandscaped their garden. It shows lush and full planting across several zones and seating areas surrounded by mature trees. Highlights incl herbaceous borders, fountain, rose garden, shady garden and terraced raised beds, with interest from spring through to late autumn. Large, homegrown plant sale in front garden.

🐖 ✳ ☕

GROUP OPENING

40 NEW DEAKS LANE GARDENS

Deaks Lane, Ansty, Haywards Heath, RH17 5AS. *3m W of Haywards Heath on A272. 1m E of A23. At r'about at junction of A272 & B2036 take exit A272 Bolney, then take immed R onto Deaks Lane. Park at Ansty Village Centre on R, where tickets for gardens are purchased.* **Sun 13 June (12-4). Combined adm £6, chd free. Light refreshments at Ansty Village Centre.**

NEW **THE BARN HOUSE**
Michael & Jackie Dykes.

NEW **BUTTERFLY HOUSE**
Cindy & Roger Edmonston.

3 LAVENDER COTTAGES
Dorry Baillieux.

NEW **NUTBOURNE**
David Miller.

NEW **THICKETS**
Becky & Matt Morgan.

Four of the gardens are close together on Deaks Lane, each bringing different designs and feel, but all united in attracting bees and wildlife. A short walk brings you to The Barn House with its formal walled garden, informal planting and fine views. Plenty of flowers to see and enjoy.

✳ ☕

41 ◆ DENMANS GARDEN

Denmans Lane, Fontwell, BN18 0SU.
Gwendolyn van Paasschen,
01243 278950, office@denmans.org,
www.denmans.org. *5m from
Chichester & Arundel. Off A27, ½ m
W of Fontwell r'about.* For NGS:
Sat 13 Feb, Sat 24 Apr, Sat 30
Oct (10-4). Adm £9, chd £7. Light
refreshments. For other opening
times and information, please phone,
email or visit garden website.
Founded by Joyce Robinson, a
pioneer in gravel gardens and
redesigned by influential landscape
designer, John Brookes MBE,
Denmans is a 4 acre garden
renowned for its curvilinear layout,
unusual plantings and yr-round
beauty. It features gravel gardens, dry
river beds, a walled garden, ponds, a
conservatory, architectural plantings
and subtropical plants. Gift shop,
plant centre with unusual plants.

42 47 DENMANS LANE

Lindfield, Haywards Heath,
RH16 2JN. Sue & Jim
Stockwell, 01444 459363,
jamesastockwell@aol.com, www.
lindfield-gardens.co.uk/47denmans-
lane. *Approx 1½ m NE of Haywards
Heath town centre. From Haywards
Heath train station follow B2028
signed Lindfield & Ardingly for 1m.
At T-lights turn L into Hickmans
Lane, then after 100 metres take
1st R into Denmans Lane.* Visits
by arrangement Mar to Sept for
groups of 5+. Adm £8, chd free.
Home-made teas included. Call to
discuss wine & canapé options.
Visits can be combined with
Lindfield Jungle.
This beautiful and tranquil 1 acre
garden was described by Sussex
Life as a 'Garden Where Plants Star'.
Created by the owners, Sue & Jim
Stockwell, over the past 20 yrs, it is
planted for interest throughout the yr.
Spring bulbs are followed by azaleas,
rhododendrons, roses and herbaceous
perennials. The garden also has
ponds, vegetable and fruit gardens.
Most of the garden accessible by
wheelchair, but some steep slopes.

43 DITTONS END

Southfields Road, Eastbourne,
BN21 1BZ. Mrs Frances
Hodkinson, 01323 647163,
franceshodkinson@yahoo.co.uk.
*Town centre, ⅓ m from train station.
Off A259 in Southfields Rd. House*
*directly opp Dittons Rd. 3 doors from
NGS open garden Hardwycke.* Sun
30 May (11-4). Combined adm
with Hardwycke £5, chd free.
Home-made teas at Hardwycke.
Visits also by arrangement Apr to
Sept for groups of 5 to 10.
Lovely well maintained, small town
garden. At the back, a very pretty
garden (35ft x 20ft) with small lawn
area, patio surrounded by a selection
of pots and packed borders with lots
of colour. In the front a compact lawn
with colourful borders (25ft x 18ft).

44 DOWN PLACE

South Harting, Petersfield,
GU31 5PN. Mrs David
Thistleton-Smith, 01730 825374,
selina@downplace.co.uk. *1m SE of
South Harting. B2141 to Chichester,
turn L down unmarked lane below
top of hill.* Sat 27, Sun 28 Mar, Sun
20, Mon 21 June (2-5.30). Adm £5,
chd free. Home-made & cream
teas. Visits also by arrangement
Apr to July for groups of 10+.
Donation to The Friends of Harting
Church.
Set on the South Downs with
panoramic views out to the undulating
wooded countryside. A garden
which merges seamlessly into its
surrounding landscape with rose and
herbaceous borders that have been
moulded into the sloping ground.
There is a well stocked vegetable
garden and walks shaded by beech
trees which surround the natural
wild meadow in which various native
orchids flourish. Substantial top
terrace and borders accessible to
wheelchairs.

45 DRIFTWOOD

4 Marine Drive, Bishopstone,
Seaford, BN25 2RS.
Geoff Stonebanks & Mark
Glassman, 01323 899296,
geoffstonebanks@gmail.com,
www.driftwoodbysea.co.uk. *A259
between Seaford & Newhaven. Turn
L into Marine Drive from Bishopstone
Rd, 2nd on R. Only park same
side as house please, not on bend
beyond the drive.* Tue 29 June,
Tue 13 July, Sun 1 Aug (11-5.30).
Adm £6, chd free. Pre-booking
essential, please visit www.ngs.
org.uk for information & booking.
Home-made teas. Visits also
by arrangement June to Aug for
groups of up to 20.
Monty Don introduced Geoff's garden
on BBC Gardeners' World saying,
a small garden by the sea, full of
character, with inspired planting and
design. Francine Raymond wrote
in her Sunday Telegraph feature,
Geoff's enthusiasm is catching, he
and his amazing garden deserve
every visitor that makes their way
up his enchanting garden path.
Over 100 5-star TripAdvisor visitors
comments, 3 successive Certificates
of Excellence and 2020 Travellers
Choice Award. Selection of Geoff's
home-made cakes available, all
served on vintage china, on trays, in
the garden. Access up steep drive,
narrow paths and many levels with
steps. Help readily available on-site or
call ahead before visit.

46 DURFORD ABBEY BARN

Petersfield, GU31 5AU. Mr & Mrs
Lund. *3m from Petersfield. Situated
on the S side of A272 between
Petersfield & Rogate, 1m from the
junction with B2072. Limited parking
by house.* Sat 5, Sun 6 June (1.30-
5.30). Adm £4, chd free. Home-
made teas.
A 1 acre plot with areas styled with
prairie, cottage and shady borders and
set in the South Downs National Park
with downs views. Plants and greeting
cards for sale. Partial wheelchair
access as some areas have quite
steep grass slopes to negotiate.

47 DURRANCE MANOR

Smithers Hill Lane, Shipley,
RH13 8PE. Gordon
Lindsay, 01403 741577,
galindsay@gmail.com. *7m SW
of Horsham. A24 to A272 (S from
Horsham, N from Worthing), turn W
towards Billingshurst. Approx 1¾ m,
2nd L Smithers Hill Lane signed to
Countryman Pub. Garden 2nd on L.*
Mon 30 Aug (12-6). Adm £6, chd
free. Home-made teas. Visits also
by arrangement Apr to Oct.
This 2 acre garden surrounding
a Medieval hall house (not open)
with Horsham stone roof, enjoys
uninterrupted views over a ha-ha of
the South Downs and Chanctonbury
Ring. There are many different
gardens here, Japanese inspired
gardens, a large pond, wildflower
meadow and orchard, colourful long
borders, hosta walk and vegetable
garden. There is also a Monet style
bridge over a pond with waterlilies.

GROUP OPENING

48 EAST GRINSTEAD GARDENS

East Grinstead, RH19 4DD.
7m E of Crawley on A264 & 14m N of Uckfield on A22. Tickets & maps at each garden. Imberhorne Lane Public Car Park 75yds from Imberhorne Allotments, disabled parking on-site. **Sun 13 June (1-5); Sun 25 July (12-5). Combined adm £5, chd free. Home-made teas at 5 Nightingale Close (13 June) & 16 Musgrave Avenue (25 July).**

NEW IMBERHORNE ALLOTMENTS

Imberhorne Allotment Association, www.imberhorneallotments.org.
Open on Sun 25 July

27 MILL WAY

Jeff Dyson.
Open on Sun 25 July

16 MUSGRAVE AVENUE

Carole & Bob Farmer.
Open on Sun 25 July

NEW 5 NIGHTINGALE CLOSE

Carole & Terry Heather.
Open on all dates

7 NIGHTINGALE CLOSE

Gail & Andy Peel.
Open on Sun 13 June

Gardens to lift the spirits and make you smile! Established gardens, displaying a mix of planting incl shrubs, perennials, dahlias and annuals, tubs and baskets. Two are past winners of East Grinstead in Bloom, Best Front Garden and the back gardens all have quite different styles. 16 Musgrave Avenue has a thriving vegetable garden. 7 Nightingale Close is on heavy clay in a frost pocket going down to a stream, plus a potager and collection of bonsai. Imberhorne Allotments consist of 80 plots with a diverse mix of planting incl grape vines, fruit, flowers and a community orchard. The enthusiastic owners are keen propagators, growing from seed, cuttings and plugs, sharing their surplus plants. Plenty of advice available and plants for sale at several gardens. The Town Council hanging baskets and planting in the High St are not to be missed and achieved a Gold Medal in 2019 from South and South East in Bloom. For steam train fans, the Bluebell Railway starts nearby. Partial wheelchair access at some gardens.

✿ ☕

Fairlight End

© Marianne Majerus

The Moongate Garden

GROUP OPENING

49 NEW EASTERGATE HOUSE GARDENS

Church Lane, Eastergate, Chichester, PO20 3UT. *Within the village of Eastergate, Church Lane can be accessed from the A29, turn at PO or from Barnham Rd. Parking is opp Eastergate House.* Sat 5 June (11-8); Sun 6 June (11-5). Combined adm £6, chd free. Home-made teas (11-5), plus music, wine, champagne & savouries (Sat only, 5-8) at Eastergate House.

NEW BRAMBLE COTTAGE
Joyce Jones.

NEW EASTERGATE HOUSE
Valerie Carter, 07545 633967, valeriesinbox@gmail.com. Visits also by arrangement June & July for groups of up to 20.

NEW THE OLD STABLES
Helen Byrne.

Three very different gardens are found within the old stone walls of Eastergate House. Eastergate House itself has much history and stories to tell. The garden retains many of the original features incl the rose garden, pond and pergola, along with wide sunny lawns ideal for an active, sporty family. Bees are busy among the flowers and hives producing the families honey. The Old Stables show what can be achieved by a wheelchair gardener, introducing smooth straight lines, wide paved areas and a plethora of colourfully planted pots, joyfully mixing vegetables among the flowers, set against the backdrop of a traditional structured garden. Bramble Cottage has a natural organic feel designed to attract wildlife. Central are the restored pools cascading into a larger pond crossed by stepping stones. Grassy pathways lead through trees and woodland planting with several quiet spots to sit and observe. Birds, bees and butterflies abound and newts, grass snake and hedgehogs can be spotted. Plants, art, crafts and books for sale. Partial wheelchair access.

✿ ☕

50 54 ELMLEIGH
Midhurst, GU29 9HA. Wendy Liddle, 07796 562275, wendyliddle@btconnect.com. *¼m W of Midhurst off A272. Reserved disabled parking at top of drive, please phone on arrival for assistance. Other visitors off road parking on marked grass area.* Sats & Suns 22, 23 May; 12, 13 June; 3, 4, 24, 25 July; 14, 15 Aug; 4, 5 Sept (10-5). Adm £4, chd free. Cream teas. Visits also by arrangement May to Sept for groups of 10 to 30. Donation to Chestnut Tree House, The King's Arms & Raynaud's Association.
⅕ acre property with terraced front garden, leading to a heavily planted rear garden with majestic 100 yr old Black Pines. Shrubs, perennials, packed with interest around every corner, providing all season colours. Many raised beds, numerous sculptures, vegetables in boxes, a greenhouse, pond, and hedgehogs in residence. A large collection of tree lilies, growing 8-10ft. Child friendly. Picnics welcome. Come and enjoy the peace and tranquility in this award-winning garden, our little bit of heaven. Not suitable for electric buggies.

♿ 🐐 ✿ ☕ 🛋

51 NEW 29 FAIRLIGHT AVENUE

Hastings, TN35 5HS. Peggy Harvey & Barbara Martin. *From Hastings seafront take A259 (Old London Rd) to Ore Village. Turn R into Fairlight Rd at Co-op store by T-lights, continue along road for 300 metres, 1st R into Fairlight Ave. Roadside parking.* Sun 18 July (11-5). Adm £3. Home-made teas.

A stunning and interesting low maintenance tropical garden newly created by the owners. Island planting with palms, yuccas and other blade shape leaf plants and shingle walkways throughout. In contrast, a smaller wildlife garden to the rear with pond and varied planting for birds and insects. The front garden is designed as a seascape with slate river run and seaside planting.

52 FAIRLIGHT END

Pett Road, Pett, Hastings, TN35 4HB. Chris & Robin Hutt, 07774 863750, chrishutt@fairlightend.co.uk, www.fairlightend.co.uk. *4m E of Hastings. From Hastings take A259 to Rye. At White Hart Beefeater turn R into Friars Hill. Descend into Pett village. Park in village hall car park, opp house.* Sun 13 June (11-5). Adm £6, chd free. Home-made teas. Also open Breanross. Visits also by arrangement May to Aug for groups of 10+. Donation to Pett Village Hall.

Gardens Illustrated said 'The 18th century house is at the highest point in the garden with views down the slope over abundant borders and velvety lawns that are punctuated by clusters of specimen trees and shrubs. Beyond and below are the wildflower meadows and the ponds with a backdrop of the gloriously unspoilt Wealden landscape' Wheelchair access with steep paths, gravelled areas and unfenced ponds.

53 FINDON PLACE

Findon, Worthing, BN14 0RF. Miss Caroline Hill, www.findonplace.com. *Directly off A24 N of Worthing. Follow signs to Findon Parish Church & park through the green gates, 1st on LH-side.* Sun 11 Apr, Sun 13 June (1-6). Adm £5, chd under 12 free. Home-made teas.

Stunning grounds and gardens surrounding a Grade II listed Georgian country house (not open), nestled at the foot of the South Downs. The most glorious setting for a tapestry of perennial borders set off by Sussex flint walls. The many charms incl a yew allee, cloud pruned trees, espaliered fruit trees, a productive ornamental kitchen garden, rose arbours and arches, and a cutting garden.

54 FITTLEWORTH HOUSE

Bedham Lane, Fittleworth, Pulborough, RH20 1JH. Edward & Isabel Braham, 01798 865074, marksaunders66.com@gmail.com, www.racingandgreen.com. *2m E, SE of Petworth. Midway between Petworth & Pulborough on the A283 in Fittleworth, turn into lane by sharp bend signed Bedham. Garden is 50yds along on the L.* Every Wed 21 Apr to 12 May (2-5). Weds 9, 16 June; 14, 21 July; 11 Aug (2-5). Adm £5, chd free. Visits also by arrangement Apr to Aug for groups of 10+.

3 acre tranquil, romantic, country garden and walled kitchen garden growing a wide range of fruit, vegetables and flowers. Large glasshouse and old potting shed, mixed flower borders, roses, rhododendrons and lawns. Magnificent 115ft tall Cedar overlooks wisteria covered Grade II listed Georgian house (not open). Wild garden and pond, new stream and rock garden. The garden sits on a gentle slope, but is accessible for wheelchairs and buggies. Non-disabled WC.

55 FIVE OAKS COTTAGE

Petworth, RH20 1HD. Jean & Steve Jackman, 07939 272443, jeanjackman@hotmail.com. *5m S of Pulborough. SatNav does not work! To ensure best route, we will provide printed directions when you book.* Visits by arrangement second week in July only for groups of up to 30. Adm £5, chd free.

An acre of delicate jungle surrounding an Arts and Crafts style cottage (not open), with stunning views of the South Downs. Our unconventional garden is designed to encourage maximum wildlife, with a knapweed and hogweed meadow on clay attracting clouds of butterflies in July, plus two small ponds and lots of seating. An award-winning, organic garden with a magical atmosphere. Home-made metal plant supports for sale. Please contact us now by phone, email or message us on Instagram @floralfringefair to arrange!

56 FOLLERS MANOR

White Way, Alfriston, BN26 5TT. Geoff & Anne Shaw, www.follersmanor.co.uk. *½m S of Alfriston. From Alfriston uphill towards Seaford. Park on L in paddock before garden. Garden next door to old Alfriston Youth Hostel, immed before road narrows. Visitors with walking difficulties drop at gate.* Sat 3, Sun 4 July (11-4). Adm £8, chd free. Home-made teas. Donation to Children with Cancer Fund, Polegate & WRAS (Local Wildlife Rescue).

Contemporary garden designed by Ian Kitson attached to C17 listed historic farmhouse. Entrance courtyard, sunken garden, herbaceous displays, wildlife pond, wildflower meadows, woodland area and beautiful views of the South Downs. Winner of Sussex Heritage Trust Award and three awards from the Society of Garden Designers; Best Medium Residential Garden, Hard Landscaping and, most prestigious, the Judges Award. A newly designed area of the garden by original designer Ian Kitson completed 2019. No dogs please.

57 THE FOLLY

Charlton, Chichester, PO18 0HU. Joan Burnett & David Ward, 07740 273603, joankcirburnett@gmail.com, www.thefollycharlton.com. *7m N of Chichester & S of Midhurst off A286 at Singleton, follow signs to Charlton. Follow NGS parking signs. No parking in lane, drop off only. Parking near pub 'Fox Goes Free'.* Sun 18 July, Sun 15 Aug (1-5). Adm £4, chd free. Home-made teas. Visits also by arrangement June to Sept for groups of 10 to 30.

Colourful cottage garden surrounding a C16 period house (not open), set in pretty downland village of Charlton, close to Levin Down Nature Reserve. Herbaceous borders well stocked with a wide range of plants. Variety of perennials, grasses, annuals and shrubs to provide long season of colour and interest. Old well. Busy bees. Partial wheelchair access with steps from patio to lawn. No dogs.

58 NEW FOREST RIDGE
Paddockhurst Lane, Balcombe, RH17 6QZ. Philip & Rosie Wiltshire, 07900 621838, rosiem.wiltshire@btinternet.com. *3m from M23, J10a. M23 J10a take B2036 to Balcombe. After ⅔m take 1st L onto B2110. After Worth School turn R into Back Lane. After 2m it becomes Paddockhurst Lane. Forest Ridge on R after 2¼m. Ignore SatNav to track!* Sat 8, Sun 9 May (2-5.30). Adm £5, chd free. Home-made teas.
A charming 4½ acre Victorian garden with far-reaching views, boasting the oldest Atlantic Cedar in Sussex. The owners themselves are currently undertaking a major restoration: felling, planting and redesigning areas. Within the garden there is formal and informal planting, woodland dell and mini arboretum. Azaleas, rhododendrons and camellias abound, rare and unusual species. A garden to explore!

59 FOXGLOVE COTTAGE
29 Orchard Road, Horsham, RH13 5NF. Peter & Terri Lefevre, 01403 256002, teresalefevre@outlook.com. *From Horsham station, over bridge, at r'about take 3rd exit (signed Crawley),1st R Stirling Way, at end turn L, 1st R Orchard Rd. From A281, take Clarence Rd, at end turn R, at end turn L. Street parking.* Sun 30 May, Wed 7, Sun 11 July (1-5). Adm £4.50, chd free. Home-made teas. Vegan, gluten & dairy free cake. Visits also by arrangement May to July for groups of 10+.
A ¼ acre plantaholic's garden, full of containers, vintage finds and quirky elements. Paths intersect both sun and shady borders bursting with colourful planting and salvias in abundance! A beach inspired summerhouse and deck are flanked by a water feature in a pebble circle. 2 small ponds encourage wildlife. The end of the garden is dedicated to plant nursery, cut flowers and fruit growing. Plenty of seating in both sun and shade throughout the garden. Member of the Hardy Plant Society. A large selection of unusual plants for sale.

60 THE GARDEN HOUSE
5 Warleigh Road, Brighton, BN1 4NT. Bridgette Saunders & Graham Lee, 07729 037182, contact@ gardenhousebrighton.co.uk, www.gardenhousebrighton.co.uk. *1½m N of Brighton pier, Garden House is 1st L after Xrds, past the open market. Paid street parking. London Road station nearby. Buses 26 & 46 stopping at Bromley Rd.* Sat 10, Sun 11 Apr, Fri 18 June (11.30-4.30). Adm £5.50, chd free. Home-made teas. Visits also by arrangement Apr to Sept for groups of 10 to 30.
One of Brighton's secret gardens. We aim to provide yr-round interest with trees, shrubs, herbaceous borders and annuals, fruit and vegetables, two glasshouses, a pond and rockery. A friendly garden, always changing with a touch of magic to delight visitors, above all it is a slice of the country in the midst of a bustling city. Garden produce and plants for sale.

61 GARDEN HOUSE, 49 GUILLARDS OAK
Midhurst, GU29 9JZ. Patty & David Christie, 01730 813323, pattychristie49@gmail.com. *Leave Midhurst on the A272 towards Petersfield. Guillards Oak is the wide opening on L, halfway up the hill before pelican crossing. Entrance to garden through green gate.* Sun 8 Aug (2-6). Adm £5, chd free. Light refreshments. Visits also by arrangement May to Aug for groups of 5 to 20.
This ¼ acre garden was changed in 2012 from an expanse of lawn under trees to a parterre with clematis, roses and trained fruit on arches to become the view from the s-facing house. We have a busy greenhouse, grand design style fruit cage, wendy house and bug palace, connected by a small woodland walk under a large Swamp Cypress, with Banksia rose and Trachelospernum climbing up the house. Refreshments will be served on the terrace overlooking the garden.

62 NEW GORSELANDS
Common Hill, West Chiltington, Pulborough, RH20 2NL. Philip Maillou. *12m N of Worthing, approx 1m N of Storrington. Take either School Hill or Old Mill Drive from Storrington. 2nd L on to Fryern Rd to West Chiltington & continue on to Common Hill. Gorselands is on the R. Roadside parking.* Sun 27 June, Sun 4 July (2-5). Adm £4, chd free. Home-made teas.
Approx ¾ acre garden featuring mature mixed borders, dahlia garden, fruit area, woodland area with giant Redwood, camellias, azaleas, rhododendrons and two ponds. There will be a small selection of oil paintings for sale by Oxfordshire artist Jackie Hughes. Wheelchair access on sloping garden with short grass.

63 ◆ GREAT DIXTER HOUSE, GARDENS & NURSERIES
Northiam, TN31 6PH. Great Dixter Charitable Trust, 01797 253107, groupbookings@greatdixter.co.uk, www.greatdixter.co.uk. *8m N of Rye. Off A28 in Northiam, follow brown signs.* For NGS: Tue 20 Apr (11-5). Adm £13.50, chd £3. Light refreshments. For other opening times and information, please phone, email or visit garden website.
Designed by Edwin Lutyens and Nathaniel Lloyd. Christopher Lloyd made the garden one of the most experimental and constantly changing gardens of our time, a tradition now being carried on by Fergus Garrett. Clipped topiary, wildflower meadows, the famous long border, pot displays, exotic garden and more. Spring bulb displays are of particular note. Regional Finalist, The English Garden's The Nation's Favourite Gardens 2019. Please see garden website for accessibility information.

64 GROVELANDS
Wineham Lane, Wineham, Henfield, BN5 9AW. Mrs Amanda Houston. *8m SW Haywards Heath. From Haywards Heath A272 W approx 6m, then L into Wineham Lane. House 1¾m on L after Royal Oak Pub. 3m NE Henfield N on A281, R onto B2116 Wheatsheaf Rd, L into Wineham Lane. House ½m on R.* Sat 29, Sun 30 May, Sat 4, Sun 5 Sept (10.30-3.30). Adm £5, chd free. Home-made teas.
A South Downs view welcomes you to this rural garden set in over an acre in the hamlet of Wineham. Created and developed by local landscape designer Sue McLaughlin and the owners, it is designed to delight throughout the seasons. Features incl mixed borders, mature shrubs and an orchard. A vegetable garden with greenhouse, cutting flower area and pond hide behind a tall clipped hornbeam hedge.

65 THE HAMBLIN CENTRE

Main Road, Bosham,
Chichester, PO18 8PJ. The
Hamblin Trust, 01243 572109,
office@thehamblinvision.org.uk,
www.hamblincentre.org.uk. *3m W
of Chichester. The Hamblin Centre is
on the R just after Bosham r'about
going W on A259, just before the
r'about on the L going E. 10 mins
walk from Bosham station. 700
Stagecoach bus stop is 2 mins walk.*
Sun 21 Mar, Sun 23 May, Sun 15
Aug (11-4.30). Adm £4.50, chd
free. Home-made teas.
An ornamental and wildlife garden in
harmony with the Hamblin Centre's
work towards health and wellbeing.
Shrubs, perennials, mature specimen
and fruit trees; including tulip tree
and mulberry, a wildlife pond with
bog garden, a wildflower meadow
and a rock garden. Many varieties
of orchids. Boundary areas are left
natural, providing a range of wildlife
habitats. Guided garden tours and
also the opportunity to learn about the
amazing bio-diversity. Plant nursery.
Wheelchair access to most areas.

♿ ✱ 🛁 ☕

66 HAMMERWOOD HOUSE

Iping, Midhurst, GU29 0PF. Mr &
Mrs M Lakin. *3m W of Midhurst.
Take A272 from Midhurst, approx
2m outside Midhurst turn R for Iping.
From A3 leave for Liphook, follow
B2070, turn L for Milland & Iping.*
Sun 4 Apr, Sun 9 May (1.30-5).
Adm £5, chd free. Home-made
teas. Donation to Iping Church.
Large s-facing garden with lots
of mature shrubs incl camellias,
rhododendrons and azaleas. An
arboretum with a variety of flowering
and fruit trees. The old yew and
beech hedges give a certain amount
of formality to this traditional English
garden. Tea on the terrace is a must
with the most beautiful view of the
South Downs. For the more energetic
there is a woodland walk. Partial
wheelchair access as garden is set
on a slope.

♿ 🐕 ✱ ☕

67 HARBOURSIDE

Prinsted Lane, Prinsted,
Southbourne, PO10 8HS.
Ann Moss, 01243 370048,
ann.moss@ngs.org.uk. *6m W of
Chichester, 1m E of Emsworth.
Turn L into Prinsted Lane off A259.
Chinese take away on corner.
Follow lane until forced to the R
past the Scout Hut car park. House*
*is adjacent. Old boat & buoy with
house name in front garden. Visits
by arrangement for groups of up
to 30. Adm £4.50, chd free. Light
refreshments.*
Award-winning coastal garden takes
you on a journey through garden
styles from around the world. Visit the
Mediterranean, France, Holland, New
Zealand and Japan. Enjoy tree ferns,
topiary, shady area, secret woodland
parlour, potager, containers, unusual
shrubs and plants, silver birch walk,
herbaceous borders, art, crafts,
wildlife, seaside garden, blue and
white area. A garden with yr-round
colour and interest. Guided and
explanatory tour of garden. Piped
music and welcoming fire. Home-
made refreshment options incl, lunch,
supper, fizz and canapés. Plants and
gardening items for sale. Wheelchair
access to most of the garden, after
10ft of gravel at entrance.

♿ ✱ ☕ 🚉

68 HARDWYCKE

Southfields Road,
Eastbourne, BN21 1BZ.
Lois Machin, 01323 729391,
loisandpeter@yahoo.co.uk. *Centre
of Eastbourne, Upperton. A259
towards Eastbourne, Southfields Rd
on R just before junction with A2270
(Upperton Rd). Limited parking,
public car park (pay) in Southfields
Rd.* Sun 30 May (11-4). Combined
adm with Dittons End £5, chd
free. Visits also by arrangement
Apr to Sept for groups of 5 to 10.
Delightful s-facing town garden
mainly of chalky soil, with many usual
and unusual plants and small new
summerhouse. Smart front garden
with sunken patio. Wide selection of
shrubs including 50 types of clematis.
Square garden at rear 70ft x 50ft.
Wheelchair access with care, two
slight steps to rear garden.

♿ 🐕 ✱ ☕

Lavender Cottage, Seaford Gardens

1 Pest Cottage

© Judi Lion

GROUP OPENING

69 HARLANDS GARDENS

Penland Road, Haywards Heath, RH16 1PH. *Follow yellow signs from Balcombe Rd or Milton Rd & Bannister Way (Sainsbury's). Bus stop on Bannister Way (Route 30, 31, 39 & 80). Gardens within 5 min easy walk from train station & bus stop.* **Sat 15, Sun 16 May (12-5). Combined adm £5, chd free. Home-made teas.**

52 PENLAND ROAD
Karen & Marcel van den Dolder.

55 PENLAND ROAD
Steve & Lisa Williams.

NEW **72 PENLAND ROAD**
Sarah Gray & Graham Delve.

5 SUGWORTH CLOSE
Lucy & Brian McCully.

27 TURNERS MILL ROAD
Sam & Derek Swanson.

We are thrilled to announce a new and additional garden into our trail.

Our eclectic mix of town gardens continue to inspire and delight our visitors. We comprise awkward shapes and differing gradients, our gardens are designed with relaxation and socialisation in mind. We offer rich cottage-style planting schemes, varied ponds, pretty courtyards, practical kitchen gardens and habitats for wildlife. Look out for interesting pots and carnivorous plants. Please check our Facebook page 'Harlands Gardens' for updated information. Our home-made cakes are always a joy including an outstanding contribution from a 'Bake-off' contestant. Roadside parking is readily available. Park once to visit all gardens. We hope Rapkyns Nursery will again be selling their beautiful plants. The quaint English villages of Cuckfield and Lindfield are worthy of a visit. Nymans (NT) and Wakehurst (Kew) are also nearby. Partial wheelchair access to most gardens, no access to No 72.

♿ ✿ ☕

GROUP OPENING

70 HASSOCKS & DITCHLING GARDEN TRAIL

Hassocks, BN6 8EY. *6m N of Brighton, off A273. Gardens are between Stonepound Xrds & the centre of Ditchling, N & S of B2116. Park on road or in free car parks in Hassocks. No parking in Parklands Rd. Use car park in Ditchling or on road.* **Sun 6 June (1-5). Combined adm £6, chd free. Home-made teas at 26 West St, Ditchling.**

13 CHANCELLORS PARK
Steve Richards & Pierre Voegeli.

PARKLANDS ROAD ALLOTMENTS
Tony Copeland & Jeannie Brooker.

NEW **26 WEST STREET**
Katja Garrood & Giles Palmer.

The two gardens on this trail are 1m apart and quite different, but each has a wide range of unusual plants and shrubs. One is well established, the other looks established, but is relatively new with an unfolding design that is organic, bee friendly and has borrowed views. Both are excellent examples of use of space to incl seating places, winding paths and wonderful selections of plants. The 55 allotments are not to be missed, a range of classic vegetables, plus some exotic ones are grown. The allotmenteers will show you their plots and answer questions. There are spectacular views across the Downs and up to Jack and Jill Windmills, and an ancient woodland lies along the north side of the allotments. One of the few remaining chalk streams, the Herring Stream, flows through the village, where flash floods have occurred. Whilst on the trail, you can visit the rain gardens in Adastra Park, BN6 8QH and find out more about them, WC available. Dogs permitted at allotments only.

✿ ☕

GROUP OPENING

71 HELLINGLY PARISH TRAIL

Brook Cottage is 500 metres from Priors Grange. Parking field opp Brook Cottage, otherwise park on the roads & lanes around church. Follow yellow signs. Bus 51 from Eastbourne. **Sun 11 July (12-5). Combined adm £5, chd free. Ploughman's lunches & afternoon tea at Brook Cottage.**

BROADVIEW
BN27 4EX. Gill Riches.

BROOK COTTAGE
BN27 4HD. Dr Colin Tourle MBE & Mrs Jane Tourle.

NEW MAY HOUSE
BN27 4FA. Lynda & David Stewart.

NEW NUTHATCH
Stone Cross, BN24 5EU. Annie Reynolds.

PRIORS GRANGE
BN27 4EZ. Sylvia Stephens.

Three gardens will open in the small village of Hellingly (Broadview, Brook Cottage and May House) and one garden will open in Stone Cross (Nuthatch). In addition a fifth garden (Priors Grange) is open for plant sales only. Brook Cottage set amongst trees and water is the refreshment venue for the trail. May House and Nuthatch are both new gardens with very different styles of planting and size. The Cuckoo Cycle Trail is nearby.

72 ◆ HERSTMONCEUX CASTLE GARDENS AND GROUNDS
Herstmonceux, BN27 1RN. Bader International Study Centre, Queen's University (Canada), 01323 833816, c_harber@bisc.queensu.ac.uk, www.herstmonceux-castle.com. *Located between Herstmonceux & Pevensey on Wartling Rd. From Herstmonceux take A271 to Bexhill, 2nd R signed Castle. Do not use SatNav.* **For NGS: Wed 28 July (10-6). Adm £7, chd free. Light refreshments. For other opening times and information, please phone, email or visit garden website.**
Herstmonceux is renowned for its magnificent moated castle set in beautiful parkland and superb Elizabethan walled gardens, leading to delightful discoveries such as our Shakespeare, rose and herb gardens and onto our woodland trails. Take a slow stroll past the lily covered lakes to the 1930s folly and admire the sheer magnificence of the castle. The gardens and grounds first opened for the NGS in 1927. Avenue of ancient sweet chestnut trees. Partial wheelchair access to formal gardens.

GROUP OPENING

73 HERSTMONCEUX PARISH TRAIL
4m NE of Hailsham. Tickets & map available at any garden. Please note this is not a walking trail. **Sun 27 June (12-5). Combined adm £6, chd free. Teas & BBQ lunch at The Windmill.**

THE ALLOTMENTS, STUNTS GREEN
BN27 4PP. Nicola Beart.

COWBEECH HOUSE
BN27 4JF. Mr Anthony Hepburn.

1 ELM COTTAGES
BN27 4RT. Audrey Jarrett.

MERRIE HARRIERS BARN
BN27 4JQ. Lee Henderson.

THE WINDMILL
BN27 4RT. Windmill Hill Windmill Trust, windmillhillwindmill.org.

Four gardens, a historic windmill and allotments will open as part of the Herstmonceux Parish Trail. Cowbeech House, a place to linger, has an exciting range of water features and sculpture in the garden dating back to 1731. Vintage car collection onsite to view. Merrie Harries Barn is a garden with sweeping lawn and open countryside beyond. Colourful herbaceous planting and a large pond with places to sit and enjoy the view. A developing garden that 5 yrs ago was agricultural land. The Allotments comprise 54 allotments growing a huge variety of traditional and unusual crops. Park in the pub across the road from 1 Elm Cottages for this garden and the Windmill. 1 Elm Cottages is a stunning cottage garden packed full of edible and flowering plants you cannot afford to miss. From here a short stroll back towards the village will take you to the historic Windmill that will be the venue for refreshments and plant sale.

74 ◆ HIGH BEECHES WOODLAND AND WATER GARDEN
High Beeches Lane, Handcross, Haywards Heath, RH17 6HQ. High Beeches Gardens Conservation Trust, 01444 400589, gardens@highbeeches.com, www.highbeeches.com. *5m NW of Cuckfield. On B2110, 1m E of A23 at Handcross.* **For NGS: Sun 6 June, Sun 3 Oct (1-5). Adm £9, chd £3. For other opening times and** information, please phone, email or visit garden website. Donation to Plant Heritage.
25 acres of enchanting landscaped woodland and water gardens with spring daffodils, bluebells and azalea walks, many rare and beautiful plants, an ancient wildflower meadow and glorious autumn colours. Picnic area. National Collection of Stewartias.

75 ◆ HIGHDOWN GARDENS
33 Highdown Rise, Littlehampton Road, Goring-by-Sea, Worthing, BN12 6FB. Worthing Borough Council, 01903 501054, highdown. gardens@adur-worthing.gov.uk, www.highdowngardens.co.uk. *3m W of Worthing. Off the A259 approx 1m from Goring-by-Sea Train Station.* **For NGS: Wed 2 June (10-7). Adm by donation. For other opening times and information, please phone, email or visit garden website.**
Curated by Sir Frederick Stern, these unique chalk gardens are home to rare plants and trees, many from seed collected by Wilson, Farrer and Kingdon Ward. New visitor centre, supported by National Lottery Heritage Fund, sharing stories of the plants and people behind the gardens. Also new accessible paths, a project to propagate and protect rare plants, and a sensory garden with a secret sea view. Accessible top pathway and lift to visitor centre. Go to garden website for further detail.

76 4 HILLSIDE COTTAGES
Downs Road, West Stoke, Chichester, PO18 9BL. Heather & Chris Lock, 01243 574802, chlock@btinternet.com. *3m NW of Chichester. From A286 at Lavant, head W for 1½m, nr Kingley Vale.* **Wed 9 June (11-3); Sun 18 July (11-4). Adm £5, chd free. Home-made teas. Visits also by arrangement June & July.**
In a rural setting this stunning garden is densely planted with mixed borders and shrubs. Large collection of roses, clematis, fuchsias and dahlias, a profusion of colour and scent in a well maintained garden. To be featured on TV, BBC Gardeners' World in 2021. Please visit www.ngs.org.uk for pop up openings in June and July.

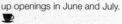

77 HOLFORD MANOR
Holford Manor Lane, North Chailey, Lewes, BN8 4DU. Martyn Price, 01444 471714, martyn@holfordmanor.com, www.chailey-iris.co.uk. *4½m SE of Haywards Heath. SE From Scaynes Hill on A272, after 1¼m turn R onto Holford Manor Lane.* **Sat 15, Sun 23 May, Sat 5 June (11-4). Adm £7, chd free. Home-made teas. Visits also by arrangement May to Aug for groups of 10+.**
5 acre garden for all seasons, surrounding a C16 Manor (not open) with far-reaching views over the ha-ha to open fields with rare breed sheep and geese. Designed and laid out by the current owners with extensive herbaceous borders, iris beds and formal parterre rose garden. Garden rooms incl a secret Chinese garden, tropical beds and wildflower meadow. Ornamental pond with water lilies and 2 acre lake and swamp garden with walking track. Brick or gravel paths give wheelchair access to most of the garden.
🚽 ✿ 🚗 🛏 🍵

78 HOLLY HOUSE
Beaconsfield Road, Chelwood Gate, Haywards Heath, RH17 7LF. Mrs Deirdre Birchell, 01825 740484, db@hollyhousebnb.co.uk, www.hollyhousebnb.co.uk. *7m E of Haywards Heath. From Nutley village on A22 turn off at Hathi Restaurant signed Chelwood Gate 2m. Chelwood Gate Village Hall on R, Holly House is opp.* **Sat 8, Sun 9 May, Fri 11 June, Sat 21, Sun 22 Aug (2-5). Adm £5, chd free. Home-made teas. Visits also by arrangement Apr to Aug for groups of up to 30.**
An acre of English garden providing views and cameos of plants and trees round every corner with many different areas giving constant interest. A fish pond and a wildlife pond beside a grassy area with many shrubs and flower beds. Among the trees and winding paths there is a cottage garden which is a profusion of colour and peace. Exhibition of paintings and cards by owner. Garden accessible by wheelchair in good weather, but it is not easy.

79 HOOPERS FARM
Vale Road, Mayfield, TN20 6BD. Andrew & Sarah Ratcliffe. *10m S of Tunbridge Wells. Turn off A267 into Mayfield. Parking is available in the village & field parking at Hoopers Farm.* **Sat 31 July, Sun 1 Aug (11-5). Adm £5, chd free. Home-made teas. Opening with Mayfield Gardens on Sat 12, Sun 13 June.**
Large s-facing garden with colour-themed mixed herbaceous planting. Mature trees, flowering shrubs, rose arbour, rock garden, secret garden and vegetable plot. New planted area with lots of late season colour developed from existing sand school. Lovely views. Plant sales by Rapkyns Nursery. Wheelchair access to most of the garden.
🚽 🐑 ✿ 🍵

80 JACARANDA
Chalk Road, Ifold, RH14 0UE. Brian & Barbara McNulty, 01403 751532, bmcn0409@icloud.com. *1m S of Loxwood. From A272/A281 take B2133 (Loxwood). ½m S of Loxwood take Plaistow Rd, then 3rd R into Chalk Rd. Follow signs for parking & garden. Wheelchair users can park in driveway.* **Visits by arrangement May to Oct for groups of 5 to 20. Adm £7.50, chd free. Tea & cake included.**
A garden for plant lovers created from scratch over the past 22 yrs. Curved borders incl unusual shrubs, trees with interesting bark, roses, climbers and perennials. Hostas in pots are a favourite feature and are displayed in many places. In the vegetable/cutting area there is a large raised bed, a greenhouse and a hanging potting bench. Our garden is for all seasons!
🚽 🐑 ✿ 🍵

81 ♦ KING JOHN'S LODGE
Sheepstreet Lane, Etchingham, TN19 7AZ. Jill Cunningham, 01580 819220, harry@kingjohnsnursery.co.uk, www.kingjohnsnursery.co.uk. *2m W of Hurst Green. Off A265 near Etchingham. From Burwash turn L before Etchingham Church, from Hurst Green turn R after church, into Church Lane, which leads into Sheepstreet Lane after ½m, then L after 1m.* **For NGS: Sun 21 Mar, Sat 1, Sun 2 May, Sat 19, Sun 20 June, Sat 11 Sept (10-5). Adm £5, chd free. Home-made teas. For other opening times and information, please phone, email or visit garden website.**
4 acre romantic garden for all seasons. An ongoing family project since 1987 with new areas completed in 2020. From the eclectic shop, nursery and tearoom, stroll past wildlife pond through orchard with bulbs, meadow, rose walk and fruit according to the season. Historic house (not open) has broad lawn, fountain, herbaceous border, pond and ha-ha. Explore secret woodland with renovated pond and admire majestic trees and 4 acre meadows. Garden is mainly flat. Stepped areas can usually be accessed from other areas. Disabled WC.
🚽 🐑 ✿ 🛏 🍵

82 KITCHENHAM FARM
Kitchenham Road, Ashburnham, Battle, TN33 9NP. Amanda & Monty Worssam. *S of Asburnham Place from A271 Herstmonceux to Bexhill Rd, take L turn 500 meters after Boreham St. Kitchenham Farm is 500 meters on the L.* **Thur 3, Tue 22 June (2-5). Adm £5, chd free. Home-made teas.**
1 acre country house garden set amongst traditional farm buildings with stunning views over the Sussex countryside. Series of borders around the house and Oast House (not open), and lawns and mixed herbaceous borders incl roses and delphiniums. A ha-ha separates the garden from the fields and sheep. The garden adjoins a working farm. Wheelchair access to the garden, but one step to the WC.
🚽 🐑 ✿ 🍵

83 KNIGHTSBRIDGE HOUSE
Grove Hill, Hellingly, Hailsham, BN27 4HH. Andrew & Karty Watson. *3m N of Hailsham, 2m S of Horam. From A22 at Boship r'about take A271, at 1st set of T-lights turn L into Park Rd & drive for 2⅔m, garden on R.* **Wed 21, Sun 25 July, Wed 8, Sat 11 Sept (1-4). Adm £6, chd free. Home-made teas.**
Mature landscaped garden set in 5 acres of tranquil countryside surrounding Georgian house (not open). Several garden rooms, spectacular herbaceous borders planted in contemporary style in traditional setting. Opening July and Sept for visitors to enjoy a very different experience. Lots of late season colour with grasses and some magnificent specimen trees; also partly walled garden. Wheelchair access to most of garden, gravel paths.
🚽 🐑 ✿ 🛏 🍵

84 NEW **28 LARKSPUR WAY**
Southwater, Horsham,
RH13 9GR. Peter & Marjorie
Cannadine, 07966 515993,
marjorie@cannadine.co.uk. *From
A24 take Worthing Rd signed
Southwater, at 2nd r'about take 1st
exit Blakes Farm Rd. At T-junction
turn R, follow road round into
Larkspur Way. At T-junction turn
L then R, No 28 is on the L.* **Sun
16 May, Sun 20 June (2-5). Adm
£3.50, chd free. Home-made teas.
Visits also by arrangement May
& June for groups of 10 to 20.
Refreshments included.**
Created for a new build this small
garden is packed with big ideas
for those looking for inspiration.
A small front garden with pond,
Mediterranean area and well stocked
border. The back garden being
approx 15 x 10 metres, backing onto
woodland is filled with cottage garden
plants, containers, ornaments and
more to inspire.

85 **LAROCHE, 43 COOMBE
DROVE**
Bramber, Steyning, BN44 3PW.
Lynne Broome, 01903 814170,
lynnecbroome@gmail.com. *From
Bramber Castle r'about, take Clays
Hill signed Steyning, turn 2nd L into
Maudlin Lane, after 100 metres, turn
R into Coombe Drove. Garden at
top of road.* **Visits by arrangement
14 Feb to end March & June for
groups of 10 to 30. If group is
smaller, ask to join another group.
Adm £4.50, chd free. Home-made
teas.**
⅓ acre garden situated on lower
slope of the South Downs. Very
wide variety of plants, many unusual.
Portland stone terracing, with a small
but steep woodland path with many
snowdrops, hellebores, aconites
and cyclamen in spring. In summer
the garden bursts into colour with
a white bed, hot bed, an array of
containers, hanging baskets and a
beautiful pergola covered in roses
and clematis. Due to steep slope and
steps, wheelchair access to lower
lawn only.

86 **LEGSHEATH FARM**
Legsheath Lane, nr Forest Row,
RH19 4JN. Mr & Mrs M Neal. *4m
S of East Grinstead. 2m W of Forest
Row, 1m S of Weirwood Reservoir.*
Sun 16 May (2-4.30). Adm £5, chd
free. Home-made teas. Donation
to Holy Trinity Church, Forest Row.
Legsheath was first mentioned
in Duchy of Lancaster records in
1545. It was associated with the
role of Master of the Ashdown
Forest. Set high in the Weald, with
far-reaching views of East Grinstead
and Weirwood Reservoir. The garden
covers 11 acres with a spring fed
stream feeding ponds. There is a
magnificent davidia, rare shrubs,
embothrium, and many different
varieties of meconopsis and abutilons.

87 **LIMEKILN FARM**
Chalvington Road, Chalvington,
Hailsham, BN27 3TA. Dr J
Hester & Mr M Royle. *10m N of
Eastbourne. Nr Hailsham. Turn S off
A22 at Golden Cross & follow the
Chalvington Rd for 1m. The entrance
has white gates on the LH-side.
Disabled parking space close to the
house, other parking 100 metres
further along road.* **Sat 5, Sun 6
June, Sat 11, Sun 12 Sept (2-5).
Adm £5, chd free. Home-made
teas in the oast house.**
The garden was designed in the
1930s when the house was owned
by Charles Stewart Taylor, MP for
Eastbourne. It has not changed in
basic layout since then. The planting
aims to reflect the age of the C17
property (not open) and original
garden design. The house and
garden are mentioned in Virginia
Woolf's diaries of 1929, depicting a
particular charm and peace that still
exists today. Flint walls enclose the
main lawn, herbaceous borders and
rose garden. Nepeta lined courtyard,
Physic garden with talk about
medicinal plants, informal pond,
and specimen trees including a very
ancient oak.

88 **LINDFIELD JUNGLE**
16 Newton Road, Lindfield,
Haywards Heath, RH16 2ND.
Tim Richardson & Clare
Wilson, 01444 484132,
info@lindfieldjungle.co.uk,
www.lindfieldjungle.co.uk. *Approx
1½ m NE of Haywards Heath town
centre. Take B2028 into Lindfield.
Turn onto Lewes Rd (B2111), then
Chaloner Rd & turn R into Chaloner
Close (no parking in close or at
garden). Garden is located at far
end, use postcode RH16 2NH.* **Mon
30 Aug (1-5). Combined adm with
5 Whitemans Close £6, chd free.**
Home-made teas at 5 Whitemans
Close. Visits also by arrangement
July to Sept for groups of 5 to 10.
Visit can be combined with 47
Denmans Lane.
A surprising, intimate garden, 17 x 8
metres. Transformed since 1999 into
an atmospheric jungle oasis planted
for tropical effect. Lush and exuberant
with emphasis on foliage and hot
colours. From the planter's terrace
enjoy the winding path through lillies,
cannas, ginger and bamboo, to
the tranquil sundowner's deck over
hidden pools.

89 **THE LONG HOUSE**
The Lane, Westdean, nr
Seaford, BN25 4AL. Robin &
Rosie Lloyd, 01323 870432,
rosiemlloyd@gmail.com,
www.thelonghousegarden.co.uk.
*3m E of Seaford, 6m W of
Eastbourne. From A27 follow signs
to Alfriston then Litlington, Westdean
1m on L. From A259 at Exceat, L on
Litlington Rd, ¼m on R. Free parking
in the village.* **Visits by arrangement
May to July for groups of 10+.
Adm £10, chd free. Home-made
teas.**
The Long House's 1 acre garden
has become a favourite for private
group visits, being compared for
romance, atmosphere and cottage
garden planting to Great Dixter and
Sissinghurst. Lavenders, hollyhocks,
roses, a wildflower meadow, a long
perennial border, water folly and pond
are just some of the features, and
everyone says Rosie's home-made
cakes are second to none. Situated
on the South Downs Way in the South
Downs National Park. Wheelchair
access over gravel forecourt at
entrance, some slopes and steps.

There are brilliant plant
sales at many gardens.
Look out for the symbol in
the garden description –
and don't forget to bring
a bag to carry your plants
home in

90 LORDINGTON HOUSE

Lordington, Chichester, PO18 9DX. Mr & Mrs John Hamilton, 01243 375862, hamiltonjanda@btinternet.com. *7m W of Chichester. On W side of B2146, ½ m S of Walderton, 6m S of South Harting. Enter through white railings on bend.* **Sat 5, Sun 6 June (1.30-4.30). Adm £5, chd free. Home-made teas. Visits also by arrangement June to Oct for groups of up to 30.**

Early C17 house (not open) and walled gardens in South Downs National Park. Clipped yew and box, lawns, borders and fine views. Vegetables, fruit and poultry in the kitchen garden. Carpet of daffodils in spring. 100+ roses planted since 2008. Trees both mature and young. Lime avenue planted in 1973 to replace elms. wildflowers in field outside walls (accessible on foot). Two walled gardens overlooking Ems valley, farmland and wooded slopes of South Downs, all in AONB. Wheelchair access is possible, but challenging with gravel paths, uneven paving and slopes. No disabled WC.

91 LOWDER MILL

Bell Vale Lane, Fernhurst, Haslemere, GU27 3DJ. Anne & John Denning, 01428 644822, anne@denningconsultancy.co.uk, www.lowdermill.com. *1½ m S of Haslemere. Follow A286 out of Midhurst towards Haslemere, through Fernhurst & take 2nd R after Kingsley Green into Bell Vale Lane. Lowder Mill is approx ½ m on R.* **For opening times and information, please visit www.ngs.org.uk**
C17 mill house and former mill set in 3 acre garden. The garden has been restored with the help of Bunny Guinness. Interesting assortment of container planting, forming a stunning courtyard between house and mill. Streams, waterfalls, innovative and quirky container planting around the potting shed and restored greenhouse. Raised vegetable garden. Rare breed chickens and ducks, as well as resident kingfishers. Extensive plant stall, mainly home propagated.

92 LUCTONS

North Lane, West Hoathly, East Grinstead, RH19 4PP. Drs Hans & Ingrid Sethi, 01342 810085, ingrid@sethis.co.uk. *4m SW of East Grinstead, 6m E of Crawley. In centre of West Hoathly village, near church, Cat Inn & Priest House. Car parks in village.* **Sat 26, Sun 27, Tue 29 June (1-5). Adm £5, chd free. Home-made teas. Visits also by arrangement May to Sept for groups of 10 to 30.**
A 2 acre Gertrude Jekyll style garden with box parterre, topiary, acclaimed herbaceous borders, swathes of spotted orchids, wildflower orchard, pond, chickens, greenhouses, large vegetable and fruit garden, croquet lawn, revamped herb garden, and a huge variety of plants, especially salvias. Admired by overseas garden tour visitors, many people comment on its peaceful atmosphere.

93 MALTHOUSE FARM

Street Lane, Street, Hassocks, BN6 8SA. Richard & Helen Keys, 01273 890356, helen.k.keys@btinternet.com. *2m SE of Burgess Hill. From r'about between B2113 & B2112 take Folders Lane & Middleton Common Lane E (away from Burgess Hill); after 1m R into Streat Lane, garden is ½ m on R. Please park carefully on the grass.* **Sun 22, Wed 25 Aug (2-5.30). Adm £6, chd free. Home-made teas. Visits also by arrangement Apr to Sept for groups of 10 to 30.**
Rural 5 acre garden with stunning views to South Downs. Garden divided into separate rooms; box parterre and borders with glass sculpture, herbaceous and shrub borders, mixed border for seasonal colour and kitchen garden. Orchard with newly planted wildflowers leading to partitioned areas with grass walks, snail mound, birch maze and willow tunnel. Wildlife farm pond with planted surround. Plants for sale. Wheelchair access mainly on grass, with some steps (caution if wet).

94 MANDALAY

High Street, Burwash, Etchingham, TN19 7EN. David & Vivienne Wright, 01435 883149, david.wright@ngs.org.uk, www.mandalaygarden.co.uk. *Mandalay is not accessed via the High St, the entrance to the garden is off Ham Lane, opp the Rose & Crown Pub.* **Sat 26 June (1-5). Home-made teas. Sun 27 June (1-5). Combined adm with Burwash Hidden Gardens £7, chd**

free. **Visits also by arrangement June to Sept for groups of 10 to 20. Combined visit with Linden Cottage.**
Contemporary, small cottage garden behind C18 listed village house (not open). Specimen olive trees, roses, clematis, border perennials, succulents, harmonious hard landscaping, water feature, and colourful containers complete the scene. Plenty of seating for relaxation and quiet reflection. NB No WC at garden.

95 MANOR OF DEAN

Tillington, Petworth, GU28 9AP. Mr & Mrs James Mitford, 07887 992349, emma@mitford.uk.com. *3m W of Petworth. From Petworth towards Midhurst on A272, pass Tillington & turn R onto Dean Lane following NGS signs. From Midhurst on A272 towards Petworth past Halfway Bridge, turn L following NGS signs.* **Thur 11 Feb, Tue 9 Mar (10.30-1); Sun 21 Mar (2-5); Thur 15 Apr, Tue 11 May (10.30-1). Adm £4.50, chd free. Visits also by arrangement Mar to Sept for groups of 20+.**
Approx 3 acres of traditional English garden with extensive views of the South Downs. Herbaceous borders, early spring bulbs, bluebell woodland walk, walled kitchen garden with fruit, vegetables and cutting flowers. NB some parts of the garden may be affected by building work. Regret, there will be no refreshments, you are very welcome to bring your own. Garden on many levels with old steps and uneven paths making it unsuitable for buggies or wheelchairs.

GROUP OPENING

96 MAYFIELD GARDENS

Mayfield, TN20 6AB. *10m S of Tunbridge Wells. Turn off A267 into Mayfield. Parking available in the village & field parking at Hoopers Farm, TN20 6BD. A detailed map available at each garden.* **Sat 12, Sun 13 June (11-5). Combined adm £7, chd free. Home-made teas at Hoopers Farm & The Oast.**

 **ABBOTSBURY**
Rebecca Morris.

HOOPERS FARM
Andrew & Sarah Ratcliffe.
(See separate entry)

12 Ainsworth Avenue

© Leigh Clapp

MULBERRY
M Vernon.

OAKCROFT
Nick & Jennifer Smith.

THE OAST
Mike & Tessa Crowe.

SOUTH STREET PLOTS
Val Buddle.

NEW **SUNNYBANK COTTAGE**
Eve & Paul Amans.

Mayfield is a beautiful Wealden village with tearooms, an old pub and many interesting historical connections. The gardens to visit are all within walking distance of the village centre. They vary in size and style, including colour themed, courtyard and cottage garden planting, wildlife meadows and fruit and vegetable plots. There are far-reaching, panoramic views over the beautiful High Weald. Partial wheelchair access to some gardens; see leaflet on the day for details.

🐕 ❄ ☕

97 ✦ MERRIMENTS GARDENS
Hawkhurst Road, Hurst Green, TN19 7RA. Lucy Cross, 01580 860666, bookings@merriments.co.uk, www.merriments.co.uk. *Off A21, 1m N of Hurst Green. On A229 Hawkhurst Rd. Situated between Hurst Green & Hawkhurst 300yds on R from A21/A229 junction.* **For NGS: Wed 18 Aug (9-5). Adm £9, chd free.** For other opening times and information, please phone, email or visit garden website.
Our beautiful 4 acre garden with colour themed borders is set amongst the rolling countryside of East Sussex. Seamlessly blending, its large borders of inspiring planting evolve through the seasons from spring pastels to the fiery autumn hues of the many trees, all on a gently sloping s-facing site with good parking and easy access. There are a number of benches in the garden to allow visitors to enjoy the atmosphere of this special ever-evolving garden. There are many unusual plants most of which are sold in the plant centre; there is also a great shop and restaurant serving

home-made lunches and teas. Wheelchairs available from the shop.

♿ 🐕 ❄ 🚗 ☕

98 ✦ MICHELHAM PRIORY
Upper Dicker, Hailsham, BN27 3QS. Sussex Archaeological Society, 01323 844224, propertymich@sussexpast.co.uk, www.sussexpast.co.uk. *3m W of Hailsham. A22 N from Eastbourne, exit L to Arlington Rd W. 1⅛ m turn R, Priory on R after approx 300yds.* **For opening times and information, please phone, email or visit garden website.**
The stunning 7 acre gardens at Michelham Priory (open to the public) are enclosed by England's longest water-filled moat, which teams with wildlife and indigenous waterlilies. Cloister and Physic gardens weave together features of medieval gardening. Over 40 yrs of developments have created a variety of features incl herbaceous borders, orchard, kitchen garden and tree lined Moat Walk. The gardens have 80,000 daffodils that create a blaze of colour from early spring onwards.

♿ ❄ 🚗 ☕

3 Normandy Drive

99 MILL HALL FARM

Whitemans Green, Cuckfield, Haywards Heath, RH17 5HX. Kate & Jonathan Berry, 01444 455986, katehod@gmail.com. *2½m E of A23, junction with B2115. Ignore SatNav. From S & E take Staplefield Rd. After 300 metres R into Burrell Cottages, then L. From A23 Warninglid exit follow signs to Cuckfield. After 30mph sign, L at end of hedge into Burrell Cottages.* Visits by arrangement Apr to Oct. Adm £5, chd free. Home-made teas. Donation to Plant Heritage.

A 2½ acre, n-facing expanding garden, now 8 yrs old with long view sloping down to maturing pond with lilies, irises and sanguisorba. Long border with ornamental trees and herbaceous plants incl cornus, acer, pulmonaria, daylily, phlox, hollyhock, brunnera, geum and potentilla, climbing and shrub roses and part of the National Collection of Noel Burr daffodils. Victorian underground water cistern guarded by a nymph and used for our watering system. Deep pond. Sloping bumpy lawn and no hard paths. No wheelchair access to WC.

100 THE MOONGATE GARDEN

6 Elm Avenue, East Preston, Littlehampton, BN16 1HJ. Helen & Derek Harnden, 07870 324654, derek@shiningmylight.plus.com. *Over railway crossing from A259, continue straight onto Golden Ave. Turn R into Elm Ave, within 110 metres.* Sun 18 July, Sat 18 Sept (2-5). Adm £5, chd free. Home-made teas. Visits also by arrangement June to Sept.

A stunning Purbeck stone wall with a moongate opening bisects the garden, creating 2 distinct halves. Many features have been incorporated into the design, such as circular lawns, a Hobbit house with vertical planting and a fern stumpery. Further highlights are provided by a Japanese area, 2 ponds with a contemporary stainless steel waterfall, and raised salad and herb planters. A haze of purple alliums in May, followed by verbena and persicarias in summer. Unusual grasses and cosmos carry the garden into autumn. Find us on Facebook by searching The Moongate Garden. Direct wheelchair access to garden on flat paving through large gate on left-hand side of house.

101 MOUNTFIELD COURT

Robertsbridge, TN32 5JP. Mr & Mrs Simon Fraser. *3m N of Battle. On A21 London-Hastings; ½m NW from Johns Cross.* Sun 9 May (2-5). Adm £5, chd free. Home-made teas.

3 acre wild woodland garden; bluebell lined walkways through exceptional rhododendrons, azaleas, camellias, and other flowering shrubs; fine trees and outstanding views. Stunning paved herb garden. Recently restored unique C18 walled garden.

102 NEW 3 NORMANDY DRIVE

East Preston, BN16 1LT. Sarah Chandler, 07903 400543, sarahchandler@live.co.uk. *From A280/A259 r'about, follow sign to East Preston on B2140. Over railway crossing, take immed L on to North Lane, leading into Sea Rd. Pass shops on your L, then Normandy Drive is 2nd R.* Thur 17 June, Thur 22 July (1-5). Adm £4.50, chd free. Home-made teas. Visits also by arrangement June to Sept for groups of 10 to 20. Adm incl tea & cake.

A self-confessed plantaholic's garden, 45' x 65', just 5 mins from the sea and a true delight. Still in its infancy, designed around a large pergola providing seating areas, borders are full of pretty herbaceous perennials, cutting flowers, climbers and a little bit of fruit and veg for good measure. Further interest incl a small wildlife pond, corten steel planters, obelisks, pots, water feature and greenhouse.

103 ◆ NYMANS

Staplefield Road, Handcross, RH17 6EB. National Trust, 01444 405250, nymans@nationaltrust.org.uk, www.nationaltrust.org.uk/nymans. *4m S of Crawley. On B2114 at Handcross signed off M23/A23 London-Brighton road. Metrobus 271 & 273 stop nearby.* For NGS: Sat 25 Sept (10-5). Adm £14.70, chd free. Home-made teas. For other opening times and information, please phone, email or visit garden website. Donation to Plant Heritage

One of NT's premier gardens with rare and unusual plant collections of national significance. In autumn dramatic shows of native tree colour can be seen in the adjoining woodland, where there are opportunities to spot wildlife. The comfortable yet elegant house, a partial ruin, reflects the personalities of the creative Messel family. Some level pathways. See full access statement on Nymans website.

 ♿ ✿ 🚌 NPC 🍵

104 OAKLANDS FARM
Hooklands Lane, Shipley, Horsham, RH13 8PX. Zsa & Stephen Roggendorff, 01403 741270, zedrog@roggendorff.co.uk. *S of Shipley village. Off the A272 towards Shipley, R at Countryman Pub, follow yellows signs. Or N of A24 Ashington, off Billingshurst Rd, 1st R signed Shipley, garden 2m up the lane.* **Mon 31 May, Wed 14 July (10.30-5.30). Adm £6, chd free. Home-made teas. Visits also by arrangement Apr to Oct for groups of 5 to 20.**
Country garden designed by Nigel Philips in 2010. Oak lined drive leading to the house and farm opens out to an enclosed courtyard with pleached hornbeam and yew. The herbaceous borders are colourful throughout the yr. Vegetable garden with raised beds and greenhouse with white peach and vine. Wild meadow leading to orchard and views across the fields, full of sheep and poultry. Mature trees. Lovely Louise will be here with her special perennials for sale. Picnics welcome. Wheelchair access over gravel and brick paths, large lawn area and grassy paths.

 ♿ ✿ 🏠 🚐 🍵 ⛲

105 OCKLYNGE MANOR
Mill Road, Eastbourne, BN21 2PG. Wendy & David Dugdill, 01323 734121, ocklyngemanor@hotmail.com, www.ocklyngemanor.co.uk. *Close to Eastbourne District General Hospital. Take A22 (Willingdon Rd) towards Old Town, turn L into Mill Rd by Hurst Arms Pub.* **Visits by arrangement May & June for groups of up to 20.**
A welcome return to the scheme for this hidden oasis behind an ancient, flint wall. Informal and tranquil, ½ acre chalk garden with sunny and shaded places to sit. Use of architectural and unusual trees. Rhododendrons, azaleas and acers in raised beds. Garden evolved over 20 yrs, maintained by owners. Georgian house (not open), former home of Mabel Lucie Attwell. Short gravel path before entering garden. Brick path around perimeter.

 ♿ 🐄 🚐

106 OFFHAM HOUSE
The Street, Offham, Lewes, BN7 3QE. Mr & Mrs P Carminger & Mr S Goodman. *2m N of Lewes on A275. Offham House is on the main road (A275) through Offham between the filling station & Blacksmiths Arms.* **Sun 6 June (1-5). Adm £5, chd free. Home-made teas.**
Romantic garden with fountains, flowering trees, arboretum, double herbaceous border and long peony bed. 1676 Queen Anne house (not open) with well knapped flint facade. Herb garden and walled kitchen garden with glasshouses, coldframes, chickens, guinea fowl, sheep and ducks. A selection of Pelargoniums and other plants for sale.

 🐄 ✿ 🍵

107 OLD CROSS STREET FARM
West Burton, Pulborough, RH20 1HD. Belinda & David Wilkinson, 01798 839373, belinda@westburton.com. *If travelling S on A29, continue & turn off at the signs for Bignor, Roman Villa. The house is in the centre of West Burton, not as indicated by the postcode! Parking in bottom of field.* **Visits by arrangement May to Oct for groups of 10 to 30. Adm £6, chd free. Light refreshments.**
A modern garden with a nod to traditional planting nestled in the ancient landscape of the South Downs. Despite its ancient buildings the garden was only designed and planted 14 yrs ago and enjoys many of the contemporary twists not usually found in such a landscape. An abundance of mass planting demonstrates the advantages of a limited planting palate. An enormous circular lawn, cloud hedging, an orchard, cutting garden, transformed farmyard, modern mass planting, all yr interest, a cottage garden, use of hedging, a raised formal pond, uses of different hard landscaping materials and planting of over 40 trees. Sorry, no picnics and WC.

 ✿ 🅳 🍵

108 NEW THE OLD HOUSE
Waldron, Heathfield, TN21 0QX. Jennifer Graham. *In centre of village opp The Star Inn & War Memorial.* **Fri 23 July (1-5). Adm £5, chd free. Home-made teas in Waldron Village Hall. Also open Warren Cottage.**
Total renovation was required on my arrival in 2004. Apart from a long hedge the garden was cleared

and levels changed. My plan was to conceal the ordinariness of the garden's shape through a strong landscape plan of my own design, and good structural planting. I wanted numerous pathways, plenty of seating, strong perfume and a formal water feature. Those key elements are still critical today.

🍵

109 NEW THE OLD RECTORY
97 Barnham Road, Barnham, PO22 0EQ. Peter & Alexandra Vining, theoldrectory97@gmail.com. *8m E of Chichester. At A27 Fontwell junction take A29 to Bognor Regis. Turn L at next r'about onto Barnham Rd. 30 metres after speed camera, turn R into Hartley Gardens. From Barnham Station turn L, we are 10 min walk. Parking for 4 cars only.* **Visits by arrangement June & July for groups of 5 to 10. Adm £3.50, chd free. Home-made teas.**
Scratch built in June 2019 after the 300m² garden was removed to 40cm, then new topsoil and returfed. By summer 2020 the garden was pretty well established. Some formal areas and a range of plants with 9 acers, salvias, lillies, 22 roses incl French roses, cypresses, boxes and grasses. We strive to bring a watercolour feel across the beds with changing hues rather than blocks of colour. We will have a small exhibition of local artists. No steps and wheelchair access through 90cm wide entrance, please advise when booking.

 ♿ 🍵

110 OLD STONELYNK EDGE
63 Battery Hill, Fairlight, Hastings, TN35 4AP. Joe & Kerry Gentleman. *5m E of Hastings. On the Ore to Cliff End main road at the junction with Warren Rd. 101 bus stops close by. Parking limited.* **Sun 6 June (11-4.30). Adm £5, chd free. Home-made teas.**
Large, well established garden with variety of shrubs and borders. Small wildlife pond. Various flower beds and lawn sloping steeply down to sizeable fish pond with a bridge leading to woodland walk and a variety of shady seating areas. The garden has some interesting architectural features and metal and wood sculptures. Not suitable for wheelchairs or those with reduced mobility.

 🐄 🍵 ⛲

⊞ THE OLD VICARAGE

The Street, Washington, RH20 4AS. Sir Peter & Lady Walters, 07766 761926, meryl.walters@me.com. *2½ m E of Storrington, 4m W of Steyning. From Washington r'about on A24 take A283 to Steyning. 500yds R to Washington. Pass Frankland Arms, R to St Mary's Church.* **Every Thur 11 Feb to 7 Oct (10.30-4). Prebooking essential, please visit www.ngs.org.uk for information & booking. Mons 5 Apr; 3, 31 May; 30 Aug (10-4). Adm £6, chd free. Home-made teas on Mondays only. Visits also by arrangement Mar to Oct for groups of 10 to 30.** Gardens of 3½ acres set around 1832 Regency house (not open). The front is formally laid out with topiary, wide lawn, mixed border and contemporary water sculpture. The rear features new and mature trees from C19, herbaceous borders, water garden and stunning uninterrupted views of the North Downs. The Japanese garden with waterfall and pond leads to a large copse, stream, treehouse and stumpery. 2000 tulips have been planted for spring as well as 10,000 mixed bulbs in the meadow area. Wheelchair access to front garden, but rear garden is on a slope.

& 🐄 ☕ 🪑

⊞ NEW 28 OLIVER ROAD

Horsham, RH12 1LH. Mary & Ian Sharp. *1½ m from A24 Hop Oast r'about, off Worthing Rd, 2nd L Longfield Rd, 1st R. No parking in Oliver Rd, use Park & Ride off the A24, at Hop Oast terminal, alight at Cricket Field bus stop & walk 6 mins, or 12 mins walk from Horsham town centre.* **Sat 12, Sun 27 June (1-5). Adm £4, chd free. Home-made teas.** A surprisingly large ¼ acre garden on the fringe of Horsham/Denne Hill. Secluded by mature trees and wildlife friendly hedges. A small pond in the wilderness area with bee hotel and log piles. Bespoke greenhouse, shed, rose pergola and a small cabin called Poets Corner, all made by the family. Top lawn repurposed for additional vegetable beds, greenhouses for tender vegetables and cottage garden beds.

☕

⊞ NEW PANGDEAN FARM

London Road, Pyecombe, Brighton, BN45 7FJ. Ian & Nicky Currie. *Leave A23 at Pyecombe, take A273 towards Hassocks. Pangdean signed on E of A23.* *Please use postcode BN45 7FN for entry to garden.* **Mon 31 May, Tue 1 June (12-5). Adm £10, chd £4. Children under 5 free. Home-made teas included.** This delightful walled garden dating from the C17 nestles in the South Downs National Park. Featuring a herb garden which supplies various herbs for the kitchen, an extensive herbaceous border which gives interest over a long period, roses and a 400 yr old James I mulberry tree. Home-made teas will be served in the stunning 300 yr old Grade II listed Sussex Barn.

& 🐄 ✿ ☕

GROUP OPENING

⊞ NEW PEACEHAVEN & WOODINGDEAN TRAIL

Follow yellow signs on A259 Brighton to Eastbourne road, the r'about at Peacehaven & in Woodingdean. Street parking adjoining these gardens. Not a walking trail. **Sat 17, Sun 18 July (12-5). Combined adm £6, chd free. Home-made teas.**

NEW BAGOTTS RATH
5 Crescent Drive North, Woodingdean, BN2 6SP. Stephen McDonnell & Den Daly.

BOXWORTH
21 Tor Road, Peacehaven, BN10 7SX. Gerard Rooney & Duncan Ward, www.boxworthflowers.com.

TOR COTTAGE
8 Tor Road, Peacehaven, BN10 7SX. Julie & Ron Basham.

Baggots Rath is a garden that wraps around the house (not open) and is divided into rooms. Each room has a different theme from tropical to English cottage garden. Large collection of succulents and other unusual plants provide a colourful backdrop to this garden perched at the top of Woodingdean. Objet d'art are peppered throughout the garden to provide focal points. Boxworth is a seaside garden developed over past 7 yrs, packed with colourful and interesting plants, perennial border, cutting garden, dahlias, shaded area with ferns and hostas. Displays in the greenhouse and floral art in the summerhouse. Seating areas for refreshments and distant sea views from the terrace. Tor Cottage is a tranquil mature garden with trees, shrubs, plants, bamboo, ornamental grasses, bananas and koi pond. Plenty of colour and different areas incl vegetable garden, rockeries and woodland. There are shady areas and seating in the sun or under the pergola. Artist studio in the garden and plants for sale.

☕

⊞ PEELERS RETREAT

70 Ford Road, Arundel, BN18 9EX. Tony & Lizzie Gilks, 01903 884981. *1m S of Arundel. At Chichester r'about take exit to Ford & Bognor Regis onto Ford Rd. We are situated close to Maxwell Rd, Arundel.* **Tue 20 Apr (11-4). Sun 25 Apr (2-5). Tue 27 Apr, Sun 11 May, Tue 25 May (11-4). Sun 30 May (2-5). Tue 8 June (11-4). Sun 20 June (2-5). Tues 22 June; 13, 27 July; 10, 24 Aug; 7 Sept (11-4). Sun 10 Oct (2-5). Tue 12 Oct (11-4). For Tuesday openings only, pre-booking essential, please visit www.ngs.org.uk for information & booking. Adm £4.50, chd free. Home-made teas. Visits also by arrangement Apr to Oct for groups of 5 to 20.** Stunning garden and gift shop with imaginative woodland sculptures and a flare for the unusual. This inspirational space is a delight in which to sit and relax, enjoying delicious teas. Interlocking beds packed with yr-round colour and scent, shaded by specimen trees, pebbled stream and fish pond. During Oct dates, we will be serving mulled wine and a variety of fruit pies with custard. Restricted wheelchair access due to narrow side entrance. Regret no motorised wheelchairs.

🐄 ♿ ☕

⊞ 33 PEERLEY ROAD

East Wittering, PO20 8PD. Paul & Trudi Harrison, 01243 673215, stixandme@aol.com. *7m S of Chichester. From A286 take B2198 to Bracklesham. Turn R into Stocks Lane, L at Royal British Legion into Legion Way. Follow road round to Peerley Rd. No 33 is halfway along.* **Sun 20 June (12-4). Adm £2.50, chd free. Visits also by arrangement May to Oct for groups of up to 20.** Small seaside garden 65ft x 32ft, 110yds from the sea. Packed full of ideas and interesting plants using every inch of space to create rooms and places for adults and children to play. A must for any suburban gardener. Specialising in unusual plants that grow well in seaside conditions with advice on coastal gardening.

117 PEMBURY HOUSE
Ditchling Road, Clayton, BN6 9PH. Nick & Jane Baker, 01273 842805, pembury@ngs.org.uk, www.pemburyhouse.co.uk. 6m N of Brighton, off A23. On B2112, 110yds from A273. Parking for groups at house. Please car share. Overflow parking at village green, BN6 9PJ; then enter by cinder track & back gate. Good public transport. Visits by arrangement Feb & Mar. Please see own website for available dates. Adm £5, chd free.
Depending on the vagaries of the season, hellebores and snowdrops are at their best in Feb and March. It is a country garden, tidy but not manicured. Work always in progress on new areas. Winding paths give a choice of walks through 3 acres of owner maintained garden, which is in and enjoys views of the South Downs National Park. Wellies, macs and winter woolies advised. A German visitor observed 'this is the perfect woodland garden'.

118 PENNS IN THE ROCKS
Groombridge, Tunbridge Wells, TN3 9PA. Mr & Mrs Hugh Gibson, 01892 864244, www.pennsintherocks.co.uk. 7m SW of Tunbridge Wells. On B2188 Groombridge to Crowborough road, just S of Xrd to Withyham. For SatNav use TN6 1UX which takes you to the white drive gates, through which you should enter the property. Sun 18 Apr, Sun 16 May, Sun 8 Aug (2-6). Adm £6, chd free. Home-made teas. Visits also by arrangement Mar to July for groups of 10+.
Large garden with spectacular outcrop of rocks, 140 million yrs old. Lake, C18 temple and woods. Old walled garden with herbaceous borders, roses and shrubs. Stone sculptures by Richard Strachey. Part C18 house (not open) once owned by William Penn of Pennsylvania. Restricted wheelchair access. No disabled WC.

119 1 PEST COTTAGE
Carron Lane, Midhurst, GU29 9LF. Jennifer Lewin. W edge of Midhurst behind Carron Lane Cemetery. Free parking at recreation ground at top of Carron Lane. Short walk on woodland track to garden, please follow signs. Fri 11 June (2.30-7); Sun 13 June (2-6). Adm £4, chd free. Light refreshments.
This edge of woodland, architects' studio garden of approx ¾ acre sits on a sloping sandy site. Designed to support wildlife and bio-diversity, a series of outdoor living spaces, connected with informal paths through lightly managed areas, creates a charming secret world tucked into the surrounding common land. The garden spaces have made a very small house (not open) into a hospitable family home. Exhibition of Architects projects. Track access and sloping site makes the garden unsuitable for wheelchairs or restricted mobility.

120 NEW PINE TREE COTTAGE
32 Mount Close, Pound Hill, Crawley, RH10 7EF. Zena & Barry Everest. 2m S of Gatwick. M23 J9, take A264/Copthorne Way. Straight over r'about, at 2nd r'about take 4th exit A2220/Copthorne Rd. Straight over r'about & L at T-lights. Then 1st L into Crawley Lane & 2nd L to Mount Close. Opp No 8. Sat 5, Sun 6 June (11-4). Adm £4, chd free. Home-made teas & cream teas.
A multi-levelled ¼ acre plot divided into 4 distinct areas. Planted with many colourful and unusual shrubs and perennials to complement this charming Sussex cottage (not open). Front garden enhanced by a fine-leaved lawn. Exuberant and colourful planting of the pond area. The 25ft stepped pergola clothed with wisteria and clematis sits between this area and the terraced top garden. All planted, landscaped and maintained by ourselves. Lovely vistas seen from one area of the garden to another. A real plantaholic's garden, well worth a visit. Partial wheelchair access.

"The annual donation from the National Garden Scheme to Perennial is the cornerstone of our fundraising activities and encourages many of our donors."
Perennial

121 6 PLANTATION RISE
Worthing, BN13 2AH. Nigel & Trixie Hall, 01903 262206, trixiehall008@gmail.com. 2m from seafront on outskirts of Worthing. A24 meets A27 at Offington r'about. Turn into Offington Lane, 1st R into The Plantation, 1st R again into Plantation Rise. Parking on The Plantation only, short walk up to 6 Plantation Rise. Visits by arrangement Mar to Sept for groups of up to 30. Adm £5, chd free. Light refreshments & home-made teas. Please advise of specific dietary requirements when booking.
Clever use is made of evergreen shrubs, azaleas, rhododendrons and acers enclosing our 70' x 80' garden, enhancing the flower decked pergolas, folly and summerhouse, which overlooks the pond. Planting incl 9 Tristus silver birches, which are semi pendula, plus a lovely combination of primroses, anemones and daffodils in spring and a profusion of roses, clematis and perennials in summer. Wheelchair access to patios only, with a good view of garden.

122 ♦ THE PRIEST HOUSE
North Lane, West Hoathly, RH19 4PP. Sussex Archaeological Society, 01342 810479, priest@sussexpast.co.uk, www.sussexpast.co.uk. 4m SW of East Grinstead, 6m E of Crawley. In centre of West Hoathly village, near church, the Cat Inn & Luctons. Car parks in village. For NGS: Sat 29 May, Sat 26 June (10.30-5.30). Adm £2, chd free. Home-made teas. Also open Luctons on 26 June only. For other opening times and information, please phone, email or visit garden website.
C15 timber-framed farmhouse with cottage garden on acid clay. Large collection of culinary and medicinal herbs in a small formal garden and mixed with perennials and shrubs in exuberant borders. Long established yew topiary, box hedges and espalier apple trees provide structural elements. Traditional fernery and stumpery, recently enlarged with a small secluded shrubbery and gravel garden. Be sure to visit the fascinating Priest House Museum, adm £1 for NGS visitors.

123 2 QUARRY COTTAGES

Wall Hill Road, Ashurst Wood, East Grinstead, RH19 3TQ. Mrs Hazel Anne Archibald. *1m S of East Grinstead. From N turn L off A22 from East Grinstead, garden adjoining John Pears Memorial Ground. From S turn R off A22 from Forest Row, garden on R at top of hill.* **Fri 21, Sat 22 May (2-5). Adm £3.50, chd free. Home-made teas.** Peaceful little garden that has evolved over 50 yrs by present owners. A natural sandstone outcrop hangs over an ornamental pond; mixed borders of perennials and shrubs with specimen trees. Many seating areas tucked into corners. Highly productive vegetable plot. Terrace round house (not open). Florist and gift shop in barn. **Also open Caxton Manor (separate admission).**
❀ ☕

124 NEW REDRIFF

Rannoch Road, Crowborough, TN6 1RA. John Mitchell. *1m S of Crowborough. From T-lights at Crowborough Cross, follow A26 (Beacon Rd) W uphill for ½ m, then turn R into Warren Rd. Follow road downhill to Xrd & turn R into Rannoch Rd, Redriff 6th on R.* **Sat 8, Sun 9 May (2-5). Combined adm with Silver Springs £5, chd free. Home-made teas.** ⅓ acre garden with a Japanese influence comprising dense planting of rhododendrons, azaleas, camellias, acers, wisteria, bamboo and topiary features. Elevated patio with shady pergola and summerhouse, both affording seating overlooking interesting fish pond, waterfall and views over the rear garden. Wheelchair access to some parts of garden.
&. ❀ ☕

125 RINGMER PARK

Uckfield Road, Ringmer, Lewes, BN8 5RW. Deborah & Michael Bedford, www.ringmerpark.com. *On A26 Lewes to Uckfield road. 1½ m NE of Lewes, 6m S of Uckfield.* **Sun 20 June (2-5). Adm £5, chd free. Home-made teas.** The garden at Ringmer Park has been developed over the last 35 yrs as the owner's interpretation of a classic English country house garden. It extends over approaching 8 acres and comprises 15 carefully differentiated individual gardens and borders which are presented to optimise the setting of the house (not

open), close to the South Downs.
&. 🐾 🚗 ☕

126 ROLFS FARM

Witherenden Road, Mayfield, TN20 6RP. Sara Jackson. *2m E of Mayfield (approx 5 mins). On Witherenden Rd, we are ½ m from Mayfield end of road. Opp Gillhope Farm. Do not follow SatNav to postcode. Long, narrow, uneven driveway downhill through woods.* **Wed 9, Wed 16 June (10-3). Adm £4, chd free. Home-made teas.** Very different from the usual NGS garden! Large wildlife friendly garden (no boundary fences to encourage visiting wildlife) with wildflower meadows, hazel trees and clipped box hedging. Large orchard next to the garden. Tom Stuart-Smith designed contemporary garden. Two ponds, small vegetable garden, and informal, naturalistic planting throughout. Uneven grass paths, sensible shoes please.
☕

127 NEW ROSE COTTAGE

Laughton, Lewes, BN8 6BX. Paul Seaborne & Glenn Livingstone, www.pelhamplants.co.uk. *5m E of Lewes; 6m W of Hailsham. Signed from junction of Common Lane & Shortage Lane, ½ m N of Roebuck Pub, Laughton.* **Wed 28, Fri 30 July (1-5). Adm £6. Home-made teas.** Nurseryman's private garden packed with unusual examples of herbaceous perennials and grasses. Informally planted 1 acre garden subdivided by strong structural shaped hedging and surrounding an old cottage (not open). Multiple densely planted borders in differing styles with yr-round interest. Specialist nursery forms part of the 2 acre woodland edge site.
🐾 ❀ ☕

128 RYMANS

Apuldram, Chichester, PO20 7EG. Mrs Michael Gayford, suzanna.gayford@btinternet.com. *1m S of Chichester. Take Witterings Rd, at 1½ m SW turn R signed Dell Quay. Turn 1st R, garden ½ m on L.* **Sat 17, Sun 18 Apr, Sat 19, Sun 20 June, Sat 18, Sun 19 Sept (2-5). Adm £6, chd free. Pre-booking essential, please visit www.ngs. org.uk for information & booking. Home-made teas. Visits also by arrangement Apr to Sept for groups of 20+.**

Walled and other gardens surrounding lovely C15 stone house (not open); bulbs, flowering shrubs, roses, ponds, and potager. Many unusual and rare trees and shrubs. In late spring the wisterias are spectacular. The heady scent of hybrid musk roses fills the walled garden in June. In late summer the garden is ablaze with dahlias, sedums, late roses, sages and Japanese anemones. Newly restored natural pond.
🐾 ❀ ☕

129 SAFFRONS

Holland Road, Steyning, BN44 3GJ. Tim Melton & Bernardean Carey, 01903 810082, tim.melton@btinternet.com. *6m NE of Worthing. Exit r'about on A283 at S end of Steyning bypass into Clays Hill Rd. 1st R into Goring Rd, 4th L into Holland Rd. Park in Goring Rd & Holland Rd.* **Visits by arrangement July & Aug for groups of 10 to 30. Home-made teas.** An artist's garden of textural contrasts and complementary colors. Well-furnished late summer flower beds of shrubby salvias, eryngiums, agapanthus, dahlias and lilies. The broad lawn is surrounded by borders with maples, rhododendrons, hydrangeas and mature trees interspersed with ferns and grasses. A large fruit cage and vegetable beds comprise the productive area of the garden. Wheelchair access difficult in very wet conditions.
&. 🐾 ☕

130 ST BARNABAS HOUSE

2 Titnore Lane, Goring-By-Sea, Worthing, BN12 6NZ. Cathy Stone, www.stbh.org.uk. *W of Worthing & just N of the r'about between A259 & A2032. Titnore Lane can be accessed via the A27 from the N, the A259 from the W (Littlehampton) or the A2032 from the E (Worthing).* **Sun 20 June, Sun 12 Sept (11-4). Adm £4.50, chd free. Pre-booking essential, please visit www.ngs. org.uk for information & booking. Home-made teas.** Guided tours. Our grounds have a central courtyard garden like an exotic atrium with seating, water features and abundant foliage from tree ferns, magnolias and katsura trees. Outside a large pond with fountain-aerator adds tranquility with the sound of running water. Lavender maze, meadow and productive vegetable plot, and areas depicting roundhouses which were part of a settlement dating back to 800BC. Good access to the site, central courtyard, main surrounding gardens and car park. Pond area paths can be affected by heavy rain.

& 🐄 ✿ 🍵 💁

GROUP OPENING

131 NEW **ST LEONARDS-ON-SEA GROUP**
St Leonards-on-Sea, TN38 0UU. *2m W of Hastings. From A259 turn into Harley Shute Rd B2092 towards Battle. Go up the hill, straight over mini r'about, over narrow railway bridge, then take 1st R into Fernside Ave. Park in Gillsmans Park.* **Sun 27 June (11-4). Combined adm £5, chd free. Home-made teas at Cornerways, 31 Fernside Avenue.**

NEW **CORNERWAYS, 31 FERNSIDE AVENUE**
Greta Romaine.

NEW **37 FERNSIDE AVENUE**
Barbara Smith.

NEW **10 WISHING TREE CLOSE**
Glenys Jacques.

Enjoy three very different gardens in this coastal town lying between Bexhill and Hastings. Start your garden safari at 37 Fernside Avenue, then move on to 31 Fernside Avenue, and then take a short 10 min walk to 10 Wishing Tree Close. The gardens offer a wealth of scent, colour, interest and ideas. Water features abound with fountains, waterfalls, a beach area with a cabin and even a pond with friendly fish. The group incl a Hastings in Bloom award-winning garden with a stunning, colourful front garden. There is also a garden in an elevated position with drought-tolerant planting, slate rocks and driftwood. From an intriguing small garden with lots of tiny areas to colourful gardens full of roses, clematis and fuchsias, there is much to see. You can also be sure of a warm welcome! Partial wheelchair access.

& ✿ 💁

132 NEW **ST MARY'S HOSPITAL**
St Martins Square, Chichester, PO19 1NR. St Mary's Trust, www.chichestercathedral.org.uk/about-us/st-marys-hospital.

Central Chichester. From East St, turn into St Martin's St adjacent to M&S Food, follow road round RH-bend & you will see the hospital entrance in the corner. **Sat 22 May, Sat 26 June, Sat 17 July (11-4). Adm £5, chd free. Light refreshments in hospital building.**
A large formal garden, opened by HRH Princess Alexandra, comprising large rose beds, traditional fruit trees, herb garden and large lawn. Four impressive knot gardens designed by Mr Ray Winnett, our Gardener who based them on medieval designs and planted them in 2015. Visitors can also visit the C13 hospital building, not usually open to the public. This is the only Almshouse in the country where residents still live in the ancient hospital, as well as in purpose built flats nearby. A hidden gem in the heart of Chichester. Wheelchair access to garden and hospital, but assistance may be required for small ridges on site.

& ✿ 💁

133 ❤ **ST MARY'S HOUSE GARDENS**
Bramber, BN44 3WE. Roger Linton & Peter Thorogood, 01903 816205, info@stmarysbramber.co.uk, www.stmarysbramber.co.uk. *1m E of Steyning. 10m NW of Brighton in Bramber village off A283.* **For NGS: Fri 9, Sat 10 July (2-5.30). Adm £6, chd free. Light refreshments. For other opening times and information, please phone, email or visit garden website.**
5 acres incl formal topiary, large prehistoric *Ginkgo biloba*, and magnificent *Magnolia grandiflora* around enchanting timber-framed Medieval house (not open for NGS). Victorian Secret Gardens incl splendid 140ft fruit wall with pineapple pits, Rural Museum, Terracotta Garden, Jubilee Rose Garden, King's Garden and circular Poetry Garden. Woodland walk and Landscape Water Garden. In the heart of the South Downs National Park. WC facilities. Wheelchair access with level paths throughout.

& ✿ 🚗 💁

134 **SANDHILL FARM HOUSE**
Nyewood Road, Rogate, Petersfield, GU31 5HU. Rosemary Alexander, 07551 777873, rosemary@englishgardeningschool.co.uk, www.rosemaryalexander.co.uk.

4m SE of Petersfield. From A272 Xrds in Rogate take road S signed Nyewood & Harting. Follow road for approx 1m over small bridge. Sandhill Farm House on R, over cattle grid. **Sat 18, Sun 19 Sept (2-5). Adm £5, chd free. Home-made teas. Visits also by arrangement Apr to Oct for groups of 10 to 30.**
Front and rear gardens broken up into garden rooms incl small kitchen garden. Front garden with small woodland area, planted with early spring flowering shrubs, ferns, bulbs and snowdrops. White garden, large leaf border and terraced area. Rear garden has rose borders, small decorative vegetable garden, red border and grasses border. Home of author and principal of The English Gardening School. Gravel paths and a few steps not easily negotiated in a wheelchair.

✿ D 💁 🎪

GROUP OPENING

136 **SEAFORD GARDENS**
Tickets & maps at each garden. 6 gardens open on 30 May & 9 on 13 June (2 Barons Close & High Trees open at noon). All signed from the A259. 12a bus route. Please note: this is not a walking trail. **Sun 30 May (12-5). Combined adm £6, chd free. Sun 13 June (11-5). Combined adm £7, chd free.**

2 BARONS CLOSE
BN25 2TY. Diane Hicks.
Open on Sun 13 June

NEW **BURFORD**
Cuckmere Road, BN25 4DE. Chris Kilsby.
Open on all dates

34 CHYNGTON ROAD
BN25 4HP. Dr Maggie Wearmouth & Richard Morland.
Open on all dates

NEW **5 CLEMENTINE AVENUE**
BN25 2UU. Joanne Davis.
Open on Sun 13 June

COSY COTTAGE
69 Firle Road, BN25 2JA. Ernie & Carol Arnold, 07763 196343, ernie.whitecrane@gmail.com.
Open on all dates
Visits also by arrangement Apr to July.

HIGH TREES
83 Firle Road, BN25 2JA. Tony & Sue Luckin.
Open on Sun 13 June

LAVENDER COTTAGE
69 Steyne Road, BN25 1QH.
Christina & Steve Machan.
Open on all dates

 MADEHURST
67 Firle Road, BN25 2JA.
Martin & Palo.
Open on all dates

SEAFORD ALLOTMENTS
Sutton Drove, BN25 3NQ.
Peter Sudell, www.salgs.co.uk.
Open on Sun 13 June

SEAFORD COMMUNITY GARDEN
East Street, BN25 1AD.
Seaford Community Garden,
www.seaford-sussex.co.uk/scg/.
Open on Sun 30 May

On 30 May, 6 gardens open. 34 Chyngton Road divided into garden rooms with pastels, hot beds and a small meadow. Cosy Cottage, a cottage garden over 3 levels with ponds, flowers, vegetables and shrubs. Lavender Cottage a flint walled garden with a coastal and kitchen garden. Madehurst is a garden on different levels with interesting planting and seasonal interest. Burford a mature garden with a mix of flowers, vegetables and an auricula theatre. The Community Garden provides an interesting space with flower and vegetable beds for members of the community to come together and share their gardening experience. On 13 June, 9 gardens will open incl 5 of the above, plus 2 Barons Close (open at noon), a riot of colour with beautiful roses. 5 Clementine Avenue slopes upwards over 3 levels and opens out onto the downs and a nice collection of succulents. High Trees (open at noon), a beautiful garden with plants, ferns and grasses. Seaford Allotments offer a unique chance to see a variety of planting and colour. A site of 189 well maintained plots with a wildlife area, a dye bed and a compost WC. Wheelchair access to Seaford Community Garden and partial access to 34 Chyngton Road.

136 SEDGWICK PARK HOUSE
Sedgwick Park, Horsham,
RH13 6QQ. Clare Davison,
01403 734930,
clare@sedgwickpark.com,
www.sedgwickpark.co.uk. *1m S of Horsham off A281. A281 towards Cowfold, Hillier Garden Center on R, then 1st R into Sedgwick*

Lane. At end of lane enter N gates of Sedgwick Park or W gate via Broadwater Lane, from Copsale or Southwater off A24. Visits by arrangement in May for groups of 10+. If your group is smaller, ask to join another group. Adm £6, chd free. Home-made teas.
Last chance to see before sold. Parkland, meadows and woodland. Formal gardens by Harold Peto featuring 20 interlinking ponds, impressive water garden known as The White Sea. Large Horsham stone terraces and lawns look out onto clipped yew hedging and specimen trees. One of the finest views of the South Downs, Chanctonbury Ring and Lancing Chapel. Turf labyrinth and organic vegetable garden. Garden has uneven paving, slippery when wet; unfenced ponds and swimming pool.

137 SELHURST PARK
Halnaker, Chichester,
PO18 0LZ. Richard & Sarah Green, 01243 839310,
mail@selhurstparkhouse.co.uk. *8m S of Petworth. 4m N of Chichester on A285.* Sun 6 June (2-5). Adm £4.50, chd free. Home-made teas. Visits also by arrangement June & July for groups of 10 to 20.
Come and explore the varied gardens surrounding a beautiful Georgian flint house (not open) approached by a chestnut avenue. The flint walled garden has a mature 160ft herbaceous border with unusual planting along with rose, hellebore and hydrangea beds. Pool garden with exotic palms and grasses divided from a formal knot and herb garden by Espalier apples. Kitchen and walled fruit garden. Wheelchair access to walled garden, partial access to other areas.

138 SENNICOTTS
West Broyle, Chichester,
PO18 9AJ. Mr & Mrs James Rank, 07950 324181,
ngs@sennicotts.com,
www.sennicotts.com. *2m NW of Chichester. White gates diagonally opp & W of the junction between Salthill Rd & the B2178.* Visits by arrangement in June for groups of 20 to 30.
Historic gardens set around a Regency villa (not open) with views across mature Sussex parkland to the South Downs. Working walled kitchen and cutting garden. Lots of space for

children and a warm welcome for all. Large buses cannot enter our gates. If you are an arranged group, please discuss refreshment options.

139 ♦ SHEFFIELD PARK AND GARDEN
Uckfield, TN22 3QX. National Trust, 01825 790231,
sheffieldpark@nationaltrust.org.uk, www.nationaltrust.org.uk/sheffieldpark. *10m S of East Grinstead. 5m NW of Uckfield; E of A275.* For NGS: Wed 5 May (11-5). Adm £11, chd £5.50. Light refreshments in Coach House Tearoom. For other opening times and information, please phone, email or visit garden website.
Magnificent landscaped garden laid out in C18 by Capability Brown and Humphry Repton covering 120 acres (40 hectares). Further development in the early yrs of this century by its owner Arthur G Soames. Centrepiece is original lakes with many rare trees and shrubs. Beautiful at all times of the yr, but noted for its spring and autumn colours. National Collection of Ghent azaleas. Natural play trail for families on South Park. Large number of Champion Trees, 87 in total. Garden largely accessible for wheelchairs, please call for information.

140 SHEPHERDS COTTAGE
Milberry Lane, Stoughton, Chichester, PO18 9JJ. Jackie & Alan Sherling, 07795 388047,
milberrylane@gmail.com. *9⅙m NW Chichester. Off B2146, next village after Walderton. Cottage is near telephone box & beside St Mary's Church. No parking in lane beside house.* Visits by arrangement Apr to Aug for groups of 10 to 30. Adm £5, chd free. Home-made teas.
A compact terraced garden using the borrowed landscape of Kingley Vale in the South Downs. The s-facing flint stone cottage (not open) is surrounded by a Purbeck stone terrace and numerous individually planted and styled seating areas. A small orchard (under-planted with meadow), lawns, yew hedges, ilex balls and drifts of wind grass provide structure and yr-round interest. Many novel design ideas for a small garden. Ample seating throughout the garden to enjoy the views. Not suitable for wheelchairs or people with mobility issues.

141 SIENNA WOOD

Coombe Hill Road, East Grinstead, RH19 4LY. Belinda & Brian Quarendon, 07970 707015, belinda222@hotmail.com. *1m W of East Grinstead. Off B2110 East Grinstead to Turners Hill. Garden is ½ m down Coombe Hill Rd on L.* **Sun 23 May (1-5). Adm £5, chd free. Light refreshments. Visits also by arrangement Mar to Oct for groups of 20+.**

Explore our beautiful 3½ acre garden, picturesque lakeside walk and 6 acre ancient woodland behind. Start at the herbaceous borders surrounding the croquet lawn, through the formal rose garden to the lawns and summer borders; then down through the arboretum to the lake and waterfall and back past the exotic border, orchard and vegetable garden. Many unusual trees and shrubs. Possible sighting of wild deer incl white deer. Partial wheelchair access to many parts of the garden.

142 NEW SILVER SPRINGS

Heavegate Road, Crowborough, TN6 1UA. Barbara Diamond. *1m S of Crowborough. From S on A26 turn L, or from N turn R, onto Fielden Rd. Garden on L at junction with Fielden Rd. Parking on Fielden Rd only.* **Sat 8, Sun 9 May (2-5). Combined adm with Dedriff £5, chd free. Home-made teas.**

¼ acre garden, developed by the current owner (a keen plantswoman) over the past 25 yrs from a bare site. Shrubs, perennials, rock garden and bulbs in three areas surrounding the house (not open), each with a different character. Yr round interest with emphasis on spring and a wide range of hydrangeas in the summer.

143 SKYSCAPE

46 Ainsworth Avenue, Ovingdean, Brighton, BN2 7BG. Lorna & John Davies. *From Brighton take A259 coast road E, passing Roedean School on L. Take 1st L at r'about into Greenways & 2nd R into Ainsworth Ave, Skyscape at the top on R.* **Sat 5, Sun 6 June (11-5). Combined adm with 12 Ainsworth Avenue £5, chd free. Tea & cake served on the patio.**

250ft s-facing rear garden on a sloping site with fantastic views of the South Downs and the sea. Garden created by owners over past 7 yrs. New orchard, flower beds and planting with bees in mind. Rich variety of plants on each level. Plants for sale. Supervised children welcomed. Full access to site via purpose built sloping path.

144 SOUTH GRANGE

Quickbourne Lane, Northiam, Rye, TN31 6QY. Linda & Michael Belton, 01797 252984, belton.northiam@gmail.com. *Between A268 & A28, approx ½ m E of Northiam. From Northiam centre follow Beales Lane into Quickbourne Lane, or Quickbourne Lane leaves A286 approx ½ m S of A28 & A286 junction. Disabled parking at front of house.* **Thur 20 May, Wed 28 July (1-5); Sat 4, Sun 5 Sept (11-5). Adm £6, chd free. Home-made teas (May & July). Teas & light lunches made to order (Sept).** **Visits also by arrangement Apr to Oct.**

Hardy Plant Society members' garden with wide variety of trees, shrubs, perennials, grasses, pots arranged into a complex garden display for yr-round colour and interest. Raised vegetable beds, wildlife pond, new water features, orchard with rose arbour, soft fruit cage and living gazebo. House roof runoff diverted to storage and pond. Small area of wildwood. An emphasis on planting for insects. We try to maintain nectar and pollen supplies and varied habitats for most of the creatures that we share the garden with, hoping that this variety will keep the garden in good heart. Home propagated plants for sale. Wheelchair access over hard paths through much of the garden, but steps up to patio and WC.

Merriments Gardens

145 STANLEY FARM

Highfield Lane, Liphook, GU30 7LW. Bill & Emma Mills. *For SatNav please use GU30 7LN, which takes you to Highfield Lane & then follow NGS signs. Track to Stanley Farm is 1m.* Sun 2 May (12-5). Adm £5, chd free. Home-made teas.

1 acre garden created over the last 15 yrs around an old West Sussex farmhouse (not open), sitting in the midst of its own fields and woods. The formal garden incl a kitchen garden with heated glasshouse, orchard, espaliered wall trained fruit, lawn with ha-ha and cutting garden. A motley assortment of animals incl sheep, donkeys, chickens, ducks and geese. Bluebells flourish in the woods, so feel free to bring dogs and a picnic, and take a walk after visiting the gardens. Wheelchair access via a ramp to view main part of the garden. Difficult access to woods due to muddy, uneven ground.

&. 🐄 ✳ 🍵

146 STONE CROSS HOUSE

Alice Bright Lane, nr Crowborough, TN6 3SH. Mr & Mrs D A Tate. *1½ m S of Crowborough Cross. At Crowborough T-lights (A26) turn S into High St & shortly R onto Croft Rd. Over 3 mini-r'abouts to Alice Bright Lane. Garden on L at next Xrds.* Sat 8 May (2-5). Adm £5, chd free. Home-made teas.

Beautiful 9 acre country property (not open) with gardens containing a delightful array of azaleas, acers, rhododendrons and camellias, interplanted with an abundance of spring bulbs. The very pretty cottage garden has interesting examples of topiary and unusual plants. Jacob sheep graze the surrounding pastures. Gravel drive, but mainly flat and no steps. No WC.

&. 🐄 🍵

147 SULLINGTON OLD RECTORY

Sullington Lane, Storrington, Pulborough, RH20 4AE. Oliver & Mala Haarmann. Mark Dixon, Head Gardener. *Traveling S on A24 take 3rd exit on Washington r'about. Proceed to Xrds on A283 for Sullington Lane & Water Lane. Take L onto Sullington Lane & garden located at the top.* Wed 14, Thur 15, Wed 21, Thur 22 July. Adm £7, chd free. Pre-booking essential, please phone 07749 394012 or email mark@sullingtonoldrectory.com for information & booking. Tour with Head Gardener at 10am or 2pm

included. Home-made teas.

With a backdrop of stunning views of the South Downs, the naturalistic style of this beautiful country garden sits perfectly into the surrounding landscape. The garden incl a potager, orchard, herb garden, established trees and shrubs, a pleached lime walk, new colour themed perennial borders, a profusion of grasses and an experimental planting in the moist meadow. Wheelchair access to most areas.

&. 🐄 🚗 🍵

148 ◆ SUSSEX PRAIRIES

Morlands Farm, Wheatsheaf Road (B2116), Henfield, BN5 9AT. Paul & Pauline McBride, 01273 495902, morlandsfarm@btinternet.com, www.sussexprairies.co.uk. *2m NE of Henfield on B2116 Wheatsheaf Rd (also known as Albourne Rd). Follow Brown Tourist signs indicating Sussex Prairie Garden.* For NGS: Sun 12 Sept (1-5). Adm £10, chd free. Home-made teas. For other opening times and information, please phone, email or visit garden website.

Exciting prairie garden of approx 8 acres planted in the naturalistic style using 60,000 plants and over 1,600 different varieties. A colourful garden featuring a huge variety of unusual ornamental grasses. Expect layers of colour, texture and architectural splendour. Surrounded by mature oak trees with views of Chanctonbury Ring and Devil's Dyke on the South Downs. Permanent sculpture collection and exhibited sculpture throughout the season. Rare breed sheep and pigs. New tropical entrance garden planted in 2018. Woodchip pathway at entrance. Soft woodchip paths in borders not accessible, but flat garden for wheelchairs and mobility scooters. Disabled WC.

&. 🐄 ✳ 🚗 🍵 🛏

149 THAKEHAM PLACE FARM

The Street, Thakeham, Pulborough, RH20 3EP. Mr & Mrs T Binnington. *In the village of Thakeham, 3m N of Storrington. The farm is at the E end of The Street, where it turns into Crays Lane. Follow signs down farm drive to Thakeham Place.* Wed 21, Sun 25 July (2-5). Combined adm with Cumberland House £7, chd free. Home-made teas.

Set in the middle of a working dairy farm, the garden has evolved over the last 30 yrs. Taking advantage of

its sunny position on free draining greensand, the borders are full of sun loving plants and grasses, with a more formal area surrounding the farmhouse (not open). Lovely views across the farm to Warminghurst from the orchard.

🐄 🍵

150 TOWN PLACE

Ketches Lane, Freshfield, Sheffield Park, RH17 7NR. Anthony & Maggie McGrath, 01825 790221, mcgrathsussex@hotmail.com, www.townplacegarden.org.uk. *5m E of Haywards Heath. From A275 turn W at Sheffield Green into Ketches Lane for Lindfield. 1¾ m on L.* Fri 18, Sat 19, Sun 20 June (11-5). Adm £7.50, chd free. Visits also by arrangement in June.

A stunning 3 acre garden with a growing international reputation for the quality of its design, planting and gardening. Set round a C17 Sussex farmhouse (not open), the garden has over 800 roses, herbaceous borders, herb garden, topiary inspired by the sculptures of Henry Moore, ornamental grasses, an 800 yr old oak, potager, and a unique ruined Priory Church and Cloisters in hornbeam. Regret, no coaches, refreshments or WC. There are steps, but all areas can be viewed from a wheelchair.

&. ✳ 🛏

151 TUPPENNY BARN

Main Road, Southbourne, PO10 8EZ. Maggie Haynes, 01243 377780, contact@tuppennybarn.co.uk, tuppennybarn.co.uk. *6m W of Chichester, 1m E of Emsworth. On Main Rd A259, corner of Tuppenny Lane. Disabled parking.* Visits by arrangement May to Sept for groups of 20 to 30. Adm £5, chd free. Home-made teas. Food intolerances & allergies catered for.

An iconic, organic smallholding used as an outdoor classroom to teach children about the environment, sustainability and healthy food. 2½ acres packed with a wildlife pond, orchard with heritage top fruit varieties, two solar polytunnels, fruit cages, raised vegetables, herbs and cut flower garden. Willow provides natural arches and wind breaks. Bug hotel and beehives support vital pollinators. Most of the grounds are accessible for wheelchairs, but there are undulated areas that are more difficult.

&. ✳ 🍵

152 NEW WADHURST PARK

Riseden Road, Wadhurst,
TN5 6NT. Nicky Browne,
wadhurstpark.co.uk. *6m SE of
Tunbridge Wells. Turn R along
Mayfield Lane off B2099 at NW end
of Wadhurst. L by The Best Beech
Inn, L at Riseden Rd.* **Wed 7, Wed
14 July (10-3.30). Adm £6, chd
free. Home-made teas in the
Common Room.**
The naturalistic gardens, designed
by Tom Stuart-Smith, created on
a C19 site, situated within a 2000
acre estate managed organically to
protect its wildlife, cultural heritage
and beauty. The gardens invite the
wider landscape in, while meadows
and hedgerows, woodland trees and
groundcover soften and frame views
to hills and lake. We strive to garden
with a greater respect for the natural
world. Features incl restored Victorian
orangery, naturalistic gardens planted
with native species, potager, log hives
and woodland walks. Wheelchair
access to main features of garden,
some surfaces incl grass, cobbles
and steps.

& ▣ ☕ ♿

153 THE WALLED GARDEN AT TILGATE PARK

Tilgate Drive, Tilgate, Crawley,
RH10 5PQ. Nick Hagon,
www.friendsoftilgatepark.co.uk.
*SE Crawley. Leave M23 at J11. From
r'about follow A23 Brighton Rd for
short distance. At T lights turn R
at sign to Tilgate Park onto Tilgate
Drive. Follow signs in Park for parking
incl disabled parking.* **Evening
opening Wed 12, Thur 13 May,
Wed 14, Thur 15 July (6.30-9).
Adm £5, accompanied children
under 16 free. No concessions.
Cash only. Light refreshments
included.**
There is much more to know about
the award-winning Tilgate Park than is
realised on a family visit. Why not join
a tour with Park Manager, Nick Hagon,
to discover more about this special
environment? In spring, learn about
the azaleas, camellias, candelabra
primulas and rhododendrons and
in July the walled garden and the
centuries old specimen trees. Talk
and tour begins at 7pm, near the
Walled Garden. Wheelchair access
within the walled garden only.

& ✿ ☕

154 NEW WARREN COTTAGE

Warren Lane, Cross In Hand,
Heathfield, TN21 0TB. Mr & Mrs
Allcorn. *2½ m E of Heathfield, 2m W
of Blackboys. At junction of B2102
Lewes Rd & Warren Lane. Follow
signs for parking near garden.* **Fri
23 July (1-5). Adm £5, chd free.
Home-made teas in Waldron
Village Hall. Also open The Old
House.**
½ acre mature, secluded garden
being redeveloped by the owners
and using many reclaimed materials.
Incl 60ft herbaceous border with oak
frame support, long grass area with
specimen shrubs and mown paths,
wildlife pond and borders, vegetable
patch with raised beds, greenhouses
and potting shed, a pub shed and
magical fairy/woodland garden.

& ✿ ☕

GROUP OPENING

155 WATERWORKS & FRIENDS

Broad Oak & Brede, TN31 6HG.
*4 Waterworks Cottages, Brede, off
A28 by church & opp Red Lion Pub,
¾ m at end of lane. Sculdown is on
B2089 Chitcombe Rd, W off A28 at
Broad Oak Xrds. Start at any garden,
directions given.* **Sat 5 June (10.30-
4). Combined adm £5, chd free.
Light refreshments at Sculdown.**

SCULDOWN
TN31 0EX. Mrs Christine Buckland,
christine.buckland@hotmail.com.
**Visits also by arrangement May
to July.**

4 WATERWORKS COTTAGES
TN31 6HG. Mrs Kristina
Clode, 07950 748097,
kristinaclode@gmail.com, www.
kristinaclodegardendesign.co.uk.
**Visits also by arrangement June
to Oct for groups of 10 to 20.**
▣

An opportunity to visit 2 unique
gardens and discover the Brede
Steam Giants 35ft Edwardian water
pumping engines, and Grade II
listed pump house located behind
4 Waterworks Cottages. Garden
designer Kristina Clode has created
her wildlife friendly garden at 4
Waterworks Cottages over the last
11 yrs. Delightful perennial wildflower
meadow, pond, wisteria covered
pergola and mixed borders packed
full of unusual specimens with yr-
round interest and colour. Sculdown's
garden is dominated by a very large

wildlife pond formed as a result of
iron-ore mining over 100 yrs ago.
The stunning traditional cottage (not
open) provides a superb backdrop
for several colourful herbaceous
borders and poplar trees. Plants for
sale at 4 Waterworks Cottages. At
Brede Steam Giants, WC available,
but regret no disabled facilities and
assistance dogs only (free entry,
donations encouraged). Wheelchair
access at Sculdown (park in flat area
at top of field) and in the front garden
of 4 Waterworks Cottages only.

& ☕ ✿ 🦮

156 WHITEHANGER

Marley Lane, Haslemere,
GU27 3PY. Lynn & David
Paynter, 07774 010901,
lynn@whitehanger.co.uk. *3m S of
Haslemere. Take the A286 Midhurst
road from Haslemere & after approx
2m turn R into Marley Lane (opp
Hatch Lane). After 1m turn into drive
shared with St Magnus Nursing.*
**Suns 11, 25 July; 8, Sun 22 Aug
(10.30-3.30). Adm £6.50, chd
£6.50. Pre-booking essential,
please visit www.ngs.org.uk for
information & booking. Visits also
by arrangement June to Oct for
groups of 5 to 30.**
Set in 6 acres on the edge of
the South Downs National Park
surrounded by NT woodland, this
rural garden was started in 2012
when a new Huf house was built
on a derelict site. Now there are
lawned areas with beds of perennials,
a serenity pool with Koi carp, a
wildflower meadow, a Japanese
garden, a sculpture garden, a
woodland walk, a large rockery
and new for 2020 an exotic walled
garden. Picnics welcome.

& 🦮 🪑

*The National Garden Scheme
was set up in 1927 to raise funds
for district nurses. We remain
strongly committed to nursing
and today our beneficiaries
include Macmillan Cancer
Support, Marie Curie, Hospice
UK and The Queen's Nursing
Institute*

157 3 WHITEMANS CLOSE

Cuckfield, Haywards Heath, RH17 5DE. **David & Christine Hart.** *1m N of Cuckfield. On B2036 signed Balcombe, Whitemans Close is 250yds from r'about on LH-side. Park on road, no parking in Whitemans Close. Buses stop at Whitemans Green, where there is also a large car park.* **Thur 13 May, Thur 10 June (10.30-4). Combined adm with 5 Whitemans Close £8.50, chd free. Pre-booking essential, please phone 01444 473520 or email shirleycarmanmartin@gmail.com for information & booking. Home-made teas at 5 Whitemans Close included.**

Maturing nicely this young garden is packed with many special delights, a treat for plants people. The sections of front garden incl a scree garden with huge grasses, trees and shrubs, plus a rockery full of unusual alpines. The rear garden is a feast of cottage garden specials, trees, plus many pots of treasures on the patio, enhancing the view from the lower level.

✿ ☕ ♿

158 5 WHITEMANS CLOSE

Cuckfield, Haywards Heath, RH17 5DE. **Shirley Carman-Martin.** *1m N of Cuckfield. On B2036 signed Balcombe Whitemans Close is 250yds from r'about on LH-side. Park on road, no parking in Whitemans Close. Buses stop at Whitemans Green, where there is also a large free car park.* **Sat 6, Mon 8, Thur 11, Mon 15, Tue 16, Wed 17 Feb (10.30-3.30). Adm £7.50, chd free. Thur 13 May, Thur 10 June (10.30-4). Combined adm with 3 Whitemans Close £8.50, chd free. Pre-booking essential, please phone 01444 473520 or email shirleycarmanmartin@gmail.com for information & booking. Home-made teas included. Mon 30 Aug (1-5). Combined adm with Lindfield Jungle £6, chd free. Home-made teas.**

A garden visit for snowdrop and plant lovers. A relatively small cottage garden packed full of exciting and unusual plants, plus a large snowdrop collection. Flower beds overflow with gorgeous plants in colour schemed borders, collection of echeveria, many pelargonium and salvias, and half hardy plants too. A plantsman's garden not to miss.

✿ ☕

159 WHITHURST PARK

Plaistow Road, Kirdford, near Billingshurst, RH14 0JW. **Mr Richard Taylor & Mr Rick Englert,** www.whithurst.com. *7m NW of Billingshurst. A272 to Wisborough Green, follow sign to Kirdford, turn R at 1st T-junction through village, then R again, 1m on Plaistow Rd, look for the white Whithurst Park sign at roadside.* **Sun 4 July (11-5). Adm £5, chd free.**

10 yr old walled kitchen garden, many espaliered fruit trees. Herb beds, vegetable beds, flower borders and cutting beds. Central greenhouse and potting shed with interesting behind the wall support buildings, incl extensive compost area close to beehives. Sustainability through permaculture principles. There will be 45 min time slots to visit the walled garden, but you can then stay as long as you wish until closing. We have plenty of open lawn around the walled garden and house which can be used for picnics. There are also footpaths around the woodlands surrounding the house and lake. Plants for sale will be in a self-serve area with a honesty box, please bring cash (no change available). Wheelchair access via ramp over 3' step on to garden paths.

♿ ✿ 🪑

160 NEW 26 WILBURY AVENUE

Hove, BN3 6HS. **Julia & Steve White,** 07568 538447, julia.wolage@btinternet.com. *Hove train station & no 7 bus route close by. Parking difficult.* **Visits by arrangement May to Aug for groups of up to 10. Adm £3, chd free. Tea.**

A town garden of a Victorian house (not open), designed to attract wildlife and yr-round colour, filled with shrubs, trees, perennials, annuals and grasses. The garden is small, approx 90ft long x 35ft wide, and is planted taking features from a cottage garden and adding a little bit of prairie and Japan. Ideal for small group visits. Lots of take home ideas.

✿ ☕

GROUP OPENING

161 WINCHELSEA'S SECRET GARDENS

Winchelsea, TN36 4EJ. *2m W of Rye, 8m E of Hastings. Purchase ticket for all gardens at first garden visited; a map will be provided showing gardens & location of refreshments.* **Sat 17 Apr (1-5.30);**

Sat 19 June (11-5). Combined adm £6, chd free. Tea at Winchelsea New Hall.

THE ARMOURY
Mr & Mrs A Jasper.
Open on all dates

CLEVELAND HOUSE
Mr & Mrs J Jempson.
Open on all dates

CLEVELAND PLACE
Sally & Graham Rhodda.
Open on all dates

EVENS
Judith & James Payne.
Open on Sat 19 June

NEW GILES POINT
Ant Parker & Tom Ashmore.
Open on all dates

KING'S LEAP
Philip Kent.
Open on all dates

LOOKOUT COTTAGE
Mary & Roger Tidyman.
Open on Sat 19 June

MAGAZINE HOUSE
Susan Stradling.
Open on Sat 19 June

THE ORCHARDS
Brenda & Ralph Courtenay.
Open on Sat 19 June

PERITEAU HOUSE
Dr & Mrs Lawrence Youlten.
Open on all dates

RYE VIEW
Howard Norton & David Page.
Open on all dates

SOUTH MARITEAU
Robert Holland.
Open on all dates

THE WELL HOUSE
Alice Kenyon.
Open on all dates

9 gardens will open in April & 13 gardens will open in June. Many styles, large and small, secret walled gardens, spring bulbs, tulips, roses, herbaceous borders and more, in the beautiful setting of the Cinque Port of Winchelsea. Explore the town with its magnificent church and famous medieval merchants' cellars. Pre-booked guided tours of cellars if rules allow. Check winchelsea.com for the latest information. If you are bringing a coach please contact david@ryeview.net, 01797 226524. Wheelchair access to 6 gardens in April and 8 in June; see map provided on the day for details.

♿ ✿ 🚗 ☕ 🪑

162 **NEW** **33 WIVELSFIELD ROAD**
Saltdean, Brighton, BN2 8FP.
Chris Briggs & Steve Jenner.
From A259 turn onto Arundel Drive West, continue onto Saltdean Vale. Take the 6th turning on L, Tumulus Rd. Take 1st R onto Wivelsfield Rd, continue up steep hill, 33 is on the R. **Sat 31 July, Sun 1 Aug (11-5). Adm £4, chd free. Home-made teas.**
A new garden that was cleared in 2016, landscaped during 2017 and planting started later that yr. Split over three levels filled with colourful perennials, annuals, grasses, and succulents planted in the ground and in containers. The formal garden leads onto a wildflower meadow extending onto the downs, giving spectacular views of both the ocean and South Downs National Park.

🐕 ❀ ☕

163 **WYCH WARREN HOUSE**
Wych Warren, Forest Row, RH18 5LF. Colin King & Mary Franck. *1m S of Forest Row. Proceed S on A22, track turning on L, 100 metres past 45mph warning triangle sign. Or 1m N of Wych Cross T-lights track turning on R. Go 400 metres across golf course till the end.* **Wed 14 July (2-8). Adm £5, chd free. Light refreshments.**
6 acre garden in Ashdown Forest, AONB, much of it mixed woodland. Perimeter walk around property (not open). Delightful and tranquil setting with various aspects of interest providing a sensory and relaxing visit. Lovely stonework, specimen trees, exotic bed, three ponds, herbaceous borders, greenhouse and always something new on the go. Plenty of space to roam and explore and enjoy the fresh air. Plants for sale and a great range of chutney and jams. Dogs on leads and children welcome. Partial wheelchair access by tarmac track to the kitchen side gate.

♿ 🐕 ❀ ☕ ⛱

Black Barn

WARWICKSHIRE

For Birmingham & West Midlands see Staffordshire

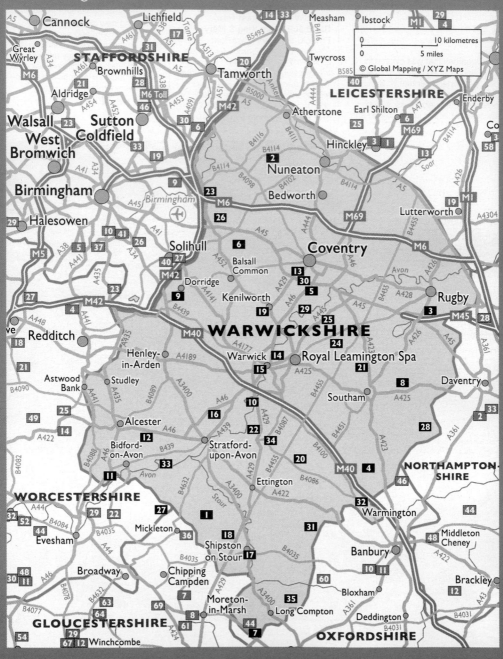

Set in the heart of England, Warwickshire is essentially a rural county, with many delightful villages and historic towns, amidst enchanting stretches of unspoilt countryside.

In the rolling hills of the south, Cotswold villages of honey-coloured stone nestle in the valleys. Stratford-upon-Avon is a mecca for tourists who flock to see Shakespeare's plays in performance, and visit his birthplace. The atmospheric county town of Warwick with its great castle, and elegant Leamington Spa lie at Warwickshire's centre, with Coventry and the more industrial landscapes to the north. The Coventry Canal runs along the county's northern border, and the towns give way once again to open countryside.

Warwickshire's gardens offer the visitor a satisfying tapestry of styles and settings, its locations ranging from picturesque villages and towns to nurseries and allotments. Our season begins in February with special Snowdrop openings and ends in October with a glow of autumn colours. Grand country house gardens and charming tiny village ones are all lovingly tended by their owners who generously open their garden gates to support the National Garden Scheme Beneficiaries in their vital work.

So do please visit us. You will be assured of a warm Warwickshire welcome by our Garden Owners who take tremendous pleasure in sharing their knowledge and enthusiasm. And, of course, there is always the promise of a glorious afternoon tea to conclude your day in our delightful gardens.

Below: 114 Hartington Crescent, Earlsdon Gardens

Volunteers

County Organiser
Liz Watson
01926 512307
liz.watson@ngs.org.uk

County Treasurer
Dee Broquard
07773 568317
dee@ngs.org.uk

Publicity
Lily Farrah
07545 560298
lily.farrah@ngs.org.uk

Social Media
Fiona Anderson
07754 943277
fiona.anderson@ngs.org.uk

Talks
Dee Broquard (as above)

Booklet Advertising
Dee Broquard (as above)

Booklet Co-ordinator
Hugh Thomas
01926 423063
hughthomas1203@gmail.com

Assistant County Organisers
Elspeth Napier
01608 666278
elspethmjn@gmail.com

Jane Redshaw
07803 234627
jane.redshaw@ngs.org.uk

David Ruffell
01926 316456
de.ruffell@btinternet.com

Isobel Somers
07767 306673
ifas1010@aol.com

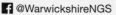 @WarwickshireNGS   @WarksNGS

OPENING DATES

All entries subject to change. For latest information check www.ngs.org.uk

Extended openings are shown at the beginning of the month

Map locator numbers are shown to the right of each garden name.

February

Snowdrop Festival

Saturday 13th
◆ Hill Close Gardens   15

April

Every Saturday and Sunday from Saturday 17th
◆ Bridge Nursery   8

Monday 5th
◆ Bridge Nursery   8

Sunday 18th
Blacksmiths Cottage   7
Broadacre   9

May

Every Saturday and Sunday
◆ Bridge Nursery   8

Monday 3rd
◆ Bridge Nursery   8
Earlsdon Gardens   13

Saturday 15th
6 Canon Price Road   10

Sunday 16th
6 Canon Price Road   10

Saturday 22nd
NEW 3 Cleeve View   11

Sunday 23rd
NEW 3 Cleeve View   11

Saturday 29th
Ilmington Gardens   18

Sunday 30th
Ilmington Gardens   18
Pebworth Gardens   27

Monday 31st
◆ Bridge Nursery   8
Pebworth Gardens   27

June

Every Saturday and Sunday
◆ Bridge Nursery   8

Saturday 5th
Tysoe Gardens   31

Sunday 6th
Lighthorne Gardens   20
Maxstoke Castle   23
Tysoe Gardens   31

Friday 11th
Priors Marston Manor   28

Sunday 13th
Packington Hall   26
Styvechale Gardens   30

Saturday 19th
6 Canon Price Road   10

Sunday 20th
6 Canon Price Road   10
Honington Gardens   17
Kenilworth Gardens   19
Whichford & Ascott Gardens   35

Saturday 26th
Welford-on-Avon Gardens   33

Sunday 27th
NEW Ansley Gardens   2
Berkswell Gardens   6
NEW Stoneleigh Village Gardens   29
Warmington Gardens   32
Welford-on-Avon Gardens   33

July

Every Saturday and Sunday
◆ Bridge Nursery   8

Saturday 3rd
Long Itchington Village Gardens   21

Sunday 4th
Avon Dassett Gardens   4
Long Itchington Village Gardens   21

Sunday 11th
NEW The Malt House   22

Saturday 17th
6 Canon Price Road   10
NEW 3 Cleeve View   11

Sunday 18th
6 Canon Price Road   10
NEW 3 Cleeve View   11

Saturday 24th
Guy's Cliffe Walled Garden   14

Sunday 25th
◆ Avondale Nursery   5

Friday 30th
Anya Court Care Home   3

Saturday 31st
Anya Court Care Home   3

August

Every Saturday and Sunday
◆ Bridge Nursery   8

Sunday 1st
Wellesbourne Allotments   34

Sunday 22nd
◆ Avondale Nursery   5

Monday 30th
◆ Bridge Nursery   8

September

Every Saturday and Sunday to Saturday 25th
◆ Bridge Nursery   8

October

Saturday 30th
◆ Hill Close Gardens   15

By Arrangement

Arrange a personalised garden visit on a date to suit you. See individual garden entries for full details.

Admington Hall   1
10 Avon Carrow, Avon Dassett Gardens   4
Broadacre   9
6 Canon Price Road   10
The Croft House   12
16 Delaware Road, Styvechale Gardens   30
Elm Close, Welford-on-Avon Gardens   33
Fieldgate, Kenilworth Gardens   19
The Hill Cottage   16
2 The Hiron, Styvechale Gardens   30
19 Leigh Crescent, Long Itchington Village Gardens   21
The Motte   24
Oak House   25
Priors Marston Manor   28
NEW Rowan House, Welford-on-Avon Gardens   33

"The support of Hospice UK and the National Garden Scheme has been invaluable to hospice nurses across the country whilst we've been battling the coronavirus crisis, helping hospices such as Derian House to continue providing vital end of life and respite care to 400 children and young adults from across the North West. Thank you." – Katie Turner, Perinatal Nurse at Derian House Children's Hospice

THE GARDENS

◻ ADMINGTON HALL
Admington, Shipston-on-Stour, CV36 4JN. Mark & Antonia Davies, 01789 450279, adhall@admingtonhall.com. *6m NW of Shipston-on-Stour. From Ilmington, follow signs to Admington. Approx 2m, turn R to Admington by Polo Ground. Continue for 1m.* Visits by arrangement May to Sept for groups of 20+. Adm £7.50, chd free. Home-made teas.
A continually evolving 10 acre garden with an established structure of innovative planning and planting. An extensive collection of fine and mature specimen trees provide the essential core structure to this traditional country garden. Features incl a lush broad lawn, orchard, water garden, large walled garden, wildflower meadows and extensive modern topiary. This is a garden in motion. Wheelchair access to most parts of garden.
&. ✿ 🚐 🛏 ☕

GROUP OPENING

◻ NEW ANSLEY GARDENS
Ansley, CV10 0QR. *Ansley is situated W of Nuneaton, adjacent to Arley. Ansley is directly off the B4114.* Sun 27 June (2-6). Combined adm £5, chd free. Light refreshments, cream teas & home-made cakes in Ansley Village Hall.

NEW **1A BIRMINGHAM ROAD**
Adrian & Heather Norgrove.

NEW **25 BIRMINGHAM ROAD**
Pat & David Arrowsmith.

NEW **188 BIRMINGHAM ROAD**
Mrs Fiona Robinson.

NEW **14 GALLEY VIEW**
Joanna Harze.

NEW **35 NUTHURST CRESCENT**
Roger & Heather Greaves.

THE OLD POLICE HOUSE
Mike & Hilary Ward.

1 PARK COTTAGES
Janet & Andy Down.

Ansley is a small ex-mining village situated in North Warwickshire. The seven gardens open are a selection of different styles and offerings. They range from a very small traditional cottage garden crammed with flowers and pots to larger gardens maximising the amazing views of the countryside. There is one with a large range of unusual plants, a country garden with mature plants and ancient roses and one new build garden. Six of the gardens are in the village with the opportunity to walk across the fields on public footpaths to the seventh garden for those who are able. The local Norman church will be open and the Morris Dancing Group will be entertaining visitors during the day. Wheelchair access to some gardens.
✿ ☕

◻ ANYA COURT CARE HOME
286 Dunchurch Road, Rugby, CV22 6JA. Karen Handley. *Opp Sainsbury's on Dunchurch Rd. If our car park is busy, please use their car park as an alternative. Please go to Anya Court reception, who will guide you to the garden.* Fri 30, Sat 31 July (2-5). Adm £5, chd free. Cream teas included.
Anya Court has a large wooded wrap around garden consisting of a number of sensory herbaceous perennial planting schemes. There is also a fruit and vegetable garden grown by the residents of the home and a colourful annual bedding and basket display. Due to the number of trees on site the garden is a good example of dry shade and drought tolerant planting for anyone requiring inspiration. Fully accessible to wheelchairs.
&. 🐕 ☕

GROUP OPENING

◻ AVON DASSETT GARDENS
Avon Dassett, CV47 2AE. *7m N of Banbury. From M40 J12 turn L & L again onto the B4100, following signs to Herb Centre & Gaydon. Take 2nd L into village (signed). Park in cemetery car park at top of hill or where signed.* Sun 4 July (2-6). Combined adm £7, chd free. Home-made teas at The Thatches.

10 AVON CARROW
Anna Prosser, 01295 690926, annaatthecarrow@btopenworld.com.
Visits also by arrangement Mar to Oct for groups of 5 to 20.
🛏

THE EAST WING, AVON CARROW
Christine Fisher & Terry Gladwin.

HILL TOP FARM
Mr D Hicks.

OLD MILL COTTAGE
Mike & Jill Lewis.

THE OLD RECTORY
Lily Hope-Frost.

POPPY COTTAGE
Audrey Butler.

THE SNUG
Mrs Deb Watts.

THE THATCHES
Trevor & Michele Gill.

Pretty Hornton stone village sheltering in the lee of the Burton Dassett hills, well wooded with parkland setting and The Old Rectory mentioned in Domesday Book. Wide variety of gardens incl kitchen gardens, cottage, gravel and tropical gardens. Range of plants incl alpines, herbaceous, perennials, roses, climbers and shrubs. The gardens are on/off the main road through the village. For 2021, we expect to run a shuttle service from top to bottom of the village. Book sale, plant sales, tombola, historic church and lunch available at the Yew Tree village pub, recently purchased by the Community, visit www.theyewtreepub.co.uk for details. Wheelchair access to most gardens.
&. 🐕 ✿ 🚐 ☕

◻ ◆ AVONDALE NURSERY
at Russell's Nursery, Mill Hill, Baginton, CV8 3AG. Gary Leaver, 07367 590620, enquiries@avondalenursery.co.uk, www.avondalenursery.co.uk. *3m S of Coventry. At junction of A45 & A46 take slip road to Baginton (Howes Lane), then 1st L to Mill Hill.* For NGS: Sun 25 July, Sun 22 Aug (11-4). Adm £3, chd free. For other opening times and information, please phone, email or visit garden website. Donation to Plant Heritage.
Vast array of flowers and ornamental grasses, incl National Collections of *Anemone nemorosa, Sanguisorba* and *Aster novae-angliae.* Choc-a-bloc with plants, our Library Garden is a well labelled reference illustrating the unusual, exciting and even some long-lost treasures. Adjacent nursery is a plantaholic's delight! Big collections of *Helenium, Crocosmia, Agapanthus, Sanguisorba* and ornamental grasses. The garden will be looking at its best in July and Aug.
&. 🐕 ✿ 🚐 NPC

GROUP OPENING

6 BERKSWELL GARDENS
Berkswell, Coventry, CV7 7BB.
7m W of Coventry. A452 to Balsall Common & follow signs to Berkswell. Tickets & maps available at each garden. Car necessary to visit all gardens. **Sun 27 June (11-6). Combined adm £6, chd free. Light refreshments at 248 Station Road & Yew Tree Barn.**

NEW 2 AGRICULTURAL COTTAGES
Mrs Jane Edwards.

BROOKSIDE HOUSE
Shirley & Frank Rounthwaite.

FAIRWAYS
Janet Lloyd-Bunt.

NEW SPENCER'S END
Gordon Clark & Nicola Content.

248 STATION ROAD
Caroline & Paul Joyner.

NEW THE TOWER HOUSE
Penny & David Stableforth.

YEW TREE BARN
Angela & Ken Shaw.

Berkswell is a beautiful village dating back to Saxon times with a C12 Norman church and has several C16 and C17 buildings including the pub. In 2014 and 2015 the village was awarded Gold in the RHS Britain in Bloom campaign, plus a special RHS award in 2014 for the Best Large Village in the Heart of England. The gardens provide great variety with fine examples of small and large, formal and informal, wild, imaginatively planted herbaceous borders and productive vegetable gardens. Something for everyone and plenty of ideas to take home. Also open to visitors is the C12 Norman church and garden. Wheelchair access to some gardens.

7 BLACKSMITHS COTTAGE
Little Compton, Moreton-in-Marsh, GL56 0SE. Mrs Andrew Lukas.
Garden entrance is opp the village hall car park. Please park at village hall or in Reed College car park nearby. **Sun 18 Apr (2-5.30). Adm £5, chd free. Home-made teas.**
This walled garden has been created in the last 6 yrs for all seasons with many fine and unusual trees, shrubs, plants and bulbs, set

around a large lawn with a beautiful Aqualens fountain at its centre. Unusual varieties of clematis climb through shrubs and up walls. There is a productive vegetable garden, greenhouse and summerhouse. All designed and made by the owners to create a sense of magic.

8 ◆ BRIDGE NURSERY
Tomlow Road, Napton, Southam, CV47 8HX. Christine Dakin & Philip Martino, 01926 812737, chris.dakin25@yahoo.com, www.bridge-nursery.co.uk. *3m E of Southam. Brown tourist sign at Napton Xrds on A425 Southam to Daventry road.* **For NGS: Mon 5 Apr (10-4). Every Sat and Sun 17 Apr to 25 Sept (10-4). Mon 3, Mon 31 May, Mon 30 Aug (10-4). Adm £3.50, chd free. For other opening times and information, please phone, email or visit garden website.**
Clay soil? Don't despair. Here is an acre of garden with an exciting range of plants which thrive in hostile conditions. Grass paths lead you round borders filled with many unusual plants, a pond and bamboo grove complete with panda! A peaceful haven for wildlife and visitors. A visitor commented 'it is garden that is comfortable with itself.' Tea or coffee and biscuits gladly provided on request.

9 BROADACRE
Grange Road, Dorridge, Solihull, B93 8QA. John Woolman, 07818 082885, jw234567@gmail.com, www.broadacregarden.org. *Approx 3m SE of Solihull. On B4101 opp The Railway Inn. Plenty of parking.* **Sun 18 Apr (2-6). Adm £5, chd free. Home-made teas. Visits also by arrangement.**
Broadacre is a semi-wild garden, managed organically. Attractively landscaped with pools, lawns and trees, beehives, and adjoining stream and wildflower meadows. Bring stout footwear to follow the nature trail. Dorridge Cricket Club is on-site (the bar will be open). Lovely venue for a picnic. Dogs and children are welcome. Excellent country pub, The Railway Inn, at the bottom of the drive.

10 6 CANON PRICE ROAD
Nursery Meadow, Barford, CV35 8EQ. Mrs Marie-Jane Roberts, 07775 584336. *From A429 turn into Barford. Park on Wellesbourne Rd & walk into Nursery Meadow by red phone box. No 6 is R in 1st close. Disabled parking by house.* **Sat 15, Sun 16 May (2-5.30). Adm £3, chd free. Sat 19, Sun 20 June, Sat 17, Sun 18 July (2-5.30). Adm £5, chd free. Light refreshments. Visits also by arrangement May to Sept.**
¾ acre mixed garden including a pond, rockery, colour themed shrubs and perennials, a nature garden and herb garden. There are areas of flowers for cutting and drying and 180 dahlias grown for exhibition, plus 7 vegetable beds and 19 types of fruit including fan trained peaches and cordon pears. 7 seating areas and colourful planted containers. Light refreshments in the garden room (please bring change). Hardy perennials for sale. Access via a single slab path that joins a wide path through the garden. No WC.

11 NEW 3 CLEEVE VIEW
Evesham Road, Salford Priors, Evesham, WR11 8UW. Ms Pip Harris. *8m from Stratford upon Avon. Enter Salford Priors from A46 r'about. The Bell will be on your L. We are a few hundred yards on the R in a red brick terrace. Small car park opp or try the church car park.* **Sat 22, Sun 23 May, Sat 17, Sun 18 July (10-4.30). Adm £3.50, chd free.**
Small cottage garden with an abundance of plants, pots and interesting metal features. Beds have been carefully planned to ensure a continuous display of colour. The beds are well stocked with a mixture of shrubs, perennials and bulbs. There is a small water feature that attracts various wildlife.

12 THE CROFT HOUSE
Haselor, Alcester, B49 6LU. Isobel & Patrick Somers, 07767 306673, ifas1010@aol.com. *6m W of Stratford-upon-Avon, 2m E of Alcester, off A46. From A46 take Haselor turn. From Alcester take old Stratford Rd, turn L signed Haselor, then R at Xrds. Garden in centre of village. Please park considerately.* **Visits by arrangement May & June for groups of 5 to 30. Adm £4, chd free. Home-made teas.**

Wander through an acre of trees, shrubs and herbaceous borders densely planted with a designer's passion for colour and texture. Hidden areas invite you to linger. Gorgeous scented wisteria on two sides of the house. Organically managed, providing a haven for birds and other wildlife. Frog pond, treehouse, small vegetable plot and a few venerable old fruit trees from its days as a market garden.

✿ ☕

GROUP OPENING

 EARLSDON GARDENS
Coventry, CV5 6FS. *Turn towards Coventry at A45 & A429 T-lights. Take 3rd L into Beechwood Ave, continue ½ m to St Barbara's Church at Xrds with Rochester Rd. Maps & tickets at St Barbara's Church Hall.* Mon 3 May (11-4). Combined adm £5, chd free. Light refreshments at St Barbara's Church Hall.

43 ARMORIAL ROAD
Gary & Jane Flanagan.

3 BATES ROAD
Victor Keene MBE.

59 THE CHESILS
John Marron & Richard Bantock.

28 CLARENDON STREET
Ruth & Symon Whitehouse.

NEW **26 CONISTON ROAD**
Judith & Colin Yates.

NEW **38 CONISTON ROAD**
Louise & Martin Prue.

27 HARTINGTON CRESCENT
David & Judith Bogle.

40 HARTINGTON CRESCENT
Viv & George Buss.

114 HARTINGTON CRESCENT
Liz Campbell & Denis Crowley.

40 RANULF CROFT
Spencer & Sue Swain.

2 SHAFTESBURY ROAD
Ann Thomson & Bruce Walker.

23 SPENCER AVENUE
Susan & Keith Darwood.

Varied selection of town gardens from small to more formal with interest for all tastes incl a mature garden with deep borders bursting with spring colour; a large garden with extensive lawns and an array of rhododendrons, azaleas and large mature trees; densely planted town garden with sheltered patio area and wilder woodland; and a surprisingly large garden offering interest to all ages! There is also a pretty garden set on several levels with hidden aspect; a large peaceful garden with water features and vegetable plot; a large mature garden in peaceful surroundings; and a plantaholic's garden with a large variety of plants, clematis and small trees.

🐕 ✿ ☕

Online booking is available for many of our gardens, at ngs.org.uk

Spencer's End, Berkswell Gardens

14 GUY'S CLIFFE WALLED GARDEN

Coventry Road, Guy's Cliffe, Warwick, CV34 5FJ. Sarah Ridgeway, www.guyscliffewalledgarden.org.uk. *Behind Hintons Nursery in Guy's Cliffe. Guy's Cliffe is on the A429, between North Warwick & Leek Wootton.* **Sat 24 July (10-3.30). Adm £3.50, chd free. Light refreshments.**
A Grade II listed garden of special historic interest, having been the kitchen garden for Guy's Cliffe House. The garden dates back to the mid-1700s. Restoration work started 7 yrs ago using plans from the early C19. The garden layout has already been reinstated and the beds, once more, planted with fruit, flowers and vegetables incl many heritage varieties. Glasshouses awaiting restoration. Original C18 walls. Exhibition of artefacts discovered during restoration. Wheelchair access through entrance of Hintons Nursery and paths through garden. WC with disabled access.

15 ◆ HILL CLOSE GARDENS

Bread and Meat Close, Warwick, CV34 6HF. Hill Close Gardens Trust, 01926 493339, centremanager@hcgt.org.uk, www.hillclosegardens.com. *Town centre. Follow signs to Warwick racecourse. Entry from Friars St onto Bread and Meat Close. Car park by entrance next to racecourse. 2 hrs free parking. Disabled parking outside the gates.* **For NGS: Sat 13 Feb, Sat 30 Oct (11-4). Adm £4.50, chd £1. Light refreshments in Visitor Centre. Gluten free options. For other opening times and information, please phone, email or visit garden website. Donation to Plant Heritage.**
Restored Grade II* Victorian leisure gardens comprising 16 individual hedged gardens, 8 brick summerhouses. Herbaceous borders, heritage apple and pear trees, C19 daffodils, over 100 varieties of snowdrops, many varieties of asters and chrysanthemums. Heritage vegetables. Plant Heritage border, auricula theatre, and Victorian style glasshouse. Children's garden. Wheelchair available, please phone to book in advance.

16 THE HILL COTTAGE

Kings Lane, Snitterfield, Stratford-upon-Avon, CV37 0QA. Gillie & Paul Waldron, 07895 369387, info@thehillcottage.co.uk, www.thehillcottage.co.uk. *5 mins from M40 J15. Take A46 Stratford, 1m take 2nd exit at r'about, ½m R into Kings Lane, over A46, 1st R, 2nd house on L. Or from village, up White Horse Hill, R at T-junction, over A46, R into Kings Lane, 2nd on L.* **Visits by arrangement Apr to Sept for groups of up to 30. Adm £5, chd free. Home-made teas.**
High on a ridge overlooking orchards and golf course with fabulous views to distant hills, this 2¼ acre garden, full of surprises, offers varied planting; sunny gravel with exotic specimens; cool, shady woodland with relaxed perennial groups; romantic green oak pond garden and stone summerhouse; pool with newts and dragonflies. Traditional glasshouse in walled kitchen garden was the big project in 2019. Bluebell wood best in late April. Gazebo, pergolas, lavender walk, pond garden, clematis, and too many pots to count!

Ascott Rise, Whichford & Ascott Gardens

GROUP OPENING

17 HONINGTON GARDENS

Shipston-on-Stour, CV36 5AA. 1½ m N of Shipston-on-Stour. Take A3400 towards Stratford-upon-Avon, then turn R signed Honington. Sun 20 June (2-5.30). Combined adm £6, chd free. Home-made teas.

HONINGTON GLEBE
Mr & Mrs J C Orchard.

HONINGTON HALL
B H E Wiggin.

MALT HOUSE RISE
Mr P Weston.

THE MALTHOUSE
Mr & Mrs R Hunt.

THE OLD HOUSE
Mr & Mrs I F Beaumont.

ORCHARD HOUSE
Mr & Mrs Monnington.

SHOEMAKERS COTTAGE
Christopher & Anne Jordan.

C17 village, recorded in Domesday, entered by old toll gate. Ornamental stone bridge over the River Stour and interesting church with C13 tower and late C17 nave after Wren. Seven super gardens. 2 acre plantsman's garden consisting of rooms planted informally with yr-round interest in contrasting foliage and texture, lily pool and parterre. Extensive lawns and fine mature trees with river and garden monuments. Secluded walled cottage garden with roses, and a structured cottage garden formally laid out with box hedging and small fountain. Small, developing garden created by the owners with informal mixed beds and borders. Wheelchair access to most gardens.

👤 🐄 ✻ 🚗 ☕

National Garden Scheme gardens are identified by their yellow road signs and posters. You can expect a garden of quality, character and interest, a warm welcome and plenty of home-made cakes!

GROUP OPENING

18 ILMINGTON GARDENS

Ilmington, CV36 4LA. 8m S of Stratford-upon-Avon. 8m N of Moreton-in-Marsh. 4m NW of Shipston-on-Stour off A3400. 3m NE of Chipping Campden. Sat 29, Sun 30 May (12.30-6). Combined adm £7, chd free. Home-made teas in Ilmington Community Shop, Upper Green (Sat) & at the village hall (Sun). Donation to Shipston Home Nursing.

THE BEVINGTONS
Mr & Mrs N Tustain.

CHERRY ORCHARD
Mr Angus Chambers.

COMPTON SCORPION FARM
Mrs Karlsen.

THE DOWER HOUSE
Mr & Mrs M Tremellen.

FOXCOTE HILL
Mr & Mrs Michael Dingley.

FROG ORCHARD
Mr & Mrs Jeremy Snowden.

GRUMP COTTAGE
Mr & Mrs Martin Underwood.

ILMINGTON MANOR
Mr Martin Taylor.

OLD FOX HOUSE
Bob & Sarah Booboo.

RAVENSCROFT
Mr & Mrs Clasper.

STUDIO COTTAGE
Sarah Hobson.

Ilmington is an ancient hillside Cotswold village 2m from the Fosse Way with two good pubs and splendid teas. Buy your ticket at Ilmington Manor (next to the Red Lion Pub); wander the 3 acre gardens with fish pond. Then walk to the upper green behind the village hall to Foxcote Hill's large gardens and Old Fox House, then tiny Grump Cottage's small stone terraced suntrap. Up Grump St to Ravenscroft's large sculpture filled sloping vistas commanding the hilltop. Walk to nearby Frog Lane, view cottage gardens of Cherry Orchard, Frog Orchard, and Studio Cottage. Then to the Bevingtons many-chambered cottage garden at the bottom of Valanders Lane near the church and manor ponds and beyond to the Dower House in Back Street. Also, visit the delightful Compton

Scorpion Farmhouse's magic garden 1m away. The Ilmington Morris Men performing round the village on Sun 30th May only.

GROUP OPENING

19 KENILWORTH GARDENS

Kenilworth, CV8 1BT. Fieldgate Lane, off A452. Parking available at Abbey Fields. Street parking on Fieldgate Lane (limited), Siddley Ave & Beehive Hill. Tickets & maps available at most gardens. Sun 20 June (12-5). Combined adm £6, chd free. Home-made teas at St Nicholas Parochial Hall from 1pm.

BEEHIVE HILL ALLOTMENTS
Kenilworth Allotment Association.

FIELDGATE
Liz & Bob Watson, 01926 512307, liz.watson@ngs.org.uk.
Visits also by arrangement May to Sept for groups of 5 to 20.

14C FIELDGATE LANE
Sandra & Bob Aulton.

65 RANDALL ROAD
Mrs Jan Kenyon.

2 ST NICHOLAS AVENUE
Mr Ian Roberts.

ST NICHOLAS PAROCHIAL HALL
St Nicholas Church.

1 SIDDELEY AVENUE
Clare Wightman.

NEW **TREE TOPS**
Joanna & George Illingworth.

Kenilworth was historically a very important town in Warwickshire. It has one of England's best castle ruins, Abbey Fields and plenty of pubs and good restaurants. The gardens open this yr are very varied. There are small and large gardens, formal, contemporary and cottage styles with trees, shrubs, herbaceous borders, ponds and more intimate, wildlife friendly areas, plus plenty of vegetables at the allotments. Many of the gardens have won Gold in the Kenilworth in Bloom garden competition. Partial wheelchair access to many of the gardens.

👤 🐄 ✻ ☕

GROUP OPENING

20 LIGHTHORNE GARDENS
Lighthorne, Warwick, CV35 0AR. *10m S of Warwick. Lighthorne will be signed from the Fosse Way & B4100.* **Sun 6 June (2-5.30). Combined adm £6, chd free. Home-made teas at the village hall.**

1 CHURCH HILL COURT
Irene Proudman.

NEW **3 CHURCH HILL COURT**
Nick & Marie.

4 CHURCH HILL COURT
Carol Schofield & Martin Preedy.

NEW **6 CHURCH HILL COURT**
Rachel Edgington.

THE OLD RECTORY
The Hon Lady Butler.

THE PADDOCK
Martin & Lesley Thornton.

NEW **ST CLEMENTS**
Mike & Carol Smith.

Lighthorne is a compact, pretty village between the Fosse Way and the B4100, with a charming church (open), pub, cafe and village hall. Featuring the spectacular walled garden of The Old Rectory to smaller, designer, contemporary and traditional gardens; there is plenty to interest the visitor. A total of seven gardens will be open. Disabled parking at the village hall. Partial wheelchair access to some of the gardens.

&. ⛟ ☕

GROUP OPENING

21 LONG ITCHINGTON VILLAGE GARDENS
Long Itchington, nr Southam, CV47 9PD. *2½m N of Southam on A423 between Coventry & Banbury, next to Grand Union Canal. Leamington 6m W, also access by A426 from Rugby & A425 from Daventry. Car parks signed at all entries to village. Maps of the village showing all open gardens available at the main car parks and church.* **Sat 3, Sun 4 July (12-5). Combined adm £7, chd free. Light refreshments in Holy Trinity Church, Church Road.**

BAKEHOUSE COURT
Martyn & Erica Smith.

1 BRAKELEY COTTAGES
Andy & Sue Jack.

CHERRYWOOD
Andy & Rosie Skilbeck.

38 DALE CLOSE
Simon & Jeni Neale.

IFFLEY LODGE
John Glare.

NEW **LAKE VIEW**
Chris & Phil Lawrence.

19 LEIGH CRESCENT
Tony Shorthouse, 01926 817192, Tonyshorthouse19@gmail.com. **Visits also by arrangement June to Sept for groups of 5 to 10.**

MEADOW COTTAGE
Charlotte Griffin, 07717 484108.
🛏

3 ODINGSELL DRIVE
Adrienne & Steve Mitchell.

20 ODINGSELL DRIVE
Jean & Gerry Bailey.

NEW **6 RUSSELL CLOSE**
Louis & Courtney Adam.

SANDY ACRE
David & Janis Tait.

NEW **6 SHORT LANE**
Alan & Jean Huitson.

8 THE SQUARE
George & Janet Powell.

NEW **THORNFIELD**
Harvey & Margaret Bailey.

THE WILLOWS
Simon & Charlotte Collyer.

Opening its gardens for the second time in 2021 Long Itchington is a large village in south Warwickshire next to the Grand Union Canal. It features a network of waymarked paths and several historic buildings including a half-timbered Tudor house where Queen Elizabeth I once stayed, a C15 Manor House and a C12/C13 church. It also boasts the biggest village pond in Warwickshire, a village green, six pubs, a Co-op and a public park and playing field with children's playground, picnic area and a wildflower meadow. Gardens of all types are featured including gravel, tropical, vegetable, cottage and contemporary. These range in size from small/courtyard gardens to an extensive landscaped garden with panoramic countryside views. Plant sales.

 ☕

22 NEW THE MALT HOUSE
Charlecote, Warwick, CV35 9EW. Katriona & Rupert Collins. *10 mins from J15 M40, take the A429. About 2m along the A429 take R turn to Charlecote. Enter the village passing telephone box on L, The Malt House is approx 100 metres past this on the L.* **Sun 11 July (12-5). Adm £5, chd free. Home-made teas & light refreshments.**
Lovely Grade II listed house (not open) set in charming hamlet of Charlecote, on the edge of Charlecote House and deer park. Beautiful borders offering interest and colour throughout the yr and featuring a small pond which attracts an abundance of wildlife. Enjoy mixed borders brimming with agapanthus and beautiful pots. Walk around the vegetable plot and cutting patch and admire a large variety of dahlias. There will be tea and delicious cake for sale, plus hand tied bunches of flowers. Access over gravel drive and 2 steps to the main border area.

🐕 ☕

23 MAXSTOKE CASTLE
Coleshill, B46 2RD. G M Fetherston-Dilke. *2½m E of Coleshill. E of Birmingham, on B4114. Take R turn down Castle Lane, Castle drive 1¼m on R.* **Sun 6 June (11-5). Adm £8, chd £5. Home-made teas.**
Approx 5 acres of garden and grounds with herbaceous, shrubs and trees in the immediate surroundings of this C14 moated castle. No wheelchair access to house.

&. ✿ ☕

24 THE MOTTE
School Lane, Hunningham, Leamington Spa, CV33 9DS. Margaret & Peter Green, 01926 632903, margaretegreen100@gmail.com. *5m E of Leamington Spa. 1m NW off B4455 (Fosse Way) at Hunningham Hill Xrds.* **Visits by arrangement May to Aug for groups of up to 20. Adm £5, chd free. Home-made teas by prior request.**
Plant lover's garden of about ⅓ acre, well stocked with unusual trees, shrubs and herbaceous plants. Set in the quiet village of Hunningham with views over the River Leam and surrounding countryside. The garden has been developed over the last 18yrs and incl woodland, exotic and herbaceous borders, raised alpine

bed, troughs, pots, wildlife pond, fruit and vegetable plot. Plant-filled conservatory. Wheelchair access to front and upper rear garden.

 ♿ 🐕 ♨

25 OAK HOUSE
Waverley Edge, Bubbenhall, Coventry, CV8 3LW. Helena Grant, 07731 419685, helena.grant@btinternet.com. *15 mins from Leamington Spa via the Oxford Rd/A423 & the Leamington Rd/A445 & the A46. Spaces for 4 cars only.* Visits by arrangement Apr to Oct for groups of up to 20. Adm £5, chd free. Tea & cake. Tucked away next to Waverley Woods, Oak House enjoys a walled garden that has been landscaped and extended over 30 yrs. The garden is split on 2 levels with 7 seating areas allowing for relaxed appreciation of every aspect of the garden, with peaceful places to sit, ponder and enjoy. A focal point is the large terracotta urn, over 60 yrs old, which delivers vertical interest and the summerhouse and arbour which face each other diagonally across the garden. The curved borders have a wide range of planting creating distinct areas which surround the lawn. There are over 80 different plant varieties giving yr-round interest which is a haven for birds and other wildlife. Tea & cake £3 per person. Wheelchair access only on level path which runs round the house.

 ♿ ✿ ♨

26 PACKINGTON HALL
Meriden, nr Coventry, CV7 7HF. Lord & Lady Guernsey, www.packingtonestate.co.uk. *Midway between Coventry & Birmingham on A45. Entrance 400yds from Stonebridge Island towards Coventry. For SatNav please use CV7 7HE.* Sun 13 June (2-5). Adm £6, chd free. Home-made teas. Packington is the setting for an elegant Capability Brown landscape. Designed from 1751, the gardens incl sweeping lawns down to a serpentine lake, impressive cedars of Lebanon, wollingtonias and various other trees, a 1762 Japanese bridge, a millennium rose garden, wildflower meadow and mixed terrace borders. Home-made teas on the terrace or in The Pompeiin Room if wet.

 🐕 ♨ ♨

GROUP OPENING

27 PEBWORTH GARDENS
Stratford-upon-Avon, CV37 8XZ. www.pebworth.org/ngs-open-gardens. *7m SW of Stratford-upon-Avon. For parking & SatNav please use CV37 8XN.* Sun 30, Mon 31 May (1-5.30). Combined adm £7, chd free.

BANK HOUSE
Clive & Caroline Warren.

FELLY LODGE
Maz & Barrie Clatworthy.

IVYBANK
Mr & Mrs R Davis.
[NPC]

THE KNOLL
Mr & Mrs K Wood.

MAPLE BARN
Richard & Wendi Weller.

MEON COTTAGE
David & Sally Donnison.

THE OLD BARN
Kevin & Tracey Morley.

PEBWORTH ALLOTMENTS
Les Madden.

PETTIFER HOUSE
Mr & Mrs Michael Veal.

4 WESLEY GARDENS
Anne Johnson.

Pebworth is a delightful village with thatched cottages and properties young and old. There are a variety of garden styles from cottage gardens to modern, walled and terraced gardens. Pebworth is topped by St Peter's Church which has a large ring of ten bells, unusual for a small rural church. This yr we have 9 gardens and the Pebworth Allotments opening, with scrumptious tea and cakes provided by the Pebworth WI in the village hall. The Pebworth Allotments have only been in existence a few yrs and residents have lovingly tended to them. They are smart and productive! Some of the Allotmenteers are real characters. Do visit and have a chat, they have a wealth of knowledge. Partial wheelchair access in some gardens. Ramp available for village hall.

 ♿ ✿ ♨ ☕ ♨

16 Delaware Road, Styvechale Gardens

The Knoll, Pebworth Gardens

28 PRIORS MARSTON MANOR
The Green, Priors Marston,
CV47 7RH. Dr & Mrs Mark
Cecil, 07375 873921,
gardenwithmatt@gmail.com.
*8m SW of Daventry. Off the A361
between Daventry & Banbury at
Charwelton. Follow signs to Priors
Marston, approx 2m. Arrive at
T-junction with a war memorial on
R. The manor will be on your L.* **Fri
11 June (11-6). Adm £5, chd free.
Home-made teas. Includes an
indoor Bonhams Valuation Day
(11-2), suggested donation £2 per
item, proceeds to NGS. Visits also
by arrangement in June for groups
of 5 to 10.**
Arrive in Priors Marston village and
explore the manor gardens. Greatly
enhanced by present owners to relate
back to a Georgian manor garden
and pleasure grounds. Wonderful
walled kitchen garden provides
seasonal produce and cut flowers for
the house (not open). Herbaceous
flower beds and a sunken terrace with
water feature by William Pye. Lawns
lead down to the lake around which
you can walk amongst the trees and
wildlife with stunning views up to the
house and garden aviary. Sculpture
on display. Partial wheelchair access.

&. ☕

GROUP OPENING

**29 NEW STONELEIGH VILLAGE
GARDENS**
Stoneleigh, CV8 3DP. *The village is
situated 5m N of Leamington Spa &
5m S of Coventry off the A46. Car
parking will be signed in the centre
of the village.* **Sun 27 June (2-6).
Combined adm £5, chd free. Tea
at Stoneleigh Village Club.**

NEW **BRIDGE END**
Tim & Nicky Sawdon.

NEW **HOLLY HOUSE**
Janet & David Gibson.

NEW **THE OLD POST OFFICE**
Peter & Jan Whitehouse.

NEW **3 WALKERS ORCHARD**
Keith & Julie Walker.

Opening its gardens for the first time
in 2021 Stoneleigh is an ancient
village. Cistercian monks founded
nearby Stoneleigh Abbey in the middle
of the C12. After the dissolution of
the monasteries, in 1561 the Abbey
passed to the Leigh family and the
Estate remained in their possession

for 400 yrs. The development of
the Estate into a thriving agricultural
community led to the need for housing
for workers, and the almshouses in
the village date back to 1594. Teas will
be served in the village club, originally
a Reading Room founded in 1856
to promote knowledge and learning
in the community. The gardens
open in the village reflect different
styles and are all interesting in their
own unique way. There are mature
shrubs, and perennials in herbaceous
borders, as well as small vegetable
plots, and some greenhouses and
summerhouses.

✿ ☕

GROUP OPENING

30 STYVECHALE GARDENS
Baginton Road, Coventry,
CV3 6FP. styvechale-gardens.
wixsite.com/ngs2020. *The
gardens are located on the s-side
of Coventry close to A45. Tickets &
map available on the day from West
Orchard United Reformed Church,
The Chesils, CV3 6FP. Advance
tickets available from suepountney@
btinternet.com.* **Sun 13 June
(11-5). Combined adm £5, chd
free. Home-made teas. Light
refreshments throughout the day
& hot bacon or sausage batches
from 11-2. Donation to Coventry
Myton Hospice.**

11 BAGINTON ROAD
Ken & Pauline Bond.

164 BAGINTON ROAD
Fran & Jeff Gaught.

16 DELAWARE ROAD
Val & Roy Howells, 02476 419485,
valshouse@hotmail.co.uk.
**Visits also by arrangement May
to Sept for groups of 10 to 30.**

2 THE HIRON
Sue & Graham Pountney,
02476 502044,
suepountney@btinternet.com.
**Visits also by arrangement May
to Sept for groups of 10 to 30.**

177 LEAMINGTON ROAD
Barry & Ann Suddens.

27 RODYARD WAY
Jon & Karen Venables.
Ⓓ

A collection of lovely, mature, suburban
gardens, each one different in style and
size. Come and enjoy the imaginatively
planted herbaceous borders,
spectacular roses, water features, fruit

and vegetable patches, cottage garden
planting and shady areas, something
for everyone and plenty of ideas for you
to take home. Relax in the gardens and
enjoy the warm, friendly welcome you
will receive from us all. Plants for sale in
some gardens. Several other gardens
will be open on the day, including 40
Ranulf Croft.

🐄 ✿ 🚗 ☕

GROUP OPENING

31 TYSOE GARDENS
Tysoe, Warwick, CV35 0SE. *W
of A422, N of Banbury (9m). E of
A3400 & Shipston-on-Stour (4m).
N of A4035 & Brailes (3m). Parking
on recreation ground CV35 0SE.
Entrance tickets & maps at village
hall.* **Free bus. Sat 5, Sun 6 June
(2-6). Combined adm £6, chd free.
Home-made teas in Tysoe Village
Hall & cold drinks at Garden
Cottage.**

DINSDALE HOUSE
Julia & David Sewell.

**GARDEN COTTAGE &
WALLED KITCHEN GARDEN**
Sue & Mike Sanderson,
www.twkg.co.uk.

IVYDALE
Sam & Malcolm Littlewood.

KERNEL COTTAGE
Christine Duke.

THE OLD POLICE HOUSE
Bridget & Digby Norton.

After a very strange 2020 and being
unable to open we are all longing to get
our gardens into shape and welcome
you in 2021. Tysoe, an original Hornton
stone village, stands on the north-east
foothills of the Cotswolds. Our gardens
and houses are diverse in character,
size and in their planting. We are
opening slightly later so have a few
extra days in early June and hope that
the weather is kind and encourages
the slothful summer plants into bloom
for us! However, you can be sure that
Tysoe will offer a happy atmosphere,
some glorious gardens, a good walk
round the village or, if you prefer, a free
bus ride, and a jolly good tea made
by the large band of WI bakers and
other cooking enthusiasts living in the
village. A warm welcome awaits you
in this buzzy, energetic, friendly village
and gardening community. Come
and check us out! Partial wheelchair
access.

&. 🐄 ✿ 🚗 ☕

GROUP OPENING

32 WARMINGTON GARDENS
Banbury, OX17 1BU. *5m NW of Banbury. Take B4100 N from Banbury, after 5m turn R across short dual carriageway into Warmington. From N take J12 off M40 onto B4100.* **Sun 27 June (1-5). Combined adm £6, chd free. Light refreshments at village hall.**

2 CHAPEL STREET
c/o Mark Broadbent, Group Coordinator.

GOURDON
Jenny Deeming.

GREENWAYS
Mark Broadbent.

NEW HILL COTTAGE, SCHOOL LANE
Mr Mike Jones.

LANTERN HOUSE
Peter & Tessa Harborne.

THE MANOR HOUSE
Mr & Mrs G Lewis.

THE ORCHARD
Mike Cable.

SPRINGFIELD HOUSE
Jenny & Roger Handscombe, 01295 690286, jehandscombe@btinternet.com.

1 THE WHEELWRIGHTS
Ms E Bunn.

Warmington is a charming historic village, mentioned in the Doomsday Book, situated at the north-east edge of the Cotswolds in a designated AONB. There is a large village green with a pond overlooked by an Elizabethan Manor House (not open). There are other historic buildings including St Michael's Church, The Plough Inn and Springfield House all dating from the C16 or before. There is a mixed and varied selection of gardens to enjoy during your visit to Warmington. These incl the formal knot gardens and topiary of The Manor House, cottage and courtyard gardens, terraced gardens on the slopes of Warmington Hill and orchards containing local varieties of apple trees. Some gardens will be selling homegrown plants. WC at village hall, along with delicious home-made cakes, and hot and cold drinks. Warmington is on a hill with many steps and gravel driveways which could be difficult for wheelchair access.

Lantern House, Warmington Gardens

GROUP OPENING

33 WELFORD-ON-AVON GARDENS

Welford-on-Avon, CV37 8PT. *5m SW of Stratford-upon-Avon. Off B439 towards Bidford-on-Avon from Stratford upon Avon.* **Sat 26, Sun 27 June (1-5). Combined adm £5, chd free. Home-made teas & WC in Village Memorial Hall.**

ASH COTTAGE
Peter & Sue Hook.

ELM CLOSE
Eric & Glenis Dyer, 01789 750793, glenisdyer@gmail.com.
Visits also by arrangement Feb to Oct for groups of 10 to 30.

NEW THE OLD FORGE
Karen Dickinson.

THE OLD RECTORY
Frank Kennedy.

6 QUINEYS LEYS
Gordon & Penny Whitehurst.

NEW ROWAN HOUSE
Lynne & Dave Smith, 01789 752676, Lynne.studley@gmail.com.
Visits also by arrangement in June.

5 WILLOWBANK
Mrs Jane Badcock.

Welford on Avon has a superb position on the river with serene swans, dabbling ducks and resident herons. It also has a beautiful C12 church and a selection of pubs serving great food. There are many different house styles and an abundance of beautiful cottages with thatched roofs, some dating from C17. The village has chocolate-box charisma! Welford also has many keen gardeners. The gardens that are opening range from small to large, from established to newly designed and planted, and incl some with fruit and vegetable plots.

34 WELLESBOURNE ALLOTMENTS

Kineton Road, Wellesbourne, Warwick, CV35 9NE. Wellesbourne Allotments, www. wellesbourneallotments.co.uk. *5m E of Stratford-upon-Avon. On LH-side of Kineton Rd (B4086) E of Wellesbourne, 400 metres from shops in precinct.* **Sun 1 Aug (2-5). Adm £4, chd free. Home-made teas.**

Garden Cottage & Walled Kitchen Garden, Tysoe Gardens

Follow the history of Wellesbourne Allotments dating from 1838 and discover the importance of the Joseph Arch story featured in BBC's Countryfile. Find out how allotments contribute to wellbeing today and can help people living with dementia. Impressive vegetables, delicious fruits and beautiful flowers abound on the site offering interest to novice and experienced gardeners of all ages. A large, lush and lovely rural site growing a huge range of crops. Meet enthusiastic gardeners and sample delicious home-made teas. Plants and produce for sale, scarecrows, children's questionnaire, bee-keepers tent, a landscaped dementia-friendly plot, music, stalls, vintage ice cream. Wheelchair access over level site with hard surface entrance and roadway.

GROUP OPENING

35 WHICHFORD & ASCOTT GARDENS

Whichford & Ascott, Shipston-on-Stour, CV36 5PG. *6m SE of Shipston-on-Stour. For parking please use CV36 5PG. We have a large car park.* **Sun 20 June (1-5). Combined adm £6, chd free. Home-made teas.**

ASCOTT LODGE
Charlotte Copley.

NEW ASCOTT RISE
Carol & Jerry Moore.

BELMONT HOUSE
Robert & Yoko Ward.

THE OLD RECTORY
Peter & Caroline O'Kane.

PLUM TREE COTTAGE
Janet Knight.

THE WHICHFORD POTTERY
Jim & Dominique Keeling, www.whichfordpottery.com.

The gardens in this group reflect many different styles. The two villages are in an AONB, nestled within a dramatic landscape of hills, pasture and woodland, which is used to picturesque effect by the garden owners. Fine lawns, mature shrub planting and much interest to plantsmen provide a peaceful visit to a series of beautiful gardens. Many incorporate the inventive use of natural springs, forming ponds, pools and other water features. Classic cottage gardens contrast with larger and more classical gardens which adopt variations on the traditional English garden of herbaceous borders, climbing roses, yew hedges and walled enclosures. Partial wheelchair access as some gardens are on sloping sites.

WILTSHIRE

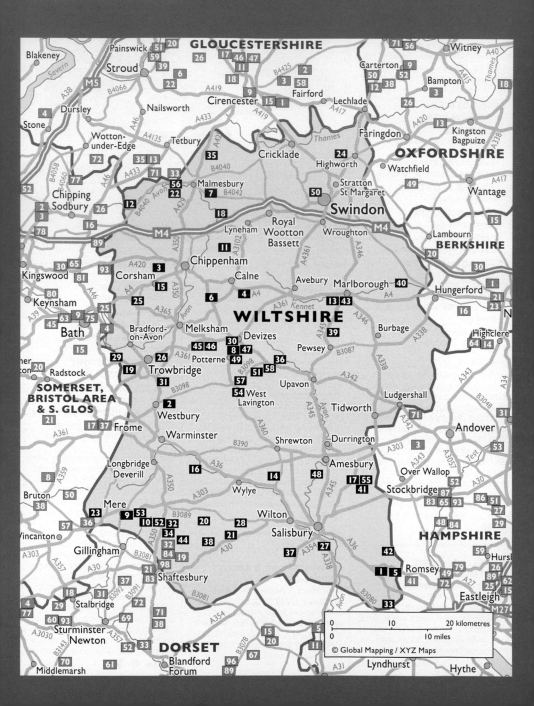

© Global Mapping / XYZ Maps

Wiltshire, a predominantly rural county, covers 1,346 square miles and has a rich diversity of landscapes, including downland, wooded river valleys and Salisbury Plain.

Chalk lies under two-thirds of the county, with limestone to the north, which includes part of the Cotswold Areas of Outstanding Natural Beauty. The county's gardens reflect its rich history and wide variety of environments.

Gardens opening for the National Garden Scheme include the grade II listed Edwardian style gardens of Hazelbury Manor and large privately owned gems such as Knoyle Place, Urchfont Manor and Cadenham Manor, together with more modest properties that are lovingly maintained by the owners, such as Cottage in the Trees in Boscombe Village and North Cottage in Tisbury.

The season opens with snowdrops at Westcroft, daffodils at Fonthill House and magnolias in spring at Corsham Court and Broadleas House Gardens.

A wide selection of gardens, large and small, are at their peak in the summer. There are also four village openings which number a total of twenty three gardens, farmhouses and a sensory garden at Julia's House Children's Hospice. There is something to delight the senses from January to September and one will always have a warm welcome wherever one goes.

Below: Urchfont Manor pool

Volunteers

County Organiser
Amelia Tester 01672 520218
amelia.tester@ngs.org.uk

County Treasurer
Tony Roper 01249 447436
tony.roper@ngs.org.uk

**Publicity
& Booklet Co-ordinator**
Tricia Duncan 01672 810443
tricia.duncan@ngs.org.uk

Social Media
Maud Peters 07595 266299
maudahpeters@gmail.com

Assistant County Organisers
Suzie Breakwell 01985 850297
suzievb@me.com

Sarah Coate 01722 782365
sarahpcoate@gmail.com

Annabel Dallas 01672 520266
annabel.dallas@btinternet.com

Alexandra Davies 01747 860351
alex@theparishhouse.co.uk

Ros Ford 01380 722778
ros.ford@ngs.org.uk

Jo Hankey 01722 742472
rbhankey@gmail.com

Alison Parker 01380 722228
alison.parker@ngs.org.uk

Diana Robertson 01672 810515
diana.robertson@ngs.org.uk

@WiltshireNGS
@WiltshireNgs
@Wiltshirengs

OPENING DATES

All entries subject to change. For latest information check www.ngs.org.uk

Map locator numbers are shown to the right of each garden name.

January

Every Thursday from Thursday 14th
Westcroft 55

February

Snowdrop Festival

Every Thursday
Westcroft 55

Sunday 7th
Westcroft 55

Friday 12th
Westcroft 55

Saturday 20th
Westcroft 55

Sunday 21st
Westcroft 55

Saturday 27th
Westcroft 55

Sunday 28th
Westcroft 55

March

Thursday 4th
Westcroft 55

Sunday 21st
◆ Corsham Court 15
Fonthill House 20

April

Sunday 18th
Broadleas House
  Gardens 8
Foxley Manor 22

Monday 19th
◆ Bowood Woodland
  Gardens 6

Thursday 22nd
Blackland House 4

Saturday 24th
◆ Iford Manor
  Gardens 29
NEW Knoyle Place 32
NEW Lower Lye 34

Sunday 25th
◆ Corsham Court 15
Cottage in the Trees 17
Oare House 39

May

Saturday 1st
Blackland House 4

Sunday 2nd
◆ Waterdale House 52

Sunday 9th
NEW Bush Farm 10
Cottage in the Trees 17

Sunday 16th
Broadleas House
  Gardens 8

Thursday 20th
Windmill Cottage 57

Friday 21st
Windmill Cottage 57

Saturday 22nd
Foxley Manor 22

Sunday 23rd
North Cottage 38
NEW Ordnance House 42
◆ Twigs Community
  Garden 50

Wednesday 26th
Hazelbury Manor
  Gardens 25

Monday 31st
NEW Tristenagh House 49

June

Thursday 3rd
Cadenham Manor 11
Windmill Cottage 57

Friday 4th
Windmill Cottage 57

Saturday 5th
Fovant House 21
West Lavington Manor 54

Sunday 6th
NEW Abbotstone House 1
Hyde's House 28

Sunday 13th
NEW Burton Grange 9
NEW Bush Farm 10
Cottage in the Trees 17
Dauntsey Gardens 18
Hannington Village
  Gardens 24
Manor Farm 35
North Cottage 38
The Old Rectory,
  Boscombe 41
NEW West Knoyle
  Gardens 53

Saturday 19th
Hilperton House 26
NEW Urchfont Manor 51

Sunday 20th
NEW Brillscote House 7
Broadleas House
  Gardens 8
Hilperton House 26
Landford Village
  Gardens 33
Manor Farm House 36

Tuesday 22nd
Whatley Manor 56

Sunday 27th
Duck Pond Barn 19
Oare House 39

July

Thursday 1st
Windmill Cottage 57

Friday 2nd
Windmill Cottage 57

Saturday 3rd
NEW Julia's House
  Children's Hospice 30
Seend House 45
Seend Manor 46

Sunday 4th
Cherry Orchard Barn 12
Hazelbury Manor
  Gardens 25
NEW Julia's House
  Children's Hospice 30
1 Southview 47
NEW Teasel 48

Sunday 18th
◆ Twigs Community
  Garden 50

Sunday 25th
The Coach House 13
The Old Mill 40
Peacock Cottage 43

Brillscote House

© Susie Bell

THE GARDENS

1 NEW **ABBOTSTONE HOUSE**
Whiteparish, Salisbury, SP5 2SH.
Mrs Andrew Lax. *On A27 in The
Street Whiteparish. 100yds from
shop and church on A27 heading
towards Salisbury. Large box hedge
with iron gates, parking in field along
driveway.* **Sun 6 June (2-6). Adm
£5, chd free. Delicious home-
made cakes and tea.**
Large garden surrounding 400 year
old Jacobean house with many roses,
herbaceous and shady borders,
pond, vegetable garden, walled
garden surrounding the swimming
pool. Variety of mature trees.

2 **BEGGARS KNOLL CHINESE
GARDEN**
Newtown, Westbury,
BA13 3ED. Colin Little &
Penny Stirling, 01373 823383,
silkendalliance@talktalk.net. *1m
SE of Westbury. Turn off B3098 at
White Horse Pottery, up hill towards
the White Horse for ³/₄ m. Parking
at end of drive for 10-12 cars.*
**Visits by arrangement June &
July for groups of 5 to 20. Adm
£6, chd free. Tea and cake by
arrangement costs £4.**
A Chinese-style garden with 12
individual garden rooms in which
mosaic paths wind past pavilions,
ponds and many rare Chinese
trees, shrubs and flowers. Garden
rooms are separated by Chinese
gateways incl moongate. Set
beneath a beechwood in 1 acre

of steep chalkland. Large potager
with chickens, a lawn for teas and
spectacular views to the Mendips.
Free garden map and plant guide.

3 **BIDDESTONE MANOR**
Chippenham Lane, Biddestone,
SN14 7DJ. Rosie Harris, Head
Gardener, 01249 713211. *On A4
between Chippenham & Corsham
turn N. from A420, 5m W of
Chippenham, turn S.* **Visits by
arrangement Apr to June for
groups of 5 to 10. Guided walk
and refreshments £10, otherwise
£5. Home-made teas or a glass of
wine for evening visits.**
Cotswold stone C17 manor house
(not open) with 5 acres of garden
to enjoy. Lake, ponds and streams,
arboretum, vegetable and cutting
gardens. Formal front garden and
natural plantings for wildlife around
watersides. Refreshments as
requested on booking. Wheelchair
access to most parts, a few steps,
help always available.

4 **BLACKLAND HOUSE**
Quemerford, Calne, SN11 8UQ.
Polly & Edward Nicholson,
www.bayntunflowers.co.uk. *Situated
just off A4. We will be operating a
one-way system, entering the grounds
through the side entrance signed St.
Peter's Church and Blackland Park
Deliveries. Opposite The Willows.*
**Thur 22 Apr, Sat 1 May (2-4.30).
Adm £10, chd free. Home-made
teas. Donation to Dorothy House
Hospice.**

A wonderfully varied 4¹/₂ acre garden
adjacent to River Marden. (House
not open). Formal walled productive
and cutting garden, traditional
glasshouses, rose garden and wide
herbaceous borders. Interesting
topiary, trained fruit trees and
specialist displays of historic tulips
and unusual spring bulbs. Hand-tied
bunches of flowers for sale. Partial
wheelchair access, steps, grass and
cobbles.

5 NEW **BLUEBELLS**
Cowesfield, Whiteparish,
Salisbury, SP5 2RB.
Hilary Mathison,
hilary.mathison@icloud.com. *SW
of Salisbury. On main A27 road
from Salisbury to Romsey. 1¹/₂ m
SW of Whiteparish, 100-200 metres
inside county boundary.* **Visits by
arrangement Mar to Sept for
groups of 10 to 30. Car parking
for 10-12 cars on site (max).
Adm £4, chd free. Cream teas
or home-made cakes can be
arranged; wine + savoury bites
for evening visits.**
Relatively new garden on established
1¹/₂ acre plot, with deciduous
woodland incl bluebells in season.
Adjoining new build contemporary
house, so rear courtyard reflects this;
other areas incl shady border and
wildlife pond. Large lawn surrounded
by differing beds and borders with
some newly planted areas. Large
feature bed planted with white
birch and cornus. Partial wheelchair
access.

6 ◆ BOWOOD WOODLAND GARDENS

Calne, SN11 9PG. **The Marquis of Lansdowne, 01249 812102, houseandgardens@bowood.org, www.bowood.org.** *3½ m SE of Chippenham. Located off J17 M4 nr Bath & Chippenham. Entrance off A342 between Sandy Lane & Derry Hill Villages. Follow brown tourist signs. For SatNav, please use SN11 9PG.* **For NGS: Mon 19 Apr (11-6). Adm £7.50, chd free. The Nosh Box offers home-made cakes and sweet treats, with teas and coffees. For other opening times and information, please phone, email or visit garden website.**
This 30 acre woodland garden of azaleas, magnolias, rhododendrons and bluebells is one of the most exciting of its type in the country. From the individual flowers to the breathtaking sweep of colour, this is a garden not to be missed. With two miles of meandering paths, you will find hidden treasures at every corner. The Woodland Gardens are 2m from Bowood House & Garden. Only partial access for wheelchairs due to rough terrain.

🐾 🐕 🚗 💷 ⛱

7 NEW BRILLSCOTE HOUSE

Lea, Malmesbury, SN16 9PF. **Mrs Simon Mounsey.** *Signs on B4042 Malmesbury to Swindon Road and road from Charlton, Milbourne & Garsdon. Please note location is Brillscote HOUSE not Farm, both on SN16 9PF. Drive has chair with eggs for sale.* **Sun 20 June (2.30-6.30). Adm £6, chd free. Home-made teas.**
The garden was created in 2005 after the house was built in 2004/5. Before that it was an agricultural field and builder's rubble. My husband's philosophy was for a garden to be 'organised chaos' - he loved wildflowers and always let them be; he loved herbaceous borders and created several here and also a wild garden. He planted over 400 trees. He created gardens all his life wherever he lived. Featured in July 2020 edition of The English Garden.

🐕 🐾 💷

8 BROADLEAS HOUSE GARDENS

Devizes, SN10 5JQ. **Mr & Mrs Cardiff, 07584 119362, robwaller76@gmail.com.** *1m S of Devizes. Turn L from Hartmoor Road onto Broadleas Park. Follow the road until you reach grassed area on R with red brick wall, stone pillars, grey gates, cattle grid. That is the entrance.* **Sun 18 Apr, Sun 16 May, Sun 20 June, Sun 15 Aug (2-5). Adm £7, chd free. Home-made teas. Visits also by arrangement Mar to Sept for groups of 10+.**
6 acre garden of hedges, perennial borders, walled rose garden, secret garden, bee garden and orchard stuffed with good plants. Well stocked kitchen and herb garden. Mature collection of specimen trees incl oaks, magnolia, handkerchief, redwood, dogwood. Overlooked by the house and arranged above the valley garden crowded with magnolias, camellias, rhododendrons, azaleas, cornus, hydrangeas etc. Pre-wrapped slices of cake with unlimited tea/coffee/ soft drink in disposable cups (please re-use your cup for refills), £5 per serving. Wheelchair access to upper garden only, some gravel and narrow grass paths.

🐾 🐕 ❀ 🚗 💷

9 NEW BURTON GRANGE

Burton, Mere, Warminster, BA12 6BR. **Sue Phipps & Paddy Sumner, www.suephipps.com.** *Take lane, signposted to Burton, on A303 just E of Mere by-pass. After 400 yards follow rd past pond and round to L. Go past wall on R. Burton Grange entrance is in laurel hedge on R.* **Sun 13 June (11.30-5.30). Adm £4, chd free. Home-made teas.**
1½ acre garden, created from scratch since 2014. Lawns, borders, large ornamental pond, some gravel planting, veg garden and pergola rose garden, together with a number of wonderful mature trees. Artist's studio open.

🐾 ❀ 💷

10 NEW BUSH FARM

West Knoyle, Warminster, BA12 6AE. **Lord & Lady Seaford, www.bisonfarm.co.uk.** *From A303 turn off at Esso petrol station (Willoughby Hedge) follow left hand lane to West Knoyle (signed Bush Farm) through to next of village entrance on bend through woods to Bush Farm.* **Sun 9 May, Sun 13 June (11-5). Adm £6, chd free. Light refreshments.**
Mature oak woodland glades jungliefied with climbing roses, clematis, honeysuckle and specimen trees. Bog garden with ferns, skunk cabbage and iris. Nearby lakeside walk with farm trail to Bison and Elk.

Wildflowers everywhere. Bluebells in May. Flat woodland walks - not paved.

🐾 💷 ⛱

11 CADENHAM MANOR

Foxham, Chippenham, SN15 4NH. **Victoria & Martin Nye, 01249 740224, garden@cadenham.com.** *B4069 from Chippenham or M4 J17 through Sutton Benger, turn R in Christian Malford and L in Foxham. From A3102 turn L from Calne or R from Lyneham (NW) at Xrds between Hilmarton and Goatacre.* **Thur 3 June (2-5.30). Adm £10, chd free. Pre-booking essential, please visit www.ngs.org.uk for information & booking. Home-made teas. Visits also by arrangement Apr to Oct for groups of 10 to 30. Home-made refreshments by agreement.**
This glorious 4 acre garden comprises a series of rooms around a listed C17 manor house and dovecote. Divided by yew hedges and moats, the rooms are furnished with specimen trees, mixed borders with stunning displays of old roses, and fountains and statues to focus the eye. There is a peony walk, a water garden in an old canal, and extensive vegetable and herb gardens.

🐾 🚗 💷

12 CHERRY ORCHARD BARN

Luckington, SN14 6NZ. **Paul Fletcher and Tim Guard.** *Cherry Orchard Barn is ¾ m before the centre of SN14 6NZ, at a T junction. Passing the Barn is ill-advised, as turning rapidly becomes difficult.* **Sun 4 July (1-5). Adm £5, chd free. Home-made teas.**
Charming 1-acre garden created over past 7 years from the corner of a field, with open views of surrounding countryside. Containing 7 rooms, 3 of which are densely planted with herbaceous perennials, each with individual identities and colour themes. The garden is described by visitors as a haven of tranquillity. Largely level access to all areas of garden. Some gravel paths.

🐾 🐕 💷

All entries are subject to change. For the latest information check ngs.org.uk

⓭ THE COACH HOUSE

Bridge Street, Manton, Marlborough, SN8 4HR. Anna Marsden. *Travelling W out of Marlborough on A4 turn L after 1m to Manton village. Garden is 50 metres from A4 by telephone pole on R. For parking continue into village and follow signs.* **Sun 25 July (2-5). Combined adm with Peacock Cottage £7, chd free. Home-made teas.**
½ - acre country garden surrounding C18 barn, comprising herbaceous border, mixed colour themed beds, pond garden and small veg/soft fruit patch. Informal lawns and mature trees provide a relaxed setting for garden art and wildflowers. C18 well used to water garden. ¾ of garden has disabled access on grass, gravel and cobble paths. Dogs on leads welcome. Plants for sale at Peacock Cottage.

&. 🐾 ✿ ☕

⓮ COCKSPUR THORNS

Berwick St James, Salisbury, SP3 4TS. Stephen & Ailsa Bush, 01722 790445, stephenjdbush@gmail.com. *8m NW of Salisbury. 1m S of A303, on B3083 at S end of village of Berwick St James.* **Visits by arrangement May to July for groups of 5 to 20. Adm £5, chd free. Home-made teas.**
2¼ acre garden, completely redesigned 20 yrs ago and developments since, featuring roses (particularly colourful in June), herbaceous border, shrubbery, small walled kitchen garden, secret pond garden, mature and new unusual small trees, fruit trees and areas of wildflowers. Beech, yew and thuja hedgings planted to divide the garden. Small number of vines planted during early 2016.

&. 🚗 ☕

⓯ ◆ CORSHAM COURT

Corsham, SN13 0BZ. Lord Methuen, 01249 701610, staterooms@corsham-court.co.uk, www.corsham-court.co.uk. *4m W of Chippenham. Signed off A4 at Corsham.* **For NGS: Sun 21 Mar, Sun 25 Apr (2-5.30). Adm £5, chd £2.50. For other opening times and information, please phone, email or visit garden website.**
Park and gardens laid out by Capability Brown and Repton. Large lawns with fine specimens of ornamental trees surround the Elizabethan mansion. C18 bath house hidden in the grounds. Spring bulbs, beautiful lily pond with Indian bean trees, young arboretum and stunning collection of magnolias. Wheelchair (not motorised) access to house, gravel paths in garden.

&.

⓰ NEW CORTINGTON MANOR

Corton, Warminster, BA12 0SY. Mr & Mrs Simon Berry. *5m S of Warminster. From Warminster, take rd thro Sutton Veny to Corton. Do not bear L into Corton but continue for ½ m. From A303, take A36, L to Boyton, cross railway and R at T jnct. Continue for ¾ m.* **Sun 12 Sept (2-5). Adm £7, chd free. Home-made teas.**
4 acres of wild and formal gardens surround rose clad C18 manor house. Herbaceous border, yew bays with Portuguese laurel line main lawn. Bank of white roses and foxgloves edges swimming pool. Sweet pea arch opens to cutting garden, veg garden and orchard divided by yew hedges. Lime avenue leads to river and pond from walled herb garden. Stable yard features pleached hornbeams and beech hedges.

✿ ☕

Ordnance House

17 COTTAGE IN THE TREES
Tidworth Rd, Boscombe Village, nr Salisbury, SP4 0AD. Karen & Richard Robertson, 01980 610921, robertson909@btinternet.com. *7m N of Salisbury. Turn L off A338 just before Social Club. Continue past church, turn R after bridge to Queen Manor, cottage 150yds on R.* **Sun 25 Apr, Sun 9 May (1.30-5). Adm £4, chd free. Sun 13 June (1.30-5). Combined adm with The Old Rectory, Boscombe £7, chd free. Home-made teas. Visits also by arrangement Mar to Sept for groups of 10+.**
Enchanting ½ acre cottage garden, immaculately planted with water feature, raised vegetable beds, small wildlife pond and gravel garden. Spring bulbs, hellebores and pulmonarias give a welcome start to the season, with pots and baskets, roses and clematis. Mixed borders of herbaceous plants, dahlias, grasses and shrubs giving all-yr interest. Large variety of cottage plants for sale, grown from the garden.

GROUP OPENING

18 DAUNTSEY GARDENS
Church Lane, Dauntsey, Malmesbury, SN15 4HT. *5m SE of Malmesbury. Approach via Dauntsey Rd from Gt Somerford, 1¼ m from Volunteer Inn Great Somerford.* **Sun 13 June (1-5). Combined adm £7.50, chd free. Home-made teas at Idover House.**

THE COACH HOUSE
Col & Mrs J Seddon-Brown.

DAUNTSEY PARK
Mr & Mrs Giovanni Amati, 01249 721777, enquiries@ dauntseyparkhouse.co.uk.

THE GARDEN COTTAGE
Miss Ann Sturgis.

IDOVER HOUSE
Mr & Mrs Christopher Jerram.

THE OLD COACH HOUSE
Tony & Janette Yates.

THE OLD POND HOUSE
Mr & Mrs Stephen Love.

This group of 6 gardens, centred around the historic Dauntsey Park Estate, ranges from the Classical C18 country house setting of Dauntsey Park, with spacious lawns, old trees and views over the River Avon, to mature country house gardens and traditional walled gardens. Enjoy the formal rose garden in pink and white, old fashioned borders and duck ponds at Idover House, and the quiet seclusion of The Coach House with its thyme terrace and gazebos, climbing roses and clematis. Here, mop-headed pruned crataegus prunifolia line the drive. The Garden Cottage has a traditional walled kitchen garden with organic vegetables, apple orchard, woodland walk and yew topiary. Meanwhile the 2 acres at The Old Pond House are both clipped and unclipped! Large pond with lilies and fat carp, and look out for the giraffe and turtle. The Old Coach House is a small garden with perennial plants, shrubs and climbers.

19 DUCK POND BARN
Church Lane, Wingfield, Trowbridge, BA14 9LW. Janet & Marc Berlin, 01225 777764, janet@berlinfamily.co.uk. *On B3109 from Frome to Bradford on Avon, turn opp Poplars PH into Church Lane. Duck Pond Barn is at end of lane. Big field for parking.* **Sun 27 June, Sun 1 Aug (10-6). Adm £5, chd free. Visits also by arrangement May to Aug.**
Garden of 1.6 acres with large duck pond, lawns, ericaceous beds, orchard, vegetable garden, big greenhouse, spinney and wild area of grass and trees with many wildflowers. Large dry stone wall topped with flower beds with rose arbour. 3 ponds linked by a rill in flower garden and large pergola in orchard. Set in farmland and mainly flat. Coach parties please ring in advance for catering purposes.

20 FONTHILL HOUSE
Tisbury, SP3 5SA. The Lord Margadale of Islay, www.fonthill.co.uk/gardens. *13m W of Salisbury. Via B3089 in Fonthill Bishop. 3m N of Tisbury.* **Sun 21 Mar (12-5). Adm £7, chd free. Sandwiches, quiches and cakes, soft drinks, wine, tea and coffee. All proceeds to NGS.**
Wonderful woodland walks with daffodils, rhododendrons, azaleas, shrubs, bulbs; magnificent views; formal gardens. The gardens have been extensively redeveloped under the direction of Tania Compton and Marie-Louise Agius. The formal gardens are being continuously improved with new designs, exciting trees, shrubs and plants. Gorgeous William Pye fountain and other sculptures. Partial wheelchair access.

21 FOVANT HOUSE
Church Lane, Fovant, Salisbury, SP3 5LA. Amanda & Noel Flint. *Fovant House is located approx. 6½ m W of Wilton and 9.3m E of Shaftesbury. Take A30 to Fovant then head N through village and follow signs to St Georges Church.* **Sat 5 June (2-5.30). Adm £6, chd free. Home-made teas.**
Fovant House is a former Rectory set in about 3 acres of formal garden. (House not open). In 2016 the garden was redesigned by Arabella Lennox Boyd. Garden incl 60m herbaceous border, terraces and parterre. A range of mature trees incl cedars, copper beech and ash. Majority of garden has easy wheelchair access subject to ground conditions being dry.

22 FOXLEY MANOR
Foxley, Malmesbury, SN16 0JJ. Richard & Louisa Turnor, lou.turnor@gmail.com. *2m W of Malmesbury. 10 mins from J17 on M4. Turn towards Malmesbury/ Cirencester, then towards Norton and follow signs for the Vine Tree, yellow signs from here.* **Sun 18 Apr, Sat 22 May (10-5). Adm £5, chd free. Home-made teas. Visits also by arrangement Apr to June for groups of up to 20.**
Yew hedges divide lawns, borders, rose garden, lily pond and a newer wild area with a natural swimming pond shaded by a liriodendron. Views through large Turkey oaks to farmland beyond. Small courtyard gravel garden. Sculptures are sited throughout the gardens. Regret steps and gravel paths make this unsuitable for wheelchairs.

23 GASPER COTTAGE
Gasper Street, Gasper Stourton, Warminster, BA12 6PY. Bella Hoare & Johnnie Gallop, bella.hoare@icloud.com. *Near Stourhead Gardens, 4m from Mere, off A303. Turn off A303 at B3092 Mere. Follow Stourhead signs. Go through Stourton. After 1m, turn R after phone box, signed Gasper. House 2nd on R*

going up hill. Parking past house on L, in field. **Sun 15 Aug (11-5). Adm £5, chd free. Visits also by arrangement June to Sept for groups of 10 to 20.** 1½ acre garden, with views to glorious countryside. Luxurious planting of dahlias, grasses, asters, cardoons and more, incl new perennial planting. Orchard with wildlife pond. Artist studio surrounded by colour balanced planting with formal pond. Pergola with herb terrace. Several seating areas. Garden model railway. Partial wheelchair access.

GROUP OPENING

24 HANNINGTON VILLAGE GARDENS

Hannington, Swindon, SN6 7RP. *Off B4019 Blunsdon to Highworth Rd by the Freke Arms. Park behind Jolly Tar PH, or opp Lushill House.* **Sun 13 June (11-5). Combined adm £8, chd free. Home-made teas in Hannington Village Hall.**

CHESTNUT HOUSE
Mary & Garry Marshall.

HANNINGTON HALL
Guillaume Molhant-Proost.

NEW LOWER FARM
Mr Piers & Mrs Jenny Martin.

LUSHILL HOUSE
John & Sasha Kennedy.

QUARRY BANK
Paul Minter & Michael Weldon.

22 QUEENS ROAD
Jan & Pete Willis.

ROSE COTTAGE
Mrs Ruth Scholes.

THE BUTLER'S COTTAGE
Mr John & Mrs Karen Mayell.

YORKE HOUSE GARDEN
Mr Miles & Mrs Cath Bozeat.

Hannington has a dramatic hilltop position on a Cotswold ridge overlooking the Thames Valley. A great variety of gardens, from large manor houses to small cottage gardens, many of which follow the brow of the hill and afford stunning views of the surrounding farm land. You will need lots of time to see all that is on offer in this beautiful historic village. Only partial wheelchair access as many gardens are on steep hills and feature steps.

25 HAZELBURY MANOR GARDENS

Wadswick, Box, Corsham, SN13 8HX. Mr L Lacroix. *5m SW of Chippenham, 5m NE of Bath. From A4 at Box, A365 to Melksham, at Five Ways junction L onto B3109 towards Corsham; 1st L in ¼m,* drive immed on R. **Wed 26 May (10-4); Sun 4 July (1-5); Wed 15 Sept (10-4). Adm £5, chd free. Home-made teas.**
The C15 Manor house comes into view as you descend along the drive and into the Grade II landscaped Edwardian gardens. The extensive plantings that surround the house are undergoing considerable redevelopment by the owners and their head gardener. A wide range of organic horticulture is practiced in 8 acres of relaxed yet playful gardens. Regretfully, very little wheelchair access.

26 HILPERTON HOUSE

The Knap, Hilperton, Trowbridge, BA14 7RJ. Chris & Ros Brown. *1½m NE of Trowbridge. Follow A361 towards Trowbridge and turn R at r'about signed Hilperton. House is next door to St Michael's Church in the Knapp off Church St.* **Sat 19, Sun 20 June (2-6.30). Adm £5, chd free. Home-made teas.**
2½ acres well stocked borders, small stream leading to large pond with fish, water lilies, waterfall and fountain. Fine mature trees incl unusual specimens. Walled fruit and vegetable garden, small woodland area. Rose walk with roses and clematis. Interesting wood carvings, mainly teak. 160yr old vine in conservatory of Grade II listed house, circa 1705 (not open). Activity sheets for children. Some gravel and lawns. Conservatory not wheelchair accessible. Path from front gate has uneven paving but can be bypassed on lawn.

27 HORATIO'S GARDEN

Duke of Cornwall Spinal Treatment Centre, Salisbury Hospital NHS Foundation Trust, Odstock Road, Salisbury, SP2 8BJ. Horatio's Garden Charity, www.horatiosgarden.org.uk. *1m from centre of Salisbury. Follow signs for Salisbury District Hospital. Please park in car park 10 (which will be free to NGS visitors on the day).* **Sun 5 Sept (2-5). Adm £5, chd free. Tea and delicious cakes - made by**

Horatio's Garden volunteers - will be served in the Garden Room.
Award winning hospital garden, opened in Sept 2012 and designed by Cleve West for patients with spinal cord injury at the Duke of Cornwall Spinal Treatment Centre. Built from donations given in memory of Horatio Chapple who was a volunteer at the centre in his school holidays. Low limestone walls, which represent the form of the spine, divide densely planted herbaceous beds. Everything in the garden designed to benefit patients during their long stays in hospital. Garden is run by Head Gardener and team of volunteers. South West Regional Winner, The English Garden's The Nation's Favourite Gardens 2019. Designer Cleve West has 8 RHS gold medals. 3pm - short talk about therapeutic gardens and the work of Horatio's Garden, Salisbury.

28 HYDE'S HOUSE

Dinton, SP3 5HH. Mr George Cruddas. *9m W of Salisbury. Off B3089 nr Dinton Church on St Mary's Rd.* **Sun 6 June (2-5). Adm £6, chd free. Home-made teas at Thatched Old School Room with outside tea tables.**
3 acres of wild and formal garden in beautiful situation with series of hedged garden rooms. Numerous shrubs, flowers and borders, all allowing tolerated wildflowers and preferred weeds, while others creep in. Large walled kitchen garden, herb garden and C13 dovecote (open). Charming C16/18 Grade I listed house (not open), with lovely courtyard. Every year varies. Free walks around park and lake. Steps, slopes, gravel paths and driveway.

> "Unfortunately, cancer was not in lockdown in 2020. The continued support of our long-standing and valued partner, the National Garden Scheme is more important than ever."
> Macmillan

29 ◆ IFORD MANOR GARDENS

Bradford-on-Avon, BA15 2BA.
Mr Cartwright-Hignett, 01225 863146,
info@ifordmanor.co.uk,
www.ifordmanor.co.uk. *7m S of
Bath. Off A36, brown tourist sign to
Iford 1m. From Bradford-on-Avon
or Trowbridge via Lower Westwood
Village (brown signs). Please note all
approaches via narrow, single track
lanes with passing places.* **For NGS:
Sat 24 Apr (11-4). Adm £7.50.
Pre-booking essential via Iford
Manor website. Cream teas. Last
garden entry at 3.30. No children
under 12 admitted. For other
opening times and information,
please phone, email or visit garden
website.**

An exemplar of Harold Peto's garden
design at his former home. This
2½ acre romantic garden provides
inspiration to many and was used as
a key location in The Secret Garden
(2020). Restored and extended by the
current owners, Troy Scott Smith was
recently appointed Head Gardener.
Steep flights of steps link the terraces
with pools, fountains, urns, colonnades
and statues. As garden is on a steeply
terraced hillside with many flights of
steps, wheelchair access is limited.
Please call for details.

30 NEW JULIA'S HOUSE CHILDREN'S HOSPICE

Bath Road, Devizes, SN10 2AT.
Nicky Clack. *Situated just off A361
Bath Road in Devizes. Please note
there is no parking available at
hospice. We kindly ask visitors to use
nearby Station Road car park. For
SatNav use SN10 1BZ.* **Sat 3, Sun 4
July (10-3.30). Adm £3, chd free.
Proceeds from refreshments and
plant sale to Julia's House.**

Julia's House Children's Hospice
provides respite care for children with
life-threatening or limiting conditions.
Our garden has been designed so
the children can experience and
enjoy different sensory elements
and is cared for by our volunteer
gardeners. Explore how our children
use the garden. Hospice tours may
be available.

31 KETTLE FARM COTTAGE

Kettle Lane, West Ashton,
Trowbridge, BA14 6AW. Tim &
Jenny Woodall, 01225 753474,
trwwoodall@outlook.com. *Kettle
Lane is halfway between West
Ashton T-lights and Yarnbrook*

r'about on S side of A350. Garden
½ m down end of lane.* **Visits by
arrangement June to Sept for
groups of up to 30. Adm £5.50,
chd free. Tea and home-made
cakes.**

Previously Priory House, Bradford
on Avon, the garden of which was on
Gardeners World, September 2017,
we have now created a new cottage
garden, full of colour and style,
flowering from June to Oct. Bring a
loved one/friend to see the garden.
One or two steps.

32 NEW KNOYLE PLACE

Holloway, East Knoyle, Salisbury,
SP3 6AF. Lizzie & Herve de la
Moriniere. *Turn off A350 into East
Knoyle and follow signs to parking
at Lower Lye. Walk 5 mins to Knoyle
Place through village following signs.*
**Sat 24 Apr (2-5). Adm £10, chd
free. Home-made teas. Also open
Lower Lye.**

Very beautiful and elegant garden
created over 60 years by previous
and current owners. Above the
house there are several acres of
mature rhododendron and magnolia
woodland planting. Among the many
different areas in this 9 acre garden
is a box parterre, rose garden,
vegetable garden and, around the
house, a recently planted formal
garden designed by Dan Combes.
Sloping lawns and woodland paths,
stone terrace.

GROUP OPENING

33 LANDFORD VILLAGE GARDENS

Landford, Salisbury, SP5 2AX.
*From Salisbury take A36 then L
through village. From S leave M27
at J2 onto A36 and R towards
Nomansland. At Xrds take Forest
Rd. Parking on L at The Gatehouse.*
**Sun 20 June (2-6). Combined
adm £6, chd free. Home-made
teas at Bentley, Whitehorn Drive.
Donation to Local Wildlife charities.**

BENTLEY
Jacky Lumby.

COVE COTTAGE
Mrs Gina Dearden.

FOREST COTTAGE
Norah Dunn.

THE GATEHOUSE
Mrs Jackie Beatham.

5 WHITEHORN DRIVE

Jackie & Barry Candler.

Landford is a small village set in the
northern New Forest, famous for the
beauty of its beech and oak trees and
freeroaming livestock. The 5 gardens
range in size, planting conditions
and setting. 3 are in the village,
adjacent to Nomansland with its green
grazed by New Forest ponies. The
Gatehouse has formal herbaceous
borders with exuberant planting and
colour schemes in mixed herbaceous
borders, looking out over paddocks to
the forest. Bentley is a small garden,
a green oasis containing a wide
range of plants and raised vegetable
borders. 5 Whitehorn Drive is a damp
garden with lush colour and surprises
in planting. A short distance away in
the forest itself are Forest Cottage, a
naturalistic garden with hidden delights
around each corner, incl field of wild
orchids, and next door Cove Cottage,
a mature garden where the owner will
be featuring her artwork.

34 NEW LOWER LYE

Holloway, East Knoyle, Salisbury,
SP3 6AQ. Belinda & Andrew Scott.
*From A350 exit at East Knoyle.
Lower Lye is at upper southern end
of village, follow signs to house.
Parking on L entrance field.* **Sat 24
Apr (2-5). Adm £20. Pre-booking
essential, please visit www.ngs.
org.uk for information & booking.
Also open Knoyle Place.**

The Lower Lye garden was completely
reconfigured 24 years ago by the
landscape architect Michael Balston.
Since then both Tania Compton and
Jane Hurst have undertaken replanting
various borders. The garden has a
wildflower area, herbaceous borders,
mature trees, both vegetable and
cutting gardens and a magnificent
view towards King Alfred's Tower. Teas
at Knoyle Place. Some steps and
slopes.

35 MANOR FARM

Crudwell, Malmesbury, SN16 9ER.
Mr & Mrs J Blanch. *4m N of
Malmesbury on A429. Heading N
on A429 in Crudwell, turn R signed
to Eastcourt. Farm entrance is on L
200m after end of speed limit sign.*
**Sun 13 June (11-5). Adm £5, chd
free. Home-made teas. Gluten
free options available.**

Set within Cotswold stone walls and

a backdrop of Crudwell church with a further 5 acres of parkland featuring Japanese maples and a Roman style summerhouse. The garden is divided by box and yew hedges to create different areas both formal and informal. Herbaceous borders with old fashioned roses are grown among fountains and extensive lawns.

 🐕 ✳ 🍵

36 MANOR FARM HOUSE
Manor Farm Lane, Patney, Devizes, SN10 3RB. Mr & Mrs Mark Alsop. *Between Pewsey and Devizes. Take 3rd entrance on L going up Manor Farm Lane from village green. Parking available in paddock.* Sun 20 June (2-5). Adm £6, chd free. Home-made teas.
2-acre plantsman's garden designed in 1980s and updated by us over the past 9 yrs. Lawned areas with borders surrounded by yew hedging, large border by tennis court, formal vegetable garden with buxus parterres and meadow with spring bulbs, box mound and walkway through moist plants garden. Limited wheelchair access due to upward slope and gravel paths. No toilets.
☕

37 MANOR HOUSE, STRATFORD TONY
Stratford Tony, Salisbury, SP5 4AT. Mr & Mrs Hugh Cookson, 01722 718496, lucindacookson@stratfordtony.co.uk, www.stratfordtony.co.uk. *4m SW of Salisbury. Take minor rd W off A354 at Coombe Bissett. Garden on S after 1m. Or take minor rd off A3094 from Wilton signed Stratford Tony and racecourse.* Visits by arrangement May to Sept for groups of 5 to 30. Adm £6, chd free. Please also look online for specific opening days.
Varied 4 acre garden with all year interest. Formal and informal areas. Herbaceous borders, vegetable garden, parterre garden, orchard, shrubberies, roses, specimen trees, lakeside planting, winter colour and structure, many original contemporary features and places to sit and enjoy the downland views. Some gravel.
♿ 🍵 🪑

38 NORTH COTTAGE
Tisbury Row, Tisbury, SP3 6RZ. Jacqueline & Robert Baker, 01747 870019, baker_jaci@yahoo.co.uk. *12m*

Burton Grange

W of Salisbury. From A30 turn N through Ansty, L at T-junction, towards Tisbury. From Tisbury take Ansty road. Car park entrance nr junction signed Tisbury flow. Sun 23 May, Sun 13 June (11.30-5). Adm £5, chd free. Home-made light lunches and teas.
Leave car park, walk past vegetables and through the paddock to reach house and gardens. Although modest, there is much variety to find. Garden is divided with lots to explore as each part differs in style and feel, augmented with pottery, woodwork handmade by the owners. From the intimacy of the garden go out to see the orchard, ponds, walk the coppice wood and see the rest of the smallholding. Ceramics, wooden furniture and handicrafts all made by garden owners, including creations from their own sheep's wool for sale. Exciting metalwork sculpture exhibition. Large variety of plants for sale.

 🐕 ✳ 🛏 🍵

39 OARE HOUSE
Rudge Lane, Oare, nr Pewsey, SN8 4JQ. Sir Henry Keswick. *2m N of Pewsey. On Marlborough Rd (A345).* Sun 25 Apr, Sun 27 June (1.30-6). Adm £6.50, chd free. Home-made teas in potting shed. Donation to The Order of St John.
1740s mansion house later extended by Clough Williams Ellis in 1920s (not open). The original formal gardens around the house have been developed over the years to create a wonderful garden full of many unusual plants. Current owner is

very passionate and has developed a fine collection of rarities. Garden is undergoing a renaissance but still maintains split compartments each with its own individual charm; traditional walled garden with fine herbaceous borders, vegetable areas, trained fruit, roses and grand mixed borders surrounding formal lawns. The magnolia garden is wonderful in spring with some trees dating from 1920s, together with strong bulb plantings. Large arboretum and woodland with many unusual and champion trees. In spring and summer there is always something of interest, with the glorious Pewsey Vale as a backdrop. Partial wheelchair access.
♿ 🐕 ✳ 🍵

40 THE OLD MILL
Ramsbury, SN8 2PN. Annabel & James Dallas. *8m NE of Marlborough. From Marlborough head to Ramsbury. At The Bell PH follow sign to Hungerford. Garden behind yew hedge on R 100yds beyond The Bell.* Sun 25 July, Sun 22 Aug (1.30-5). Adm £6, chd free. Tea.
The River Kennet runs through this garden with a millrace, a millstream and small channels to walk across and watch the reflections in. The planting is relaxed and colourful near the house, and gets more naturalistic further out into the meadows beyond. Gravel and grass paths wind through controlled wilder areas with seating for visitors. There is a more formal and productive kitchen garden. Partial wheelchair access as gravel paths and bridges and very soft ground in places.
🐕 🍵

41 THE OLD RECTORY, BOSCOMBE

Tidworth Road, Boscombe, Salisbury, SP4 0AB. Helen & Peter Sheridan, 07712 004797, helen.sheridan@uwclub.net. *7m N of Salisbury on A338. From Salisbury, 2nd L turning after Earl of Normanton PH, just S of Boscombe Social Club and Black Barn.* **Sun 13 June (1.30-5). Combined adm with Cottage in the Trees £7, chd free. Home-made teas. Visits also by arrangement June & July for groups of 10 to 30.**

The elegant gardens of this period property feature a traditional walled garden with herbaceous borders and vegetable beds, contrasting with a parkland of sweeping lawns, mature trees and flowering shrubs. Wheelchair access, soft ground in places. Gravel paths.

&♿ 🐕 ❀ 🚌 🍵

42 NEW ORDNANCE HOUSE

West Dean, Salisbury, SP5 1JE. Terry & Vanessa Winters, www.ordnancehouse.co.uk. *8m W of Romsey, 8m E of Salisbury. From A36 Southampton to Salisbury road follow signs for West Dean via West Grimstead. From Romsey, follow signs via Awbridge, Lockerley & East Dean. From Stockbridge, follow signs from Broughton.* **Sun 23 May (11-4). Adm £5, chd free. Home-made teas in aid of Salisbury Hospice.**

Only public opening of 2021.This garden was awarded a RHS Gold Medal in May 2020 in a competition held in conjunction with the BBC One Show.The garden won the 'Back Garden' category judged by a panel incl Monty Don, James Alexander-Sinclair, Sue Biggs (DG of RHS) and One Show host Alex Jones. The garden has drifts of white and purple alliums and foxgloves in spring with unusual varieties of lavender in summer. Its design features topiary balls and spheres, a vegetable garden, ornamental parterre and small orchard while seating areas offer views of the garden and to the borrowed landscape of open countryside beyond.

🐕 ❀ 🍵

43 PEACOCK COTTAGE

Preshute Lane, Manton, Marlborough, SN8 4HQ. Ian & Clare Maurice. *1m to W of Marlborough along A4, turn L at Manton Village sign. Pass signs to partner garden and follow Car Park signs. If arriving through Manton village turn R by pub.* **Sun 25 July (2-5). Combined adm with The Coach House £7, chd free.**

Set in rural surroundings, this expertly tended, undulating garden, sweeps up from the rear of the house and comprises herbaceous and mixed borders, parterre and exotic bed. Created in 1966 when 2 cottages and their gardens combined, and has been developed considerably over last 15yrs. Disabled parking places in car park. Wheelchair access to garden is via lawns and grass paths.

&♿ 🐕 ❀

44 PYTHOUSE KITCHEN GARDEN

West Hatch, Tisbury, SP3 6PA. Mr Piers Milburn, 01747 870444, info@pythousekitchengarden.co.uk, www.pythousekitchengarden.co.uk. *A350 S of E Knoyle, follow brown signs to Walled Garden approx 3m. From Tisbury take Newtown road past church, stay on it 2½m, garden on R. Check map on www. pythousekitchengarden.co.uk.* **Thur 9 Sept (10-3.30). Adm £5, chd free. Light refreshments. If hoping to have lunch in the restaurant, pre-booking required on www.pythousekitchengarden. co.uk. Visits also by arrangement May to Oct for groups of up to 30.**

3 acres of working kitchen garden, in largely continuous use since C18, with fruit-lined walls and gnarled apple trees leading down to an orchard via ravishing, rosa rugosa-edged, beds of flowers, soft fruit, vegetables and beehives. A restaurant now occupies the old potting shed and conservatory, with terraces for tea on open days. Abundant herbs, kiwis and apricots, as well as the deliciously scented 1920s HT rose, Mrs Oakley Fisher, growing by the kitchen door. Restaurant opens 10am for delicious home-made food at coffee, lunch and teatime, served both inside and out on terrace, open until 4pm on open days. Grass paths across slope.

&♿ 🐕 🚌 🍵

45 SEEND HOUSE

High Street, Seend, Melksham, SN12 6NR. Maud Peters. *In Seend village. Near church and opp post office.* **Sat 3 July (1-6). Combined adm with Seend Manor £10, chd free. Home-made teas.**

Seend House is a Georgian house with 6 acres of gardens and paddocks. Framed with yew and box.

Highlights include: cloud and rose garden, stream lavender, view of knot garden from above, fountain with grass border, walled garden as well as formal borders. Amazing view across the valley to Salisbury Plain.

🐕 🍵

46 SEEND MANOR

High Street, Seend, Melksham, SN12 6NX. Stephen & Amanda Clark. *In centre of village, opp village green with car parking for garden visitors.* **Sat 3 July (1-6). Combined adm with Seend House £10, chd free. Home-made teas.**

Created over 20 yrs, a stunning walled garden with 4 quadrants evoking important parts of the owners lives - England, China, Africa and Italy, with extensive trelliage, hornbeam hedges on stilts, cottage orne, temple, Chinese ting, grotto, fern walk, fountains, parterres, stone loggia and more. Wonderful views. Kitchen garden. Folly ruin in woods. Courtyard garden. Extensive walled gardens, water features, garden structures, topiary and hedging, fountains and one of the best views in Wiltshire! Many gravel paths, so wheelchairs with thick tyres are best.

&♿ 🐕 🍵

47 1 SOUTHVIEW

Wick Lane, Devizes, SN10 5DR. Teresa Garraud, 01380 722936, tl.garraud@hotmail.co.uk. *From Devizes Mkt Pl go S (Long St). At r'about go straight over, at mini r'about turn L into Wick Lane. Continue to end of Wick Lane. End of terrace on L. Park in road or roads nearby.* **Sun 4 July, Sun 12 Sept (2-5). Adm £4, chd free. Visits also by arrangement May to Sept for groups of up to 20.**

An atmospheric and very long town garden, full of wonderful planting surprises at every turn. Densely planted with both pots near the house and large borders further up, it houses a collection of beautiful and often unusual plants, shrubs and trees many with striking foliage. Colour from seasonal flowers is interwoven with this textural tapestry. 'Truly inspirational' is often heard from visitors.

❀

Online booking is available for many of our gardens, at ngs.org.uk

Knoyle Place

Lower Lye

Delightful 2 acre community garden, created and maintained by volunteers. Features incl 7 individual display gardens, ornamental pond, plant nursery, Iron Age round house, artwork, fitness trail, separate kitchen garden site, Swindon beekeepers and the haven, overflowing with wildflowers. Featured in Garden Answers & Wiltshire Life magazine and on Great British Gardens website. In Top 100 Attractions in SW England. Most areas wheelchair accessible. Disabled WC.

51 NEW URCHFONT MANOR

Urchfont, Devizes, SN10 4RF. Chris Legg. *Main entrance from gate on Blackboard Lane. Signed parking 250 metres walk from Blackboard Lane entrance. Entrance for disabled from gate off B3098.* **Sat 19 June (11-5.30). Adm £10, chd free. Light refreshments, also available to purchase for picnics.**
Urchfont Manor was built in 1685 in 10 acres. DelBuono Gazerwitz designed the formal gardens with a modern twist. Maze-like beech spiral garden, cloud box, lime allee, reflective pool. Extensive kitchen garden with herb and passion flower arch. Rose garden with pergola. Gravel garden irises and euphorbia. Ferns and foxgloves in woodland. Parkland of fine specimen trees, wildflower meadow. Orchard. Most areas accessible by wheelchair.

48 NEW TEASEL

Wilsford, Amesbury, Salisbury, SP4 7BL. Ray Palmer. *2m SW of Amesbury in the Woodford Valley on western banks of River Avon which runs through the gardens.* **Sun 4 July (11-5). Adm £5, chd free. Home-made teas.**
The gardens at Teasel were originally laid out in the 1970's by James Mitchell the publisher who made his fortune from the worldwide best seller ' The Joy of Sex '. The gardens have been the subject of extensive restoration and new landscaping during 2020. This includes tropical gardens around the swimming pool terrace, long herbaceous borders and a ½ m riverside walk.

49 NEW TRISTENAGH HOUSE

Devizes Road, Potterne, Devizes, SN10 5LW. Ros Ford. *lm S of Devizes on A360. Turn down byway by postbox and cottage (27 Devizes Rd) and follow yellow signs to car park and entrance to the garden.* **Mon 31 May (1.30-5.30). Adm £5, chd free. Home-made teas. Refreshments in aid of The Wiltshire Bobby Van Trust.**
Garden of approx 2 acres, created over the past 20 years. Island beds with a mixture of herbaceous plants, bulbs, annuals and shrubs. All year round structure provided by beech and yew hedges, box topiary and by mature beech and Scots pine trees. Some gravelled areas with containers. Good views of valley from terrace. Wheelchair access to most parts of garden on grass paths. Access to terrace limited due to gravel. Some steps.

50 ◆ TWIGS COMMUNITY GARDEN

Manor Garden Centre, Cheney Manor, Swindon, SN2 2QJ. TWIGS, 01793 523294, twigs.reception@gmail.com, www. twigscommunitygardens.org.uk. *From Gt Western Way, under Bruce St Bridges onto Rodbourne Rd. 1st L at r'about, Cheney Manor Industrial Est. Through estate, 2nd exit at r'about. Opp Pitch & Putt. Signs on R to Manor Garden Centre.* **For NGS: Sun 23 May, Sun 18 July (12-4). Adm £3.50, chd free. Home-made teas. Excellent hot and cold lunches available at Olive Tree café within Manor Garden centre adj to Twigs (pre booking required) 01793 533152. For other opening times and information, please phone, email or visit garden website.**

52 ◆ WATERDALE HOUSE

East Knoyle, SP3 6BL. Mr & Mrs Julian Seymour, 01747 830262. *8m S of Warminster. N of East Knoyle, garden signed from A350. Do not use SatNav.* **For NGS: Sun 2 May (2-5). Adm £5, chd free. Home-made teas. For other opening times and information, please phone.**
In the event of inclement weather, before visiting, please check www. ngs.org.uk. 4 acre mature woodland garden with rhododendrons, azaleas, camellias, maples, magnolias, ornamental water, bog garden, herbaceous borders. Bluebell walk. Shrub border created by storm damage, mixed with agapanthus and half-hardy salvias. Difficult surfaces, sensible footwear essential, parts of garden very wet. Please keep to raked and marked paths in woodland. Partial wheelchair access.

GROUP OPENING

53 NEW WEST KNOYLE GARDENS

West Knoyle, Warminster, BA12 6AJ. *From E on A303 take road signed West Knoyle at petrol station. Continue down hill and past church into village. Follow signs. To Barrowstreet take turning to Charnage off A303 and follow signs.* Sun 13 June (2-5). Combined adm £8, chd free. Also open Bush Farm (11-5), separate adm. Home-made teas at Barrowstreet House.

NEW BARROWSTREET HOUSE
Sarah & Peter Fineman.

NEW BROADMEAD FARM
Brigit & Brian Wessely.

THE PARISH HOUSE
Philip & Alex Davies.

The 3 very different gardens in West Knoyle are wonderful examples of what can be achieved on very different sites. Barrowstreet House has an elegant formal garden surrounded by yew hedges with beautiful views beyond. There is a wildlife pond and smallholding with chickens, ducks and lambs. Broadmead Farm also has a large pond beautifully planted around the edges, a woodland walk and large perennial and grasses border. The gravel garden at The Parish House is small and detailed, planted with self-seeding perennials, roses and evergreen topiary. There is a long grass area beyond the perimeters of the gravel garden with orchids (we hope) and mown paths. At Parish House most of level ground is gravel but there is also lawn. At Broadmead Farm garden is gently sloping.

🔥 ✻ ☕ 🪑

54 WEST LAVINGTON MANOR

1 Church Street, West Lavington, SN10 4LA. Andrew Doman, andrewdoman01@gmail.com. *6m S of Devizes, on A360. House opp White St, where parking is available.* Sat 5 June (11-6). Adm £10, chd free. Lunch available all day along with a wide range of cakes, scones and refreshments. Visits also by arrangement Feb to Dec. Donation to West Lavington Youth Club.

5 acre walled garden first established in C17 by John Danvers who brought Italianate gardens to the UK. Herbaceous border, Japanese garden, rose garden, orchard and arboretum with some outstanding specimen trees all centred around a trout stream and duck pond. White Birch grove and walk along southern bank of the stream. Picnickers welcome. Partial wheelchair access.

🔥 🐑 ☕ 🪑 ☕ 🪑

55 WESTCROFT

Boscombe Village, nr Salisbury, SP4 0AB. Lyn Miles, 01980 610877, lynmiles@icloud.com, www.westcroftgarden.co.uk. *7m N/E Salisbury. On A338 from Salisbury, just past Boscombe & District Social Club, park there or in field opp house, signed on day. Disabled parking only on drive.* Every Thur 14 Jan to 4 Mar (11-4). Sun 7, Fri 12, Sat 20, Sun 21, Sat 27, Sun 28 Feb (11-4). Adm £3, chd free. Home-made soups, scrummy teas. Visits also by arrangement Jan to Mar for groups of 20+.

Whilst overflowing with roses in June, in Jan and Feb the bones of this ²⁄₃ acre galanthophile's garden on chalk are on show. Brick and flint walls, terraces, rustic arches, gates and pond add character. Drifts of snowdrops carpet the floor whilst throughout is a growing collection of over 400 named varieties. Many holleboros, pulmonarias, grasses and seedheads add interest. Snowdrops (weather dependent) and snowdrop sundries for sale incl greetings cards, mugs, bags, serviettes. Partial wheelchair access to lower levels although not easy when the ground is soft.

🔥 ✻ ☕ ☕

56 WHATLEY MANOR

Easton Grey, Malmesbury, SN16 0RB. Christian & Alix Landolt, 01666 822888, reservations@whatleymanor.com, www.whatleymanor.com. *4m W of Malmesbury. From A429 at Malmesbury take B4040 signed Sherston. Manor 2m on L.* Tue 22 June (2-7). Adm £6.50, chd free. Home-made teas in The Loggia Garden. Hotel also open for lunch and full afternoon tea.

12 acres of English country gardens with 26 distinct rooms each with a unique theme based on colour, scent or style. Original 1920s Arts & Crafts plan inspired the design and combines classic style with more contemporary touches; incl specially commissioned sculpture. Dogs must be on a lead at all times.

🔥 🐑 ☕ ☕

57 WINDMILL COTTAGE

Kings Road, Market Lavington, SN10 4QB. Rupert & Gill Wade, 01380 813527. *5m S of Devizes. Turn E off A360 1m N of W. Lavington, signed Market Lavington & Easterton. At top of hill turn L into Kings Rd, L into Windmill Lane after 200yds. Limited parking on site, more parking nearby.* Thur 20, Fri 21 May, Thur 3, Fri 4 June, Thur 1, Fri 2 July (2-5). Adm £4, chd free. Home-made teas. Visits also by arrangement May & June for groups of 5+. Refreshments must be arranged in advance.

1 acre cottage style, wildlife friendly garden on greensand. Mixed beds and borders with long season of interest. Roses on pagoda, large vegetable patch for kitchen and exhibition at local shows, greenhouse, polytunnel and fruit cage. Whole garden virtually pesticide free for last 20 yrs. Small bog garden by wildlife pond. Secret glade with prairie. Grandchildren's little wood and wild place. Mostly accessible, some soft and gravel paths.

🔥 🐑 ✻ ☕ ☕

58 WUDSTON HOUSE

High Street, Wedhampton, Devizes, SN10 3QE. David Morrison, 07881 943213, djm@plml.co.uk. *Wedhampton lies on N side of A342 approx 4m E of Devizes. House is set back on E side of village street at end of drive with beech hedge on either side.* Visits by arrangement June to Sept for groups of 10+. Light refreshments.

The garden of Wudston House was started in 2010, following completion of the house. It consists, inter alia, of formal gardens round the house, perennial meadow, pinetum and arboretum. Nick Macer and James Hitchmough, who has pioneered the concept of perennial meadows, have been extensively involved in aspects of the garden, which is still developing. Partial wheelchair access.

✻ ☕

WORCESTERSHIRE

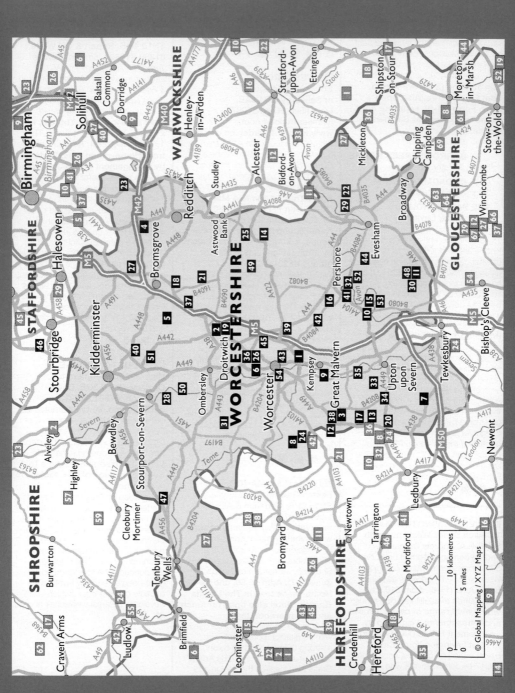

Worcestershire has something to suit every taste, and the same applies to its gardens.

From the magnificent Malvern Hills, the inspiration for Edward Elgar, to the fruit orchards of Evesham which produce wonderful blossom trails in the spring, and from the historic city of Worcester, with its 11th century cathedral and links to the Civil War, to the numerous villages and hamlets that are scattered throughout, there is so much to enjoy in this historic county.

Worcestershire is blessed with gardens created by both amateur and professional gardeners, and the county can boast properties with grounds of many acres to small back gardens of less than half an acre, but all have something special to offer.

There are gardens with significant historical interest and some with magnificent views and a few that are not what you might consider to be a "typical" National Garden Scheme garden! We also have a number of budding artists within the Scheme, and a few display their works of art on garden open days.

Worcestershire's garden owners guarantee visitors beautiful gardens, some real surprises and a warm welcome.

Left: The Lodge

Volunteers

County Organiser
David Morgan
01214 453595
meandi@btinternet.com

County Treasurer
Doug Bright
01886 832200
doug.bright@ngs.org.uk

Publicity
Pamela Thompson
01886 888295
peartree.pam@gmail.com

Social Media
Brian Skeys
01684 311297
brian.skeys@ngs.org.uk

Booklet Co-ordinator
Steven Wilkinson & Linda Pritchard
01684 310150
steven.wilkinson48412@gmail.com

Assistant County Organisers
Andrea Bright
01886 832200
andrea.bright@ngs.org.uk

Brian Bradford
07816 867137
brianbradford101@outlook.com

Philippa Lowe
01684 891340
philippa.lowe@ngs.org.uk

Stephanie & Chris Miall
0121 445 2038
stephaniemiall@hotmail.com

Alan Nokes
alan.nokes@ngs.org.uk

@WorcestershireNGS
@WorcsNGS

OPENING DATES

All entries subject to change. For latest information check **www.ngs.org.uk**

Map locator numbers are shown to the right of each garden name.

February

Snowdrop Festival

Sunday 7th
Brockamin 9

March

Sunday 7th
Bridges Stone Mill 8

Friday 12th
◆ Little Malvern Court 20

Wednesday 17th
The Dell House 13

Friday 19th
◆ Little Malvern Court 20

Sunday 21st
Brockamin 9

Friday 26th
◆ Little Malvern Court 20

April

Saturday 3rd
◆ Whitlenge Gardens 51

Sunday 4th
◆ Whitlenge Gardens 51

Friday 9th
◆ Spetchley Park
Gardens 39

Sunday 11th
White Cottage &
Nursery 49

Wednesday 14th
The Dell House 13

Sunday 18th
The Dell House 13

Saturday 24th
The River School 36

May

Saturday 1st
NEW 35 Alexandra
Road 3

Sunday 2nd
NEW 35 Alexandra
Road 3
Pear Tree Cottage 31
White Cottage &
Nursery 49

Monday 3rd
◆ Little Malvern Court 20
White Cottage &
Nursery 49

Sunday 9th
Hiraeth 19

Wednesday 12th
The Dell House 13

Sunday 16th
The Dell House 13
White Cottage &
Nursery 49
The White House 50

Saturday 22nd
1 Church Cottage 10
Oak Tree House 27
Ravelin 33
NEW Rhydd Gardens 35

Sunday 23rd
1 Church Cottage 10
Ravelin 33
Rothbury 38
Warndon Court 45

Saturday 29th
Oak Tree House 27

Sunday 30th
1 Church Cottage 10
White Cottage &
Nursery 49
68 Windsor Avenue 54

Monday 31st
1 Church Cottage 10
Rothbury 38
White Cottage &
Nursery 49
68 Windsor Avenue 54

June

Saturday 5th
Eckington Gardens 15
Pershore Gardens 32

Sunday 6th
NEW Acorns Children's
Hospice 1
Brockamin 9
Eckington Gardens 15
Pershore Gardens 32

Wednesday 9th
The Dell House 13

Sunday 13th
Cowleigh Lodge 12
Warndon Court 45
White Cottage &
Nursery 49
NEW Wick Village 52

Saturday 19th
Alvechurch Gardens 4
North Worcester
Gardens 26
NEW Rest Harrow 34

Sunday 20th
Alvechurch Gardens 4
Birtsmorton Court 7
North Worcester
Gardens 26
NEW Rest Harrow 34
Walnut Cottage 44

Saturday 26th
Long Hyde House 22
The Walled Garden 43
◆ Whitlenge Gardens 51

Sunday 27th
Hiraeth 19
Long Hyde House 22
3 Oakhampton Road 28
White Cottage &
Nursery 49
◆ Whitlenge Gardens 51

July

Saturday 3rd
The Lodge 21
The Walled Garden 43
Wharf House 47

Sunday 4th
The Lodge 21
◆ Spetchley Park
Gardens 39
Walnut Cottage 44
Wharf House 47
The White House 50

Saturday 10th
Westacres 46

Sunday 11th
Westacres 46

July (cont.)

Wednesday 14th
The Dell House 13

Sunday 18th
The Firs 16
White Cottage &
Nursery 49

Thursday 22nd
NEW 12 Three Springs
Road 41

Saturday 24th
Millbrook Lodge 24
Rothbury 38

Sunday 25th
Millbrook Lodge 24
Rothbury 38

Saturday 31st
5 Beckett Drive 6
3 Oakhampton Road 28

August

Sunday 1st
5 Beckett Drive 6
The Firs 16
Hiraeth 19

Saturday 7th
Offenham Gardens 29
The River School 36

Sunday 8th
Offenham Gardens 29

Wednesday 11th
The Dell House 13

Sunday 15th
Bridges Stone Mill 8
Cowleigh Lodge 12

Saturday 21st
NEW Rest Harrow 34

Sunday 22nd
NEW Rest Harrow 34

Saturday 28th
NEW The Granary 18
Morton Hall Gardens 25

Sunday 29th
1 Church Cottage 10
NEW The Granary 18
3 Oakhampton Road 28
Pear Tree Cottage 31

Monday 30th
1 Church Cottage 10
3 Oakhampton Road 28

THE GARDENS

1 NEW **ACORNS CHILDREN'S HOSPICE**
350 Bath Road, Worcester,
WR5 3EZ. *On A38 approx 1¼m S
of Worcester. From M5 J7 towards
Worcester then L at first island onto
A4440. R at 2nd island onto A38
towards city centre for approx ½m.*
**Sun 6 June (11-4). Adm £4, chd
free. Light refreshments. Soup
and roll available at lunchtime.**
1½ acres designed for the benefit of
children and young people with life
limiting illnesses and their families.
Incl play areas, raised beds for
young people to plant vegetables,
tranquil memorial garden with gentle
stream and memorial pebbles for
children who have passed away and
an enclosed bedroom garden set
aside for recently bereaved families.
Acers, silver birch and other trees and
shrubs feature. This is not a traditional
National Garden Scheme garden but
one that shows how gardens can
have a positive effect on peoples'
wellbeing. All areas wheelchair
accessible.
&. ✿ ☕

2 **24 ALEXANDER AVENUE**
Droitwich Spa, WR9 8NH. Malley
& David Terry, 01905 774907,
terrydroit@aol.com. *1m S of
Droitwich. Droitwich Spa towards
Worcester A38. Or from M5 J6 to*
Droitwich Town centre. **Visits by
arrangement Apr to Sept. Adm
£4, chd free. Refreshments
available by prior arrangement.**
Beautifully designed, now with mature
plantings and giving a feeling of
space and tranquillity. Many varieties
of clematis growing through shrubs
and high hedges. Borders with
herbaceous perennials and shrubs,
some unusual. Swooping curves of
lawns and paths to woodland area
with shade-loving plants. Drought-
tolerant plants in S-facing gravel
front garden. Alpine filled troughs. In
April spring bulbs incl good variety of
snowdrops, erythroniums, species
tulips and alpines. In July/August
clematis, colourful mixed shrub
and herbaceous borders. Partial
wheelchair access.
&. 🚜 ☕

3 NEW **35 ALEXANDRA ROAD**
Malvern, WR14 1HE. Margaret
& David Cross, 01684 569019,
margaret.cross@ifdev.net. *From
A449. From Malvern Link, pass train
stn, ahead at T-lights, 1st R into
Alexandra Rd. From Great Malvern,
go through Link Top T-lights (junction
with B4503), 1st L.* **Sat 1, Sun 2
May (10.30-5.30). Adm £4.50, chd
free. Light refreshments. Visits
also by arrangement Apr to Oct for
groups of up to 10.**
Overlooked by the Malvern Hills,
this garden is on 2 levels with
mature horse chestnuts, western
red cedars, wildlife pond, small
wildflower meadow, shady woodland
area, cottage garden, Japanese
garden, rookeries and a New Zealand
influenced area. An arbour inspired
by Geoff Hamilton, also a treehouse
and den for accompanied children.
Elgar wrote some of the Enigma
variations in a bell tent here! With
help, the garden can be accessed by
wheelchair, there are few steps but
gravel paths and gradients. If advised
we can offer closer parking.
&. 🚜 ☕

"I love the National Garden
Scheme which has been the
most brilliant supporter of
Queen's Nurses like me. It
was founded by the Queen's
Nursing Institute which
makes me very proud. As we
battle Coronavirus on the
front line in the community,
knowing we have their
support is a real comfort." –
Liz Alderton, Queen's Nurse

GROUP OPENING

⁴ ALVECHURCH GARDENS

Alvechurch, B48 7LF. Group Co-ordinator Martin & Janet Wright. *3m N of Redditch, 3m NE of Bromsgrove. Alvechurch is on the B4120 close to J2 of M42. Gardens are signed from all roads into the village. Pick up your map when you pay for your ticket at the first garden visited.* **Sat 19, Sun 20 June (1-6). Combined adm £6, chd free. Light refreshments at The Shrubbery, Bear Hill, Alvechurch, B48 7JX.**

THE ALLOTMENTS
Eileen McHugh.

18 BEAR HILL
Alexandra & Nicholas Wood.

69 BIRMINGHAM ROAD
Anna & Andy Ingram.

NEW **43 BLYTHESWAY**
Tony & Cheryl Davis-Wetton.

NEW **11 CALLOW HILL ROAD**
Sheila Dickinson.

28 CALLOW HILL ROAD
Janet & Martin Wright.

NEW **33 CALLOW HILL ROAD**
Mrs Chris Grainger.

CORNER HOUSE
Janice Wiltshire.

NEW **41 HINTON AVENUE**
Lindsay & Jules Bissell.

THE MOAT HOUSE
Mike & Tracy Fallon.

RECTORY COTTAGE
Celia Hitch, 0121 445 4824, celia@rectorycottage-alvechurch.co.uk.
🛏

THE SHRUBBERY
Chris Thompson & Liz Fox.

NEW **WYCHWOOD HOUSE**
Chris & Stephanie Miall.

Large village with new development and historically interesting core with buildings spanning medieval to Edwardian and St Laurence Church dedicated in 1239. There is a selection of lovely open gardens ranging from a riverside former rectory with waterfall to a corner plot gardened for wildlife. There are professionally landscaped terraced gardens and cottage gardens with lots of colour. Also a large garden on the old site of the Bishop of Worcester's Summer Palace. Around the gardens there are mature trees, rose beds, shrubberies, herbaceous and fruit and vegetable beds. Some gardens include sculptures. The village allotments will also be open for viewing. Not all gardens are wheelchair accessible or they may only have partial access.
 ✿ ☕

⁵ BADGE COURT

Purshull Green Lane, Elmbridge, Droitwich, WR9 0NJ. Stuart & Diana Glendenning, 01299 851216, dianaglendenning1@gmail.com. *5m N of Droitwich Spa. 2½m from J5 M5. Turn off A38 at Wychbold down side of the Swan Inn. Turn R into Berry Ln. Take next L into Cooksey Green Ln. Turn R into Purshull Green Ln. Garden is on L.* **Visits by arrangement May to July for groups of 5 to 30. Smaller groups Mon, Wed or Fri during day. Larger groups on Wed evenings. Adm £5, chd free. Home-made teas.**
The 2½ acre garden is divided into a series of uniquely different areas and contains a large selection of specimen trees and over 80 varieties of roses and clematis. Areas within the garden include a large pool with fish, stumpery, walled garden with aviary, Mediterranean garden (new), large vegetable garden, orchard, cherry tunnel, herbaceous and specialist borders.
♿ ✿ ☕

5 Becket Drive

© Michael Warren

6 5 BECKETT DRIVE

Northwick, Worcester, WR3 7BZ. Jacki & Pete Ager, 01905 451108, agers@outlook.com. *1½ m N of Worcester city centre. A cul-de-sac off the A449 Ombersley Road directly opp Grantham's Autocare Garage. 1m S of the Claines r'about on A449 & just N of Northwick Cinema on Ombersley Rd.* **Sat 31 July, Sun 1 Aug (2-5). Adm £4, chd free. Home-made teas. Opening with North Worcester Gardens on Sat 19, Sun 20 June. Visits also by arrangement May to Aug for groups of 10 to 20.**
An extraordinary town garden on the northern edge of Worcester packed with different plants and year-round interest guaranteed to give visitors ideas and inspiration for their own gardens. Over 15 years visitors have enjoyed the unique and surprising features of this garden which has many planting schemes for a variety of situations. Plants at bargain prices and delicious home-made teas.

7 BIRTSMORTON COURT

Birtsmorton, nr Malvern, WR13 6JS. Mr & Mrs N G K Dawes. *7m E of Ledbury. Off A438 Ledbury/Tewkesbury rd.* **Sun 20 June (2-5.30). Adm £7. Home-made teas provided by Castlemorton School.**
10 acre garden surrounding beautiful medieval moated manor house (not open). White garden, built and planted in 1997 surrounded on all sides by old topiary. Potager, vegetable garden and working greenhouses, all beautifully maintained. Rare double working moat and waterways including Westminster Pool laid down in Henry VII's reign to mark the consecration of the knave of Westminster Abbey. Ancient yew tree under which Cardinal Wolsey reputedly slept in the legend of the Shadow of the Ragged Stone. No dogs.

8 BRIDGES STONE MILL

Alfrick Pound, WR6 5HR. Sir Michael & Lady Perry. *6m NW of Malvern. A4103 from Worcester to Bransford r'about, then Suckley Rd for 3m to Alfrick Pound.* **Sun 7 Mar, Sun 15 Aug (2-5.30). Adm £6, chd free. Home-made teas.**
Once a cherry orchard adjoining the mainly C19 flour mill, this is now a 2½ acre all-year-round garden

laid out with trees, shrubs, mixed beds and borders. The garden is bounded by a stretch of Leigh Brook (an SSSI), from which the mill's own weir feeds a mill leat and small lake. A rose parterre and a traditional Japanese garden complete the scene. Wheelchair access by car to courtyard.

9 BROCKAMIN

Old Hills, Callow End, Worcester, WR2 4TQ. Margaret Stone, 01905 830370, stone.brockamin@btinternet.com. *5m S of Worcester. ½ m S of Callow End on the B4424, on an unfenced bend, turn R into the car park signed Old Hills. Walk towards the houses keeping R.* **Sun 7 Feb (11-4); Sun 21 Mar, Sun 6 June, Sun 19 Sept (2-5). Adm £4, chd free. Home-made teas. Visits also by arrangement Feb to Oct for groups of 10+. Donation to Plant Heritage.**
This is a plant specialist's 1½ acre informal working garden, parts of which are used for plant production rather than for show. Situated next to common land. Mixed borders with wide variety of hardy perennials where plants are allowed to self seed. Includes Plant Heritage National Collections of Symphyotrichum (Aster) novae-angliae and some Hardy Geraniums. Unusual plants for sale. Open for snowdrops in Feb, daffodils in March, geraniums in June and asters in September. Good collection of pulmonarias. Seasonal pond/bog garden and kitchen garden. Teas with home-made cakes from Malvern Country Market. An access path reaches a large part of the garden.

> "The annual donation from the National Garden Scheme to Perennial is the cornerstone of our fundraising activities and encourages many of our donors." Perennial

---

10 1 CHURCH COTTAGE

Church Road, Defford, WR8 9BJ. John Taylor & Ann Sheppard, 01386 750863, ann98sheppard@btinternet.com. *3m SW of Pershore. A4104 Pershore to Upton Rd, turn into Harpley Rd, Defford. Don't go up Bluebell Lane as directed by sat nav, black & white cottage at side of church. Parking in village hall car park.* **Sat 22, Sun 23, Sun 30, Mon 31 May, Sun 29, Mon 30 Aug (11-5). Adm £4, chd free. Home-made teas. Visits also by arrangement May to Sept for groups of 10 to 30.**
True countryman's ⅓ acre cottage garden. Japanese style feature with 'dragons den'. Specimen trees, water features, perennial garden, vegetable garden, poultry, streamside bog garden. New features in progress. Wheelchair access to most areas.

11 CONDERTON MANOR

Conderton, nr Tewkesbury, GL20 7PR, Mr & Mrs W Carr, 01386 725389, carrs@conderton.com. *5½ m NE of Tewkesbury. From M5 - A46 to Beckford - I for Overbury/ Conderton. From Tewkesbury B4079 to Bredon - then follow signs to Overbury. Conderton from B4077 follow A46 directions from Teddington r'about.* **Visits by arrangement Mar to Nov for groups of up to 30. Light refreshments by prior arrangement. Tea/coffee and biscuits. Wine/snacks may be offered on an evening visit.**
7 acre garden with magnificent views of Cotswolds. Flowering cherries and bulbs in spring. Formal terrace with clipped box parterre; huge rose and clematis arches, mixed borders of roses and herbaceous plants, bog bank and quarry garden. Many unusual trees and shrubs make this a garden for all seasons. Visitors are particularly encouraged to visit in spring and autumn when the trees are at their best. This is a garden/small arboretum of particular interest for tree lovers. The views towards the Cotswolds escarpment are spectacular and it provides a peaceful walk of about an hour. Some gravel paths and steps - no disabled WC.

12 COWLEIGH LODGE

16 Cowleigh Bank, Malvern, WR14 1QP. Jane & Mic Schuster, 01684 439054, dalyan@hotmail.co.uk. *7m SW from Worcester. From Worcester or Ledbury follow the A449 to Link Top. Take North Malvern Rd (behind Holy Trinity Church), follow yellow signs. From Hereford take B4219 after Storridge church, follow yellow signs.* **Sun 13 June, Sun 15 Aug (11-5). Adm £5, chd free. Home-made teas. Visits also by arrangement May to Aug for groups of 10 to 30.** The garden on the slopes of the Malvern Hills has been described as 'quirky'. Formal rose garden, grass beds, bamboo walk, colour themed beds, nature path leading to a pond, acer bank, chickens. Large vegetable plot and orchard with views overlooking the Severn Valley. Explore the polytunnel and then relax with a cuppa and slice of home-made cake served with a smile. This is the 7th year of opening of this developing and expanding garden - visitors from previous years will be able to see the difference. Lots of added interest with staddle stones, troughs, signs and other interesting artefacts. The garden is definitely now a mature one! Slopes and steps throughout the garden. WC.

13 THE DELL HOUSE

2 Green Lane, Malvern Wells, WR14 4HU. Kevin & Elizabeth Rolph, 01684 564448, stay@thedellhouse.co.uk, www.dellhousemalvern.uk. *2m S of Gt Malvern. Behind former church on corner of A449 Wells Rd & Green Ln. NB Satnavs/Google don't work with postcode. Small car park.* **Wed 17 Mar (1-5); Wed 14, Sun 18 Apr (1-6); Wed 12, Sun 16 May (1-7); Wed 9 June, Wed 14 July, Wed 11 Aug (1-8); Sun 5, Wed 8 Sept (1-7); Wed 13 Oct (1-6). Adm £5, chd free. Pre-booking essential, please visit www.ngs. org.uk for information & booking. Light refreshments. Visits also by arrangement Mar to Nov for groups of up to 20.** Two acre wooded hillside garden of this 1820s former rectory, now a B&B. Peaceful and natural, the garden contains many magnificent specimen trees including a Wellingtonia Redwood. Informal in style with meandering bark paths, historic garden buildings, garden railway and

a paved terrace with outstanding views. Spectacular tree carvings by Steve Elsby, and other sculptures by various artists. Featured in 'A survey of Historic Parks & Gardens in Worcestershire'. Partial wheelchair access but good views from the level paved terrace. Parking is on gravel. Sloping bark paths, some quite steep.

14 6 DINGLE END

Inkberrow, WR7 4EY. Mr & Mrs Glenn & Gabriel Allison, 01386 792039. *12m E of Worcester. A422 from Worcester. At the 30 sign in Inkberrow turn R down Appletree Ln then 1st L up Pepper St. Dingle End is 4th on R of Pepper St. Limited parking in Dingle End but street parking on Pepper St.* **Visits by arrangement Apr to Sept for groups of 5 to 20. Tea and cakes or soup and sandwiches by prior arrangement. Adm £4, chd free.** Over 1 acre garden with formal area close to the house opening into a flat area featuring a large pond, stream and weir with apple orchard and woodland area. Large vegetable garden incl an interesting variety of fruits. Garden designed for wildlife and attractive to birds on account of water and trees. Named varieties of apple and pear trees. Wheelchair access - slopes alongside every terrace.

GROUP OPENING

15 ECKINGTON GARDENS

Hilltop, Nafford Road, Eckington, WR10 3DH. Group Coordinator Richard Bateman. *3 gardens - 1 in Upper End, 1 close by in Nafford Rd, 3rd about 1 mile along Nafford Rd. A4104 Pershore to Upton & Defford, L turn B4080 to Eckington. In centre, by war memorial turn L into New Rd (becomes Nafford Rd).* **Sat 5, Sun 6 June (11-5). Combined adm £6, chd free. Home-made teas at Mantoft.**

HILLTOP FARM

Richard & Margaret Bateman, 01386 750667, richard. bateman111@btinternet.com. **Visits also by arrangement May to Sept for groups of 5 to 30.**

MANTOFT

Mr & Mrs M J Tupper, 01386 750819.

Visits also by arrangement May to Sept for groups of 5 to 30.

NAFFORD HOUSE

George & Joanna Stylianou.

3 very diverse gardens set in/ close to lovely village of Eckington. Hilltop – 1 acre garden with sunken garden/pond, rose garden, herbaceous borders, interesting topiary incl cloud pruning and formal hedging to reduce effect of wind and having 'windows' for views over the beautiful Worcestershire countryside. Sculptures made by owner. Vegetable yurt (added 2018) has been successful in allowing pollination and preventing damage to young plants. Mantoft - Wonderful ancient thatched cottage with 1½ acres of magical gardens. Fishpond with ghost koi, Cotswold and red brick walls, large topiary, treehouse with seating, summer house and dovecote, pathways, vistas and stone statues, urns and herbaceous borders. Featured in Cotswold Life - should not be missed. Nafford House is 2 acre mature natural garden, wood with walk and slopes to R Avon, formal gardens around the house/ magnificent wisteria. Some wheelchair access issues, particularly at Nafford House.

16 THE FIRS

Brickyard Lane, Drakes Broughton, Pershore, WR10 2AH. Ann & Ken Mein. *From J7 M5 take B4084 towards Pershore. At Drakes Broughton turn L to Stonebow Rd. First R into Walcot Ln then second R into Brickyard Ln. The Firs is 200 yds on L.* **Sun 18 July, Sun 1 Aug (12.30-4.30). Adm £4, chd free. Cream teas.** Two acre garden - constantly evolving as new ideas and plans are introduced. Over 200 trees have been planted and mixed borders have been established around the house and barn. Birch trees, roses and hydrangeas are a passion! Open Studio with Ceramics.

The National Garden Scheme searches the length and breadth of England and Wales for the very best private gardens

Wharf House

ⓘ7 THE FOLLY

87 Wells Road, Malvern,
WR14 4PB. David & Lesley
Robbins, 01684 567253,
lesleycmedley@btinternet.com.
*1½m S of Great Malvern & 9m S
of Worcester. Approx 8m from M5
via J7 or J8. Situated in Malvern
Wells on A449, 0.7m N of B4209
and 0.2m S of Malvern Common.
Parking off A449 0.25m N on lay-by
or side road, or in 0.1m N turn E on
Peachfield Rd which runs by Malvern
Common.* **Visits by arrangement
Apr to June for groups of 5 to
20. Tea, coffee and biscuits by
arrangement.**
Steeply sloping garden on Malvern
Hills with views over Severn Vale. 3
levels accessed by steps, paved/
gravel paths and ramps. Potager
and greenhouse, courtyard, formal
terrace and lawn, pergola, mature
cedars, ornaments and sculpture
in landscaped beds and borders.
Climbers, small trees, shrubs, hostas,
grasses and ferns, with new areas
developing. Seating on each level.
Gravel and rockery gardens, small
cottage garden, stumpery and
shrubbery linked by winding paths
with an intimate atmosphere as views
are concealed and revealed.

ⓘ8 NEW THE GRANARY

Buntsford Hill, Stoke Heath,
Bromsgrove, B60 3AP.
Mr & Mrs Jonathan &
Lorna Hill, 01527 879401,
lornashill0468@gmail.com. *1m SW
of Bromsgrove. Follow tourist signs
to Avoncroft Museum. From A38 take
Buntsford Dr to mini r'about (opp
museum) then L down Buntsford
Hill. After 200 metres R into very
wide drive. Parking after 50 metres.*
**Evening opening Sat 28, Sun 29
Aug (7.30-10); Fri 19, Sat 20 Nov (4-
8). Adm £4, chd free. Pre-booking
essential, please phone 01527
879401 or email lornashill0468@
gmail.com for information &
booking. Light refreshments. Visits
also by arrangement Aug to Nov for
groups of 5 to 20. Best seen dusk
to dark.**
A hidden away ⅓ acre traditional
garden with extensive views.
Sweeping lawn down to pond area,
mature trees, various borders with
mixed flowers and shrubs, pots,
seating, patios and sculptures. It is
in the dark that this garden reveals
its other identity through creative
lighting. Come and experience
the transition into darkness. Enjoy
homemade refreshments, music
and art in the summer house. The
November opening will have a festive
theme. Please Note - visitor numbers
restricted due to limited local parking
so visitors must book in advance with
garden owners.

ⓘ9 HIRAETH

30 Showell Road, Droitwich,
WR9 8UY. Sue & John Fletcher,
07752 717243 or 01905 778390,
sueandjohn99@yahoo.com. *1m
S of Droitwich. On The Ridings
estate. Turn off A38 r'about into
Addyes Way, 2nd R into Showell
Rd, 500yds on R. Follow the yellow
signs.* **Sun 9 May, Sun 27 June,
Sun 1 Aug (2-5). Adm £3.50, chd
free. Home-made teas. Visits also
by arrangement May to Aug for
groups of 10 to 30.**
⅓ acre gardens, front and rear,
containing many plant species,
cottage, herbaceous, hostas, ferns,
acer trees, 300yr old olive tree, pool,
waterfall, oak sculptures, metal
animals incl giraffes, elephant, birds.
New patio in 2019 and rear lawn
removed but still an oasis of colours in
a garden not to be missed described
by visitor as 'A haven on the way
to heaven'. Excellent tea, coffee,
cold drinks, home-made cakes and
scones served with china cups,
saucers, plates, tea-pots and coffee-
pots. Partial wheelchair access.

ⓘ20 ◆ LITTLE MALVERN COURT

Little Malvern, WR14 4JN. Mrs
T M Berington, 01684 892988,
littlemalverncourt@hotmail.com,
www.littlemalverncourt.co.uk.
*3m S of Malvern. On A4104 S of
junction with A449.* **For NGS: Fri 12,
Fri 19, Fri 26 Mar (2-5). Adm £5,
chd free. Tea. Mon 3 May (2-5).
Adm £8, chd £1. Home-made
teas. For other opening times and
information, please phone, email or
visit garden website.**
10 acres attached to former
Benedictine Priory, magnificent views
over Severn Valley. Garden rooms
and terrace around house designed
and planted in early 1980s; chain of
lakes; wide variety of spring bulbs,
flowering trees and shrubs. Notable
collection of old-fashioned roses.
Topiary hedge and fine trees. Regional
Finalist, The English Garden's The
Nation's Favourite Gardens 2019. The
May Bank Holiday - Flower Festival in
the Priory Church. Partial wheelchair
access.

Rhydd Gardens

21 THE LODGE

off Holmes Lane, Dodderhill
Common, Hanbury, Bromsgrove,
B60 4AU. Mark & Lesley Jackson.
*1m N of Hanbury Village. 3½m E
from M5 J5, on A38 at the Hanbury
Turn Xrds take A4091 S towards
Hanbury. After 2½m turn L into
Holmes Ln, then immediately L
again into dirt track. Car Park (Worcs
Woodland Trust) on L.* **Sat 3, Sun
4 July (11-4). Adm £4, chd free.
Cream teas.**
1½ acre garden attached to a
part C16 black and white house
(not open) with 3 lawned areas,
traditional greenhouse, various beds
plus stunning views over the North
Worcestershire countryside. Small
display of classic cars plus other
cars of interest. Garden is wheelchair
accessible, disabled parking within
grounds, toilets not wheelchair
accessible. No dogs.

  ♿ ☕

22 LONG HYDE HOUSE

Long Hyde Road, South Littleton,
Evesham, WR11 8TH. David
& Linda Lamb, 01386 834697,
davidlamb1943@gmail.com.
*From Badsey r'about on A46,
follow B4035 towards Bretforton,
then L on B4085 to The Littletons
approx 1.7m. L onto Long Hyde Rd.
Garden on L opp playing field.* **Sat
26, Sun 27 June (1-5). Adm £5,
chd free. Cream teas. Visits also
by arrangement June & July for
groups of 10 to 30.**
Beautiful 1 acre traditional garden
in Vale of Evesham with stunning
views of Cotswolds. Formal rose
garden, clipped box hedging, parterre
garden, herb garden with 2 ponds,
large vegetable area, extensive
borders, giant chess set with dark
planting, lavender bed and a variety of
baskets and tubs. Honey locust tree
dominates main lawn, raised patios
and seating areas. Wide paths and
disabled parking.

  ♿ ☕ 🪑

23 74 MEADOW ROAD

Wythall, B47 6EQ. Joe
Manchester, 01564 829589,
joe@cogentscreenprint.co.uk.
*4m E of Alvechurch. 2m N from J3
M42. On A435 at Becketts Farm
r'about take rd signed Earlswood/
Solihull. Approx 250 metres turn L
into School Dr, then L into Meadow
Rd.* **Visits by arrangement. Adm
£3.50, chd free.**
Has been described as one of
the most unusual urban gardens
dedicated to woodland, shade-loving
plants. 'Expect the unexpected' in
a few tropical and foreign species.
Meander through the garden under
the majestic pine, eucalyptus and
silver birch. Sit and enjoy the peaceful
surroundings and see how many
different ferns and hostas you can
find. As seen on BBC Gardeners'
World.

 ✿ 🚗 ☕

24 MILLBROOK LODGE

Millham Lane, Alfrick, Worcester,
WR6 5HS. Andrea & Doug Bright,
andreabright@hotmail.co.uk. *11m
from M5 J7. A4103 from Worcester
to Bransford r'about, then Suckley
Rd for 3m to Alfrick Pound to find car
park on L for Nature Reserve.* **Sat
24, Sun 25 July (1-5). Adm £5, chd
free. Home-made teas. Visits also
by arrangement Mar to Oct for
groups of 10+.**
3½ acre garden and woodland
developed by current owners over
25 years. Large informal flowerbeds
planted for all year round interest,
from spring flowering bulbs, camellias
and magnolias through to asters and
dahlias. Pond with stream and bog
garden. Fruit and vegetable garden
and gravel garden. Peaceful setting in
an area of outstanding natural beauty
and opposite a nature reserve. Slopes
and grass/woodland paths.

✿ ☕

25 MORTON HALL GARDENS

Morton Hall Lane,
Holberrow Green, Redditch,
B96 6SJ. Mrs A Olivieri,
www.mortonhallgardens.co.uk.
*In centre of Holberrow Green, at a
wooden bench around a tree, turn
up Morton Hall Ln. Follow NGS signs
to gate opposite Morton Hall Farm.*
**Sat 28 Aug (10-4.30). Adm £9, chd
free. Home-made teas.**
Perched atop an escarpment with
breathtaking views, hidden behind
a tall hedge, lies a Georgian country
house with a unique garden of
outstanding beauty. A garden for
all seasons, it features one of the
country's largest fritillary spring
meadows, sumptuous herbaceous
summer borders, a striking potager,
a majestic woodland rockery and an
elegant Japanese Stroll Garden with
tea house. For May Tulip Festival
and other visiting options please see
website.

☕

GROUP OPENING

26 NORTH WORCESTER GARDENS

Northwick, Worcester, WR3 7BZ.
Jacki & Pete Ager. *Five town
gardens off A449, Ombersley Rd, S
of Claines r'about. Look for Yellow
signs from the main rd.* **Sat 19, Sun
20 June (11-5). Combined adm
£5, chd free. Teas and home-
made cakes at 27 Sheldon Park
Road. Home-made ice cream at
5 Beckett Drive. Cold drinks at 14
Beckett Road.**

5 BECKETT DRIVE
Jacki & Pete Ager.
(See separate entry)
NEW 14 BECKETT ROAD
Mr & Mrs Mike & Julia Roberts.
NEW 19 BEVERE CLOSE
Mr & Mrs Malcolm & Diane Styles.
10 LUCERNE CLOSE
Mark & Karen Askwith.
**NEW 27 SHELDON PARK
ROAD**
Mr & Mrs Alan & Helen Kirby.

Five town gardens on the northern
edge of Worcester each having its own
identity. The diversity of this quintet of
gardens would satisfy anyone from a
dedicated plantsman to an enthusiastic
amateur while also offering plenty of
ideas for landscaping, planting and
structures. The landscaped garden
at 5 Beckett Drive includes borders
planted at different levels using palettes
of harmonising or contrasting colours
complemented by innovative features.
10 Lucerne Close is a cottage-style
garden literally packed with a wide
variety of plants with secluded seating
areas. 27 Sheldon Park Road is a
contemporary style garden planned
with low maintenance in mind and
expertly planted with many different
foliage plants. The hard landscaping
set amongst colourful borders with
vegetables planted amongst the
flowers are significant features at 14
Beckett Road. And by no means
least, 19 Bevere Drive with a stunning
collection of bonsai trees planted in
an oriental setting. Tickets can be
purchased from any of the gardens.
Maps showing driving and walking
routes available. Wheelchair access
at Beckett Drive and Lucerne Close is
restricted.

✿ ☕

27 OAK TREE HOUSE

504 Birmingham Road, Marlbrook, Bromsgrove, B61 0HS. Di & Dave Morgan, 0121 445 3595, meandi@btinternet.com. *On main A38 midway between M42 J1 & M5 J4. Park in old A38 - R fork 250 yds N of garden or small area in front of Miller & Carter Pub car park 200 yds S or local roads.* **Sat 22, Sat 29 May (1-5.30). Adm £4, chd free. Pre-booking essential, please phone 0121 445 3595 or email meandi@btinternet.com for information & booking. Home-made teas. Visits also by arrangement May to Aug for groups of 10 to 30.**

Plantswoman's cottage garden overflowing with plants, pots and interesting artifacts. Secluded patio with plants and shrubs for spring, small pond and waterfall. Plenty of seating, separate wildlife pond, water features, alpine area, rear open vista. Scented plants, hostas, dahlias and lilies. Conservatory with art by owners. Visitor HS said: "Such a wonderful peaceful oasis". Also: 'Wynn's Patch' - part of next door's garden being maintained on behalf of the owner. Please Note - visitor numbers managed due to limited local parking so visitors must book in advance with garden owners.

28 3 OAKHAMPTON ROAD

Stourport-On-Severn, DY13 0NR. Sandra & David Traynor, 07970 014295, traynor007@btinternet.com. *Between Astley Cross & Kings Arms PHs. From Stourport take A451 Dunley Rd towards Worcester. 1600 yds turn L into Pearl Ln. 4th R into Red House Rd, past the Kings Arms Pub, and next L to Oakhampton Rd. Extra parking at Kings Arms Pub.* **Sun 27 June, Sat 31 July, Sun 29, Mon 30 Aug (10-6.30). Adm £4.50, chd free. Home-made teas. Visits also by arrangement June to Aug for groups of 5 to 20.**

Beginning in March 2016 our plan was to create a garden with a decidedly tropical feel to include palms from around the world, with tree ferns, bananas and many other strange and unusual plants from warmer climes that would normally be considered difficult to grow here as well as a pond and small waterfall. Not a large garden but you'll be surprised what can be done with a small space. Some narrow paths.

GROUP OPENING

29 OFFENHAM GARDENS

Main Street, Offenham, WR11 8QD. *Approaching Offenham on B4510 from Evesham, L into village signed Offenham & ferry ¾ m. Follow road round into village. Parking in Village Hall car park opp Church. Walk to gardens from car park.* **Sat 7, Sun 8 Aug (11-5). Combined adm £5, chd free. Home-made teas. Light savoury snacks, strawberries & cream and alcohol free Pimms.**

BROADWAY VIEW
Brett Pillinger & Rachel Bates.

DECHMONT
Angela & Paul Gash.

LANGDALE
Sheila & Adrian James, www.langdalegarden.uk.

Offenham is a picturesque village in the heart of the Vale of Evesham, with thatched cottages and traditional maypole. Three interesting & attractive gardens of diverse styles including Langdale a Daily Mail 2019 National Garden Competition Finalist. Broadway View has a formal front garden. In the rear courtyard garden peaceful water features together with lush big leaf plants, olives and lemon trees make for a very tranquil space. With a mature walnut tree Dechmont features box topiary, conifers, shrubs and acers, colour from bulbs, perennials, annuals, clematis and roses set within curved borders. Langdale is a plant lover's garden designed for all year round interest. Surrounding a tranquil rill & pergola are garden rooms in a variety of styles including cottage, woodland, naturalistic and exotic, with many unusual plants.

30 OVERBURY COURT

Overbury, GL20 7NP. Sir Bruce & Lady Bossom, 01386 725111(office), pa@overburyenterprises.co.uk. *5m NE of Tewkesbury. Overbury signed off A46. Turn off village road beside the church. Park by the gates & walk up the drive. What3words app - cars.blurs.crunches.* **Visits by arrangement Mar to Sept for groups of 10+. Adm £5, chd free. There are local pubs and**

Fantastically Fresh's The Eatery in Beckford GL20 7AU.

Georgian house 1740 (not open); landscaped garden of same date with stream and pools, daffodil bank and grotto. Plane trees, yew hedges, shrubs, cut flowers, coloured foliage, gold and silver, shrub rose borders. Norman church adjoins garden. Close to Whitcombe and Conderton Manor. Some slopes, while all the garden can be viewed, parts are not accessible to wheelchairs.

31 PEAR TREE COTTAGE

Witton Hill, Wichenford, Worcester, WR6 6YX. Pamela & Alistair Thompson, 01886 888295, peartree.pam@gmail.com, www.peartreecottage.me. *13m NW of Worcester & 2m NE of Martley. From Martley, take B4197. Turn R into Horn Lane then 2nd L signed Witton Hill. Keep L & Pear Tree Cottage is on R at top of hill.* **Sun 2 May (11-5); Sun 29 Aug (2-9.30). Adm £5, chd free. Home-made teas. Sun 30 Aug - wine & Pimms served after 6pm. Visits also by arrangement May to Sept. Refreshments by prior arrangement.**

A Grade II listed black and white cottage (not open) SW-facing gardens with far reaching views across orchards to Abberley Clock Tower. The ¾ acre garden comprises gently sloping lawns with mixed and woodland borders, shade and plenty of strategically placed seating. The garden exudes a quirky and humorous character with the odd surprise and even includes a Shed of the Year Runner Up 2017. 'Garden by Twilight' evenings are very popular. Trees, shrubs and sculptures are softly uplit and the garden is filled with 100s of candles and nightlights (weather permitting). Visitors are invited to listen to the owls and watch the bats whilst enjoying a glass of wine. Partial wheelchair access.

GROUP OPENING

32 PERSHORE GARDENS

Pershore, WR10 1BG. Group Co-ordinator Jan Garratt, www.visitpershore.co.uk. *On B4084 between Worcester & Evesham, & 6m from J7 on M5. There is also a train station to N of town.* **Sat 5, Sun 6 June (1-5).**

Whitcombe House

Combined adm £6, chd free. Also refreshments at some individual gardens. These will be indicated on the map/description sheet. Most years about twenty gardens open in Pershore. This small town has been opening gardens as part of the NGS for 50 years, almost continuously. In those days the open gardens were in the Georgian heart of the town but now, gardens open from all over the town. Some gardens are surprisingly large, well over an acre, while others are courtyard gardens. All have their individual appeal and present great variety. The wealth of pubs, restaurants and cafes offer ample opportunities for refreshment while the Abbey and the River Avon are some of the many points of interest in this market town. Tickets with a map are valid for both days and are available in advance from the Tourist Information in the library

and 'Blue' in Broad Street and on the day at Number 8 Community Arts Centre in the High Street and any open garden. Refreshments available in pubs, hotels and cafes in the town, at Holy Redeemer School and in Number 8 Community Arts Centre. Wheelchair access to some gardens.

🐃 ✳ 🚗 🛏 ☕

33 RAVELIN

Gilberts End, Hanley Castle, WR8 0AS. Mrs Christine Peer, 01684 310215, cvpeer55@btinternet.com. *From Worcester/Callow End B4424 or from Upton B4211 to Hanley Castle. Then B4209 to Hanley Swan. From Malvern B4209 to Hanley Swan. At pond/ Xrds turn to Welland. ½ m turn L opp Hall to Gilberts End.* **Sat 22, Sun 23 May (1-5); Sat 18, Sun 19 Sept,**

Sun 3 Oct (12-4). Adm £5, chd free. Light refreshments. **Visits also by arrangement Apr to Sept.** ½ acre mature yet ever changing garden with a wide range of unusual plants full of colour and texture. Of interest to plant lovers and flower arrangers alike with views overlooking the fields and the Malvern Hills. Seasonal interest provided by a wide variety of holleboree, hardy geraniums, aconitums, heucharas, Michaelmas daisies, grasses and dahlias and a fifty-year-old silver pear tree. Thought to be built on medieval clay works in the royal hunting forest. Garden containing herbaceous and perennial planting with gravel garden, woodland area, pond, summer house and plenty of seating areas around the garden. A quiz for children. Largely flat, partial access for disabled - regret no mobility scooters.

🐃 ✳ 🚗 ☕

34 NEW **REST HARROW**
California Lane, Welland, Malvern, WR13 6NQ. Mr Malcolm & Mrs Anne Garner. *4.6m S of Gt Malvern In un-adopted California Lane off B4208 Worcester Rd. From Gt Malvern A449 towards Ledbury, L onto Hanley Rd/B4209 signed Upton. After about 1m R (Blackmore Park Rd/B4209). After 1m R onto B4208. After ⅓m R (California Lane) - garden 300 yds on L.* **Sat 19, Sun 20 June, Sat 21, Sun 22 Aug (2-5.30). Adm £5, chd free. Pre-booking essential, please phone 01684 310503 or email anne.restharrow@gmail.com for information & booking. Light refreshments.**
1½ acres developed over 14 yrs with 5 acre field, woodland & stunning views of Malvern Hills. Colourful & diverse flower beds, unusual plants, roses, alstroemeria, stocks & shrubs. Potager kitchen garden with veg & flowers, large fruit cage, fruit trees and rustic trellis made from our own pollarded trees. Sit, relax, enjoy the views or stroll down through the field to the wooded wetland border. Wheelchair access in garden but not down in field.

♿ 🍷 🪑

35 NEW **RHYDD GARDENS**
Worcester Road, Hanley Castle, Worcester, WR8 0AB. Bill Bell & Sue Brooks, 01684 311001, NGS@Rhyddgardens.co.uk. *2Km N of Hanley Castle. Gates 200m N of layby on B4211.* **Sat 22 May, Sat 18 Sept (12-4). Adm £5, chd free. Cream teas.**
Two walled gardens and a 60 foot greenhouse from the early 1800s set in 6 acres with wonderful views of the entire length of the Malvern ridge. One Walled garden is set out with formal paths and borders bounded by box hedging. We are planting fruit trees in espaliers and cordons as they would have been when the garden was first set out and have a nature area with walks and some woodland. Wheelchair access to the main walled garden with grass and paving paths. Parking near gates can be arranged in advance.

♿ 🏠 🍷

36 **THE RIVER SCHOOL**
Oakfield House, Droitwich Road, Worcester, WR3 7ST. Christian Education Trust -Worcester, 01905 451309, lacerta@btinternet.com. *2.4m N*

of Worcester City Centre on A38 towards Droitwich. At J6 M5, take A449 signed for Kidderminster, turn off at 1st turning marked for Blackpole. Turn R to Fernhill Heath & at T- junction with A38 turn L. The school is ½m on R. **Sat 24 Apr, Sat 7 Aug (10.30-3.30). Adm £5, chd free. Light refreshments in the Lewis Room near garden entrance. Visits also by arrangement Apr to Aug for groups of 10 to 30.**
A former Horticultural College garden being brought back to life. For 35 years after WW2 it was known as Oakfield Teacher Training College for Horticulture. With its reputation visitors came from 58 countries and at least 8 other Horticultural Colleges were founded by people inspired by it. The Estate features many less common shrubs and trees as well as a Forest School pond area.

37 ◆ **RIVERSIDE GARDENS AT WEBBS**
Wychbold, Droitwich, WR9 0DG. Webbs of Wychbold, 01527 860000, www.webbsdirect.co.uk. *2m N of Droitwich Spa. 1m N of M5 J5 on A38. Follow tourism signs from M5.* **For opening times and information, please phone or visit garden website.**
2½ acres. Themed gardens incl colour spectrum, tropical and dry garden, rose garden, vegetable garden, National Collection of Harvington Hellebores, seaside garden, bamboozeleum and self-sufficient garden. The New Wave garden is a natural wildlife area and includes seasonal interest with grasses and perennials. There are willow wigwams and wooden tepees made for children to play in, a bird hide, the Hobbit House and beehives which produce honey for our own food hall. Our New Wave Garden has grass paths which are underlaid with mesh and wheelchair accessible.

♿ ❀ 🚌 NPC 🍷

38 **ROTHBURY**
5 St Peter's Road, North Malvern, WR14 1QS. John Bryson, Philippa Lowe & David, www.facebook.com/RothburyNGS. *7m W of M5 J7 (Worcester). Turn off A449 Worcester to Ledbury Rd at B4503, signed Leigh Sinton. Almost immed take the middle rd (Hornyold Rd). St Peter's Rd is ¼m uphill, 2nd*

R. **Sun 23 May (1-5.30); Mon 31 May, Sat 24 July (10-5.30); Sun 25 July (1-5.30). Adm £4, chd free. Home-made teas. Tea in teapots and fresh coffee in cafetieres. Gluten free options available.**
Set on slopes of Malvern Hills, ⅓ acre plant-lovers' garden surrounding Arts and Crafts house (not open), created by owners since 1999. Herbaceous borders, rockery, pond, vegetables, small orchard. Siberian irises in May and magnificent Eucryphia glutinosa in July. A series of hand-excavated terraces accessed by sloping paths and steps. Views and seats. Partial wheelchair access. One very low step at entry, one standard step to main lawn and one to WC. Decking slope to top lawn. Dogs on leads.

♿ 🐕 ❀ 🍷

39 ◆ **SPETCHLEY PARK GARDENS**
The Estate Office, Spetchley Park, Worcester, WR5 1RS. Mr Henry Berkeley, 01905 345106, enquiries@spetchleygardens.co.uk, www.spetchleygardens.co.uk. *2m E of Worcester. On A44, follow brown signs.* **For NGS: Fri 9 Apr, Sun 4 July (10.30-5). Adm £8, chd £3.50. Light refreshments. For other opening times and information, please phone, email or visit garden website.**
Surrounded by glorious countryside lays one of Britain's best-kept secrets. Spetchley is a garden for all tastes and ages, containing one of the biggest private collections of plant varieties outside the major botanical gardens and weaving a magical trail for younger visitors. Spetchley is not a formal paradise of neatly manicured lawns or beds but rather a wondrous display of plants, shrubs and trees woven into a garden of many rooms and vistas. Plant sales, gift shop and tea room. Gravel paths.

♿ ❀ 🚌 🍷

40 ◆ **STONE HOUSE COTTAGE GARDENS**
Church Lane, Stone, DY10 4BG. Louisa Arbuthnott, 07817 921146, louisa@shcn.co.uk, www.shcn.co.uk. *2m SE of Kidderminster. Via A448 towards Bromsgrove, next to church, turn up drive.* **For opening times and information, please phone, email or visit garden website.**
A beautiful and romantic walled garden adorned with unusual brick

follies. This acclaimed garden is exuberantly planted and holds one of the largest collections of rare plants in the country. It acts as a shop window for the adjoining nursery. Open Wed to Sat late March to late August 10-5. Partial wheelchair access.

& ✿ ☕

41 NEW **12 THREE SPRINGS ROAD**
Pershore, WR10 1HH. Mr & Mrs Roger Smith. *From M5 J6 take B4084 to Pershore. At bottom of hill turn R onto A4104. Garden is 300 yards on R - opposite BP garage. 12 is on pillar. Please park on local roads ensuring no obstruction.* **Thur 22 July (10-3). Adm £3, chd free. Light refreshments.**
Garden has well defined areas with mature and young trees, perennial shrubs, 1 formal colour themed parterres and informal planting. Major part of garden is behind house and is gently sloping from top to bottom, about 25 by 45 yards. New projects most years. Enthusiastic amateur. Paving slabs (and lawn) allow easy access for wheelchairs.

& 🐐 ☕

42 **THE TYNINGS**
Church Lane, Stoulton, Worcester, WR7 4RE. John & Leslie Bryant, 01905 840189, johnlesbryant@btinternet.com. *5m S of Worcester; 3m N of Pershore. On the B4084 between M5 J7 & Pershore. The Tynings lies beyond the church at the extreme end of Church Lane. Ample parking.* **Visits by arrangement June to Sept. Admission incl light refreshments. Adm £5, chd free.**
Acclaimed plantsman's ½ acre garden, generously planted with a large selection of rare trees and shrubs. Features incl specialist collection of lilies, many unusual climbers and rare ferns. The colour continues into late summer with cannas, dahlias, berberis, euonymus and tree colour. Surprises around every corner. Lovely views of adjacent Norman Church and surrounding countryside. Plants labelled and plant lists available.

& 🐐 ✿ ☕ ☕

43 **THE WALLED GARDEN**
6 Rose Terrace, off Fort Royal Hill, Worcester, WR5 1BU. William & Julia Scott. *Close to City centre. ½m from Cathedral. Via Fort Royal Hill, off London Rd (A44). Park on 1st section of Rose Terrace & walk 20yds down track.* **Sat 26 June, Sat 3 July (1-5). Adm £4, chd free. Tea in the herb room.**
This C19 Walled Kitchen Garden, reawakened in 1995, is formal in layout, with relaxed chemical free planting. An outer path connects the camellia walk, soft fruit cordons, compost heaps, an area awaiting archeological investigation, bees and chickens. The inner walk, leads to herbs, topiary, vegetables, flower gardens, cook's garden, old and new fruit trees, medlar, mulberry and quince trees. Historic garden with a focus on herbs and their uses.

✿ ☕ 🛋

44 **WALNUT COTTAGE**
Lower End, Bricklehampton, Pershore, WR10 3HL. Mr & Mrs Richard & Janet Williams. *2½ m S of Pershore on B4084, then R into Bricklehampton Lane to T-junction, then L. Cottage is on R.* **Sun 20 June, Sun 4 July (3-7). Adm £6, chd free. Wine.**
1½ acre garden with views of Bredon Hill, designed into rooms, many created with high formal hedging of beech, hornbeam, copper beech and yew. There is a small 'front garden' with circular gravel path, well-stocked original garden area with pond and arches to the side of the house. Magnolia garden with several species and magnificent tree garden with specimens from around the world. Over 200 roses, in colour themed beds, climbing over a pergola or up trees, greenhouse, raised Koi Carp pond and metal stairway leading to roof-based viewing platform surrounded by roses. The garden continues to evolve and a Japanese garden is planned. Plenty of seating and interesting artefacts.

☕

In our first year 609 private gardens opened their gates to all, for the modest sum of one shilling. Today the National Garden Scheme retains that combination of inclusivity and affordability

45 **WARNDON COURT**
St Nicholas Lane, Worcester, WR4 0SL. Drs Rachel & David Pryke, rachelgpryke@btinternet.com. *½m from J6 of M5, Worcester North. St Nicholas Lane is off Hastings Drive.* **Sun 23 May, Sun 13 June (12.30-5.30). Adm £5, chd free. Home-made teas in St Nicholas Church barn, next door. Visits also by arrangement May to Sept for groups of 10 to 30. Diane Cooksey garden/church visits 01905 611268 diane_cooksey@hotmail.com.**
Warndon Court is a 2 acre family garden surrounding a Grade 2 listed farmhouse (not open) featuring a circular route taking in formal rose gardens and terraces, two ponds, pergolas, topiary (incl a dragon dressed as The Gruffalo!) a potager and woodland walk along the dry moat and through the secret garden. It has bee friendly wildlife areas and is home to great-crested newts and slow-worms. Grade 1 listed St Nicholas Church will also be open to visitors. There will be an exhibition of original oil paintings and display of vintage cars. The gardens around the house can be accessed across the lawn. The potager is accessible but the woodland walk is bumpy with slopes at each end.

& 🐐 ✿ ☕ 🛋

46 **WESTACRES**
Wolverhampton Road, Prestwood, Stourbridge, DY7 5AN. Mrs Joyce Williams, 01384 877496, Koijoy62@yahoo.co.uk. *3m W of Stourbridge. A449 in between Wall Heath (2m) & Kidderminster (6m). Ample parking Prestwood Nurseries (next door).* **Sat 10, Sun 11 July (11-4). Adm £5, chd free. Light refreshments. Visits also by arrangement June to Sept for groups of 5+.**
¾ acre plant collector's garden with unusual plants and many different varieties of acers, hostas, shrubs. Woodland walk, large fish pool. Covered tea area with home-made cakes. Come and see for yourselves, you won't be disappointed. Described by a visitor in the visitors book as 'A garden which we all wished we could have, at least once in our lifetime'. Garden is flat. Disabled parking.

& 🐐 ✿ ☕ ☕

47 WHARF HOUSE
Newnham Bridge, Tenbury
Wells, WR15 8NY. Gareth
Compton & Matthew Bartlett,
01584 781966, gco@no5.com,
www.wharfhousegardener.blog.
*Off A456 in hamlet of Broombank,
between Mamble & Newnham
Bridge. Follow signs. Do not rely on
SatNav.* **Sat 3, Sun 4 July (10-5).
Adm £5, chd free. Home-made
teas. Visits also by arrangement
for groups of 10+.**
2 acre country garden, set around
an C18 house and outbuildings (not
open). Mixed herbaceous borders
with colour theming: White Garden,
Bright Garden, Spring Garden,
Canal Garden, long double borders,
several courtyards, scented border,
stream with little bridge to an island,
vegetable garden. The garden is on
several levels, with limited wheelchair
access and some uneven paths.

48 WHITCOMBE HOUSE
Overbury, Tewkesbury,
GL20 7NZ. Faith & Anthony
Hallett, 01386 725206,
faith.hallett1@gmail.com. *9m S of
Evesham, 5m NE Tewkesbury. Leave
A46 at Beckford to Overbury (2m).
Or B4080 from Tewkesbury through
Bredon/Kemerton (5m). Or small lane
signed Overbury at r'about junction
A46, A435 & A4077. Approx 5m
from J9 M5.* **Visits by arrangement
Apr to Aug for groups of 5 to 30.
Adm £4, chd free.**
Stone walled, English cottage garden,
(1 acre) set in a Cotswold village.
Long borders of herbaceous, climbers
and shrubs, gravel stream flanked by
primula, astilbe, hydrangea, lavender
and rose. Pastel colours merge
with cool blue and late hot planting
framed under canopy of acer, beech
and catalpa. Nooks and crannies, a
bridge and vegetable parterre adjoin
surrounding yew and box.

49 WHITE COTTAGE & NURSERY
Earls Common Road, Stock
Green, Inkberrow, B96 6SZ.
Mr & Mrs S M Bates, 01386 792414,
smandjbates@aol.com.
whitecottage.garden. *2m W
of Inkberrow, 2m E of Upton
Snodsbury. A422 Worcester to
Alcester, turn at sign for Stock
Green by Red Hart PH, 1½m to
T- junction, turn L 500 yds on the
L.* **Sun 11 Apr, Sun 2, Mon 3, Sun
16, Sun 30, Mon 31 May, Sun 13,**

**Sun 27 June, Sun 18 July, Sun
26 Sept (11-4.30). Adm £4.50,
chd free. Refreshments are on a
commercial basis.**
2 acre garden with large herbaceous
and shrub borders, island beds,
stream and bog area. Spring meadow
with 1000s of snakes head fritillaries.
Formal area with lily pond and circular
rose garden. Alpine rockery and
new fern area. Large collection of
interesting trees incl Nyssa Sylvatica,
Parrotia persica, and Acer 'October
Glory' for magnificent Autumn colour
and many others. Nursery and Garden
open most Thursdays (10.30-5),
please check before visiting. For
further information please phone, email
or visit garden website and Facebook.

50 THE WHITE HOUSE
Seedgreen Lane, Astley Burf,
DY13 0SA. John & Joanna Daniels,
www.talesfromacountrygarden.
co.uk. *3m SW of Stourport on
Severn. Off B4196 Holt Heath/
Stourport Rd. From Stourport L
towards Larford Fishing Lakes,
garden 1.8m. From Holt Heath R
onto Crundles Lane towards Astley
Burf, R in village then 1st L into
Seedgreen Lane.* **Sun 16 May, Sun
4 July (2-5). Adm £3.50, chd free.
Home-made teas.**
Informal 1 acre constantly evolving
country garden and orchard in
peaceful rural surroundings. Large
mixed borders with shrubs, roses
and cottage garden plants. Orchard
has ornamental and productive trees
and veg garden. Large Victorian style
greenhouse with scented leaved
pelargoniums. New features in the
orchard are a wildlife pond and an
'autumn bed' planted with perennial
helianthus, rudbeckia and echinops.
A small collection of vintage garden
tools is exhibited in an outbuilding.
Gardened with wildlife in mind leaving
areas of long grass in the orchard with
paths mown through. Car parking
is in adjacent field, disabled drop off
point at garden gate.

51 ◆ WHITLENGE GARDENS
Whitlenge Lane, Hartlebury,
DY10 4HD. Mr & Mrs K J
Southall, 01299 250720, keith.
southall@creativelandscapes.
co.uk, www.whitlenge.co.uk. *5m
S of Kidderminster, on A442. A449
Kidderminster to Worcester L at
T-lights, A442 signed Droitwich, over
island, ¼m, 1st R into Whitlenge Ln.*

Follow brown signs. **For NGS: Sat
3, Sun 4 Apr, Sat 26, Sun 27 June,
Sat 4, Sun 5 Sept (10-5). Adm
£4.50, chd £2. Light refreshments
in the adjacent tea rooms.
For other opening times and
information, please phone, email or
visit garden website.**
3 acre show garden of professional
garden designer incl large variety
of trees, shrubs etc. Features incl a
twisted brick pillar pergola, 2½ metre
high solid oak moon gate set into
reclaimed brickwork, 4 cascading
waterfalls, ponds and streams. Mystic
features of the Green Man, 'Sword
in the Stone' and cave fernery, deck
walk through giant gunnera leaves,
herb garden. Walk the 400 sq metre
turf labyrinth and the new standing
stone circle plus children's play/
pet corner. Extensive plant nursery.
Wheelchair access but mix of hard
paths, gravel paths and lawn.

GROUP OPENING

52 NEW WICK VILLAGE
School Lane, Wick, Pershore,
WR10 3PD. Co-ordinator Mark
Heath. *1m E of Pershore on B4084.
Sign to Wick is almost opp Pershore
Horticultural College. Tickets for
gardens at car park next to St
Marys.* **Sun 13 June (1.30-6.30).
Combined adm £6, chd free. Ice
cream/soft drinks only.**

LAMBOURNE HOUSE
Mr & Mrs G Power.

NEW MERRYWAY
Mrs Clair Meikle-Taylor.

THE OLD FORGE
Sean & Elaine Young.

**NEW THE OLD STABLES,
YOCK LANE**
Mr & Mrs Tony & Amalia Knight.

NEW 13 TIMBERDOWN
Jo Sainsbury & Denis Rock.

NEW 13 WICK HOUSE CLOSE
Mr & Mrs Nigel & Gill Barker.

Wick Village gardens offers the visitor
a range of extraordinary garden
visiting experiences. We can show you
gardens that are low maintenance,
small and large gardens planted with
an array of plants which are trendy
modern day, in vogue ornamentals, as
well as the more unusual enthusiast
and specialist plants. As well as
ornamental gardens we can show
you the keen vegetable growers

gardens containing vegetables both everyday and the more unusual. We last opened in 2015 and since then gardens in the village have developed and new gardens are still emerging. Wick is an historic village on the edge of Pershore. Ice creams and cold drinks will be available. Teas may be available please check online. Many of the gardens are suitable for wheelchair access.

53 NEW WILLOW POND

Pass Street, Eckington, Pershore, WR10 3AX. James Field & Mike Washbourne, 07970 962842, james.field@washbournefield. co.uk, www.willow-pond.co.uk. *Black and white cottage about half way down Pass St. 4m S of Pershore; 7m N of Tewkesbury. In Eckington, 2nd R off New Rd.* **Visits by arrangement Apr to Sept for groups of 5 to 20. Adm £5. Home-made teas.**
Colourful and informal village garden of approx 1 acre with a listed black and white cottage (not open). Gardened for wildlife, there are areas of lawn and long grass with well-stocked, herbaceous borders and a vegetable plot. Several water features incl a pond. Rose covered pergola, terrace garden and cottage garden along with large planted pots and choice plants.

54 68 WINDSOR AVENUE

St Johns, Worcester, WR2 5NB. Roger & Barbara Parker. *W area of Worcester, W side of R Severn. Off A44 to Bromyard. Into Comer Rd, 3rd L into Laugherne Rd, 3rd L into Windsor Ave, at bottom in cul-de-sac. Limited parking, please park courteously on road sides, car share if possible.* **Sun 30, Mon 31 May (1-5). Adm £5, chd free. Light refreshments.**
Our 10th year of opening. Almost one acre garden divided into three areas, situated behind a 1930s semi detached house in a cul-de-sac. Visitors are amazed and comment on size of garden and the tranquility. The garden includes bog gardens, flowerbeds, 'oriental' area, vegetable patch, five greenhouses and a Koi pond plus two other ponds each in very different styles. We also have ornamental pheasants and other birds. Gravel paths.

27 Sheldon Park Road

YORKSHIRE

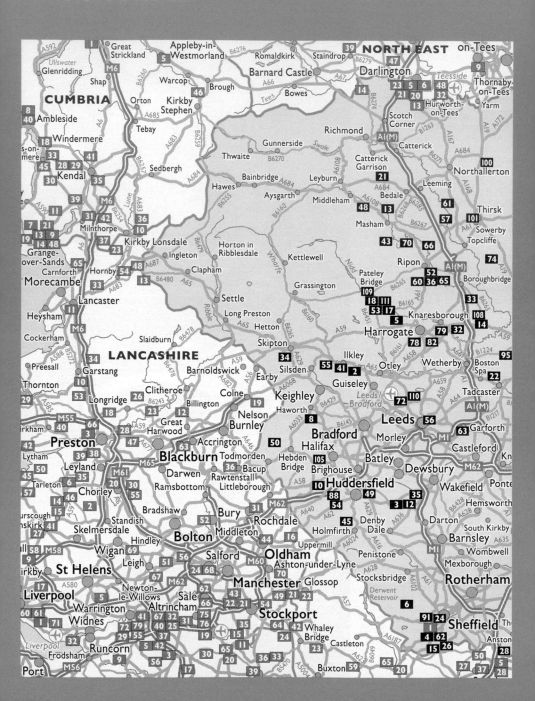

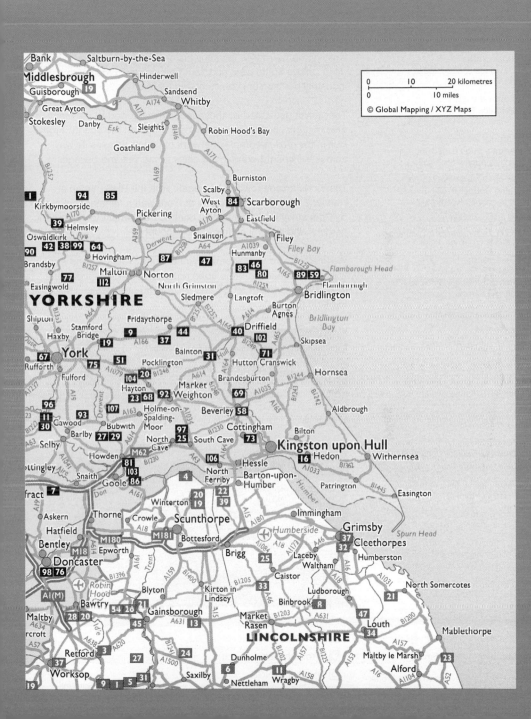

Volunteers

County Organisers

East Yorks
Helen Marsden
01430 860222
helen.marsden@ngs.org.uk

North Yorks
David Lis 01439 788846
david.lis@ngs.org.uk

South & West Yorks
Veronica Brook 01423 340875
veronica.brook@ngs.org.uk

County Treasurer
Angela Pugh 01423 330456
angela.pugh@ngs.org.uk

Publicity & Social Media
Jane Cooper 01484 604232
jane.cooper@ngs.org.uk

Booklet Advertising
Sally Roberts 01423 871419
sally.roberts@ngs.org.uk

By Arrangement Visits, Clubs & Societies
Penny Phillips 01937 834970
penny.phillips@ngs.org.uk

Assistant County Organisers

East Yorks
Ian & Linda McGowan
01482 896492
ianandlinda.mcgowan@ngs.org.uk

Hazel Rowe 01430 861439
hazel.rowe@ngs.org.uk

Natalie Verow 01759 368444
natalieverow@aol.com

North Yorks
David Morgan 07827 958103
david.j.morgan@ngs.org.uk

John Plant 01347 888125
rewelacottage@gmail.com

Judi Smith 01845 567518
judi.smith@ngs.org.uk

South & West Yorks
Felicity Bowring 01729 823551
felicity.bowring@ngs.org.uk

Jane Hudson 01484 866697
jane.hudson@ngs.org.uk

Chris & Fiona Royffe
01937 530306
plantsbydesign@btinternet.com

Elizabeth & David Smith
01484 644320
elizabethanddavid.smith@ngs.org.uk

Yorkshire, England's largest county, stretches from the Pennines in the west to the rugged coast and sandy beaches of the east: a rural landscape of moors, dales, vales and rolling wolds.

Nestling on riverbanks lie many historic market towns, and in the deep valleys of the west and south others retain their 19th century industrial heritage of coal, steel and textiles.

The wealth generated by these industries supported the many great estates, houses and gardens throughout the county. From Hull in the east, a complex network of canals weaves its way across the county, connecting cities to the sea and beyond.

The Victorian spa town of Harrogate with the RHS garden at Harlow Carr, or the historic city of York with a minster encircled by Roman walls, are both ideal centres from which to explore the gardens and cultural heritage of the county.

We look forward to welcoming you to our private gardens – you will find that many of them open not only on a specific day, but also by arrangement for groups and individuals - we can help you to get in touch.

f @YorkshireNGS
@YorkshireNGS
@YorkshireNGS

Above: The Terrace Gardens

OPENING DATES

All entries subject to change. For latest information check **www.ngs.org.uk**

Extended openings are shown at the beginning of the month.

Map locator numbers are shown to the right of each garden name.

February

Snowdrop Festival

Sunday 21st

Devonshire Mill 20

March

Sunday 21st

Fawley House 25

Sunday 28th

Goldsborough Hall 32

April

Sunday 11th

Clifton Castle 13
Ellerker House 23

Saturday 17th

249 Barnsley Road 3

Sunday 18th

249 Barnsley Road 3
Fawley House 25

May

Saturday 1st

Linden Lodge 51

Sunday 2nd

Highfield Cottage 40
Linden Lodge 51
Scape Lodge 88

Monday 3rd

Linden Lodge 51
Whixley Gardens 108

Saturday 8th

NEW The Cottage 16

Sunday 9th

NEW The Cottage 16
Low Hall 53
Millrace Garden 63
Woodlands Cottage 109

Tuesday 11th

Havoc Hall 38
◆ Himalayan Garden & Sculpture Park 43

Saturday 15th

NEW Ferry House 27

Sunday 16th

◆ RHS Garden Harlow Carr 78
◆ Stillingfleet Lodge 96
Warley House Garden 105

Tuesday 18th

Warley House Garden 105

Wednesday 19th

Land Farm 50
Primrose Bank Garden and Nursery 75

Friday 21st

NEW Thimbleby Hall 100

Saturday 22nd

The Red House 76
Tamarind 98

Sunday 23rd

Bramblewood Cottage 6
◆ Jackson's Wold 47
The Orchard 72
The Red House 76
The Ridings 80
NEW Rudding Park 82
Tamarind 98

Wednesday 26th

5 Hill Top 41

Saturday 29th

NEW Lumb Beck Farmhouse 55

Sunday 30th

5 Hill Top 41
Rewela Cottage 77
1 School Lane 89
The Villa 103

Monday 31st

Bridge Farm House 7
Holmfield 44

June

Wednesday 2nd

NEW Lumb Beck Farmhouse 55

Thursday 3rd

NEW South Park 95

Friday 4th

◆ Shandy Hall Gardens 90

Saturday 5th

Old Sleningford Hall 70
Shiptonthorpe Gardens 92

Sunday 6th

◆ Norton Conyers 66
Old Sleningford Hall 70
Shiptonthorpe Gardens 92
◆ The Yorkshire Arboretum 112

Wednesday 9th

Brookfield 8
Glencoe House 31

Thursday 10th

Skipwith Hall 93

Saturday 12th

NEW Frog Hall Barn 29
NEW Maggie's at St James' Hospital 56
NEW Thirsk Hall 101

Sunday 13th

Brookfield 8
Clifton Castle 13
NEW Frog Hall Barn 29
The Old Vicarage 71
Tythe Farm House 102

Wednesday 16th

Holmfield 44
Primrose Bank Garden and Nursery 75

Thursday 17th

NEW Jervaulx Hall 48

Friday 18th

NEW Arden Hall 1

Sunday 20th

Birstwith Hall 5
3 Embankment Road 24
◆ Jackson's Wold 47
Millrace Garden 63
NEW The Old Priory 68
White Wynn 107
Yorke House & White Rose Cottage 111

Wednesday 23rd

NEW The Cottage 15
Hillside 42
Land Farm 50
Ness Hall 64

Friday 25th

◆ Shandy Hall Gardens 90

Sunday 27th

Fernleigh 26
Havoc Hall 38
115 Millhouses Lane 62
Ridgefield Cottage & Nursery 79

Wednesday 30th

The Grange 34

July

Sunday 4th

Bugthorpe Gardens 9
Cobble Cottage 14
NEW Grafton Gardens 33
Highfield Cottage 40
NEW Maunby Hall 61
The Ridings 80
Sleightholmedale Lodge 94

Tuesday 6th

◆ Scampston Walled Garden 87

Wednesday 7th

NEW 3 The Barns 2

Saturday 10th

Cawood Gardens 11
NEW Galehouse Barn 30
NEW St Catherine's Hospice 84

Sunday 11th

Bramblewood Cottage 6
Cawood Gardens 11
Dacre Banks & Summerbridge Gardens 18
NEW Galehouse Barn 30
Honey Head 45
NEW 19 Manor Road 58
23 The Paddock 73

Tuesday 13th

Havoc Hall 38

Wednesday 14th

NEW The Cottage 15

Saturday 17th

The Manor House 57
Stonefield Cottage 97

Sunday 18th

Brookfield 8
Daneswell House 19
East Wing, Newton Kyme Hall 22
Goldsborough Hall 32
The Nursery 67
Stonefield Cottage 97
NEW The Vines 104

Rosemary Cottage

THE GARDENS

SPECIAL EVENT

1 [NEW] **ARDEN HALL**
Hawnby, York, YO62 5LS. Victoria
Savile. *Turn left at the shop in
Hawnby and follow the road up
past a church and over a bridge.*
Fri 18 June (10.30-1). Adm £25.
Pre-booking essential, please
visit www.ngs.org.uk/special-
events for information & booking.
Light refreshments.
Historic gardens featuring a two
acre walled garden with formal stone
terraces providing exceptional views
over Hawnby Hill and the National
Park. The planting is an informal mix
of shrubs and perennials. A spring-fed
stream flows through the garden in a
series of rills and formal ponds. Other
striking features include magnificent
yew topiary, clipped box, a laburnum
walk and a sunken greenhouse.

2 [NEW] **3 THE BARNS**
Ben Rhydding Drive, Ilkley,
LS29 8BG. Miss Linda Jones & Mr
Allan Shaw. *2m from centre of Ilkley
in Ben Rhydding on edge of Ilkley
moor. From A65 turn at T-lights on
to Wheatley Lane to Ben Rhydding.
After going under railway bridge as
road bears R on the corner turn L &
then immed L through 4 stone pillars
on to The Drive.* Wed 7 July (11-
4.30). Adm £3, chd free.
Charming small garden on the edge
of Ilkley Moor with mixed herbaceous
borders, pond, oak summerhouse,
greenhouse and raised bed
area. Roses, hostas, geraniums,
alstroemerias, perennials, trees and
shrubs. Lovely view of Wharfe Valley
can be seen on approach to the
garden. Limited parking nearby. Coffee,
lunch and afternoon tea available in the
restaurant or on the terrace at Raven
Audley. 10% discount for NGS garden
visitors with their ticket.

3 **249 BARNSLEY ROAD**
Flockton, Wakefield,
WF4 4AL. Nigel & Anne
Marie Booth, 01924 848967,
nigel.booth1@btopenworld.com.
*On A637 Barnsley Road. M1 J38 or
39 follow the signs for Huddersfield.
Parking on Manor House Rd (WF4*
4AL for Sat Nav) & Hardcastle Lane.
Sat 17, Sun 18 Apr, Sat 31 July,
Sun 1 Aug (1-5). Adm £4, chd
free. Light refreshments. Visits
also by arrangement Apr to Sept.
An elevated garden with fantastic
panoramic views. ⅓ acre south
facing garden packed with an
abundance of spring colour, created
from 1000's of bulbs, perennials,
shrubs and trees. Make a return
visit in the summer to view the
transformation, displaying up to 60
hanging baskets and over 150 pots,
creating the 'wow factor' garden.
Limited wheelchair access. Ukulele
group playing live in the garden at
the summer opening. Massive plant
sale at both the spring and summer
openings. Disabled drop off point at
the bottom of the drive.

4 **90 BENTS ROAD**
Bents Green, Sheffield, S11 9HL.
Mrs Hilary Hutson, 0114 225 8570,
h.hutson@paradiseregained.net.
*3m SW of Sheffield. From ring road
in Sheffield Centre (nr Waitrose)
follow A625. After approx 3m
turn R on to Bents Rd.* Visits by
arrangement July & Aug for
groups of up to 30. Adm £4, chd
free. Light refreshments.
Plantswoman's NE facing garden with
many unusual and borderline-hardy
species. Patio with alpine troughs
for year round interest and pots of
colourful tropical plants in summer.
Mixed borders surround a lawn which
leads to mature trees underplanted
with shade-loving plants at end of
garden. Front garden peaks in summer
with hot-coloured blooms. Front
garden and patio flat and accessible.
Remainder of back garden accessed
via 6 steps with handrail, so unsuitable
for wheelchairs.

5 **BIRSTWITH HALL**
High Birstwith, Harrogate,
HG3 2JW. Sir James & Lady
Aykroyd, 01423 770250,
ladya@birstwithhall.co.uk.
*5m NW of Harrogate. Between
Hampsthwaite & Birstwith villages,
close to A59 Harrogate/Skipton
road.* Sun 20 June (2-5). Adm £5,
chd free. Home-made teas. Visits
also by arrangement Apr to Aug
for groups of 10 to 30.
Charming and varied 4 acre garden
nestling in secluded Yorkshire dale.
Formal garden and ornamental orchard,
extensive lawns leading to picturesque
stream and large pond. Walled garden
and Victorian greenhouse.

SPECIAL EVENT

6 **BRAMBLEWOOD COTTAGE**
Old Coach Road, Bradfield,
Sheffield, S6 6HX. Nigel Dunnett.
*6m from Sheffield City centre. From
Sheffield follow Loxley Road (B6077)
turn L onto New Road just before
Damflask reservoir. Follow alongside
reservoir until Old Coach Road.*
Sun 23 May, Sun 11 July, Sun 15
Aug (11-4). Adm £15. Pre-booking
essential, please visit www.
ngs.org.uk/special-events for
information & booking. Home-
made teas.
Guided tour and lecture at the
experimental, hillside home garden
of Nigel Dunnett, which is based on
strong ecological and sustainable
principles, with extensive areas of
naturalistic perennial planting. The
front garden is designed around a
pool, rain gardens and bioswales
that collect rainwater runoff. To the
rear, a range of different designed
annual and perennial meadows,
large-scale log-pile and habitat
sculptures, bluebell woodland and
lockdown project vegetable garden.
No wheelchair access possible.

7 **BRIDGE FARM HOUSE**
Long Lane, Great Heck, Selby,
DN14 0BE. Barbara & Richard
Ferrari. *6m S of Selby, 3m E M62
J34. At M62 J34 take A19 to Selby,
at r'about turn E towards Snaith on
A645. After level crossing turn R at
T-lights, L at T-junction onto Main St,
past Church, to T-junction, cross to
car park.* Mon 31 May (12-4). Adm
£4, chd free. Home-made teas in
church (opp).
2 acre garden, developed from green
field over 18 years by a plantaholic
and a conservationist in a spirit of
compromise (usually). Divided by
hedges into separate areas planted
with unusual and interesting plants for
all year interest. Long double mixed
borders, bog, gravel and woodland
gardens; interesting trees, hens,
compost heaps and wildlife areas.
Wheelchair access easiest by front
gate, please ask.

8 BROOKFIELD

Jew Lane, Oxenhope, Keighley, BD22 9HS. Mrs R L Belsey, 01535 643070. *5m SW of Keighley. From Keighley take A629 (Halifax) Fork R A6033 towards Haworth & Oxenhope turn L at Xrds into village. Turn R (Jew Lane) at bottom of hill.* Wed 9, Sun 13 June, Sun 18 July (1.30-5). Adm £4, chd free. Home-made teas. **Visits also by arrangement June to Aug for groups of up to 20.**

An intimate 1 acre sloping garden with steps and paths leading down to a large pond with an island, mallards, and wild geese. Many varieties of primula and dactylorrhiza which have seeded into the lower lawn around a small pond with stream; azaleas and rhododendrons. Unusual trees and shrubs in a series of island beds, screes, greenhouses and conservatory. 'Round and round the garden' children's quiz. Steep narrow paths, not suitable for wheelchairs.

GROUP OPENING

9 BUGTHORPE GARDENS

Bugthorpe, York, YO41 1QG. *4m E of Stamford Bridge, A166, village of Bugthorpe.* Sun 4 July (10-4). Combined adm £6, chd free. Light refreshments. cakes/scones/tea/coffee/wine available at 3 Church Walk.

3 CHURCH WALK
Barrie Creaser & David Fielding.

THE OLD RECTORY
Dr & Mrs P W Verow.

Two contrasting gardens situated in the small village of Bugthorpe. 3 Church Walk: a garden created in 2000, surprisingly mature, with mixed borders and trees, water feature and pond. Raised vegetable garden and greenhouse. The lawn leads onto a paddock with views of open countryside. The Old Rectory: ¾ acre country garden with views of the Yorkshire Wolds. Mixed borders, ponds, terrace, summerhouse, courtyard and many mature trees. Raised vegetable beds. Gravel drive with lawn access to the garden.

10 NEW CALF HEY FARM

Barkisland, Halifax, HX4 0ET. Anthea Thornber, 07771 357752, calfhey@icloud.com. *Between Barkisland, Krumlin & Stainland villages. Set Sat Nav to Barkisland Village FIRST, then use garden post code. Do not approach from Sonoco Paper Mill. Not suitable for buses.* Visits by arrangement 13-19 Sept, for groups of up to 20. Adm £5, chd free. Home-made teas.

A young, evolving garden around a West Yorkshire farmhouse with fine views. The owners are creating a relaxing space to enjoy an eclectic mix of colourful scented borders and the surrounding views of Krumlin. The garden features 320ft of stone walls, small trees, stone sculptures, raised beds, wildlife areas, lawns, native hedges and a young cut flower paddock.

GROUP OPENING

11 CAWOOD GARDENS

Cawood, nr Selby, YO8 3UG. 01757 268571, davidjones051946@gmail.com. *On B1223 5m N of Selby & 7m SE of Tadcaster. Between York & A1 on B1222. Village maps at all gardens.* Sat 10, Sun 11 July (12-5). Combined adm £5, chd free. Home-made teas. Also open Galehouse Barn.

9 ANSON GROVE
Brenda Finnigan, 01757 268888, beeart@ansongrove.co.uk. Visits also by arrangement June & July for groups of up to 20.

21 GREAT CLOSE
David & Judy Jones, 01757 268571, davidjones051946@gmail.com. Visits also by arrangement June & July for groups of up to 20.

Two contrasting gardens in an attractive historic village are linked by a pretty riverside walk to the C11 church and memorial garden and across the Castle Garth to the remains of Cawood Castle. 9 Anson Grove is a small garden with tranquil pools and secluded seating areas. Narrow winding paths lead to views of oriental-style pagoda, bridge and Zen garden. 21 Great Close is a flower arranger's garden, designed and built by the owners. Interesting trees and shrubs combine with herbaceous

borders incl many grasses. Two ponds are joined by a stream, winding paths lead to the vegetable garden and summerhouse, then back to the colourful terrace for views across the garden and countryside beyond. Arts and crafts on sale. Partial wheelchair access.

12 THE CIRCLES GARDEN

8 Stocksmoor Road, Midgley, nr Wakefield, WF4 4JQ. Joan Gaunt, penny.phillips@ngs.org.uk. *Equidistant from Huddersfield, Wakefield & Barnsley, W of M1. Turn off A637 in Midgley at the Black Bull Pub (sharp bend) onto B6117 (Stocksmoor Rd). Please park adjacent to houses.* Visits by arrangement Apr to Sept for groups of 5 to 30. Adm £4, chd free. Light refreshments.

An organic and self-sustaining plantswoman's ½ acre garden on gently sloping site overlooking fields, woods and nature reserve opposite. Designed and maintained by owner. Herbaceous, bulb and shrub plantings linked by grass and gravel paths, woodland area with mature trees, meadows, fernery, greenhouse, fruit trees, viewing terrace with pots. About 100 hellebores propagated from owners seed. South African plants, hollies, and small bulbs of particular interest.

13 CLIFTON CASTLE

Ripon, HG4 4AB. Lord & Lady Downshire. *2m N of Masham. On rd to Newton-le-Willows & Richmond. Gates on L next to red telephone box.* Sun 11 Apr, Sun 13 June (2-5). Adm £5, chd free. Home-made teas.

Breathtaking views over lower Wensleydale; the gardens include formal walks through the wooded 'pleasure grounds' with abundant wildflowers, affording views over the river from the top path and returning at river level. The walled kitchen garden is similar to how it was set out in the C19. New wildflower meadows have been laid out with modern sculptures. Gravel paths and steep slopes to river.

Jervaulx Hall

14 COBBLE COTTAGE

Rudgate, Whixley, YO26 8AL. John Hawkridge & Barry Atkinson, 01423 331419, cobblecottage@outlook.com. *8m W of York, 8m E of Harrogate, 6m N of Wetherby. From High St, L at Anchor Inn onto Station Rd, on L.* **Sun 4 July (11-5). Combined adm with Grafton Gardens £7, chd free. Opening with Whixley Gardens on Mon 3 May. Visits also by arrangement May to July for groups of 20+.** Imaginatively designed, constantly changing, small cottage garden full of decorative architectural plants and old family favourites. Interesting water garden, containers and use of natural materials. Black and white courtyard garden and Japanese-style garden with growing willow screen.

✿ 🚗

15 NEW THE COTTAGE

55 Bradway Road, Bradway, Sheffield, S17 4QR. Jane Holbrey. *6m S Sheffield City Centre. Follow Greenhill Parkway (B6054) towards Derbyshire, cottage set back just before Dore & Totley Golf Club.* **Wed 23 June, Wed 14 July (12-5). Adm £3.50, chd free. Light refreshments.** Contemporary, stylish cottage garden with herbaceous borders, vegetable plot, wild flower area, and auricula theatre. Modern shepherd's hut and pond. Specimen fruit trees including step-over apples. Many seating areas. No wheelchair access possible due to steps and uneven York stone paving.

✿ ☕

16 NEW THE COTTAGE

3 Fletcher Gate, Hedon, Hull, HU12 8ET. Mr & Mrs Ian & Yvonne Mcfarlane. *6m E of Hull. Take A1033 to Hedon, L at r'about (Hungry Horse) onto Hull Rd which joins Fletcher Gate. Garden is on R after 2nd zebra crossing. No parking at property. 2 free car parks 2 mins walk away.* **Sat 8, Sun 9 May (1-4). Adm £4, chd free. Light refreshments.** The Cottage is grade 2 listed. Mature 0.3 acre suburban secret garden with small pond, summerhouse, unusual planting in hidden areas accessed via steps, sloping and grassed paths. Lots of seating. Small stream originally constructed in early 1900s in a gravel garden. Sitting area in reclaimed palm house under construction. Many farm and garden implements. Some small areas may not be accessible to wheelchairs.

♿ 🐑 ✿ ☕ ☕

17 COW CLOSE COTTAGE

Stripe Lane, Hartwith, Harrogate, HG3 3EY. William Moore & John Wilson, 01423 779813, cowclose1@btinternet.com. *8m NW of Harrogate. From A61(Harrogate-Ripon) at Ripley take B6165 to Pateley Bridge. 2m beyond Burnt Yates turn R, signed Hartwith/ Brimham Rocks onto Stripe Lane. Parking available.* **Visits by arrangement July & Aug for groups of 10 to 30. Adm £5, chd free. Tea.** $2/3$ acre country garden on sloping site with stream and far reaching views. Large borders with drifts of interesting, well-chosen, later flowering summer perennials and some grasses contrasting with woodland shade and streamside plantings. Gravel path leading to vegetable area. Courtyard area, terrace and seating with views of the garden. Orchard and ha-ha with steps leading to wildflower meadow. The lower part of the garden can be accessed via the orchard.

♿ 🐑 ✿ 🚗 ☕

GROUP OPENING

18 DACRE BANKS & SUMMERBRIDGE GARDENS

Nidderdale, HG3 4EW. www.yorkehouse.co.uk. *4m SE Pateley Bridge, 10m NW Harrogate,10m N Otley, on B6451 & B6165. Parking at each garden. Maps available for garden locations.* **Sun 11 July (12-5). Combined adm £10, chd free. Home-made teas at Yorke House, Low Hall and Woodlands Cottage.**

LOW HALL
Mrs P A Holliday.
(See separate entry)

RIVERSIDE HOUSE
Joy Stanton.

WOODLANDS COTTAGE
Mr & Mrs Stark.
(See separate entry)

YORKE HOUSE & WHITE ROSE COTTAGE
Tony & Pat Hutchinson & Mark & Amy Hutchinson.
(See separate entry)

Dacre Banks and Summerbridge Gardens are situated in the beautiful countryside of Nidderdale and designed to take advantage of the scenic Dales landscape. The gardens are linked by attractive walks along the valley, but each may be accessed individually by car. Low Hall has a romantic walled garden set on different levels around the historic C17 family home (not open) with herbaceous borders, shrubs, climbing roses and tranquil water garden. Riverside House is an atmospheric waterside garden on many levels, supporting shade-loving plants and features a Victorian folly, fernery, courtyard and naturalistic riverside plantings. Woodlands Cottage is a garden of many rooms, with exquisite formal and informal plantings, and an attractive wildflower meadow which harmonises with mature woodland. Yorke House has extensive colour-themed borders and water features with beautiful waterside plantings. The newly developed garden at White Rose Cottage is specifically designed for wheelchair users. Visitors welcome to use orchard picnic area at Yorke House. Partial wheelchair access at some gardens. Full wheelchair access at White Rose Cottage.

🐑 ✿ 🚗 ☕ 🏕

19 DANESWELL HOUSE

35 Main Street, Stamford Bridge, YO41 1AD. Brian & Pauline Clayton, 01759 371446, pauline-clayton44@outlook.com. *7m E of York. On A166 at the E end of the village. Garden on your L as you leave the village towards Bridlington.* **Sun 18 July (10-4). Adm £4, chd free. Cream teas. Visits also by arrangement July & Aug for groups of 5+.** A $3/4$ acre secluded garden that sweeps down to the R Derwent. It has tiered terraces, ponds, including a new pond with bridge, water feature, shrubs, borders. The garden attracts abundant wildlife. Teas are served at the top of the garden. This is a sloping garden down to river, however wheelchair access is possible with care.

♿ 🚗 ☕

20 DEVONSHIRE MILL

Canal Lane, Pocklington, York, YO42 1NN. Sue & Chris Bond, 01759 302147, chris.bond.dm@btinternet.com, www.devonshiremill.co.uk. *1m S of Pocklington. Canal Lane, off A1079 on the opposite side of the road from the canal towards Pocklington.* **Sun 21 Feb (11-4.30). Adm £4, chd free. 2022: Sun 13 Feb. Visits also by arrangement.**

Drifts of double snowdrops, hellebores and ferns surround the historic grade II listed water mill. Explore the two acre garden with mill stream, orchards, woodland, herbaceous borders, hen run and greenhouses. The old mill pond is now a vegetable garden with raised beds and polytunnel. Over the past twenty years the owners have developed the garden on organic principles to encourage wildlife.
🐑 🚘

21 NEW **DIAMOND HILL FARM**
Bedale, DL8 1LS. David & Joan Ford, 01677 450885, joan@diamondhillfarm.co.uk, www.diamondhillfarm.co.uk. *Leave A1 at J51, Leeming Bar. Take A684 towards Leyburn. After 6m (through Crakehall & Patrick Brompton), turn R towards Catterick Garrison. Diamond Hill Farm is 1m on L.* Sun 12 Sept (1-5). Adm £5, chd free. Cream teas.
Developed since 2008. Half acre, south facing open site. Multiple garden areas including many colourful and interesting herbaceous beds, shrubs, ornamental trees and vegetable garden. Also wildlife pond, stream and woodland walk. Level hill top location with stunning views and secluded vistas. Seating. Partial wheelchair access.
🐑 🚘 ☕

22 **EAST WING, NEWTON KYME HALL**
Croft Lane, Newton Kyme, nr Boston Spa, LS24 9LR. Fiona & Chris Royffe, plantsbydesign@btinternet.com. www.plantsbydesign.info. *2m from Tadcaster or Boston Spa. Follow directions for Newton Kyme Village from A659.* Sun 18 July (11-5). Adm £4, chd free. Home-made teas.
Contemporary designed garden, dramatic setting. Views of Kyme Castle, St Andrews Church, C18 Newton Kyme Hall (not open). Sculptural planting, herb & cutting garden, small meadow. As well as emphasising the broad scene and setting for the East Wing, planted areas provide for the rich variety of visual interest through the seasons as well as featured to attract wildlife. Garden design exhibition.
✹ ☕

23 **ELLERKER HOUSE**
Everingham, York, YO42 4JA. Mr & Mrs M Wright, www.ellerkerhouse.weebly.com. *15m SE of York. 5½m from Pocklington. Just out of the village towards Harswell on R.* Sun 11 Apr (1-5). Adm £6, chd free. Home-made teas served all day. Savouries served at lunch time.
5 acre garden with many fine old trees, lawns surrounded by colour themed herbaceous borders planted with old roses, unusual shrubs and herbaceous plants for all year colour. Daffodils, spring bulbs and alpines planted around the lake in a stumpery. 11 acres of woodland, thatched oak hut and several sitting areas. Rare Plant Fair with many different plant stalls. See website for details. Most of the garden is accessible by wheelchair.
♿ ✹ 🚘 ☕

24 **3 EMBANKMENT ROAD**
Broomhill, Sheffield, S10 1EZ. Charlotte Cummins. *1½m W of city centre, 7½m from J33 M1. A61 ring rd follow A57 to Manchester. In Broomhill turn R onto Crookes Rd, 1st R on to Crookesmoor Rd. Embankment Rd is 2nd on L.* Sun 20 June (10-4.30). Adm £4, chd free. Light refreshments. Good selection of home-made cakes with tea and coffee available.
Compact, city garden with cottage style planting with well stocked borders of interesting, carefully selected herbaceous perennials. Front garden features clipped box, hostas and lavender bed with alliums. Rear garden with steep steps to elevated lawn, herbaceous garden, raised vegetable beds and many pots. Collection of astrantias. Over 100 different varieties of hostas. Home grown plants for sale.
✹ ☕

25 **FAWLEY HOUSE**
7 Nordham, North Cave, nr Brough, Hull, HU15 2LT. Mr & Mrs T Martin, 01430 422266, louisem200@hotmail.com, www.nordhamcottages.co.uk. *15m W of Hull. L at J38 on M62E. L at '30' & signs. Wellands & Pub. At L bend, R into Nordham. Fawley's on R. From Boverloy, B1230 to N Cave. R after church. Over bridge.* Sun 21 Mar, Sun 18 Apr (12-5). Adm £5, chd free. Please check website availability and venue for refreshments. Visits also by arrangement Feb to Oct for groups of 10+. Refreshments can be arranged.
Tiered, 2½ acre garden with lawns, mature trees, formal hedging and gravel pathways. Lavender beds, mixed shrub/herbaceous borders, and hot double herbaceous borders. Apple espaliers, pears, soft fruit, produce and herb gardens. Terrace with pergola and vines. Sunken garden with white border. Woodland with naturalistic planting and spring bulbs. Quaker well, stream and spring area with 3 bridges, ferns and hellebores near mill stream. Beautiful snowdrops and aconites early in year. Treasure hunt for children. Self catering accommodation at Nordham Cottages - see website. Refreshments - please check the website. Partial wheelchair access to top of garden and terrace on pea gravel, sloping paths thereafter.
♿ 🐑 ✹ 🚘 🚘 ☕

26 **FERNLEIGH**
9 Meadowhead Avenue, Meadowhead, Sheffield, S8 7RT. Mr & Mrs C Littlewood, 01142 747234, littlewoodchristine@gmail.com. *4m S of Sheffield city centre. From Sheffield city centre. A61, A6102, B6054 r'about, exit B6054. 1st R Greenhill Ave, 2nd R. From M1 J33, A630 to A6102, then as above.* Sun 27 June, Sun 25 July, Sun 29 Aug (11-5). Adm £3.50, chd free. Light refreshments. Visits also by arrangement Apr to Aug for groups of 10 to 30.
Plantswoman's ⅓ acre cottage style suburban garden. Large variety of unusual plants set in different areas provide all year interest. Several seats to view different aspects of garden. Auricula theatre, patio, gazebo and greenhouse. Miniature log cabin with living roof and cobbled area with unusual plants in pots. Sempervivum, alpine displays, collection of epimedium and wildlife 'hotel'. Wide selection of home grown plants for sale. Animal Search for children.
✹ ☕

Online booking is available for many of our gardens, at ngs.org.uk

27 NEW **FERRY HOUSE**
Breighton, Selby, YO8 6DH. Group
Captain Neil & Mrs Claire Bale.
*4m NW of Howden. From A63 take
rd from Loftsome Bridge towards
Wressle & Breighton. From Bubwith
take Church St then Breighton Rd to
Breighton.* **Sat 15 May (1-5). Adm
£4, chd free. Home-made teas in
a nearby garden.**
Newly landscaped country garden
with informal cottage style planting.
Stunning displays of tulips, daffodils
and spring flowers. Colourful
herbaceous borders with variety of
perennials. Large pond with Monet
style bridge, interesting paths and
seating from which to enjoy the
garden. Gate through to woodland
area with wildflower walk. Greenhouse
and vegetable beds. Most of garden
accessible for wheelchairs.

28 **FIRVALE ALLOTMENT
GARDEN**
Winney Hill, Harthill, nr
Worksop, S26 7YN. Don &
Dot Witton, 01909 771366,
donshardyeuphorbias@
btopenworld.com,
www.euphorbias.co.uk. *12m SE of
Sheffield, 6m W of Worksop. M1 J31
A57 to Worksop. Turn R to Harthill.
Allotments at S end of village, 26
Casson Drive at N end on Northlands
Estate.* **Visits by arrangement Apr
to June. Adm £3.50, chd free.
Home-made teas.**
Large allotment with 13 island
beds displaying 500+ herbaceous
perennials including the National
Collection of hardy euphorbias with
124 varieties flowering between
March and October. Organic
vegetable garden. Refreshments,
WC, plant sales at 26 Casson Drive,
a small garden with mixed borders,
shade and seaside garden.

29 NEW **FROG HALL BARN**
Breighton, Selby, YO8 6DH. Mrs
Sarah Clarke. *When coming from
Bubwith. Pass the sign for Breighton,
we are the 3rd house on the R.* **Sat
12, Sun 13 June (10.30-4). Adm
£4, chd free. Home-made teas.**
Bee-friendly flower and veg garden in
³/₄ acre. To the front, walled courtyard
garden of perennials and annuals. To
the rear, garden is anchored by horse
chestnut trees. Borders full of cottage
garden flowers selected to encourage
wildlife. A small meadow, fruit trees,
beehives and chickens. The kitchen
garden has a traditional greenhouse,
with berry fruits.

30 NEW **GALEHOUSE BARN**
Bishopdyke Road, Cawood,
Selby, YO8 3UB. Mr & Mrs
P Lloyd, 07768 405642,
junelloyd042@gmail.com.
*On B1222 1m out of Cawood
towards Sherburn in Elmet.* **Sat**

Paddock House, Grafton Gardens

10, Sun 11 July (12-5). Adm £3, chd free. Also open Cawood Gardens. Home-made teas at Cawood Gardens. **Visits also by arrangement Apr to Sept for groups of 5 to 10.**
The Barn: A plantaholic's informal cottage garden, created in 2015, to encourage birds and insects. Raised beds with tranquil seating area. The Farm: South facing partly shaded varied herbaceous border. North facing exposed shaded border redeveloped 2017 and ongoing for spring and autumn interest. Small new experimental 50 shades of white garden. Raised beds for vegetables. Partial wheelchair access.

31 GLENCOE HOUSE
Main Street, Bainton, Driffield, YO25 9NE. Liz Dewsbury, 01377 217592, efdewsbury@gmail.com. *6m SW of Driffield on A614. 10m N of Beverley on B1248 Malton Rd. The house is on the W side of A614 in centre of the village. Parking is around village or in lay-by 280m N of house towards the Bainton r'about.* **Wed 9 June (12-5). Adm £4, chd free. Home-made teas. Visits also by arrangement in June for groups of 5 to 30.**
A tranquil 3 acre garden developed from a small-holding over 40+ years and continually evolving. Trees, shrubs, roses, clematis and masses of herbaceous perennials grow in the cottage garden, which is next to an area of parkland planted with specimen trees and an orchard. Furthest from the house, there is a naturalistic established planting of native and unusual trees plus large wildlife pond. Wheelchair access to paved area in cottage garden but difficult elsewhere.

32 GOLDSBOROUGH HALL
Church Street, Goldsborough, HG5 8NR. Mr & Mrs M Oglesby, 01423 867321, info@goldsboroughhall.com, www.goldsboroughhall.com. *2m SE of Knaresborough. 3m W of A1M. Off A59 (York-Harrogate). Car parking E of village in field off Midgley Lane. Disabled parking only at the Hall.* **Sun 28 Mar (11-4); Sun 18 July (11-5). Adm £5, chd free. Light refreshments. Donation to St Mary's Church.**
Previously opened for NGS from 1928-30 and now beautifully restored by present owners and re-opened in 2010. 12 acre garden and formal landscaped grounds in parkland setting and Grade II*, C17 house, former residence of HRH Princess Mary, daughter of George V and Queen Mary. Gertrude Jekyll inspired 120ft double herbaceous borders and rose garden. Quarter-mile Lime Tree Walk planted by royalty in the 1920s, orchard, restored kitchen garden and new glasshouse, flower borders featuring 'Yorkshire Princess' rose, named after Princess Mary. Gravel paths and some steep slopes.

GROUP OPENING

33 NEW GRAFTON GARDENS
Marton Cum Grafton, York, YO51 9QJ. Mrs Glen Garnett. *2½m S of Boroughbridge. Turn off the A168 or B6265 to Marton or Grafton (South of Boroughbridge).* **Sun 4 July (11-5). Combined adm with Cobble Cottage £7, chd free. Tea at The Punch Bowl PH, Marton. Tel: 01423 322519. Booking preferred. Lunches and teas.**

NEW PADDOCK HOUSE
Tim & Jill Smith.

WELL HOUSE
Glen Garnett.

These two gardens in adjacent rural villages are also connected by a public footpath. Overlooking open fields, close to The Punch Bowl PH in Marton, Paddock House on an elevated site with extensive views down a large sloping lawn to a wildlife pond and the Hambleton Hills. A plant lover's garden where the house is encircled by a profusion of pots and extensive plant collections combining cottage gardening with the Mediterranean and Tropical. A curved terrace of York stone and steps using gravel and wood sleepers leads to many seating areas culminating in a cutting garden and small greenhouse. Well House in Grafton nestles under the hillside, also with long views to the White Horse. This 1½ acre garden was begun 39 years ago and is under constant change. A traditional English cottage garden with herbaceous borders, climbing roses and ornamental shrubs with a variety of interesting species. Paths meander through the borders to an orchard with geese and chickens.

34 THE GRANGE
Carla Beck Lane, Carleton in Craven, Skipton, BD23 3BU. Mr & Mrs R N Wooler, 07740 639135, margaret.wooler@hotmail.com. *1½m SW of Skipton. Turn off A56 (Skipton-Clitheroe) into Carleton. Keep L at Swan Pub, continue to end of village then turn R into Carla Beck Lane.* **Wed 30 June, Wed 28 July (12-4.30). Adm £5, chd free. Cream teas. Visits also by arrangement July & Aug for groups of 30+. (min charge). Donation to Sue Ryder Care Manorlands Hospice.**
Over 4 acres of garden set in the grounds of Victorian house (not open) with mature trees and panoramic views towards The Gateway to the Dales. The garden has been restored by the owners over the last 3 decades with many areas of interest being added to the original footprint. Bountiful herbaceous borders with many unusual species, rose walk, parterre, mini-meadows and water features. Large greenhouse and raised vegetable beds. Oak seating placed throughout the garden invites quiet contemplation - a place to 'lift the spirits'. Gravel paths and steps.

35 GREAT CLIFF EXOTIC GARDEN
Cliff Drive, Crigglestone, Wakefield, WF4 3EN. Kristofer Swaine, www.facebook.com/yorkshirekris. *1m from J39 M1. From M1 (J39) take A636 towards Denby Dale, past Cedar Court Hotel then L at British Oak pub onto Blacker Lane. Parking on Cliff Road motorway bridge.* **Every Sun 8 Aug to 22 Aug (1-5). Adm £4, chd free.**
An exotic garden on a long narrow plot. Possibly the largest collection of palm species planted out in Northern England including a large Chilean wine palm. Colourful and exciting borders with zinnias, cannas, ensete, bananas, tree ferns, agaves, aloes, colcasias and bamboos. Jungle hut, winding paths, pond that traverses the full width of the garden and children's area. Unsuitable for wheelchairs and those dependent on a walking frame due to narrow paths.

All entries are subject to change. For the latest information check ngs.org.uk

36 GREENCROFT

Pottery Lane, Littlethorpe, Ripon, HG4 3LS. David & Sally Walden, 01765 602487, s-walden@outlook.com. *1½m SE of Ripon town centre. Off A61 Ripon bypass follow signs to Littlethorpe, turn R at Church. From Bishop Monkton take Knaresborough Road towards Ripon then R to Littlethorpe.* Sun 1 Aug (12-4). Adm £5, chd free. Home-made teas. **Visits also by arrangement July & Aug.**
½ acre informal garden made by the owners with long herbaceous borders packed with colourful late summer perennials, annuals and exotics culminating in a circular garden with views through to large wildlife pond and surrounding countryside. Special ornamental features incl gazebo, temple pavilions, formal pool, stone wall with mullions and gate to pergola and cascade water feature.

37 GREENWICK FARM

Huggate, York, YO42 1YR. Fran & Owen Pearson, 01377 288122, greenwickfarm@hotmail.com. *2m W of Huggate. From York on A166, turn R Im after Garrowby Hill, at brown sign for picnic area & scenic route. White wind turbine on drive.* Sun 15 Aug (12-5). Adm £4.50, chd free. Home-made teas. Tea tables in conservatory & outside. **Visits also by arrangement July & Aug.**
1 acre woodland garden created in 2010 from disused area of the farm. Set in a large dell with mature trees. Paths up the hillside through borders lead to terrace and woodland planting. Many seating areas with spectacular views across wooded valley and the Wolds. Stumpery and hot border. New summerhouse, water feature and brushed steel sculpture. Described by guests as a graceful and tranquil garden. Access for wheelchairs difficult, but good view of garden from hard standing outside house/tea area.

SPECIAL EVENT

38 HAVOC HALL

York Rd, Oswaldkirk, York, YO62 5XY. David & Maggie Lis, 01439 788846, davidglis@me.com, www.havochall.co.uk. *21m N of York. On B1363, 1st house on R as you enter Oswaldkirk from S & last house on L as you leave village from N.* Sun 27 June

(1-5). Adm £6.50, chd free. Home-made teas. Special Event Tue 11 May & Tue 13 July (10.30-1). Adm £25. Pre-booking essential, please visit www.ngs.org.uk/special-events for information & booking. **Visits also by arrangement Apr to Sept for groups of 20+.**
12 areas incl knot, herbaceous, mixed shrub and flower gardens, courtyard, vegetable area and orchard, woodland walk and large lawned area with hornbeam trees and hedging. New in 2021: two prairie beds on the south lawn. To the S is a 2 acre wild flower meadow and small lake. Extensive collection of roses, herbaceous perennials and grasses. See website for more detail. Wheelchair access: some steps but these can be avoided.

SPECIAL EVENT

39 NEW ◆ HELMSLEY WALLED GARDEN

Cleveland Way, Helmsley, YO62 5AH. Helmsley Walled Garden Ltd, 01439 772314, info@helmsleywalledgarden.org.uk, www.helmsleywalledgarden.org.uk. *Helmsley is on A170, 25m from York. 14m from Thirsk & 16m from Malton. The Garden is a short walk from Cleveland Way car park in the centre of Helmsley.* For NGS: Thur 26 Aug (6-8.30). Adm £25. Pre-booking essential, please visit www.ngs.org.uk/special-events for information & booking. **For other opening times and information, please phone, email or visit garden website.**
A tranquil and historic five acre garden with over one hundred apple varieties, medicinal herb garden, white garden and kitchen garden. Special collections of clematis, iris, peony. Volunteers maintain the garden and help make this a special place of beauty for all. Double herbaceous borders planted with hot colours of summer. Special collections of Yorkshire apple varieties, peony and clematis, Victorian glass houses in use for growing salads, vines and tender perennials and annuals for display. Full access to garden, Plant Centre, Gift Shop and Vine House Café. Disabled toilet in the café.

40 HIGHFIELD COTTAGE

North Street, Driffield, YO25 6AS. Debbie Simpson, 01377 256562, debbie@simpsonhighfield.karoo.co.uk. *30m E of York, 29m E of M62. Exit at A614/A166 r'about onto York Rd into Driffield. Straight on until you reach the park with the Indian takeaway opp. Highfield Cottage is the white detached house opposite the park.* Sun 2 May, Sun 4 July (10.30-3.30). Adm £4, chd free. Cream teas are weather dependent. **Visits also by arrangement Mar to Sept for groups of 30+. Cream teas.**
A ¾ acre suburban garden bordered by mature trees and stream. Structure is provided by numerous yew and box topiary, a pergola and sculptures. Constantly evolving, the garden has something for everyone; lawns with island beds, mixed shrubs, fruit trees and herbaceous borders. The garden has been described as 'magical' and a 'hidden gem' by NGS visitors. Refreshments weather permitting, if raining there are alternative cafes in town centre about 600 metres from garden. No WC. The garden is not suited to wheelchairs and access is uneven.

41 5 HILL TOP

Westwood Drive, Ilkley, LS29 9RS. Lyn & Phil Short. *½m S of Ilkley town centre, steep uphill. Turn S at town centre T-lights up Brook St, cross The Grove taking Wells Rd up to the Moors & follow NGS signs.* Wed 26, Sun 30 May (11-4). Adm £4, chd free. Home-made teas.
Delightful ⅔ acre steep garden on edge of Ilkley Moor. Sheltered woodland underplanted with naturalistic, flowing tapestry of foliage, shade-loving flowers, shrubs and ferns amongst large moss covered boulders. Many Japanese maples. Natural stream, bridges, meandering gravel paths and steps lend magic to 'Dingley Dell'. Lawns, large rockery and summerhouse with stunning views. Some steep steps.

42 HILLSIDE

West End, Ampleforth, York, YO62 4DY. Sue Shepherd & Jon Borgia, 01439 788993, sue@sueandjon.net. *West End of Ampleforth village, 4m S of Helmsley. From A19, follow brown sign to Byland Abbey & continue to Ampleforth. From A170 take B1257 to Malton, after 1m turn R to*

The Old Priory

Ampleforth. Roadside parking only. Wed 23 June, Wed 21 July, Wed 25 Aug, Wed 22 Sept (1-5). Adm £4, chd free. Light refreshments. Visits also by arrangement June to Oct for groups of 10 to 20.
Half acre garden on a south facing slope, and half acre field. The design is evolving, based on informal planting and a wildlife friendly approach. Woodland and meadow areas. Ponds and bog garden. Lawn rising up to summerhouse and deck with fine views of the Coxwold - Gilling Gap. Fruit trees and kitchen garden. Wild garden in field. All year round interest with an emphasis on autumn colour.

🐕 ♿ ☕ 🪑

43 ◆ HIMALAYAN GARDEN & SCULPTURE PARK
The Hutts, Hutts Lane, Grewelthorpe, Ripon, HG4 3DA. Mr & Mrs Roberts, 01765 658009, info@himalayangarden.com, www.himalayangarden.com. 5m NW of Ripon. From N: A1(M) J51 A684 (Bedale) then B6268 (Masham). From S: A1M J50 (Ripon) then A6108 (Masham) after North Stainley turn L (Mickley & Grewelthorpe). Follow AA & garden signs. For NGS: Evening opening Tue 11 May (4-8.30). Adm £9, chd £4. Light refreshments. For other opening times and information, please phone, email or visit garden website.

Winner of Yorkshire in Bloom Tourist Attractions Award 2018 & 2019. 45 acres of garden inspired by the Himalayas. Widely considered to have the North's largest collection of rhododendrons, azaleas and magnolias, and a haven for woodland plants. There are almost 20,000 plants including some 1,400 rhododendron varieties, 250 azalea varieties and 150 different magnolias; and a new 20-acre arboretum. 90 Contemporary sculptures, Himalayan shelter, lakeside pagoda, thatched summerhouse, Buddha garden, Norse shelter, contemplation circle. Sturdy footwear required. The topography of the gardens means it is not suitable for mobility scooter and wheelchair users.

🐕 ♿ 🚐 NPC 🪑 ☕

44 HOLMFIELD
Fridaythorpe, YO25 9RZ. Susan & Robert Nichols, 01377 236627, susan@wiresculptures.net. 9m W of Driffield. From York A166 through Fridaythorpe. 1m turn R signed Holmfield. 1st house on lane. Mon 31 May, Wed 16 June (12-5). Adm £4.50, chd free. Visits also by arrangement May to July for groups of 10+.
Informal 2 acre country garden on gentle S-facing slope. Developed from a field in 1988. Large mixed borders, bespoke octagonal gazebo, family friendly garden with 'Hobbit House',

sunken trampoline, large lawn, tennis court, hidden paths for hide and seek. Productive fruit cage, vegetable and cut flower area. Collection of phlomis. Display of wire sculptures. Bee friendly planting. Some gravel areas, sloping lawns. Wheelchair access possible with help.

♿ ♿ 🚐 ☕

45 HONEY HEAD
33 Wood Nook Lane, Meltham, Holmfirth, HD9 4DU. Susan & Andrew Brass. 6m S of Huddersfield. Turn from A616 to Honley, through village then follow Meltham rd for 1m, turn L on Wood Nook Lane. Sun 11 July (10-4). Adm £4, chd free. Home-made teas.
Set high on a Pennine hillside with panoramic views, Honey Head aspires to provide year round interest whilst attempting to be self sufficient in fruit, vegetables, cut flowers and plants. Formal gardens with interconnecting ponds lead to extensive kitchen gardens with greenhouses complemented by areas planted to encourage wildlife. Weather permitting we will have "The Saxpots" playing an assortment of music, a local ensemble who are keen to support our event.

🐕 ♿ ☕ 🪑

46 HUNMANBY GRANGE
Wold Newton, Driffield,
YO25 3HS. Tom & Gill Mellor,
enquiries@woldtopbrewery.co.uk.
*01723 892222. 12½m SE of
Scarborough. Hunmanby Grange
home of Wold Top Brewery, between
Wold Newton & Hunmanby on rd
from Burton Fleming to Fordon.*
**Visits by arrangement June to
Aug for groups of 10+. Light
refreshments in Wold Top
Brewery bar area.**
Hunmanby Grange sits high on the
Yorkshire Wolds where the power
of the wind and the shallow chalk
wolds soils have dictated the garden
design. A series of gardens surround
the house giving scope for changes
in design and planting. In the Brewery
courtyard is the water feature from
The Welcome to Yorkshire Chelsea
Garden - The Brewers Yard. The
Wold Top Brewery will be open for
refreshments. Steps can be avoided
by using grass paths and lawns. Pond
garden not completely accessible to
wheelchairs but can be viewed from
gateway.

⚧ 🐄 🚗 ☕

47 ◆ JACKSON'S WOLD
Sherburn, Malton,
YO17 8QJ. Mr & Mrs Richard
Cundall, 07966 531995,
jacksonswoldgarden@gmail.com,
www.jacksonswoldgarden.com.
*11m E of Malton, 10m SW of
Scarborough. Signs only from A64.
A64 Eastbound to Scarborough.
R at T-lights in Sherburn, take the
Weaverthorpe rd, after 100 metres
R fork to Helperthorpe & Luttons.
1m to top of hill, turn L at garden
sign.* For NGS: Sun 23 May, Sun
20 June (1-5). Adm £5, chd free.
Home-made teas. **For other
opening times and information,
please phone, email or visit garden
website.**
2 acre garden with stunning views of
the Vale of Pickering. Walled garden
with mixed borders, numerous
old shrub roses underplanted with
unusual perennials. Woodland paths
lead to further shrub and perennial
borders. Lime avenue with wildflower
meadow. Traditional vegetable garden
with roses, flowers and framed by
a Victorian greenhouse. Adjoining
nursery. Tours by appointment.

⚧ ✿ 🚗 ☕

48 NEW JERVAULX HALL
Jervaulx, Ripon,
HG4 4PH. Mr & Mrs Phillip
Woodrow, 01677 460008,
phillip@nowtryus.net. *Parking
will be in the grounds of Jervaulx
Abbey and is accessed via The East
Lodge on Kilgram Lane, which is
just after Brymor Ice Cream (from
the South). Access from the A6108
is not permitted.* **Thur 17 June,
Wed 21 July, Wed 11 Aug (12-5).
Adm £7, chd free. Pre-booking
essential, please visit www.ngs.
org.uk for information & booking.
Home-made teas at Jervaulx
Abbey tearooms. Visits also by
arrangement June to Aug for
groups of 10+. Refreshments will
be available at Jervaulx Abbey
tearooms.**
Eight acre garden undergoing
renovation. Mixed borders and beds,
croquet lawn, parterre, glasshouse,
small vegetable garden and fernery.
Woodland garden areas with choice
trees and shrubs planted in last
six years, including magnolia, acer,
sorbus and betula. Magnificent older
trees. Contemporary sculpture.
Adjacent to ruins of Jervaulx Abbey.
Views of river Ure. Gardens include
Abbey Mill ruins.

🚗 ☕

49 NEW KIRKWOOD HOSPICE
21 Albany Road,
Dalton, Huddersfield,
HD5 9UY. Susan Wood,
www.kirkwoodhospice.co.uk. *2m
from Huddersfield town centre along
A629. Wakefield Rd. Follow Kirkwood
Hospice signs L onto Dalton Green
Rd then R onto Albany Rd.* **Sat 31
July (10-3). Adm £4, chd free. Pre-
booking essential, please visit
www.ngs.org.uk for information &
booking. Baked products on sale.**
Large, mature formal summer garden,
with seasonal interest. Level access
with composite pathways draws
the visitor around sweeping bed
with prairie planting. Lawns leading
to mixed beds, labyrinth and pond.
Mature trees and shrubs enclose the
garden boundaries. Extensive patio
and viewing area. Specimen plants
as well as favourites, new for 2021 a
small wild flower area and allotment.
Complete wheelchair accessibility.

⚧ 🐄 ✿ ☕

50 LAND FARM
Edge Lane, Colden, Hebden
Bridge, HX7 7PJ. Mr J Williams.
*8m W of Halifax. At Hebden Bridge
(A646) after 2 sets of T-lights take
turning circle to Heptonstall &
Colden. After 2¾m in Colden village
turn R at Edge Lane 'no through rd'.
In ¾m turn L down lane.* **Wed 19
May, Wed 23 June (10-5). Adm £5,
chd free. Home-made teas.**
An intriguing 6 acre upland garden
within a sheltered valley, created by
the present owner over the past 40
years. In that time the valley has also
been planted with 20,000 trees which
has encouraged a habitat rich in bird
and wildlife. Within the garden vistas
have been created around thought-
provoking sculptures. Moss garden
and well established meconopsis and
cardiocrinum lilies. Partial wheelchair
access, please telephone 01422
842260.

⚧ 🐄 ☕

51 LINDEN LODGE
Newbridge Lane, nr Wilberfoss,
York, YO41 5RB. Robert Scott &
Jarrod Marsden, 07900 003538,
rdsjsm@gmail.com. *Equidistant
between Wilberfoss, Bolton &
Fangfoss. From York on A1079, ignore
signs for Wilberfoss, take next turn to
Bolton village. After 1m at the Xrds,
turn L onto Newbridge Lane.* **Sat
1, Sun 2, Mon 3 May (12-5). Adm
£4.50, chd free. Light lunches,
refreshments, cakes & cream
teas in the Bothy. Visits also by
arrangement in May for groups
of 20+.**
6 acres in all. 1 acre garden, owner
designed and constructed since 2000.
Gravel paths edged with brick or
lavender, many borders with unusual
mixed herbaceous perennials, shrubs
and feature trees. A wildlife pond,
summerhouse, nursery, glasshouse
and fruit cage. Orchard and woodland
area. Formal garden with pond/
water feature. 5 acres of developing
meadow, trees, pathways, hens and
vegetable garden. Gravel paths and
shallow steps.

⚧ 🐄 ✿ 🚗 ☕

52 LITTLETHORPE MANOR
Littlethorpe Road,
Littlethorpe, Ripon,
HG4 3LG. Mrs J P Thackray,
www.littlethorpemanor.com.
*Outskirts of Ripon nr racecourse.
Ripon bypass A61. Follow
Littlethorpe Rd from Dallamires Lane
r'about to stable block with clock*

tower. Map supplied on application.
Sun 12 Sept (1-5). Adm £7, chd free. Home-made teas served in a marquee. £3/head.
11 acres. Walled garden with, herbaceous, roses, gazebo. Sunken garden with ornamental planting and herbs. Brick pergola with wisteria, blue and yellow borders. Formal lawn with fountain pool, hornbeam towers, yew hedging. Box headed hornbeam drive with Aqualens. Large pond with classical pavilion and boardwalk. Winter garden. New in 2021, Physic garden with rill, raised beds and medicinal plants. Cut flower garden. Canal walk. Extensive perennial borders. Parkland with mature trees, autumn and winter walks and late summer plantings. Spring bulbs, annual and perennial meadows. Wheelchair access - gravel paths, some steep steps.

  ❖ 🍵

53 LOW HALL
Dacre Banks, Nidderdale, HG3 4AA, Mrs P A Holliday, 01423 780230, 1pamelaholliday@gmail.com. *10m NW of Harrogate. On B6451 between Dacre Banks & Darley.* **Sun 9 May (1-5). Adm £4.50, chd free. Home-made teas. Also open Woodlands Cottage. Opening with Dacre Banks & Summerbridge Gardens on Sun 11 July. Visits also by arrangement May to Sept for groups of 5 to 30.**
Romantic walled garden set on differing levels designed to complement historic C17 family home (not open). Spring bulbs, rhododendrons; azaleas round tranquil water garden. Asymmetric rose pergola underplanted with auriculas and lithodora links orchard to the garden. Extensive herbaceous borders, shrubs and climbing roses give later interest. Bluebell woods, lovely countryside and farmland all around, overlooking the R Nidd. 80% of garden can be seen from a wheelchair but access involves three stone steps.

  🐕 ❖ 🚗 🍵

There are brilliant plant sales at many gardens. Look out for the symbol in the garden description – and don't forget to bring a bag to carry your plants home in

54 7 LOW WESTWOOD
Golcar, Huddersfield, HD7 4ER. Craig Limbert. *3½m W of Huddersfield off A62. R at T-lights in Linthwaite signed 'Titanic Spa'. Park on road nr Titanic Spa. Garden is over canal bridge.* **Sat 31 July (10.30-2). Adm £4, chd £1.**
With 110yds of canal frontage, this landscaped garden of 1½ acres has both flat and steeply sloping aspects with views across the Colne Valley. Mature lime tree walk, terraced herbaceous beds, pond and vegetable plot, contrasting shady and sunny sites. Kniphofia, astrantia, agapanthus and hydrangeas. Late summer colour is plentiful. Partial wheelchair access - into lower garden only.

55 NEW LUMB BECK FARMHOUSE
Moorside Lane, Addingham, Ilkley, LS29 9JX. Mrs F W Crott, 01943 830400, lumbbeckfarmhouse@btinternet.com, www.lumbbeckfarmhouse.co.uk. *A65 to Addingham r'about A6034 towards Silsden 1st L to Xrds, straight ahead 1m along Straight Lane.* **Sat 29 May, Wed 2 June (11-5). Adm £4, chd free. Home-made teas. Visits also by arrangement May & June for groups of 5 to 30.**
This ½ acre hillside country garden has an abundance of features. Meander along the cobbled paths and herbaceous-edged lawns to find a productive vegetable garden, lily ponds, interesting stone troughs, an enchanting summerhouse and babbling brook with a natural waterfall and bridge. Surrounding this plant-lovers paradise are awe inspiring views adding to the tranquil and immersive experience.

🐕 ❖ 🍵

56 NEW MAGGIE'S AT ST JAMES' HOSPITAL
Alma Street, Leeds, LS9 7BE. Laura Riach, www.maggies.org/our-centres/maggies-leeds/. *Please park in the multi-storey car park next to Maggie's. Take the lift to level B2 & exit to the L.* **Sat 12 June (11-4). Adm by donation. Home-made teas.**
The garden at Maggie's surrounds the centre and incls rooftop gardens. Designed by award winning Landscape Architect Balston Agius, the garden environment sets out to be in sympathy with the organic

form of the building and seeks to provide physical and psychological shelter. The development of low-key woodland will be welcoming, grounding and will celebrate the cycle of the seasons. The gardens are a calming oasis in the concrete desert. Wheelchair access is available at the main entrance to the centre. There is a lift inside the centre.

  🍵

57 THE MANOR HOUSE
Holme-On-Swale, Thirsk, YO7 4JE. Steve & Judi Smith, 01845 567518, judiandsteve@outlook.com. *7m S/W from Thirsk. From A1 J50 onto A6055 N, onto B6267 E from Ripon/Thirsk A61 onto B6267.* **Sat 17 July (1-5). Adm £5, chd free. Home-made teas. Variety of home baked refreshments and drinks. Visits also by arrangement May to Sept for groups of up to 30.**
Almost two acres of mature, well established gardens with level lawns, a variety of specimen trees and shrubs, magnolias, rhododendrons, camellias. Sweeping herbaceous beds with many seasonal bulbs and perennial planting, peonies, roses, rambling clematis. Restored walled kitchen garden with vegetables, herbs and fruit, greenhouse and potting shed, newly planted orchard and woody wild areas. Access by gravelled path onto large level lawns.

  🐕 ❖ 🚗 🍵

58 NEW 19 MANOR ROAD
Beverley, HU17 7AR. Rosemary Dyason, 01482 866085, rosemary@hogwashhouse.karoo.co.uk. *From E on A1174 to T-lights (Rose & Crown on L) straight over onto A164 at r'about L onto Manor Rd. No 19 on L Park in side streets.* **Sun 11 July (10-5). Adm £4, chd free. Visits also by arrangement June & July for groups of 5 to 10.**
1930s house, side entrance to garden, passing many varieties of hostas and ferns leading to garden with established mature shrubs, evergreens, perennials and annuals. Collection of washday buckets, tubs and washtubs planted. Mangle and a small museum of laundry items. One step at gate.

  🐕 ❖

59 MANSION COTTAGE
8 Gillus Lane, Bempton, Bridlington, YO15 1HW. Polly & Chris Myers, 01262 851404, chrismyers0807@gmail.com. *2m NE of Bridlington. From Bridlington take B1255 to Flamborough. 1st L at T-lights - Bempton Lane, turn 1st R into Short Lane then L at end. Continue - L fork at Church. Sat 7, Sun 8 Aug (10-4). Adm £4.50, chd free. Light refreshments.* Visits also by arrangement June to Sept for groups of 10 to 30.
Exuberant, lush, vibrant borders with late perennial planting in this peaceful, surprising, hidden garden. Visitors' book says 'A veritable oasis', 'The garden is inspirational'. Globe garden, mini hosta walk, Japanese themed area, 100ft border, summerhouse, and Art studio, vegetable plot, cuttery, late summer hot border, bee and butterfly border, deck and lawns. Produce, plants and home made soaps. No wheelchair access.

🐐 ✿ ☕

60 MARKENFIELD HALL
Ripon, HG4 3AD. Lady Deirdre & Mr Ian Curteis, 01765 692303, wehaveamoat@gmail.com, www.markenfield.com. *A61 between Ripon & Ripley. Turning between two low stone gateposts west of main road. Beware Sat Navs.* Visits by arrangement May to Sept for groups of 5-10 as a guided tour. Adm £4, chd free. Home-made teas in the Hall. Refreshments to be requested at the time of booking.
The work of the Hall's owner Lady Deirdre Curteis and gardener Giles Gilbey. Mature planting combines with newly-designed areas, where walls with espaliered apricots and figs frame a mix of hardy perennials. The final phase of restoration started in 2017 when the Farmhouse-wing's garden was re-planted to eventually blend seamlessly with the Hall's main East Border. The gardens surround Markenfield Hall - a moated, medieval manor house - one of the oldest, continuously inhabited houses in the country. Partial wheelchair access.

✿ 🚗 ☕

61 NEW MAUNBY HALL
Maunby, Thirsk, YO7 4HA. Mr Peter Hill Walker. *5m NW of Thirsk, 5m S of Northallerton. From Northallerton follow A167 S. From A1(M) J50 take A61 towards Thirsk. Turn L at r'about (A167-Northallerton). Parking at farm opp.*

Sun 4 July (2-5). Adm £5, chd free. Home-made teas.
An established country garden overlooking parkland with fine trees. Long mixed herbaceous border, box parterre, lawns and ha-ha. Large shrubs, nut walk, rose arches and clipped yews. Partial wheelchair access.

♿ 🐐 ✿ ☕

62 115 MILLHOUSES LANE
Sheffield, S7 2HD. Sue & Phil Stockdale. *Approx 4m SW of Sheffield City Centre. Follow A625 Castleton/Dore Road, 4th L after Prince of Wales Pub, 2nd L. OR take A621 Baslow Rd; after Tesco garage take 2nd R, then 1st L.* Sun 27 June (11-5). Adm £3.50, chd free. Light refreshments.
Plantswoman's $\frac{1}{3}$ acre south-facing level cottage style garden with many choice and unusual perennials and bulbs, providing year round colour and interest. Large collection of 50+ hostas, roses, peonies, iris and clematis, with unusual, tender and exotic plants incl echeverias, bananas and echiums. Collection of specimen aeoniums. Seating areas around the garden. Many home propagated plants for sale.

♿ 🐐 ✿ 🚗 ☕

63 MILLRACE GARDEN
84 Selby Road, Garforth, Leeds, LS25 1LP. Mr & Mrs Carthy, 0113 2869233, carolcarthy.millrace@gmail.com, www.millrace-plants.co.uk. *5m E of Leeds. On A63 in Garforth. 1m from M1 J46, 3m from A1.* Sun 9 May, Sun 20 June, Sun 1 Aug (1-5). Adm £5, chd free. Home-made teas. Visits also by arrangement May to Aug. Groups are welcome to use their visit as a propagation opportunity.
Overlooking a secluded valley, garden incl large herbaceous borders containing over 3000 varieties of perennials, shrubs and trees, many of which are unusual and drought tolerant. Ornamental pond, vegetable garden and walled terraces leading to wild flower meadow, small woodland, bog garden and wildlife lakes. Propagation opportunity depending on season (cuttings, seeds, divisions). Art exhibition and The Illuminate Choir at some of our openings. Most of the garden is accessible for wheelchairs. Although there are steps in places there is generally an alternative ramp.

♿ 🐐 ✿ 🚗 ☕

64 NESS HALL
East Ness, Nunnington, YO62 5XD. Mr Richard & the Hon Mrs Murray Wells. *6m E of Helmsley, 22m N of York. From B1257 Helmsley-Malton rd turn L at Slingsby signed Kirkbymoorside, 3m to Ness.* Wed 23 June (10.30-1). Adm £25. Pre-booking essential, please visit www.ngs.org.uk/special-events for information & booking.
Ness is a romantic English flower garden created by three generations of keen gardeners. Surrounded by parkland it has views to the North Yorkshire Moors. A large walled garden with mixed and herbaceous borders as well as self seeding beds, and a water garden, vegetable, cutting garden, rockery, rose, woodland area and orchard. There are steps and slopes in the walled garden although some of the garden is accessible for wheelchairs.

♿ 🐐

65 ◆ NEWBY HALL & GARDENS
Ripon, HG4 5AE. Mr R C Compton, 01423 322583, info@newbyhall.com, www.newbyhall.com. *4m SE of Ripon. (HG4 5AJ for Sat Nav). Follow brown tourist signs from A1(M) or from Ripon town centre.* For opening times and information, please phone, email or visit garden website.
40 acres of extensive gardens and woodland laid out in 1920s. Full of rare and beautiful plants. Formal seasonal gardens, stunning double herbaceous borders to R Ure and National Collection of Cornus. Miniature railway and adventure gardens for children. Sculpture exhibition (open June - Sept). Free parking. Licensed restaurant. Shop and plant nursery. Wheelchair map available. Disabled parking. Manual and electric wheelchairs available on loan, please call to reserve.

♿ ✿ 🚗 NPC ☕ 🍴

66 ◆ NORTON CONYERS
Wath, Ripon, HG4 5EQ. Sir James & Lady Graham, 01765 640333, info@nortonconyers.org.uk, www.nortonconyers.org.uk. *4m NW of Ripon. Take Melmerby & Wath sign off A61 Ripon-Thirsk. Go through both villages to boundary wall. Signed entry 300 metres on R, follow track to car park.* For NGS: Sun 6 June (2-4.30).

Adm £6, chd free. Pre-booking essential, please visit www.ngs. org.uk for information & booking. Home-made teas. For other opening times and information, please phone, email or visit garden website.

Romantic mid C18 walled garden of interest to garden historians. Lawns, herbaceous borders, yew hedges, and Orangery with attractive pond. The garden retains essential features of its original C18th design with sympathetic planting in the English style. There are borders of gold and silver plants, of old fashioned peonies and irises in season. Visitors frequently comment on the tranquil and romantic atmosphere. For House opening dates and times see website. Unusual hardy plants for sale. Most areas wheelchair accessible, gravel paths.

67 THE NURSERY

15 Knapton Lane, Acomb, York, YO26 5PX. Tony Chalcraft & Jane Thurlow, 01904 781691,

janeandtonyatthenursery@hotmail. co.uk. 2½m W of York. From A1237 take B1224 direction Acomb. At r'about turn L (Beckfield Ln.), after 150 metres Turn L. Sun 18 July (1-5); Mon 19, Tue 20 July (2-7). Adm £3, chd free. Home-made teas. Visits also by arrangement Apr to Oct for groups of 10+. Able to give daytime/evening talks to groups in addition to visits.

A former suburban commercial nursery, now an attractive and productive 1 acre organic, private garden. Wide range of fruit with over 100 fruit trees, many in trained form. Many different vegetables grown both outside and under cover in 20m greenhouse. Productive areas interspersed with informal ornamental plantings provide colour and habitat for wildlife. The extensive planting of different forms and varieties of fruit trees make this an interesting garden for groups to visit by appointment at blossom and fruiting times in addition to the main summer openings.

68 NEW THE OLD PRIORY

Everingham, YO42 4JD. Dr J D & Mrs H J Marsden, 01430 860222, helen.marsden@ngs.org.uk. 15m SE of York, 6m from Pocklington. 2m S of A1079. Garden is on E side of village. Parking at village hall opposite. Drop off in garden if necessary. Adm £5, chd free. Home-made teas in Everingham Village Hall. Visits also by arrangement May to July. Wine & beer offered for evening visits.

2 acre rural garden reflecting surrounding countryside. Created in 1990s. Walled veg garden. Polytunnel. Borders and lawn near house are on dry sandy loam. Garden slopes down to bog. Roughly mown pathway through woodland, along lakeside and lightly grazed pasture. Refreshments in aid of RNLI. Gravel driveway. Please drive up if necessary.

St Mary's

69 THE OLD RECTORY

Arram Road, Leconfield, Beverley, HU17 7NP. David Baxendale, 01964 502037, davidbax@newbax.co.uk. *On entering Arram Rd you will see a double bend sign approx 80yds on L. The entrance to the Old Rectory is by the sign. If you reach the church you have missed it.* **Visits by arrangement Jan to May for groups of up to 10. Adm £4, chd free. No refreshments, so if you would like to picnic please bring your own.**
Approx 3 acres of garden and paddock. The garden is particularly attractive from early spring until mid summer. Notable for aconites, snowdrops, crocuses, daffodils and bluebells. Later hostas, irises, lilies and roses. There is a small wildlife pond with all the usual residents incl grass snakes. Well established trees and shrubs, with new trees planted when required.

70 OLD SLENINGFORD HALL

Mickley, nr Ripon, HG4 3JD. Jane & Tom Ramsden. *5m NW of Ripon. Off A6108. After N Stainley turn L, follow signs to Mickley. Gates on R after 1½m opp cottage.* **Sat 5, Sun 6 June (12-4). Adm £5, chd free.**

Home-made teas.
A large English country garden and award winning permaculture forest garden. Early C19 house (not open) and garden with original layout; wonderful mature trees, woodland walk and Victorian fernery; romantic lake with islands, watermill, walled kitchen garden; beautiful long herbaceous border, yew and huge beech hedges. Several plant and other stalls. Picnics very welcome. Reasonable wheelchair access to most parts of garden. Disabled WC at Old Sleningford Farm next to the garden.

71 THE OLD VICARAGE

North Frodingham, Driffield, YO25 8JT. Professor Ann Mortimer. *6m E of Driffield on B1249. From Driffield take B1249 E for approx 6m. The church is on L. Garden is opp. Entrance is on T-junction of rd to Emmotland & B1249. From North Frodingham take the B1249 W for ½m.* **Sun 13 June (10.30-4.30). Adm £4.50, chd free. Light refreshments. Also open Tythe Farm House.**
1½ acre plantsman's garden, developed by owner over 24 years. Many themed areas e.g. rose garden, jungle, desert, fountain, scented,

kitchen gardens, glasshouses. Numerous classical statues, unusual trees and shrubs, large and small ponds, orchard, nuttery. Children's interest with 'Jungle Book' and wild animal statues. Neo-Jacobean revival house, built 1837, mentioned in Pevsner (not open). The land now occupied by the house and garden was historically owned by the family of William Wilberforce. Visitors are welcome to pick produce from the fruit cage and kitchen garden. Parking in farmyard opposite. Parts of the garden are inaccessible but most can be viewed from above (not the jungle).

72 THE ORCHARD

4A Blackwood Rise, Cookridge, Leeds, LS16 7BG. Carol & Michael Abbott, 0113 2676764, michael.john.abbott@hotmail.co.uk. *5m N of Leeds centre, 5 mins from York Gate garden. Off A660 (Leeds-Otley) N of A6120 Ring Rd. Turn L up Otley Old Rd. At top of hill turn L at T-lights (Tinshill Lane). Please park in Tinshill Lane.* **Sun 23 May (12-5). Adm £3.50, chd free. Light refreshments. Pop up cafe and cover for inclement weather. Visits also by arrangement May & June for groups of 10+.**

Low Hall, Dacre Banks & Summerbridge Gardens

⅓ acre plantswoman's hidden oasis. A wrap around garden of differing levels made by owners using stone found on site, planted for yr-round interest. Extensive rockery, unusual fruit tree arbour, oriental style seating area and tea house, linked by grass paths, lawns and steps. Mixed perennials, hostas, ferns, shrubs, bulbs and pots amongst paved and pebbled areas.

✿ ☕ 🍷

🟥 23 THE PADDOCK

Cottingham, HU16 4RA. Jill & Keith Stubbs. *Half way between Beverley & Hull. From Humber Bridge signs to Beverley. Castle Rd on R past hospital. From Beverley signs to Humber Bridge. Harland Way on L at 1st r'about. Garden at S end of village.* **Sun 11 July (10-4.30). Adm £4, chd free.**

A secret garden found behind a small mixed frontage. Archway to themed areas within garden created and maintained by current owners - Japanese, Mediterranean, fairy, mixed herbaceous, patios and lawns. Also two differing ponds, ornamental and tree sculptures. New water feature. Plants for sale along with blacksmith's garden ornaments and sculptures. Assisted access for disabled through side gate.

♿ 🐐 ✿ ☕

🟥 PILMOOR COTTAGES

Pilmoor, nr Helperby, YO61 2QQ. Wendy & Chris Jakeman, 01845 501848, cnjakeman@aol.com. *20m N of York. From A1M J48. N end B'bridge follow rd towards Easingwold. From A19 follow signs to Hutton Sessay then Helperby. Garden next to mainline railway.* **Sun 29 Aug (11-5). Adm £4, chd free. Light refreshments. Visits also by arrangement May to Sept.**

A year round garden for rail enthusiasts and garden visitors. A ride on the 7¼ " gauge railway runs through 2 acres of gardens and gives you the opportunity to view the garden from a different perspective. The journey takes you across water, through a little woodland area, past flower filled borders, and through a tunnel behind the rockery and water cascade. 1½ acre wildflower meadow and pond. Clock-golf putting green. Refreshments by two local WIs with donation to NGS.

♿ 🐐 ✿ ☕ 🍴

🟥 PRIMROSE BANK GARDEN AND NURSERY

Dauby Lane, Kexby, York, YO41 5LH. Sue Goodwill & Terry Marran, 07774 944447, suegoodwill@yahoo.co.uk, www.primrosebank.co.uk. *4m E of York just off A1079. Well signposted.* **Wed 19 May, Wed 16 June, Wed 21 July (11-4). Adm £4, chd free. Home-made teas. Includes designated area for dogs outside the Tearoom. Visits also by arrangement.**

Over an acre of rare and unusual plants, shrubs and trees. Bulbs, hellebores and flowering shrubs in spring, followed by planting for yr round interest. Courtyard garden, mixed borders, summer house and pond. Lawns, contemporary rock garden, shade and woodland garden with pond, stumpery, and shepherd's hut. Poultry and Hebridean sheep. Dogs allowed in car park and at designated tables outside the tearoom. Extensive eranthis collection. CL Caravan Site (CAMC members only) adjoining the nursery. Most areas of the garden are level and are easily accessible for wheelchairs. Accessible WC.

♿ ✿ ☕ 🍷

🟥 THE RED HOUSE

17 Whin Hill Road, Bessacarr, Doncaster, DN4 7AF. Rosie Hamlin. *2m S of Doncaster. A638 South, L at T-lights for B1396, Whin Hill Rd is 2nd R. A638 North, R signed Branton B1396 onto Whin Hill.* **Sat 22, Sun 23 May (1-5). Combined adm with Tamarind £5, chd free. Home-made teas.**

Mature ⅔ acre garden. Dry shade a challenge but acid loving plants a joy. Fine acers, camellia, daphne, rhododendrons, skimmia and eucryphia. Terrace, rockery, small shady garden through to lawn with modern rotating summerhouse, shrubs and young trees. White border conceals pond, compost and hens. Woodland new planting large leaved rhododendrons and trees.

✿ ☕ LH

"Unfortunately, cancer was not in lockdown in 2020. The continued support of our long-standing and valued partner, the National Garden Scheme is more important than ever." Macmillan

🟥 REWELA COTTAGE

Skewsby, YO61 4SG. John Plant & Daphne Ellis, 01347 888125, rewelacottage@gmail.com. *4m N of Sheriff Hutton, 15m N of York. After Sheriff Hutton, towards Terrington, turn L towards Whenby & Brandsby. Turn R just past Whenby to Skewsby. Turn L into village. 400yds on R.* **Sun 30 May, Sun 25 July (11-5). Adm £5, chd free. Cream teas and BBQ. Visits also by arrangement May to July for groups of 20+.**

Situated in a lovely quiet country village this is one of Yorkshire's little hidden treasures. Rewela Cottage was designed from an empty paddock, to be as labour saving as possible, using unusual trees and shrubs to offer interest all year. Their foliage, bark and berries, enhance the well designed structure of the garden. Unusual trees and shrubs have labels giving full descriptions, a picture and cultivation notes incl propagation. Plant sales are specimens from garden. Many varieties of heuchera, heucherella and tiarellas, penstemon, hostas, ferns and herbs for sale. Some gravel paths.

♿ 🐐 ✿ ☕ 🍷 🍴

🟥 ♦ RHS GARDEN HARLOW CARR

Crag Lane, Harrogate, HG3 1QB. Royal Horticultural Society, 01423 565418, harlowcarr@rhs.org.uk, www.rhs.org.uk/harlowcarr. *1½m W of Harrogate town centre. On B6162 (Harrogate - Otley).* **For NGS: Sun 16 May (9.30-5). Adm £12.15, chd £6.10. For other opening times and information, please phone, email or visit garden website.**

One of Yorkshire's most relaxing yet inspiring locations. Highlights include spectacular herbaceous borders, streamside garden, alpines, scented and kitchen gardens. Lakeside gardens, woodland and wildflower meadows. Betty's Cafe Tearooms, gift shop, plant centre and children's play area incl tree house and log ness monster. Wheelchairs and mobility scooters available, advanced booking recommended.

♿ ✿ ☕ 🍷

79 RIDGEFIELD COTTAGE & NURSERY
Forest Moor Road,
Knaresborough, HG5 8JP.
Tony & Jo Pickering. *Between Knaresborough & Harrogate. From A1(J47) take A59 then A658 towards Harrogate & Bradford over 3 r'abouts. After 1½ m turn R (B6163) to Knaresborough & Calcutt. In ¾ m turn L opposite The Cricketers Pub.* Sun 27 June (11-4). Adm £4, chd free. Home-made teas.
½ acre garden evolved over 20yrs around Victorian mound planted orchard. Rocks and dry stone walling surround island beds. Unusual trees, shrubs and shade loving perennials are planted amongst old apple, damson and plum trees. Small fish pond and raised alpine bed. The plant nursery will be open. Half refreshments money donated to St Michael's Hospice, Harrogate. Car parking in field. For disabled some flat grass and gravel, Dogs on a lead only as we have sheep.

80 THE RIDINGS
South Street, Burton Fleming, Driffield, YO25 3PE. **Roy & Ruth Allerston, 01262 470489.** *11m NE of Driffield. 11m SW of Scarborough. 7m NW of Bridlington. From Driffield B1249, before Foxholes turn R to Burton Fleming. From Scarborough A165 turn R to Burton Fleming.* Sun 23 May, Sun 4 July (1-5). Adm £4, chd free. Home-made teas. Visits also by arrangement Apr to July.
Secluded cottage garden with colour-themed borders surrounding neat lawns. Grass and paved paths lead to formal and informal areas through rose and clematis covered pergolas and arbours. Box hedging defines well stocked borders with roses, herbaceous plants and trees. Seating in sun and shade offer vistas and views. Potager, greenhouse and summerhouse. Terrace with water feature and farming bygones. Terrace, tea area and main lawn accessible via ramp.

82 NEW RUDDING PARK
Follifoot, Harrogate, HG3 1JH. **Mr & Mrs Simon Mackaness, 01423 871350, www.ruddingpark.co.uk.** *3m S of Harrogate off the southern bypass A658. Follow brown tourist signs. Use hotel entrance.* Sun 23 May

(2-5). Adm £5, chd free. Light refreshments. Visits also by arrangement.
20 acres of attractive formal gardens, kitchen gardens and lawns around a Grade 1 Regency House extended and used as an Hotel. Humphry Repton parkland. Formal gardens designed by Jim Russell with extensive rhododendron and azalea planting. Recent designs by Matthew Wilson, contemporary look with grasses and perennials.

83 RUSTIC COTTAGE
Front Street, Wold Newton, nr Driffield, YO25 3YQ. **Jan Joyce, 01262 470710, janetmjoyce@icloud.com.** *13m N of Driffield. From Driffield take B1249 to Foxholes (12m), take R turning signed Wold Newton. Turn L onto Front St, opp village pond, continue up hill, garden on L.* Visits by arrangement Mar to Aug for groups of up to 10. Adm £4, chd free.
Plantswoman's cottage garden of much interest with many choice and unusual plants. Hellebores and bulbs are treats for colder months. Old-fashioned roses, fragrant perennials, herbs and wildflowers, all grown together provide habitat for birds, bees, butterflies and small mammals. It has been described as 'organised chaos'! The owner's 2nd NGS garden. Small dogs only.

84 NEW SAINT CATHERINE'S HOSPICE
Throxenby Lane, Scarborough, YO12 5RE. **Susan Stephenson, Communications & Marketing Manager, 01723 351421, susan. stephenson@saintcatherines.org. uk, www.saintcatherines.org.uk.** *On N side of Scarborough, just off A171 Scarborough to Whitby road. Travelling N along Scalby Rd (A171), turn L onto Throxenby Lane & Saint Catherine's is located at the top of the hill.* Sat 10 July (10-3). Adm £3, chd free. Home-made teas. Visits also by arrangement June to Aug.
The award-winning gardens at Saint Catherine's provide a tranquil, relaxing and beautiful environment for patients, visitors, staff and volunteers to enjoy. All of the patients' rooms have double doors which open out onto the gardens and many describe the joy of seeing birds feeding, flowers in bloom and the freedom of

being able to spend time outdoors. Wheelchair access fully accessible.

85 NEW ST MARY'S
Anserdale Lane, Lastingham, York, YO62 6TN. **Mr Clemens & Mrs Johanna Heinrichs.** *Located close to St. Mary's church & to Blacksmith Arms pub on route towards Hutton Le Hole. Parking in adjacent field available.* Sat 14 Aug (11-4). Adm £6, chd free. Home-made teas.
Located within peaceful Lastingham Village with views towards the church. The garden has landscaped areas with grasses and perennials, meadows and lawns as well as formal structures and mature trees. There is a bog garden next to the beck with a natural waterfall. Newer additions are productive areas and a gravel garden. Only upper part of garden accessible to wheelchairs. Stream areas are inaccessible to visitors with mobility issues.

86 NEW SALTMARSHE HALL
Howden, DN14 7RX. **Claire Connely, 01430 434920, info@saltmarshehall.com, www.saltmarshehall.com/.** *6m E of Goole. From Howden (M62, J37) follow signs to Howdendyke & Saltmarshe. House in park W of Saltmarshe village.* Wed 4 Aug (12-4). Adm £4.50, chd free. Light refreshments in the Hall. Tea, coffee, cream teas and alcoholic beverages available. Afternoon teas and picnic hampers available via pre-booking. Visits also by arrangement.
17 acre estate with Regency House (open for refreshments and pre-booked afternoons teas) set on the banks of the River Ouse with a walled garden, herbaceous borders brimming with shrubs, climbers, herbaceous plants and roses, expansive lawns, fine old trees, pond, orchard, courtyards, river and countryside views. Wheelchair to the house via a ramp. Garden is accessed via gravel paths, narrow walkways and shallow steps.

SPECIAL EVENT

87 ♦ SCAMPSTON WALLED GARDEN
Scampston Hall, Scampston, Malton, YO17 8NG. **The Legard Family, 01944 759111,**

info@scampston.co.uk, www.scampston.co.uk/gardens. *5m E of Malton. ½m N of A64, nr the village of Rillington & signed Scampston only.* **For NGS: Tue 6 July (4-6.30). Adm £25. Pre-booking essential, please visit www.ngs.org.uk/special-events for information & booking.** **For other opening times and information, please phone, email or visit garden website.**
An exciting modern garden designed by Piet Oudolf. The 4-acre walled garden contains a series of hedged enclosures designed to look good throughout the year. The garden contains many unusual species and is a must for any keen plant lover. The Walled Garden is set within the grounds and parkland surrounding Scampston Hall. The Hall opens to visitors for a short period during the summer months. A restored Richardson conservatory at the heart of the Walled Garden is used as a Heritage and Learning Centre. The Walled Garden, cafe and facilities are accessible by wheelchair. Some areas of the parkland and the first floor of the Hall are harder to access.
♿ ❀ 🚗 ☕ 🏕

88 SCAPE LODGE
11 Grand Stand, Scapegoat Hill, Golcar, Huddersfield, HD7 4NQ. Elizabeth & David Smith, 01484 644320, elizabethanddavid. smith@ngs.org.uk. *5m W of Huddersfield. From J23 or 24 M62, follow signs to Rochdale. After Outlane village, 1st L. At top of hill, 2nd L. Park at Scapegoat Hill Baptist Church (HD7 4NU) or in village. 5 mins walk to garden. 303/304 bus.* **Sun 2 May, Sun 22 Aug (1.30-4.30). Adm £5, chd free. Home-made teas. Visits also by arrangement May to Aug. Donation to Mayor of Kirklees Charity Appeal.**
⅓ acre contemporary country garden at 1000ft in the Pennines on a steeply sloping site with far-reaching views. Gravel paths lead between mixed borders on many levels. Colour themed informal planting chosen to sit comfortably in the landscape and give year round interest. Steps lead to terraced kitchen and cutting garden. Gazebo, pond, shade garden, collection of pots and tender plants. Regional finalist, The English Garden's "The Nation's Favourite Gardens 2019"
❀ 🚗 ☕

89 1 SCHOOL LANE
Bempton, Bridlington, YO15 1JA. Robert & Elizabeth Tyas, rktyas10@gmail.com. *Close to the centre of village. 3m N of Bridlington. From Brid take B1255 toward Flamborough. L at 1st T-lights. Follow Bempton Lane out of residential area then 2nd R onto Bolam Lane. Continue to end and garden is facing.* **Sun 30 May (10-4). Adm £4, chd free. Home-made teas. Visits also by arrangement May to July for groups of 5 to 20. Weekdays only.**
A short distance from award winning Bempton Cliffs this village garden provides year round interest. Mixed planting in beds and borders offers a cottage garden feel enhanced by container planting, a crevice rockery, troughs planted with unusual alpines, a small collection of bonsai, miniature hostas, fruit trees, and a growing display of lewisia. Many additional garden features. All areas of the garden can be viewed from a wheelchair.
♿ 🐕 ❀ ☕

90 ◆ SHANDY HALL GARDENS
Thirsk Bank, Coxwold, York, YO61 4AD. The Laurence Sterne Trust, 01347 868465, www. laurencesternetrust.org.uk/ shandy-hall-garden.php. *N of York. From A19, from both Easingwold & Thirsk, turn E signed Coxwold. Park on road.* **For NGS: Evening opening Fri 4, Fri 25 June (6.30-8). Adm £3, chd free. For other opening times and information, please phone or visit garden website.**
Home of C18 author Laurence Sterne. 2 walled gardens, 1 acre of unusual perennials interplanted with tulips and old roses in low walled beds. In old quarry, another acre of trees, shrubs, bulbs, climbers and wildflowers encouraging wildlife, incl over 440 recorded species of moths. Moth demonstration.
♿ 🐕 ❀ 🏕

GROUP OPENING

91 SHEFFIELD GARDENS
Crookes, Sheffield, S10 1UZ. *Two city gardens in Crookes. Fir Street lies between Northfield Rd & South Rd / Howard Rd in Crookes. Nr High Street. Tasker Road is 1st turning L off Mulehouse Rd. The garden is behind a tall hedge.* **Sat 24 July (12-**

5). Combined adm £8, chd free. Home-made teas at 68 Tasker Road.

19 FIR STREET
S6 3TG. James Hitchmough.

68 TASKER ROAD
S10 1UZ. Andy Clayden.

Two inspiring, small, urban gardens designed and owned by members of The Department of Landscape Architecture, The University of Sheffield. The garden owners are at the forefront of current ideas in naturalistic planting design, creating sustainable landscapes and gardening for a changing planet. The owners will be available to discuss their gardens. 68 Tasker Road is a family garden using reclaimed materials, managing water through green roofs, ponds and soakaways. 19 Fir Street explores a meadow aesthetic using many plants usually considered too tender for a northern city, to maintain colour and foliage interest from February to November.

GROUP OPENING

92 SHIPTONTHORPE GARDENS
Shiptonthorpe, York, YO43 3PQ. *2m NW of Market Weighton. Both gardens are in the main village on the N of A1079.* **Sat 5, Sun 6 June (11-5). Combined adm £5, chd free.**

6 ALL SAINTS DRIVE
Di Thompson.

WAYSIDE
Susan Sellars.

Two contrasting gardens offering different approaches to gardening style. 6 All Saints Drive, 'Langdale End' is an eclectic 'maze-like' garden with a mix of contemporary and cottage garden styles features; a hidden corner with tropical plants and greenhouse, a Japanese /Asian inspired garden with water features and pond. Wayside has interesting planting in different areas of the garden. Developing vegetable and fruit growing areas with greenhouse. Proceeds for teas to go to the village hall and church. Wheelchairs possible with help at Wayside.

93 SKIPWITH HALL

Skipwith, Selby, YO8 5SQ.
Mr & Mrs C D Forbes Adam,
rosalind@escrick.com,
www.escrick.com/hall-gardens. *9m
S of York, 6m N of Selby. From York
A19 Selby, L in Escrick, 4m to Skipwith.
From Selby A19 York, R onto A163 to
Market Weighton, then L after 2m to
Skipwith.* Thur 10 June (1-4). Adm
£5, chd free. Home-made teas.
Visits also by arrangement in June
for groups of 10 to 30.
4-acre walled garden of Queen Anne
house (not open). Mixture of historic
formal gardens (in part designed by
Cecil Pinsent), wildflower walks and
lawns. 'No-dig' kitchen garden with
herb maze, Italian garden, collection
of old-fashioned shrub roses and
climbers. Orchard with espaliered and
fan-trained fruit and small arboretum.
Gravel paths.

94 SLEIGHTHOLMEDALE LODGE

Fadmoor, YO62 7JG. Patrick &
Natasha James. *6m NE of Helmsley.
Parking can be limited in wet
weather. Garden is 1st property in
Sleightholmedale, 1m from Fadmoor.*
Sun 4 July (2-6). Adm £5, chd
free. Home-made teas.
A south facing, 3 acre hillside garden
with views over a peaceful valley in
the North York Moors. Cultivated for
over 100 years, wide borders and
descending terraces lead down the
valley with beautiful, informal planting
within the formal structure of walls
and paths. In July and August, the
garden features roses, delphiniums
and verbascums.

SPECIAL EVENT

95 NEW SOUTH PARK

Hutton Wandesley, York, YO26 7LL.
Mrs Sasha York. *From B1224 in
Long Marston, follow signs to Hutton
Wandesley, South Park is the last
house on the L as you leave the
village towards Angram.* Thur 3 June
(10.30-1). Adm £25. Pre-booking
essential, please visit www.
ngs.org.uk/special-events for
information & booking.
A 2 acre garden created from a patch
of grass, started in 2008 and still
evolving. A garden of rooms which
include herbaceous borders, a cutting
garden, vegetable area, wildflower
meadow and orchard. Features include
topiary, pleached hornbeam and

espaliered fruit trees. South Park is a
good source of inspiration for anyone
just creating a garden. Most of the
garden is accessible for wheelchairs.

96 ♦ STILLINGFLEET LODGE

Stewart Lane, Stillingfleet, York,
YO19 6HP. Mr & Mrs J Cook,
01904 728506, vanessa.cook@
stillingfleetlodgenurseries.co.uk,
www.stillingfleetlodgenurseries.
co.uk. *6m S of York. From A19 York-
Selby take B1222 towards Sherburn
in Elmet. In village turn opp church.*
For NGS: Sun 16 May, Sun 12 Sept
(1-5). Adm £6, chd £1. Home-made
teas. For other opening times and
information, please phone, email or
visit garden website.
Organic, wildlife garden subdivided
into smaller gardens, each based on
a colour theme with emphasis on use
of foliage plants. Wild flower meadow,
natural pond, 55yd double herbaceous
borders and modern rill garden. Rare
breeds of poultry wander freely in
garden. Adjacent nursery. Garden
courses run all summer - see website.
Art exhibitions in the cafe. Gravel paths
and lawn. Ramp to cafe if needed. No
disabled WC.

97 STONEFIELD COTTAGE

27 Nordham, North Cave, Brough,
HU15 2LT. Nicola Lyte. *15m W of
Hull. M62 E, J38 towards N Cave.
Turn L towards N Cave Wetlands, then
R at LH bend. Stonefield Cottage is on
R, ¼ m along Nordham.* Sat 17, Sun
18 July (10-5). Adm £4.50, chd free.
Home-made teas.
A hidden and surprising 1-acre
garden, with an emphasis throughout
on strong, dramatic colours and
sweeping vistas. Rose beds, mixed
borders, vegetables, a riotous hot
bed, boggy woodland, wildlife pond
and jacquemontii under-planted
with red hydrangeas. Collections
of hellebores, primulas, ferns,
astilbes, hostas, heucheras, dahlias,
hemerocallis and hydrangeas.

98 TAMARIND

2 Whin Hill Road, Bessacarr,
Doncaster, DN4 7AE. Ken & Carol
Kilvington. *2m S of Doncaster.
Through Lakeside, pass Dome on R
turn R at Bawtry Rd T-lights, through
pedestrian crossing & T-lights
then 1st L. Or - A638 - North from
Bawtry to Doncaster turn R signed
Cantley-Branton (B1396).* Sat 22,
Sun 23 May (1-5). Combined
adm with The Red House, chd
free. Home-made teas on Sat at
Tamarind; Sun at The Red House.
The gardens are within easy
walking distance of each other.
A ⅔ acre garden, changing each
year, level at the front with acers and
varied planting. Shaped lawn leads
to steep steps and terraced rear
garden full of colour and differing
styles. White border with dovecote
and doves. Japanese garden, hot
border, rose garden, herbaceous
embankment, fern garden and
rhododendron garden. Stream with
waterfalls and ponds. Rockery,
thatched summerhouse, patio. The
front garden and rear lower patio are
accessible to wheelchairs, from which
most of the rear garden can be seen.
Steps to the rest of the garden.

GROUP OPENING

99 NEW THE TERRACE GARDENS

Oswaldkirk, York, YO62 5XZ.
Bridget Hannigan. *20m N of York,
4m S of Helmsley. Single track road
on bend of B1363 in the center of
Oswaldkirk. Parking in Main St opp.*
Sun 15 Aug (1-5). Combined adm
£6, chd free. Home-made teas in
the Village Hall.
Collection of four gardens nestled
along a south-facing hillside known
as The Hag. All with stunning views
of the Coxwold-Gilling Gap and
Howardian Hills beyond. Bramleys:
plantswoman's garden with winding
paths through naturalistic planting
and wildflower orchard. Orchard
House: beekeepers' garden with an
emphasis on wildlife and fruit and
vegetable production. Ewe Cote:
traditional cottage garden full of roses
and perennial favourites and a small
orchard. Pavilion House: terraced
garden, creating different 'rooms' and
planting themes. There is no vehicular
access to The Terrace but parking is
available nearby. All gardens are on a
slope; Bramleys and Pavilion House
in particular have steps and paths
unsuitable for wheelchairs and those
with mobility issues.

100 NEW THIMBLEBY HALL

Thimbleby, Northallerton, DL6 3PY. **Mr & Mrs A Shelley.** *From A19, take signs for Osmotherly then turn R in the village centre down South End Rd towards Thimbleby.* **Fri 21 May (1-4.30). Adm £7, chd free. Prebooking essential, please visit www.ngs.org.uk for information & booking. Tea.**

Country estate on the western edge of the NY Moors. A number of formal garden areas include terraced gardens, herbaceous garden and lake all within an extensive parkland setting with sweeping views of the surrounding countryside. The gardens have been substantially extended by the owners with formal areas and a large banked area to the west of the house, planted with spring flowering shrubs featuring azaleas and cornus. Mature trees are another feature within the estate and especially impressive is an avenue of redwoods framing views across the lake to the south of the Hall.

101 NEW THIRSK HALL

Kirkgate, Thirsk, YO7 1PL. **Willoughby & Daisy Gerrish, 01845523420, Info@thirskhall.com, www.thirskhall.com.** *Kirkgate is off the main market square. Thirsk Hall sits in the town centre of Thirsk on Kirkgate, next to St Mary's Church.* **Sat 12 June (2-5.30) Adm £6, chd free. Home-made teas.**

Thirsk Hall is a Grade II* listed townhouse completed by John Carr in 1777 (not open). Behind the house the unexpected 20 acre grounds include lawns, herbaceous borders, kitchen gardens, walled paddocks and parkland beyond a ha ha. The present layout and planting are the result of a sensitive restoration blending formal and informal beds, shrubbery, and mature trees. Wheelchair access. Sculpture park.

102 TYTHE FARM HOUSE

Carr Lane, Wansford, Driffield, YO25 8NP. **Terry & Susanne Hardcastle, 07951 126588, susanannhardcastle@gmail.com, tythefarmgardens.com.** *Wansford is 3m E of Driffield on B1249. Turn L at mini r'about onto Nafferton Rd. Carr Lane is opposite the Church. Approx 600 yds on R.* **Sun 13 June (10-5). Adm £5, chd free. Light refreshments. Visits also by arrangement Feb to Oct for groups of 5+.**

Terry and Susanne welcome you to their secret garden that extends to 10 acres. Deciduous woodlands, orchards, lake, courtyard garden, herbaceous, rose garden and Italian sculptures. Formerly a working farm with traditional buildings now renovated. Mature garden developed over 25 years. Excavation of the lake and landscaping projects have created something very special. Ample parking space on site and easy access to most areas. An optional short hike through the woods will prepare you for home-made refreshments in the courtyard garden or the garden room as the weather dictates. Access to many parts of the garden is via a few paved paths and level lawns that can become soggy in wet weather.

103 THE VILLA

High Street, Hook, Goole, DN14 5PJ. **Penny & John Settle.** *Close to M62 J36 & J37. Approach Hook from A614 Boothferry Road along Westfield Lane around 'z' bend to Xrds where the village hall stands. Turn L, garden approx. 300yds down on R. Park on High St.* **Sun 30 May, Sun 12 Sept (11.30-4.30). Adm £4, chd free. Light refreshments.**

Acquired by the current owners in 2011 in a derelict state, the garden now offers a stunning variety of different areas incl. Mediterranean, an extensive Bonsai display, an oriental area featuring Asian sculptures, a pergola walkway, a paved terrace and a large lawned area. There will be a display by local artists, with some works being offered for sale. The garden also features a collection of statues, alongside a large variety of perennial shrubs, bamboos and ornamental grasses, bearded irises and hostas. The garden offers plenty of seating for visitors to sit and take in the tranquillity of this hidden gem. Sorry, no wheelchair access.

104 NEW THE VINES

Waplington Hall, Allerthorpe, York, YO42 4RS. **Penny & Bill Simmons.** *Leave A1079 at Pocklington r'about, SW through Allerthorpe, turn R signed 'Waplington only'. Coming from Melbourne turn L at start of Allerthorpe village.* **Sun 18 July (12-4.30). Adm £4.50, chd free. Home-made teas at Allerthorpe Village Hall.**

This spacious 1¾ acre English country garden is traditionally planted with varied herbaceous borders of perennials, shrubs and climbers. Walk through an unusual vinehouse full of fruiting grapevines to a thriving walled kitchen garden with fruit trees, vegetables, soft fruit and productive greenhouse. An informal garden featuring extensive grassed areas with mature and ornamental trees. Most areas of the garden are grassed and mainly level and so accessible for wheelchairs, apart from the greenhouses.

Kirkwood Hospice

105 WARLEY HOUSE GARDEN

Stock Lane, Warley, Halifax, HX2 7RU. Dr & Mrs P J Hinton, 07915 550994, warleyhousegardens@yahoo.com, www.warleyhousegardens.com. *2m W of Halifax. Take A646 (towards Burnley) from Halifax centre. Go through large intersection after approx 1m. Approx 1m, turn R up Windle Royd Lane. Disabled parking on site for 4/5 vehicles.* **Sun 16, Tue 18 May (1-5). Adm £5, chd free. Home-made teas. Local hospice volunteers supply home-made teas, drinks and variety of cakes. Visits also by arrangement May to July for groups of 10 to 30.**

Partly walled 2½ acre garden of demolished C18 House, renovated by the present owners. Rocky paths and Japanese style planting lead to lawns and lovely S-facing views. Alpine ravine planted with ferns and fine trees give structure to the woodland area. Drifts of shrubs, herbaceous plantings, wild flowers and heathers maintain constant seasonal interest. This is an historic garden, renovated after total neglect from 1945 to 1995. Spring planting is enhanced by many new rhododendrons and woodland. A lengthy rockery, alpine ravine and Japanese garden. Most of the garden is accessible to wheelchairs. Lawns usually suitable unless very wet. Disabled access to WCs and tea-room.

106 NEW WELTON LODGE

Dale Road, Welton, Brough, HU15 1PE. Brendon Swallow, 07792 345858, manager@weltonlodge.co.uk. *From A63 E bound take exit for Welton, Elloughton & Brough. Immed turn after R after junction. Garden on R just after Xrds.* **Sun 5 Sept (10-4). Adm £4, chd free. Cream teas & home-made cakes.**

Welton Lodge is a beguiling Grade II listed house occupying 1.5 acres of land in the beautiful village of Welton and features a series of tiered gardens. The recently renovated walled kitchen gardens have espaliers of fruit trees and vines around formal lawns and flower beds leading to a 20m Orangery. Block paved ramp leading to gravelled access to lower and upper gardens without steps. Steep upper garden.

107 WHITE WYNN

Ellerton, York, YO42 4PN. Cindy & Richard Hutchinson, 01757 289495, richardahutchinson@btinternet. com. *Garden NOT in Ellerton Village. It is W of B1228, ½m up Shortacre Lane.* **Sun 20 June (12-5). Adm £4, chd free. Visits also by arrangement Apr to Aug for groups of 5 to 30.**

1½ acres developed over 11 years. Front garden and gravel areas. Lawns and patios adjoining 60m long border, wildflower meadow, heather bed, orchard, rose garden with arbour, vegetable beds, woodland and shade beds. Secret garden contains less usual plantings to give a tropical feel. Wildlife friendly with 2 ponds, blackthorn and damson copses, stored logs provide shelter for hedgehogs and toads. Very good, level access, mostly on grass.

Littlethorpe Manor

GROUP OPENING

108 WHIXLEY GARDENS
York, YO26 8AR. *8m W of York, 8m E of Harrogate, 6m N of Wetherby. 3m E of A1(M) off A59 York-Harrogate. Signed Whixley.* Mon 3 May (11-5). Combined adm £6, chd free. Light refreshments at The Old Vicarage.

COBBLE COTTAGE
John Hawkridge & Barry Atkinson.
(See separate entry)

THE OLD VICARAGE
Mr & Mrs Roger Marshall,
biddymarshall@btinternet.com.
Visits also by arrangement Mar to Aug.

Attractive rural yet accessible village nestling on the edge of the York Plain with beautiful historic church and Queen Anne Hall (not open). The gardens are at opposite ends of the village with good footpaths. A plantsman's and flower arranger's garden at Cobble Cottage, has views to the Hambleton Hills. Close to the church, The Old Vicarage, with a ¾-acre walled flower garden, overlooks the old deer park. The walls, house and various structures are festooned with climbers. Gravel and old brick paths lead to hidden seating areas creating the atmosphere of a romantic English garden. Wheelchair access only to The Old Vicarage.
🚗 🐴 ✿ ♿ 🍵

"I was amazed to discover that the National Garden Scheme is Marie Curie's largest single funder and has given the charity nearly £10 million over 25 years. Their continued support makes such a difference to me and all Marie Curie Nurses on the frontline of the coronavirus crisis, as we continue to provide expert care and support to people at end of life." – Tracy McWilliams, Marie Curie Nurse

109 WOODLANDS COTTAGE
Summerbridge, HG3 4BT. Mr & Mrs Stark, 01423 780765, annstark@btinternet.com, www.woodlandscottagegarden.co.uk. *10m NW of Harrogate. On the B6165 W of Summerbridge.* Sun 9 May (1-5). Adm £4, chd free. Home-made teas. Also open Low Hall. Opening with Dacre Banks & Summerbridge Gardens on Sun 11 July. Visits also by arrangement May to Aug.

A 1 acre country garden in Nidderdale, created by its owners and making full use of its setting, which includes natural woodland with wild bluebells and gritstone boulders. There are several gardens within the garden, from a wildflower meadow and woodland rock-garden to a formal herb garden and herbaceous areas. Also a productive fruit and vegetable garden. Gravel paths and grassy slopes, with a few steps, but these can be avoided.
🚗 🐴 ✿ ♿ 🍵

110 ◆ YORK GATE
Back Church Lane, Adel, Leeds, LS16 8DW. Perennial, 0113 267 8240, yorkgate@perennial.org.uk, www.yorkgate.org.uk. *5m N of Leeds. A660 from to Adel, turn R at lights a few meters before 'Divino' restaurant, L on to Church Lane. After the church, turn R onto Back Church Lane then entrance to the garden is on the left.* For opening times and information, please phone, email or visit garden website.

An internationally acclaimed one acre jewel. A garden of 14 individual garden rooms. Many unusual plants and architectural topiary. 2020 saw an extension to the garden with new facilities incl a new cafe and gift shop as well as exciting new gardens, plant nursery and heritage exhibition. Owned by Perennial, the charity that looks after horticulturists and their families in times of need. Narrow gravel and cobbled paths make most of the original garden inaccessible to wheelchairs. The new gardens, tea room, shop and WC are accessible.
✿ ♿ 🍵

111 YORKE HOUSE & WHITE ROSE COTTAGE
Dacre Banks, Nidderdale, HG3 4EW. Tony & Pat Hutchinson & Mark & Amy Hutchinson, 01423 780456, pat@yorkehouse.co.uk, www.yorkehouse.co.uk. *4m SE of Pateley Bridge, 10m NW of Harrogate, 10m N of Otley. On B6451 near centre of Dacre Banks. Car park.* Sun 20 June (11-5). Adm £5, chd free. Cream teas. Opening with Dacre Banks & Summerbridge Gardens on Sun 11 July. Visits also by arrangement June to Aug for groups of 10+.
Award winning English country garden in the heart of Nidderdale. Designed as a series of distinct areas which flow through 2 acres of ornamental garden. Colour-themed borders, natural pond & stream with delightful waterside plantings. Secluded seating areas & attractive views. Adjacent cottage has newly developed garden designed for wheelchair access. Large collection of hostas. Orchard picnic area. Winner Harrogate's Glorious Gardens. All main features accessible to wheelchair users.
♿ 🐴 ✿ ♿ 🍵 🪑

112 ◆ THE YORKSHIRE ARBORETUM
Castle Howard, York, YO60 7BY. The Castle Howard Arboretum Trust, 01653 648598, marketing@yorkshirearboretum.org, www.yorkshirearboretum.org. *15m NE of York. Off A64. Follow signs to Castle Howard then look for Yorkshire Arboretum signs at the obelisk r'about.* For NGS: Sun 6 June (10-3.30). Adm £7, chd free. For other opening times and information, please phone, email or visit garden website. Donation to Plant Heritage.
A glorious, 120 acre garden of trees from around the world set in a stunning landscape of parkland, lakes and ponds. Walks and lake-side trails, tours, family activities. We welcome visitors of all ages wanting to enjoy the space, serenity and beauty of this sheltered valley as well as those interested in our extensive collection of trees and shrubs. Internationally renowned collection of trees in a beautiful setting, accompanied by a diversity of wildflowers, birds, insects and other wildlife. Children's playground, cafe and gift shop. Dogs on leads welcome. Not suitable for wheelchairs. Motorised all-terrain buggies are available on loan, please book 24hrs in advance on 01653 648598.
🐴 ♿ NPC 🍵 🪑

Esgair Angell

WALES

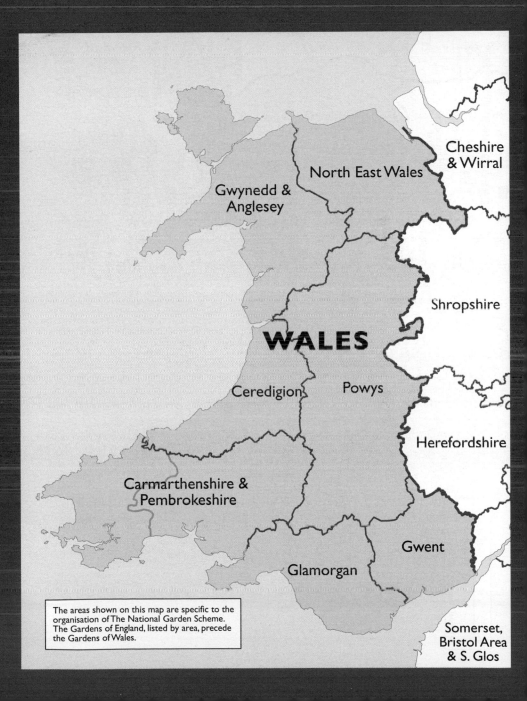

Cheshire & Wirral

North East Wales

Gwynedd & Anglesey

Shropshire

WALES

Ceredigion

Powys

Herefordshire

Carmarthenshire & Pembrokeshire

Gwent

Glamorgan

The areas shown on this map are specific to the organisation of The National Garden Scheme. The Gardens of England, listed by area, precede the Gardens of Wales.

Somerset, Bristol Area & S. Glos

CARMARTHENSHIRE & PEMBROKESHIRE

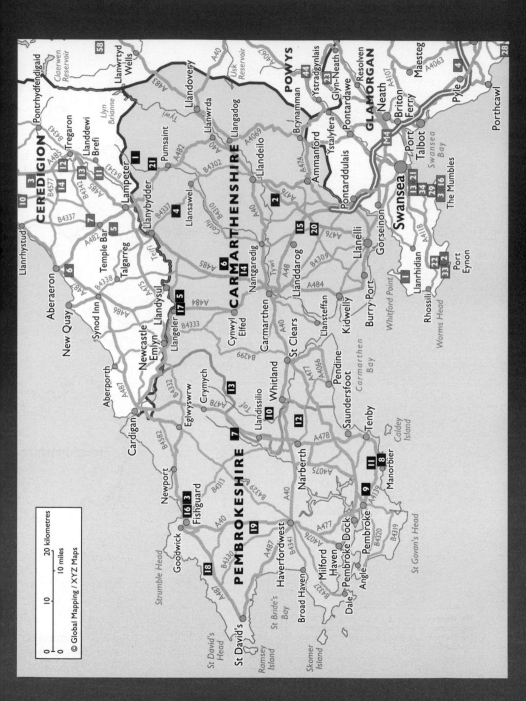

From the rugged Western coast and beaches to the foothills of the Brecon Beacons and Black Mountain, these counties offer gardens as varied as the topography and weather.

In gardens such as Llwyngarreg and Treffgarne Hall tender plants flourish. Look out for new gardens this year: The Grange, Dwynant, Skanda Vale Hospice Garden, and Tyn y Waun. Each of our gardens has something different to offer the visitor and many welcome dogs as well as their owners!

Several fascinating gardens have restricted parking areas and so are only open by pre-booking; the owners will be delighted to see you, but please do telephone first to arrange a visit. Most of our gardens also offer teas or light refreshments and what better way to enjoy an afternoon in a garden, where there is no weeding or washing up for you to do! We look forward to welcoming you to our gardens.

Volunteers

County Organisers
Jackie Batty
01437 741115
jackie.batty@ngs.org.uk

County Treasurer
Brian Holness
01437 742048
brian.holness@ngs.org.uk

Assistant County Organisers
Elena Gilliatt
01558 685321
elenamgilliatt@hotmail.com

Liz and Paul O'Neill
01994 240717
lizpaulfarm@yahoo.co.uk

Brenda Timms
01558 650187
brendatimmsuk@gmail.com

Social Media
Mary-Ann Nossent
07985077022
maryann.nossent@ngs.org.uk

 @CarmsandPembsNGS

 carmsandpembsngs

Left: Norchard

OPENING DATES

All entries subject to change. For latest information check www.ngs.org.uk
Map locator numbers are shown to the right of each garden name.

April

Monday 5th
Llwyngarreg   10

May

Sunday 2nd
Treffgarne Hall   19
NEW Tyn y Waun   20

Sunday 9th
NEW Dwynant   2
◆ Dyffryn Fernant   3

Sunday 23rd
NEW Dwynant   2

Wednesday 26th
Pont Trecynny   16

Saturday 29th
NEW Skanda Vale
Hospice Garden   17

Sunday 30th
Norchard   11
NEW Tyn y Waun   20

June

Sunday 6th
Llwyngarreg   10

Sunday 13th
NEW Dwynant   2

Saturday 19th
Pen-y-Garn   15

Sunday 20th
Pentresite   14
Pen-y-Garn   15

Wednesday 23rd
Pont Trecynny   16

Saturday 26th
◆ Lamphey Walled
Garden   9

Sunday 27th
◆ Lamphey Walled
Garden   9

July

Saturday 3rd
Glangwili Lodges   6

Sunday 4th
Glangwili Lodges   6
Treffgarne Hall   19

Saturday 10th
Glandwr   5

Sunday 11th
Glandwr   5

August

Wednesday 4th
Glyn Bach Gardens   7

Saturday 7th
Pen-y-Garn   15
NEW Skanda Vale
Hospice Garden   17

Sunday 8th
Pen-y-Garn   15

Sunday 15th
◆ Dyffryn Fernant   3

Monday 30th
Llwyngarreg   10

September

Saturday 4th
◆ Lamphey Walled
Garden   9

Sunday 5th
◆ Lamphey Walled
Garden   9
Pentresite   14

By Arrangement

Arrange a personalised garden visit on a date to suit you. See individual garden entries for full details.

Bwlchau Duon   1
Gelli Uchaf   4
Glandwr   5
NEW The Grange   8
Llwyngarreg   10
Norchard   11
The Old Rectory   12
Pencwm   13
Pentresite   14
Pen-y-Garn   15
Pont Trecynny   16
Stable Cottage   18
Treffgarne Hall   19
Ty'r Maes   21

Dyffryn Fernant

© Carole Drake

THE GARDENS

1 BWLCHAU DUON

Ffarmers, Llanwrda,
Carmarthenshire,
SA19 8JJ. Brenda & Allan
Timms, 01558 650187,
brendatimmsuk@gmail.com. *7m
SE Lampeter, 8m NW Llanwrda.
From A482 turn to Ffarmers. In
Ffarmers, take lane opp Drovers
Arms PH. After caravan site on
L, turn L at small Xrds into single
track lane & follow NGS arrows.*
**Visits by arrangement June to
Aug for groups of up to 30. Teas
on request when booking. Adm
£4.50, chd free.**
A 1 acre, ever evolving garden
challenge, set in the foothills of the
Cambrian Mountains at 1100ft. This
is a plantaholics haven where borders
are full of many unusual plants and
lots of old favourites. There are
raised vegetable gardens, a 100ft
herbaceous border, natural bog areas,
winding pathways through semi-
woodland and magnificent views over
the Cothi valley. Rare breed turkeys,
chickens, geese.

🐕 ✳ 🍵 🪑

2 NEW DWYNANT

Golden Grove, Carmarthen,
SA32 8LT. Mrs Sian
Griffiths, 01558 668727,
sian.41@btinternet.com,
www.airbnb.co.uk/rooms/31081717.
*14m east of Carmarthen, 3m from
Llandeilo. Take B4300 Llandeilo to
Carmarthen. Take L turning to Gelli
Aur, pass church & vicarage then
1st R onto Old Coach Rd. Dwynant
is approximately ¼m on R.* **Sun
9, Sun 23 May, Sun 13 June
(10.30-5). Adm £4, chd free. Pre-
booking essential, please phone
01558 668727 or email sian.41@
btinternet.com for information &
booking. Teas on request when
booking.**
A ¾ acre garden set on a steep
slope designed to sit comfortably
within a verdant countryside
environment with beautiful scenery
and tranquil woodland setting. A
spring garden with lily pond, selection
of plants and shrubs incl. azaleas,
rhododendrons, rambling roses
set amongst a carpet of bluebells.
Seating in appropriate areas to enjoy
the panoramic view and flowers.

✳ 🚌 🍵

3 ♦ DYFFRYN FERNANT

Llanychaer, Fishguard,
SA65 9SP. Christina Shand &
David Allum, 01348 811282,
christina@dyffrynfernant.co.uk,
www.dyffrynfernant.co.uk. *3m E
of Fishguard, then ½m inland. A487
east, 2m from Fishguard turn R
towards Llanychaer. Follow lane for
½m, entrance on L. Look for yellow
garden signs.* **For NGS: Sun 9 May,
Sun 15 Aug (11-5). Adm £7, chd
free. For other opening times and
information, please phone, email or
visit garden website.**
A modern, 6-acre garden, which,
over 2 decades, has grown out of
the ancient landscape. Distinctly
different areas, including a lush bog
garden, exotic courtyard, and a field
of ornamental grasses combine to
create a garden that unfolds as you
journey though it. Abundant sitting
places and a garden library invite
visitors to take their time. RHS Partner
Garden. Scented azaleas and peonies
in May. Dahlias, ornamental grasses
and exotics in August.

🐕 🚌 🚍 🪑

4 GELLI UCHAF

Rhydcymerau, Llandeilo,
SA19 7PY. Julian & Fiona
Wormald, 01558 685119,
thegardenimpressionists@
gmail.com,
www.thegardenimpressionists.com.
*5m SE of Llanybydder. 1m NW of
Rhydcymerau. See our website for
detailed directions to the garden.
Parking v. limited: essential to phone
or email first.* **Visits by arrangement
Jan to Oct for groups of up to 20.
Adm £5, chd free.**
Complementing a C17 Longhouse
and 11 acre smallholding this 1½
acre garden is mainly organic. Trees &
shrubs are underplanted with hundreds
of thousands of snowdrops, crocus,
cyclamen, daffodils, woodland shrubs,
clematis, rambling roses, hydrangeas &
autumn flowering perennials. Extensive
views, shepherd's hut & seats to
enjoy them. Year-round flower interest
with naturalistic plantings. 6 acres of
wildflower meadows, 2 ponds and
stream.

✳ 🪑

5 GLANDWR

Pentrecwrt, Llandysul, SA44 5DA.
Mrs Jo Hicks, 01559 363729,
leehicks@btinternet.com. *15m N of
Carmarthen, 2m S of Llandysul, 7m
E of Newcastle Emlyn. On A486. At
Pentrecwrt village, take minor rd opp*
Black Horse PH. After bridge keep
L for ¼m. Glandwr is on R. **Sat 10,
Sun 11 July (11-5). Adm £3.50,
chd free. Home-made teas. Visits
also by arrangement June to Aug.
Teas on request when booking.**
Delightful 1 acre enclosed mature
garden, with a natural stream.
Includes a rockery, various flower
beds (some colour themed), many
clematis and shrubs. Do spend time
in the woodland with interesting trees,
shade loving plants and ground cover,
surprise paths, and secluded places
to hear the bird song.

🐕 🍵

6 GLANGWILI LODGES

Llanllawddog, Carmarthen,
SA32 7JE. Chris & Christine Blower.
*7m NE of Carmarthen. Take A485
from Carmarthen. ¼m after Gwili
Pottery on L in Pontarsais, turn R for
Llanllawddog and Brechfa. ½m after
Llanllawddog Chapel, rd bears sharply
R, Glangwili Lodges 100yds on R.* **Sat
3, Sun 4 July (11-5). Adm £4.50,
chd free. Light refreshments.**
Our 16 acre estate has a one
acre enclosed walled garden, with
rockeries, water features, flower and
shrub beds, a wisteria arbour, maples,
magnolias and espalier fruit trees.
Areas outside the wall incl one acre of
woodland, with stream, an orchard, 2
acre wildlife area and a hedge tunnel
containing a diverse collection of
trees. A productive vegetable garden,
with polytunnels and a fruit cage.
Wheelchair access outside walled
garden is limited.

♿ 🐕 ✳ 🍵 🪑

> "The support of Hospice UK
> and the National Garden
> Scheme has been invaluable
> to hospice nurses across the
> country whilst we've been
> battling the coronavirus
> crisis, helping hospices such
> as Derian House to continue
> providing vital end of life and
> respite care to 400 children
> and young adults from across
> the North West. Thank you."
> – Katie Turner, Perinatal Nurse
> at Derian House Children's
> Hospice

7 GLYN BACH GARDENS

Efailwen, Pembrokeshire,
SA66 7JP. Peter & Carole Whittaker,
www.glynbachgardens.co.uk.
Efailwen 8m N of Narberth, 15m S
of Cardigan. About 1m N of Efailwen
turn W off A478 at Glandy Cross
garage, follow signs for 1m towards
Llangolman & Pont Hywel Bridge. Glyn
Bach is situated on L before 'road
narrows' sign. **Wed 4 Aug (10-6).**
Adm £5, chd free. Pre-booking
essential, please phone 01994
419104 or email carole_whittaker@
hotmail.com for information &
booking. Home-made teas on
request when booking. Donation to
Plant Heritage.
3 acres of garden with numerous
perennial borders, alpine walls,
tropical beds, large pond, bog
garden, rose garden, grass beds,
cottage garden, polytunnels,
greenhouse and succulent bed,
surrounded by 3 acres of mixed
woodland and grassland. Holders
of a National Collection of Monarda.
Wheelchair access on grass
pathways.

8 NEW THE GRANGE

Manorbier, Tenby, SA70 7TY.
Joan Stace, 01834 871311,
joanstace@btinternet.com. *4m*
W of Tenby. From Tenby, take the
A4239 for Pembroke. ½m after
Lydstep, The Grange is on L, at
the Xrds. Parking limited. **Visits by**
arrangement June to Aug. Teas
on request when booking. Adm
£4, chd free.
5 acre country garden of two parts.
Older established garden around
Grade II listed house (not open)
has colourful herbaceous borders,
rose garden, outdoor chess set and
swimming pool (private). Newer
garden was reclaimed from a heavy
clay bog field which was drained and
now consists of a lake with islands,
ponds and dry river bed, newly-
planted with wide variety of grasses
as a 'prairie' garden.

9 ◆ LAMPHEY WALLED GARDEN

Lamphey, Pembroke, SA71 5PD.
Mr Simon Richards, 07503 976766,
ullapoolsi@hotmail.co.uk. *½m*
from Lamphey village off the
Ridgeway at Lower Lamphey Park.
Turn off A477 at Milton towards
Lamphey for 1m. At T-junction, turn

R to Lamphey (signposted), after 1m
take R up farm track signed Lower
Lamphey Park. Parking behind
cottages. **Sat 26, Sun 27 June, Sat**
4, Sun 5 Sept (10-4). Adm £4, chd
free. Home-made teas. For other
opening times and information,
please phone or email. Donation
to The Bumblebee Conservation
Trust.
Built in late 1700's an acre with
12 ft-high walls. Renovated 2005,
extensive plantings from 2015.
Several hundred different species, a
plantsperson's garden with a natural
feel. Over 40 Salvia species, old
roses, lavender, sunflowers, grasses
and ferns. Espalier apples and pears
plus fan-trained stone fruit. Heritage
vegetable garden. Organic and
wildlife-friendly planting. Grassed
paths slope gently uphill and allow
wheelchair access to the majority of
garden.

10 LLWYNGARREG

Llanfallteg, Whitland, SA34 0XH.
Paul & Liz O'Neill, 01994 240717,
lizpaulfarm@yahoo.co.uk,
www.llwyngarreg.co.uk. *19m*
W of Carmarthen. A40 W from
Carmarthen, turn R at Llandewi
Velfrey, 2½m to Llanfallteg. Go
through village, garden ½m further
on: 2nd farm on R. Disabled car
park in bottom yard on R. **Mon 5**
Apr, Sun 6 June, Mon 30 Aug
(1-6). Adm £5, chd free. Home-
made teas. Refreshments on
fixed Open Days and for pre-
booked groups. Visits also by
arrangement: open most days,
please phone ahead.
Llwyngarreg is always changing,
delighting plant lovers with its many
rarities incl species Primulas, many
huge bamboos with Roscoeas,
Hedychiums and Salvias extending
the season through to riotous autumn
colour. Trees and rhododendrons
are underplanted with perennials.
The exotic sunken garden and gravel
gardens continue to mature. Springs
form a series of linked ponds across
the main garden, providing colourful
bog gardens. Spot the subtle mobiles
and fun constructions hidden around
the 4 acre garden. Wildlife ponds,
fruit and veg, composting, twig piles,
numerous living willow structures,
mobiles, swings, chickens, goldfish.
Partial wheelchair access.

11 NORCHARD

The Ridgeway, Manorbier,
Tenby, SA70 8LD. Ms H
Davies, 07790 040278,
h.norchard@hotmail.co.uk. *4m W*
of Tenby. From Tenby, take A4139 for
Pembroke. ½m after Lydstep, take R
at Xrds. Proceed down lane for ¾m.
Norchard on R. **Sun 30 May (1-5).**
Adm £5, chd free. Check website
for refreshment info. Visits also
by arrangement Apr to July for
groups of 10+.
Historic gardens at medieval
residence. Nestled in tranquil and
sheltered location with ancient oak
woodland backdrop. Strong structure
with formal and informal areas incl
early walled gardens with restored
Elizabethan parterre and potager. 1½
acre orchard with old (many local)
apple varieties. Mill and millpond.
Range of wildlife habitats. Extensive
collections of roses, daffodils and
tulips. Partial wheelchair access.
Access to potager via steps only.

12 THE OLD RECTORY

Lampeter Velfrey, Narberth,
SA67 8UH. Jane & Stephen
Fletcher, 01834 831444,
jane_e_fletcher@hotmail.com.
3m E of Narbeth. 3m E of Narbeth,
next to church in Lampeter Velfrey.
Parking in church car park. **Visits by**
arrangement Feb to Sept. Teas
on request when booking. Adm
£3.50, chd free.
Historic approx 2 acre garden,
sympathetically redesigned and
replanted since 2009 and still
being restored. Many unique trees,
some over 300yrs old, wide variety
of planting and several unique,
architecturally designed buildings.
Formal beds, mature woodland
surrounding an old quarry with
recently planted terraces and
rhododendron bank. Meadow,
fernery and orchard.

The National Garden
Scheme searches the
length and breadth of
England and Wales for the
very best private gardens

Skanda Vale Hospice Garden

13 PENCWM

Hebron, Whitland, SA34 0JP. Lorna Brown, 07967274830, lornambrown@hotmail.com. *10m N of Whitland. From A40 take St Clears exit & head N to Llangynin then on to Blaenwaun. Through village, after speed limit signs take 1st L. Over Xrds, 1¼m then 2nd lane on L marked Pencwm.* **Visits by arrangement Mar to Oct for groups of up to 20. Adm £4, chd free. Teas on request when booking.**

A secluded garden of about one acre set among large native trees, designed for year round interest and for benefit of wildlife. A wide variety of exotic specimen trees and shrubs incl magnolias, rhododendrons, hydrangeas, bamboos and acers. Drifts of bluebells and other spring bulbs and good autumn colour. Incl boggy area and pond with appropriate planting. Wellies recommended at most times.

14 PENTRESITE

Rhydargaeau Road, Carmarthen, SA32 7AJ. Gayle & Ron Mounsey, 01267 253928, gayle.mounsey@gmail.com. *4m N of Carmarthen. Take A485 heading N out of Carmarthen, once out of village of Peniel take 1st R to Horeb & cont for 1m. Turn R at NGS sign, 2nd house down lane.* **Sun 20 June, Sun 5 Sept (11-5). Adm £4, chd free. Home-made teas. Visits also by arrangement Apr to Sept for groups of up to 20.**

Approx. 2 acre garden developed over the last 14 yrs with extensive lawns, colour filled herbaceous and mixed borders, on several levels. A bog garden and magnificent views of the surrounding countryside. There is now a new area planted with trees and herbaceous plants. This garden is south facing and catches the south westerly winds from the sea.

15 PEN-Y-GARN

Foelgastell, Cefneithin, Carmarthen, SA14 7EU. Mary-Ann Nossent & Mike Wood, 07985077022, maryann.nossent@ngs.org.uk, shorturl.at/hsABI. *10m SE Carmarthen. A48 N from Cross Hands take 1st L to Foelgastell R at T junction, 300m sharp L, 300m 1st gateway on L. A48 S Botanic Gdns turning r'about R to Porthythyd 1st L before T junction. 1m 1st R & 300m on L.* **Sat 19, Sun 20 June, Sat 7, Sun 8 Aug (10-5). Adm £4, chd free. Home-made teas. Visits also by arrangement June to Sept for groups of up to 10. Teas on request when booking.**

1⅓ acres unusual setting within a former old limestone quarry, the garden is on several levels with slopes and steps. Sympathetically developed to sit within the landscape, there are 5 distinct areas with a mixture of wild and cultivated plants. A shady area with woodland planting & wild ponds; no dig kitchen garden; terraced borders with shrubs and herbaceous planting; lawns and pond; and a wild garden. Steep in places.

16 PONT TRECYNNY

Garn Gelli Hill, Fishguard, SA65 9SR. Wendy Kinver, 01348873040, wendykinver@icloud.com. *1½m N of Fishguard. Driving up the hill from Fishguard to Dinas turn R ½way up the rd & follow signs. Satnav will take you to a lay-by opp the garden. Parking limited.* **Wed 26 May, Wed 23 June (11-4). Adm £5, chd free. Pre-booking essential, please phone 01348873040 or email wendykinver@icloud.com for information & booking. Visits also by arrangement May to July for groups of 5 to 10.**

A diverse garden of 3½ acres. Meander through the meadow planted with native trees, pass the pond and over a bridge which takes you along a path, through an arboretum, orchard and gravel garden and into the formal garden full of cloud trees, exotic plants and pots,which then leads you to the stream and vegetable garden.

17 NEW SKANDA VALE HOSPICE GARDEN

Saron, Llandysul, Carmarthen, SA44 5DY. Skanda Vale Hospice, 07967245912, brotherfrancis@skandavale.org, www.skandavalehospice.org. *On A484, in village of Saron, 13m N of Carmarthen. Between Carmarthen and Cardigan.* **Sat 29 May, Sat 7 Aug (11-5). Adm £3.50, chd free. Light refreshments.**

A tranquil garden of approx.1acre, built for therapy, relaxation and fun. Maintained by volunteers, the lawns and glades link garden buildings with willow spiral, wildlife pond, colourful borders, sculptures and stained glass. Planting schemes of blue and gold at entrance inspire calm and confidence, leading to brighter red, orange and white. Hospice facilities open for visitors to view. A garden for quiet contemplation, wheelchair friendly, with lots of small interesting features.

Llwyngarreg

Craft stall. For accommodation please visit: www.skanda-hafan.com or phone 01559 384566.

♿ 🐐 ✿ 🛏 ☕

18 STABLE COTTAGE
Rhoslanog Fawr, Mathry, Haverfordwest, SA62 5HG. Mr Michael & Mrs Jane Bayliss, 01348 837712, michaelandjane1954@michaelandjane.plus.com. *Between Fishguard & St David's. Head W on A487 turn R at Square & Compass sign. ½ m, at hairpin take track L. Stable Cottage on L with block paved drive.* **Visits by arrangement May to Aug for groups of up to 20. Teas on request when booking. Adm £3, chd free.** Garden extends to approx ⅓ of an acre. It is divided into several smaller garden types, with a seaside garden, small orchard and wildlife area, scented garden, small vegetable/kitchen garden, and two Japanese areas - a stroll garden and courtyard area.

🐐 ✿ ☕ 🛏 🏠

19 TREFFGARNE HALL
Treffgarne, Haverfordwest, SA62 5PJ. Martin & Jackie Batty, 01437 741115, jmv.batty@gmail.com. *7m N of Haverfordwest, signed off A40. Proceed up through village & follow rd round sharply to L, Hall ¼ m further on L.* **Sun 2 May, Sun 4 July (1-5). Adm £5, chd free. Picnic boxes (with cake and savouries) only on sale. Please reserve by phone or email. Visits also by arrangement.**
Stunning hilltop location with panoramic views: handsome Grade II listed Georgian house (not open) provides formal backdrop to garden of 4 acres with wide lawns and themed beds. A walled garden, with double rill and pergolas, is planted with a multitude of borderline hardy exotics. Also large scale sculptures, summer broadwalk, meadow patch, gravel garden, heather bed and stumpery. Planted for yr-round interest. The planting schemes seek to challenge the boundaries of what can be grown in Pembrokeshire.

🐐 ✿ 🚗 ☕ 🛏

20 NEW TYN Y WAUN
Bethesda Road, Pontyberem, Llanelli, SA15 5LN. Ms Emma Hipkiss. *2½ m SW of Cross Hands. From Cross Hands take the A476 towards Llanelli. In Tumble turn R onto the B4310 for 0.7m. Turn L onto Bethesda Rd and continue for 0.6m. Ty'n Y Waun on R. Limited parking.* **Sun 2, Sun 30 May (1.30-6). Adm £3, chd free.**
¾ acre garden with organic ethos, developed since 2007 in the Gwendraeth Valley with stunning views and borrowed landscape. Specimen trees and shrubs. Largely informal planting with an abundance of wildlife. Recently added more formal raised beds for vegetables, and many spring flowering perennials and bulbs. Wheelchair users please email emmahipkiss@hotmail.com to reserve parking.

♿ ✿

21 TY'R MAES
Ffarmers, Llanwrda, Carmarthenshire, SA19 8JP. John & Helen Brooks, 01558 650541, johnhelen140@gmail.com. *7m SE of Lampeter. 8m NW of Llanwrda. 1½ m N of Pumsaint on A482, opp turn to Ffarmers (please ignore satnav).* **Visits by arrangement Apr to Oct. Adm £5, chd free. Teas on request when booking**
4 acre garden with splendid views. Herbaceous and shrub beds – formal design, exuberantly informal planting, full of cottage garden favourites and many unusual plants. Woodland garden with over 200 types of tree; wildlife and lily ponds; pergola, gazebos, post and rope arcade covered in climbers. Gloriously colourful from early spring till late autumn. Wheelchair note: Some gravel paths.

♿ 🐐 ✿ 🚗 ☕

Ty'r Maes

CEREDIGION

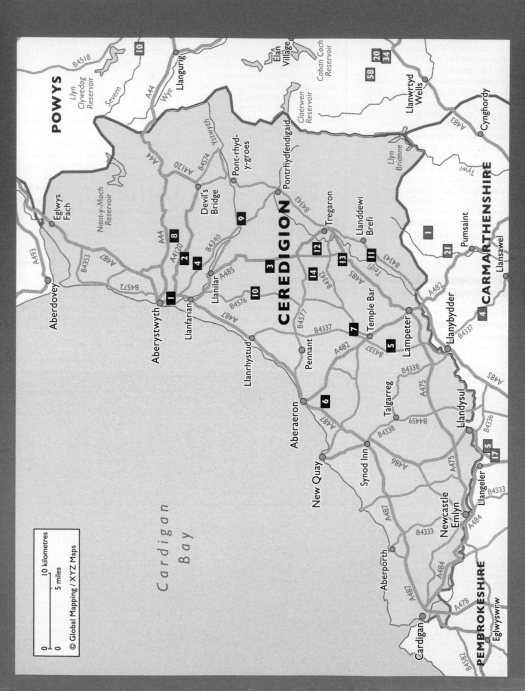

Ceredigion is essentially a rural county, the second most sparsely populated in Wales devoid of any large commercial area.

Much of the land Is elevated, particularly towards the east of the county. There are steep sided wooded valleys, fast flowing rivers and streams, acres of moorland and a dramatic coastline with some lovely sandy beaches. From everywhere in the county there are breath taking views of the Cambrian Mountains and from almost everywhere glimpses of the stunning Cardigan Bay are visible. The gardens in Ceredigion reflect this natural beauty and sit comfortably in the rugged scenery.

As you would expect from this varied landscape, the gardens that open for the National Garden Scheme are equally diverse. Bwlch y Geuffordd is particularly family friendly and will provide children all ages and adults too with hours of interest, adventure and fun. There are a number of dedicated vegetable growers and you can see the fruits of their labours at Trewern Fawr, Yr Efail, Aberystwyth Allotments and Melindwr Valley Bees – which as its name implies, is also a bee farm. Whether your preference is for traditional rolling acres around a grand manor house, cottage gardens, woodland gardens or water gardens you will find something to inspire you, and you can be sure of receiving a warm welcome from the garden owners, and some delicious home-made refreshments to complete a perfect day out.

Volunteers

County Organiser
Pat Causton
01974 272619
pat.causton@ngs.org.uk

County Treasurer
Elaine Grande
01974 261196
e.grande@zoho.com

Booklet Co-ordinator
Shelagh Yeomans
01974 299370
shelagh.yeomans@ngs.org.uk

Assistant County Organisers
Gay Acres
01974 251559
gayacres@aol.com

Brenda & Norman Jones
01974 261737
tobrenorm@gmail.com

 @Ceredigion Gardens

Below: Penybont

OPENING DATES

All entries subject to change. For latest information check **www.ngs.org.uk**

Map locator numbers are shown to the right of each garden name.

All entries are subject to change. For the latest information check ngs.org.uk

By Arrangement

Arrange a personalised garden visit on a date to suit you. See individual garden entries for full details.

Rhos Villa

THE GARDENS

1 ABERYSTWYTH ALLOTMENTS
5th Avenue, Penparcau,
Aberystwyth, SY23 1QT.
Aberystwyth Town Council. *On S
side of R Rheidol on Aberystwyth
by-pass. From N or E, take A4120
between Llanbadarn & Penparcau.
Cross bridge then take 1st R into
Minyddol. Allotments ¼ m on R.* **Sun
18 July (1-5). Adm £3.50, chd free.
Home-made teas.**
There are 37 plots in total on 2 sites
just a few yards from each other. The
allotments are situated in a lovely
setting alongside River Rheidol close
to Aberystwyth. Wide variety of
produce grown, vegetables, soft fruit,
top fruit, flowers, herbs. Car parking
available. For more information
contact Brian Heath 01970 617112.
Sample tastings from allotment
produce. Grass and gravel paths.

2 BRYNGWYN
Capel Seion, Aberystwyth,
SY23 4EE. Mr Terry & Mrs
Sue Reeves, 01970 880760,
sueterr02@btinternet.com. *On
A4120 between the villages of Capel
Seion & Pant y Crug.* **Sat 8, Mon 10
May, Sat 12, Mon 14 June (2-5.30).
Adm £4, chd free. Home-made
teas. Visits also by arrangement
May & June for groups of up to 20.**
Bluebells in May. Traditional wildflower
rich hay meadows managed for
wildlife. As a result of conservation
grazing, hedgerow renovation and tree
planting, habitat has been restored
such that numbers and diversity of
wildflowers and wildlife have increased.
Pond for wildlife. Meander along mown
paths through the meadows. Small
orchard containing Welsh heritage
apples and pears. Wildflower seed
sales. Mown grass paths. Partly
accessible to wheelchair users.

3 BWLCH Y GEUFFORDD
GARDENS
Bronant, Aberystwyth, SY23 4JD.
Mr & Mrs J Acres, 01974 251559,
gayacres@aol.com, bwlch-y-
geuffordd-gardens.myfreesites.net.
*12m SE of Aberystwyth, 6m NW of
Tregaron off A485. Take turning opp
village school in Bronant for 1½ m
then L up ½ m uneven track.* **Visits by
arrangement Apr to Dec. Adm £5,**

chd £2. Tea. Please call beforehand
to book and arrange cake. For a
Mad Hatter's Tea Party, please see
details on website.
1000ft high, 3 acre, constantly
evolving wildlife and water garden. An
adventure garden for children. There
are a number of themed gardens,
incl Mediterranean, cottage garden,
woodland, oriental, memorial and
jungle. Plenty of seating. Unique
garden sculptures and buildings,
incl a cave, temple, gazebo, jungle
hut, tree house and willow den.
Children's adventure garden, musical
instruments, pond dipping, treasure
hunt, beautiful lake, temple and
labyrinth, tree house, sculptures,
wildlife rich, particularly insects and
birds. Paths are gravel, and there are
some steps. There is a shorter route
covering the main features, without
steps.

4 BWLCH Y GEUFFORDD
New Cross, Aberystwyth,
SY23 4LY. Manuel & Elaine
Grande. *5m SE of Aberystwyth. Off
A487, take B4340 to New Cross.
Garden on R at bottom of small dip.
Parking in lay-bys opp house.* **Sun
16 May (10.30-4.30). Adm £4, chd
free. Home-made teas.**
Landscaped hillside 1½ acre garden,
fine views of Cambrian mountains.
Embraces its natural features with
different levels, 3 ponds, mixed borders
merging into carefully managed informal
areas. Banks of rhododendrons,
azaleas, bluebells in spring. Full of
unusual shade & damp-loving plants,
flowering shrubs, mature trees, clematis
& climbing roses scrambling up the
walls of the old stone buildings. Partial
wheelchair access only to lower levels
around house. Some steps and steep
paths further up.

5 ♦ CAE HIR GARDENS
Cribyn, Lampeter, SA48 7NG. Julie
& Stuart Akkermans, 01570 471116,
info@caehirgardens.com,
caehirgardens.com. *5m W of
Lampeter. Take A482 from Lampeter
towards Aberaeron. After 5m turn S on
B4337. Garden on N side of village of
Cribyn.* **For NGS: Sun 18 July (10-5).
Adm £5, chd free. Home-made
teas. For other opening times and
information, please phone, email or
visit garden website.**
A Welsh Garden with a Dutch History,
Cae Hir is a true family garden of
unassuming beauty, made tenable by

its innovative mix of ordinary garden
plants and wildflowers growing in
swathes of perceived abandonment.
At Cae Hir the natural meets the
formal and riotous planting meets
structure and form. A garden not
just for plant lovers, but also for
design enthusiasts. 5 acres of fully
landscaped gardens. Tea room
serving a selection of homemade
cakes and scones and a 'Soup of the
Day'. Limited wheelchair access.

6 NEW FFYNNON LAS
Ffosyffin, Aberaeron, SA46 0HB.
Liz Roberts, 01545 571687,
lizhomerent@hotmail.co.uk. *A
short distance off the A487. 2m S
of Aberaeron. Turn off A487 opp
The Forge Garage in Ffosyffin,
300m up the rd take the L turn at
the T junction.* **Sat 29 May, Sun 13
June (11-5). Adm £5, chd free.
Home-made teas. Visits also
by arrangement May & June for
groups of 5 to 20.**
A 2 acre garden that has been in the
making for over 15 years, Ffynnonlas
is a beautiful area that delivers on
many different aspects of gardening.
There are large lawns, several beds of
mature shrubs and flowers. There is
a small lake and 2 smaller ponds that
are separated by a Monet style bridge
with lilies. A wildflower meadow as a
work in progress that has wild orchids
in spring. Lake with water lillies and
other aquatic plants. Grass paths,
level ground.

7 LLANLLYR
Talsarn, Lampeter, SA48 8QB.
Mrs Loveday Gee, 01570 470900,
lgllanllyr@aol.com. *6m NW of
Lampeter. On B4337 to Llanrhystud.
From Lampeter, entrance to garden
on L just before village of Talsarn.*
**Sun 20 June (2-6). Adm £4.50, chd
free. Home-made teas. Visits also
by arrangement Apr to Oct.**
Large early C19 garden on site of
medieval nunnery, renovated and
replanted since 1989. Large pool,
bog garden, formal water garden,
rose and shrub borders, gravel
gardens, rose arbour, allegorical
labyrinth and mount, all exhibiting fine
plantsmanship. Year-round appeal,
interesting and unusual plants.
Spectacular rose garden planted
with fragrant old fashioned shrub and
climbing roses. Specialist Plant Fair
by Ceredigion Growers Association.

8 NEW **MELINDWR VALLEY BEES, TYNYFFORDD ISAF**
Capel Bangor, Aberystwyth, SY23 3NW. Vicky Lines & Jim Palmer, 01970 880534, vickysweetland@googlemail.com. *Capel Bangor. From Aberystwth take L turn off A44 at E end of Capel Bangor. Follow this rd round to R until you see sign with Melindwr Valley Bees.* **Visits by arrangement July & Aug for groups of up to 15. Suitable footwear advised as we are a working holding. Adm £3.50, chd free. Light refreshments.**
A bee farm dedicated to wildlife with permaculture and forest garden ethos. Fruit, vegetables, culinary and medicinal herbs and bee friendly plants within our wildlife zones. Beehives are situated in dedicated areas and can be observed from a distance. Ornamental and wildflower areas in a cottage garden theme. Formal and wildlife ponds. Grass paths, not suitable for wheelchairs. Vintage tractors and other machinery. Subtropical greenhouse with carnivorous plants, orchids, exotic fruits.

9 **PENYBONT**
Llanafan, Aberystwyth, SY23 4BJ. Norman & Brenda Jones, 01974 261737, tobrenorm@gmail.com. *9m SE of Aberystwyth. Ystwyth Valley B4340 from Aberystwyth. Stay on B4340 for 9m via Trawscoed towards Pontrhydfendigaid. Curve R over stone bridge. ¼m up hill, R by row of cream houses.* **Sun 25 July, Sun 1 Aug (11-5). Adm £4, chd free. Home-made teas. Visits also by arrangement May to Sept. If we are home, we're open.**
A stroll around Penybont in Wild Wales shows what can be achieved from a green hillside in just a few years. This exciting garden (about an acre) designed from scratch, complements the modern building, its forest backdrop and panoramic views. Rhododendrons, azaleas, acers. Fabulous colour all summer, lots of lavender, roses and hydrangeas. Original design, maturing fast. Seats with stunning views over the borrowed landscape, forest, valley, hills. Kites & buzzards. Sloping ground, gravel paths. Partial wheelchair access.

10 **PLAS TREFLYS**
Llangwyryfon, Aberystwyth, SY23 4HD. Mrs Pat Causton, 01974 272619, pat.causton@ngs.org.uk. *11m SE of Aberystwyth on B4576 between Llangwyryfon and Bethania. From Llangwyryfon, continue on B4576 for 1m, take 2nd turning on R after staggered Xrd. Garden at bottom of track (about ⅓m).* **Visits by arrangement Apr & May for groups of up to 20. Adm £3.50, chd free. Home-made teas.**
Tranquil 1 acre garden in rural situation with a variety of habitats incorporating lawns, exuberant flower borders, wildlife pond, shaded stream fringed with irises, marsh marigolds and ferns, bog, dry and herb borders, terraced banks. Woodland areas and productive ornamental kitchen garden. Daffodils, bluebells and tulips in the springtime. Grass and gravel paths, steps and slopes.

11 **RHOS VILLA**
Llanddewi Brefi, Tregaron, SY25 6PA. Andrew & Sam Buchanan, 01570 493787, sam@sabuchanan.plus.com. *From Lampeter take the A485 towards Tregaron. After Llangybi turn R opp junction for Olmarch, The property can be found on the RHS after 1½m.* **Sun 6 June (11-4.30). Adm £3.50, chd free. Home-made teas. Visits also by arrangement Apr to Aug for groups of up to 15.**
A ¾ acre garden creatively utilising local materials. Secret pathways meander through sun and shade, dry and damp. A variety of perennials and shrubs are inter-planted to create interest throughout the year. A productive vegetable and fruit garden with semi formal structure contrasts the looser planting through the rest of the garden.

12 NEW **TREWERN FAWR**
Llangeitho, Tregaron, SY25 6TX. The Richards family. *3 miles NW of Tregaron. From Aberystwyth take A485 to Tyncelyn to join B4578. After 2km turn l at bottom of steep hill beside post box. From Lampeter take A485 to Llanio, turn l onto B4578 for 5km. Turn r beside post box.* **Sun 8 Aug (11-5). Adm £3.50, chd free. Home-made teas.**
Smallholding with 2 large vegetable gardens, polycrub and solar tunnel. Orchard with 20 apple tree varieties, 7 acres of wildflower meadow, large pond. Home of Huw Richards (gardening youtuber and author of 2 gardening books). Innovative and expert vegetable growing.

13 **YR EFAIL**
Llanio Road, Tregaron, SY25 6PU. Mrs Shelagh Yeomans, 01974 299370, shelagh.yeomans@ngs.org.uk. *3m SW of Tregaron. On B4578 between Llanio and Stag's Head.* **Visits by arrangement Feb to Oct for groups of up to 30. Adm £4, chd free. Home-made teas.**
Savour one of the many quiet spaces to sit and reflect amongst the informal gardens of relaxed perennial planting, incl a wildlife pond, shaded areas, bog and gravel gardens. Be inspired by the large productive vegetable plots, three polytunnels and greenhouse. Wander along grass paths through the maturing, mostly native, woodland. Enjoy home-made teas incorporating homegrown fruit and veg. Bilingual quiz sheet for children. Seasonal vegetables and plants for sale. Gravel and grass paths accessible to wheelchairs with pneumatic wheels.

14 **YSGOLDY'R CWRT**
Llangeitho, Tregaron, SY25 6QJ. Mrs Brenda Woodley, 01974 821542. *1½m N of Llangeitho. From Llangeitho, turn L at school signed Penuwch. Garden 1½m on R. From Cross Inn take B4577 past Penuwch Inn, R after brown sculptures in field. Garden ¾m on L.* **Sun 13 June (11-5). Adm £4, chd free. Home-made teas. Visits also by arrangement Apr to Aug.**
One acre hillside garden, with 4 natural ponds which are a magnet for wildlife plus a fish pond. Areas of wildflower meadow, bog, dry and woodland gardens. Established rose walk. Rare trees, large herbaceous beds with ornamental grasses, acer collection, bounded by a mountain stream, with 2 natural cascades and magnificent views. Shade bed with acers and azaleas. Rockery. Large Iris ensata and Iris laevigata collections in a variety of colours. Regional Finalist in The English Garden's competition 'The Nation's Favourite Gardens 2019'. Unsuitable for wheelchairs.

Melindwr Valley Bees, Tynyffordd Isaf

GLAMORGAN

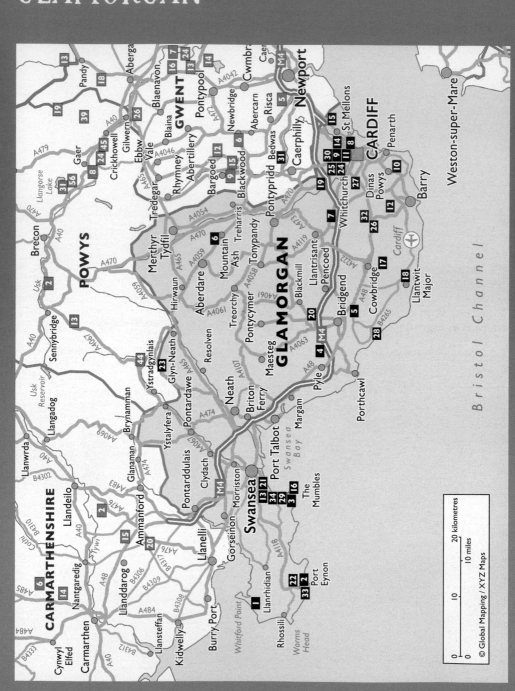

Glamorgan is a large county stretching from the Brecon Beacons in the north to the Bristol Channel in the south, and from the city of Cardiff in the east to the Gower Peninsula in the west. The area has a natural divide where the hills rise from the vale in a clear line of demarcation.

There are gardens opening for the National Garden Scheme throughout the county, and in recent years the number of community openings has greatly increased and have been very successful.

A number of gardens open in villages or suburbs, often within walking distance of each other, providing a very pleasant afternoon for the visitors. Each garden has its own distinct character and the locality is full of hospitality and friendliness.

Gardens range from Mediterranean-style to gardens designed to encourage wildlife. Views from our coastal gardens are truly spectacular.

Our openings start around Easter with a woodland and spring bulbs garden and continue through to mid-September.

So just jump in the car – *Garden Visitors Handbook* book in hand – and head west on the M4. The gardens in Wales are waiting for you!

Below: St Peter's Community Garden

Volunteers

County Organiser
Rosamund Davies
01656 880048
rosamund.davies@ngs.org.uk

County Treasurer
Steven Thomas
01446 772339
steven.thomas@ngs.org.uk

Publicity
Rhian James
rhian.james@ngs.org.uk
07802 438299

Social Media
Rhian Rees
rhian.rees@ngs.org.uk
01446 774817

Booklet Co-ordinator
Lesley Sherwood
02920 890055
lesley.sherwood@ngs.org.uk

Talks Co-ordinator
Frances Bowyer
02920 892264
frances.bowyer@ngs.org.uk

Health and Gardens Co-ordinator
Miranda Workman 02920 766225
miranda.parsons@talktalk.net

Assistant County Organisers
Trevor Humby 02920 512709
trevor.humby@ngs.org.uk

Sol Blytt Jordens 01792 391676
sol.blyttjordens@ngs.org.uk

Tony Leyshon 07896 799378
anthony.leyshon@icloud.com

Ceri Macfarlane
01792 404906
ceri@mikegravenor.plus.com

f @GlamorganNGS
@GlamorganNGS
@ngsglamorgan

OPENING DATES

All entries subject to change. For latest information check www.ngs.org.uk

Map locator numbers are shown to the right of each garden name.

Feburary

Snowdrop Festival

Sunday 14th
Slade   28

Sunday 21st
Slade   28

April

Saturday 10th
Slade   28

Sunday 11th
Slade   28

Saturday 24th
Llandough Castle   17

Sunday 25th
Llandough Castle   17

May

Saturday 8th
9 Willowbrook
   Gardens   34

Sunday 9th
9 Willowbrook
   Gardens   34
105 Heath Park
   Avenue   11

Saturday 15th
Plas y Coed   26

Sunday 16th
Plas y Coed   26

Saturday 29th
17 Maes y Draenog   19

Sunday 30th
17 Maes y Draenog   19

June

Saturday 5th
Sunny Cottage   31

Sunday 6th
6 The Boarlands   2
Llantwit Garden   18
NEW 3 Monksland
   Road   22
Sunny Cottage   31

Sunday 13th
22 Dan-y-Coed Road   9

Saturday 19th
Big House Farm   1
NEW Uplands   32

Sunday 20th
Creigiau Village
   Gardens   7
NEW 28 Slade
   Gardens   29
NEW Uplands   32

Saturday 26th
Corntown Gardens   5
St Peter's Community
   Garden   27

Sunday 27th
Cefn Cribwr Garden
   Club   4
NEW 185 Pantbach
   Road   24
St Peter's Community
   Garden   27

July

Saturday 3rd
38 South Rise   30

Sunday 4th
Hen Felin & Swallow
   Barns   12
38 South Rise   30

Saturday 10th
Dinas Powys   10
110 Heritage Park   15

Sunday 11th
Dinas Powys   10
110 Heritage Park   15

Saturday 17th
Westcliffe   33

Sunday 18th
The Cedars   3
50 Pen y Dre   25

Sunday 25th
Maes-y-Wertha Farm   20

August

Sunday 1st
7 Cressy Road   8
4 Hillcrest   16

September

Saturday 11th
NEW Uplands   32

Sunday 12th
16 Hendy Close   13
NEW Uplands   32

February 2022

Sunday 13th
Slade   28

Sunday 20th
Slade   28

By Arrangement

Arrange a personalised garden visit on a date to suit you. See individual garden entries for full details.

77 Cefn Road, Cefn
   Cribwr Garden Club   4
The Cottage   6
7 Cressy Road   8
22 Dan-y-Coed Road   9
23 Henllys Road   14
110 Heritage Park   15
Maggie's Swansea   21
NEW Nant y Melyn
   Farm   23

Online booking is available for many of our gardens, at ngs.org.uk

Ashgrove

THE GARDENS

1 BIG HOUSE FARM
Llanmadoc, Gower, Swansea, SA3 1DE. Mark & Sheryl Mead.
15m W of Swansea. M4 J47, L A483 for Swansea, 2nd r'about R, A484 Llanelli 3rd r'about L, B4296 Gowerton T-lights, R B4295, pass Bury Green R to Llanmadoc. **Sat 19 June (1-5.30). Adm £5, chd free. Home-made teas.**
Multi awarded inspirational garden of around 1 acre at this lovely listed property combines colour form and texture, described by visitors as 'the best I've seen', 'a real gem'. 'better than Chelsea & Gardeners World, should visit'. Large variety of interesting plants and shrubs, with ambient cottage garden feel, Victorian glasshouse with rose garden potager. Beautiful views over sea and country. Located on the Gower Peninsular, Britain's first designated Area of Outstanding Natural Beauty. Majority of garden accessible to wheelchairs.

2 6 THE BOARLANDS
Porteynon, Swansea, SA3 1NX. Robert & Annette Dyer. *16.8m W of Swansea on Gower Peninsula. Follow A4118 until it starts to descend towards the sea. Take R turn signposted Overton & then L turn over cattlegrid into small estate of bungalows. Proceed to bottom of the estate to No 6.* **Sun 6 June (2-5). Adm £4, chd free. Home-made teas.**
Small plantsmans garden with enchanting views over Port Eynon Bay. Rare and unusual plants and bulbs with many from the Southern Hemisphere, incl proteas,banksias and eucalyptus. Formal pond with wildlife area Decking area overlooking view, small woodland area, many climbers, herbaceous/shrub borders and scree/rockery. Two front gardens with rare bulbs and alpine plants. Several seating areas with panoramic views of Port Eynon Bay. Rough surfaces and a number of steps make wheelchair access difficult.

3 THE CEDARS
20A Slade Road, Newton, Swansea, SA3 4UF. Mr Ian & Mrs Madelene Scott. *Take the A4067 to Oystermouth, turn R at White Rose PH continue along Newton Rd keeping R at fork to T-junction, turn R & 1st R into Slade Rd, follow yellow signs.* **Sun 18 July (2-5). Adm £4, chd free. Home-made teas.**
South facing garden on a sloping site consisting of rooms subdivided by large shrubs and trees. Small kitchen garden with greenhouse. Ornamental pond. Herbaceous perennials, number of fruit trees. It is a garden which affords all year interest with azalea camellias rhododendrons magnolias hydrangeas.

GROUP OPENING

4 CEFN CRIBWR GARDEN CLUB
Cefn Cribwr, Bridgend, CF32 0AP.
www.cefncribwrgardeningclub.com.
5m W of Bridgend on B4281. **Sun 27 June (11-5). Combined adm £5, chd free. Home-made teas.**

6 BEDFORD ROAD
Carole & John Mason.

13 BEDFORD ROAD
Mr John Loveluck.

2 BRYN TERRACE
Alan & Tracy Birch.

CEFN CRIBWR GARDEN CLUB ALLOTMENTS
Cefn Cribwr Garden Club.

CEFN METHODIST CHURCH
Cefn Cribwr Methodist Church.

77 CEFN ROAD
Peter & Veronica Davies & Mr Fai Lee, 07879671326.
Visits also by arrangement May to Sept for groups of up to 10.

25 EAST AVENUE
Mr & Mrs D Colbridge.

6 TAI THORN
Mr Kevin Burnell.

Cefn Cribwr is an old mining village atop a ridge with views of Swansea to the west, Somerset to the south and home to Bedford Park and the Cefn Cribwr Iron Works. The village hall is at the centre with teas, cakes and plants for sale. The allotments are to be found behind the hall. Children, art and relaxation are just some of the themes to be found in the gardens besides the flower beds and vegetables. There are also water features, fish ponds, wildlife ponds, summerhouses and hens adding to the diverse mix in the village. Themed colour borders, roses, greenhouses, recycling, composting and much more. The chapel grounds are peaceful with a meandering woodland trail.

GROUP OPENING

5 CORNTOWN GARDENS
Corntown, Bridgend, CF35 5BB.
Take B4265 from Bridgend to Ewenny. Take L in Ewenny on B4525 to Corntown follow yellow NGS signs. From A48 take B4525 to Corntown. **Sat 26 June (11-4). Combined adm £4, chd free. Home-made teas at Y Bwythyn.**

RHOS GELER
Bob Priddle & Marie D Robson.

Y BWYTHYN
Mrs Joyce Pegg.

Y Bwythyn has had over 30yrs of hard labour, some guesswork and considerable good luck resulting in a delightful garden. The area at the front of the house is a mixture of hot colour combinations whilst at the rear of this modest sized garden the themes are of a more traditional cottage garden style which incl colour themed borders as well as soft fruit, herbs and vegetables. Rhos Geler's garden has a lavender hedge at the front and subjects to attract butterflies. The main garden area at the back of the house is a long narrow garden that is in a series of themed areas. These incl an herbaceous border, shade loving plants, an Elizabethan style knot garden and a Japanese influenced area. Containers hold a range of subjects incl a collection of sempervivums.

"The annual donation from the National Garden Scheme to Perennial is the cornerstone of our fundraising activities and encourages many of our donors." Perennial

6 THE COTTAGE

Cwmpennar, Mountain Ash, CF45 4DB. Helen & Hugh Jones, 01443 472784, hhjones1966@yahoo.co.uk. *18m N of Cardiff. A470 from N or S. Then follow B4059 to Mountain Ash. Follow signs for Cefnpennar then turn R before bus shelter into village of Cwmpennar & to top of lane at end of Middle Row.* **Visits by arrangement May to July for groups of up to 20. Limited parking. Adm £4, chd free. Home-made teas.**

50 years of amateur muddling have produced this 4 acre garden, which, while scarcely a designer's dream, is at least well stocked. An area of natural woodland adjoins a colourful camellia, rhododendron and azalea shrubbery, whilst the main garden contains shrub and herbaceous borders, some majestic trees and a rose garden with some 80 varieties. Since the garden slopes from N/E-S/W unfortunately only the central part is suitable for wheelchairs.

GROUP OPENING

7 CREIGIAU VILLAGE GARDENS

Maes Y Nant, Creigiau, CF15 9EJ. *W of Cardiff (J34 M4). From M4 J34 follow A4119 to T-lights, turn R by Castell Mynach Pub, pass through Groes Faen & turn L to Creigiau. Follow NGS signs.* **Sun 20 June (11-5). Combined adm £5, chd free. Home-made teas at 28 Maes y Nant.**

28 MAES Y NANT
Mike & Lesley Sherwood.

WAUNWYLLT
John Hughes & Richard Shaw.

On the NW side of Cardiff and with easy access from the M4 J34, Creigiau Village Gardens incl two vibrant and innovative gardens. Each quite different, they combine some of the best characteristics of design and planting for modern town gardens as well as cottage gardens. Each has its own forte; Waunwyllt has incorporated next door's garden which is an ongoing development. 28 Maes y Nant is surrounded on three sides by cottage style planting, with a large area to the side of the house surrounded by native hedging which incl a small uncultivated wildlife patch

and an area of informal mixed annual planting. Anyone looking for ideas for a garden in an urban setting will not go away disappointed; enjoy a warm welcome, plant sales available.

8 7 CRESSY ROAD

Penylan, Cardiff, CF23 5BE. Victoria Thornton, 02920 311215, penylanhillbilys@hotmail.com. *Approaching from Wellfield Rd, turn R at the T-lights into Marlborough Rd & Cressy Rd is the 3rd turning on the L. From Newport Rd follow Marlborough Rd to Thomas Court, turn L.* **Sun 1 Aug (2-6). Adm £3.50, chd free. Home-made teas. Visits also by arrangement July & Aug for groups of 5 to 10.**

Enjoy a warm welcome at No 7. Small subtropical garden, creating an illusion of a much larger space. Lush tropical planting, alive with colour and texture. With an emphasis on wildlife, nectar rich planting, barrel pond, attracting frogs, dragonflies and damselflies. Bug hotel. Ever changing,new plantings to be seen every year. Silver medal, Cardiff in Bloom 2014.

9 22 DAN-Y-COED ROAD

Cyncoed, Cardiff, CF23 6NA. Alan & Miranda Workman, 029 2076 6225, miranda.parsons@talktalk.net. *Dan y Coed Rd leads off Cyncoed Rd at the top and Rhydypenau Rd at the bottom. No 22 is at the bottom of Dan y Coed Rd. There is street parking and level access to the R of the property.* **Adm £4, chd free. Home-made teas. Visits also by arrangement Apr to Sept for groups of 5 to 20.**

A medium sized, much loved garden. Owners share a passion for plants and structure, each year the lawn gets smaller to allow for the acquisition of new plants and features. Hostas, ferns, Acers and other trees form the central woodland theme as a backdrop is provided by the Nant Fawr woods. Year round interest has been created for the owner's and visitor's greater pleasure. There is a wildlife pond, many climbing plants and a greenhouse with a cactus and succulent collection. There is wheelchair access to the garden patio area only.

GROUP OPENING

10 DINAS POWYS

Dinas Powys, CF64 4TL. *Approx 6m SW of Cardiff. Exit M4 at J33, follow A4232 to Leckwith, onto B4267 & follow to Merry Harrier T-lights. Turn R & enter Dinas Powys. Follow yellow NGS signs.* **Sat 10, Sun 11 July (11-5). Combined adm £5, chd free. Home-made teas at Ashgrove & the Community Gardens, incl gluten free and vegan cakes. Donation to Dinas Powys Voluntary Concern and Dinas Powys Community Library.**

1 ASHGROVE
Sara Bentley.
Open on all dates

BROOKLEIGH
Melanie Syme.
Open on all dates

32 LONGMEADOW DRIVE
Julie & Nigel Barnes.
Open on Sun 11 July

30 MILLBROOK ROAD
Mr & Mrs R Golding.
Open on all dates

32 MILLBROOK ROAD
Mrs G Marsh.
Open on all dates

NIGHTINGALE COMMUNITY GARDENS
Keith Hatton.
Open on all dates

There are six gardens to visit in this small friendly village, all with something different to offer, plus the village church. The garden at 1 Ashgrove, has a large variety of different plants and trees in sweeping beds, with many hidden nooks and places to sit. The community garden features displays of vegetable and fruit. 30 Millbrook Road, a large and beautiful garden, established over 25 years, with lovely pond and waterfall. 32 Millbrook Road is a lovely family garden with interesting touches. 32 Longmeadow is full of gorgeous and colourful plants with an unusual water feature. Brookleigh an interesting garden with a good variety of perennial plants and trees. Our village church, St Peters, is also opening with displays of flowers, classic cars and other attractions. Home-made teas will be served, incl vegan and gluten free options, and there will be plants for sale. There are many restful and beautiful areas to sit and relax. Good wheelchair access at the community gardens. Partial access elsewhere.

105 HEATH PARK AVENUE
Heath, Cardiff, CF14 3RG.
Mr & Mrs Gambles,
beverly_gambles@hotmail.com.
Approx 1m from University Hospital. From Gabalfa R'about into Birchgrove turn R into Heathwood Rd. At top of Rd turn R at T-lights into Heath Park Ave. Follow NGS signs. Sun 9 May (2-6). Adm £4, chd free. Home-made teas.
A South facing family garden that is constantly evolving. Perennial borders and plenty of pots surrounding a central lawn give lots of colour throughout the year. A small vegetable plot and greenhouse to grow a variety of plants and vegetables. Small fishpond with lots of frogs and newts.

GROUP OPENING

HEN FELIN & SWALLOW BARNS
Vale of Glamorgan, Dyffryn, CF5 6SU. 02920 593082,
rozanne.lord@blueyonder.co.uk.
3m from Culverhouse. Cross r'about. Sun 4 July (10-5). Combined adm £5, chd free. Home-made teas.
Dyffryn hamlet is a hidden gem in the Vale of Glamorgan; despite the lack of a PH there is a fantastic community spirit. Yr Hen Felin: Beautiful cottage garden with stunning borders, breathtaking wildflower meadows, oak tree with surrounding bench, 200yr old pig sty, wishing well, secret garden with steps to river, lovingly tended vegetable garden and chickens. Mill stream running through garden with adjacent wildflowers. Swallow Barns: Cross the bridge over the river and pass under the weeping ash to enter Swallow Barns garden, a mature garden with packed herbaceous borders - formal and informal, orchard, hens, herb garden, lavender patio, willow arches leading to woodland walk and deep secluded pond. We welcome all visitors - old and new. Wheelchair access possible to most of the two gardens. Unprotected river access, children must be supervised in both gardens. Home-made jams, Pimm's at the pond, cheese and wine, artists' stall, face painting, white elephant stall, jewellery stall and plant stall.

77 Cefn Road

13 16 HENDY CLOSE

Derwen Fawr, Swansea, SA2 8BB.
Peter & Wendy Robinson. *Approx 3m W of Swansea. A4067 Mumbles Rd follow sign for Singleton Hospital. Then R onto Sketty Lane at mini r'about, turn L then 2nd R onto Saunders Way. Follow yellow NGS signs. Please park on Saunders Way if possible.* **Sun 12 Sept (2-5.30). Adm £6, chd free. Light refreshments.**
Originally the garden was covered with 40ft conifers. Cottage style, some unusual and mainly perennial plants which provide colour in spring, summer and autumn. Hopefully the garden is an example of how to plan for all seasons. Visitors say it is like a secret garden because there are a number of hidden places. Plants to encourage all types of wildlife in to the garden.

14 23 HENLLYS ROAD

Cyncoed, Cardiff, CF23 6NL.
**Mrs Jill Evans, 07870654205,
jilltreenut@uwclub.net.** *Turn off Cyncoed Rd, in Cyncoed Village down Bettws y Coed Rd. Take 2nd L onto Llangorse Rd and 2nd R into Henllys Rd. Garden is halfway down on the R.* **Visits by arrangement May to Sept for groups of 5 to 10. Adm £4, chd free. Home-made teas. Open by arrangement to Health groups/local charities.**
Urban garden with cottage style planting, with some unusual shrubs, perennials, and bulbs, which try to provide all year round colour. The garden has various seating areas with different aspects incl a summerhouse. There are two lawns, a pergola with wisteria, roses and clematis, with garden sculptures and water features. In the Spring the many containers are filled with dozens of varieties of tulips, which are followed by Agapanthus and Dahlias. Autumn colour is provided by late flowering perennials and shrubs incl Acers, Viburnams, and Cercis. The garden is constantly evolving with new plants and ideas. There are some steps and gravel paths but most of the garden can be viewed from the terrace which is wheelchair friendly. Garden entered via side gate on tarmac drive. Paved ramp down to terrace. Elsewhere there are gravel paths.

15 110 HERITAGE PARK

St Mellons, Cardiff, CF3 0DS.
**Sarah Boorman, 07969499967,
sarahbooam@me.com.** *Leave A48 at St Mellons junction, take 2nd exit at r'about. Turn R to Willowdene Way & R to Willowbrook Drive. Heritage Park is 1st R. Park outside the cul-de-sac.* **Sat 10, Sun 11 July (1-5). Adm £4, chd free. Home-made teas. Gluten free options available. Visits also by arrangement May to Sept for groups of 5 to 10. Suggested £5 donation adm. Home-made tea and cake incl.**
An unexpected gem within a modern housing estate. Opening in July instead of May. Evergreen shrubs, herbaceous borders, box topiary and terracotta pots make for a mix between cottage garden and Italian style. With numerous seating areas and a few quirky surprises this small garden is described by neighbours as a calm oasis. The garden would be able to be viewed from the patio area.

16 4 HILLCREST

Langland, Swansea, SA3 4PW.
Mr Gareth & Mrs Penny Cross. *Approach from L, turn at Langland corner on Langland Rd Mumbles; take first L. off Higher Lane; take 3rd L. up Worcester Rd and bungalow faces you at top.* **Sun 1 Aug (1-5). Adm £5, chd free.**
Redesigned and developed over the last 7 years, our hilltop garden backs onto woodland overlooking Swansea bay and attracts many birds. We have planted large herbaceous borders with mainly perennial plants/shrubs to provide interest and colour through the seasons. A wildlife pond and fruit/veg garden add to interest.

17 LLANDOUGH CASTLE

Llandough, Cowbridge, CF71 7LR.
Mrs Rhian Rees. *1½m outside Cowbridge. At the T- lights in Cowbridge turn onto the St Athan Rd. Continue then turn R to Llandough. Drive into the village follow the car park signs. Walk up a short lane and into the gardens.* **Sat 24, Sun 25 Apr (10.30-4). Adm £5, chd free. Light refreshments.**
Set within castle grounds and with a backdrop of an ancient monument, the 3½ acres of garden incl a potager with a hint of the Mediterranean, formal lawns and herbaceous beds, a wildlife pond with waterfall and a woodland garden with stumpery and sculpture. Over 60,000 Spring bulbs have been planted incl snowdrops, narcissi and tulips. Potager, wildlife pond, formal borders, stumpery.

Although some of the gardens are flat, areas like the woodland and gravel drive may be difficult for wheelchair access.

18 LLANTWIT GARDEN

21 Monmouth Way, Boverton, Llantwit Major, CF61 2GT. **Don & Ann Knight.** *At Llanmaes Rd, T-lights turn onto Eagleswell Rd, next L into Monmouth Way, garden ½ way down on R.* **Sun 6 June (11-5). Adm £4, chd free. Home-made teas.**
This is a Japanese style garden with a Zen gate, Torri gate and Japanese lanterns featuring a large collection of Japanese style trees, a pagoda and 3 water features which incl the great Amazon waterfall along with large Buddha's head and Koi pond. Oriental themed garden with a collection of Bonsai and Koi.

19 17 MAES Y DRAENOG

Maes Y Draenog, Tongwynlais, Cardiff, CF15 7JL. **Mr Derek Price.** *N of M4. From S: M4 , J32, take A4054 into village. R at Lewis Arms pub, up Mill Rd. 2nd R into Catherine Drive, park in signed area (no parking in Maes y draenog). Follow signs to 17 Maes y draenog.* **Sat 29, Sun 30 May (12.30-5.30). Adm £4. Light refreshments.**
A hidden gem of a garden, in the shadow of Castell Coch, fed by a mountain stream having a footbridge to naturalised areas and small woodland area. Set against a woodland backdrop. Developed over 12 yrs with a good variety of plants, lavender beds and two main herbaceous borders, with plants for spring and summer displays. Garden structures, summer house, patios, greenhouse and small veg area. Natural mountain stream with wooden footbridge, Summer house, greenhouse and a good variety of plants in different borders around house. Rear of house is set against woodland and fields. Not suitable for young children and partial access for wheelchair users.

20 MAES-Y-WERTHA FARM

Bryncethin, CF32 9YJ. **Stella & Tony Leyshon.** *3m N of Bridgend. Follow sign for Bryncethin, turn R at Masons Arms. Follow sign for Heol-y-Cyw garden about 1m outside Bryncethin on R.* **Sun 25 July (12-7). Adm £5, chd free. Home-made teas.**

A 3 acre hidden gem outside Bridgend. Entering the garden you find a small Japanese garden fed by a stream, this leads you to informal mixed beds & enclosed herbaceous borders. Ponds & rill are fed by a natural spring. A meadow with large lawns under new planting gives wonderful vistas over surrounding countryside. Mural in the summerhouse by contemporary artist Daniel Llewelyn Hall. His work is represented in the Royal Collection and House of Lords. Fresh hand made sandwiches available.

21 MAGGIE'S SWANSEA

Singleton Hospital, Sketty Lane, Sketty, Swansea, SA2 8QL. Miss Leanne Jennett, www.maggies.org/swansea. *On the grounds of Singleton Hospital. They are next to the Genetic Building & close to the chemotherapy day unit at the back of the main hospital. Please follow orange signs.* **Visits by arrangement Mar to Oct for groups of up to 10.**
Maggie's Swansea's gardens wrap around the building, and overlook into Swansea Bay. The garden, designed by Kim Wilkie, attracts wildlife, heightening the natural and tranquil feel, and there is also a fully functional allotment. The Centre sits among a small wooded area, and the wings of the design also help to shelter the outside seating areas, meaning that visitors can enjoy sitting out for as much of the year as possible. Decking all the way around the building.

22 NEW 3 MONKSLAND ROAD

Scurlage, Reynoldston, Swansea, SA3 1AY. Mrs Vhairi Cotter, vhairi.cotter@btinternet.com. *From Swansea follow A4118 signed to Port Eynon. Continue until turning R onto B4247 signed to Rhossili. At Medical Centre on the L, turn into Monksland Rd, continue to 3rd house along on R.* **Sun 6 June (1.30-6). Adm £5, chd free. Home-made teas.**
This small garden, formerly mostly grass, is just 6 years old and still evolving. However it has surprisingly mature and diverse planting that, within seven distinct areas, features everything from coastal varieties and grasses to roses, perennials and shrubs. Water features encourage wildlife, while plants and shrubs attract bees and butterflies as well as

providing year-round interest. Pergola, several seating areas. An ornamental bee hunt for children.

23 NEW NANT Y MELYN FARM

Seven Sisters, Neath, SA10 9BW. Mr Craig Pearce, cgpearce@hotmail.co.uk. *From J43 on M4 take A465 towards Neath. Exit for Seven Sisters at r'about take 3rd exit, 6m for Seven Sisters you come to Pantyffordd sign, turn L under low bridge.* **Visits by arrangement June to Sept for groups of up to 20. Adm £3.50, chd free. Light refreshments.**
A spacious interesting garden with many features which include a stunning natural waterfall, a meandering woodland stream covered by a canopy of entwined trees, a picturesque Japanese garden, beautiful lawned areas with winding pathways.

185 Pantbach Road

Waunwyllt

24 NEW 185 PANTBACH ROAD
Rhiwbina, Cardiff, CF14 6AD. Kate & Glynn Canning. *185 Pantbach Road. Situated 0.5m from Heol-y -Deri in the heart of Rhiwbina.* Sun 27 June (12-5). Adm £4, chd free. Cream teas. A selection of refreshments available including home-made scones and cakes. An Urban Retreat...Quite literally a brand new garden in the heart of Cardiff. The owners, Kate and Glynn always dreamt of a large outdoor space with a central winding path and numerous seating areas. Although in its infancy, the garden features six seating areas, a rose garden and a central water feature.. The winding path is framed by standard roses and interesting shaped borders and lawns.
👨‍🦽 ☕

25 50 PEN Y DRE
Rhiwbina, Cardiff, CF14 6EQ. Ann Franklin. *N Cardiff. M4 J32, A470 to Cardiff, 1st L to mini r'about, turn R. At T-lights in village, turn R to Pen-y-Dre.* Sun 18 July (11-4.30). Adm £3, chd free. Home-made teas. Situated in the heart of the conservation area of Rhiwbina Garden Village, this north facing garden features deep herbaceous borders with a variety of cottage garden style

plants and shrubs, some rare and unusual. Small pond for wildlife and veg plot. It provides a peaceful haven with year round interest. Home-made refreshments are available in the conservatory and plants for sale by the veg plot.
✿ ☕

26 PLAS Y COED
Bonvilston, CF5 6TR. Hugh & Gwenda Child. *Plas y Coed is just off the A48 at Bonvilston. Find the Church & you will find us. Parking will be clearly marked at the Reading Rooms.* Sat 15, Sun 16 May (10-6). Adm £5, chd free.
The garden at Plas y Coed has become more than 'just a garden' in the fifteen years we have lived here. The field that extends beyond the pond and folly is now an arboretum with close on fifty trees - trees chosen for their form, bark and colour. The garden is an interesting marriage of architecture, artwork and landscape. Partial wheelchair access.
☕

27 ST PETER'S COMMUNITY GARDEN
St Fagans Road, Fairwater, Cardiff, CF5 3DW. St Peter's Church, www. stpeterscommunitygarden.org.

uk. *On the St Fagans Rd opp Gorse Place. Next Door to Church. A48 to Culverhousecross r'bout take A48 Cowbridge Rd West to Ely r'bout 1st L. At T-lights go L B4488 to Fairwater Green, follow yellow NGS signs.* Sat 26, Sun 27 June (10-5). Adm £5, chd free. Home-made teas. Secret garden in city suburb. Unusual combination of flower beds, raised vegetable beds and nature reserve, all created by volunteers. Features incl 2 large natural ponds surrounded by wild plants, Welsh heritage apple trees, long herb border and wildflower meadow, Quiet Garden with a zen feel - water features and monoliths and 2 beehives. Book browsing area, Honey raffle. All day refreshments, mostly home-made. Disabled WC available. Most of the Garden wheelchair access friendly.
👨‍🦽 ✿ 🚗 ☕

28 SLADE
Southerndown, CF32 0RP. Rosamund & Peter Davies, 01656 880048, rosamund.davies@ngs.org.uk, www.sladeholidaycottages.co.uk. *5m S of Bridgend. M4 J35 Follow A473 to Bridgend. Take B4265 to St. Brides Major. Turn R in St. Brides Major for Southerndown, then follow yellow NGS signs.* Sun 14, Sun 21

Feb (2-4.30), Sat 10, Sun 11 Apr (2-5). Adm £5, chd free. Home-made teas. 2022: Sun 13, Sun 20 Feb.

Hidden away Slade garden is an unexpected jewel to discover next to the sea with views overlooking the Bristol Channel. The garden tumbles down a valley protected by a belt of woodland. In front of the house are delightful formal areas a rose and clematis pergola and herbaceous borders. From terraced lawns great sweeps of grass stretch down the hill enlivened by spring bulbs and fritillaries. Heritage Coast wardens will give guided tours of adjacent Dunraven Gardens with slide shows every hour from 2pm (Apr opening only). Partial wheelchair access.

&. ❀ 🚗 🛏 ☕ 🎋 🪑

29 NEW 28 SLADE GARDENS
West Cross, Swansea, SA3 5QP. Peter & Helen Shoterline. *At the end of Slade Gardens facing Oystermouth Cemetery. Wheelchair drop off only in Slade Gdns. Parking in cemetery carpark at top of lane from Newton Rd indicated with white sign. Approx 150m level walk on tarmac. Additional parking on Bellevue Rd. Sun 20 June* (1-5.30). Adm £5, chd free. Home-made teas.

This is a recent (2 years) renovation of the old garden of the original Lodge to Oystermouth Cemetery. It is tucked under the steep limestone woodland at the eastern end of the cemetery valley and faces south overlooking the Victorian part of the cemetery. It is planted in cottage garden style with perennials and selected shrubs on the side of the hill separated from a lawn by a limestone wall. The garden can be enjoyed from the terrace of the house which is reached without climbing steps and where tea and cakes can be enjoyed. The view over the garden and 'hidden Valley' of the cemetery is a lovely surprise even for those who know Mumbles well. The scented roses should be in full bloom.

☕

30 38 SOUTH RISE
Lanishen, Cardiff, CF14 0RH. Dr Khalida Hasan. *N of Cardiff, from Llanishen Village Station Rd past Train Stn go R down The Rise or further down onto S Rise directly. Following yellow signs.* Sat 3, Sun 4 July (11-5). Adm £5, chd free. Home-made teas. South Asian savouries (e.g. samosa, chick pea chaat, pakora) and ice cream available.

A relatively new garden backing on to Llanishen Reservoir gradually establishing with something of interest and colour all yr round. Herbaceous borders, vegetables and fruit plants surround central lawn. Wildlife friendly; variety of climbers and exotics. In front shrubs and herbaceous borders to a lawn. Stepping stones leading to children's play area and vegetable plot also at the back. Variety of home made cakes and Asian savouries such as samosas, yoghurt and chick peas chaat. Wheelchair access to rear from the side of the house.

&. ❀ 🚗 ☕

31 SUNNY COTTAGE
Mountain Road, Bedwas, Caerphilly, CF83 8ES. Mr Paul & Mrs Carol Edwards. *Head E from Caerphilly to Bedwas on A468. Turn L at T-lights into village onto Church St follow rd up to St Barrwgs Church. Turn R after bridge then 1st l onto Mountain Rd. Follow NGS signs.* Sat 5, Sun 6 June (11-4). Adm £3.50, chd free. Home-made teas.

Terraced garden surrounded by wildlife friendly hedges. Hidden areas on the terraces are linked by paths and arches. Each level is very well stocked with a multitude of flowers, shrubs & trees. Something of interest for every season. 2 lawn areas provide a welcome splash of green. One is shaded by 2 mature apple trees. Seating areas to relax and enjoy the vistas can be found on most levels.

❀ ☕

32 NEW UPLANDS
Peterston-Super-Ely, Cardiff, CF5 6LG. David Richmond. *From the A48 between St Nicholas & Bonvilston take the Peterston Super Ely turning & follow yellow NGS arrow, park at small green near Gwern-y-Steeple sign.* Sat 19, Sun 20 June, Sat 11, Sun 12 Sept (12-5). Adm £4, chd free.

Uplands has a rustic heart within a cottage garden design. The front is vegetables and currant bushes. The back was planted in 2019, on once all grass. There are four main herbaceous borders, young fruit trees, ornamental grasses, ferns, shrubs and roses, with many places to relax and enjoy the garden. Medium sized greenhouse with citrus and fig tree. Several sculptures located around the garden.

33 WESTCLIFFE
Overton Lane, Porteynon, Swansea, SA3 1NR. David Carlsen-Browne. *Westcliffe House. Make for Porteynon. When nearly at Porteynon ½ way down hill turnoff on R, signed Overton, drive to the end of lane, Westcliffe House on your L. Last house in the lane.* Sat 17 July (1.30-5.30). Adm £5, chd free. Light refreshments. Pimms available.

Westcliffe House has a garden of interest, diversity and surprise. It has been created despite salt laden winds at its cliff top situation. There is a Mediterranean atmosphere inspired by the gardener's admiration for Italy. The garden is intimate with division into several areas incl two water features. Each turn in your walk brings a surprise.

☕

34 9 WILLOWBROOK GARDENS
Mayals, Swansea, SA3 5EB. Gislinde Macpherson. *Nr Clyne Gardens. Go along Mumbles Rd to Blackpill. Turn R at Texaco garage up Mails Rd. 1st R along top of Clyne Park, at mini r'about into Westport Ave. 1st L into Willowbrook gardens.* Sat 8, Sun 9 May (1-5). chd free.

Informal ½ acre mature garden on acid soil, designed to give natural effect with balance of form and colour between various areas linked by lawns; unusual trees suited to small suburban garden, especially conifers and maples; rock and water garden. Sculptures, ponds and waterfall.

 ❀

"I love the National Garden Scheme which has been the most brilliant supporter of Queen's Nurses like me. It was founded by the Queen's Nursing Institute which makes me very proud. As we battle Coronavirus on the front line in the community, knowing we have their support is a real comfort." – Liz Alderton, Queen's Nurse

GWENT

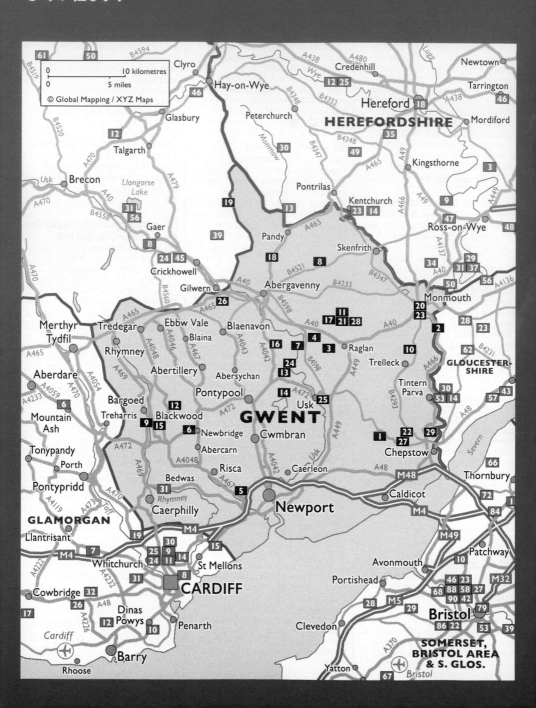

© Global Mapping / XYZ Maps

0 10 kilometres
0 5 miles

HEREFORDSHIRE

GLOUCESTER-SHIRE

GWENT

GLAMORGAN

SOMERSET, BRISTOL AREA & S. GLOS.

Newtown
Tarrington
Mordiford
Credenhill
Hereford
Kingsthorne
Clyro
Hay-on-Wye
Peterchurch
Ross-on-Wye
Glasbury
Pontrilas
Kentchurch
Talgarth
Skenfrith
Brecon
Llangorse Lake
Gaer
Pandy
Monmouth
Crickhowell
Gilwern
Abergavenny
Merthyr Tydfil
Tredegar
Ebbw Vale
Blaenavon
Raglan
Trelleck
Rhymney
Blaina
Aberdare
Abertillery
Abersychan
Tintern Parva
Mountain Ash
Pontypool
Usk
Chepstow
Tonypandy
Treharris
Blackwood
Porth
Newbridge
Cwmbran
Thornbury
Pontypridd
Abercarn
Caerleon
Risca
Bedwas
Caldicot
Caerphilly
Newport
Llantrisant
Whitchurch
St Mellons
Avonmouth
Patchway
Cowbridge
Portishead
CARDIFF
Dinas Powys
Penarth
Bristol
Cardiff
Barry
Clevedon
Rhoose
Yatton

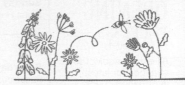

Welcome to Gwent! It is a county of valleys and hills, of castles and farms, of country lanes yet very accessible main roads, easily reached from Cardiff, Bristol or Hereford.

Some of our gardens are clustered around the delightful market towns of Monmouth and Abergavenny, many with breathtaking panoramic views. The historic small town of Usk opens around 20 varied gardens over a June weekend. Our Gwent gardens range from small jewels of town gardens to gracious estates, from manicured lawns to hillside gardens that blend into the landscape. Look out for some lovely gardens opening for the first time - marked clearly in the book as New. For a few of the gardens (Highfield Farm, Nant y Bedd, Trengrove House, Woodlands Farm and Wyndcliffe) pre-booking via the NGS website is required.

A very warm welcome to our gardens and thank you for your support.

Left: High House

Volunteers

County Organiser
Cathy Davies
01291 672625 / 07976 633743
cathydavies127@gmail.com

County Treasurer
Ian Mabberley
01873 890219
ian.mabberley@ngs.org.uk

Publicity
Cathy Davies
(as above)

Social Media
Roger Lloyd 01873 880030
droger.lloyd@btinternet.com

Assistant County Organiser
Cherry Taylor
07803 853681
cherry.taylor@dynamicmarkets.co.uk

Sue Torkington
01873 890045
sue.torkington@ngs.org.uk

Veronica Ruth
07967 157806 / 01873 859757
veronica.ruth@ngs.org.uk

Jenny Lloyd
01873 880030 / 07850 949209
jenny.lloyd@ngs.org.uk

Booklet Co-Ordinator
Veronica Ruth (as above)

f @gwentngs
🐦 @GwentNGS
📷 @gwentngs

OPENING DATES

All entries subject to change. For latest information check **www.ngs.org.uk**

Map locator numbers are shown to the right of each garden name.

March

Sunday 28th
Llanover 16

April

Every Monday and Sunday from Monday 5th
Woodlands Farm 28

Saturday 24th
Glebe House 7

Sunday 25th
Glebe House 7

May

Saturday 1st
Park House 22
Woodhaven 27

Sunday 2nd
High Glanau Manor 10
High House 11
The Old Vicarage 21

Sunday 9th
The Alma 1

Monday 10th
The Alma 1

Sunday 23rd
Wenallt Isaf 26

Saturday 29th
Hillcrest 15
Rockfield Park 23

Sunday 30th
Croesllanfro Farm 5
Hillcrest 15

◆ Nant y Bedd 19
Rockfield Park 23

Monday 31st
Hillcrest 15
◆ Nant y Bedd 19

June

Every Thursday and Sunday from Thursday 17th
Trengrove House 24

Friday 4th
◆ Wyndcliffe Court 29

Sunday 6th
NEW Cwmdows Farm 6
Highfield Farm 13
NEW North Parade House 20

Saturday 12th
Longhouse Farm 17

Sunday 13th
Longhouse Farm 17

Sunday 20th
Croesllanfro Farm 5
NEW Glen Trothy 8
Mione 18

Friday 25th
Mione 18

Saturday 26th
Usk Open Gardens 25

Sunday 27th
Usk Open Gardens 25

July

Every Sunday
Hillcrest 15

Every Thursday and Sunday to Sunday 4th
Trengrove House 24

Friday 2nd
Mione 18

Saturday 3rd
Hill House 14

Sunday 4th
Birch Tree Well 2
Hill House 14
Mione 18

Saturday 10th
14 Gwerthonor Lane 9

Sunday 11th
14 Gwerthonor Lane 9

Sunday 18th
Clytha Park 4
Highfield Farm 13

Sunday 25th
Croesllanfro Farm 5

August

Every Sunday
Hillcrest 15

Saturday 7th
32/33 High Street 12

Sunday 8th
32/33 High Street 12

Sunday 15th
NEW Cwmdows Farm 6
Highfield Farm 13

Sunday 22nd
Croesllanfro Farm 5

Sunday 29th
Wenallt Isaf 26

September

Every Sunday
Hillcrest 15

Sunday 5th
Highfield Farm 13

By Arrangement

Arrange a personalised garden visit on a date to suit you. See individual garden entries for full details.

The Alma

THE GARDENS

1 THE ALMA

Bully Hole Bottom, Usk Road, Shirenewton, NP16 6SA. Dr Pauline Ruth, 01291 641902, pmruth@hotmail.co.uk. *S-west facing slope overlooking valley. B4235 Usk to Chepstow signposted Bully Hole Bottom. Down hill over bridge up to T junction. Drive straight ahead along track signposted The Alma. Parking in meadow on L.* **Sun 9, Mon 10 May (12-5). Adm £5, chd free. Home-made teas. Visits also by arrangement May to Sept for groups of up to 30.**
Large and beautiful sheltered SW facing garden, uncommon trees, wisteria, roses and acid loving shrubs. Long border, hot border, productive vegetable garden, old brick outbuildings, fruit cage and vines, wildlife pond, sunset arbour and stream side walk. Drive packed with native daffodils, snowdrops and bluebells in the spring. Wildflower meadow and orchard in development. Sunny terrace for teas. Children's play area and picnics welcome. Wheelchair access to level terrace.
&. 🐄 ☕ 🪑

2 BIRCH TREE WELL

Upper Ferry Road, Penallt, Monmouth, NP25 4AN. Jill Bourchier, 01600 775327, gillian.bourchier@btinternet.com. *4m SW of Monmouth. Approx 1m from Monmouth on B4293, turn L for Penallt & Trelleck. After 2m turn L to Penallt. On entering village turn L at Xrds & follow yellow signs.* **Sun 4 July (2-5.30). Adm £4, chd free. Home-made teas. Visits also by arrangement May to Sept for groups of 10 to 20.**
Situated in the heart of the Lower Wye Valley, amongst the ancient habitat of woodland, rocks and streams. These 3 acres are shared with deer, badgers and foxes. A woodland setting with streams and boulders which can be viewed from a lookout tower and a butterfly garden planted with specialist hydrangeas incl many plants to also attract bees and insects. Live music will be played (harp and cello). Children are very welcome (under supervision) with plenty of activities in the form of treasure hunts. Not all areas of garden suitable for wheelchairs but refreshments certainly are!
&. ✿ ☕

3 BRYNGWYN MANOR

Raglan, NP15 2JH. Peter & Louise Maunder, 01291 691485, louiseviola@live.co.uk. *2m W of Raglan. Turn S (between the two garden centres) off B4598 (old A40) Abergavenny-Raglan road at Croes Bychan. House 1/4 m up lane on left.* **Visits by arrangement. Light refreshments or afternoon tea can be provided. Picnics are welcome if preferred. Please contact us to discuss your group's requirements.**
3 acres. Winter snowdrops, daffodil walk, mature trees, walled parterre garden, mixed borders, lawns, ponds and shrubbery. All areas of the garden can be accessed without using the steps, however, ground is uneven and consist mainly of grass paths.
&. 🐄 ☕ 🪑

4 CLYTHA PARK

Abergavenny, NP7 9BW. Jack & Susannah Tenison. *Between Abergavenny (5m) & Raglan (3m). On old A40 signed Clytha at r'abouts either end.* **Sun 18 July (2-5). Adm £5, chd free. Home-made teas.**
Large C18/19 garden around lake with wide lawns and specimen trees, original layout by John Davenport, with C19 arboretum, and H Avray Tipping influence. Visit the 1790 walled garden and the newly restored greenhouses. Due to current restrictions, the usual features and attractions are no longer available. Gravel and grass paths.
&. 🐄 ☕

5 CROESLLANFRO FARM

Groes Road, Rogerstone, Newport, NP10 9GP. Barry & Liz Davies, 07957694230, lizplants@gmail.com. *3m W of Newport. From M4 J27 take B4591 towards Risca. Take 3rd R, Cefn Walk (also signed 14 Locks Canal Centre). Proceed over bridge, cont 1/2 m to island in middle of lane.* **Sun 30 May, Sun 20 June, Sun 25 July, Sun 22 Aug (1.30-5). Adm £5, chd free. Visits also by arrangement May to Sept for groups of 5+.**
An informal two acre garden featuring mass planted perennials, grasses, wildflower meadow and exotic garden. Spring and early summer is a tapestry of green concentrating on leaf form and texture. Late summer, early autumn brings the the garden to a finale with an explosion of colour. A formal garden designed on 6 different levels for easy maintenance. Large barn

open to the public, 'fabulous folly' and grotto. Owner is a Garden Designer and co-author of 'Designing Gardens on Slopes'. Some gravel paths and shallow steps to main area of garden.
&. ✿ 🚗 ☕

6 NEW CWMDOWS FARM

New Bryngwyn Road, Newbridge, Newport, NP11 4NE. Dewi Reynolds and Dan Pearce, 07874 204034. *Cwmdows Farm is situated just off New Bryngwyn Rd, Newbridge.* **Sun 6 June, Sun 15 Aug (10-5). Adm £4, chd free. Tea. Visits also by arrangement May to Aug for groups of up to 20. Sundays only.**
We started creating our garden from scratch in 2017. Our aim was to create a number of individual rooms in the garden so that each area delivered separate interest to complement the backdrop of our medieval farmhouse. The garden has a lawn, herbaceous borders, pergola, pots, rose arches, courtyard, water features and seating areas. The garden is constantly evolving with new plants and ideas. Garden houses a small collection of very rare Auricula plants.
✿ 🚗 ☕

7 GLEBE HOUSE

Llanvair Kilgeddin, Abergavenny, NP7 9DE. Mr & Mrs Murray Kerr, 01873 840422, joanna@amknet.com. *Midway between Abergavenny (5m) & Usk (5m) on B4598.* **Sat 24, Sun 25 Apr (2-6). Adm £5, chd free. Home-made teas. Visits also by arrangement Apr to July.**
Borders bursting with spring colour incl tulips, narcissi and camassias. South facing terrace with wisteria and honeysuckle, decorative veg garden and orchard underplanted with succession of bulbs. Some topiary and formal hedging in 1 1/2 acre garden set in AONB in Usk valley. Old rectory of St Mary's, Llanfair Kilgeddin which will also be open to view famous Victorian Scraffito Murals. Some gravel and gently sloping lawns.
&. ✿ ☕ 🪑

The National Garden Scheme searches the length and breadth of England and Wales for the very best private gardens

8 NEW GLEN TROTHY

Llanvetherine, Abergavenny, NP7 8RB. Mr & Mrs Ben Herbert. *5m NE of Abergavenny. 6m from Abergavenny off B4521 (Old Ross Rd).* **Sun 20 June (2-5.30). Adm £6.50, chd free. Home-made teas.** Victorian house (not open) in the Scottish Baronial style, set in mature parkland with a small pinetum and arboretum. The walled garden has been renovated over the past 12yrs, incorporating blue and white herbaceous borders, a rose garden and ornamental vegetable garden with pear tunnel as well as an Italianate loggia.

9 14 GWERTHONOR LANE

Gilfach, Bargoed, CF81 8JT. **Suzanne & Philip George.** *8m N of Caerphilly. A469 to Bargoed, through the T-lights next to school then L filterlane at next T-lights onto Cardiff Rd. First L into Gwerthonor Rd, 4th R into Gwerthonor Lane.* **Sat 10, Sun 11 July (11-6). Adm £4, chd free. Light refreshments.** The garden has a beautiful panoramic view of the Rhymney Valley. A real plantswoman's garden with over 800 varieties of perennials, annuals, bulbs, shrubs and trees. There are numerous rare, unusual and tropical plants combined with traditional and well loved favourites (many available for sale). A pond with a small waterfall adds to the tranquil feel of the garden.

10 HIGH GLANAU MANOR

Lydart, Monmouth, NP25 4AD. Mr & Mrs Hilary Gerrish, 01600 860005, helenagerrish@gmail.com, www.highglanaugardens.com. *4m SW of Monmouth. Situated on B4293 between Monmouth & Chepstow. Turn R into private rd, ¼m after Craig-y-Dorth turn on B4293.* **Sun 2 May (2-6). Adm £6, chd free. Home-made teas. Visits also by arrangement May to July for groups of 20+.** Listed Arts and Crafts garden laid out by H Avray Tipping in 1922. Original features incl impressive stone terraces with far reaching views over the Vale of Usk to Blorenge, Skirrid, Sugar Loaf and Brecon Beacons. Pergola, herbaceous borders, Edwardian glasshouse, rhododendrons, azaleas, tulips, orchard with wildflowers. Originally open for the NGS in 1927. Garden guidebook by owner, Helena

Gerrish, available to purchase. Gardens lovers cottage to rent.

11 HIGH HOUSE

Penrhos, NP15 2DJ. Mr & Mrs R Cleeve. *4m N of Raglan. From r'about on A40 at Raglan take exit to Clytha. After 50yds turn R at Llantilio Crossenny. Follow NGS signs, 10mins through lanes.* **Sun 2 May (2-6). Combined adm with The Old Vicarage £6.50, chd free. Home-made teas.** 3 acres of spacious lawns and trees surrounding C16 house (not open) in a beautiful, hidden part of Monmouthshire. South facing terrace and extensive bed of old roses. Swathes of grass with tulips, camassias, wildflowers and far reaching views. Espaliered cherries, pears and scented evergreens in courtyard. Large extended pond, orchard with chickens and ducks, large vegetable garden. Partial wheelchair access, some shallow steps, sloping lawn, gravel courtyard.

12 32/33 HIGH STREET

Argoed, Blackwood, NP12 0HG. **Graeme & Sue Moore.** *2m N of Blackwood on A4048 Blackwood to Tredegar Rd, turn E into High St, Argoed.* **Sat 7, Sun 8 Aug (10-5). Adm £4, chd free. Home-made teas.** This lovely little garden in the village high street, crowded with flowering plants, has an Italianate air with a little gravel terrace and a sunken garden below. Inspiration for anyone with a really small plot. Visitors have asked about the Irish yews which are usually seen as multi-stemmed specimens and in old age often need wiring in. I have raised these from cuttings and trained them up on a single leader and they continue to grow up on a single stem just like Italian cypresses. Steps throughout make garden unsuitable for people with limited mobility.

13 HIGHFIELD FARM

Penperlleni, Goytre, NP4 0AA. Dr Roger & Mrs Jenny Lloyd, 01873 880030, jenny.plants@btinternet.com, www.instagram.com/jenny.plants/. *4m W of Usk, 6m S of Abergavenny. Turn off the A4042 at the Goytre Arms, over railway bridge, bear L.*

Garden ½m on R. From Usk off B4598, turn L after Chain Bridge, then L at crossroads. Garden 1m on L. **Sun 6 June, Sun 18 July, Sun 15 Aug, Sun 5 Sept (11-5). Adm £5, chd free. Pre-booking essential, please visit www.ngs.org.uk/ events for information & booking. Visits also by arrangement May to Sept for groups of 5 to 30.** This is a garden defined by its plants. There are over 1200 cultivars, with many rarities, densely planted over 2 acres to generate an exuberant display across the seasons. It provides an intimate immersive experience and a real tactile and spiritual connection with this diverse array of herbaceous, shrubs and trees. Huge sale of plants from the garden. Access to almost all garden without steps.

14 HILL HOUSE

Church Lane, Glascoed, Pontypool, NP4 0UA. Susan & John Wright. *Between Usk & Pontypool. Approx 2m W of Usk via A472 turn L before Beaufort Arms into Glascoed Lane. Bear L up hill, Church Lane 1st turn on R. House at the very end of narrow lane, do not turn off.* **Sat 3, Sun 4 July (2-5.30). Adm £5, chd free. Home-made teas.** The garden surrounds a modest 17thC farmhouse with fine views. Small orchard, reflecting pond, herbaceous borders, grasses, farmyard garden with standard parrotia and massed crocosmia and picket beds. Many seats. 30 mins walk through fields, steep return climb. Restricted parking, marshals in attendance. Some gravel paths. No disabled WC.

15 HILLCREST

Waunborfa Road, Cefn Fforest, Blackwood, NP12 3LB. Mr M O'Leary, 01443 837029, olearymichael18@gmail.com. *3m W of Newbridge. B4254/ A469 at T/L head to B'wood. At X take lane ahead, 1st on L at top of hill. A4048/B4251 cross Chartist Bridge, 2nd exit on next 2 r'abouts. End of road turn L & immed R onto Waunborfa. 400m on R.* **Sat 29, Sun 30, Mon 31 May (11-6). Light refreshments. Every Sun 4 July to 26 Sept (11-5). Adm £5, chd free. No refreshments on Sunday openings. Visits also by** arrangement Apr to Oct for groups of up to 30.

A cascade of secluded gardens of distinct character, all within 1½ acres. Magnificent, unusual trees with interesting shrubs, ferns and perennials. With choices at every turn, visitors exploring the gardens are well rewarded as hidden delights and surprises are revealed. Well placed seats encourage a relaxed pace to fully appreciate the garden's treasures. Tulips in April, glorious blooms of the Chilean Firebushes, Handkerchief Tree and cornuses in May and many trees in their autumnal splendour in October. Lowest parts of garden not accessible to wheelchairs.

& 🐕 ✿ ♨ 🏕

LLANOVER
nr Abergavenny, NP7 9EF. Mr & Mrs M H Murray, 07753 423635, elizabeth@llanover.com, www.llanovergarden.co.uk, *4m S of Abergavenny, 15m N of Newport, 20m SW Hereford. On A4042 Abergavenny - Cwmbran Rd, in village of Llanover.* Sun 28 Mar (11-4). Adm £6, chd free. Visits also by arrangement Mar to Oct for groups of 20+.

Benjamin Waddington, the direct ancestor of the current owners, purchased the house and land in 1792. Subsequently he created a series of ponds, cascades and rills which form the backbone of the 15 acre garden as the stream winds its way from its source in the Black Mountains to the River Usk. There are herbaceous borders, a drive lined with narcissi, spring bulbs and wildflowers, a water garden, champion trees and two arboreta. The house (not open) is the birthplace of Augusta Waddington, Lady Llanover, C19 patriot, supporter of the Welsh language and traditions. Gwerinyr Gwent will be performing Welsh folk dances during the afternoon in traditional Welsh costume. Lawns, spring bulbs, shrubs and trees. Gravel and grass paths and lawns. No disabled WC.

& 🐕 ✿ 🏠 ♨ 🏕

LONGHOUSE FARM
Penrhos, Raglan, NP15 2DE. Mr & Mrs M H C Anderson, 01600 780389, m.anderson666@btinternet.com. *Midway between Monmouth & Abergavenny. 4m from Raglan. Off Old Raglan/Abergavenny rd signed Clytha. At Bryngwyn/Great Oak Xrds turn towards Great Oak - follow*

yellow NGS signs from red phone box down narrow lane. Sat 12, Sun 13 June (2-6). Adm £5, chd free. Home-made teas. Visits also by arrangement June to Sept for groups of 5+.

Over the past 10 years this hidden garden with extensive views and spacious lawns has matured. The woodland walk and its series of ponds and stream will continue to develop. The productive vegetable garden has an additional fruit cage and potting shed. The avenue of malus trees seen from the south-facing terrace are still a feature supported by borders planted with year-round colourful plants.

& ✿ ♨

MIONE
Old Hereford Road, Llanvihangel Crucorney, Abergavenny, NP7 7LB. Yvonne & John O'Neil. *5m N of Abergavenny. From Abergavenny take A465 to Hereford. After 4.8m turn L - signed Pantygelli. Mione is ½m on L.* Sun 20, Fri 25 June, Fri 2, Sun 4 July (10.30-6). Adm £4, chd free. Home-made teas.

Beautiful garden with a wide variety of established plants, many rare and unusual. Pergola with climbing roses and clematis. Wildlife pond with many newts, insects and frogs. Numerous containers with diverse range of planting. Several seating areas, each with a different atmosphere. Enjoy our new benches in a secret hideaway under the pergola. Lovely home-made cakes, biscuits and scones to be enjoyed sitting in the garden or pretty summerhouse.

✿ ♨

◆ NANT Y BEDD
Grwyne Fawr, Fforest Coal Pit, Abergavenny, NP7 7LY. Sue & Ian Mabberley, 01873 890219, garden@nantybedd.com, www.nantybedd.com. *In Grwyne Fawr valley. From A465 Llanvihangel Crucorney, direction Llanthony, then L to Fforest Coal Pit. At grey telephone box cont for 4½m towards Grwyne Fawr Reservoir.* For NGS: Sun 30, Mon 31 May (10-5). Adm £7, chd free. Pre-booking essential, please visit www.ngs. org.uk/events for information & booking. Home-made teas. For other opening times and information, please phone, email or visit garden website.

Blending wild and tame, 10 acres described as 'Absolutely enchanting', 'of the place'. Set high in the Black Mountains by the Grwyne Fawr river with places to sit & enjoy the tranquility. An 'amazing mixture and riot of interest' with vegetables and fruit, mature trees and water. 'A lesson in working with the land and listening to what she tells you'. Productive organic vegetable and fruit gardens, stream, forest and river walk, wildflowers, natural swimming pond, treehouse, tree sculpture, shepherd's hut and eco-features. Ducks, chickens, sheep, pigs, cats. See www.nantybedd.com for details., Wales' Favourite Garden in The English Garden's The Nation's Favourite Gardens 2019. Access for visitors with mobility issues is limited by topography and terrain. See www. nantybedd.com for further details.

♨ 🏕

Rockfield Park

20 NEW NORTH PARADE HOUSE

12 Hereford Road, Monmouth, NP25 3PB. **Tim Haynes & Lisa O'Neill.** *5mins walk from the centre of town. Take the sign to Monmouth off the A40 r'about by the town. At the T-lights, turn R on to the A466 & the house is on the R after about 300 yds. Parking is on the street.* **Sun 6 June (2-6). Adm £5, chd free. Home-made teas in the garden.**
A secluded town walled garden of ⅔ acre which is in the process of restoration. It includes a kitchen garden, mature specimen trees, shrubs, spring bulbs and herbaceous border.

21 THE OLD VICARAGE

Penrhos, Raglan, Usk, NP15 2LE. **Mrs Georgina Herrmann.** *3m N of Raglan. From A449 take Raglan exit, join A40 & move immed into R lane & turn R across dual carriageway. Follow yellow NGS signs. One way traffic system between Old Vicarage & High House.* **Sun 2 May (2-6). Combined adm with High House £6.50, chd free.**
Set in rolling Monmouthshire countryside, the Old Vicarage, a 154yr old (1867) Victorian Gothic house, surrounded by sweeping lawns with young, mature and unusual trees. In addition to traditional and kitchen gardens, there is a parterre with a charming gazebo and a pond area. Range of spring flowers, wildflower meadow and bee friendly areas. This gem gets better each year.

22 PARK HOUSE

School Lane, Itton, Chepstow, NP16 6BZ. **Professor Bruce & Dr Cynthia Matthews.** *From M48 take A466 Tintern. At 2nd r'about turn L B4293 After blue sign Itton turn R Park House is at end of lane. Parking 200m before house. From Devauden B4293 1st L in Itton.* **Sat 1 May (10-5). Combined adm with Woodhaven £6, chd free.**
Approx one acre garden with large vegetable areas and many mature trees, rhododendrons, azaleas, camellias in a woodland setting. Bordering on Chepstow Park Wood. Magnificent views over open country. A few small steps. and irregular paths too narrow for wheelchairs. Disabled parking adjacent to house.

23 ROCKFIELD PARK

Rockfield, Monmouth, NP25 5QB. **Mark & Melanie Molyneux.** *On arriving in Rockfield village from Monmouth, turn R by phone box. After approx 400yds, church on L. Entrance to Rockfield Park on R, opp church, via private bridge over river.* **Sat 29, Sun 30 May (10.30-4). Adm £5, chd free. Home-made teas. We will be serving light lunches as well.**
Rockfield Park dates from C17 and is situated in the heart of the Monmouthshire countryside on the banks of the River Monnow. The extensive grounds comprise formal gardens, meadows and orchard, complemented by riverside and woodland walks. Possible to picnic on riverside walks. Main part of gardens can be accessed by wheelchair but not steep garden leading down to river.

24 TRENGROVE HOUSE

Nantyderry, Abergavenny, NP7 9DP. **Guin Vaughan & Chris Jofeh.** *Approx 4m N of Usk , 6m SE of Abergavenny. From Abergavenny, L off the A4042 after Llanover, signed to Nantyderry. From Usk, B4598 N to Chainbridge, then immediately L after bridge.* **Every Thur and Sun 17 June to 4 July (11-5). Adm £5, chd free. Pre-booking essential, please visit www.ngs.org.uk/**

Cwmdows Farm

events for information & booking. Home-made teas.

Garden designer's own 'single-handed' 2½ acre country garden developed over 18 yrs along 'right plant, right place' lines. Informal borders planted for a range of conditions with interesting and some unusual shrubs, trees, perennials and grasses. 1 acre meadow, managed to encourage only naturally occurring species with many wildflowers and grasses. Tranquil and atmospheric. Follow the garden on Instagram #trengrovegarden or on the Trengrove Garden Facebook page.

GROUP OPENING

25 USK OPEN GARDENS
Twyn Square, Usk, NP15 1BH. 07944 616448, UskOpenGardens@gmail.com, www.uskopengardens.com. *From M4 J24 take A449, proceed 8m N to Usk exit. Free parking signposted in town. Blue badge car parking in main car parks & at Usk Castle. Map of gardens provided with ticket.* **Sat 26, Sun 27 June (10-5). Combined adm £7.50, chd free.**
Winner of Wales in Bloom for 40 years, Usk's floral colour is a wonderful backdrop to the gardens, ranging from small cottages packed with unusual plants to large gardens with brimming herbaceous borders. Romantic garden around the ramparts of Usk Castle. Gardeners' Market with interesting plants. Great day out for all the family with lots of places to eat and drink incl places to picnic. Sorry but NGS discretionary entry tickets are not accepted at this event. Various cafes, PH and restaurants available for refreshments, plus volunteer groups offering teas and cakes; usually one of the gardens has a pop up Pimms, Prosecco and ice cream bar with a picnic and children's play area by their lake. Gardens allowing well-behaved dogs on leads noted on ticket/map. Usk Castle and School fully accessible but some gardens/areas of gardens are partially wheelchair accessible. Accessibility noted on ticket/map.

26 WENALLT ISAF
Twyn Wenallt, Gilwern, Abergavenny, NP7 0HP. Tim & Debbie Field, 01873 832753, wenalltisaf@gmail.com. *3m W of Abergavenny. Between Abergavenny & Brynmawr. Leave the A465 at Gilwern & follow yellow NGS signs through the village. Sat navs may not be accurate due to recent road changes.* **Sun 23 May, Sun 29 Aug (2-5.30). Adm £5, chd free. Home-made teas. Gluten free and lactose free cakes available. Visits also by arrangement May to Sept for groups of 10+.**
An everchanging garden of nearly 3 acres designed in sympathy with its surroundings and the challenges of being 650ft up on a N facing hillside. Far reaching views of the magnificent Black Mountains, mature trees, rhododendrons, viburnum, hydrangeas, borders, vegetable garden, small polytunnel, orchard, pigs (alternate years), chickens, beehives. Child friendly with plenty of space to run about.

27 WOODHAVEN
Itton, Chepstow, NP16 6BX. Mr & Mrs Kelly, 01291 641219, lesley@classics.co.uk. *Take the B4293 at Chepstow Racecourse r'about at Itton Common triangle, turn L at red telephone box, and then L along the the rd to Shirenewton, Woodhaven is on the L.* **Sat 1 May (10-5). Combined adm with Park House £6, chd free. Visits also by arrangement Apr to Sept for groups of 10 to 30.**
A modern house built on the site of a former sawmills for the Itton Court Estate. Garden of two-thirds of an acre developed over the last twenty years for all yr round colour and interest. Level front garden and gently sloping rear garden with extensive views over the valley. Lots of seating areas to enjoy the views. Meadow area with bulbs, display of tulips in spring, fruit trees and wildflowers. Level access to the front garden and the rear terrace. No wheelchair access to the toilet.

28 WOODLANDS FARM
Penrhos, NP15 2LE. Craig Loane & Charles Horsfield, 01600 780203, Woodlandsfarmwales@gmail.com, www.woodlandsfarmwales.com. *3m N of Raglan. From A449 take Raglan exit, join A40 & move immed into R lane & turn R across dual*
carriageway towards Tregare. Follow NGS signs or check website for directions. **Every Mon and Sun 5 Apr to 26 Apr (11-5). Adm £5, chd free. Pre-booking essential, please visit www.ngs.org.uk/ events for information & booking. Home-made teas in the Pavilion at Woodlands Farm Barns. Visits also by arrangement May to Sept for groups of 10+.**
A design led garden that's built to entertain, even your reluctant partner. The garden has opened for 12 years under the NGS. In 2021 we're opening Easter Monday and three other Sundays and Mondays in April (booking online essential). The 4 acre garden contains rooms and hidden spaces which draw you in, with buildings, ponds, viewing platform, jetty, hard landscaping, sculptures and Pygmy goats in their own palace. Beginner pottery classes and floristry classes available by arrangement. See our website for details.

29 ✦ WYNDCLIFFE COURT
St Arvans, NP16 6EY. Mr & Mrs Anthony Clay, 07710 138972, sarah@wyndcliffecourt.com, www.wyndcliffecourt.com. *3m N of Chepstow. Off A466, turn at Wyndcliffe signpost coming from the Chepstow direction.* **For NGS: Fri 4 June (2-6). Adm £10. Pre-booking essential, please visit www.ngs.org.uk for information & booking. Home-made teas. For other opening times and information, please phone, email or visit garden website.**
Exceptional and unaltered garden designed by H. Avray Tipping and Sir Eric Francis in 1922. Arts and Crafts 'Italianate' style. Stone summerhouse, terracing and steps with lily pond. Yew hedging and topiary, sunken garden, rose garden, bowling green and woodland. Walled garden and double tennis court lawn under refurbishment. Rose garden completely replanted to a new design by Sarah Price in 2017. Not suitable for children under 12.

> "Unfortunately, cancer was not in lockdown in 2020. The continued support of our long-standing and valued partner, the National Garden Scheme is more important than ever." Macmillan

GWYNEDD & ANGLESEY

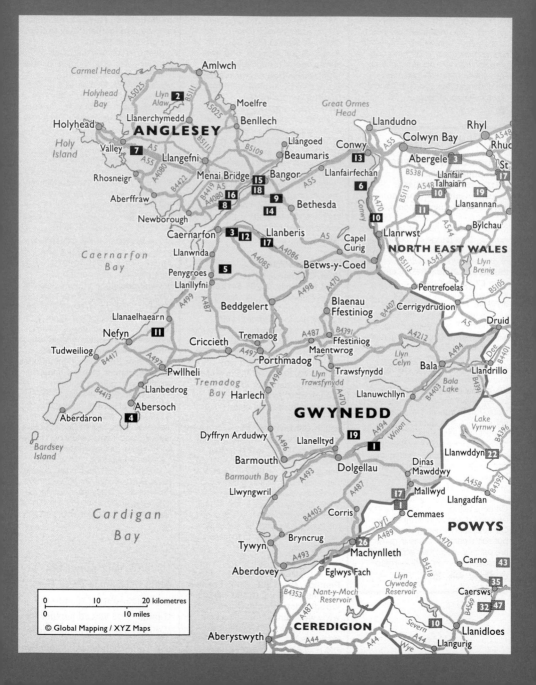

Gwynedd is a county rich in history and outstanding natural beauty. Bordered by the Irish Sea and home to Snowdonia National Park, Gwynedd can boast some of the most impressive landscapes in the UK.

The mountains in Gwynedd are world famous, and have attracted visitors for hundreds of years – the most famous perhaps, was Charles Darwin in 1831. As well as enjoying the tallest peaks in the UK, Gwynedd has fine woodland – from hanging oak forests in the mountains to lush, riverside woods.

Holiday-makers flock to Gwynedd and Anglesey to take advantage of the sandy beaches, and many can enjoy sightings of dolphins and porpoises off the coast.

The gardens of Gwynedd and Anglesey are just as appealing an attraction for visitors. A variety of gardens open for Gwynedd National Garden Scheme, ranging from Crowrach Isaf a two-acre garden with views of Snowdonia and Cardigan Bay to Ty Capel Ffrwd, a true cottage garden in the Welsh mountains.

So why not escape from the hustle and bustle of everyday life and relax in a beautiful garden? You will be assured of a warm welcome at every garden gate.

Below: Cae Newydd

Volunteers

Gwynedd & Anglesey County Organiser
Position Vacant – for details please contact hello@ngs.org.uk or 01483 211535

County Treasurer
Nigel Bond
01407 831354
nigel.bond@ngs.org.uk

Assistant County Organisers
Hazel Bond
07378 844295
hazelcaenewydd@gmail.com

Janet Jones
01758 740296
janetcoron@hotmail.co.uk

Delia Lanceley
01286 650517
delia@lanceley.com

 @gwyneddandangleseyngs

OPENING DATES

All entries subject to change. For latest information check **www.ngs.org.uk**

Extended openings are shown at the beginning of the month.

Map locator numbers are shown to the right of each garden name.

April

Saturday 17th
NEW Plas Llwynonn　16

Sunday 25th
Maenan Hall　10

Wednesday 28th
◆ Plas Cadnant Hidden
　Gardens　15

May

Sunday 2nd
Gilfach　6

Saturday 15th
Mynydd Heulog　11

Sunday 16th
Bryn Gwern　1
Mynydd Heulog　11

Saturday 22nd
Ty Capel Ffrwd　19

Sunday 23rd
◆ Gardd y Coleg　5
Llys-y-Gwynt　9
Swn-Y-Gwynt　17
Ty Capel Ffrwd　19

Saturday 29th
Gwaelod Mawr　7

Sunday 30th
Gwaelod Mawr　7
Pant Ifan　12

June

Sunday 6th
◆ Pensychnant　13

Saturday 12th
Cae Newydd　2

Sunday 13th
Cae Newydd　2

Saturday 19th
Crowrach Isaf　4
Llanidan Hall　8

Sunday 20th
Crowrach Isaf　4

July

Saturday 3rd
Llanidan Hall　8
NEW Plas Llwynonn　16

Sunday 4th
Gilfach　6

Saturday 10th
NEW Cae Rhydau　3

Sunday 11th
NEW Cae Rhydau　3

Saturday 17th
Pentir Gardens　14

Sunday 18th
Bryn Gwern　1
◆ Pensychnant　13

Sunday 25th
Maenan Hall　10

September

Saturday 11th
Treborth Botanic Garden,
　Bangor University　18

Sunday 19th
◆ Gardd y Coleg　5
Llys-y-Gwynt　9

By Arrangement

Arrange a personalised garden visit on a date to suit you. See individual garden entries for full details.

Bryn Gwern　1
Crowrach Isaf　4
Gilfach　6
Llys-y-Gwynt　9
Mynydd Heulog　11
Ty Capel Ffrwd　19

Crowrach Isaf

Open your garden with the National Garden Scheme and join a community of like-minded individuals, all passionate about gardens, and raise money for nursing and health charities. Big or small, if your garden has quality, character and interest we'd love to hear from you. Call us on 01483 211535 or email hello@ngs.org.uk

THE GARDENS

🚻 BRYN GWERN

Llanfachreth, Dolgellau, LL40 2DH.
H O & P D Nurse, 01341 450 255,
antique_pete@btinternet.com.
*5m NE of Dolgellau. Do not go to
Llanfachreth, stay on Bala rd, A 494,
5m from Dolgellau, 14m from Bala.* Sun
16 May, Sun 18 July (10-5). Adm £4,
chd free. Cream teas. Visits also by
arrangement May to Sept.
Sloping 2 acre garden in the hills
overlooking Dolgellau with views to
Cader Idris, originally wooded but
redesigned to enhance its natural
features with streams, ponds and
imaginative and extensive planting
and vibrant colour. The garden is now
a haven for wildlife with hedgehogs
and 26 species of birds feeding last
winter as well as being home to
ducks, dogs and cats. Dogs must be
on a lead. Stone mason at work and
items for sale or orders taken.
🐂 ✿ 🚗 🍵

🔢 CAE NEWYDD

Rhosgoch, Anglesey, LL66 0BG.
Hazel & Nigel Bond. *3m SW of
Amlwch. A5025 from Benllech to
Amlwch, follow signs for leisure
centre & Lastra Farm. Follow
yellow NGS signs (approx 3m), car
park on L.* Sat 12, Sun 13 June
(11-4). Adm £4, chd free. Light
refreshments.
A mature country garden of 2½ acres
which blends seamlessly into the
open landscape with stunning views
of Snowdonia and Llyn Alaw. Variety
of shrubs, trees and herbaceous
areas, large wildlife pond, polytunnel,
greenhouses, raised vegetable beds.
Collections of fuchsia, pelargonium,
cacti & succulents. An emphasis on
gardening for wildlife throughout the
garden. Hay meadow. Lots of seating
throughout the garden, visitors are
welcome to bring a picnic. Garden
area closest to house suitable for
wheelchairs.
♿ 🐂 ✿ 🍵 🍽

🔢 NEW CAE RHYDAU

Caeathro, Caernarfon, LL55 2TN
Ms Lieneke van der Veen &
Mr Victor van Daal. *4m E of
Caernarfon. Take A4086 from
Caernarfon to Llanberis. After narrow
bridge take R to Bontnewydd and
then immediately L. Follow signs
Lieneke's Flowers. Take the 2nd*
track on R. Sat 10, Sun 11 July
(11-4.30). Adm £3.50, chd free.
Home-made teas.
Formerly a dairy mill. A Dutch couple
started in 2000 redesigning the
gardens of a 14 acre smallholding.
One acre garden contains a variety
of plants, shrubs, trees and grasses.
Also a vegetable and a cut flower
garden. Gardens accessed along an
800 metres footpath. The garden is
bordered on one side by a stream.
Very peaceful, no through road. Free
workshop by award-winning florist to
make a buttonhole using flowers from
the garden.
✿ 🍵

🔢 CROWRACH ISAF

Bwlchtocyn, LL53 7BY. Margaret
& Graham Cook, 01758 712860,
crowrach_isaf@hotmail.com.
*1½m SW of Abersoch. Follow road
through Abersoch & Sarn Bach, L
at sign for Bwlchtocyn for ½m until
junction & no-through road sign.
Turn R, parking 50 metres on R.* Sat
19, Sun 20 June (1-4.30). Adm £4,
chd free. Cream teas. Visits also
by arrangement May to Sept for
groups of 20+.
2 acre plot developed from 2000,
incl island beds, windbreak hedges,
vegetable garden, wildflower area
and wide range of geraniums,
unusual shrubs and herbaceous
perennials. Views over Cardigan Bay
and Snowdonia. Grass and gravel
paths, some gentle slopes. Parking at
garden for disabled visitors.
♿ ✿ 🚗 🍵

🔢 ◆ GARDD Y COLEG

Carmel, LL54 7RL. Pwyllgor
Pentref Carmel Village Committee.
*Garden at Carmel village centre.
Parking on site.* For NGS: Sun 23
May, Sun 19 Sept (11-3). Adm £3,
chd free. Tea.
Approx ½ acre featuring raised
beds planted with ornamental and
native plants mulched with local
slate. Benches and picnic area, wide
pathways suitable for wheelchairs.
Spectacular views. Garden created
and maintained by volunteers.
Development of the garden is
ongoing. Ramped access.
♿ ✿ 🍵

🔢 GILFACH

Rowen, Conwy, LL32 8TS.
James & Isoline Greenhalgh,
01492 650216, isolinegreenhalgh@
btinternet.com. *4m S of Conwy. At*
Xrds 100yds E of Rowen S towards
Llanrwst, past Rowen School on L,
turn up 2nd drive on L.* Sun 2 May,
Sun 4 July (2-5.30). Adm £3.50,
chd free. Home-made teas. Visits
also by arrangement Apr to Aug.
1 acre country garden on S-facing
slope with magnificent views of the
River Conwy and mountains; set
in 35 acres of farm and woodland.
Collection of mature shrubs is
added to yearly; woodland garden,
herbaceous border and small pool.
Spectacular panoramic view of the
Conwy Valley and the mountain range
of the Carneddau. Classic cars. Large
coaches can park at bottom of steep
drive, disabled visitors can be driven
to garden by the owner.
♿ ✿ 🚗 🍵

🔢 GWAELOD MAWR

Caergeiliog, Anglesey, LL65 3YL.
Tricia Coates. *6m E of Holyhead.
½m E of Caergeiliog. From A55 J4,
r'about 2nd exit signed Caergeiliog.
300yds, Gwaelod Mawr is 1st
house on L.* Sat 29, Sun 30 May
(11-4). Adm £4, chd free. Light
refreshments.
2 acre garden created by owner over
20 yrs with lake, large rock outcrops
and palm tree area. Spanish style
patio and laburnum arch lead to
sunken garden and wooden bridge
over lily pond with fountain and
waterfall. Peaceful Chinese orientated
garden offering contemplation.
Separate Koi carp pond. Abundant
seating throughout. Mainly flat, with
gravel and stone paths, no wheelchair
access to sunken lily pond area.
♿ 🐂 🚗 🍵

🔢 LLANIDAN HALL

Brynsiencyn, LL61 6HJ. Mr J W
Beverley (Head Gardener). *5m E of
Llanfair Pwll. From Llanfair PG follow
A4080 towards Brynsiencyn for 4m.
After Hooton's farm shop on R take
next L, follow lane to gardens.* Sat
19 June (10-4). Sat 3 July (10-4),
also open Plas Llwynonn. Adm
£4, chd free. Donation to CAFOD.
Walled garden of 1¾ acres. Physic
and herb gardens, ornamental
vegetable garden, herbaceous
borders, water features and many
varieties of old roses. Sheep, rabbits
and hens to see. Children must be
kept under supervision. Well behaved
dogs on leads welcome. Llanidan
Church will be open for viewing. Hard
gravel paths, gentle slopes.
♿ 🐂

9 LLYS-Y-GWYNT

Pentir Road, Llandygai, Bangor, LL57 4BG. Jennifer Rickards & John Evans, 01248 353863, mjrickards@gmail.com. *3m S of Bangor. 300yds from Llandygai r'about at J11, A5 & A55, just off A4244. Follow signs for services (Gwasanaethau). Turn off at No Through Rd sign, 50yds beyond. Do not use SatNav.* Sun 23 May, Sun 19 Sept (11-4). Adm £5, chd free. Cream teas. Visits also by arrangement. Tea/coffee, biscuits available.

Interesting, harmonious and very varied 2 acre garden incl magnificent views of Snowdonia. An exposed site incl Bronze Age burial cairn. Winding paths and varied levels planted to create shelter, yr-round interest, microclimates and varied rooms. Ponds, waterfall, bridge and other features use local materials and craftspeople. Wildlife encouraged, well organised compost. Good family garden. Come and help us celebrate 25 yrs with NGS.

  🚻 🚐 ☕

10 MAENAN HALL

Maenan, Llanrwst, LL26 0UL. The Hon Mr & Mrs Christopher Mclaren. *2m N of Llanrwst. On E side of A470, ¼m S of Maenan Abbey Hotel.* Sun 25 Apr, Sun 25 July (10.30-5). Adm £4, chd free. Light refreshments. Donation to Ogwen Valley Mountain Rescue Organisation.

A superbly beautiful 4 hectares on the slopes of the Conwy Valley, with dramatic views of Snowdonia, set amongst mature hardwoods. Both the upper part, with sweeping lawns, ornamental ponds and retaining walls, and the bluebell carpeted woodland dell contain copious specimen shrubs and trees, many originating at Bodnant. Magnolias, rhododendrons, camellias, pieris, cherries and hydrangeas, amongst many others, make a breathtaking display. Upper part of garden accessible but with fairly steep slopes.

  🐾 ❄ ☕

11 MYNYDD HEULOG

Llithfaen, Pwllheli, LL53 6PA. Mrs Christine Jackson, 01758 750400, christine.jackson007@btinternet. com. *From A499 take B4417 road at r'about signed Nefyn, approx 3m enter Llithfaen, 1st R turn opp chapel. Follow NGS signs, garden last property on R, limited parking.* Sat 15, Sun 16 May (10-4). Adm £3.50, chd free. Home-made teas.

Visits also by arrangement May & June. Donation to Dog's Trust.

Mynydd Heulog is an C18 stone cottage set in approx. 1 acre of sloping garden with amazing views over the Lleyn and Cardigan Bay. Gradually being developed over 25 years, the garden is now an eclectic mix of mature trees, shrubs, perennials and exotics. Features incl arches, statues, bridges, summer house and shepherds hut. Large terrace and verandah with views and secret seating areas. Narrow paths.

  🐾 🚐 ☕

12 PANT IFAN

Ceunant, Llanrug, Caernarfon, LL55 4HX. Mrs Delia Lanceley. *2m E of Caernarfon. From Llanrug take rd opp old PO between Convenience Store & Monumental Mason. Straight across at next Xrds. Then 3rd turn on L at Xrds. Pant Ifan 2nd house on L.* Sun 30 May (11-5). Adm £4, chd free. Home-made teas.

2 acre mix of formal and wildlife garden set around farmhouse and yard. Herbaceous borders, shrubs, vegetables, fruit, ponds and recently planted woodland. Field walks, sitting areas in the sun or shade. Poultry, ducks, geese, donkeys, horse and two ponies, greenhouses and poly tunnel. Deep water, children must be supervised at all times. Wheelchair users can access yard and teas.

  ❄ ☕ 🏕

13 ♦ PENSYCHNANT

Sychnant Pass, Conwy, LL32 8BJ. Pensychnant Foundation; Warden Julian Thompson, 01492 592595, jpt.pensychnant@btinternet.com, www.pensychnant.co.uk. *2½m W of Conwy at top of Sychnant Pass. From Conwy: L into Upper Gate St; after 2½m Pensychnant's Dr signed on R. From Penmaenmawr: fork R, up Sychnant Pass; after walls at top of Pass, U turn L into drive.* For NGS: Sun 6 June, Sun 18 July (11-5). Adm £3.50, chd free. Light refreshments. For other opening times and information, please phone, email or visit garden website.

Wildlife Garden. Diverse herbaceous cottage garden borders surrounded by mature shrubs, banks of rhododendrons, ancient and Victorian woodlands. 12 acre woodland walks with views of Conwy Mountain and Sychnant. Woodland birds. Picnic tables, archaeological trail on mountain. A peaceful little gem. Large

Victorian Arts and Crafts house (open) with art exhibition. Partial wheelchair access, please phone for advice.

  ❄ ☕

GROUP OPENING

14 PENTIR GARDENS

Pentir, Bangor, LL57 4YA. *Pentir 3m S of Bangor, Gwynedd. A4244 from J11 of A55/A5, continue for 1.7m to Pentir. Either turn L, follow car park signs to field (LL574YB), or proceed, turn R signed Caerhun to car park opp church (LL574EA).* Sat 17 July (12-5). Combined adm £5, chd free. Home-made teas at Bryn Meddyg. Donation to Ogwen Valley Mountain Rescue, Llanberis Mountain Rescue.

NEW BRYN HYFRYD
Mr John Turner.

BRYN MEDDYG
Mr Wyn James.

TAN RALLT
John Lewis & Gary Carvalho.

TY UCHAF
Sian Lewis.

Starting from Pentir square, Ty Uchaf: a small densely packed garden with a wide variety of cottage favourites, plus raised beds and willow lattice. It enjoys a romantic-feel, prioritising colour and texture. The secret gate takes visitors to neighbouring Bryn Meddyg, serving refreshments within a large garden comprising mature shrubs, with several island beds, vegetable garden, soft fruit and secluded seating. A 600-yard walk takes you to Bryn Hyfryd, a front garden set to lawn, borders and pond, and via archway to rear garden with stunning views of Snowdonia and owner's champion Shetland sheep, with kitchen garden and mixed planting. Continue along the lane, a few minutes' walk, to Tan Rallt, backing onto Moel y Ci, the one acre stunningly diverse garden includes a range of mature trees and shrubs, vegetable and soft fruit, herbaceous borders and areas of lawn. Diversely planted pond with tree ferns, bog plants, and giant Gunnera. Adjacent to the Field car park (LL574YB), there is an attractive large lake (with safety restrictions). There is an alternative access to Bryn Meddyg for wheelchair access. There is restricted disabled access at Ty Uchaf and Bryn Hyfryd.

  🐾 ❄ ☕

Treborth Botanic Garden

15 ◆ PLAS CADNANT HIDDEN GARDENS

Cadnant Road, Menai Bridge, LL59 5NH. Mr Anthony Tavernor, 01248 717174, plascadnantgardens@gmail.com, www.plascadnantgardens.co.uk. *½m E of Menai Bridge. Take A545 & leave Menai Bridge heading for Beaumaris, then follow brown tourist information signs. Sat Nav not always reliable.* **For NGS: Wed 28 Apr (12-5). Adm £8, chd free. Light refreshments in traditional Tea Room. For other opening times and information, please phone, email or visit garden website. Donation to Wales Air Ambulance; Anglesey Red Squirrel Trust; Menai Bridge Community Heritage Trust.** Early C19 picturesque garden undergoing restoration since 1996.

Valley gardens with waterfalls, large ornamental walled garden, woodland and early pit house. Recently created Alpheus water feature and Ceunant (Ravine) which gives visitors a more interesting walk featuring unusual moisture loving Alpines. Restored area following flood damage. Guidebook available. Visitor centre open. Partial wheelchair access to parts of gardens. Some steps, gravel paths, slopes. Access statement available. Accessible Tea Room and WC.

&♿ ✳ 🚌 🚐 ☕

16 NEW ▶ PLAS LLWYNONN

Llanedwen, Llanfairpwllgwyngyll, LL61 6DQ. Dominique Carpenter. *2m E of Llanfairpwll. From Llanfairpwll follow A4080 towards Brynsiencyn for 2m. 100yds after Plas Newydd National Trust entrance, turn R at lodge house. Over cattle grid and follow track through parkland for 500yds.* **Sat 17 Apr, Sat 3 July (11-4). Adm £4, chd free.**

A Plas Newydd Estate grade 2 listed building. 3 acres semi formal and wild gardens. Espalier pears, grapes, kiwi and Welsh Heritage apple orchard. Surrounded by 12 acres of woodland - snowdrops, wild garlic, bluebells and daffodils until late spring. Picnic areas and benches throughout the gardens and woods. Abundance of birdlife and red squirrels. Walled kitchen garden with Victorian Peach House. Beehives. If wet, ground can be very soft.

&♿ 🐕 ✳ 🛋

Llys y gwynt

Bryn Gwern

☑7 SWN-Y-GWYNT

High Street, Llanberis, Caernarfon, LL55 4EN. Keith & Olwen Chadwick. *Llanberis, at the foot of Snowdon. From all directions head towards the village of Llanberis. Turn off the A4086, follow NGS sign onto Llanberis High St then a green sign for Swn-y-Gwynt with the driveway entrance and private car park.* **Sun 23 May (11-4.30). Adm £3, chd free.**

A S-facing terraced garden with dramatic views of Snowdon and surrounding hills. Steep slate steps lead to the garden comprising of a variety of conifers, mature acers, azaleas, rose bed and herbaceous perennial border with a variety of spring bulbs complemented by hellebores and ferns. There is also a secluded small pond area leading onto slate steps towards a Scots pine terraced area. Children must be supervised at all times.

☑8 TREBORTH BOTANIC GARDEN, BANGOR UNIVERSITY

Treborth, Bangor, LL57 2RQ. Natalie Chivers, *treborth.bangor.ac.uk. On the outskirts of Bangor towards Anglesey. Approach Menai Bridge either from Upper Bangor on A5 or leave A55 J9 & travel towards Bangor for 2m. At Antelope Inn r'bout turn L just before entering Menai Bridge.* **Sat 11 Sept (10-1). Adm £4, chd free. Home-made teas including vegan and gluten free options.**

Owned by Bangor University and used as a resource for teaching, research, public education and enjoyment. Treborth comprises planted borders, species rich natural grassland, ponds, arboretum, Chinese garden, ancient woodland, and a rocky shoreline habitat. Six glasshouses provide specialised environments for tropical, temperate, orchid and carnivorous plant collections. Partnered with National Botanic Garden of Wales to champion Welsh horticulture, protect wildlife and extol the virtues of growing plants for food, fun, health and wellbeing. Glasshouse and garden Q&As. Wheelchair access to some glasshouses and part of the garden. Woodland path is surfaced but most of the borders only accessed over grass.

☑9 TY CAPEL FFRWD

Llanfachreth, Dolgellau, LL40 2NR. Revs Mary & George Bolt, 01341 422006, maryboltminstrel@gmail.com. *4m NE of Dolgellau, 18m SW of Bala. From Dolgellau 4m up hill to Llanfachreth. Turn L at War Memorial. Follow lane ½ m to chapel on R. Park & walk down lane past chapel to cottage.* **Sat 22, Sun 23 May (12-5.30). Adm £4, chd free. Home-made teas. Visits also by arrangement May to Sept for groups of up to 10.**

True cottage garden in Welsh mountains. Azaleas, rhododendrons, acers; large collection of aquilegia. Many different hostas give added strength to spring bulbs and corms. Stream flowing through the garden, 10ft waterfall and on through a small woodland bluebell carpet. For summer visitors there is a continuous show of colour with herbaceous plants, roses, clematis and lilies, incl Cardiocrinum giganteum. Harp will be played in the garden.

NORTH EAST WALES

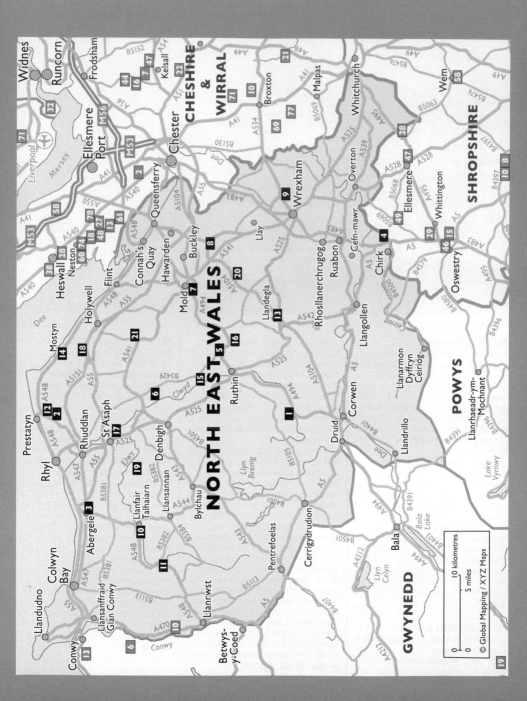

With its diversity of countryside from magnificent hills, seaside vistas and rolling farmland, North East Wales offers a wide range of gardening experiences.

Our gardens offer a wealth of designs and come in all shapes and sizes, ranging from large old parks offering a good walk, to the compact town gardens. Visitors will have something to see from the frost-filled days of February through till the magnificent colourful days of autumn.

The majority of our gardens are within easy reach of North West England, and being a popular tourist destination make an excellent day out for all the family.

Come and enjoy the beauty and the variety of the gardens of North East Wales with the added bonus of a delicious cup of tea and a slice of cake. Our garden owners await your visit.

Volunteers

County Organiser
Jane Moore
07769 046317
jane.moore@ngs.org.uk

County Treasurer
Iris Dobbie
01745 886730
iris.dobbie@ngs.org.uk

Booklet Co-ordinator
Roy Hambleton
01352 740206
roy.hambleton@ngs.org.uk

Assistant County Organisers
Fiona Bell
07813 087797
bell_fab@hotmail.com

Kate Bunning
01978 262855
kate.bunning@ngs.org.uk

Lesley Callister
01824 705444
lesley.callister@ngs.org.uk

Suzanne Hamer
01492640843
hamersuzanne@yahoo.co.uk

Anne Lewis
01352 757044
anne.lewis@ngs.org.uk

Pat Pearson
01745 813613
pat.pearson@ngs.org.uk

Helen Robertson
01978 790666
helen.robertson@ngs.org.uk

© Joe Wainwright

Left: Aberclwyd Manor

OPENING DATES

All entries subject to change. For latest information check www.ngs.org.uk

Map locator numbers are shown to the right of each garden name.

February

Snowdrop Festival

Wednesday 10th
Aberclwyd Manor 1

Wednesday 17th
Aberclwyd Manor 1
Clwydfryn 6

March

Wednesday 3rd
Aberclwyd Manor 1

Wednesday 17th
Aberclwyd Manor 1

Wednesday 31st
Aberclwyd Manor 1

April

Wednesday 14th
Aberclwyd Manor 1

Sunday 25th
NEW ◆ St Kentigern
Hospice 17

Wednesday 28th
Aberclwyd Manor 1

May

Wednesday 12th
Aberclwyd Manor 1

Sunday 16th
Mostyn Hall 14

Sunday 23rd
Brynkinalt Hall 4

Wednesday 26th
Aberclwyd Manor 1

Saturday 29th
Hafodunos Hall 11

Monday 31st
Garthewin 10

June

Saturday 5th
Plas Coch 15

Sunday 6th
Plas Coch 15

Wednesday 9th
Aberclwyd Manor 1

Wednesday 23rd
Aberclwyd Manor 1

Saturday 26th
33 Bryn Twr and Lynton 3

Sunday 27th
33 Bryn Twr and Lynton 3

July

Wednesday 7th
Aberclwyd Manor 1

Sunday 11th
The Cottage Nursing
Home 7

Sunday 18th
NEW Llandegla Village
Gardens 13

Wednesday 21st
Aberclwyd Manor 1

August

Sunday 1st
White Croft 21

Wednesday 4th
Aberclwyd Manor 1

Sunday 8th
NEW Tyn Rhos 20

Sunday 15th
NEW ◆ St Kentigern
Hospice 17

Wednesday 18th
Aberclwyd Manor 1

Sunday 22nd
NEW Cedar Gardens 5

September

Wednesday 1st
Aberclwyd Manor 1

Thursday 2nd
Brynkinalt Hall 4

Sunday 12th
◆ Erlas Victorian Walled
Garden 9

Wednesday 15th
Aberclwyd Manor 1

Wednesday 29th
Aberclwyd Manor 1

By Arrangement

Arrange a personalised garden visit on a date to suit you. See individual garden entries for full details.

Aberclwyd Manor 1
5 Birch Grove 2
33 Bryn Twr and Lynton 3
Dove Cottage 8
Garthewin 10
Hilbre, Manor Close 12
Plas Y Nant 16
NEW Saith Ffynnon
Farm 18
Tal-y-Bryn Farm 19
White Croft 21

Coreopsis tinctoria
American wildflower that gives reds and orang...

Saith Ffynnon Farm

THE GARDENS

The Gate House, Ruthin Road

1 ABERCLWYD MANOR

Derwen, Corwen, LL21 9SF.
Miss Irene Brown & Mr G
Sparvoli, 01824 750431,
irene662010@live.com. *7m from
Ruthin. Travelling on A494 from
Ruthin to Corwen. At Bryn SM
Service Station turn R, follow sign
to Derwen. Aberclwyd gates on L
before Derwen. Do not follow sat nav
directions .* Wed 10, Wed 17 Feb,
Wed 3, Wed 17, Wed 31 Mar, Wed
14, Wed 28 Apr, Wed 12, Wed 26
May, Wed 9, Wed 23 June, Wed 7,
Wed 21 July, Wed 4, Wed 18 Aug,
Wed 1, Wed 15, Wed 29 Sept (10-
4). Adm £3.50, chd free. Cream
teas. **Visits also by arrangement.**
4 acre garden on a sloping hillside
overlooking the Upper Clwyd Valley.
The garden has many mature trees
underplanted with snowdrops, fritillaries
and cyclamen. An Italianate garden
of box hedging lies below the house
and shrubs, ponds, perennials, roses
and an orchard are also to be enjoyed
within this cleverly structured area.
Mass of cyclamen in Sept. Abundance
of spring flowers. Cyclamen in August/
September. Snowdrops in February.
Mostly flat with some steps and slopes.

2 5 BIRCH GROVE

Woodland Park, Prestatyn,
LL19 9RH. Mrs Iris
Dobbie, 01745 886730,
iris.dobbie@ngs.org.uk. *A547
from Rhuddlan turn up The Avenue,
Woodland Park after railway bridge
1st R into Calthorpe Dr, 1st L Birch
Gr. 10 mins walk from town centre
- at top of High St, turn R & L onto
The Avenue. Visits by arrangement
May to Sept for groups of up to
20.* Adm £3, chd free.
The gardens consist of a variety
of borders incl woodland, grass,
herbaceous, alpine, shrub, drought,
tropical and a simulated bog garden
with a small pond. The more formal
front lawned garden has borders
of mixed colourful planting and box
balls. A small greenhouse is fully
used for propagation. Gravel drive
with few steps. Plenty of parking and
not far from the beach, Offa's Dyke,
town centre. A variety of borders and
planting. Gravel drive, few steps.

3 33 BRYN TWR AND LYNTON

Lynton, Highfield Park, Abergele,
LL22 7AU. Mr & Mrs Colin
Knowlson & Bryn Roberts & Emma
Knowlson-Roberts, 07712 623836,
apk@slaters.ltd. *From A55 heading
W take slip rd into Abergele town
centre. Turn L at 2nd set of T-lights
signed Llanfair TH, 3rd rd on L. For
SatNav use LL22 8DD.* Sat 26, Sun
27 June (1-5). Adm £5, chd free.
**Home-made teas. Visits also by
arrangement June & July.**
Bryn Twr is a family garden with
chickens, shrubs, roses, pots & Lawn.
Lynton completely different, intense
cottage style garden: Veg, Trees,
shrub, roses, ornamental grasses &
lots of pots. Garage with interesting
fire engine; classic cars and
memorabilia; greenhouse over large
water capture system that was part
of an old swimming pool. For 2021
new water feature. Partial wheelchair
access.

Online booking is available
for many of our gardens, at
ngs.org.uk

4 BRYNKINALT HALL

Brynkinalt, Chirk, Wrexham,
LL14 5NS. Iain & Kate Hill-Trevor,
www.brynkinalt.co.uk. *6m N of
Oswestry, 10m S of Wrexham.
Come off A5/A483 & take B5070
into Chirk village. Turn into Trevor Rd
(beside St Mary's Church). Continue
past houses on R. Turn R on bend
into Estate Gates. N.B. Do not use
postcode with SatNav.* Sun 23 May,
Thur 2 Sept (12-4). Adm £5, chd
free. **Home-made teas.**
5 acre ornamental woodland
shrubbery, overgrown until recently,
now cleared and replanted,
rhododendron walk, historic ponds,
well, grottos, ha-ha and battlements,
new stumpery, ancient redwoods
and yews. Also 2 acre garden beside
Grade II* house (see website for
opening), with modern rose and formal
beds, deep herbaceous borders, pond
with shrub/mixed beds, pleached limes
and hedge patterns. Home of the first
Duke of Wellington's grandmother and
Sir John Trevor, Speaker of House of
Commons. Stunning Rhododendrons
and formal West Garden. Partial
wheelchair access. Gravel paths in
West Garden and grass paths and
slopes in shrubbery.

5 NEW **CEDAR GARDENS**
Llanbedr Hall, Llanbedr Dyffryn
Clwyd, Ruthin, LL15 1YD. Mr Peter
Jones. *For directions please go to
cedargardens.co.uk & click map to
download a copy. Also there will be
the usual yellow signs to follow.* Sun
22 Aug (11-4). Adm £3.50, chd
free.
This tranquil garden is made up
of two gardens with the top one
featuring a 300 year old Cedar of
Lebanon tree. As you wander along
to the lower secret garden, there is a
rill running through the lower part of
the large lawn. At the end of the path
is a garden room where refreshments
may be served, and adjoining this
there are two patios on different levels
with beds full of colourful planting.

6 **CLWYDFRYN**
Bodfari, LL16 4HU. Keith & Susan
Watson. *5m outside Denbigh. ½ way
between Bodfari & Llandyrnog on
B5429. Yellow signs at bottom of
lane.* Wed 17 Feb (11-3). Adm £4,
chd free.
¾ acre garden developed from a
field, which has had many changes
over the past 30 years. Extensive
collection of snowdrops. Orchard
with various fruit trees now well
established. Vegetable and colourful
garden planting below the orchard.
Alpine house with sand plunge beds
for alpines and bulbs. Greenhouse
with many interesting succulents.
Garden access to a paved area at
back of house for wheelchair users.
&

7 **THE COTTAGE NURSING
HOME**
54 Hendy Road, Mold, CH7 1QS.
Blue Ocean Care Group. *12m W of
Chester. From Mold town centre take
A494 towards Ruthin then follow
yellow NGS signs.* Sun 11 July
(2-5). Adm £2, chd £0.50. Light
refreshments. Donation to British
Heart Foundation.

Beautiful garden set in approx 1 acre.
Well-established shrubs, herbaceous
plants and abundance of colourful
window boxes and tubs. Recently remodelled garden we have now added
a waterfall feature incl a fish pond
and small area to grow vegetables.
The central courtyard has two water
features. At the bottom of the garden
lies a wooden summer house and
extended seating.
& 🐕 ✿ ☕ 🍴

8 **DOVE COTTAGE**
Rhos Road, Penyffordd,
Chester, CH4 0JR. Chris &
Denise Wallis, 01244 547539,
dovecottage@supanet.com. *6m
SW of Chester. Leave A55 at J35
take A550 to Wrexham. Drive 2m,
turn R onto A5104. From A541
Wrexham/Mold Rd in Pontblyddyn
take A5104 to Chester. Garden opp
train stn.* Visits by arrangement
June to Aug for groups of 5+. Adm
£4, chd free. Home-made teas.

Hilbre, Manor Close

Approx 1½ acre garden, shrubs and herbaceous plants set informally around lawns. Established vegetable area, 2 ponds (1 wildlife), summerhouse and woodland planted area. Gravel paths.

♿ ✱ 🚗 🚌 ☕

9 ◆ ERLAS VICTORIAN WALLED GARDEN

Bryn Estyn Road, Wrexham, LL13 9TY. Erlas Victorian Walled Garden, 01978265058, info@erlas.org, www.erlas.org. *From A483 follow signs for the Wrexham Ind Est. From A5156 follow signs to Wrexham on A534 (Holt Rd). At 2nd r'about on Holt Rd take 1st L on to Brynestyn Rd, for ½m. For NGS: Sun 12 Sept (10.30-3.30). Adm £4, chd free. Home-made teas. For other opening times and information, please phone, email or visit garden website.*
Home of the Erlas Victorian Walled Garden charity, our garden is a place of work, solace & inspiration for adults of all abilities. A garden of 4 parts. The Walled Garden has many delights incl a centuries-old Mulberry tree; The West Garden is full of fruit, veg, herbs & our Roundhouse; The Orchard has a mixtura of Apple & Pear varieties. Our ecology area provides a haven for flora & fauna. We have wheelchair access throughout the garden, howovor it is on a slopo, but the gradiont is not too extreme.

♿ 🐕 ✱ ☕

10 GARTHEWIN

Llanfairtalhaiarn, LL22 8YR. Mr Michael Grime, 01745 720288, michaelgrime12@btinternet.com. *6m S of Abergele & A55. From Abergele take A548 to Llanfair TH & Llanrwst. Entrance to Garthewin 300yds W of Llanfair TH on A548 to Llanrwst. SatNav misleading. Mon 31 May (2-6). Adm £5, chd free. Home-made teas. Visits also by arrangement Apr to Oct for groups of up to 30.*
Valley garden with ponds and woodland areas. Much of the 8 acres has been reclaimed and redesigned providing a younger garden with a great variety of azaleas, rhododendrons and young trees, all within a framework of mature shrubs and trees. Teas in old theatre. Chapel open. Some stalls to promote local arts, crafts and foods.

🐕 ☕

11 HAFODUNOS HALL

Llangernyw, Abergele, Conwy, LL22 8TY. Dr Richard Wood, www.hafodunoshall.co.uk. *1m W of Llangernyw. ½ way between Abergele & Llanrwst on A548. Signed from opp Old Stag Public House. Parking available on site. Sat 29 May (11-5). Adm £5, chd free. Home-made teas in Victorian conservatory.*
Historic garden undergoing restoration after 30years of neglect surrounds a Sir George Gilbert Scott Grade I listed Hall, derelict after arson attack. Unique setting. ½m tree-lined drive, formal terraces, woodland walks with ancient redwoods, laurels, yews, lake, streams, waterfalls and a gorge. Wonderful rhododendrons. Uneven paths, steep steps. Children must have adult supervision. Most areas around the hall accessible to wheelchairs by gravel pathways. Some gardens are set on slopes.

♿ 🐕 ✱ 🚗 ☕

12 HILBRE, MANOR CLOSE

Manor Close, Bishops Wood Road, Prestatyn, LL19 9PH. Robert & Margaret Smith, 01745853628, margaret.smith8101@gmail.com. *From A458 turn N at T-lights up Ffordd Las, turn 2nd R before top of the hill into Stoneby Dr. 1st L onto Orme View Dr then first L Manor Close. Park anywhere near here. Walk up grass track. Visits by arrangement Mar to Sept for groups of up to 20. Adm £3.50, chd free. Home-made teas. Teas may be arranged when booking*
Hilbre is surrounded by a cottage garden with fish pond and soft fruit. One side has hostas, heucheras and geraniums, the other side has helleborcs, and fuchsias. In spring the garden is full of daffodils and tulips. In the last few yrs it has been steeply terraced with different narrow levels featuring a woodland walk, orchard, wildflower garden and climbing roses. Sea views from top. Wheelchair access to front garden. Steep steps lead to the terraces and are unsuitable for people with walking difficulties.

🐕 ✱ ☕

The National Garden Scheme searches the length and breadth of England and Wales for the very best private gardens

GROUP OPENING

13 NEW LLANDEGLA VILLAGE GARDENS

Llandegla, LL11 3AW. *Please follow NGS signs for parking in Llandegla village. Mini bus available from car park to take visitors to out-lying gardens. Sun 18 July (11-5). Combined adm £6, chd free. Light refreshments.*

BRYN EITHIN
Mr & Mrs A Fife.
Open on Sun 18 July

NEW BRYNIAU MANOR
Mrs Hilary Berry.

THE GATE HOUSE, RUTHIN ROAD
Rod & Shelagh Williams.
Open on Sun 18 July

Llandegla is a small Welsh village on the banks of the River Alyn in North East Wales. Nestling below the Clwydian Hills, the village lies in an Area of Outstanding Natural Beauty. Bryn Eithin is a large garden with variety of colourful bedding, perennial plants and extensive views over Horseshoe Pass and Clwydian Range. Bryniau Manor is a variety of mature trees, shrubs and herbaceous borders. The Gate House is a garden with a wide variety of features - colourful shrubs and borders, water features, orchards and native trees, vegetable and fruit plots.

🐕 ☕

14 MOSTYN HALL

Mostyn, Holywell, CH8 9HN. Lord Mostyn, www.mostynestates.co.uk. *Use J31 A55 towards Tre Mostyn or A458 turning through Rhewl Mostyn. Sun 16 May (10-4). Adm £5, chd free. Light refreshments in Mostyn Hall.*
Mostyn Hall is set in approx 25 acres of gardens overlooking the Dee estuary. Large lawns lead to camellia and rhododendron lined walks among Victorian specimen trees. Other areas incl the rose garden, Japanese dell, herbaceous borders, orchard and 3 acre walled Victorian kitchen garden. Generally accessible to wheelchairs with some gravel areas and slight gradients.

♿ 🐕 ☕

15 PLAS COCH
Llanychan, Ruthin, LL15 1UF.
Sir David & Lady Henshaw,
Enquiries@annedd.com. *Situated
on B5429 between villages of
Llandyrnog & Llanbedr DC.* Sat 5,
Sun 6 June (11-4). Adm £3, chd
free. Home-made teas.
Well established country garden with
deep and varied herbaceous borders,
vegetable garden and fruit trees with
recently planted heritage variety small
orchard. Other sections incl small
yard garden, pond areas, three seater
tybach (outside privy) MG TC 1949,
all in the centre of the vale of Clwyd
with extensive views towards the
Clwydian Hills. Wheelchair access but
gravel paths.

16 PLAS Y NANT
Llanbedr Dyffryn Clwyd,
Ruthin, LL15 1YF. Lesley
& Ian, 01824 705444,
lesleycallister@icloud.com. *From
A494 turn onto B5429 Graigfechan,a
pprox 1m, 4th turning L private rd. If
using Satnav follow postcode LL15
2YA. Proceed through farmyard
continue & keep L, uphill to cottage
on R, go straight onto forest track.
Travel 0.8m on track, drive on R.*
Visits by arrangement Apr to Sept.
Adm £4, chd free. Refreshments to
be discussed when booking.
Listed Gothic Villa in a serene upland
valley (AONB) amid seven acres
of gardens, bluebell woods and
stream. Rhododendrons, magnolia
and specimen trees abound, formal
parterre of clipped box and yew.
Embryonic Dragon's Head rose
garden and trelliage. Procession of
seasonal colour led by snowdrops,
primroses and daffodils. Other
aspects incl water features, loggia,
summer and greenhouses, beehives.
Dragon Head carved into fallen tree
trunk that has formed an archway
over Ha-Ha.

17 NEW ◆ ST KENTIGERN
HOSPICE
Upper Denbigh Road, St
Asaph, LL17 0RS. Ms Laura
Ellis-Bartlett, 01745585221,
mail@stkentigernhospice.org.uk.
*Exit the A55 at J27 & follow the A525
through St Asaph High St, turn R at
the mini r'about, passed the High
School, turn L into the Pure housing
estate & follow signs to the Hospice.*
For NGS: Sun 25 Apr, Sun 15 Aug
(11-3). Adm £2.50. With breakfast,

lunch and afternoon teas
available. A variety of hot drinks
can be purchased and a daily
selection of cakes made by our
in-house team. For other opening
times and information, please
phone or email.
Recently updated as part of
a major expansion project, St
Kentigern Hospice has a number
of gardens built into the Hospice
grounds including a number of open
courtyards available to view during
our Open Days. Our onsite cafe
will be available for visitors to enjoy
tea, coffee and cakes during the
event. Tours around the Hospice can
be arranged with prior notice and
depending on availability.

18 NEW SAITH FFYNNON FARM
Downing Road, Whitford, Holywell,
CH8 9EN. Mrs Jan Miller-Klein,
01352 711198, Jan@7wells.org,
www.7wells.co.uk. *1m outside
Whitford village. from Holywell
follow signs for Pennant Park Golf
Course. Turn R just past the Halfway
House. Take 2nd lane on R (signed
for Downing & Trout Farm) & Saith
Ffynnon Farm is the 1st house on
R.* Visits by arrangement Sept &
Oct for groups of up to 10. During
Covid-19 restrictions no groups
larger than 4 allowed. Adm £4, chd
£2. Donation to Plant Heritage.
Wildlife garden 1 acre and re-wilded
meadows of 8 acres, incl ponds,
woods, wildflower meadows,
butterfly and bee gardens, Medieval
herbal, natural dye garden and the
Plant Heritage National Collection
of Eupatorium. Guided walks with
the owner. Other displays may
be possible depending on Covid
restrictions at the time of opening.
Hard surface from gate to patio and 2
sections of the garden. No wheelchair
access on the damp meadows.

19 TAL-Y-BRYN FARM
Llannefydd, Denbigh,
LL16 5DR. Mr & Mrs Gareth
Roberts, 01745 540256,
falmai@villagedairy.co.uk,
www.villagedairy.co.uk. *3m W of
Henllan. From Henllan take rd signed
Llannefydd. After 2½m turn R signed
Llaeth y Llan. Garden ½m on L.*
Visits by arrangement Mar to Oct.
Adm £4, chd free. Refreshments to
be discussed when booking.
Medium sized working farmhouse
cottage garden. Ancient farm

machinery. Incorporating ancient
privy festooned with honeysuckle,
clematis and roses. Terraced arches,
sunken garden pool and bog garden,
fountains and old water pumps. Herb
wheels, shrubs and other interesting
features. Lovely views of the Clwydian
range. Water feature, new rose tunnel,
vegetable tunnel and small garden
summer house.

20 NEW TYN RHOS
Ffordd Y Rhos, Treuddyn,
Mold, CH7 4NJ. Karen & Robert
Waight. *4m S of Mold. On A541
at Pontblyddyn, turn onto A5104
toward Corwen. After 2.3m turn
R onto Ffordd-y-Llan, signed for
Treuddyn. After 0.4m turn L at Xrds
onto Ffordd-y-Rhos. After 0.2m take
farm track on R.* Sun 8 Aug (1-5).
Adm £5, chd free.
5 acre garden, Waights wood,
orchard, vegetable plot. Main garden
planning by Jenny Hendy, laid to
herbaceous planting, enhanced
during summer with home grown
annuals. Path winds past the natural
pond and under planting to the British
woodland containing native bulbs,
leading to paddocks and hives.
Rare breed pigs roam part of the
wood and paddock. Newly planted
orchard contains Denby plum trees.
A graveled yard and small steps (2
inch) need to be negotiated. For the
intrepid wheelchair user the woodland
walk is a possibility.

21 WHITE CROFT
19 Ffordd Walwen, Lixwm,
Holywell, CH8 8LW. Mr John &
Mrs Mary Jones, 01352781829,
fforddwalwen1@talktalk.net. *From
A541 turn onto the B5121 for Lixwm,
From A55 come off at J32 or J32A.
Drive South, follow Lixwm signs.* Sun
1 Aug (1-5). Adm £4, chd free.
Light refreshments. Visits also by
arrangement in Aug for groups of
5 to 10.
Wrap around garden incl Cottage
style planting, Dahlia bed, Rock
garden, Vegetable plot, Fruit trees.
Greenhouse, Stream and Fish pond,
Pergola, Balcony views over Clwydian
Range and Japanese style garden.
The garden has been established
over the last 8 yrs. It was originally
just lawn and conifer trees. A double
garage containing a Westfield Kit Car
and a sports car. Car park available.

Erlas Victorian Walled Garden

POWYS

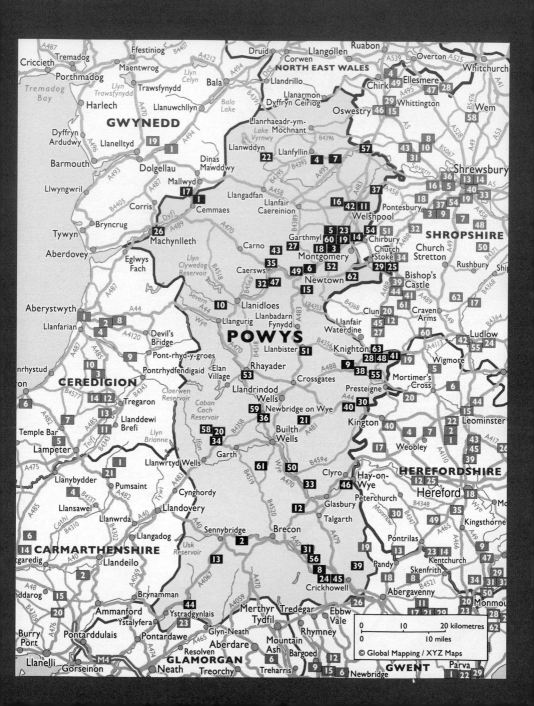

A three hour drive through Powys takes you through the spectacular and unspoilt landscape of Mid Wales, from the Berwyn Hills in the north to south of the Brecon Beacons.

Through the valleys and over the hills, beside rippling rivers and wooded ravines, you will see a lot of sheep, pretty market towns, half-timbered buildings and houses of stone hewn from the land.

The stunning landscape is home to many of the beautiful National Garden Scheme gardens of Powys. Gardens nestling in valleys, gardens high in the hills, wild-life gardens, riverside gardens, walled gardens, grand gardens, cottage gardens, gardens in picturesque villages, gardens in towns... they can all be found in Powys!

Here in Powys all is the spectacular, the unusual, the peaceful and the enchanting all opened by generous and welcoming garden owners.

Below: Cwm Farm

Volunteers

North Powys County Organiser
Susan Paynton 01686 650531
susan.paynton@ngs.org.uk

County Treasurer
Jude Boutle 07702 061623
jude.boutle@ngs.org.uk

Publicity
Helen Anthony 01686 941795
helen.anthony@ngs.org.uk

Social Media
Nikki Trow 07958 958382
nikki.trow@ngs.org.uk

Booklet Co-ordinator
Carole Jones 01650 511176
jonesey200@gmail.com

Assistant County Organisers
Simon Cain 07958 915115
simon.cain@ngs.org.uk

Sue McKillop 01650 511333
sue.mckillop@ngs.org.uk

Lucy Price 07538 038401
lucy.price@ngs.org.uk

South Powys County Organiser
Christine Carrow 01591 620461
christine.carrow@ngs.org.uk

County Treasurer
Steve Carrow 01591 620461
steve.carrow@ngs.org.uk

Publicity Officer
Gail Jones 07974 103692
gail.jones@ngs.org.uk

Assistant County Organisers
Bob & Andrea Deakin
01982 551718
bandeak@googlemail.com

Chris Harris 01982 551418
chris.h48.ch@gmail.com

Sadie Woods 07932 324101
sadie_wc@hotmail.co.uk

f @powysngs
🐦 @PowysNGS
📷 @powysngs

OPENING DATES

All entries subject to change. For latest information check **www.ngs.org.uk**
Map locator numbers are shown to the right of each garden name.

March

Sunday 28th
Oak Cottage 42

April

Sunday 4th
Gorsty House 25

Monday 5th
Oak Cottage 42

Wednesday 7th
Bryngwyn Hall 7

Sunday 11th
Oak Cottage 42

Saturday 24th
Little House 32

Sunday 25th
Oak Cottage 42

May

Sunday 2nd
NEW Cwm Farm 14
NEW Pontsioni House 50

Saturday 8th
◆ Dingle Nurseries & Garden 16

Sunday 9th
◆ Dingle Nurseries & Garden 16
NEW Pontsioni House 50

Saturday 15th
Rock Mill 52

Sunday 16th
◆ Gregynog Hall & Garden 27
Penmyarth House - Glanusk Estate 45
Rock Mill 52

Sunday 23rd
Gorsty House 25
Llysdinam 36

Wednesday 26th
◆ Grandma's Garden 26

Monday 31st
Llanstephan House 33

Llwyn Madoc 34

June

Saturday 5th
Tranquillity Haven 55

Sunday 6th
The Neuadd 39
The Rock House 51
Tranquillity Haven 55

Friday 11th
The Rock House 51

Saturday 12th
NEW No 2 The Old Coach House 41
Tranquillity Haven 55
1 Ystrad House 63

Sunday 13th
Hurdley Hall 29
NEW No 2 The Old Coach House 41
Tranquillity Haven 55
1 Ystrad House 63

Wednesday 16th
NEW The Kitchen Garden 31

Saturday 19th
Tremynfa 57

Sunday 20th
Gorsty House 25

Pen-y-Maes 46
Tremynfa 57

Friday 25th
White Hopton Farm 62

Sunday 27th
Esgair Angell 17
NEW New Radnor Gardens 40

July

Every Friday from Friday 9th
◆ Welsh Lavender 61

Saturday 3rd
NEW Abercamlais House 2
The Hymns 30

Sunday 4th
Cwm-Weeg 15
Hurdley Hall 29
The Hymns 30
NEW South Street Gardens 53

Monday 5th
Garthmyl Hall 19

Saturday 10th
Bachie Uchaf 4
Stockton Mill 54

Sunday 11th
Bachie Uchaf 4

Vaynor Park

© Simon Morgan

Garthmyl Hall

THE GARDENS

1 ABERANGELL, THE OLD COACH HOUSE

Dolcorsllwyn, Aberangell, Machynlleth, SY20 9AB. Sue McKillop, www. theoldcoachhousecottage.co.uk. *On A470 midway beween Dolgellau & Machynlleth. From Mallwyd r'about to Cemmaes Rd, turn R after 3m just after turn for Aberangell village. From Machynlleth, come through Cemmaes and Cwm Llinau. 1m on L.* **Sun 22 Aug (11-4). Adm £4, chd free. Home-made teas.**
Nestled in heart of Dyfi Valley this small, cottage style garden is haven for birds and pollinating insects. Narrow paths take you around the flower and vegetable beds to a little pond. Planting is informal with mostly perennials and shrubs. Secluded seating areas allow the visitor to relax and enjoy different aspects of the garden with views down to R Dyfi. Home made preserves and conserves for sale. A gravel drive and cobbled area in front of the house leads to a grassy slope into the garden. There are a few low steps within the garden.
♿ ✿ ☕

2 NEW ABERCAMLAIS HOUSE

Abercamlais, Brecon, LD3 8EY. Mr Anthony & Mrs Andrea Ballance, www.abercamlais.com. *6m W of Brecon. Take the A40 from Brecon towards Sennybridge. After 6m turn R immed after the bus stop & follow yellow NGS signs to the car park.* **Sat 3 July (11-4). Adm £7.50, chd free. Home-made teas.**
Abercamlais is a splendid Grade 1 listed mansion house set in extensive grounds. At the entrance is a lodge and a beautiful drive contains a Wellingtonia, a Metasequoia and some fine Oak trees. There is a fine octagonal dovecote which used to be a privy. The grounds also contain a unique Victorian suspension bridge that leads to a large walled garden which is currently being restored. Easy access to some parts of the garden. No wheelchair access over the suspension bridge, however, there is an alternative route via a stone bridge.
♿ 🐐 🚏 ☕

3 ABERNANT

Garthmyl, SY15 6RZ. Mrs B M Gleave, 01686 640494. *1½m S of Garthmyl. On A483 midway between Welshpool & Newtown (both 8m). Approached over steep humpback bridge with wooden statue of workman. Straight ahead to house & parking.* **Visits by arrangement Apr to July for groups of up to 20. Adm £4, chd free. Light refreshments.**
3 acres incl large cherry blossom orchard (85 trees). Knot garden, box hedging, formal rose garden, rockery, pond, shrubs, ornamental trees, archaic sundials, fossilised wood and stone heads. Additional woodland of 9 acres borrowed views of the Severn Valley. 85 cherry trees in blossom in late April. Roses in late June. Picnics welcome from 12pm.
🐐 ☕

4 BACHIE UCHAF

Bachie Road, Llanfyllin, SY22 5NF. Glyn & Glenys Lloyd. *S of Llanfyllin. Going towards Welshpool on A490 turn R onto Bachie Rd after Llanfyllin primary school. Keep straight for 0.8m. Take drive R uphill at cottage on L.* **Sat 10, Sun 11 July (12.30-5). Adm £4.50, chd free. Home-made teas.**
Inspiring, colourful hillside country garden. Gravel paths meander around extensive planting & over streams cascading down into ponds. Specimen trees, shrubs & vegetable garden. Enjoy the wonderful views from one of the many seats; your senses will be rewarded. Winner Best Flower & Shrub Garden, Llanfyllin Show and Winner Montgomeryshire Agricultural Assoc Best Garden Competition 2019. Finalist Montgomeryshire Best Garden Competition.
🐐 ✿ ☕

GROUP OPENING

5 BERRIEW VILLAGE GARDENS

Berriew, Welshpool, SY21 8BA. *A483 5m S Welshpool. From Welshpool take A483 S for approx 4m. Turn R onto B4390. Continue through village and turn R to Car Park at school. Map and admission tickets at car park. All within walking distance of car park.* **Sun 8 Aug (12-5). Combined adm £6, chd free. Home-made teas in the Old School in centre of village.**

NEW BRONAFON

Gill Evans.

3 CHURCH TERRACE

Mrs Jane Hancock.

1 GLAN YR AFON

Ms Lesley Ellis.

THE OLD COURT HOUSE

Michael Davis & Andrew Logan.

NEW RHIEW HOUSE

Mr Richard & Mrs Fiona Noyce.

The picturesque village of Berriew is on the Montgomeryshire Canal with the R Rhiew flowing through its heart. Black & white cottages, church, two pubs, shops, William O'Brien artist/blacksmith forge and Andrew Logan Museum of Sculpture. Five very different gardens: 1 Glan yr Afon colourful small garden with bark paths, veg plot and pergola covered with grape vine: The Old Court House idyllic situation on banks of R Rhiew terrace with metal staircase by Berriew sculptor/metalworker William O'Brien, raised beds for soft fruits & vegetables: Rhiew House combines a degree of order and colour with a sense of wildness: Bronafon large lawn sweeping down to R Rhiew with colourful mixed beds inc beautiful cercis (forest pansy tree): 3 Church Terrace ¼-acre plantsman's garden with river below and diverse range of rare and unusual plants to create interest throughout yr. Partial wheelchair access.
♿ ✿ ☕

6 BRYN TEG

Bryn Lane, Newtown, SY16 2DP. Novlet Childs. *N side of Newtown. Take Llanfair Caereinion Rd towards Bettws Cedewain and turn L before hospital. Up hill on L.* **Sat 24 July (10-5), also open Ponthafren. Sun 25 July (11-3). Adm £4, chd free.**
Amazing exotic secret Caribbean garden in centre of Newtown planted to remind me of my childhood in Jamaica. An exciting walk through the jungle. High above the head are banana leaves & colourful climbers. A winding path takes you on a journey through another land. Sounds of water fill the air; explosion of colourful intermingling flowers & plants with giant leaves -paulownia, bamboos, tetrapanex. A huge number of plants on many levels. All shapes, sizes and colours mixed together as found in tropical jungles. Feel transported to another continent. Even in the rain you will get that jungle experience. Manual wheelchair access to most of the garden. Not suitable for mobility scooters.
♿ 🐐 ✿

7 NEW **BRYNGWYN HALL**
Bwlch-y-Cibau, Llanfyllin, SY22 5LJ.
Auriol Marchioness of Linlithgow,
www.bryngwyn.com. *3m SE
Llanfyllin. From Llanfyllin take A490
towards Welshpool for 3m turn L up
drive just before Bwlch-y-Cibau.*
**Wed 7 Apr (10-4). Adm £6,
Chd free. Home-made Teas.**
Stunning grade II* listed 9 acre
garden with 60 acres parkland
design inspired by William Emes.
Prunus subhirtella 'Autumnalis',
varieties of hamamelis, mahonia,
early flowering daphnes, corylopsis
and chimonanthus. Woodland
garden carpeted with snowdrops
then stunning show of thousands of
daffodils, camassias and fritillaries in
the long grass down to serpentine
lake. Unusual trees and shrubs &
unique Poison Garden.
  ♿ 🐕 🚗 ☕ ⚲

8 **BRYNTEG**
Old Road, Bwlch, Brecon, LD3 7NJ.
Maria Pritchard, 01874 730582,
m-pritchard@live.co.uk. *4m NW
of Crickhowell. Access to Old Rd
entrance situated next to the only
village public house, The New Inn.
Brynteg is situated 300m from the
main A40.* **Visits by arrangement Mar
to Oct for groups of up to 10. Adm
£4, chd free. Home-made teas.**
Brynteg, built 1887 with an interesting
background & unique garden,
situated in the Brecon Beacons
National Park with unobstructed
views of the Black Mountains.
Herbaceous Borders, herb beds fruit
& specimen trees. Spend the day
at a creative or wellbeing workshop
offered to small groups within the
beautiful setting of Brynteg. Contact
me for info and booking. We look
forward to welcoming you. An
established Magnolia welcomes all
who visit Brynteg.
  ♿ ✿ 🏠 ☕ ⚲

9 NEW **CEFNSURAN FARM**
Llangunllo, Knighton, LD7 1SL.
Gill & Gordon Morgan,
cefnsuran@hotmail.com. *6m west
Knighton. A488 from Knighton
towards Llandrindod Wells. After 4m
turn R B4356. After 1m turn R sign
post Cefnsuran/Rally School follow
track 3/4 m.* **Sat 14, Sun 15 Aug
(1-5). Adm £4, chd free. Home-
made teas.**
1 acre sheltered farmhouse garden
set in hollow at 1000 ft in beautiful
secluded location. Surrounded by

fields, ponds and woodland with
wonderful selection of shrubs, trees
& flowers plus a wild flower area &
ancient hollow oak. Easy access to
all of the garden. Walks can be taken
around the two large ponds, along
farm trails and also the Glyndwr Way
which passes through our garden.
Wildlife haven with bees, house
martins and swallows. sloping paths
mainly level lawns.
  ♿ 🏠 ⚲

10 **CEUNANT**
Old Hall, Llanidloes, SY18 6PW.
Sharon McCready, 01686 412345,
sharon.mccready@yahoo.co.uk.
*2m W of Llanidloes. Leave Llanidloes
over Shortbridge St, turn L along
Pen y Green Rd, approx ½m turn L
signposted Llangurig, Glyn Brochan.
Follow signs property is on R.* **Sun
11 July, Sun 1 Aug (1-5). Adm £5,
chd free. Home-made teas.**
4 acre riverside garden started in
2013 in a beautiful setting on the
River Severn just 7m from the source.
Wooded wildlife area and 2 wildlife
ponds, orchard, ornamental garden,
herbaceous borders, scented seating
areas, Hobbit House, riverside
path, meadow & veg plot. Full of
unusual features. Planting aimed at
pollinators & encouraging wildlife.
Sleeping mud maid, clay oven and
brook home to Indian Runner ducks,
lots of seating areas and recycled
features. Accommodation available
in shepherd's hut and Hobbit
house. www.holidaycottages.co.uk/
cottage/56492-severn-way. Some
rough terrain and lots of slopes.
  🐕 ✿ 🏠 ☕ ⚲

11 **1 CHURCH BANK**
Welshpool, SY21 7DR. Mel &
Heather Parkes, 01938 559112,
melandheather@live.co.uk. *Centre
of Welshpool. Church Bank leads
onto Salop Rd from Church St.
Follow one way system, use main car
park then short walk. Follow yellow
NGS signs.* **Visits by arrangement
Apr to Aug for groups of 5 to 30.
Adm £3.50, chd free. Home-made
teas.**
An intimate jewel in the town. C17
barrel maker's cottage with museum
of tools & motor memorabilia. Gothic
arch leads to shell grotto with mystic
pool of smoke & sounds, bonsai
garden, fernery, many interesting
plants & unusual features.
  🐕 ☕

12 **CHURCH HOUSE**
Llandefalle, Brecon, LD3 0ND. Chris
& Anne Taylor. *Between Brecon
& Llyswen, off A470, signposted
Llandefalle, then single track lane for
½m. Parking is next to Church.* **Sat
11, Sun 12 Sept (2-5.30). Adm £5,
chd free. Home-made teas.**
One acre garden, subdivided into
different areas, each with its own
character incl orchard. Skilfully
landscaped in the 1980s to provide
terraces, generous borders and year
round interest. New owners have
added small pond, herb garden,
Stumpery and rill. Tranquil places to
sit with fine views to Black Mts. Next
to St Matthew's Church, 15th century
with notable earlier features. Some
steps. Wheelchair access to terraces
over sloping lawns but only if ground
is dry.
  ♿ ✿ ☕ 🏠

GROUP OPENING

13 **CRAI GARDENS**
Crai, LD3 8YP. *13m SW of Brecon.
Turn W off A4067 signed 'Crai'
Village hall is 50yds straight ahead.
Park here for admission, information
about gardens, map, teas.* **Sun 15
Aug (2-5). Combined adm £5, chd
free. Home-made teas in Village
Hall. Gluten free & wheat free
refreshments available.**
Set against the backdrop of Fan
Gyhirych and Fan Brycheiniog, at
1000 ft above sea level the Crai valley
is a hidden gem, off the beaten track
between Brecon and Swansea. Those
in the know have long enjoyed visiting
our serene valley, with its easy access
to the hills and its fabulous views. In
difficult climatic conditions - fierce
wind and heavy rain are frequent
visitors - the Crai Gardens reflect a true
passion for gardening. The gardens
come in a variety of size, purpose and
design, and include: a range of shrubs,
perennials and annuals; vegetables
and fruits; raised beds; water features;
patio containers and window boxes,
manicured lawns. Not to forget the
chickens and ducks. And to complete
your Sunday afternoon, come and
enjoy the renowned hospitality of the
Crai ladies by sampling their delicious
home-made cakes in the village hall,
also ice creams for sale. Some of
the gardens are fully, some partly,
accessible by wheelchair. Some have
gravel paths or steep grassy banks.
  ✿ ☕

Tynrhos

14 NEW ◆ CWM FARM

Forden, Welshpool, SY21 8NB. Mr Michael & Mrs Gemma Hughes, 07957 810225. *3m S Welshpool. from A483 turn L A490 for 2.4m then ahead on B4388 for 1.1m.* **Sun 2 May (12-5). Adm £5, chd free. Home-made teas. Gluten and dairy free options available.**
Charming C19th farmhouse down wooded cwm with impressive crag. 5 acre garden in stunning location with panoramic view of Corndon & Roundton Hills. Bluebells, herbaceous beds, Welsh apple orchard, pond, stream, wildflower meadow, wide lawns and paths to wander and places to sit. Engaging mix of wild and cultivated. Gemma's studio will be open, showing her abstract sculptures that re-reflect daylight and colour. Live light classical music in the garden, weather permitting. Wheelchair drop off point by farmhouse. Highest view point from top meadow has steep steps so not accessible.

15 CWM-WEEG

Dolfor, Newtown, SY16 4AT. Dr W Schaefer & Mr K D George, 01686 628992, wolfgang@cwmweeg.co.uk, www.cwmweeg.co.uk. *4½m SE of Newtown. Off Bypass, take A489 E from Newtown for 1½m, turn R towards Dolfor. After 2m turn L down asphalted farm track, signed at entrance. Do not rely on SatNav. Also signed from Dolfor village on NGS days.* **Sun 4 July (2-5). Adm £6, chd free. Home-made teas.**
2½ acre garden set within 24 acres of wildflower meadows and bluebell woodland with stream centred around C15 farmhouse (open by prior arrangement). Formal garden in English landscape tradition with vistas, grottos, sculptures, stumpery, lawns and extensive borders terraced with stone walls. Translates older garden vocabulary into an innovative C21 concept. Extensive under cover seating area in the new Garden Pavilion. Partial wheelchair access.

For further information please see garden website.

16 ◆ DINGLE NURSERIES & GARDEN

Welshpool, SY21 9JD. Mr & Mrs D Hamer, 01938 555145, info@ dinglenurseriesandgarden.co.uk, www.dinglenurseries.co.uk. *2m NW of Welshpool. Take A490 towards Llanfyllin & Guilsfield. After 1m turn L at sign for Dingle Nurseries & Garden. Follow signs and enter the Garden from adjacent plant centre.* **For NGS: Sat 8, Sun 9 May, Sat 16, Sun 17 Oct (9-5). Adm £3.50, chd free. Tea. For other opening times and information, please phone, email or visit garden website.**
4½ acre internationally acclaimed RHS partner garden on S-facing site, sloping down to lakes. Huge variety of rare and unusual trees, ornamental shrubs & herbaceous plants give yr-round interest. Set in the hills of

mid Wales this beautiful well known garden attracts visitors from Britain & abroad. Open all yr except 24 Dec - 2 Jan. Partial access to top areas of garden over gravel path for those with mobility issues.

🐕 ❄ 🚗 ☕

17 ESGAIR ANGELL
Aberangell, Machynlleth, SY20 9QG. Carole Jones, 01650 511176, jonesey200@gmail.com, www.upperbarncottage.co.uk. *Midway between Dolgellau & Machynlleth just inside Dyfi Forest. Turn off A470 towards village of Aberangell, then signed.* **Sun 27 June (12.30-4). Adm £6, chd free. Home-made teas.**
25 acres with 3 acre garden in beautiful setting in Dyfi Forest. Focusing on wildlife - incl otters. 3 large trout ponds, half acre birdseed field, bird box sculptures and unusual log bee hives, manicured lawns, raised beds, wild flowers. Aviaries housing rescue owls. Various stalls incl street food, plants etc. Partial wheelchair access, mainly laid to lawn. Access on gravelled area above lake affording excellent views.

🐕 🐕 ❄ 🚐 ☕

18 FRON HEULOG
Pentre Llifior, Berriew, Welshpool, SY21 8QJ. Annie Bratt & Tim Ward, 01686 640544, FronHeulogGardens@outlook.com. *8 miles SW from Welshpool. Map with directions to garden will be provided on confirmation of 'by arrangement' booking.* **Visits by arrangement Apr to Sept for groups of 10 to 20. Restricted parking available. Max 8 cars. Adm £6, chd free.**
Stunning 1½ acre landscaped garden in beautiful setting within 8 acres managed for wildlife. Divided into different contrasting areas with exceptional range of plants for all season colour and unusual shrubs & trees. Incl terraced rockery, dingle and potager. Adjoining wildflower meadow can be explored via network of paths & includes wildlife pond, shepherd's hut and magnificent views.

19 GARTHMYL HALL
Garthmyl, Montgomery, SY15 6RS. Julia Pugh, 07716 763567, hello@garthmylhall.co.uk, www.garthmylhall.co.uk. *On A483 midway between Welshpool & Newtown (both 8m). Turn R 200yds*

S of Nag's Head Pub. **Mon 5 July (12-5). Adm £5, chd free. Home-made teas.**
Grade II listed Georgian manor house (not open) surrounded by 5 acres of grounds. 100 metre herbaceous borders, newly restored 1 acre walled garden with gazebo, circular flowerbeds, lavender beds, wildflower meadow, pond, two fire pits and gravel paths. Fountain, 3 magnificent Cedar of Lebanon and giant redwood. Partial wheelchair access. Accessible WC.

♿ 🐕 🚗 🚐 ☕

20 GILWERN BARN
Beulah, LD5 4YG. Mrs Penelope Bourdillon, 01591 620203, pbourdillon@gmail.com, www.gilwerngarden.co.uk. *2m N of Beulah. Turn L exactly ¾m from Beulah on B4358. Follow rd for over 1m. Past Cefnhafdref Farm. L at T junction. In 600 yds fork L through stone gateposts.* **Visits by arrangement May to Aug. Open Thursdays only from 20th May until 12th August. Call to pre-book. Adm £6, chd free. Home made teas.**
Terraced garden on very challenging site situated on steep rocky hillside in the beautiful and secluded Cammarch Valley. Roses, herbaceous and shrub borders. Fine walling and gate posts making use of stone and slate found in the garden. Man-made waterfall and rill. The situation, on a steep rocky slope, is fairly dramatic and makes it interesting. The waterfall is quite a feature. Late flowering roses. The peonies should be worth seeing in late May. A new rill, recently planted.

♿ 🐕 ☕

21 NEW GLANOER
Bettws, Hundred House, Llandrindod Wells, LD1 5RP. Dave & Sue Stone, 07771767246, stone@glanoer.co.uk. *Centre of postcode area. Use postcode for satnav. From A481 take turning to Bettws & Franksbridge by Hundred House Inn. Take L fork signposted Bettws. Glanoer is 1m further along lane on L (past St Mary's Church).* **Sun 25 July (10.30-5). Adm £5, chd free. Home-made teas. Visitors welcome to have their picnics on the field within the grounds.**
Just over an acre in an alder valley with running brook. Numerous 'rooms' throughout providing seating,

vistas and interest. Includes an allotment, bluebell wood, wildlife pond, secret garden and water features. Planting for colour & pollination throughout the year for resident bees with over 50 varieties of rose. A young garden created since 2015 on boulder clay from open space. Free-range chickens, bee hives, allotment, terraces. Sorry not suitable for wheelchairs.

🐕 ❄ 🚐 ☕ 🪑

22 NEW 1 GLANRAFON
Llanwddyn, Oswestry, SY10 0LU. Margaret Herbert. *24m W of Oswestry. From Llanfyllin, follow brown signs to Lake Vyrnwy. Through village of Llanwddyn & turn L across dam then L past Artisans Cafe & park in public car park by playground. Follow yellow signs to garden.* **Sat 17, Sun 18 July (11-4.30). Adm £4.50, chd free. Home-made teas.**
1 acre secluded garden under development. As a keen plantsperson. I have been establishing this steeply sloping, densely planted wildlife garden with hundreds of varieties of shrubs & perennials over the past 5 years. Very productive raised veg beds, fruit cages and polytunnels, large pond. Large greenhouse with ornamental & edible crops. Fruit trees, bushes and vegetables, many of which are unusual varieties. Yard with chickens and ducks. Wheelchairs can access the upper & lower levels via the access road & separate gates. Disabled car park down access road by house.

♿ 🚗 ☕

"I was amazed to discover that the National Garden Scheme is Marie Curie's largest single funder and has given the charity nearly £10 million over 25 years. Their continued support makes such a difference to me and all Marie Curie Nurses on the frontline of the coronavirus crisis, as we continue to provide expert care and support to people at end of life." - Tracy McWilliams, Marie Curie Nurse

24 GLIFFAES COUNTRY HOUSE HOTEL

Gliffaes Rd, Crickhowell, NP8 1RH. Mrs N Brabner & Mr & Mrs J C Suter, 01874 730371, calls@gliffaeshotel.com, www.gliffaes.com. *3½ m W of Crickhowell. From Crickhowell, drive W for 2½ m, L off A40 and continue for 1m.* **Sun 24 Oct (1.30-4.30). Adm £5, chd free. Cream teas.** Gliffaes Hotel has, within its 33 acres of grounds, one of the best small arboretums in Wales with an enviable collection of specimen trees from around the globe planted by far sighted Victorian collectors. It also has numbers of much older trees shedding light on how the woodlands were used and managed in past times. There will be a guided Autumn colours Tree Walk at 2pm. Gliffaes is a country house hotel and is open for lunch, bar snacks, afternoon tea and dinner to non residents and garden visitors. Wheelchair ramp to the west side of the hotel. In dry weather main lawns accessible, but more difficult if wet.

25 GORSTY HOUSE

Hyssington, Montgomery, SY15 6AT. Gary & Annie Frost. *A488 N from Bishop's Castle. Approx 3½ m, turn L to Hyssington, follow NGS signs. From Churchstoke towards Bishop's Castle 1m then turn L to Hyssington, follow NGS signs. 250yd walk from Village carpark, disabled parking at house.* **Sun 4 Apr, Sun 23 May, Sun 20 June, Sun 15 Aug (1.30-5.30). Adm £5, chd free. Home-made teas in Village Hall for April opening (or in poor weather), otherwise served in garden. Gluten free and vegan cakes available.** Quintessential country garden, abundantly planted with cottage garden perennials that follow the seasons, magnet for bees and butterflies, & haven for wildlife. 2 acres of garden includes deep borders, pond, young orchard underplanted with wildflowers, and acre of flowering hay meadow. Thousands of daffodils in spring, 130+ roses in June. Wonderful views. Beautiful cottage garden planting. A haven for wildlife. Sorry no dogs.

26 ◆ GRANDMA'S GARDEN

Dolguog Estates, Felingerrig, Machynlleth, SY20 8UJ. Richard Rhodes, 01654 702244, info@plasdolguog.co.uk, www.plasdolguog.co.uk/. *1½ m E of Machynlleth. Turn L off A489 Machynlleth to Newtown rd. Follow brown tourist signs to Plas Dolguog Hotel.* **For NGS: Wed 26 May (10.30-4.30). Adm £4, chd free. Home-made teas. For other opening times and information, please phone, email or visit garden website.** Inspiration for the senses, unique, fascinating, educational and fun. Strategic seating, wildlife abundant, 9 acres of peace. Sculptures, poetry arboretum. Seven sensory gardens, wildlife pond, riverside boardwalk, stone circle, labyrinth. Azaleas and bluebells in May. Children welcome. Plas Dolguog Hotel open their café in the conservatory - the hotel is the admission point - serving inside and out on patio overlooking gardens.

27 ◆ GREGYNOG HALL & GARDEN

Tregynon, Newtown, SY16 3PL. Gregynog Trust, 01686 650224, enquiries@gregynog.org, www.gregynog.org. *5m N of Newtown. From main A483, take turning for Berriew. In Berriew follow sign for Bettws then for Tregynon (£2.50 car parking charge applies).* **For NGS: Sun 16 May (10-4). Adm £5, chd free. Light refreshments at Courtyard Cafe. For other opening times and information, please phone, email or visit garden website.** Gardens are Listed Grade I and some features of William Emes, 18th Century landscape architect, still remain. Set within 750 acres of Gregynog Estate a designated National Nature Reserve in 2013. Parkland with fountains, lily lake and water garden. A mass display of rhododendrons & azaleas & unique yew hedge create a spectacular backdrop to the sunken lawns. Unusual trees. David Austin roses. Courtyard cafe serving morning coffee, light lunches and Welsh afternoon teas. Some gravel paths.

28 HEBRON

Ludlow Road, Knighton, LD7 1HP. Kevin Collins & Anita Lewis, 01547 529576, kd.collins@hotmail.co.uk. *From Knighton, take A4113 towards Ludlow, house on R opp playing fields where parking is available.* **Visits by arrangement Mar to Oct for groups of up to 20. Adm £4, chd free.** ½ acre Victorian garden full of surprises. Divided into distinct rooms by yew & beech hedges with box topiary, willow arches & sculptures. Variety of trees incl large silver birches, magnolia, weeping willow and cherry; shrubs provide colour throughout the season. Heathers, herbs, box edged vegetable beds and flower beds. Plenty of seating. Beehives and garden art add extra interest.

29 HURDLEY HALL

Hurdley, Churchstoke, SY15 6DY. Simon Cain & Simon Quin. *2m from Churchstoke. Take turning for Hurdley off A489, 1m E of Churchstoke. Garden is a further 1m up the lane.* **Sun 13 June, Sun 4 July, Sun 1 Aug (11-5). Adm £6, chd free. Home-made teas.** Regional Finalist in The English Garden's Favourite Garden Competition 2019. 2 acre garden incl herbaceous & mixed borders leading to 18 acres of Coronation Meadows, woodland & orchard. Visit June for meadow flowers & roses; July for lavender & herbaceous borders; or August for hot & white borders. Uneven ground & steep slopes give wide ranging views but may restrict access.

30 THE HYMNS

Walton, Presteigne, LD8 2RA. E Passey, 07958 762362, thehymns@hotmail.com, www.thehymns.co.uk. *5m W of Kington. Take A44 W, then 1st R for Kinnerton. After approx 1m, at the top of small hill, turn L (W).* **Sat 3, Sun 4 July (11-4). Adm £5, chd free. Home-made teas. Visits also by arrangement Apr to Oct.** In a beautiful setting in the heart of the Radnor valley, the garden is part of a restored C16 farmstead, with long views to the hills, and The Radnor Forest. It is a traditional garden reclaimed from the wild, using locally grown plants and seeds, and with a herb patio, wildflower meadows and a short woodland walk. It is designed for all the senses: sight, sound and smell.

31 NEW THE KITCHEN GARDEN
Ty Mawr, Llangasty, Brecon, LD3 7PJ. Rae Gervis, 07949396589, rae.gervis@lime.org.uk. *2½ m from Brecon off A40. ¼ m from Treberfydd House.* **Wed 16 June, Wed 15 Sept (11-4). Adm £10, chd free. Pre-booking essential, please visit www.ngs.org.uk/events for information & booking. Light refreshments. Visits also by arrangement June to Sept for groups of 5 to 10.**
Admission to include talk and tour. Regenerative vegetable garden. Uses no-dig practices. Growing edibles sustainably & regeneratively. Preserving & enhancing the soil is key to all we do. Overlooking Llangorse lake & at the foot of the Allt we grow, vegetables, herbs & flowers which are sold through the shop in the 18c. Granary along with produce & products which have the environment

at their core. Selling fresh vegetables, perennial herbs, organic produce and all things sustainable. Homemade, cakes, teas, cordials, ice lollies, wines & beers available.

32 LITTLE HOUSE
Llandinam, Newtown, SY17 5BH. Peter & Pat Ashcroft, www.littlehouse1692.uk. *1m from Llandinam Village Hall. Cross river at statue of David Davies on A470 in Llandinam. Follow rd for just under 1m (ignore GPS), Little House is black & white cottage on roadside. Limited parking.* **Sat 24 Apr, Sat 17 July, Sat 4 Sept (1-4.30). Adm £4, chd free. Home-made teas.**
⅓ acre plantswoman's garden on a quiet lane surrounded by fields, woodland & stream. Slate & bark paths give access to the many features incl fish & wildlife ponds, conifer, azalea, grass & mixed beds, vegetable garden, woodland, mini meadow, sensory garden, carnivorous plant & grotto water features, auricula theatre & alpine/cactus house. 500+ different plants, many propagated for sale. Spring; auricula theatre will be

at its best with other primulas, winter heathers & hellebores. Summer; the garden in full bloom & the sensory garden alive with colour & wonderful scents. Autumn; cyclamen, grasses, crocosmia, michaelmas daisies & interesting seed heads on show.

33 LLANSTEPHAN HOUSE
Llanstephan, Llyswen, LD3 0YR. Lord & Lady Milford. *10m SW of Builth Wells. Leave A470 at Llyswen onto B4350. 1st L after crossing river in Boughrood. From Builth Wells leave A470, Erwood Bridge, 1st L. Follow signs.* **Mon 31 May (1-5). Adm £5, chd free. Tea.**
20 acre garden, first laid out almost 200 years ago and in the present owner's family for more than a century, featuring a Victorian walled kitchen garden and greenhouses, 100 year old wisteria, woodland walks punctuated by azaleas and rhododendrons, specimen trees and immaculate lawns. Beautiful and celebrated views of Wye Valley and Black Mountains.

New Radnor Gardens

Monaughty Mill

34 LLWYN MADOC

Beulah, Llanwrtyd Wells, LD5 4TT. Patrick & Miranda Bourdillon, 01591 620564, miranda.bourdillon@gmail.com. *8m W of Builth Wells. On A483 at Beulah take rd towards Abergwesyn for 1m. Drive on R. Parking on field below drive - follow signs.* **Mon 31 May (1.30-5.30). Adm £5, chd free. Cream teas.**
Terraced garden in attractive wooded valley overlooking lake; yew hedges; rose garden with pergola; kitchen garden and small orchard; azaleas and rhododendrons. As the garden is situated on a slope, it is quite steep in parts.

35 NEW LLYS CELYN

Llanwnog, Caersws, SY17 5JG. Lesley & Tony Geary. *9m W of Newtown. From Newtown A489 W to A470, turn R to Machynlleth & Caersws. After 3m turn R on B4589. Follow signs. Alternative route: from Newtown B4589 from McDonalds*

direct to Llanwnog. **Sat 24, Sun 25 July (2-5). Adm £4, chd free. Home-made teas.**
An evolving one acre garden renovated and developed over the last 9 years. Extensive vegetable garden with fruit cage, polytunnel and greenhouse. Newly created herbaceous borders and bog garden. Wildlife areas incl ponds and habitats, rockery and courtyard garden. Fabulous views of the surrounding countryside and village church steeple.

36 LLYSDINAM

Newbridge-on-Wye, LD1 6NB. Sir John & Lady Venables-Llewelyn & Llysdinam Charitable Trust, 01597 860190/07748492025, llysdinamgardens@gmail.com, llysdinamgardens.co.uk. *5m SW of Llandrindod Wells. Turn W off A470 at Newbridge-on-Wye; turn R immed after crossing R Wye; entrance up hill.* **Sun 23 May (2-5). Adm £5, chd free. Cream teas. Visits also by arrangement.**

Llysdinam Gardens are among the loveliest in mid Wales, especially noted for a magnificent display of rhododendrons and azaleas in May. Covering some 6 acres in all, they command sweeping views down the Wye Valley. Successive family members have developed the gardens over the last 150yrs to incl woodland with specimen trees, large herbaceous and shrub borders and a water garden, all of which provide varied and colourful planting throughout the yr. The Victorian walled kitchen garden and extensive greenhouses grow a wide variety of vegetables, hothouse fruit and exotic plants. Gravel paths.

37 MAESFRON HALL AND GARDENS

Trewern, Welshpool, SY21 8EA. Dr & Mrs TD Owen, 01938 570600, maesfron@aol.com, www.maesfron.co.uk. *4m E of Welshpool. On North side of A458 Welshpool to Shrewsbury Rd. Visits by arrangement Mar to Sept for* groups of 10+. Adm £5, chd free. Home-made teas. Wider range of refreshments available by prior arrangement.

Largely intact Georgian estate of 6 acres. House (partly open) built in Italian villa style set in South-facing gardens on lower slopes of Moel-y-Golfa with panoramic views of The Long Mountain. Terraces, Walled Kitchen Garden, Chapel, Display Beds, Restored Victorian Conservatories, Tower, Shell Grotto and gardens below tower. Explore ground floor, wine cellar, old kitchen and servants quarters. Parkland walks with wide variety of trees. Refreshments served in reception rooms or on terrace. Some gravel, steps and slopes.

38 NEW MONAUGHTY MILL
Monaughty, Knighton, LD7 1SH. Peter & Shelley Lane. *5m SW Knighton. From Knighton take A488 towards Llandrindod Wells. After 4m turn R B4356 Garden 400m on L.* Sat 31 July, Sun 1 Aug (1-5). Adm £4.50, chd free. Home-made teas.

C17th converted mill (not open) with approx 2 acre garden set on banks of R. Lugg in stunning position with lovely views of Glogg Hill. Informal herbaceous & shrub borders, pond, small woodland areas, riverside, shady & bog plantings, unusual rusted iron 'church' sculpture laden with climbers & shrubs.

39 THE NEUADD
Llanbedr, Crickhowell, NP8 1SP. Robin & Philippa Herbert, 01873 812164, philippahherbert@gmail.com. *1m NE of Crickhowell. Leave Crickhowell by Llanbedr Rd. At junction with Great Oak Rd bear L, cont up hill for approx 1m, garden on L. Ample parking.* Sun 6 June (2-6). Adm £5, chd free. Home-made teas. Visits also by arrangement June to Sept for groups of 5 to 10.

Robin and Philippa Herbert have worked on the restoration of the garden at The Neuadd since 1999 and have planted many unusual trees and shrubs in the dramatic setting of the Brecon Beacons National Park. One of the major features is the walled garden, which has both traditional and decorative planting of fruit, vegetables and flowers. There is also a woodland walk with ponds and streams and a formal garden with flowering terraces. Spectacular views, water feature, rare trees and shrubs,

plant stall and teas. The owner uses a wheelchair and most of the garden is accessible, but some steep paths.

GROUP OPENING

40 NEW NEW RADNOR GARDENS
New Radnor, New Radnor, Presteigne, LD8 2SS. *New Radnor LD8 2SS Powys. On the A44, 6m W of Kington & 12m E from Llandrindod Wells. Tickets & Map from The Hub in School Lane, (near the monument.) Parking at the end of School Lane.* Sun 27 June (10.30-4.30). Combined adm £6, chd free. Light refreshments at The Old School Hub and Garden.

The Conservation Village of New Radnor, formally the county town of Radnorshire and strategic site of a Motte and Bailey Castle, whose mound now dominates the village, lies in the beautiful Radnor Valley, cradled between the two high hills of The Smatcher and The Wimble. Wheelchair access to most gardens. One garden unsuitable for visitors with limited mobility.

41 NEW NO 2 THE OLD COACH HOUSE
Church Road, Knighton, LD7 1ED. Mr Richard & Mrs Jenny Vaughan, 01547 520246. *Next door to the town's Bowling Green. Out of Knighton on the A400 road to Clun, take 1st L into Church Rd, in the courtyard of Ystrad House.* Sat 12, Sun 13 June, Sat 31 July, Sun 1 Aug (2-5). Combined adm with 1 Ystrad House £4.50, chd free. Home-made teas. Visits also by arrangement June to Aug.

The garden at the Old Coach House is a green haven of rooms as you move along a winding path edged by the top borders of shrubs, colourful perennials & circular lawns. An archway of Clematis & Passiflora leads to the lower lawn, flanked by beech hedging, apple trees, a greenhouse, shrubs, hostas & ferns, arriving at a private quiet riverside & woodland glade. The garden can be accessed through the side gate for wheelchair users, where the lawns are generally level.

42 OAK COTTAGE
23 High Street, Welshpool, SY21 7JP. Tony Harvey. *Entered from back of house via Bowling Green Lane parallel to the High Street in centre of Welshpool.* Sun 28 Mar, Mon 5, Sun 11, Sun 25 Apr (2-5). Adm £3.50, chd free. Home-made teas.

An oasis of green in the town centre. Gravel paths and stepping stones meander through a wide variety of plants, incl unusual species, and make the garden seem much larger than it is. Alpines are still a favourite as are the insectivorous plants. Modern representation of a Wardian cabinet, with hepaticas. Places to sit and enjoy garden views. Gravel paths and steps or steep slope at entrance.

44 OSPREY STUDIOS
57 Ynyswen, Penycae, Swansea, SA9 1YT. Rebecca Buck, 07913743457, osprey.studios@btinternet.com, www.ospreystudios.org. *20m from Swansea. M4 exit 45, 15m on A4067 towards Brecon. L after Ynyswen rd sign. L at T junction, Blue house on R, no. 57. 13m from A40/A4067 junction, Sennybridge.* Visits by arrangement May to Sept for groups of up to 10. Adm by donation. Light refreshments.

Wild-life, sculpture and fruit in compact front and back gardens. Blending into the wet meadow at the base of Cribarth, Sleeping Giant Mountain on the SW corner of the Brecon Beacons National Park. Specialising in slug-proof planting! Ceramic sculpture studio on site, sculptures and planters for sale. Teas and home-made cake. Toilet. Easy parking in front. Unique Sculpture for sale or commission. A varied selection of 'Seconds' starting at £1! Lovely walks and great Pubs in this area. Access to front garden and the back garden patio that offers good views.

National Garden Scheme gardens are identified by their yellow road signs and posters. You can expect a garden of quality, character and interest, a warm welcome and plenty of home-made cakes!

45 PENMYARTH HOUSE - GLANUSK ESTATE

The Glanusk Estate, Crickhowell, NP8 1SH. Mrs Harry Legge-Bourke, 01873 810414, info@glanuskestate.com, www.glanuskestate.com. *2m NW of Crickhowell. Please access the Glanusk Estate via the Main Entrance (NP8 1SH) on the A40 and follow signs to car park. There is no access from the B4558 Cwm Crawnon Road.* **Sun 16 May (11-4). Adm £7.50. Light refreshments. Food and Beverage Village onsite for the Day with wide selection for many tastes during the Estate Fayre and NGS Open Garden.**
The garden is adorned with many established plant species such as rhododendrons, azaleas, acers, camelia, magnolia, prunus and dogwood giving a vast array of colour in the spring and summer months and over 300 cultivars of Oaks. Alongside the Open Garden, we will be holding the annual Estate Fayre, showcasing over 30 artisans with exhibits of works for sale. Penmyarth Church, a short distance from the gardens, will be open. For more information see our website and social media channels. Historic gardens, Oaks, Cottage Orné, Artisan stands, home-made cakes, coffee, tea and gourmet catering, licensed bar, talks, garden tours, plant sale, activities and demonstrations. The gardens contain some historic features, incl paths and steps which are not suitable for access by wheelchairs or pushchairs.

&. ﬁ ❀ 🚗 NPC 🚌 ☕ ⛲

46 PEN-Y-MAES

Hay-on-Wye, HR3 5PP. Shân Egerton, 01497 820423, penymaes.hay@gmail.com. *1m SW of Hay-on-Wye. On B4350 towards Hay from Brecon. 2½ m from Glasbury.* **Sun 20 June, Sun 19 Sept (2-5). Adm £5, chd free. Home-made teas. Visits also by arrangement June & July for groups of 10 to 30.**
2 acre garden incl mixed and herbaceous borders; orchard, topiary; walled formal kitchen garden; shrub, modern and climbing roses, peony borders, clematis, espaliered pears. Fine mulberry. Beautiful dry stone walling and mature trees. Great double view of Black Mountains and the Brecon Beacons. Emphasis on foliage and shape. Artist's garden.

47 PLAS DINAM

Llandinam, SY17 5DQ. Eldrydd Lamp, 07415 503554, eldrydd@plasdinam.co.uk, www.plasdinamcountryhouse.co.uk. *7½ m SW Newtown. on A470.* **Visits by arrangement Apr to Nov for groups of 10+. Adm £8 incl tea, coffee and cake.**
12 acres of parkland, gardens, lawns and woodland set at the foot of glorious rolling hills with spectacular views across the Severn Valley. A host of daffodils followed by one of the best wildflower meadows in Montgomeryshire with 36 species of flowers and grasses incl hundreds of wild orchids; Glorious autumn colour with parrotias, liriodendrons, cotinus etc. Millennium wood. From 1884 until recently the home of Lord Davies and his family (house not open).

&. ﬁ 🚗 🚌 ☕

48 PONT FAEN HOUSE

Farrington Lane, Knighton, LD7 1LA. Mr John & Mrs Brenda Morgan, 01547 520847. *S of Knighton off Ludlow Rd. W from Ludlow on A4113 into Knighton. 1st L after 20mph sign before school.* **Visits by arrangement May to Sept for groups of up to 30. Adm £4, chd free.**
Colourful ¾ acre flat garden brimming with flowers on edge of town. Arches lead from shady ferny corners to deep borders filled with large range of colourful perennials, annuals & fish pond. Stunning wisteria,10 varieties of alstromeria, rhododendrons & azaleas followed by colourful perennials, shrubs, and 40 rose varieties. Good late summer colour. Enjoy views of hills from the many seats. Year round colour: tulips, clematis, rudbeckias, inulas. Disabled parking in garden.

&. ﬁ ☕

49 PONTHAFREN

Long Bridge Street, Newtown, SY16 2DY. www.ponthafren.org.uk. *Park in main car park in town centre, 5 mins walk. Turn L out of car park, turn L over bridge, garden on L. Limited disabled parking, please phone for details.* **Sat 24 July (10-5). Adm by donation. Home-made teas. Also open Bryn Teg.**
Ponthafren is a registered charity that provides a caring community to promote positive mental health and well-being for all. Open door policy so everyone is welcome. Interesting community garden on banks of R

Severn run and maintained totally by volunteers: sensory garden with long grasses, herbs, scented plants and shrubs, quirky objects. Productive vegetable plot. Lots of plants for sale. Covered seating areas positioned around the garden to enjoy the views. Partial wheelchair access.

&. ﬁ ❀ 🚗 ☕ ⛲

50 NEW PONTSIONI HOUSE

Aberedw, Builth Wells, LD2 3SQ. Mr & Mrs Jonathan Reeves. *5m SE of Builth Wells. On B4567 between Erwood Bridge & Aberedw on Radnorshire side of R Wye.* **Sun 2, Sun 9 May (2.30-6). Adm £5, chd free. Home-made teas.**
With a background of old ruins and steep rocky woodland, this Wye Valley garden with herbaceous, shrub borders, terraces and natural rockery merge with lawns. Recently constructed small walled vegetable and fruit garden. Walks through wildflower meadow along a mile of old railway line with bluebell woods and walks up to the Aberedw Rocks. Dogs welcome along old railway line. Spectacular rocky and woody situation. Extensive bluebells.

&. ﬁ ❀ ☕ ⛲

51 THE ROCK HOUSE

Llanbister, LD1 6TN. Jude Boutle & Sue Cox. *10m N of Llandrindod Wells. Off B4356 just above Llanbister village.* **Sun 6 June (1-5). Adm £4.50, chd free. Home-made teas. Evening opening Fri 11 June (6-9). Adm £6, chd £1.50. Light refreshments.**
About an acre of informal hillside garden, 1000ft up with views over the Radnorshire Hills. Wildlife ponds, bluebell meadow and a laburnum arch. It's a bit of a battle with nature so come and visit and see who you think is winning! New for 2021 Bee hives and an epic summer house. Minimal wheelchair access.

❀ ⛲

52 ROCK MILL

Abermule, Montgomery, SY15 6NN. Rufus & Cherry Fairweather, 01686 630664, fairweathers66@btinternet.com. *1m S of Abermule on B4368 towards Kerry. Best approached from Abermule village as there is an angled entrance into field for parking.* **Sat 15, Sun 16 May (2-5). Adm £6, chd free. Home-made teas. Visits also by arrangement in May for** groups of 10 to 30.

2 acre riverside garden on different levels in wooded valley. Colourful borders & shrubberies, specimen trees, terraces, woodland walks, bridges, extensive lawns, fishponds, orchard, herb & vegetable gardens, beehives, dovecote, heather thatch roundhouse, remnants of industrial past (corn mill & railway line). Child and adult friendly activities (supervision required) incl sunken trampoline, croquet and badminton, animal treasure hunt, interactive quiz, wilderness trails, cockleshell tunnel. Sensible shoes and a sense of fun/adventure recommended.

GROUP OPENING

53 NEW SOUTH STREET GARDENS

Beechcroft, South Street, Rhayader, LD6 5BH. Gwyneth Rose & Steve Harley, 01597 811868, info@penralleyhouse.com. *All gardens are 100 metres S of the centre of Rhayader on A470. Parking in public car park in St Harmon Rd, by sports centre. Penralley House & Beechcroft gardens are in South St on the RHS & the Toll house is opp, approx 100 metres from town clock.* **Sun 4 July (1-4.30). Combined adm £5, chd free. Light refreshments in Penralley garden.**
Three very different gardens close to the centre of town. The Toll House shows what can be done in a tiny space, with raised beds of annuals, troughs and traditional roses around the door. Penralley House has a large terraced garden with flower borders, mature trees, orchard, chicken coop, fruit garden and lawns. Beechcroft, the largest of the three, has an herbaceous border, fruit garden with green house and compost area, raised vegetable garden, pond and wild flower meadow. Below is a alopod woodland garden with zig zag paths and natural spring at the bottom. All gardens have views of the lovely Gwastedyn Hill at the edge of town. Tea and cake available in Penralley garden. Plant sales and open textile studio in Beechcroft garden. Both Beechcroft and Penralley gardens are accessible. The Toll house is restricted. Blue badge only Disabled parking in Beechcroft.

54 STOCKTON MILL

Marton, Welshpool, SY21 8JL. Stephen & Caroline Cox, 01938 561990, carolinecox74@gmail.com, www.stocktonmill.wordpress.com. *6m SE Welshpool. From Welshpool take A483 S. Turn L on A490 for 3.3m (rd takes L turn 200m after The Cock Inn, Forden) then turn L signed Marton. Continue 1.2m & turn R before pink building. Park in field.* **Sat 10 July (10-5). Adm £4, chd free. Home-made teas.**
Peaceful 3 acre garden by R Camlad in Vale of Montgomery with lovely views. Trees mainly planted 30 yrs ago now matured into striking woodland including Himalayan Birch, Himalayan Cherry, Dogwoods and varieties of unusual conifers. Extensive shrub & herbaceous borders, lawns, veg patch & walks by the river. Two wildlife ponds. C19 Grade II listed Old Mill (holiday let not open).

"The support of Hospice UK and the National Garden Scheme has been invaluable to hospice nurses across the country whilst we've been battling the coronavirus crisis, helping hospices such as Derian House to continue providing vital end of life and respite care to 400 children and young adults from across the North West. Thank you."
— Katie Turner, Perinatal Nurse at Derian House Children's Hospice

55 TRANQUILLITY HAVEN

7 Lords Land, Whitton, Knighton, LD7 1NJ. Val Brown, 01547 560070, valerie.brown1502@gmail.com. *approx 3m from Knighton & 5m Presteigne. From Knighton take B4355 after approx. 2m turn R on B4357 to Whitton. Car park on L by yellow NGS signs.* **Sat 5, Sun 6, Sat 12, Sun 13 June (2-5); Sat 2, Sun 3, Sat 16, Sun 17 Oct (2-4). Adm £4, chd free. Home-made teas. Visits also by arrangement Apr to Oct for groups of 5 to 20.**
Amazing Japanese Stroll Garden with borrowed views to Offa's Dyke. Winding paths pass small pools and lead to Japanese bridges over natural stream with dippers and kingfishers. Sounds of water fill the air. Enjoy peace and tranquillity from one of the seats or the Japanese Tea House. Dense oriental planting with Cornus kousa satomi, acers, azaleas, unusual bamboos and wonderful cloud pruning,.

56 TREBERFYDD HOUSE

Llangasty, Bwlch, Brecon, LD3 7PX. David Raikes & Carla Rapoport, www.treberfydd.com. *6m E of Brecon. From Abergavenny on A40, turn R in Bwlch on B5460. Take 1st turning L towards Pennorth & cont 2m along lane. From Brecon, turn L off A40 towards Pennorth in Llanhamlach.* **Sun 18 July (1-5.30). Adm £5.50, chd free. Tea.**
Grade I listed Victorian Gothic house with 10 acres of grounds designed by W A Nesfield. Magnificent Cedar of Lebanon, avenue of mature Beech, towering Atlantic Cedars, Victorian rockery, herbaceous border and manicured lawns ideal for a picnic. Wonderful views of the Black Mountains. Plants available from Commercial Nursery in grounds - Walled Garden Treberfydd. Easy wheelchair access to areas around the house, but herbaceous border only accessible via steps.

57 TREMYNFA

Carreghofa Lane,
Llanymynech, SY21 6LA. Jon
& Gillian Fynes, 01691 839471,
gillianfynes@btinternet.com. *Edge
of Llanymynech village. From N leave
Oswestry on A483 to Welshpool.
In Llanymynech turn R at Xrds (car
wash on corner). Take 2nd R then
follow yellow NGS signs. 300yds
park signed field, limited disabled
parking nr garden.* **Sat 19, Sun 20
June (1-5). Adm £4.50, chd free.
Home-made teas. Gluten free
cakes available. Visits also by
arrangement 14th-25th June for
groups of 10 to 30. Weekdays,
afternoon/evening only.**
S-facing 1 acre garden developed
over 15yrs. Old railway cottage set
in herbaceous and raised borders,
patio with many pots of colourful
and unusual plants. Garden slopes
to productive fruit and vegetable
area, ponds, spinney, unusual trees,
wild areas & peat bog. Patio &
seats to enjoy extensive views incl
Llanymynech Rocks. Pet ducks on
site, Montgomery canal close by. 100s
of home grown plants and home-
made jams for sale. Wheelchairs can
access most areas but users will need
to walk up/down 2 or 3 shallow steps,
please ring to discuss if concerned.

58 TYN Y CWM

Beulah, Llanwrtyd Wells,
LD5 4TS. Steve & Christine
Carrow, 01591 620461,
steve.carrow@ngs.org.uk. *10m W
of Builth Wells. On A483 at Beulah
take rd towards Abergwesyn for 2m.
Drive drops down to L.* **Visits by
arrangement May to Aug for groups
of up to 30. Adm £5, chd free.**
Garden mainly started 19yrs ago, lower
garden has spring/woodland area,
raised beds mixed with vegetables,
fruit trees, fruit and flowers. Perennial
borders, summer house gravel paths
through rose and clematis pergola.
Upper garden, partly sloped, incl bog
and water gardens. Perennial beds
with unusual slate steps. Beautiful
views. Property bounded by small river.

59 NEW TYNRHOS

Newbridge-On-Wye, Llandrindod
Wells, LD1 6ND. Clare Wilkinson.
*1 mile NW Newbridge-on-Wye. From
A470 take B4518 (signed Beulah)
after 400yds cross bridge, turn R for
Llysdinam. Proceed up the hill for
1m, Tynrhos is on L.* **Sun 11 July**

**(1.30-5.30). Adm £5, chd free.
Home-made teas.**
A ¾ acre family cottage garden lying
at approx 800' with views over the
Wye Valley. Mainly herbaceous planting
with summer annuals and pots.
Vegetable plot with small greenhouse
leading to orchard, fields and wildlife
pond. Cobbled area to terrace. Access
to rear garden area over grass. Paths
are gravel/grass, maybe slippery if wet.

60 NEW VAYNOR PARK

Berriew, Welshpool, SY21 8QE.
Mr & Mrs William Corbett-Winder.
*Leave Berriew going over the bridge
& straight up the hill on the Bettws rd.
The entrance to Vaynor Park is on the
R ¼ m from the speed derestriction
sign.* **Sun 5 Sept (12-5). Adm £6,
chd free. Home-made teas.**
Beautiful C17 house (not open) with
5 acre garden. Stunning herbaceous
borders with late flowering salvias,
penstemons & dahlias, box edged
rose parterre, banks of hydrangeas;
topiary yew birds, box buttresses
and spires bring formality. Courtyard
with lime green hydrangea paniculata
and Annabelle. Woodland garden.
Orangery. Spectacular views. Home to
the Corbett-Winder family since 1720.

61 ♦ WELSH LAVENDER

Cefnperfedd Uchaf,
Maesmynis, Builth Wells,
LD2 3HU. Nancy Durham & Bill
Newton-Smith, 01982 552467,
farmers@welshlavender.com,
www.welshlavender.com. *Approx
4½ m S of Builth Wells & 13m north
from Brecon Cathedral off B4520. The
farm is 1⅓ m from turn signed Farmers'
Welsh Lavender.* **For NGS: Every Fri
9 July to 20 Aug (11-4). Adm £5,
chd free. Pre-booking essential,
please visit www.ngs.org.uk/events
for information & booking. Light
refreshments. Coffee, tea, wine and
light refreshments available. For
other opening times and information,
please phone, email or visit garden
website.**
Welsh Lavender's fields of blue will be
at their peak from mid July through
August. The surrounding gardens
continue to develop with Jeni Arnold's
stylish wild planting of the steep bank
above the ever popular wild swimming
pond. Walk in the lavender fields, learn
how the distillation process works, and
visit the farm shop to try body creams
and balms made with lavender oil
distilled on the farm. Swim in the pond

before enjoying coffee, tea, wine and
light refreshments which are available.
Partial wheelchair access. Large paved
area adjacent to teas and shop area
easy to negotiate.

62 WHITE HOPTON FARM

Wern Lane, Sarn, Newtown,
SY16 4EN. Claire Austin, www.
claireaustin-hardyplants.co.uk.
*From Newtown A489, E towards
Churchstoke for 7m & turn R in Sarn
follow yellow NGS signs.* **Fri 25 June
(10-4). Adm £5, chd free. Light
refreshments at nursery or plants
and light lunches at The Sarn, Nr
Newtown, SY16 4EJ.**
Horticulturist and author Claire
Austin's private 1½ acre plant
collector's garden designed along
the lines of a cottage garden with
hundreds of different perennials. Front
garden mainly full of May blooming
perennials, back garden, which is
split into various areas, has a June
and July garden, a small woodland
walk, a Victorian fountain and mixed
rose borders. Fabulous views over
the Kerry Vale. Wide range of Claire
Austin's Hardy Plants for sale at The
Sarn or on the nursery.

63 1 YSTRAD HOUSE

1 Church Road, Knighton,
LD7 1EB. John & Margaret
Davis, 01547 528154,
jamdavis@ystradhouse.plus.com.
*At junction of Church Rd & Station Rd.
Take the turning opp Knighton Hotel
(A488 Clun) travel 225yds along Station
Rd. Yellow House, red front door, at
junction with Church Rd.* **Sat 12, Sun
13 June, Sat 31 July, Sun 1 Aug
(2-5). Combined adm with No 2 The
Old Coach House £4.50, chd free.
Home-made teas. incl gluten free
cakes. Visits also by arrangement
June to Aug for groups of up to 30.**
A town garden behind a Regency Villa
of earlier origins. A narrow entrance
door opens revealing unexpected
calm & timelessness. A small
walled garden with box hedging &
greenhouse lead to broad lawns, wide
borders with soft colour schemes &
mature trees. More intimate features:
pots, urns & pools add interest &
surprise. The formal areas merge with
wooded glades leading to a riverside
walk. croquet on request. Lawns and
gravelled paths mostly flat, except
access to riverside walk.

Bryn Teg

Early Openings 2022

Plan your garden visiting well ahead – put these dates in your diary!

Gardens across the country open early – before the next year's guide is published – with glorious displays of colour including hellebores, aconites, snowdrops and carpets of spring bulbs.

Cheshire & Wirral
Sun 27 February (1–4)
Bucklow Farm

By arrangement in February
Rosewood
The Well House

Devon
Fri 4, Fri 11, Sat 19 February
(2–5)
Higher Cherubeer

Essex
Sun 23 January (10–5)
Green Island

Glamorgan
Sun 13, Sun 20 February (2–5)
Slade

Gloucestershire
Sun 30 January,
Sun 13 February (11–4)
Home Farm

Hampshire
Sun 13, Mon 14, Sun 20,
Mon 21 February (2–4.30)
Little Court

Kent
Sat 5, Sun 6, Mon 7 February
(11–3.30)
Knowle Hill Farm

Sat 12, Sun 20 February (12–4)
Copton Ash

By arrangement in February
The Old Rectory

Northamptonshire
Sun 27 February (11–3)
67-69 High Street

Oxfordshire
Sun 6 February (2–5)
23 Hid's Copse Road

Sun 6 February (2–5)
Stonehaven

Sun 13 February (1.30–4.30)
Hollyhocks

**Somerset, Bristol &
South Gloucestershire**
Sun 30 January,
Sun 6 February (11–4)
Rock House

Thur 10 February (10–5)
East Lambrook Manor Gardens

**Staffordshire, Birmingham
& West Midlands**
Sun 13 February (11–3)
5 East View Cottages

Suffolk
Sun 20 February (11–4)
The Laburnums

Surrey
Wed 12, Wed 19, Wed 26
January, Mon 7, Mon 14,
Mon 21 February (12–2)
Timber Hill

Sussex
By arrangement in February
Pembury House

Yorkshire
Sun 13 February (11–4.30)
Devonshire Mill

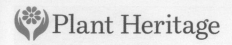 Plant Heritage

Over 70 gardens that open for the National Garden Scheme are holders of a Plant Heritage National Plant Collection although this may not always be noted in the garden description. These gardens carry the NPC symbol.

Plant Heritage, 12 Home Farm, Loseley Park, Guildford, Surrey GU3 1HS.
01483 447540 www.plantheritage.org.uk

Acer (excl. palmatum cvs.)
Blagdon, North East

Alnus
Stone Lane Gardens, Devon
Blagdon, North East

Anemone nemorosa
Avondale Nursery, Warwickshire

Anemone nemorosa cvs.
Kingston Lacy, Dorset

Araliaceae (excl. Hedera)
Meon Orchard, Hampshire

Aspidistra elatior & sichuanensis cvs.
12 Woods Ley, Kent

Aster & related genera (autumn flowering)
The Picton Garden, Herefordshire

Aster (Symphyotrichum) novae-angliae
Avondale Nursery, Warwickshire
Brockamin, Worcestershire

Astilbe
Holehird Gardens, Cumbria
Marwood Hill Garden, Devon

Astrantia
Norwell Nurseries, Norwell Gardens, Nottinghamshire

Betula
Ness Botanic Gardens, Cheshire
Stone Lane Gardens, Devon

Buddleja davidii cvs. & hybrids
Shapcott Barton Knowstone Estate, Devon

Camellia (autumn and winter flowering)
Green Island, Essex

Camellias & Rhododendrons introduced to Heligan pre-1920
The Lost Gardens of Heligan, Cornwall

Carpinus
Sir Harold Hillier Gardens, Hampshire

Carpinus betulus cvs.
West Lodge Park, London

Ceanothus
Eccleston Square, London

Cercidiphyllum
Sir Harold Hillier Gardens, Hampshire
Hodnet Hall Gardens, Shropshire

Chlorophytum comosum cvs.
52 Cobblers Bridge Road, Kent

Chrysanthemum (Hardy) Dispersed
Norwell Nurseries, Norwell Gardens, Nottinghamshire

Clematis viticella
Longstock Park, Hampshire

Clematis viticella cvs.
Roseland House, Cornwall

Codonopsis & related genera
Woodlands, Lincolnshire

Colchicum
East Ruston Old Vicarage, Norfolk

Cornus
Sir Harold Hillier Gardens, Hampshire

Cornus (excl. C. florida cvs.)
Newby Hall & Gardens, Yorkshire

Corokia
33 Wood Vale, London

Corylus
Sir Harold Hillier Gardens, Hampshire

Cotoneaster
Sir Harold Hillier Gardens, Hampshire

Cyclamen (excl. persicum cvs.)
Higher Cherubeer, Devon

Daboecia
Holehird Gardens, Cumbria

Daffodil Dispersed Noel Burr Cultivars
Mill Hall Farm, Sussex

Dierama spp.
Yew Tree Cottage, Staffordshire

Erica & Calluna - Sussex heather cvs
Nymans, Sussex

Erythronium
Greencombe Gardens, Somerset

Eucalyptus
Meon Orchard, Hampshire

Eucalyptus spp.
The World Garden at Lullingstone Castle, Kent

Eucryphia
Whitstone Farm, Devon

Euonymus (deciduous)
The Place for Plants, East Bergholt Place Garden, Suffolk

Eupatorium
Saith Ffynnon Farm, North East Wales

Euphorbia (hardy)
Firvale Allotment Garden, Yorkshire

Fuchsia (hardy spp. & cvs.)
Croxteth Hall Walled Garden, Lancashire

Galanthus
127 Stoke Road, Bedfordshire

Gaultheria (incl Pernettya)
Greencombe Gardens, Somerset

Geranium sanguineum,
macrorrhizum & x cantabrigiense
Brockamin, Worcestershire

Geranium sylvaticum
& renardii - forms, cvs.
& hybrids
Wren's Nest, Cheshire

Geum
1 Brickwall Cottages, Kent

Hakonechloa macra & cvs.
12 Woods Ley, Kent

Hamamelis
Sir Harold Hillier Gardens, Hampshire

Hamamelis cvs.
Green Island, Essex

Hedera
Ivybank, Pebworth Gardens,
Warwickshire

Heliotropium
Hampton Court Palace, London

Helleborus (Harvington hybrids)
Riverside Gardens at Webbs,
Worcestershire

Hilliers (Plants raised by)
Sir Harold Hillier Gardens, Hampshire

Hoheria
Abbotsbury Gardens, Dorset

Hypericum
Sir Harold Hillier Gardens, Hampshire

Iris (Bearded)
White Hopton Farm, Powys

Iris ensata
Marwood Hill Garden, Devon

Juglans
Upton Wold, Gloucestershire

Lapageria rosea (& named cvs)
Roseland House, Cornwall

Leucanthemum x superbum
(Chrysanthemum maximum)
Shapcott Barton Knowstone Estate,
Devon

Ligustrum
Sir Harold Hillier Gardens, Hampshire

Lithocarpus
Sir Harold Hillier Gardens, Hampshire

Malus (ornamental)
Barnards Farm, Essex

Meconopsis (large flowered blue
spp. & cvs.)
Holehird Gardens, Cumbria

Metasequoia
Sir Harold Hillier Gardens, Hampshire

Monarda
Glyn Bach Gardens, Camarthenshire
Hole's Meadow, Devon

Nepeta
Hole's Meadow, Devon

Ophiopogon japonicus cvs.
12 Woods Ley, Kent

Paeonia (hybrid herbaceous)
White Hopton Farm, Powys

Pelargonium
Ivybank, Pebworth Gardens,
Warwickshire

Pennisetum spp. & cvs. (hardy)
Knoll Gardens, Dorset

Photinia
Sir Harold Hillier Gardens, Hampshire

Picea spp.
The Yorkshire Arboretum, Yorkshire

Pinus (excl dwarf cvs.)
Sir Harold Hillier Gardens, Hampshire

Pinus spp.
The Lovell Quinta Arboretum,
Cheshire

Podocarpaceae
Meon Orchard, Hampshire

Polystichum
Holehird Gardens, Cumbria
Greencombe Gardens, Somerset

Primula auricula (Border)
12 Meres Road, Staffordshire

Pterocarya
Upton Wold, Gloucestershire

Quercus
Chevithorne Barton, Devon
Sir Harold Hillier Gardens, Hampshire

Rhododendron (Ghent Azaleas)
Sheffield Park and Garden, Sussex

Rhododendron (Kurume Azalea
Wilson 50)
Trewidden Garden, Cornwall

Rhododendron spp.
Himalayan Garden & Sculpture Park,
Yorkshire

Rhus
The Place for Plants, East Bergholt
Place Garden, Suffolk

Rosa - Hybrid Musk intro
by Pemberton & Bentall
1912–1939
Dutton Hall, Lancashire

Rosa (rambling)
Moor Wood, Gloucestershire

Sanguisorba
Avondale Nursery, Warwickshire

Santolina
The Walled Gardens of Cannington,
Somerset

Sarracenia
Beaufort, Shropshire

Saxifraga sect. Ligulatae: spp. &
cvs.
Waterperry Gardens, Oxfordshire

Saxifraga sect. Porphyrion
subsect. Porophyllum
Waterperry Gardens, Oxfordshire

Sorbus
Ness Botanic Gardens, Cheshire

Sorbus (British endemic spp.)
Blagdon, North East

Stewartia - Asian spp.
High Beeches Woodland and Water
Garden, Sussex

Styracaceae (incl Halesia,
Pterostyrax, Styrax, Sinojackia)
Holker Hall Gardens, Cumbria

Taxodium spp. & cvs.
West Lodge Park, London

Toxicodendron
The Place for Plants, East Bergholt
Place Garden, Suffolk

Tulbaghia spp & subsp
Marwood Hill Garden, Devon

Vaccinium
Greencombe Gardens, Somerset

Yucca
Renishaw Hall & Gardens, Derbyshire

Proud to Be A Shed

Not all sheds are the same. Posh Sheds are designed with care and built to last. Perfect for the discerning gardener.

- Unique triple layer walls with plywood lining
- Choice of 3 build options
- Option of flat pack delivery or full installation service
- Range of sizes & designs
- Felt shingles, cedar shingles or slate effect roof

THE POSHSHED® COMPANY

01544 387 101
info@theposhshedcompany.co.uk
theposhshedcompany.co.uk

Society of Garden Designers

The 🄳 symbol at the end of a garden description indicates that the garden has been designed by a Fellow, Member, Pre-Registered Member, Student or Friend of the Society of Garden Designers.

Fellow of the Society of Garden Designers (FSGD) is awarded to Members for exceptional contributions to the Society or to the profession

Rosemary Alexander FSGD
Christopher Bradley-Hole FSGD
Roderick Griffin FSGD
Sarah Massey FSGD
Nigel Philips FSGD
David Stevens FSGD
Julie Toll FSGD

Member of the Society of Garden Designers (MSGD) is awarded after passing adjudication

Tommaso Del Buono MSGD
Peter Eustance MSGD
Jill Fenwick MSGD
Paul Gazerwitz MSGD
Fiona Harrison MSGD
Joanna Herald MSGD
Nic Howard MSGD
Barbara Hunt MSGD (retired)
Ian Kitson MSGD
Arabella Lennox-Boyd MSGD
Emma Mazzullo MSGD
Robert Myers MSGD
Chris Parsons MSGD
Dan Pearson MSGD
Emma Plunket MSGD
Paul Richards MSGD
Jilayne Rickards MSGD
Debbie Roberts MSGD

Libby Russell MSGD
Charles Rutherfoord MSGD
Ian Smith MSGD
Lorenzo Soprani-Volpini MSGD
Tom Stuart-Smith MSGD
Joe Swift MSGD
Jo Thompson MSGD
Sue Townsend MSGD
Cleve West MSGD
Matthew Wilson MSGD
Rebecca Winship MSGD

Pre-Registered Member is a member working towards gaining Registered Membership

Tamara Bridge
Alasdair Cameron
Kristina Clode
Linsey Evans
Anoushka Feiler
Claire Merriman
Sarah Naybour
Anne-Marie Powell
Faith Ramsay
Caz Renshaw
Judy Shardlow
Helen Thomas
Fiona Cadwallader

Students
Julianne Fernandez
Tom Gadsby
Will Jennings
Paul Kimberley
Jonathan Venables
Adam Vetere

Acknowledgements

Each year the National Garden Scheme receives fantastic support from the community of garden photographers who donate and make available images of gardens. We would like to thank them for their generous donations. Our thanks also to our wonderful garden owners who kindly submit images of their gardens.

Unless otherwise stated, photographs are kindly reproduced by permission of the garden owner.

The 2021 Production Team: Elna Broe, Jean Crockford, Vicky Flynn, Louise Grainger, Vince Hagan, Joanne McGowan, Kay Palmer, Elena Pearce, Helena Pretorius, George Plumptre, Jane Sennett, Linda Shelton, Catherine Swan, Georgina Waters, Anna Wili.

A CIP catalogue record for this book is available from the British Library.

ISBN: 978-1-4087-1526-0

Designed by Level Partnership
Maps by Mary Spence © Global Mapping and XYZ Maps
Typeset in Helvetica Neue
Printed and bound in Italy by Rotolito S.p.A.

Constable
An imprint of Little, Brown Book Group
Carmelite House, 50 Victoria Embankment,
London EC4Y 0DZ

An Hachette UK Company
www.hachette.co.uk www.littlebrown.co.uk

If you require this information in alternative formats, please telephone 01483 211535 or email hello@ngs.org.uk